Denis Grebenkov · Ralf Metzler · Gleb Oshanin
Editors



# Target Search Problems

Springer

*Editors*
Denis Grebenkov
CNRS—Ecole Polytechnique
Palaiseau, France

Ralf Metzler
University of Potsdam
Potsdam, Brandenburg, Germany

Gleb Oshanin
Sorbonne University—CNRS
Paris, France

ISBN 978-3-031-67801-1 ISBN 978-3-031-67802-8 (eBook)
https://doi.org/10.1007/978-3-031-67802-8



This Springer imprint is published by the registered company Springer Nature Switzerland AG
The registered company address is: Gewerbestrasse 11, 6330 Cham, Switzerland

# Preface

Recent years have witnessed a substantial interest in the so-called 'Target Problem' that encompasses a wide range of different situations in which some 'agents', either living organisms or inanimate ones, perform a deterministic or stochastic motion searching for a certain immobile or mobile 'target'. Such problems emerge across many disciplines, including computer science, evolution of stock markets, biochemistry, biomedicine, evolutionary games, as well as diverse areas of physics. Respectively, a 'target' can be a file in a database, a website or specific information in the internet, a prescribed threshold value that a given stock price is supposed to reach, a binding site or a reactive center in a chemical container, or a specific site on a DNA, a prey hunted by predators, a bacterium chased by a phage, a nucleus of a new phase, targeted regions in a human body to which a drug is to be delivered and so on. Depending on the system at hand, the dynamics of the search process can also be very different: the motion can be intermittent with active slow search phases alternating with inactive long-range relocations, as performed by some living creatures, standard or active Brownian motion, or anomalous diffusion, almost instantaneous relocations like in an eye-guided or computer search, advected motion, random motion assisted by smell or facilitated by a long-ranged perception (vision) and so on. Searchers may act individually or operate in swarms, thus exchanging information between each other.

For each situation, one may concentrate on different properties that characterize the efficiency of the search for a target. It can be the mean first-passage time to the target or, more generally, the mean first-reaction time, if the target is not certain to be detected, or a chemical reaction initiated, upon the first encounter with it. As both the first-passage and the first-reaction times are random variables, it is desirable to know their full probability density functions to access the fluctuations around the mean values. In some instances, a salient property can be the area covered by the searchers or the target survival probability. Last but not least, one may seek different optimization schemes with respect to the searchers' motion, in order to device the most efficient search strategies.

The Target Problem has been studied in different areas of science, which resulted in numerous seminal papers focusing on its particular aspects and instances, and introducing various concepts, notations and approaches. The literature is abundant but, due to the evidently multidisciplinary character of the Target Problem, is scattered among many journals and also devoted to different subjects and focuses on different audiences. As a consequence, a researcher working in a particular domain cannot easily embrace and grasp all the achievements made. Aiming to provide a unifying picture, in this volume we propose an up-to-date collection of 31 chapters, contributed by leading experts in their respective fields. Each of the chapters concentrates on some particular aspect of the general Target Problem, and presents a comprehensive, contemporary overview of previous work, new results and perspectives for further research, written in a widely accessible language.

We believe that this edition is timely and will be beneficial for established and early career researchers working in related fields. This volume will serve as a source of inspiration for further research and permit a cross-fertilization of approaches, ideas and research directions.

Palaiseau, France — Denis Grebenkov
Potsdam, Germany — Ralf Metzler
Paris, France — Gleb Oshanin

# Contents

1 **Target Search Problems** ........................................ 1
Denis Grebenkov, Ralf Metzler, and Gleb Oshanin

2 **Concentration of Empirical First-Passage Times** ................. 31
Rick Bebon and Aljaž Godec

3 **Defocusing of First-Reaction Times** ............................ 55
Denis Grebenkov, Ralf Metzler, and Gleb Oshanin

4 **Encounter-Based Approach to Target Search Problems** ........... 77
Denis Grebenkov

5 **Multi-target Search in Bounded and Heterogeneous Environments: A Lattice Random Walk Perspective** .............. 107
Luca Giuggioli, Seeralan Sarvaharman, Debraj Das, Daniel Marris, and Toby Kay

6 **Effects of Molecular Crowding and Obstacles on a Target Search** ........................................................ 135
Kazuhiko Seki

7 **Search and Escape of Mortal Random Walkers** .................. 155
Enrique Abad and Santos Bravo Yuste

8 **Semi-permeable Interfaces and the Target Problem** .............. 177
Paul C. Bressloff

9 **Big Jump Principle for First Passage Times** ..................... 209
Marc Höll, Alon Nissan, Brian Berkowitz, and Eli Barkai

10 **Optimization of Trap Locations for Narrow Capture Problems** .... 225
Alexei Cheviakov and Michael Ward

11 **Boundary Homogenization for Target Search Problems** ........... 247
Denis Grebenkov and Alexei Skvortsov

**12 Competition of Many Searchers** ... 281
Sean D. Lawley

**13 First-Passage Times for Random Diffusivity Models** ... 305
Yann Lanoiselée and Vittoria Sposini

**14 Random Resetting in Search Problems** ... 323
Arnab Pal, Viktor Stojkoski, and Trifce Sandev

**15 The Distribution of the Maximum of Independent Resetting Brownian Motions** ... 357
Alexander K. Hartmann, Satya N. Majumdar, and Grégory Schehr

**16 Target Search Kinetics for Random Walkers with Memory** ... 391
Olivier Bénichou, Thomas Guérin, Nicolas Levernier, and Raphaël Voituriez

**17 Non-Markovian and Collective Search Strategies** ... 413
Hugues Meyer and Heiko Rieger

**18 Modeling Ionic Flow Between Small Targets: Insights from Diffusion and Electro-Diffusion Theory** ... 433
Frédéric Paquin-Lefebvre and David Holcman

**19 Target Search by Active Particles** ... 463
Urna Basu, Sanjib Sabhapandit, and Ion Santra

**20 Deterministic Pursuit of Randomly Moving Targets** ... 489
Davide Bernardi and Benjamin Lindner

**21 Conservative Versus Non-conservative Diffusion Toward a Target in a Networked Environment** ... 511
Ernesto Estrada

**22 Searching Through Cellular Landscapes** ... 541
Elena F. Koslover

**23 How to Find Targets That Are Always Hidden: Nucleosome-Covered DNA and Pioneer Transcription Factors** ... 579
Anupam Mondal and Anatoly B. Kolomeisky

**24 Exploring the Benefits of DNA-Target Search With Antenna** ... 601
Lucas Hedström and Ludvig Lizana

**25 Target Search on DNA—Effect of Coiling** ... 617
Michael A. Lomholt and Ralf Metzler

**26 First-Reaction Times at the Terminal Target Site in Molecular Relay Races** ... 639
Denis Grebenkov, Ralf Metzler, and Gleb Oshanin

**27 Size-Distribution of Polymer-Free Voids in Frozen Polymer Solutions** ........ 659
Sergei F. Burlatsky and Gleb Oshanin

**28 Target Search Processes During Horizontal Gene Transfer in Bacteria** ........ 675
Rinat Arbel-Goren and Joel Stavans

**29 Optimal Random Searches: Lévy Walks and Animal Movement** ........ 687
Marcos G. E. da Luz, Gandhimohan M. Viswanathan, and Ernesto P. Raposo

**30 Olfactory Search** ........ 711
Antonio Celani and Emanuele Panizon

**31 From Features to Foraging: Current Developments in Visual Search by Human Observers** ........ 733
Todd S. Horowitz

# Chapter 1
# Target Search Problems

**Denis Grebenkov, Ralf Metzler, and Gleb Oshanin**

**Abstract** This chapter presents an introduction to the Target Problem. It contains a brief description of the origins of its name, an exposition of a historical background and an overview of different applications of this problem, settings, notations and most important concepts and results. We also briefly describe the problems studied in the subsequent chapters, putting their contributions in a general context.

## 1.1 The Target Problem and Its Applications

Searching is one of the most fundamental processes in nature and in human life. At the microscopic scale, many biochemical reactions can be described in terms of "searchers" and "targets". Inanimate entities such as molecules typically rely on random search in the search space, similar to conscious beings without prior knowledge about the target location. On a larger scale, animals search for food, mates, and shelters. Humans face various target problems every day, searching for a lost key, information, or the meaning of life. As a consequence, target problems have been thoroughly investigated in vastly diverse disciplines such as applied mathematics, computer sciences, physics, engineering, chemistry, biology, economics, and social sciences. The applications of the resulting theories and models are equally diversified, ranging from efficient search algorithms in web search engines to targeted drug delivery and missile interception.

---

D. Grebenkov (✉)
Laboratoire de Physique de la Matière Condensée, CNRS/Ecole Polytechnique/Institut Polytechnique de Paris, 91120 Palaiseau, France
e-mail: denis.grebenkov@polytechnique.edu

R. Metzler
Institute of Physics and Astronomy, University of Potsdam, 14476 Potsdam-Golm, Germany
e-mail: rmetzler@uni-potsdam.de

G. Oshanin
Laboratoire de Physique Théorique de la Matière Condensée, CNRS/Sorbonne Université, 4 Place Jussieu, 75252 Paris, France
e-mail: gleb.oshanin@sorbonne-universite.fr

 
D. Grebenkov et al. (eds.), *Target Search Problems*,
https://doi.org/10.1007/978-3-031-67802-8_1

### 1.1.1 Historical Background

The idea that the phase of search in space of reactive species for each other can be a rate-controlling factor in various chemical processes is due to Smoluchowski. Prior to his seminal paper [1, 2], a kinetic description of chemical reactions was believed to be entirely governed by the so-called law of mass action. For example, for simple bimolecular reaction schemes of the form $A + B \to C$, where $A$ and $B$ are reactive species and $C$ are the desired reaction products, textbooks suggest to write

$$-\dot{C}_C(t) = \dot{C}_A(t) = \dot{C}_B(t) = -K C_A(t) C_B(t), \tag{1.1}$$

where the dot denotes the time derivative, $C_A(t)$, $C_B(t)$ and $C_C(t)$ are the concentrations of the reactive species and the reaction product at time instant $t$, and $K$ is the intrinsic reaction rate dependent on the activation energy barrier against reaction [3]. In the simple case when $C_A(0) = C_B(0) = C(0)$, e.g., for the process of recombination of radicals, the decay of concentrations of $A$s and $B$s therefore obeys

$$C_A(t) = C_B(t) = \frac{C(0)}{1 + C(0)Kt}. \tag{1.2}$$

A legitimate question addressed by Smoluchowski is what happens if the reaction is perfect. That means that when an $A$ and a $B$ are in immediate contact, a reaction happens with probability 1, i.e., there is no barrier against the reaction and therefore the intrinsic reaction rate $K$ is infinite. For such a situation, Eq. (1.2) predicts that $C_A(t)$ and $C_B(t)$ instantaneously drop down to zero, once $t > 0$, which does not appear to make sense from a physicist's point of view: unless the particles are introduced into the system in geminate $A$-$B$ pairs, they first have to meet each other within their random motion, in order to have the possibility to react. For the reaction of coagulation of colloids taking place in three-dimensional reaction baths, Smoluchowski put forth a scheme that permits to incorporate the random search phase into a theoretical description and therefore to quantify the "resistance" offered by the diffusive motion of reactants toward each other with respect to the reaction process. In essence, using an auxiliary diffusion-reaction search problem, he showed that the infinitely large rate constant $K$ in Eq. (1.1) is to be replaced by a finite quantity, nowadays called the Smoluchowski constant $K_S = 4\pi DR$, where $D$ is the sum of the diffusion coefficients and $R$ is the sum of the radii of the species involved. Since then it was understood that, indeed, the random search represents a stage of utmost importance for many chemical reactions which makes the insight provided by Smoluchowski's work conceptually fundamental.

Analogues of the Smoluchowski constant have been derived for various reaction schemes encountered in diverse chemical, biochemical and biophysical systems under a variety of different physical conditions [4–15]. Subsequently, it was realized that the Smoluchowski approach, based on the formal-kinetic Eq. (1.1) with an appropriately calculated reaction rate, describes the kinetics of some reaction scenarios adequately well, albeit sometimes with incorrect values for the numerical factors. Concurrently, it was shown that the formal-kinetic Eq. (1.1) fails to predict the correct temporal evolution of some systems with chemical reactions, because some other effects come into play—namely, the effects of fluctuations in spatial distributions of reactants which render mean-field-like descriptions incorrect. Examples of such systems are the above-mentioned bimolecular reactions $A + B \to C$ with $C_A(0) = C_B(0) = C(0)$ [16–19], reversible diffusion-limited reactions (see, e.g., [20–25]), spatially inhomogeneous reactions with propagating fronts [26–28] and pattern formation [29], and eventually the trapping reactions $A + B \to B$ with immobile, randomly placed traps $B$ (see below).

The term "Target Problem" was apparently first introduced in the context of chemical kinetics in the analysis of fluctuation phenomena in the simplest reaction-diffusion processes of the form $A + B \to B$ (see, e.g., early papers [30–33]). The purpose was to distinguish between two extreme cases, contrasting the Target Problem with the above mentioned Trapping Problem. In the former, one considers a single immobile target $A$ placed in a sea of mobile diffusive particles (scavengers or searchers) $B$, any of which annihilates the target upon the first encounter with it. In the latter case, on the contrary, a single particle $A$ diffuses in the presence of immobile, randomly scattered (throughout an infinitely large volume) traps $B$, and the first encounter of $A$ with any of the $B$s results in an annihilation of $A$. In both problems, one focuses on the evolution with time $t$ of the survival probability $S(t)$ of the particle $A$ (see Sect. 1.2.2.2 below). The precise modeling of this time-evolution turns out to be remarkably different depending on the considered setting.

For the Trapping Problem—mathematically a very complicated problem which generally cannot be solved in explicit form—the long-time behavior has been scrutinized through a variety of different approaches, including, to name but a few, the optimal-fluctuation method [34], a derivation of rigorous lower and upper bounds on $S(t)$ [35, 36], perturbation theory analysis and diagrammatic expansions [33], or renormalized instanton technique [37]. It was realized that the long-time behavior of the survival probability is dominated by the randomness of the spatial distribution of the traps,[1] namely, by the possibility of having arbitrarily large regions of space which are completely devoid of immobile $B$s. In such regions, the $A$ particle survives for anomalously long times and the long-$t$ decay of $S(t)$, averaged over all possible spatial arrangements of traps $B$, is described by a stretched-exponential function of time (see, e.g., the seminal works [34, 35, 38]), $\ln S(t) \sim -c\,t^z$ with $z = d/(d+2)$, in any spatial dimension $d$.[2] The dimensional prefactor $c$ in this dependence is a non-analytic (in general, sublinear) function of the mean density $n_B$ of the $B$s and the diffusion coefficient of $A$. Such a behavior is strikingly different from the exponential time dependence predicted by the formal-kinetic Eq. (1.1) or

[1] For non-random placement of traps, i.e., when they form a regular lattice, the decay of $S(t)$ is exponential in any spatial dimension (see, e.g., [33]).

[2] Note that such a value of $z$ is specific to a random Poissonian placement of traps. The exponent $z$ may have a different value if the spatial distribution of traps is correlated, e.g., the traps are placed on polymer chains in solution [39].

the Smoluchowski approaches. The survival probability in the Trapping Problem therefore becomes an exponential function of time, i.e., exhibits the fastest decay, only in the limit $d \to \infty$ such that $z \to 1$. Note that this stretched-exponential form is intrinsically linked to the celebrated Lifshitz singularity of the density of states of an electron in a random array of immobile scatterers [40].

For the Target Problem, one usually assumes that the initial positions of the searchers (i.e., $B$ particles) are uniformly distributed in the reaction bath and that all of them diffuse *independently* of each other. Under such assumptions, a calculation of the survival probability of the target (i.e., the $A$ particle) reduces to an exactly-solvable two-body problem [30–33]. In this case, the target survival probability attains the form

$$S(t) = \exp\left(-n_B v_t\right) , \tag{1.3}$$

in which the argument of the exponential function is linear with $n_B$, and $v_t$ is a particular characteristic property of the trajectories of a single $B$ particle: it shows how "compactly" (see, e.g., [41]) each of these trajectories explores the space, how often it overlaps with itself, and how many islands of unvisited space it leaves in the course of time. More specifically, $v_t = \overline{\mathcal{V}_t}$, where the overbar denotes averaging over all possible realizations of trajectories of a single $B$ particle, while $\mathcal{V}_t$ is a functional of a given trajectory which defines the volume of the so-called Wiener sausage.[3] The latter is formally defined as a neighborhood of the trace of a given Brownian trajectory up to a time $t$, obtained by taking all points within some fixed distance of this given trajectory (see, e.g., [35, 42] and also Chap. 27). If the model is defined in discrete space, e.g., on a regular lattice, the meaning of $v_t$ becomes most apparent—it represents the mean number of *distinct* sites visited up to time $t$, i.e., each site which the trajectory visits more than once is counted only once [43]. The situation becomes slightly more complicated for *imperfect* reactions with the target, such that the annihilation of the target occurs only with a finite probability upon encounter with it. In this case, other "geometrical" properties of random walk trajectories come into play, such as, e.g., the mean number of sites visited a prescribed number of times [44] (see also Chap. 4).

The mean volume $v_t$ of the Wiener sausage is evidently a growing function of time: as time evolves, each trajectory visits progressively more and more previously unvisited areas. It appears that the asymptotic, long-time behavior of $v_t$ depends essentially on the spatial dimension $d$, i.e., on whether $d$ is smaller, equal or greater than 2, the fractal dimension of Brownian trajectories. In one-dimensional continuous- or discrete-space systems, $v_t$ grows sub-linearly with time, $v_t \sim \sqrt{t}$, such that $S(t)$ is a stretched-exponential function of time, $\ln S(t) \sim -\sqrt{t}$. This "slower-than-exponential" decay is due to the fact that in one-dimensional systems each Brownian trajectory too thoroughly explores ("oversamples") the space revisiting many times already visited sites. In two dimensions, i.e., in case when the spatial dimension is equal to the fractal dimension of Brownian motion, each trajectory again revisits

[3] In two dimensions, $\mathcal{V}_t$ defines an area covered by a finite-size particle, while in one dimension it is simply the range (or the span) of the Brownian motion [43].

to some extent already visited areas but here such an "oversampling" is marginal, resulting only in a logarithmic correction to the linear growth, $\upsilon_t \sim t/\ln(t)$. Lastly, in three and higher dimensions, $\upsilon_t$ exhibits a linear growth with time, $\upsilon_t \sim t$. In other words, for the Target Problem in unbounded systems with a finite concentration of searchers $S(t)$ decays *slower* in the asymptotic limit $t \to \infty$ in low-dimensional systems ($d = 1$ and $d = 2$) than in high-dimensional ones ($d \geq 3$), such that the target has less chances to survive in systems with $d \geq 3$, than in low-dimensional systems. Some other interesting effects of the spatial dimension will be discussed below on example of the so-called "reduction-of-dimensionality" scenario in biophysical systems.

In more general settings, when both species move diffusively, the problem of determining the target survival probability is even more complicated than the Trapping Problem with immobile traps and clearly, $S(t)$ does not obey the simple form in Eq. (1.3). While this problem has been rather extensively studied in the past, we are only aware of a single exact result on the long-time decay of $S(t)$. In [45] rigorous lower and upper bounds on $S(t)$ were derived, which both exhibit the same temporal behavior. For $D_B > 0$, it was shown that $S(t)$ exhibits exactly the same dependence on time as the one encountered in the Target Problem via the function $\upsilon_t$: $S(t)$ has the above-mentioned stretched-exponential form in one dimension, an exponential function with a logarithmic correction in two dimensions, and $S(t)$ is a simple exponential function in three and higher dimensions [45]. In turn, the prefactor in front of $\upsilon_t$ is an unknown complicated function of all parameters. The effect of confinement onto encounters between diffusing $A$ and $B$ particles was also investigated, both analytically and numerically (see [46, 47] and references therein).

The bounds derived in [45] are not sharp in the sense that they do not converge to each other in the limit $t \to \infty$, which means that the characteristic decay time—a function of both diffusion coefficients and $n_B$—remains unknown, except for some limiting situations. For example, the form of the characteristic decay times has been determined exactly for low-dimensional systems [48–52]. Remarkably, these low-dimensional characteristic decay times appear to be independent of the diffusion coefficient of the $A$ particle, such that $S(t)$ has exactly the same long-time asymptotic behavior as the solution of the Target Problem. Lastly, it was shown that if the $B$s perform a sufficiently confined subdiffusive motion, the $A$ particle "sees" the traps as essentially immobile, and in these situations $S(t)$ shows exactly the same behavior as the one observed in the Trapping Problem [53].

### 1.1.2 Diverse Applications and Research Directions

The Target Problem can be evidently viewed from a much more general perspective, well-beyond the above-mentioned simple reaction scheme. Indeed, different search processes for different "targets" are ubiquitous in the human or animal worlds and, in general, in nature. The target can be a bacterium, a file in a database, a book on a shelf, a lost key, food, a global minimum in a high-dimensional energy landscape, a mating partner, etc. Respectively, the $B$s can be bacteriophages or other predators, human beings or foraging animals, birds or insects. In biophysics, different kinds of ligands search for receptors or some specific binding sites, or eventually, for the entrance to or the exit from a nuclear pore (see, e.g., [54–56]).

Depending on the case at hand, there may be a single target or a single searcher, or targets as well as the searchers may be present in some amounts (see Chaps. 5 and 12). The target and the searchers may have their own life-times or not, such that the search process may last only for some finite time (fixed or random), or conversely, may continue indefinitely (see [57] and Chap. 7). The searchers may move diffusively in continuous space or in networked environments (see Chap. 21), or undergo some kind of anomalous diffusion, at least in the short-time regimes, as it happens with proteins moving in crowded cellular environments or in tortuous labyrinths created by impermeable organelles (see Chaps. 6 and 22). Some aspects of search processes in the complex interior of a bacterial cell, within the context of horizontal gene transfer in bacteria are discussed in Chap. 28. In many instances, the searchers can undergo *active* Brownian motion (see Chap. 19), non-Gaussian diffusion with "diffusing diffusivity" (see Chap. 13) or perform random walks with memory (see Chap. 16). In cell-to-cell communications, in which "messengers" are seeking special targets at intermediate stages of the signal transduction process, the search at different stages may proceed in spaces with different spatial dimensions—effectively one-dimensional ionic channels, nuclear pores, or along DNA (Chap. 25), two-dimensional membranes of the cell or the nucleus, and three-dimensional extracellular medium, the cytoplasm or the nucleoplasm [54, 55]. More ample discussion of such a multi-stage Target Problem is presented in Chap. 26.

In turn, the targets may be located within the bulk of the searched volume, being completely exposed to the searchers (see Chap. 8), or may be localized on the (smooth or rough) boundary of this volume, which can effectively screen the target [58]. The searchers may move "blindly", recognizing the target only upon an encounter with it or, on the contrary, may perceive the targets in some way, assisted by either smell or another kind of a distant perception (see, e.g., [59, 60] and Chap. 30), and change accordingly the mode of their motion from a random search to a direct drift toward the target [61]. Lastly, in the animal and human worlds, one often uses a visual search to localize a desired target (see Chap. 31). In summary, the Target Problem manifests itself as richly-faceted from such a perspective, regarding a wealth of applications, possible conditions and geometries, perception and transport mechanisms.

### *1.1.3 Optimal Search Strategies*

Yet another important aspect of the Target Problem consists in finding the most efficient (or optimal) search strategies with respect to the searchers' motion and concrete geometric setting. Below we name just a few such examples. Predating or foraging animals in their search for prey or food often adapt their dynamics: they swarm, communicate information and eventually coordinate their efforts (see, e.g.,

[62–65] and Chaps. 17 and 29). Foraging birds, insects or animals, in competition for scarce randomly distributed resources, typically move in their environment along apparently erratic trajectories which are often described as superdiffusive Lévy flights or Lévy walks, thus ensuring the maximal spread. Such a behavior was evidenced, e.g., in analyses of recorded trajectories of albatrosses [66, 67], jackals [68] or spider monkeys [69].[4] In cellular systems, a protein, i.e., the Lac repressor, that searches for a specific binding site on a coiled DNA finds it in remarkably short times, much faster than one can expect by assuming that it moves via a simple three-dimensional diffusion through the solution to the target site (see, e.g., [70, 71]). In fact, the protein takes advantage of the "coiled" structure of the DNA and also of the non-specific DNA-protein interactions: it finds first any segment of the DNA, performs a one-dimensional "sliding" motion along non-specific DNA, detaches and occasionally re-attaches to the DNA to parts which are spatially close but appear at large distances along the chain itself. In other words, the searching process can be viewed as a combination of diffusive sliding alternated with effectively long jumps along the DNA, the so-called "facilitated" diffusion, a process that appears to be very robust and results in accelerated targeting [75–85].

During the last decades much effort has been invested in modeling, theoretically or numerically, various search processes in different settings, in order to develop efficient strategies and to gain a conceptual understanding of their essential features and underlying mechanisms (see, e.g., [86] for an overview). Historically, apparently the first attempt in this direction was performed by Adam and Delbrück [87], who put forth an idea how organisms may resolve some problems of efficiency and timing, being limited by molecular diffusion. Namely, Adam and Delbrück suggested that the organisms may do it by subsequently reducing the dimensionality in which the diffusion takes place, even despite the fact that the diffusivities on low-dimensional substrates may have significantly lower values than in the three-dimensional bulk. For instance, for a simple geometrical setting with a target localized on a two-dimensional surface being in contact with a bulk solution, in which the particle is initially located, an efficient search scenario should be that the particle first finds any point on the surface (not necessarily the target itself), non-specifically binds to the surface and diffuses further along the surface until it locates the target. Such a "reduction-of-dimensionality" scenario has been verified and substantiated for several situations, and it was shown that, indeed, it appears advantageous over the direct search through the bulk when some physical conditions are met (see, e.g., [88] and references therein). There are, however, some indications (see, e.g., [89]) that such a strict scenario in which the particle cannot detach from the surface, may not be the most efficient strategy; instead, a scenario in which a particle has some finite probability of detachment from the surface, such that, overall, the search process will consist of a series of alternating tours of decorrelating bulk and surface diffusion,

[4] It is worth noting that a re-analysis of albatross flights showed that they generally are not Lévy flights [72]; however, strong evidence was presented according to which Lévy flights are indeed a search pattern for individuals [73]. See also discussions on optimality of Lévy flights strategy in [74] and in Chap. 29.

may outperform in efficiency the process suggested by Adam and Delbrück (see [89] and references therein).

Another direction of research was devoted to modeling different prey-predator contests, formulated as differential games with an emphasis put on devising optimal pursuit strategies (see, e.g., [90–104] and references therein). Numerous advances in this field are summarized in Chaps. 17 and 20. Moreover, efficient escape strategies permitting the prey to avoid a meeting with the predators as long as possible can be found in, e.g., [61, 105]. The above-mentioned different foraging strategies used by animals or birds are discussed in Chap. 29.

An intriguing question of the facilitated search of a protein for a specific binding site on a coiled DNA received a lot of attention within the last decades. Various aspects of this complicated problem have been scrutinized in a series of papers (see, e.g., [76, 78–85] and reviews [70, 71]). A detailed account of different achievements can be found in Chaps. 23, 24 and 25.

In parallel, considerable interest has been devoted to understanding more fundamental, model-independent aspects of stochastic search processes. Motivated by the analysis of trajectories of foraging animals, Bénichou et al. [106] (see also [107] for an extensive review) proposed the idea of the so-called *intermittent* search, in which the searching trajectory consists of fast, blind relocation stages alternated with stages of a thorough diffusive search. While during the diffusive stages, which have a random duration, the searcher is capable to identify the target, the relocation stages (also of a random duration) do not allow to perceive the target and the only purpose of such stages is to bring the searcher away from already well-explored areas in order to reduce the above-mentioned oversampling. Focusing on the mean first-passage time as an optimizable parameter (and therefore, formally allowing the search process to last for an infinite time), it was shown that the latter reaches a minimum at a certain frequency of relocation/diffusive stages and therefore, upon a fine-tuning of this frequency, one can achieve the most pronounced outcome. A somewhat different line of thought concerning the optimization of intermittent search processes was pursued in [108–110]. Here, instead, it was assumed that the searchers have some finite life-time $\tau$ such that the search process is allowed to run only over a finite time-interval. Therefore, one was seeking to minimize the target's survival probability $S(t)$ calculated at the time moment $\tau$ of eventual death of the searcher. Optimal conditions corresponding to such a setting were defined in [108–110]. Concurrently, the reliability, efficiency and other important features of intermittent search with Lévy-distributed relocation lengths were examined in [111–113].

Finally, a conceptually different idea of stochastic search was put forth by Evans and Majumdar [114]. The point is that in high-dimensional spaces, e.g., in the search for a global minimum of a multi-dimensional energy landscape, the searching trajectories can escape to infinity and therefore will never arrive to the target. To this end, it might sometimes be more reasonable to restart the search process than to let it continue. Consequently, Evans and Majumdar [114] suggested a search algorithm with a spontaneous reset, at random time instants, of the searcher position to its starting point (or any randomly chosen position), which appeared to be very efficient. A salient feature of such a scenario is that the search process under the stochastic reset condition becomes *certain*, i.e., it finds the target with probability 1, even in unbounded high-dimensional ($d \geq 3$) spaces, unlike Brownian or intermittent search processes in such systems which find the target only with a finite probability. Moreover, all moments of the first-passage time are well-defined. This algorithm received a lot of attention and has been extended in various direction (see also [115, 116] for a review). In this volume, different aspects of search with a stochastic reset are discussed in Chaps. 14 and 15.

## 1.2 Mathematical Background

In this section, we briefly describe the mathematical background for target search problems, which is intensively employed and further detailed in the various chapters. While notations vary from one chapter to another, many underlying concepts are common. This section aims therefore to guide the reader through different notions such as Markovian/non-Markovian dynamics, anomalous scaling, non-Gaussian diffusion, search in random and disordered media, as well as in networks. As target search problems appear in such diverse disciplines and contexts, it is hard to elaborate a unique mathematical language to describe all of them. Without pretending neither for mathematical rigor, nor for generality, we aim at providing a systematic description that encompasses the majority of settings described in this book (see also the mathematical introduction to Chap. 2). General concepts and approaches can be also found in [117–119].

From the mathematical point of view, a target search process can be understood as a "trajectory" $\boldsymbol{X}_t$ in a phase space $\Omega$, i.e., a collection of successive "states enumerated by time" $t$, until reaching a target $\Gamma$ (a specific state or a subset of states). In the most basic case, $\boldsymbol{X}_t$ can be the position of a single searcher at time $t$; in general, however, $\boldsymbol{X}_t$ characterizes all degrees of freedom in the system that are relevant for the considered target search problem; for instance, $\boldsymbol{X}_t$ can represent the positions of many searchers (and targets or obstacles if they are also moving); $\boldsymbol{X}_t$ can combine the position and the velocity of a massive particle, or the position and the orientation of an anisotropic object (e.g., a Janus particle or a bacterium); $\boldsymbol{X}_t$ can be a conformational state of a protein, etc.

One of the central quantities that is commonly employed to characterize the search process is the above-mentioned *first-passage time* (FPT)

$$\mathcal{T} = \inf\{t > 0 \; : \; \boldsymbol{X}_t \in \Gamma\}, \tag{1.4}$$

i.e., it is the first time instant when the process $\boldsymbol{X}_t$ finds the target $\Gamma$ [117–119]. Determining the dependence of the probability distribution of this random variable on both the dynamics and the target is a challenging problem that will be addressed in this book. As the time $t$ can be either continuous or discrete, one often distinguishes *discrete-time* and *continuous-time* search processes; however, this distinction is not significant as one can standardly discretize a continuous time variable or, conversely, transform the discrete time into a continuous time by employing random waiting times (see below). In turn, the distinction between *discrete-space* and *continuous-space* search is crucial, as it settles the mathematical tools to be employed.

### 1.2.1 Discrete-Space Dynamics

When the phase space $\Omega$ (e.g., a lattice, a network or a graph) is countable (finite or infinite), all the states, denoted as $\boldsymbol{x}_k$, can be enumerated by a positive integer index $k$. The diffusive dynamics is then defined by introducing the conditional probabilities [120, 121]

$$\mathbb{P}\{\boldsymbol{X}_{t+1} = \boldsymbol{x}_{k_{t+1}} | \boldsymbol{X}_t = \boldsymbol{x}_{k_t}, \ldots, \boldsymbol{X}_0 = \boldsymbol{x}_{k_0}\} = p_t(k_{t+1}|k_t, \ldots, k_0)$$

of finding the process at time $t+1$ in the state $\boldsymbol{x}_{k_{t+1}}$ given that its previous states were $\boldsymbol{x}_{k_t}, \ldots, \boldsymbol{x}_{k_0}$ (here we assume for simplicity that $t$ is a discrete variable). A typical example is a self-avoiding random walk, for which the probability of the next move to a neighboring site depends on whether this site was earlier visited or not. In such *non-Markovian* settings, when the probability of the future state depends on (some) states in the past, finding the probability distribution of FPTs is an extremely difficult and mostly unsolved task. In turn, many advanced tools are available for *Markovian* dynamics when the probability of the future state (at time $t+1$) depends only on the present state (at time $t$), i.e.,

$$\mathbb{P}\{\boldsymbol{X}_{t+1} = \boldsymbol{x}_{k_{t+1}} | \boldsymbol{X}_t = \boldsymbol{x}_{k_t}, \ldots, \boldsymbol{X}_0 = \boldsymbol{x}_{k_0}\} = p_t(k_{t+1}|k_t).$$

Averaging over intermediate times yields the *propagator*, i.e., the probability of finding the process in a prescribed state $\boldsymbol{x}_{k_t}$ at time $t$:

$$\begin{aligned} G(k_t, t|k_0) &= \mathbb{P}\{\boldsymbol{X}_t = \boldsymbol{x}_{k_t} | \boldsymbol{X}_0 = \boldsymbol{x}_{k_0}\} \\ &= \sum_{k_1,\ldots,k_{t-1}} p_{t-1}(k_t|k_{t-1})\, p_{t-2}(k_{t-1}|k_{t-2}) \ldots p_0(k_1|k_0), \end{aligned} \tag{1.5}$$

given the initial state $\boldsymbol{x}_{k_0}$ of the system. Interpreting the transition probabilities $p_t(k_{t+1}|k_t)$ as the elements of a matrix, $(\Pi_t)_{k_{t+1},k_t} = p_t(k_{t+1}|k_t)$, the above sums over indices $k_1, \ldots, k_{t-1}$ turn into matrix products, yielding a compact, though still formal expression for the propagator:

$$G(k_t, t|k_0) = \left[\Pi_{t-1}\Pi_{t-1} \ldots \Pi_0\right]_{k_t,k_0}. \tag{1.6}$$

Moreover, the dynamics is called *homogeneous in time*, if the transition probability $p_t(k_{t+1}|k_t)$ does not depend explicitly on $t$, i.e., $\Pi_t = \ldots = \Pi_0 = \Pi$, in which case the propagator admits a simpler form, involving the $t$-th power of the transition matrix $\Pi$:

$$G(k_t, t|k_0) = \left[\Pi^t\right]_{k_t,k_0}. \tag{1.7}$$

Dividing $\Pi$ by the duration $\delta$ of one step gives the matrix of transition rates, $\Pi/\delta$, which stands in the master equation that is often used for describing the discrete-space dynamics (see Chap. 5). In the case of random walks, the matrix $\Pi$ can be constructed from the adjacency matrix of the underlying lattice or graph that encodes their structure. It can also be related to the discrete Laplace operator.

The matrix representation (1.7) allows for efficient numerical computations, while the diagonalization of the transition matrix $\Pi$ gives access to a spectral decomposition of the propagator and further analytical insights onto its properties. If the target plays the role of a sink for the searcher (i.e., the search is stopped after hitting the target), summing the propagator over all states $\boldsymbol{x}_{k_t}$ yields the *survival probability*

$$S(t|k_0) = \mathbb{P}\{\mathcal{T} > t|\boldsymbol{X}_0 = \boldsymbol{x}_{k_0}\} = \sum_{k_t} G(k_t, t|k_0), \tag{1.8}$$

i.e., the probability of not finding the target up to time $t$. In turn, the difference between $S(t-1|k_0)$ and $S(t|k_0)$ is the probability of finding the target at time $t$,

$$\mathbb{P}\{\mathcal{T} = t|\boldsymbol{X}_0 = \boldsymbol{x}_{k_0}\} = S(t-1|k_0) - S(t|k_0) \quad (t = 1, 2, \ldots). \tag{1.9}$$

As a consequence, the knowledge of the propagator determines the distribution of the first-passage time $\mathcal{T}$ in the Markovian setting. Many former studies on target search problems with discrete-space dynamics were focused on this setting (e.g., most random walk models) and its extensions (e.g., heterogeneous continuous-time random walks on graphs [122]). It is worth stressing that the apparent simplicity of Eq. (1.7) is deceptive. In fact, the relation between the search process on the phase space $\Omega$, encoded via the matrix $\Pi$, and its spectral properties, is sophisticated [123, 124]. Further details and examples of discrete-space search processes are presented in Chaps. 5 and 21.

### *1.2.2 Continuous-Space Dynamics*

Since the points of a continuum space cannot be enumerated, the continuous-space dynamics requires different concepts and tools such as stochastic differential equations and Itô calculus [125, 126]. In the context of target search problems, a stochastic process $\boldsymbol{X}_t$ usually represents the position of a particle (or multiple particles) diffusing in a Euclidean space, inside a confining domain, or on a manifold (e.g., a plasma membrane or the Earth surface) representing the available region.

#### 1.2.2.1 Markovian Case

In the Markovian case, a stochastic process $\boldsymbol{X}_t$ can be defined by describing its infinitesimal increment over an infinitesimal time step $dt$, i.e., $\boldsymbol{X}_{t+dt} = \boldsymbol{X}_t + d\boldsymbol{X}_t$. Knowing the starting position $\boldsymbol{X}_0$, one can rigorously construct $\boldsymbol{X}_t$ by summing the increments. It is common to separate *deterministic* and *random* contributions to the increment $d\boldsymbol{X}_t$ by writing

$$d\boldsymbol{X}_t = \boldsymbol{\mu}(\boldsymbol{X}_t, t)dt + \sigma(\boldsymbol{X}_t, t)d\boldsymbol{W}_t, \tag{1.10}$$

where the vector $\boldsymbol{\mu}(\boldsymbol{x}, t)$ is a drift term, which can account for an external force or a potential, a flow field, or interactions between particles, while $\sigma(\boldsymbol{x}, t)$ is the $n \times m$ volatility matrix, which characterizes the amplitudes of random fluctuations $d\boldsymbol{W}_t$ (here both $\boldsymbol{X}_t$ and $d\boldsymbol{W}_t$ are vectors in $\mathbb{R}^n$ and $\mathbb{R}^m$ respectively, where $n$ is the number of degrees of freedom of the considered system, and $m$ is the number of independent noises that drive the stochastic process $\boldsymbol{X}_t$). The random fluctuations $d\boldsymbol{W}_t$ are defined as increments of the standard Wiener process $\boldsymbol{W}_t$, i.e., a centered Gaussian process with independent increments and the covariance function $\langle \boldsymbol{W}_{t_1} \boldsymbol{W}_{t_2} \rangle = \min\{t_1, t_2\}$, where $\langle \ldots \rangle$ denotes the expectation (i.e., ensemble-average). In physical literature, Eq. (1.10) is often written in terms of the "velocity" of the particle, $\dot{\boldsymbol{X}}_t = d\boldsymbol{X}_t/dt$, as an overdamped Langevin equation [127]:

$$\dot{\boldsymbol{X}}_t = \boldsymbol{\mu}(\boldsymbol{X}_t, t) + \sigma(\boldsymbol{X}_t, t)\eta_t. \tag{1.11}$$

Here $\eta_t = d\boldsymbol{W}_t/dt$ is called Gaussian white noise, which has zero mean and no correlation: $\langle \eta_{t_1} \eta_{t_2} \rangle = \delta(t_1 - t_2)$, with $\delta(t)$ being the Dirac distribution. Since the time derivative $\dot{\boldsymbol{X}}_t$ of a stochastic process is generally ill-defined, Eq. (1.11) should be understood as a physically appealing form of Eq. (1.10). We also mention the units of different quantities: if $\boldsymbol{X}_t$ is the position of a particle and thus has units of meters (m), then $\boldsymbol{\mu}(\boldsymbol{x}, t)$ has units of velocity (m/s), $\boldsymbol{W}_t$ has units $\mathrm{s}^{1/2}$, and $\sigma(\boldsymbol{x}, t)$ has units $\mathrm{m}/\mathrm{s}^{1/2}$.

The drift and volatility functions allow one to describe a very broad class of stochastic processes and to cover many applications in physics and other disciplines. When there are no random fluctuations ($\sigma \equiv 0$), or their effect is negligible, one retrieves the framework of classical mechanics and ordinary differential equations. In this setting, the FPT is not random, and many challenges of the target problem become trivial (e.g., if the drift is constant, the FPT is simply equal to the distance to the target divided by the drift). This setting will be excluded from our discussion.

Basic examples of *Gaussian* stochastic processes are isotropic Brownian motion ($\boldsymbol{\mu} = 0$, $\sigma_{ij} = \sqrt{2D}\,\delta_{ij}$) with a constant diffusion coefficient $D$, and the Ornstein-Uhlenbeck process ($\boldsymbol{\mu} = -(k/\gamma)\boldsymbol{x}$, $\sigma_{ij} = \sqrt{2D}\,\delta_{ij}$), which corresponds to diffusion with the drag coefficient $\gamma$ in a harmonic potential $k|\boldsymbol{x}|^2/2$ (or, equivalently, with a Hookian restoring force $-k\boldsymbol{x}$) with stiffness $k$. In the former case, the mean-squared displacement (MSD) of the particle, started from the origin, $\boldsymbol{X}_0 = 0$, grows linearly with time,

$$\langle |\boldsymbol{X}_t|^2 \rangle = 2nDt, \tag{1.12}$$

which is a reminiscent feature of ordinary diffusion since Einstein's works. In turn, the harmonic potential effectively traps the particle near the origin, yielding an exponential saturation of the MSD at long times:

$$\langle |\boldsymbol{X}_t|^2 \rangle = n\frac{\gamma D}{k}\left(1 - e^{-2kt/\gamma}\right) \sim \begin{cases} 2nDt & (t \ll \gamma/k), \\ n\gamma D/k & (t \gg \gamma/k). \end{cases} \tag{1.13}$$

In general, however, nonlinear drift or a spatially dependent volatility matrix render the stochastic process $\boldsymbol{X}_t$ defined by Eq. (1.10) to be *non-Gaussian*, even though the random fluctuations $d\boldsymbol{W}_t$ are still Gaussian. Various anomalous features of spatially heterogeneous diffusions were investigated (see [128, 129] and references therein). For instance, one often observes the anomalous, power-law behavior, $\langle |\boldsymbol{X}_t|^2 \rangle \propto t^\alpha$, with the scaling exponent $\alpha$ distinguishing *subdiffusion* ($0 < \alpha < 1$), diffusion ($\alpha = 1$), and *superdiffusion* ($\alpha > 1$) [130, 131].

In most applications, it is sufficient to characterize the random position $\boldsymbol{X}_t$ via the propagator $G(\boldsymbol{x}, t|\boldsymbol{x}_0)$, i.e., *the probability density* of finding the stochastic process $\boldsymbol{X}_t$ in a vicinity of a point $\boldsymbol{x}$ at a given time $t$: $\mathbb{P}_{\boldsymbol{x}_0}\{\boldsymbol{X}_t \in (\boldsymbol{x}, \boldsymbol{x} + d\boldsymbol{x})\} = G(\boldsymbol{x}, t|\boldsymbol{x}_0)d\boldsymbol{x}$, with $\boldsymbol{x}_0$ being the starting point. The Markovian character of Eq. (1.10) and the Gaussian nature of fluctuations allow one to deduce the partial differential equation (PDE), known as the Fokker-Planck equation, for the propagator [132]

$$\partial_t G(\boldsymbol{x}, t|\boldsymbol{x}_0) = -\nabla \cdot \boldsymbol{J}(\boldsymbol{x}, t|\boldsymbol{x}_0) = \mathcal{L} G(\boldsymbol{x}, t|\boldsymbol{x}_0), \tag{1.14}$$

where

$$\mathcal{L} = -\sum_{i=1}^{n} \partial_{x_i} \boldsymbol{\mu}_i(\boldsymbol{x}, t) + \sum_{i,j=1}^{n} \partial_{x_i}\partial_{x_j} D_{ij}(\boldsymbol{x}, t) \tag{1.15}$$

is the Fokker-Planck operator, $D(\boldsymbol{x}, t) = \frac{1}{2}\sigma^\dagger\sigma$ is the matrix of diffusion coefficients, and $\boldsymbol{J}(\boldsymbol{x}, t|\boldsymbol{x}_0)$ is the probability density flux with components

$$J_i(\boldsymbol{x}, t|\boldsymbol{x}_0) = \boldsymbol{\mu}_i(\boldsymbol{x}, t)G(\boldsymbol{x}, t|\boldsymbol{x}_0) - \sum_{j=1}^{n} \partial_{x_j}\left(D_{ij}(\boldsymbol{x}, t)G(\boldsymbol{x}, t|\boldsymbol{x}_0)\right) \quad (i = 1, \ldots, n).$$

In other words, the Fokker-Planck equation incorporates the continuity equation ensuring the conservation of probability. The initial condition $G(\boldsymbol{x}, 0|\boldsymbol{x}_0) = \delta(\boldsymbol{x} - \boldsymbol{x}_0)$ determines the *fixed* starting point $\boldsymbol{x}_0$. In turn, if there is a prescribed initial concentration $c_0(\boldsymbol{x}_0)$ of particles, the propagator yields their concentration at time $t$:

$$c(\boldsymbol{x}, t) = \int d\boldsymbol{x}_0\, G(\boldsymbol{x}, t|\boldsymbol{x}_0)\, c_0(\boldsymbol{x}_0). \tag{1.16}$$

In the simplest case of isotropic Brownian motion ($\boldsymbol{\mu}(\boldsymbol{x},t) = 0$, $D_{ij}(\boldsymbol{x},t) = D\delta_{ij}$ with a constant $D$), the Fokker-Planck operator $\mathcal{L} = D\nabla^2 = D\Delta$ is proportional to the Laplace operator $\Delta = \partial^2_{x_1} + \ldots + \partial^2_{x_n}$, while Eq. (1.14) is reduced to the diffusion (or heat) equation,

$$\partial_t G(\boldsymbol{x},t|\boldsymbol{x}_0) = D\Delta G(\boldsymbol{x},t|\boldsymbol{x}_0), \tag{1.17}$$

whose solution is Gaussian,

$$G(\boldsymbol{x},t|\boldsymbol{x}_0) = \frac{\exp\left(-\frac{|\boldsymbol{x}-\boldsymbol{x}_0|^2}{4Dt}\right)}{(4\pi Dt)^{n/2}}, \tag{1.18}$$

as expected.

When the particle diffuses in a confining domain $\Omega$ (e.g., in a porous medium, in the cytoplasm of a living cell, in a chemical reactor), the restricting effect of impermeable walls should be taken into account in both stochastic and PDE descriptions. Let us briefly mention two common implementations of this effect, referred to as "soft" and "hard" walls.[5] A soft wall is an effective virtual confinement implemented by adding a long-range potential. A common choice is the aforementioned harmonic potential $k|\boldsymbol{x}|^2/2$, which confines the diffusing particle near the origin. A similar confining mechanism is used in optical tweezers that confine a dielectric particle in all three dimensions by a laser beam (for the effect of such confinement on first-passage times, see the review [133]). The long-time asymptotic value of the MSD from Eq. (1.13) determines the typical size $\sqrt{n\gamma D/k}$ of the induced confinement.

Despite the relative simplicity of the above description, the position of a soft wall is not sharp, while the dynamics of the particle is strongly affected by the artificially imposed harmonic potential. As a consequence, a more common setting is the hard wall, which affects the motion of particles only locally and prohibits their diffusion outside the confining domain. The restricting effect of a hard wall is usually incorporated *directly* via a reflecting boundary condition to the Fokker-Planck equation that imposes no probability density flux $\boldsymbol{J}(\boldsymbol{x},t|\boldsymbol{x}_0)$ in the normal direction to the wall $\partial\Omega$:

$$\left(\boldsymbol{n}_{\boldsymbol{x}} \cdot \boldsymbol{J}(\boldsymbol{x},t|\boldsymbol{x}_0)\right) = 0 \qquad (\boldsymbol{x} \in \partial\Omega), \tag{1.19}$$

where $\boldsymbol{n}_{\boldsymbol{x}}$ is the unit vector orthogonal to $\partial\Omega$ at each point $\boldsymbol{x}$. In turn, its implementation to the stochastic description was mostly ignored in physical literature. Indeed, one would need to include into Eq. (1.10) an infinitely short-ranged but infinitely strong force to repel the particle from the hard wall, without affecting its dynamics in the bulk. A rigorous implementation of such a repelling force is well established in mathematical literature [125, 134] and involves the so-called boundary local time (see SM I in [135] for more details). The underlying stochastic description has numerous advantages as discussed in Chap. 4.

[5] We stress that these mathematical notions have no relation to the physical softness of the wall.

A geometric confinement or the presence of impenetrable obstacles (e.g., molecular crowding) considerably alters the dynamics of diffusing particles and may lead to anomalous features such as subdiffusive scaling, non-Gaussian diffusion, various nonlinear effects due to interactions with the crowders, etc. [136–146]. Some of these anomalous features are discussed in Chap. 6. In turn, in biological systems one often deals with *partially-permeable walls* (e.g., plasma membranes or capillary walls) that can be crossed by a diffusing particle after a number of repetitive attempts. A mathematical description of these crossing events and their effect on the search process are discussed in Chap. 8.

#### 1.2.2.2 Mathematical Framework for Target Search Problems

Once the stochastic process $\boldsymbol{X}_t$ is defined, one can formulate various target search problems. Despite the vast variety of such problems raised in different disciplines (see Sects. 1.1.2 and 1.1.3), we focus on the archetypical problem of searching for a target $\Gamma$, which is either a subset of the confining domain $\Omega$, or a subset of its boundary $\partial\Omega$. For the sake of clarity, we focus on the most studied setting of a *perfect* target, which is found upon the first arrival on it. The first-passage time $\mathcal{T}$ to the target, defined via Eq. (1.4), is a random variable, whose realized value depends on the stochastic trajectory $\boldsymbol{X}_t$ followed by the diffusing particle. In other words, if the search process was repeated a number of times, $\mathcal{T}$ would take different (random) values. Its probability distribution is characterized by the cumulative distribution function, $\mathbb{P}_{\boldsymbol{x}_0}\{\mathcal{T} < t\}$ or, equivalently, by the above-mentioned survival probability

$$S(t|\boldsymbol{x}_0) = 1 - \mathbb{P}_{\boldsymbol{x}_0}\{\mathcal{T} < t\} = \mathbb{P}_{\boldsymbol{x}_0}\{\mathcal{T} > t\}, \tag{1.20}$$

which is here conditioned on the fixed starting point $\boldsymbol{x}_0$. We recall that the term "survival" originates from the common situation when the particle disappears or dies after hitting the target. From the mathematical point of view, however, it does not matter what in reality happens to the particle after its arrival onto the target, given that the search process is stopped at $\mathcal{T}$. The stopping condition plays the central role in the stochastic description of the search process.

In analogy to the Fokker-Planck equation (1.14) for the propagator, the survival probability satisfies the *backward* Fokker-Planck (or Kolmogorov) equation, subject to the so-called terminal condition [147–149]. To ease the presentation, we restrict the following discussion to stochastic processes that are *homogeneous in time* (i.e., the drift and volatility functions do not depend on time), so that the backward equation can be written in the forward form

$$\partial_t S(t|\boldsymbol{x}_0) = \mathcal{L}^\dagger S(t|\boldsymbol{x}_0) \qquad (\boldsymbol{x}_0 \in \Omega), \tag{1.21}$$

where $\mathcal{L}^\dagger$ is the adjoint Fokker-Planck operator (compare with Eq. (1.15)),

$$\mathcal{L}^\dagger = \sum_{i=1}^{n} \mu_i(\boldsymbol{x}_0)\partial_{x_{0,i}} + \sum_{i,j=1}^{n} D_{ij}(\boldsymbol{x}_0)\partial_{x_{0,i}}\partial_{x_{0,j}}, \tag{1.22}$$

acting on the starting point $\boldsymbol{x}_0$. The initial condition $S(0|\boldsymbol{x}_0) = 1$ simply states that the particle is surely present at the beginning ($t = 0$). If the confining domain $\Omega$ has an impenetrable boundary $\partial\Omega$, the above equation (1.21) should be completed by the *Neumann boundary condition*:

$$\partial_n S(t|\boldsymbol{x}_0) \equiv \big(\boldsymbol{n}_{\boldsymbol{x}_0} \cdot \nabla S(t|\boldsymbol{x}_0)\big) = 0 \qquad (\boldsymbol{x}_0 \in \partial\Omega), \tag{1.23}$$

which is compatible with Eq. (1.19) for the forward Fokker-Planck equation (here $\partial_n$ denotes the normal derivative). In turn, the stopping condition on the perfect target $\Gamma$ yields the *Dirichlet boundary condition*:

$$S(t|\boldsymbol{x}_0) = 0 \qquad (\boldsymbol{x}_0 \in \Gamma). \tag{1.24}$$

This condition states that if the particle started on the target, it finds the target instantly, so that $\mathcal{T} = 0$, and the probability of having $\mathcal{T} > t$ for any $t > 0$ is zero. From the "forward" point of view, the *absorbing* boundary condition (1.24) qualifies the target as a *sink* (or a perfect trap) for diffusing particles. Expectedly, the survival probability $S(t|\boldsymbol{x}_0)$ is therefore a monotonously decreasing function of time $t$. In this scenario, the solution of the Fokker-Planck equation, $G(\boldsymbol{x}, t|\boldsymbol{x}_0)$, is no longer normalized and thus not a probability density, strictly speaking.

The survival probability $S(t|\boldsymbol{x}_0)$ depends, in general, on the mechanism underlying the stochastic dynamics (represented by the operator $\mathcal{L}^\dagger$), the geometry of the confining domain $\Omega$, the shape and location of the target $\Gamma$, and the position of the starting point $\boldsymbol{x}_0$. The survival probability is known explicitly only for few simple cases [117, 142, 150–152], some of which were mentioned above. In general, one needs to rely either on numerical techniques (e.g., finite-element methods or Monte Carlo simulations), or on approximate and asymptotic tools. A considerable progress in optical single-molecule experiments opened a promising way to estimate the survival probability from the recorded trajectories of individual molecules [153, 154]. Moreover, as the survival probability of a single particle is generally proportional to the concentration of particles, it can also be observed in some macroscopic experiments such as fluorescence recovery after photobleaching [155–158] or diffusion-weighted nuclear magnetic resonance [159–161].

The time derivative of the survival probability, $H(t|\boldsymbol{x}_0) = -\partial_t S(t|\boldsymbol{x}_0)$, determines the probability density function (PDF) of the FPT: $\mathbb{P}_{\boldsymbol{x}_0}\{\mathcal{T} \in (t, t + dt)\} = H(t|\boldsymbol{x}_0)dt$. The FPT PDF contains the same information but provides a complementary insight onto the FPT $\mathcal{T}$ as a "histogram" of its values. Yet another equivalent representation is the moment-generating function,

$$\tilde{H}(p|\boldsymbol{x}_0) = \langle e^{-p\mathcal{T}} \rangle = \int\limits_0^\infty dt\, e^{-pt}\, H(t|\boldsymbol{x}_0), \tag{1.25}$$

which turns out to be the Laplace transform of the FPT PDF. Its derivatives with respect to the parameter $p$, evaluated at $p = 0$, yield the positive-integer moments of the FPT:

$$\langle \mathcal{T}^k \rangle = (-1)^k \lim_{p\to 0} \partial_p^k \tilde{H}(p|\boldsymbol{x}_0) \qquad (k = 1, 2, \ldots). \tag{1.26}$$

Moreover, the inverse Laplace transform allows one to reconstruct the FPT PDF $H(t|\boldsymbol{x}_0)$ from the moment-generating function. The latter can be obtained by solving the boundary value problem:

$$(p - \mathcal{L}^\dagger)\tilde{H}(p|\boldsymbol{x}_0) = 0 \quad (\boldsymbol{x}_0 \in \Omega), \qquad \begin{cases} \tilde{H}(p|\boldsymbol{x}_0) = 1 \quad (\boldsymbol{x}_0 \in \Gamma), \\ \partial_n \tilde{H}(p|\boldsymbol{x}_0) = 0 \quad (\boldsymbol{x}_0 \in \partial\Omega), \end{cases} \tag{1.27}$$

which is deduced from Eq. (1.21). As a consequence, one also gets the boundary value problems for positive-integer moments of the FPT. For instance, if the *mean FPT* exists, it satisfies

$$\mathcal{L}^\dagger \langle \mathcal{T} \rangle = -1 \quad (\boldsymbol{x}_0 \in \Omega), \qquad \begin{cases} \langle \mathcal{T} \rangle = 0 \quad (\boldsymbol{x}_0 \in \Gamma), \\ \partial_n \langle \mathcal{T} \rangle = 0 \quad (\boldsymbol{x}_0 \in \partial\Omega). \end{cases} \tag{1.28}$$

Many former studies were focused exclusively on the analysis of the mean FPT, partly due to the simpler character of the underlying mathematical problem. It is worth noting, however, that finding an exact analytical solution of this problem even for the mean FPT is still challenging due to the *mixed* Dirichlet-Neumann boundary condition [89, 162–168]. For this reason, many efficient approximate and asymptotic methods were developed, including matching asymptotic techniques (Chap. 10) and boundary homogenization (Chap. 11).

In many settings, the mean FPT is the only relevant timescale, so that its knowledge may be sufficient for a complete characterization and comprehension of the search process. For instance, this is often the case when the starting point $\boldsymbol{x}_0$ is uniformly distributed over the confinement. However, if the starting point is close to the target, or if the confining domain has irregular structure at multiple scales, the PDF of the FPT can exhibit various asymptotic regimes so that the whole distribution is necessary for understanding the search process (see Chap. 3). It is also the case if many particles search simultaneously for the target. If $N$ particles diffuse independently from each other, the search is terminated when the target is found by the *fastest* particle. In other words, we are interested in the minimum among $N$ independent FPTs $\mathcal{T}_1, \ldots, \mathcal{T}_N$, which has the following distribution:

$$\mathbb{P}_{x_0}\{\min\{\mathcal{T}_1, \ldots, \mathcal{T}_N\} > t\} = \prod_{j=1}^{N} \mathbb{P}_{x_0}\{\mathcal{T}_j > t\} = [S(t|\boldsymbol{x}_0)]^N. \tag{1.29}$$

Even if only the *mean* fastest FPT is needed, its computation and analysis require the knowledge of the survival probability $S(t|\boldsymbol{x}_0)$ for a single particle. Many properties of the fastest FPT were studied (see [169–172]), with a number of implications in biology (e.g., the redundancy principle) [173, 174]. Recent advances in this direction are presented in Chap. 12. Similar questions arise in the analysis of the inference of the FPT from a limited amount of empirical data (see Chap. 2). Even in the alternative setting of *sequential* search in the so-called molecular relay race, when the arrival of one searcher triggers the release of the next one, the relevant timescale is not the sum of the mean FPTs of intermediate steps, as one might intuitively expect (see further discussions in Chap. 26).

For the sake of clarity, we restricted the above discussion to perfect targets which are found upon the first arrival. In most applications, however, this idealized setting is not sufficient. For instance, the diffusing particle and/or the target should be in active states (e.g., an open ion channel, a proper conformation of a protein, etc.) to trigger an event of interest (e.g., a chemical reaction, a vesicle release, etc.); a molecule has to overcome an energy activation barrier to react on a substrate or with an enzyme; a particle searching to escape needs to overcome an entropic barrier when moving through a small hole; more generally, the target may need to be recognized (see, e.g., [44]). The completion of such a particle-target reaction event is not granted, and if it fails, the particle resumes its stochastic motion and restarts the search process. In this more general scenario of a *partially (re)active* target, the successful reaction is usually preceded by a sequence of failed reaction attempts, intervened by diffusive explorations of the confining domain. In other words, the first arrival onto the target is just the very first step of the search process. The effect of partial reactivity of the target has been thoroughly investigated (see [14, 15] and references therein). In the stochastic description, the first-arrival stopping condition $\boldsymbol{X}_t \in \Gamma$ should be replaced by another condition that accounts for the number of failed reaction attempts (see Chap. 4 for the related encounter-based approach). In the PDE description, the Dirichlet boundary condition (1.24) is usually replaced by the *Robin boundary condition* (also known as Fourier or radiation boundary condition)

$$\partial_n S(t|\boldsymbol{x}_0) + qS(t|\boldsymbol{x}_0) = 0 \qquad (\boldsymbol{x}_0 \in \Gamma), \tag{1.30}$$

where the parameter $q = \kappa/D$ accounts for the partial reactivity $0 \le \kappa \le +\infty$ and may range from 0 for passive targets to $+\infty$ for perfectly reactive targets [175]. Note that target search problems with more sophisticated particle-target reaction events can be analyzed within the encounter-based approach [135, 176]. One can also consider temporal binding or adsorption of particles to the boundary [177, 178], as well as intermittent dynamics with alternating phases of bulk and surface diffusion (see [179] and references therein).

### 1.2.3 Further Generalizations

The above description of the diffusive dynamics via the stochastic equation (1.10) or the Fokker-Planck equation (1.14) is indeed quite general. In the most basic case, one considers a single massless particle in the $n$-dimensional space $\mathbb{R}^n$ whose interactions with the environment are represented by given functions $\boldsymbol{\mu}(\boldsymbol{x}, t)$ and $D(\boldsymbol{x}, t)$ (or $\sigma(\boldsymbol{x}, t)$). For instance, one can consider heterogeneous diffusion processes with, e.g., $D(\boldsymbol{x}) \propto |\boldsymbol{x}|^\alpha$ [128, 129] or scaled Brownian motion with $D(t) \propto t^{\alpha-1}$ [180]. The Fokker-Planck equation may also be coupled to the Poisson equation for the electric voltage in order to model electro-diffusion of charged ions that diffuse, interact and generate an electric field (Chap. 18). Moreover, one can even describe the dynamics of a massive particle by representing Newton's second equation $m\,\ddot{\boldsymbol{x}}_t = F(\boldsymbol{x}_t, t)$ as the system of two equations: $\boldsymbol{v}_t = \dot{\boldsymbol{x}}_t$ and $m\,\dot{\boldsymbol{x}}_t = F(\boldsymbol{x}_t, t)$ and then combining the position $\boldsymbol{x}_t$ and velocity $\boldsymbol{v}_t$ as components of a higher-dimensional vector $\boldsymbol{X}_t$ (here the force $F(\boldsymbol{x}, t)$ includes both deterministic and stochastic contributions for brevity). In this way, one can study the stochastic dynamics of active particles such as Janus particles, swimming bacteria with propulsive motion, or fishes, birds and other animals. Search processes with such a kind of motion are amply discussed in Chap. 19, which also presents some other examples of active random motion. Similarly, this framework is suitable for dealing with $N$ particles that interact via a prescribed potential (e.g., Coulombic interactions between charged particles), or the statistics of encounters between the particle and moving targets [46, 47, 181, 182]. Moreover, one can even study the deterministic pursuit of randomly moving targets (Chap. 20).

Despite this generality, many further extensions to the stochastic equation (1.10) and its PDE counterpart were proposed; here we just mention some of them.

(i) both drift and volatility can be treated as random fields in order to investigate the effect of *disordered or random environments* [43]. In this case, the average over random realizations of $\boldsymbol{\mu}$ or/and $\sigma$ bring new features into the problem and widen the class of described phenomena. For instance, one often considers diffusion in a random energy landscape that allows for temporal trapping of the particle in local minima (see the review [183] and references therein); such spatially heterogeneous landscapes appear for electron diffusion in amorphous semiconductors or can be realized for colloidal particles by laser speckles [184, 185]. Alternatively, a diffusion coefficient varying as a function of time can be assigned to be a stochastic process, resulting in a class of diffusing diffusivity models [186–190]. Such random fluctuations of $D(t)$ can account for conformational changes of a diffusing polymer, the effect of dynamic heterogeneous media rapidly evolving in time, assembling-disassembling kinetics, etc. (see Chap. 13). Yet another mechanism is random switching between a discrete set of diffusion coefficients due to binding-unbinding of the diffusing particle to mobile or immobile buffer molecules [23, 191].

(ii) In the above description, the particle could search for the target arbitrarily long, as if the particle was "immortal". In many applications, however, the search process is limited by the particle lifetime or other temporal constraints. For instance, a radioactive nucleus can disintegrate, an excited state of a spin or a fluorophore can be spontaneously relaxed, a protein can be disassembled, an active particle can be trapped or transformed into a passive form, a bacterium or an animal can be killed or die on starvation, etc. Whatever the microscopic origin of these mechanisms, the diffusing particle may spontaneously become disabled to find the target, and thus to fail its search. Such a particle disappears from the target search. In the simplest setting, a spontaneous disappearance of the particle is characterized by its random, exponentially distributed lifetime $\tau$, with a prescribed mean lifetime $\langle\tau\rangle$ or by the decay rate $\gamma = 1/\langle\tau\rangle$. In the stochastic description, it simply means that the stochastic process $\boldsymbol{X}_t$ is stopped at $\tau$, and one may wonder whether the search is completed before $\tau$ or not. Equivalently, such a first-order kinetics can be included by adding the reaction term $-\gamma G(\boldsymbol{x}, t|\boldsymbol{x}_0)$ to the right-hand side of the Fokker-Planck equation (1.14). The effect of *mortality* onto the search efficiency was thoroughly investigated (see [192–194] and Chap. 7). More sophisticated reaction mechanisms can also be explored. A totally different scenario is a *random resetting* of the particle to its initial (or another) position, from which diffusion restarts anew [114, 115, 195–198]. In the context of target search problems, resetting helps to terminate pointless explorations far from the target and thus potentially improves the search efficiency. Different resetting protocols and their impact onto the target problem were investigated (see Chaps. 14 and 15 for more details).

(iii) A *subordination* technique, a sort of stochastic time change, is often employed to incorporate subtle temporal features into the stochastic dynamics; for instance, a very popular model of anomalous diffusion with a heavy-tailed distribution of waiting times [199–201] can be directly introduced by means of this technique. In the original microscopic model of continuous-time random walks (CTRW), a particle executes a sequence of random jumps, after each jump pausing for a random time drawn from a prescribed distribution. When the variance of the jump and the mean sojourn time are finite, such a CTRW converges to Brownian motion in the macroscopic limit, as standard lattice random walks do. However, if the distribution of sojourn times is heavy-tailed, anomalously long pausing periods make the mean sojourn time infinite. In this anomalous setting, the Fokker-Planck (or diffusion) equation becomes fractional and thus nonlocal in time, whereas the stochastic dynamics becomes non-Markovian. Such CTRWs exhibit various anomalous features such as subdiffusive scaling and *aging* [131]. For instance, a single jump may provide the dominant contribution to the target search efficiency (the so-called *big jump principle*, see Chap. 9). Another application of the subordination technique was demonstrated for diffusing-diffusivity dynamics in [187–189].

(iv) One can also relax the assumption of the uncorrelated (white) noise. As a linear combination of Gaussian variables is still a Gaussian variable, one can construct various Gaussian but correlated noises and then build more elaborate stochastic processes. An archetypical example is the fractional Brownian motion (fBm) [42, 202, 203] (see also [204, 205] for more details), i.e., a centered Gaussian process with the covariance

$$\langle B_H(t_1) B_H(t_2)\rangle = D_H\big(t_1^{2H} + t_2^{2H} - |t_1 - t_2|^{2H}\big), \tag{1.31}$$

where $D_H$ is the generalized diffusion coefficient (with units $m^2/s^{2H}$), and $0 < H < 1$ is the Hurst exponent (here we consider only fBm in one dimension). Since the MSD of $B_H(t)$ is simply $2D_H t^{2H}$, the fBm can be subdiffusive or superdiffusive for $H < 1/2$ and $H > 1/2$, respectively, while one-dimensional Brownian motion is retrieved for $H = 1/2$. Note that fBm can be constructed as a weighted combination of white noises [206, 207]. More generally, one can use colored noises to generate a broad class of Gaussian non-Markovian processes with temporal correlations from the generalized or fractional Langevin equation based on memory kernels [208–213]. (see also [214–216] for other examples). By construction, these stochastic processes are non-Markovian that prohibits using PDE description such as the Fokker-Planck equation. As a consequence, the analysis of target search problems turns out to be much more sophisticated [217, 218]. Despite these challenges, some progress in this setting is presented in Chaps. 16 and 17.

(v) Finally, Gaussian fluctuations $d\boldsymbol{W}_t$ that drive the stochastic dynamics in Eq. (1.10) can be replaced by more general Lévy $\alpha$-stable noises in order to model large displacements with infinite variance. The resulting Lévy processes have found numerous applications in physics and finances, including efficient search strategies [66, 67]. The underlying mathematical description is more elaborate. For instance, the Laplace operator, which represented local displacements due to Gaussian fluctuations, is now replaced by fractional spatial derivatives, i.e., by spatially nonlocal operators. The consequences for search strategies are discussed in Chap. 29.

In summary, this book covers very distinct aspects and applications of target search problems and presents its current state of the art. Despite the impressive progress over the past twenty years, many open questions and unsolved problems wait to be challenged. We hope that this book will stimulate further developments in this active field and build new bridges and collaborations across multiple disciplines studying search processes.

## References

1. M. Smoluchowski, Versuch einer mathematischen Theorie der Koagulationskinetik kolloider Lösungen. Z. Phys. Chem. **92U**, 129–168 (1918)
2. E. Gudowska-Nowak, K. Lindenberg, R. Metzler, Preface: Marian Smoluchowski's 1916 paper - a century of inspiration. J. Phys. A: Math. Theor. **50**, 380301 (2017)
3. P. Hänggi, P. Talkner, M. Borkovec, Reaction-rate theory: fifty years after Kramers. Rev. Mod. Phys. **62**, 251 (1990)
4. F.C. Collins, G.E. Kimball, Diffusion-controlled reaction rates. J. Colloid Sci. **4**, 425 (1949)
5. A. Szabo, K. Schulten, Z. Schulten, First passage time approach to diffusion controlled reactions. J. Chem. Phys. **72**, 4350 (1980)
6. D. Shoup, G. Lipari and A. Szabo, Diffusion-controlled bimolecular reaction rates. The effect of rotational diffusion and orientation constraints. Biophys. J. **36**, 697 (1981)
7. D.F. Calef, J.M. Deutch, Diffusion-controlled reactions. Annu. Rev. Phys. Chem. **34**, 493 (1983)
8. R. Zwanzig, A. Szabo, Time dependent rate of diffusion-influenced ligand binding to receptors on cell surfaces. Biophys. J . **60**, 671 (1991)
9. G.H. Weiss, Overview of theoretical models for reaction rates. J. Stat. Phys. **42**, 3 (1986)

10. G. Oshanin, M. Moreau, S. Burlatsky, Models of chemical reactions with participation of polymers. Adv. Colloid Interface Sci. **49**, 1 (1994)
11. G. Oshanin, A. Stemmer, S. Luding, A. Blumen, Smoluchowski approach for three-body reactions in one dimension. Phys. Rev. E **52**, 5800 (1995)
12. G. Oshanin, A. Blumen, Kinetic description of diffusion-limited reactions in random catalytic media. J. Chem. Phys. **108**, 1140 (1998)
13. O.A. Vasilyev, L. Lizana, G. Oshanin, Smoluchowski rate for diffusion-controlled reactions of molecules with antenna. J. Phys. A: Math. and Theor. **50**, 264004 (2017)
14. K. Lindenberg, R. Metzler, G. Oshanin (eds.), *Chemical Kinetics: Beyond the Textbook* (World Scientific Publishers Europe, London, 2019)
15. D.S. Grebenkov, Diffusion-controlled reactions: an overview. Molecules **28**, 7570 (2023)
16. A.A. Ovchinnikov, Ya.. B. Zeldovich, Role of density fluctuations in bimolecular reaction kinetics. Chem. Phys. **28**, 215 (1978)
17. S.F. Burlatskii, The effect of the nonuniform distribution of reactants on the kinetics of bimolecular recombination. Theor. Exp. Chem. **14**, 373 (1979). [Translated from Teoreticheskaya i eksperimentalnaya khimiya, Vol. 14, No. 4, pp. 483-487, July-August 1978]
18. D. Toussaint, F. Wilczek, Particle-antiparticle annihilation in diffusive motion. J. Chem. Phys. **78**, 2642 (1983)
19. K. Kang, S. Redner, Fluctuation-dominated kinetics in diffusion-controlled reactions. Phys. Rev. A **32**, 435 (1985)
20. G.S. Oshanin, S.F. Burlatsky, Fluctuation-induced kinetics of reversible coagulation. J. Phys. A: Math. and Gen. **22**, L973 (1989)
21. G.S. Oshanin, A.A. Ovchinnikov, S.F. Burlatsky, Fluctuation-induced kinetics of reversible reactions. J. Phys. A: Math. and Gen. **22**, L977 (1989)
22. R. Voituriez, M. Moreau, G. Oshanin, Corrections to the law of mass action and properties of the asymptotic $t = \infty$ state for reversible diffusion-limited reactions. J. Chem. Phys. **122**, 084103 (2005)
23. M. Reva, D.A. DiGregorio, D.S. Grebenkov, A first-passage approach to diffusion-influenced reversible binding: insights into nanoscale signaling at the presynapse. Sci. Rep. **11**, 5377 (2021)
24. D.S. Grebenkov, A. Kumar, Reversible target-binding kinetics of multiple impatient particles. J. Chem. Phys. **156**, 084107 (2022)
25. Y. Scher, S. Reuveni, D.S. Grebenkov, Escape of a sticky particle. Phys. Rev. Res **5**, 043196 (2023)
26. L. Gálfi, Z. Rácz, Properties of the reaction front in an $A+B \to C$ type reaction-diffusion process. Phys. Rev. A **38**, 3151(R) (1988)
27. R. Goodman, D. Graff, L.M. Sander, P. Leroux-Hugon, E. Clement, Trigger waves in a model for catalysis. Phys. Rev. E **52**, 5904 (1995)
28. J. Mai, I.M. Sokolov, A. Blumen, Front propagation and local ordering in one-dimensional irreversible autocatalytic reactions. Phys. Rev. Lett. **77**, 4462 (1996)
29. J.D. Murray, *Mathematical Biology* (Springer, Berlin, 1993)
30. A. Blumen, G. Zumofen, J. Klafter, Target annihilation by random walkers. Phys. Rev. B **30**, 5379(R) (1984)
31. M. Tachiya, Theory of diffusion-controlled reactions: formulation of the bulk reaction rate in terms of the pair probability. Radiat. Phys. Chem. **21**, 167 (1983)
32. S. Redner, K. Kang, Kinetics of the 'scavenger' reaction. J. Phys. A: Math. Gen. **17**, L451 (1984)
33. S.F. Burlatskii, A.A. Ovchinnikov, Effect of reactant-fluctuation density on the kinetics of recombination, multiplication, and trapping processes. Sov. Phys. JETP **65**, 908 (1987)
34. B. Va. Balagurov, V.G. Vaks, Random walks of a particle on lattices with traps. Sov. Phys. JETP **38**, 968 (1974)
35. M.D. Donsker, S.R.S. Varadhan, Asymptotics of the wiener sausage. Commun. Pure Appl. Math. **28**, 525 (1975)

36. M.D. Donsker, S.R.S. Varadhan, On the number of distinct sites visited by a random walk. Commun. Pure Appl. Math. **32**, 721 (1979)
37. Th.M. Nieuwenhuizen, Trapping and Lifshitz tails in random media, self-attracting polymers, and the number of distinct sites visited: a renormalized instanton approach in three dimensions. Phys. Rev. Lett. **62**, 357 (1989)
38. R.F. Kayser, J.B. Hubbard, Diffusion in a medium with a random distribution of static traps. Phys. Rev. Lett. **51**, 79 (1983)
39. S.F. Burlatsky, G.S. Oshanin, Diffusion-controlled reactions with polymers. Phys. Lett. A **145**, 61 (1990)
40. I.M. Lifshitz, S.A. Gredescul, L.A. Pastur, *An Introduction to the Theory of Disordered Systems* (Wiley, New York, 1988)
41. P.G. de Gennes, Kinetics of diffusion-controlled processes in dense polymer systems. J. Chem. Phys. **76**, 3316 (1982)
42. A.N. Kolmogorov, Wienersche Spiralen und einige andere interessante Kurven im Hilbertschen Raum. C. R. (Doklady) Acad. Sci. URSS (N. S.) **26**, 115 (1940)
43. B.D. Hughes, *Random Walks and Random Environments*, Volume 1 (Clarendon Press, 1995)
44. O. Bénichou, M. Moreau, G. Oshanin, Kinetics of stochastically gated diffusion-limited reactions and geometry of random walk trajectories. Phys. Rev. E **61**, 3388 (2000)
45. M. Bramson, J.L. Lebowitz, Asymptotic behavior of densities in diffusion-dominated annihilation reactions. Phys. Rev. Lett. **61**, 2397 (1988)
46. F. Le Vot, S.B. Yuste, E. Abad, D.S. Grebenkov, First-encounter time of two diffusing particles in confinement. Phys. Rev. E **102**, 032118 (2020)
47. F. Le Vot, S.B. Yuste, E. Abad, D.S. Grebenkov, First-encounter time of two diffusing particles in two- and three-dimensional confinement. Phys. Rev. E **105**, 044119 (2022)
48. A.J. Bray, R.A. Blythe, Exact asymptotics for one-dimensional diffusion with mobile traps. Phys. Rev. Lett. **89**, 150601 (2002)
49. A.J. Bray, R.A. Blythe, Survival probability of a diffusing particle in the presence of Poisson-distributed mobile traps. Phys. Rev. E **67**, 041101 (2003)
50. A.J. Bray, S.N. Majumdar, R.A. Blythe, Formal solution of a class of reaction-diffusion models: Reduction to a single-particle problem. Phys. Rev. E **67**, 060102 (2003)
51. G. Oshanin, O. Bénichou, M. Coppey, M. Moreau, Trapping reactions with randomly moving traps: exact asymptotic results for compact exploration. Phys. Rev. E **66**, 060101R (2002)
52. M. Moreau, G. Oshanin, O. Bénichou, M. Coppey, Lattice theory of trapping reactions with mobile species. Phys. Rev. E **69**, 046101 (2004)
53. S.B. Yuste, G. Oshanin, K. Lindenberg, O. Bénichou, J. Klafter, Survival probability of a particle in a sea of mobile traps: a tale of tails. Phys. Rev. E **78**, 021105 (2008)
54. D.P. Snustad, M.J. Simmons, *Principles of Genetics* (Wiley, New York, 2015)
55. R.A. Bradshaw, E.A. Dennis (eds.), *Handbook of Cell Signaling* (Academic, New York, 2010)
56. H.C. Berg, E.M. Purcell, Physics of chemoreception. Biophys. J. **20**, 193 (1977)
57. S.B. Yuste, J.J. Ruiz-Lorenzo, K. Lindenberg, The target problem with evanescent subdiffusive traps. Phys. Rev. E **74**, 046119 (2006)
58. C. Guerrier, D. Holcman, Brownian search for targets hidden in cusp-like pockets: progress and applications. Eur. Phys. J. Special Topics **223**, 3273 (2014)
59. M. Vergassola, E. Villermaux, B.I. Shraiman, "Infotaxis" as a strategy for searching without gradients. Nature **445**, 406 (2007)
60. A. Sengupta, T. Kruppa, H. Loewen, Chemotactic predator-prey dynamics. Phys. Rev. E **81**, 031914 (2011)
61. G. Oshanin, O. Vasilyev, P.L. Krapivsky, J. Klafter, Survival of an evasive prey. Proc. Natl. Acad. Sci. U.S.A. **106**, 13696 (2009)
62. J.K. Parrish, L. Edelstein-Keshet, Complexity, pattern, and evolutionary trade-offs in animal aggregation. Science **284**, 99 (1999)
63. I.D. Couzin, Collective cognition in animal groups. Trends Cognit. Sci. **13**, 35 (2009)
64. F. Bartumeus, M.G.E. da Luz, G.M. Viswanathan, J. Catalan, Animal search strategies: a quantitative random-walk analysis. Ecology **86**, 3078 (2005)

65. L.C.M. Salvador, F. Bartumeus, S.A. Levin, W.S. Ryu, Mechanistic analysis of the search behaviour of Caenorhabditis elegans. J. R. Soc. Interface **11**, 20131092 (2014)
66. G.M. Viswanathan, V. Afanasyev, S.V. Buldyrev, E.J. Murphy, P.A. Prince, H.E. Stanley, Lévy flight search patterns of wandering albatrosses. Nature **381**, 413–415 (1996)
67. G.M. Viswanathan, S.V. Buldyrev, S. Havlin, M.G.E. da Luz, E.P. Raposok, H.E. Stanley, Optimizing the success of random searches. Nature **401**, 911 (1999)
68. R.P.D. Atkinson, C.J. Rhodes, D.W. Macdonald, R.M. Anderson, Scale-free dynamics in the movement patterns of jackals. Oikos **98**, 134 (2002)
69. G. Ramos-Fernández, J.L. Mateos, O. Miramontes, H. Larralde, G. Cocho, B. Ayala-Orozco, Lévy walk patterns in the foraging movements of spider monkeys (Ateles geoffroyi). Behav. Ecol. Sociobiol. **55**, 223 (2004)
70. S.E. Halford, J.F. Marko, How do site-specific DNA-binding proteins find their targets? Nucleic Acids Res. **32**, 3040 (2004)
71. M. Sheinman, O. Bénichou, Y. Kafri, R. Voituriez, Classes of fast and specific search mechanisms for proteins on DNA. Rep. Prog. Phys. **75**, 026601 (2012)
72. A.M. Edwards et al., Revisiting Lévy flight search patterns of wandering albatrosses, bumblebees and deer. Nature **449**, 1044–1048 (2007)
73. N.E. Humphries, H. Weimerskirch, N. Queiroz, E.J. Southall, D.W. Sims, Foraging success of biological Lévy flights recorded in situ. Proc. Natl. Acad. Sci. U.S.A. **109**, 7169–7174 (2012)
74. N. Levernier, J. Textor, O. Bénichou, R. Voituriez, Inverse square Lévy walks are not optimal search strategies for $d \geq 2$. Phys. Rev. Lett. **124**, 080601 (2020)
75. J. Elf, G.-W. Li, X.S. Xie, Probing transcription factor dynamics at the single-molecule level in a living cell. Science **316**, 1191–1194 (2007)
76. M.A. Lomholt, B. van den Broek, S-M.J. Kalisch, G.J.L. Wuite R. Metzler, Facilitated diffusion with DNA coiling. Proc. Natl Acad. Sci. USA **106**, 8204 (2009)
77. O. Pulkkinen, R. Metzler, Distance matters: the impact of gene proximity in bacterial gene regulation. Phys. Rev. Lett. **110**, 198101 (2013)
78. O.G. Berg, R.B. Winter, P.H. von Hippel, Diffusion-driven mechanisms of protein translocation on nucleic acids: 1. Models and theory, Biochem. **20**, 6929 (1981)
79. P.H. von Hippel, O.G. Berg, Facilitated target location in biological systems. J. Biol. Chem. **264**, 675 (1989)
80. M. Coppey, O. Bénichou, R. Voituriez, M. Moreau, Kinetics of target site localization of a protein on DNA: a stochastic approach. Biophys. J. **87**, 1640 (2004)
81. M.A. Lomholt, T. Ambjornsson, R. Metzler, Optimal target search on a fast-folding polymer chain with volume exchange. Phys. Rev. Lett. **95**, 260603 (2005)
82. T. Hu, A. Yu, Grosberg, B.I. Shklovskii, How proteins search for their specific sites on DNA: the role of DNA conformation. Biophys. J. **90**, 2731 (2006)
83. T. Hu, B.I. Shklovskii, How does a protein search for the specific site on DNA: the role of disorder. Phys. Rev. E **74**, 021903 (2006)
84. L. Mirny, M. Slutsky, Z. Wunderlich, A. Tafvizi, J. Leith, A. Kosmrlj, How a protein searches for its site on DNA: the mechanism of facilitated diffusion. J. Phys. A: Math. Theor. **42**, 434013 (2009)
85. C. Loverdo, O. Bénichou, R. Voituriez, A. Biebricher, I. Bonnet, P. Desbiolles, Quantifying hopping and jumping in facilitated diffusion of DNA-binding proteins. Phys. Rev. Lett. **102**, 188101 (2009)
86. M.G.E. da Luz, A. Yu. Grosberg, E.P. Raposo G.M. Viswanathan, eds., The random search problem: trends and perspectives. J. Phys. A: Math. Theor. **42**, 430301 (2009)
87. G. Adam, M. Delbrück, Reduction of dimensionality in biological diffusion processes, in *Structural Chemistry and Molecular Biology* (Freeman, San Francisco, 1968)
88. D.S. Grebenkov, R. Metzler, G. Oshanin, Search efficiency in the Adam-Delbrück reduction-of-dimensionality scenario versus direct diffusive search. New J. Phys. **24**, 083035 (2022)
89. D.S. Grebenkov, G. Oshanin, Diffusive escape through a narrow opening: new insights into a classic problem. Phys. Chem. Chem. Phys. **19**, 2723 (2017)

90. Y. Chen, T. Kolokolnikov, A minimal model of predator-swarm interactions. J. R. Soc. Interface **11**, 20131208 (2014)
91. R.S. Olson, A. Hintze, F.C. Dyer, D.B. Knoester, C. Adami, Predator confusion is sufficient to evolve swarming behaviour. J. R. Soc. Interface **10**, 20130305 (2013)
92. D. Campos, E. Abad, V. Mendez, S.B. Yuste, K. Lindenberg, Optimal search strategies of space-time coupled random walkers with finite lifetimes. Phys. Rev. E **91**, 052115 (2015)
93. J. Wang, W. Li, Motion patterns and phase-transition of a defender-intruder problem and optimal interception strategy of the defender. Commun. Nonlinear Sci. Numer. Simul. **27**, 294 (2015)
94. M. Sato, Chasing and escaping by three groups of species. Phys. Rev. E **85**, 066102 (2012)
95. A. Gabel, S.N. Majumdar, N.K. Panduranga, S. Redner, Can a lamb reach a haven before being eaten by diffusing lions? J. Stat. Mech. P05011 (2012)
96. R. Martínez-García, J.M. Calabrese, T. Mueller, K. Olson, C. López, Optimizing the search for resources by sharing information: Mongolian gazelles as a case study. Phys. Rev. Lett. **110**, 248106 (2013)
97. S. Yang, S. Jiang, L. Jiang, G. Li, Z. Han, Aggregation increases prey survival time in group chase and escape. New J. Phys. **16**, 083006 (2014)
98. A. Kamimura, T. Ohira, Group chase and escape. New J. Phys. **12**, 053013 (2010)
99. T. Iwama, M. Sato, Group chase and escape with some fast chasers. Phys. Rev. E **86**, 067102 (2012)
100. L. Angelani, Collective predation and escape strategies. Phys. Rev. Lett. **109**, 118104 (2012)
101. R. Nishi, A. Kamimura, K. Nishinari, T. Ohira, Group chase and escape with conversion from targets to chasers. Phys. A **391**, 337 (2012)
102. A. Ramanantoanina, C. Hui, A. Ouhinou, Effects of density-dependent dispersal behaviours on the speed and spatial patterns of range expansion in predator-prey metapopulations. Ecol. Model. **222**, 3524 (2011)
103. T. Vicsek, Statistical physics: closing in on evaders. Nature **466**, 43 (2010)
104. M. Schwarzl, A. Godec, G. Oshanin, R. Metzler, A single predator charging a herd of prey: effects of self volume and predator-prey decision-making. J. Phys. A: Math. Theor. **49**, 225601 (2016)
105. M. Moreau, O. Bénichou, G. Oshanin, R. Voituriez, The shadow principle: An optimal survival strategy for a prey chased by random predators. Phys. A **392**, 2837 (2013)
106. O. Bénichou, M. Coppey, M. Moreau, P.-H. Suet, R. Voituriez, Optimal search strategies for hidden targets. Phys. Rev. Lett. **94**, 198101 (2005)
107. O. Bénichou, C. Loverdo, M. Moreau, R. Voituriez, Intermittent search strategies. Rev. Mod. Phys. **83**, 81 (2011)
108. G. Oshanin, H.S. Wio, K. Lindenberg, S.F. Burlatsky, Intermittent random walks for an optimal search strategy: one-dimensional case. J. Phys.: Condens. Matter **19**, 065142 (2007)
109. G. Oshanin, K. Lindenberg, H.S. Wio, S.F. Burlatsky, Efficient search by optimized intermittent random walks. J. Phys. A: Math. Theor. **42**, 434008 (2009)
110. F. Rojo, J. Revelli, C.E. Budde, H.S. Wio, G. Oshanin, K. Lindenberg, Intermittent search strategies revisited: effect of the jump length and biased motion. J. Phys. A: Math. Theor. **43**, 345001 (2010)
111. M.A. Lomholt, K. Tal, R. Metzler, J. Klafter, Lévy strategies in intermittent search processes are advantageous. Proc. Natl. Acad. Sci. U.S.A. **105**, 11055 (2008)
112. V.V. Palyulin, A.V. Chechkin, R. Metzler, Lévy flights do not always optimize random blind search for sparse targets. Proc. Natl. Acad. Sci. U.S.A. **111**, 2931 (2014)
113. V.V. Palyulin, A.V. Chechkin, R. Klages, R. Metzler, Search reliability and search efficiency of combined Lévy-Brownian motion: long relocations mingled with thorough local exploration. J. Phys. A: Math. Theor. **49**, 394002 (2016)
114. M.R. Evans, S.N. Majumdar, Diffusion with Stochastic Resetting. Phys. Rev. Lett. **106**, 160601 (2011)
115. M.R. Evans, S.N. Majumdar, G. Schehr, Stochastic resetting and applications. J. Phys. A: Math. Theor. **53**, 193001 (2020)

116. S. Gupta, A.M. Jayannavar, Stochastic resetting: a (very) brief review. Front. Phys. **10**, 789097 (2022)
117. S. Redner, *A Guide to First Passage Processes*. (Cambridge University Press, 2001)
118. R. Metzler, G. Oshanin, S. Redner (eds.), *First-Passage Phenomena and Their Applications* (World Scientific, Singapore, 2014)
119. J. Masoliver, *Random Processes: First-passage And Escape*. (World Scientific Publishing, 2018)
120. W. Feller, *An Introduction to Probability Theory and Its Applications* (Wiley, New York, 1968)
121. F. Spitzer, *Principles of Random Walk* (Springer, New York, 1976)
122. D.S. Grebenkov, L. Tupikina, Heterogeneous continuous-time random walks. Phys. Rev. E **97**, 012148 (2018)
123. J.W. Haus, K.W. Kehr, Diffusion in regular and disordered lattices. Phys. Rep. **150**, 263 (1987)
124. L. Giuggioli, Exact spatiotemporal dynamics of confined lattice random walks in arbitrary dimensions: a century after Smoluchowski and Pólya. Phys. Rev. X **10**, 021045 (2020)
125. K. Itô, H.P. McKean, *Diffusion Processes and Their Sample Paths* (Springer, Berlin, 1965)
126. P. Mörters, Y. Peres, *Brownian Motion*. Cambridge Series in Statistical and Probabilistic Mathematics. (Cambridge University Press, New York, 2010)
127. W.T. Coffey, Y.P. Kalmykov, J.T. Waldron, *The Langevin Equation: With Applications to Stochastic Problems in Physics, Chemistry and Electrical Engineering*, 2nd edn. (World Scientific, Singapore, 2004)
128. A.G. Cherstvy, A.V. Chechkin, R. Metzler, Anomalous diffusion and ergodicity breaking in heterogeneous diffusion processes. New J. Phys. **15**, 083039 (2013)
129. A.G. Cherstvy, R. Metzler, Ergodicity breaking and particle spreading in noisy heterogeneous diffusion processes. J. Chem. Phys. **142**, 144105 (2015)
130. J. Klafter, I.M. Sokolov, *First Steps in Random Walks: From Tools to Applications* (Oxford University Press, Oxford, 2011)
131. R. Metzler, J.-H. Jeon, A.G. Cherstvy, E. Barkai, Anomalous diffusion models and their properties: non-stationarity, non-ergodicity, and ageing at the centenary of single particle tracking. Phys. Chem. Chem. Phys. **16**, 24128 (2014)
132. H. Risken, *The Fokker-Planck Equation: Methods of Solution and Applications*, 3rd edn. (Springer, Berlin, 1996)
133. D.S. Grebenkov, First exit times of harmonically trapped particles: a didactic review. J. Phys. A: Math. Theor. **48**, 013001 (2015)
134. M. Freidlin, *Functional Integration and Partial Differential Equations* (Princeton University Press, Annals of Mathematics Studies, Princeton, NJ, 1985)
135. D.S. Grebenkov, Paradigm shift in diffusion-mediated surface phenomena. Phys. Rev. Lett. **125**, 078102 (2020)
136. D. ben-Avraham, S. Havlin, *Diffusion and Reactions in Fractals and Disordered Systems*. (Cambridge University Press, 2000)
137. P.E. Levitz, D.S. Grebenkov, M. Zinsmeister, K. Kolwankar, B. Sapoval, Brownian flights over a fractal nest and first-passage statistics on irregular surfaces. Phys. Rev. Lett. **96**, 180601 (2006)
138. S. Condamin, O. Bénichou, V. Tejedor, R. Voituriez, J. Klafter, First-passage time in complex scale-invariant media. Nature **450**, 77 (2007)
139. A. Rozanova-Pierrat, D.S. Grebenkov, B. Sapoval, Faster diffusion across an irregular boundary. Phys. Rev. Lett. **108**, 240602 (2012)
140. O. Bénichou, R. Voituriez, From first-passage times of random walks in confinement to geometry-controlled kinetics. Phys. Rep. **539**, 225–284 (2014)
141. S.K. Ghosh, A.G. Cherstvy, D.S. Grebenkov, R. Metzler, Anomalous, non-Gaussian tracer diffusion in heterogeneously crowded environments. New J. Phys. **18**, 013027 (2016)
142. L. Dagdug, J. Peña and I. Pompa-García, *Diffusion Under Confinement. A Journey Through Counterintuition*. (Springer, 2024)

143. O. Bénichou, G. Oshanin, Ultraslow vacancy-mediated tracer diffusion in two dimensions: the Einstein relation verified. Phys. Rev. E **66**, 031101 (2002)
144. O. Bénichou, P. Illien, G. Oshanin, A. Sarracino, R. Voituriez, Microscopic theory for negative differential mobility in crowded environments. Phys. Rev. Lett. **113**, 268002 (2014)
145. O. Bénichou, P. Illien, G. Oshanin, A. Sarracino, R. Voituriez, Tracer diffusion in crowded narrow channels. J. Phys.: Condens. Matter **30**, 443001 (2018)
146. O. Bénichou, P. Illien, G. Oshanin, A. Sarracino, R. Voituriez, Nonlinear response and emerging nonequilibrium microstructures for biased diffusion in confined crowded environments. Phys. Rev. E **93**, 032128 (2016)
147. C.W. Gardiner, *Handbook of Stochastic Methods for Physics, Chemistry and the Natural Sciences* (Springer, Berlin, 1985)
148. N.G. Van Kampen, *Stochastic Processes in Physics and Chemistry* (Elsevier, Amsterdam, 1992)
149. Z. Schuss, *Brownian Dynamics at Boundaries and Interfaces in Physics, Chemistry and Biology* (Springer, New York, 2013)
150. J. Crank, *The Mathematics of Diffusion*. (Oxford University Press, 1956)
151. H.S. Carslaw, J.C. Jaeger, *Conduction of Heat in Solids*, 2nd Ed. (Oxford University Press, 1959)
152. D.S. Grebenkov, B.-T. Nguyen, Geometrical structure of Laplacian eigenfunctions. SIAM Rev. **55**, 601–667 (2013)
153. S. Condamin, V. Tejedor, R. Voituriez, O. Bénichou, J. Klafter, Probing microscopic origins of confined subdiffusion by first-passage observables Proc. Nat. Acad. Sci. USA **105**, 5675 (2008)
154. D.A. Kenwright, A.W. Harrison, T.A. Waigh, P.G. Woodman, V.J. Allan, First-passage-probability analysis of active transport in live cells. Phys. Rev. E **86**, 031910 (2012)
155. D.M. Soumpasis, Theoretical analysis of fluorescence photobleaching recovery experiments. Biophys. J. **41**, 95–97 (1983)
156. A. Szabo, Theory of diffusion-influenced fluorescence quenching. J. Phys. Chem. **93**, 6929 (1989)
157. M.J. Saxton, Anomalous subdiffusion in fluorescence photobleaching recovery: a Monte Carlo study. Biophys. J. **81**, 2226 (2001)
158. S.B. Yuste, E. Abad, K. Lindenberg, A reaction-subdiffusion model of fluorescence recovery after photobleaching (FRAP). J. Stat. Mech. **2014**, P11014 (2014)
159. K.R. Brownstein, C.E. Tarr, Importance of classical diffusion in NMR studies of water in biological cells. Phys. Rev. A **19**, 2446 (1979)
160. P.P. Mitra, P.N. Sen, Effects of microgeometry and surface relaxation on NMR pulsed-field-gradient experiments: Simple pore geometries. Phys. Rev. B **45**, 143 (1992)
161. D.S. Grebenkov, NMR survey of reflected Brownian motion. Rev. Mod. Phys. **79**, 1077 (2007)
162. I.N. Sneddon, *Mixed Boundary Value Problems in Potential Theory* (Wiley, NY, 1966)
163. A. Singer, Z. Schuss, D. Holcman, R.S. Eisenberg, Narrow escape, Part I. J. Stat. Phys. **122**, 437 (2006)
164. A. Singer, Z. Schuss, D. Holcman, Narrow escape, Part II The circular disk. J. Stat. Phys. **122**, 465 (2006)
165. D. Holcman, Z. Schuss, The narrow escape problem. SIAM Rev. **56**, 213–257 (2014)
166. G. Oshanin, M. Tamm, O. Vasilyev, Narrow-escape times for diffusion in microdomains with a particle-surface affinity: mean-field results. J. Chem. Phys. **132**, 235101 (2010)
167. D.S. Grebenkov, Universal formula for the mean first passage time in planar domains. Phys. Rev. Lett. **117**, 260201 (2016)
168. J.S. Marshall, Analytical solutions for an escape problem in a disc with an arbitrary distribution of exit holes along its boundary. J. Stat. Phys. **165**, 920–952 (2016)
169. G.H. Weiss, K.E. Shuler, K. Lindenberg, Order statistics for first passage times in diffusion processes. J. Stat. Phys. **31**, 255–278 (1983)
170. S.D. Lawley, Universal formula for extreme first passage statistics of diffusion. Phys. Rev. E **101**, 012413 (2020)

171. S.D. Lawley, Distribution of extreme first passage times of diffusion. J. Math. Biol. **80**, 2301–2325 (2020)
172. D.S. Grebenkov, R. Metzler, G. Oshanin, From single-particle stochastic kinetics to macroscopic reaction rates: fastest first-passage time of *N* random walkers. New J. Phys. **22**, 103004 (2020)
173. K. Reynaud, Z. Schuss, N. Rouach, D. Holcman, Why so many sperm cells? Commun. Integr. Biol. **8**, e1017156 (2015)
174. Z. Schuss, K. Basnayake, D. Holcman, Redundancy principle and the role of extreme statistics in molecular and cellular biology. Phys. Life Rev. **28**, 52–79 (2019)
175. F.C. Collins, G.E. Kimball, Diffusion-controlled reaction rates. J. Coll. Sci. **4**, 425 (1949)
176. D.S. Grebenkov, Encounter-based approach to the escape problem. Phys. Rev. E **107**, 044105 (2023)
177. Y. Scher, O.L. Bonomo, A. Pal, S. Reuveni, Microscopic theory of adsorption kinetics. J. Chem. Phys. **158**, 094107 (2023)
178. Y. Scher, S. Reuveni, D.S. Grebenkov, Escape from textured adsorbing surfaces. J. Chem. Phys. **160**, 184105 (2024)
179. O. Bénichou, D.S. Grebenkov, P.E. Levitz, C. Loverdo, R. Voituriez, Optimal reaction time for surface-mediated diffusion. Phys. Rev. Lett. **105**, 150606 (2010)
180. J.-H. Jeon, A.V. Chechkin, R. Metzler, Scaled Brownian motion: a paradoxical process with a time dependent diffusivity for the description of anomalous diffusion. Phys. Chem. Chem. Phys. **16**, 15811 (2014)
181. A. Szabo, R. Zwanzig, N. Agmon, Diffusion-controlled reactions with mobile traps. Phys. Rev. Lett. **61**, 2496 (1988)
182. S. Redner, P. Krapivsky, Capture of the lamb: diffusing predators seeking a diffusing prey. Am. J. Phys. **67**, 1277 (1999)
183. J.-P. Bouchaud, A. Georges, Anomalous diffusion in disordered media: statistical mechanisms, models and physical applications. Phys. Rep. **195**, 127 (1990)
184. R.D.L. Hanes, C. Dalle-Ferrier, M. Schmiedeberg, M.C. Jenkins, S.U. Egelhaaf, Colloids in one dimensional random energy landscapes. Soft Matter **8**, 2714 (2012)
185. J. Bewerunge, I. Ladadwa, F. Platten, C. Zunke, A. Heuer, S.U. Egelhaaf, Time- and ensemble-averages in evolving systems the case of Brownian particles in random potentials. Phys. Chem. Chem. Phys. **18**, 18887 (2016)
186. M.V. Chubynsky, G.W. Slater, Diffusing diffusivity: a model for anomalous, yet Brownian, diffusion. Phys. Rev. Lett. **113**, 098302 (2014)
187. A.V. Chechkin, F. Seno, R. Metzler, I.M. Sokolov, Brownian yet non-gaussian diffusion: from superstatistics to subordination of diffusing diffusivities. Phys. Rev. X **7**, 021002 (2017)
188. Y. Lanoiselée, N. Moutal, D.S. Grebenkov, Diffusion-limited reactions in dynamic heterogeneous media. Nat. Comm. **9**, 4398 (2018)
189. Y. Lanoiselée, D.S. Grebenkov, A model of non-Gaussian diffusion in heterogeneous media. J. Phys. A: Math. Theor. **51**, 145602 (2018)
190. D.S. Grebenkov, V. Sposini, R. Metzler, G. Oshanin, F. Seno, Exact first-passage time distributions for three random diffusivity models. J. Phys. A: Math. Theor. **54**, 04LT01 (2021)
191. D.S. Grebenkov, A unifying approach to first-passage time distributions in diffusing diffusivity and switching diffusion models. J. Phys. A **52**, 174001 (2019)
192. S.B. Yuste, E. Abad, K. Lindenberg, Exploration and trapping of mortal random walkers. Phys. Rev. Lett. **110**, 220603 (2013)
193. B. Meerson, S. Redner, Mortality, redundancy, and diversity in stochastic search. Phys. Rev. Lett. **114**, 198101 (2015)
194. D.S. Grebenkov, J.-F. Rupprecht, The escape problem for mortal walkers. J. Chem. Phys. **146**, 084106 (2017)
195. A. Pal, S. Reuveni, First passage under restart. Phys. Rev. Lett. **118**, 030603 (2017)
196. A.V. Chechkin, I.M. Sokolov, Random search with resetting: a unified renewal approach. Phys. Rev. Lett. **121**, 050601 (2018)

197. O. Tal-Friedman, Y. Roichman, S. Reuveni, Diffusion with partial resetting. Phys. Rev. E **106**, 054116 (2022)
198. C. Di Bello, A.V. Chechkin, A.K. Hartmann, Z. Palmowski, R. Metzler, Time-dependent probability density function for partial resetting dynamics. New J. Phys. **25**, 082002 (2023)
199. H. Scher, E.W. Montroll, Anomalous transit-time dispersion in amorphous solids. Phys. Rev. B **12**, 2455 (1975)
200. R. Metzler, J. Klafter, The random walk's guide to anomalous diffusion: a fractional dynamics approach. Phys. Rev. **339**, 1 (2000)
201. R. Metzler, J. Klafter, The restaurant at the end of the random walk: recent developments in the description of anomalous transport by fractional dynamics. J. Phys. A: Math. Theor. **37**, R161 (2004)
202. P. Lévy, Random functions; general theory with special reference to Laplacian random functions. Univ. California Publ. in Statist. **1**, 331–390 (1953)
203. B. Mandelbrot, J.W. Van Ness, Fractional Brownian motions, fractional noises and applications. SIAM Rev. **10**, 422 (1968)
204. B. Meerson, O. Bénichou, G. Oshanin, Path integrals for fractional Brownian motion and fractional Gaussian noise. Phys. Rev. E **106**, L062102 (2022)
205. O. Bénichou, G. Oshanin, A unifying representation of path integrals for fractional Brownian motions. J. Phys. A: Math. Theor. **57**, 225001 (2024)
206. L. Decreusefond, A. Üstünel, Stochastic analysis of the fractional Brownian motion. Poten. Anal. **10**, 177–214 (1999)
207. D.S. Grebenkov, D. Beliaev, P.W. Jones, A multiscale guide to Brownian motion. J. Phys. A: Math. Theor. **49**, 043001 (2016)
208. K.-G. Wang, C.W. Lung, Long-time correlation effects and fractal Brownian motion. Phys. Lett. A **151**, 119 (1990)
209. J.M. Porrá, K.-G. Wang, J. Masoliver, Generalized Langevin equations - Anomalous diffusion and probability distributions. Phys. Rev. E **53**, 5872 (1996)
210. E. Lutz, Fractional Langevin equation. Phys. Rev. E **64**, 051106 (2001)
211. N. Pottier, Aging properties of an anomalously diffusing particule. Phys. A (Amsterdam) **317**, 371 (2003)
212. A.D. Vinales, M.A. Desposito, Anomalous diffusion - Exact solution of the generalized Langevin equation for harmonically bounded particle. Phys. Rev. E **73**, 016111 (2006)
213. D.S. Grebenkov, M. Vahabi, Analytical solution of the generalized Langevin equation with hydrodynamic interactions: subdiffusion of heavy tracers. Phys. Rev. E **89**, 012130 (2014)
214. W. Wang, F. Seno, I.M. Sokolov, A.V. Chechkin, R. Metzler, Unexpected crossovers in correlated random-diffusivity processes. New J. Phys. **22**, 083041 (2020)
215. J. Slezak, R. Metzler, Minimal model of diffusion with time changing Hurst exponent. J. Phys. A: Math. Theor. **56**, 35LT01 (2023)
216. W. Wang, M. Balcerek, K. Burnecki, A.V. Chechkin, S. Janusonis, J. Slezak, T. Vojta, A. Wylomanska, R. Metzler, Memory-multi-fractional Brownian motion with continuous correlations. Phys. Rev. Res **5**, L032025 (2023)
217. T. Guérin, N. Levernier, O. Bénichou, R. Voituriez, Mean first-passage times of non-Markovian random walkers in confinement. Nature **534**, 356–359 (2016)
218. N. Levernier, T.V. Mendes, O. Bénichou, R. Voituriez, T. Guérin, Everlasting impact of initial perturbations on first-passage times of non-Markovian random walks. Nat. Comm. **13**, 5319 (2022)

# Chapter 2
# Concentration of Empirical First-Passage Times

**Rick Bebon and Aljaž Godec**

**Abstract** First-passage properties are central to the kinetics of target-search processes. Theoretical approaches so far primarily focused on predicting first-passage statistics for a given process or model. In practice, however, one faces the reverse problem of inferring first-passage statistics from, typically sub-sampled, experimental or simulation data. Obtaining trustworthy estimates from undersampled data and unknown underlying dynamics remains a daunting task, and the assessment of the uncertainty is imperative. In this chapter, we highlight recent progress in understanding and controlling finite-sample effects in empirical first-passage times of reversible Markov processes. Precisely, we present concentration inequalities bounding from above the deviations of the sample mean for any sample size from the true mean first-passage time and construct *non-asymptotic* confidence intervals. Moreover, we present two-sided bounds on the range of fluctuations, i.e., deviations of the expected maximum and minimum from the mean in any given sample, which control uncertainty even in situations where the mean is a priori not a sufficient statistic.

## 2.1 Introduction

In the past decades, there has been a surge of interest in target-search problems, typically devoted to determining the time it takes a random searcher to find a target. This stochastic time, canonically referred to as *first-passage* or *first-hitting time*, has proven to be invaluable in understanding kinetic properties of various physical, chemical, and biological processes [1–5].

As a pioneering example consider diffusive barrier crossing, when a thermally driven searcher (e.g., a reactant molecule) must overcome an energy barrier in order

R. Bebon · A. Godec (✉)
Mathematical bioPhysics Group, Max-Planck-Institute for Multidisciplinary Sciences, 37077 Göttingen, Germany
e-mail: agodec@mpinat.mpg.de

R. Bebon
e-mail: rick.bebon@mpinat.mpg.de



D. Grebenkov et al. (eds.), *Target Search Problems*,
https://doi.org/10.1007/978-3-031-67802-8_2

to "collide" with its reaction or binding partner. This situation arises naturally in the classical picture of Kramers' reaction-rate theory [6, 7], according to which kinetic rates of chemical reactions can be understood in terms of inverse mean first-crossing times. Building on the success of these ideas, first-passage theory remained an effective paradigm for investigating (bio-) chemical kinetics ever since [8–27].

In a similar vein, numerous events in biological settings are initiated when a diffusing searcher, for example, a messenger ligand, first binds to its target, its receptor. Evidently, this prompts the application of first-passage ideas to study the kinetics of molecular and cellular processes such as cell signaling and gene regulation [28–36, 38], intracellular transport [13], RNA biosynthesis [14], stochastic protein accumulation [15, 16], DNA-binding [17], virus uptake [19], cell adhesion [24–26], kinetochore capture [39], or gating of ion channels [27]. Notably, going beyond the molecular and cellular scope, first-passage principles have also found biological relevance in exploring the emergence of drug resistance [18], spreading dynamics of diseases [20, 21], and foraging behavior of bacteria and animals [22].

It should thus come to no surprise that addressing target-search questions through the lens of first-passage properties led to an immense body of literature with successful applications spanning virtually all scientific domains, including mathematics, physics, chemistry, biology, geology, economics, and finance, among others (see e.g. [1–3, 40–43] and references therein).

First-passage concepts have further been applied in more abstract settings to characterize persistence properties [44–50], as well as diffusion through interfaces [51] and across phase boundaries [52]. Recently, first-passage problems arose in the field of stochastic thermodynamics [53] where they give insight into the statistics of stochastic currents [54, 55], thermodynamic entropy production [56–59], and dynamical activity [60, 61] in systems driven far from equilibrium. Lastly, we remark that first-passage phenomena are intimately tied to the statistics of extremes [62–65], and their mathematical description has been extended to capture quantum systems [66, 67], additive functionals of stochastic paths [68–73], intermittent targets [74–78], active particles [79–81], non-Markovian dynamics [82–84], and resetting processes [85–95], including soft resets [96].

Unfortunately, the exact first-passage time density and thus the full statistics are typically known only for rather elementary examples and simple geometries (e.g., see Ref. [1]), and therefore remains a grand theoretical challenge on its own. The long-time behavior of the first-passage time density is generally somewhat better accessible but still requires substantial efforts [65, 97, 98]. A more accessible, albeit still demanding, statistic is the *mean first-passage time*. Remarkably, this simplest statistic is often a highly non-trivial quantity, especially in cases where the dynamics span a broad range of timescales. Here, some trajectories almost directly reach the target and contribute to the short-time behavior, whereas others survive much longer and feed into the long-time behavior [36]. This makes an estimation of the mean first-passage time or its inverse (i.e., the kinetic rate constant) from experimental or simulation data on a complex problem. The central theme of this chapter is thus how to efficiently control the uncertainty of such empirically derived estimates of the mean first-passage time, especially in the context of small sample sizes.

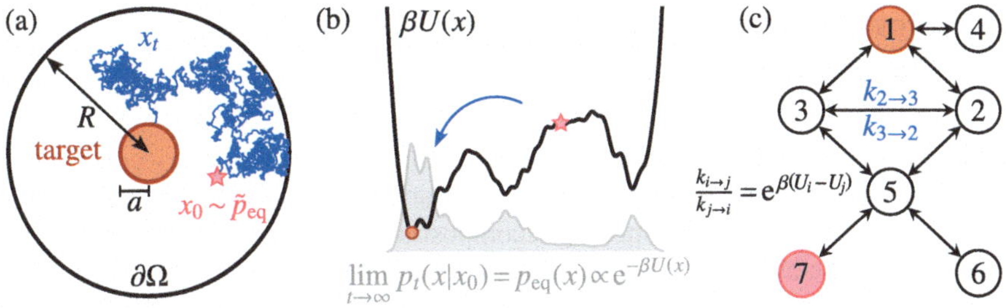


**Fig. 2.1** Schematic of target search processes for ergodic reversible Markov dynamics. Diffusive dynamics in **a** $d$ dimensional spherical domains (here $d = 2$) with reflecting boundary $\partial\Omega$ and **b** arbitrary one-dimensional confining potential landscapes $U(x)$. **c** Markov jump dynamics on a discrete network state-space with transition rates that obey detailed balance. Search processes are initialized from $x_{t=0}$ (magenta), which is drawn from the stationary density $\tilde{p}_{eq}(x)$ and we consider the first-passage time $\tau$ to reach the target (red)

We will focus on the first-passage properties of general target searches described by reversible Markovian dynamics in discrete or continuous state-spaces (see Fig. 2.1). In particular, we will discuss Markov jump processes with arbitrary transition-rate matrices and effectively one-dimensional diffusion processes in arbitrary confining potentials $U(x)$ that also include radial representations of hyper-spherically symmetric diffusion processes in $d$ dimensions where the spurious drift in radial direction allows the mapping to a one-dimensional geometric free energy of purely entropic origin (e.g., see [36]).

The chapter is structured as follows. We begin in Sect. 2.2.1 with an overview of general first-passage concepts. In Sect. 2.2.2, we outline the basics of spectral theory of first-passage processes. Two illustrative examples are worked out in Sects. 2.2.3 and 2.2.4. Section 2.3 introduces the challenges associated with estimating mean first-passage times from observations, and Sect. 2.3.1 explores *concentration inequalities* for bounding deviations of inferred estimates from the true mean first-passage time, applicable for all sample sizes. Building on these results, Sect. 2.3.2 demonstrates how they can be applied to construct performance guarantees that do *not* hinge on asymptotic assumptions. In Sect. 2.4, we move beyond mean first-passage times and showcase results that bound deviations of minimum and maximum first-passage times in samples of arbitrary sizes. These findings are relevant in situations when the mean is a priori an insufficient statistic and problems involving multiple searchers. The chapter concludes in Sect. 2.5 with a discussion and an outlook.

## 2.2 First-Passage Fundamentals: A Spectral-Theoretic Perspective

In this section we present some background, by introducing elementary concepts of first-passage theory of target search. Of particular importance is hereby the mathematical structure encoded in a spectral decomposition of first-passage time densities, which will be the essential building block of derivations throughout the chapter.

### 2.2.1 Generic Concepts in First-Passage Theory

Let $x_t$ denote the abstract trajectory of a "searcher" in a state space $\Omega$; for instance, it may correspond to the position coordinate of a diffusive particle confined within a spherical domain (see Fig. 2.1a), a particle exploring an arbitrary confining potential landscape $U(x)$ (see Fig. 2.1b), or a network-based representation of protein configurations in the search of the native conformational state during folding (see Fig. 2.1c). We suppose that $x_t$ follows a reversible Markovian equation of motion. A more precise formulation of reversible dynamics will be given at the end of this section. The description will be kept "light" and informal.

The time required for the searcher to make its *first* encounter with some predefined target $a \in \Omega$ starting from an initial condition distributed according to the density $p_0$ then corresponds to the first-passage time $\tau$, formally defined as,

$$\tau = \inf_t \left[ t | x_t = a, \, p_0(x_0) \right]. \tag{2.1}$$

The stochastic nature of the underlying dynamics inherently renders the first-passage time a random variable that not only depends on the location of the target $a$ but also the initial position $x_{t=0}$ (drawn from $p_0(x)$). The complete statistics of $\tau$ is encoded in the so-called survival probability

$$S_a(t|x_0) \equiv \mathbb{P}(\tau \geq t), \tag{2.2}$$

the probability that the searcher has *not* found the target by time $t$, i.e., the first-passage event has not yet occurred by $t$. The normalized probability flux into the target $a$ corresponds to the first-passage time density $\wp_a(t|x_0) = \langle \delta(t - \tau[\{x_{t'}\}]) \rangle$, where $\langle \cdot \rangle$ represents the average over all first-passage paths $\{x_{t'}\}_{0 \leq t' \leq \tau}$, i.e., paths that hit $a$ only once. By the fundamental theorem of calculus, the survival probability $S_a(t|x_0)$ in Eq. (2.2) and the first-passage time density $\wp_a(t|x_0)$ are related via

$$\wp_a(t|x_0) = -\frac{\partial}{\partial t} S_a(t|x_0). \tag{2.3}$$

**Assumptions about the dynamics**

We assume $x_t$ to be a strongly ergodic time-homogeneous Markov process on a continuous or discrete state space $\Omega$ with invariant (equilibrium) density $p_{\rm eq}(x)$. The transition probability density $p_t(x|x_0)$ to find $x_t$ at $x$ at time $t$, given that it started from $x_0$, is given by $p_t(x|x_0) = e^{\hat{L}t}\delta_{x_0}(x)$ where $\delta_{x_0}(x)$ denotes either the Dirac or Kronecker delta, depending on the specific state-space being considered, and the generator $\hat{L}$ is a linear reversible (dissipative) operator that is essentially self-adjoint (i.e., it is self-adjoint in some basis; e.g., the operator $\sqrt{p_{\rm eq}(x)}^{-1}\hat{L}\sqrt{p_{\rm eq}(x)}$ is self-adjoint).

In our context, $\hat{L}$ can take the form of a Markov-rate matrix or, equivalently, an effectively one-dimensional Fokker–Planck operator with discrete spectrum and real non-positive eigenvalues; obeying detailed balance in both cases. The effectively one-dimensional character of the process is crucial, as the theory in Sect. 2.2.2 builds on the renewal theorem [37, 97, 98]. For the latter to hold, a process starting from some $x_0$ *must*, before reaching the final state $x$, pass through the point-like target set $a$ that, however, has zero measure in $d \geq 2$. If we consider absorbing sets with positive measure, the renewal structure breaks down unless the symmetry allows a reduction to an effectively one-dimensional Markov dynamics. The results may even hold without this requirement, but are not covered by our proof. As we assume $x_t$ to be ergodic, $p_t$ reaches Boltzmann–Gibbs equilibrium for long times $\lim_{t\to\infty} p_t(x|x_0) = p_{\rm eq} \propto \exp(-\beta U)$ with thermal energy $\beta^{-1} = k_{\rm B}T$. Lastly, a discrete and real spectrum of eigenvalue is ensured by assuming that $\hat{L}$ is (i) either bounded (Fig. 2.1c) (ii) $\Omega$ is finite with reflecting boundary $\partial\Omega$ (Fig. 2.1a) or (iii) that $\beta U(x)$ is sufficiently confining (Fig. 2.1b).

The processes discussed above are *relaxation* processes that conserve probability. In the first-passage setting one has to additionally introduce the absorbing target $x = a$, which formally modifies the generator $\hat{L} \to \hat{L}_a$. For discrete-state dynamics the generator is obtained by removing all transitions out of the absorbing target, i.e., $\hat{L}_a = \hat{L} - \hat{L}|a\rangle\langle a|$. Here $|a\rangle$ is a vector containing zero entries except an entry of unity at the $a$th position. In a continuous state-space $L_a$ remains the Fokker–Planck operator, however, the subscript $a$ now denotes an absorbing (Dirichlet) boundary condition $p_t(a, t|x_0) = 0$.

### 2.2.2 *Spectral Decomposition of First-Passage Processes*

The reversibility of the underlying Markov process and the finite/confining domain ensure that the first-passage generator $\hat{L}_a$ has a spectral expansion in a bi-orthogonal eigenbasis

$$\hat{L}_a = -\sum_{k\geq 1} \mu_k |\phi_k^{\rm R}\rangle\langle\phi_k^{\rm L}| \tag{2.4}$$

in terms of the eigenvalues $\mu_k > 0$ and right $|\phi_k^{\rm R}\rangle$ and left $\langle\phi_k^{\rm L}|$ real eigenfunctions, respectively. Without loss of generality we assume the ordering $\mu_k \leq \mu_{k+1}$, where $\mu_1 > 0$. For more details on the spectral-theoretic description of relaxation and first passage see [97, 98].

The first-passage time density $\wp_a(t|x_0)$ to reach the absorbing target at $x = a$ starting from $x_0$ in turn also admits a spectral representation of the form [97, 98]

$$\wp_a(t|x_0) = \sum_{k\geq 1} w_k(x_0)\mu_k e^{-\mu_k t}. \tag{2.5}$$

Each eigenvalue $\mu_k$ corresponds to a distinct first-passage time-scale via $1/\mu_k$, and the contribution of each time-scale is governed by the corresponding (generally not necessarily positive) first-passage weight $w_k(x_0)$. The weights are normalized according to $\sum_k w_k(x_0) = 1$. Moreover, first-passage weights explicitly depend on the initial condition $x_0$ and for pseudo-equilibrium initial conditions $x_0 \sim \tilde{p}_{\text{eq}}(x)$ (i.e., equilibrium but excluding—effectively disregarding—the target; see below) one always finds $w_k^{\text{eq}} \geq 0, \forall k$ [99]. Furthermore, it is important to highlight that both, $\mu_k(a)$ and $w_k(x_0, a)$, depend on the location of the absorbing target $a$, which we strictly account for, but omit in the notation for the sake of readability. The survival probability Eq. (2.3) can equally be expressed in a spectral form,

$$S_a(t|x_0) = \sum_{k\geq 1} w_k(x_0)\mathrm{e}^{-\mu_k t}, \tag{2.6}$$

and the $m$th first-passage moment is computed via $\langle \tau^m \rangle = m! \sum_{k\geq 1} w_k(x_0)/\mu_k^m$.

**! Pseudo-equilibrium initial conditions**

We stress that we will in all subsequent discussions assume that $x_{t=0}$ is drawn from the pseudo-equilibrium density $p_0(x) = \tilde{p}_{\text{eq}}(x)$. This has important consequences for $w_k(x_0)$ (and hence $\wp_a(t|x_0)$ and $S_a(t|x_0)$). These pseudo-equilibrium initial conditions apply in a large number of settings [34, 35, 106–111] since, for example, $x_0$ usually cannot be controlled in practice and is thus effectively sampled from the stationary density. For a discussion of general initial conditions see [99].

In contrast to the equilibrium density $p_{\text{eq}}(x)$ for standard relaxation processes (see Fig. 2.1b in gray), the absorption process will reach the target with probability 1 as $t \to \infty$. A corresponding definition of a non-trivial stationary *pseudo-equilibrium* density $\tilde{p}_{\text{eq}}(x)$ as an initial condition for the first-passage process (i.e., absorption) therefore becomes more nuanced by the presence of the absorbing target $a$. For discrete state spaces $\Omega_{\text{ds}}$ with states $x \in \Omega_{\text{ds}}$ the pseudo-equilibrium density is obtained via a renormalization of $p_{\text{eq}}(x)$ by simply excluding the target in the sense of $\tilde{p}_{\text{eq}}(x) \equiv p_{\text{eq}}(x)/\sum_{k\neq a} p_{\text{eq}}(k)$ for $x \neq a$ and $\tilde{p}_{\text{eq}}(x) = 0$ when $x = a$. The resulting pseudo-equilibrium density therefore adopts the target as perfectly reflecting, effectively disregarding its presence. This ensures $S_a(t = 0|x_0 \sim \tilde{p}_{\text{eq}}) = 1$. Note that by contrast, in this case the *true invariant distribution of the absorption process* is trivially $p_{\text{inv}}^{\text{abs}}(x) = \delta_{x,a}$. The situation is straightforward for continuous state-spaces, where we have $\tilde{p}_{\text{eq}}(x) = p_{\text{eq}}(x)$, since the target $a$ has zero measure.

### 2.2.3 Example 1: Markov Jump Network

For illustrative purposes we first consider Markov process on a discrete state space (see Fig. 2.1c) as a model of a conformational search of a protein for the native folded state $a = 1$ (red), with the initial condition $x_0$ drawn from the pseudo-equilibrium $\tilde{p}_{\text{eq}}(x)$. The dynamics are encoded in the transition rates $k_{i\to j}$ from state $i$ to $j$ and enter the transition-rate matrix $\hat{L}$ via $L_{ji} = k_{i\to j}$ and $L_{ii} = -\sum_{j\neq i} L_{ji}$. Since we consider reversible dynamics, $\hat{L}$ obeys detailed balance $p_{\text{eq},j}/p_{\text{eq},i} = L_{ji}/L_{ij} = \exp[\beta(U_i - U_j)]$, and the transition rates are related to the (free) energy of states $U_i$. Computing the weights is readily achieved by diagonalization of $\hat{L}_a$ and using $w_k(x_0) = -\langle a|\phi_k^{\text{R}}\rangle\langle\phi_k^{\text{L}}|x_0\rangle$ [97, 98]. Recall that $w_k^{\text{eq}} \geq 0, \forall k$ as a result of the pseudo-equilibrium initial condition [99].

### 2.2.4 Example 2: Diffusive Target Search in a Confined Spherical Domain

As a second example we analyze a confined diffusive search of a Brownian particle in a $d = 3$ dimensional unit sphere with reflecting boundary $\partial\Omega$ at $R = 1$ (see Fig. 2.1a). A perfectly absorbing spherical target with radius $0 < a < 1$ is placed in the center (we take $a = 0.1$). The time-evolution of the distance to the absorbing sphere $x_t$ is a confined Bessel process obeying the stochastic Skorokhod equation [100–105]

$$dx_t = \frac{2}{x_t}dt + \sqrt{2}dW_t + \mathbb{1}_{\partial\Omega}(x_t)dL_t, \tag{2.7}$$

where $dW_t$ denotes the Wiener increment, i.e., Gaussian white noise with $\langle dW_t\rangle = 0$, $\langle dW_t dW_{t'}\rangle = \delta(t - t')dtdt'$, and we set the diffusion constant to unity $D = 1$ without loss of generality. The second term explicitly accounts for (normal) reflections on the boundary $\partial\Omega$, i.e., that $x_t$ does not leave the domain $\Omega$. Here, the boundary local time $L_t$ is a non-decreasing process with $L_0 = 0$, that by construction only increases when $x_t \in \partial\Omega$, i.e., the indicator function $\mathbb{1}_{\partial\Omega}(x_t) = 1$ if $x_t \in \partial\Omega$ and 0 otherwise. (For a more detailed discussion see Chap. 4). Note that the general case of a sphere with radius $R$ with any $0 < D < \infty$ is easily recovered by expressing time in units $R^2/D$. It is not difficult to show that the pseudo-equilibrium weights are given by

$$w_k^{\text{eq}} = \frac{2}{\mu_k}\frac{3a^2}{1 - a^3}\frac{\tan[(1 - a)\sqrt{\mu_k}] + \frac{1}{\sqrt{\mu_k}}}{(1 - a)\tan[(1 - a)\sqrt{\mu_k}] - \frac{a}{\sqrt{\mu_k}}}, \tag{2.8}$$

and the first-passage rates $\mu_k$ are solutions of the transcendental equation $\sqrt{\mu_k} = \tan([1 - a]\sqrt{\mu_k})$.

## 2.3 Inference of the Mean First-Passage Time

From a practical point of view, the inference of the mean first-passage time $\langle \tau \rangle$ from experimental or simulation data via the direct *unbiased*[1] sample-mean estimator that we call *empirical first-passage time* $\overline{\tau}_n \equiv n^{-1} \sum_i \tau_i$, generally poses a serious challenge, especially in situations where only a limited number $n$ of first-passage events $\tau_i$ are available. The sub-sampling may be due to, e.g., practical constraints in experimental setups or limited computational resources. Hence, sample numbers in the range from 1 to 10 [113–118] or sometimes up to 100 [119] are quite common. Clearly, when dealing with sub-sampling issues, the inferred estimate is afflicted by large uncertainties, especially when first-passage times are distributed across many time-scales and extreme values substantially contribute to (and potentially skew) the sample mean. Understanding the fluctuations of $\overline{\tau}_n$ for any $n$ is thus crucial but challenging, since limited sample-sizes lead to non-Gaussian errors in turn rendering uncertainty quantification not amendable to standard error-analysis techniques.

As detailed in the following section, we recently addressed the statistical deviations of the inferred empirical $\overline{\tau}_n$ from the actual mean first-passage time $\langle \tau \rangle$ by bounding the probability of such deviations in the small-sample regime [99]. Our approach rests on a non-asymptotic description of the *concentration-of-measure phenomenon*.

### 2.3.1 Bounding the Probability of Deviations

The study of the concentration behavior of the sample-mean (here $\overline{\tau}_n$) around the true mean (here $\langle \tau \rangle$) has a long tradition in probability theory. For example, in the asymptotic limit $n \to \infty$ the (weak or strong) law of large numbers establishes that $\overline{\tau}_n$ converges to $\langle \tau \rangle$ (in probability or almost surely) and the central limit theorem asserts that $\sqrt{n}(\overline{\tau}_n - \langle \tau \rangle)$ converges to a normal distribution with mean 0 and variance $\sigma^2$. Similar statements for finite $n$ pose a much greater challenge that requires a more intricate analysis. On a related note, we remark that understanding finite-sample effects is important beyond statistical inference, as these manifest in diverse contexts such as, e.g., molecule-number fluctuations in chemical reactions [120].

Returning to statistics, to assess the fluctuations of the inferred $\overline{\tau}_n$ around the true mean $\langle \tau \rangle$ in more general situations (i.e., small $n$) where the sample-average $\overline{\tau}_n$ is *not yet* tightly concentrated around $\langle \tau \rangle$ and its probability density[2] is not close to a Gaussian (see Fig. 2.2a), we examine deviation probabilities $\mathbb{P}(\overline{\tau}_n \geq \langle \tau \rangle + t)$ and $\mathbb{P}(\overline{\tau}_n \leq \langle \tau \rangle - t)$, respectively. These reflect the probability that the empirical first-passage time $\overline{\tau}_n$, estimated from a sample of $n \geq 1$ recorded first-passage events,

---

[1] The bias of an estimator (e.g., the sample-mean) is defined as the difference between its expectation and the true value. Correspondingly, estimators with zero bias are said to be unbiased [112].

[2] Recall that the sample-mean $\overline{\tau}_n$ is a random variable on its own as every set of $n$ i.i.d. recorded first-passage events $\{\tau_1, \tau_2, \ldots, \tau_n\}$ gives rise to a different estimate $\overline{\tau}_n$.

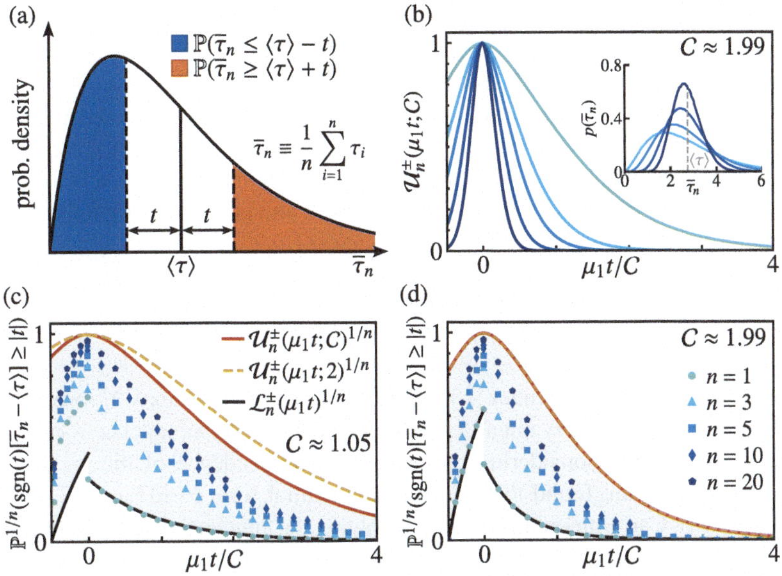


**Fig. 2.2** Non-asymptotic concentration and sample-size effects of empirical first-passage times $\overline{\tau}_n$ around the true mean $\langle\tau\rangle$. **a** Schematic probability density of $\overline{\tau}_n$ inferred from a (small) sample of $n$ realizations. Fluctuations are quantified by tail probabilities of deviations of $\overline{\tau}_n$ from $\langle\tau\rangle$ by more than $t$ toward the right $\mathbb{P}(\overline{\tau}_n \geq \langle\tau\rangle + t)$ or the left $\mathbb{P}(\overline{\tau}_n \leq \langle\tau\rangle - t)$ and are shown in orange and blue, respectively. **b** Dependence of the upper bounds $\mathcal{U}_n^{\pm}(\mu_1 t; C)$ on sample size $n$ for the confined Brownian search in $d = 3$. Bounds quantify how the probability of deviations from the sample mean drastically diminishes as $n$ increases (bright to dark; see (**d**)). Inset: corresponding empirical histogram of the sample mean $\overline{\tau}_n$ for different $n$ values. **c**, **d** Deviation probabilities and corresponding bounds for **c** a Markov network representation of protein folding ($C \approx 1.05$) and **d** a spatially confined Brownian search ($C \approx 1.99$) as introduced in Fig. 2.1. Probabilities are scaled as $\mathbb{P}^{1/n}(\mathrm{sgn}(t)[\overline{\tau}_n - \langle\tau\rangle] \geq |t|)$; right tail areas are shown for $t > 0$ and left for $t < 0$, respectively. Lower $\mathcal{L}_n^{\pm}(\mu_1 t)^{1/n}$ and upper bounds $\mathcal{U}_n^{\mp}(\mu_1 t; C)^{1/n}$ are depicted as black and red lines, respectively, and the model-free bounds $\mathcal{U}_n^{\mp}(\mu_1 t; 2)^{1/n}$ are shown as yellow dashed line. Corresponding deviation probabilities obtained by numerical simulations as a function of $t$ for different fixed $n$ are denoted by symbols. Yellow and red curves coincide in (**d**) since $C \approx 1.99$

deviates from the true mean $\langle\tau\rangle$ by more than some value $t$ in either direction, as illustrated in Fig. 2.2a by the colored regions. These probabilities gauge the likelihood and magnitude of fluctuations in the estimate $\overline{\tau}_n$ regardless of sample-size; however they cannot be obtained in a generic fashion since the probability density of $\overline{\tau}_n$ is not known.

Here, we describe a solution of this problem in terms of the recently derived general upper bounds, so-called concentration inequalities, $\mathbb{P}(\pm[\overline{\tau}_n - \langle\tau\rangle] \geq t) \leq \mathcal{U}_n^{\pm}(t)$, which do *not* require the knowledge of the distribution of $\overline{\tau}_n$. Notably, these bounds are valid for all values $n \geq 1$ and $t \geq 0$ and their most conservative version is independent of any specific details about the underlying dynamics. From a technical perspective, they are derived from a bounding technique referred to as the *Cramér–Chernoff method* [121] which is summarized briefly at the end of this section.

A complete derivation with all mathematical details is beyond the scope of this chapter (see [99]), we therefore only outline the key concepts. The procedure starts from Chernoff's inequality $\mathbb{P}(\pm[\tau - \langle\tau\rangle] \geq t) \leq \mathrm{e}^{-\lambda t + \psi_{\pm(\tau-\langle\tau\rangle)}(\lambda)}$, where $\psi_X(\lambda)$ denotes the cumulant generating function of a random variable $X$ (see below), and essentially involves two main steps. First, we prove an upper bound on the cumulant generating function $\ln\langle\exp[\pm\lambda(\tau - \langle\tau\rangle)]\rangle \leq \phi_\pm(\lambda)$, $\lambda \in \mathbb{R}^+$, in the form of

$$\phi_+(\lambda) = \frac{\lambda^2}{2}\frac{\mu_1^2\langle\tau^2\rangle}{1-\lambda/\mu_1} \quad \text{and} \quad \phi_-(\lambda) = \frac{\lambda^2}{2}\frac{\mu_1^2\langle\tau^2\rangle}{1-(\lambda/\mu_1)^2}, \tag{2.9}$$

for the right tail ($\tau \geq \langle\tau\rangle$; index +) and left tail ($\tau < \langle\tau\rangle$; index −), respectively. In the second step, we carry out a Legendre transform $\phi_\pm^*(t) \equiv \sup_\lambda[\lambda t - \phi_\pm(\lambda)]$ to optimize the bound. Considering the statistical independence of the $\tau_i$, the concentration bounds are found that have an exponential form $\mathbb{P}(\pm[\overline{\tau}_n - \langle\tau\rangle] \geq t) \leq \exp(-n\phi_\pm^*(t))$. Carrying out these steps we obtain the announced upper bounds [99]

$$\begin{aligned} \mathbb{P}(\overline{\tau}_n - \langle\tau\rangle \geq t) &\leq \exp(-nCh_+(\mu_1 t/C)) \equiv \mathcal{U}_n^+(\mu_1 t; C) \qquad 0 \leq t \leq \infty \\ \mathbb{P}(\langle\tau\rangle - \overline{\tau}_n \geq t) &\leq \exp(-nCh_-(\mu_1 t/C)) \equiv \mathcal{U}_n^-(\mu_1 t; C) \qquad 0 \leq t \leq \langle\tau\rangle, \end{aligned} \tag{2.10}$$

where we introduced $C \equiv \mu_1^2\langle\tau^2\rangle$ and defined the auxiliary functions

$$h_+(u) \equiv 1 + u - \sqrt{1+2u} \quad \text{and} \quad h_-(u) \equiv \Lambda(u)u - \frac{1}{2}\frac{\Lambda(u)^2}{1-\Lambda(u)^2}, \tag{2.11}$$

with

$$\Lambda(u) \equiv \frac{1}{2}\left[g(u) - \sqrt{4 + 2/g(u) - g(u)^2}\right] \tag{2.12}$$

$$g(u) \equiv \frac{2}{\sqrt{3}}\left\{1 + 2\cosh\left[\frac{1}{3}\operatorname{arccos}\left(1 + \frac{3^3}{2^7 u^2}\right)\right]\right\}^{1/2}. \tag{2.13}$$

It is worth noting that, for the bounds presented in Eq. (2.10), the underlying detailed dynamics manifest solely via the system-dependent constant $C$. Remarkably, however, it can be shown that for pseudo-equilibrium initial conditions ($x_0 \sim \tilde{p}_{\mathrm{eq}}$; see end of Sect. 2.2.2) this constant falls within the range $0 \leq C \leq 2$ [99]. This insight can be leveraged to directly establish

$$\mathcal{U}_n^\pm(\mu_1 t; C) \leq \mathcal{U}_n^\pm(\mu_1 t; 2) \equiv \mathcal{U}_n^\pm(\mu_1 t). \tag{2.14}$$

While these bounds tend to be more conservative (i.e., when the actual value of $C$ significantly deviates from 2 one may still take $C = 2$ but the bound becomes less sharp), they are entirely model-independent. In the case of general initial conditions $p_0(x) \neq \tilde{p}_{\rm eq}$, $C$ becomes replaced with a different constant (for details see [99]).

At this point we emphasize that the magnitude of deviations, in other words, the "error", is naturally parameterized by the dimensionless variable $\mu_1 t$ as evident from the argument of $h_\pm(\mu_1 t/C)$ in Eq. (2.10). Therefore, considering errors in units $1/\mu_1$, $\pm\mu_1(\overline{\tau}_n - \langle\tau\rangle)$, there is no need to specify or know $\mu_1$ (i.e., the longest time-scale) itself. This relative error normalized by the "observable largest value" of first passage events $1/\mu_1$ thus quantifies the deviation of the inferred empirical mean first-passage time $\overline{\tau}_n$ from the true mean $\langle\tau\rangle$ relative to the characteristic (maximal, "extreme-case-scenario"[3]) time-scale of the system. In this context deviation bounds (2.10) state that, for any sample size $n$, the probability $\mathbb{P}(\mu_1[\overline{\tau}_n - \langle\tau\rangle] \geq \mu_1 t)$ of observing a relative error larger than a specified (dimensionless) value $\mu_1 t$ (e.g., say $\mu_1 t = 0.1$ or 10%) is lower that the corresponding upper bound $\mathcal{U}_n^\pm(\mu_1 t)$.

The validity of concentration bounds (2.10) and sample-size effects are illustrated in Fig. 2.2b–d for the target-search examples introduced earlier. To make the comparison, we evaluated "exact" empirical deviation probabilities (symbols) obtained from extensive sampling of $\overline{\tau}_n$ ($10^{11}$ repetitions) for various fixed values of $n$ (see inset Fig. 2.2b for histograms of $\overline{\tau}_n$). For visualization purposes, we formally let $t \to -t$ for the left tails, allowing us to depict both, the right and left deviation bounds, in a single plot over the support $[-\langle\tau\rangle, \infty)$. Illustrated in Fig. 2.2b, deviation bounds quantify an exponential concentration rate of $\overline{\tau}_n$ around $\langle\tau\rangle$ as $n$ increases. In panels (c-d) we additionally employed the scaling $\mathbb{P}^{1/n}$ to collapse the results onto a single master curve for all values of $n$. Data points approach the upper bound as $n$ increases and as anticipated, the model-independent bound $\mathcal{U}_n(\mu_1 t; 2)$ (yellow) holds universally but tends to be more conservative.

Finally, we mention that in [99] we also derived corresponding lower bounds $\mathcal{L}_n^\pm$, for the sake of completeness shown in black,

$$\begin{aligned}\mathbb{P}(\overline{\tau}_n - \langle\tau\rangle \geq t) &\geq \left(w_1 \mathrm{e}^{-\mu_1(\langle\tau\rangle + t)}\right)^n \equiv \mathcal{L}_n^+(\mu_1 t)\\ \mathbb{P}(\langle\tau\rangle - \overline{\tau}_n \geq t) &\geq \left(1 - \mathrm{e}^{-\mu_1(\langle\tau\rangle - t)}\right)^n \equiv \mathcal{L}_n^-(\mu_1 t).\end{aligned} \tag{2.15}$$

by again leveraging the spectral analysis discussed in Sect. 2.2.2 in combination with ideas from extreme-value theory. Specifically, the searcher is assumed to start at $x_0$ drawn from pseudo-equilibrium $\tilde{p}_{\rm eq}(x)$. A detailed discussion of lower bounds (2.15)—explicitly demonstrating the existence of an intrinsic noise-floor—is beyond the scope of this chapter, as these do not play a role in the construction of performance guarantees presented in the following section. Nevertheless, these non-trivial lower bounds underscore the critical importance of a robust assessment of uncertainty of the observable $\overline{\tau}_n$, which we explore next.

[3] The time-scale $1/\mu_1$ sets a "cut-off" on the range of observable first-passage events; events longer than $O(1) \times 1/\mu_1$ are exponentially unlikely.

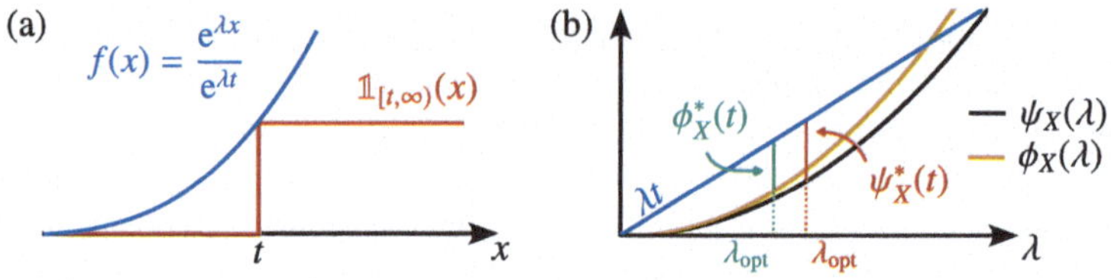


**Fig. 2.3** Sketch of the Cramér–Chernoff method. **a** Schematic of the inequality in Eq. (2.16). **b** Illustration of the Legendre transform of $\psi_X(\lambda)$ (black) and $\phi_X(\lambda)$ (yellow). The optimal $\lambda_{\text{opt}}$ maximizes the difference with $\lambda t$ (blue); $\psi_X^*(t)$ (red) and $\phi_X^*(t)$ denoting the corresponding Legendre transform (Eq. (2.18)), respectively. Note that $\phi_X(\lambda) \geq \psi_X(\lambda)$ implies $\phi_X^*(t) \leq \psi_X^*(t)$

**Cramér–Chernoff method in a nutshell**

Here we present a short overview of the Cramér–Chernoff method, for more details see, e.g., [121]. Consider a real-valued positive random variable $X$, in our context either $\tau - \langle\tau\rangle$ for the right tail or $\langle\tau\rangle - \tau$ for the left tail. For any $\lambda \in \mathbb{R}^+$, we begin with the obvious inequality (see Fig. 2.3a):

$$\mathbb{1}_{[t,\infty)} \leq \frac{e^{\lambda X}}{e^{\lambda t}} \quad \forall X \geq 0, \tag{2.16}$$

where $\mathbb{1}_A$ denotes the indicator function of set $A$. Taking expectation over $X$, $\langle\cdot\rangle$, on both sides, we immediately obtain

$$\mathbb{P}(X \geq t) \leq e^{-\lambda t}\langle e^{\lambda X}\rangle = e^{-\lambda t + \psi_X(\lambda)}, \tag{2.17}$$

where $\psi_X(\lambda) \equiv \ln\langle e^{\lambda X}\rangle$ denotes the cumulant generating function of $X$. Since this inequality holds for all values $\lambda \geq 0$, one can select $\lambda$ in a way that optimizes the upper bound, achieved by computing the Cramér-Legendre transform (see Fig. 2.3b)

$$\psi_X^*(t) = \sup_{\lambda\in\mathbb{R}^+} [\lambda t - \psi_X(\lambda)]. \tag{2.18}$$

By Eq. (2.17) we thus obtain an upper bound on the tail probability according to

$$\mathbb{P}(X \geq t) \leq e^{-\psi_X^*(t)}. \tag{2.19}$$

The strength of the method lies in its extension to sums $\overline{X} = X_1 + \ldots + X_n$ of $n$ i.i.d. real-valued random variables $X_1, \ldots X_n$. Due to independence, we have $\psi_{\overline{X}}(\lambda) = n\psi_X(\lambda)$, leading to $\psi_{\overline{X}}^*(t) = n\psi_X^*(t/n)$, in turn allowing one to write

$$\mathbb{P}(\overline{X}/n \geq t) \leq e^{-n\psi_X^*(t)}, \tag{2.20}$$

which underscores the evident connection to sample-averages like $\overline{\tau}_n$. We remark that in our case, the Legendre transform was computed not directly for $\psi_X(\lambda)$ but rather for $\phi_X(\lambda) \geq \psi_X(\lambda)$ (Eq. (2.9)), implying that $\psi_X^*(t) \geq \phi_X^*(t)$ (see Fig. 2.3b).

### 2.3.2 *Uncertainty Quantification*

Having established the importance of assessing uncertainty in estimates derived from undersampled data, we are now confronted with the challenge of constructing reliable performance guarantees for the kinetic inference of empirical first-passage times $\overline{\tau}_n$. Unfortunately, for the task at hand many standard procedures fall short as they are grounded in asymptotic theory that necessitates a large sample size ($n \to \infty$). This requirement, clearly, is practically unattainable, making their applicability in subsampled regimes, by definition, problematic. Likewise, complementary approaches to statistical inference via Bayesian methods (see e.g., Ref. [122]) face challenges when dealing with limited data, as the influence of prior distributions must be carefully considered. A sensitivity to the choice of a prior potentially results in "biased" prior-dependent uncertainty estimates, even in the large-sample limit.

This apparent lack of accurate finite-sample guarantees highlights the pressing need for reliable non-asymptotic error estimation techniques—e.g., in the form of confidence intervals—with correct finite-sample coverage probabilities to assess the quality of inferred estimates. Luckily, the concentration bounds (2.10) hold for any $n$, thus providing a well-suited candidate for addressing the challenges of the small-sample regime. As we will outline next, they provide a framework to quantify the discrepancy between the estimate $\overline{\tau}_n$ and the true parameter $\langle\tau\rangle$ in terms of non-asymptotic "with high probability" guarantees.

A construction of such performance guarantees is straightforward and easily implementable. By setting $\mathcal{U}_n^{\pm}(\mu_1 t^{\pm}_{\alpha_\pm,n}) = \alpha_\pm$ for a user-defined acceptable left and right error probability of $\alpha_\pm$, we promptly obtain the implicit definition of a confidence interval $[-t^-_{\alpha_-,n}, t^+_{\alpha_+,n}]$ in the form

$$\mathbb{P}(-t^-_{\alpha_-,n} \leq \overline{\tau}_n - \langle\tau\rangle \leq t^+_{\alpha_+,n}) \geq 1-(\alpha_- + \alpha_+) \equiv 1-\alpha \tag{2.21}$$

by performing an ordinary union bound. In other words, Eq. (2.21) asserts that with a high probability of at least $1-\alpha$ the inferred sample-average $\overline{\tau}_n$ falls within the interval $[\langle\tau\rangle - t^-_{\alpha_-,n}, \langle\tau\rangle + t^+_{\alpha_+,n}]$ (see green region in Fig. 2.4a). For illustrative purposes we present in Fig. 2.3b the most conservative, universal 90% confidence region (i.e., $\alpha = 0.1$ and $\alpha_\pm = \alpha/2$) for the relative error as a function of the given sample-size $n$ for the model systems previously introduced in Sects. 2.2.3 and 2.2.4.

The non-asymptotic confidence intervals (2.21) are further valuable in addressing the equally important question of how large the minimal number of realizations $n_{\min}$ has to be if one wants to ensure that the error of the estimate $\overline{\tau}_{n_{\min}}$ does *not* exceed a specified threshold with a desired level of confidence—an essential consideration in the planning of experimental or computational studies. Correspondingly, this number is implicitly defined via

$$\mathcal{U}^-_{n_{\min}}(\mu_1 t^-_{\alpha_-,n}; C) + \mathcal{U}^+_{n_{\min}}(\mu_1 t^+_{\alpha_+,n}; C) = \alpha. \tag{2.22}$$

Using our simple examples, we explicitly show in Fig. 2.4c the minimal sample-size $n_{\min}$ necessary to guarantee a (dimensionless) relative error $\mu_1(\overline{\tau}_n - \langle\tau\rangle)$ of at most $\pm 10\%$ as a function of the confidence level $1-\alpha$ for the Markov network model (blue; $C \approx 1.05$) and compare them with the most conservative model-independent version (black line). Note that the conservative bound holds for *any* setup that is described by Markovian dynamics as we introduced in Sect. 2.2.1 and a reliable inference of the mean (i.e., 90% confidence level; dashed line in Fig. 2.4c) typically requires hundreds to thousands of samples.

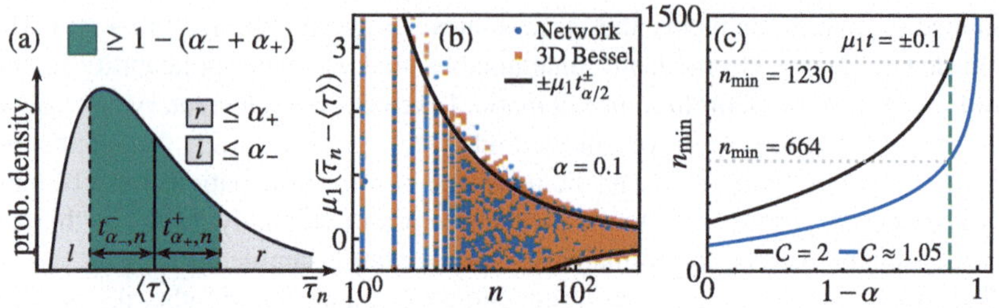


**Fig. 2.4** Uncertainty quantification of the sample-mean $\overline{\tau}_n$ for any sample-size $n$. **a** Probability that the sample-mean $\overline{\tau}_n$ lies within an interval of $[-t^-_{\alpha_-,n}, t^+_{\alpha_+,n}]$ around $\langle\tau\rangle$ with a probability of at least $1-(\alpha_- + \alpha_+)$. **b** Model-free 90% confidence region (i.e., $\alpha = 0.1$) that the error in units $1/\mu_1$ remains within ±10% as a function of sample-size $n$. **c** Minimal-sample size $n_{\mathrm{min}}$ required to ensure that the relative error $\mu_1(\overline{\tau}_n - \langle\tau\rangle)$ does not exceed ±10% as a function of confidence level $1-\alpha$; a confidence of at least 90% is shown as dashed line

Collectively, Eqs. (2.21) and (2.22) provide a framework for evaluating uncertainties that arise in the kinetic inference of $\overline{\tau}_n$ when dealing with undersampled data. The required values of $t^\pm_{\alpha_\pm,n}$ and $n_{\mathrm{min}}$ can readily be determined using standard root-finding techniques (e.g., bisection method). We remark that in our discussion, we have chosen equal tail probabilities $\alpha_\pm = \alpha/2$, resulting in what are commonly referred to as "central" confidence intervals. However, in general, two-sided confidence regions are not uniquely defined by their confidence level, and various other choices exist (e.g., symmetric intervals). Expanding on this, there are situations where only one-sided confidence intervals are needed [99]. In this case we even explicitly obtain that

$$\overline{\tau}_n - \langle\tau\rangle \leq -\frac{\ln(\alpha)}{\mu_1 n} + \frac{\sqrt{2}\sqrt{-\ln(\alpha)}}{\mu_1\sqrt{n/C}} \tag{2.23}$$

with probability of $1-\alpha$. Correspondingly, one obtains the explicit bound

$$n^\pm_{\mathrm{min}} \geq -\frac{\ln(\alpha_\pm)}{Ch_\pm(\mu_1 t/C)} \tag{2.24}$$

where $n^\pm_{\mathrm{min}}$ now denotes the required sample-sizes to ensure that $\overline{\tau}_n - \langle\tau\rangle \leq t$ with a probability of at least $1-\alpha_\pm$.

## 2.4 Going Beyond the Mean First-Passage Time

Thus far we reviewed recent advances in quantifying uncertainty of the sample-mean $\overline{\tau}_n$ inferred from a (small) finite number $n$ of experimental or simulated trajectories, and explored how many samples are needed to ensure that the uncertainty of $\overline{\tau}_n$ falls below a predefined threshold value. However, the fact that possibly multiple relevant time-scales are at play (e.g., in "compact" search processes [34–36, 123]) introduces an additional layer of complexity to the uncertainty of $\overline{\tau}_n$. In such situations the mean on its own may not be a sufficient statistic (i.e., as the "characteristic" time-scale) and a priori may not faithfully characterize the entire distribution.[4] This calls for a deeper understanding of the first-passage time density $\wp_a(t)$, including aspects such as its behavior in the short and long-time regimes [11, 34–36, 123].

Interestingly, assuming that a reliable estimate of the mean is available (which can be achieved using the outlined procedure designed to handle uncertainty in $\overline{\tau}_n$ even where estimation becomes challenging, including scenarios involving multiple time-scales), we can apply the upper bounds $\mathcal{U}_n^{\pm}(t)$ to gain insight. By inspecting the case $n = 1$, i.e., $\overline{\tau}_n = \tau$ and $\mathbb{P}(\pm[\tau - \langle\tau\rangle] \geq t) \leq \mathcal{U}^{\pm}(t)$, the bounds directly provide knowledge about the tail regions of the first-passage time density $\wp_a(t)$ (see Fig. 2.2a) since, in this scenario, the mean first-passage time $\langle\tau\rangle$ acts as a "reference point". Understanding these tail regions is crucial, as trajectories that survive for long times (right tail) explore their environment extensively, and thus retain information about structural properties. Conversely, shorter trajectories (left tail) typically reach their target more directly, often following the shortest direct path and reveal intricate features such as the number of energy minima [124]. On top of that, in cases where the mean is not a priori representative the spread of extreme (i.e., minimal and maximal) first-passage times provides insight about the range of first-passage events (see below).

Another intriguing and related challenge emerges when moving beyond the "single searcher setting" to the first-passage properties of multiple searchers—the fastest search time [97, 125–128]. Here, the statistics may be dominated by the time it takes the fastest (minimal $\tau$) or slowest (maximal $\tau$) out of many searchers to reach a target for the first time. This phenomenon is particularly evident in various biological systems, such as the search process of sperm cells seeking an egg cell [129].

Motivated by these challenges, we now summarize recent progress in understanding first-passage properties in scenarios where the mean first-passage time $\langle\tau\rangle$ alone falls short in capturing the complete range of the dynamics or when dealing with multiple searchers [99].

### *2.4.1 Bounding Extreme Deviations*

The presence of many different distinct time-scales is already included in the spectral expansion of the first-passage time density $\wp_a(t) = \sum_k w_k^{\mathrm{eq}} \mu_k \exp(-\mu_k t)$; recall that $w_k^{\mathrm{eq}} \geq 0, \forall k$ for pseudo-equilibrium initial conditions. Evidently, whenever more than a single weight $w_k^{\mathrm{eq}}$ contributes, multiple time-scales become relevant. In such cases, relying solely on the knowledge of $\langle\tau\rangle$ may not suffice for an accurate characterization of the entire dynamics, as $\wp_a(t)$ significantly deviates from a single

[4] Clearly, for an exponentially distributed $\tau$ the mean $\langle\tau\rangle$ gives complete knowledge since $\wp_a(t) = 1/\langle\tau\rangle e^{-t/\langle\tau\rangle}$.

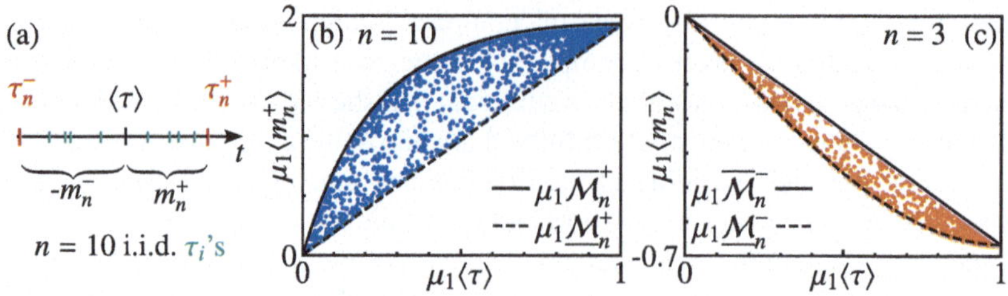


**Fig. 2.5** Extreme deviations from the mean first-passage time. **a** Maximal $\tau_n^+ \equiv \max_{i\in[1,n]} \tau_i$ and minimal $\tau_n^- \equiv \min_{i\in[1,n]} \tau_i$ first-passage time in a sample of $n$ i.i.d. realizations. **b, c** Average maximal (+) and minimal (−) deviation from the mean $\langle m_n^\pm\rangle \equiv \langle \tau_n^\pm - \langle\tau\rangle\rangle$ from extensive computer simulations (symbols) with lower $\underline{\mathcal{M}}_n^\pm$ (dashed black lines) and upper $\overline{\mathcal{M}}_n^\pm$ bound (black lines) shown for $n = 10$ and $n = 3$, respectively. Quantities are expressed in units of $1/\mu_1$ and shown as a function of $\mu_1\langle\tau\rangle$

exponential. For the following discussion, we will make use of the fact that the presence of multiple time-scales is reflected in the dimensionless factor $\mu_1\langle\tau\rangle \in (0, 1]$, i.e., it substantially differs from 1, and the dimensionless parameter $C = \mu_1^2\langle\tau^2\rangle$ becomes substantially smaller than 2.

To address the range of these sample-to-sample fluctuations of the first-passage time $\tau$ we consider *extreme deviations from the mean*, i.e., we are interested how much the maximal $\tau_n^+ \equiv \max_{i\in[1,n]} \tau_i$ and minimal $\tau_n^- \equiv \min_{i\in[1,n]} \tau_i$ in a sample of $n$ i.i.d. observed first-passage times $\tau_i$ deviate from the mean (see Fig. 2.5a) on average. This expected range of extreme deviations $\langle m_n^\pm\rangle \equiv \langle \tau_n^\pm - \langle\tau\rangle\rangle$ is the appropriate observable controlling uncertainty when multiple time-scales are concerned, especially in the small-sample setting, as it provides information regarding the spread (i.e., uncertainty) of observations. To this end, we recently derived sharp two-sided "squeeze" bounds [99]

$$\underline{\mathcal{M}}_n^\pm \leq \langle m_n^\pm\rangle \leq \overline{\mathcal{M}}_n^\pm, \quad n \geq 1 \tag{2.25}$$

in the form of

$$\begin{aligned}
\overline{\mathcal{M}}_n^+ &\equiv \sum_{k=1}^{n} \binom{n}{k} (-1)^{k+1} \frac{(\mu_1\langle\tau\rangle)^k}{\mu_1 k} - \langle\tau\rangle, \quad & \overline{\mathcal{M}}_n^- &\equiv \langle\tau\rangle\left[\frac{1}{n} - 1\right] \\
\underline{\mathcal{M}}_n^+ &\equiv \langle\tau\rangle \sum_{k=2}^{n} \frac{1}{k}, & \underline{\mathcal{M}}_n^- &\equiv \langle\tau\rangle\left[\frac{1}{n}(\mu_1\langle\tau\rangle)^{n-1} - 1\right].
\end{aligned} \tag{2.26}$$

Notably, in dimensionless units of $1/\mu_1$, these bounds are functions of $\mu_1\langle\tau\rangle$ only. In other words, the mean $\langle\tau\rangle$ sharply bounds the expected range of extreme first-passage times even when it is a priori not a representative statistic due to the presence of multiple time-scales. Therefore, these results motivate an estimation of $\langle\tau\rangle$ via $\overline{\tau}_n$ even in the presence of many time-scales. The validity and sharpness is depicted in Fig. 2.5b, c for expected deviations of the maximum (blue) and the minimum (orange) for the Markov jump network from Fig. 2.1c. Each data point corresponds to one particular realization of the jump process (see Sect. 2.2.3) where transition rates are drawn randomly under the constraint of detailed balance. Note that the "squeeze" bounds (2.26) can be saturated (for a discussion see [99]).

Considering that the maximal and minimal first-passage time in a sample of $n$ i.i.d. realizations can be viewed as representing the fastest and slowest first-passage times among the $n$ independent searchers, respectively, these results provide insight into the slowest and fastest first-passage events of $n$ independent searchers. Therefore, it is anticipated that these bounds will prove to be relevant in target-searches with multiple searches and in the few-encounter limit.

## 2.5 Discussion

To robustly assess the uncertainty of inferred empirical first-passage times in the small-sample regime in a general setting, one needs to go beyond conventional approaches relying on asymptotic reasoning or require prior belief. In this chapter, we presented one approach that achieves reliable error estimates by tackling the issue from a non-asymptotic concentration-of-measure perspective that builds on spectral properties of the first-passage problem.

In particular, we put forward a framework that allows to bound uncertainties associated with empirical first-passage times in terms of deviation probabilities that hold regardless of sample size. Subsequently, we showcased how these bounds can be applied by constructing non-asymptotic confidence intervals, which allowed us to tackle the critical question of what sample size is required to ensure tolerable errors.

In the last part, we addressed the challenging situation of many relevant time-scales, where it is no longer adequate to rely solely on the mean first-passage time for an appropriate description. In doing so, we presented sharp non-asymptotic bounds on average minimal and maximal deviations around the mean, and argued that these results have direct implications for target-search processes with multiple searchers.

To conclude, the presented framework of non-asymptotic uncertainty quantification is applicable to the broad class of reversible Markov dynamics that include discrete Markovian jump-processes in any dimension and Markovian diffusion in effectively one-dimensional potential landscapes. Accordingly, it will be interesting to see how the developed theory can be turned into new statistical tests in the future. Moreover, extensions to, for example, irreversible (i.e., driven) dynamics and more complicated observables will be carried out in the future.

**Acknowledgements** Financial support from Studienstiftung des Deutschen Volkes (to R. B.) and the German Research Foundation (DFG) through the Emmy Noether Program GO 2762/1-2 (to A. G.) is gratefully acknowledged.

## References


1. S. Redner, *A Guide to First-Passage Processes* (Cambridge University Press, Cambridge, 2001)
2. R. Metzler, S. Redner, G. Oshanin, *First-Passage Phenomena and their Applications* (World Scientific, Singapore, 2014)
3. K. Lindenberg, R. Metzler, G. Oshanin, *Chemical Kinetics: Beyond the Textbook* (World Scientific, New Jersey, 2019)
4. S. Iyer-Biswas, A. Zilman, First-passage processes in cellular biology. Adv. Chem. Phys. **160**, 261–306 (2016)
5. Y. Zhang, O.K. Dudko, First-passage processes in the genome. Annu. Rev. Biophys. **45**, 117–134 (2016)
6. H. Kramers, Brownian motion in a field of force and the diffusion model of chemical reactions. Physica **7**, 284–304 (1940)
7. P. Hänggi, P. Talkner, M. Borkovec, Reaction-rate theory: fifty years after Kramers. Rev. Mod. Phys. **62**, 251–341 (1990)
8. A. Szabo, K. Schulten, Z. Schulten, First passage time approach to diffusion controlled reactions. J. Chem. Phys. **72**, 4350–4357 (1980)
9. E. Ben-Naim, S. Redner, F. Leyvraz, Decay kinetics of ballistic annihilation. Phys. Rev. Lett. **70**, 1890–1893 (1993)
10. D.S. Grebenkov, R. Metzler, G. Oshanin, Towards a full quantitative description of single-molecule reaction kinetics in biological cells. Phys. Chem. Chem. Phys. **20**, 16393–16401 (2018)
11. D.S. Grebenkov, R. Metzler, G. Oshanin, Strong defocusing of molecular reaction times results from an interplay of geometry and reaction control. Commun. Chem. **1**, 1 (2018)
12. D.S. Grebenkov, Universal formula for the mean first passage time in planar domains. Phys. Rev. Lett. **117**, 260201 (2016)
13. P.C. Bressloff, J.M. Newby, Stochastic models of intracellular transport. Rev. Mod. Phys. **85**, 135–196 (2013)
14. É. Roldán, A. Lisica, D. Sánchez-Taltavull, S.W. Grill, Stochastic resetting in backtrack recovery by RNA polymerases. Phys. Rev. E **93**, 062411 (2016)
15. K.R. Ghusinga, J.J. Dennehy, A. Singh, First-passage time approach to controlling noise in the timing of intracellular events. Proc. Natl. Acad. Sci. **114**, 693–698 (2017)
16. K. Rijal, A. Prasad, A. Singh, D. Das, Exact distribution of threshold crossing times for protein concentrations: Implication for biological timekeeping. Phys. Rev. Lett. **128**, 048101 (2022)
17. J.J. Parmar, D. Das, R. Padinhateeri, Theoretical estimates of exposure timescales of protein binding sites on DNA regulated by nucleosome kinetics. Nucleic Acids Res. **44**, 1630–1641 (2015)
18. D.A. Charlebois, N. Abdennur, M. Kaern, Gene expression noise facilitates adaptation and drug resistance independently of mutation. Phys. Rev. Lett. **107**, 218101 (2011)
19. F. Frey, F. Ziebert, U.S. Schwarz, Stochastic dynamics of nanoparticle and virus uptake. Phys. Rev. Lett. **122**, 088102 (2019)
20. A.L. Lloyd, R.M. May, How viruses spread among computers and people. Science **292**, 1316–1317 (2001)
21. L. Hufnagel, D. Brockmann, T. Geisel, Forecast and control of epidemics in a globalized world. Proc. Natl. Acad. Sci. **101**, 15124–15129 (2004)
22. O. Bénichou, C. Loverdo, M. Moreau, R. Voituriez, Intermittent search strategies. Rev. Mod. Phys. **83**, 81–129 (2011)
23. F. Boccardo, O. Pierre-Louis, Controlling the shape of small clusters with and without macroscopic fields. Phys. Rev. Lett. **128**, 256102 (2022)
24. T. Erdmann, U.S. Schwarz, Stability of adhesion clusters under constant force. Phys. Rev. Lett. **92**, 108102 (2004)
25. S. Chakrabarti, M. Hinczewski, D. Thirumalai, Plasticity of hydrogen bond networks regulates mechanochemistry of cell adhesion complexes. Proc. Natl. Acad. Sci. **111**, 9048–9053 (2014)

26. K. Blom, A. Godec, Criticality in cell adhesion. Phys. Rev. X **11**, 031067 (2021)
27. I. Goychuk, P. Hänggi, Ion channel gating: a first-passage time analysis of the kramers type. Proc. Natl. Acad. Sci. **99**, 3552–3556 (2002)
28. O.G. Berg, R.B. Winter, P.H. Von Hippel, Diffusion-driven mechanisms of protein translocation on nucleic acids. 1. models and theory. Biochem. **20**, 6929–6948 (1981)
29. E. Koslover, M. Díaz de la Rosa, A. Spakowitz, Theoretical and computational modeling of target-site search kinetics in vitro and in vivo. Biophys. J. **101**, 856–865 (2011)
30. D. Holcman, Z. Schuss, Time scale of diffusion in molecular and cellular biology. J. Phys. A: Math. Theor. **47**, 173001 (2014)
31. O. Bénichou, C. Chevalier, B. Meyer, R. Voituriez, Facilitated diffusion of proteins on chromatin. Phys. Rev. Lett. **106**, 038102 (2011)
32. E.G. Marklund, A. Mahmutovic, O.G. Berg, P. Hammar, D. van der Spoel, D. Fange, J. Elf, Transcription-factor binding and sliding on DNA studied using micro- and macroscopic models. Proc. Natl. Acad. Sci. **110**, 19796–19801 (2013)
33. M. Bauer, R. Metzler, In vivo facilitated diffusion model. PLoS ONE **8**, e53956 (2013)
34. O. Bénichou, C. Chevalier, J. Klafter, B. Meyer, R. Voituriez, Geometry-controlled kinetics. Nat. Chem. **2**, 472–477 (2010)
35. O. Bénichou, R. Voituriez, From first-passage times of random walks in confinement to geometry-controlled kinetics. Phys. Rep. **539**, 225–284 (2014)
36. A. Godec, R. Metzler, Universal proximity effect in target search kinetics in the few-encounter limit. Phys. Rev. X **6**, 041037 (2016)
37. A.F. Siegert, On the first passage time probability problem. Phys. Rev E **81**, 617–623 (1951)
38. J. Newby, J. Allard, First-passage time to clear the way for receptor-ligand binding in a crowded environment. Phys. Rev. Lett. **116**, 128101 (2016)
39. I. Nayak, D. Das, A. Nandi, Comparison of mechanisms of kinetochore capture with varying number of spindle microtubules. Phys. Rev. Res. **2**, 013114 (2020)
40. Z. Schuss, *Theory and Applications of Stochastic Processes: An Analytical Approach* (Springer Science & Business Media, New York, 2009)
41. J.-P. Bouchaud, M. Potters, *Theory of Financial Risk and Derivative Pricing: From Statistical Physics to Risk Management* (Cambridge University Press, Cambridge, 2003)
42. W. Paul, J. Baschnagel, *Stochastic Processes: From Physics to Finance* (Springer International Publishing, Cham, 2013)
43. C.W. Gardiner, *Handbook of Stochastic Methods for Physics, Chemistry and the Natural Sciences* (Springer, Berlin, 1985)
44. D.B. Dougherty, I. Lyubinetsky, E.D. Williams, M. Constantin, C. Dasgupta, S. Sarma, Experimental persistence probability for fluctuating steps. Phys. Rev. Lett. **89**, 136102 (2002)
45. M. Constantin, S.D. Sarma, C. Dasgupta, O. Bondarchuk, D.B. Dougherty, E.D. Williams, Infinite family of persistence exponents for interface fluctuations. Phys. Rev. Lett. **91**, 086103 (2003)
46. J. Merikoski, J. Maunuksela, M. Myllys, J. Timonen, M.J. Alava, Temporal and spatial persistence of combustion fronts in paper. Phys. Rev. Lett. **90**, 024501 (2003)
47. M. Constantin, C. Dasgupta, P.P. Chatraphorn, S.N. Majumdar, S.D. Sarma, Persistence in nonequilibrium surface growth. Phys. Rev. E **69**, 061608 (2004)
48. D.B. Dougherty, C. Tao, O. Bondarchuk, W.G. Cullen, E.D. Williams, M. Constantin, C. Dasgupta, S.D. Sarma, Sampling-time effects for persistence and survival in step structural fluctuations. Phys. Rev. E **71**, 021602 (2005)
49. C. Godrèche, S.N. Majumdar, G. Schehr, Longest excursion of stochastic processes in nonequilibrium systems. Phys. Rev. Lett. **102**, 240602 (2009)
50. A.J. Bray, S.N. Majumdar, G. Schehr, Persistence and first-passage properties in nonequilibrium systems. Adv. Phys. **62**, 225–361 (2013)
51. T. Kay, L. Giuggioli, Diffusion through permeable interfaces: Fundamental equations and their application to first-passage and local time statistics. Phys. Rev. Res. **4**, 032039 (2022)
52. S. Bo, L. Hubatsch, J. Bauermann, C.A. Weber, F. Jülicher, Stochastic dynamics of single molecules across phase boundaries. Phys. Rev. Res. **3**, 043150 (2021)

53. U. Seifert, Stochastic thermodynamics, fluctuation theorems and molecular machines. Rep. Prog. Phys. **75**, 126001 (2012)
54. T.R. Gingrich, J.M. Horowitz, Fundamental bounds on first passage time fluctuations for currents. Phys. Rev. Lett. **119**, 170601 (2017)
55. S. Singh, P. Menczel, D.S. Golubev, I.M. Khaymovich, J.T. Peltonen, C. Flindt, K. Saito, É. Roldán, J.P. Pekola, Universal first-passage-time distribution of non-gaussian currents. Phys. Rev. Lett. **122**, 230602 (2019)
56. E. Roldán, I. Neri, M. Dörpinghaus, H. Meyr, F. Jülicher, Decision making in the arrow of time. Phys. Rev. Lett. **115**, 250602 (2015)
57. I. Neri, E. Roldán, F. Jülicher, Statistics of infima and stopping times of entropy production and applications to active molecular processes. Phys. Rev. X **7**, 011019 (2017)
58. G. Falasco, M. Esposito, Dissipation-time uncertainty relation. Phys. Rev. Lett. **125**, 120604 (2020)
59. I. Neri, Estimating entropy production rates with first-passage processes. J. Phys. A: Math. Theor. **55**, 304005 (2022)
60. J.P. Garrahan, Simple bounds on fluctuations and uncertainty relations for first-passage times of counting observables. Phys. Rev. E **95**, 032134 (2017)
61. K. Hiura, S. Ichi Sasa, Kinetic uncertainty relation on first-passage time for accumulated current. Phys. Rev. E **103**, 050103 (2021)
62. M. Kac, On some connections between probability theory and differential and integral equations, *Proceedings of the Second Berkeley Symposium on Mathematical Statistics and Probability*, pp. 189–215, (University of California Press, Berkeley, 1951). D. Hartich, A. Godec, Reaction Kinetics in the Few-Encounter Limit, in *Chemical Kinetics* (World Scientific, New Jersey, 2019), Chap. 11, pp. 265–283
63. G. Schehr, S.N. Majumdar, Exact record and order statistics of random walks via first-passage ideas, in *First-Passage Phenomena and Their Applications* (World Scientific, Singapore, 2014), Chap. 1, pp. 226–251
64. S.N. Majumdar, G. Schehr, G. Wergen, Record statistics and persistence for a random walk with a drift. J. Phys. A: Math. Theor. **45**, 355002 (2012)
65. D. Hartich, A. Godec, Extreme value statistics of ergodic Markov processes from first passage times in the large deviation limit. J. Phys. A: Math. Theor. **52**, 244001 (2019)
66. H. Friedman, D.A. Kessler, E. Barkai, Quantum walks: the first detected passage time problem. Phys. Rev. E **95**, 032141 (2017)
67. F. Thiel, E. Barkai, D.A. Kessler, First detected arrival of a quantum walker on an infinite line. Phys. Rev. Lett. **120**, 040502 (2018)
68. M.J. Kearney, S.N. Majumdar, On the area under a continuous time Brownian motion till its first-passage time. J. Phys. A: Math. Gen. **38**, 4097–4104 (2005)
69. M.J. Kearney, S.N. Majumdar, R.J. Martin, The first-passage area for drifted Brownian motion and the moments of the airy distribution. J. Phys. A: Math. Theor. **40**, F863–F869 (2007)
70. M.J. Kearney, S.N. Majumdar, Statistics of the first passage time of Brownian motion conditioned by maximum value or area. J. Phys. A: Math. Theor. **47**, 465001 (2014)
71. M.J. Kearney, R.J. Martin, Statistics of the first passage area functional for an Ornstein–Uhlenbeck process. J. Phys. A: Math. Theor. **54**, 055002 (2021)
72. S.N. Majumdar, B. Meerson, Statistics of first-passage Brownian functionals. J. Stat. Mech.: Theory Exp. 023202 (2020)
73. P. Singh, A. Pal, First-passage Brownian functionals with stochastic resetting. J. Phys. A: Math. Theor. **55**, 234001 (2022)
74. G. Mercado-Vásquez, D. Boyer, First hitting times to intermittent targets. Phys. Rev. Lett. **123**, 250603 (2019)
75. A. Kumar, A. Zodage, M.S. Santhanam, First detection of threshold crossing events under intermittent sensing. Phys. Rev. E **104**, 052103 (2021)
76. J.L. Spouge, A. Szabo, G.H. Weiss, Single-particle survival in gated trapping. Phys. Rev. E **54**, 2248–2255 (1996)

77. Y. Scher, S. Reuveni, Unified approach to gated reactions on networks. Phys. Rev. Lett. **127**, 018301 (2021)
78. A. Kumar, Y. Scher, S. Reuveni, M.S. Santhanam, Inference from gated first-passage times. Phys. Rev. Res. **5** (2023)
79. E. Woillez, Y. Zhao, Y. Kafri, V. Lecomte, J. Tailleur, Activated escape of a self-propelled particle from a metastable state. Phys. Rev. Lett. **122**, 258001 (2019)
80. F. Mori, P.L. Doussal, S.N. Majumdar, G. Schehr, Universal survival probability for a $n$-dimensional run-and-tumble particle. Phys. Rev. Lett. **124**, 090603 (2020)
81. F.D. Trapani, T. Franosch, M. Caraglio, Active Brownian particles in a circular disk with an absorbing boundary. Phys. Rev. E **107** (2023)
82. P. Hänggi, P. Talkner, Memory index of first-passage time: a simple measure of non-Markovian character. Phys. Rev. Lett. **51**, 2242–2245 (1983)
83. P. Hänggi, P. Talkner, First-passage time problems for non-Markovian processes. Phys. Rev. A **32**, 1934–1937 (1985)
84. T. Guérin, N. Levernier, O. Bénichou, R. Voituriez, Mean first-passage times of non-Markovian random walkers in confinement. Nature **534**, 356–359 (2016)
85. M.R. Evans, S.N. Majumdar, Diffusion with stochastic resetting. Phys. Rev. Lett. **106**, 160601 (2011)
86. L. Kusmierz, S.N. Majumdar, S. Sabhapandit, G. Schehr, First order transition for the optimal search time of Lévy flights with resetting. Phys. Rev. Lett. **113**, 220602 (2014)
87. S. Reuveni, Optimal stochastic restart renders fluctuations in first passage times universal. Phys. Rev. Lett. **116**, 170601 (2016)
88. A. Pal, S. Reuveni, First passage under restart. Phys. Rev. Lett. **118**, 030603 (2017)
89. A. Pal, I. Eliazar, S. Reuveni, First passage under restart with branching. Phys. Rev. Lett. **122**, 020602 (2019)
90. M.R. Evans, S.N. Majumdar, G. Schehr, Stochastic resetting and applications. J. Phys. A: Math. Theor. **53**, 193001 (2020)
91. B. Besga, A. Bovon, A. Petrosyan, S.N. Majumdar, S. Ciliberto, Optimal mean first-passage time for a brownian searcher subjected to resetting: Experimental and theoretical results. Phys. Rev. Res. **2**, 032029 (2020)
92. O. Tal-Friedman, A. Pal, A. Sekhon, S. Reuveni, Y. Roichman, Experimental realization of diffusion with stochastic resetting. J. Phys. Chem. Lett. **11**, 7350–7355 (2020)
93. B.D. Bruyne, J. Randon-Furling, S. Redner, Optimization in first-passage resetting. Phys. Rev. Lett. **125**, 050602 (2020)
94. B.D. Bruyne, S.N. Majumdar, G. Schehr, Optimal resetting Brownian bridges via enhanced fluctuations. Phys. Rev. Lett. **128**, 200603 (2022)
95. V. Stojkoski, P. Jolakoski, A. Pal, T. Sandev, L. Kocarev, R. Metzler, Income inequality and mobility in geometric Brownian motion with stochastic resetting: theoretical results and empirical evidence of non-ergodicity. Phil. Trans. R. Soc. A. **380**, 20210157 (2022)
96. P. Xu, T. Zhou, R. Metzler, W. Deng, Stochastic harmonic trapping of Lévy walk: transport and first-passage dynamics under soft resetting strategies. New J. Phys. **24**, 033003 (2022)
97. D. Hartich, A. Godec, Duality between relaxation and first passage in reversible markov dynamics: rugged energy landscapes disentangled. New J. Phys. **20**, 112002 (2018)
98. D. Hartich, A. Godec, Interlacing relaxation and first-passage phenomena in reversible discrete and continuous space Markovian dynamics. J. Stat. Mech. **2019**, 024002 (2019)
99. R. Bebon, A. Godec, Controlling uncertainty of empirical first-passage times in the small-sample regime. Phys. Rev. Lett. **131**, 237101 (2023)
100. A.V. Skorokhod, Stochastic equations for diffusion processes in a bounded region. Theory Probab. its Appl. **6**, 264–274 (1961)
101. K. Ito, H.P. McKean, *Diffusion Processes and Their Sample Paths* (Springer, Berlin, 1965)
102. H. Tanaka, Stochastic differential equations with reflecting boundary condition in convex regions. Hiroshima Math. J. **9**, 163–177 (1979)
103. P.-L. Lions, A.-S. Sznitman, Stochastic differential equations with reflecting boundary conditions. Commun. Pure Appl. Math. **37**, 511–537 (1984)

104. D.S. Grebenkov, Probability distribution of the boundary local time of reflected Brownian motion in Euclidean domains. Phys. Rev. E **100**, 062110 (2019)
105. D.S. Grebenkov, Paradigm shift in diffusion-mediated surface phenomena. Phys. Rev. Lett. **125**, 078102 (2020)
106. T.G. Mattos, C. Mejía-Monasterio, R. Metzler, G. Oshanin, First passages in bounded domains: when is the mean first passage time meaningful? Phys. Rev. E **86**, 031143 (2012)
107. C. Chevalier, O. Bénichou, B. Meyer, R. Voituriez, First-passage quantities of Brownian motion in a bounded domain with multiple targets: a unified approach. J. Phys. A: Math. Theor. **44**, 025002 (2010)
108. B. Meyer, C. Chevalier, R. Voituriez, O. Bénichou, Universality classes of first-passage-time distribution in confined media. Phys. Rev. E **83**, 051116 (2011)
109. G.R. Bowman, V.S. Pande, F. Noé, *An Introduction to Markov State Models and their Application to Long Timescale Molecular Simulation* (Springer Science & Business Media, Dordrecht, 2013)
110. F. Noé, S. Olsson, J.Köhler, H. Wu, Boltzmann generators: sampling equilibrium states of many-body systems with deep learning. Science **365** (2019)
111. E. Braun, J. Gilmer, H.B. Mayes, D.L. Mobley, J.I. Monroe, S. Prasad, D.M. Zuckerman, Best practices for foundations in molecular simulations [article v1.0]. Living J. Comp. Mol. Sci. **1**, 1 (2019)
112. R.V. Hogg, J.W. McKean, A.T. Craig, *Introduction to Mathematical Statistics* (Pearson, Boston, 2018)
113. K. Lindorff-Larsen, S. Piana, R.O. Dror, D.E. Shaw, How fast-folding proteins fold. Science **334**, 517–520 (2011)
114. J.L. Adelman, M. Grabe, Simulating rare events using a weighted ensemble-based string method. J. Chem. Phys. **138**, 044105 (2013)
115. V. Gapsys, B.L. de Groot, On the importance of statistics in molecular simulations for thermodynamics, kinetics and simulation box size. eLife **9**, e57589 (2020)
116. B. Mostofian, D.M. Zuckerman, Statistical uncertainty analysis for small-sample, high log-variance data: Cautions for bootstrapping and Bayesian bootstrapping. J. Chem. Theory Comput. **15**, 3499–3509 (2019)
117. R. Mehra, K.P. Kepp, Cell size effects in the molecular dynamics of the intrinsically disordered Aβ peptide. J. Chem. Phys. **151**, 085101 (2019)
118. A. Militaru, M. Innerbichler, M. Frimmer, F. Tebbenjohanns, L. Novotny, C. Dellago, Escape dynamics of active particles in multistable potentials. Nat. Commun. **12**, 1 (2021)
119. L. Rondin, J. Gieseler, F. Ricci, R. Quidant, C. Dellago, L. Novotny, Direct measurement of Kramers turnover with a levitated nanoparticle. Nat. Nanotechnol. **12**, 1130–1133 (2017)
120. O. Pulkkinen, R. Metzler, Variance-corrected Michaelis-Menten equation predicts transient rates of single-enzyme reactions and response times in bacterial gene-regulation. Sci. Rep. **5**, 17820 (2015)
121. S. Boucheron, G. Lugosi, P. Massart, *Concentration Inequalities: A Nonasymptotic Theory of Independence* (Oxford University Press, New York, 2013)
122. A. Gelman, J.B. Carlin, H.S. Stern, D.B. Rubin, *Bayesian Data Analysis* (Chapman and Hall/CRC, London, 1995)
123. A. Godec, R. Metzler, First passage time distribution in heterogeneity controlled kinetics: going beyond the mean first passage time. Sci. Rep. **6** (2016)
124. A.L. Thorneywork, J. Gladrow, Y. Qing, M. Rico-Pasto, F. Ritort, H. Bayley, A.B. Kolomeisky, U.F. Keyser, Direct detection of molecular intermediates from first-passage times. Sci. Adv. **6**, 1 (2020)
125. D. Hartich, A. Godec, Reaction kinetics in the few-encounter limit, in *Chemical Kinetics* (World Scientific, New Jersey, 2019), Chap. 11, pp. 265–283
126. S.D. Lawley, J.B. Madrid, First passage time distribution of multiple impatient particles with reversible binding. J. Chem. Phys. **150**, 214113 (2019)
127. S.D. Lawley, J.B. Madrid, A probabilistic approach to extreme statistics of Brownian escape times in dimensions 1, 2, and 3. J. Nonlin. Sci. **30**, 1207–1227 (2020)

128. S.D. Lawley, Universal formula for extreme first passage statistics of diffusion. Phys. Rev. E **101**, 012413 (2020)
129. J. Yang, I. Kupka, Z. Schuss, D. Holcman, Search for a small egg by spermatozoa in restricted geometries. J. Math. Biol. **73**, 423–446 (2015)

# Chapter 3
# Defocusing of First-Reaction Times

**Denis Grebenkov, Ralf Metzler, and Gleb Oshanin**

**Abstract** In this chapter, we focus on the full probability density function $H(\mathbf{x}_0, t)$ of the first-reaction time (FRT) in a geometrically simple model, in which a diffusive particle, starting at a fixed position $\mathbf{x}_0$, searches for and eventually binds to an immobile, *partially reactive* (that is, having a finite intrinsic reactivity $\kappa$) spherical binding site, located in the center of a spherical domain. The entire domain is enclosed by an impermeable boundary. For such a model, we determine an exact expression for $H(\mathbf{x}_0, t)$ for arbitrary values of $\kappa$ and $\mathbf{x}_0$. Inspecting its functional form, we identify four distinct temporal regimes: a rise to the maximal value, a power-law descent, an extended in time plateau-like region and the terminal exponential tail, as well as the corresponding regime-delimiting time-scales. We show that the commonly studied *mean* FRT is associated with the large-$t$ exponential tail and hence, its value is supported by anomalously long searching trajectories. In turn, the most probable FRTs, that should be observed for a substantial amount of realizations, can be several orders of magnitude shorter than the mean FRT, revealing a large defocusing of the first-reaction times. We also define the effective broadness of $H(\mathbf{x}_0, t)$ by calculating the corresponding coefficient of variation and specify the reaction depths at each of the temporal stages of the binding process. In particular, we demonstrate that a large fraction of trajectories react before the mean FRT. Finally, we present a simple approximate expression for $H(\mathbf{x}_0, t)$ which may be useful for fitting data garnered in experiments.

D. Grebenkov (✉)
Laboratoire de Physique de la Matière Condensée (PMC),
CNRS/Ecole Polytechnique/Institut Polytechnique de Paris, 91120 Palaiseau, France
e-mail: denis.grebenkov@polytechnique.edu

R. Metzler
Institute of Physics and Astronomy, University of Potsdam, 14476 Potsdam-Golm, Germany
e-mail: rmetzler@uni-potsdam.de

G. Oshanin
Laboratoire de Physique Théorique de la Matière Condensée, CNRS/Sorbonne Université,
4 Place Jussieu, 75252 Paris, France
e-mail: gleb.oshanin@sorbonne-universite.fr



D. Grebenkov et al. (eds.), *Target Search Problems*,
https://doi.org/10.1007/978-3-031-67802-8_3

## 3.1 Introduction

The binding of a diffusive particle (e.g., a ligand) to an immobile, small target (binding) site located within some spatially bounded domain represents an essential step in a variety of processes encountered in molecular biology [1, 2]. In particular, such events control different intermediate stages of the multi-stage cell-to-cell communication processes, or of the intra-cellular signal transduction; in the latter, a binding event often accomplishes a preceding stage and triggers the launch of the next one in the transduction cascade. Similarly, diverse processes underlying the immune reactions, regulation, infection, metabolism or transmitter release in neurons [1–3] (see also Chap. 26) are entirely controlled by such events.

Depending on the case at hand, this particle can be an ion or a protein, or simply a chemically active molecule. The binding site can be a receptor (see, e.g., [4, 5]), or, in some applications, a small aperture in, e.g., the nuclear envelope through which the particle can enter or conversely, exit the nucleus [6–11]. The bounded domain can be a cell or its nucleus, a compartment, a vesicle, etc. In turn, a binding event may occur instantaneously upon the first encounter of the particle with the target site, in which case the reaction is called *perfect* and one is naturally interested in determining the *first-passage time* to the binding site from the initial particle's position. These are random variables whose properties are amply discussed in the present book, for different geometrical settings and physical conditions.

Concurrently, if a successful binding to the target site requires overpassing an energy barrier against the reaction [12–17], or an entropy barrier in case of an exit from a bounded domain or of an entrance to a nuclear pore (see, e.g., [18]), one deals with an *imperfect* reaction occurring at a partially reactive site. In this latter situation, the first encounter with the binding site may not necessarily lead to the binding event–the particle can be reflected back to the medium and may attempt to bind again and again, when it arrives at the site at some random instants of time after a series of unsuccessful binding attempts. A binding event may therefore necessitate many repeated attempts interspersed with diffusive excursions in the bulk. Evidently, the crucial random variable is the *first-reaction time* (FRT). Of course, typical values of such random variables (which are also often called the first-passage times to the reaction event) may by far exceed the first-passage time to the target site.

The standard line of thought often relies only on the *mean* values of these random variables–of the first-passage times and/or of the FRTs. Indeed, in a bounded domain a diffusive particle is *certain* to find the target site, and eventually, to react with it even for imperfect reactions. In confinement, the probability density function (PDF) of the FRT should be characterized by an exponential long-time tail which will ensure the existence of all moments of the FRT (see, e.g., [19]). In particular, it was argued in [20] that for a perfectly reactive target site and in situations when the ligand's starting point is random–that being, the particle may start at arbitrary point within the bounded domain of volume $V$ with uniform probability density $1/V$—the PDF $H(t)$ of the first-passage time $t$ is very accurately approximated by

$$H(t) = q\delta(t) + \frac{(1-q)}{\langle t \rangle} \exp\left(-\frac{t}{\langle t \rangle}\right), \tag{3.1}$$

where $\delta(t)$ is the Dirac delta-distribution, $\langle t \rangle$ is here the volume-averaged mean first-passage time[1] to the target site and $q$ represents the fraction of trajectories that arrive at the target site (almost) immediately—the so-called "direct" trajectories [21–23] which go straight from the starting point to the target site. On this basis, it is often argued that the first moment of the PDF is a meaningful characteristic time-scale which suffices to comprehend how much time indeed is necessary for the desired binding event to take place. As a consequence, most of the available analyses dealing with reactions in bounded domains concentrate exclusively on the calculation of the mean FRT.

However, while the mean $\langle t \rangle$ is an important time-scale (whose derivation already often is quite elaborate), one should not forget that we deal with random variables and $\langle t \rangle$ is just the first moment of the PDF. In this way, it defines only a mean value in an ensemble of observed FRTs and it is legitimate to inquire how different these observed values can be and how strongly they can deviate from $\langle t \rangle$. In particular, if one takes two different realizations of ligand-binding processes out of many, how likely is it that the observed values of the FRTs $t_1$ and $t_2$ in two realizations will be close to each other and, respectively, to $\langle t \rangle$? In other words, is the PDF centered at some value close to $\langle t \rangle$, or is $\langle t \rangle$ rather associated with the tail of the PDF and therefore a somewhat misleading measure, largely overestimating the typical FRTs observed in experiments?

As we already remarked above, the PDF in Eq. (3.1)—in which $\langle t \rangle$ indeed appears to be the unique characteristic time-scale—applies to the somewhat simplified situation when a) the reaction is *perfect* and takes place instantaneously upon the first arrival to the binding site, and b) the starting point $x_0$ is *uniformly* distributed within the bounded domain. In many ligand-binding processes, however, the starting point is fixed to a prescribed location, e.g., it may be at the entrance of (or exit from) a nuclear pore or of a gated ion-channel. Concurrently, the binding events are often conditioned on the passage through a barrier against the reaction and therefore, may require multiple reaction attempts. In fact, there is ample evidence that even in rather geometrically simple systems the first-passage times may exhibit significant sample-to-sample fluctuations—their values observed in different realizations of the binding process can be disproportionally different, such that the coefficient of variation of the PDF, defined as the ratio of the standard deviation of the first-passage time and of its first moment, can exceed unity. This means that fluctuations around the mean value are bigger than the mean value itself [24–26]. In more complicated geometries and for imperfect reactions, a defocusing of the FRTs—a scatter of FRTs observed in different realizations of the process—can be strongly amplified [27–33]. Note that even the PDF in Eq. (3.1) is, in fact, a monotonically decreasing function of $t$ and hence, is not centered at the mean first-passage time. Therefore, smaller values of $t$,

[1] The triangular brackets $\langle \cdot \rangle$ here and henceforth denotes averaging with respect to the FRT PDF.

as compared to $\langle t \rangle$, are more probable.[2] Overall, this suggests that the mean FRT may not be a unique time-scale, fluctuations of the FRTs may be important and therefore, a knowledge of the full FRT PDF is indispensable in order to get a fully comprehensive picture of the time-scales characterizing the ligand-binding processes.

In this chapter, going well-beyond the picture based on the mean FRT, we address the question of the full PDF of this random variable. To this end, following our earlier paper [34], we concentrate here on the case (depicted in Fig. 3.1) of a spherical bounded domain with an immobile, partially reactive spherical target site placed at the center of the domain. For such a model we evaluate an exact expression for the full PDF $H(\mathbf{x}_0, t)$ of the FRT, for fixed starting point $\mathbf{x}_0$ and for arbitrary values of the intrinsic reactivity. Capitalizing on the exact expression for $H(\mathbf{x}_0, t)$, we identify four distinct temporal regimes and the regime-delimiting time-scales, as well as the reaction depths (i.e., the relative weights) corresponding to different temporal stages. We will show, in particular, that the mean FRT is associated with the large-$t$ tail of the PDF and hence, its value is supported by anomalously long searching trajectories. In turn, the most probable FRTs, that should be observed for a large fraction of realizations, can be several orders of magnitude shorter than the mean FRT, revealing a large defocusing of the FRTs. We also discuss a generalization of the approximate expression (3.1) for the case of a fixed starting point and finite intrinsic reactivity.

We note finally that despite the geometrical simplicity of the model (e.g., in realistic scenarios the binding site may not necessarily be placed in the center of the domain, only have a finite reactive surface patch, or be attached to some structure that screens it [19, 20, 27–33]), it takes into account explicitly the two most important aspects: a diffusive search for the target site and its finite reactivity, allowing us to unveil in a very direct way several generic features without resorting to any approximation. The identified temporal regimes are also present in geometrically more complicated situations, but to make them apparent would require a much more mathematically involved analysis or simulations (see [27–33, 35, 36]).

## 3.2 The Model and Basic Equations

Consider a pair of reactive molecules (see Fig. 3.1): a partially reactive immobile binding site of radius $\rho$ placed at the center of a three-dimensional spherical domain of radius $R$ enclosed by an impermeable boundary, and a (point-like) particle initially placed at some prescribed position within the bounded domain and diffusing with the diffusion coefficient $D$. Whenever the molecule hits the surface of the binding site, it may react with (or irreversibly bind to) it with a finite probability $p$ ($0 \le p \le 1$), which defines the so-called intrinsic reaction rate $\kappa \propto p/(1-p)$ (see, e.g., [15, 16]

---

[2] Note that the coefficient of variation of the random variable in (3.1) is always larger than unity for all $q$ values.

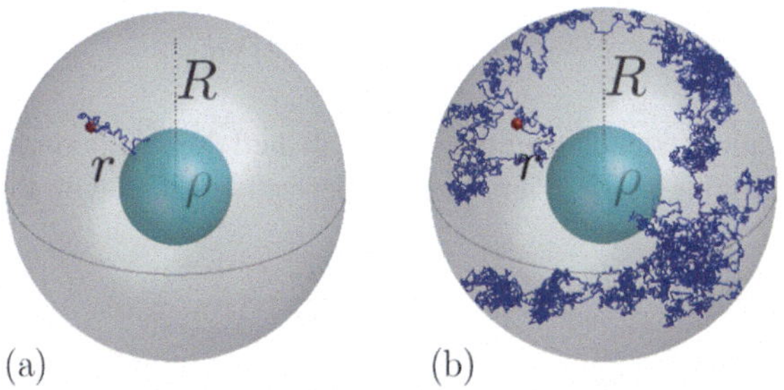


**Fig. 3.1** A sketch of the target search problem: a molecule (small red ball) starts a distance $r$ away from the origin and diffusively searches for an immobile spherical target of radius $\rho$ (light blue), which is surrounded by a concentric impenetrable sphere of radius $R$ (gray). Its random trajectory (in blue) can either go directly to the target (panel **a**) or explore the confining domain (panel **b**), resulting in a small or large value of the random first-reaction time, respectively

for more details). Evidently, the limiting case $p = 1$ (and hence, $\kappa = \infty$) corresponds to a *perfect* reaction in which the FRT is just the first-passage time to the target site, while in the opposite limit $p = 0$ (implying that $\kappa = 0$) the reaction is completely inhibited, i.e., the target site is perfectly reflecting and the binding never takes place.

Let $P(\mathbf{x}, t|\mathbf{x}_0)$ denote the position PDF at time instance $t$ for a particle started from $\mathbf{x}_0$ at time 0. This propagator obeys the diffusion equation within a spherical-shell domain, bounded by the surface of the target site and by the outer impermeable boundary,

$$\frac{\partial}{\partial t} P(\mathbf{x}, t|\mathbf{x}_0) = D \Delta P(\mathbf{x}, t|\mathbf{x}_0) , \tag{3.2}$$

where $\Delta$ is the Laplace operator. Equation (3.2) is to be complemented by the initial condition:

$$P(\mathbf{x}, t = 0|\mathbf{x}_0) = \delta(\mathbf{x} - \mathbf{x}_0) , \tag{3.3}$$

and two boundary conditions

$$\partial_n P(\mathbf{x}, t|\mathbf{x}_0)|_{\|\mathbf{x}\|=R} = 0, \tag{3.4a}$$

$$-D\partial_n P(\mathbf{x}, t|\mathbf{x}_0)|_{\|\mathbf{x}\|=\rho} = \kappa \; P(\mathbf{x}, t|\mathbf{x}_0)|_{\|\mathbf{x}\|=\rho} , \tag{3.4b}$$

where $\partial_n$ stands for the normal derivative, directed outwards from the domain. The first boundary condition describes that the particle is perfectly reflected by the boundary at $\|\mathbf{x}\| = R$ and therefore cannot leave the bounded domain, while the second is the so-called partially reactive boundary condition, which takes into account the imperfect reaction occurring at the surface $\|\mathbf{x}\| = \rho$ of the target site (see [15, 16]).

The properties of focal interest here is the probability that the particle, starting at $\mathbf{x}_0$, does not react with the target site up to time $t$, i.e., the so-called survival probability,

$$S(\mathbf{x}_0, t) = \int d^3\mathbf{x}\, P(\mathbf{x}, t|\mathbf{x}_0), \tag{3.5}$$

where the integration extends over the spherical-shell domain bounded by the target site and the enveloping surface, and its time-derivative, taken with the sign minus,

$$H(\mathbf{x}_0, t) = -\frac{\partial S(\mathbf{x}_0, t)}{\partial t}, \tag{3.6}$$

which defines the desired PDF of the time instant $t$ at which the first-reaction event takes place.

The rotational symmetry of the problem at hand greatly simplifies the analysis. Skipping the details of intermediate calculations (see [34] for these details), one finds that the Laplace-transformed FRT PDF,

$$\tilde{H}(r_0, p) = \int_0^\infty dt \exp(-pt)\, H(r_0, t), \tag{3.7}$$

with $r_0$ being the radial distance to the starting point, obeys

$$\tilde{H}''(r_0, p) + \frac{2}{r_0}\tilde{H}'(r_0, p) - \frac{p}{D}\tilde{H}(r_0, p) = 0, \tag{3.8a}$$

$$\tilde{H}'(r_0 = R, p) = 0, \tag{3.8b}$$

$$\left(\tilde{H}(r_0, p) - \frac{D}{\kappa}\tilde{H}'(r_0, p)\right)\Big|_{r_0=\rho} = 1, \tag{3.8c}$$

where the prime denotes a derivative with respect to $r_0$. To somewhat ease the notations, we drop the subscript 0 such that the initial radial distance to the starting point will be simply denoted by $r$ in what follows.

The solution of Eqs. (3.8a) to (3.8c) can be found by standard means [37–39] and reads [34]

$$\tilde{H}(r, p) = \frac{\kappa g(r)}{\kappa g(\rho) - D g'(\rho)}, \tag{3.9}$$

where

$$g(r) = \frac{R\sqrt{p/D}\cosh\xi - \sinh\xi}{r\sqrt{p/D}}, \tag{3.10}$$

with $\xi = (R - r)\sqrt{p/D}$. Expression (3.9) permits to determine the PDF in the time-domain.

## 3.3 The PDF of the First-Reaction Times: Temporal Regimes and Characteristic Time-Scales

This section is outlined as follows: We first briefly discuss the steps involved in the calculation of the inverse Laplace transform of $\tilde{H}(r, p)$. Then, we analyze the behavior of $H(r, t)$ as function of $t$ in the case of a small target site for two different starting positions—away from the target site and close to its surface, and for four different values of the intrinsic reaction rate $\kappa$ (see Fig. 3.3). This will permit us to get a qualitative understanding of how the value of $\kappa$ affects the shape of the PDF. Next, in Fig. 3.2, we depict the PDF as function of $t$ for two values of the intrinsic reaction rate and four different starting positions, highlighting in this way the crucial importance of the initial distance to the target. These figures will also allow us to identify four different temporal stages and the three corresponding characteristic time-scales. Next, we briefly quantify the effective broadness of $H(r, t)$ by analyzing the coefficient of variation of the PDF. Lastly, we discuss the asymptotic behavior of $H(r, t)$ in the short- and long-time limits and propose a generalization of the approximate expression (3.1), valid for arbitrary intrinsic reactivity and arbitrary fixed initial position.

### 3.3.1 The PDF in the Time-Domain

The Laplace inversion of expression (3.9) is standardly performed by identifying the poles of the function $\tilde{H}(r, p)$ in the complex plane, which amounts to finding the zeros of the denominator in (3.9). Introducing the dimensionless Laplace variable $s = (R - \rho)^2 p/D$, one thus searches for the zeros of the function

$$F(s) = \frac{1}{\rho^2\sqrt{s}}\Big(\big(\rho R + \mu(R-\rho)^2\big)\sqrt{s}\cosh\sqrt{s} \\ - \big(\rho(R-\rho) + \mu(R-\rho)^2 - \mu R\rho s\big)\sinh\sqrt{s}\Big), \tag{3.11}$$

where the parameter $\mu$ is given by

$$\mu = \frac{D}{\kappa(R-\rho)}. \tag{3.12}$$

We remark that $\mu$ is an important dimensionless parameter: in the case of a perfectly reactive target when $\kappa = \infty$, one evidently has $\mu = 0$, which corresponds, in terms of the classic theory of diffusion-limited reactions (see, e.g., [40, 41]), to the regime of the so-called "diffusion" control. When $\mu$ is equal to zero or is close to zero, reactions occur more likely upon the first encounter with the target site. The opposite regime of the "kinetic" control is realized when $\mu$ is large, which happens when

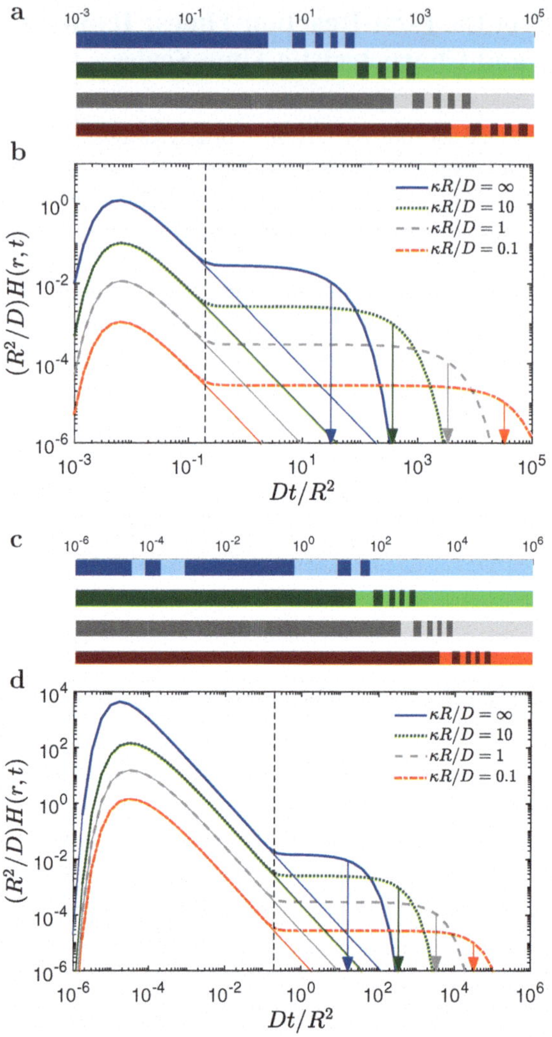


**Fig. 3.2** Effect of the distance to the starting point. **b**, **d** The rescaled PDF, $R^2 H(r, t)/D$, as a function of the dimensionless time $Dt/R^2$, for a target with radius $\rho/R = 0.01$, for the different initial radii $r$ indicated in the panels ($r$ increasing from top to bottom). The values of the reactivity are (**b**) $\kappa = \infty$ (perfectly reactive) and (**d**) $\kappa R/D = 1$ (partially reactive). The length and time units are fixed by setting $R = 1$ and $R^2/D = 1$. The colored vertical arrows indicate the corresponding mean FRTs for these cases (note that some arrows coincide). The vertical black dashed line indicates the crossover time $t_c$. Thin colored lines show the reaction time density $H_\infty(r, t)$ from Eq. (3.29) for the unbounded case ($R \to \infty$). **a, c** Colored bar codes indicate the cumulative depths corresponding to the four considered values of $r/R$ in increasing order from top to bottom. Each bar code is split into ten regions of alternating brightness, representing ten 10%-quantiles of the distribution. Despite distinctions in the probability densities in panel (**d**), the corresponding cumulative distributions are close to each other and result in very similar reaction depths [Adapted from [34]]

either $\kappa$ is small, or when the target site is large, $\rho \sim R$. In this case, it is most likely that multiple encounters with the target site are required before a successful reaction takes place.

A direct analysis shows that all poles of the Laplace-transformed PDF $\tilde{H}(r, p)$, i.e., the zeros of $F(p)$, lie on the negative real axis and can be simply related to the solutions of the transcendental equation

$$\tan\alpha = \frac{\alpha\big(\rho R + \mu(R-\rho)^2\big)}{\rho(R-\rho) + \mu(R-\rho)^2 + \mu R\rho\alpha^2}\,. \tag{3.13}$$

This equation has infinitely many positive solutions, which we denote as $\alpha_n$, $n = 0, 1, 2, \ldots$, organizing them in ascending order. A straightforward analysis [34] shows that the smallest solution $\alpha_0$ is always located within the interval $(0, \pi/2)$. More generally, the $n$-th order solution lies within the interval $(\pi n, \pi(n+1/2))$ and tends, for any fixed $\kappa$, to the left boundary of this interval as $n \to \infty$. Note that $\alpha = 0$ (or $p = 0$) is not a pole of the function $\tilde{H}(r, p)$.

Once the poles are identified, we calculate the residues by taking the derivative of $F(p)$ at the poles. Applying next the residue theorem, we arrive at the exact expression for the desired FRT PDF $H(r, t)$, for a particle starting at a distance $r$ from the origin,

$$H(r,t) = \sum_{n=0}^{\infty} u_n(r)\, e^{-Dt\lambda_n}, \tag{3.14}$$

where

$$\lambda_n = \alpha_n^2/(R-\rho)^2, \tag{3.15}$$

$$u_n(r) = c_n \frac{D}{(R-\rho)^2} \frac{R\alpha_n \cos\big(\alpha_n \frac{R-r}{R-\rho}\big) - (R-\rho)\sin\big(\alpha_n \frac{R-r}{R-\rho}\big)}{r\alpha_n}, \tag{3.16}$$

while the expansion coefficients $c_n$ are given explicitly by

$$c_n = \frac{2\rho^2\alpha_n^2}{(\rho R + \mu(R^2+\rho^2))\alpha_n \sin\alpha_n + \rho(\mu R\alpha_n^2 - \rho)\cos\alpha_n}\,. \tag{3.17}$$

Therefore, the form of the PDF is now completely defined in the time-domain.

### 3.3.2 Temporal Regimes and Effects of the Intrinsic Reactivity and of the Initial Position

In Figs. 3.3 and 3.2 we depict the rescaled FRT PDF $R^2 H(r,t)/D$, defined in Eqs. (3.14) to (3.17), as function of the dimensionless time $Dt/R^2$ for $\rho/R = 0.01$, and different values of the intrinsic reactivity and of the starting point.

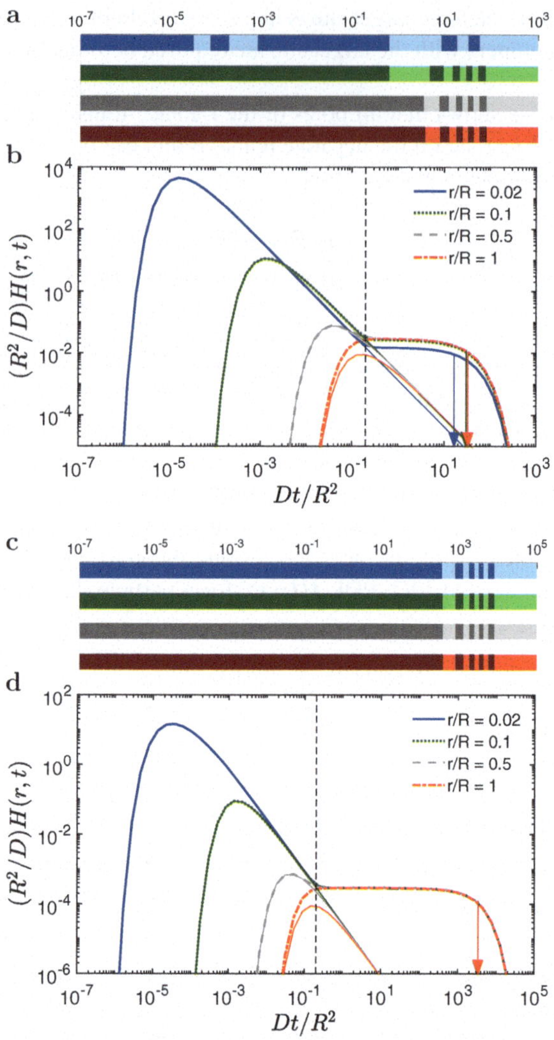


**Fig. 3.3** Effect of the intrinsic reactivity $\kappa$. **b, d** The rescaled PDF, $R^2H(r,t)/D$, as a function of the dimensionless time $Dt/R^2$, for a target site with radius $\rho/R = 0.01$, the starting point (**b**) $r/R = 0.2$ and (**d**) $r/R = 0.02$, and four progressively decreasing (from top to bottom) values of the dimensionless reactivity $\kappa' = \kappa R/D$. Length and time units are fixed by setting $R = 1$ and $R^2/D = 1$. The colored vertical arrows show the corresponding mean FRTs, while the vertical black dashed line indicates the crossover time $t_c$. Thin colored lines show the reaction time density $H_\infty(r,t)$ from Eq. (3.29) for the unbounded case ($R \to \infty$). **a, c** Colored bar codes indicate the cumulative depths corresponding to the four considered values of $\kappa'$ in decreasing order from top to bottom (see Sect. 3.4). Each bar code is split into ten regions of alternating brightness, representing ten 10%-quantiles of the distribution (e.g., the first dark blue region of the top bar code in panel (**c**) indicates that 10% of reaction events occur till $Dt/R^2 \simeq 1$) [Adapted from [34]]

We note first that the range of relevant reaction times (the horizontal axis) is very broad; namely, it spans over eight decades for panel **(b)** and over 12 decades for panel **(d)**. Second, we observe that the PDFs $H(r, t)$ in both figures look quite similar and qualitatively, the curves can be divided into four apparent temporal regimes:

(i) The first regime (at very short times) is characterized by an abrupt rise to the maximal value attained at the most probable FRT $t_{\mathrm{mp}}$.

(ii) In the second regime, $H(r, t)$ descends to a constant value, which is reached at the crossover time $t_c$. Below we will show that overall this hump-like region comprising both regimes is well described by a Lévy-Smirnov density (thin colored curves) specific to unbounded systems ($R \to \infty$). Therefore, the crossover time $t_c$ is roughly the time instant at which the particle first engages with the boundary, realizing eventually that it moves in a bounded domain.

(iii) Thirdly, we observe an extended plateau-like regime, in which, remarkably, all values of the FRT are equally probable. This regime terminates at the time instant which roughly equals the mean FRT, given by (3.23), and designated by colored arrows.

(iv) Finally, for $t > \langle t \rangle$, $H(r, t)$ rapidly (exponentially) decays to zero.

Next, still at a qualitative level, we discuss the effects of the intrinsic reactivity and of the initial position on the PDF. We focus first on Fig. 3.3 and observe that upon diminishing $\kappa$, the hump-like region remains unaffected (except that the most probable value decreases)—the values of $t_{\mathrm{mp}}$ and $t_c$ remain nearly the same. This is quite plausible on physical grounds because the former time $t_{\mathrm{mp}}$ is controlled by the "direct" trajectories which go straight from the initial position to the target and react with it upon the first encounter, while the latter time $t_c$ is dominated by the size of the system and the diffusion coefficient. The major effect of lowering $\kappa$ is that the extended plateau-like region stretches over progressively longer times. Respectively, the mean FRT increases by orders of magnitude such that the ultimate exponential stage is delayed to much longer times. The same behavior takes place regardless of the actual starting position (compare panels **(b)** and **(d)**).

In turn, from Fig. 3.2 we infer that the major effect of the initial distance to the target is the duration of the hump-like region. The latter becomes progressively extended in time the closer the initial position to the target is. In particular, the hump-like region stretches over more than four decades when the particle starts very close to the target site, and it is almost absent in case when the particle starts right at the outer boundary of the domain. Intuitively, this can be understood as follows: once a particle starts very close to the target, some of the trajectories can reach it by going straight to the target but at the same time, much more trajectories go away from the target exploring the bulk for a long time, until their eventual arrival to the target and reaction with it. On the contrary, when a particle starts far away from the target, the fraction of trajectories arriving directly to the target becomes very low, such that the hump-like region shrinks. In turn, the plateau-like regime and the ultimate exponential decay are almost unaffected by the value of $\kappa$. In Sect. 3.4 below we will also discuss the reaction depths corresponding to each temporal stage, which show the fraction of trajectories reacting at each stage.

### 3.3.3 Three Characteristic Time-Scales

The shortest time-scale is the most probable FRT $t_{\mathrm{mp}}$ which defines the position of the maximum of $H(r,t)$. In [34] it was demonstrated that $t_{\mathrm{mp}}$ obeys

$$t_{\mathrm{mp}} = \frac{(r-\rho)^2}{6D} z^2 , \tag{3.18}$$

where $z$ is a dimensionless function of system parameters which is defined implicitly as the solution of the following, rather complicated transcendental equation

$$\beta^2 z^4 - 3(1+\beta)z^2 + 9 - \sqrt{\pi/6}\,\beta^3 z^5 \mathrm{erfcx}\left(\frac{\sqrt{3/2}}{z} + \frac{\beta z}{\sqrt{6}}\right) = 0, \tag{3.19}$$

with $\mathrm{erfcx(x)} = \exp(\mathrm{x}^2)\mathrm{erfc(x)}$ and $\mathrm{erfc(x)}$ being the complementary error function, and

$$\beta = \frac{r-\rho}{\rho}\left(1 + \frac{\kappa\rho}{D}\right). \tag{3.20}$$

We denote the solution of this equation as $z_\beta$. When $\beta$ tends to 0 (i.e., when $r$ is very close to $\rho$), a Taylor expansion of the left-hand side of Eq. (3.19) yields $z^2 - 9 + O(\beta)$, and hence, $z_0 = \sqrt{3}$. In the opposite limit $\beta \to \infty$, (i.e., when $\kappa \to \infty$), one uses the asymptotic behavior of the function $\mathrm{erfcx}(x)$ to get

$$z_\beta \simeq 1 + \frac{3}{2\beta} + O(\beta^{-2}). \tag{3.21}$$

With some technical efforts [34], one can prove that $z_\beta$ is a monotonously decreasing function of $\beta$, which is bounded between $\sqrt{3}$ and 1 so that the most probable time $t_{\mathrm{mp}}$ lies between $(r-\rho)^2/(6D)$ (for $\kappa\rho \gg 1$) and $(r-\rho)^2/(2D)$ (for $\kappa\rho \ll 1$). This suggests that, not counter-intuitively, the only important parameters are the distance to the target and the diffusion coefficient, and hence, emphasizes the role of the direct trajectories. Indeed, we observe in Figs. 3.3 and 3.2 that $t_{\mathrm{mp}}$ shows very little variation with the reactivity $\kappa$.

The second characteristic time-scale is the crossover time $t_c$ from the hump-like Lévy-Smirnov region (see below) specific to an unbounded system, to the plateau region. Hence, $t_c$ can be interpreted as the time at which a molecule starts to feel the confinement and obeys [34]

$$t_c \approx \frac{2(R-\rho)^2}{\pi^2 D} , \tag{3.22}$$

showing that $t_c$ is independent of the starting point and of the reactivity $\kappa$, being entircly dominated by the diffusivity and the difference between the sizes of the domain and of the target. Considering the ratio $t_{\mathrm{mp}}/t_c \sim (r-\rho)^2/(R-\rho)^2$, one realizes that the crossover time $t_c$ may be comparable to the most probable time (in which case the hump-like region shrinks), but may also become much larger than the latter when $r$ is close to $\rho$, which we indeed observe in Fig. 3.2. In this case, of course, the hump-like region should be most pronounced.

Finally, the onset of the right exponential shoulder at long times coincides with the mean FRT, as indicated by the arrows in Figs. 3.3 and 3.2. The latter can be straightforwardly obtained by differentiating $\tilde{H}(r, p)$ in Eq. (3.9) with respect to the Laplace parameter $p$ and setting $p = 0$. This gives the following exact formula

$$\langle t \rangle = \frac{(r-\rho)(2R^3 - \rho r(r+\rho))}{6Dr\rho} + \frac{R^3-\rho^3}{3\kappa\rho^2}. \tag{3.23}$$

The first term in Eq. (3.23) is the standard mean first-passage time to a perfectly reactive target [39]. The second contribution is proportional to $\kappa^{-1}$ and hence accounts for the imperfect reaction with finite reactivity. This contribution is independent of the particle's starting point and the diffusion coefficient. Therefore, the contribution due to a diffusive search and the intrinsic reactivity to the mean FRT are additive and can be regarded as an analogue of the Collins-Kimball relation in the classic theory of diffusion controlled imperfect reactions [42]. As evidenced in Figs. 3.3 and 3.2, this ultimate time-scale may exceed by several orders of magnitude the shortest time-scale $t_{mp}$.

### 3.3.4 *Effective Broadness of the PDF*

Due to the exponential cut-off, the FRT PDF $H(r, t)$ in Eq. (3.14) has finite moments of arbitrary order. In standard nomenclature, such distributions are usually considered as "narrow", in contrast to PDFs with power-law tails which therefore do not possess all the moments. Concurrently, the fact that $H(r, t)$ has moments of arbitrary order does not imply that there is no significant spread of FRTs. Usually, the actual spread of FRTs and its effective broadness are probed by the so-called coefficient of variation of the PDF (see, e.g., [24–26]), defined as the ratio

$$\gamma = \frac{\sqrt{\langle t^2 \rangle - \langle t \rangle^2}}{\langle t \rangle}. \tag{3.24}$$

Therefore, $\gamma$ quantifies the relative scatter of values of the FRT around the mean value and points to a possible lack of self-averaging or defocusing of the FRTs, if its value exceeds unity.

The second moment of the FRT is obtained by differentiating $H(r, p)$ with respect to $p$ twice and setting $p = 0$ afterwards. In doing so, one gets the variance of the FRT [34]

$$\langle t^2\rangle - \langle t\rangle^2 = \frac{1}{90D^2r^2\rho^4}\Big\{10r^2(R^3-\rho^3)^2(D/\kappa)^2+ \\ +4\rho r^2(5R^3+6R^2\rho+3R\rho^2+\rho^3)(R-\rho)^3(D/\kappa)+ \\ +\rho^2(r-\rho)\big(2R^3(5R^3\rho+5R^3r+10r^2\rho^2-18R^2r\rho) \\ -\rho^2r^2(\rho+r)(r^2+\rho^2)\big)\Big\}, \tag{3.25}$$

which together with expression (3.23) defines the coefficient of variation.

The resulting expression for $\gamma$ appears to be quite cumbersome and depends on all model parameters. For our purpose here it will suffice to illustrate that it can exceed unity by far, by considering its behavior just in two particular limits. For example, for a small ($\rho/R \ll 1$) and perfectly reactive target, and with a particle starting close to the target surface, i.e. for $r/\rho \to 1$, one finds to leading order that

$$\gamma^2 \simeq \frac{2\rho}{r-\rho} + O(1), \tag{3.26}$$

meaning that $\gamma$ actually diverges when $r$ approaches $\rho$! Such a behavior is, in fact, not counter-intuitive–this is precisely the limit in which the hump-like region has a maximal extent, as evidenced in Fig. 3.2. We will comment more on this issue in the next subsection. Next, for a partially reactive target, the squared coefficient of variation is finite in the limit $r \to \rho$ and for a small target reads

$$\gamma^2 \simeq 1 + \frac{2\rho\kappa}{D}. \tag{3.27}$$

Here, as well, $\gamma$ turns out to be *generically* bigger than unity.

Capitalizing on these two examples we may conclude that, at least in some relevant cases, the PDF can be effectively "broad" and therefore, one should expect to observe a significant scatter of the values of the FRTs recorded in different realizations of the process.

### 3.3.5 *Asymptotic Behavior of the PDF and a Generalization of the Approximation in Eq.* (3.1)

The short-time asymptotic behavior of $H(r,t)$ in Eq. (3.14) can be accessed by using the following argument: at sufficiently short times, until the particle engages for the first time with the boundary, it is unaware that it diffuses in a bounded domain. Therefore, to determine the behavior of $H(r,t)$ in this limit it is legitimate to set $R \to \infty$, e.g., in Eq. (3.9), to get

$$\tilde{H}_\infty(r, p) = \frac{(\rho/r)\, e^{-(r-\rho)\sqrt{p/D}}}{1 + \left(1 + \rho\sqrt{p/D}\right)D/(\kappa\rho)}\,. \tag{3.28}$$

Inverting this expression, we find the PDF in the time-domain for arbitrary $\kappa$ and $r$:

$$\begin{aligned} H_\infty(r, t) = \frac{\kappa}{r} \exp\left(-\frac{(r-\rho)^2}{4Dt}\right)\Bigg\{ & \frac{\rho}{\sqrt{\pi Dt}} \\ & - \left(1 + \frac{\kappa\rho}{D}\right) \mathrm{erfcx}\left(\frac{r-\rho}{\sqrt{4Dt}} + \left(1 + \frac{\kappa\rho}{D}\right)\frac{\sqrt{Dt}}{\rho}\right)\Bigg\}, \end{aligned} \tag{3.29}$$

where $\mathrm{erfcx}(x)$ is the scaled complementary error function defined in the text after Eq. (3.19). We note next that due to the transient character of three-dimensional diffusion, the above PDF is not normalized to 1, i.e.,

$$\int_0^\infty dt\, H_\infty(r, t) = \tilde{H}_\infty(r, p = 0) = \frac{\rho/r}{1 + D/(\kappa\rho)} < 1\,. \tag{3.30}$$

The expression on the right-hand side defines the probability of reacting with the target before escaping to infinity. As demonstrated in Figs. 3.3 and 3.2, Eq. (3.29), which was derived using very heuristic arguments, reproduces remarkably well the behavior of $H(r, t)$ within the hump-like region (see the thin colored curves) for any choice of the system parameters.

Two following remarks are in order: We note first that in the limit of a perfectly absorbing target, i.e., for $\kappa \to \infty$, the expression (3.29) reduces to

$$H_\infty(r, t) = \frac{\rho}{r} \frac{r-\rho}{\sqrt{4\pi Dt^3}} \exp\left(-\frac{(r-\rho)^2}{4Dt}\right), \tag{3.31}$$

which has the form (apart from the factor $\rho/r \le 1$ which reflects the transient nature of diffusion in three dimensions) of the celebrated Lévy-Smirnov density describing the PDF of the first-passage times in *one-dimensional* systems (see, e.g., [39]). The "direct" trajectories which define the hump-like region in three-dimensional systems are also essentially "one-dimensional". This is precisely the reason why we referred above to such a behavior as the Lévy-Smirnov regime. The second remark concerns the divergence of the coefficient of variation in the case when $r \to \rho$ (see Eq. (3.26)) for which the hump-like region becomes most pronounced. The point is that the descent of $H(r, t)$ from its maximal value is dictated by the Lévy-Smirnov form, i.e., obeys a power-law $t^{-3/2}$ (see Eqs. (3.29) and (3.31)) and therefore contributes most to the values of the moments of the FRTs. In consequence, the more extended the hump-like region is, the higher are the values of the moments.

In turn, the long-time asymptotic behavior of the PDF $H(r, t)$ in Eq. (3.14) is determined by the smallest eigenvalue $\lambda_0$. For the spherical domain, one can accurately compute this eigenvalue [34] to find that, for a sufficiently small target with a low reactivity $\kappa$, one has $\lambda_0 \approx \kappa S_\rho/(DV)$, where $S_\rho = 4\pi\rho^2$ is the surface area of the target while $V$ is the volume of the domain, $V \approx 4\pi R^3/3$. According to Eq. (3.23), in this limit $\langle t \rangle \approx 1/(D\lambda_0)$, i.e., the mean first-reaction time is dominated by multiple returns to the target until the reaction occurs. As the target shrinks ($\rho$ vanishes), the smallest eigenvalue tends to zero. In turn, the other eigenvalues $\lambda_n$, corresponding to rotation-invariant eigenfunctions of the Laplace operator in the spherical domain, are bounded from below $\lambda_n > \pi^2 n^2/R^2$ for $n = 1, 2, \ldots$. As a consequence, there is an intermediate range of times, $1/(D\lambda_1) \ll t \ll 1/(D\lambda_0)$, for which the contribution of all higher-order eigenmodes vanishes, that is, $e^{-Dt\lambda_n} \ll 1$, whereas the contribution of the lowest eigenmode is almost constant in time, $e^{-Dt\lambda_0} \approx 1$. This is precisely the reason why the intermediate, plateau-like region emerges (see Figs. 3.3 and 3.2). Note that this region protrudes over an increasing range of time-scales when either the reactivity $\kappa$, or the target radius $\rho$, or both, decrease.

Capitalizing on the above results on the asymptotic short- and long-$t$ behavior, we arrive at the following approximate formula for the full PDF $H(r, t)$:

$$H(r,t) \approx H_\infty(r,t) + (1-q)\frac{e^{-t/\langle t \rangle}}{\langle t \rangle}, \tag{3.32}$$

where $\langle t \rangle \approx 1/(D\lambda_0)$ and

$$q = \int_0^\infty dt\, H_\infty(r,t) = \frac{\rho/r}{1 + D/(\kappa\rho)} < 1 \tag{3.33}$$

is the hitting probability of the target. The correct normalization of $H(r, t)$ is ensured by the prefactor in front of the second term. Result (3.32) is more general than the simple form (3.1) suggested in [20]. The form (3.32) not only extends expression (3.1) to the partially reactive case, i.e., for arbitrary finite values of $\kappa$, but also provides an explicit form for the contribution from the hump-like region around $t_{\mathrm{mp}}$, which is most relevant for reactions in which the molecule starts close to the target.

Figure 3.4 illustrates the quality of this approximation, showing that it becomes most accurate when the target radius $\rho$ or reactivity $\kappa$ is small. One observes that it accurately captures both the maximum, the plateau, and the exponential cut-off of the first-reaction time distribution. In turn, the transition between the maximum and the plateau region is less sharp than in the exact form. A minor inaccuracy of approximation (3.32) is that it reaches a constant—set by the second term—in the short time limit while the exact distribution vanishes as $t \to 0$. This feature can be simply removed by multiplying the second term by a Heaviside step function $\Theta(t - t_c)$ and re-evaluating the normalization constant. But even in the present form, approximation (3.32) provides a remarkably good insight into the behavior of the FRT density and can thus be used as an efficient and easy-to-handle fit formula for data analysis or for explicit analytical derivations of follow-up processes.

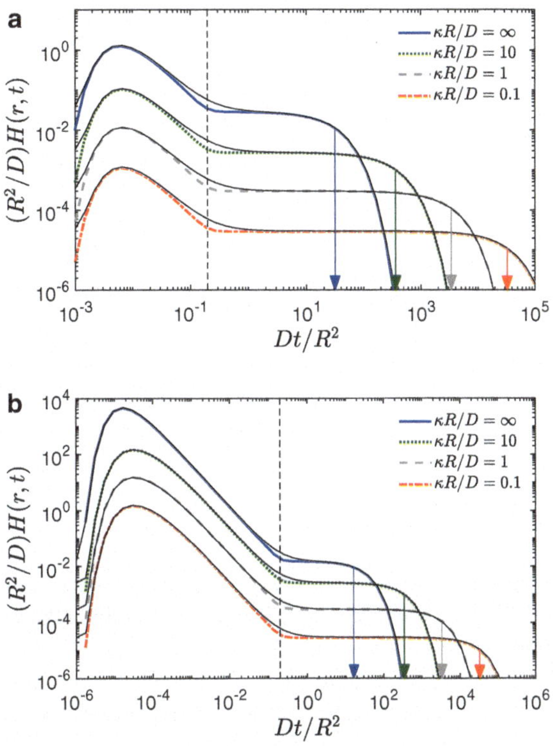


**Fig. 3.4** Explicit approximation for the first-reaction time PDF $H(r,t)$. It is evaluated for a reaction with an inner target of radius $\rho/R = 0.01$ with starting point (**a**) $r/R = 0.2$ and (**b**) $r/R = 0.02$, and four values of the dimensionless reactivity $\kappa' = \kappa R/D$ (decreasing from top to bottom). The colored vertical arrows indicate the respective mean FRT. The black vertical dashed line shows the crossover time $t_c$. Thin black lines present the approximation (3.32). Length and time units are fixed by setting $R = 1$ and $R^2/D = 1$ [Adapted from [34]]

## 3.4 Cumulative Distribution and Reaction Depths

We finally address the question about the relative importance of different regimes for the overall reaction process. Namely, what is the fraction of trajectories that result in reaction events up to times $t_{\rm mp}$, $t_c$ and $\langle t \rangle$? To this end, we focus on the cumulative distribution function (CDF) of the FRTs, defined as

$$F(r,t) = \int_0^t dt' H(r,t') = 1 - S(r,t), \tag{3.34}$$

where $S(r,t)$ is the survival probability introduced in Sect. 3.2. The CDF is an increasing function of $t$ which has the evident property $F(r,\infty) = 1$. This function thus shows explicitly which fraction of trajectories have reacted up to time $t$. Integrating (3.14), one gets an explicit spectral expansion for $F(r,t)$. For the unbounded case ($R = \infty$), one also has the explicit relation

$$F_\infty(r,t) = \frac{R - R_\kappa}{r}\left\{\mathrm{erfc}\left(\frac{r-R}{\sqrt{4Dt}}\right) + e^{-(r-R)^2/(4Dt)}\mathrm{erfcx}\left(\frac{r-R}{\sqrt{4Dt}} + \frac{\sqrt{Dt}}{R_\kappa}\right)\right\}, \tag{3.35}$$

**Table 3.1** Impact of the target reactivity and of the initial distance onto the reaction depth. Relative weights (in per cents) of three characteristic regions of the FRT density for $\rho/R = 0.01$: the hump-like region around the most probable reaction time $t_{mp}$, extending from 0 till $t_c$ (and thus merging two subregions discussed in the text: the exponential tail to the left of $t_{mp}$ and the power-law decay to the right of $t_{mp}$); the plateau-like region stretching from $t_c$ to the mean FRT $\langle t \rangle$; and the exponential tail which persists beyond $\langle t \rangle$. Two starting points $r/R$ and four values of the dimensionless reactivity $\kappa' = \kappa R/D$ are used, corresponding to Figs. 3.3 and 3.2

| $r/R$ | Region | $\kappa' = \infty$ | $\kappa' = 10$ | $\kappa' = 1$ | $\kappa' = 0.1$ |
|---|---|---|---|---|---|
| 0.2 | Hump-like | 3.8 | 0.34 | 0.04 | 0.004 |
| | Plateau-like | 59.4 | 62.9 | 63.2 | 63.2 |
| | Exponential tail | 36.8 | 36.8 | 36.8 | 36.8 |
| 0.02 | Hump-like | 49.4 | 4.4 | 0.5 | 0.05 |
| | Plateau-like | 20.0 | 58.8 | 62.7 | 63.15 |
| | Exponential tail | 30.6 | 36.8 | 36.8 | 36.8 |

where $R_\kappa = R/(1 + \kappa R/D)$. In the limit of a perfectly reactive target ($\kappa \to \infty$), this expression, which was given by Collins and Kimball [42], is reduced to the seminal result by Smoluchowski: $F_\infty(r, t) = \frac{R}{r}\mathrm{erfc}\big((r - R)/\sqrt{4Dt}\big)$ [43].

In panels **a** and **c** in Figs. 3.3 and 3.2 we present the FRT CDF by colored bar codes corresponding to four considered values of $\kappa$ (Fig. 3.3) and four values of $r/R$ (Fig. 3.2) in increasing order from top to bottom. Each bar code is split into ten regions of alternating brightness, representing ten 10%-quantiles of the distribution. This presentation permits to observe very directly how many trajectories up to a fixed moment of $t$ result in successful reactive events, for different values of the reactivity and of the starting position.

Further on, Table 3.1summarizes the values of the reaction depths of the three characteristic regions of the PDF $H(r, t)$: the hump-like region around $t_{mp}$, the plateau region, and the exponential tail. We realize that for $r/R = 0.2$ the least amount of the reaction events happens within the hump-like region: it is of the order of just 4% for perfect reactions, and this fraction rapidly diminishes upon a decrease of $\kappa$. In turn, a much larger amount of the first-reaction events is collected within the final exponential region. It is typically of the order of 37%, almost independently of the value of $\kappa$, meaning that for such a value of the ratio $r/R$ roughly one third of all realizations remain unreacted at time $t = \langle t \rangle$. However, most of the realizations of the first-reaction events occur within the plateau-like regime–it amounts to roughly 59% for perfect reactions, and becomes even bigger for smaller values of $\kappa$. The situation becomes different for a smaller release radius: $r/R = 0.02$. Here, for perfect reactions the majority of trajectories (49% such that $t_c$ should be close to the median time) react within the hump-like region, while the plateau region and the final exponential tail contribute only 20% and 30%, respectively. Upon lowering $\kappa$, the hump-like region is no longer representative, and more reaction events take place during the exponential tail ($\sim 37\%$) and the plateau-like regions ($\sim 63\%$), respectively. In conclusion, the plateau region appears to be the most important part of $H(r, t)$ which contributes most to the overall number of reaction events, except for the case $r/R \ll 1$ and $\kappa R/D \gg 1$, for which the hump-like region becomes the dominant one. Concurrently, this plateau is the region of the strongest defocusing effect.

## 3.5 Conclusions

In this chapter, we focused on a simple geometric setting allowing to exactly solve the underlying target search problem and thus to explore a broad range of time-scales. At the same time, the qualitative conclusions of our study remain valid in much more general settings. Thus, the FRT PDF does not look like a narrow Gaussian curve centered around the mean FRT; even though the latter presents a relevant time-scale of the target search problem, it is usually not unique, whereas the scatter around the mean can be large or even diverging. A short initial distance to the target or its low reactivity can dramatically widen the PDF and further enhance the defocusing of the FRTs. In the considered spherical setting, we identified four distinct regimes: a rise to the maximum, a descent from the maximum, a plateau-like region, and an exponential decay. We also characterized their relevant time-scales: the most probable FRT $t_{mp}$, the crossover time $t_c$, and the exponential cut-off time, which is usually close to the mean FRT. It is worth noting that other regimes can emerge in more sophisticated settings. For instance, if the outer impenetrable spherical boundary contains a very small hole, the aforementioned exponential regime will last up to a new time-scale, which is inversely proportional to the hole size, i.e., until the mean escape time from the confining sphere of radius $R$. If there is a bigger outer confinement, another exponential regime will then be established. This scenario can be relevant, e.g., for the target search inside the nucleus (here modeled as a sphere of radius $R$), from which the searcher can escape through nuclear pores into the cytoplasm, the latter being surrounded by the cell membrane. In contrast, if the escaped molecule starts to diffuse in an unbounded (or extremely large) space, the ultimate regime would actually be described by a function $H_\infty(r, t)$, in which $\rho$ is replaced by $R$. In this case, the mean FRT is infinite and thus totally uninformative, regardless of the smallness of the hole. This example further highlights that rare, anomalously long trajectories provide the dominant contribution to the mean. Yet another example is diffusion towards an array of absorbing nano-pillars, for which the Lévy-Smirnov regime is replaced by another intermediate regime, $H(r, t) \propto 1/(t \ln^2(t))$, due to effectively two-dimensional search outside a cylindrical pillar [35].

In order to reveal the rich structure of the PDF with distinct regimes, the log-log scale was employed in Figs. 3.3 and 3.2. In turn, if linear or semi-logarithmic scales were used, all 12 decades on the time axis would collapse into a single point, keeping emphasis on the exponential tail; all shown curves would become identical and thus boring. To some extent, this collapse is quintessentialized by (3.1), in which all interesting regimes at short and intermediate time-scales are represented via the Dirac distribution $\delta(t)$. Our approximation (3.32) aims to grasp more features in the PDF. In other words, plotting the PDF of the FRTs (or their histogram) at a linear scale would eliminate most of its features. It is therefore recommended to employ the logarithmic time-scale to present both theoretical results and empirical histograms obtained from experimental or numerically generated FRTs. From a practical point of view, if $\{t_1, \ldots, t_N\}$ denotes the sequence of such empirical FRTs, one can produce a histogram of $\ln(t_k)$, from which the PDF of the FRT can be easily constructed at the logarithmic scale. This also suggests the potential importance of more elaborate "means" such as the harmonic mean $\langle 1/t \rangle$ or the typical value $\langle \ln(t) \rangle$ (see [34] for further discussions).

The main take-home message of this chapter is that the distribution of the FRT in general has a rich structure and is not reducible to the mean FRT, on which most former contributions were focused. As a consequence, the analysis of target search problems should not stop at the first moment of the search time and go beyond this common time-scale. At the same time, it is worth stressing that finding the mean FRT alone can be a difficult problem even for simple search processes [44–56]. Moreover, getting the mean FRT for more sophisticated diffusive dynamics (e.g., non-Markovian) becomes a real challenge [57, 58] (see also Chaps. 16 and 17). It is therefore not surprising that the full distribution of the FRTs is still unaccessible in many relevant scenarios and urging for future research.

## References

1. B. Alberts et al., *Molecular Biology of the Cell*, 6th edn. (Garland, New York, 2015)
2. D.P. Snustad, M.J. Simmons, *Principles of Genetics*, 7th edn. (John Wiley, New York, NY, 2015)
3. P.C. Bressloff, *Stochastic Processes in Cell Biology* (Springer, Cham, Switzerland, 2014)
4. H.C. Berg, E.M. Purcell, Physics of chemoreception. Biophys. J. **20**, 193–219 (1977)
5. J.J. Lindemann, D.A. Lauffenburger, Analysis of intracellular receptor/ligand sorting. Biophys. J. **50**, 295–305 (1986)
6. I.V. Grigoriev, Y.A. Makhnovskii, A.M. Berezhkovskii, V.Y. Zitserman, Kinetics of escape through a small hole. J. Chem. Phys. **116**, 9574 (2002)
7. D. Holcman, Z. Schuss, Escape through a small opening: receptor trafficking in a synaptic membrane. J. Stat. Phys. **117**, 975 (2004)
8. O. Bénichou, R. Voituriez, Narrow-escape time problem: time needed for a particle to exit a confining domain through a small window. Phys. Rev. Lett. **100**, 168105 (2008)
9. A.F. Cheviakov, A.S. Reimer, M.J. Ward, Mathematical modeling and numerical computation of narrow escape problems. Phys. Rev. E **85**, 021131 (2012)
10. G. Oshanin, M. Tamm, O. Vasilyev, Narrow-escape times for diffusion in microdomains with a particle-surface affinity: mean-field results. J. Chem. Phys. **132**, 235101 (2010)
11. Z. Yang, E.F. Koslover, Diffusive exit rates through pores in membrane-enclosed structures. Phys. Biol. **20**, 026001 (2023)
12. K. Lindenberg, R. Metzler, G. Oshanin (eds.), *Chemical Kinetics: Beyond the Textbook* (World Scientific Publishers Europe, London, 2019)
13. E. Gudowska-Nowak, K. Lindenberg, R. Metzler, Preface: Marian Smoluchowski's 1916 paper—a century of inspiration. J. Phys. A Math. Theor. **50**, 380301 (2017)

14. O. Bénichou, M. Moreau, G. Oshanin, Kinetics of stochastically gated diffusion-limited reactions and geometry of random walk trajectories. Phys. Rev. E **61**, 3388 (2000)
15. D.S. Grebenkov, Searching for partially reactive sites: analytical results for spherical targets. J. Chem. Phys. **132**, 034104 (2010)
16. D.S. Grebenkov, Imperfect diffusion-controlled reactions, in Ref. [6], Ch. 8 (2019)
17. D.S. Grebenkov, Diffusion-controlled reactions: an overview. Molecules **28**, 7570 (2023)
18. D.S. Grebenkov, G. Oshanin, Diffusive escape through a narrow opening: new insights into a classic problem. Phys. Chem. Chem. Phys. **19**, 2723 (2017)
19. O. Bénichou, R. Voituriez, From first-passage times of random walks in confinement to geometry-controlled kinetics. Phys. Rep. **539**, 225 (2014)
20. O. Bénichou, C. Chevalier, J. Klafter, B. Meyer, R. Voituriez, Geometry-controlled kinetics. Nat. Chem. **2**, 472 (2010)
21. O. Pulkkinen, R. Metzler, Distance matters: the impact of gene proximity in bacterial gene regulation. Phys. Rev. Lett. **110**, 198101 (2013)
22. A. Godec, R. Metzler, Universal proximity effect in target search kinetics in the few encounter limit. Phys. Rev. X **6**, 041037 (2016)
23. A. Godec, R. Metzler, First passage time distribution in heterogeneity controlled kinetics: going beyond the mean first passage time. Sci. Rep. **6**, 20349 (2016)
24. C. Mejía-Monasterio, G. Oshanin, G. Schehr, First passages for a search by a swarm of independent random searchers. J. Stat. Mech. P06022 (2011)
25. T. Mattos, C. Mejía-Monasterio, R. Metzler, G. Oshanin, First passages in bounded domains: when is the mean first passage time meaningful? Phys. Rev. E **86**, 031143 (2012)
26. T. Mattos, C. Mejía-Monasterio, R. Metzler, G. Oshanin, G. Schehr, Trajectory-to-trajectory fluctuations in first-passage phenomena in bounded domains, in *First-Passage Phenomena and Their Applications*, ed. by R. Metzler, G. Oshanin, S. Redner (Singapore, World Scientific, 2014)
27. D.S. Grebenkov, R. Metzler, G. Oshanin, Full distribution of first exit times in the narrow escape problem. New J. Phys. **21**, 122001 (2019)
28. D.S. Grebenkov, R. Metzler, G. Oshanin, From single-particle stochastic kinetics to macroscopic reaction rates: fastest first-passage time of $N$ random walkers. New J. Phys. **22**, 103004 (2020)
29. D.S. Grebenkov, R. Metzler, G. Oshanin, Towards a full quantitative description of single-molecule reaction kinetics in biological cells. Phys. Chem. Chem. Phys. **20**, 16393 (2018)
30. S.D. Lawley, Distribution of extreme first passage times of diffusion. J. Math. Biol. **80**, 2301 (2020)
31. S.D. Lawley, Universal formula for extreme first passage statistics of diffusion. Phys. Rev. E **101**, 012413 (2020)
32. D.S. Grebenkov, R. Metzler, G. Oshanin, Distribution of first-reaction times with target regions on boundaries of shell-like domains. New J. Phys. **23**, 123049 (2021)
33. D.S. Grebenkov, R. Metzler, G. Oshanin, Search efficiency in the Adam-Delbrück reduction-of-dimensionality scenario versus direct diffusive search. New J. Phys. **24**, 083035 (2022)
34. D.S. Grebenkov, R. Metzler, G. Oshanin, Strong defocusing of molecular reaction times results from an interplay of geometry and reaction control. Commun. Chem. **1**, 96 (2018)
35. D.S. Grebenkov, A.T. Skvortsov, Survival in a nanoforest of absorbing pillars. J. Phys. A Math. Theor. **56**, 165002 (2023)
36. J. Ma, M. Do, M.A. Le Gros, C.S. Peskin, C.A. Larabell, Y. Mori et al., Strong intracellular signal inactivation produces sharper and more robust signaling from cell membrane to nucleus. PLOS Comput. Biol. **16**, e1008356 (2020)
37. J. Crank, *The Mathematics of Diffusion* (Oxford University Press, 1956)
38. H.S. Carslaw, J.C. Jaeger, *Conduction of Heat in Solids*, 2nd edn. (Oxford University Press, 1959)
39. S. Redner, *A Guide to First Passage Processes* (Cambridge University Press, Cambridge, 2001)
40. D.F. Calef, J.M. Deutch, Diffusion-controlled reactions. Ann. Rev. Phys. Chem. **34**, 493–524 (1983)

41. G.H. Weiss, Overview of theoretical models for reaction rates. J. Stat. Phys. **42**, 3–36 (1986)
42. F.C. Collins, G.E. Kimball, Diffusion-controlled reaction rates. J. Colloid Sci. **4**, 425–437 (1949)
43. M. Smoluchowski, Versuch einer mathematischen Theorie der Koagulationskinetic kolloider Lösungen. Z. Phys. Chem. **92U**, 129–168 (1918)
44. I.V. Grigoriev, Y.A. Makhnovskii, A.M. Berezhkovskii, V.Y. Zitserman, Kinetics of escape through a small hole. J. Chem. Phys. **116**, 9574 (2002)
45. A. Singer, Z. Schuss, D. Holcman, R.S. Eisenberg, Narrow escape, part I. J. Stat. Phys. **122**, 437–463 (2006)
46. A. Singer, Z. Schuss, D. Holcman, Narrow escape, part II: the circular disk. J. Stat. Phys. **122**, 465 (2006)
47. A. Singer, Z. Schuss, D. Holcman, Narrow escape, part III: non-smooth domains and Riemann surfaces. J. Stat. Phys. **122**, 491 (2006)
48. Z. Schuss, A. Singer, D. Holcman, The narrow escape problem for diffusion in cellular microdomains. Proc. Nat. Acad. Sci. USA **104**, 16098–16103 (2007)
49. S. Pillay, M.J. Ward, A. Peirce, T. Kolokolnikov, An asymptotic analysis of the mean first passage time for narrow escape problems: part I: two-dimensional domains. Multiscale Model. Simul. **8**, 803–835 (2010)
50. A.F. Cheviakov, M.J. Ward, R. Straube, An asymptotic analysis of the mean first passage time for narrow escape problems: part II: the sphere. Multiscale Model. Simul. **8**, 836–870 (2010)
51. A.F. Cheviakov, A.S. Reimer, M.J. Ward, Mathematical modeling and numerical computation of narrow escape problems. Phys. Rev. E **85**, 021131 (2012)
52. D. Holcman, Z. Schuss, The narrow escape problem. SIAM Rev. **56**, 213–257 (2014)
53. O. Bénichou, R. Voituriez, From first-passage times of random walks in confinement to geometry-controlled kinetics. Phys. Rep. **539**, 225–284 (2014)
54. J.S. Marshall, Analytical solutions for an escape problem in a disc with an arbitrary distribution of exit holes along its boundary. J. Stat. Phys. **165**, 920–952 (2016)
55. D.S. Grebenkov, Universal formula for the mean first passage time in planar domains. Phys. Rev. Lett. **117**, 260201 (2016)
56. A.J. Bernoff, A.E. Lindsay, Numerical approximation of diffusive capture rates by planar and spherical surfaces with absorbing pores. SIAM J. Appl. Math. **78**, 266–290 (2018)
57. T. Guérin, N. Levernier, O. Bénichou, R. Voituriez, Mean first-passage times of non-Markovian random walkers in confinement. Nature **534**, 356–359 (2016)
58. I.M. Sokolov, Cyclization of a polymer: a first passage problem for a non-Markovian process. Phys. Rev. Lett. **90**, 080601 (2003)

# Chapter 4
# Encounter-Based Approach to Target Search Problems

**Denis Grebenkov**

**Abstract** In this chapter, we present the encounter-based approach to target search problems, in which the diffusive dynamics is described by the joint probability of the position of the particle and the number of its encounters with a given target set. The knowledge of the statistics of encounters allows one to implement various mechanisms of reactions on the target set, beyond conventional reaction schemes. We formulate this approach for three relevant settings: discrete random walks, Brownian motion with bulk reactions, and reflected Brownian motion with surface reactions. In all the cases, we discuss the advantages of this approach, its recent applications and possible extensions.

## 4.1 Introduction

Marian von Smoluchowski formulated and solved one of the first target search problems in chemical physics [1]. He studied the dynamics of colloidal particles diffusing in a solvent and coagulating upon their encounters. At the level of single-particle events, his solution can be interpreted in terms of the survival probability that two particles have not coagulated up to time $t$, i.e., the random first-encounter time $\mathcal{T}$ does not exceed $t$. Assuming that both particles are spheres of radii $R_1$ and $R_2$ that undergo independent Brownian motions with diffusion coefficients $D_1$ and $D_2$, Smoluchowski associated the origin of the coordinate system with the center of one particle and thus mapped the original first-encounter problem for two moving particles onto an equivalent target problem when a single point-like particle with diffusion coefficient $D = D_1 + D_2$ searches for a target—the static sphere of radius $R = R_1 + R_2$. As ordinary diffusion is governed by the Laplace operator $\Delta$, Smoluchowski solved the diffusion equation $\partial_t S_\infty(t|\boldsymbol{x}_0) = D\Delta S_\infty(t|\boldsymbol{x}_0)$ and found the survival probability:

D. Grebenkov (✉)
Laboratoire de Physique de la Matière Condensée, CNRS – Ecole Polytechnique, Institut Polytechnique de Paris, 91120 Palaiseau, France
e-mail: denis.grebenkov@polytechnique.edu

 
D. Grebenkov et al. (eds.), *Target Search Problems*,
https://doi.org/10.1007/978-3-031-67802-8_4

$$S_\infty(t|\boldsymbol{x}_0) = \mathbb{P}_{\boldsymbol{x}_0}\{\mathcal{T} > t\} = 1 - \frac{R}{|\boldsymbol{x}_0|}\mathrm{erfc}\left(\frac{|\boldsymbol{x}_0| - R}{\sqrt{4Dt}}\right), \tag{4.1}$$

where $\mathrm{erfc}(z)$ is the complementary error function, $\boldsymbol{x}_0$ is the initial position of the searcher in the target problem, while $|\boldsymbol{x}_0|$ can be interpreted as the initial distance between the centers of two particles in the original problem. Expectedly, the survival probability is higher when the particles are initially farther from each other, and decreases with time. Curiously, the survival probability does not vanish in the limit $t \to \infty$ because the particle may escape to infinity without encountering the target due to the transient character of Brownian motion in three dimensions.

The seminal work by Smoluchowski emphasized the importance of diffusion in chemical reactions that are nowadays called diffusion-limited, diffusion-controlled, diffusion-influenced, diffusion-assisted or diffusion-mediated reactions [2, 3]. This field goes far beyond the original coagulation dynamics and finds applications in physics, chemistry, biology, ecology and social sciences [4–9]. The particle can be an atom, an ion, a molecule, a protein, a virus, a bacterium, or even an animal or a human. The target can be another particle, an enzyme, a receptor, a specific region on the DNA, on the plasma membrane or inside the nucleus, a channel, a catalytic site, a magnetic impurity, a trap, a hole, a pray, or a website. The dynamics can be an ordinary diffusion in the bulk, a continuous-time random walk, a Lévy flight, a biased spatially heterogeneous diffusion, diffusion in a dynamic heterogeneous medium, or any other stochastic process. Finally, the reaction event can be a chemical transformation, a binding to the adsorbing surface, an association to another molecule, an escape or an entrance, a level crossing, a relaxation of an excited state (e.g., loss of fluorescence or transverse spin magnetization), eating, knowledge acquisition, or death. The influence of the dynamics onto the target problem is discussed in several chapters of this book. In contrast, we keep the diffusive dynamics as simple as possible in this chapter and focus on different mechanisms of reaction events.

In the original work by Smoluchowski, the coagulation of two particles was assumed to occur immediately upon their first encounter. This condition was implemented via Dirichlet boundary condition to the diffusion equation: $S_\infty(t|\boldsymbol{x}_0) = 0$ on the contact surface (i.e., on the target). In probabilistic terms, it simply states that if the particle starts on the target, the target is immediately found so that $\mathcal{T} = 0$ and the survival probability for any $t > 0$ is zero. In many applications, however, the first encounter is not sufficient for the reaction event [3]. In fact, a diffusing molecule and/or a target may need to be in appropriate conformational states to bind or to overcome an energy activation barrier to react; an ion channel needs to be in active (open) state to ensure the ion transfer; the diffusing particle has to arrive onto an active catalytic germ on the catalytic surface to be chemically transformed, etc. If such a condition is not fulfilled at the first arrival of the particle onto the target, its diffusion is resumed until the next arrival onto the target, and so on. As a consequence, the successful reaction event is generally preceded by a sequence of failed reaction attempts and the consequent diffusive explorations of the bulk around the target. This issue was recognized in 1949 by Collins and Kimball who suggested to account for partial reactivity of the target via Robin boundary condition (also known as radiation or third boundary condition):

$$-D\partial_n S_q(t|\boldsymbol{x}_0) = \kappa S_q(t|\boldsymbol{x}_0), \tag{4.2}$$

where $\partial_n$ is the normal derivative oriented outwards the confining domain, $\kappa$ is the reactivity of the boundary, and $q = \kappa/D$ [10]. This condition *postulates* that the diffusive flux of particles from the bulk (the left-hand side) is proportional to their reactive flux on the target (the right-hand side) at each boundary point. The reactivity $\kappa$ has units of meter per second and ranges from 0 for an inert impermeable target (no reaction) to $+\infty$ for a perfectly reactive target, in which case Dirichlet boundary condition is retrieved. The emergence of Robin boundary condition was further rationalized via various microscopic mechanisms, including stochastic gating or rapid switching between active and passive states in time [11–13], homogenization of micro-heterogeneity of reactive patches [14–20], overcoming energetic or entropic barriers or interactions [21–25]. Probabilistic interpretations of Robin boundary condition via random walks and reflected Brownian motions were discussed [26–32]. The role of reactivity onto diffusion-controlled reactions and related target problems has been thoroughly investigated [33–48].

The conventional approach for studying such diffusion-controlled reactions relies on the analysis of the Laplace operator $\Delta$ that governs diffusion in a given confining domain $\Omega$ with boundary $\partial\Omega$. For instance, when the confinement $\Omega$ is bounded, the Laplace operator is known to have a discrete spectrum that allows one to expand the survival probability and related quantities as [5, 49]

$$S_q(t|\boldsymbol{x}_0) = \sum_{k=0}^{\infty} e^{-Dt\lambda_k^{(q)}} u_k^{(q)}(\boldsymbol{x}_0) \int_{\Omega} d\boldsymbol{x}\, u_k^{(q)}(\boldsymbol{x}), \tag{4.3}$$

where $\lambda_k^{(q)}$ and $u_k^{(q)}(\boldsymbol{x})$ are the eigenvalues and normalized eigenfunctions satisfying

$$-\Delta u_k^{(q)} = \lambda_k^{(q)} u_k^{(q)} \quad \text{in } \Omega, \qquad \partial_n u_k^{(q)} + q u_k^{(q)} = 0 \quad \text{on } \partial\Omega. \tag{4.4}$$

The superscript $q = \kappa/D$ highlights the dependence of eigenvalues and eigenfunctions on the reactivity through Robin boundary condition. In this way, one can investigate the short-time and long-time behavior of the survival probability in general domains, and find explicit solutions in simple domains such as an interval, a rectangle, a disk, a sphere, a circular annulus and a spherical shell [5, 49, 50]. Despite the long history of successful applications of this spectral method, it has several drawbacks that hindered further advances in this field: (i) both $\lambda_k^{(q)}$ and $u_k^{(q)}$ exhibit *implicit* dependence on $q$ that hides the impact of the reactivity $\kappa$ on diffusion-controlled reactions; for instance, if these quantities have to be found numerically, the computation should be repeated for each value of $q$; (ii) as the reactivity enters through Robin boundary condition, the diffusive dynamics is tightly coupled to imposed surface reactions in $S_q(t|\boldsymbol{x}_0)$; in other words, it is hard to disentangle these effects; (iii) at the microscopic level, Robin boundary condition describes surface reaction, which may occur at each encounter with the target with a constant probability (see below); this description does not allow to deal with more sophisticated reaction mechanisms such as progressive passivation or activation of the target by its interactions with the diffusing particle; (iv) when diffusion occurs in an unbounded domain (e.g., in the original Smoluchowski problem with $\Omega$ being the exterior of a ball of radius $R$), the spectrum of the Laplace operator is continuous, and the spectral expansion (4.3) is not applicable.

In this chapter, we describe an alternative approach to target search problems that resolves all these limitations and provides complementary insights onto diffusion-controlled reactions. This so-called encounter-based approach was first introduced in [30] and then further elaborated and extended in [51–65]. In Sect. 4.2, we formulate this approach for three settings: discrete random walks, Brownian motion with bulk reactions, and reflected Brownian motion with surface reactions. The first setting is intuitively more appealing, even though exact computations turn out to be the most difficult. In turn, the formulation in the last setting that relies on the notion of the boundary local time is more subtle but computations are usually easier. Section 4.3 briefly describes these computations via spectral expansions based on the Dirichlet-to-Neumann operator. In Sect. 4.4, we discuss several applications and extensions of the encounter-based approach, whereas Sect. 4.5 concludes the chapter.

## 4.2 Probabilistic Insights

In this section, we formulate the encounter-based approach for three related settings and give some examples to illustrate the introduced concepts.

### *4.2.1 Discrete Random Walks*

To convey the main idea of the encounter-based approach, we start with a simpler setting of a symmetric discrete-time random walk on a square (or hyper-cubic) lattice with the spacing $a$ between the closest nodes (Fig. 4.1a). A walker starts from a point $\boldsymbol{x}_0$ and performs consecutive random jumps from its current position to one of four neighboring sites with probability $1/4$. For a given set $\Gamma$ of target sites, we count the number of encounters of the walker with the target set $\Gamma$ after $n$ steps,

$$L_n = \sum_{i=1}^{n} \mathbb{I}_\Gamma(\boldsymbol{x}_i), \qquad (4.5)$$

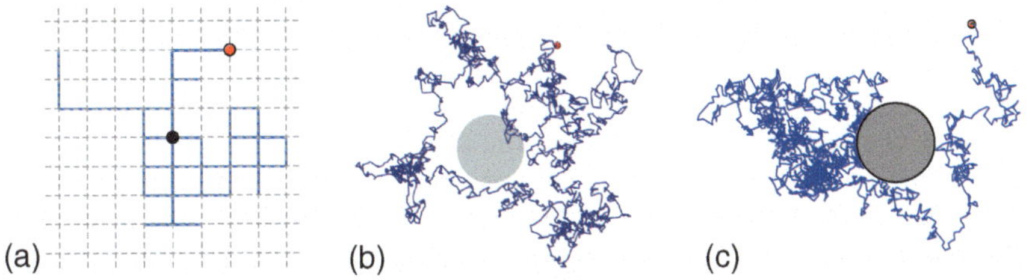


**Fig. 4.1** A random trajectory (in blue) of a particle that started from $\boldsymbol{x}_0$ (small red disk) and searching for a target set. **a** A random walk on the square lattice visiting a single target site at the origin (small black disk): $\Gamma = \{0\}$; **b** Brownian motion that may react when diffusing through the target set $\Gamma = \{|\boldsymbol{x}| < R\}$ (gray disk); here, the particle dynamics is not affected by the target; **c** reflected Brownian motion that may react when hitting the target set $\Gamma = \{|\boldsymbol{x}| = R\}$ (in black) on the boundary of an impenetrable obstacle (gray disk); the latter modifies the trajectory of the particle

where $\boldsymbol{x}_i$ is the random position of the walker after $i$ steps, and $\mathbb{I}_\Gamma(\boldsymbol{x})$ is the indicator function of $\Gamma$: $\mathbb{I}_\Gamma(\boldsymbol{x}) = 1$ if $\boldsymbol{x} \in \Gamma$ and 0 otherwise. In other words, at each intermediate step $i = 1, 2, \ldots, n$, the counter $L_n$ is incremented by 1 if the walker's position $\boldsymbol{x}_i$ belongs to the target set $\Gamma$ (i.e., the walker visits any site of $\Gamma$). The random variable $L_n$ plays the central role in the encounter-based approach. We describe the diffusive dynamics of this walker by the so-called encounter propagator

$$p(\boldsymbol{x}, L, n|\boldsymbol{x}_0) = \mathbb{P}_{\boldsymbol{x}_0}\{\boldsymbol{x}_n = \boldsymbol{x},\ L_n = L\}, \tag{4.6}$$

i.e., the joint probability distribution of the walker's position $\boldsymbol{x}_n$ and of the number of encounters $L_n$. Rubin and Weiss proposed a formal method to determine this quantity for any finite configuration of target sites [66]. To avoid technical issues, we skip the description of their method and just provide an example below. In turn, we discuss the advantages of the encounter propagator $p(\boldsymbol{x}, L, n|\boldsymbol{x}_0)$ in describing target search problems.

In this framework, the walker merely counts its encounters with the target set during the diffusive exploration of the space, i.e., the encounter propagator $p(\boldsymbol{x}, L, n|\boldsymbol{x}_0)$ does not include any "trigger" to stop this search process and to say that the target is found. For instance, if the target set can trap or bind the walker after some encounters, such a stopping condition has to be added explicitly. The most common assumption is that the walker reacts with the target set with a given probability $\sigma$ at each visit, and each reaction attempt is independent of the dynamics and other reaction attempts. The probability of finding the *survived* walker in $\boldsymbol{x}$ after $n$ steps, i.e., *without* reaction on the target set, follows immediately as

$$g_\sigma(\boldsymbol{x}, n|\boldsymbol{x}_0) = \sum_{L=0}^{\infty} (1-\sigma)^L\, p(\boldsymbol{x}, L, n|\boldsymbol{x}_0). \tag{4.7}$$

In fact, this sum classifies the contributions from different random trajectories from $\boldsymbol{x}_0$ to $\boldsymbol{x}$ according to the number of encounters with the target set. For instance, $p(\boldsymbol{x}, 0, n|\boldsymbol{x}_0)$ is the probability of moving from $\boldsymbol{x}_0$ to $\boldsymbol{x}$ in $n$ steps without any encounter with the target set. The next term, $(1-\sigma)p(\boldsymbol{x}, 1, n|\boldsymbol{x}_0)$, is the contribution of trajectories that experienced only one encounter with $\Gamma$, and this reaction attempt fails with probability $1-\sigma$. Similarly, $(1-\sigma)^L p(\boldsymbol{x}, L, n|\boldsymbol{x}_0)$ is the contribution of trajectories that experienced $L$ encounters with $\Gamma$, and each of these reaction attempts fails, yielding the factor $(1-\sigma)^L$. Note that the sum in Eq. (4.7) has actually a finite number of terms because $p(\boldsymbol{x}, L, n|\boldsymbol{x}_0) = 0$ for any $L > n$. While $p(\boldsymbol{x}, L, n|\boldsymbol{x}_0)$ characterized the search process alone (without any reaction), the propagator $g_\sigma(\boldsymbol{x}, n|\boldsymbol{x}_0)$ incorporates reactions on the target set and thus implements the survival of the walker. The expression (4.7) relates these two fundamental quantities and allows one to translate one from the other.

Conventional approaches to target search problems with random walks deal directly with the propagator $g_\sigma(\boldsymbol{x}, n|\boldsymbol{x}_0)$ (or related quantities), which is often easier to access than $p(\boldsymbol{x}, L, n|\boldsymbol{x}_0)$. In fact, the propagator $g_\sigma(\boldsymbol{x}, n|\boldsymbol{x}_0)$ satisfies the master equation

$$g_\sigma(\boldsymbol{x}, n+1|\boldsymbol{x}_0) = \frac{1}{2d}\left(\sum_{|\boldsymbol{x}'-\boldsymbol{x}|=a} g_\sigma(\boldsymbol{x}', n|\boldsymbol{x}_0)\right) - \sigma\, \mathbb{I}_\Gamma(\boldsymbol{x}) g_\sigma(\boldsymbol{x}, n|\boldsymbol{x}_0), \tag{4.8}$$

with the initial condition $g_\sigma(\boldsymbol{x}, 0|\boldsymbol{x}_0) = \delta_{\boldsymbol{x},\boldsymbol{x}_0}$, which can be solved by standard tools [67–69]. One may thus wonder what the purpose of introducing a less accessible quantity $p(\boldsymbol{x}, L, n|\boldsymbol{x}_0)$ is? *Its disentanglement from the reaction mechanism* is the fundamental advantage of the encounter propagator, allowing one to consider more sophisticated reactions. To illustrate this idea, let us assume that reaction on the target set occurs after a prescribed number of encounters $L_{\max}$ (for instance, a cleaning procedure or a security update is triggered when the number of authorized visits exceeds $L_{\max}$). The conventional propagator $g_\sigma(\boldsymbol{x}, n|\boldsymbol{x}_0)$ does not allow one to describe the position of the walker that survives upon such a stopping condition. In turn, it is enough to sum the encounter propagator $p(\boldsymbol{x}, L, n|\boldsymbol{x}_0)$ over $L$ from 0 to $L_{\max} - 1$ to determine the probability of finding the walker at $\boldsymbol{x}$ at time $n$ before the random walk is stopped at the $L_{\max}$th encounter. More generally, one can replace a fixed threshold $L_{\max}$ for the number of encounters $L_n$ by a random threshold $\hat{L}$ that is defined by a given probability law $\Psi(L) = \mathbb{P}\{\hat{L} > L\}$. In this case, the reaction on the target set occurs at a random moment $n$ when the number of encounters $L_n$ reaches the random threshold $\hat{L}$. In other words, we employ the stopping condition $L_n = \hat{L}$ to define the reaction mechanism, whereas the condition $L_n < \hat{L}$ describes the survival of the walker in such reactions. As the threshold $\hat{L}$ itself is random, the probability of finding the survived walker at $\boldsymbol{x}$ after $n$ steps is

$$g_\Psi(\boldsymbol{x}, n|\boldsymbol{x}_0) = \mathbb{P}_{\boldsymbol{x}_0}\{\boldsymbol{x}_n = \boldsymbol{x},\ L_n < \hat{L}\} = \sum_{L=0}^{\infty} \Psi(L)\, p(\boldsymbol{x}, L, n|\boldsymbol{x}_0). \tag{4.9}$$

Setting $\Psi(L) = \mathbb{I}_{[0,L_{\max}-1]}(L)$ yields the fixed threshold at $L_{\max}$, whereas $\Psi(L) = (1-\sigma)^L$ implies the earlier discussed conventional setting with a constant reaction probability $\sigma$; see Eq. (4.7). One sees that the introduction of a random threshold opens the possibility to describe much more general reaction mechanisms. For instance, the constant reaction probability $\sigma$ can be replaced by a sequence of reaction probabilities $\sigma_1, \sigma_2, \ldots$, in which case the contribution of the $L$th term with $p(\boldsymbol{x}, L, n|\boldsymbol{x}_0)$ is weighted not by $(1-\sigma)^L$, as in Eq. (4.7), but by $(1-\sigma_1)(1-\sigma_2)\ldots(1-\sigma_L)$. In other words, such a model of encounter-dependent reactivity corresponds to

$$\Psi(L) = \prod_{l=1}^{L} (1-\sigma_l). \tag{4.10}$$

The controlled dependence of the reaction probability $\sigma_l$ on the number of encounters allows one to describe activation or passivation of the target set by its interactions with the random walker. Alternatively, one can rewrite this relation as $\Psi(L)/\Psi(L-1) = 1-\sigma_L$ to determine the sequence of reaction probabilities $\sigma_l$ that represents a given probability law $\Psi(L)$. These generalized reaction mechanisms were introduced in [30] for the encounter-based approach in the continuous setting (see Sect. 4.2.3). We stress that such mechanisms require the knowledge of the encounter propagator and therefore are not accessible in conventional approaches.

### Example of a Single Target Site

When the target set $\Gamma$ contains a single site, located at the origin, the probabilities $p(\boldsymbol{x}, L, n|\boldsymbol{x}_0)$ can be determined from the generating function

$$\tilde{p}(\boldsymbol{x}, L, z|\boldsymbol{x}_0) = \sum_{n=0}^{\infty} z^n \, p(\boldsymbol{x}, L, n|\boldsymbol{x}_0). \tag{4.11}$$

For a symmetric random walk on the square lattice, one gets [66, 70]

$$\tilde{p}(\boldsymbol{x}, L, z|\boldsymbol{x}_0) = \begin{cases} \dfrac{\tilde{q}(\boldsymbol{x}, z)\tilde{q}(\boldsymbol{x}_0, z)}{[\tilde{q}(0, z)]^2}\left(1 - \dfrac{1}{\tilde{q}(0, z)}\right)^{L-1} & (L \geq 1), \\ \tilde{q}(\boldsymbol{x}-\boldsymbol{x}_0, z) - \dfrac{\tilde{q}(\boldsymbol{x}, z)\tilde{q}(\boldsymbol{x}_0, z)}{\tilde{q}(0, z)} & (L = 0), \end{cases} \tag{4.12}$$

where

$$\tilde{q}(\boldsymbol{x}, z) = \frac{1}{(2\pi)^2} \int_{-\pi}^{\pi} dk_1 \int_{-\pi}^{\pi} dk_2 \, e^{-i(\boldsymbol{x}\cdot\boldsymbol{k})/a} \left(1 - z\frac{\cos(k_1) + \cos(k_2)}{2}\right)^{-1} \tag{4.13}$$

is the lattice Green's function, i.e., the generating function of the probabilities $q(\boldsymbol{x}, n)$ of finding the walker at $\boldsymbol{x}$ after $n$ steps on the square lattice without any target (see also Chap. 5 for more details). Note that $\tilde{q}(0, z)$ can be expressed as $\tilde{q}(0, z) = \frac{2}{\pi} K(z)$, where $K(z)$ is the complete elliptic integral of the first kind [71]. From the propagator $p(\boldsymbol{x}, L, n|\boldsymbol{x}_0)$, one can deduce other quantities of interest; e.g., Eq. (4.7) yields the generating function for the propagator $g_\sigma(\boldsymbol{x}, n|\boldsymbol{x}_0)$:

$$\tilde{g}_\sigma(\boldsymbol{x}, z|\boldsymbol{x}_0) = \sum_{n=0}^{\infty} z^n \, g_\sigma(\boldsymbol{x}, n|\boldsymbol{x}_0) = \tilde{q}(\boldsymbol{x} - \boldsymbol{x}_0, z) - \frac{\sigma \, \tilde{q}(\boldsymbol{x}, z)\tilde{q}(\boldsymbol{x}_0, z)}{1 - (1-\sigma)\tilde{q}(0, z)} . \tag{4.14}$$

In turn, the sum over $\boldsymbol{x}$ allows one to determine the statistics of encounters, $\rho(L, n|\boldsymbol{x}_0) = \sum_{\boldsymbol{x}} p(\boldsymbol{x}, L, n|\boldsymbol{x}_0)$, for which we get for $L \geq 1$

$$\tilde{\rho}(L, z|\boldsymbol{x}_0) = \sum_{n=0}^{\infty} z^n \, \rho(L, n|\boldsymbol{x}_0) = \sum_{\boldsymbol{x}} \tilde{p}(\boldsymbol{x}, L, z|\boldsymbol{x}_0) = \frac{\big(1 - 1/\tilde{q}(0, z)\big)^{L-1} \tilde{q}(\boldsymbol{x}_0, z)}{(1-z)[\tilde{q}(0, z)]^2} , \tag{4.15}$$

where we used the normalization of $q(\boldsymbol{x}, n)$ to evaluate the sum of $\tilde{q}(\boldsymbol{x}, z)$ over $\boldsymbol{x}$. After finding this generating function, one can access these probabilities:

$$\rho(L, n|\boldsymbol{x}_0) = \mathbb{P}_{\boldsymbol{x}_0}\{L_n = L\} = \frac{1}{n!} \lim_{z \to 0} \frac{d^n}{dz^n} \tilde{\rho}(L, z|\boldsymbol{x}_0). \tag{4.16}$$

The same solution holds for the symmetric random walk on higher dimensional lattices with an appropriate modification of the lattice Green's function $\tilde{q}(\boldsymbol{x}, z)$ [66]. For instance, Joyce obtained an explicit representation for $\tilde{q}(0, z)$ for the simple cubic lattice [72]:

$$\tilde{q}(0, z) = \frac{(1 - 9\xi^4)\big[\frac{2}{\pi} K(\eta)\big]^2}{(1-\xi)^3(1+3\xi)}, \quad \eta = \sqrt{\frac{16\xi^3}{(1-\xi)^3(1+3\xi)}}, \quad \xi = \sqrt{\frac{1 - \sqrt{1 - z^2/9}}{1 + \sqrt{1 - z^2}}} . \tag{4.17}$$

Despite the explicit character of Eq. (4.12), a practical implementation of this solution to get the probabilities $p(\boldsymbol{x}, L, n|\boldsymbol{x}_0)$ is rather tedious even numerically. In fact, one needs to compute the double integral in Eq. (4.13) and then to evaluate the $n$th order derivative $\frac{d^n}{dz^n} \tilde{p}(\boldsymbol{x}, L, z|\boldsymbol{x}_0)$ at $z = 0$. Even the inversion of the fully explicit relation (4.15) with $\boldsymbol{x}_0 = 0$ to get $\rho(L, n|0)$ is not straightforward, especially for large $n$. In turn, one can employ this formalism to study the asymptotic behavior.[1]

[1] For instance, one can easily compute the generating function for the mean number of encounters

$$f(z|\boldsymbol{x}_0) = \sum_{n=0}^{\infty} z^n \, \mathbb{E}_{\boldsymbol{x}_0}\{L_n\} = \sum_{L=1}^{\infty} L \, \tilde{\rho}(L, z|\boldsymbol{x}_0)$$

Moreover, one easily gets the first values of $\rho(L, n|0)$; e.g., for the square lattice,

| $\rho(L, n\|0)$ | $n = 0$ | $n = 2$ | $n = 4$ | $n = 6$ |
|---|---|---|---|---|
| $L = 1$ | 1 | 3/4 | 43/64 | 161/256 |
| $L = 2$ | 0 | 1/4 | 17/64 | 69/256 |
| $L = 3$ | 0 | 0 | 4/64 | 22/256 |
| $L = 4$ | 0 | 0 | 0 | 4/256 |

and for the simple cubic lattice:

| $\rho(L, n\|0)$ | $n = 0$ | $n = 2$ | $n = 4$ | $n = 6$ |
|---|---|---|---|---|
| $L = 1$ | 1 | 5/6 | 57/72 | 2995/3888 |
| $L = 2$ | 0 | 1/6 | 13/72 | 731/3888 |
| $L = 3$ | 0 | 0 | 2/72 | 144/3888 |
| $L = 4$ | 0 | 0 | 0 | 18/3888 |

Each column gives the distribution of the number of encounters $L_n$ for even $n$ (for odd $n$, the distribution is the same as for $n - 1$ because the walker needs an even number of steps to return to the target site and to change $L_n$). For instance, after two steps, the walker who started from the target site can either return to it (with probability $(2d)^{-1}$) and thus increase the number of encounters to 2, or stay away from it and keep the initial number of encounters $L = 1$ (with probability $1 - (2d)^{-1}$).

The above example illustrates that, even in the simplest case of a single target site on the square lattice, the computation of the probabilities $p(\boldsymbol{x}, L, n|\boldsymbol{x}_0)$ and $\rho(L, n|\boldsymbol{x}_0)$ is feasible but rather sophisticated, even numerically. When the number of target sites increases, the computation is still feasible in a matrix form (see [66]), but the dependence of the propagators $p(\boldsymbol{x}, L, n|\boldsymbol{x}_0)$ and $g_\sigma(\boldsymbol{x}, n|\boldsymbol{x}_0)$ on the starting and arrival points and on the arrangement of the target sites becomes hidden behind matrix operations. Moreover, even though infinite lattices can be replaced by general graphs to describe, e.g., random walks in confined environments, the obtained results would be rather formal.

In the next section, we discuss how to generalize the encounter-based approach to continuous diffusion, for which computations become simpler.

---

$$= \frac{1}{1-z} \frac{\tilde{q}(\boldsymbol{x}_0, z)}{[\tilde{q}(0, z)]^2} \sum_{L=1}^{\infty} L\big(1 - 1/\tilde{q}(0, z)\big)^{L-1} = \frac{\tilde{q}(\boldsymbol{x}_0, z)}{1-z} .$$

From the behavior of $f(z|\boldsymbol{x}_0)$ near $z = 1$, one can determine the large-$n$ asymptotic behavior of the mean with the help of Tauberian theorems. For the simple cubic lattice, one has in the leading order $f(z|\boldsymbol{x}_0) \approx \tilde{q}(\boldsymbol{x}_0, 1)/(1 - z)$ as $z \to 1$, so that $\mathbb{E}_{\boldsymbol{x}_0}\{L_n\} \to \tilde{q}(\boldsymbol{x}_0, 1)$ as $n \to \infty$ (e.g., one has $\tilde{q}(0, 1) \approx 1.5164$ at $\boldsymbol{x}_0 = 0$). Accounting for the next-order correction allows one to show that the approach to this limit is rather slow, as $1/\sqrt{n}$. Note that the situation is different for the square lattice, for which $\tilde{q}(0, 1) \propto \ln(1 - z)$ so that $\mathbb{E}_{\boldsymbol{x}_0}\{L_n\}$ exhibits a logarithmic growth with $n$, due to the recurrent nature of the random walk on the square lattice.

### 4.2.2 Brownian Motion with Bulk Reactions

In the continuous limit $a \to 0$, the discrete-step random walk $\boldsymbol{x}_n$ is known to approach Brownian motion $\boldsymbol{X}_t$, which is continuous in time and space. To extend the conventional approach to the continuous setting, we consider now the target set $\Gamma \subset \mathbb{R}^d$ as a region in the bulk, and the particle diffusing inside $\Gamma$ can undertake a first-order bulk reaction with the rate $Q$ (Fig. 4.1b). The propagator $G_Q(\boldsymbol{x}, t|\boldsymbol{x}_0)$ satisfies then the standard diffusion-reaction equation:

$$\partial_t G_Q(\boldsymbol{x}, t|\boldsymbol{x}_0) = D\Delta G_Q(\boldsymbol{x}, t|\boldsymbol{x}_0) - Q\,\mathbb{I}_\Gamma(\boldsymbol{x})\,G_Q(\boldsymbol{x}, t|\boldsymbol{x}_0), \tag{4.18}$$

subject to the initial condition $G_Q(\boldsymbol{x}, 0|\boldsymbol{x}_0) = \delta(\boldsymbol{x} - \boldsymbol{x}_0)$ with the Dirac distribution $\delta(\boldsymbol{x} - \boldsymbol{x}_0)$ that fixes its starting point $\boldsymbol{x}_0$. Here the time evolution of $G_Q(\boldsymbol{x}, t|\boldsymbol{x}_0)$ is governed by diffusion with a constant diffusion coefficient $D$ (the first term) and eventual disappearance due to bulk reactions inside $\Gamma$ (the second term). In contrast to the discrete setting, $G_Q(\boldsymbol{x}, t|\boldsymbol{x}_0)$ is not the probability of arriving at $\boldsymbol{x}$ (which is strictly zero for continuous diffusion), but the *probability density* of finding the particle in a vicinity of $\boldsymbol{x}$ at time $t$, which survives against bulk reactions (see below).

It is worth noting that Szabo et al. proposed a more direct "translation" from the discrete to continuous setting by incorporating point-like partial traps, localized at points $\boldsymbol{x}_1, \ldots, \boldsymbol{x}_J$, via Dirac peaks into the diffusion equation [73]. In other words, the reactivity profile $\mathbb{I}_\Gamma(\boldsymbol{x})$ in Eq. (4.18) was replaced by $\sum_j \delta(\boldsymbol{x} - \boldsymbol{x}_j)$. In this case, the propagator $G_Q(\boldsymbol{x}, t|\boldsymbol{x}_0)$ in the presence of partial traps can be expressed in terms of the propagator $G_0(\boldsymbol{x}, t|\boldsymbol{x}_0)$ without traps in a way, which is similar to the discrete solution (4.14). However, this approach is only applicable in the one-dimensional case[2] because a point is not accessible to Brownian motion in higher dimensions. In fact, the probability of visiting a given point (or even a countable set of points) is strictly zero. As a consequence, a target site cannot be treated as point-like anymore but rather as a small ball (or any open set $\Gamma \subset \mathbb{R}^d$ as we did above), to make it accessible by Brownian motion.

To proceed toward an encounter-based approach, one can replace the number of encounters $L_n$, introduced in Eq. (4.5) for a random walk, by the residence time of Brownian motion inside $\Gamma$ up to time $t$:

$$L_t = \int\limits_0^t dt'\,\mathbb{I}_\Gamma(\boldsymbol{X}_{t'}). \tag{4.19}$$

[2] The problem of localized partial traps was formulated in [73] in a way that might wrongly suggest that its solution is valid in dimensions higher than 1 (e.g., bold notation $\boldsymbol{r}$ for the position was employed). This is not true, as the authors stated on page 234: "...but that in higher dimensions one must have a reactive surface in order to have a properly posed problem."

One sees that the residence time is a specific functional of Brownian trajectory $\boldsymbol{X}_t$. According to the Feynman–Kac formula, the propagator $G_Q(\boldsymbol{x}, t|\boldsymbol{x}_0)$ satisfying Eq. (4.18) admits the following probabilistic interpretation:

$$G_Q(\boldsymbol{x}, t|\boldsymbol{x}_0) = \mathbb{E}_{\boldsymbol{x}_0}\left\{e^{-QL_t}\delta(\boldsymbol{X}_t - \boldsymbol{x})\right\} = \int_0^\infty dL\, e^{-QL}\, P(\boldsymbol{x}, L, t|\boldsymbol{x}_0). \tag{4.20}$$

The first equality states that the probability density $G_Q(\boldsymbol{x}, t|\boldsymbol{x}_0)$ accounts for all trajectories from $\boldsymbol{x}_0$ to $\boldsymbol{x}$ (of duration $t$), whose contribution is weighted by their residence time $L_t$ spent in $\Gamma$, during which the particle might react and thus disappear with probability $1 - e^{-QL_t}$. In turn, the second equality spells out the definition of this average in terms of the encounter propagator $P(\boldsymbol{x}, L, t|\boldsymbol{x}_0)$, i.e., the joint probability density of finding the particle, started from $\boldsymbol{x}_0$ at time 0, in a vicinity of $\boldsymbol{x}$ at time $t$, with the residence time in a vicinity of $L$:

$$\mathbb{P}_{\boldsymbol{x}_0}\{\boldsymbol{X}_t \in (\boldsymbol{x}, \boldsymbol{x} + d\boldsymbol{x}), L_t \in (L, L + dL)\} = P(\boldsymbol{x}, L, t|\boldsymbol{x}_0)d\boldsymbol{x}dL. \tag{4.21}$$

If one manages to solve Eq. (4.18) for a given target set $\Gamma$, the encounter propagator $P(\boldsymbol{x}, L, t|\boldsymbol{x}_0)$ can be formally obtained via the inverse Laplace transform with respect to $Q$. Alternatively, as multiplication by $Q$ in Eq. (4.18) becomes the derivative by $L$ in the dual space with respect to the Laplace transform, one can write directly the partial differential equation for the encounter propagator:

$$\partial_t P(\boldsymbol{x}, L, t|\boldsymbol{x}_0) = D\Delta P(\boldsymbol{x}, L, t|\boldsymbol{x}_0) - \mathbb{I}_\Gamma(\boldsymbol{x})\partial_L P(\boldsymbol{x}, L, t|\boldsymbol{x}_0), \tag{4.22}$$

subject to the initial condition $P(\boldsymbol{x}, L, 0|\boldsymbol{x}_0) = \delta(\boldsymbol{x} - \boldsymbol{x}_0)\delta(L)$. The second term can be interpreted as the probability flux due to the increase of the residence time when the particle diffuses inside the target set $\Gamma$. Multiplying this equation by $e^{-QL}$ and evaluating its integral over $L$ from 0 to $+\infty$ via integration by parts, one retrieves Eq. (4.18) for the propagator $G_Q(\boldsymbol{x}, t|\boldsymbol{x}_0)$, given that $P(\boldsymbol{x}, +\infty, t|\boldsymbol{x}_0) = 0$, and $P(\boldsymbol{x}, 0, t|\boldsymbol{x}_0) = 0$ for any $\boldsymbol{x} \in \Gamma$.

Let us look again at Eq. (4.20), in which the factor $e^{-QL}$ can be interpreted as the survival condition $\{L_t < \hat{L}\}$, where $\hat{L}$ is the exponentially distributed random threshold for the residence time $L_t$ that determines the bulk reaction. The exponential law for this threshold is the consequence of choosing the first-order bulk reaction inside the target set $\Gamma$. In analogy to Eq. (4.9), one can introduce an arbitrary probability law $\Psi(L) = \mathbb{P}\{\hat{L} > L\}$ for the random threshold $\hat{L}$, in which case the propagator $G_Q(\boldsymbol{x}, t|\boldsymbol{x}_0)$ is substituted by the generalized propagator

$$\begin{aligned} G_\Psi(\boldsymbol{x}, t|\boldsymbol{x}_0) &= \mathbb{E}_{\boldsymbol{x}_0}\{\delta(\boldsymbol{X}_t - \boldsymbol{x}) \cdot \mathbb{I}_{L_t < \hat{L}}\} = \mathbb{E}_{\boldsymbol{x}_0}\{\delta(\boldsymbol{X}_t - \boldsymbol{x})\, \Psi(L_t)\} \\ &= \int_0^\infty dL\, \Psi(L)\, P(\boldsymbol{x}, L, t|\boldsymbol{x}_0), \end{aligned} \tag{4.23}$$

where $\mathbb{I}_{L_t<\hat{L}}$ is the indicator function of the probabilistic event $L_t < \hat{L}$. When $\Psi(L) \neq e^{-QL}$, the bulk reaction is not of the first-order, and the generalized propagator does not satisfy Eq. (4.18). One deals therefore with non-Markovian bulk reactions introduced via non-exponential random threshold $\hat{L}$ and its distribution $\Psi(L)$.

### Example of a Ball as the Target Set

To illustrate this approach, let us consider a ball of radius $R$ as the target set: $\Gamma = \{\boldsymbol{x} \in \mathbb{R}^3 \; : \; |\boldsymbol{x}| < R\}$. While the propagator $G_Q(\boldsymbol{x}, t|\boldsymbol{x}_0)$ can be found in this case, we focus on the survival probability,

$$S_Q(t|\boldsymbol{x}_0) = \int_{\mathbb{R}^3} d\boldsymbol{x}\, G_Q(\boldsymbol{x}, t|\boldsymbol{x}_0), \tag{4.24}$$

i.e., the probability that the particle has not reacted up to time $t$. Note that the integral of Eq. (4.20) over $\boldsymbol{x}$ implies that

$$S_Q(t|\boldsymbol{x}_0) = \int_0^\infty dL\, e^{-QL}\, \rho(L, t|\boldsymbol{x}_0), \tag{4.25}$$

i.e., the inverse Laplace transform of $S_Q(t|\boldsymbol{x}_0)$ with respect to $Q$ determines the statistics of the residence time $L_t$. As the solution is fairly standard, we only sketch the main steps. The survival probability satisfies backward diffusion-reaction equation, alike Eq. (4.18), but with respect to the starting point $\boldsymbol{x}_0$. In turn, its Laplace transform with respect to time $t$,

$$\tilde{S}_Q(p|\boldsymbol{x}_0) = \int_0^\infty dt\, e^{-pt}\, S_Q(t|\boldsymbol{x}_0), \tag{4.26}$$

satisfies

$$\big(p - D\Delta_{\boldsymbol{x}_0} + Q\,\mathbb{I}_\Gamma(\boldsymbol{x}_0)\big)\tilde{S}_Q(p|\boldsymbol{x}_0) = 1. \tag{4.27}$$

The rotational symmetry of $\Gamma$ implies that $\tilde{S}_Q(p|\boldsymbol{x}_0)$ depends only on the radial coordinate $r = |\boldsymbol{x}_0|$ that considerably simplifies its computation. Substituting $\Delta = \partial_r^2 + \frac{2}{r}\partial_r$ into Eq. (4.27) and writing its solutions separately for $r < R$ and for $r > R$, one then matches two solutions to ensure the continuity of $\tilde{S}_Q(p|\boldsymbol{x}_0)$ and of its radial derivative at $r = R$. After simplifications, one gets

$$\tilde{S}_Q(p|\boldsymbol{x}_0) = \begin{cases} \dfrac{1}{p'} - (C-1)\dfrac{p'-p}{pp'}\dfrac{R\,\sinh(|\boldsymbol{x}_0|\sqrt{p'/D})}{|\boldsymbol{x}_0|\,\sinh(R\sqrt{p'/D})} & (|\boldsymbol{x}_0| < R), \\ \dfrac{1}{p} - C\dfrac{p'-p}{pp'}\dfrac{R}{|\boldsymbol{x}_0|}e^{-(|\boldsymbol{x}_0|-R)\sqrt{p/D}} & (|\boldsymbol{x}_0| \geq R), \end{cases} \tag{4.28}$$

where

$$p' = p + Q, \qquad C = \frac{1 - \tanh(R\sqrt{p'/D})/(R\sqrt{p'/D})}{1 + \sqrt{p/p'}\tanh(R\sqrt{p'/D})}. \tag{4.29}$$

As a consequence, the inverse Laplace transform of $\tilde{S}_Q(p|\boldsymbol{x}_0)$ with respect to $p$ yields $S_Q(t|\boldsymbol{x}_0)$, whereas another inverse Laplace transform with respect to $Q$ gives access to $\rho(L,t|\boldsymbol{x}_0)$:

$$\rho(L,t|\boldsymbol{x}_0) = \mathcal{L}^{-1}_{Q,L}\{\mathcal{L}^{-1}_{p,t}\{\tilde{S}_Q(p|\boldsymbol{x}_0)\}\}. \tag{4.30}$$

Despite the explicit form of $\tilde{S}_Q(p|\boldsymbol{x}_0)$ in Eq. (4.28), the above representation of $\rho(L,t|\boldsymbol{x}_0)$ is rather formal because it involves two inverse Laplace transforms. Since $\tilde{S}_Q(p|\boldsymbol{x}_0)$ in Eq. (4.28) depends on $p$ and $p' = p + Q$, it is convenient to shift $Q$ by $p$ and write

$$\rho(L,t+L|\boldsymbol{x}_0) = \mathcal{L}^{-1}_{L,Q}\{L^{-1}_{p,t}\{\tilde{S}_{Q-p}(p|\boldsymbol{x}_0)\}\}. \tag{4.31}$$

This shift allows one to evaluate the inverse Laplace transform with respect to $p$ explicitly. For instance, for $\delta = |\boldsymbol{x}_0| - R > 0$, one gets

$$\begin{aligned} \rho(L,t+L|\boldsymbol{x}_0) &= \left[1 - \frac{R}{|\boldsymbol{x}_0|}\mathrm{erfc}\left(\frac{\delta}{\sqrt{4Dt}}\right)\right]\delta(L) + \frac{\sqrt{D}}{|\boldsymbol{x}_0|}\mathrm{erfc}\left(\frac{\delta}{\sqrt{4Dt}}\right)\mathcal{L}^{-1}_{L,Q}\left\{\frac{1}{A(Q)}\right\} \\ &+ \frac{R}{|\boldsymbol{x}_0|}e^{-\delta^2/(4Dt)}\mathcal{L}^{-1}_{L,Q}\left\{\left(1 - \frac{\sqrt{D}}{RA(Q)}\right)\left[\frac{A(Q)}{Q\sqrt{\pi t}}\right.\right. \\ &+ \left.\left.\left(1 - \frac{A^2(Q)}{Q}\right)\mathrm{erfcx}\left(\frac{\delta}{\sqrt{4Dt}} + \sqrt{t}A(Q)\right)\right]\right\}, \end{aligned} \tag{4.32}$$

where $A(Q) = \sqrt{Q}/\tanh(R\sqrt{Q/D})$, and $\mathrm{erfcx}(z) = e^{z^2}\mathrm{erfc}(z)$ is the scaled complementary error function. Expectedly, the first term accounts for trajectories that never arrive on the target set $\Gamma$ so that the residence time remains 0; the coefficient in front of it is simply $S_\infty(t|\boldsymbol{x}_0)$, as if the (perfectly reactive) target set was on the boundary. The other terms can be evaluated numerically or used to investigate the asymptotic behavior or the moments of the residence time.[3]

---

[3] In practice, it is more convenient to evaluate the derivatives of $\tilde{S}_Q(p|\boldsymbol{x}_0)$ with respect to $Q$ at $Q = 0$. For instance, to compute the mean residence time, one finds

$$-\left.\frac{\partial \tilde{S}_Q(p|\boldsymbol{x}_0)}{\partial Q}\right|_{Q=0} = \frac{\sqrt{D}}{|\boldsymbol{x}_0|p^{5/2}}\left(\frac{1 + R\sqrt{p/D}}{1 + \tanh(R\sqrt{p/D})} - 1\right),$$

so that

In the above discussion, the particle was assumed to diffuse in the whole space $\mathbb{R}^d$. The description can be easily adapted to consider diffusion in a confined medium or a reservoir $\Omega \subset \mathbb{R}^d$ by imposing Neumann boundary condition on the reflecting wall $\partial\Omega$, i.e., by requiring that the diffusive flux of particles across the wall vanishes. The inclusion of this boundary condition does not change the above discussion but may render the computation of the encounter propagator $P(\boldsymbol{x}, L, t|\boldsymbol{x}_0)$ more sophisticated.

### 4.2.3 Reflected Brownian Motion with Surface Reactions

In the above description of bulk reactions, the particle can freely diffuse through the target set $\Gamma$, i.e., the particle dynamics is not affected by $\Gamma$. For some target search problems, this description naturally represents the physical process; for instance, a laser beam can relax the excited state of fluorescent particles in a given spot, without alternating their diffusion. However, there are many other applications when reaction occurs on (a subset of) the boundary of an impenetrable obstacle or on the solid wall of the confining domain. For instance, this is the case of heterogeneous catalysis when the particle has to arrive onto a catalytic germ located on the surface of a solid catalyst. Alternatively, the particle can bind to or adsorb on adherent patches on the boundary but cannot go through it. Similarly, if there is an ion channel in the otherwise reflecting wall, the ion may exit or escape the confining domain through this target set on the boundary, or be reflected back. In all these cases, the target set is located on the boundary, which strongly affects the particle dynamics due to eventual reflections. We speak therefore about *surface reactions* that will be the main focus in the rest of this chapter. More precisely, we consider ordinary diffusion of a point-like particle inside a confining domain $\Omega \subset \mathbb{R}^d$ with a boundary $\partial\Omega$, and assume that the target set $\Gamma$ is located on the boundary, i.e., $\Gamma = \partial\Omega$ or $\Gamma \subset \partial\Omega$ (Fig. 4.1c).

The standard diffusion-reaction equation (4.18) is not applicable in this setting. In fact, this equation holds only for the interior (bulk) points $\boldsymbol{x} \in \Omega$, for which $\mathbb{I}_\Gamma(\boldsymbol{x}) \equiv 0$. From the probabilistic point of view, as the target set $\Gamma$ is a $(d-1)$-dimensional surface, the fraction of time that Brownian motion spent on it is strictly zero, i.e., $L_t \equiv 0$ for any $t$. A simple way to overcome this problem is to introduce a thin layer near $\Gamma$ of thickness $a$, $\Gamma_a = \{\boldsymbol{x} \in \Omega \,:\, |\boldsymbol{x} - \Gamma| < a\}$, where $|\boldsymbol{x} - \Gamma|$ is the Euclidean distance between $\boldsymbol{x}$ and the target set $\Gamma$. For this target set, one can still apply Eq. (4.18) to get the propagator $G_Q^{(a)}(\boldsymbol{x}, t|\boldsymbol{x}_0)$ (the superscript $(a)$ highlights its dependence on $a$). If we keep the reaction rate $Q$ fixed and set $a \to 0$, the effect of

$$\mathbb{E}_{\boldsymbol{x}_0}\{L_t\} = \frac{\sqrt{D}}{|\boldsymbol{x}_0|}\left(-\frac{t^{3/2}}{\Gamma(5/2)} + \mathcal{L}^{-1}_{p,t}\left\{\frac{R\sqrt{p/D}+1}{p^{5/2}(1+\tanh(R\sqrt{p/D}))}\right\}\right).$$

At short times, one has $\mathbb{E}_{\boldsymbol{x}_0}\{L_t\} \approx \frac{Rt}{2|\boldsymbol{x}_0|}\big(1 - \frac{4\sqrt{Dt}}{3\sqrt{\pi}R}\big)$, with exponentially small corrections. In turn, at long times, one gets $\mathbb{E}_{\boldsymbol{x}_0}\{L_t\} \approx \frac{R^3}{3D|\boldsymbol{x}_0|}\big(1 - \frac{R}{\sqrt{\pi Dt}} + O(1/t)\big)$, i.e., the mean residence time reaches a constant because the particle can escape to infinity and thus never return to the target set.

the reactive layer $\Gamma_a$ will disappear, as explained above. In order to preserve its effect, one has to enhance the reaction rate as the layer gets thinner. Setting $Q = qD/a$ with a constant $q \geq 0$ and taking the limit $a \to 0$, one can achieve a nontrivial limit

$$G_q(\boldsymbol{x}, t|\boldsymbol{x}_0) = \lim_{a\to 0} G^{(a)}_{qD/a}(\boldsymbol{x}, t|\boldsymbol{x}_0). \tag{4.33}$$

One can show (see [30]) that $G_q(\boldsymbol{x}, t|\boldsymbol{x}_0)$ satisfies the diffusion equation,

$$\partial_t G_q(\boldsymbol{x}, t|\boldsymbol{x}_0) = D\Delta G_q(\boldsymbol{x}, t|\boldsymbol{x}_0) \quad (\boldsymbol{x} \in \Omega), \tag{4.34}$$

subject to the initial condition $G_q(\boldsymbol{x}, 0|\boldsymbol{x}_0) = \delta(\boldsymbol{x} - \boldsymbol{x}_0)$ to fix the starting point $\boldsymbol{x}_0$, Robin boundary condition on the target set $\Gamma$, and Neumann boundary condition on the remaining part of the boundary:

$$-\partial_n G_q(\boldsymbol{x}, t|\boldsymbol{x}_0) = q\, G_q(\boldsymbol{x}, t|\boldsymbol{x}_0) \quad (\boldsymbol{x} \in \Gamma), \qquad \partial_n G_q(\boldsymbol{x}, t|\boldsymbol{x}_0) = 0 \quad (\boldsymbol{x} \in \partial\Omega\backslash\Gamma). \tag{4.35}$$

Comparing with Eq. (4.2), one sees that the parameter $q$ determines the reactivity $\kappa = qD$ of the target set. Most importantly, the conventional propagator $G_q(\boldsymbol{x}, t|\boldsymbol{x}_0)$ depends on $q$ *implicitly*, through Robin boundary condition.

In the same vein, one can rescale the residence time $L_t^{(a)}$ inside the layer $\Gamma_a$ by $a$ to ensure the existence of a nontrivial limit of $L_t^{(a)}/a$. In this way, the *boundary local time* $\ell_t$ is introduced:

$$\ell_t = \lim_{a\to 0} \frac{D}{a} L_t^{(a)} = \lim_{a\to 0} \frac{D}{a} \int\limits_0^t dt'\, \mathbb{I}_{\Gamma_a}(\boldsymbol{X}_{t'}) \tag{4.36}$$

(here the multiplication by $D$ is done for convenience). Despite its name, $\ell_t$ has units of length (we also stress that $\ell_t$ should not be confused with a bulk point local time; see [74, 75]). In practice, one can think of $\ell_t$ as the rescaled residence time in a thin layer near the target set, i.e., $L_t^{(a)} \approx a\ell_t/D$, for small enough $a$. Since the time needed to cross a thin layer of width $a$ is of the order of $a^2/D$, the ratio $L^{(a)}/(a^2/D) \approx \ell_t/a$ can be interpreted as the number of such crossings, which represents the number of encounters between the particle and the target set $\Gamma$ for reflected Brownian motion.

As the representation (4.20) is valid for the target set $\Gamma_a$, one can rewrite it by changing the integration variable $\ell = DL/a$ and taking the limit $a \to 0$ as

$$G_q(\boldsymbol{x}, t|\boldsymbol{x}_0) = \int\limits_0^\infty d\ell\, e^{-q\ell}\, P(\boldsymbol{x}, \ell, t|\boldsymbol{x}_0), \tag{4.37}$$

where the encounter propagator $P(\boldsymbol{x}, \ell, t|\boldsymbol{x}_0)$ is introduced as the rescaled limit of the former encounter propagator $P(\boldsymbol{x}, L, t|\boldsymbol{x}_0)$ discussed in Sect. 4.2.2:

$$P(\boldsymbol{x}, \ell, t|\boldsymbol{x}_0) = \lim_{a\to 0} \frac{a}{D} P_{\Gamma_a}(\boldsymbol{x}, \ell a/D, t|\boldsymbol{x}_0). \tag{4.38}$$

As previously mentioned, the encounter propagator is the joint probability density of finding the particle that started from $\boldsymbol{x}_0$ at time 0, in a vicinity of $\boldsymbol{x}$ at time $t$, with the boundary local time $\ell_t$:

$$\mathbb{P}_{\boldsymbol{x}_0}\{\boldsymbol{X}_t \in (\boldsymbol{x}, \boldsymbol{x} + d\boldsymbol{x}),\ \ell_t \in (\ell, \ell + d\ell)\} = P(\boldsymbol{x}, \ell, t|\boldsymbol{x}_0) d\boldsymbol{x} d\ell. \tag{4.39}$$

As earlier, this is the main building block of the encounter-based approach. In particular, the encounter propagator determines the conventional propagator $G_q(\boldsymbol{x}, t|\boldsymbol{x}_0)$ via the Laplace transform in Eq. (4.37) with respect to the boundary local time $\ell$, in direct analogy with Eq. (4.20). The major advantage of the encounter propagator is that it describes the diffusive motion *alone*, whereas the effect of surface reactions is taken into account via the *explicit* factor $e^{-q\ell}$ in Eq. (4.37). The disentanglement of diffusive dynamics from surface reactions opens efficient ways for solving optimization problems and introducing new surface reaction mechanisms. In fact, as in Sects. 4.2.1 and 4.2.2, one can assume that surface reaction occurs at time

$$\mathcal{T} = \inf\{t > 0\ :\ \ell_t > \hat{\ell}\} \tag{4.40}$$

when the number of encounters, here represented by $\ell_t$, exceeds a random threshold $\hat{\ell}$ with a given probability law $\Psi(\ell) = \mathbb{P}\{\hat{\ell} > \ell\}$. As a consequence, the generalized propagator for the survived particle reads

$$G_\Psi(\boldsymbol{x}, t|\boldsymbol{x}_0) = \mathbb{E}_{\boldsymbol{x}_0}\{\delta(\boldsymbol{X}_t - \boldsymbol{x}) \cdot \mathbb{I}_{t<\mathcal{T}}\} = \int\limits_0^\infty d\ell\, \Psi(\ell)\, P(\boldsymbol{x}, \ell, t|\boldsymbol{x}_0), \tag{4.41}$$

in analogy to Eqs. (4.9), (4.23). Comparison with Eq. (4.37) indicates that an exponentially distributed threshold with $\Psi(\ell) = e^{-q\ell}$ describes the conventional setting when a particle attempts to react at each encounter with the target set $\Gamma$ with a constant reactivity $\kappa = qD$. In turn, other choices of $\Psi(\ell)$ correspond to encounter-dependent reactivity and allow one to describe activation or passivation of the target set (see further discussions in [30]). Note that the generalized propagator $G_\Psi(\boldsymbol{x}, t|\boldsymbol{x}_0)$ does not satisfy Robin boundary condition.

We complete this section by clarifying the following point. To provide more intuitive views on the encounter-based approach, we started this section from the discrete case, for which the problem could be easily formulated but getting its explicit solution was very difficult. Its extension to the continuous case was straightforward for bulk reactions but less intuitive for surface reactions. In particular, we had to rescale the residence time to characterize the encounters with the boundary target set $\Gamma$ via the boundary local time $\ell_t$. Such an indirect introduction of the encounter-based approach for surface reactions was meant to ease its presentation. At the same time, it is important to stress that the encounter propagator $P(\boldsymbol{x}, \ell, t|\boldsymbol{x}_0)$ emerges naturally and straightforwardly in the mathematical theory of stochastic differential equations in confined domains. In fact, the mathematical definition of reflected Brownian motion relies on the Skorokhod equation [76, 77],

$$dX_t = \sqrt{2D}\, dW_t - n(X_t)\, d\ell_t, \tag{4.42}$$

where $W_t$ is the standard Wiener process, and $n(x)$ is the unit normal vector at a boundary point $x$ oriented outwards the domain $\Omega$. The first term determines random displacements inside $\Omega$, whereas the second term, which is nonzero only when $X_t$ is on the boundary $\partial\Omega$, describes normal reflections on $\partial\Omega$ and ensures that the process $X_t$ does not leave the domain. Curiously, this single equation determines simultaneously two tightly related stochastic processes, the position $X_t$ and the boundary local time $\ell_t$. In this setting, the encounter propagator $P(x, \ell, t|x_0)$ naturally describes the pair $\{X_t, \ell_t\}$. This construction and other advantages of the encounter-based approach were discussed in [30].

## 4.3 Spectral Insights

While the encounter propagator $P(x, \ell, t|x_0)$ plays the central role in the encounter-based approach, its computation is a difficult task. For instance, the standard inversion of the Laplace transform in Eq. (4.37) employs the Bromwich integral,

$$P(x, \ell, t|x_0) = \frac{1}{2\pi i} \int\limits_{\gamma - i\infty}^{\gamma + i\infty} dq\, e^{q\ell}\, G_q(x, t|x_0), \tag{4.43}$$

where $\gamma$ is a vertical contour in the complex plane $q \in \mathbb{C}$ chosen so that all singularities of $G_q(x, t|x_0)$ are to the left of it. In other words, it requires the knowledge of the conventional propagator $G_q(x, t|x_0)$ even for complex values of the parameter $q$ (which was supposed to be positive throughout this chapter). In practice, one needs other representations to access the encounter propagator more efficiently. The conventional approach employs spectral expansions, alike Eq. (4.3), based on the eigenfunctions and eigenvalues of the Laplace operator that governs the diffusive dynamics [5, 49]. In turn, the encounter-based approach was shown to rely on another mathematical operator, known as the Dirichlet-to-Neumann operator $\mathcal{M}_p$ [30].

To introduce this operator, let us consider an auxiliary steady-state diffusion-reaction problem when particles diffuse inside a confining domain $\Omega$ and can undergo first-order bulk reactions with a rate $p$. In the steady-state regime, their concentration $u(x)$ obeys the modified Helmholtz equation: $D\Delta u - pu = 0$. If a prescribed concentration of particles $f$ is maintained on the target set $\Gamma \subset \partial\Omega$, one can think of $\Gamma$ as a permanent source of particles, which diffuse in $\Omega$ and disappear due to bulk reactions. The diffusive flux of particles starting from $\Gamma$, $g = D(\partial_n u)|_\Gamma$, is fully determined by the solution $u$ of the modified Helmholtz equation, which, in turn, is determined by the imposed concentration $f$. In other words, to a given concentration $f$ on $\Gamma$, one can associate the diffusive flux density $g$ on $\Gamma$. This is precisely the action of the Dirichlet-to-Neumann operator $\mathcal{M}_p$. More formally, for a given function $f$ on $\Gamma$, one defines $\mathcal{M}_p f = (\partial_n u)|_{\partial\Omega}$, where $u$ is the solution of the boundary value problem

$$(p - D\Delta)u = 0 \quad (\boldsymbol{x} \in \Omega), \qquad u|_\Gamma = f, \qquad (\partial_n u)|_{\partial\Omega\setminus\Gamma} = 0. \tag{4.44}$$

When $p \geq 0$ and the target set $\Gamma$ is bounded and smooth enough, $\mathcal{M}_p$ is a self-adjoint elliptic pseudodifferential operator with a discrete spectrum (see technical details and mathematical restrictions in [78] and references therein). In other words, there is a countable sequence of eigenvalues $\mu_k^{(p)}$ and eigenfunctions $v_k^{(p)}$, enumerated by $k = 0, 1, 2, \ldots$, that satisfy $\mathcal{M}_p v_k^{(p)} = \mu_k^{(p)} v_k^{(p)}$. Each eigenfunction $v_k^{(p)}$ living on $\Gamma$ can be extended into the whole domain $\Omega$ as the unique solution $V_k^{(p)}$ of the boundary value problem:

$$(p - D\Delta)V_k^{(p)} = 0 \quad (\boldsymbol{x} \in \Omega), \qquad V_k^{(p)}|_\Gamma = v_k^{(p)}, \qquad (\partial_n V_k^{(p)})|_{\partial\Omega\setminus\Gamma} = 0. \tag{4.45}$$

In mathematics, $V_k^{(p)}$ are called the eigenfunctions of the Steklov spectral problem. In analogy to the eigenfunctions $u_k^{(q)}$ of the Laplace operator, one can use $V_k^{(p)}$ in spectral expansions. In particular, these eigenfunctions allowed us to determine the encounter propagator [30]:

$$P(\boldsymbol{x}, \ell, t|\boldsymbol{x}_0) = G_\infty(\boldsymbol{x}, t|\boldsymbol{x}_0)\delta(\ell) + \mathcal{L}_{p,t}^{-1}\left\{\frac{1}{D}\sum_{k=0}^{\infty} V_k^{(p)}(\boldsymbol{x}_0)V_k^{(p)}(\boldsymbol{x})e^{-\mu_k^{(p)}\ell}\right\}, \tag{4.46}$$

where $\mathcal{L}_{p,t}^{-1}$ denotes the inverse Laplace transform with respect to time $t$. The first term describes the contribution of trajectories of duration $t$ that moved from $\boldsymbol{x}_0$ to $\boldsymbol{x}$ without hitting the target set $\Gamma$. The probability of this event is determined by the conventional propagator $G_\infty(\boldsymbol{x}, t|\boldsymbol{x}_0)$ with Dirichlet boundary condition $G_\infty(\boldsymbol{x}, t|\boldsymbol{x}_0)|_{\boldsymbol{x}\in\Gamma} = 0$ that prohibits any arrival onto $\Gamma$. As these trajectories never encountered $\Gamma$, the boundary local time remains 0, as expressed by the factor $\delta(\ell)$. In turn, the second term accounts for all other trajectories that could encounter the target set $\Gamma$ and thus correspond to positive values $\ell$ of the boundary local time $\ell_t$.

The spectral expansion (4.46) is the cornerstone of the encounter-based approach. Many other characteristics of search processes and diffusion-controlled reactions can be expressed with the help of this relation, including the conventional propagator $G_q(\boldsymbol{x}, t|\boldsymbol{x}_0)$ and its extension $G_\Psi(\boldsymbol{x}, t|\boldsymbol{x}_0)$, the surface-hopping propagator [52], the distribution of the boundary local time [27, 51], the survival probability, the probability density of various first-passage times [53], the particle concentration, the spread harmonic measure [41, 79], the overall reaction rate, etc. The intricate connection to the Dirichlet-to-Neumann operator opens promising opportunities to translate the knowledge on this operator from spectral geometry [78] to diffusion-controlled reactions and related target search problems.

### Example of a Spherical Target Set

The spectral expansion (4.46) is also an efficient practical tool for computing the encounter propagator (see, e.g., [52, 53]). For instance, if the target set $\Gamma$ is the boundary of a ball of radius $R$, $\Gamma = \{\boldsymbol{x} \in \mathbb{R}^3 \ : \ |\boldsymbol{x}| = R\}$, whereas the particle diffuses in the exterior of this ball, $\Omega = \{\boldsymbol{x} \in \mathbb{R}^3 \ : \ |\boldsymbol{x}| > R\}$, the eigenvalues $\mu_k^{(p)}$ and eigenfunctions $V_k^{(p)}$ have a relatively simple form. To skip technical details, let us focus again on the statistics of encounters by integrating Eq. (4.46) over $\boldsymbol{x} \in \Omega$. In this case, the first term in Eq. (4.46) is just the survival probability of a particle outside a perfected reactive sphere, given by Eq. (4.1), while the sum in the second term contains only one term due to the rotational invariance of the problem. Surprisingly, the inverse Laplace transform of this single term can be computed explicitly, yielding[4] [54]

$$\rho(\ell, t|\boldsymbol{x}_0) = \left(1 - \frac{R}{|\boldsymbol{x}_0|}\mathrm{erfc}\left(\frac{|\boldsymbol{x}_0| - R}{\sqrt{4Dt}}\right)\right)\delta(\ell) + \frac{e^{-\ell/R}}{|\boldsymbol{x}_0|}\left(\mathrm{erfc}\left(\frac{|\boldsymbol{x}_0| - R + \ell}{\sqrt{4Dt}}\right) + \frac{Re^{-(|\boldsymbol{x}_0|-R+\ell)^2/(4Dt)}}{\sqrt{\pi Dt}}\right). \tag{4.47}$$

To appreciate the remarkable simplicity of this result, it is enough to compare it with similar quantities $\rho(L, n|\boldsymbol{x}_0)$ and $\rho(L, t|\boldsymbol{x}_0)$ given by Eqs. (4.16), (4.30). Unfortunately, this is an exception: even for the circular target set in the plane, the expression for $\rho(\ell, t|\boldsymbol{x}_0)$ is much more involved [54].

We complete this discussion by stressing that the three encounter-based approaches presented in Sects. 4.2.1, 4.2.2, and 4.2.3 are similar but not fully equivalent. Indeed, despite different units, the number of encounters $L_n$, the residence time $L_t$ and the boundary local time $\ell_t$ convey essentially the same information about the interaction of the diffusing particle with the target set. For instance, the means of these random variables, $\mathbb{E}_{\boldsymbol{x}_0}\{L_n\}$, $\mathbb{E}_{\boldsymbol{x}_0}\{L_t\}$ and $\mathbb{E}_{\boldsymbol{x}_0}\{\ell_t\}$, reach constant values in the long time limit for the search in three dimensions (or higher). This is the consequence of the transient character of the diffusive motion, which can escape to infinity and never return to the target set. Moreover, this approach is quite slow,[5] typically as $n^{-1/2}$ or $t^{-1/2}$. At the same time, the diffusive motion of the particle is hindered by

---

[4] Note that there is a misprint in Eq. (5) of [54]: erf$(z)$ should be replaced by erfc$(z)$, as in our Eq. (4.47).

[5] To get the mean boundary local time, it is easier to start from the known expression for the survival probability

the target set due to reflections in the case of reflected Brownian motion, in sharp contrast to bulk reactions. As a consequence, the statistics of random trajectories can be considerably different between the cases of bulk and surface reactions. Even if the target set is small, the distributions of $L_t$ and $\ell_t$ differ. In fact, as the residence time $L_t$ is a fraction of time that Brownian motion spends on the target set, it cannot exceed $t$, so that the distribution of $L_t$ is supported on a finite interval $(0, t)$. In turn, the boundary local time $\ell_t$ can take any positive value on $\mathbb{R}_+$, even though the probability of very high $\ell_t$ is extremely small; see Eq. (4.47).

## 4.4 Applications and Extensions

The encounter-based approach provided complementary insights onto diffusion-controlled reactions and stimulated new developments in first-passage phenomena in statistical physics. In this section, we list some of these achievements and highlight the related open problems.

### A Variety of First-Passage Times

In previous sections, we showed that the successful realization of a reaction event occurs when the number of encounters exceeds some threshold. In particular, the random time of the reaction event can be defined via Eq. (4.40) as the first-crossing time of a random threshold $\hat{\ell}$. The distribution of reaction times, which characterizes the efficiency of diffusion-controlled reactions, was thoroughly investigated for conventional surface reactions with a constant reactivity (see [5, 8, 9, 44–47] and references therein; see also Chap. 3). As the boundary local time $\ell_t$ is a nondecreasing process, the survival condition $\mathcal{T} > t$ is identical to $\ell_t < \hat{\ell}$, i.e., the particle survives up to time $t$ if its boundary local time $\ell_t$ has not yet crossed the threshold $\hat{\ell}$. As a consequence, the survival probability, i.e., the integral of the propagator, determines the distribution of the reaction time $\mathcal{T}$:

---

$$\tilde{S}_q(p|\boldsymbol{x}_0) = \frac{1}{p}\left(1 - \frac{qR^2 e^{-(|\boldsymbol{x}_0|-R)\sqrt{p/D}}}{|\boldsymbol{x}_0|(1 + qR + R\sqrt{p/D})}\right).$$

Since this is the double Laplace transform of Eq. (4.47), the mean boundary local time can be found by evaluating the derivative of this expression with respect to $q$ at $q = 0$ and then computing the inverse Laplace transform with respect to $p$:

$$\mathbb{E}_{\boldsymbol{x}_0}\{\ell_t\} = \frac{R^2 e^{-(|\boldsymbol{x}_0|-R)^2/(4Dt)}}{|\boldsymbol{x}_0|}\left[\mathrm{erfcx}\left(\frac{|\boldsymbol{x}_0| - R}{\sqrt{4Dt}}\right) - \mathrm{erfcx}\left(\frac{|\boldsymbol{x}_0| - R}{\sqrt{4Dt}} + \sqrt{Dt}/R\right)\right].$$

The mean boundary local time vanishes in the short-time limit and approaches the constant $R^2/|\boldsymbol{x}_0|$ in the long-time limit because the particle can escape to infinity and never return to the target set with the probability $R/|\boldsymbol{x}_0|$. More precisely, one has $\mathbb{E}_{\boldsymbol{x}_0}\{\ell_t\} = R^2/|\boldsymbol{x}_0| - R^2/\sqrt{\pi Dt} + O(1/t)$ as $t \to \infty$. If the particle starts on the target set, $|\boldsymbol{x}_0| = R$, one finds $\mathbb{E}_{\boldsymbol{x}_0}\{\ell_t\} = R(1 - \mathrm{erfcx}(\sqrt{Dt}/R))$. Similarly, one can find the variance and higher order moments of the boundary local time.

$$\mathbb{P}_{x_0}\{\mathcal{T} > t\} = \int\limits_{\Omega} dx\, G_\Psi(x, t|x_0) = \int\limits_0^\infty d\ell\, \Psi(\ell) \int\limits_{\Omega} dx\, P(x, \ell, t|x_0)$$
$$= \int\limits_0^\infty d\ell\, \Psi(\ell) \rho(\ell, t|x_0). \tag{4.48}$$

In other words, the knowledge of the probability density $\rho(\ell, t|x_0)$ of the boundary local time $\ell_t$ determines the distribution of reaction times. For instance, setting $\Psi(\ell) = e^{-q\ell}$, one can express the survival probability $S_q(t|x_0)$, which is a cornerstone of the conventional approach to first-passage times, as the moment-generating function of the boundary local time:

$$S_q(t|x_0) = \mathbb{P}_{x_0}\{\mathcal{T}_q > t\} = \int\limits_0^\infty d\ell\, e^{-q\ell} \rho(\ell, t|x_0) = \mathbb{E}_{x_0}\{e^{-q\ell_t}\}. \tag{4.49}$$

The latter can be obtained by integrating Eq. (4.46) over the confining domain $\Omega$:

$$\rho(\ell, t|x_0) = S_\infty(t|x_0)\delta(\ell) + \mathcal{L}^{-1}_{p,t}\left\{\frac{1}{D}\sum_{k=0}^{\infty} e^{-\mu_k^{(p)}\ell}\, V_k^{(p)}(x_0) \int\limits_{\Omega} dx\, V_k^{(p)}(x)\right\}. \tag{4.50}$$

While this analysis can be extended in several directions, we mention only two of them.

(i) One can consider $N$ independent particles searching simultaneously for the target set $\Gamma$, and investigate various extreme value statistics for this ensemble, such as the minimum of their reaction times $\mathcal{T}_q^1, \ldots, \mathcal{T}_q^N$. Since the seminal work by Weiss et al. [80], the distribution of the *fastest* first-passage time and its mean were actively studied for conventional surface reactions [81–89]. The encounter-based approach offers an alternative view onto this problem [58]. If $\ell_t^1, \ldots, \ell_t^N$ denote the boundary local times of each of $N$ particles, one can define the first crossing-time $\mathcal{T}_{\ell,N} = \inf\{t > 0 : \ell_t^1 + \ldots + \ell_t^N > \ell\}$ when the total number of encounters exceeds a given threshold $\ell$. If each encounter is associated with consumption of resources on the target set $\Gamma$, the random variable $\mathcal{T}_{\ell,N}$ determines the moment when the initial amount $\ell$ of resources is depleted by a population of diffusing species. The distribution of $\mathcal{T}_{\ell,N}$ can be determined in terms of the survival probability $S_q(t|x_0)$ for a single particle: $\mathbb{P}_{x_0}\{\mathcal{T}_{\ell,N} > t\} = \mathcal{L}^{-1}_{q,\ell}\{[S_q(t|x_0)]^N/q\}$, from which many asymptotic results can be deduced [58]. Choosing the threshold $\ell = 0$ implies that none of the particles has reached the target set, so that $\mathcal{T}_{0,N} = \min\{\mathcal{T}_\infty^1, \ldots, \mathcal{T}_\infty^N\}$, i.e., one retrieves the conventional fastest first-passage time to the perfectly reactive target studied earlier. Replacing $\ell$ by a random exponentially distributed threshold $\hat{\ell}$, one gets $\mathcal{T}_{\hat{\ell},N} = \min\{\mathcal{T}_q^1, \ldots, \mathcal{T}_q^N\}$ for a partially reactive target (see [58] for more details). In turn, the choice of a random threshold $\hat{\ell}$ with an arbitrary distribution $\Psi(\ell)$ may yield more sophisticated extreme value statistics.

(ii) Alternatively, one can consider many target sets, $\Gamma_1, \ldots, \Gamma_n$, and investigate the statistics of their encounters by a single particle. For each target set $\Gamma_i$, one can introduce its own boundary local time $\ell_t^i$ and then study the encounter propagator $P(\boldsymbol{x}, \ell^1, \ldots, \ell^n, t|\boldsymbol{x}_0)$ as the joint probability density of the position $\boldsymbol{x}$ and boundary local times $\ell_t^1, \ldots, \ell_t^n$. In this setting, different kinds of first-passage times emerge, e.g., the first instance when the maximum or the minimum of $\ell_t^1, \ldots, \ell_t^n$ exceeds some threshold, or the first instance when one boundary local time exceeds the other, and so on. These questions naturally generalize the notion of splitting probabilities that characterize diffusional screening or diffusion interactions between the targets that compete for the diffusing particle. Some of these first-passage problems were investigated in simple geometric settings with two target sets [53]. Moreover, if one of the target sets is perfectly reactive, one deals with the escape problem [62]. For instance, one can investigate the statistics of encounters of a protein with its receptor (a target set $\Gamma_1$) in the presence of a protein channel (another target set $\Gamma_2$) on the plasma membrane through which the protein can leave the living cell [90]. Despite recent progress in this field, many problems with multiple target sets are still open. For instance, a spectral expansion for the encounter propagator $P(\boldsymbol{x}, \ell^1, \ldots, \ell^n, t|\boldsymbol{x}_0)$ is missing.

### Small Target Limit

According to the spectral expansion (4.50), the dependence of $\rho(\ell, t|\boldsymbol{x}_0)$ on the shape of the confining domain $\Omega$ and the target set $\Gamma$ is captured implicitly via the spectral properties of the Dirichlet-to-Neumann operator $\mathcal{M}_p$. In general, it is very hard to get the statistics of encounters explicitly. However, if the target set is small and located far away from the remaining confining boundary $\partial\Omega\backslash\Gamma$, some approximations can be derived. This setting, known as the narrow escape or narrow capture problem, was thoroughly investigated for conventional surface reactions (see [91–93] and references therein). Bressloff extended previous works by employing the matched asymptotic analysis to obtain an approximate form of the encounter propagator [59]. These results were compared to two other approximations in [60]. In particular, a fully explicit approximation was derived in three dimensions for $\rho(\ell, t|\boldsymbol{x}_0)$ averaged over the uniformly distributed starting point $\boldsymbol{x}_0$:

$$\overline{\rho(\ell, t)} = \frac{1}{|\Omega|} \int\limits_{\Omega} d\boldsymbol{x}_0 \, \rho(\ell, t|\boldsymbol{x}_0)$$

$$\approx e^{-t/t_0}\delta(\ell) + \sqrt{\frac{t/t_0}{\ell\ell_0}} e^{-\ell/\ell_0 - t/t_0} I_1\Big(2\sqrt{(t/t_0)(\ell/\ell_0)}\Big), \qquad (4.51)$$

where $|\Omega|$ is the volume of the confining domain $\Omega$, $I_\nu(z)$ is the modified Bessel function of the first kind, $\ell_0 = |\Gamma|/C$, $t_0 = |\Omega|/(DC)$, $|\Gamma|$ is the surface area of the target set, and $C$ is the harmonic capacity (or capacitance) of the target set (e.g., $C = 4\pi R$ is the capacity of a ball of radius $R$) [60]. Moreover, when $t\ell \gg t_0\ell_0$, the asymptotic behavior of the modified Bessel function yields a remarkably simple expression:

$$\overline{\rho(\ell, t)} \approx e^{-t/t_0}\delta(\ell) + \frac{(t/t_0)^{\frac{1}{4}}}{2\sqrt{\pi}(\ell/\ell_0)^{\frac{3}{4}}\ell_0} e^{-\left(\sqrt{\ell/\ell_0}-\sqrt{t/t_0}\right)^2} . \tag{4.52}$$

As time increases, the particle experiences more encounters with the target set, so that the maximum of this density is shifted to larger $\ell$. Despite the simple form of this expression, the dependence on $\ell$ is quite nontrivial. One also concludes that when the target set is small, the statistics of encounters is essentially determined by two scales, $t_0$ and $\ell_0$, that are formed by three geometric characteristics: the surface area and the capacity of the target set, and the volume of the domain.

#### Reversible Binding

The spectral expansion (4.46) is particularly well suited for describing diffusive explorations of the confining domain $\Omega$ between reflections on the target set $\Gamma$. In particular, the notion of the surface-hopping propagator, that characterizes the position of the particle after a prescribed number of encounters, was introduced in [52]. Moreover, an accurate description of repeated returns to the target set allows one to consider reversible binding of the particle on $\Gamma$. A general theory of diffusion-controlled reactions with non-Markovian binding/unbinding kinetics was developed in [32]. The binding events were described via a given distribution $\Psi(\ell)$ of the threshold $\hat{\ell}$ for the boundary local time $\ell_t$. When $\Psi(\ell)$ is heavy tailed, i.e., $\Psi(\ell) \propto (\ell_0/\ell)^\alpha$ at large $\ell$ with an exponent $0 < \alpha < 1$, one deals with anomalous, non-Markovian binding kinetics. Its competition with non-Markovian unbinding kinetics and their consequences on diffusion-controlled reactions were described in [32].

#### Diffusion Across Permeable Barriers

Bressloff employed the encounter-based approach to describe diffusion across permeable barriers [63–65]. In this case, the surface reaction is understood as permeation through the target set $\Gamma$, and one can employ again the stopping condition $\ell_t < \hat{\ell}$ as a trigger. The subtle point here is that, after permeation, the particle starts diffusing in the adjacent domain (on the other side of $\Gamma$), whose encounter-based description involves another boundary local time. Indeed, after a number of encounters with $\Gamma$ on the other side, the particle can permeate back into the initial domain, and so on (see more details in Chap. 8). Note that Bressloff also studied different functionals of reflected Brownian motion in the presence of generalized surface reactions [61].

**Extensions**

The encounter-based approach can be naturally extended in many ways. As stated from the beginning, we focused on the simplest diffusive dynamics in order to highlight the role of surface reactions. The main concepts of the encounter-based approach are expected to be valid and valuable for much more general stochastic processes. However, a practical realization of such extensions requires considerable mathematical efforts. For instance, an extension to diffusion with a gradient drift was discussed in [55]. Further extensions to anomalous diffusions such as continuous-time random walks [94–97], Lévy flights/walks [98], diffusing-diffusivity processes [99–101], and non-Markovian dynamics [102, 103] present interesting perspectives. Similarly, the encounter-based approach suggests an alternative way to deal with intermittent diffusion with alternating phases of bulk and surface diffusion [104–113] (see a review [114] on intermittent search strategies). Another extension consists in stochastic resetting of the position or the boundary local time [56, 57].

## 4.5 Conclusion

In this chapter, we presented the basic ideas and applications of the encounter-based approach. Unlike earlier publications on this topic, we started from the discrete setting of a random walk on a lattice, for which the formulation of the encounter-based approach is simple and intuitively appealing. In fact, the conventional propagator $g_\sigma(\boldsymbol{x}, n|\boldsymbol{x}_0)$ that characterizes the random position of the survived walker is substituted by the joint distribution $p(\boldsymbol{x}, L, n|\boldsymbol{x}_0)$ of the position and the number of encounters $L_n$ of the particle with a given target set $\Gamma$. This discrete setting was then extended to Brownian motion that undergoes a first-order bulk reaction inside the target set. The number of encounters is naturally replaced by the residence time $L_t$ in a bulk region $\Gamma$. In the next step, the target set is put on the boundary of an impenetrable obstacle, in which case the residence time $L_t$ has to be rescaled and replaced by the boundary local time $\ell_t$. In all three settings, the reaction event occurs when the (rescaled) number of encounters ($L_n$, $L_t$ or $\ell_t$) exceeds a prescribed random threshold. The use of such a stopping condition generalizes conventional reaction events with a constant reactivity, allowing one to describe much more sophisticated processes such as activation or passivation of the target set by interactions with the diffusing particle. The disentanglement of the diffusive dynamics from reaction events is one of the crucial advantages of the encounter-based approach.

Other advantages were rapidly illustrated by considering several applications, including reversible binding/unbinding kinetics, the escape problem, resource depletion by a population of diffusing species, the statistics of various first-passage times for multiple targets, etc. We also showed how the Dirichlet-to-Neumann operator can substitute the Laplace operator to compute the encounter propagator and related quantities. This spectral tool is particularly suitable for describing diffusive explorations in the bulk between reflections on the target set.

Despite impressive progress in this direction, many problems remain open. In particular, the numerical computation of the encounter propagator in complex media is challenging that prohibits further understanding of the impact of the geometrical complexity onto diffusion-controlled reactions. In the same vein, the disentanglement between the diffusive dynamics and reactions event may help in solving optimization problems such as finding an optimal structure of the target set to minimize or maximize the moments of the reaction time, or to reshape its distribution. Such problems may appear in the design of chemical reactors or in a programmable drug release in pharmaceutical industry. This research direction remains unexplored yet. Further developments of the encounter-based approach should bring new insights onto the intricate diffusive dynamics in complex environments and provide complementary views onto diffusion-controlled reactions and other target problems.

## References

1. M. Smoluchowski, Versuch einer mathematischen Theorie der Koagulationskinetik kolloider Lösungen. Z. Phys. Chem. **92U**, 129–168 (1918)
2. S.A. Rice, *Diffusion-Limited Reactions* (Elsevier, Amsterdam, 1985)
3. D.S. Grebenkov, Diffusion-Controlled Reactions: An Overview. Molecules **28**, 7570 (2023)
4. N.G. Van Kampen, *Stochastic Processes in Physics and Chemistry* (Elsevier, Amsterdam, 1992)
5. S. Redner, *A Guide to First Passage Processes* (Cambridge University Press, Cambridge, 2001)
6. J.E. House, *Principles of Chemical Kinetics* (Academic, 2007)
7. Z. Schuss, *Brownian Dynamics at Boundaries and Interfaces in Physics, Chemistry and Biology* (Springer, New York, 2013)
8. R. Metzler, G. Oshanin, S. Redner (eds.), *First-Passage Phenomena and Their Applications* (World Scientific, Singapore, 2014)
9. K. Lindenberg, R. Metzler, G. Oshanin (eds.), *Chemical Kinetics: Beyond the Textbook* (World Scientific, New Jersey, 2019)
10. F.C. Collins, G.E. Kimball, Diffusion-controlled reaction rates. J. Colloid Sci. **4**, 425–437 (1949)
11. O. Bénichou, M. Moreau, G. Oshanin, Kinetics of stochastically gated diffusion-limited reactions and geometry of random walk trajectories. Phys. Rev. E **61**, 3388–3406 (2000)
12. J. Reingruber, D. Holcman, Gated narrow escape time for molecular signaling. Phys. Rev. Lett. **103**, 148102 (2009)
13. S.D. Lawley, J.P. Keener, A new derivation of Robin boundary conditions through homogenization of a stochastically switching boundary. SIAM J. Appl. Dyn. Syst. **14**, 1845–1867 (2015)
14. H.C. Berg, E.M. Purcell, Physics of chemoreception. Biophys. J. **20**, 193–219 (1977)
15. A.M. Berezhkovskii, Y.A. Makhnovskii, M.I. Monine, V.Y. Zitserman, S.Y. Shvartsman, Boundary homogenization for trapping by patchy surfaces. J. Chem. Phys. **121**, 11390 (2004)
16. A.M. Berezhkovskii, M.I. Monine, C.B. Muratov, S.Y. Shvartsman, Homogenization of boundary conditions for surfaces with regular arrays of traps. J. Chem. Phys. **124**, 036103 (2006)

17. C. Muratov, S. Shvartsman, Boundary homogenization for periodic arrays of absorbers. Multiscale Model. Simul. **7**, 44–61 (2008)
18. A. Bernoff, A. Lindsay, D. Schmidt, Boundary homogenization and capture time distributions of semipermeable membranes with periodic patterns of reactive sites. Multiscale Model. Simul. **16**, 1411–1447 (2018)
19. D.S. Grebenkov, Spectral theory of imperfect diffusion-controlled reactions on heterogeneous catalytic surfaces. J. Chem. Phys. **151**, 104108 (2019)
20. B. Punia, S. Chaudhury, A.B. Kolomeisky, Understanding the reaction dynamics on heterogeneous catalysts using a simple stochastic approach. J. Phys. Chem. Lett. **12**, 11802–11810 (2021)
21. G.H. Weiss, Overview of theoretical models for reaction rates. J. Stat. Phys. **42**, 3–36 (1986)
22. P. Hänggi, P. Talkner, M. Borkovec, Reaction-rate theory: fifty years after Kramers. Rev. Mod. Phys. **62**, 251–341 (1990)
23. H.-X. Zhou, R. Zwanzig, A rate process with an entropy barrier. J. Chem. Phys. **94**, 6147–6152 (1991)
24. D. Reguera, G. Schmid, P.S. Burada, J.-M. Rubí, P. Reimann, P. Hänggi, Entropic transport: kinetics, scaling, and control mechanisms. Phys. Rev. Lett. **96**, 130603 (2006)
25. S.J. Chapman, R. Erban, S. Isaacson, Reactive boundary conditions as limits of interaction potentials for Brownian and Langevin dynamics. SIAM J. Appl. Math **76**, 368–390 (2016)
26. D.S. Grebenkov, Partially reflected brownian motion: a stochastic approach to transport phenomena, in *Focus on Probability Theory*, ed. by L.R. Velle (Nova Science Publishers, New York, 2006), pp. 135–169
27. D.S. Grebenkov, Residence times and other functionals of reflected Brownian motion. Phys. Rev. E **76**, 041139 (2007)
28. R. Erban, S.J. Chapman, Reactive boundary conditions for stochastic simulations of reaction-diffusion processes. Phys. Biol. **4**, 16–28 (2007)
29. A. Singer, Z. Schuss, A. Osipov, D. Holcman, Partially reflected diffusion. SIAM J. Appl. Math. **68**, 844–868 (2008)
30. D.S. Grebenkov, Paradigm shift in diffusion-mediated surface phenomena. Phys. Rev. Lett. **125**, 078102 (2020)
31. F. Piazza, The physics of boundary conditions in reaction-diffusion problems. J. Chem. Phys. **157**, 234110 (2022)
32. D.S. Grebenkov, Diffusion-controlled reactions with non-Markovian binding/unbinding kinetics. J. Chem. Phys. **158**, 214111 (2023)
33. H. Sano, M. Tachiya, Partially diffusion-controlled recombination. J. Chem. Phys. **71**, 1276–1282 (1979)
34. K.R. Brownstein, C.E. Tarr, Importance of classical diffusion in NMR studies of water in biological cells. Phys. Rev. A **19**, 2446–2453 (1979)
35. J.G. Powles, M.J.D. Mallett, G. Rickayzen, W.A.B. Evans, Exact analytic solutions for diffusion impeded by an infinite array of partially permeable barriers. Proc. R. Soc. London A **436**, 391–403 (1992)
36. B. Sapoval, General formulation of Laplacian transfer across irregular surfaces. Phys. Rev. Lett. **73**, 3314–3316 (1994)
37. B. Sapoval, M. Filoche, E. Weibel, Smaller is better - but not too small: a physical scale for the design of the mammalian pulmonary acinus. Proc. Nat. Ac. Sci. USA **99**, 10411–10416 (2002)
38. D.S. Grebenkov, M. Filoche, B. Sapoval, M. Felici, Diffusion-reaction in branched structures: theory and application to the lung acinus. Phys. Rev. Lett. **94**, 050602 (2005)
39. S.D. Traytak, W. Price, Exact solution for anisotropic diffusion-controlled reactions with partially reflecting conditions. J. Chem. Phys. **127**, 184508 (2007)
40. P.C. Bressloff, B.A. Earnshaw, M.J. Ward, Diffusion of protein receptors on a cylindrical dendritic membrane with partially absorbing traps. SIAM J. Appl. Math. **68**, 1223–1246 (2008)

41. D.S. Grebenkov, Analytical representations of the spread harmonic measure density. Phys. Rev. E **91**, 052108 (2015)
42. A.S. Serov, C. Salafia, D.S. Grebenkov, M. Filoche, The role of morphology in mathematical models of placental gas exchange. J. Appl. Physiol. **120**, 17–28 (2016)
43. D.S. Grebenkov, G. Oshanin, Diffusive escape through a narrow opening: new insights into a classic problem. Phys. Chem. Chem. Phys. **19**, 2723–2739 (2017)
44. D.S. Grebenkov, R. Metzler, G. Oshanin, Strong defocusing of molecular reaction times results from an interplay of geometry and reaction control. Commun. Chem. **1**, 96 (2018)
45. D.S. Grebenkov, R. Metzler, G. Oshanin, Towards a full quantitative description of single-molecule reaction kinetics in biological cells. Phys. Chem. Chem. Phys. **20**, 16393–16401 (2018)
46. D.S. Grebenkov, R. Metzler, G. Oshanin, Full distribution of first exit times in the narrow escape problem. New J. Phys. **21**, 122001 (2019)
47. T. Guérin, M. Dolgushev, O. Bénichou, R. Voituriez, Universal kinetics of imperfect reactions in confinement. Commun. Chem. **4**, 157 (2021)
48. F. Piazza, D.S. Grebenkov, Diffusion-controlled reaction rate on non-spherical partially absorbing axisymmetric surfaces. Phys. Chem. Chem. Phys. **21**, 25896–25906 (2019)
49. C.W. Gardiner, *Handbook of Stochastic Methods for Physics, Chemistry and the Natural Sciences* (Springer, Berlin, 1985)
50. D.S. Grebenkov, B.-T. Nguyen, Geometrical structure of Laplacian eigenfunctions. SIAM Rev. **55**, 601–667 (2013)
51. D.S. Grebenkov, Probability distribution of the boundary local time of reflected Brownian motion in Euclidean domains. Phys. Rev. E **100**, 062110 (2019)
52. D.S. Grebenkov, Surface hopping propagator: an alternative approach to diffusion-influenced reactions. Phys. Rev. E **102**, 032125 (2020)
53. D.S. Grebenkov, Joint distribution of multiple boundary local times and related first-passage time problems with multiple targets. J. Stat. Mech. 103205 (2020)
54. D.S. Grebenkov, Statistics of boundary encounters by a particle diffusing outside a compact planar domain. J. Phys. A.: Math. Theor. **54**, 015003 (2021)
55. D.S. Grebenkov, An encounter-based approach for restricted diffusion with a gradient drift. J. Phys. A: Math. Theor. **55**, 045203 (2022)
56. P.C. Bressloff, Diffusion-mediated surface reactions and stochastic resetting. J. Phys. A: Math. Theor. **55**, 275002 (2022)
57. Z. Benkhadaj, D.S. Grebenkov, Encounter-based approach to diffusion with resetting. Phys. Rev. E **106**, 044121 (2022)
58. D.S. Grebenkov, Depletion of resources by a population of diffusing species. Phys. Rev. E **105**, 054402 (2022)
59. P.C. Bressloff, Narrow capture problem: an encounter-based approach to partially reactive targets. Phys. Rev. E **105**, 034141 (2022)
60. D.S. Grebenkov, Statistics of diffusive encounters with a small target: three complementary approaches. J. Stat. Mech. 083205 (2022)
61. P.C. Bressloff, Diffusion-mediated absorption by partially-reactive targets: Brownian functionals and generalized propagators. J. Phys. A: Math. Theor. **55**, 205001 (2022)
62. D.S. Grebenkov, Encounter-based approach to the escape problem. Phys. Rev. E **107**, 044105 (2023)
63. P.C. Bressloff, A probabilistic model of diffusion through a semipermeable barrier. Proc. Roy. Soc. A **478**, 20220615 (2022)
64. P.C. Bressloff, Renewal equation for single-particle diffusion through a semipermeable interface. Phys. Rev. E. **107**, 014110 (2023)
65. P.C. Bressloff, Renewal equations for single-particle diffusion in multilayered media. SIAM J. Appl. Math. **83**, 1518–1545 (2023)
66. R.J. Rubin, G.H. Weiss, Random walks on lattices. The problem of visits to a set of points revisited. J. Math. Phys. **23**, 250–253 (1982)

67. W. Feller, *An Introduction to Probability Theory and Its Applications* (Wiley, New York, 1968)
68. F. Spitzer, *Principles of Random Walk* (Springer, New York, 1976)
69. B.D. Hughes, *Random Walks and Random Environments* (Clarendon Press, Oxford, 1995)
70. E.W. Montroll, G.H. Weiss, Random Walks on Lattices. II. J. Math. Phys. **6**, 167–181 (1965)
71. A.J. Guttmann, Lattice Green's functions in all dimensions. J. Phys. A: Math. Theor. **43**, 305205 (2010)
72. G.S. Joyce, Exact evaluation of the simple cubic lattice Green function for a general lattice point. J. Phys. A: Math. Gen. **31**, 5105–5115 (1998)
73. A. Szabo, G. Lamm, G.H. Weiss, Localized partial traps in diffusion processes and random walks. J. Stat. Phys. **34**, 225–238 (1984)
74. A.N. Borodin, P. Salminen, *Handbook of Brownian Motion: Facts and Formulae* (Birkhäuser Verlag, Basel-Boston-Berlin, 1996)
75. S.N. Majumdar, Brownian functionals in physics and computer science. Curr. Sci. **88**, 2076–2092 (2005)
76. K. Itô, H.P. McKean, *Diffusion Processes and Their Sample Paths* (Springer, Berlin, 1965)
77. M. Freidlin, *Functional Integration and Partial Differential Equations* (Annals of Mathematics Studies). (Princeton University Press, Princeton, NJ, 1985)
78. M. Levitin, D. Mangoubi, I. Polterovich, *Topics in spectral geometry*. Graduate studies in mathematics **237** (American Mathematical Society, Providence, RI, 2023)
79. D.S. Grebenkov, Scaling properties of the spread harmonic measures. Fractals **14**, 231–243 (2006)
80. G.H. Weiss, K.E. Shuler, K. Lindenberg, Order statistics for first passage times in diffusion processes. J. Stat. Phys. **31**, 255–278 (1983)
81. A.J. Bray, S.N. Majumdar, G. Schehr, Persistence and first-passage properties in nonequilibrium systems. Adv. Phys. **62**, 225–361 (2013)
82. Z. Schuss, K. Basnayake, D. Holcman, Redundancy principle and the role of extreme statistics in molecular and cellular biology. Phys. Life Rev. **28**, 52–79 (2019)
83. K. Basnayake, Z. Schuss, D. Holcman, Asymptotic formulas for extreme statistics of escape times in 1, 2 and 3-dimensions. J. Nonlinear Sci. **29**, 461–499 (2019)
84. S.D. Lawley, Distribution of extreme first passage times of diffusion. J. Math. Biol. **80**, 2301–2325 (2020)
85. S.D. Lawley, J.B. Madrid, A probabilistic approach to extreme statistics of Brownian escape times in dimensions 1, 2, and 3. J. Nonlinear Sci. **30**, 1207–1227 (2020)
86. S.D. Lawley, Universal formula for extreme first passage statistics of diffusion. Phys. Rev. E **101**, 012413 (2020)
87. J.B. Madrid, S.D. Lawley, Competition between slow and fast regimes for extreme first passage times of diffusion. J. Phys. A: Math. Theor. **53**, 335002 (2020)
88. D.S. Grebenkov, R. Metzler, G. Oshanin, From single-particle stochastic kinetics to macroscopic reaction rates: fastest first-passage time of $N$ random walkers. New J. Phys. **22**, 103004 (2020)
89. D.S. Grebenkov, R. Metzler, G. Oshanin, Search efficiency in the Adam-Delbrück reduction-of-dimensionality scenario versus direct diffusive search. New J. Phys. **24**, 083035 (2022)
90. D.A. Lauffenburger, J. Linderman, *Receptors: Models for Binding, Trafficking, and Signaling* (Oxford University Press, Oxford, 1993)
91. Z. Schuss, A. Singer, D. Holcman, The narrow escape problem for diffusion in cellular microdomains. Proc. Nat. Acad. Sci. USA **104**, 16098–16103 (2007)
92. O. Bénichou, R. Voituriez, Narrow-escape time problem: time needed for a particle to exit a confining domain through a small window. Phys. Rev. Lett. **100**, 168105 (2008)
93. D. Holcman, Z. Schuss, The narrow escape problem. SIAM Rev. **56**, 213–257 (2014)
94. R. Metzler, J. Klafter, The random walk's guide to anomalous diffusion: a fractional dynamics approach. Phys. Rep. **339**, 1–77 (2000)
95. J. Klafter, I.M. Sokolov, *First Steps in Random Walks: From Tools to Applications* (Oxford University Press, 2011)

96. D.S. Grebenkov, Searching for partially reactive sites: analytical results for spherical targets. J. Chem. Phys. **132**, 034104 (2010)
97. D.S. Grebenkov, Subdiffusion in a bounded domain with a partially absorbing-reflecting boundary. Phys. Rev. E **81**, 021128 (2010)
98. V. Zaburdaev, S. Denisov, J. Klafter, Lévy walks. Rev. Mod. Phys. **87**, 483–530 (2015)
99. M.V. Chubynsky, G.W. Slater, Diffusing diffusivity: a model for anomalous, yet Brownian, diffusion. Phys. Rev. Lett. **113**, 098302 (2014)
100. A.V. Chechkin, F. Seno, R. Metzler, I.M. Sokolov, Brownian yet non-Gaussian diffusion: from superstatistics to subordination of diffusing diffusivities. Phys. Rev. X **7**, 021002 (2017)
101. Y. Lanoiselée, N. Moutal, D.S. Grebenkov, Diffusion-limited reactions in dynamic heterogeneous media. Nature Commun. **9**, 4398 (2018)
102. O. Bénichou, R. Voituriez, From first-passage times of random walks in confinement to geometry-controlled kinetics. Phys. Rep. **539**, 225–284 (2014)
103. T. Guérin, N. Levernier, O. Bénichou, R. Voituriez, Mean first-passage times of non-Markovian random walkers in confinement. Nature **534**, 356–359 (2016)
104. P. Levitz, M. Zinsmeister, P. Davidson, D. Constantin, O. Poncelet, Intermittent Brownian dynamics over a rigid strand: heavily tailed relocation statistics in a simple geometry. Phys. Rev. E **78**, 030102(R) (2008)
105. A.V. Chechkin, I.M. Zaid, M. Lomholt, I.M. Sokolov, R. Metzler, Bulk-mediated surface diffusion along a cylinder: propagators and crossovers. Phys. Rev. E **79**, 040105(R) (2009)
106. O. Bénichou, D.S. Grebenkov, P. Levitz, C. Loverdo, R. Voituriez, Optimal reaction time for surface-mediated diffusion. Phys. Rev. Lett. **105**, 150606 (2010)
107. O. Bénichou, D.S. Grebenkov, P. Levitz, C. Loverdo, R. Voituriez, Mean first-passage time of surface-mediated diffusion in spherical domains. J. Stat. Phys. **142**, 657–685 (2011)
108. F. Rojo, C.E. Budde, Enhanced diffusion through surface excursion: a master-equation approach to the narrow-escape-time problem. Phys. Rev. E **84**, 021117 (2011)
109. A.V. Chechkin, I.M. Zaid, M. Lomholt, I.M. Sokolov, R. Metzler, Effective surface motion on a reactive cylinder of particles that perform intermittent bulk diffusion. J. Chem. Phys. **134**, 204116 (2011)
110. A.V. Chechkin, I.M. Zaid, M. Lomholt, I.M. Sokolov, R. Metzler, Bulk-mediated diffusion on a planar surface: Full solution. Phys. Rev. E **86**, 041101 (2012)
111. J.-F. Rupprecht, O. Bénichou, D.S. Grebenkov, R. Voituriez, Exact mean exit time for surface-mediated diffusion. Phys. Rev. E **86**, 041135 (2012)
112. J.-F. Rupprecht, O. Bénichou, D.S. Grebenkov, R. Voituriez, Kinetics of active surface-mediated diffusion in spherically symmetric domains. J. Stat. Phys. **147**, 891–918 (2012)
113. A.M. Berezhkovskii, L. Dagdug, S.M. Bezrukov, A new approach to the problem of bulk-mediated surface diffusion. J. Chem. Phys. **143**, 084103 (2015)
114. O. Bénichou, C. Loverdo, M. Moreau, R. Voituriez, Intermittent search strategies. Rev. Mod. Phys. **83**, 81–130 (2011)

# Chapter 5
# Multi-target Search in Bounded and Heterogeneous Environments: A Lattice Random Walk Perspective

**Luca Giuggioli, Seeralan Sarvaharman, Debraj Das, Daniel Marris, and Toby Kay**

**Abstract** For more than a century lattice random walks have been employed ubiquitously, both as a theoretical laboratory to develop intuition about more complex stochastic processes and as a tool to interpret a vast array of empirical observations. Recent advances in lattice random walk theory in bounded and heterogeneous environments have opened up opportunities to cope with the finely resolved spatio-temporal nature of modern movement data. We review such advances and their formalisms to represent analytically the walker spatio-temporal dynamics in arbitrary dimensions and geometries. As new findings, we derive the exact spatio-temporal representation of biased walks in a periodic hexagon, we use the discrete Feynman–Kac equation to describe a walker's interaction with a radiation boundary and we unearth a disorder indifference phenomenon. To demonstrate the power of the formalism for first-passage processes we uncover the appearance of multiple first-passage peaks with biased walkers in a periodic hexagon, we display the dependence of the first-transmission probability on the proximity transfer efficiency between two resetting walkers in a one-dimensional periodic lattice, we present an example of spatial disorder in a two-dimensional square lattice that strongly affects the splitting probabilities to either of two targets and we study the first-reaction dynamics to a single lattice site in an unbounded one-dimensional lattice.

L. Giuggioli (✉) · S. Sarvaharman · D. Marris · T. Kay
School of Engineering Mathematics and Technology, University of Bristol, Bristol, UK
e-mail: luca.giuggioli@bristol.ac.uk

S. Sarvaharman
e-mail: seeralan.sarvaharman@mpinat.mpg.de

D. Marris
e-mail: daniel.marris@bristol.ac.uk

T. Kay
e-mail: t.kay@imperial.ac.uk

D. Das
Quantitative Life Sciences, ICTP – The Abdus Salam International Centre for Theoretical Physics, 34151 Trieste, Italy
e-mail: ddas1@ictp.it

 
D. Grebenkov et al. (eds.), *Target Search Problems*,
https://doi.org/10.1007/978-3-031-67802-8_5

## 5.1 Introduction

The dynamics of reaction-diffusion processes in heterogeneous space is a topic with vast applicability from transport processes in physical and chemical systems [1] to animal ecology and disease spread [2]. While the literature encompasses both continuous and discrete variable descriptions, we focus here on the latter providing an overview of the discrete space-time formalism. Such formalism is the lattice random walk (LRW) whose history (see, e.g. [3] for an historical account of probability theory) dates back to the early works by de Moivre [4] and Laplace [5] and to seminal contributions, during the last century, by Smoluchowski [6], Polya [7], Chandrasekhar [8], Kac [9], Feller [10], Erdös [11], Sparre Andersen [12] and Spitzer [13], with more recent monographs by Weiss [14] and Hughes [15].

The studies of LRW dynamics in spatially heterogeneous space can be traced back to the work by Montroll in the 60's to study trapping reactions in photosynthetic units [16–18]. Since then a large literature on the transport properties of random walks both in continuous and discrete time emerged [19–28], but it mainly focused on translationally invariant one-dimensional systems. Interest for theoretical studies in non-periodic bounded domains have surfaced more recently (see, e.g. [29–31]). Part of the more recent surge of interest stems from the technological progress in various types of sensors and instruments that allow to record with unprecedented resolution and quality the spatio-temporal dynamics of physical particles and biological organisms as well as of the environment where motion occurs.

These advances demand the development of models that account in an explicit manner, rather than simply averaging over the disorder, for how random movement is affected by the presence of spatial heterogeneities. More broadly, there is a need for models capable of describing how the underlying space affect motion. Very recently a series of exact analytic studies on the dynamics of (Markov) LRWs in arbitrary dimensions [32–37] have opened up this possibility. Here we review this recent literature, what knowledge gaps it has filled and what it has accomplished. We also present some new findings and we point to open problems where LRW can provide some useful contribution.

### *5.1.1 The Missing Links*

While mathematical convenience has often favoured studies of diffusive transport processes with continuous space description—no need for spatial discretisation—the benefit of discrete space models is that the dynamics of many relevant quantities can now be found in closed form. This benefit is especially relevant in dimensions greater than one where finding the spatio-temporal dependence of the diffusion equation, except in very symmetric geometries, requires the numerical solution of complicated boundary value problems. In addition, when time is discrete, that is when dealing with LRWs, the inversion to time of closed form expressions of generating functions is computationally more convenient when compared to a numerical Laplace inversion to continuous time.

Given these advantages, one may wonder why LRWs have not been employed more extensively to study first-passage processes in heterogeneous space. The answer to this question lies in the multiple technical challenges, our missing links, that needed to be overcome, starting from the analytic knowledge of the LRW spatio-temporal dependence of the occupation probability $P_{\boldsymbol{n}_0}(\boldsymbol{n}, t)$, where $\boldsymbol{n}$ represents the lattice coordinate, $\boldsymbol{n}_0$ the localised initial lattice and $t$ the discrete time variable. In unbounded hypercubic lattices the exact solution of the LRW Master equation, the so-called propagator or Green's function, for nearest-neighbour jumps can be easily found for each Fourier component $\left(\widehat{f}(k_i) = \sum_{n_i=-\infty}^{+\infty} e^{ik_i n_i} f(n_i)\right)$ and as a generating function $\left(\widetilde{f}(z) = \sum_{t=0}^{\infty} f(t) z^t\right)$, namely in the form $\widetilde{\widehat{P}}_{\boldsymbol{n}_0}(\boldsymbol{k}, z)$. The same could not be said, instead, about finite lattices of arbitrary dimensions for which exact propagators have remained unknown for over a century [32], except for periodic domains [38]. In Sect. 5.2, we present the development that led to the exact spatio-temporal dependence of $P_{\boldsymbol{n}_0}(\boldsymbol{n}, t)$ in hypercubic lattices with arbitrary boundaries, i.e. absorbing, periodic, reflecting and mixed absorbing/reflecting.

Knowledge of the propagators has allowed us to study in closed form the LRW dynamics in the presence of spatial disorder [37, 39]. For the latter, and to make a distinction with the dynamics in the presence of reactive sites, we have used the name inert spatial heterogeneities to indicate that the LRW occupation probability is conserved over time. Such heterogeneities consist, e.g. of permeable and reflecting barriers, regions with different diffusivities, local biases as well as long-range connections. We show this advance in Sect. 5.2.1. A special type of inert heterogeneity, that affects the movement dynamics globally, occurs when a LRW jumps with some probability from anywhere on the lattice to a specific site. These dynamics represent the so-called resetting LRW in discrete space and discrete time, originally introduced in the literature for Brownian walks in Ref. [40]. The form of the resetting discrete Master equation, derived and solved in Ref. [34], is presented in Sect. 5.2.2. Use and extension of the above approaches to other geometries has also been possible, one such important endeavour has been the exact derivation of the LRW dynamics in finite periodic and reflecting hexagonal and honeycomb domains in Ref. [35]. In Sect. 5.2.3 we report on these latter findings, but we also extend them and find the propagator when an hexagonal LRW is subject to global biases.

In Sect. 5.3, we present the mathematical formalism [36] that allows us to determine the splitting probabilities when the environment contains multiple partially or fully absorbing targets. As studies of radiation boundary conditions for LRWs have been very limited, we have dedicated Sect. 5.4 to introduce the formalism and show the link to the discrete Feynman–Kac equation [41–43], which allows to analyse functionals of random processes. In Sect. 5.5, we cover four topics and we do so by applying the formal theories presented in the preceding sections. We study the dynamics of the first-passage probability with a single target on a periodic hexagonal geometry, the splitting probabilities to two targets in a heterogeneous 2D reflecting square lattice, the transmission probability between two resetting LRWs in a 1D periodic domain, and the first-reaction probability in a 1D unbounded domain. The analytical nature of the formalism is exploited in discovering an interesting disorder indifference phenomenon in Sect. 5.6. While for a semi-bounded 1D domain a disorder indifference phenomenon has already been presented in the literature [36], here we show another example, which occurs in a 2D domain. Finally, in the concluding Sect. 5.7, we point to research areas where our formalism could be further exploited as well as to areas where LRWs can provide fruitful approaches.

## 5.2 Dynamics in Bounded Hypercubic Lattices

The first step in identifying the missing links has been to devise a hierarchical procedure that exploits the analytic solution in a lower dimension to build the solution in a higher one [32]. Such procedure, which involves a symmetrisation step in the presence of global biases [33], allows to find the exact spatio-temporal solution of the nearest-neighbour LRW Master equation in hypercubic lattices of $d$ dimensions, namely,

$$P(\boldsymbol{n}, t+1) = \left(1 - \frac{1}{d}\sum_{i=1}^{d} q_i\right) P(\boldsymbol{n}, t) + \sum_{i=1}^{d}\left\{\frac{q_i}{2d}(1-g_i)P(\boldsymbol{n}-\boldsymbol{e}_i, t) + \frac{q_i}{2d}(1+g_i)P(\boldsymbol{n}+\boldsymbol{e}_i, t)\right\}, \quad (5.1)$$

where $\boldsymbol{e}_i$ is the $i$th basis vector and where, at each time step, $-1 \leq g_i \leq 1$ represents the bias along the $i$th axis ($g_i > 0$ indicates a bias in the negative direction), and with $0 \leq q_i \leq 1$ for all $i$ yielding $0 \leq 1 - d^{-1}\sum_{i=1}^{d} q_i \leq 1$ the probability of not moving. Given that the $q_i$'s are proportional to the diffusion constant along each direction in the continuous space-time limit of Eq. (5.1), we refer to them as the LRW diffusivities.

In bounded hypercubic lattices, once the types of domain boundary along each axis are specified, the propagator of Eq. (5.1) for the initial condition $P_{\boldsymbol{n}_0}(\boldsymbol{n}, 0) = \delta_{\boldsymbol{n},\boldsymbol{n}_0}$, with $\delta_{\boldsymbol{a},\boldsymbol{b}}$ representing the Kronecker delta function for each axis, is given by [32]

$$P_{\boldsymbol{n}_0}^{(\gamma)}(\boldsymbol{n}, t) = \sum_{k_1=w^{(\gamma_1)}}^{W^{(\gamma_1)}} \cdots \sum_{k_d=w^{(\gamma_d)}}^{W^{(\gamma_d)}} \prod_{j=1}^{d} h_{k_j}^{(\gamma_j)}(n_j, n_{0_j}) \left[1 + \frac{s_{k_1}^{(\gamma_1)}}{d} + \cdots + \frac{s_{k_d}^{(\gamma_d)}}{d}\right]^t, \quad (5.2)$$

where the superscript $\gamma$, accounting along each axis for domains that are periodic ($\gamma = p$), absorbing ($\gamma = a$), reflecting ($\gamma = r$) or a mixture of reflecting on one side and absorbing on the other ($\gamma = m$), specifies the spatial and temporal dependence. While we refer the reader to Refs. [32, 33] for the explicit expressions for $h_k^{(\gamma)}(n, n_0)$, $s_k^{(\gamma)}$, $w^{(\gamma)}$ and $W^{(\gamma)}$, we plot in Fig. 5.1 an illustrative example, that of a LRW in a square domain periodic and reflecting along two orthogonal directions, globally biased towards the bottom left corner. At intermediate times the occupation probability, $P_{n_0}(\boldsymbol{n}, t)$, displays some non-monotonicity in space along the horizontal axis due to reflection, while a non-zero probability appearing close to the top boundary indicates the presence of the periodic boundary conditions along the vertical axis. At step $t = 200$ we observe the peak of the probability at coordinates (13,13), which is 16 lattice sites away from the starting location in the South West direction. To explain this observation we calculate the ensemble average, indicated with the symbol $\langle\rangle$, using the unbounded propagator and obtain the mean displacement $\langle n_i \rangle = \sum_{\boldsymbol{n}} n_i P(\boldsymbol{n}, \ \mathrm{t}) = n_{0_i} - g_i q_i t/2$, where $i = 1$ and 2 denotes, respectively, the horizontal and vertical directions. To determine the time dependent mean position of the probability we simply subtract the initial position along each axis and we have $\langle (n_i - n_{0_i}) \rangle = -g_i q_i t/2$, which is exactly -8 along each axis for the parameters used in Fig. 5.1. An analogous calculation for the mean square displacement (MSD) extracted from the unbounded propagator along each axis gives $\langle (n_i - \langle n_i \rangle)^2 = q_i t \left(1 - g_i^2 q_i/2\right)/2$, whose square root gives a distance of around 9 for the parameters of Fig. 5.1. With the square domain of width 31 and the mode centred at (13, 13) at time $t = 200$, it becomes clear why the effects of the boundaries start to be conspicuous at such time scale, and are particularly visible along the horizontal axis because of the probability accumulation due to the reflecting boundary.

Differently from the corresponding cases for random walks in continuous time as well as Brownian walks, one may notice that Eq. (5.2) is not the product of 1D propagators, which ultimately has made identifying the exact spatio-temporal solution of the Master equation challenging. Another notable characteristic of Eq. (5.1) and its solution is that the dynamics include the chance at each time step of staying at a site with some arbitrary probability. In the mathematics literature a symmetric nearest-neighbour LRW that stays put with probability 1/2 and moves in all allowed directions with probability $1/(4d)$ is called a lazy random walk [44–46]. In our Eqs. (5.1) and (5.2), we obtain such kind of transition probability by setting $g_i = 0$ and $q_i = 1/2$. This implies that our formalism generalises the lazy LRW, allowing to describe its occupation probability when the chance of not moving at each time step can be chosen arbitrarily along each direction, a practical convenient aspect since it avoids the so-called parity issues [47] that affect the spatio-temporal dependence of the LRW occupation probability dynamics when $q_i = 1$. From now on we refer more broadly to a lazy LRW whenever a walker at each time step has a non-zero chance of staying.

### 5.2.1 *Inert Heterogeneities*

In the presence of $M$ inert spatial heterogeneities the transition probabilities between any two sites can be modified, leading to the Master equation [37]

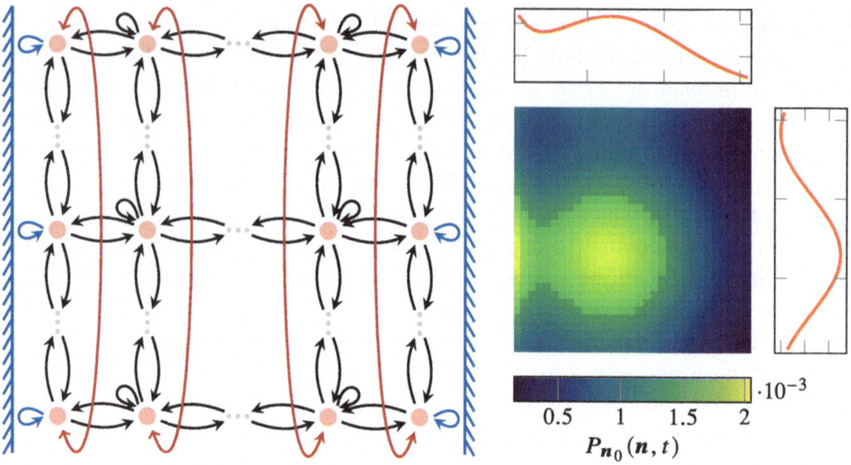


**Fig. 5.1** Left panel: Schematic representation of the transition probabilities for a biased LRW on a 2D square lattice limited by reflecting and periodic boundaries along the horizontal and vertical axes, respectively. The black arrows denote transition probabilities in the bulk with the following values: $q_1(1+g_1)/2$ for horizontal left steps, $q_1(1-g_1)/2$ for horizontal right steps, $q_2(1+g_2)/2$ for vertical steps down, $q_2(1-g_2)/2$ for vertical steps up and $1-(q_1+q_2)/2$ for the slanted oval, representing the probability of staying. Two-headed long red arrows depict transition probabilities across the periodic boundary and therefore have the same values as the vertical black arrows with corresponding directions. The blue oval arrows at the left and right extremes denote the modified probabilities of staying because of the reflecting boundaries along the horizontal directions. The values of these arrows on the left and right extremes are given by $1+q_1g_1/2-q_2/2$ and $1-q_1g_1/2-q_2/2$, respectively. Right panel: Corresponding occupation probability and its marginals (top and side panels) at time $t=200$ of the biased LRW with $q_1=q_2=0.8$ and $g_1=g_2=0.1$, starting at $\boldsymbol{n}_0=(21,21)$, in a square domain of size $N=31$ with reflecting and periodic boundaries along, respectively, the horizontal and vertical axis. With the definitions $f=(1-g)/(1+g)$ and $\sigma=(1-g^2)^{-1/2}$, the temporal dependence in Eq. (5.2) is given by $s_k^{(p)}=q\cos(2\pi k/N)+iqg\sin(2\pi k/N)-q$ for the periodic axis, while for the reflective axis we have $s_0^{(r)}=0$ or $s_k^{(r)}=q\cos(\pi k_i/N)/\sigma-q$ when $k>0$. For the spatial dependence we have $h_k^{(p)}(n,n_0)=\exp[2\pi ik(n-n_0)/N]/N$, while $h_0^{(r)}(n,n_0)=f^{n-1}(1-f)/(1-f^N)$ and $h_k^{(r)}(n,n_0)=N^{-1}f^{(n-n_0-1)/2}g(n)g(n_0)/[\sigma-\cos(\pi k/N)]$ with $g(m)=\sqrt{f}\sin(\pi km/N)-\sin[\pi k(m-1)/N]$ when $k>0$. The finite series for the $k_i$ indices start with $\omega=0$ and terminates at $W=N-1=30$

$$\mathcal{P}(\boldsymbol{n},t+1)=\sum_{\boldsymbol{m}}\underline{\boldsymbol{A}}_{\boldsymbol{n},\boldsymbol{m}}\mathcal{P}(\boldsymbol{m},t) + \sum_{k=1}^{M}\left(\delta_{\boldsymbol{n},\boldsymbol{u}_k}-\delta_{\boldsymbol{n},\boldsymbol{v}_k}\right)\left[\lambda_{\boldsymbol{v}_k,\boldsymbol{u}_k}\mathcal{P}(\boldsymbol{u}_k,t)-\lambda_{\boldsymbol{u}_k,\boldsymbol{v}_k}\mathcal{P}(\boldsymbol{v}_k,t)\right], \tag{5.3}$$

where $\underline{\boldsymbol{A}}_{\boldsymbol{n},\boldsymbol{m}}$ is a compact notation for the tensor representing the transition probabilities between any pair of sites, e.g. Eq. (5.1) if the lattice is unbounded, or Eq. (5.1) together with the appropriate boundary constraints if the domain is finite. The role of

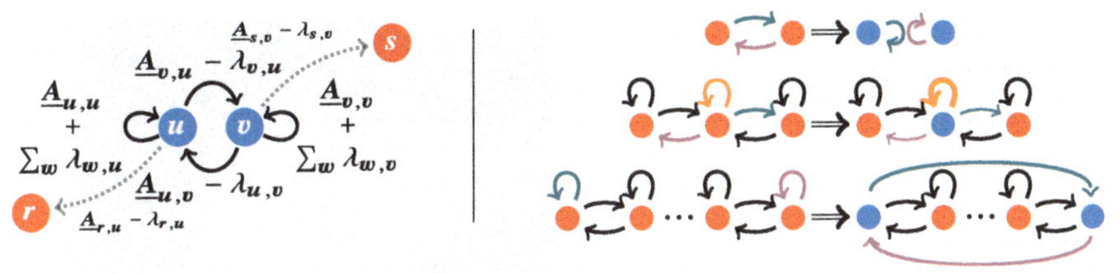


**Fig. 5.2** Schematic representation of the transition probabilities after the introduction of spatial heterogeneities or disorder between a pair of sites subject to the constraints $\lambda_{\boldsymbol{w},\boldsymbol{u}} \leq \underline{A}_{\boldsymbol{w},\boldsymbol{u}}$ for all neighbouring sites $\boldsymbol{w}$ with a (modified) heterogeneous connection in the direction $\boldsymbol{u}$ to $\boldsymbol{w}$, and $0 \leq \underline{A}_{\boldsymbol{u},\boldsymbol{u}} + \sum_{\boldsymbol{w}} \lambda_{\boldsymbol{w},\boldsymbol{u}}$ to ensure that the heterogeneous transition probabilities are positive. The left panel shows that the probability of hopping from site $\boldsymbol{u}$ to $\boldsymbol{v}$ is given by $\underline{A}_{\boldsymbol{v},\boldsymbol{u}}$. When $\lambda_{\boldsymbol{v},\boldsymbol{u}}$ is positive, the probability of jumping from $\boldsymbol{u}$ to $\boldsymbol{v}$ decreases, while the probability of staying put increases. When $\lambda_{\boldsymbol{v},\boldsymbol{u}}$ is negative, the opposite effect occurs with a decrease in the probability of staying, while increasing the jump probability from $\boldsymbol{u}$ to $\boldsymbol{v}$. The parameter $\lambda_{\boldsymbol{u},\boldsymbol{v}}$ affects the transition probability from $\boldsymbol{v}$ to $\boldsymbol{u}$ and the probability of remaining at $\boldsymbol{v}$ in an equivalent manner. On the right panel we show three specific modifications: the introduction of a reflecting barrier (top), of a sticky site (middle), and of a long-range connection (bottom)

the right elements inside the summation can be evinced by looking at Fig. 5.2 where we illustrate how the transition probabilities between pairs of lattice sites are being modified whenever the $\lambda$ parameters are non-zero.

By generalising the so-called Montroll's defect technique for reactive sites to the case of inert heterogeneities [37], we have obtained the exact propagator generating function of Eq. (5.3) as

$$\widetilde{\mathcal{P}}_{\boldsymbol{n}_0}(\boldsymbol{n}, z) = \widetilde{P}_{\boldsymbol{n}_0}(\boldsymbol{n}, z) - 1 + \frac{\left|\underline{\boldsymbol{H}}(\boldsymbol{n}, \boldsymbol{n}_0)\right|}{\left|\underline{\boldsymbol{H}}\right|}, \tag{5.4}$$

where $\widetilde{P}_{\boldsymbol{n}_0}(\boldsymbol{n}, z)$ is the propagator generating function of the defect free problem, while $\left|\underline{\boldsymbol{H}}\right|$ and $\left|\underline{\boldsymbol{H}}(\boldsymbol{n}, \boldsymbol{n}_0)\right|$ are determinants with matrices of size $M \times M$ built from the defect-free propagator dynamics and the heterogeneity properties as follows:

$$\underline{\boldsymbol{H}}_{i,j} = \lambda_{\boldsymbol{v}_i,\boldsymbol{u}_i} \widetilde{P}_{\langle \boldsymbol{u}_j - \boldsymbol{v}_j \rangle}(\boldsymbol{u}_i, z) - \lambda_{\boldsymbol{u}_i,\boldsymbol{v}_i} \widetilde{P}_{\langle \boldsymbol{u}_j - \boldsymbol{v}_j \rangle}(\boldsymbol{v}_i, z) - z^{-1}\delta_{i,j}, \tag{5.5}$$

$$\underline{\boldsymbol{H}}(\boldsymbol{n}, \boldsymbol{n}_0)_{i,j} = \underline{\boldsymbol{H}}_{i,j} - \widetilde{P}_{\langle \boldsymbol{u}_j - \boldsymbol{v}_j \rangle}(\boldsymbol{n}, z) \left[\lambda_{\boldsymbol{v}_i,\boldsymbol{u}_i} \widetilde{P}_{\boldsymbol{n}_0}(\boldsymbol{u}_i, z) - \lambda_{\boldsymbol{u}_i,\boldsymbol{v}_i} \widetilde{P}_{\boldsymbol{n}_0}(\boldsymbol{v}_i, z)\right], \tag{5.6}$$

where $\widetilde{P}_{\langle \boldsymbol{u}-\boldsymbol{v} \rangle}(\boldsymbol{n}, z) = \widetilde{P}_{\boldsymbol{u}}(\boldsymbol{n}, z) - \widetilde{P}_{\boldsymbol{v}}(\boldsymbol{n}, z)$ represents the difference between defect-free propagators evaluated at pairs of sites.

We consider a disordered lattice represented in the left panel of Fig. 5.3, and through a straightforward numerical $z$-inversion [48, 49] of Eq. (5.4), we plot $\mathcal{P}_{\boldsymbol{n}_0}(\boldsymbol{n}, t)$ at times $t = 50$, 100 and 200 in the right panels of Fig. 5.3. It is evident that the local biases and the sticky sites force the occupation probability to deviate quite considerably from a uniform spreading around the starting site even at relatively early times.

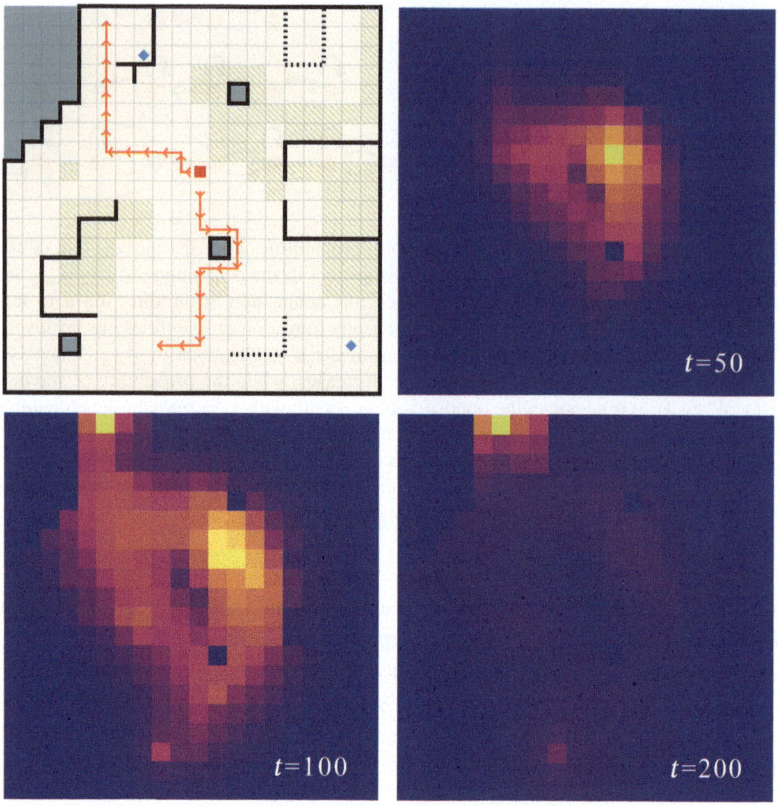


**Fig. 5.3** Spatial dependence of the propagator $\mathcal{P}_{n_0}(\boldsymbol{n}, t)$ at different times for a diffusive LRW starting from $\boldsymbol{n}_0 = (11, 12)$ in a square disordered lattice of size $N = 20$ indicated in the left panel. The diffusivity along each axis is set to $q_i = 0.2$. The inert spatial heterogeneities are as follows: arrows indicate a local bias, solid and dotted lines are, respectively, fully reflecting or permeable barriers, while shaded areas represent sticky sites. The permeable barriers are constructed such that 80% of the time they reflect the walkers allowing the walker to pass through the remaining 20% of the time, the sticky sites results in a reduction of movement by a factor of 0.2, finally, the local biases are constructed to reduce the sojourning probability by 20% whilst increasing the probability of moving in the indicated biased direction by the equivalent probability. The blue marked diamond at coordinates (8, 18) and (19, 3) represent target sites used later in Sect. 5.3

In some instances, when the number of spatial heterogeneities is large, we have devised a technique that avoids to calculate the determinants in Eq. (5.4). This aspect is particularly convenient to study LRW dynamics in a space with multiple domains. For want of space we refer the reader to the publication [39] for further details. That mathematical technique has also allowed us to clarify the microscopic limiting procedure to obtain the so-called leather or Tanner boundary condition [50] in continuous space, something relevant to study transport processes in porous materials.

### 5.2.2 Resetting Dynamics

When a LRW is subject to resetting to a given site $\boldsymbol{n}_c$, its dynamics is as follows: at each time step it either resets to $\boldsymbol{n}_c$ with a given probability $r$ $(0 \leq r \leq 1)$, or performs its underlying Markov walk with the complementary probability $1 - r$. The Master equation of the occupation probability of such a process, ${}_rP_{\boldsymbol{n}_0}(\boldsymbol{n}, t)$ (the left subscript $r$ expresses the fact that the LRW is subject to resetting), can be written [34] in terms of the LRW dynamics in the absence of resetting. The renewal form of the Master equation is given by

$$
{}_rP_{\boldsymbol{n}_0}(\boldsymbol{n}, t) = r \sum_{t'=0}^{t-1} (1-r)^{t'} P_{\boldsymbol{n}_c}(\boldsymbol{n}, t') + (1-r)^t P_{\boldsymbol{n}_0}(\boldsymbol{n}, t) . \tag{5.7}
$$

The structure of Eq. (5.7) is advantageous as it connects directly the resetting propagator ${}_rP_{\boldsymbol{n}_0}(\boldsymbol{n}, t)$ to the reset-free propagator $P_{\boldsymbol{n}_0}(\boldsymbol{n}, t)$, thus allowing to express various quantities of the resetting dynamics in terms of the reset-free propagator.

In $z$-domain, one may write Eq. (5.7) as

$$
{}_r\widetilde{P}_{\boldsymbol{n}_0}(\boldsymbol{n}, z) = \frac{rz}{1-z} \widetilde{P}_{\boldsymbol{n}_c}(\boldsymbol{n}, (1-r)z) + \widetilde{P}_{\boldsymbol{n}_0}(\boldsymbol{n}, (1-r)z) . \tag{5.8}
$$

The form of Eq. (5.8) is particularly useful to determine the steady-state probability in the presence of resetting. In the limit $t \to \infty$, using the final value theorem for the $z$-transform, we obtain the steady-state probability ${}_rP^{\text{ss}}(\boldsymbol{n})$ as

$$
{}_rP^{\text{ss}}(\boldsymbol{n}) \equiv \lim_{z \to 1} [(1-z)\, {}_r\widetilde{P}_{\boldsymbol{n}_0}(\boldsymbol{n}, z)] = r\, \widetilde{P}_{\boldsymbol{n}_c}(\boldsymbol{n}, (1-r)) , \tag{5.9}
$$

which indicates, as expected, that the steady-state probability depends on the resetting site $\boldsymbol{n}_c$ and is independent of the initial site $\boldsymbol{n}_0$. For an unbiased $1d$ LRW with diffusivity $q$ in unbounded domain subject to resetting at site $n_c$, the steady-state probability obtained using Eq. (5.9) is given by

$$
{}_rP^{\text{ss}}(n) = \frac{\sqrt{\chi_r - 1}}{\sqrt{\chi_r + 1}} \exp\left\{ -|n - n_c| \ln\left( \chi_r + \sqrt{\chi_r^2 - 1} \right) \right\} , \tag{5.10}
$$

where $\chi_r \equiv 1 + r/(q(1-r)) \geq 1$ is a renormalized parameter dependent only on the LRW diffusivity and the resetting probability.

We emphasise that Eqs. (5.7), (5.8), and (5.9) hold for any reset-free propagator and its generating function, and can thus be employed in arbitrary dimensions with arbitrary boundary conditions, e.g. for any choice of $\gamma$ in Eq. (5.2), or with any spatial heterogeneity as dictated by Eq. (5.4), as well as with arbitrary geometries, an example of which is presented in the next subsection.

### *5.2.3 Hexagonal and Honeycomb Domains*

It is well-known that it is not only the dimensionality of the lattice but its coordination number, that is the number of nearest neighbours of each site, which affects the properties of the walk [51]. This is exemplified in finite lattices where the combination of the shape or outer boundaries of the spatial domain and the coordination number have critical implications on the LRW dynamics. An important non-hypercubic lattice is the six neighbour hexagonal lattice employed in models ranging from the movement of mobile phone users [52] through to the formation of territories in scent-marking organisms [53–57]. In this section we introduce the periodic propagators for bounded hexagonal lattices and describe the wealth of quantities that can be obtained from them.

When considering hexagonal geometries, the non-orthogonality of the lattice sites makes it convenient to use the so-called Her's three-axis coordinate system [58, 59], where each coordinate is given by three linearly dependent integer coordinates $(n_1, n_2, n_3)$ such that $n_1 + n_2 + n_3 = 0$ as displayed in Fig. 5.4.

Since the coordinates are linearly dependent, we require only two of them to uniquely describe a coordinate. In unbounded space, the Master equation governing the evolution of the occupation probability, $Q^{\mathcal{H}}(n_1, n_2, t)$ (the superscript $\mathcal{H}$ stands for hexagonal), is

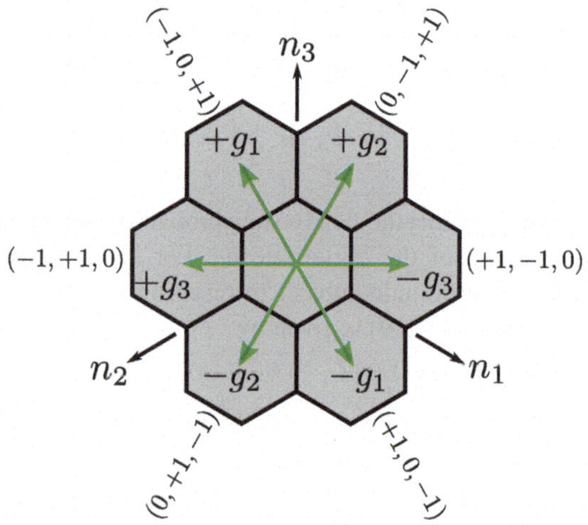


**Fig. 5.4** Schematic representation of the hexagonal lattice and the LRW movement steps. The $n_i$ unit vectors of the Her's coordinate system are shown next to the sites, while permissible movement directions between lattice sites are shown with the arrows emerging from the centre. The three outer numbers indicate the increase, decrease or no change of the $n_i$ coordinate along the specified direction. The magnitude of the transition probability along each direction, $(q/6)$, is modified when either of the three biases is non-zero, namely by the multiplicative factor $1 \pm g_i$. For clarity, we omit arrows depicting the option of staying on lattice sites, which is given by $1 - q$, and we also do not display the modifications that occur when periodic or reflecting boundaries are present

$$
\begin{aligned}
Q^{\mathcal{H}}(n_1, n_2, t+1) = \frac{q}{6}\Big[&(1+g_1)Q^{\mathcal{H}}(n_1+1, n_2, t) + (1-g_1)Q^{\mathcal{H}}(n_1-1, n_2, t) \\
&+ (1+g_2)Q^{\mathcal{H}}(n_1, n_2+1, t) + (1-g_2)Q^{\mathcal{H}}(n_1, n_2-1, t) \\
&+ (1+g_3)Q^{\mathcal{H}}(n_1+1, n_2-1, t) + (1-g_3)Q^{\mathcal{H}}(n_1-1, n_2+1, t)\Big] \\
&+ (1-q)Q^{\mathcal{H}}(n_1, n_2, t),
\end{aligned}
\tag{5.11}
$$

where $-1 \leq g_i \leq 1, i \in \{1, 2, 3\}$ governs the independent bias along the three movement directions as shown in Fig. 5.4. Following standard procedures [15], we obtain the unbounded propagator

$$
\widetilde{Q}^{\mathcal{H}}_{\boldsymbol{n}_0}(n_1, n_2, z) = \frac{1}{(2\pi)^2}\int_{-\pi}^{\pi}\int_{-\pi}^{\pi}\frac{\mathrm{e}^{i[(\boldsymbol{n}-\boldsymbol{n}_0)\cdot\boldsymbol{k}]}}{1 - z\mu(k_1, k_2)}\mathrm{d}k_1\mathrm{d}k_2, \tag{5.12}
$$

where $\boldsymbol{k} = (k_1, k_2)^{\mathsf{T}}$ (the symbol $^{\mathsf{T}}$ indicating the transpose operation), $\boldsymbol{n} - \boldsymbol{n}_0 = (n_1 - n_{0_1}, n_2 - n_{0_2})$ and

$$
\begin{aligned}
\mu(k_1, k_2) = 1 - q + \frac{q}{3}\Big[&\cos(k_1) + \cos(k_2) + \cos(k_1 - k_2) \\
&- ig_1\sin(k_1) - ig_2\sin(k_2) - ig_3\sin(k_1 - k_2)\Big],
\end{aligned}
\tag{5.13}
$$

is the so-called structure or characteristic function [17, 60], or discrete Fourier transform of the step probabilities. The quantity $\widetilde{Q}^{\mathcal{H}}_{\boldsymbol{n}_0}(n_1, n_2, z)$ reduces to known results [18] when $q = 1$ and $g_i = 0, i \in \{1, 2, 3\}$. We note here that in the other limit viz. $g_i = \pm 1, i \in \{1, 2, 3\}$, the walker is confined to three choices of movement direction. One can obtain a fully ballistic walker by solving a slightly modified Master equation whereby individual $q_i$ values are assigned to each unit vector such that $(q_1 + q_2 + q_3)/3 = q$. For example, in this case by setting $g_1 = 1, q_2 = q_3 = 0$ one obtains ballistic motion along the $(-1, 0, +1)$ direction.

In order to find the exact dynamics in finite hexagonal domains, we make use of the so-called method of images, a technique exploited extensively for LRW in hypercubic lattices. For the technique to be applied to hexagonal domains, a generalisation to non-orthogonal coordinate systems is necessary. In doing so for periodic hexagonal domains, as a result of the zig-zag nature of the boundaries there are two unique image sets, which we refer to as the left and right shift (see Figs. 2 and 8 in Ref. [35] for pictorial representations). As the two image sets create two unique periodic propagators with differing transport statistics, we denote them below with the superscripts $\ell$ and $\mathfrak{r}$, respectively.

In applying the generalised method of images we take a finite hexagon with a circumradius of size $R$, containing $\Omega = 3R^2 + 3R + 1$ lattice sites, and we construct the periodic propagator via [35] $\widetilde{P}^{\mathcal{H}^{(p)}}_{\boldsymbol{n}_0}(n_1, n_2, z) = \sum_{m_1=-\infty}^{\infty}\sum_{m_2=-\infty}^{\infty}\widetilde{Q}^{\mathcal{H}}_{\boldsymbol{n}_0}(n_1 + \hat{n}_1^{[i]}, n_2 + \hat{n}_2^{[i]}, z)$ with the $[i]$ shift using

$$\begin{bmatrix} \widehat{n}_1^{[\ell]} \\ \widehat{n}_2^{[\ell]} \end{bmatrix} = \begin{bmatrix} (2R+1)m_1 - Rm_2 \\ -(R+1)m_1 + (2R+1)m_2 \end{bmatrix}; \quad \begin{bmatrix} \widehat{n}_1^{[\mathfrak{r}]} \\ \widehat{n}_2^{[\mathfrak{r}]} \end{bmatrix} = \begin{bmatrix} -Rm_1 + (2R+1)m_2 \\ -(R+1)m_1 - Rm_2 \end{bmatrix}. \tag{5.14}$$

Using (5.14), we find the periodic propagator in closed form as

$$P_{\boldsymbol{n}_0}^{\mathcal{H}^{(p)[i]}}(n_1, n_2, t) = \frac{1}{\Omega} + \frac{1}{\Omega}\sum_{r=0}^{R-1}\sum_{s=0}^{3r+2}\Big\{ e^{i[(\boldsymbol{n}-\boldsymbol{n}_0)\cdot \boldsymbol{k}^{[i]}]}\left[\mu\left(k_1^{[i]}(r,s), k_2^{[i]}(r,s)\right)\right]^t + e^{-i[(\boldsymbol{n}-\boldsymbol{n}_0)\cdot \boldsymbol{k}^{[i]}]}\left[\mu\left(-k_1^{[i]}(r,s), -k_2^{[i]}(r,s)\right)\right]^t \Big\}, \tag{5.15}$$

where $k_1^{[\mathfrak{r}]}(r,s) = k_2^{[\ell]}(r,s) = 2\pi\Omega^{-1}\left[R(s+1)+s-r\right]$ and $k_2^{[\mathfrak{r}]}(r,s) = k_1^{[\ell]}(r,s) = 2\pi\Omega^{-1}[R(2-s+3r)+r+1]$. Equation (5.15) reduces to Eq. (12) of Ref. [35] in the absence of any bias. Note that using Eq. (5.15) the LRW propagator in a finite hexagons with reflecting boundaries can also be found by using the general framework for spatial heterogeneities presented in Sect. 5.2.1. To do that one needs to insert $6R+3$ pairs of defective sites (3 pairs when the site is a corner site, and 2 pairs otherwise) to ensure that a walker cannot jump to periodically-linked hexagonal boundary sites on the other side of the lattice. Such an approach was used for the unbiased case and can be found in Ref. [35].

The creation of the hexagonal periodic image set in Eq. 5.14 has also opened up the ability to represent, entirely analytically, random motion in six-sided hexagonal geometry. We note in fact that Eq. (5.15) is valid with any normalised structure-function, and the inclusion of internal states [61, 62] into the random walk allows the analytic representation of the three-neighbour, non-Bravais honeycomb lattice [35], which finds use in transport processes over carbon structures such as graphite [63] and single-walled carbon nanotubes [64, 65]. We omit the details here, but we point the interested reader once again to Ref. [35].

## 5.3 First Passage Processes to Multiple Targets

The LRW renewal equation, first introduced by Erdös, Feller and Pollard [66], and used extensively since (see, e.g. [29–31]), has a fundamental role in first-passage processes since it allows to link the generating functions of the occupation probability with the first-passage probability to reach the target site $\boldsymbol{n}$ from the initial site $\boldsymbol{n}_0$:

$$\widetilde{F}_{\boldsymbol{n}_0\to\boldsymbol{n}}(z) = \frac{\widetilde{P}_{\boldsymbol{n}_0}(\boldsymbol{n}, z)}{\widetilde{P}_{\boldsymbol{n}}(\boldsymbol{n}, z)}. \tag{5.16}$$

Generalisations of relation (5.16) to multiple targets have appeared, e.g. [32, 67, 68], but the proposed relations have been constructed in terms of the solution of an algebraic set of coupled probability equations. In [21] Rubin and Weiss developed a LRW formalism to compute the number of visits to an arbitrary set of lattice sites in unbounded or periodic domains. The generalisation of that method to arbitrary geometry, devised in Ref. [36], has led recently to the sought-after general theory of first-passage processes to a set of multiple (partially) absorbing targets.

With a set of targets at coordinates $\boldsymbol{m}_i$ $(i = 1, ..., M)$ and absorbing probability $\boldsymbol{\rho} = \left(\rho_{\boldsymbol{m}_1}, \rho_{\boldsymbol{m}_2}, ..., \rho_{\boldsymbol{m}_M}\right)$ with $(0 < \rho_i \le 1)$, the multi-target theory allows to determine first-passage probabilities to any of the absorbing sites as well as the so-called splitting probabilities to reach only one, that is the probability to reach target $j$ and none of the remaining ones, $\widetilde{F}_{\boldsymbol{n}_0 \to (\boldsymbol{m}_j|\boldsymbol{m}_1;...;\boldsymbol{m}_{j-1};\boldsymbol{m}_{j+1};...;\boldsymbol{m}_M)}(\boldsymbol{\rho}, z)$. That quantity is given in terms of individual first-passage and first-return probability as follows

$$\widetilde{F}_{\boldsymbol{n}_0 \to (\boldsymbol{m}_j|\boldsymbol{m}_1;...;\boldsymbol{m}_{j-1};\boldsymbol{m}_{j+1};...;\boldsymbol{m}_M)}(\boldsymbol{\rho}, z) = \frac{|\mathbb{F}^{(j)}(\boldsymbol{\rho}, z)|}{|\mathbb{F}(\boldsymbol{\rho}, z)|}, \tag{5.17}$$

where the $M \times M$ matrix $\mathbb{F}(\boldsymbol{\rho}, z)$ has diagonal components equal to $\mathbb{F}_{k,k}(\boldsymbol{\rho}, z) = 1 + \frac{1-\rho_{\boldsymbol{m}_k}}{\rho_{\boldsymbol{m}_k}}\left[1 - \widetilde{\mathcal{R}}_{\boldsymbol{m}_k}(z)\right]$, with $\widetilde{\mathcal{R}}_{\boldsymbol{m}_k}(z)$ the generating function of the (first-)return probability to site $\boldsymbol{m}_k$, and the off-diagonal terms $\mathbb{F}_{j,k}(\boldsymbol{\rho}, z) = \widetilde{F}_{\boldsymbol{m}_k \to \boldsymbol{m}_j}(z)$. The matrix $\mathbb{F}^{(j)}(\boldsymbol{\rho}, z)$ is the same as $\mathbb{F}(\boldsymbol{\rho}, z)$, but with the $j$th column replaced by $\mathcal{G}(z) = \left(\widetilde{F}_{\boldsymbol{n}_0 \to \boldsymbol{m}_1}(z), \widetilde{F}_{\boldsymbol{n}_0 \to \boldsymbol{m}_2}(z), \ldots, \widetilde{F}_{\boldsymbol{n}_0 \to \boldsymbol{m}_M}(z)\right)^{\mathsf{T}}$.

If one were interested in the first-passage over a given subset of the $M$ targets, given the mutually exclusive nature of the first-passage and first-return trajectories, it suffices to sum Eq. (5.17) over the appropriate $j$ elements. When the interest instead lies in knowing the first-passage probability to any of the $M$ targets it is trivial to show that the multi-linearity of the determinant allows to write the first-passage probability to any of the targets as

$$\widetilde{F}_{\boldsymbol{n}_0 \to (\boldsymbol{m}_1;...;\boldsymbol{m}_M)}(\boldsymbol{\rho}, z) = 1 - \frac{|\overline{\mathbb{F}}(\boldsymbol{\rho}, z)|}{|\mathbb{F}(\boldsymbol{\rho}, z)|}, \tag{5.18}$$

where $\overline{\mathbb{F}}_{i,j}(\boldsymbol{\rho}, z) = \mathbb{F}_{i,j}(\boldsymbol{\rho}, z) - \mathcal{G}_i(z)$.

Using Eq. (5.18) the mean first-passage, or mean first-absorption time when $\rho < 1$, to any of the $M$ targets can be conveniently obtained by standard procedure, namely $T_{\boldsymbol{n}_0 \to (\boldsymbol{m}_1;...;\boldsymbol{m}_M)} = (\mathrm{d}/\mathrm{d}z)\, \widetilde{F}_{\boldsymbol{n}_0 \to (\boldsymbol{m}_1;...;\boldsymbol{m}_M)}(\boldsymbol{\rho}, z)\big|_{z=1}$. The resulting expression is given by

$$T_{\boldsymbol{n}_0 \to (\boldsymbol{m}_1;...;\boldsymbol{m}_M)} = \frac{|\mathbb{T}_0|}{|\mathbb{T}_1| - |\mathbb{T}|}. \tag{5.19}$$

The $M \times M$ elements of the matrix $\mathbb{T}$ are constructed by calculating individual mean first-passage and first-return quantities, namely, $\mathbb{T}_{i,i} = -\frac{1-\rho_{\boldsymbol{m}_i}}{\rho_{\boldsymbol{m}_i}} R_{\boldsymbol{m}_i}$ for the diagonal terms, with $R_{\boldsymbol{m}_i}$ the mean return time to $\boldsymbol{m}_i$, and the off-diagonal terms $\mathbb{T}_{i,j} = T_{\boldsymbol{m}_j \to \boldsymbol{m}_i}$, with the mean first-passage time from $\boldsymbol{m}_j$ to $\boldsymbol{m}_i$. The other matrices, $\mathbb{T}_0$ and $\mathbb{T}_1$, are constructed as follows: $\mathbb{T}_{0_{i,j}} = \mathbb{T}_{i,j} - T_{\boldsymbol{n}_0 \to \boldsymbol{m}_i}$ and $\mathbb{T}_{1_{i,j}} = \mathbb{T}_{i,j} - 1$.

As a transmission event of a token of information, e.g. the passing of a pathogen or an internal degree of freedom, or its limiting case an encounter event when $\rho = 1$, is exactly equivalent to a first-absorption event at any of the locations where two LRW may meet [19], an important achievement of Eq. (5.18) is that it allows to compute the time dependence of the probability of encounter and transmission [36]. The necessary ingredients consist of the joint time-dependent occupation probability in the $2d$-dimensional space for the two independent walkers that move in $d$ dimensions, which are now available in hypercubic lattices for simple [32] and biased walks [33] with resetting dynamics [34, 69] as well as in triangular and hexagonal lattices [35]. Using the knowledge of the joint propagators, the evaluation of mean first-passage times between lattice points and mean return times is also possible, leading to the evaluation of the mean transmission or mean encounter time via Eq. (5.19).

## 5.4 Dynamics in the Presence of Radiation Boundaries

The so-called radiation boundary condition, introduced in LRW theory by Weiss [14], consists of a lattice site at which a walker may be reflected with probability $\eta$ and absorbed with probability $1-\eta$. If we consider for simplicity a 1D unbounded LRW, described by Eq. (5.1) in the absence of any bias, starting to the right of the origin and such that a radiation boundary is present at the origin, the boundary constraint imposes the following flux condition [70]:

$$J^{+}(0,t)=\eta J^{-}(0,t), \tag{5.20}$$

where the $-$ and $+$ superscripts indicate the flux in and out of the origin, respectively, while the total flux is given by $J(0,t)=J^{+}(0,t)-J^{-}(0,t)$. If we take a nearest-neighbour 1$d$ LRW, with lattice points $n=0,1,...,$ the fluxes in and out of the origin are, $J^{-}(0,t)=qP(1,t)/2$ and $J^{+}(0,t)=qP(0,t)/2$. These relations lead to the discrete version of the so-called radiation (also referred to as Robin or Fourier) boundary condition

$$J(0,t)=\frac{q(\eta-1)}{2\eta}P(0,t), \tag{5.21}$$

which is a generalisation to the case of lazy LRWs, and are well-known in the literature (see, e.g. [71]). For $\eta=1$ we recover from Eq. (5.21) the reflecting boundary condition, whereas for $\eta=0$ we have the absorbing boundary condition.

To find the first-reaction (or first-absorption) probability when the LRW Master equation is supplemented by the boundary condition (5.21), we need to formulate the dynamics in terms of the number of times, $\mathcal{S}_t$, the LRW visits the origin [72–75]. Using $\mathcal{N}_t$ as the stochastic variable that represents the location of the LRW, with $\mathcal{N}_0\neq 0$, we define $\mathcal{S}_t=\sum_{i=1}^{t}\delta_{\mathcal{N}_i,0}$ and write a Master equation describing the joint probability of $\mathcal{N}_t$ and $\mathcal{S}_t$, $\Lambda(n,s,t)=\mathbb{P}[\mathcal{N}_t=n,\mathcal{S}_t=s]$, as

$$\Lambda(n,s,t+1) = \left[\frac{q}{2}\Lambda(n-1,s,t) + \frac{q}{2}\Lambda(n+1,s,t) + (1-q)\Lambda(n,s,t)\right](1-\delta_{n,0})$$
$$+ \left[\frac{q}{2}\Lambda(n-1,s-1,t) + \frac{q}{2}\Lambda(n+1,s-1,t) + (1-q)\Lambda(n,s-1,t)\right]\delta_{n,0}. \quad (5.22)$$

Taking the generating function of Eq. (5.22) with respect to $s$, i.e. $\sum_{s=0}^{\infty}\eta^s\Lambda(n,s,t) = P^{(\eta)}(n,t)$, we find

$$P^{(\eta)}(n,t+1) = \left[\frac{q}{2}P^{(\eta)}(n-1,t) + \frac{q}{2}P^{(\eta)}(n+1,t) + (1-q)P^{(\eta)}(n,t)\right](1-\delta_{n,0})$$
$$+ \eta\left[\frac{q}{2}P^{(\eta)}(n-1,t) + \frac{q}{2}P^{(\eta)}(n+1,t) + (1-q)P^{(\eta)}(n,t)\right]\delta_{n,0}$$
$$= \eta^{\delta_{n,0}}\left[\frac{q}{2}P^{(\eta)}(n-1,t) + \frac{q}{2}P^{(\eta)}(n+1,t) + (1-q)P^{(\eta)}(n,t)\right], \quad (5.23)$$

which is the discrete time Feynman–Kac equation [41–43] for the functional that indicates the number of visit to the origin. Through a simple rearrangement of the terms on the right-hand side of the top equation in (5.23), one realises that we have written the Master equation of a lazy symmetric unbounded LRW with a partially absorbing defect at $n = 0$ with absorption probability $\rho = 1 - \eta$.

Since $\Lambda(n,s,t) = \left\langle \delta_{\mathcal{N}_t,n}\,\delta_{\mathcal{S}_t,s}\right\rangle$, we have that the $\eta$-transform is $P^{(\eta)}(n,t) = \sum_{s=0}^{\infty}\eta^s\Lambda(n,s,t) = \left\langle \delta_{\mathcal{N}_t,n}\,\eta^{\mathcal{S}_t}\right\rangle$, which leads to

$$P^{(\eta)}(n,t) = \sum_{s=0}^{\infty}\eta^s\Big(\mathbb{P}[\mathcal{N}_t = n, \mathcal{S}_t < s+1] - \mathbb{P}[\mathcal{N}_t = n, \mathcal{S}_t < s]\Big). \quad (5.24)$$

As $\mathbb{P}[\mathcal{N}_t = n, \mathcal{S}_t < 0] = 0$, we may write

$$P^{(\eta)}(n,t) = \sum_{s=1}^{\infty}\mathbb{P}[\mathcal{N}_t = n, \mathcal{S}_t < s](\eta^{s-1} - \eta^s) = \sum_{s=0}^{\infty}\mathbb{P}[\mathcal{N}_t = n, \mathcal{S}_t \le s](1-\eta)\eta^s. \quad (5.25)$$

Identifying $(1-\eta)\eta^s$ as the probability distribution of a geometrically distributed random variable, $\hat{s}$, i.e. $\mathbb{P}[\hat{s} = s] = \langle\delta_{\hat{s},s}\rangle = (1-\eta)\eta^s$, we obtain

$$P^{(\eta)}(n,t) = \left\langle \sum_{s=0}^{\infty}\mathbb{P}[\mathcal{N}_t = n, \mathcal{S}_t \le s]\delta_{\hat{s},s}\right\rangle = \mathbb{P}[\mathcal{N}_t = n, \mathcal{S}_t \le \hat{s}]. \quad (5.26)$$

Thus, $P^{(\eta)}(n,t)$, the solution of the Feynman–Kac equation (5.23), represents the LRW occupation probability at $n$, whilst having not visited the origin more times than a geometrically distributed random variable, $\hat{s}$, with mean $\eta/(1-\eta)$.

Now, we return to the radiation boundary. Consider a random walker that visits the origin $m$ times without being absorbed, but being reflected at each visit, then the probability of this is simply $\eta^m$, $\eta$ being the reflection probability. As the number of visits to the origin is a random variable, $\mathcal{S}_t$, we must average over all possible realisations of $\mathcal{S}_t$. The probability for the random walker to be located at the lattice point $n$ at time $t$, whilst having not been absorbed by the radiation boundary, is $\left\langle \delta_{\mathcal{N}_t,n}\, \eta^{\mathcal{S}_t} \right\rangle$, which is simply the solution to Eq. (5.23), with a reflecting boundary condition at the origin. Thus we have established that the radiation boundary can be formulated in terms of a reflected random walk that is absorbed once the number of visits to the origin exceed a geometrically distributed random variable with mean $\eta/(1-\eta)$.

Following the arguments of Ref. [74], as $\mathcal{S}_t$ is a monotonically non-decreasing process, the event $\left\{\mathcal{S}_t > \hat{s}\right\}$ is equivalent to $\{t > \mathcal{T}\}$, where $\mathcal{T}$ is the time of reaction (absorption), therefore

$$\mathcal{T} = \min\left\{t > 0 : \mathcal{S}_t > \hat{s}\right\}. \tag{5.27}$$

Equation (5.27) represents the spatio-temporal discrete form of the reaction time, which was found in the continuous paradigm in Ref. [74]. From Eq. (5.27) one obtains

$$P^{(\eta)}(n,t) = \mathbb{P}\left[\mathcal{N}_t = n, t < \mathcal{T}\right]. \tag{5.28}$$

One is then able to study the first-reaction time probability, $\mathcal{F}^{(\eta)}_{n_0\to 0}(t) = \mathbb{P}[\mathcal{T} = t|\mathcal{N}_0 = n_0]$, for a random walker in the presence of a radiation boundary. Due to the connection between the radiation boundary and discrete Feynman–Kac equation explained above, we solve Eq. (5.23), with a reflecting boundary at the origin and the localised initial condition at $n_0$, using the defect technique. The resulting probability is the propagator of Eq. (5.1) for a $1d$ unbiased walker that satisfy the boundary condition (5.21).

Then marginalisation over $n$ leads to the survival probability, $\mathbb{P}[t < \mathcal{T}]$, which then gives the first-reaction probability as

$$\widetilde{\mathcal{F}}^{(\eta)}_{n_0\to 0}(z) = \frac{\widetilde{P}_{n_0}(0,z) + \widetilde{P}_{-n_0}(0,z)}{\frac{\eta}{1-\eta} + 2\widetilde{P}_0(0,z)}, \tag{5.29}$$

where $\widetilde{P}_{n_0}(n,z)$ is the generating function of the lazy $1d$ symmetric unbounded propagator [32], i.e.

$$\widetilde{P}_{n_0}(n,z) = \frac{\left(\frac{1}{\beta(z)} + \frac{1}{\beta(z)}\sqrt{1-\beta^2(z)}\right)^{-|n-n_0|}}{[1-z(1-q)]\sqrt{1-\beta^2(z)}}, \tag{5.30}$$

where $\beta(z) = zq/[1-z(1-q)]$ $|\beta(z)| \le 1$ (see Sect. 5.5 for the time dependence of $\mathcal{F}^{(\eta)}_{n_0\to 0}(t)$).

## 5.5 Dynamics of First-Passage, First-Encounter and First-Reaction Processes

To show the general applicability of the multi-target formalism described in Sect. 5.3, we use it to plot the time dependent probability of various target hitting processes, namely, a first-passage to a single target in a hexagonal domain, a first-passage to either of two targets in a disordered lattice and a first-transmission between two resetting walkers in a periodic domain. We also exploit the formalism introduced in Sect. 5.4 to display the dynamics of first-reaction to a single target in a 1D unbounded domain. We present these four examples in the four panels of Fig. 5.5.

The top left panel displays a very rich first-passage dynamics with multimodal shapes, a feature rarely observed with Markov LRW, but present in non-Markov [76] and quantum walks [77]. This top left panel represents the first passage of a biased periodic hexagonal LRW from the origin to a target located towards the bottom right of the domain for different bias values. As the level of bias towards the target increases we see that the mode of $F^{\mathcal{H}}_{\boldsymbol{n}_0\to\boldsymbol{n}}(t)$ increases in magnitude and moves towards shorter times. With high bias, exemplified by the case with $g_i = 0.8$, the initial peak is followed by a region of near zero probability, corresponding to the timescale at which the walkers travelling towards the target are likely to have passed it and the walkers who originally wandered in the opposite direction have yet to reach the target. In general we observe that higher biases lead to sustained oscillations at intermediate times in $F^{\mathcal{H}}_{\boldsymbol{n}_0\to\boldsymbol{n}}(t)$, with troughs corresponding to timescales in which heavily biased trajectories are away from the target. Oscillations are also sustained for longer times for larger domain sizes as the chance of missing the target increases with the circumradius $R$. Eventually, once the indirect trajectories dominate the dynamics, the oscillations dampen and $F^{\mathcal{H}}_{\boldsymbol{n}_0}(\boldsymbol{n}, t)$ drops to zero, as shown in the inset of the top left panel.

In the top right panel, we plot the the first passage and time-dependent splitting probabilities to either of the targets with the environment and initial condition shown in the left-most panel of Fig. 5.2 where the two targets, are indicated using blue diamonds. Due to the local biases, the walker starting at $\boldsymbol{n}_0 = (11, 12)$ is carried quickly to the target $\boldsymbol{n}_1 = (8, 18)$ resulting in an increase of the splitting probability $F_{\boldsymbol{n}_0\to(\boldsymbol{n}_1|\boldsymbol{n}_2)}(t)$ at earlier times. The local biases instead do not facilitate the movement towards the target at $\boldsymbol{n}_2 = (19, 3)$, which is reached mainly diffusively from $\boldsymbol{n}_0$. The reflecting barriers also act to confine the walker temporarily nearby $\boldsymbol{n}_1$ further increasing the likelihood of absorption at $\boldsymbol{n}_1$, while much of the heterogeneities present do not assist in reaching $\boldsymbol{n}_2$, and the walker must explore a much larger proportion of the domain before reaching $\boldsymbol{n}_2$, which results in a mode at a later time for $F_{\boldsymbol{n}_0\to(\boldsymbol{n}_2|\boldsymbol{n}_1)}(t)$. Beyond the delay in the mode of the splitting probabilities, another interesting feature is that the trajectories that contribute to the first-passage probability to either of the targets is mostly made up of trajectories that reach $\boldsymbol{n}_1$, as one may evince by the small difference in the curves of $F_{\boldsymbol{n}_0\to(\boldsymbol{n}_1|\boldsymbol{n}_2)}(t)$ and $F_{\boldsymbol{n}_0\to(\boldsymbol{n}_1;\boldsymbol{n}_2)}(t)$. More quantitatively, one has that the relative ratio between the integrated splitting probabilities, i.e. $\sum_{t=0}^{\infty} F_{\boldsymbol{n}_0\to(\boldsymbol{n}_1|\boldsymbol{n}_2)}(t) \approx 0.74$ and $\sum_{t=0}^{\infty} F_{\boldsymbol{n}_0\to(\boldsymbol{n}_2|\boldsymbol{n}_1)}(t) \approx 0.26$. Given that

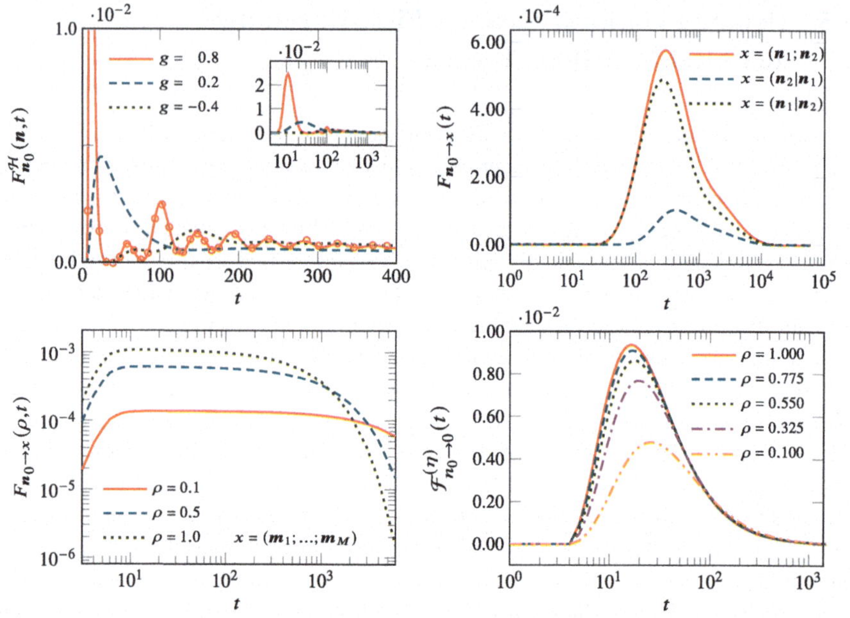


**Fig. 5.5** Top left panel: First passage time from the origin to a single target at $(n_1, n_2, n_3) = (7, -2, -5)$ in a left shift periodic hexagon with $R = 14$ for different bias strength $g$ with $g_i = g$ and diffusivity $q = 0.9$. The solid lines is obtained by numerically inverting Eq. (5.16) and making use of Eq. (5.15), while the open-circle markers are results from $10^6$ stochastic simulations. Inset shows the same information plotted up to $t = 3000$ on a semilog scale. Top right panel: First-passage probabilities to either of two targets as well as their splitting probabilities for the 2D LRW moving in the disordered domain shown in the left panel of Fig. 5.3. The LRW parameters are the same as those in that panel, and the targets, indicated in blue, are located, respectively, at $\boldsymbol{n}_1 = (8, 18)$ and $\boldsymbol{n}_2 = (19, 3)$. The splitting probability to reach $\boldsymbol{n}_1$ is the curve with the higher mode. The propagator in the disordered space, shown in Eq. (5.4), is used in Eqs. (5.17) and (5.18) to plot, respectively, the splitting probabilities or the first-passage probability to either target. Bottom left panel: First-transmission probability for two independent resetting 1D LRW without bias in a periodic lattice of size $N = 21$ $(1 \leq n \leq 21)$. The joint propagator for the two resetting walkers used in Eq. (5.18) is obtained by taking the generating function of the product of 1D time-dependent resetting propagators through Eq. (5.7). The first and the second walker starts from their corresponding resetting sites set at $n_{c_1} = 8$ and $n_{c_2} = 14$, respectively. The diffusivities and resetting probabilities of the walkers are taken to be $q_1 = q_2 = 0.8$ and $r_1 = r_2 = 0.4$, respectively. The walkers can interact with probability $\rho$ when they simultaneously occupy a site $n$ in the 1D domain, which corresponds to a partially absorbing site $\boldsymbol{m}_i = (i, i)$, where $i \in [1, M = 21]$ in the 2D domain. Bottom right panel: First-reaction probability, $\mathcal{F}^{(\eta)}_{n_0 \to 0}(t)$, for a radiation boundary at the origin, for varying reaction probabilities $\rho = 1 - \eta$, for $n_0 = 5$ and $q = 0.5$. The case $\rho = 1$ or $\eta = 0$ corresponds to the fully absorbing case in the flux and boundary condition, respectively, Eqs. (5.20) and (5.21), transforming the first-reaction probability into a first-passage probability. The expression for $\widetilde{\mathcal{F}}^{(\eta)}_{n_0 \to 0}(z)$ can be found in Eq. (5.29)

in the absence of any heterogeneities, the proportion of trajectories that reach either $\boldsymbol{n}_1$ or $\boldsymbol{n}_2$ from $\boldsymbol{n}_0$ is more evenly split, these findings suggest empirical approaches to favour the selection of certain targets versus others based on the type and strength of spatial disorder that one may introduce in a given spatial domain.

In the bottom panels we show the first-transmission probability between two resetting LRWs (left) and the first-reaction probability at a lattice site for an individual LRW (right). In the left panel we show the effects of the co-location or proximity transfer efficiency, $\rho$, on the first-transmission probability (at any of the sites). The transport limited regime, which corresponds to the case $\rho = 1$, is simply the encounter probability. It displays the highest mode and at the earliest time. The reaction limited regime is particularly evident when $\rho = 0.1$, for which the curve remains essentially constant after the mode, with the flat profile lasting longer the smaller is $\rho$. Given that the system is finite eventually a fast decay to zero characterises all of the curves for any $\rho > 0$. In the right panel, we present the first-reaction probability, $\mathcal{F}^{(\eta)}_{n_0 \to 0}(t)$, for different reactivity, $\rho$, at the origin. One can clearly see that, as the reactivity gets reduced, the mode shifts to longer times as more and more visits to the origin are necessary before the walker reacts. The widening of the shape of the probability and the appearance of longer tails is also a result of the need for multiple return to the origin before a reaction takes place.

## 5.6 Disorder Indifference Phenomena

When a LRW moves on a disordered lattice, the type and location of the spatial heterogeneities heavily affect the spatio-temporal dependence of the walker's occupation probability. But in some cases, certain quantities, surprisingly, are not affected at all by the disorder. We have uncovered two such examples and we report them below.

The first example is the mean first-passage time from site $n_0$ to site $n$ in a semi-bounded 1D domain in presence of a symmetric permeable barrier at lattice site $u$ in between $n$ and $n_0$. To be more precise consider a left reflecting barrier between site $n = 0$ and site $n = 1$, then take the LRW initial position at $n_0 \geq 1$, and to the right of $n_0$ we place the barrier at $u$ and further to the right we have the site $n$, that is $1 \leq n_0 < u < n$. In this arrangement, while the time dependence of the first-passage probability from $n_0$ to $n$ does depend on the location $u$ and the permeability of the barrier, which is proportional to the $\lambda$ parameter in Sect. 5.2.1, the mean first-passage time is independent of both. This surprising effect, reported in Ref. [37], which was shown to appear even in the presence of a global bias, is lost when the barrier permeability is made asymmetric. An analogous effect was also shown for Brownian walks [78] in 1D and in radially symmetric geometries [79].

The second disorder indifference phenomenon we have observed manifests when one is interested in counting the number of visits or more generally the number of detectable visits. By detectable it is meant that there is a parameter governing the probability that the visit is accounted for. To define the problem more precisely, we consider the number of detectable visits that occurs at the set of sites $S_v$ before being absorbed at any of the set of sites $S_a$. Let the set $S_a = \boldsymbol{r}_1, \cdots, \boldsymbol{r}_{M_a}$ be the destination points where the process terminates, i.e. the targets, and let $S_v = \boldsymbol{r}_{M_a+1}, \cdots, \boldsymbol{r}_{M_a+M_v}$ be the set of sites or points of interest where one is interested in counting the number of detections. We refer to sites in $S_a$ as the destination sites and sites in $S_v$ as the visitation sites. The sets $S_a$ and $S_v$ are mutually exclusive and the set of defects is given by $S = S_a \cup S_v$ with the total number of defects being $|S| = M_a + M_v = M$. Note that each site has its own independent probability of, respectively, detectability and absorption, that is we have $\boldsymbol{\rho} = (\rho_1, \cdots, \rho_M)$ with $\rho_i$ for $1 \le i \le M_a$ being the partial absorption probability at site $\boldsymbol{r}_i$, while $\rho_j$ for $M_a + 1 \le j \le M$ being the detection probability of a visit at site $\boldsymbol{r}_j$. With this set-up, by generalising the formalism shown in [21], it is possible to show that the generating function of the probability of (detected) visits at the sites of interest is given by

$$\overline{V}_{\boldsymbol{n}_0}(\boldsymbol{x}, S, \boldsymbol{\rho}) = 1 - \frac{\left|C^{(d_1)} - C^{(1)}\right| - \left|C^{(d_2)} - C^{(2)}\right|}{\left|C^{(d_1)}\right| - \left|C^{(d_2)}\right|}, \tag{5.31}$$

where

$$C_{i,j}^{(d_1)} = \begin{cases} \rho_j T_{\boldsymbol{r}_j \to \boldsymbol{r}_i}(1 - \delta_{i,j}) + (\rho_j - 1) R_{\boldsymbol{r}_i} \delta_{i,j}, & 1 \le j \le M_a, \\ \rho_j (1 - x_j) T_{\boldsymbol{r}_j \to \boldsymbol{r}_i}(1 - \delta_{i,j}) + \left[\rho_j (1 - x_j) - 1\right] R_{\boldsymbol{r}_i} \delta_{i,j}, & M_a + 1 \le j \le M, \end{cases} \tag{5.32}$$

$$C_{i,j}^{(d_2)} = C_{i,j}^{(d_1)} - \begin{cases} \rho_j, & 1 \le j \le M_a, \\ \rho_j (1 - x_j), & M_a + 1 \le j \le M, \end{cases} \tag{5.33}$$

$$C_{i,j}^{(1)} = \begin{cases} \rho_j T_{\boldsymbol{n}_0 \to \boldsymbol{r}_i}, & 1 \le j \le M_a, \\ 0, & M_a + 1 \le j \le M, \end{cases} \tag{5.34}$$

$$C_{i,j}^{(2)} = \begin{cases} \rho_j (T_{\boldsymbol{n}_0 \to \boldsymbol{r}_i} - 1) & 1 \le j \le M_a, \\ 0, & M_a + 1 \le j \le M, \end{cases} \tag{5.35}$$

with the notation $\overline{V}_{\boldsymbol{n}_0}(\boldsymbol{x}, S, \boldsymbol{\rho}) = \sum_{k_1} \cdots \sum_{k_{M_v}} V_{\boldsymbol{n}_0}(k_1, \cdots, k_{M_v}, S, \boldsymbol{\rho}) x^{k_1} \cdots x^{k_{M_v}}$ representing the multi-dimensional generating function over the visits $k_1, \cdots, k_{M_v}$ to the sites $\boldsymbol{r}_1, \cdots, \boldsymbol{r}_{M_v}$.

From Eqs. (5.32)–(5.35) it is easy to realise that $\overline{V}_{\boldsymbol{n}_0}(\boldsymbol{x}, S, \boldsymbol{\rho})$ depends upon mean first-passage times and mean return times between the initial condition and the defective sites as well as between the defective sites, both the lattice of interest (visitation sites) and the absorbing lattice sites (terminating sites). When the lattice contains inert spatial heterogeneities, the disorder is expected to affect $\overline{V}_{\boldsymbol{n}_0}(\boldsymbol{x}, S, \boldsymbol{\rho})$ through the mean first-passage and mean return times dependence on the locations and strengths of such heterogeneities in Eqs. (5.32)–(5.35).

It turns out that when the disorder is due to sticky sites, $\overline{V}_{\boldsymbol{n}_0}(\boldsymbol{x}, S, \boldsymbol{\rho})$ as well as $V_{\boldsymbol{n}_0}(\boldsymbol{k}, S, \boldsymbol{\rho})$, that is the probability of the number of $k$ visits before being absorbed at any of the reactive sites, is independent of the numbers, strength and location of such heterogeneities. We show this indifference phenomenon in the simpler scenario

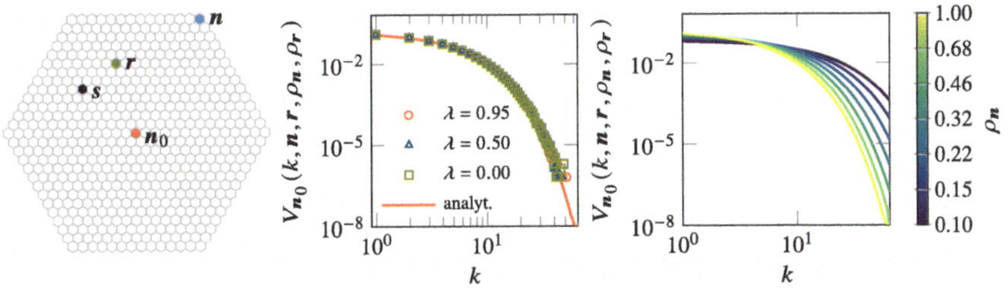


**Fig. 5.6** The probability of the number of visits at site $\boldsymbol{r}$ before being absorbed at $\boldsymbol{n}$ in the presence of a sticky site $\boldsymbol{s}$. Left most panel shows a schematic of the spatial arrangement. We use a periodic hexagonal propagator of circumradius $R = 13$, yielding a total of 547 sites, a LRW diffusivity of $q = 1$, and perfect detection at site $\boldsymbol{r}$, i.e. $\rho_r = 1.0$. The sticky heterogeneity is constructed such that the movement probability at the site is $q_s = q(1-\lambda)$. Centre panel, probability of $k$ number of visits, when $\rho_n = 1.0$, for different values of $\lambda$. The marks, indicating $10^6$ simulations, all collapse onto the the analytic curve obtained by numerically inverting Eq. (5.31). Right most panel shows how the probability of $k$ visits depends on the reaction probability at the target site, i.e. $\rho_n$

of one sticky site at coordinate $\boldsymbol{s}$, one visitation site at $\boldsymbol{r}$ and one absorbing site at $\boldsymbol{n}$ in a periodic hexagonal domain. In Fig. 5.6 we plot $V_{n_0}(k, \boldsymbol{n}, \boldsymbol{r}, \rho_n, \rho_r)$, that is the probability of $k$ visits before being absorbed at $\boldsymbol{r}$ as a function of $k$ for different choices of the heterogeneous parameter $\lambda$ and absorption probability $\rho_n$ and detection probability $\rho_r$. The left most panel of Fig. 5.6 shows the geometrical arrangement of the problem. In the centre most panel we show that for different properties of the sticky site, all of the simulations collapse onto the analytic prediction, while the right most panel shows how the $k$-dependent probability changes as one reduces the reaction probability $\rho_n$ of the target site $\boldsymbol{n}$. This latter dependence is expected since a decrease in $\rho_n$ maintains the walkers in the system for a longer period of time, which in turn allows for increased number visits to the site $\boldsymbol{r}$.

While surprising at first, the disorder indifference phenomenon shown in the central panel of Fig. 5.6 has an intuitive explanation. It is the result of the isotropic nature of the sticky heterogeneity, which slows down the walker in a symmetric fashion, that is in an equal manner irrespective of the direction with which it moves away from the heterogeneous site. In other words, the sticky site acts to slow down all trajectories in all directions, lengthening their time before being absorbed even though their actual spatial path remains unchanged. Hence, the fraction of trajectories that visit $\boldsymbol{r}$ $k$-times before being absorbed at $\boldsymbol{n}$ also remains unchanged.

Given that the distribution is independent of the sticky heterogeneity, it is expected that the mean is also independent of the heterogeneity, i.e.

$$\mathcal{V}_{n_0}(\boldsymbol{n}, \boldsymbol{r}, \rho_n, \rho_r) = \frac{\rho_r}{R_r}\left[T_{n_0\to n} + T_{n\to r} - T_{n_0\to r} + R_n\left(\frac{1}{\rho_n} - 1\right)\right]. \tag{5.36}$$

Using the exact expressions for the mean first-passage time $T_{u \to v}$ and the mean return time $R_n$ when sticky heterogeneities are present (see Eqs. (8) and (9) in Ref. [37]) one can show that indeed $\mathcal{V}_{n_0}(\boldsymbol{n}, \boldsymbol{r}, \rho_n, \rho_r)$ is independent of the location and strength of the isotropic sticky site.

## 5.7 Discussions, Open Problems and Future Directions

While we have included a few new findings, we have focused on presenting a series of quantitative tools that have recently appeared in the LRW literature and that we believe could be exploited in many scenarios both to understand empirical observations and to make theoretical predictions.

In relation to the former, note that the challenge in defining the appropriate jump probabilities of the transition matrix is not different from the one faced by using models with continuous variables when extracting information about diffusion constants or other transport parameters. When the dynamics is relatively complex, e.g. in the presence of spatial disorder, a convenient approach is often that of comparing the dynamics observed in the absence of the underlying heterogeneities to theoretically obtained outputs, one prototypical example being the MSD in an unbounded domain from which the system diffusivity can be extracted. The analysis of the change in dynamics when the disorder is present then allows to identify the elements of the transition matrix relative to the homogeneous environment.

In regards to the theoretical advantages note that there is a computational gain in employing the analytical exact formulation as compared to a numerical iteration of the Master equation [80], e.g. in finding the first-passage probability and the mean first passage time to a single target in a hypercubic lattice of $d$ dimensions and length $N$ there is a factor of $N^d$ reduction in the time complexity [36]. More generally in the context of spatially disordered lattices, the advantage of our framework rests on being able to compute determinants of size equal to the number of spatial defects versus the numerical evaluation of eigenvalues and eigenvectors of a large sparse matrix whose size corresponds to the number of lattice sites, an approach suggested in Ref. [81].

Many areas in probability and combinatorics could benefit from these advances. For example, the analysis of mixing time [83] and cut-off phenomena [84] in finite domains may find convenient to use the analytical expressions of the propagators and provide more rigorous bounds on known time scales.

The generality of the approach to study the dynamics on disordered lattice could also be exploited to study the dynamics of LRWs on networks [85]. By adding, cutting or modifying the link strengths of regular lattices, we may build an arbitrary network. Starting from a lattice of which a propagator generating function is known, it would thus be possible to determine analytically the LRW propagator on the network by employing the formalism for inert spatial heterogeneities of Sect. 5.2.1.

There exist classical problems with Brownian walks that account for diffusion in some representative force field, e.g. a quadratic potential or the so-called V-potential. With the exception of Ref. [47] there has not been many attempts in the LRW literature to study these types of problems, but it would be beneficial to add them to the arsenal of tools in LRW theory.

While our theory can be easily applied to study interacting LRWs for dilute systems, in light of the already existing link between many-particle stochastic systems and quantum many-body dynamics, there are potential avenues to look at some fundamental collective phenomena with some new techniques. One example is the classic asymmetric exclusion process on a $1D$ line [86] for which the exclusion interaction can be formulated by inserting reflecting barriers in the higher dimensional space spanned by the walkers.

We have discussed so far Markov problems, but there are plenty of movement processes for which the non-Markov nature of the process cannot be neglected [87–89]. While the amount of Markov literature would dwarf the non-Markov one if compared, there is great interest in developing a formalism for non-Markov LRWs. The majority of the work on the latter has been computational, while advances on the analytical front has been scarce, but see Refs. [90–94] for some exceptions. For the one-step non-Markovian LRW, the correlated LRW, analytic results have appeared in the past for the full first-passage probability in unbounded domains [95] and for periodically bounded domains only in terms of the mean first-passage time [96]. Recently, a general framework has been developed [97], which allows for the exact determination of the full first-passage probability of a bounded correlated LRW on both the hypercubic and the hexagonal lattice.

We conclude by thanking the book editors for the opportunity to write about the topic and with the hope that this short review will help raise awareness of the strengths and possibilities when one study transport processes using LRWs.

**Acknowledgements** LG acknowledges funding from the Biotechnology and Biological Sciences Research Council (BBSRC) Grant No. BB/T012196/1, and the National Environment Research Council (NERC) Grant No. NE/W00545X/1. All authors would like to thank the Isaac Newton Institute for Mathematical Sciences for support and hospitality during the programme 'Mathematics of Movement: an interdisciplinary approach to mutual challenges in animal ecology and cell biology', when the work on this chapter was undertaken, supported by the EPSRC Grant Number EP/R014604/1. All authors wish to thank the book editors, Denis Grebenkov, Ralf Metzler and Gleb Oshanin, for the invitation to contribute.

## References

1. D. Ben-Avraham, S. Havlin, *Diffusion and Reactions in Fractals and Disordered Systems* (Cambridge University Press, Cambridge, 2004)
2. V.M. Kenkre, L. Giuggioli, *Theory of the Spread of Epidemics and Movement Ecology of Animals: An Interdisciplinary Approach Using Methodologies of Physics and Mathematics* (Cambridge University Press, Cambridge, 2021)

3. I. Todhunter, *A History of the Mathematical Theory of Probability: From the Time of Pascal to That of Laplace* (Salzwasser Verlag, Frankfurt, 1865)
4. A. De Moivre, *The Doctrine of Chances* (Millar, London, 1756)
5. P.S. Laplace, *Théorie analytique des probabilités* (Courcier, Paris, 1820)
6. M.V. Smoluchowski, Zur kinetischen Theorie der Brownschen Molekularbewegung und der Suspensionen. Ann. Phys. **326**, 756–780 (1906)
7. G. Pólya, Über eine Aufgabe der Wahrscheinlichkeitsrechnung betreffend die Irrfahrt im Straßennetz. Math. Ann. **84**, 149–160 (1921)
8. S. Chandrasekhar, Stochastic problems in physics and astronomy. Rev. Mod. Phys. **15**, 1–89 (1943)
9. M. Kac, On the notion of recurrence in discrete stochastic processes. Bull. Amer. Math. Soc. **53**, 1002–1010 (1947)
10. W. Feller, *An Introduction to Probability Theory and Its Applications* (Wiley, New Jersey, 1950)
11. A. Dvoretzky, P. Erdös, Some problems on random walk in space, in *Proceedings of the 2nd Berkeley Symposium* (1951), pp. 353–367
12. E. Sparre Andersen, On the fluctuations of sums of random variables II. Math. Scand. 195–223 (1954)
13. F. Spitzer, *Principles of Random Walk* (Springer, New York, 1976)
14. G.H. Weiss, *Aspects and Applications of the Random Walk* (North Holland, Amsterdam, 1994)
15. B.D. Hughes, *Random Walks and Random Environments: Volume 1: Random Walks* (Oxford University Press, Oxford, 1995)
16. E.W. Montroll, Random walks on lattices, in *Proceedings of Symposia in Applied Mathematics*, pp. 193–220 (1964)
17. E.W. Montroll, G.H. Weiss, Random walks on lattices. II. J. Math. Phys. **6**, 167–181 (1965)
18. E.W. Montroll, Random walks on lattices. III. Calculation of first-passage times with application to exciton trapping on photosynthetic units. J. Math. Phys. **10**, 753–765 (1969)
19. V.M. Kenkre, Theory of exciton annihilation in molecular crystals. Phys. Rev. B **22**, 2089–2098 (1980)
20. W.T.F. Den Hollander, P.W. Kasteleyn, Random walks with spontaneous emission on lattices with periodically distributed imperfect traps. Physica A **112**, 523–543 (1982)
21. R.J. Rubin, G.H. Weiss, Random walks on lattices. The problem of visits to a set of points revisited. J. Math. Phys. **23**, 250–253 (1982)
22. V.M. Kenkre, A theoretical approach to exciton trapping in systems with arbitrary trap concentration. Chem. Phys. Lett. **93**, 260–263 (1982)
23. B. Derrida, Velocity and diffusion constant of a periodic one-dimensional hopping model. J. Stat. Phys. **31**, 433–450 (1983)
24. A. Szabo, G. Lamm, G.H. Weiss, Localized partial traps in diffusion processes and random walks. J. Stat. Phys. **34**, 225–238 (1984)
25. J.W. Haus, K.W. Kehr, Diffusion in regular and disordered lattices. Phys. Rep. **150**, 263–406 (1987)
26. K.P.N. Murthy, K.W. Kehr, Mean first-passage time of random walks on a random lattice. Phys. Rev. A **40**, 2082–2087 (1989)
27. B.D. Hughes, *Random Walks and Random Environments: Volume 2: Random Environments.* (Oxford University Press, Oxford, 1996)
28. H. Cordes, S.D. Baranovskii, K. Kohary, P. Thomas, S. Yamasaki, F. Hensel, J.H. Wendorff, One-dimensional hopping transport in disordered organic solids. I. Analytic calculations. Phys. Rev. B **63**, 094201 (2001)
29. S. Redner, *A Guide to First-Passage Processes* (Cambridge University Press, Cambridge, 2001)
30. O. Bénichou, R. Voituriez, From first-passage times of random walks in confinement to geometry-controlled kinetics. Phys. Rep. **539**, 225–284 (2014)
31. R. Metzler, G. Oshanin, S. Redner, *First-Passage Phenomena and Their Applications* (World Scientific, Singapore, 2014)
32. L. Giuggioli, Exact spatiotemporal dynamics of confined lattice random walks in arbitrary dimensions: a century after Smoluchowski and Pólya. Phys. Rev. X **10**, 021045 (2020)

33. S. Sarvaharman, L. Giuggioli, Closed-form solutions to the dynamics of confined biased lattice random walks in arbitrary dimensions. Phys. Rev. E **102**, 062124 (2020)
34. D. Das, L. Giuggioli, Discrete space-time resetting model: application to first-passage and transmission statistics. J. Phys. A: Math. Theor. **55**, 424004 (2022)
35. D. Marris, S. Sarvaharman, L. Giuggioli, Exact spatiotemporal dynamics of lattice random walks in hexagonal and honeycomb domains. Phys. Rev. E. **107**, 054139 (2023)
36. L. Giuggioli, S. Sarvaharman, Spatio-temporal dynamics of random transmission events: from information sharing to epidemic spread. J. Phys. A: Math. Theor. **55**, 375005 (2022)
37. S. Sarvaharman, L. Giuggioli, Particle-environment interactions in arbitrary dimensions: a unifying analytic framework to model diffusion with inert spatial heterogeneities. Phys. Rev. Res. **5**, 043281 (2023)
38. E.W. Montroll, Random walks in multidimensional spaces, especially on periodic lattices. J. Soc. Ind. Appl. Math. **4**, 241–260 (1956)
39. D. Das, L. Giuggioli, Dynamics of lattice random walk within regions composed of different media and interfaces. J. Stat. Mech.: Theory Exp. **2023**, 013201 (2023)
40. M.R. Evans, S.N. Majumdar, Diffusion with optimal resetting. J. Phys. A: Math. Theor. **44**, 435001 (2011)
41. M. Kac, On distributions of certain wiener functionals. T. Am. Math. Soc. **65**, 1–13 (1949)
42. M. Kac, On some connections between probability theory and differential and integral equations, in *Proceedings of the Second Berkeley Symposium on Mathematical Statistics and Probability*, pp. 189–216. University of California Press (1951)
43. E. Csáki, A discrete Feynman-Kac formula. J. Stat. Plan. Inference **34**, 63–73 (1993)
44. M. Hildebrand, Random lazy random walks on arbitrary finite groups. J. Theor. Probab. **14**, 1019–1034 (2001)
45. J. Miller, Y. Peres, Uniformity of the uncovered set of random walk and cutoff for lamplighter chains. Ann. Probab. **40**, 535–577 (2012)
46. V. Kovač, On binomial sums, additive energies, and lazy random walks. J. Math. Anal. **528**, 127510 (2023)
47. M. Kac, Random walk and the theory of Brownian motion. Am. Math. Mon. **54**, 369–391 (1947)
48. J. Abate, W. Whitt, Numerical inversion of probability generating functions. Oper. Res. Lett. **12**, 245–251 (1992)
49. J. Abate, G.L. Choudhury, W. Whitt, An introduction to numerical transform inversion and its application to probability models, in *Computational Probability* (Springer, 2000), pp. 257–323
50. J.E. Tanner, Transient diffusion in a system partitioned by permeable barriers. Application to NMR measurements with a pulsed field gradient. J. Chem. Phys. **69**, 1748–1754 (1978)
51. G. Zumofen, A. Blumen, Energy transfer as a random walk II. Two-dimensional regular lattices. J. Chem. Phys. **76**, 3713–3731 (1982)
52. I.F. Akyildiz, Y. Lin, W. Lai, R. Chen, A new random walk model for pcs networks. IEEE J. Sel. Area. Comm. **18**, 1254–1260 (2000)
53. L. Giuggioli, J.R. Potts, S. Harris, Animal interactions and the emergence of territoriality. PLoS Comp. Bio. **7**, e1002008 (2011)
54. L. Giuggioli, J.R. Potts, D.I. Rubenstein, S.A. Levin, Stigmergy, collective actions, and animal social spacing. Proc. Natl. Acad. Sci. USA **110**, 16904–16909 (2013)
55. L. Giuggioli, V.M. Kenkre, Consequences of animal interactions on their dynamics: Emergence of home ranges and territoriality. Mov. Ecol. **2**, 1–22 (2014)
56. A. Heiblum Robles, L. Giuggioli, Phase transitions in stigmergic territorial systems. Phys. Rev. E **98**, 042115 (2018)
57. S. Sarvaharman, A. Heiblum Robles, L. Giuggioli, From Micro-to-Macro: how the movement statistics of individual walkers affect the formation of segregated territories in the territorial random walk model. Front. Phys. **7**, 129 (2019)
58. I. Her, C.T. Yuan, Resampling on a pseudohexagonal grid. CVGIP-Graph Model Im. **56**, 336–347 (1994)

59. I. Her, Geometric transformations on the hexagonal grid. IEEE Trans. Imag. Process. **4**, 1213–1222 (1995)
60. J. Klafter, I.M. Sokolov, *First Steps in Random Walks: From Tools to Applications* (Oxford University Press, Oxford, 2011)
61. G.H. Weiss, R.J. Rubin, Random walks: theory and selected applications. Adv. Chem. Phys **52**, 363–505 (1983)
62. J.B.T.M. Roerdink, K.E. Shuler, Asymptotic properties of multistate random walks I. Theory. J. Stat. Phys. **40**, 205–240 (1985)
63. B. Bercu, F. Montégut, Asymptotic analysis of random walks on ice and graphite. J. Math. Phys. **62**, 10 (2021)
64. N. Cotfas, Random walks on carbon nanotubes and quasicrystals. J. Phys. A: Math. Theor. **33**, 2917 (2000)
65. M.T. Batchelor, B.I. Henry, Exact solution for random walks on the triangular lattice with absorbing boundaries. J. Phys. A: Math. Gen. **35**, 5951–5959 (2002)
66. P. Erdös, W. Feller, H. Pollard, A property of power series with positive coefficients. Bull. Amer. Math. Soc. **55**, 201–204 (1949)
67. H. Larralde, G.H. Weiss, A generating function for the second moment of the distinct number of sites visited by an n-step lattice random walk. J. Phys. A: Math. Gen. **28**, 5217 (1995)
68. C. Chevalier, O. Bénichou, B. Meyer, R. Voituriez, First-passage quantities of Brownian motion in a bounded domain with multiple targets: a unified approach. J. Phys. A: Math. Theor. **44**, 025002 (2011)
69. D. Das, V.M. Kenkre, R. Nathan, L. Giuggioli, Misconceptions about quantifying animal encounter and interaction processes. Front. Ecol. Evol. **11**, 1230890 (2023)
70. T. Kay, L. Giuggioli, Subdiffusion in the presence of reactive boundaries: a generalized Feynman-Kac approach. J. Stat. Phys. **190**, 92 (2023)
71. D.S. Grebenkov, M. Filoche, B. Sapoval, Spectral properties of the Brownian self-transport operator. Eur. Phys. J. B **36**, 221–231 (2003)
72. K. Itô, P. Henry Jr., *Diffusion Processes and Their Sample Paths* (Springer Science & Business Media, Berlin, 1965)
73. M.I. Freidlin, *Functional Integration and Partial Differential Equations* (Princeton University Press, Princeton N.J., 1985)
74. D.S. Grebenkov, Paradigm shift in diffusion-mediated surface phenomena. Phys. Rev. Lett. **125**, 078102 (2020)
75. D.S. Grebenkov, Imperfect diffusion-controlled reactions, in *Chemical Kinetics: Beyond the Textbook* (World Scientific, 2020), pp. 191–219
76. T. Verechtchaguina, I.M. Sokolov, L. Schimansky-Geier, First passage time densities in non-Markovian models with subthreshold oscillations. Europhys. Lett. **73**, 691 (2006)
77. M. Kulkarni, S.N. Majumdar, First detection probability in quantum resetting via random projective measurements. J. Phys. A: Math. Theor. **56**, 385003 (2023). https://doi.org/10.1088/1751-8121/acf103
78. T. Kay, L. Giuggioli, Diffusion through permeable interfaces: fundamental equations and their application to first-passage and local time statistics. Phys. Rev. Res. **4**, L032039 (2022)
79. A. Godec, R. Metzler, Optimization and universality of Brownian search in a basic model of quenched heterogeneous media. Phys. Rev. E **91**, 052134 (2015)
80. S. Condamin, O. Bénichou, M. Moreau, First-exit times and residence times for discrete random walks on finite lattices. Phys. Rev. E **72**, 016127 (2005)
81. D.S. Grebenkov, L. Tupikina, Heterogeneous continuous-time random walks. Phys. Rev. E **97**, 012148 (2018)
82. J.H.P. Schulz, E. Barkai, R. Metzler, Aging renewal theory and application to random walks. Phys. Rev. X **4**, 011028 (2014)
83. D.A. Levin, Y. Peres, *Markov Chains and Mixing Times* (American Mathematical Society, Rhode Island, 2017)
84. P. Diaconis, The mathematics of mixing things up. J. Stat. Phys. **144**, 445–458 (2011)

85. N. Masuda, M.A. Porter, R. Lambiotte, Random walks and diffusion on networks. Phys. Rep. **716**, 1–58 (2017)
86. R. Stinchcombe, Stochastic non-equilibrium systems. Adv. Phys. **50**, 431–496 (2001)
87. O. Vilk, Y. Orchan, M. Charter, N. Ganot, S. Toledo, R. Nathan, M. Assaf, Ergodicity breaking in area-restricted search of avian predators. Phys. Rev. X **12**, 031005 (2022)
88. P. Bovet, S. Benhamou, Spatial analysis of animals' movements using a correlated random walk model. J. Theor. Biol. **131**, 419–433 (1988)
89. V.M. Kenkre, V. Ern, A. Fort, Coherence effects in triplet-exciton transport via time-dependent delayed fluorescence. Phys. Rev. B **28**, 598 (1983)
90. G.M. Schütz, S. Trimper, Elephants can always remember: Exact long-range memory effects in a non-markovian random walk. Phys. Rev. E **70**, 045101 (2004)
91. D. Boyer, C. Solis-Salas, Random walks with preferential relocations to places visited in the past and their application to biology. Phys. Rev. Lett. **112**, 240601 (2014)
92. A. Falcón-Cortés, D. Boyer, L. Giuggioli, S.N. Majumdar, Localization transition induced by learning in random searches. Phys. Rev. Lett. **119**, 140603 (2017)
93. D. Boyer, A. Falcón-Cortés, L. Giuggioli, S.N. Majumdar, Anderson-like localization transition of random walks with resetting. J. Stat. Mech.: Theory Exp. **2019**, 053204 (2019)
94. H. Meyer, H. Rieger, Optimal non-Markovian search strategies with n-step memory. Phys. Rev. Lett. **127**, 070601 (2021)
95. H. Larralde, First-passage probabilities and mean number of sites visited by a persistent random walker in one-and two-dimensional lattices. Phys. Rev. E **102**, 062129 (2020)
96. V. Tejedor, R. Voituriez, O. Bénichou, Optimizing persistent random searches. Phys. Rev. Lett. **108**, 088103 (2012)
97. D. Marris, L. Giuggioli, Persistent and anti-persistent motion in bounded and unbounded space: resolution of the first-passage problem. New J. Phys. **26**, 073020 (2024)

# Chapter 6
# Effects of Molecular Crowding and Obstacles on a Target Search

**Kazuhiko Seki**

**Abstract** We are reviewing diffusion-controlled target reactions, where an immobile target is surrounded by diffusing reactants. The reaction occurs between the target and one of the reactants. We summarize the conventional results for the decay of the target survival probability in the dilute limit. In the dilute limit, the diffusion of each reactant is considered as an independent event. Subsequently, we present recent results obtained at high reactant concentrations, where the migration of each reactant is influenced by the configurations of the other reactants, and the diffusion is correlated. Using lattice models, we demonstrate that the search for a target by surrounding reactants can be expedited by prohibiting simultaneous double occupancy of a site. This is achieved by blocking the migration of a reactant to a site that is already occupied by another reactant, resulting in a site-blocking effect. This effect proves advantageous for efficiently searching the target. We present a detailed analytical method to account for site-blocking effects in a simple case, instead of providing numerous examples.

## 6.1 Introduction

The conventional theory of diffusion-controlled reactions assumes that the migration of reactants is independent of the configurations of other reactants [1–4]. In the dilute limit, the diffusion of each reactant can be considered an independent event. However, the migration of reactants is influenced by the configurations of other reactants when the concentration is high. The excluded volume interactions among reactants result in correlated diffusion, where the diffusive migration of a reactant is influenced by the correlations among the reactant configurations [5–9]. Even without a reaction event, calculating the tracer diffusion constant is complicated due to the influence of correlations [5–9]. Correlated diffusion arises from interactions among multiple particles, and the tracer diffusion constant depends on both the concentration and

K. Seki (✉)
National Institute of Advanced Industrial Science and Technology (AIST), Tsukuba Onogawa 16-1 AIST West, Ibaraki 305-8569, Japan
e-mail: k-seki@aist.go.jp



D. Grebenkov et al. (eds.), *Target Search Problems*,
https://doi.org/10.1007/978-3-031-67802-8_6

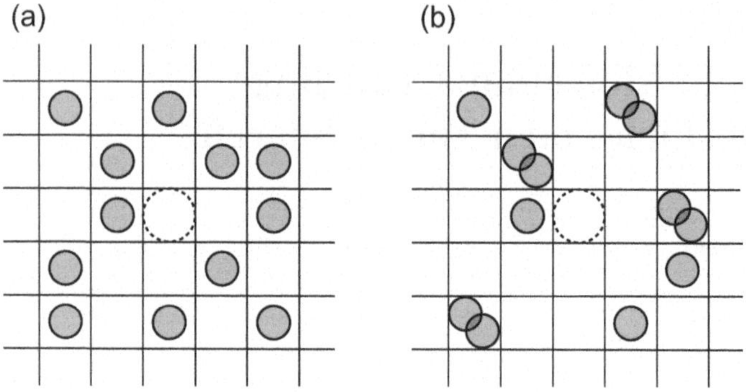


**Fig. 6.1** **a** Double occupancy is not allowed by the site-blocking effect. **b** Double occupancy is allowed. The target, denoted by a circle with a dashed line, is located at the origin (target site). The reactants are denoted by circles with solid lines. Reaction occurs when one of the reactants occupies the target site. The initial number of occupied sites by reactants is $N = 12$. In (**a**), the number of occupied sites by reactants is always the same ($N = 12$) until the reaction occurs. In (**b**), the number of occupied sites by reactants is decreased to $N = 8$ by allowing double occupancy, though the number of reactants is 12. As a result, the probability that the occupied sites of reactants include the target site is higher under the site-blocking effect

lattice structure. Here, we present an approach to account for such correlations in diffusion-controlled reactions.

We are considering a target problem involving diffusion-controlled reactions [10–12]. In the target problem, an immobile target is surrounded by numerous mobile reactants, and a reaction takes place between the target and one of the mobile reactants. The opposite case, where a mobile target reacts with one of the immobile reactants, is referred to as the trapping problem. Using lattice models for a target problem, correlated diffusion can be accounted for by prohibiting simultaneous double occupancy of a site as shown schematically in Fig. 6.1. Excluded volume interactions are considered by prohibiting the double occupancy of reactants due to site-blocking effects. Reactants can only move to adjacent empty sites. In this way, the migration of reactants depends on the configuration of reactants surrounding each reactant. The kinetics of the survival probability of the target are well-known in the dilute limit. We present an approximate theory for the kinetics of the target survival probability under the effects of site-blocking. The theory interpolates between high and low limits of reactant concentrations. We focus on lattice models, although various approaches have been adopted to consider molecular crowding encountered in living cells [13–25]. Possible applications of lattice models include the study of porous materials and surface reactions [8, 12, 26–28].

## 6.2 Conventional Theory

Here, we present the known results of diffusion-mediated reactions for the target problem in the dilute limit [1–4, 29]. We consider an immobile target located at the origin. The target reacts with one of the $N$ mobile reactants and becomes inert. In this section, we consider the movement of each reactant as an independent event, disregarding correlated diffusion by assuming the dilute limit.

We consider that $N$ reactants are placed on the lattice with $M$ sites, including the origin where the target is located. The configuration occupied by reactants is denoted by $(\mathbf{r}_1, \mathbf{r}_2, \ldots, \mathbf{r}_N)$. The reaction rate constant between the target and a reactant placed at $\mathbf{r}$ is denoted by $k(\mathbf{r})$. Later, we will consider the case where the reaction occurs exclusively at the target site. The reaction rate constant for this case can be expressed as $k(\mathbf{r}) = k_0\delta_{\mathbf{r},0}$.

The survival probability of the target at time $t$ denoted by $P_s(t)$ is related to the probability of taking the configuration $(\mathbf{r}_1, \mathbf{r}_2, \ldots, \mathbf{r}_N)$ at time $t$ denoted by $P(\mathbf{r}_1, \mathbf{r}_2, \ldots, \mathbf{r}_N; t)$; the relation is given by

$$P_s(t) = \sum_{\{\mathbf{r}_i\}} P(\mathbf{r}_1, \mathbf{r}_2, \ldots, \mathbf{r}_N; t), \tag{6.1}$$

where the summation indicates the sum over all possible reactant configurations. Equation (6.1) in general holds even under site-blocking effects.

In the dilute limit, the movement of each reactant can be independent and we obtain

$$\sum_{\{\mathbf{r}_i\}} P(\mathbf{r}_1, \mathbf{r}_2, \ldots, \mathbf{r}_N; t) = \left[\sum_{\ell=1}^{M} P_i(\mathbf{r}_\ell, t)/M\right]^N, \tag{6.2}$$

where $P_i(\mathbf{r}_\ell, t)$ indicates the pair correlation function; the fraction of a reactant at $\mathbf{r}_\ell$ from the target site at time $t$ divided by the bulk fraction. We consider the initial condition corresponds to a uniform initial configuration for $P_i(\mathbf{r}_\ell, t)$; the initial condition is given by $P_i(\mathbf{r}_\ell, 0) = 1$ and the normalization can be confirmed by $\sum_{\ell=1}^{M} P_i(\mathbf{r}_\ell, 0)/M = 1$. By substituting Eq. (6.2) into Eq. (6.1), we obtain the conventional expression in the dilute limit as [2, 3, 30]

$$\begin{aligned} P_s(t) &= \left(\sum_{\ell=1}^{M} P_i(\mathbf{r}_\ell, t)/M\right)^N \\ &= \left(1 - \frac{1}{M}\sum_{\ell=1}^{M}[1 - P_i(\mathbf{r}_\ell, t)]\right)^N \\ &\simeq \exp\left(-c\sum_{\mathbf{r}}[1 - P_i(\mathbf{r}, t)]\right), \end{aligned} \tag{6.3}$$

where the limit of $M \to \infty$ is taken while keeping the fraction of reactants fixed, $c = \lim_{M\to\infty} N/M$; $\sum_{\mathbf{r}}$ indicates the sum over all lattice sites. The relation given by Eq. (6.3) holds without specifying the lattice structure and the spatial dependence of the reaction rate constant $[k(\mathbf{r})]$; the only condition required is that the reactant concentration is dilute. However, in a crowded situation, we need to take into account that each reactant is surrounded by other reactants differently; Eq. (6.2) does not hold because each reactant migrates differently due to the site-blocking effects. In the conventional theory, Eq. (6.2) is adopted by assuming that reactants move independently of each other.

The rate constant of a target reaction can be defined by

$$k_{\mathrm{tg}}(t) = -\frac{\frac{d}{dt}P_{\mathrm{s}}(t)}{P_{\mathrm{s}}(t)} = -c\sum_{\mathbf{r}}\frac{d}{dt}P_{\mathrm{i}}(\mathbf{r},t). \tag{6.4}$$

The probability of finding a reactant at any site on the lattice $[\sum_{\mathbf{r}} P_{\mathrm{i}}(\mathbf{r},t)]$ decays by reaction with the rate given by $\sum_{\mathbf{r}} k(\mathbf{r})P_{\mathrm{i}}(\mathbf{r},t)$. Therefore, Eq. (6.4) can be expressed as

$$k_{\mathrm{tg}}(t) = c\sum_{\mathbf{r}} k(\mathbf{r})P_{\mathrm{i}}(\mathbf{r},t). \tag{6.5}$$

We can rewrite Eq. (6.4) as

$$P_{\mathrm{s}}(t) = \exp\left[-\int_0^t dt_1 k_{\mathrm{tg}}(t_1)\right], \tag{6.6}$$

where $P_{\mathrm{s}}(0) = 1$ is used. Equation (6.6) can be expressed by substituting Eq. (6.5) as

$$P_{\mathrm{s}}(t) = \exp\left[-c\int_0^t dt_1 \sum_{\mathbf{r}} k(\mathbf{r})P_{\mathrm{i}}(\mathbf{r},t_1)\right]. \tag{6.7}$$

At this stage, it is interesting to compare Eq. (6.3) with Eq. (6.7). Equation (6.7) is expressed using both $k(\mathbf{r})$ and $P_{\mathrm{i}}(\mathbf{r},t)$, while Eq. (6.3) is expressed using $P_{\mathrm{i}}(\mathbf{r},t)$ alone. In this sense, Eq. (6.3) is conceptually superior over Eq. (6.7).

For a localized reaction $[k(\mathbf{r}) = k_0\delta_{\mathbf{r},0}]$, Eq. (6.7) reduces to

$$P_{\mathrm{s}}(t) = \exp\left[-c\int_0^t dt_1 k_0 P_{\mathrm{i}}(0,t_1)\right]. \tag{6.8}$$

The probability distribution $P_{\mathrm{i}}(\mathbf{r},t)$ decays solely due to reactions occurring at the origin, which can be expressed using the free Green's function $[G_0(\mathbf{r},0,t)]$ for the time-evolution equation of $P_{\mathrm{i}}(\mathbf{r},t)$ described by $\mathcal{L}$ [see Eq. (6.47) below]. $\mathcal{L}$ is an operator for hopping transitions alone, without any reaction. The explicit equation for the free Green's function is given by Eq. (6.51). Using the free Green's function, $P_{\mathrm{i}}(\mathbf{r},t)$ can be expressed formally as $P_{\mathrm{i}}(\mathbf{r},t) = 1 - \int_0^t dt_1 k_0 G_0(\mathbf{r},0,t-t_1)P_{\mathrm{i}}(0,t_1)$. Throughout the manuscript, we denote the Laplace transform of any time-dependent function $f(t)$ as $\hat{f}(s)$. In the Laplace domain, we have $\hat{P}_{\mathrm{i}}(0,s) =$

$\left[1 + k_0 \hat{G}_0 (0, 0, s)\right]^{-1}$. Equation (6.8) yields the conventional expression for the survival probability of the target when the reaction occurs at the target site. The conventional expression is derived by ignoring a site-blocking effect among reactants and is given by [3, 4, 31]

$$P_s(t) = \exp\left(-c \int_0^t dt_1 k_{cv}(t_1),\right), \tag{6.9}$$

where $\hat{k}_{cv}(s)$ can be expressed using the free Green's function [Eq. (6.52)] in the Laplace domain as

$$s\hat{k}_{cv}(s) = \frac{k_0}{1 + k_0 \hat{G}_0 (0, 0, s)}. \tag{6.10}$$

It has been recognized that the conventional expression of the survival probability overestimates the experimental results at high reactant concentrations [32, 33].

## 6.3 Static Target Reaction Without Double Occupancy

For simplicity, we consider static reactions without diffusion but take into account the hard-core repulsion by prohibiting the double occupancy of the same site by reactants as shown in Fig. 6.1. Even without diffusion, the acceleration of decay can be found by preventing double occupancy. The derivation also illustrates a part of our method that is unrelated to diffusion.

We consider $N$ reactants on the lattice with $M$ sites. The lattice sites occupied by reactants are denoted by $(\mathbf{r}_1, \mathbf{r}_2, \ldots, \mathbf{r}_N)$. The probability of taking the configuration $(\mathbf{r}_1, \mathbf{r}_2, \ldots, \mathbf{r}_N)$ at time $t$ is denoted by $P(\mathbf{r}_1, \mathbf{r}_2, \ldots, \mathbf{r}_N; t)$. We consider the case that the reactants are initially uniformly distributed without double occupancy on $M$ lattice sites,

$$P(\mathbf{r}_1, \mathbf{r}_2, \ldots, \mathbf{r}_N; 0) = 1/\binom{M}{N}, \tag{6.11}$$

where $\binom{M}{N} = M!/[N!(M-N)!]$ indicates the $N$-combination of $M$ members. When the target reacts with one of the $N$-reactants with the reaction rate constant given by $k(\mathbf{r}_\ell)$, the survival probability of the target can be calculated by applying Cauchy's integral theorem [34],

$$P_s(t) = \frac{1}{2\pi i} \int dx \frac{1}{x^{N+1}} \frac{1}{\binom{M}{N}} \prod_{\ell=1}^{M} (1 + x \exp(-k(\mathbf{r}_\ell) t)), \tag{6.12}$$

where the integration path encircles the origin on the complex plane; the coefficient in front of $x^N$ of $\prod_{\ell=1}^{M}(1+x\exp(-k(\mathbf{r}_\ell)t))$ represents the survival probability of the target, where all possible reactant configurations are taken into account with the rate constants of the sites for each configuration. By Cauchy's integral theorem, the coefficient can be obtained. We rewrite Eq. (6.12) as

$$P_s(t) = \frac{1}{2\pi i}\int dx \frac{(1+x)^M}{x^{N+1}}\frac{1}{\binom{M}{N}}\exp\left[\sum_{\ell=1}^{M}\ln\left(1+\frac{x}{1+x}\left(\exp(-k(\mathbf{r}_\ell)t)-1\right)\right)\right], \tag{6.13}$$

and take the thermodynamic limit using the Stirling formula $n! \approx \sqrt{2\pi n}\exp(-n)n^n$. In Eq. (6.13), we introduce the Stirling formula obtaining

$$\frac{1}{\binom{M}{N}} = \frac{2\pi\sqrt{(M-N)N}}{\sqrt{2\pi M}}\frac{(M-N)^{M-N}N^N}{M^M} \tag{6.14}$$

$$= \sqrt{2\pi Mc(1-c)}(1-c)^M\left(\frac{c}{1-c}\right)^{Mc} \tag{6.15}$$

$$= \sqrt{2\pi Mc(1-c)}\exp\left[M\left((1-c)\ln(1-c)+c\ln(c)\right)\right], \tag{6.16}$$

where $c = N/M$ is the concentration. Therefore, we obtain

$$\frac{1}{x^{N+1}}\frac{(1+x)^M}{\binom{M}{N}} = \sqrt{2\pi Mc(1-c)}\exp\left[Mf(x)\right]/x, \tag{6.17}$$

where $f(x)$ is given by

$$f(x) = (1-c)\ln(1-c) + c\ln(c) + \ln(1+x) - c\ln(x). \tag{6.18}$$

Now, we consider the thermodynamic limit ($M \to \infty$) with $c$ being fixed; the number of reactants is constant for the reactant configurations which survive reaction. In this limit, we can apply the steepest descent method by considering the form of $f(x)$. We define $x^*$ using $f'(x^*) = 0$ and find $x^* = c/(1-c)$. By applying the steepest descent method to function $\varphi(x)$, we obtain

$$\frac{1}{2\pi i}\int dx\exp\left[Mf(x)\right]\varphi(x)/x = \frac{1}{\sqrt{2\pi Mf''(x^*)}}\exp\left[Mf(x^*)\right]\varphi(x^*)/x^*. \tag{6.19}$$

We find $f''(x^*) = (1-c)^3/c$, $f(x^*) = 0$ and

$$\sqrt{2\pi Mc(1-c)}\frac{\exp\left[Mf(x^*)\right]}{x^*}\frac{1}{\sqrt{2\pi Mf''(x^*)}} = 1. \tag{6.20}$$

By applying the Stirling formula and the steepest descent method using $1 + x^* = 1/(1-c)$, Eq. (6.13) can be expressed as

$$P_s(t) = \exp\left[\sum_{\ell=1}^{M} \ln\left\{1 + c\left(\exp\left(-k\left(\mathbf{r}_\ell\right)t\right) - 1\right)\right\}\right]. \tag{6.21}$$

This is the expression for the static reaction obtained using a different method by Allinger and Blumen [35]. Allinger and Blumen derived Eq. (6.21) by giving a concentration $c$, whereas we derived Eq. (6.21) by assuming $N$ reactants on $M$-sites, which becomes equal to $c$ in the thermodynamic limit.

The limiting expressions can be obtained as

$$P_s(t) = \exp\left(-\sum_{\ell=1}^{M} k\left(\mathbf{r}_\ell\right)t\right) \quad \text{for } c = 1, \tag{6.22}$$

$$P_s(t) \simeq \exp\left[-\sum_{\ell=1}^{M} c\left(1 - \exp\left(-k\left(\mathbf{r}_\ell\right)t\right)\right)\right] \quad \text{for } c \sim 0. \tag{6.23}$$

The latter limit is well recognized [2, 35]; $\exp\left(-k\left(\mathbf{r}_\ell\right)t\right)$ can be interpreted as the pair survival probability of the reactant at $\mathbf{r}_\ell$ from the target site at time $t$ in the static limit. If a long-range reaction is not considered and the reaction is assumed only at a target site at the origin, $k\left(\mathbf{r}_\ell\right) = k_0\delta_{\mathbf{r}_\ell,0}$, Eq. (6.21) reduces to

$$P_s(t) = 1 - c + c\exp(-k_0 t). \tag{6.24}$$

We find $P_s(t) \approx \exp(-k_0 t)$ for $c \sim 1$ and $P_s(t) \approx \exp\left[-c(1 - \exp(-k_0 t))\right]$ for $c \sim 0$. When a reaction occurs only at the target site in the static limit, the target site should initially be occupied by one of the reactants. In the presence of diffusion, the reactant that initially occupies the target site may escape the reaction by diffusing away from the target site. On the other hand, the vacant target site can be filled by a reactant through diffusion. In the upcoming section, we will analyze these contradictory effects.

## 6.4 Target Reaction Under Correlated Diffusion with Site-Blocking Effect

Here, we consider the effect of the random walks of reactants in the presence of site-blocking effects. Reactants can only move to an empty neighboring site, and their movement to an occupied site is blocked. This phenomenon is known as the site-blocking effect. The movement of a reactant is influenced by the configuration of the other reactants. For simplicity, we assume an immobile target and that the target site can be occupied by a reactant, similar to the other sites. We apply Nakazato-Kitahara's theory of tracer diffusion to account for the site-blocking effects on reactant migration [5].

We consider $N$ reactants on $M$ sites. A target is located at the origin of the above lattice. We consider the initial random occupancy of the site given by Eq. (6.11). The survival probability can be expressed by Eq. (6.1).

We follow Nakazato and Kitahara to consider the site-blocking effect of a random walk [5]. The self-diffusion constant is influenced by the correlation among the configuration of the other random walkers and the effect is called correlated random walk.

Following the Nakazato-Kitahara theory, we define the ket vector $|\mathbf{r}, \bullet\rangle$ to indicate the occupation of site $\mathbf{r}$ by a reactant. We also define the ket vector $|\mathbf{r}, \phi\rangle$ to indicate that site $\mathbf{r}$ is empty [5–7]. The probability of finding a configuration $(\mathbf{r}_1, \ldots \mathbf{r}_N)$ at time $t$ can be expressed as

$$P(\mathbf{r}_1, \ldots, \mathbf{r}_N; t) = \left(\prod_{\ell=1}^{N} \langle \mathbf{r}_\ell, \bullet|\right) \left(\prod_{k=N+1}^{M} \langle \mathbf{r}_k, \phi|\right) e^{Ht} \sum_{\{i\}} \frac{1}{\binom{M}{N}} \left(\prod_{\ell=1}^{N} |\mathbf{r}_\ell, \bullet\rangle\right) \left(\prod_{k=N+1}^{M} |\mathbf{r}_k, \phi\rangle\right), \tag{6.25}$$

where the sum indicates the summation over all possible configurations of $N$ occupied sites on the $M$ sites. $H$ corresponds to the Hamiltonian in the second quantization representation and can be divided into two parts, $H = H_{\mathrm{w}} + H_{\mathrm{rc}}$; $H_{\mathrm{w}}$ represents diffusion [5–7],

$$H_{\mathrm{w}} = \sum_{\langle n,m\rangle} \frac{\Gamma}{2d} \left(|\mathbf{r}_n, \bullet\rangle\langle \mathbf{r}_n, \phi| \cdot |\mathbf{r}_m, \phi\rangle\langle \mathbf{r}_m, \bullet| - |\mathbf{r}_n, \bullet\rangle\langle \mathbf{r}_n, \bullet| \cdot |\mathbf{r}_m, \phi\rangle\langle \mathbf{r}_m, \phi|\right), \tag{6.26}$$

where $2d$ is the coordination number of the $d$-dimensional hyper-cubic lattice, $\Gamma$ is the sum of the transition rates from a site to all accessible nearest neighbor sites assuming them being empty, and the sum with $\langle n, m\rangle$ indicates double summation; the sum is over all sites with the subsequent summation over all neighbor sites of each site in the first summation.

We understand that $H_{\mathrm{w}}$ represents diffusion by considering the following transition processes. A transition from the state $|\mathbf{r}_m, \bullet\rangle|\mathbf{r}_n, \phi\rangle$ to the state $|\mathbf{r}_n, \bullet\rangle|\mathbf{r}_m, \phi\rangle$ is possible because the site $n$ is empty and is accessible for a reactant occupying a neighboring site. If the site $n$ is occupied by a reactant, an inverse transition is possible for the state, $|\mathbf{r}_n, \bullet\rangle|\mathbf{r}_m, \phi\rangle$; the occupancy probability of the state $|\mathbf{r}_n, \bullet\rangle|\mathbf{r}_m, \phi\rangle$ decreases by the inverse transition with the overall transition rate denoted by $\Gamma$.

$H_{\mathrm{rc}}$ represents a reaction with the rate constant $k(\mathbf{r}_n)$ from an occupied site $\mathbf{r}_n$ [36–39],

$$H_{\rm rc} = -\sum_{n=1}^{M} k\left(\mathbf{r}_n\right) |\mathbf{r}_n, \bullet\rangle \langle \mathbf{r}_n, \bullet|. \tag{6.27}$$

We introduce the generating function to calculate the survival probability [40], [Eq. (6.1)],

$$G(x,t) = \sum_{N=0}^{M} \sum_{\{\mathbf{r}_i\}} x^N P(\mathbf{r}_1, \mathbf{r}_2, \ldots, \mathbf{r}_N; t). \tag{6.28}$$

Using the generating function, the survival probability can be obtained from [40],

$$P_{\rm s}(t) = \frac{1}{2\pi i} \int dx \frac{1}{x^{N+1}} G(x,t). \tag{6.29}$$

The explicit expression of the generating function is given by [5],

$$G(x,t) = \frac{1}{\binom{M}{N}} \prod_{\ell=1}^{M} \left(\langle \mathbf{r}_\ell, \phi| + \sqrt{x} \langle \mathbf{r}_\ell, \bullet|\right) e^{Ht} \prod_{k=1}^{M} \left(|\mathbf{r}_k, \phi\rangle + \sqrt{x} |\mathbf{r}_k, \bullet\rangle\right). \tag{6.30}$$

Equation (6.12) in the static limit is generalized to Eq. (6.29) with Eq. (6.30) by including the correlated diffusion with site-blocking effect.

We rewrite Eq. (6.30) by introducing the operator

$$S \equiv \sum_{\ell=1}^{M} \left(|\mathbf{r}_\ell, \bullet\rangle \langle \mathbf{r}_\ell, \phi| - |\mathbf{r}_\ell, \phi\rangle \langle \mathbf{r}_\ell, \bullet|\right) \tag{6.31}$$

and use the identity

$$\exp(\theta S) = E \cos\theta + S \sin\theta + 1 - E, \tag{6.32}$$

where $E$ is the identity operator given by

$$E = \sum_{n=1}^{M} \left[|\mathbf{r}_n, \bullet\rangle \langle \mathbf{r}_n, \bullet| + |\mathbf{r}_n, \phi\rangle \langle \mathbf{r}_n, \phi|\right]. \tag{6.33}$$

The identity [Eq. (6.32)] can be confirmed by power series expansion in terms of $\theta$, $\exp(\theta S) = 1 + E(-\theta^2/2! + \theta^4/4! - \cdots) + S(\theta - \theta^3/3! \cdots)$. We choose the rotating angle $\theta$ to satisfy $x = \tan^2\theta$ and find

$$\prod_{k=1}^{M} |\mathbf{r}_k, \phi\rangle + \sqrt{x} |\mathbf{r}_k, \bullet\rangle = \prod_{k=1}^{M} \left(1 + \tan(\theta) |\mathbf{r}_k, \bullet\rangle \langle \mathbf{r}_k, \phi|\right) |\mathbf{r}_k, \phi\rangle. \tag{6.34}$$

By noticing $\sqrt{x} = \tan\theta$, we divide both sides of Eq. (6.32) by $\cos\theta$ and obtain

$$\prod_{k=1}^{M} (1 + \tan(\theta)|\mathbf{r}_k, \bullet\rangle\langle\mathbf{r}_k, \phi|)\, |\mathbf{r}_k, \phi\rangle = \prod_{k=1}^{M} \frac{1}{\cos\theta} \exp(\theta S)|\mathbf{r}_k, \phi\rangle. \tag{6.35}$$

Using $1 + x = 1/\cos^2\theta$, we find $(1/\cos\theta)^{2M} = (1 + x)^M$, which is the same factor in Eq. (6.13). Using Eq. (6.35) and the conjugate of it, Eq. (6.30) can be expressed as

$$G(x, t) = \frac{(1 + x)^M}{\binom{M}{N}} \left(\prod_{\ell=1}^{M} \langle\mathbf{r}_\ell, \phi|\right) \exp(-\theta S) \exp(Ht) \exp(\theta S) \left(\prod_{\ell=1}^{M} |\mathbf{r}_\ell, \phi\rangle\right). \tag{6.36}$$

We apply the saddle point method given by Eq. (6.19) when considering the limit of $M \to \infty$ while keeping the fraction of reactants fixed, $c = N/M$, and find $x^* = c/(1 - c)$ from $f'(x^*) = 0$ using Eq. (6.18); we can define $\tan\theta^* = \sqrt{c/(1 - c)}$ from $x^* = \tan^2\theta^*$. Equation (6.29) with Eq. (6.36) can be evaluated by noticing Eq. (6.20) as

$$P_\mathrm{s}(t) = \left(\prod_{\ell=1}^{M} \langle\mathbf{r}_\ell, \phi|\right) \exp\left(\tilde{H}t\right) \left(\prod_{\ell=1}^{M} |\mathbf{r}_\ell, \phi\rangle\right), \tag{6.37}$$

where we define $\tilde{H} = \exp(-\theta^* S)\, H \exp(\theta^* S)$ by noticing the relation given by $\exp(-\theta^* S) \exp(Ht) \exp(\theta^* S) = \exp\left(\tilde{H}t\right)$; $\tilde{H}$ is obtained using Eq. (6.32) as $\tilde{H} = \tilde{H}_0 + \tilde{H}_1$, where $\tilde{H}_0$ conserves the particle number and is given by

$$\tilde{H}_0 = H_\mathrm{w} - \sum_{n=1}^{M} k(\mathbf{r}_n) [(1 - c)|\mathbf{r}_n, \bullet\rangle\langle\mathbf{r}_n, \bullet| + c|\mathbf{r}_n, \phi\rangle\langle\mathbf{r}_n, \phi|] \tag{6.38}$$

and

$$\tilde{H}_1 = -\sum_{n=1}^{M} k(\mathbf{r}_n) \sqrt{c(1 - c)}\, (|\mathbf{r}_n, \bullet\rangle\langle\mathbf{r}_n, \phi| + |\mathbf{r}_n, \phi\rangle\langle\mathbf{r}_n, \bullet|). \tag{6.39}$$

The time-evolution equation for the survival probability of a target can be obtained using time differentiation of Eq. (6.37) as

$$\frac{\partial}{\partial t} P_\mathrm{s}(t) = -c\kappa P_\mathrm{s}(t) - \sqrt{c(1 - c)} \sum_{j=1}^{M} k(\mathbf{r}_j)\, q(\mathbf{r}_j, t), \tag{6.40}$$

where $\kappa$ denotes the sum of reaction rate constants

$$\kappa = \sum_{j=1}^{M} k\left(\mathbf{r}_j\right). \tag{6.41}$$

In Eq. (6.40), $q\left(\mathbf{r}_j, t\right)$ is defined by

$$q\left(\mathbf{r}_j, t\right) \equiv \left(\prod_{\ell=1}^{M'} \langle \mathbf{r}_\ell, \phi | \langle \mathbf{r}_j, \bullet | \right) \exp\left(\tilde{H}t\right) \left(\prod_{\ell=1}^{M} |\mathbf{r}_\ell, \phi\rangle\right), \tag{6.42}$$

where $M'$ indicates the product of all sites except the site denoted by $\mathbf{r}_j$. The initial condition of Eq. (6.40) is given by $P(0) = 1$. By Laplace transform of Eq. (6.40), we obtain

$$\hat{P}_s(s) = \frac{1}{s + \sum_{j=1}^{M} ck\left(\mathbf{r}_j\right)} \left[1 - \sqrt{c(1-c)} \sum_{j=1}^{M} k\left(\mathbf{r}_j\right) \hat{q}\left(\mathbf{r}_j, s\right)\right]. \tag{6.43}$$

$q\left(\mathbf{r}_j, t\right)$ can be calculated by the perturbation expansion of $\exp\left(\tilde{H}t\right)$ in terms of $\tilde{H}_1$, which is proportional to $\sqrt{c(1-c)}$. By further considering $\sqrt{c(1-c)}$ in Eq. (6.40), the expansion parameter is proportional to $c(1-c)$.

We take into account the lowest order perturbation, which can be expressed as $\hat{q}\left(\mathbf{r}_j, s\right) \approx -\hat{q}_0(1 + c(1-c)\hat{q}_1)$, where $\hat{q}_0$ and $\hat{q}_1$ indicate the non-perturbative and perturbative term, respectively. Using the Páde approximation, $\hat{q}\left(\mathbf{r}_j, s\right) \approx -\hat{q}_0/(1 - c(1-c)\hat{q}_1)$, we obtain

$$\hat{q}\left(\mathbf{r}_j, s\right) = -\frac{\sum_{\ell=1}^{M} \hat{G}\left(\mathbf{r}_j, \mathbf{r}_\ell, s\right) k\left(\mathbf{r}_\ell\right) \sqrt{c(1-c)} \frac{1}{s + c\kappa}}{1 - c(1-c) \sum_{v=1}^{M} \sum_{w=1}^{M} \frac{k\left(\mathbf{r}_v\right) \hat{G}\left(\mathbf{r}_v, \mathbf{r}_w, s\right) k\left(\mathbf{r}_w\right)}{s + c\kappa}}, \tag{6.44}$$

where $\kappa$ is given by Eq. (6.41) and $\hat{G}\left(\mathbf{r}_i, \mathbf{r}_j, s\right)$ indicates the Laplace transform of

$$G\left(\mathbf{r}_i, \mathbf{r}_j, t\right) \equiv \left(\prod_{\ell=1}^{M'} \langle \mathbf{r}_\ell, \phi | \langle \mathbf{r}_i, \bullet | \right) \exp\left(\tilde{H}_0 t\right) \left(\prod_{\ell=1}^{M'} |\mathbf{r}_\ell, \phi\rangle |\mathbf{r}_j, \bullet\rangle\right). \tag{6.45}$$

The time differentiation of Eq. (6.45) yields

$$\frac{\partial}{\partial t} G\left(\mathbf{r}_i, \mathbf{r}_j, t\right) = \mathcal{L} G\left(\mathbf{r}_i, \mathbf{r}_j, t\right) - (1-2c)k\left(\mathbf{r}_i\right) G\left(\mathbf{r}_i, \mathbf{r}_j, t\right) - c\kappa G\left(\mathbf{r}_i, \mathbf{r}_j, t\right), \tag{6.46}$$

where the initial condition is given by $G\left(\mathbf{r}_i, \mathbf{r}_j, 0\right) = \delta_{i,j}$. In Eq. (6.46), $\mathcal{L}$ describes hopping transitions,

$$\mathcal{L}G\left(\mathbf{r}_i, \mathbf{r}_j, t\right) = \sum_{k=1}^{2d} \Gamma/(2d)\left[G\left(\mathbf{b}_k + \mathbf{r}_i, \mathbf{r}_j, t\right) - G\left(\mathbf{r}_i, \mathbf{r}_j, t\right)\right], \tag{6.47}$$

where $\mathbf{b}_k + \mathbf{r}_i$ indicates a nearest neighbor of the site $\mathbf{r}_i$ and the sum over all nearest neighbor sites given by $2d$ should be taken.

By substituting Eq. (6.44), the survival probability of a target [Eq. (6.43)] is expressed in the Laplace domain as

$$\hat{P}_s(s) = \frac{1}{s + c\kappa - c(1-c)\sum_{v=1}^{M}\sum_{w=1}^{M} k\left(\mathbf{r}_v\right)\hat{G}\left(\mathbf{r}_v, \mathbf{r}_w, s\right)k\left(\mathbf{r}_w\right)}. \tag{6.48}$$

The term with $c(1-c)$ originates from the correlated diffusion. We can obtain the survival probability, $P_s(t)$ by inverse Laplace transform of Eq. (6.48), by introducing the solution of Eq. (6.46).

## 6.5 Localized Reaction Sink Term

Here, we consider the simplest case that reaction takes place only at a target site, $k\left(\mathbf{r}_\ell\right) = k_0\delta_{\mathbf{r}_\ell,0}$. For this case, the survival probability in the Laplace domain is given by

$$\hat{P}_s(s) = \frac{1}{s + ck_0 - c(1-c)k_0\hat{G}\left(\mathbf{r}_R, \mathbf{r}_R, s\right)k_0}. \tag{6.49}$$

Equation (6.49) represents the approximate solution in the Laplace domain, which interpolates between the cases of low and high reactant concentrations. By substituting $k\left(\mathbf{r}_\ell\right) = k_0\delta_{\mathbf{r}_\ell,0}$, the solution of Eq. (6.46) in the Laplace domain is obtained as

$$\hat{G}(\mathbf{r}_R, \mathbf{r}_R, s) = \frac{\hat{G}_0(0, 0, s + ck_0)}{1 + (1-2c)\,k_0\hat{G}_0(0, 0, s + ck_0)}, \tag{6.50}$$

where the free Green's function in the Laplace domain denoted by $\hat{G}_0(\mathbf{r}_i, \mathbf{r}_j, s)$ satisfies

$$\frac{\partial G_0\left(\mathbf{r}_i, \mathbf{r}_j, t\right)}{\partial t} = \mathcal{L}G_0\left(\mathbf{r}_i, \mathbf{r}_j, t\right) + \delta_{\mathbf{r}_i,\mathbf{r}_j}\delta(t), \tag{6.51}$$

where $\mathcal{L}G_0\left(\mathbf{r}_i, \mathbf{r}_j, t\right)$ describes the overall gain and loss rate by hopping transitions associated with $\mathbf{r}_i$, which is given by Eq. (6.47). Green's function for free random walks, $\hat{G}_0(0, 0, s)$, is used in Eq. (6.50), where the Laplace variable is replaced from $s$ to $s + ck_0$ owing to the site-blocking effect.

We need to know the expression of Green's function for free random walks in the absence of site-blocking effect and the reaction. In the Laplace domain, the free Green's function can be expressed as

$$\hat{G}_0\left(0, 0, s\right) = \frac{1 - \hat{\psi}(s)}{s} U(s), \tag{6.52}$$

where the target site is allowed to be occupied by a reactant as the other sites. $U(s)$ represents the $d$-dimensional hyper-cubic lattice Green's function in the Laplace domain defined by

$$U(s) = \frac{1}{(2\pi)^d} \int \cdots \int_{-\pi}^{\pi} d^d\mathbf{k} \frac{1}{1 - \hat{\psi}(s)\tilde{\lambda}(\mathbf{k})}, \tag{6.53}$$

where we introduced $\hat{\psi}(s) = \Gamma/(s + \Gamma)$, and the structure factor given by $\tilde{\lambda}(\mathbf{k}) \equiv \frac{1}{2d}\sum_{j=1}^{2d} \cos\left(\mathbf{k} \cdot \mathbf{b}_j/b\right)$. $b$ is the lattice constant. Equation (6.49) can be expressed as [10, 11],

$$\hat{P}_s(s) = \left[s + \left(\frac{1}{ck_0} + \frac{(1-c)}{c}\left(1/\hat{G}_0\left(0, 0, s + ck_0\right) - ck_0\right)^{-1}\right)^{-1}\right]^{-1}. \tag{6.54}$$

The site-blocking effect gives rise to a factor of $c/(1-c)$ in front of the free Green's function. This factor indicates that the rate of diffusional encounter is not proportional to $c$, but is actually enhanced by the factor of $1/(1-c)$.

The analytical expressions of the Lattice Green's function are known for hyper-cubic lattices including the simple cubic lattice [41]. As an example, we consider the lattice Green's function for the simple cubic lattice. The lattice Green's function for the simple cubic lattice is shown in Eqs. (4) and (5) of Ref. [11]. As a reference, we show the lattice Green's function for the 1-dimensional lattice given by [42] $U(s) = 1/\sqrt{1 - \hat{\psi}(s)^2}$. We also studied some other lattices [10–12]. The survival probability is obtained from Eq. (6.54) by introducing these lattice Green's functions in Eq. (6.52).

It turned out that Eq. (6.54) provides a lower bound for the survival probability, which closely matches the exact numerical results in two and three dimensions [10, 11]. In the limit of $k_0 \to \infty$ (diffusion-controlled limit) Eq. (6.54) reduces to exponential decay

$$P_s(t) = (1 - c)\exp\left(-c\Gamma t\right) \tag{6.55}$$

for any $d$ in the dilute limit. In the case of $d = 1$ (1 dimension) in the dilute limit, the survival probability is known to exhibit a non-exponential decay if the model does not take into account the effect of site-blocking. However, this particular case is not replicated using Eq. (6.55). The reason behind the discrepancy is not clear, but it could be related to the fact that diffusion under site-blocking effect should occur within a finite interval in one dimension. If we consider the target reaction occurring under one-dimensional diffusion within a finite interval, the survival probability of the target decays exponentially. In addition to the thermodynamic limit, we used the Páde approximation. A more rigorous theory is required to study the dilute limit of the survival probability of the target in one dimension.

## 6.6 Comparison to Numerical Simulation

In the analytical calculation, we utilized the steepest descent method and the Páde approximation. We compare analytical results with those obtained using a kinetic Monte Carlo method. The simulations are performed for the simple cubic lattice ($d = 3$). A reactant is able to move to an empty nearest neighbor site with the transition rate constant $\Gamma/(2d)$. The target is located at the origin of the lattice. The reaction occurs with the rate constant $k_0$ when one of the reactants occupies the target site. The initial condition is that $M$ lattice sites including the target site are occupied by $N$ reactants according to the even probability. We used the variable step size method (VSSM) [49]; the event is selected in proportion to the probability that the selected event will occur against all possible events of all reactants. Ensemble averaging is performed using 100,000 independent runs.

In Fig. 6.2, the survival probability is shown against time for the simple cubic lattice for $c = 0.5$. Theoretical results [Eq. (6.54)] are in agreement with the simulation results. We compare the result in the diffusion-controlled limit ($\Gamma/k_0 = 0.01$) and $\Gamma/k_0 = 1$. The initial decay is insensitive to the values of $\Gamma/k_0$ and proceeds mainly without movement of reactants; the initial decay follows the result of the static limit, Eq. (6.24) [11]. The long-time decay depends on the values of $\Gamma/k_0$, where the site-blocking effect is larger by decreasing $\Gamma/k_0$. When $\Gamma/k_0 = 0.01$, the survival probability of the target exhibits a two-step decay. The initial decay follows the result of the static limit, Eq. (6.24); the long-time decay follows the result of the hopping-controlled limit, Eq. (6.55) [11]. We can define an interpolating function of Eqs. (6.24) and (6.55) as

$$P_s(t) \approx (1-c)\exp(-c\Gamma t) + c\exp(-k_0 t). \tag{6.56}$$

The results of Eq. (6.56) are shown by red dots in Fig. 6.2; a two-step decay is well reproduced. The same type of equation has been applied to analyze quenching kinetics of excited probes in zeolites [12].

Previously, the effect of concentrations was investigated; the results of the site-blocking effect taken into account by Eq. (6.54) reproduce the numerical results for

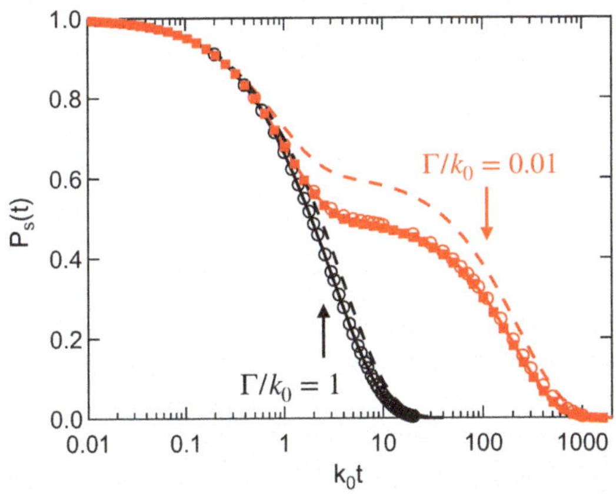


**Fig. 6.2** The survival probability is shown against normalized time, $k_0t$ for the simple cubic lattice for $c = 0.5$. Red lines indicate $\Gamma/k_0 = 0.01$ and black lines indicate $\Gamma/k_0 = 1$. The solid lines are obtained using a numerical Laplace inversion of Eq. (6.54), where the site-blocking effect is considered. The circles represent the simulation results. The dashed lines represent the conventional expression without site-blocking effect, Eqs. (6.9) with (6.10). The red dots indicate the results of Eq. (6.56) with $\Gamma/k_0 = 0.01$. The original data are taken from Ref. [11]. The initial decay is insensitive to the values of $\Gamma/k_0$ and follows the decay without movement of reactants (static limit). The long-time components of decay depend strongly on the values of $\Gamma/k_0$, where the decay is diffusion controlled. The decay is accelerated by prohibiting double occupancy in the diffusion-controlled limit

all concentrations, whereas the results of the conventional expression [Eqs. (6.9) with (6.10)] are close to the simulation results only in the dilute limit [11].

In all cases, the survival probability decreases more rapidly when site-blocking is in effect compared to when it is not [50, 51]. The results are consistent with those of other models on one- and two-dimensional lattices [45–47, 52]. The reason for the accelerated decay of the target survival probability is as follows [10–12], see Fig. 6.1. The initially placed $N$ reactants change their configuration over time until one of the reactants reacts at the target site. When the site-blocking effect is absent, multiple reactants are allowed to occupy the same site simultaneously. If double occupancy is allowed, the number of sites occupied by reactants is less than $N$, while the number of sites occupied by reactants is always $N$ under the site-blocking effect before the target reaction occurs. In general, the number of sites occupied by reactants (searchers of the target) is higher under the site-blocking effects, as shown in Fig. 6.1. Prohibiting double occupancy at any given time increases the probability that the occupied sites of reactants include the target site. A similar effect has also been found by other approaches [53–55].

The acceleration of the decay of the survival probability under site-blocking effect can be characterized by the reaction time distribution given by

$$f_{\mathrm{rt}}(t) = -\frac{\partial}{\partial t} P_{\mathrm{s}}(t). \tag{6.57}$$

The reaction time distribution for the perfectly absorbing boundary condition can be interpreted as the first passage time distribution to the absorbing boundary. The mean reaction time can be defined as

$$T_{\mathrm{mrt}} = \int_0^\infty dt t f_{\mathrm{rt}}(t) \tag{6.58}$$

$$= \int_0^\infty dt P_{\mathrm{s}}(t), \tag{6.59}$$

where Eq. (6.57) and integration by parts are used to obtain the last equation.

In Fig. 6.3, we show the reaction time distribution for $\Gamma/k_0 = 0.01$ and $\Gamma/k_0 = 1$. In both cases, the reaction time distribution increases at short times and decreases at long times by imposing site-blocking effect. Highly non-exponential decay of the reaction time distribution is obtained in the diffusion-controlled limit of $\Gamma/k_0 = 0.01$, whereas a simple decay close to an exponential decay is obtained for $\Gamma/k_0 = 1$. Correspondingly, the mean reaction time for $\Gamma/k_0 = 0.01$ is too large to characterize the main part of the reaction time distribution, while the mean reaction time for $\Gamma/k_0 = 1$ characterizes the reaction time distribution.

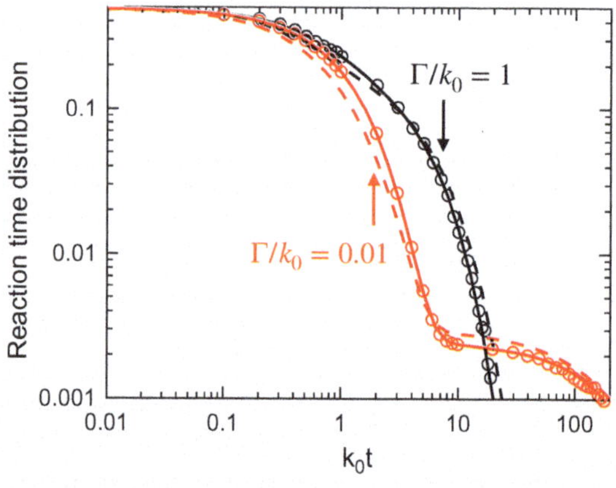


**Fig. 6.3** The reaction time distribution is shown against normalized time, $k_0 t$ for the simple cubic lattice for $c = 0.5$. Red lines and symbols indicate $\Gamma/k_0 = 0.01$ and black lines and symbols indicate $\Gamma/k_0 = 1$. The solid lines are obtained using a numerical Laplace inversion of $1 - s\hat{P}_{\mathrm{s}}(s)$, where $\hat{P}_{\mathrm{s}}(s)$ is obtained from Eq. (6.54) derived under the site-blocking effect. The circles represent $-\partial P_{\mathrm{s}}(t)/(\partial t)$, where $P_{\mathrm{s}}(t)$ is obtained by simulations. The dashed lines represent the conventional expression without site-blocking effect obtained by $ck_{\mathrm{cv}}(t)P_{\mathrm{s}}(t)$, where Eqs. (6.9) and (6.10) are used. The mean reaction time is 102.3 (with site-blocking effect) and 143.6 (without site-blocking effect) for $\Gamma/k_0 = 0.01$. The mean reaction time is 3.15 (with site-blocking effect) and 3.87 (without site-blocking effect) for $\Gamma/k_0 = 1.0$. Because of the highly non-exponential decay of the survival probability, the mean reaction time for $\Gamma/k_0 = 0.01$ fails to characterize the main part of the reaction time distribution. By imposing site-blocking effect, the reaction time distribution increases at short times and decreases at long times

Note that the decay of the survival probability is accelerated by a site-blocking effect for the target reaction, in which a target reacts with one of the many reactants surrounding it. For the site-blocking effect on a separated geminate pair, the decrease in relative distance between a pair for the reaction to occur can be delayed in the presence of inert particles due to the hindrance effect [56, 57]. Strictly speaking, the hindrance effect was shown for a pair of a mobile reactant and an immobile reactant. When both moves, numerical calculations were performed without taking into account the site-blocking effect [58], where the hindrance effect was absent. A less explored but interesting area of research for studying the effect of molecular crowding on geminate reactions might be the Adam-Delbrück reduction-of-dimensionality scenario [59, 61–63]. This scenario involves the transition from bulk diffusion to surface diffusion. Another interesting area to explore is facilitated diffusion; a one-dimensional search is facilitated by a three-dimensional search in the bulk [60], where some of the reactants diffusing along an entangled polymer may have a shorter real distance to the target in three dimensions compared to the distance along the polymer. In both cases, the impact of molecular crowding on three-dimensional bulk diffusion and diffusion in reduced dimensionality could vary, potentially affecting the time required to search for the target. We have not explored the target reaction under the presence of inert particles which cause a hindrance effect to the diffusion of reactants as the concentration of inert particles increases [18, 21]. The Nakazato-Kitahara theory was originally developed to study correlated diffusion, where the reaction is absent. Focused on transport without taking into account reactions, the passive transport of molecules in crowded environments has been studied [64]. The Nakazato-Kitahara theory is potentially applicable to such problems.

## 6.7 Conclusions

We have reviewed a detailed derivation of analytical results on the target reaction problem under site-blocking effects among reactants. Reactants migrate on any lattice sites until a reaction takes place. For simplicity, we consider the case where a reaction occurs when a reactant occupies the target site. Once the reaction occurs, the target becomes inert. The method can be generalized for the case that the reaction rate constant depends on the distance between the reactant and the target.

In conventional theory, reactants are assumed to move independently of the configurations of other reactants. At high reactant concentrations, reactant migration is affected by site-blocking caused by the presence of other reactants, specifically their configurations. In the absence of a reaction, the migration is termed as correlated diffusion. Because of the correlation among reactants, the analytical method for calculating the survival probability of the target is technically complicated. Here, we focus on the simplest model and provide a detailed derivation, which was given concisely in the original Nakazato-Kitahara theory of correlated diffusion.

Although technically difficult, the reason for the accelerated decay of the target survival probability is straightforward. In the presence of site-blocking effects, $N$

reactants always occupy $N$ lattice sites until a reaction occurs at the origin, where the target is located. However, if double occupancy is allowed, $N$ reactants can occupy at most $N$ lattice sites. Therefore, the search for the target by reactants is less efficient in the absence of a site-blocking effect. The conventional theory on the target reaction assumes that reactants move independently, allowing for double occupancy implicitly. However, the number of sites occupied by reactants is reduced by allowing double occupancy at any given time. The target survival probability should be higher in the conventional theory, as it implicitly ignores site-blocking effects.

**Acknowledgements** The author would like to thank Professor M. Tachiya and Professor M. Wojcik for their collaboration on the original works.

## References

1. S.A. Rice, Diffusion-limited reactions, in *Comprehensive Chemical Kinetics*, ed. by C.H. Bamford, C.F.H. Tipper, R.G. Compton, vol. 25 (Elsevier, Amsterdam, 1985) and references cited therein
2. M. Tachiya, A. Mozumder, Decay of trapped electrons by tunnelling to scavenger molecules in low-temperature glasses. Chem. Phys. Lett. **28**, 87 (1974)
3. M. Tachiya, Theory of diffusion-controlled reactions: Formulation of the bulk reaction rate in terms of the pair probability. Radiat. Phys. Chem. **21**, 167 (1983)
4. A. Szabo, Theory of diffusion-influenced fluorescence quenching. J. Phys. Chem. **93**, 6929 (1989)
5. K. Nakazato, K. Kitahara, Site blocking effect in tracer diffusion on a lattice. Prog. Theor. Phys. **64**, 2261 (1980)
6. Y. Suzuki, K. Kitahara, Y. Fujitani, S. Kinouchi, Vacancy-assisted diffusion in a honeycomb lattice and in a diamond lattice. J. Phys. Soc. Jpn. **71**, 2936 (2002)
7. R. Okamoto, Y. Fujitani, Group-theoretical calculation of the diffusion coefficient via the vacancy-assisted mechanism. J. Phys. Soc. Jpn. **74**, 2510 (2005)
8. O. Bénichou, G. Oshanin, Ultraslow vacancy-mediated tracer diffusion in two dimensions: The Einstein relation verified. Phys. Rev. E **66**, 031101 (2002)
9. H. van Beijeren, R. Kutner, Mean square displacement of a tracer particle in a hard-core lattice gas. Phys. Rev. Lett. **55**, 238 (1985)
10. K. Seki, M. Tachiya, Reaction under vacancy-assisted diffusion at high quencher concentration Phys. Rev. E **80**, 041120 (2009)
11. K. Seki, M. Wojcik, M. Tachiya, Effects of excluded volume interaction and dimensionality on diffusion-mediated reactions. J. Chem. Phys. **134**, 094506 (2011)
12. K. Seki, A. Ballal, M. Tachiya, Site blocking effect on diffusion-mediated reactions in porous media. J. Phys. Chem. C **116**, 22086 (2012)
13. R.J. Ellis, A.P. Minton, Join the crowd. Nature **425**, 27 (2003)
14. S. Schnell, T.E. Turner, Reaction kinetics in intracellular environments with macromolecular crowding: simulations and rate laws. Prog. Biophys. Mol. Biol. **85**, 235 (2004)
15. A.P. Minton, The influence of macromolecular crowding and macromolecular confinement on biochemical reactions in physiological media. J. Biol. Chem. **276**, 10577 (2001)
16. G. Rivas, A.P. Minton, Macromolecular crowding in vitro, in vivo, and in between. Trends Biochem. Sci. **41**, 970 (2016)
17. S. Nakano, D. Miyoshi, N. Sugimoto, Effects of molecular crowding on the structures, interactions, and functions of nucleic acids. Chem. Rev. **114**, 2733 (2014)

18. C. Echevería, K. Tucci, R. Kapral, Diffusion and reaction in crowded environments. J. Phys. Condens. Matter **19**, 065146 (2007)
19. S.B. Zimmerman, S.O. Trach, Estimation of macromolecule concentrations and excluded volume effects for the cytoplasm of Escherichia coli. J. Mol. Biol. **222**, 599 (1991)
20. H.-X. Zhou, G. Rivas, A.P. Minton, Macromolecular crowding and confinement: biochemical, biophysical, and potential physiological consequences. Annu. Rev. Biophys. **37**, 375 (2008)
21. J.S. Kim, A. Yethiraj, Crowding effects on association reactions at membranes. Biophys. J. **96**, 1333 (2009)
22. J.S. Schreck, J. Bridstrup, J.-M. Yuan, Investigating the effects of molecular crowding on the kinetics of protein aggregation. J. Phys. Chem. B **124**, 9829 (2020)
23. D. Gnutt, M. Gao, O. Brylski, M. Heyden, S. Ebbinghaus, Excluded-volume effects in living cells. Angew. Chem. Int. Ed. **54**, 2548 (2015)
24. S. DasGupta, Molecular crowding and RNA catalysis. Org. Biomol. Chem. **18**, 7724 (2020)
25. D. Tsiapalis, D.I. Zeugolis, It is time to crowd your cell culture media - Physicochemical considerations with biological consequences. Biomater. **275**, 120943 (2021)
26. O. Bénichou, G. Oshanin, Atomic slide puzzle: Self-diffusion of an impure atom. Phys. Rev. E **64**, 020103 (2001)
27. R. van Gastel, E. Somfai, W. van Saarloos, J.W.M. Frenken, Vacancy mediated surface diffusion. Nature (London) **408**, 665 (2000)
28. A.V. Barzykin, S. Hashimoto, Reaction kinetics in zeolites as a random walk problem. J. Chem. Phys. **113**, 2841 (2000)
29. M. Inokuti, F. Hirayama, Influence of Energy transfer by the exchange mechanism on donor luminescence. J. Chem. Phys. **43**, 1978 (1965)
30. H. Scher, E.W. Montroll, Anomalous transit-time dispersion in amorphous solids. Phys. Rev. B **12**, 2455 (1975)
31. A. Szabo, G. Lamm, G.H. Weiss, Localized partial traps in diffusion processes and random walks. J. Stat. Phys. **34**, 225 (1984)
32. S.F. Swallen, K. Weidemaier, H.L. Tavernier, M.D. Fayer, Experimental and theoretical analysis of photoinduced electron transfer: including the role of liquid structure. J. Phys. Chem. **100**, 8106 (1996)
33. A. Rosspeintner, D.R. Kattnig, G. Angulo, S. Landgraf, G. Grampp, A. Cuetos, On the coherent description of diffusion influenced fluorescence quenching experiments. Chem. Eur. J. **13**, 6474 (2007)
34. G.B. Arfken, H.J. Weber, *Mathematical Methods for Physicists*, 4th edn. (Academic Press, San Diego, 1995)
35. K. Allinger, A. Blumen, On the direct energy transfer to moving acceptors. J. Chem. Phys. **72**, 4608 (1980)
36. M. Doi, Stochastic theory of diffusion-controlled reaction. J. Phys. A **9**, 1479 (1976)
37. V. Kuzovkov, E. Kotomin, Kinetics of bimolecular reactions in condensed media. Rep. Prog. Phys. **51**, 1479 (1988)
38. L. Peliti, Path integral approach to birth-death processes on a lattice. J. Physique **46**, 1469 (1985)
39. V.N. Kuzovkov, E.A. Kotomin, Theory of diffusion-controlled colloid formation in irradiated solids. J. Chem. Phys. **98**, 9107 (1993)
40. N.G. Van Kampen, *Stochastic Processes in Physics and Chemistry*, 2nd edn. (North-Holland, Amsterdam, 1992)
41. G.S. Joyce, I.J. Zucker, on the evaluation of generalized Watson integrals. Proc. Am. Math. Soc. **133**, 71 (2004)
42. B.D. Hughes, *Random Walks and Random Environments*, vol. 1 (Clarendon Press, Oxford, 1995) and references cited therein
43. A. Szabo, R. Zwanzig, N. Agmon, Diffusion-controlled reactions with mobile traps. Phys. Rev. Lett. **61**, 2496 (1988)
44. M. Bramson, J.L. Lebowitz, Asymptotic behavior of densities in diffusion-dominated annihilation reactions. Phys. Rev. Lett. **61**, 2397 (1988)

45. D. Arora, D.P. Bhatia, M.A. Prasad, Survival probability in one dimension for the A+B → B reaction with hard-core repulsion. J. Stat. Phys. **84**, 697 (1996)
46. S.F. Burlatsky, M. Moreau, G. Oshanin, A. Blumen, Comment on pair and triple correlations in the A + B → B diffusion-controlled reaction. Phys. Rev. Lett. **75**, 585 (1995)
47. D.P. Bhatia, M.A. Prasad, D. Arora, Comment on pair and triple correlations in the A + B → B diffusion-controlled reaction. Phys. Rev. Lett. **75**, 586 (1995)
48. I.M. Sokolov, R. Metzler, K. Pant, M.C. Williams, First passage time of *N* excluded-volume particles on a line. Phys. Rev. E **72**, 041102 (2005)
49. R.J. Gelten, A.P.J. Jansen, R.A. van Santen, J.J. Lukkien, J.P.L. Segers, P.A.J. Hilbers, Monte Carlo simulations of a surface reaction model showing spatio-temporal pattern formations and oscillations. J. Chem. Phys. **108**, 5921 (1998)
50. J. Lee, J. Sung, S. Lee, Excluded volume effects on the diffusion-influenced reaction: the many particle kernel approach. J. Chem. Phys. **113**, 8686 (2000)
51. J. Park, H. Kim, K.J. Shin, Excluded volume effect on diffusion-influenced reactions in one dimension. J. Chem. Phys. **118**, 9697 (2003)
52. G. Zumofen, A. Blumen, J. Klafter, Excluded volume effects in annihilation reactions. Chem. Phys. Lett. **117**, 340 (1985)
53. J. Dzubiella, J.A. McCammon, Substrate concentration dependence of the diffusion-controlled steady-state rate constant. J. Chem. Phys. **122**, 184902 (2005)
54. N. Dorsaz, C. De Michele, F. Piazza, P. De Los Rios, G. Foffi, Diffusion-limited reactions in crowded environments. Phys. Rev. Lett. **105**, 120601 (2010)
55. H.X. Zhou, Speeding up in a crowd. Physics **3**, 77 (2010)
56. J.D. Schmit, E. Kamber, J. Kondev, Lattice model of diffusion-limited bimolecular chemical reactions in confined environments. Phys. Rev. Lett. **102**, 218302 (2009)
57. K. Seki, M. Wojcik, M. Tachiya, Diffusion-mediated geminate reactions under excluded volume interactions. Phys. Rev. E **85**, 011131 (2012)
58. F. Le Vot, S.B. Yuste, D.S. Grebenkov, First-encounter time of two diffusing particles in two- and three-dimensional confinement. Phys. Rev. E **105**, 044119 (2022)
59. G. Adam, M. Delbrück, Reduction of dimensionality in biological diffusion processes, in *Structural Chemistry and Molecular Biology*, ed. by A. Rich, N. Davidson (W H Freeman and Co., San Francisco, 1968)
60. H.C. Berg, E.M. Purcell, Physics of chemoreception. Biophys. J. **20**, 193 (1977)
61. D.S. Grebenkov, R. Metzler, G. Oshanin, Search efficiency in the Adam-Delbrück reduction-of-dimensionality scenario versus direct diffusive search. New J. Phys. **24**, 083035 (2022)
62. D. Krepel, D. Gomez, S. Klumpp, Y. Levy, Mechanism of facilitated diffusion during a DNA search in crowded environments. J. Phys. Chem. B **120**, 11113 (2016)
63. D. Krepel, Y. Levy, Protein diffusion along DNA: on the effect of roadblocks and crowders. J. Phys. A Math. Theor. **49**, 494003 (2016)
64. S.K. Ghosh, A.G. Cherstvy, D.S. Grebenkov, R. Metzler, Anomalous, non-Gaussian tracer diffusion in crowded two-dimensional environments. New J. Phys. **18**, 013027 (2016)

# Chapter 7
# Search and Escape of Mortal Random Walkers

**Enrique Abad and Santos Bravo Yuste**

**Abstract** We review some representative results for first-passage problems involving so-called mortal or evanescent walkers, i.e., walkers with a finite lifetime. The mortality constraint plays a key role in the modeling of many real scenarios, as it filters out long Brownian trajectories, thereby drastically modifying space exploration properties. Among such scenarios, we consider here different first-passage problems, including one or many searchers, resetting, anomalous diffusion, evolving domains, and the narrow escape problem. In spite of the different physics, the mathematical treatment draws strongly on the formalism for standard (i.e., immortal) walkers.

## 7.1 Introduction

Mortal or evanescent walkers "die" at a certain rate as they move. In the context of search and escape problems, the death process should be understood as any constraint leading to a premature end of a diffusive trajectory, or as a change of identity which prevents the walker to find a target or to exit a domain. Beyond the not-so-frequent cases of literal death (e.g., a prey hunted by a predator), other constraints that can be regarded as death processes in a broad sense include degradation, inactivation, premature detachment from a substrate, decay into an inert species, or absorption/trapping of any kind. In spite of the ubiquity of such mechanisms and of their drastic effects, they are not always incorporated in random walk models, or not always in the correct way. For example, in models of anomalous diffusion, mortality typically leads to a transport term that includes the mortality rate (see, e.g., Eq. (7.28) below); this peculiarity has been sometimes overlooked, as it is at first glance counterintuitive.

E. Abad (✉)
Universidad de Extremadura; Departamento de Física Aplicada and Instituto de Computación Científica Avanzada (ICCAEx), Mérida, Spain
e-mail: eabad@unex.es

S. B. Yuste
Universidad de Extremadura, Departamento de Física and Instituto de Computación Científica Avanzada (ICCAEx), Badajoz, Spain
e-mail: santos@unex.es

 
D. Grebenkov et al. (eds.), *Target Search Problems*,
https://doi.org/10.1007/978-3-031-67802-8_7

An example of mortal walks often invoked in the literature concerns the diffusion of radioactive particles, which decay according to Rutherford-Soddy's exponential law [1, 2]. Other phenomena which can be modeled as diffusion-decay processes include motility of spermatozoa [3], surface diffusion on catalytic substrates [4], scavenging reactions [5], photoluminescence decay due to electron-hole recombination [6], diffusion of defects [7], mRNA translation [8], light absorption in tissues [9], morphogen gradient formation [10, 11], transport and detachment of molecular motors from polymer tracks [12], virus trafficking subject to killing [13], intracellular signaling [14], and channel transport of degrading information molecules [15], to name but a few. Even the random walk corresponding to the classical problem of the gambler's ruin can be modified by allowing for a finite probability to quit the game before each bet [4], resulting in a strong shortening of the game duration.

A fast mortality results in a drastic decrease of long trajectories; consequently, first-passage properties conditioned on survival may be very different from their counterparts for immortal walkers. As pointed out by Lawley [16], "if inactivation is fast, then the conditional first-passage time (FPT) compared to the FPT without inactivation is (i) much faster, (ii) much less affected by spatial heterogeneity, and (iii) much less variable". Thus, mortality selects the most direct trajectories to a target and shortens the measured search time at the expense of having a poorer long-time statistics; in this sense, it is different from other mechanisms such as, e.g., chemotaxis or the Adam-Delbrück reduction-of-dimensionality scenario [17, 18].

On a more technical note, we mention that death processes can also be used to "heal" some pathologies of certain equations, e.g., the lack of robustness of subdiffusive fractional equations as far as their stationary distribution is concerned [19]. There are also some surprising connections between the mortal walk formalism and averaging dynamics on graphs [20], as well as between lattice Green functions of immortal walkers [21] and space exploration properties of mortal ones [7, 21, 22].

Our aim is to give a short overview of a few representative problems of interest. For reasons of space, it is unfortunately not possible to describe (even minimally) all the problems with mortal walkers studied so far by different authors (and by ourselves). Moreover, the description of those presented here necessarily remains schematic. A previous book chapter [22] coauthored by two of us also deals with mortal random walkers, although mainly on discrete supports and for a different set of problems. For the sake of continuity with Ref. [22], we begin our presentation with the discrete case, but then focus exclusively on continuum diffusion.

## 7.2 Basic Properties of Mortal Brownian Motion

### *7.2.1 Probability to Be at a Certain Location*

We consider a mortal walker performing a discrete-time Markovian random walk consisting of equiprobable jumps to nearest-neighbor sites on a connected graph.

This walker has a probability $q$ of surviving at each time step; thus, its probability of demise at each time step is $1 - q$. Standard ("immortal") random walks are recovered for $q = 1$.

We will devote the remainder to the continuum or diffusion limit of a mortal random walk, as this has been extensively studied in the literature. The lack of explicit spatial structure (apart from dimensionality and boundary conditions) makes this limit generally easier to analyze. We note, however, that discrete geometries are important to mimic the details of real systems, and a number of works address this case by making use of lattice Green functions [4, 7, 23–27], as well as of recursive relations in the case of hierarchical lattices [17] and fractal supports [17, 28].

For concreteness, we focus on the case of an infinite one-dimensional lattice and label its sites with integers. We respectively denote by $a$ and $\tau$ the lattice constant and time step. The probability $p_j(n)$ that the walker is at site $j$ at time $n\tau$ reads

$$p_j(n+1) = \tfrac{1}{2}q\,[p_{j-1}(n) + p_{j+1}(n)]. \tag{7.1}$$

We build the continuum limit of this difference equation in the usual way, i.e., by letting $a \to 0$ and $\tau \to 0$. Additionally we let $q \to 1$ in such a way that [17]

$$\lim qa^2/(2\tau) = D, \qquad \lim (1-q)/\tau = \mu, \tag{7.2}$$

where $D$ stands for the walker's diffusivity and $\mu$ for the mortality rate. Replacing $ja$ by $x$ and $n\tau$ by $t$ in Eq. (7.1), and retaining the symbol $p(x,t)$ for the positional red probability density function (pdf), we obtain the following diffusion-decay equation:

$$\frac{\partial p}{\partial t} = D\,\frac{\partial^2 p}{\partial x^2} - \mu\, p. \tag{7.3}$$

The solution $p(x,t|\,x_0)$ of Eq. (7.3) satisfying the initial condition $p(x,0) = \delta(x - x_0)$ reads

$$p(x,t|\,x_0) = e^{-\mu t}\, p_I(x,t|\,x_0), \tag{7.4}$$

where

$$p_I(x,t|\,x_0) = (4\pi Dt)^{-1/2}\, e^{-(x-x_0)^2/(4Dt)} \tag{7.5}$$

denotes the propagator for immortal Brownian motion. Equation (7.4) is just the standard Gaussian solution with the extra exponentially decaying factor $e^{-\mu t}$. A similar decomposition holds true in $d$ dimensions, so the corresponding $d$-dimensional solution $p(\mathbf{x},t|\,\mathbf{x}_0)$ is given by (7.4) and (7.5) replacing $(4\pi Dt)^{-1/2}$ by $(4\pi Dt)^{-d/2}$ and $x - x_0$ by $|\mathbf{x} - \mathbf{x}_0|$. Note that now $p(\mathbf{x},t|\,\mathbf{x}_0)$ is no longer normalized to unity, as $\int d\mathbf{x}\, p(\mathbf{x},t|\,\mathbf{x}_0) = e^{-\mu t}$.

### 7.2.2 Probability to Die at a Certain Location

The probability distribution $p_\mu(x,t|\,x_0)$ for a mortal Brownian particle starting at position $x_0$ on the real line to die inside an interval $(x, x+dx)$ between the times $t$ and $t+dt$ is easily obtained by exploiting the statistical independence of the Brownian motion and the death process. Thus, $p_\mu(x,t|\,x_0) = p_\mu(t)p_I(x,t|\,x_0)$, where $p_\mu(t) = \mu e^{-\mu t}$ is the death time distribution. The probability to die at any time at a given location is then obtained by integration [29]:

$$p_\mu(x|\,x_0) = \int_0^\infty p_\mu(x,t|\,x_0)\,dt = \sqrt{\frac{\mu}{4D}}e^{-\sqrt{\mu/D}|x-x_0|}. \tag{7.6}$$

This result can be generalized to higher Euclidean dimensions. We quote the two- and three-dimensional results [29]:

$$p_\mu(\mathbf{x}|\,\mathbf{x}_0) = \begin{cases} \dfrac{\mu}{2\pi D}K_0\left(\sqrt{\dfrac{\mu}{D}}|\mathbf{x}-\mathbf{x}_0|\right), & d=2, \quad (7.7)\\[2ex] \dfrac{\mu}{4\pi D|\mathbf{x}-\mathbf{x}_0|}e^{-\sqrt{\frac{\mu}{D}}|\mathbf{x}-\mathbf{x}_0|}, & d=3, \quad (7.8)\end{cases}$$

where $K_0$ is the zeroth-order modified Bessel function of the second kind. These results differ strongly from the Gaussian profile of the propagator for an immortal walker. As noted by Meerson [29], the expressions given by Eqs. (7.7) and (7.8) both diverge at $\mathbf{x}_0$ (but are of course integrable), reflecting the strong localization about the starting point induced by mortality.

## 7.3 First-Passage Properties of a Single Searcher

### 7.3.1 Survival Probability, Conditional First-Passage Time, and Territory Explored

Consider a mortal walker (searcher) diffusing from the starting point $x_0$ to an arbitrary target point $x$ on the real line (we take $x < x_0$ without loss of generality). In the remainder, we will occasionally refer to this setting as the one-sided target problem (OSTP). The corresponding FPT density $f(t, x|x_0)$ follows from the Markovian renewal equation

$$p(x_1, t\,|\,x_0) = \int_0^t dt'\, f(t', x\,|\,x_0)\, p(x_1, t-t'\,|\,x), \tag{7.9}$$

where $x_1 < x < x_0$. With the help of Laplace transforms, we find

$$f(t, x \,|\, x_0) = \frac{e^{-\mu t}}{(4\pi D t^3)^{1/2}} (x_0 - x)\, e^{-(x_0-x)^2/(4Dt)}. \tag{7.10}$$

The attenuation factor $e^{-\mu t}$ forces the first moment of $f(t, x \,|\, x_0)$ to become finite, whence the non-recurrence of this one-dimensional mortal walk follows. Consequently, first-passage properties are strongly modified, notably the FPT density and the volume explored by the walker (see below). For completeness, we note that Eq. (7.10) also follows from the method of images upon use of Eq. (7.4) (see, e.g., Eqs. (1) and (2) of Ref. [3]). The probability $F(t, x \,|\, x_0) = \int_0^t dt' f(t', x \,|\, x_0)$ that the searcher has reached point $x$ by time $t$ is [3, 30]

$$\begin{aligned} F(t, x \,|\, x_0) &= \tfrac{1}{2}\, e^{\sqrt{\frac{\mu}{D}}(x_0-x)} \mathrm{erfc}\big([(x_0 - x)/\sqrt{4Dt}] + \sqrt{\mu t}\big) \\ &\quad + \tfrac{1}{2}\, e^{-\sqrt{\frac{\mu}{D}}(x_0-x)} \mathrm{erfc}\big([(x_0 - x)/\sqrt{4Dt}] - \sqrt{\mu t}\big) \end{aligned} \tag{7.11}$$

with $\mathrm{erfc}(z) = 1 - \mathrm{erf}(z)$, where $\mathrm{erf}(z) = (2/\sqrt{\pi}) \int_0^z e^{-u^2} du$ is the error function. The probability that a searcher starting from $x_0$ ever reaches $x$ is

$$\mathcal{F}(x \,|\, x_0) = \int_0^\infty dt\; f(t, x \,|\, x_0) = \exp\left[-(\mu/D)^{1/2} (x_0 - x)\right], \tag{7.12}$$

which exhibits exponential attenuation with increasing distance $x_0 - x$. Correspondingly, the (conditional) mean first-passage time (MFPT) reads

$$\langle T(x \,|\, x_0)\rangle = \frac{\int_0^\infty dt\, t\, f(t, x \,|\, x_0)}{\int_0^\infty dt\, f(t, x \,|\, x_0)} = \frac{x_0 - x}{\sqrt{4D\mu}}. \tag{7.13}$$

As in ballistic motion, the mean travel time is thus proportional to the covered distance. Here, the effective speed is $(4D\mu)^{1/2}$. However, one should bear in mind that the conditional MFPT is an average over a finite fraction of trajectories only, namely, those involving a first passage from $x_0$ to $x$. This fraction decreases exponentially with increasing distance $x_0 - x$, implying a loss of statistical accuracy in a hypothetical experimental measurement of the MFPT.

The calculation of the variance of the FPT is also straightforward. We skip the details and give directly the final result (see, e.g., Eq. (5b) of Ref. [3]):

$$\sigma_T^2 \equiv \langle T^2(x \,|\, x_0)\rangle - \langle T(x \,|\, x_0)\rangle^2 = \frac{2D(x_0 - x)}{(4D\mu)^{3/2}}. \tag{7.14}$$

The non-recurrence of mortal Brownian motion reduces strongly the explored territory; the latter would be infinite for immortal particles, but turns out to be finite for mortal ones. Formally, the explored volume in the continuum is known as "the Wiener sausage". The characterization of its statistical properties is key for the understanding of diffusion-controlled kinetics and of the so-called target problem [7] in particular. The average volume $v_t$ of the Wiener sausage generated by a mortal

Brownian particle of radius $R$ up to time $t$ has been recently determined [7] under the assumption that $t = n\ell^2/(2dD)$ is long enough ($n \gg 1$), thus ensuring that the volume explored by the particle is much larger than its own volume. As it turns out, $v_t \sim \gamma_d(\ell/R)^2 R^d S_n$ for $d \ge 2$; here, $\gamma_d$ denotes a constant that depends on dimension (e.g., $\gamma_2 = 1$, $\gamma_3 \simeq 2.917$ [31]), and $S_n$ is the territory covered by a particle that evolves in a $d$-dimensional simple cubic lattice with period $\ell \ll R$. Explicit expressions for $S_n$ and its limiting value $S_\infty$ (the latter being of special relevance for applications) can be found in Ref. [7]. E.g., for $d = 1$ one finds

$$S_\infty = \left[\frac{1+e^{-\tilde{\mu}}}{1-e^{-\tilde{\mu}}}\right]^{1/2}, \tag{7.15}$$

where $\tilde{\mu} = [\ell^2/(2D)]\mu$. For $d \ge 2$ and small $\mu$, $S_\infty$ goes as $1/\mu$, albeit with a logarithmic correction in the two-dimensional case [7].

### 7.3.2 *First Passage with Resetting*

Resetting changes drastically first-passage properties [32] and has a large number of applications [33]. Following [34], let us discuss the OSTP (defined in Sect. 7.3.1) with a mortal searcher starting from $x_0 \ge 0$. The searcher is subject to stochastic reset to the initial position $x_0$ at a rate $r$. The search process ends either with the detection of a point target at $x = 0$, or else with the searcher's death. The searcher's pdf $p(x, t|\, x_0)$ satisfies the equation [34, Supplementary material]

$$\frac{\partial p}{\partial t} = D\frac{\partial^2 p}{\partial x^2} - (\mu + r)p + r\,\delta\,(x - x_0)\int_0^\infty dy\; p(y, t|\, x_0), \tag{7.16}$$

which must be solved with the initial condition $p(x, 0|\, x_0) = \delta\,(x - x_0)$ and the boundary condition $p(0, t|\, x_0) = 0$. The second term on the right-hand side (rhs) takes into account the negative probability flux $(\mu + r)p$ out of each point $x$ due to both resetting and particle death. The third term gives the probability influx at $x_0$ due to resetting of the particles that have survived by time $t$. For the Laplace transform $\tilde{p}(x, s|\, x_0) = \int_0^\infty dt\, e^{-st}\, p(x, t|\, x_0)$, we obtain

$$s\tilde{p} - \delta\,(x - x_0) = D\frac{\partial^2 \tilde{p}}{\partial x^2} - (\mu + r)\tilde{p} + r\,\delta\,(x - x_0)\int_0^\infty dy\; \tilde{p}(y, s|\, x_0). \tag{7.17}$$

The pdf $p(x, t|\, x_0)$ must vanish both for $x = 0$ and for $x \to +\infty$. Together with the continuity condition at $x_0$, this yields [34]

$$\tilde{p}(x, s|\, x_0) = \begin{cases} A \sinh \gamma_s x, & x \le x_0, \\ A e^{\gamma_s (x_0 - x)} \sinh \gamma_s x_0, & x \ge x_0, \end{cases} \tag{7.18}$$

where $\gamma_s = \sqrt{(s + r + \mu)/D}$. The unknown coefficient $A$ is obtained from the jump of the derivative $\partial_x \tilde{p}$ at $x = x_0$, which gives $A = \gamma_s/[(s + \mu)\exp(\gamma_s x_0) + r]$.

We are now in the position to compute the Laplace transform of the flux $j(t|\, x_0) = D\partial_x p(x, t|\, x_0)$ to the target, which reads

$$\tilde{j}(s|\, x_0) = D\left.\frac{\partial \tilde{p}}{\partial x}\right|_{x=0} = \frac{s + r + \mu}{(s + \mu)e^{\gamma_s x_0} + r}. \tag{7.19}$$

From here, the probability to ever reach the target follows as

$$p_r(x_0) = \int_0^{\infty} j(t|\, x_0)dt = \tilde{j}(0|\, x_0) = \frac{r + \mu}{\mu e^{w} + r}, \tag{7.20}$$

where $w = \sqrt{(r + \mu)/D}x_0$. One then concludes from Eq. (7.20) that $p_r$ decreases monotonically with increasing $r$ if $\mu \geq \mu_0 = (z^*)^2\, D/x_0^2$, where $z^* \approx 1.59362426\ldots$ is the solution to $z/2 = 1 - e^{-z}$. In contrast, for $\mu < \mu_0$, the probability $p_r$ takes its maximum at the nonvanishing restart rate $r_0 = \mu_0 - \mu$.

As explained in Ref. [34], the observed behavior of $p_r$ can be qualitatively understood as follows. When $\mu$ is small compared to $D/x_0^2$, the searcher lives long enough to perform a typical diffusive path and reach the target. On the other hand, the searcher can still make distant excursions into empty regions of the search space. In this context, a nonvanishing restart rate (resetting) censors fatal paths and increases the chance of finding the target. However, an excessively large resetting rate may hinder target detection, as the searcher has less time between restarts to reach the origin while facing the same mortality rate. Hence, there exists an optimal non-zero resetting rate $r^*$ which maximizes the probability of finding the target before death. At this point, it is worth noting that a recent work by Radice [35] extends the analysis to other quantities, including explicit expressions for the mean and the variance of the search time.

According to Belan [34], the above problem of target detection by a mortal walker subject to resetting can be viewed as a specific type of Bernoulli experiment with restart (target detection is identified with "success", whereas non-detection due to the searcher's premature death is identified with "failure"). However, Belan's discussion goes beyond this particular example; he concludes that in a generic first-passage process for which an optimal resetting rate $r^*$ exists [in the sense that $r^*$ maximizes (minimizes) the probability of success (failure)], then for $r = r^*$ the completion time of the process conditioned on a specific outcome (no matter whether success or failure) equals the completion time over both successful and failed trials (unconditional MFPT).

## 7.4 $N$-Walker Problem: Order Statistics

### 7.4.1 Shortest First-Passage Time

Let us consider a collection of $N$ (non-interacting) diffusing particles (also termed indistinctly "walkers" or "searchers" hereafter) that compete to reach a target. In this context, the FPT statistics of the fastest searcher (first searcher to reach the target) is of particular interest. Beyond biological problems like gene regulation [36], calcium signaling [37], or oocyte fertilization [3], the order statistics of random walkers also plays a key role in other fields, e.g., for information spreading in networks [38].

We consider the simplest initial condition, i.e., $N$ mortal searchers all starting at the same distance $x_0 > 0$ from a target placed at $x = 0$. We ask about the FPT of the first one to reach the target. To reduce the problem to a single-particle one (cf. Sect. 7.3), we use the simplified notation $f_1(t) \equiv f(t, 0 \mid x_0)$ and $F_1(t) \equiv F(t, 0 \mid x_0)$.

In order to exploit previous results, we will first consider the above problem for $\mu = 0$, i.e., for a collection of $N$ immortal walkers (quantities with hats will hereafter refer to immortal walkers). The probability that the target is annihilated by any of the walkers at time $t$ is characterized by the density $\hat{f}_N(t)$, which is in turn expressible in terms of quantities for the one-walker problem [3, 39, 40]:

$$
\begin{aligned}
\hat{f}_N(t) &= N \hat{f}_1(t)\left[1 - \hat{F}_1(t)\right]^{N-1} \\
&= N \frac{x_0}{\sqrt{4\pi D t^3}}\, e^{-x_0^2/4Dt} \left[\operatorname{erf}\left(x_0/\sqrt{4Dt}\right)\right]^{N-1} \\
&\simeq N \frac{x_0}{\sqrt{4\pi D t^3}} \left(\frac{x_0}{\sqrt{\pi D t}}\right)^{N-1}, \qquad t \to \infty,
\end{aligned}
\tag{7.21}
$$

where the third line follows from the $z \ll 1$ asymptotics of $\operatorname{erf}(z)$. A consequence of the fact that $\hat{f}_N(t)$ exhibits the algebraic long-time tail $\hat{f}_N(t) \sim t^{-(N+2)/2}$ is that the average search time

$$
\langle \hat{T}_N \rangle = \int_0^\infty dt\, t\, \hat{f}_N(t) \tag{7.22}
$$

(time taken by the fastest walker to reach the target) is divergent for $N \leq 2$, but finite for $N \geq 3$ [41].

At this stage, it is worth noting that the case $N \geq 3$ has attracted much attention over the last decades (see, e.g., [39–46]). In Ref. [39], the statistics of first-arrival times of $N \gg 1$ immortal walkers to the target was studied in detail. In particular, the first two moments of $\hat{T}_N$ were computed, and an asymptotic expansion with a (not fully correct) first subdominant term was provided. This problem was revisited in Ref. [42]; the first subdominant term was amended and the next one (needed to estimate the variance) was also evaluated. To first order in the corrective terms, the final expression for $\langle \hat{T}_N \rangle$ reads [42, 43]

$$\langle \hat{T}_N \rangle \approx \frac{x_0^2}{4D\ln(\lambda_0 N)} \left\{ 1 + \frac{\frac{1}{2}\ln\ln(\lambda_0 N) - \gamma}{\ln(\lambda_0 N)} \right\}, \tag{7.23}$$

where $\lambda_0 = 1/\sqrt{\pi}$ and $\gamma \approx 0.5772$ is the Euler-Mascheroni constant. Note that the corrective term remains non-negligible even for very large values of $N$. The approximate expression given by the leading term in Eq. (7.23) only,

$$\langle \hat{T}_N \rangle \simeq \frac{x_0^2}{4D\ \ln N}, \tag{7.24}$$

agrees with the one obtained in Ref. [3] by a different method.

Turning now to the case of mortal walkers, for $N$ identical searchers all starting from $x_0$, the probability $p$ that at least one of them eventually hits the target is given by the interplay between the intrinsic mean lifetime $1/\mu$ and the diffusive time scale $\tau_D = x_0^2/(4D)$. Meerson and Redner [3] conclude that if $N$ is large enough, $p$ becomes non-negligible even for a relatively high mortality. Reaching the target is no longer guaranteed, but when $M \equiv \sqrt{\mu\tau_D} \gg 1$, one has for the conditional MFPT [3]

$$\langle T_N \rangle = \frac{\int_0^\infty t\ f_N(t)\,dt}{\int_0^\infty f_N(t)\,dt} \simeq \frac{\tau_D}{M} = \frac{x_0}{\sqrt{4D\mu}}. \tag{7.25}$$

The result (7.25) is consistent with Eq. (7.13) for the MFPT of a *single* mortal searcher. Thus, in the high-mortality regime, the MFPT is predominantly influenced by the fastest searcher only, which renders the other searchers superfluous unless their number is extremely large. In this latter case, when $N \gg \sqrt{M}e^{2M}$ (but with $M$ also large), one recovers Eq. (7.24) [3]. Therefore, in the large-$N$ limit, even a relatively high mortality becomes ultimately irrelevant for the computation of the MFPT. Thus, $\langle T_N \rangle \approx \langle \hat{T}_N \rangle$ in this limit. Consequently, although the search time (7.24) decays logarithmically with $N$, in the problem of oocyte fertilization it still entails a reduction by a factor of around 20 with respect to the characteristic diffusion time $\tau_D$ because of the typically large number of human sperm ($N = 3 \times 10^8$) [3].

Finally, we note that an extension of the above $N$-walker problem to the case of a position-dependent death rate $\mu(x)$ was recently studied by Toste and Holcman [47].

### 7.4.2 Longest First-Passage Time

The problem of estimating the moments of the time $\hat{T}_{j,N}$ taken by the $j$-th fastest searcher among a collection of $N \gg 1$ *immortal* searchers to reach a predefined set of targets was addressed in Refs. [39, 42, 43] for several initial distributions of searchers and different target arrangements in $d = 1$. Variants of the same problem with different embedding geometries and target distributions were later considered, e.g., $d$-dimensional Euclidean lattices with absorbing hyperspherical surface [48,

49], $d$-dimensional Euclidean media filled with random traps [50], disordered media filled with random traps [51], disordered media with an absorbing hyperspherical (Euclidean or chemical) surface in [52], and fractal lattices [53, 54]. In particular, for $N$ immortal walkers starting all at $x_0$ and a point target at $x = 0$, one has [42]

$$\langle \hat{T}_{j,N} \rangle = \langle \hat{T}_N \rangle + \frac{x_0^2}{4D \ln^2(\lambda_0 N)} H_{j-1}. \tag{7.26}$$

$H_k = \sum_{r=1}^{k} r^{-1}$ is the $k$-th harmonic number, and the expression for $\langle \hat{T}_N \rangle \equiv \langle \hat{T}_{1,N} \rangle$ is given by Eq. (7.23).

Beyond a number of biological systems where the shortest FPT plays a central role, Lawley and Johnson [55] point out that in other instances the quantity of interest is the search time of the *slowest* walker(s), i.e., the longest FPT, "as it can define the termination of a process or perhaps the exhaustion of a supply". In this context, Lawley and Johnson deal extensively with the onset of menopause, which occurs when the number of primordial follicles in the ovarian reserve drops from an initial value of around $5 \times 10^5$ to approximately $10^3$ (see [55] and references therein).

Using extreme value statistics, Lawley and Johnson recently proved [55] that when the survival probability $S(t)$ of an individual *immortal* searcher exhibits the behavior $S(t) \sim (\zeta t)^{-\eta}$ as $t \to \infty$ (with $\zeta > 0$ and $\eta > 0$), then for $N \to \infty$ *mortal* searchers and a fixed $k$, the MFPT of the $N - k$-th searcher behaves as follows:

$$\langle T_{N-k,N} \rangle = \mu^{-1} \big( \ln N - (\eta + 1) \ln \ln N + \ln \mathcal{A} + \gamma - H_k + o(1) \big), \tag{7.27}$$

where $\mathcal{A} = \eta(\mu/\zeta)^\eta / \int_0^\infty [1 - S(t')] \mu e^{-\mu t'} dt'$. In particular, for our canonical example of $N$ mortal walkers starting at $x_0$ with a target at $x = 0$, one has $S(t) = \mathrm{erf}[x_0/(4Dt)]$, $\zeta = \pi D / x_0^2$, $\eta = 1/2$, and $\mathcal{A} = \sqrt{\mu x_0^2/(4\pi D)} \exp\left(\sqrt{\mu x_0^2/D}\right)$.

## 7.5 The Target Problem with Anomalous Diffusion

Problems with anomalous diffusion and mortality have been addressed by many authors. Theoretical approaches rely often on (separable) CTRW models and the corresponding fractional diffusion equations, as they lend themselves well to the inclusion of chemical reactions in general and of decay processes in particular [56–68]. We note, however, that the case of non-separable walks [69, 70] has also been addressed, as well as that of a decay process coupled to the statistics of jumps [24, 71, 72] or to the territory explored [73, 74].

In the remainder of this section, we focus on the survival probability of an immobile target surrounded by one or more subdiffusively moving evanescent searchers. These searchers follow the statistics of an uncoupled CTRW with a jump length pdf of finite variance and a waiting-time pdf displaying the long-time behavior $t^{-1-\gamma}$

(with $0 < \gamma < 1$) [30, 62, 63]. On sufficiently long spatial and temporal time scales, this yields a fractional diffusion equation with anomalous diffusion coefficient $D_\gamma$ ($D_1 \equiv D$ recovers the case of normal diffusion) [75].

### 7.5.1 The One-Searcher Problem

Our setting consists of an impenetrable, hyperspherical target with radius $R$ placed at the origin $\mathbf{x} = 0$ of a $d$-dimensional volume $V$, and an evanescent subdiffusive point searcher located inside $V$ at an initial position $\mathbf{x}_0$ ($|\mathbf{x}_0| > R$). The target is spherical in three dimensions, circular in two dimensions, and a segment in one dimension. Unless stated otherwise, the searcher's survival probability is assumed to decrease exponentially when the target is absent, i.e., $p_\mu(t) = \mu e^{-\mu t}$. Since the searcher may evanesce as it moves, its pdf $p(\mathbf{x}, t|\mathbf{x}_0)$ follows the subdiffusion-decay equation [11, 62]

$$\frac{\partial p(\mathbf{x}, t|\mathbf{x}_0)}{\partial t} = e^{-\mu t} D_\gamma \, {}_0\mathcal{D}_t^{1-\gamma} e^{\mu t} \nabla_{\mathbf{x}}^2 p(\mathbf{x}, t|\mathbf{x}_0) - \mu p(\mathbf{x}, t|\mathbf{x}_0) \tag{7.28}$$

where ${}_0\mathcal{D}_t$ denotes the Riemann-Liouville fractional derivative[1]

$${}_0\mathcal{D}_t^{1-\gamma} p(\mathbf{x}, t|\mathbf{x}_0) \equiv \frac{1}{\Gamma(\gamma)} \frac{\partial}{\partial t} \int_0^t dt' \, \frac{p(\mathbf{x}, t|\mathbf{x}_0)}{(t-t')^{1-\gamma}}. \tag{7.29}$$

While the presence of exponential functions $e^{\pm\mu t}$ in the first term on the rhs of Eq. (7.28) may at first glance seem counterintuitive, they arise from a rigorous derivation of this equation from the underlying CTRW integral equation. These exponentials ensure that Eq. (7.28) reduces to a pure subdiffusion equation for the normalized density $e^{\mu t} p(\mathbf{x}, t|\mathbf{x}_0)$ in a similar way as the density $p_I(x, t|x_0)$ defined in Sect. 7.2.1 fulfills the standard diffusion equation $\partial_t p_I = D\partial_x^2 p_I$. Failure to include these exponentials may result in non-physical solutions for the problem at hand, including spurious effects such as negative values of $p(\mathbf{x}, t|\mathbf{x}_0)$.

We assume that the searcher destroys the target instantaneously upon contact and seek to compute the survival probability $Q_1(\mathbf{x}_0, t; R)$ of the target. Since the subsequent fate of the searcher is immaterial for this purpose, we will assume for convenience that the target is *f*ully absorbing, i.e., the searcher also dies instantly upon hitting the target (the particular case $d = 1$, $R = 0$, and $\gamma = 1$ recovers the OSTP studied in Sect. 7.3.1). The searcher's survival probability $Q_{1,s}(\mathbf{x}_0, t; R) = \int p(\mathbf{x}, t|\mathbf{x}_0) d\mathbf{x}$ is related to $Q_1$ via the equation [30, 62]:

[1] Strictly speaking, the derivation of Eq. (7.28) involves the so-called Grünwald-Letnikov derivative, which is equivalent to the Riemann-Liouville one in the present case of a sufficiently smooth function at $t = 0$ (see, e.g., Eqs. (2.255), (2.248), and (2.240) in Ref. [76]).

$$Q_1(\mathbf{x}_0, t; R) = Q_{1,s}(\mathbf{x}_0, t; R) + \mu \int_0^t Q_{1,s}(\mathbf{x}_0, t'; R) dt'. \tag{7.30}$$

Thus, the target has a higher survival probability than the searcher at any time, since the latter may evanesce spontaneously before being able to hit the target [cf. Eq. (7.28)]. The second term on the rhs of Eq. (7.30) quantifies the difference by adding up the infinitesimal probabilities for the searcher to decay spontaneously between $t'$ and $t' + dt'$ at *any* previous time $0 < t' < t$, thereby increasing the target's chance of survival $Q_1$ by time $t$. This quantity is easily obtained in the limit $V \to \infty$ of an infinite volume, as in this limit $Q_1$ and $Q_{1,s}$ do not depend on the full vector $\mathbf{x}_0$, but only on its modulus $|\mathbf{x}_0|$. First, to compute $Q_{1,s}(|\mathbf{x}_0|, t; R)$ in this case, one must solve the boundary value problem

$$\frac{\partial Q_{1,s}(|\mathbf{x}_0|, t; R)}{\partial t} = e^{-\mu t} D_\gamma \, {}_0\mathcal{D}_t^{1-\gamma} e^{\mu t} \nabla^2_{\mathbf{x}_0} Q_{1,s}(|\mathbf{x}_0|, t; R) - \mu Q_{1,s}(|\mathbf{x}_0|, t; R), \tag{7.31a}$$

$$Q_{1,s}(|\mathbf{x}_0|, 0; R) = 1, \tag{7.31b}$$

$$Q_{1,s}(R, t; R) = 0, \tag{7.31c}$$

$$Q_{1,s}(|\mathbf{x}_0| \to \infty, t; R) = e^{-\mu t}. \tag{7.31d}$$

For $d = 1, 3$ one obtains the following long-time expression [30]:

$$Q_1(|\mathbf{x}_0|, \infty; R) - Q_1(|\mathbf{x}_0|, t; R) \sim -\left(\frac{R}{|\mathbf{x}_0|}\right)^{(d-1)/2} \frac{1}{\Gamma(1-\frac{\gamma}{2})} \frac{|\mathbf{x}_0| - R}{\mu^{1+\frac{\gamma}{2}} \ell_\gamma} t^{-1-\frac{\gamma}{2}} e^{-\mu t}, \tag{7.32}$$

where $\ell_\gamma = \left(4D_\gamma/\mu^\gamma\right)^{1/2}$ can be regarded as the typical distance explored by the searcher before dying spontaneously and

$$Q_1(|\mathbf{x}_0|, \infty; R) = 1 - \left(\frac{R}{|\mathbf{x}_0|}\right)^{(d-1)/2} e^{-2(|\mathbf{x}_0| - R)/\ell_\gamma}. \tag{7.33}$$

In $d = 2$, the long-time decay takes a more complicated form:

$$Q_1(|\mathbf{x}_0|, t; R) \sim Q_1(|\mathbf{x}_0|, \infty; R) + \frac{2}{\gamma} \ln\left(\frac{|\mathbf{x}_0|}{R}\right) \frac{e^{-\mu t}}{\mu t \ln^2(\alpha_\gamma t)} \tag{7.34}$$

with $\alpha_\gamma = (4D_\gamma/R^2)^{1/\gamma}$ and

$$Q_1(|\mathbf{x}_0|, \infty; R) = 1 - \frac{K_0(2|\mathbf{x}_0|/\ell_\gamma)}{K_0(2R/\ell_\gamma)}. \tag{7.35}$$

Thus, in $d = 1, 2$ the target has a non-zero probability of eternal survival, as opposed to the standard case of an immortal searcher ($\mu = 0$). In this latter case, the target death would be certain because of the recurrence of the searcher's walk.

### 7.5.2 The Many-Searcher Problem

The results for the one-searcher target problem provide the starting point to address the many-searcher problem. Specifically, consider a hyperspherical target of radius $R$ centered at $\mathbf{x} = 0$ and surrounded by a collection of $N_0$ non-interacting evanescent point searchers in a volume $V$ around $\mathbf{x} = 0$ (ultimately, the precise form of $V$ does not matter for our purposes, but it will be assumed to be hyperspherical). In each ensemble realization of the system, the initial positions of these searchers are drawn from a uniform pdf, and each of them satisfies the diffusion-decay equation (7.28). As in the one-searcher problem, we assume that if any searcher hits the target, both are instantaneously annihilated. We again focus on the survival probability $Q(t;\, R)$ of the target, i.e., the probability that it has not been hit by any searcher up to time $t$.

Because of the statistical independence of the searchers' trajectories, the solution can be expressed in terms of its counterpart for the one-searcher problem:

$$Q(t;\, R) = \left[\frac{1}{V}\int_{\mathbf{x}_0 \in V\setminus S_d(R)} Q_1(|\mathbf{x}_0|, t;\, R)\, d\mathbf{x}_0\right]^{N_0}, \tag{7.36}$$

where the above volume integral excludes a hypersphere $S_d(R)$ of radius $R$ centered at the origin. Now, we take the thermodynamic limit, i.e., the double limit $N_0 \to \infty$, $V \to \infty$ in such a way that the initial global density of searchers $\rho_0 \equiv \rho(0) = \lim_{N_0, V\to\infty} N_0/V$ remains finite. In this limit, from the solution for $Q_1$ in an infinite volume, one finds

$$Q(t;\, R) = \exp\left\{-\rho_0 R^d \sigma(t,\, R)\right\} \tag{7.37}$$

with

$$\sigma(t;\, R) \equiv \frac{1}{R^d}\int_{|\mathbf{x}_0|>R} \left[1 - Q_1(|\mathbf{x}_0|, t;\, R)\right] d\mathbf{x}_0. \tag{7.38}$$

As pointed out in Refs. [24, 63], the quantity $R^d\sigma(t,\, R)$ is directly related to the average volume swept by a *single* mortal walker up to time $t$. Long-time asymptotic expressions for $\sigma(t,\, R)$ were obtained in Refs. [30, 62]. The general expression for its limiting value in $d$ dimensions is

$$\sigma(\infty;\, R) = S_d \frac{\ell_\gamma}{2R} \frac{K_{d/2}(2R/\ell_\gamma)}{K_{d/2-1}(2R/\ell_\gamma)}, \tag{7.39}$$

where $S_d = 2\pi^{d/2}/\Gamma(d/2)$ stands for the surface of a $d$-dimensional hypersphere of unit radius and $K_\nu$ is the $\nu$th-order modified Bessel function of the second kind. Thus, even though we have an infinite initial number of searchers, the target has a non-zero chance of eternal survival in all dimensions. Thus, in a finite fraction of realizations, a sufficiently fast evanescence process may kill all the searchers before any of them is able to hit the target. In the absence of such a process, the target is of course eventually killed with certainty in all spatial dimensions [56].

At this stage, a few closing remarks are in order. We first note that the problem with a power-law decaying density of searchers $\rho(t) \propto t^{-\beta}$ is also interesting; depending on the values of $\beta$ and $\gamma$, the final survival probability of the target $Q(\infty; R)$ is finite or not. We will not deal with this problem here for reasons of space, but the details can be found in Ref. [30]. Finally, we note that the important case of a *partially absorbing target* has been addressed in Refs. [77, 78] for settings with one and many searchers, respectively.

## 7.6 The Target Problem in an Evolving Domain

A rubber band that stretches/shrinks or the surface of a balloon being inflated/deflated are elementary examples of evolving media. Some motivation for studying diffusion in expanding media comes, e.g., from the fields of developmental biology (diffusion-mediated formation of growing biological structures) and cosmology (diffusion of cosmic rays in the expanding universe), see, e.g., [79–81] and references therein. In particular, the search problems addressed in Sect. 7.5 can now be considered in an expanding (or contracting) domain.

In the case of a shrinking domain, the searcher concentration grows as the volume decreases, and this may compensate for the particle loss due to evanescence. Therefore, a competition between both mechanisms settles. For simplicity, we illustrate this competition via the normal diffusive case ($\gamma = 1$) on the real line ($d = 1$).

### *7.6.1 The One-Searcher Problem*

As before, we consider the target problem with a single evanescent point searcher, but now each length element $dy$ of the real line grows/shrinks (homogeneously) in the course of time as $dy = a(t)\,dx$, where $dx$ denotes its initial length and $a(t)$ is the so-called scale factor, which takes the initial value $a(0) = 1$ [79]. Thus, a point on the line with initial coordinate $x$ (comoving coordinate) is located at the physical coordinate $y = a(t)\,x$ after a time $t$.

We now place a target (i.e., a segment with absorbing endpoints) centered at $y = x = 0$. We assume that the target size follows the evolution law of the one-dimensional domain, i.e., $R(t) = a(t)R$, where $R$ denotes the initial "radius" (half-length of the target). The searcher starts at $y_0 = x_0$ with $x_0 > R$ and performs random jumps drawn from a stationary pdf $\lambda(y)$ with finite variance. In $x$-space, the corresponding jump length pdf becomes time-dependent $\lambda'(x, t)$, and so does the diffusivity. Between consecutive jumps, the searcher sticks rigidly to the expanding/shrinking length element on which it rests; thus, in $y$-space, the searcher experiences a deterministic drift while still waiting to jump. On the other hand, between $t$ and $t + dt$, the searcher has a non-zero probability $\mu dt$ to spontaneously disappear. The corresponding reaction-diffusion equation in $x$-space then is [81]:

$$\frac{\partial p(x,t|\,x_0)}{\partial t} = \frac{D}{a^2(t)}\frac{\partial^2 p(x,t|\,x_0)}{\partial x^2} - \mu p(x,t|\,x_0). \tag{7.40}$$

From here, one can easily compute the searcher's survival probability $Q_{1,s}(x_0,t) = \int p(x,t|\,x_0)dx$. The solution of the corresponding boundary value problem is

$$Q_{1,s}(x_0,t) = e^{-\mu t}\hat{Q}_{1,s}(x_0,t) = e^{-\mu t}\text{erf}\left(\frac{x_0 - R}{\sqrt{4D\tau(t)}}\right) \tag{7.41}$$

with the so-called Brownian conformal time $\tau(t) = \int_0^t dt'/a^2(t')$ [79, 81]. The conformal time $\tau$ accounts for the effect of the domain deformation; it grows superlinearly (sublinearly) in time in a shrinking (expanding) domain.

In the present case of an evolving domain, the survival probability of the target $Q_1(x_0,t) > Q_{1,s}(x_0,t)$ is still given by Eq. (7.30). Depending on the form $a(t)$, the integral on the rhs (which quantifies the difference between both survival probabilities) may or may not be explicitly evaluated. In some cases, the asymptotic value $Q_1(x_0,\infty)$ and the long-time approach to it can be computed via Laplace transforms and Tauberian theorems. In any case, for fixed $x_0$ and $t$, $Q_1(x_0,t)$ is larger (smaller) on the expanding (shrinking) line than its counterpart for a static domain.

### 7.6.2 The Many-Searcher Problem

Equations (7.36), (7.37), and (7.38) still hold in $x$-space. From Eqs. (7.38) and (7.30), one finds

$$\frac{\partial\sigma(t;\,R)}{\partial t} = -\frac{1}{R}\int_{|x_0|>R}\left[\frac{\partial Q_{1,s}}{\partial t} + \mu e^{-\mu t}Q_{1,s}\right]dx_0. \tag{7.42}$$

We now use Eq. (7.40) and change the "volume integral" into a "surface integral"

$$\frac{\partial\sigma(t;\,R)}{\partial t} = \frac{2D}{a^2(t)}\frac{e^{-\mu t}}{R}\left.\frac{\partial\hat{Q}_{1,s}(x_0,t;\,R)}{\partial x_0}\right|_{x_0=R} \tag{7.43}$$

via Gauss's theorem [62]. Integration from 0 to $t$ yields

$$\sigma(t;\,R) = \frac{2D}{R}\int_0^t \frac{1}{a^2(t')}\left.\frac{\partial\hat{Q}_{1,s}\left(x_0,t';\,R\right)}{\partial x_0}\right|_{x_0=R} e^{-\mu t'}\,dt' \tag{7.44}$$

(recall that $\sigma(0;\,R) = 0$ $[Q(0;\,R) = 1]$). Finally, the explicit form of $\hat{Q}_{1,s}$ gives

$$\sigma(t;\,R) = R^{-1}\sqrt{\frac{4D}{\pi}}\int_0^t \frac{1}{a^2(t')}\frac{e^{-\mu t'}}{\sqrt{\tau(t')}}dt'. \tag{7.45}$$

In particular, for an exponentially shrinking domain $a(t) = e^{Ht}$ with $H < 0\,[\tau(t) = H^{-1}e^{-Ht}\sinh(Ht)]$, one sees that $Q(\infty, R)$ remains finite as long as $\mu > |H|$, but tends to zero otherwise. For an expanding line ($H > 0$), the target has a finite chance of eternal survival even when the searchers are immortal (see Sect. 3.3.2 in [80]).

## 7.7 Escape

Escape from a domain with reflecting boundaries through a narrow target is an important topic in biology [13, 82]. In this context, inactivation or degradation of the entity diffusing toward the target [83] often competes on time scales comparable or shorter than those of pure diffusion and must therefore be accounted for to compute rates of binding, etc. In Ref. [8], Grebenkov showed that the mean escape time (here identified as the MFPT) and the target absorption probabilities can be related to their counterparts for immortal walkers. Below we sketch his derivation.

Let $\hat{Q}(\mathbf{x}_0, t)$ denote the survival probability, i.e., the probability that an *immortal* walker started at a point $\mathbf{x}_0 \in \Omega$ has not left a confining domain $\Omega \subset \mathbb{R}^d$ through an escape region $\Gamma$ on the boundary $\partial\Omega$. The corresponding FPT density is

$$\hat{\rho}(\mathbf{x}_0, t) = -\frac{\partial \hat{Q}(\mathbf{x}_0, t)}{\partial t}. \tag{7.46}$$

Note that if $\tilde{\hat{Q}}(\mathbf{x}_0, s)$ is the Laplace-transformed survival probability, then the MFPT is simply $\tilde{\hat{Q}}(\mathbf{x}_0, 0)$. For *mortal* walkers, assuming as before that the walker's mortality and the diffusion process are independent, one has that its survival probability (i.e., the probability that the walker has neither evanesced nor left the domain $\Omega$ up to time $t$) is simply $Q(\mathbf{x}_0, t) = e^{-\mu t}\,\hat{Q}(\mathbf{x}_0, t)$. Thus, in Laplace space, $\tilde{Q}(\mathbf{x}_0, s) = \tilde{\hat{Q}}(\mathbf{x}_0, s + \mu)$, and the MFPT for mortal walkers then takes the form

$$\langle T(\mathbf{x}_0)\rangle = \tilde{Q}(\mathbf{x}_0, 0) = \tilde{\hat{Q}}(\mathbf{x}_0, \mu). \tag{7.47}$$

There is a connection between the MFPT and the probability $H(\mathbf{x}_0)$ that a mortal walker started from $\mathbf{x}_0$ leaves the confining domain before dying, namely,

$$H(\mathbf{x}_0) = \int_0^\infty dt\, e^{-\mu t}\,\hat{\rho}(\mathbf{x}_0, t) = 1 - \mu\tilde{\hat{Q}}(\mathbf{x}_0, \mu) = 1 - \mu\langle T(\mathbf{x}_0)\rangle. \tag{7.48}$$

For low mortality ($\mu\langle\hat{T}(\mathbf{x}_0)\rangle \ll 1$), the Taylor expansion of $\tilde{\hat{Q}}(\mathbf{x}_0, s)$ yields

$$\langle T(\mathbf{x}_0)\rangle = \langle\hat{T}(\mathbf{x}_0)\rangle - \frac{1}{2}\mu\langle\hat{T^2}(\mathbf{x}_0)\rangle + O(\mu^2). \tag{7.49}$$

From Eq. (7.48), we then find

$$H(\mathbf{x}_0) = 1 - \mu \langle \hat{T}(\mathbf{x}_0) \rangle + \frac{1}{2} \mu^2 \langle \hat{T}^2(\mathbf{x}_0) \rangle + O(\mu^3). \tag{7.50}$$

We are thus led to the analysis of moments of the FPT for immortal walkers:

$$\langle \hat{T}^n(\mathbf{x}_0) \rangle = (-1)^n \left. \frac{\partial^n}{\partial \mu^n} H(\mathbf{x}_0) \right|_{\mu=0}. \tag{7.51}$$

Exact asymptotic formulas for the MFPT of immortal walkers are known for some simple domains [82]. E.g., for $d = 3$, $\langle \hat{T}(\mathbf{x}_0) \rangle \simeq |\Omega|/(4DR\epsilon) + O(\ln \epsilon)$, where $R = \sqrt{|\partial\Omega|}$ is the characteristic size ("radius") of the domain, and $\epsilon = (|\Gamma|/|\partial\Omega|)^{1/2} \ll 1$ is the square root of the area of the escape region normalized by the area of the boundary. In Ref. [8], a thorough study of the exit probability and the MFPT for small, moderate, and large death rates can be found.

A variant of the above problem concerns the statistics of the escape time for $N$ walkers initially uniformly distributed throughout the confining domain [84]. Another situation of interest refers to diffusion processes switching between $d = 3$ and $d = 2$ (Adam-Delbrück scenario) [see Ref. [18] for a recent treatment with one and $N$ walkers]. The effect of mortality in these two settings is still an open question.

## 7.8 Summary and Conclusion

Mortality changes drastically the properties of Brownian motion inasmuch as it induces a stronger localization and promotes quasiballistic behavior. Only sufficiently short trajectories have a chance to reach a target, as illustrated by the canonical example of one or many particles diffusing on the real half-line with a target placed at the origin: the conditional MFPT is finite and linear in the initial target-searcher separation; mortal random walks become non-recurrent already in one dimension, and quantities like the mean number of distinct sites visited converge to a finite value.

Resetting of an individual mortal walker eliminates long trajectories which lead to premature death before finding the target. On the other hand, an excessively large resetting decreases the search efficiency. The trade-off between these two effects leads to an optimal resetting rate induced by the mortality constraint.

The order statistics of a collection of $N$ mortal walkers searching for the target depends strongly on how the mortality time scale compares to the diffusive one. For high mortality, the MFPT of the fastest searcher is identical to the one obtained in the one-searcher problem; for low mortality, the leading term of this quantity is the same as for immortal walkers and thus proportional to $1/\ln N$. In contrast, the MFPT of the slowest searchers turns out to grow as $\ln N$ in the large-$N$ limit.

The target problem for the half-line can also be solved in the case of subdiffusive searchers. Both for a single searcher and for many (randomly distributed) searchers, the target has (as in the case of normal diffusion) a finite probability of eternal survival. The asymptotic decay of the target's survival probability to this finite value is given by the intrinsic survival probability of the mortal searcher $e^{-\mu t}$ multiplied by slower decaying terms (power-law, logarithmic) that depend on the searcher's anomalous diffusion exponent.

Another variant of the one-dimensional target problem is obtained by introducing a homogeneous expansion (contraction) of the real line. For an exponential contraction, if the searchers' mortality rate is larger than the contraction rate, the target may survive forever even if the number of (randomly distributed) searchers is infinite.

Finally, we have also reproduced some results for the conditional MFPT and the exit probability in the narrow escape problem for mortal walkers. Explicit expressions follow from the formalism for immortal walkers in the low-mortality limit.

Many of the results presented here are recent, which shows that research on mortal walkers is very much alive. In view of the number of possible extensions and the wealth of applications, we expect that a good deal of new findings relevant to both theory and experiments will be reported in the near future.

**Acknowledgements** We acknowledge financial support from Grant PID2020-112936GB-I00 funded by MCIN/AEI/10.13039/501100011033 and from Grant IB20079 funded by Junta de Extremadura (Spain) and by ERDF "A way of making Europe".

## References

1. L.O. Nicolaysen, Solid diffusion in radioactive minerals and the measurement of absolute age. Geochim. Cosmochim. Acta **11**, 41–59 (1957)
2. M.P. Silverman, Brownian motion of decaying particles: Transition Probability, computer simulation, and first-passage times. J. Mod. Phys. **8**, 1809–1849 (2017)
3. B. Meerson, S. Redner, Mortality, redundancy, and diversity in stochastic search. Phys. Rev. Lett. **114**, 198101 (2015)
4. E. Abad, J.J. Kozak, Competing reaction processes on a lattice as a paradigm for catalyst deactivation. Phys. Rev. E **91**, 022106 (2015)
5. H. Sano, M. Tachiya, Partially diffusion controlled recombination. J. Chem. Phys. **71**, 1276–1282 (1979)
6. K. Seki, M. Murayama, M. Tachiya, Dispersive photoluminescence decay by geminate recombination in amorphous semiconductors. Phys. Rev. B **71**, 235212 (2005)
7. S.B. Yuste, E. Abad, K. Lindenberg, Exploration and trapping of mortal random walkers. Phys. Rev. Lett. **110**, 220603 (2013)
8. D.S. Grebenkov, J.-F. Rupprecht, The escape problem for mortal walkers. J. Chem. Phys. **146**, 084106 (2017)
9. R.F. Bonner, R. Nossal, S. Havlin, G.H. Weiss, Model for photon migration in turbid biological media. J. Opt. Soc. Am. A **4**, 423–432 (1987)
10. O. Wartlick, A. Kicheva, M. González-Gaitán, Morphogen gradient formation. Cold Spring Harb. Perspect. Biol. **1**, a001255 (2009)
11. S.B. Yuste, E. Abad, K. Lindenberg, A reaction-subdiffusion model of morphogen gradient formation. Phys. Rev. E **82**, 061123 (2010)
12. A.B. Kolomeisky, M.E. Fisher, Periodic sequential kinetic models with jumping, branching and deaths. Physica A **279**, 1–20 (2000)

13. D. Holcman, Z. Schuss, *Stochastic Narrow Escape in Molecular and Cellular Biology* (Springer, New York, 2015)
14. J. Ma, M. Do, M.A. Le Gros, C.S. Peskin, C.A. Larabell, Y. Mori, S.A. Isaacson, Strong intracellular signal inactivation produces sharper and more robust signaling from cell membrane to nucleus. PLoS Comput. Biol. **16**, e1008356 (2020)
15. N. Briantceva, L. Chouhan, M. Parsani, M.-S. Alouini, Impact of evanescence process on three-dimensional sub-diffusion-based molecular communication channel. IEEE Trans. Nanobioscience **22**, 923–932 (2023)
16. S.D. Lawley, The effects of fast inactivation on conditional first passage times of mortal diffusive searchers. SIAM J. Appl. Math. **81**, 1–24 (2021)
17. V. Balakrishnan, E. Abad, T. Abil, J.J. Kozak, First-passage properties of mortal random walks: Ballistic behavior, effective reduction of dimensionality, and scaling functions for hierarchical graphs. Phys. Rev. E **99**, 062110 (2019)
18. D.S. Grebenkov, R. Metzler, G. Oshanin, Search efficiency in the Adam-Delbrück reduction-of-dimensionality scenario versus direct diffusive search. New J. Phys. **24**, 083035 (2022)
19. S. Fedotov, S. Falconer, Random death process for the regularization of subdiffusive fractional equations. Phys. Rev. E **87**, 052139 (2013)
20. O. Sikder, Averaging dynamics, mortal random walkers and information aggregation on graphs. J. Phys. Complex. **2**, 045005 (2021)
21. B.D. Hughes, *Random Walks and Random Environments. Volume 1: Random Walks* (Clarendon Press, Oxford, 1995), pp. 123–124
22. S.B. Yuste, E. Abad, K. Lindenberg, Arrival statistics and exploration properties of mortal walkers, in *First-Passage Phenomena and their Applications*, ed. by R. Metzler, G. Oshanin, S. Redner (World Scientific, Singapore, 2014), pp. 1–20
23. I. Lohmar, J. Krug, Diffusion-limited reactions and mortal random walkers in confined geometries. J. Stat. Phys. **134**, 307–336 (2009)
24. E. Abad, S.B. Yuste, K. Lindenberg, Evanescent continuous time random walks. Phys. Rev. E **88**, 062110 (2013)
25. M.A. Ré, N.C. Bustos, Reaction rate in an evanescent random walkers system. Pap. Phys. **7**, 070003 (2015)
26. J.P. Esguerra, J. Reyes, Exploration properties of biased evanescent random walkers on a one-dimensional lattice. AIP Conf. Proc. **1871**, 050002 (2017)
27. E. Abad, T. Abil, A. Santos, J.J. Kozak, Random walks on lattices. Influence of competing reaction centers on diffusion-controlled processes. Physica A **511**, 336–357 (2018)
28. Z. Wu, G. Xu, Mortal random walks on a family of treelike regular fractals with a deep trap. Int. J. Mod. Phys. B **34**, 2050031 (2020)
29. B. Meerson, Mortal Brownian motion: Three short stories. Int. J. Mod. Phys. B **33**, 1950172 (2019)
30. E. Abad, S.B. Yuste, K. Lindenberg, Elucidating the role of subdiffusion and evanescence in the target problem: some recent results. Math. Model. Nat. Phenom. **8**, 100–113 (2013)
31. A.M. Berezhkovskii, Y.A. Makhnovskii, R.A. Suris, Wiener sausage volume moments. J. Stat. Phys. **57**, 333–346 (1989)
32. M.R. Evans, S.N. Majumdar, Diffusion with Stochastic Resetting. Phys. Rev. Lett. **106**, 160601 (2011)
33. M.R. Evans, S.N. Majumdar, G. Schehr, Stochastic resetting and applications. J. Phys. A Math. Theor. **53**, 193001 (2020)
34. S. Belan, Restart could optimize the probability of success in a Bernoulli trial. Phys. Rev. Lett. **120**, 080601 (2018)
35. M. Radice, Effects of mortality on stochastic search processes with resetting. Phys. Rev. E **107**, 024136 (2023)
36. C.T. Harbison et al., Transcriptional regulatory code of a eukaryotic genome. Nature **431**, 99–104 (2004)
37. Z. Schuss, K. Basnayake, D. Holcman, Redundancy principle and the role of extreme statistics in molecular and cellular biology. Phys. Life Rev. **28**, 52–79 (2019)

38. S.P. Wang, W.J. Pei, First passage time of multiple Brownian particles on networks with applications. Physica A **387**, 4699–4708 (2008)
39. G.H. Weiss, K.E. Shuler, K. Lindenberg, Order statistics for first passage times in diffusion processes. J. Stat. Phys. **31**, 255–278 (1983)
40. C. Mejía-Monasterio, G. Oshanin, G. Schehr, First passages for a search by a swarm of independent random searchers. J. Stat. Mech. P06022 (2011)
41. K. Lindenberg, V. Seshadri, K.E. Shuler, G.H. Weiss, Lattice random walks for sets of random walkers. First passage times. J. Stat. Phys. **23**, 11–25 (1980)
42. S.B. Yuste, K. Lindenberg, Order statistics for first passage times in one-dimensional diffusion processes. J. Stat. Phys. **85**, 501–512 (1996)
43. S.B. Yuste, L. Acedo, Multiparticle trapping problem in the half-line. Physica A **297**, 321–336 (2001)
44. S.D. Lawley, Distribution of extreme first passage times of diffusion. J. Math. Biol. **80**, 2301–2325 (2020)
45. S.D. Lawley, J.B. Madrid, A probabilistic approach to extreme statistics of Brownian escape times in dimensions 1, 2, and 3. J. Nonlinear Sci. **30**, 1207–1227 (2020)
46. S.D. Lawley, Universal formula for extreme first passage statistics of diffusion. Phys. Rev. E **101**, 012413 (2020)
47. S. Toste, D. Holcman, Extreme diffusion with point-sink killing fields: application to fast calcium signaling at synapses. SIAM J. Appl. Math. **84**, S269–S296 (2024)
48. S.B. Yuste, L. Acedo, Diffusion of a set of random walkers in Euclidean media, first passage times. J. Phys. A **33**, 507–512 (2000)
49. S.B. Yuste, L. Acedo, K. Lindenberg, Order statistics for $d$-dimensional diffusion processes. Phys. Rev. E **64**, 052102 (2001)
50. S.B. Yuste, L. Acedo, Order statistics of the trapping problem. Phys. Rev. E **64**, 061107 (2001)
51. S.B. Yuste, L. Acedo, Order statistics of Rosenstock's trapping problem in disordered media. Phys. Rev. E **68**, 036134 (2003)
52. L. Acedo, S.B. Yuste, Survival probability and order statistics of diffusion on disordered media. Phys. Rev. E **66**, 011110 (2002)
53. S.B. Yuste, Escape times of $N$ random walkers from a fractal labyrinth. Phys. Rev. Lett. **79**, 3565–3568 (1997)
54. S.B. Yuste, Order statistics of diffusion on fractals. Phys. Rev. E **57**, 6327–6333 (1998)
55. S.D. Lawley, J. Johnson, Slowest first-passage times, redundancy, and menopause timing. J. Math. Biol. **86**, 90 (2023)
56. S.B. Yuste, J.J. Ruiz-Lorenzo, K. Lindenberg, Target problem with evanescent subdiffusive traps. Phys. Rev. E **74**, 046119 (2006)
57. I.M. Sokolov, M.G.W. Schmidt, F. Sagués, Reaction-subdiffusion equations. Phys. Rev. E **73**, 031102 (2006)
58. B.I. Henry, T.A.M. Langlands, S.L. Wearne, Anomalous diffusion with linear reaction dynamics. Phys. Rev. E **74**, 031116 (2006)
59. A. Zoia, Continuous-time random-walk approach to normal and anomalous reaction-diffusion processes. Phys. Rev. E **77**, 041115 (2008)
60. D. Campos, V. Méndez, Nonuniversality and the role of tails in reaction-subdiffusion fronts. Phys. Rev. E **80**, 021133 (2009)
61. E. Abad, S.B. Yuste, K. Lindenberg, Reaction-subdiffusion and reaction-superdiffusion equations for evanescent particles performing continuous-time random walks. Phys. Rev. E **81**, 031115 (2010)
62. E. Abad, S.B. Yuste, K. Lindenberg, Survival probability of an immobile target in a sea of evanescent diffusive or subdiffusive traps: A fractional equation approach. Phys. Rev. E **86**, 061120 (2012)
63. E. Abad, S.B. Yuste, K. Lindenberg, Fractional reaction-transport equations arising from evanescent continuous time random walks, in *Fractional Calculus: Theory*, ed. by R.A.Z. Daou, X. Moreau (Nova Science Publishers, New York, 2015), pp. 183–202

64. C.N. Angstmann, I.C. Donnelly, B.I. Henry, Continuous time random walks with reactions forcing and trapping. Math. Model. Nat. Phenom. **8**, 17–27 (2013)
65. H. Stage, Aging in mortal superdiffusive Lévy walkers. Phys. Rev. E **96**, 062150 (2017)
66. S. Lawley, Anomalous reaction-diffusion equations for linear reactions. Phys. Rev. E **102**, 032117 (2020)
67. P.C. Bresloff, Encounter-based reaction-subdiffusion model I: surface adsorption and the local time propagator. J. Phys. A Math. Theor. **56**, 435004 (2023)
68. P.C. Bresloff, Encounter-based reaction-subdiffusion model II: partially absorbing traps and the occupation time propagator. J. Phys. A Math. Theor. **56**, 435005 (2023)
69. D. Campos, E. Abad, V. Méndez, S.B. Yuste, K. Lindenberg, Optimal search strategies for space-time coupled random walkers with finite lifetimes. Phys. Rev. E **91**, 052115 (2015)
70. S. Fedotov, A. Tan, A. Zubarev, Persistent random walk of cells involving anomalous effects and random death. Phys. Rev. E **91**, 042124 (2015)
71. G. Hornung, B. Berkowitz, N. Barkai, Morphogen gradient formation in a complex environment: an anomalous diffusion model. Phys. Rev. E **72**, 041916 (2005)
72. V.P. Shkilev, Subdiffusion with the disappearance of particles at the time of a jump. J. Exp. Theor. Phys. **112**, 1071–1076 (2011)
73. O. Bénichou, S. Redner, Depletion-controlled starvation of a diffusing forager. Phys. Rev. Lett. **113**, 238101 (2014)
74. O. Bénichou, M. Chupeau, S. Redner, Role of depletion on the dynamics of a diffusing forager. J. Phys. A Math. Theor. **49**, 394003 (2016)
75. R. Metzler, J. Klafter, The random walk's guide to anomalous diffusion: a fractional dynamics approach. Phys. Rep. **339**, 1–77 (2000)
76. I. Podlubny, *Fractional Differential Equations: An Introduction to Fractional Derivatives, Fractional Differential Equations, to Methods of Their Solution and Some of Their Applications* (Academic Press, San Diego, 1999)
77. D. Grebenkov, Subdiffusion in a bounded domain with a partially absorbing-reflecting boundary. Phys. Rev. E **81**, 021128 (2010)
78. D. Grebenkov, Searching for partially reactive sites: analytical results for spherical targets. J. Chem. Phys. **132**, 034104 (2010)
79. S.B. Yuste, E. Abad, C. Escudero, Diffusion in an expanding medium: Fokker-Planck equation. Green's function, and first-passage properties. Phys. Rev. E **94**, 032118 (2016)
80. E. Abad, C. Escudero, F. Le Vot, S.B. Yuste, First-passage processes and encounter-controlled reactions, in *Growing Domains*, in *Chemical Kinetics: Beyond the Textbook*, ed. by K. Lindenberg, R. Metzler, G. Oshanin (World Scientific, New Jersey, 2019), pp. 409–433
81. E. Abad, C.N. Angstmann, B.I. Henry, A.V. McGann, F. Le Vot, S.B. Yuste, Reaction-diffusion and reaction-subdiffusion equations on arbitrarily evolving domains. Phys. Rev. E **102**, 032111 (2020)
82. D. Holcman, Z. Schuss, The narrow escape problem. SIAM Rev. **56**, 213–257 (2014)
83. D. Holcman, A. Marchewka, Z. Schuss, Survival probability of diffusion with trapping in cellular neurobiology. Phys. Rev. E **72**, 031910 (2005)
84. D.S. Grebenkov, R. Metzler, G. Oshanin, From single-particle stochastic kinetics to macroscopic reaction rates: fastest first-passage time of $N$ random walkers. New J. Phys. **22**, 103004 (2020)

# Chapter 8
# Semi-permeable Interfaces and the Target Problem

**Paul C. Bressloff**

**Abstract** Diffusion through semi-permeable interfaces arises in a wide range of processes, including reverse osmosis for desalination, molecular transport in biological cells, porous media, and animal migration in heterogeneous landscapes. In this chapter, we review recent work on first passage time (FPT) problems for a particle to be absorbed within some target domain $\mathcal{U}$ whose boundary $\partial\mathcal{U}$ acts as a semi-permeable interface. We begin by considering an interface with a constant permeability $\kappa_0$ and a target with a constant rate of absorption $\gamma$. Quantities of interest such as the survival probability, absorption flux, and the FPT density can be determined by solving a Robin boundary value problem for the forward diffusion equation in Laplace space. We present a general method of solution based on Green's functions and the spectral decomposition of certain Dirichlet-to-Neumann (D-to-N) operators. We then extend the theory by considering modifications in either the absorption process or the transport process across the interface, both of which are non-Markovian. First, we introduce an encounter-based model of partial absorption within the target $\mathcal{U}$. Encounter-based models consider the joint probability density or generalized propagator for particle position and the amount of particle-target contact time prior to absorption. Absorption occurs when the contact time exceeds a random threshold. If the probability distribution of the latter is an exponential function, then one recovers the Markovian example of absorption at a constant rate, whereas a non-exponential distribution signifies non-Markovian absorption. In the case of a partially absorbing target $\mathcal{U}$ (surface $\partial\mathcal{U}$) the contact time is given by a Brownian functional known as the occupation time (boundary local time). Second, we describe a more general probabilistic model of single-particle diffusion through a semi-permeable interface, by combining the encounter-based approach with so-called snapping out Brownian motion (BM). Snapping out BM sews together successive rounds of partially absorbing BMs that are restricted to either the interior or the exterior of $\partial\mathcal{U}$. The rule for terminating each round is implemented using the encounter-based model. We show that this results in a time-dependent permeability that can be heavy-tailed.

---

P. C. Bressloff (✉)
Department of Mathematics, Imperial College London, London SW7 2AZ, UK
e-mail: pbresslo@ic.ac.uk



D. Grebenkov et al. (eds.), *Target Search Problems*,
https://doi.org/10.1007/978-3-031-67802-8_8

## 8.1 Introduction

A classical example of a target problem is shown in Fig. 8.1. A Brownian particle or searcher is confined within some bounded domain $\Omega \subset \mathbb{R}^d$ that contains an interior target $\mathcal{U} \subset \Omega$. Assuming that the exterior boundary $\partial\Omega$ is totally reflecting, one is typically interested in calculating the statistics of the first passage time (FPT) for the particle to be absorbed by the target. The details will depend on the nature of the target and its surface interface $\partial\mathcal{U}$. The most common scenario is shown in Fig. 8.1(a), where $\partial\mathcal{U}$ is a partially reactive surface. That is, whenever the particle encounters the target boundary, it is absorbed at some constant rate $\kappa_0$ or reflected back into the domain $\Omega\backslash\mathcal{U}$. (In the limit $\kappa_0 \to \infty$, the interface $\partial\mathcal{U}$ becomes totally absorbing.) An alternative scenario is illustrated in Fig. 8.1(b), where $\partial\mathcal{U}$ now acts as a semi-permeable membrane surrounding a partially absorbing target $\mathcal{U}$. The particle flux across the interface is continuous but there is a jump discontinuity in the density. Whenever the particle diffuses within $\mathcal{U}$, it is absorbed at some constant rate $\gamma$. One notable example of the latter scenario is the lateral diffusion of neurotransmitter receptors within the plasma membrane of a neuron. The partially absorbing traps correspond to local synaptic trapping regions that bind receptors to scaffolding proteins, followed by internalization of the receptors via endocytosis [1–6]. Treating the synaptic interfaces as semi-permeable membranes is motivated by the so-called partitioned fluid-mosaic model of the plasma membrane [7], in which confinement domains are formed by a fluctuating network of cytoskeletal fence proteins combined with transmembrane picket proteins that act as fence posts.

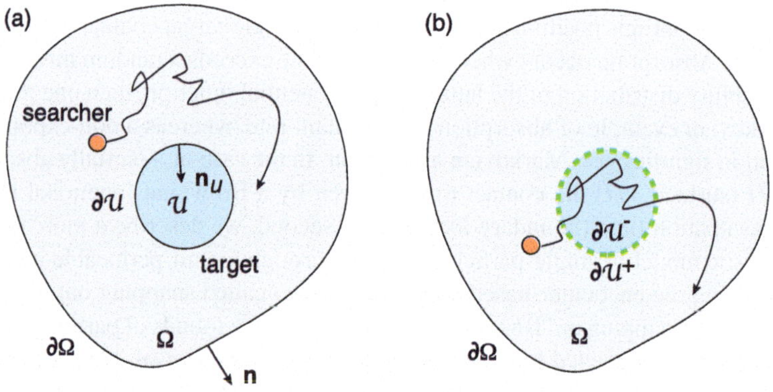


**Fig. 8.1** Diffusion of a particle (searcher) in a bounded domain $\Omega$ with $\mathcal{U} \subset \Omega \subset \mathbb{R}^d$. Partially reactive interface $\partial\mathcal{U}$ in which the probability of absorption depends on the particle-surface encounter time (boundary local time). (The outward unit normal at a point on $\partial\Omega$ is denoted by $\mathbf{n}$ and the inward unit normal at a point on $\partial\mathcal{U}$ is denoted by $\mathbf{n}_u$.) **b** Semi-permeable interface $\partial\mathcal{U}$ with a partially absorbing target interior or trap $\mathcal{U}$. The particle can now enter and exit the target. The probability flux across $\partial\mathcal{U}$ is continuous, but there is a jump discontinuity in the probability density that depends on the permeability of $\partial\mathcal{U}$. The probability of particle absorption depends on the amount of time spent within $\mathcal{U}$ (occupation time)

In this chapter, we review our recent work on FPT problems for absorption by a target $\mathcal{U}$ whose interface $\partial\mathcal{U}$ is semi-permeable. For pedagogical reasons, we focus on a single Brownian particle searching for a single target in a bounded domain. We begin by writing down the forward diffusion equation for the target problem shown in Fig. 8.1b, and define various quantities of interest such as the survival probability, absorption flux, and the FPT density (see Sect. 8.2). We then present a general method of solution based on Green's functions and the spectral decomposition of so-called Dirichlet-to-Neumann (D-to-N) operators. Such methods have previously been applied to the target problem of Fig. 8.1a [8, 9] and the target problem of Fig. 8.1b when the interface $\partial\mathcal{U}$ is transparent [10]. In the multidimensional case ($d > 1$), $\partial\mathcal{U}$ is a finite-dimensional compact surface and the D-to-N operators have countably infinite spectra. This means that the solution of the diffusion equation takes the form of an infinite series that requires inverting an infinite-dimensional matrix. In Sect. 8.4, we consider the simpler problem of diffusion in the interval $\Omega = [-L', L]$ with a partially absorbing subinterval $\mathcal{U} = [-L', 0)$ and a semi-permeable membrane at $x = 0$. The D-to-N operators reduce to scalar multipliers, which allows us to derive an explicit formula for the mean FPT (MFPT) as a function of the interface permeability.

In Sect. 8.5, we use an encounter-based approach to extend the theory of partial absorption within the target $\mathcal{U}$. Encounter-based models consider the joint probability density or generalized propagator for particle position and the amount of particle-target contact time prior to absorption [9–12]. Absorption occurs when the contact time exceeds a random threshold. If the probability distribution of the latter is an exponential function, then one recovers the Markovian example of absorption at a constant rate, whereas a non-exponential distribution signifies non-Markovian absorption. In the case of a partially absorbing target $\mathcal{U}$ (surface $\partial\mathcal{U}$) the contact time is given by a Brownian functional known as the occupation time (boundary local time). We illustrate the theory using the 1D example of Sect. 8.4. In particular, we derive an explicit expression for the MFPT that depends on various moments of the occupation time threshold distribution. Finally, in Sect. 8.6, we develop a more general probabilistic model of single-particle diffusion through semi-permeable interfaces, by combining the encounter-based approach with so-called snapping out Brownian motion (BM) [13–15]. The latter was originally formulated for 1D single-particle diffusion through a semi-permeable barrier [16–18], but has recently been extended to higher spatial dimensions [14]. Snapping out BM sews together successive rounds of partially absorbing BMs that are restricted to either the interior or the exterior of $\partial\mathcal{U}$. The rule for terminating each round is implemented using the encounter-based model of partially absorbing BM introduced in Ref. [9]. We show that this results in a time-dependent permeability that can be heavy-tailed.

## 8.2 Single Target with a Semi-permeable Interface

Consider the single target problem shown in Fig. 8.1(b), in which a semi-permeable interface $\partial\mathcal{U}$ surrounds a partially absorbing interior $\mathcal{U}$, with $\partial\mathcal{U}^+$ ($\partial\mathcal{U}^-$) denoting the side approached from outside (inside) $\mathcal{U}$. Let $p(\mathbf{x}, t|\mathbf{x}_0)$ denote the probability density that the particle position $\mathbf{X}_t$ is in a neighborhood of $\mathbf{x} \in \Omega\backslash\mathcal{U}$ at time $t$, given that it started at $\mathbf{x}_0$. That is, $p(\mathbf{x}, t|\mathbf{x}_0)d\mathbf{x} = \mathbb{P}[\mathbf{x} < \mathbf{X}_t < \mathbf{x} + d\mathbf{x}|\mathbf{X}_0 = \mathbf{x}_0]$. Denote the corresponding probability density within $\mathcal{U}$ by $q(\mathbf{x}, t|\mathbf{x}_0)$. The forward diffusion equation takes the form

$$\frac{\partial p(\mathbf{x}, t|\mathbf{x}_0)}{\partial t} = D\nabla^2 p(\mathbf{x}, t|\mathbf{x}_0), \quad \mathbf{x} \in \Omega\backslash\mathcal{U}, \quad \nabla p \cdot \mathbf{n} = 0, \quad \mathbf{x} \in \partial\Omega, \tag{8.1a}$$

$$\frac{\partial q(\mathbf{x}, t|\mathbf{x}_0)}{\partial t} = D\nabla^2 q(\mathbf{x}, t|\mathbf{x}_0) - \gamma q(\mathbf{x}, t|\mathbf{x}_0), \quad \mathbf{x} \in \mathcal{U}, \tag{8.1b}$$

where $\mathbf{n}$ is the outward unit normal at a point on the boundary $\partial\Omega$ and $\gamma$ is the rate at which the particle is absorbed within the target $\mathcal{U}$. These are supplemented by the semi-permeable boundary conditions

$$D\nabla p(\mathbf{x}^+, t|\mathbf{x}_0) \cdot \mathbf{n}_u = D\nabla q(\mathbf{x}^-, t|\mathbf{x}_0) \cdot \mathbf{n}_u \equiv -\mathcal{J}(\mathbf{x}, t|\mathbf{x}_0) \tag{8.1c}$$

$$\mathcal{J}(\mathbf{x}, t|\mathbf{x}_0) = \kappa\left[(1-\alpha)p(\mathbf{x}^+, t|\mathbf{x}_0) - \alpha q(\mathbf{x}^-, t|\mathbf{x}_0)\right], \quad \mathbf{x}^\pm \in \partial\mathcal{U}^\pm, \tag{8.1d}$$

where $\mathbf{n}_u$ denotes the inward unit normal at a point on $\partial\mathcal{U}$, $\mathcal{J}(\mathbf{x}, t|\mathbf{x}_0)$ is the continuous inward flux across the point $\mathbf{x} \in \partial\mathcal{U}$, $\kappa$ is the permeability of the interface $\partial\mathcal{U}$, and $\alpha \in [0, 1]$ specifies a directional bias with $\alpha = 1/2$ the unbiased case. (One could also take the diffusivities within $\Omega\backslash\mathcal{U}$ and $\mathcal{U}$ to be different.) Equations (8.1c) and (8.1d) are one version of the well-known Kedem-Katchalsky (KK) equations [19–21]. Note that in the limit $\kappa \to \infty$ with $\alpha = 1/2$, the interface is transparent and we obtain the pair of continuity equations

$$D\nabla p(\mathbf{x}^+, t|\mathbf{x}_0) \cdot \mathbf{n}_u = D\nabla q(\mathbf{x}^-, t|\mathbf{x}_0) \cdot \mathbf{n}_u, \quad p(\mathbf{x}^+, t|\mathbf{x}_0) = q(\mathbf{x}^-, t|\mathbf{x}_0) \tag{8.2}$$

for $\mathbf{x}^\pm \in \partial\mathcal{U}^\pm$. On the other hand, if $\kappa = 0$, then the interface is totally reflecting on both sides. Finally, if the particle started outside the domain $\mathcal{U}$, then in the limit $\gamma \to \infty$ the particle is absorbed as soon as it hits the target boundary, see Fig. 8.1a. Hence, we recover a totally absorbing target with $p$ evolving according to Eq. (8.1a) such that $p(\mathbf{x}, t|\mathbf{x}_0) = 0$ for all $\mathbf{x} \in \partial\mathcal{U}$.

Consider the survival probability that the particle hasn't been absorbed by the target in the time interval $[0, t]$, having started at $\mathbf{x}_0$ [22]:

$$S(\mathbf{x}_0, t) = \int_{\Omega\backslash\mathcal{U}} p(\mathbf{x}, t|\mathbf{x}_0)d\mathbf{x} + \int_{\mathcal{U}} q(\mathbf{x}, t|\mathbf{x}_0)d\mathbf{x}. \tag{8.3}$$

Differentiating both sides of this equation with respect to $t$ and using Eqs. (8.1a)–(8.1d) together with the divergence theorem gives

$$\begin{aligned}\frac{\partial S(\mathbf{x}_0,t)}{\partial t} =& D\int_{\partial\mathcal{U}^+} \nabla p(\mathbf{x},t|\mathbf{x}_0)\cdot\mathbf{n}_u d\sigma - D\int_{\partial\mathcal{U}^-} \nabla q(\mathbf{x},t|\mathbf{x}_0)\cdot\mathbf{n}_u d\sigma \\ &-\gamma\int_{\mathcal{U}} q(\mathbf{x},t|\mathbf{x}_0)d\mathbf{x}. \end{aligned} \tag{8.4}$$

Continuity of the flux across the interface implies that the first two terms on the right-hand side cancel, so that

$$\frac{\partial S(\mathbf{x}_0,t)}{\partial t} = -\gamma\int_{\mathcal{U}} q(\mathbf{x},t|\mathbf{x}_0)d\mathbf{x} \equiv -J(\mathbf{x}_0,t), \tag{8.5}$$

where $J(\mathbf{x}_0,t)$ is the total absorption flux within $\mathcal{U}$. It follows that we can identify $J(\mathbf{x}_0,t)$ with the FPT density $f(\mathbf{x}_0,t)$. In particular, the moments of the FPT density can be written as

$$\begin{aligned}\mathbb{E}[\mathcal{T}^n] &= \int_0^\infty J(\mathbf{x}_0,t)t^n dt = \lim_{s\to 0}\left(-\frac{\partial}{\partial s}\right)^n \int_0^\infty \mathrm{e}^{-st}J(\mathbf{x}_0,t)dt \\ &= \lim_{s\to 0}\left(-\frac{\partial}{\partial s}\right)^n \widetilde{J}(\mathbf{x}_0,s), \end{aligned} \tag{8.6}$$

where $\widetilde{J}(\mathbf{x}_0,s)$ is the Laplace transformed flux. In other words, the latter acts as the moment generating function for the FPT density. Finally, the Laplace transformed fluxes $\widetilde{J}(\mathbf{x}_0,s)$ and $\widetilde{\mathcal{J}}(\mathbf{x},s|\mathbf{x}_0)$ can be related as follows. First, equation (8.5) implies

$$\begin{aligned}\frac{\partial J(\mathbf{x}_0,t)}{\partial t} &= \gamma D\int_{\mathcal{U}} \nabla^2 q(\mathbf{x},t|\mathbf{x}_0)d\mathbf{x} - \gamma J(\mathbf{x}_0,t) \\ &= \gamma\int_{\partial\mathcal{U}} \mathcal{J}(\mathbf{x},t|\mathbf{x}_0)d\mathbf{x} - \gamma J(\mathbf{x}_0,t), \end{aligned} \tag{8.7}$$

where $\mathcal{J}(\mathbf{x},t|\mathbf{x}_0) = -\nabla q(\mathbf{x},t|\mathbf{x}_0)\cdot\mathbf{n}_u$. Laplace transforming Eq. (8.7) with respect to $t$ and using the initial condition $J(\mathbf{x}_0,0)=0$ for $\mathbf{x}_0\notin\mathcal{U}$, we have

$$(s+\gamma)\widetilde{J}(\mathbf{x}_0,s) = \gamma\int_{\partial\mathcal{U}} \widetilde{\mathcal{J}}(\mathbf{x},s|\mathbf{x}_0)d\mathbf{x}. \tag{8.8}$$

In particular, for all $\mathbf{x}_0\notin\mathcal{U}$

$$T(\mathbf{x}_0) \equiv -\left.\frac{\partial}{\partial s}\widetilde{J}(\mathbf{x}_0,s)\right|_{s=0}. \tag{8.9}$$

Finally, in order to determine $\widetilde{J}(\mathbf{x}_0, s)$, it is necessary to solve the forward diffusion equation in Laplace space:

$$D\nabla^2\widetilde{p}(\mathbf{x}, s|\mathbf{x}_0) - s\widetilde{p}(\mathbf{x}, s|\mathbf{x}_0) = -\delta(\mathbf{x}-\mathbf{x}_0), \quad \mathbf{x}, \mathbf{x}_0 \in \Omega\backslash\mathcal{U}, \tag{8.10a}$$

$$-\nabla\widetilde{p}(\mathbf{x}, s|\mathbf{x}_0)\cdot\mathbf{n} = 0, \quad \mathbf{x} \in \partial\Omega, \tag{8.10b}$$

$$D\nabla^2\widetilde{q}(\mathbf{x}, s|\mathbf{x}_0) - (s+\gamma)\widetilde{q}(\mathbf{x}, s|\mathbf{x}_0) = 0, \quad \mathbf{x} \in \mathcal{U} \tag{8.10c}$$

$$D\nabla\widetilde{p}(\mathbf{x}^+, s|\mathbf{x}_0)\cdot\mathbf{n}_u = D\nabla\widetilde{q}(\mathbf{x}^-, s|\mathbf{x}_0)\cdot\mathbf{n}_u \equiv -\widetilde{\mathcal{J}}(\mathbf{x}, s|\mathbf{x}_0) \tag{8.10d}$$

$$\widetilde{\mathcal{J}}(\mathbf{x}, s|\mathbf{x}_0) = \kappa\left[(1-\alpha)\widetilde{p}(\mathbf{x}, s|\mathbf{x}_0) - \alpha\widetilde{q}(\mathbf{x}, s|\mathbf{x}_0)\right], \quad \mathbf{x} \in \partial\mathcal{U}. \tag{8.10e}$$

For the sake of illustration, we have taken $\mathbf{x}_0 \in \Omega\backslash\mathcal{U}$.

## 8.3 Green's Functions and Spectral Decompositions

As highlighted in the previous section, one way to calculate the moments of the FPT density for a partially absorbing target $\partial\mathcal{U}$ is to solve the forward diffusion equation in Laplace space, which yields the Laplace transformed target flux $\widetilde{J}(\mathbf{x}_0, s)$. In the case of a partially absorbing interface $\partial\mathcal{U}$, a general method for solving the corresponding Robin boundary value problem (BVP) is based on a spectral decomposition of a so-called Dirichlet-to-Neumann (D-to-N) operator [9]. The analogous spectral analysis for a semi-permeable interface $\partial\mathcal{U}$ is considerably more involved when $d \geq 2$. This is true even in the infinite permeability limit $\kappa \to \infty$ with $\alpha = 1/2$, for which the interface $\mathcal{U}$ becomes completely transparent. The latter example was analyzed in Ref. [10] by replacing the continuity equations (8.2) with the inhomogeneous Dirichlet conditions (in Laplace space) $\widetilde{p}(\mathbf{x}, s|\mathbf{x}_0) = \widetilde{q}(\mathbf{x}, s|\mathbf{x}_0) = f(\mathbf{x}, s)$ for all $\mathbf{x} \in \partial\mathcal{U}$. Here we consider the case of finite $\kappa$. The first step is to replace the semi-permeable boundary conditions (8.1d,e) with a pair of Dirichlet conditions $\widetilde{p}(\mathbf{x}, s|\mathbf{x}_0) = f(\mathbf{x}, t)$ and $\widetilde{q}(\mathbf{x}, s|\mathbf{x}_0) = \overline{f}(\mathbf{x}, s)$ for the unknown functions $f, \overline{f}$. The general solution of Eqs. (8.10a)–(8.10c) can then be written in the form

$$\widetilde{p}(\mathbf{x}, s|\mathbf{x}_0) = \mathcal{F}(\mathbf{x}, s) + G(\mathbf{x}, s|\mathbf{x}_0), \quad \mathbf{x} \in \Omega\backslash\mathcal{U}, \tag{8.11a}$$

$$\widetilde{q}(\mathbf{x}, s|\mathbf{x}_0) = \overline{\mathcal{F}}(\mathbf{x}, s), \quad \mathbf{x} \in \mathcal{U}, \tag{8.11b}$$

where

$$\mathcal{F}(\mathbf{x}, s) = -D\int_{\partial\mathcal{U}} \partial_{\sigma'}G(\mathbf{x}', s|\mathbf{x})f(\mathbf{x}', s)d\mathbf{x}', \tag{8.12a}$$

$$\overline{\mathcal{F}}(\mathbf{x}, s) = D\int_{\partial\mathcal{U}} \partial_{\sigma'}\overline{G}(\mathbf{x}', s+\gamma|\mathbf{x})\overline{f}(\mathbf{x}', s)d\mathbf{x}', \tag{8.12b}$$

and $\partial_{\sigma'} = \nabla_{\mathbf{x}'} \cdot \mathbf{n}_u$. We have introduced the modified Helmholtz Green's functions $G$ and $\overline{G}$ for the two domains $\mathcal{U}^c = \Omega\backslash\mathcal{U}$ and $\mathcal{U}$, respectively:

$$D\nabla^2 G(\mathbf{x}, s|\mathbf{x}') - sG(\mathbf{x}, s|\mathbf{x}') = -\delta(\mathbf{x} - \mathbf{x}'), \quad \mathbf{x}, \mathbf{x}' \in \Omega\backslash\mathcal{U}, \tag{8.13a}$$

$$G(\mathbf{x}, s|\mathbf{x}') = 0, \ \mathbf{x} \in \partial\mathcal{U}, \quad \nabla G(\mathbf{x}, s|\mathbf{x}') \cdot \mathbf{n} = 0, \quad \mathbf{x} \in \partial\Omega, \tag{8.13b}$$

$$D\nabla^2 \overline{G}(\mathbf{x}, s|\mathbf{x}') - s\overline{G}(\mathbf{x}, s|\mathbf{x}') = -\delta(\mathbf{x} - \mathbf{x}'), \quad \mathbf{x}, \mathbf{x}' \in \mathcal{U}, \tag{8.13c}$$

$$\overline{G}(\mathbf{x}, s|\mathbf{x}') = 0, \quad \mathbf{x} \in \partial\mathcal{U}. \tag{8.13d}$$

The Green's functions have dimensions of [time]/[length]$^d$

The unknown functions $f, \overline{f}$ are determined by substituting the solutions (8.11a, b) into Eq. (8.10d, e):

$$\mathbb{L}_s[f](\mathbf{x}, s) + \partial_\sigma G(\mathbf{x}, s|\mathbf{x}_0) = -\overline{\mathbb{L}}_{s+\gamma}[\overline{f}](\mathbf{x}, s), \tag{8.14a}$$

$$\mathbb{L}_s[f](\mathbf{x}, s) + \partial_\sigma G(\mathbf{x}, s|\mathbf{x}_0) = -\frac{\kappa}{D}\left[(1-\alpha)f(\mathbf{x}, s) - \alpha\overline{f}(\mathbf{x}, s)\right], \quad \mathbf{x} \in \partial\mathcal{U}, \tag{8.14b}$$

where $\mathbb{L}_s$ and $\overline{\mathbb{L}}_s$ are the D-to-N operators

$$\mathbb{L}_s[f](\mathbf{x}, s) = -D\partial_\sigma \int_{\partial\mathcal{U}} \partial_{\sigma'} G(\mathbf{x}', s|\mathbf{x}) f(\mathbf{x}', s) d\mathbf{x}', \tag{8.15a}$$

$$\overline{\mathbb{L}}_s[\overline{f}](\mathbf{x}, s) = -D\partial_\sigma \int_{\partial\mathcal{U}} \partial_{\sigma'} \overline{G}(\mathbf{x}', s|\mathbf{x}) \overline{f}(\mathbf{x}', s) d\mathbf{x}'. \tag{8.15b}$$

acting on the space $L_2(\partial\mathcal{U})$. The D-to-N operators $\mathbb{L}_s$ and $\overline{\mathbb{L}}_s$ both have discrete spectra. That is, there exist countable sets of eigenvalues $\lambda_n(s), \overline{\lambda}_n(s)$ and eigenfunctions $v_n(\mathbf{x}, s), \overline{v}_n(\mathbf{x}, s)$ satisfying (for fixed $s$)

$$\mathbb{L}_s v_n(\mathbf{x}, s) = \lambda_n(s) v_n(\mathbf{x}, s), \quad \overline{\mathbb{L}}_s \overline{v}_n(\mathbf{x}, s) = \overline{\lambda}_n(s) \overline{v}_n(\mathbf{x}, s). \tag{8.16}$$

We can now solve Eq. (8.14) by introducing the eigenfunction expansions

$$f(\mathbf{x}, s) = \sum_{m=0}^{\infty} f_m(s) v_m(\mathbf{x}, s), \quad \overline{f}(\mathbf{x}, s) = \sum_{m=0}^{\infty} \overline{f}_m(s) v_m(\mathbf{x}, s). \tag{8.17}$$

Substituting Eq. (8.17) into (8.14) and taking the inner product with the adjoint eigenfunctions $v_n^*(\mathbf{x}, s)$ yields the following matrix equations for the coefficients $f_m, \overline{f}_m$:

$$\lambda_n(s) f_n(s) - g_n(s) = -\sum_{m\geq 1} H_{nm}(s+\gamma)\overline{f}_m(s), \tag{8.18a}$$

$$\lambda_n(s) f_n(s) - g_n(s) = -\kappa\left[(1-\alpha) f_n(s) - \alpha\overline{f}_n(s)\right], \tag{8.18b}$$

where

$$g_n(s) = -\int_{\partial\mathcal{U}} v_n^*(\mathbf{x}, s)\partial_\sigma G(\mathbf{x}, s|\mathbf{x}_0) d\mathbf{x} = \frac{1}{D}\mathcal{V}_n^*(\mathbf{x}_0, s), \tag{8.19a}$$

$$\mathcal{F}_n(s) = D\int_{\partial\mathcal{U}} v_n^*(\mathbf{x}, s)\mathcal{V}_n(\mathbf{x}, s) d\mathbf{x}, \tag{8.19b}$$

$$\overline{\mathcal{F}}_n(s) = D\int_{\partial\mathcal{U}} v_n^*(\mathbf{x}, s)\overline{\mathcal{V}}_n(\mathbf{x}, s) d\mathbf{x}, \tag{8.19c}$$

$$H_{nm}(s) = D\int_{\partial\mathcal{U}} v_n^*(\mathbf{x}, s)\partial_\sigma\mathcal{V}_m(\mathbf{x}, s) d\mathbf{x}. \tag{8.19d}$$

Here $\mathcal{V}_n$ and $\overline{\mathcal{V}}_n$ are defined according to

$$\mathcal{V}_n(\mathbf{x}, s) = -D\int_{\partial\mathcal{U}} v_n(\mathbf{x}', s)\partial_{\sigma'} G(\mathbf{x}', s|\mathbf{x}) d\mathbf{x}', \tag{8.20a}$$

$$\overline{\mathcal{V}}_n(\mathbf{x}, s) = D\int_{\partial\mathcal{U}} v_n(\mathbf{x}', s)\partial_{\sigma'}\overline{G}(\mathbf{x}', s|\mathbf{x}) d\mathbf{x}'. \tag{8.20b}$$

Equation (8.18b) implies that

$$\overline{f}_n(s) = \frac{1}{\kappa\alpha}\left[(\lambda_n(s) + \kappa(1-\alpha)) f_n(s) - g_n(s)\right]. \tag{8.21}$$

Introducing the vectors $\mathbf{f}(s) = (f_n(s), n \geq 0)$ and $\mathbf{g}(s) = (g_n(s), n \geq 0)$, we can now formally write the solution of Eq. (8.18a) as

$$\begin{aligned}\mathbf{f}(s) &= \left[\mathbf{M}(s) + \frac{1}{\kappa\alpha}\mathbf{H}(s+\gamma)\left[\mathbf{M}(s) + \kappa(1-\alpha)\mathbf{I}\right]\right]^{-1}\left[\mathbf{I} + \frac{1}{\kappa\alpha}\mathbf{H}(s+\gamma)\right]\mathbf{g}(s)\\ &\equiv \mathbf{\Lambda}_1^{-1}(s)\mathbf{\Lambda}_2(s)\mathbf{g}(s),\end{aligned} \tag{8.22}$$

where $\mathbf{H}(s)$ is the matrix with elements $H_{nm}(s)$ and $\mathbf{M}(s) = \text{diag}(\lambda_1(s), \lambda_2(s)\ldots)$. Finally, substituting Eq. (8.22) into Eq. (8.11) gives

$$\widetilde{p}(\mathbf{x}, s|\mathbf{x}_0) = G(\mathbf{x}, s|\mathbf{x}_0) + \frac{1}{D}\sum_{n,m}\mathcal{V}_n(\mathbf{x}, s)\left[\mathbf{\Lambda}_1^{-1}(s)\mathbf{\Lambda}_2(s)\right]_{nm}\mathcal{V}_m^*(\mathbf{x}_0, s), \quad \mathbf{x}\in\Omega\backslash\mathcal{U}, \tag{8.23a}$$

$$\widetilde{q}(\mathbf{x}, s|\mathbf{x}_0) = \frac{1}{D\kappa\alpha}\sum_{n,m}\overline{\mathcal{V}}_n(\mathbf{x}, s+\gamma)\Big(\lambda_n(s)+\kappa(1-\alpha)\Big)\Big[\mathbf{\Lambda}_1^{-1}(s)\mathbf{\Lambda}_2(s)\Big]_{nm}\mathcal{V}_m^*(\mathbf{x}_0, s)$$
$$-\frac{1}{D\kappa\alpha}\sum_n \overline{\mathcal{V}}_n(\mathbf{x}, s+\gamma)\mathcal{V}_n^*(\mathbf{x}_0, s), \quad \mathbf{x}\in\mathcal{U}. \tag{8.23b}$$

There are two distinct challenges in using the spectral decompositions (8.23) to determine the flux $\widetilde{J}(\mathbf{x}_0, s)$, and hence the FPT statistics, when $d \geq 2$:

(i) Obtaining the eigenvalues and eigenfunctions of the D-to-N operators. One higher-dimensional example where the spectral decompositions of $\mathbb{L}_s$ and $\overline{\mathbb{L}}_s$ are known exactly is a partially absorbing sphere [8]. The rotational symmetry of $\mathcal{U}$ means that if $\mathbb{L}_s$ and $\overline{\mathbb{L}}_s$ are expressed in spherical polar coordinates $(r, \theta, \phi)$, then the eigenfunctions are given by spherical harmonics, and are independent of the Laplace variable $s$ and the radius $r$:

$$v_{nm}(\theta, \phi) = \overline{v}_{nm}(\theta, \phi) = \frac{1}{R}Y_n^m(\theta, \phi), \quad n \geq 0, \ |m| \leq n. \tag{8.24}$$

From orthogonality, it follows that the adjoint eigenfunctions are

$$v_{nm}^*(\theta, \phi) = \overline{v}_{nm}^*(\theta, \phi) = (-1)^m\frac{1}{R}Y_n^{-m}(\theta, \phi). \tag{8.25}$$

(Note that eigenfunctions are labeled by the pair of indices $(nm)$.) The corresponding eigenvalues are [8]

$$\lambda_n(s) = -\beta(s)\frac{k_n'(\beta(s)R)}{k_n(\beta(s)R)}, \quad \overline{\lambda}_n(s) = \beta(s)\frac{i_n'(\beta(s)R)}{i_n(\beta(s)R)}, \tag{8.26}$$

where $\beta(s) = \sqrt{s/D}$, and $i_n(x), k_n(x)$ are modified spherical Bessel functions of the first and second kind, respectively. Since the $n$th eigenvalue is independent of $m$, it has a multiplicity $2n+1$. It is also possible to compute the projections of the boundary fluxes in (8.20) by using appropriate series expansions of the corresponding Green's functions [9].

(ii) Numerically truncating the infinite series expansions in Eq. (8.23) and inverting the matrix $\mathbf{\Lambda}_1(s)$.

## 8.4 Partially Absorbing Interval with a Semi-permeable Barrier

In the case of a one-dimensional (1D) substrate, the D-to-N operators reduce to scalar multipliers so that the difficulties of higher dimensional interfaces are avoided. As a simple illustration of this, consider a particle diffusing in the interval $\Omega = [-L', L]$ with a partially absorbing subinterval $\mathcal{U} = [-L', 0)$ and a semi-permeable

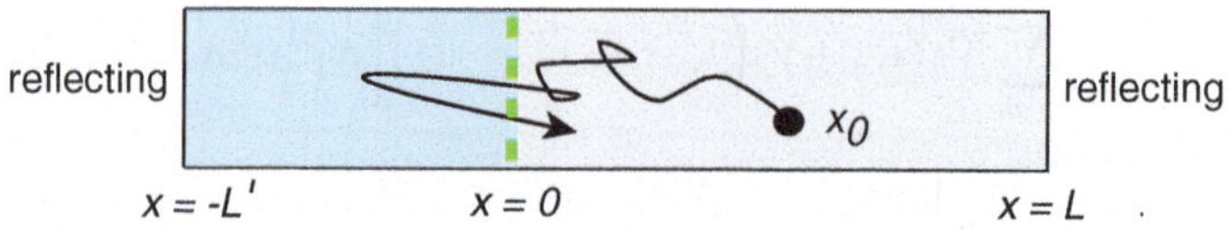


**Fig. 8.2** Partially absorbing substrate in 1D with $\mathcal{U} = [-L', 0]$ and $\Omega\backslash\mathcal{U} = [0, L]$

membrane at $x = 0$, see Fig. 8.2. It follows that $\partial\Omega = \{-L', L\}$ and $\partial\mathcal{U} = \{0\}$. The 1D version of Eq. (8.1) takes the form

$$\frac{\partial p(x,t|x_0)}{\partial t} = D\frac{\partial^2 p(x,t|x_0)}{\partial x^2}, \quad x \in (0, L), \quad \partial_x p(L, t|x_0) = 0, \tag{8.27a}$$

$$\frac{\partial q(x,t|x_0)}{\partial t} = D\frac{\partial^2 q(x,t|x_0)}{\partial x^2} - \gamma q(x,t|x_0), \quad x \in (-L', 0), \tag{8.27b}$$

together with the semi-permeable boundary conditions

$$D\partial_x p(0^+, t|x_0) = D\partial_x q(0^-, t|x_0) \equiv -\mathcal{J}(x_0, t) \tag{8.27c}$$

$$\mathcal{J}(x_0, t) = \kappa\left[(1-\alpha)p(0^+, t|x_0) - \alpha q(0^-, t|x_0)\right]. \tag{8.27d}$$

The general solution (8.11) in Laplace space becomes

$$\widetilde{p}(x, s|x_0) = D\ \partial_{x'} G(x', s|x)\big|_{x'=0} f(s) + G(x, s|x_0), \quad x, x_0 \in (0, L], \tag{8.28a}$$

$$\widetilde{q}(x, s|x_0) = -D\ \partial_{x'} \overline{G}(x', s+\gamma|x)\big|_{x'=0} \overline{f}(s), \quad x \in [-L', 0), \tag{8.28b}$$

for the unknown functions $f, \overline{f}$. Note that $\partial_{\sigma'} = -\partial_{x'}$. Equation (8.13a) for the modified Helmholtz Green's function $G(x, s|x_0)$ reduces to

$$D\frac{d^2}{dx^2}G(x, s|x_0) - sG(x, s|x_0) = -\delta(x - x_0), \quad 0 < x, x_0 < L, \tag{8.29a}$$

$$G(0, s|x_0) = 0, \quad \frac{d}{dx}G(L, s|x_0)\bigg|_{x=L} = 0. \tag{8.29b}$$

The explicit solution is $G = G_L$, where

$$G_L(x, s|x_0) = \frac{\Theta(x_0 - x)g(x, s)\widehat{g}(x_0, s) + (x - x_0)g(x_0, s)\widehat{g}(x, s)}{\sqrt{sD}\cosh(\beta(s)L)}, \tag{8.30}$$

$\beta(s) = \sqrt{s/D}$ and

$$g(x, s) = \sinh \beta(s)x, \quad \text{and} \quad \widehat{g}(x, s) = \cosh[\beta(s)(L - x)]. \tag{8.31}$$

Similarly, $\overline{G}(x, s|x_0) = G_{L'}(-x, s|-x_0)$ for $x, x_0 \in [-L', 0)$. It also follows that

$$D\, \partial_{x'} G(x', s|x)\big|_{x'=0} = \frac{\cosh(\beta(s)[L-x]}{\cosh(\beta(s)L)}, \tag{8.32a}$$

$$-D\, \partial_{x'} \overline{G}(x', s+\gamma|x)\big|_{x'=0} = \frac{\cosh(\beta(s+\gamma)[L'+x]}{\cosh(\beta(s)L')}. \tag{8.32b}$$

Substituting for the Green's functions into Eq. (8.15) and setting $\partial_\sigma = -\partial_x$ etc., we find that

$$\mathbb{L}_s[f](s) \equiv -Df(s)\, \partial_x\partial_{x'} G(x', s|x)\big|_{x=x'=0} = f(s)\sqrt{\frac{s}{D}} \tanh(\sqrt{s/D}L), \tag{8.33a}$$

$$\overline{\mathbb{L}}_s[f](s) \equiv -Df(s)\, \partial_x\partial_{x'} \overline{G}(x', s|x)\big|_{x=x'=0} = f(s)\beta(s) \tanh(\beta(s)L'). \tag{8.33b}$$

We deduce that for 1D diffusion, the D-to-N operators reduce to scalars with single eigenvalues $\lambda(s) = \beta(s)\tanh(\beta(s)L)$ and $\overline{\lambda}(s) = \beta(s)\tanh(\beta(s)L')$. The unknown functions $f, \overline{f}$ are then determined from the 1D version of Eq. (8.14):

$$f(s)\beta(s)\tanh(\beta(s)L) - \partial_x G(0, s|x_0) = -\overline{f}(s)\beta(s+\gamma)\tanh(\beta(s+\gamma)L'), \tag{8.34a}$$

$$f(s)\beta(s)\tanh(\beta(s)L) - \partial_x G(0, s|x_0) = -\frac{\kappa}{D}\left[(1-\alpha)f(s) - \alpha\overline{f}(s)\right]. \tag{8.34b}$$

After some algebra, we find that

$$\overline{f}(s) = \Lambda(s)f(s), \quad f(s) = \frac{1}{D\Phi(s)} \frac{\cosh\beta(s)(L-x_0)}{\cosh\beta(s)L}, \tag{8.35}$$

where

$$\Lambda(s,\gamma) = \frac{1-\alpha}{\alpha} \frac{1}{1+\kappa^{-1}[D/\alpha]\beta(s+\gamma)\tanh(\beta(s+\gamma)L')}, \tag{8.36a}$$

$$\Phi(s,\gamma) = \beta(s)\tanh(\beta(s)L) + \frac{1-\alpha}{\alpha} \frac{\beta(s+\gamma)\tanh(\beta(s+\gamma)L')}{1+\kappa^{-1}[D/\alpha]\beta(s+\gamma)\tanh(\beta(s+\gamma)L')}. \tag{8.36b}$$

In the specific case $x_0 = 0^+$, the full solution has the particularly simple form

$$\widetilde{p}(x, s|0^+) = \frac{1}{\Phi(s,\gamma)D} \frac{\cosh\beta(s)(L-x)}{\cosh\beta(s)L}, \quad x \in [0, L], \tag{8.37a}$$

$$\widetilde{q}(x, s|0^+) = \frac{\Lambda(s,\gamma)}{\Phi(s,\gamma)D} \frac{\cosh\beta(s+\gamma)(L'+x)}{\cosh\beta(s+\gamma)L'}, \quad x \in [-L', 0]. \tag{8.37b}$$

From the 1D version of Eq. (8.9), the MFPT for absorption is

$$T(0^+) = \frac{1}{\gamma} - \partial_s \widetilde{\mathcal{J}}(0), \tag{8.38}$$

with

$$\widetilde{\mathcal{J}}(s) = \kappa \left[ (1-\alpha)\widetilde{p}(0, s|0^+) - \alpha \widetilde{q}(0, s|0^+) \right] = \frac{\kappa}{D\Phi(s,\gamma)} \left[ (1-\alpha) - \alpha \Lambda(s,\gamma) \right] \tag{8.39}$$

Substituting for $\Phi(s,\gamma)$ and $\Lambda(s,\gamma)$ gives

$$\widetilde{\mathcal{J}}(s) = \frac{1-\alpha}{\alpha} \frac{1}{\beta(s)\tanh(\beta(s)L)\left[\frac{1}{\beta(s+\gamma)\tanh(\beta(s+\gamma)L')} + \frac{D}{\alpha\kappa}\right] + \frac{1-\alpha}{\alpha}}. \tag{8.40}$$

Hence, the MFPT for finite $\kappa$ is

$$T(0^+) = \frac{\alpha}{1-\alpha} \frac{L}{\sqrt{\gamma D}\tanh(\sqrt{\gamma/D}L')} + \frac{L}{(1-\alpha)\kappa} + \frac{1}{\gamma}. \tag{8.41}$$

The final term on the right-hand side is the expected time for absorption when the particle is within the target, whereas the first two terms are the mean time spent outside the target, where no absorption can occur. The latter includes the contribution $L/(1-\alpha)\kappa$ associated with the effective "resistance" of the semi-permeable membrane to particle influx.

## 8.5 Encounter-Based Model of a Partially Absorbing Target

So far we have assumed that the absorption rate $\gamma$ is a constant. However, various absorption-based reactions are better modeled in terms of a reactivity that is a function of the amount of contact time between a particle and a target [23, 24]. That is, the substrate may need to be progressively activated by repeated encounters with a diffusing particle, or an initially highly reactive substrate may become less active due to multiple interactions with the particle (passivation). Recently, a so-called encounter-based approach has been developed for analyzing a more general class of partially absorbing surfaces $\partial\mathcal{U}$ [9, 11], and partially absorbing interiors $\mathcal{U}$ with totally permeable interfaces [10, 12]. Here we extend the latter to the case of a semi-permeable interface $\partial\mathcal{U}$.

### 8.5.1 Occupation Time Propagator

The basic idea of the encounter-based approach is to consider the joint probability density or generalized propagator $P(\mathbf{x}, a, t)$ for the pair $(\mathbf{X}_t, \mathcal{A}_t)$, where $\mathcal{A}_t$ is a Brownian functional that specifies the amount of contact time with the target over the time interval $[0, t]$ (in the absence of absorption). In the case of an interior target or trap $\mathcal{U}$, the functional $\mathcal{A}_t$ is identified with the occupation time [25]

$$\mathcal{A}_t = \int_0^t I_{\mathcal{U}}(\mathbf{X}_\tau)d\tau. \tag{8.42}$$

Here $I_{\mathcal{U}}(\mathbf{x})$ denotes the indicator function of the set $\mathcal{U} \subset \Omega$, that is, $I_{\mathcal{U}}(\mathbf{x}) = 1$ if $\mathbf{x} \in \mathcal{U}$ and is zero otherwise. The effects of partial absorption are then incorporated by introducing the stopping time $\mathcal{T} = \inf\{t > 0 : \ \mathcal{A}_t > \widehat{\mathcal{A}}\}$, with $\widehat{\mathcal{A}}$ a randomly distributed occupation time threshold. Given the probability distribution $\Psi(a) = \mathbb{P}[\widehat{\mathcal{A}} > a]$, the marginal probability density for particle position is defined according to

$$p^{\Psi}(\mathbf{x}, t)d\mathbf{x} = \mathbb{P}[\mathbf{X}_t \in (\mathbf{x}, \mathbf{x} + d\mathbf{x}), \ t < \mathcal{T}].$$

Since $\mathcal{A}_t$ is a nondecreasing process, the condition $t < \mathcal{T}$ is equivalent to the condition $\mathcal{A}_t < \widehat{\mathcal{A}}$. This implies that

$$\begin{aligned} p^{\Psi}(\mathbf{x}, t)d\mathbf{x} &= \mathbb{P}[\mathbf{X}_t \in (\mathbf{x}, \mathbf{x} + d\mathbf{x}), \ \mathcal{A}_t < \widehat{\mathcal{A}}] \\ &= \int_0^\infty da \ \psi(a)\mathbb{P}[\mathbf{X}_t \in (\mathbf{x}, \mathbf{x} + d\mathbf{x}), \ \mathcal{A}_t < a] \\ &= \int_0^\infty da \ \psi(a) \int_0^a da'[P(\mathbf{x}, a', t)d\mathbf{x}], \end{aligned}$$

where $\psi(a) = -\Psi'(a)$ and $P(\mathbf{x}, a, t)$ denotes the joint probability density for the pair $(\mathbf{X}_t, \mathcal{A}_t)$. Using the identity

$$\int_0^\infty dv \ f(v) \int_0^v dv' \ g(v') = \int_0^\infty dv' \ g(v') \int_{v'}^\infty dv \ f(v)$$

for arbitrary integrable functions $f, g$, it follows that

$$p^{\Psi}(\mathbf{x}, t) = \int_0^\infty \Psi(a)P(\mathbf{x}, a, t)da. \tag{8.43}$$

Let $P(\mathbf{x}, a, t|\mathbf{x}_0)$ denote the occupation time propagator under the initial conditions $\mathbf{X}_0 = \mathbf{x}_0$ and $\mathcal{A}_0 = 0$. It follows that

$$P(\mathbf{x}, a, t|\mathbf{x}_0) = \left\langle \delta\left(a - \mathcal{A}_t\right) \right\rangle_{\mathbf{X}_0 = \mathbf{x}_0}^{\mathbf{X}_t = \mathbf{x}}, \tag{8.44}$$

where expectation is taken with respect to all random paths realized by $\mathbf{X}_\tau$ between $\mathbf{X}_0 = \mathbf{x}_0$ and $\mathbf{X}_t = \mathbf{x}$. Using the Feynman-Kac formula, it can be shown that away from the boundaries $\partial\Omega$ and $\partial\mathcal{U}^\pm$, the propagator satisfies a BVP of the form [10]

$$\frac{\partial P(\mathbf{x}, a, t|\mathbf{x}_0)}{\partial t} = D\nabla^2 P(\mathbf{x}, a, t|\mathbf{x}_0) - I_{\mathcal{U}}(\mathbf{x})\frac{\partial P}{\partial a}(\mathbf{x}, a, t|\mathbf{x}_0) \; - \delta(a) I_{\mathcal{U}}(\mathbf{x}) P(\mathbf{x}, 0, t|\mathbf{x}_0), \tag{8.45a}$$

for all $\mathbf{x} \in \Omega$. Since the reflecting boundary condition on $\partial\Omega$ and the semi-permeable boundary conditions across $\partial\mathcal{U}$ are independent of the occupation time, they also hold for the propagator. We thus have the following BVP for the occupation time propagator:

$$\frac{\partial P(\mathbf{x}, a, t|\mathbf{x}_0)}{\partial t} = D\nabla^2 P(\mathbf{x}, a, t|\mathbf{x}_0), \quad \mathbf{x} \in \Omega\backslash\mathcal{U}, \tag{8.46a}$$

$$\nabla P(\mathbf{x}, a, t|\mathbf{x}_0) \cdot \mathbf{n} = 0, \quad \mathbf{x} \in \partial\Omega, \tag{8.46b}$$

$$\frac{\partial Q(\mathbf{x}, a, t|\mathbf{x}_0)}{\partial t} + \frac{\partial Q(\mathbf{x}, a, t|\mathbf{x}_0)}{\partial a} = D\nabla^2 Q(\mathbf{x}, a, t|\mathbf{x}_0) - \delta(a) Q(\mathbf{x}, 0, t|\mathbf{x}_0) \tag{8.46c}$$

for $\mathbf{x} \in \mathcal{U}$, and

$$D\nabla P(\mathbf{x}, a, t|\mathbf{x}_0) \cdot \mathbf{n}_u = D\nabla Q(\mathbf{x}, a, t|\mathbf{x}_0) \cdot \mathbf{n}_u \equiv -\mathcal{J}(\mathbf{x}, a, t|\mathbf{x}_0) \tag{8.46d}$$

$$\mathcal{J}(\mathbf{x}, a, t|\mathbf{x}_0) = \kappa\left[(1-\alpha)P(\mathbf{x}, a, t|\mathbf{x}_0) - \alpha Q(\mathbf{x}, a, t|\mathbf{x}_0)\right], \quad \mathbf{x} \in \partial\mathcal{U}. \tag{8.46e}$$

We now denote the propagator within the target $\mathcal{U}$ by $Q$. The initial conditions are $P(\mathbf{x}, a, 0|\mathbf{x}_0) = \delta(\mathbf{x} - \mathbf{x}_0)\delta(a)$, $Q(\mathbf{x}, a, 0|\mathbf{x}_0) = 0$, assuming that the particle starts in the non-absorbing region. (The analysis is easily modified if $\mathbf{x}_0 \in \mathcal{U}$.) One interesting observation is that Eq. (8.46c) takes the form of an age-structured model, reflecting the fact that whenever the particle is within the target interior $\mathcal{U}$, the occupation time $\mathcal{A}_t$ increases at the same rate as the absolute time $t$. (Age-structured models are typically found within the context of birth-death processes in ecology and cell biology, where the birth and death rates of individual organisms and cells depend on their age [26–28].) Finally, note that the term involving the Dirac delta function $\delta(a)$ in Eq. (8.46c) ensures that the probability of being in the boundary layer is zero if the occupation time is zero.

### *8.5.2 Marginal Probability Density and Flux*

Laplace transforming Eq. (8.46a, c) with respect to $a$ and setting

$$\widetilde{P}(\mathbf{x}, z, t|\mathbf{x}_0) = \int_0^\infty \mathrm{e}^{-za} P(\mathbf{x}, a, t|\mathbf{x}_0) da, \quad \widetilde{Q}(\mathbf{x}, z, t|\mathbf{x}_0) = \int_0^\infty \mathrm{e}^{-za} Q(\mathbf{x}, a, t|\mathbf{x}_0) da, \tag{8.47}$$

yields

$$\frac{\partial \widetilde{P}(\mathbf{x}, z, t|\mathbf{x}_0)}{\partial t} = D\nabla^2 \widetilde{P}(\mathbf{x}, z, t|\mathbf{x}_0), \quad \mathbf{x} \in \Omega \backslash \mathcal{U}, \tag{8.48a}$$

$$\frac{\partial \widetilde{Q}(\mathbf{x}, z, t|\mathbf{x}_0)}{\partial t} = D\nabla^2 \widetilde{Q}(\mathbf{x}, z, t|\mathbf{x}_0) - z\widetilde{Q}(\mathbf{x}, z, t|\mathbf{x}_0), \quad \mathbf{x} \in \mathcal{U}, \tag{8.48b}$$

together with the Laplace transformed versions of the boundary conditions (8.46b, d, e). We thus recover the BVP (8.1) for diffusion in a domain with a partially absorbing trap $\mathcal{U}$ with a constant rate of absorption $z$. This establishes that the original BVP for a partially absorbing trap with a constant absorption rate $z$ is recovered by taking the occupation time threshold to be an exponential random variable. That is, $\Psi(a) = \mathrm{e}^{-za}$. In other words, the marginal density $p(\mathbf{x}, t|\mathbf{x}_0)$ for a constant absorption rate $z$ is equivalent to the Laplace transform $\widetilde{P}(\mathbf{x}, z, t|\mathbf{x}_0)$ of the occupation time propagator. Assuming that the inverse Laplace transform exists, we have the general result

$$p^\Psi(\mathbf{x}, t) = \int_0^\infty \Psi(a)\, \mathrm{LT}^{-1}[\widetilde{P}](\mathbf{x}, a, t)] da. \tag{8.49}$$

The general probabilistic framework for analyzing single-particle diffusion in partially absorbing media is summarized in the commutative diagram of Fig. 8.3. One of the challenges of implementing the encounter-based method is that solutions of the classical BVP with a constant rate of absorption $z$ tend to have a non-trivial parametric dependence on the Laplace variable $z$, which makes it difficult to calculate the inverse transform. This is clear from the spectral decomposition given by Eq. (8.23) on replacing $\gamma$ by the Laplace variable $z$. (In contrast, solving the Robin BVP for a

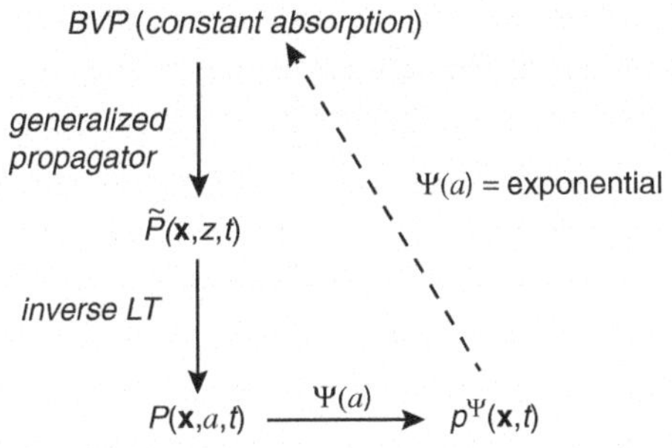


**Fig. 8.3** Diagram illustrating the encounter-based framework for diffusion in a domain with a partially absorbing target. The solution of the BVP in the case of a constant absorption rate $\gamma$ generates the Laplace transform $\widetilde{P}(\mathbf{x}, z, t)$ of the occupation time propagator $P(\mathbf{x}, a, t)$. The inverse LT determines the marginal probability density $p^\Psi(\mathbf{x}, t)$ according to Eq. (8.49)

reactive surface in terms of the spectrum of an associated D-to-N operator yields a series expansion that is easily inverted with respect to the Laplace variable $z$ conjugate to the boundary contact time or local time [9, 11].) In order to invert the $z$-Laplace transforms term by term in Eq. (8.23), we require these infinite series to be uniformly convergent. Assuming that this is the case, one then has to determine how many terms in the series are required in order to obtain a given level of accuracy for quantities of interest such as the MFPT. After taking the $s \to 0$ limit, accuracy will depend on the choice of the stopping time distribution $\Psi$. That is, although the spectral decomposition of the propagator is independent of $\Psi$, the numerical truncation of the corresponding expansion of the MFPT will be $\Psi$-dependent.

Finally, note that the statistics of the FPT density for non-exponential $\Psi$ proceeds along analogous lines to the exponential case. In particular, the FPT moment generator is given by the Laplace transform of the target flux $J^{\Psi}(\mathbf{x}_0, t)$. The generalized survival probability is

$$S^{\Psi}(\mathbf{x}_0, t) = \int_{\Omega\backslash\mathcal{U}} p^{\Psi}(\mathbf{x}, t|\mathbf{x}_0) d\mathbf{x} + \int_{\mathcal{U}} q^{\Psi}(\mathbf{x}, t|\mathbf{x}_0) d\mathbf{x}, \tag{8.50}$$

with

$$p^{\Psi}(\mathbf{x}, t) = \int_0^{\infty} \Psi(a) P(\mathbf{x}, a, t) da, \quad q^{\Psi}(\mathbf{x}, t) = \int_0^{\infty} \Psi(a) Q(\mathbf{x}, a, t) da. \tag{8.51}$$

Differentiating with respect to $t$ and using Eq. (8.46a) and (8.46c) gives

$$\begin{aligned} \frac{\partial S^{\Psi}(\mathbf{x}_0, t)}{\partial t} =& D \int_{\Omega\backslash\mathcal{U}} \nabla^2 p^{\Psi}(\mathbf{x}, t|\mathbf{x}_0) d\mathbf{x} + D \int_{\mathcal{U}} \nabla^2 q^{\Psi}(\mathbf{x}, t|\mathbf{x}_0) d\mathbf{x} \\ &- \int_{\mathcal{U}} \int_0^{\infty} \psi(a) Q(\mathbf{x}, a, t|\mathbf{x}_0) da\, d\mathbf{x}. \end{aligned} \tag{8.52}$$

Applying the divergence theorem to the first two integrals on the right-hand side, imposing the Neumann boundary condition on $\partial\Omega$ and flux continuity at $\partial\mathcal{U}$ shows that these two integrals cancel. The result is then

$$\frac{\partial S^{\Psi}(\mathbf{x}_0, t)}{\partial t} = - \int_{\mathcal{U}} \int_0^{\infty} \psi(a) Q(\mathbf{x}, a, t|\mathbf{x}_0) da\, d\mathbf{x} \equiv -J^{\Psi}(\mathbf{x}_0, t). \tag{8.53}$$

In the exponential case, $\psi(a) = \gamma \mathrm{e}^{-\gamma a}$, we recover Eq. (8.5).

### *8.5.3 One-Dimensional Substrate*

Let us return to the 1D example of Sect. 8.4. Equation (8.37b) implies that

$$\widetilde{Q}(x,z,s|0^+) = \frac{\Lambda(s,z)}{\Phi(s,z)D}\frac{\cosh\beta(s+z)(L'+x)}{\cosh\beta(s+z)L'}, \quad x\in[-L',0]. \tag{8.54}$$

In order to determine the flux $\widetilde{J}^{\Psi}(0^+,s)$ given by Laplace transforming the 1D version of Eq. (8.53) with respect to $t$, we need to calculate the inverse Laplace transform of Eq. (8.54) with respect to $z$. This is relatively straightforward in the limit $L'\to\infty$, since $\tanh(\sqrt{(s+z)/D}L')\to 1$ and

$$\widetilde{Q}(x,z,s|0^+) = \frac{[(1-\alpha)/\alpha D]\mathrm{e}^{\beta(s+z)x}}{\beta(s)\tanh(\beta(s)L)\left[1+\frac{D\beta(s+z)}{\kappa\alpha}\right]+\frac{1-\alpha}{\alpha}\beta(s+z)} \tag{8.55}$$

for $x\in(-\infty,0]$. Integrating with respect to $x$ then gives

$$\int_{-\infty}^{0}\widetilde{Q}(x,z,s|0^+)dx = \frac{[(1-\alpha)/\alpha]}{D\beta(s)\tanh(\beta(s)L)\left[\beta(s+z)+\frac{s+z}{\kappa\alpha}\right]+\frac{1-\alpha}{\alpha}(s+z)}. \tag{8.56}$$

We thus find that

$$T(0^+) = -\partial_s\widetilde{J}^{\Psi}(0^+,0) = -\int_0^{\infty}da\,\psi(a)\mathrm{LT}^{-1}\left[\int_{-\infty}^{0}\partial_s\widetilde{Q}(x,z,0|0^+)dx\right]. \tag{8.57}$$

Differentiating Eq. (8.56) with respect to $s$, we find that

$$\begin{aligned}T(0^+) &= \int_0^{\infty}da\,\psi(a)\mathrm{LT}^{-1}\left[\frac{1}{z^2}\left(1+\frac{\alpha L}{1-\alpha}\left(\sqrt{\frac{z}{D}}+\frac{z}{\kappa\alpha}\right)\right)\right]\\ &= \int_0^{\infty}da\,\psi(a)\left[a+\frac{\alpha L}{1-\alpha}\left(2\sqrt{\frac{a}{\pi D}}+\frac{1}{\kappa\alpha}\right)\right]\\ &= \frac{2\alpha L\mathbb{E}[\sqrt{a}]}{[1-\alpha]\sqrt{\pi D}}+\frac{L}{[1-\alpha]\kappa}+\mathbb{E}[a].\end{aligned} \tag{8.58}$$

A few comments are in order. First, in the case of the exponential density $\psi(a)=\gamma\mathrm{e}^{-\gamma a}$, we have $\mathbb{E}[a]=\gamma^{-1}$ and $\mathbb{E}[\sqrt{a}]=\sqrt{\pi/\gamma}/2$. Hence, Eq. (8.58) reduces to Eq. (8.41) in the limit $L'\to\infty$. Second, the $\kappa$-dependent term is independent of the occupation time distribution $\Psi$. Finally, in the case of a non-exponential distribution $\Psi$, the MFPT only exists if the corresponding moments $\mathbb{E}[a]$ and $\mathbb{E}[\sqrt{a}]$ are finite.

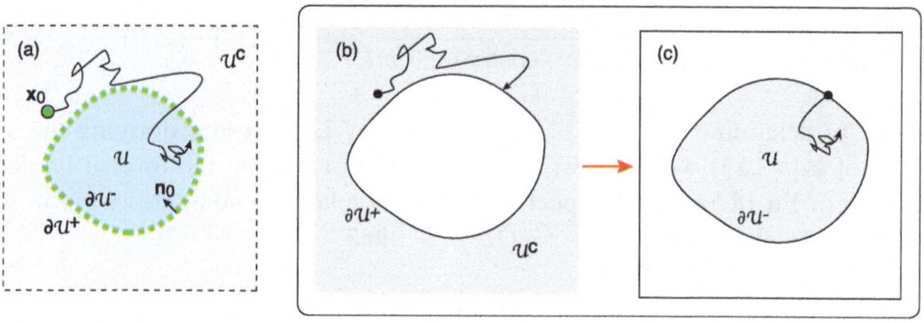


**Fig. 8.4** Decomposition of **a** snapping out BM into two partially reflected BMs corresponding to **b** $\mathbf{X}_t \in \mathcal{U}^c$ and **c** $\mathbf{X}_t \in \mathcal{U}$, respectively

## 8.6 Snapping Out Brownian Motion for Semi-permeable Interfaces

The encounter-based framework for absorbing targets can also be used to develop a more general probabilistic model of single-particle diffusion through semi-permeable interfaces, by combining it with so-called snapping out Brownian motion (BM) [13–15]. The latter was originally formulated for 1D single-particle diffusion through a semi-permeable barrier [16–18], but has recently been extended to higher spatial dimensions [14]. In order to present the basic theory, we ignore the effects of absorption by setting $\gamma = 0$. Snapping out BM sews together successive rounds of partially reflecting BM that are restricted to either the interior or the exterior of $\partial\mathcal{U}$, see Fig. 8.4. Suppose that the particle starts in the domain $\mathcal{U}^c = \Omega\backslash\mathcal{U}$ ($\mathcal{U}$). It realizes reflected BM until it is killed when its local time on $\partial\mathcal{U}^+$ ($\partial\mathcal{U}^-$) is greater than an exponentially distributed random threshold. (This is analogous to the killing of BM when the occupation time $\mathcal{A}_t$ spent within a target interior $\mathcal{U}$ exceeds an exponentially distributed random threshold, signifying an absorption event, see Sect. 8.5.) Let $\mathbf{y} \in \partial\mathcal{U}$ denote the point on the boundary where killing occurs. The stochastic process immediately restarts as a new round of partially reflected BM, either from $\mathbf{y}^+$ into $\mathcal{U}^c$ or from $\mathbf{y}^-$ into $\mathcal{U}$. These two possibilities occur with the probabilities $\alpha$ and $1-\alpha$, respectively. Subsequent rounds of partially reflected BM are generated in the same way. We thus have a stochastic process on the set $\mathbb{G} = \overline{\mathcal{U}} \cup \overline{\mathcal{U}^c}$. It can be proven that the probability density of sample paths generated by snapping out BM evolves according to equations (8.1a) and (8.1b) for $\gamma = 0$, together with the semi-permeable boundary conditions (8.1c) and (8.1d) [13, 14, 16]. (One version of the proof for the 1D case is given in Sect. 8.6.3.) For simplicity, we set $\alpha = 1/2$ in the following.

### 8.6.1 Partially Reflected BMs in $\mathcal{U}$ and $\mathcal{U}^c$

Consider a Brownian particle diffusing in the bounded domain $\mathcal{U}^c$, see Fig. 8.4b with $\partial\mathcal{U}^+$ totally reflecting. Let $\mathbf{X}_t$ denote the position of the particle at time $t$. In order to write down a stochastic differential equation (SDE) for $\mathbf{X}_t$, we introduce the boundary local time [25, 29–32]

$$\ell_t^+ = \lim_{\epsilon\to 0}\frac{D}{\epsilon}\int_0^t \Theta(\epsilon - \text{dist}(\mathbf{X}_\tau, \partial\mathcal{U}^+))d\tau, \tag{8.59}$$

where $\Theta$ is the Heaviside function, and $\text{dist}(\mathbf{X}_\tau, \partial\mathcal{U}^+)$ denotes the shortest Euclidean distance of $X_\tau$ from the boundary $\partial\mathcal{U}^+$. The corresponding SDE takes the form

$$d\mathbf{X}_t = \sqrt{2D}d\mathbf{W}_t - \mathbf{n}_u(\mathbf{X}_t)d\ell_t^+, \tag{8.60}$$

where $\mathbf{W}_t$ is a $d$-dimensional Brownian motion and $\mathbf{n}_u(\mathbf{X}_t)$ is the inward unit normal at the point $\mathbf{X}_t \in \partial\mathcal{U}$. The differential $d\ell_t^+$ can be expressed in terms of a Dirac delta function:

$$d\ell_t^+ = Ddt\left(\int_{\partial\mathcal{U}^+}\delta(\mathbf{X}_t - \mathbf{y})d\mathbf{y}\right). \tag{8.61}$$

Partially reflected BM in $\mathcal{U}$ is then obtained by stopping the stochastic process $\mathbf{X}_t$ when the local time $\ell_t^+$ exceeds a random exponentially distributed threshold $\widehat{\ell}^+$ [9]. That is, the particle is absorbed somewhere on $\partial\mathcal{U}^+$ at the stopping time

$$\mathcal{T}^+ = \inf\{t > 0:\ \ell_t^+ > \widehat{\ell}^+\},\quad \mathbb{P}[\widehat{\ell}^+ > \ell] = \mathrm{e}^{-\kappa_0\ell/D}. \tag{8.62}$$

Consider the local time propagator $P(\mathbf{x}, \ell, t|\mathbf{x}_0)$ for the pair $(\mathbf{X}_t, \ell_t)$, which evolves according to [9]

$$\frac{\partial P(\mathbf{x}, \ell, t|\mathbf{x}_0)}{\partial t} = D\nabla^2 P(\mathbf{x}, \ell, t|\mathbf{x}_0),\quad \mathbf{x}\in\Omega\backslash\mathcal{U}, \tag{8.63a}$$

$$\nabla P(\mathbf{x}, \ell, t|\mathbf{x}_0)\cdot\mathbf{n} = 0,\quad \mathbf{x}\in\partial\mathcal{U}^c, \tag{8.63b}$$

$$-D\nabla P(\mathbf{x}, \ell, t|\mathbf{x}_0)\cdot\mathbf{n}_u = DP(\mathbf{x}, \ell = 0, t|\mathbf{x}_0)\,\delta(\ell) + D\frac{\partial}{\partial\ell}P(\mathbf{x}, \ell, t|\mathbf{x}_0),\quad \mathbf{x}\in\partial\mathcal{U}. \tag{8.63c}$$

This can be derived using a Feynman-Kac formula along analogous lines to the occupation time propagator of Sect. 8.5, see Ref. [10]. Equations (8.63) are supplemented by the initial condition $P(\mathbf{x}, \ell, 0|\mathbf{x}_0) = \delta(\mathbf{x} - \mathbf{x}_0)\delta(\ell)$. Laplace transforming the local time BVP (8.63) with respect to $\ell$ and setting

$$\widetilde{P}(\mathbf{x}, \omega, t|\mathbf{x}_0) = \int_0^\infty \mathrm{e}^{-\omega\ell}P(\mathbf{x}, \ell, t|\mathbf{x}_0)d\ell \tag{8.64}$$

yields

$$\frac{\partial \widetilde{P}(\mathbf{x}, \omega, t|\mathbf{x}_0)}{\partial t} = D\nabla^2 \widetilde{P}(\mathbf{x}, \omega, t|\mathbf{x}_0), \quad \mathbf{x} \in \Omega \backslash \mathcal{U}, \tag{8.65a}$$

$$\nabla \widetilde{P}(\mathbf{x}, \omega, t|\mathbf{x}_0) \cdot \mathbf{n} = 0, \quad \mathbf{x} \in \partial\Omega, \tag{8.65b}$$

$$-\nabla \widetilde{P}(\mathbf{x}, \omega, t|\mathbf{x}_0) \cdot \mathbf{n}_u = \omega \widetilde{\mathcal{P}}(\mathbf{x}, \omega, t|\mathbf{x}_0), \quad \mathbf{x} \in \partial\mathcal{U}^+, \tag{8.65c}$$

and $\widetilde{P}(\mathbf{x}, \omega, 0|\mathbf{x}_0) = \delta(\mathbf{x} - \mathbf{x}_0)$. We see that Eq. (8.65b) is a classical Robin boundary condition on $\partial\mathcal{U}^+$ with a constant reactivity $\kappa_0 = \omega D$. Hence, the Robin boundary condition is equivalent to an exponential law for the local time threshold $\widehat{\ell}^+$. Following Ref. [9], we now modify the rule for killing each round of partially reflected BM by taking $\widehat{\ell}^+$ to have a non-exponential distribution $\Psi^+(\ell)$. We then define the corresponding marginal probability density according to

$$p_+^{\Psi}(\mathbf{x}, t|\mathbf{x}_0) = \int_0^\infty \Psi^+(\ell) P(\mathbf{x}, \ell, t|\mathbf{x}_0) d\ell, \quad \mathbf{x} \in \mathcal{U}^c. \tag{8.66}$$

Multiplying both sides of the boundary condition (8.63b) by $\Psi(\ell)$ and integrating by parts with respect to $\ell$ shows that

$$j_+^{\Psi}(\mathbf{x}, t|\mathbf{x}_0) \equiv -D\nabla p_+^{\Psi}(\mathbf{x}, t|\mathbf{x}_0) \cdot \mathbf{n}_u = D \int_0^\infty \psi(\ell) P(\mathbf{x}, \ell, t|\mathbf{x}_0) d\ell, \quad \mathbf{x} \in \partial\mathcal{U}^+, \tag{8.67}$$

with $\psi^+(\ell) = -\Psi^{+'}(\ell)$. We have used Eq. (8.63b) and the identity $\Psi^+(0) = 1$. Integrating with respect to points on the boundary then gives

$$J_+^{\Psi}(\mathbf{x}_0, t) = \int_{\partial\mathcal{U}} j_+^{\Psi}(\mathbf{x}, t|\mathbf{x}_0) d\sigma \equiv D \int_{\partial\mathcal{U}} \left[ \int_0^\infty \psi^+(\ell) P(\mathbf{x}, \ell, t|\mathbf{x}_0) d\ell \right] d\sigma. \tag{8.68}$$

Finally, we can determine $P(\mathbf{x}, \ell, t|\mathbf{x}_0)$ by inverting the solution $\widetilde{P}(\mathbf{x}, \omega, t|\mathbf{x}_0)$ to the Robin BVP with respect to $\omega$, which is the local time analog of Eq. (8.49). In other words, a commutative diagram of the form shown in Fig. 8.3 also applies to the local time propagator.

An analogous construction holds for partially reflected BM in $\mathcal{U}$, see Fig. 8.4(c). Given the local time

$$\ell_t^- = \lim_{\epsilon \to 0} \frac{D}{\epsilon} \int_0^t H(\epsilon - \text{dist}(\mathbf{X}_\tau, \partial\mathcal{U}^-)) d\tau, \tag{8.69}$$

and stopping time

$$\mathcal{T}^- = \inf\{t > 0 : \ \ell_t^- > \widehat{\ell}^-\}, \quad \mathbb{P}[\widehat{\ell}^- > \ell] = \mathrm{e}^{-\kappa_0 \ell / D}, \tag{8.70}$$

we introduce the local time propagator $Q(\mathbf{x}, \ell, t|\mathbf{x}_0)$, which evolves according to

$$\frac{\partial Q(\mathbf{x}, \ell, t|\mathbf{x}_0)}{\partial t} = D\nabla^2 Q(\mathbf{x}, \ell, t|\mathbf{x}_0), \quad \mathbf{x} \in \mathcal{U}, \tag{8.71a}$$

$$D\nabla Q(\mathbf{x}, \ell, t|\mathbf{x}_0) \cdot \mathbf{n}_u = DQ(\mathbf{x}, \ell = 0, t|\mathbf{x}_0)\, \delta(\ell) + D\frac{\partial}{\partial \ell} Q(\mathbf{x}, \ell, t|\mathbf{x}_0), \quad \mathbf{x} \in \partial\mathcal{U}. \tag{8.71b}$$

Laplace transforming with respect to $\ell$ yields the following Robin BVP:

$$\frac{\partial \widetilde{Q}(\mathbf{x}, \omega, t|\mathbf{x}_0)}{\partial t} = D\nabla^2 \widetilde{Q}(\mathbf{x}, \omega, t|\mathbf{x}_0), \quad \mathbf{x} \in \mathcal{U}, \tag{8.72a}$$

$$\nabla \widetilde{Q}(\mathbf{x}, \omega, t|\mathbf{x}_0) \cdot \mathbf{n}_u = \omega \widetilde{Q}(\mathbf{x}, \omega, t|\mathbf{x}_0), \quad \mathbf{x} \in \partial\mathcal{U}, \tag{8.72b}$$

and $\widetilde{Q}(\mathbf{x}, \omega, 0|\mathbf{x}_0) = \delta(\mathbf{x} - \mathbf{x}_0)$ assuming that $\mathbf{x}_0 \in \mathcal{U}$. Finally, given a local time threshold distribution $\Psi^-(\ell)$, the generalized marginal density is

$$p_-^\Psi(\mathbf{x}, t|\mathbf{x}_0) = \int_0^\infty \Psi^-(\ell) Q(\mathbf{x}, \ell, t|\mathbf{x}_0) d\ell, \quad \mathbf{x} \in \mathcal{U}. \tag{8.73}$$

### 8.6.2 Renewal Equations for Snapping Out BM

The crucial step in formulating snapping out BM is sewing together successive rounds of partially reflected BM. As we have recently shown [12, 14], this can be achieved by constructing renewal equations that relate the full probability density $\rho^\Psi$ of snapping out BM to the corresponding probability densities $p_\pm^\Psi$. First, it is convenient to consider a distribution of initial conditions by setting

$$p_+^\Psi(\mathbf{x}, t) = \int_{\mathcal{U}^c} \left[ \int_0^\infty \Psi^+(\ell) P(\mathbf{x}, \ell, t|\mathbf{x}_0) d\ell \right] h(\mathbf{x}_0) d\mathbf{x}_0, \quad \mathbf{x} \in \mathcal{U}^c, \tag{8.74a}$$

$$p_-^\Psi(\mathbf{y}, t) = \int_{\mathcal{U}} \left[ \int_0^\infty \Psi^-(\ell) Q(\mathbf{y}, \ell, t|\mathbf{y}_0) d\ell \right] \overline{h}(\mathbf{y}_0) d\mathbf{y}_0, \quad \mathbf{y} \in \mathcal{U}, \tag{8.74b}$$

with

$$\int_{\mathcal{U}^c} h(\mathbf{x}_0) d\mathbf{x}_0 + \int_{\mathcal{U}} \overline{h}(\mathbf{y}_0) d\mathbf{y}_0 = 1. \tag{8.75}$$

Denote the probability density of generalized snapping out BM given $\mathbf{x}_0 \in \mathbb{G}$ by $\rho^\Psi(\mathbf{x}, t|\mathbf{x}_0)$ and set

$$\rho^\Psi(\mathbf{x}, t) = \int_{\mathcal{U}^c} \rho^\Psi(\mathbf{x}, t|\mathbf{x}_0) h(\mathbf{x}_0) d\mathbf{x}_0 + \int_{\mathcal{U}} \rho^\Psi(\mathbf{x}, t|\mathbf{y}_0) \overline{h}(\mathbf{y}_0) d\mathbf{y}_0. \tag{8.76}$$

It can then be shown that $\rho^\Psi(\mathbf{x}, t)$ satisfies the first renewal equation [14]

$$\rho^{\Psi}(\mathbf{x},t) = I_{\mathcal{U}^c}(\mathbf{x})p_+^{\Psi}(\mathbf{x},t) + I_{\mathcal{U}}(\mathbf{x})p_-^{\Psi}(\mathbf{x},t) + \frac{\kappa}{2}\int_0^t d\tau \int_{\partial\mathcal{U}} d\mathbf{z} \times \left[\rho^{\Psi}(\mathbf{x},t-\tau|\mathbf{z}^+) + \rho^{\Psi}(\mathbf{x},t-\tau|\mathbf{z}^-)\right] f^{\Psi}(\mathbf{z},\tau) \tag{8.77}$$

for $\mathbf{x} \in \mathbb{G}$. In addition, $f^{\Psi}(\mathbf{z},\tau)$ is the FPT density for the particle to be killed at time $\tau$ and a point $\mathbf{z} \in \mathcal{U}$ for the given distribution of initial conditions. That is,

$$f^{\Psi}(\mathbf{z},\tau) = \int_{\mathcal{U}^c} j_+^{\Psi}(\mathbf{z}^+,t|\mathbf{x}_0)h(\mathbf{x}_0)d\mathbf{x}_0 + \int_{\mathcal{U}} j_-^{\Psi}(\mathbf{z}^-,t|\mathbf{y}_0)\overline{h}(\mathbf{y}_0)d\mathbf{y}_0. \tag{8.78}$$

The first two terms on the right-hand side of Eq. (8.77) represent all sample trajectories that have never been absorbed by the boundaries $\partial\mathcal{U}^+$ and $\partial\mathcal{U}^-$, respectively. The integrand for a given $\mathbf{z} \in \partial\mathcal{U}$ represents all trajectories that were first absorbed (stopped) at time $\tau$ and position $\mathbf{z}$, and then switched to either the domain $\mathcal{U}^c$ or $\mathcal{U}$ with probability 1/2, after which multiple killing events can occur before reaching $\mathbf{x}$ at time $t$. The probability that the first stopping event occurred at $\mathbf{z}$ in the interval $(\tau, \tau + d\tau)$ is $f^{\Psi}(\mathbf{z},\tau)d\tau$. Finally, it is necessary to integrate with respect to all first stopping positions $\mathbf{z}$.

Laplace transforming the renewal Eq. (8.77) with respect to time $t$ and using the convolution theorem gives

$$\widetilde{\rho}^{\Psi}(\mathbf{x},s) = I_{\mathcal{U}^c}(\mathbf{x})\widetilde{p}_+^{\Psi}(\mathbf{x},s) + I_{\mathcal{U}}(\mathbf{x})\widetilde{p}_-^{\Psi}(\mathbf{x},s) + \frac{\kappa}{2}\int_{\partial\mathcal{U}} d\mathbf{z} \times \left[\widetilde{\rho}^{\Psi}(\mathbf{x},s|\mathbf{z}^+) + \widetilde{\rho}^{\Psi}(\mathbf{x},s|\mathbf{z}^-)\right]\widetilde{f}^{\Psi}(\mathbf{z},s). \tag{8.79}$$

In order to determine the factor $\widetilde{\rho}^{\Psi}(\mathbf{x},s|\mathbf{z}^+) + \widetilde{\rho}^{\Psi}(\mathbf{x},s|\mathbf{z}^-)$ we set $g(\mathbf{x}_0) = \delta(\mathbf{x}_0 - \mathbf{z}^+)$ and $\overline{g}(\mathbf{y}_0) = \delta(\mathbf{y}_0 - \mathbf{z}^-)$ in Eqs. (8.74), (8.76) and (8.78). This gives

$$\widetilde{\rho}^{\Psi}(\mathbf{x},s|\mathbf{z}^+) + \widetilde{\rho}^{\Psi}(\mathbf{x},s|\mathbf{z}^-) = I_{\mathcal{U}^c}(\mathbf{x})\widetilde{p}_+^{\Psi}(\mathbf{x},s|\mathbf{z}^+) + I_{\mathcal{U}}(\mathbf{x})\widetilde{p}_-^{\Psi}(\mathbf{x},s|\mathbf{z}^-) + \frac{\kappa}{2}\int_{\partial\mathcal{U}} d\mathbf{z}\left[\widetilde{\rho}^{\Psi}(\mathbf{x},s|\mathbf{z}^+) + \widetilde{\rho}^{\Psi}(\mathbf{x},s|\mathbf{z}^-)\right]\left[\widetilde{j}_+^{\Psi}(\mathbf{z}^+,s|\mathbf{z}^+) + \widetilde{j}_-^{\Psi}(\mathbf{z}^-,s|\mathbf{z}^-)\right]. \tag{8.80}$$

For general geometries, solving this implicit integral equation for $\widetilde{\rho}^{\Psi}(\mathbf{x},s|\mathbf{z}^+) + \widetilde{\rho}^{\Psi}(\mathbf{x},s|\mathbf{z}^-)$ is nontrivial. Therefore, we will illustrate the theory using a 1D interface.

### *8.6.3 Snapping Out BM in an Interval*

Let us return to the 1D example shown in Fig. 8.2 with zero absorption ($\gamma = 0$). After Laplace transforming with respect to $t$, the 1D version of Eq. (8.65) becomes

$$D\frac{\partial^2 \widetilde{P}(x,\omega,s|x_0)}{\partial x^2} - s\widetilde{P}(x,\omega,s|x_0) = -\delta(x-x_0), \tag{8.81a}$$

$$\partial_x \widetilde{P}(0,\omega,s|x_0) = \omega\widetilde{P}(0,\omega,s|x_0), \quad \partial_x \widetilde{P}(L,\omega,s|x_0) = 0, \tag{8.81b}$$

with $0 < x, x_0 < L$. We can identify $\widetilde{P}(x,\omega,s|x_0)$ as a Green's function of the modified Helmholtz equation on $[0, L]$, similar to $G$ of Eq. (8.29) but with different boundary conditions

$$\widetilde{P}(x,\omega,s|x_0) = \begin{cases} A_L(\omega,s)g(x,\omega,s)\widehat{g}_L(x_0,\omega,s), & 0 \le x \le x_0 \\ A_L(\omega,s)g(x_0,\omega,s)\widehat{g}(x,\omega,s), & x_0 \le x \le L \end{cases} \tag{8.82}$$

with

$$g(x,\omega,s) = \frac{\sqrt{sD}\cosh(\beta(s)x) + \omega D\sinh(\beta(s)x)}{\sqrt{sD}+\omega D}, \quad \widehat{g}_L(x,s) = \cosh(\beta(s)(L-x)) \tag{8.83}$$

and

$$A_L(\omega,s) = \frac{1}{\sqrt{sD}}\frac{\sqrt{sD}+\omega D}{\sqrt{sD}\sinh(\beta(s)L) + \omega D\cosh(\beta(s)L)}. \tag{8.84}$$

Similarly, $\widetilde{Q}(x,\omega,s|x_0)$ satisfies Eq. (8.81) for $L \to L'$ and $x, x_0 \to -x, -x_0$. Hence,

$$\widetilde{Q}(x,\omega,s|x_0) = \begin{cases} A_{L'}(\omega,s)g(-x,\omega,s)\widehat{g}_{L'}(-x_0,\omega,s), & x_0 \le x \le 0 \\ A_{L'}(\omega,s)g(-x_0,\omega,s)\widehat{g}_{L'}(-x,\omega,s), & x \le x_0 \le 0 \end{cases} \tag{8.85}$$

The Laplace transformed propagators have simple poles in the complex $\omega$-plane and can thus be inverted straightforwardly. For the sake of illustration, suppose that $x_0 = 0$. Then

$$\widetilde{P}(x,\omega,s|0) = \frac{\cosh(\beta(s)(L-x))}{\sqrt{sD}\sinh(\beta(s)L) + \omega D\cosh(\beta(s)L)}, \tag{8.86a}$$

$$\widetilde{Q}(x,\omega,s|0) = \frac{\cosh(\beta(s)(L'+x))}{\sqrt{sD}\sinh(\beta(s)L') + \omega D\cosh(\beta(s)L')}, \tag{8.86b}$$

and

$$\widetilde{P}(x,\ell,s|0) = \frac{\cosh(\beta(s)(L-x))}{D\cosh(\beta(s)L)}\exp\left[-\ell\beta(s)\tanh(\beta(s)L)\right], \tag{8.87a}$$

$$\widetilde{Q}(x,\ell,s|0) = \frac{\cosh(\beta(s)(L'+x))}{D\cosh(\beta(s)L')}\exp\left[-\ell\beta(s)\tanh(\beta(s)L')\right]. \tag{8.87b}$$

The corresponding marginal probability densities are thus

$$\widetilde{p}_{+}^{\Psi}(x,s|0^{+})=\frac{\cosh(\beta(s)(L-x))}{D\cosh(\beta(s)L)}\widetilde{\Psi}^{+}\left[\beta(s)\tanh(\beta(s)L)\right], \tag{8.88a}$$

$$\widetilde{p}_{-}^{\Psi}(x,s|0^{-})=\frac{\cosh(\beta(s)(L'+x))}{D\cosh(\beta(s)L')}\widetilde{\Psi}^{-}\left[\beta(s)\tanh(\beta(s)L')\right]. \tag{8.88b}$$

Similarly, the flux densities $\widetilde{j}_{\pm}^{\Psi}(x,s|0)$ are obtained by replacing $\widetilde{\Psi}^{\pm}$ with $D\widetilde{\psi}^{\pm}$, where

$$\widetilde{\psi}^{\pm}(s)=1-s\widetilde{\Psi}^{\pm}(s). \tag{8.89}$$

The 1D version of Eq. (8.76) for $p^{\Psi}(x,t)$ on $x\in\mathbb{G}$ is

$$\begin{aligned} p^{\Psi}(x,t)&=\Theta(x)\int_0^L\left[\int_0^{\infty}\Psi^{+}(\ell)P(x,\ell,t|x_0)d\ell\right]h(x_0)dx_0\\ &\quad+\Theta(-x)\int_{-\infty}^0\left[\int_0^{\infty}\Psi^{-}(\ell)Q(x,\ell,t|y_0)d\ell\right]\overline{g}(y_0)dy_0. \end{aligned} \tag{8.90}$$

Similarly, the 1D version of the first renewal Eq. (8.77) takes the form

$$\rho^{\Psi}(x,t)=p^{\Psi}(x,t)+\frac{1}{2}\int_0^t[\rho^{\Psi}(x,t-\tau|0^{+})+\rho^{\Psi}(x,t-\tau|0^{-})]f^{\Psi}(\tau)d\tau \tag{8.91}$$

for $x\in\mathbb{G}$. Laplace transforming with respect to time $t$ then gives

$$\widetilde{\rho}^{\Psi}(x,s)=\widetilde{p}^{\Psi}(x,s)+\frac{1}{2}[\widetilde{\rho}^{\Psi}(x,s|0^{+})+\widetilde{\rho}^{\Psi}(x,s|0^{-})]\widetilde{f}^{\Psi}(s),\quad x\in\mathbb{G}. \tag{8.92}$$

The factor $\widetilde{\rho}^{\Psi}(x,s|0^{+})+\widetilde{\rho}^{\Psi}(x,s|0^{-})$ can now be determined by setting $g(x_0)=[\delta(x_0-0^{+})+\delta(x-0^{-})]/2$ in Eq. (8.92):

$$\begin{aligned} &\widetilde{\rho}^{\Psi}(x,s|0^{+})+\widetilde{\rho}^{\Psi}(x,s|0^{-})\\ &\quad=\Theta(x)\widetilde{p}_{+}^{\Psi}(x,s|0^{+})+\Theta(-x)\widetilde{p}_{-}^{\Psi}(x,s|0^{-})\\ &\quad\quad+\frac{1}{2}[\widetilde{\rho}^{\Psi}(x,s|0^{+})+\widetilde{\rho}^{\Psi}(x,s|0^{-})]\left[\widetilde{j}_{+}^{\Psi}(0^{+},s|0^{+})+\widetilde{j}_{-}^{\Psi}(0^{-},s|0^{-})\right] \end{aligned}$$

which can be arranged to yield the result

$$\widetilde{\rho}^{\Psi}(x,s|0^{+})+\widetilde{\rho}^{\Psi}(x,s|0^{-})=\frac{\Theta(x)\widetilde{p}_{+}^{\Psi}(x,s|0^{+})+\Theta(-x)\widetilde{p}_{-}^{\Psi}(x,s|0^{-})}{1-[\widetilde{j}_{+}^{\Psi}(0^{+},s|0^{+})+\widetilde{j}_{-}^{\Psi}(0^{-},s|0^{-})]/2}.$$

Substituting back into Eq. (8.80) yields the explicit solution

$$\widetilde{\rho}^{\Psi}(x,s) = \widetilde{p}^{\Psi}(x,s) + \frac{\Theta(x)\widetilde{p}_{+}^{\Psi}(x,s|0^{+}) + \Theta(-x)\widetilde{p}_{-}^{\Psi}(x,s|0^{-})}{1-[\widetilde{j}_{+}^{\Psi}(0^{+},s|0^{+}) + \widetilde{j}_{-}^{\Psi}(0^{-},s|0^{-})]/2}\widetilde{f}^{\Psi}(s), \ x \in \mathbb{G}. \tag{8.93}$$

Further simplification occurs if we take $\Psi^{+} = \Psi^{-}$ and $L, L' \to \infty$, such that

$$\widetilde{p}_{\pm}^{\Psi}(x,s|0) = \frac{1}{D}\mathrm{e}^{-\beta(s)|x|}\widetilde{\Psi}(\beta(s)), \quad \widetilde{j}_{\pm}^{\Psi}(x,s|0) = \mathrm{e}^{-\beta(s)|x|}\widetilde{\psi}(\beta(s)). \tag{8.94}$$

We then find that

$$\widetilde{\rho}^{\Psi}(x,s) = \widetilde{p}^{\Psi}(x,s) + \frac{\mathrm{e}^{-\beta(s)|x|}}{2\sqrt{sD}}\Gamma^{\Psi}(s), \quad x \in \mathbb{G}, \tag{8.95}$$

where

$$\begin{aligned}\Gamma^{\Psi}(s) &\equiv \widetilde{f}^{\Psi}(s) = D\partial_x\widetilde{p}^{\Psi}(0^{+},s) - D\partial_x\widetilde{p}^{\Psi}(0^{-},s)\\ &= \widetilde{\psi}(\beta(s))\left[\int_0^{\infty}\mathrm{e}^{-\beta(s)x_0}h(x_0)dx_0 + \int_{-\infty}^{0}\mathrm{e}^{\beta(s)x_0}\overline{h}(x_0)dx_0\right].\end{aligned} \tag{8.96}$$

Since the propagator satisfies the diffusion equation in the bulk of the domain, the density $\rho^{\Psi}(x,t)$ does too. The remaining issue concerns the boundary condition at the interface. We proceed along the lines of Ref. [12]. First, it follows from Eq. (8.95) that

$$\widetilde{\rho}^{\Psi}(x,s) + \widetilde{\rho}^{\Psi}(-x,s) = \widetilde{p}^{\Psi}(x,s) + \widetilde{p}^{\Psi}(-x,s) + \frac{\mathrm{e}^{-\beta(s)|x|}}{\sqrt{sD}}\Gamma^{\Psi}(s), \tag{8.97a}$$

$$\widetilde{\rho}^{\Psi}(x,s) - \widetilde{\rho}^{\Psi}(-x,s) = \widetilde{p}^{\Psi}(x,s) - \widetilde{p}^{\Psi}(-x,s), \tag{8.97b}$$

and

$$D\partial_x\widetilde{\rho}^{\Psi}(0^{+},s) - D\partial_x\widetilde{\rho}^{\Psi}(0^{-},s) = D\partial_x\widetilde{p}^{\Psi}(0^{+},s) - D\partial_x\widetilde{p}^{\Psi}(0^{-},s) - \Gamma^{\Psi}(s), \tag{8.98a}$$

$$D\partial_x\widetilde{\rho}^{\Psi}(0^{+},s) + D\partial_x\widetilde{\rho}^{\Psi}(0^{-},s) = D\partial_x\widetilde{p}^{\Psi}(0^{+},s) + D\partial_x\widetilde{p}^{\Psi}(0^{-},s). \tag{8.98b}$$

Equations (8.96) and (8.98a) establish that $\partial_x\widetilde{\rho}^{\Psi}(0^{+},s) = \partial_x\widetilde{\rho}^{\Psi}(0^{-},s)$. In other words, the flux through the membrane is continuous, as it is in the standard permeable boundary conditions (8.27c,d). Equation (8.98b) then implies that

$$\begin{aligned}2D\partial_x\widetilde{\rho}^{\Psi}(0^{\pm},s) &= \widetilde{\psi}(\beta(s))\left[\int_0^{\infty}\mathrm{e}^{-\beta(s)x_0}h(x_0)dx_0 - \int_{-\infty}^{0}\mathrm{e}^{\beta(s)x_0}\overline{h}(x_0)dx_0\right]\\ &= \frac{D\widetilde{\psi}(\beta(s))}{\widetilde{\Psi}(\beta(s))}[\widetilde{p}^{\Psi}(0^{+},s) - \widetilde{p}^{\Psi}(0^{-},s)]\end{aligned}$$

$$= \frac{D\widetilde{\psi}(\beta(s))}{\widetilde{\Psi}(\beta(s))}[\widetilde{\rho}^{\Psi}(0^+,s) - \widetilde{\rho}^{\Psi}(0^-,s)]. \tag{8.99}$$

The final line follows from Eq. (8.97b).

In the exponential case $\Psi(\ell) = e^{-\omega_0 \ell}$, we have $\psi(\ell) = \omega_0 \Psi(\ell)$, and we recover the semi-permeable boundary conditions of Eq. (8.27c, d) with permeability $\kappa = \omega_0 D$ and $\alpha = 1/2$. For non-exponential distributions, the boundary condition involves a time-dependent permeability. More specifically, setting

$$\widetilde{\kappa}(s) = \frac{D\widetilde{\psi}(\beta(s))}{\widetilde{\Psi}(\beta(s))} \tag{8.100}$$

and using the convolution theorem, the boundary condition in the time domain takes the form

$$2D\partial_x \rho^{\Psi}(0^{\pm}, t) = \int_0^t \kappa(\tau)[\rho^{\Psi}(0^+, t-\tau) - \rho^{\Psi}(0^-, t-\tau)]d\tau. \tag{8.101}$$

For the sake of illustration, suppose that $\psi(\ell)$ is given by the gamma distribution:

$$\psi(\ell) = \frac{\omega_0(\omega_0 \ell)^{\mu-1} e^{-\omega_0 \ell}}{\Gamma(\mu)}, \quad \mu > 0, \tag{8.102}$$

where $\Gamma(\mu)$ is the gamma function. Here $\omega_0$ determines the effective absorption rate and $\mu$ characterizes the deviation of $\psi(\ell)$ from the exponential case $\mu = 1$. The corresponding Laplace transforms are

$$\widetilde{\psi}(z) = \left(\frac{\omega_0}{\omega_0 + z}\right)^{\mu}, \quad \widetilde{\Psi}(z) = \frac{1 - \widetilde{\psi}(z)}{z}. \tag{8.103}$$

If $\mu < 1$ ($\mu > 1$) then $\psi(\ell)$ decreases more rapidly (slowly) as a function of the local time $\ell$. Substituting the gamma distribution into Eq. (8.100) yields

$$\widetilde{\kappa}(s) = \frac{\sqrt{sD}\,\omega_0^{\mu}}{(\omega_0 + \beta(s))^{\mu} - \omega_0^{\mu}}. \tag{8.104}$$

If $\mu = 1$ then $\widetilde{\kappa}(s) = \omega_0 D = \kappa_0$ and $\kappa(\tau) = \omega_0 \delta(\tau)$. An example of $\mu \neq 1$ that has a simple inverse Laplace transform is $\mu = 2$:

$$\widetilde{\kappa}(s) = \frac{D\omega_0^2}{2\omega_0 + \beta(s)} = \frac{\kappa_0^2}{2\kappa_0 + \sqrt{sD}} \tag{8.105}$$

and

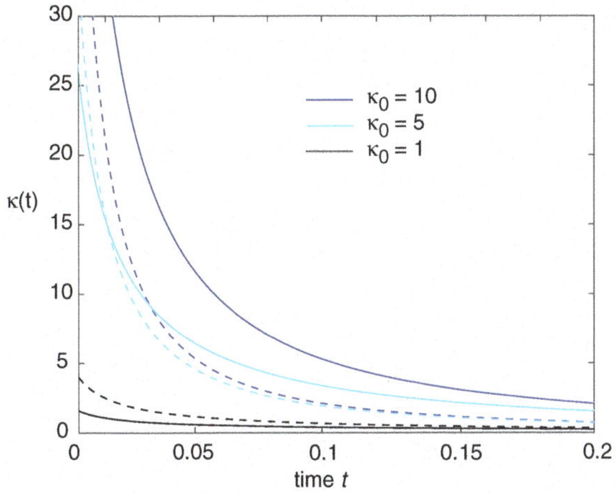


**Fig. 8.5** Plot of permeability function $\kappa(t)$ as a function of time $t$ for various values of $\omega_0$ with $D = 10$ (solid curves) and $D = 1$ dashed curves

$$\kappa(\tau) = \frac{\kappa_0^2}{\sqrt{D}}\left[\frac{1}{\sqrt{\pi\tau}} - \frac{2\kappa_0}{\sqrt{D}}e^{4\kappa_0^2\tau/D}\mathrm{erfc}(2\kappa_0\sqrt{\tau/D})\right], \tag{8.106}$$

where $\mathrm{erfc}(x) = (2/\sqrt{\pi})\int_x^\infty e^{-y^2}dy$ is the complementary error function. Example plots of $\kappa(\tau)$ are shown in Fig. 8.5. It can be seen that $\kappa$ is a monotonically decreasing function of time whose rate of decay depends on $\kappa_0$ and $D$. Asymptotically expanding $\mathrm{erfc}(x)$ in Eq. (8.106) using the formula

$$\mathrm{erfc}(x) \sim \frac{1}{\sqrt{\pi}}e^{-x^2}\sum_{k=0}^{\infty}(-1)^k\frac{(2k)!}{2^{2k}k!}\frac{1}{x^{2k+1}}, \tag{8.107}$$

shows that $\kappa(t)$ is heavy-tailed with

$$\kappa(t) \sim \sqrt{\frac{D}{\pi}}\frac{1}{8t^{3/2}}, \quad t \to \infty. \tag{8.108}$$

### 8.6.4 *Absorbing Target*

So far in our discussion of snapping out BM we have ignored absorption within the target $\mathcal{U}$. If absorption is included, then each round of partially reflected BM within $\mathcal{U}$ has two distinct killing events: either the particle is first absorbed on $\partial\mathcal{U}$, after which snapping out BM continues in the normal fashion, or the particle is first absorbed within the interior $\mathcal{U}$, after which the stochastic process is terminated. A single round of BM within $\mathcal{U}$ thus becomes a competition between two absorbing targets, namely, the boundary $\partial\mathcal{U}$ and the interior $\mathcal{U}$. Certain care has to be taken in treating these two targets as independent, since $\mathcal{U}$ is an open set whose closure includes $\partial\mathcal{U}$. One way to deal with this situation would be to introduce a boundary layer around $\partial\mathcal{U}$, within which the particle can only be absorbed by $\partial\mathcal{U}$. This is also consistent with how one would numerically calculate the boundary local time. Here we ignore such details, and simply consider the dual-aspect propagator $Q(\mathbf{x}, \ell, a, t|\mathbf{x}_0)$ for the pair $(\ell_t^-, \mathcal{A}_t)$ with $\mathcal{A}_t$ the occupation time within $\mathcal{U}$, see Eq. (8.42).

The evolution equation for $Q(\mathbf{x}, \ell, a, t|\mathbf{x}_0)$ is obtained by combining Eqs. (8.71a) and (8.46c):

$$\begin{aligned}&\frac{\partial Q(\mathbf{x}, \ell, a, t|\mathbf{x}_0)}{\partial t} + \frac{\partial Q(\mathbf{x}, \ell, a, t|\mathbf{x}_0)}{\partial a}\\ &\quad = D\nabla^2 Q(\mathbf{x}, \ell, a, t|\mathbf{x}_0) - \delta(a) Q(\mathbf{x}, \ell, a = 0, t|\mathbf{x}_0), \quad \mathbf{x} \in \mathcal{U}, \qquad (8.109a)\\ &D\nabla Q(\mathbf{x}, \ell, a, t|\mathbf{x}_0) \cdot \mathbf{n}_u\\ &\quad = D Q(\mathbf{x}, \ell = 0, a, t|\mathbf{x}_0)\, \delta(\ell) + D\frac{\partial}{\partial \ell} Q(\mathbf{x}, \ell, a, t|\mathbf{x}_0), \quad \mathbf{x} \in \partial\mathcal{U}. \qquad (8.109b)\end{aligned}$$

Let $\Psi_1(\ell)$ and $\Psi_2(a)$ denote the threshold distributions for the local time and occupation time, respectively. Following along similar lines to our previous examples, the associated marginal probability density is

$$\begin{aligned} q^{\Psi_1,\Psi_2}(\mathbf{x}, t|\mathbf{x}_0) &= \int_0^\infty d\ell\, \Psi_1(\ell) \int_0^\infty da\, \Psi_2(a) Q(\mathbf{x}, \ell, a, t|\mathbf{x}_0)\\ &= \int_0^\infty d\ell\, \Psi_1(\ell) \int_0^\infty da\, \Psi_2(a) \mathrm{LT}_\omega^{-1} \mathrm{LT}_z^{-1} \widetilde{Q}(\mathbf{x}, \omega, z, t|\mathbf{x}_0), \end{aligned} \qquad (8.110)$$

with

$$\frac{\partial \widetilde{Q}(\mathbf{x}, \omega, z, t|\mathbf{x}_0)}{\partial t} = D\nabla^2 \widetilde{Q}(\mathbf{x}, \omega, z, t|\mathbf{x}_0) - \gamma \widetilde{Q}(x, \omega, z, t|x_0), \qquad (8.111a)$$

$$\nabla \widetilde{Q}(\mathbf{x}, \omega, z, t|\mathbf{x}_0) \cdot \mathbf{n}_u = \omega \widetilde{Q}(\mathbf{x}, \omega, z, t|\mathbf{x}_0), \quad \mathbf{x} \in \partial\mathcal{U}. \qquad (8.111b)$$

The corresponding marginal fluxes are as follows:

$$\begin{aligned} J_1^{\Psi_1,\Psi_2}(\mathbf{x}_0, t) &\equiv D \int_{\partial\mathcal{U}} \nabla q^{\Psi_1,\Psi_2}(\mathbf{x}, t|\mathbf{x}_0) \cdot \mathbf{n}_u d\mathbf{x}\\ &= D \int_{\partial\mathcal{U}} \left\{ \int_0^\infty d\ell\, \psi_1(\ell) \int_0^\infty da\, \Psi_2(a) Q(\mathbf{x}, \ell, a, t|\mathbf{x}_0) \right\} d\mathbf{x}, \end{aligned} \qquad (8.112)$$

$$J_2^{\Psi_1,\Psi_2}(\mathbf{x}_0, t) = D \int_{\mathcal{U}} \left\{ \int_0^\infty d\ell\, \Psi_1(\ell) \int_0^\infty da\, \psi_2(a) Q(\mathbf{x}, \ell, a, t|\mathbf{x}_0) \right\} d\mathbf{x}. \qquad (8.113)$$

The stochastic process is killed as soon as one of the contact times exceeds its corresponding threshold, which occurs at the stopping time $\mathcal{T} = \min\{\tau_1, \tau_2\}$ with

$$\tau_1 = \inf\{t > 0 : \ell_t > \widehat{\ell}\}, \quad \tau_2 = \inf\{t > 0 : \mathcal{A}_t > \widehat{\mathcal{A}}\}. \tag{8.114}$$

Since there are now two effective targets, we have to introduce the associated splitting probabilities $\pi_j^{\Psi_1,\Psi_2}(\mathbf{x}_0)$ and conditional FPT densities $f_j^{\Psi_1,\Psi_2}(\mathbf{x}_0, t)$ for $j = 1, 2$. These are defined according to

$$\pi_j^{\Psi_1,\Psi_2}(\mathbf{x}_0) = \int_0^\infty J_j^{\Psi_1,\Psi_2}(\mathbf{x}_0, t)dt, \quad f_j^{\Psi_1,\Psi_2}(\mathbf{x}_0, t) = \frac{J_j^{\Psi_1,\Psi_2}(\mathbf{x}_0, t)}{\pi_j^{\Psi_1,\Psi_2}(\mathbf{x}_0)}. \tag{8.115}$$

We can now define snapping out BM with absorption using the conditional FPTs. In particular, the first renewal equation still holds with $\Psi \to (\Psi_1, \Psi_2)$ and the FPT density $f^\psi$ replaced by $\pi_j^{\Psi_1,\Psi_2} f_j^{\Psi_1,\Psi_2}$. Note that an analogous FPT problem has been studied elsewhere within the context of two impenetrable targets [9].

## 8.7 Discussion

In this paper we considered the FPT problem for a single target $\mathcal{U}$ in a bounded domain whose interior is partially absorbing and whose boundary $\partial\mathcal{U}$ is a semi-permeable interface, see Fig. 8.1(b). We described several scenarios of increasing complexity.

1. A classical semi-permeable membrane $\mathcal{U}$ with permeability $\kappa$ and bias $\alpha$, and a constant rate of absorption $\gamma$ within $\mathcal{U}$. We showed that one way to solve the FPT problem was in terms of the spectral properties of a pair of D-to-N operators.
2. A classical semi-permeable membrane $\mathcal{U}$ and a non-Markovian process of absorption within $\mathcal{U}$. We used an encounter-based method to formulate the absorption process in terms of a random threshold-crossing condition for the occupation time within $\mathcal{U}$. If the probability density of the occupation time threshold is an exponential, then one recovers the case of a constant rate of absorption. On the other hand, a non-exponential density leads to a non-Markovian form of absorption. The resulting MFPT depends on various moments of the occupation time threshold.
3. It is also possible to generalize the classical model of a semi-permeable membrane by formulating single-particle diffusion in terms of snapping out BM. Snapping out BM latter sews together successive rounds of partially reflecting BM that are restricted to either the interior or the exterior of $\partial\mathcal{U}$. Each round of partially reflected BM is killed when the boundary local time on the current side of the semi-permeable interface exceeds a randomly generated local time threshold. If the probability density of the latter is exponential, then the classical case of constant

permeability is recovered. On the other hand, a non-exponential density leads to a time-dependent permeability that tends to be heavy-tailed. It is also possible to combine snapping out BM with a non-Markovian absorption mechanism by keeping track of both the boundary local time on $\partial\mathcal{U}^-$ and the occupation time within the interior $\mathcal{U}$.

There are a number of natural generalizations of the single target problem considered here.

(i) *Multiple targets with semi-permeable interfaces in* $\mathbb{R}^d$. As we highlighted in this paper, solving the FPT problem for a single target in 2D or 3D is non-trivial even for simplified geometries. The analysis becomes even more difficult in the case of multiple targets, where one has to calculate splitting probabilities and conditional FPTs. However, considerable simplification occurs in the small-target limit, since one can then use matched asymptotic expansions and Green's function methods. More specifically, an inner or local solution is constructed in an $O(\epsilon)$ neighborhood of each target, where $\epsilon$ characterizes the relative size of each target compared to the size of the search domain. (The inner solution ignores the effects of other targets and treats the search domain as $\mathbb{R}^d$.) The inner solution is then matched to an outer or global solution that is valid away from each neighborhood. For details see Ref. [6] for 2D and Ref. [34] for 3D. A comparative analysis of different approaches to the single small target problem are presented in Ref. [35].

(ii) *Numerical methods.* In this paper, we focused on analytical methods for solving the target problem with semi-permeable interfaces. If one is also interested in studying single-particle trajectories, then it is necessary to construct efficient numerical schemes for simulating snapping out BM. Along these lines, we have recently developed a fast Monte Carlo algorithm for solving multi-dimensional snapping out BM for multiple interfaces, which combines a walk-on-spheres method with an efficient numerical scheme for calculating boundary local times [36]. The numerical methods were shown to have high accuracy when compared to solutions obtained from matched asymptotic analysis in the small-target regime. Note that there are also a number of alternative computational schemes for solving 1D diffusion problems in heterogeneous media with semi-permeable interfaces [37–39]. However, these do not generate exact sample trajectories of snapping out BM.

(iii) *Biophysical mechanisms.* Finally, from a modeling perspective, it would be interesting to identify plausible biophysical mechanisms underlying non-Markovian models of semi-permeable membranes. It is known that various surface-based reactions are better modeled in terms of a reactivity that is a function of the local time. For example, the surface may become progressively activated by repeated encounters with a diffusing particle, or an initially highly reactive surface may become less active due to multiple interactions with the particle (passivation) [23, 24]. One potential application is synaptic receptor trafficking in neurons [6], where the clustering of receptors within postsynaptic domains can be modeled in terms of a diffusion-trapping model. In this example, the boundary of the postsynaptic domain could be treated as an asymmetric semi-permeable membrane that is likely to involve non-Markovian components due to the complexity of the crowded molecular environment.

## References

1. P.C. Bressloff, B.A. Earnshaw, A biophysical model of AMPA receptor trafficking and its regulation during LTP/LTD. J. Neurosci. **26**, 12362–12373 (2006)
2. D. Holcman, A. Triller, Modeling synaptic dynamics driven by receptor lateral diffusion. Biophys. J. **91**, 2405–2415 (2006)
3. P.C. Bressloff, B.A. Earnshaw, M.J. Ward, Diffusion of protein receptors on a cylindrical dendritic membrane with partially absorbing traps. SIAM J. Appl. Math. **68**, 1223–1246 (2008)
4. K. Czondor, M. Mondin, M. Garcia, M. Heine, R. Frischknecht, D. Choquet, J.B. Sibarita, O.R. Thoumine, A unified quantitative model of AMPA receptor trafficking at synapses. Proc. Nat. Acad. Sci. USA **109**, 3522–3527 (2012)
5. R.D. Schummand, P.C. Bressloff, Local accumulation times in a diffusion-trapping model of synaptic receptor dynamics. Phys. Rev. E **105**, 064407 (2022)
6. P.C. Bressloff, 2D interfacial diffusion model of inhibitory synaptic receptor dynamics. Proc. Roy. Soc. A **479**, 20220831 (2023)
7. A. Kusumi, C. Nakada, K. Ritchie, K. Murase, K. Suzuki, H. Murakoshi, R.S. Kasai, J. Kondo, T. Fujiwara, Paradigm shift of the plasma membrane concept from the two-dimensional continuum fluid to the partitioned fluid: high-speed single-molecule tracking of membrane molecules. Annu. Rev. Biophys. Biomol. Struct. **34**, 351 (2005)
8. D.S. Grebenkov, Spectral theory of imperfect diffusion-controlled reactions on heterogeneous catalytic surfaces. J. Chem. Phys. **151**, 104108 (2019)
9. D.S. Grebenkov, Paradigm shift in diffusion-mediated surface phenomena. Phys. Rev. Lett. **125**, 078102 (2020)
10. P.C. Bressloff, Spectral theory of diffusion in partially absorbing media. Proc. Roy. Soc. A **478**, 20220319 (2022)
11. D.S. Grebenkov, An encounter-based approach for restricted diffusion with a gradient drift. J. Phys. A. **55**, 045203 (2022)
12. P.C. Bressloff, Diffusion-mediated absorption by partially reactive targets: Brownian functionals and generalized propagators. J. Phys. A. **55**, 205001 (2022)
13. P.C. Bressloff, A probabilistic model of diffusion through a semi-permeable barrier. Proc. R. Soc. A **478**, 20220615 (2022)
14. P.C. Bressloff, Renewal equations for single-particle diffusion through a semi-permeable interface. Phys. Rev. E **107**, 014110 (2023)
15. P.C. Bressloff, Renewal equations for single-particle diffusion in multi-layered media. SIAM J. Appl. Math. **83**, 1518–1545 (2023)
16. A. Lejay, The snapping out Brownian motion. Ann. Appl. Prob. **26**, 1727–1742 (2016)
17. A. Lejay, Monte Carlo estimation of the mean residence time in cells surrounded by thin layers. Math. Comp. Simul. **143**, 65–77 (2018)
18. A. Bobrowski, Semigroup-theoretic approach to diffusion in thin layers separated by semi-permeable membranes. J. Evol. Equ. **21**, 1019–1057 (2021)
19. O. Kedem, A. Katchalsky, Thermodynamic analysis of the permeability of biological membrane to non-electrolytes. Biochim. Biophys. Acta **27**, 229–246 (1958)
20. A. Katchalsky, O. Kedem, Thermodynamics of flow processes in biological systems. Biophys. J. **2**, 53–78 (1962)
21. A. Kargo, M. Kargol, S. Przestalski, The Kedem-Katchalsky equations as applied for describing substance transport across biological membranes. Cell. Mol. Biol. Lett. **2**, 117–124 (1996)

22. S. Redner, *A Guide to First-Passage Processes* (Cambridge University Press, Cambridge UK, 2001)
23. C.H. Bartholomew, Mechanisms of catalyst deactivation. Appl. Catal., A: Gen. **212**, 17–60 (2001)
24. M. Filoche, D.S. Grebenkov, J.S. Andrade Jr., B. Sapoval, Passivation of irregular surfaces accessed by diffusion. Proc. Natl. Acad. Sci. **105**, 7636–7640 (2008)
25. S.N. Majumdar, Brownian functionals in physics and computer science. Curr. Sci. **89**, 2076–2092 (2005)
26. A.G. McKendrick, Applications of mathematics to medical problems. Proc. Edinb. Math. Soc. **44**, 98 (2025)
27. H. Von Foerster, Some remarks on changing populations, in *The Kinetics of Cellular Proliferation*. ed. by F. Stohlman Jr. (Grune and Stratton, New York 1959)
28. M. Iannelli, F. Milner, *The Basic Approach to Age-structured Population Dynamics: Models, Methods and Numerics*. Lecture Notes on Mathematical Modelling in the Life Sciences. (Springer, New York, 2017)
29. P. Lévy, Sur certains processus stochastiques homogénes. Compos. Math. **7**, 283 (1940)
30. K. Itô, H.P. McKean, Brownian motions on a half line. Illinois J. Math. **7**, 181–231 (1963)
31. E.B. Dynki, *Markov Processes I and II* (Springer, Berlin, 1965)
32. H.P. McKean, Brownian local time. Adv. Math. **15**, 91–111 (1975)
33. D.S. Grebenkov, Joint distribution of multiple boundary local times and related first-passage time problems with multiple targets. J. Stat. Mech. **103205** (2020)
34. P.C. Bressloff, The 3D narrow capture problem for traps with semi-permeable interfaces. Multiscale Model. Simul. **21**, 1268–1298 (2023)
35. D.S. Grebenkov, Statistics of diffusive encounters with a small target: three complementary approaches, J. Stat. Mech. **083205** (2022)
36. R.D. Schumm, P.C. Bressloff, A numerical method for solving snapping out Brownian motion in 2D bounded domains. J. Comp. Phys. **493**, 112479 (2023)
37. S. Regev, O. Farago, Application of underdamped Langevin dynamics simulations for the study of diffusion from a drug-eluting stent. Phys. A Stat. Mech. Appl. **507**, 231–239 (2018)
38. O. Farago, Algorithms for Brownian dynamics across discontinuities. J. Chem. Phys. **423**, 109802 (2020)
39. N. Moutal, D.S. Grebenkov, Diffusion across semi-permeable barriers: spectral properties, efficient computation, and applications. J. Sci. Comput. **81**, 1630–1654 (2019)

# Chapter 9
# Big Jump Principle for First Passage Times

**Marc Höll, Alon Nissan, Brian Berkowitz, and Eli Barkai**

**Abstract** First passage times in disordered media, such as those involving charge carriers in amorphous semiconductors, contaminants in hydrogeological systems, or single macro-molecules in the the cell environment, are often described by time scale-free processes. We study the statistical properties of the first passage time of biased processes in different models, and employ the big jump principle that shows the dominance of the maximum trapping time on the first passage time. This leads to a scenario unlike that found for ordered systems, where many small short lived jumps lead to the completion of the first passage process. Inspired by the restart paradigm, we demonstrate that the removal of the maximum significantly expedites the first passage process. As the disorder increases, the system enters a phase where the removal shows a dramatic effect, the system is shown to exhibit a dynamical phase transition.

## 9.1 Introduction

Tracer pathways in disordered environments are typically seen to be similar over a large range of length and time scales, being observed ubiquitously in nature. In these systems, the waiting time distribution describing the transition times between micro-states is far from a simple exponential [1–4]. It is worth mentioning that even for Brownian motion in a confined space, many different time scales are relevant, while

M. Höll · E. Barkai (✉)
Bar-Ilan University, Ramat Gan, Israel
e-mail: Eli.Barkai@biu.ac.il

M. Höll
e-mail: marc-holl@outlook.com

A. Nissan
ETH Zürich, Zurich, Switzerland

B. Berkowitz
Weizmann Institute of Science, Rehovot, Israel
e-mail: Brian.Berkowitz@weizmann.ac.il

 
D. Grebenkov et al. (eds.), *Target Search Problems*,
https://doi.org/10.1007/978-3-031-67802-8_9

the distribution of the first passage time is broad [5, 6]. In many cases, the probability density function (PDF) of the transition times is heavy-tailed $\psi(\tau) \sim A\tau^{-(1+\alpha)}$ with some amplitude $A$ and the scaling exponent $\alpha > 0$. A critical consequence is that extremely long trapping times occur in deep traps, narrow passages, or at major obstacles, hence, the motion slows down dramatically. As a rule of thumb, as the disorder is increased, the first passage times from a given starting point to the target state are typically very long, and not surprisingly are also fat-tailed distributed, at least over a significant amount of time, before cutoffs become important.

To characterize the first passage time process in these systems we use the big jump principle [7–10]. The name might be a bit confusing for a start because the jumps we consider are not the spatial jumps found in random walk processes, but rather refer to the waiting times along the path. More specifically, consider a particle in a disordered system, and assume that the motion can be modeled by a series of waiting/transition/sojourn times and jumps. The particle starting at some location eventually reaches a fixed target at some random time. This is the first arrival or passage time studied in this book. Along the path, we may rank the waiting times from the smallest to the largest. To do this in practice demands some coarse graining of the system, to define the meaning of trapping times. We also require information on the path, which is now routine with single particle tracking techniques. A key theoretical topic is to find the relation between the longest trapping time in the sequence and the first passage time. It may be natural, at least for large systems, that this maximum is very small compared to the first passage time, and it is obvious that the latter is always the larger of the two. This scenario holds true when the disorder is weak. In contrast, for strong disorder, related to the exponent $\alpha$, the largest waiting time can be of the order of the first passage time. We will refine this statement below, and show that it holds mainly for the statistically large first passage times.

The study of the big jump principle thus belongs to the domain of extreme value statistics [11–13]. These extreme events might be rare in general, however their importance in practice cannot be underestimated. For example, consider a system to which, whether intentionally or not, contaminating particles are added. The absorption of the particles is on the boundary, where the particles are removed from the system. The largest first passage time is of interest because it signals the time when the system is clean and free from the intruders. We will focus on these long first passage times. It should be noted that also the shortest first passage times play a special role and they are also defined as extremes, in the sense that they are far from the typical or averaged first passage time [14]. For example, consider a study of the time for a single particle, diffusion-controlled reaction to take place, where the target is unique, say a single lattice point in a finite system. The first passage time stemming from a large number of searchers [15–18], when the process of search ends when the fastest searcher hits the target for the first time, is dominated under certain conditions by the short time statistics of the single particle distribution of first passage times, at least when the particles are non-interacting. It follows that extreme value statistics are of special importance in the study of first passage times.

Focusing on systems with long trapping times, we can ask at least academically how to exploit the dominance of one long sticking time in these processes. Specifically what will be the effect of the removal of the longest trapping time on the first passage time? Will this significantly speed up the first passage times, and if so how? It is then natural to ponder how to perform this task in practice. Here one must distinguish between two classes of models. The basic issue is whether there is some spatial region in the system that is repeatedly responsible for the largest trapping times in the process. For example, considering a bottleneck in the system like a very deep trap, will its removal be beneficial in terms of speeding up the first passage time? This might be possible in systems where the disorder is quenched, namely fixed in time, while for time varying disorder which is homogeneous in space, the same scenario seems unlikely. Here we study these new ideas briefly. The derivation of the main results in this text and further relevant literature can be found in our recent paper [19].

## 9.2 Big Jump Principle

We start with a simple model, which still can capture some of the complexities in the following more challenging and realistic approaches. The transport is unidirectional on a one-dimensional lattice with length $L$ where the lattice spacing is unity, so that $L$ is a positive integer. Of course, most disordered systems are not lattices. Hence in this pedagogical example, we mean that we coarse grain the system with each unit cell representing a region that already contains many molecules and atoms. At each lattice point, the particle waits randomly according to $\psi(\tau)$. The transition times $(\tau_1, \ldots, \tau_L)$ are independent and identically distributed, their PDF is $\psi(\tau)$. The first passage time is denoted as $t_f$ and it is simply the sum of the transition times because the process is unidirectional $t_f = \sum_{x=1}^{L} \tau_x$. This time is related to the maximum transition time $\tau_{\max} = \max(\tau_1, \ldots, \tau_L)$ via the big jump principle, which here could also be called the big transition time principle. This gives the probabilities [7, 8]

$$\mathrm{Prob}(t_f > t) \sim \mathrm{Prob}(\tau_{\max} > t). \tag{9.1}$$

Thus in a statistical sense and for large $t$ and any $L$, the first passage time is solely determined by a single summand, the maximum transition time.

This statement holds for any PDF $\psi(\tau)$ that is decaying slower than exponential for large $\tau$. These are called sub-exponential distributions. A special but important class of functions is when the PDF of the waiting time $\tau$ is described by the mentioned power law and then

$$\begin{aligned} \mathrm{Prob}(t_f > t) &\sim \mathrm{Prob}(\tau_{\max} > t) \\ &\sim L\frac{A}{\alpha}t^{-\alpha} \end{aligned} \tag{9.2}$$

which holds for any $\alpha$. Thus, surprisingly or not, the condition for the validity of the big jump principle is different from the familiar problem of summation of many random variables which converge in the limit of a large number of summands to either Gaussian or Lévy statistics, i.e., the central limit theorem and its generalization. Then the anomalous, non-Gaussian statistics hold when the variance of the waiting time diverges, namely, when $0 < \alpha < 2$. While mathematically the big jump theory is valid for any sub-exponential PDF, in practice its applicability is expected to be important for power laws with $\alpha$ being not too large. The reason is simple: when $\alpha$ is not big, sampling of large first passage times is easier. In other words, when $\alpha$ is small the extreme events, while rare, are typically relevant and sampling problems are not too bad. This statement must be tested case by case, as also parameters like $L$ and the small $\tau$ limit of $\psi(\tau)$ are important for this convergence issue. We also emphasize that the principle holds for any $N$, and in that sense it differs from Fréchet's law, from extreme value statistics, which holds as a limit for $N \to \infty$.

What is the effect of the elimination of the maximum transition time $\tau_{\max}$ from the sequence $(\tau_1, \ldots, \tau_L)$? We define this time as

$$t_r = t_f - \tau_{\max} \tag{9.3}$$

where the index "$r$" stands for *removal*. The idea is to observe if the removal of the maximum can have a large impact on the statistics of first passage times. The asymptotic behavior of the distribution of $t_r$ is

$$\text{Prob}(t_r > t) \sim \frac{1}{2} L(L-1) \left(\frac{A}{\alpha}\right)^2 t^{-2\alpha}, \tag{9.4}$$

see Fig. 9.1. The remarkable issue is the doubling effect of the exponent to $-2\alpha$, whereas previously we had the $-\alpha$ decay in Eq. (9.2). Thus, the probability of large $t_r$ is decreased compared to the probability of large $t_f$. In particular, if $\alpha < 1$, as is found in many disordered systems, the mean first passage time $\langle t_f \rangle$ diverges, but once we eliminate $\tau_{\max}$ from each trajectory, the mean $\langle t_r \rangle$ will diverge only if $\alpha < 1/2$. Furthermore, $t_r$ can be thought of as a first passage time, modified compared to $t_f$, though currently the modification can be made only after the completion of the path, because determination of the maximum demands the knowledge of all waiting times. We will treat this issue soon, when presenting models with quenched disorder. For now, Eq. (9.4) is a further indication that the removal of the maximum trapping time can have a significant effect.

Using scale invariance, one continues to wonder whether the first passage time after removal of $t_r$ is still related to the extreme events of the sequence, in a way similar to the big jump principle of the original $t_f$. This is of importance if we wish to expedite transport, not by simply removing the largest waiting time from the path, but by, e.g., removing two or more of them. One can show that [19].

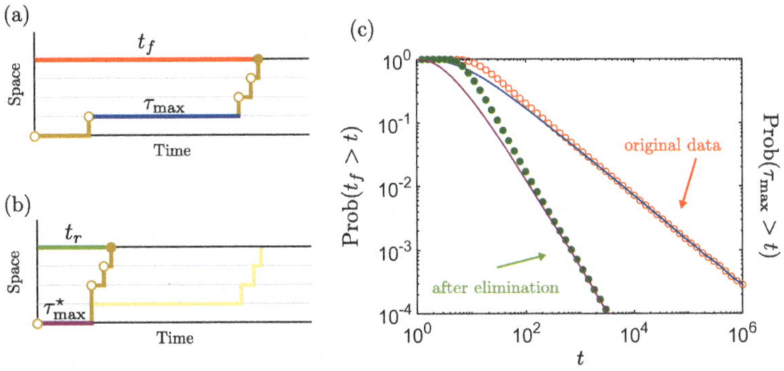


**Fig. 9.1** **a** Conceptual figure of the unidirectional transport model on a lattice. The first passage time $t_f$ (red) to reach some boundary is the sum of the transition times between lattice points. The maximum transition time $\tau_{\max}$ (blue) is also marked. **b** Conceptual figure of the same transport model as in **a** but $\tau_{\max}$ has been removed from the trajectory. This shows the modified first passage time $t_r = t_f - \tau_{\max}$ (green) and the second maximum $\tau^{\star}_{\max}$ (purple). **c** The distributions of $t_f$ (red circles), $\tau_{\max}$ (blue line), $t_r$ (green filled circles) and $\tau^{\star}_{\max}$ (purple line) obtained from Monte Carlo simulations for the unidirectional model. We used the Pareto distribution with $\alpha = 0.7$ for the transition times with $L = 5$ and $10^6$ trajectories. As shown, the right tails before and after the elimination of $\tau_{\max}$ from each trajectory match for the maximum and the first passage time distributions, as predicted by the principle of the single long transition time Eq. (9.1) and the relationship after elimination Eq. (9.5). Importantly, the elimination decreases the power law of the $t_f$ distribution Eqs. (9.2) to (9.4). This shows clearly that the elimination of the maximum transition time dramatically modifies the statistics in the tail distribution by orders of magnitude. This figure is reprinted with permission from [19]

$$\begin{aligned}\text{Prob}(t_r > t) &\sim \text{Prob}(\tau^{\star}_{\max} > t) \\ &\sim \frac{1}{2}L(L-1)\left(\frac{A}{\alpha}\right)^2 t^{-2\alpha},\end{aligned} \tag{9.5}$$

see Fig. 9.1. Here, $\tau^{\star}_{\max}$ is the second longest transition time of the original sequence of transition times. This shows that the principle holds also after the first removal. Removing also $\tau^{\star}_{\max}$ would result in more speedup $\text{Prob}(t_r > t) \sim t^{-3\alpha}$, where $t_r$ is the first passage time without both $\tau_{\max}$ and $\tau^{\star}_{\max}$. Further removals of the third, fourth, ... longest transition times result in an even faster speedup of $\text{Prob}(t_r > t)$, see details in [19].

As mentioned in the introduction, one of our goals is to check if we can in principle speed up first passage times by the manipulation of a single waiting time. To quantify the effect we define a measure of gain

$$G = \frac{\langle t_r \rangle}{\langle t_f \rangle}, \tag{9.6}$$

where a small value indicates significant impact of the removal. When the disorder becomes stronger, namely when the mean transition time diverges, so $\alpha < 1$, the

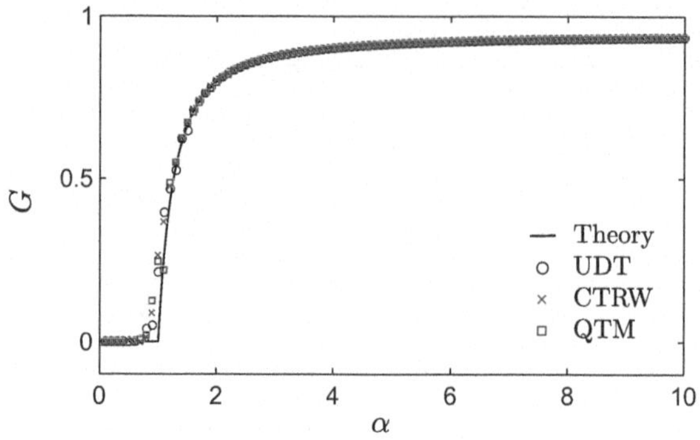


**Fig. 9.2** The quantifier $G$, which is a measure of the transport improvement Eq. (9.6), versus the parameter $\alpha$, which is a measure of disorder. When $G = 0$, the benefit from removal of the maximum transition time is maximal, in contrast with $G = 1$. A transition is found at $\alpha = 1$ for three models of transport considered in this work: the unidirectional model (UDT) (black circles), the continuous time random walk (red crosses), and the quenched trap model (QTM) averaged over the disorder (blue squares), see below. We used Pareto distributed transition times which cannot be smaller than $t_p = 1$, namely $\psi(\tau) = 0$ for $\tau < 1$. Each point in the plot used $10^5$ trajectories in the Monte Carlo simulation. For the UDT, we have $L = 20$, for continuous time random walk (CTRW) $L = 10$ and $p = 0.75$ so that $\langle N \rangle = 20$, and for QTM $L = 20$ and $p = 1$. See details in [19]. This figure is reprinted with permission from [19]

ratio $G$ is zero, which indicates a significant effect upon elimination. In Fig. 9.2 we present $G$ of Eq. (9.6) versus $\alpha$, for Pareto distributed transition times, the circle symbols for unidirectional transport (UDT) are those we have discussed so far. We see a sharp transition in the behavior of $G$ in Eq. (9.6) at the critical point $\alpha = 1$ similar to a dynamical phase transition.

## 9.3 Biased Continuous Time Random Walk

We now consider a basic model for anomalous transport, the continuous time random walk, which has found applications in a vast number of systems [1, 4]. As for the unidirectional model from above, at each lattice point a particle takes a transition time according to $\psi(\tau) \sim A\tau^{-(1+\alpha)}$. But now jumping to the left with probability $1 - p$ is permitted where $p$ is the probability of jumping to the right. We consider $p > 1/2$ hence the bias is driving the particle to the right. The unbiased case $p = 1/2$ is discussed in [19]. A particle that starts at $x = 1$ has a random number of jumps $N$, up to the target state, namely the absorption point $x = L + 1$. Therefore, to find the distribution of the first passage times we have to average over $N$ with a technique called subordination [20–22].

The idea is to consider a discrete time random walk with the same bias, initial conditions, and boundary conditions, as for the original model. Let $\phi_{\text{dis}}(n)$ be the probability that a particle performing this walk made $N = n$ jumps before its absorption, and the subscript "dis" stands for discrete time. Clearly, $n$ is the discrete time version of the first passage time. For the biased case $p > 1/2$ the mean $\langle N \rangle = \sum_{n=1}^{\infty} n\phi_{\text{dis}}(n) < \infty$ is finite. We obtain the big jump principle

$$\begin{aligned} \text{Prob}(t_f > t) &\sim \text{Prob}(\tau_{\max} > t) \\ &\sim \langle N \rangle \frac{A}{\alpha} t^{-\alpha}. \end{aligned} \tag{9.7}$$

The difference compared to Eq. (9.2) is that we replace $L$ by $\langle N \rangle$. Thus Eq. (9.7) shows again the dominance of the largest trapping time, and it is clear that the fluctuations in the number of trapping events, before the completion of the first passage process, do not alter the main results of the theory. We calculate $\phi_{\text{dis}}(n)$ using standard techniques and find $\langle N \rangle = L/(p-q)$ where $q = 1-p$ is the probability to jump left. Then Eq. (9.7) reads

$$\begin{aligned} \text{Prob}(t_f > t) &\sim \text{Prob}(\tau_{\max} > t) \\ &\sim \frac{L}{p-q} \frac{A}{\alpha} t^{-\alpha}. \end{aligned} \tag{9.8}$$

The modified first passage time upon removal behaves as

$$\begin{aligned} \text{Prob}(t_r > t) &\sim \text{Prob}(\tau^{\star}_{\max} > t) \\ &\sim \frac{1}{2}\langle N(N-1) \rangle \left(\frac{A}{\alpha}\right)^2 t^{-2\alpha} \end{aligned} \tag{9.9}$$

with $\langle N(N-1) \rangle = \sum_{n=1}^{\infty} n(n-1)\phi_{\text{dis}}(n)$. Compared to the result of the unidirectional model Eq. (9.4), we find again the same power law decay $-2\alpha$ but the $N$-dependent pre-factor must be averaged. The explicit expression for the pre-factor depends on whether or not we add a reflecting wall at $x = 0$; see further details in [19]. In the examples we use here, we employ a free boundary condition on the left, though recall that because $p > 1/2$ the left boundary condition is not of major concern. The insight gained is that the big jump principle works even when the particles can retract and when the number of jumps in a given time interval is random.

For the example of the Pareto distributed transition times, the dimensionless gain $G$ of Eq. (9.6) is

$$G = \begin{cases} 0, & 0 < \alpha < 1, \\ 1 - h_\alpha \sum\limits_{n=1}^{\infty} \phi_{\text{dis}}(n)(-1)^n n! \Gamma\left(-n + \frac{1}{\alpha}\right), & 1 < \alpha \end{cases} \tag{9.10}$$

with $h_\alpha = [(p-q)(\alpha-1)]/[\alpha L \Gamma(1/\alpha)]$. The phase transition-like in $G$ is presented in Fig. 9.2. Importantly, the exponent $\alpha$, which is the control parameter in this figure, is related to microscopic parameters, usually using mean field arguments. In particular for the quenched trap model, $\alpha = T/T_g$ where $T$ is the thermal temperature of the system, and $T_g$ is a measure of the disorder defined in the next section. The subscript $g$ is for glass, as this model describes certain properties of glassy dynamics like aging, anomalous diffusion, and weak ergodicity breaking.

## 9.4 Quenched Trap Model

In the unidirectional transport model and the continuous time random walk, the transition times are spatially homogeneous, i.e., independent of the lattice points. These models are thus semi-Markovian in the sense that the process is renewed after a jump is made. We remove this simplifying property now, considering the quenched trap model [23, 24]. In this approach, the disorder is spatially fixed, more specifically the particle is performing a biased random walk with $1/2 < p < 1$ in a random energy landscape. The energetic traps $E_x$ on the lattice points $x$ are mutually independent, identically distributed exponential random variables; their distribution is $\text{Prob}(E_x > E) = \exp(-E/T_g)$, where $T_g$ is a measure of disorder, specifically the average depth of the traps. The Boltzmann constant is considered as $k_B = 1$. As before, the particle starts at $x = 1$ and the absorption is at $x = L + 1$. According to the basic theory of activation, the distribution of the transition times is exponential $\text{Prob}(\tau_x > t) = \exp(-t/\bar{\tau}_x)$ with the well-known Arrhenius time $\bar{\tau}_x = t_0 \exp(E_x/T)$ to escape from an energy trap, used in many activation processes. Here, $T$ is the temperature of the system and $t_0$ is a well-studied timescale for dynamics. Averaging $\text{Prob}(\tau_x > t)$ over the disorder gives the fat-tailed $\psi(\tau)$ with $\alpha = T/T_g$ [23].

We consider now the occupation time $\hat{\tau}_x$, which is the sum of all transition times $\tau_x$ at a single trap $E_x$. The full derivation of the asymptotic behaviors of the distribution of first passage times is non-trivial and can be found in [19]. Briefly, the theory starts with the study of the joint probability $\phi(n_x, k)$ with $k$ visited traps and $n_x$ visits at trap $x$. The number of visits per trap is important because it is needed to estimate the total time spent in each trap. For the average over the energy landscape, we obtain

$$\begin{aligned} \langle \text{Prob}(t_f > t) \rangle_{\text{en}} &\sim \langle \text{Prob}(\hat{\tau}_{\max} > t) \rangle_{\text{en}} \\ &\sim (t_0)^\alpha M_\alpha t^{-\alpha} \end{aligned} \tag{9.11}$$

with the function

$$\begin{aligned} M_\alpha &= \sum_{x=-\infty}^{L} \left\langle \frac{\Gamma(N_x + \alpha)}{\Gamma(N_x)} \right\rangle \\ &= \sum_{x=-\infty}^{L} \sum_{k=1}^{\infty} \sum_{n_x=0}^{\infty} \phi(n_x, k) \frac{\Gamma(n_x + \alpha)}{\Gamma(n_x)} \end{aligned} \tag{9.12}$$

The non-trivial function $M_\alpha$ is obtained from the simulation of a discrete time and space random walk, i.e., without using any disorder. The strong bias case $p = 1$ gives $M_\alpha = \Gamma(1+\alpha)L$.

The idea is to now remove the deepest visited trap of each particle. The pair of probability distributions of the first passage time and of the maximum occupation time, both after the removal, behave as

$$\begin{aligned} \langle \text{Prob}(t_r > t)\rangle_{\text{en}} &\sim \langle \text{Prob}(\hat{\tau}^{\star}_{\max} > t)\rangle_{\text{en}} \\ &\sim \frac{(t_0)^{2\alpha}}{2} M^{\star}_{\alpha} t^{-2\alpha} \end{aligned} \tag{9.13}$$

with the function

$$\begin{aligned} M^{\star}_{\alpha} &= \sum_{x=-\infty}^{L} \left\langle (K-1)\frac{\Gamma(N_x+2\alpha)}{\Gamma(N_x)} \right\rangle \\ &= \sum_{x=-\infty}^{L} \sum_{k=1}^{\infty} \sum_{n_x=0}^{\infty} \phi(n_x,k)(k-1)\frac{\Gamma(n_x+2\alpha)}{\Gamma(n_x)}, \end{aligned} \tag{9.14}$$

Both probabilities are related and experience a doubling effect in the power law exponent; compared with Eq. (9.11). We obtain $M^{\star}_{\alpha}$ similar to the method of finding $M_\alpha$. For $p = 1$, the function is $M^{\star}_{\alpha} = \Gamma(1+2\alpha)L(L-1)$.

In Fig. 9.3, we present the theoretical predictions and compare them to theory for the case $T = T_g$ and $L = 10$. We present the distribution of first passage times both before and after the removal. The modification of the first passage time by removal

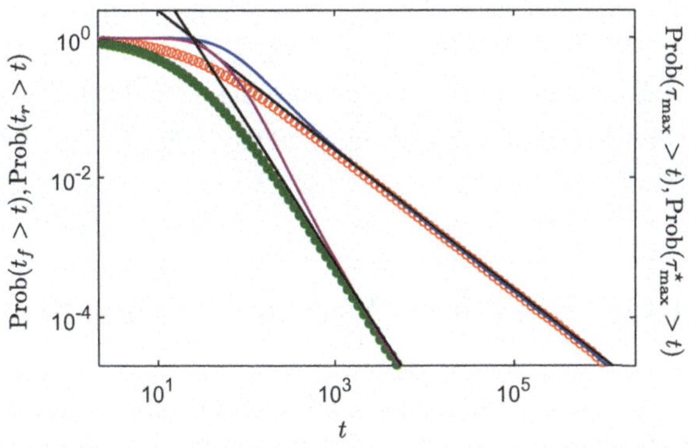


**Fig. 9.3** The distributions of $t_f$ (red circles), $\hat{\tau}_{\max}$ (blue line), $t_r$ (green filled circles) and $\hat{\tau}^{\star}_{\max}$ (purple line) compared with the theories Eq. (9.11) (upper black line) and Eq. (9.13) (lower black line) for the weakly biased quenched trap model with $L = 10$ and $p = 0.7$. We used $10^6$ particles and the parameters $T = T_g = 2$ (so that $\alpha = 1$) and $t_0 = 1$. We see the long transition time principle at work, the distributions of the maximum transition time and the first passage time follow the same laws for large arguments; moreover, the effect of removal is clearly strong. This figure is reprinted with permission from [19]

of one trap is clearly very significant, in particular for the large first passage times, where the big jump principle works. Also the distributions clearly follow power law statistical behaviors as the theory predicts. To verify the big jump principle, we also plot the distribution of the maximum occupation time, both before and after the removal. They match the corresponding first passage distributions, as they should. Note again that we are removing the trap in the sense that we remove the total occupation time in the deepest trap. This is not exactly the same as for the continuous time random walk model, where the maximum of the waiting times is removed. We also use the gain factor $G = \langle \bar{t}_r \rangle_{\text{en}} / \langle \bar{t}_f \rangle_{\text{en}}$ where averages over an ensemble of disordered systems are included. For the case $p = 0.7$, $L = 10$ and $\alpha = 1$, the gain quantifier $G$ has the numerical value 0.151, showing once again the impact of elimination of the maximum contribution; here the deepest trap yields a big gain. A semi-analytical expression for $G$ can be found for the case $p = 1$ which represents the strong bias limit. This theoretical prediction is compared to simulation in Fig. 9.2, showing excellent agreement. Not surprisingly, $G$ in this high field limit resembles the one found with the simpler models under study.

So far, we have considered the first passage time statistics when two types of averages are made. These are over stochastic paths, in a given realization of disorder, and over the disorder itself. This averaging is important in many applications and used extensively. However, in an experiment, with a single realization of the disorder, the results can be different than what was presented so far. In particular, imagine a finite system, and in the spirit of our discussion, the deepest trap is dominating the maximum of the occupation time. It is clear that this maximum is roughly $t_0 \exp(E_m/T)$ (at least for strong bias) where $E_m$ is the maximal energy. It follows that for this specific system, the first passage time distribution will not decay like a power. In other words, we have a natural cutoff on the first passage distribution, arising from the deepest trap in the system. This is a non-self-averaging effect. We studied the big jump principle for the quenched trap model in a single realization of disorder in our recent paper [19]. Here we will present in the next section simulations of transport in porous media, where similar, non-self-averaging and cutoff issues arise. The basic question: is the big jump principle at all valid in the absence of averaging over disorder?

## 9.5 First Passage in a Porous Medium: A Simulation Study

Using extensive numerical simulations, we discuss the advection-dominated transport behavior in a porous medium, thus bringing our insights closer to truly complex systems. Over twenty years of field and laboratory experiments [25], numerical simulations [26] and theoretical studies [27] have shown how advective-dispersive tracer motion in hydrogeological systems is characterized by many time scales. In particular, the distribution of trapping times in these systems is very broad; for example, for biased transport in a porous medium, the continuous time random walk with power law sticking times is an appropriate model.

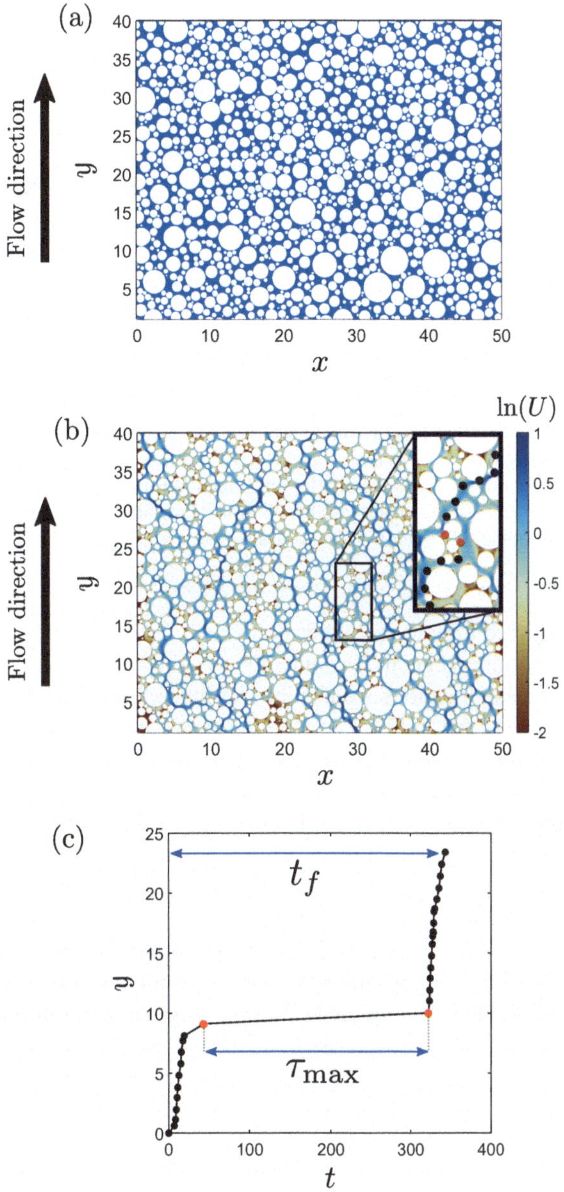


**Fig. 9.4** **a** The generated porous medium, with a porosity of 0.35, and mean grain diameter of $\lambda = 1$ [mm]. **b** Fluid velocity field in the generated pore-scale system; the color bar shows the normalized velocity field (divided by the mean value of the field, $\overline{U}$) with a logarithmic scale. The inset shows an example of the location of the longest transition time from a single particle trajectory within a scale of $\lambda$. Transitions occur between two successive dots (black marker), where the longest transition time is marked in red. **c** A single trajectory showing the vertical axis $y(t)$ with $N = 30$ transitions. Each time point of a transition is represented by a dot. The $X$-axis is shown in a non-dimensional time domain, i.e., $t = t_{\text{dim}}\lambda/\overline{U}$ where the dimensional time $t_{\text{dim}}$ has the unit $(\overline{U}/\lambda)^{-1}$, and $\overline{U}$ is the mean velocity. For the presented trajectory, we find $\tau_{\max} \approx 279$ and $t_f \approx 343$ where $t_f$ is the time it takes the particles to traverse the typical distance $L \approx \lambda N$ from bottom to top. This figure is reprinted with permission from [19]

Briefly, we simulated particle transport in a two-dimensional heterogeneous system, based on the flow solution of the Navier-Stokes Equations and use of a particle tracking technique along streamlines [26]. In Fig. 9.4a the porous medium model is presented; see [19] for the details of how it is constructed. The flow of the particles is from bottom to top, and releasing particles on the bottom we can find the first

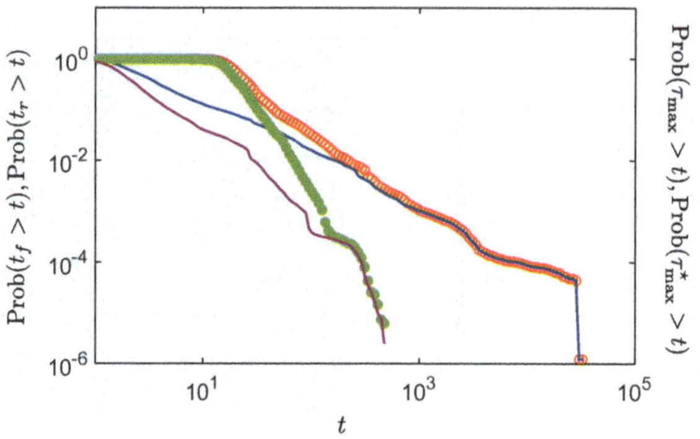


**Fig. 9.5** **a** The distributions of the first passage time $t_f$ (red circles), the maximum of the transition times $\tau_{max}$ (blue line), the modified first passage time $t_r$ (green filled circles) and the second longest transition time $\tau^\star_{max}$ (purple line) for the simulation of transport in a pore-scale system, where we followed $N = 30$ transition times of $10^6$ particles. The matching of the right tails of the respective distributions shows the single long transition time principle, which is non-trivial if compared to Eq. (9.2), as we do not have here a power law tail. The removal yields a significant decrease in the tail of the distribution of first passage times. This figure is reprinted with permission from [19]

passage time to the upper boundary. The flow field presented in Fig. 9.4b is highly non-homogeneous. In some regions, the local velocity field is very small, and hence in these locations we expect and indeed find very long sticking times. This effect is presented in Fig. 9.4c where now we may identify a long sticking event with the spatial $y$ coordinate.

More specifically, we coarse grain the system into patches, such that the sticking times are well defined, with which we can identify the maximum of the sticking times, see details [19]. In Fig. 9.5, we present the distributions for $t_f$ and $\tau_{max}$. These distributions are very wide (note the log-log scale) but unlike the previous models, we do not observe power laws because we do not average over disorder. The striking behavior is that we find excellent matching in the right part of the tails of the distributions, i.e., when $t_f$ and the maximum are large. Thus, we find the principle of the single big jump applied to the long transition times in a non-trivial system. This indicates that the longest first passage times are dominated by the longest transition times, and not by a sequence of many relatively short sticking times where particles simply wandered far from the absorbing boundary. An interesting effect is found in Fig. 9.5: the tails of both distributions, of $t_f$ and $\tau_{max}$, are not smooth, but exhibit step-like structures. The transport is dominated by the disorder configuration of the system. Namely, while the specific pattern of the distributions depends on the details of the porous medium model at hand, the single long transition principle holds also in this fixed background system. As the tail of the distribution of first passage times $t_f$ and, similarly, the extreme $\tau_{max}$, are sensitive to the fine details of the disorder, these observables are called non-self-averaging [28].

We now eliminate $\tau_{\max}$ from each trajectory and study $t_r$. In Fig. 9.5, we present the distribution of $t_r$ (green). Clearly, we see a dramatic improvement in the transport of $t_f$ (red). In the same figure, we also show the distribution for $\tau^\star_{\max}$ after elimination (purple). The figure illustrates the second level of the long big jump/transition time principle, namely, that the tails of the distribution of $t_r$ and of $\tau^\star_{\max}$ match, where the global maximum distribution is compared with the distribution of $t_f$; compare also to Eq. (9.4). The matching of the two distributions for $t_r$ and $\tau^\star_{\max}$ is valid even for a relatively complicated structure of the right tail.

Practically the most important observation is that the elimination leads to a significant truncation of the first passage times. Viewing the tracers as contamination, clearly, the removal or treatment of the long transition time has a dramatic effect on the leaching of the system (see the original data in red and the data after removal in green). Comparing the mean first passage times before and after elimination, as in Eq. (9.6), we obtain the value $G \approx 0.6085$; thus the elimination of $\tau_{\max}$ leads to expedited transport by about 39%, at least for our model system.

## 9.6 Comparison with the Restart Paradigm

Extensive literature exists on the influence of restart on the first passage time. In that paradigm, the particle is initialized at random or non-random times. A particle searching for a target can under certain conditions benefit from the restart, in the sense that one can reduce the mean first passage time, in some cases significantly [29–31]. In our case, because we deal with biased systems, the restart of the process will not in general speed up the process of search, at least not for strong bias.

We avoid the restart completely. Instead we aim to reduce the first passage time using a *single* move, but an educated one. This means that it is enough to remove the maximum of the sticking times to gain a dramatic improvement of the mean first passage time. We showed that this effect is described by a dynamical phase transition, the gain $G$ is non-analytical.

Of course, the removal of the maximum transition time can be done only after the completion of the process. Hence, in practical applications one will need to consider the following protocol. For a system with quenched disorder, we start with a learning session. This means that we follow particles in the system and detect the bottlenecks, for example, the location of the deepest trap in the quenched trap model. Then we have a spatial location where the removal must be applied. For example, the slowest regions in the flow field in our porous medium system that still have a significant impact on many particles. The removal is then, roughly speaking, a fixing of a local bottleneck. Of course, there is nothing new in this idea. Any engineer who deals with traffic flow in a city has to estimate the location where public resources for by-passes or bridges will have some strong impact on the arrival times of persons to different destinations. The new aspect is that in strongly disordered systems, we can use the big jump principle, find a dynamical phase transition to characterize this goal, and quantify the effect of removal in statistical terms.

**Acknowledgements** M.H. is funded by the Deutsche Forschungsgemeinschaft (DFG, German Research Foundation), Grant No. 436344834. The support of Israel Science Foundation Grant No. 1614/21 (E.B.) and ViTamins project funded by the Volkswagen Foundation grant AZ 9B192 (B.B.) is acknowledged. A.N. is supported by an ETH Zurich Postdoctoral Fellowship. B.B. holds the Sam Zuckerberg Professorial Chair in Hydrology. M.H. and A.N. contributed equally to this work

## References

1. R. Metzler, J. Klafter, The random walk's guide to anomalous diffusion: a fractional dynamics approach. Phys. Rep. **339**, 1 (2000)
2. R. Klages, G. Radons, I.M. Sokolov, Anomalous Transport. (Wiley Online Library, 2008)
3. F. Höfling, T. Franosch, Anomalous transport in the crowded world of biological cells. Rep. Prog. Phys. **76**, 046602 (2013)
4. R. Metzler, J.-H. Jeon, A.G. Cherstvy, E. Barkai, Anomalous diffusion models and their properties: non-stationarity, non-ergodicity, and ageing at the centenary of single particle tracking. Phys. Chem. Chem. Phys. **16**, 24128 (2014)
5. Denis S. Grebenkov, Ralf Metzler, Gleb Oshanin, Strong defocusing of molecular reaction times results from an interplay of geometry and reaction control. Commun. Chem. **1**(1), 96 (2018)
6. A. Godec, R. Metzler, Universal proximity effect in target search kinetics in the few-encounter limit. Phys. Rev. X **6**(4), 041037 (2016)
7. V.P. Chistyakov, A theorem on sums of independent positive random variables and its applications to branching random processes. Theory Prob. Appl. **9**, 640 (1964)
8. A. Vezzani, E. Barkai, R. Burioni, Single-big-jump principle in physical modeling. Phys. Rev. E **100**, 012108 (2019)
9. W. Wang, A. Vezzani, R. Burioni, E. Barkai, Transport in disordered systems: the single big jump approach. Phys. Rev. Res. **1**, 033172 (2019)
10. R.K. Singh, S. Burov, Universal to Non-Universal Transition of the statistics of Rare Events During the Spread of Random Walks. Phys. Rev. E. **108**, L052102 (2023). https://doi.org/10.1103/PhysRevE.108.L052102
11. S.N. Majumdar, A. Pal, G. Schehr, Extreme value statistics of correlated random variables: a pedagogical review. Phys. Rep. **840**, 1–32 (2020)
12. L. Zarfaty, E. Barkai, D.A. Kessler, Accurately approximating extreme value statistics. J. Phys. A: Math. Theor. **54**(31) (2021)
13. S.D. Lawley, J. Johnson, Slowest first passage times, redundancy, and menopause timing. J. Math. Biol. **86**(6) (2023)
14. A. Vezzani, R. Burioni, Fast rare events in exit times distributions of jump processes. Phys. Rev. Lett. **132**, 187101 (2024). https://doi.org/10.1103PhysRevLett.132.187101
15. D.S. Grebenkov, Searching for partially reactive sites: analytical results for spherical targets. J. Chem. Phys. **132**(3) (2010)
16. C. Mejía-Monasterio, G. Oshanin, G. Schehr, First passages for a search by a swarm of independent random searchers. J. Stat. Mech. P06022 (2011)
17. D.S. Grebenkov, R. Metzler, G. Oshanin, From single-particle stochastic kinetics to macroscopic reaction rates: fastest first-passage time of N random walkers. New J. Phys. **22**(10), 103004 (2020)
18. S.D. Lawley, Chapter 12 of this book
19. M. Höll, A. Nissan, B. Berkowitz, E. Barkai, Controls that expedite first passage times in disordered systems. Phys. Rev. E. **108**, 034124 (2023)
20. H.C. Fogedby, Langevin equations for continuous time Lévy flights. Phys. Rev. E **50**, 1657 (1994)

21. E. Barkai, Fractional Fokker-Planck equation, solution and application. Phys. Rev. E **63**, 046118 (2001)
22. S. Condamin, O. Benichou, J. Klafter, First-passage time distributions for sub-diffusion in confined geometry. Phys. Rev. Lett. **98**, 250602 (2007)
23. C. Monthus, J.-P. Bouchaud, Models of traps and glass phenomenology. J. Phys. A Math. Theor. **29**, 3847 (1996)
24. S. Burov, E. Barkai, Time transformation for random walks in the quenched trap model. Phys. Rev. Lett. **106**, 140602 (2011)
25. B. Berkowitz, H. Scher, Exploring the nature of non-Fickian transport in laboratory experiments. Adv. Water Resour. **32**(5), 750 (2009)
26. A. Nissan, B. Berkowitz, Anomalous transport dependence on Péclet number, porous medium heterogeneity, and a temporally varying velocity field. Phys. Rev. E **99**, 033108 (2019)
27. B. Berkowitz, A. Cortis, M. Dentz, H. Scher: Modeling non-Fickian transport in geological formations as a continuous time random walk. Rev. Geophys. **44**(2) ( 2006)
28. T. Akimoto, E. Barkai, K. Saito, Universal fluctuations of single-particle diffusivity in quenched environment. Phys. Rev. Lett. **117**, 180602 (2016)
29. G. Bel, B. Munsky, I. Nemenman, The simplicity of completion time distributions for common complex biochemical processes. Phys. Biol. **7**, 016003 (2009)
30. M.R. Evans, S.N. Majumdar, Diffusion with stochastic resetting. Phys. Rev. Lett. **106**, 160601 (2011)
31. A. Pal, S. Reuveni, First passage under restart. Phys. Rev. Lett. **118**, 030603 (2017)

# Chapter 10
# Optimization of Trap Locations for Narrow Capture Problems

**Alexei Cheviakov and Michael Ward**

**Abstract** The determination of the mean first passage time (MFPT) for a Brownian particle in a domain that contains a collection of small absorbing traps in its interior is a narrow capture problem with many biophysical applications. The average MFPT is the expected capture time assuming a uniform distribution of starting points for the random walk. We survey some results for determining the spatial locations of the traps that minimize the average MFPT for certain 2-D and 3-D domains. In the limit of small trap radii, the optimization of the average MFPT over the trap locations can be reduced to the problem of seeking global minima of a discrete energy for an interacting particle system that is encoded by the Neumann Green's matrix. Open problems for the optimization of the average MFPT are discussed.

## 10.1 Introduction

Narrow capture problems arise naturally in models of diffusive transport in cellular biology owing to the highly heterogeneous spatial environment that diffusing signaling molecules typically encounter before reaching a specific target site to initiate some biological function [1–4]. Modeling these diffusing signaling molecules as Brownian particles, a narrow capture problem involves calculating the mean time, and other related averaged quantities, that such a particle takes in order to reach a small target when it is confined within a multi-dimensional domain. More generally, an overview of results for a wide variety of first passage-problems and their applications are given in [5, 6].

In this article, we exploit the connection between mean first passage time problems and PDE problems that have origins in electrostatics. In a bounded domain

A. Cheviakov
Department of Mathematics and Statistics, University of Saskatchewan, Saskatoon, SK S7N 5E6, Canada
e-mail: shevyakov@math.usask.ca

M. Ward (✉)
Departemt of Mathematics, University of British Columbia, Vancouver, BC V6T 1Z2, Canada
e-mail: ward@math.ubc.ca

 
D. Grebenkov et al. (eds.), *Target Search Problems*,
https://doi.org/10.1007/978-3-031-67802-8_10

$\Omega$ containing $N$ small traps, the mean first passage time $u(\mathbf{x})$ for a narrow capture problem is the solution to the singularly perturbed PDE

$$\begin{aligned} D\,\Delta u &= -1\,, \quad \mathbf{x} \in \bar{\Omega}\,; \qquad \bar{\Omega} \equiv \Omega \setminus \cup_{j=1}^{N} \Omega_{\varepsilon j}\,, \\ \partial_n u &= 0\,, \quad \mathbf{x} \in \partial\Omega\,; \qquad u = 0\,, \quad \mathbf{x} \in \partial\Omega_{\varepsilon j}\,, \quad j = 1, \ldots, N\,. \end{aligned} \tag{10.1}$$

The small traps $\Omega_{\varepsilon j}$, each of radius $\mathcal{O}(\varepsilon) \ll 1$, are centered at locations $\mathbf{x}_1, \ldots, \mathbf{x}_N$ within $\Omega$. We assume that the traps are well-separated in the sense that $|\mathbf{x}_i - \mathbf{x}_j| = \mathcal{O}(1)$ for $i \neq j$ and that $\text{dist}(\partial\Omega, \mathbf{x}_j) = \mathcal{O}(1)$ as $\varepsilon \to 0$. The prototypical case in 2-D and 3-D is where the traps are all either disks or spheres, respectively, of a common radius $\varepsilon$. The average MFPT $\overline{u}$, which assumes that the starting point for Brownian motion is uniformly distributed over the domain, is the spatial average

$$\overline{u} \equiv \frac{1}{|\bar{\Omega}|} \int_{\bar{\Omega}} u \, d\mathbf{x}\,. \tag{10.2}$$

Our goal for a fixed $N$ is to identify the locations $\mathbf{x}_1, \ldots, \mathbf{x}_N$ of the traps that minimizes $\overline{u}$. Such an optimal trap configuration is the one that most rapidly captures, on average, a Brownian particle when the initial starting point for the random walk is uniformly distributed in the domain.

With regard to full numerical PDE simulations, the identification of optimal trap configurations for (10.1) is a challenging shape optimization problem since the trap locations must be adjusted and the PDE re-meshed and re-solved at each iterate of an optimization procedure. For an arbitrary 2-D domain with a collection of circular traps of small radius $\varepsilon \ll 1$, a **closest-point method (CPM)** coupled to a particle-swarm optimizer was developed in [7] to compute optimal trap patterns.

From the viewpoint of asymptotic analysis, in the limit $\varepsilon \to 0$ strong localized perturbation theory [8, 9] can be used to derive an approximation $\overline{u}_0$ for the average MFPT. For a survey of strong localized perturbation for MFPT problems and in other contexts see [10]. In the limit $\varepsilon \to 0$ for a 2-D or 3-D domain, we obtain

$$\overline{u} \sim \overline{u}_0 + \mathcal{O}(\varepsilon^m)\,, \qquad \overline{u}_0 \equiv \frac{1}{|\Omega|} \int_{\Omega} u_0 \, d\mathbf{x}\,, \tag{10.3}$$

where $m = 2$ (2-D) or $m = 3$ (3-D). Here, for $\varepsilon \to 0$, $u_0$ is the outer approximation for the solution to (10.1) when the traps are replaced by effective point singularities at the "particles" $\mathbf{x}_j \in \Omega$ for $j = 1, \ldots, N$. In this limiting problem, traps of different shapes are encoded by their logarithmic capacitances or capacitances in a 2-D or 3-D setting, respectively. These trap-shape dependent terms can be calculated separately from certain **local problems** for Laplace's equation that are valid in the near-field region of a trap. The explicit solution to this limiting system in terms of the Neumann Green's function determines a discrete energy for our effective "interacting-particle" system that is defined in terms of the Neumann Green's matrix, which depends on the trap locations. By applying particle-swarm optimization software to this discrete energy the asymptotic prediction for optimal trap configurations can be readily computed for moderately small $N$. However, as is typically the case with optimizing the discrete energy associated with other interacting-particle systems, as $N$ increases it becomes increasingly more challenging to numerically identify globally optimal trap patterns owing to the existence of many local minima [12].

As shown in Appendix C of [11] the problem of identifying optimal trap configurations that minimize the average MFPT is closely related to a classical problem in spectral theory of identifying trap locations that maximize the fundamental Neumann eigenvalue for the Laplacian in the domain with traps. For this class of eigenvalue problem, optimal trap locations have been analyzed in [13–19].

The outline of this brief survey is as follows. In Sects. 10.2 and 10.3, we highlight some previous results obtained for certain specific 2-D and 3-D domains, respectively. For a 2-D or 3-D domain with small traps, in Sect. 10.4 we relate the problem of minimizing the average MFPT to the spectral problem of maximizing the fundamental Neumann eigenvalue for the Laplacian. In Sect. 10.5 we discuss some related narrow capture and escape problems, and we suggest a few open problems.

## 10.2 2-D Domains

A hybrid asymptotic-numerical method was implemented for (10.1) to identify optimal trap configurations that minimize the average MFPT for disk-shaped domains [17], for near-disk domains [7], and for elliptical-shaped domains [11, 20]. For a class of dumbbell-shaped domains, the optimal location of a single trap has been identified in [17]. A full numerical approach to compute optimal trap configurations for (10.1) using the closest point methodology (CPM) of [21] was developed in [7]. In this way, approximate optimization results obtained from the hybrid asymptotic-numerical theory, which is based on the assumption of small trap radii, were validated.

In a 2-D domain, strong localized perturbation theory can be applied to (10.1) in the limit of small trap radii to derive a linear algebraic system (LAS) that determines the average MFPT in terms of the spatial locations of $N$ traps. In this asymptotic limit, the spatial interaction between the traps is mediated by the Neumann Green's matrix, which is defined in terms of the Neumann Green's function for the domain.

We now highlight the derivation of this LAS. Under the assumption that the traps $\Omega_{\varepsilon j}$ have a common shape, the outer problem $u_0$ satisfies [11, 22]

$$\begin{aligned} &D\Delta u_0 = -1\,, \quad \mathbf{x} \in \Omega \setminus \{\mathbf{x}_1, \ldots, \mathbf{x}_N\}\,; \quad \partial_n u_0 = 0\,, \quad \mathbf{x} \in \partial\Omega\,; \\ &u_0 \sim A_j \log|\mathbf{x} - \mathbf{x}_j| + A_j/\nu \quad \text{as} \quad \mathbf{x} \to \mathbf{x}_j\,, \qquad j = 1, \ldots, N\,, \end{aligned} \tag{10.4}$$

where $\nu \equiv -1/\log(\varepsilon d)$ and $A_j$ for $j = 1, \ldots, N$ are to be determined. Here $d$ is the logarithmic capacitance associated with the assumed common trap shape, which is independent of the orientation of the traps in the domain. It is defined by the solution to the following local problem near a trap formulated in terms of $\mathbf{y} = \varepsilon^{-1}(\mathbf{x} - \mathbf{x}_j)$:

$$\Delta_{\mathbf{y}} v = 0\,, \quad \mathbf{y} \notin \Omega_j\,; \quad v = 0\,, \quad \mathbf{y} \in \partial\Omega_j\,,$$
$$v \sim \log|\mathbf{y}| - \log d + O(|\mathbf{y}|^{-1})\,, \quad \text{as} \quad |\mathbf{y}| \to \infty\,, \tag{10.5}$$

where $\Omega_j = \Omega_{\varepsilon j}/\varepsilon$. If $\Omega_{\varepsilon j}$ are disks of a common radius $\varepsilon$ then $\Omega_j$ is the unit disk and $d = 1$ from (10.5). Table 1 of [22] gives $d$ for traps of other shapes.

The solution to (10.4) is

$$u_0 = -2\pi \sum_{k=1}^{N} A_k G(\mathbf{x}; \mathbf{x}_k) + \overline{u}_0\,, \qquad \text{where} \qquad \overline{u}_0 = \frac{1}{|\Omega|} \int_\Omega u_0 \, d\mathbf{x}\,. \tag{10.6}$$

Here $\overline{u}_0$ is the approximation to the average MFPT (see (10.3)) and $G(\mathbf{x}; \mathbf{x}_j)$ is the unique Neumann Green's function satisfying

$$\Delta G = \frac{1}{|\Omega|} - \delta(\mathbf{x} - \mathbf{x}_j)\,, \quad \mathbf{x} \in \Omega\,; \quad \partial_n G = 0\,, \quad \mathbf{x} \in \partial\Omega\,; \quad \int_\Omega G \, d\mathbf{x} = 0\,,$$
$$G \sim -\frac{1}{2\pi} \log|\mathbf{x} - \mathbf{x}_j| + R_j + \nabla_{\mathbf{x}} R_j \cdot (\mathbf{x} - \mathbf{x}_j) + \cdots\,, \quad \text{as} \quad \mathbf{x} \to \mathbf{x}_j\,. \tag{10.7}$$

Moreover, $R_j \equiv R(\mathbf{x}_j)$ is the regular part of the Green's function at $\mathbf{x} = \mathbf{x}_j$. By enforcing that $u_0$ in (10.6) satisfies the singularity conditions in (10.4), we obtain that $\overline{u}_0$ is given in terms of a LAS and a discrete energy $q(\mathbf{x}_1, \ldots, \mathbf{x}_N)$ defined by

$$\overline{u}_0 = \frac{|\Omega|}{2\pi D \nu N} + \frac{2\pi}{N} q(\mathbf{x}_1, \ldots, \mathbf{x}_N)\,,$$
$$\text{where} \quad q \equiv \mathbf{e}^T \mathcal{G} \mathcal{A} \quad \text{and} \quad \Big[ I + 2\pi\nu(I - E)\mathcal{G} \Big] \mathcal{A} = \frac{|\Omega|}{2\pi D N} \mathbf{e}\,. \tag{10.8}$$

Here $E \equiv \mathbf{e}\mathbf{e}^T/N$, $\mathbf{e} \equiv (1, \ldots, 1)^T$, $\mathcal{A} \equiv (A_1, \ldots, A_N)^T$, $I$ is the identity matrix, and $\nu \equiv -1/\log(\varepsilon d)$. The $N \times N$ symmetric Neumann Green's matrix $\mathcal{G}$, which depends on the trap locations $\mathbf{x}_1, \ldots, \mathbf{x}_N$, has matrix entries

$$(\mathcal{G})_{jj} = R_j\,, \quad \text{and} \quad (\mathcal{G})_{ij} = (\mathcal{G})_{ji} = G(\mathbf{x}_i; \mathbf{x}_j) \ \text{ for } i \neq j\,. \tag{10.9}$$

For $\nu \ll 1$, for which we can estimate $\mathcal{A} \sim |\Omega|\mathbf{e}/(2\pi D N) + O(\nu)$, (10.8) yields the following two-term approximation for $\overline{u}_0$ in terms of $\nu \ll 1$:

$$\overline{u}_0 = \tilde{u}_0 + O(\nu^2)\,, \quad \text{with} \quad \tilde{u}_0 \equiv \frac{|\Omega|}{2\pi D \nu N} \left[ 1 + \frac{2\pi\nu}{N} p(\mathbf{x}_1, \ldots, \mathbf{x}_N) \right], \quad p \equiv \mathbf{e}^T \mathcal{G} \mathbf{e}\,. \tag{10.10}$$

When $G(\mathbf{x}; \mathbf{x}_j)$ is known analytically as either an explicit formula or as a rapidly converging infinite series, the optimal trap locations in the limit $\varepsilon \to 0$ can be calculated numerically by applying particle swarm optimization schemes [23–25] to the LAS (10.8) to minimize $q$ and hence $\overline{u}_0$. Domains for which $G$ is readily available are disks, rectangles (see Sect. 4.2 of [26] and Sect. 3.2 of [27]), and ellipses.

In Sects. 10.2.1 and 10.2.2 we highlight some results for optimal trap patterns for the disk and the ellipse. For a spatially periodic arrangement of traps, in §10.2.3 we discuss results from [28] for the optimization of $\overline{u}_0$ over the class of Bravais lattices with a fixed area of the primitive cell.

### 10.2.1 The Unit Disk

For the unit disk, the Neumann Green's function and its regular part for (10.7) are [17]

$$
\begin{aligned}
G(\mathbf{x};\mathbf{x}_j) &= -\frac{1}{2\pi}\log|\mathbf{x}-\mathbf{x}_j| - \frac{1}{4\pi}\log\left(|\mathbf{x}|^2|\mathbf{x}_j|^2 + 1 - 2\mathbf{x}\cdot\mathbf{x}_j\right) + \frac{|\mathbf{x}|^2}{4\pi} + \frac{|\mathbf{x}_j|^2}{4\pi} - \frac{3}{8\pi}\,,\\
R(\mathbf{x}_j) &= -\frac{1}{2\pi}\log\left(1-|\mathbf{x}_j|^2\right) + \frac{|\mathbf{x}_j|^2}{2\pi} - \frac{3}{8\pi}\,.
\end{aligned}
\tag{10.11}
$$

In this way, the Neumann Green's matrix $\mathcal{G}$ in (10.8) and (10.10) is readily evaluated.

In [17] a restricted optimization of the discrete energy $p(\mathbf{x}_1,\ldots,\mathbf{x}_N)$ for $\tilde{u}_0$ in (10.10) was undertaken for $2 \leq N \leq 25$ where the traps were forced to lie on concentric rings about the origin, with possibly a center trap. For such two- and three-ring patterns analytical expressions for $p$ were derived in [17] in terms of the ring radii and the number of traps on each ring. The results of this restricted optimization process for $p$ for $6 \leq N \leq 25$ are shown in Fig. 10.1 (see also Table 2 of [17]). With the restricted optimal values of $p$ determined in this way, the two-term approximation $\tilde{u}_0$ in (10.10) can be estimated for a given $\varepsilon$ and $D$.

In [20] the results of the restricted optimization in [17] were compared with corresponding results computed by optimizing $\overline{u}_0$ in the LSA (10.8) by using a particle swarm optimizer on $q$ in (10.8) over the $2N$ variables $\mathbf{x}_j = (x_j, y_j)$ with $j = 1,\ldots,N$ and $|\mathbf{x}_j| < 1$ (see [20] for details). In Fig. 10.2, we show a reasonably close comparison between the optimal $\tilde{u}_0$ and $\overline{u}_0$ for $6 \leq N \leq 25$ when $\varepsilon = 0.05$ and $D = 1$. In [20] it was shown that as $N$ increases the optimal trap configuration begins to deviate significantly from a ring-type structure (see Figs. 4a)–7a) of [20]).

Finally, we show that when $N$ traps are equally spaced on one ring of radius $r$ concentric within the unit disk, the approximate result $\tilde{u}_0$ in (10.10) is an exact solution to the LSA (10.8). For such a ring pattern, the Neumann Green's matrix is cyclic and symmetric and has the eigenpair (see Principal Result 4.3 of [17])

$$
\mathcal{G}\mathbf{e} = \sigma\mathbf{e}\,, \quad \sigma = \sigma(r) \equiv \frac{1}{2\pi}\left(-\log\left(Nr^{N-1}\right) - \log\left(1-r^{2N}\right) + r^2 N - \frac{3N}{4}\right). \tag{10.12}
$$

Since $(I-E)\mathbf{e} = 0$, the solution to (10.8) for $\mathcal{A}$ is $\mathcal{A} = |\Omega|\mathbf{e}/(2\pi DN)$, so that

$$
\overline{u}_0 = \tilde{u}_0 = \frac{|\Omega|}{2\pi DN\nu}\left(1 + 2\pi\sigma\nu\right). \tag{10.13}
$$

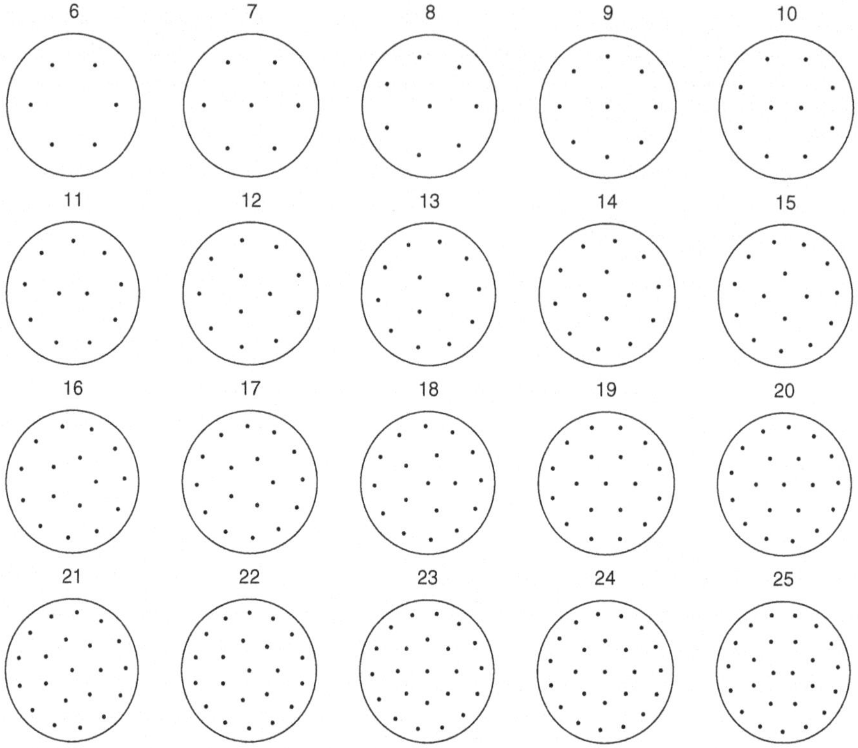


**Fig. 10.1** The optimum configurations for $N = 6$ to $N = 25$ traps for the discrete energy $p$ in (10.10) for the class of two-ring and three-ring patterns, with possibly a center trap

The optimal ring radius $r_c$ is found by setting $d\sigma/dr = 0$ and solving for $r = r_c$. For $D = 1$ and $\varepsilon = 0.003$, and for $2 \leq N \leq 10$, in Fig. 10.3 we show a very close comparison between $r_c$ obtained from (10.13) with that computed by the CPM from the full PDE (10.1) for the restricted optimization problem where $N$ equally spaced traps are restricted to one ring that is concentric within the unit disk (see Fig. 4 of [7]).

### 10.2.2 The Ellipse

A new rapidly converging infinite series representation for the Neumann Green's function $G(\mathbf{x}; \mathbf{x}_0)$ and its regular part $R(\mathbf{x}_0)$ for an elliptical-shaped domain $\Omega \equiv \{\mathbf{x} = (x, y) \,|\, x^2/a^2 + y^2/b^2 \leq 1\}$ with $a > b$ was derived in Sect. 5 of [11].

In the analysis of [11], the ellipse was first mapped to a rectangular domain $0 \leq \xi \leq \xi_b$ and $0 \leq \eta \leq 2\pi$ using the elliptical coordinates $(\xi, \eta)$ defined by

| $N$ | $\tilde{u}_0$ | $\overline{u}_0$ |
|---|---|---|
| 6 | 0.11648 | 0.11648 |
| 7 | 0.09299 | 0.09297 |
| 8 | 0.07660 | 0.07660 |
| 9 | 0.06518 | 0.06512 |
| 10 | 0.05653 | 0.05624 |
| 11 | 0.04920 | 0.04900 |
| 12 | 0.04291 | 0.04278 |
| 13 | 0.03805 | 0.03796 |
| 14 | 0.03380 | 0.03375 |
| 15 | 0.03042 | 0.03038 |
| 16 | 0.02747 | 0.02745 |
| 17 | 0.02502 | 0.02499 |
| 18 | 0.02286 | 0.02280 |
| 19 | 0.02078 | 0.02076 |
| 20 | 0.01909 | 0.01907 |
| 21 | 0.01756 | 0.01755 |
| 22 | 0.01626 | 0.01624 |
| 23 | 0.01512 | 0.01510 |
| 24 | 0.01411 | 0.01403 |
| 25 | 0.01314 | 0.01307 |

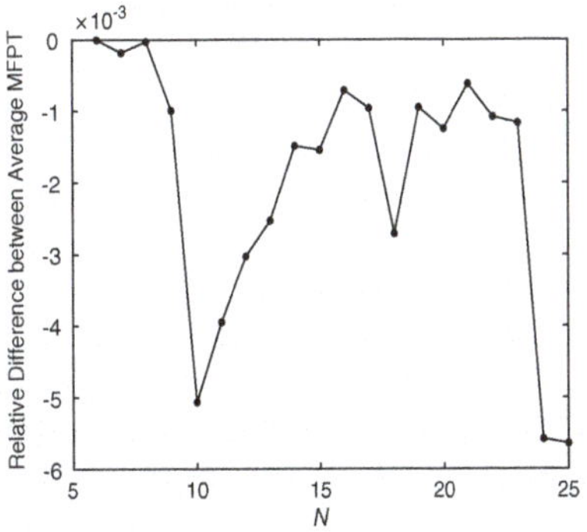


**Fig. 10.2** Left: The optimal average MFPT $\tilde{u}_0$ in the unit disk for a restricted optimization with traps equally spaced on rings (see [17]) is compared with $\overline{u}_0$ from the LSA (10.8) for optimal trap locations. Right: The relative difference $(\overline{u}_0 - \tilde{u}_0)/\tilde{u}_0$ versus $N$

| $N$ | $\tilde{u}_0$ | $\overline{u}$ (CPM) |
|---|---|---|
| 2 | 0.4536 | 0.4533 |
| 3 | 0.5517 | 0.5480 |
| 4 | 0.5985 | 0.5987 |
| 5 | 0.6251 | 0.6275 |
| 6 | 0.6417 | 0.6411 |
| 7 | 0.6527 | 0.6467 |
| 8 | 0.6604 | 0.6609 |
| 9 | 0.6662 | 0.6689 |
| 10 | 0.6706 | 0.6708 |

(a) Optimal ring radius $r_c$ for $N$ traps.

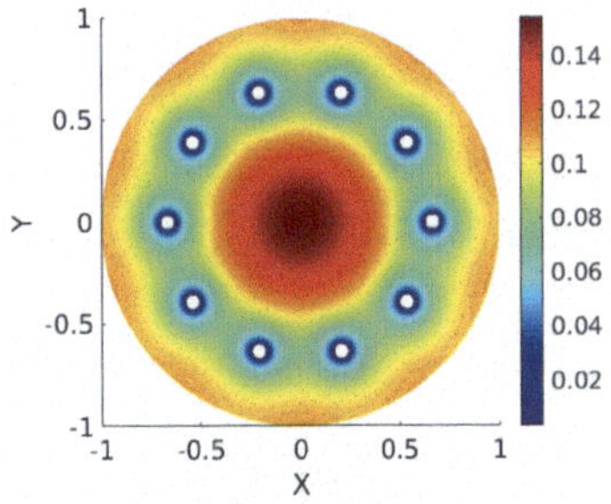


(b) MFPT for the optimal 10 trap ring.

**Fig. 10.3** The optimal ring radius $r_c$ for $N$ circular traps of radius $\varepsilon = 0.003$ that are equally spaced on a ring concentric within the unit disk. Left: Comparison of asymptotic results obtained by minimizing (10.13) with full numerical results computed from the MFPT-PDE (10.1) by the CPM. Right: The optimal MFPT $u(\mathbf{x})$ in the unit disk computed from (10.1) from the CPM with $N = 10$ traps on a ring

$$x = f \cosh \xi \cos \eta \,, \quad y = f \sinh \xi \sin \eta \,, \quad f = \sqrt{a^2 - b^2} \,, \tag{10.14}$$

where

$$\xi_b = \tanh^{-1}\left(\frac{b}{a}\right) = -\frac{1}{2}\log\beta\,, \quad \text{with} \quad \beta \equiv \left(\frac{a-b}{a+b}\right)\,. \tag{10.15}$$

For a given pair $(x, y)$, $(\xi, \eta)$ is determined by inverting the transformation (10.14) as

$$\xi = \frac{1}{2}\log\left(1 - 2s + 2\sqrt{s^2 - s}\right)\,, \quad s \equiv \frac{-\mu - \sqrt{\mu^2 + 4f^2y^2}}{2f^2}\,, \quad \mu \equiv x^2 + y^2 - f^2\,,$$
$$\eta = \begin{cases} \eta_\star, & \text{if } x \geq 0,\ y \geq 0, \\ \pi - \eta_\star, & \text{if } x < 0,\ y \geq 0, \\ \pi + \eta_\star, & \text{if } x \leq 0,\ y < 0, \\ 2\pi - \eta_\star, & \text{if } x > 0,\ y < 0, \end{cases} \quad \text{with } \eta_\star \equiv \sin^{-1}\left[\left(\frac{-\mu + \sqrt{\mu^2 + 4f^2y^2}}{2f^2}\right)^{1/2}\right]. \tag{10.16}$$

In this way, the Dirac point at $\mathbf{x}_0 = (x_0, y_0)$ is mapped to $(\xi_0, \eta_0)$.

The analysis in Sect. 5 of [11] determined $G$ as

$$\begin{aligned} G(\mathbf{x}; \mathbf{x}_0) = &\frac{1}{4|\Omega|}\left(|\mathbf{x}|^2 + |\mathbf{x}_0|^2\right) - \frac{3}{16|\Omega|}(a^2 + b^2) - \frac{1}{4\pi}\log\beta - \frac{1}{2\pi}\xi_> \\ &- \frac{1}{2\pi}\sum_{n=0}^{\infty}\log\left(\prod_{j=1}^{8}|1 - \beta^{2n}z_j|\right)\,, \quad \text{for} \quad \mathbf{x} \neq \mathbf{x}_0\,, \end{aligned} \tag{10.17}$$

where $|\Omega| = \pi ab$, $\xi_> \equiv \max(\xi, \xi_0)$, and the complex constants $z_1, \ldots, z_8$ are defined in terms of $(\xi, \eta)$, $(\xi_0, \eta_0)$ and $\beta$ by

$$\begin{aligned} &z_1 \equiv e^{-|\xi-\xi_0|+i(\eta-\eta_0)}\,, \quad z_2 \equiv \beta^2 e^{|\xi-\xi_0|+i(\eta-\eta_0)}\,, \quad z_3 \equiv \beta e^{-(\xi+\xi_0)+i(\eta-\eta_0)}\,, \\ &z_4 \equiv \beta e^{\xi+\xi_0+i(\eta-\eta_0)}\,, \quad z_5 \equiv \beta^2 e^{\xi+\xi_0+i(\eta+\eta_0)}\,, \quad z_6 \equiv e^{-(\xi+\xi_0)+i(\eta+\eta_0)}\,, \\ &z_7 \equiv \beta e^{|\xi-\xi_0|+i(\eta+\eta_0)}\,, \quad z_8 \equiv \beta e^{-|\xi-\xi_0|+i(\eta+\eta_0)}\,. \end{aligned} \tag{10.18}$$

By letting $(\xi, \eta) \to (\xi_0, \eta_0)$, the regular part of the Neumann Green's function is

$$\begin{aligned} R(\mathbf{x}_0) = &\frac{|\mathbf{x}_0|^2}{2|\Omega|} - \frac{3(a^2 + b^2)}{16|\Omega|} + \frac{1}{2\pi}\log(a + b) + \frac{1}{4\pi}\log\left(\cosh^2\xi_0 - \cos^2\eta_0\right) \\ &- \frac{\xi_0}{2\pi} - \frac{1}{2\pi}\sum_{n=1}^{\infty}\log(1 - \beta^{2n}) - \frac{1}{2\pi}\sum_{n=0}^{\infty}\log\left(\prod_{j=2}^{8}|1 - \beta^{2n}z_j^0|\right)\,, \end{aligned} \tag{10.19}$$

where $z_j^0$ for $j = 2, \ldots, 8$ are given by

$$\begin{aligned} &z_2^0 = \beta^2\,, \quad z_3^0 = \beta e^{-2\xi_0}\,, \quad z_4^0 = \beta e^{2\xi_0}\,, \quad z_5^0 = \beta^2 e^{2\xi_0+2i\eta_0}\,, \\ &z_6^0 = e^{-2\xi_0+2i\eta_0}\,, \quad z_7^0 = \beta e^{2i\eta_0}\,, \quad z_8^0 = \beta e^{2i\eta_0}\,. \end{aligned} \tag{10.20}$$

By using (10.17) and (10.19), the Neumann Green's matrix $\mathcal{G}$ in (10.9), as needed in the LAS (10.8), is readily evaluated. As a result, a particle swarm optimization procedure was used in [11, 20] to identify how the optimal trap configurations depend on the aspect ratio of the ellipse when its area is fixed as $|\Omega| = \pi ab = \pi$. In the rapidly converging infinite series (10.17) and (10.19), the aspect ratio of the ellipse is determined by the parameter $\beta$, as defined in (10.15), which satisfies $0 \leq \beta < 1$.

In particular, when $N = 4$, $\varepsilon = 0.05$ and $D = 1$ we show in Figs. 10.4 and 10.5 that four optimally located traps become colinear with the semi-major axis of the ellipse when the ellipse becomes sufficiently elongated (see Fig. 10 and 11 of [11]). From our numerical computations using (10.8), we find that this transition to colinearity occurs for an ellipse of area $\pi$ when $a = a_c \approx 1.7$. It is an open problem to calculate the critical threshold $a_c$ analytically.

As motivated by the colinear trap pattern in Fig. 10.5, for a long and thin ellipse where $a = 1/\delta \gg 1$, $b = \delta \ll 1$, and $|\Omega| = \pi$, we use thin domain asymptotics to

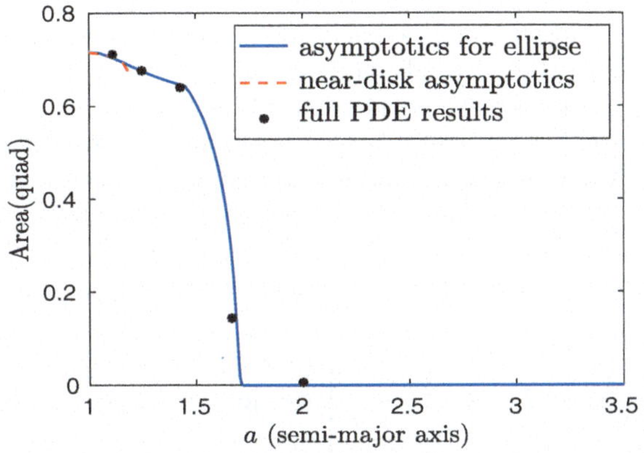


**Fig. 10.4** Area of the quadrilateral formed by four optimally located traps of a common radius $\varepsilon = 0.05$ with $D = 1$ in a deforming ellipse of area $\pi$ and semi-major axis $a$. The optimal traps become colinear as $a$ increases. Solid curve: results from (10.8). Discrete points: full numerical CPM results. Dashed line near $a = 1$: near-disk asymptotics of [11]

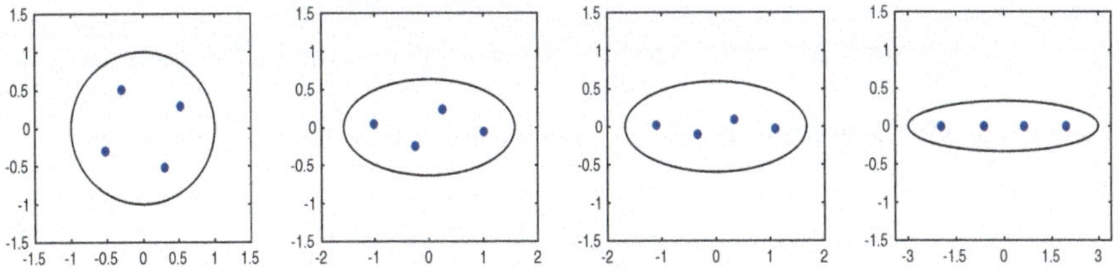

**Fig. 10.5** Optimal four-trap patterns for $D = 1$ in a deforming ellipse of area $\pi$ with semi-major axis $a$ and a common trap radius $\varepsilon = 0.05$. Left: $a = b = 1$. Middle Left: $a = 1.577$, $b \approx 0.634$. Middle Right: $a = 1.675$, $b \approx 0.597$. Right: $a = 3.0$, $b = 1/3$. The optimal trap pattern is colinear along the major axis when $a > a_c \approx 1.75$

formulate, as done in [11], a 1-D optimization problem for the optimal trap locations and the optimal average MFPT. Thin domain asymptotics for MFPT problems with a single trap have also been developed in a 2-D and 3-D setting in [29, 30].

From Appendix A of [11], for the thin ellipse the MFPT is approximated as $u(x,y) \sim \delta^{-2} D^{-1} U_0(\delta x) + \mathcal{O}(\delta^{-1})$, where the 1-D profile $U_0(X)$ satisfies

$$\left[F(X)\, U_0'\right]' = -F(X)\,, \quad \text{on} \quad |X| \leq 1\,, \quad \text{with} \quad F(X) \equiv \sqrt{1-X^2}\,, \tag{10.21}$$

where $X = \delta x$ and with $U_0$ and $U_0'$ bounded as $X \to \pm 1$. Assuming that $U_0(X) = U_0(-X)$, the average MFPT for the thin ellipse with $|\Omega| = \pi$ is estimated as

$$\overline{u}_0 \sim \frac{1}{\pi} \int_{-1/\delta}^{1/\delta} \int_{-\delta\sqrt{1-\delta^2 x^2}}^{\delta\sqrt{1-\delta^2 x^2}} u\, dx dy \sim \frac{4}{D\pi\delta^2} \int_0^1 F(X)\, U_0(X)\, dX\,, \tag{10.22}$$

where $b = \delta \ll 1$. For the approximating BVP (10.21), the colinear circular traps of a common radius $\varepsilon$ centered on the semi-major axis are approximated by zero point constraints for $U_0$ at locations on the interval $|X| \leq 1$. Therefore, (10.21) is a multi-point BVP, whose solution depends on the locations of the zero point constraints. The optimal trap locations and optimal average MFPT are estimated by minimizing the 1-D integral (10.22) for $\overline{u}_0$ with respect to the locations of these point constraints.

Following Sect. 4.2 of [11], we illustrate this approach for $N = 4$ colinear traps in a thin ellipse. We assume that two optimal traps are located on either side of the origin and that $U_0(X) = U_0(-X)$. By symmetry, we first solve (10.21) with $U_0'(0) = 0$, $U_0(d_1) = 0$, and $U_0(d_2) = 0$, where $0 < d_1 < d_2$, to obtain

$$U_0(X) = -\frac{1}{4}\left[\left(\sin^{-1} X\right)^2 + X^2\right] - \frac{1}{4} \begin{cases} \left(-\pi \sin^{-1} X + c_2\right)\,, & \text{if } d_2 \leq X \leq 1\,, \\ \left(b_1 \sin^{-1} X + b_2\right)\,, & \text{if } d_1 \leq X \leq d_2\,, \\ c_1\,, & \text{if } 0 \leq X \leq d_1\,, \end{cases} \tag{10.23}$$

where $c_1$, $c_2$, $b_1$ and $b_2$ are given in terms of $d_1$ and $d_2$ by

$$\begin{aligned} c_1 &= -d_1^2 - \left(\sin^{-1} d_1\right)^2\,, \quad b_1 = \frac{\left(\sin^{-1} d_1\right)^2 - \left(\sin^{-1} d_2\right)^2 + d_1^2 - d_2^2}{\sin^{-1} d_2 - \sin^{-1} d_1}\,, \\ c_2 &= -d_2^2 + \pi \sin^{-1} d_2 - \left(\sin^{-1} d_2\right)^2\,, \quad b_2 = -b_1 \sin^{-1} d_1 - d_1^2 - \left(\sin^{-1} d_1\right)^2\,. \end{aligned} \tag{10.24}$$

The average MFPT, defined in (10.22), is determined in terms of $d_1$ and $d_2$ as

$$\overline{u}_0(d_1, d_2) \sim -\frac{1}{\pi D \delta^2}\left[J_0 + \mathcal{H}(d_1, d_2)\right]\,, \tag{10.25}$$

where the two integrals $J_0$ and $\mathcal{H}(d_1, d_2)$ are given by

$$J_0 \equiv \int_0^1 F(X)\left[\left(\sin^{-1} X\right)^2 + X^2 - \pi \sin^{-1}(X)\right] dX \approx -0.703\,,$$

$$\mathcal{H}(d_1, d_2) \equiv c_2 \int_{d_2}^1 F(X)\, dX + \int_{d_1}^{d_2} F(X)\, \left((b_1+\pi)\sin^{-1}(X) + b_2\right) dX$$
$$+ \pi \int_0^{d_1} F(X)\, \left(\pi \sin^{-1}(X) + c_1\right) dX\,. \tag{10.26}$$

To minimize $\overline{u}_0$, we use a grid search to maximize the function $\mathcal{H}(d_1, d_2)$ on $0 < d_1 < d_2 < 1$. As derived in [11], this yields $d_{1\text{opt}} \approx 0.215$ and $d_{2\text{opt}} \approx 0.656$. In this way, and in terms of the original variables, the thin domain asymptotics predicts that the optimal trap locations and the optimal average MFPT have the scaling law

$$x_{1\text{opt}} \sim \frac{0.215}{b}\,, \quad x_{2\text{opt}} \sim \frac{0.656}{b}\,, \quad \overline{u}_{0\min} \equiv \overline{u}_0(d_{1\text{opt}}, d_{2\text{opt}}) \sim \frac{0.0179}{b^2 D}\,, \tag{10.27}$$

for $b \ll 1$. In Fig. 10.6 (left) (reproduced from Fig. 12 of [11]) we observe that the scaling laws for the optimal distances on $b < 0.5$ almost exactly reproduce results obtained from the CPM computed from the PDE (10.1) and from the hybrid asymptotic-numerical theory based on the LAS (10.8). From Fig. 10.6 (right), we observe that the scaling law for the optimal average MFPT closely approximates that from the CPM and the LAS (10.8) only when $b \leq 0.25$.

For $N \leq 50$, in [20] optimum trap configurations and the optimum average MFPT were computed numerically from a particle swarm optimization of the LAS (10.8) for various domain eccentricities $\kappa = \sqrt{1-(b/a)^2}$, where $b = 1/a$. In Fig. 10.7 (left) we plot the discrete energy $q(\mathbf{x}_1, \ldots, \mathbf{x}_N)$ in (10.8) versus $N$ for $\kappa = 0$, 0.472, 0.802, and 0.995. Interestingly, we observe that as $N$ increases, the optimal $q$ seems

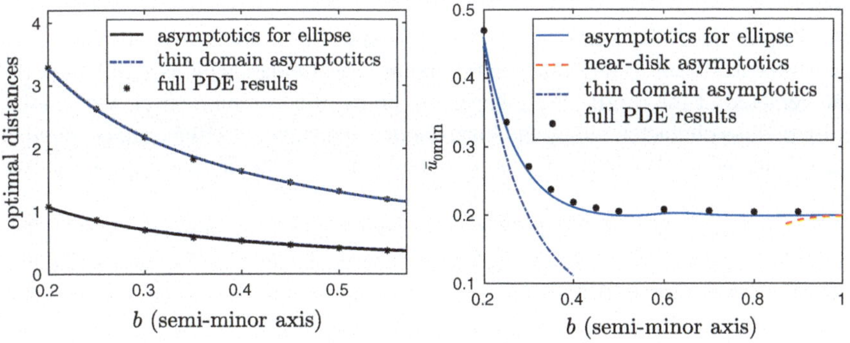


**Fig. 10.6** Left: Optimal distances from the origin for a colinear four-trap pattern on the major-axis of an ellipse of area $\pi$ and semi-minor axis $b$. When $b \leq 0.57$ the optimal pattern has two pairs of traps symmetrically located on either side of the origin. Right: optimal average MFPT $\overline{u}_{0\min}$ versus $b$. Solid curves: optimal results from the LAS (10.8). Discrete points: full PDE results for (10.1) computed from the CPM. Dashed-dotted: thin-domain asymptotics (10.27). Dashed: near-disk results of [11]

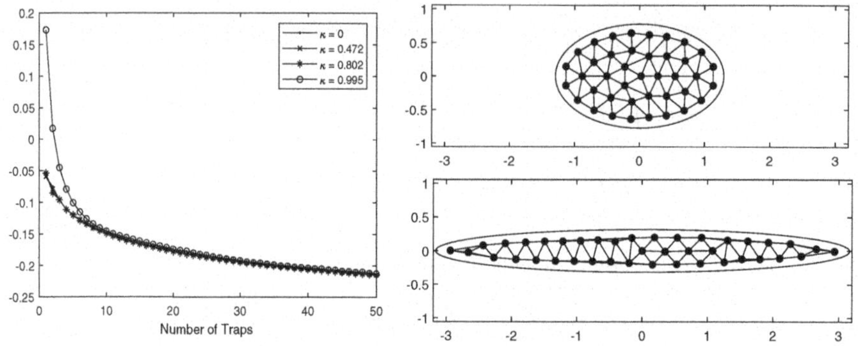


Fig. 10.7 Left: For an ellipse of area $\pi$, the optimal values of the discrete energy $q(\mathbf{x}_1, \ldots, \mathbf{x}_N)$ in (10.8) are plotted versus the number $N$ of traps for a few ellipse eccentricities $\kappa$. Right: optimal trap pattern for $N = 40$ when $\kappa = 0.802$ (top) and $\kappa = 0.995$ (bottom). The Delaunay triangulations are shown

to approach a universal curve that is independent of $\kappa$. In Fig. 10.7 (right) we show the computed optimal trap configurations when $\kappa = 0.802$ and $\kappa = 0.995$.

### 10.2.3 Periodic Trap Arrangements

Next we consider, as in [28], a periodic arrangement of traps, where the traps are centered at the lattice points of a Bravais lattice

$$\Lambda \equiv \left\{ m\boldsymbol{l}_1 + n\boldsymbol{l}_2 \,\middle|\, m,\, n \in \mathbb{Z} \right\}. \tag{10.28}$$

The *Wigner-Seitz (WS)* cell centered at a fixed $\boldsymbol{l} \in \Lambda$ is the set of all points that are closer to $\boldsymbol{l}$ than to any other lattice point. The fundamental WS cell $\Omega$ is the one centered at the origin. We assume that traps of a common shape with radius $O(\varepsilon) \ll 1$, as characterized by the logarithmic capacitance $d$ (10.5), are centered at each lattice point. The WS cells for an oblique lattice with circular traps are shown in Fig. 10.8.

Fixing the area $|\boldsymbol{l}_1 \times \boldsymbol{l}_2|$ of the WS cells to unity, and formulating the MFPT PDE in the fundamental WS cell, we seek to determine the specific Bravais lattice that minimizes the average MFPT for

$$D\Delta u = -1\,, \quad \mathbf{x} \in \Omega \backslash \Omega_\varepsilon\,; \quad u \in \mathcal{P}\,, \quad \mathbf{x} \in \partial\Omega\,; \quad u = 0\,, \quad \mathbf{x} \in \partial\Omega_\varepsilon\,, \tag{10.29}$$

where the operator $\mathcal{P}$ denotes periodic boundary conditions on $\partial\Omega$.

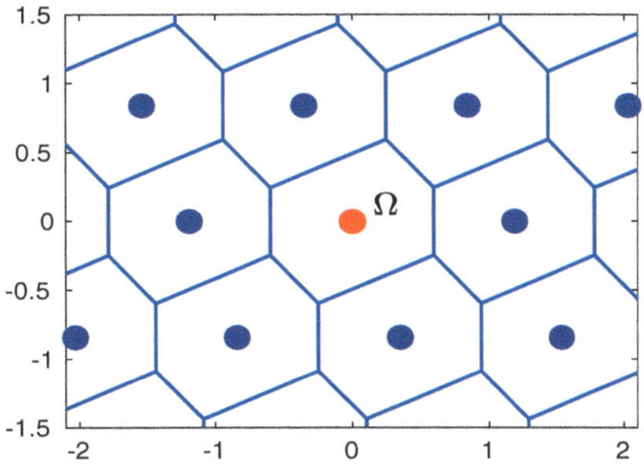


**Fig. 10.8** Wigner Seitz (WS) cells for an oblique Bravais lattice with generators $l_1 = (2^{1/4}, 0)^T$ and $l_2 = 2^{-1/4}(1, 1)^T$ with circular traps (blue dots) of a common radius $\varepsilon$. The fundamental WS cell $\Omega$ of unit area is centered at the origin and contains the red trap

The asymptotic analysis needed to analyze (10.29) is analogous to that done in (10.4)–(10.8), except that now we must use the periodic source-neutral Green's function $G_p(\mathbf{x})$, defined by

$$\Delta G_p = \frac{1}{|\Omega|} - \delta(\mathbf{x}), \quad \mathbf{x} \in \Omega; \quad G_b \in \mathcal{P}, \quad \mathbf{x} \in \partial\Omega; \quad \int_\Omega G_p \, d\mathbf{x} = 0,$$
$$G_p \sim -\frac{1}{2\pi} \log|\mathbf{x}| + R_p + \frac{|\mathbf{x}|^2}{4} + o(|\mathbf{x}|^2), \quad \text{as} \quad \mathbf{x} \to \mathbf{0}, \tag{10.30}$$

instead of the Neumann Green's function. Setting $N = 1$ and $|\Omega| = 1$ in (10.8), we obtain in terms of $R_p$ and the logarithmic capacitance $d$ of the trap that

$$\overline{u}_0 \sim \frac{1}{2\pi D}(-\log[\varepsilon d]) + \frac{R_p}{D}, \quad \text{as} \quad \varepsilon \to 0. \tag{10.31}$$

In [31] a formula for the regular part $R_p$ was derived. By identifying a point $\mathbf{x}$ as a complex number $z = x + iy$ and by writing the Bravais lattice using the generators $\alpha \in \mathbb{C}$ and $\beta \in \mathbb{C}$ as $\Lambda \equiv \left\{ m\alpha + n\beta \;\middle|\; m, n \in \mathbb{Z} \right\}$, with $\text{Im}(\beta/\alpha) > 0$ and $\text{Im}(\bar{\alpha}\beta) = 1$ to fix $|\Omega| = 1$, it was derived in [31] that, with $\zeta \equiv \beta/\alpha$ and $e(w) \equiv e^{2\pi i w}$,

$$R_p = -\frac{1}{2\pi} \log(2\pi) - \frac{1}{2\pi} \log \left| \sqrt{\text{Im}(\zeta)}\, e\left(\frac{\zeta}{12}\right) \prod_{n=1}^{\infty} (1 - e(n\zeta))^2 \right|. \tag{10.32}$$

For a Bravais lattice with $|\Omega| = 1$, Theorem 2 of [31] proved that $R_p$ in (10.32) is minimized for a regular hexagonal lattice. From (10.31), this establishes that for Bravais lattices, the average MFPT as $\varepsilon \to 0$ is smallest for the hexagonal lattice. A plot of $R_p$ is shown in Fig. 10.9 for the one-parameter family of lattices $\Lambda$ with $\alpha = 1/\sqrt{\sin\theta}$ and $\beta = \alpha e^{i\theta}$. For the hexagonal and square lattices, where $\theta = \pi/3$ and $\theta = \pi/2$, we calculate $R_p \approx -0.210262$ and $R_{p0} \approx -0.208578$, respectively. From

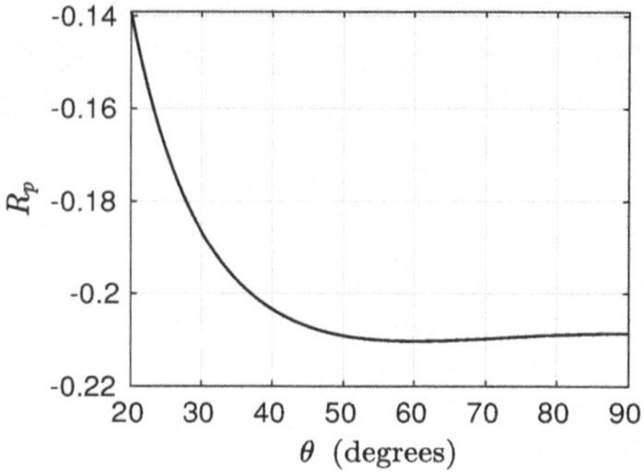


**Fig. 10.9** Plot of $R_p$ in (10.32) for oblique lattices with unit area of the primitive cell for which $l_1 = (1/\sqrt{\sin(\theta)}, 0)^T$ and $l_2 = (\cos(\theta)/\sqrt{\sin(\theta)}, \sqrt{\sin(\theta)})^T$. This yields $\alpha = 1/\sqrt{\sin\theta}$ and $\beta = \alpha e^{i\theta}$ in (10.32). The minimum occurs for the hexagon where $\theta = \pi/3$

Fig. 10.9 we observe that $R_{p0}$ varies only slightly on the range $\theta > \pi/4$. As a result, the average MFPT is rather insensitive to the choice of lattice when $\theta > \pi/4$.

In [28], a higher order approximation for $\overline{u}_0$ was derived for a circular trap by using the fact that, since $\Omega$ has two lines of symmetry that intersect at the origin, the usual gradient term $\nabla_{\mathbf{x}} G_p|_{\mathbf{x}=\mathbf{0}} \cdot \mathbf{x}$ is absent in the local behavior of $G_p$ in (10.30) as $\mathbf{x} \to \mathbf{0}$. From Principal Result 4.1 of [28], for the special case where $\Omega_\varepsilon$ is a circular disk of radius $\varepsilon$ the following improved asymptotic approximation was derived:

$$\overline{u}_0 \sim \frac{1}{2\pi D}(-\log\varepsilon) + \frac{R_p}{D} + \frac{\varepsilon^2}{D}\left[-\frac{1}{2}\log\varepsilon + \pi R_p + \frac{1}{2}\right] + o(\varepsilon^2), \quad \text{as} \quad \varepsilon \to 0. \tag{10.33}$$

For a square or hexagonal lattice, for which $R_p$ was evaluated above, (10.33) yields

$$\overline{u}_0 \sim \frac{-1}{2\pi D}(\log\varepsilon + 1.310533) - \frac{\varepsilon^2}{D}\left(\frac{1}{2}\log\varepsilon + 0.155266\right), \quad \text{(square)}$$
$$\overline{u}_0 \sim \frac{-1}{2\pi D}(\log\varepsilon + 1.321117) - \frac{\varepsilon^2}{D}\left(\frac{1}{2}\log\varepsilon + 0.160559\right), \quad \text{(hexagon)}. \tag{10.34}$$

We remark that the leading order result for the square lattice, without the $O(\varepsilon^2)$ correction, was derived previously in [32] by using the method of pseudo-potentials together with a numerical evaluation of certain lattice sums.

## 10.3 3-D Domains

For a bounded 3-D domain containing $N$ small traps, strong localized perturbation theory was used in [33] to approximate $\overline{u}_0$ in (10.3) for the MFPT-PDE (10.1). Related results for the MFPT and the splitting probability were derived in [34] using the method of pseudo-potentials. An extension of this narrow capture analysis has been developed recently in [35, 36] to calculate the MFPT, the splitting probability, and conditional moments of the first passage time distribution in terms of the solution to a single modified narrow capture PDE in a 3-D and 2-D setting, respectively.

Focusing on the MFPT, for the 3-D narrow capture problem we obtain, in place of (10.5), that the local problem defined in the near-field region of a trap is to solve

$$\Delta_{\mathbf{y}} v = 0\,, \quad \mathbf{y} \notin \Omega_j\,; \quad v = 0\,, \quad \mathbf{y} \in \partial\Omega_j\,; \quad v \sim \frac{C_j}{|\mathbf{y}|} + O(|\mathbf{y}|^{-2}) + \cdots \text{ as } |\mathbf{y}| \to \infty\,. \tag{10.35}$$

Here $C_j$ is the capacitance of the trap and $\Omega_j = \Omega_{\varepsilon j}/\varepsilon$. The capacitance has two key properties: it is invariant under rotations of the trap shape and, over all trap shapes $\Omega_j$ of the same volume, $C_j$ is minimized for a spherical trap. The capacitance is known analytically for some simple trap shapes (see Table 1 of [33]).

For the 3-D problem, the Neumann Green's function $G(\mathbf{x}; \mathbf{x}_j)$ is defined by

$$\begin{aligned} \Delta G &= \frac{1}{|\Omega|} - \delta(\mathbf{x} - \mathbf{x}_j)\,, \quad \mathbf{x} \in \Omega\,; \quad \partial_n G = 0\,, \quad \mathbf{x} \in \partial\Omega\,; \quad \int_\Omega G \, d\mathbf{x} = 0\,, \\ G &\sim \frac{1}{4\pi|\mathbf{x} - \mathbf{x}_j|} + R_j + \nabla_{\mathbf{x}} R_j \cdot (\mathbf{x} - \mathbf{x}_j) + \cdots\,, \quad \text{as } \mathbf{x} \to \mathbf{x}_j\,, \end{aligned} \tag{10.36}$$

where $R_j \equiv R(\mathbf{x}_j)$ is the regular part of the Green's function at $\mathbf{x} = \mathbf{x}_j$.

As derived in Principal Result 3.1 of [33], the average MFPT $\overline{u}_0$ for $\varepsilon \to 0$ is

$$\overline{u}_0 \sim \frac{|\Omega|}{D} \left( \frac{1}{4\pi\varepsilon N\bar{C}} + \frac{1}{N^2} p_c(\mathbf{x}_1, \ldots, \mathbf{x}_N) + O(\varepsilon) \right), \quad \text{where} \quad p_c \equiv \frac{\mathbf{C}^T \mathcal{G} \mathbf{C}}{(\bar{C})^2}\,. \tag{10.37}$$

Here $\mathbf{C} \equiv (C_1, \ldots, C_N)^T$, $\bar{C} \equiv N^{-1}(C_1 + \ldots C_N)$, and $\mathcal{G}$ is the Neumann Green's matrix in (10.9) now given in terms of the 3-D Green's function (10.36). For identically-shaped traps with a common capacitance $C = C_j$, for $j = 1, \ldots, N$, (10.37) becomes

$$\overline{u}_0 \sim \frac{|\Omega|}{D} \left[ \frac{1}{4\pi\varepsilon NC} + \frac{1}{N^2} p(\mathbf{x}_1, \ldots, \mathbf{x}_N) + O(\varepsilon) \right], \quad \text{where} \quad p \equiv \mathbf{e}^T \mathcal{G} \mathbf{e}\,, \tag{10.38}$$

with $\mathbf{e} \equiv (1, \ldots, 1)^T$. As a result, for $N$ identical traps, to minimize the average MFPT in (10.38), we must determine the global minimum of the discrete energy $p(\mathbf{x}_1, \ldots, \mathbf{x}_N)$.

For the unit sphere, we can evaluate the Neumann Green's matrix by using an analytical result for the 3-D Neumann Green's function and its regular part (see [33])

$$\begin{aligned} G(\mathbf{x};\mathbf{x}_j) &= \frac{1}{4\pi|\mathbf{x}-\mathbf{x}_j|} + \frac{1}{4\pi|\mathbf{x}||\mathbf{x}'-\mathbf{x}_j|} + \frac{1}{4\pi}\log\left(\frac{2}{1-\mathbf{x}\cdot\mathbf{x}_j+|\mathbf{x}||\mathbf{x}'-\mathbf{x}_j|}\right) \\ &\quad + \frac{1}{6|\Omega|}\left(|\mathbf{x}|^2+|\mathbf{x}_j|^2\right) - \frac{7}{10\pi}, \\ R(\mathbf{x}_j) &= \frac{1}{4\pi\left(1-|\mathbf{x}_j|^2\right)} + \frac{1}{4\pi}\log\left(\frac{1}{1-|\mathbf{x}_j|^2}\right) + \frac{|\mathbf{x}_j|^2}{4\pi} - \frac{7}{10\pi}, \end{aligned} \tag{10.39}$$

where $|\Omega| = 4\pi/3$ and $\mathbf{x}' = \mathbf{x}/|\mathbf{x}|^2$ is the image point to $\mathbf{x}$ outside the unit sphere. With the Neumann Green's matrix easily calculated, a particle swarm optimizer can be used to optimize $\overline{u}_0$ from the discrete energies in (10.38) and (10.37).

For $N \leq 20$ in [33] it was shown that the optimal trap locations lie, approximately, on a concentric shell inside the unit sphere, with possibly a center trap. The optimal trap patterns for $N \leq 100$ identical traps and the optimal average MFPT were computed numerically in [37] by optimizing the discrete energy $p$ in (10.38).

## 10.4 Neumann Eigenvalue Problem for Laplacian

The problem of minimizing the average MFPT is closely related to the problem of identifying the trap configuration that maximizes the lowest, or fundamental, eigenvalue $\lambda_0$ of

$$\begin{aligned} &\Delta u + \lambda u = 0\,, \quad x \in \Omega \setminus \cup_{j=1}^{N}\Omega_{\varepsilon j}\,; \qquad \partial_n u = 0\,, \quad x \in \partial\Omega\,, \\ &u = 0\,, \quad x \in \partial\Omega_{\varepsilon j}\,, \quad j = 1, \ldots, N\,. \end{aligned} \tag{10.40}$$

When $\Omega$ is a 2-D bounded domain that contains traps of a common shape, it was shown in [17] that $\lambda_0 \ll 1$ is the smallest positive root of

$$\mathcal{F}(\lambda) = 0\,, \qquad \text{where} \qquad \mathcal{F}(\lambda) \equiv \det\left(I + 2\pi\nu\mathcal{G}_H\right)\,. \tag{10.41}$$

Here $\nu = -1/\log(\varepsilon d)$, $d$ is the logarithmic capacitance of the identical traps, and $\mathcal{G}_H$ is the $N \times N$ Helmholtz Green's matrix with matrix entries

$$(\mathcal{G})_{Hjj} = R_{Hj} \quad \text{and} \quad (\mathcal{G})_{Hij} = (\mathcal{G})_{Hji} = G_H(\mathbf{x}_i;\mathbf{x}_j) \ \text{ for } \ i \neq j\,, \tag{10.42}$$

where $G_H(\mathbf{x};\mathbf{x}_j)$ and its regular part $R_{Hj}$ satisfy

$$\begin{aligned} &\Delta G_H + \lambda G_H = -\delta(\mathbf{x}-\mathbf{x}_j)\,, \quad \mathbf{x} \in \Omega\,; \qquad \partial_n G_H = 0\,, \quad \mathbf{x} \in \partial\Omega\,; \\ &G_H \sim -\frac{1}{2\pi}\log|\mathbf{x}-\mathbf{x}_j| + R_{Hj} + o(1)\,, \quad \text{as} \quad \mathbf{x} \to \mathbf{x}_j\,. \end{aligned} \tag{10.43}$$

By expanding $G_H$ for $\lambda \ll 1$ and substituting the resulting expression into (10.41), it was shown in Appendix C of [11] that

$$\lambda_0 \sim \frac{2\pi N \nu}{|\Omega|} - \frac{4\pi^2 \nu^2}{|\Omega|} p(\mathbf{x}_1, \ldots, \mathbf{x}_N) + O(\nu^3), \quad \text{with} \quad p(\mathbf{x}_1, \ldots, \mathbf{x}_N) \equiv \mathbf{e}^T \mathcal{G} \mathbf{e}, \tag{10.44}$$

where $\mathbf{e} \equiv (1, \ldots, 1)^T$ and $\mathcal{G}$ is the Neumann Green's matrix. For an arbitrary bounded 2-D domain, it follows by comparing (10.44) with (10.10) that, up to terms of $O(\nu^2)$, the trap locations that maximize the rate at which a Brownian particle is captured to provide the minimizing trap configuration for the average MFPT.

A similar correspondence between the average MFPT and the fundamental Neumann eigenvalue holds for a bounded 3-D domain. For a configuration of $N$ traps centered at $\mathbf{x}_j \in \Omega$ with capacitances $C_j$ for $j = 1, \ldots, N$, it was shown in [33] that

$$\lambda_0 \sim \frac{4\pi \varepsilon N}{|\Omega|} \bar{C} - \frac{16\pi^2 \varepsilon^2}{|\Omega|} p_c(\mathbf{x}_1, \ldots, \mathbf{x}_N) + O(\varepsilon^3), \quad \text{where} \quad p_c \equiv \mathbf{C}^t \mathcal{G} \mathbf{C}. \tag{10.45}$$

Here $\bar{C} \equiv N^{-1}(C_1 + \ldots + C_N)$, $\mathbf{C} \equiv (C_1, \ldots, C_N)^T$, and $\mathcal{G}$ is the 3-D Neumann Green's matrix. By comparing (10.45) with (10.37) we conclude that, for $\varepsilon \to 0$, the trap locations that minimize the average MFPT also maximize $\lambda_0$.

## 10.5 Discussion and Open Challenges

We have surveyed some previous results for the optimization of the average MFPT for various narrow capture problems in a 2-D or 3-D domain. In the limit of small trap radii, we have shown that these optimization problems typically involve minimizing certain discrete energies for interacting particle systems, which are encoded by Green's matrices. These discrete energies that are derived from the strong localized perturbation analysis of the MFPT PDE are all closely related to the classical Fekete point energies [38, 39] of interacting Coulombic particle systems. When the Green's matrix can be calculated either explicitly or as a rapidly converging infinite series, a particle swarm optimizer can be readily used on the discrete energies to numerically determine the optimal trap set for a moderately small number of traps.

Our brief survey has focused on narrow capture problems with interior traps. Related narrow escape problems where the traps are located on the boundary of a domain have been analyzed for 2-D domains in [27, 40, 41] and for the sphere in [42] (see also the references therein). For the narrow escape problem from a sphere, optimal boundary trap configurations together with a scaling law have been computed in [12, 43] from minimizing the appropriate discrete energy, which is defined in terms of an explicit analytical formula for the surface Neumann Green's function. A survey of results for the narrow escape problem is given in [4].

A related narrow capture problem is the Berg-Purcell [44] problem that involves calculating the effective capacitance and effective trapping rate of a sphere that has a large number $N$ of small nanotraps on its otherwise reflecting boundary. Strong localized perturbation theory has been used to derive a discrete energy, involving the exterior surface Neumann Green's function, that determines the effective capacitance [45, 46]. For equally-distributed points, and in the limit $N \to \infty$, a scaling law for the discrete energy has been used to derive an explicit formula for the effective trapping rate in the dilute trap fraction limit that improves upon the heuristic Berg-Purcell formula. This "first principles" result for the homogenized trapping rate has a similar form to some empirically postulated trapping rates used in [47, 48]. Related work that determines the effective trapping rate for a simple model of semi-permeable traps is given in [49, 50]. More biologically elaborate boundary trapping models that include the effect of patchy particles and or finite receptor kinetics have been analyzed recently in [51–53]. Based on an integral equation formulation of the relevant PDE, a fast full numerical solver for narrow capture and narrow escape problems in the sphere, which can effectively treat 2000 or more nanotraps, has been developed in [54].

Various extensions of the narrow capture modeling framework discussed in this survey can be incorporated into our hybrid asymptotic-numerical theory. For a collection of traps that are only partially absorbing, we need only replace the boundary condition $v = 0$ on each trap by the Robin condition $\partial_n v + bv = 0$. From a numerical boundary integral solver on the local problems (10.5) and (10.35), the constants $d_j(b)$ and $C_j(b)$ can be computed for the 2-D and 3-D cases, respectively. A corresponding modified discrete energy can then be readily derived and optimized.

As a further extension of the methodology, in the 2-D case it was shown in [28] how to use a complex variable approach together with a least squares fitting to numerically calculate the logarithmic capacitance for a cluster of non-overlapping traps that are centered at some point in $\Omega$. Such trap clusters are characterized by inter-trap separations of $O(\varepsilon)$ in a confining domain of radius $O(1)$.

The numerical treatment for problems involving clusters of traps is closely related to the challenging problem of calculating the MFPT in the non-perturbative setting where the traps do not have asymptotically small radii. In this context, an exact solution involving bipolar coordinates was found in [55] for calculating the MFPT in a circular domain containing one non-concentric disk of arbitrary radius. For a spherical domain containing non-overlapping spheres of arbitrary radii, a semi-analytical method based on the generalized method of separation of variables (GMSV) was developed in [56] to calculate Green's function for the Laplacian in the perforated domain. In this approach, the global approximate solution is represented as a linear combination of local solutions involving solid harmonics centered near at each inclusion. Translation addition theorems for solid harmonics are then used for deriving a linear algebraic system that allows all the boundary conditions to be satisfied. This semi-analytical approximation of Green's function is key for estimating various stationary quantities such as the escape probability, harmonic measure, and the MFPT. This GMSV approach of [56] should be ideal for estimating the capacitance of a cluster of non-overlapping spherical traps in $\mathbb{R}^3$. For the same class of domains, an extension of this GMSV methodology was developed in [57] to approximate Green's function for the reduced-wave equation. This Green's function arises in calculating the Laplace transform for the time-dependent probability distribution.

We now discuss a few open problems related to narrow capture problems.

To easily implement a particle swarm optimizer on the discrete energy associated with the Neumann Green's matrix it is essential to efficiently calculate the Neumann Green's function and its regular part. For the 2-D case, this can be done for a disk, for a rectangle [26, 27], and for an ellipse. For 3-D domains, the Neumann Green's function can be readily calculated for a sphere. For a triclinic cell in 3-D, the source neutral periodic Green's function can be evaluated in terms of rapidly converging infinite series [58]. For arbitrary 2-D or 3-D domains, a key open problem is to develop a fast multipole algorithm, such as done in [59] for the reduced-wave equation, to rapidly calculate the Neumann Green's function and its regular part for (10.7) and (10.36). By using such a fast numerical evaluation of the Neumann Green's matrix, it should be numerically tractable to use a particle swarm optimizer on either the LAS (10.8) for the 2-D domain or on (10.37) for a 3-D domain so as to numerically identify optimal trap configurations.

For 2-D or 3-D domains with a collection of interior traps, it would be interesting to derive an approximate scaling law for the optimal average MFPT when $N \to \infty$. Intuitively, one might expect that the optimal trap configuration for large $N$ has a near lattice structure in the interior of the domain with only a small modification needed near the domain boundary. It would be interesting to investigate this conjecture and to relate the large $N$ limiting problem, but with $N \sum_j |\Omega_{\varepsilon j}| \ll 1$, to results that can be derived by homogenization theory in the dilute trap fraction limit.

Another open problem is to determine the optimal lattice arrangement for a 3-D periodic lattice with identical shaped traps centered at the lattice points. The main challenge here is to provide a tractable analytical formula for the regular part of the periodic source-neutral Green's function for the 14 distinct Bravais lattices.

Finally, it would also be worthwhile to investigate optimizing MFPT problems for some highly challenging situations where the traps are not stationary, such as in [60], in Sect. 4.3 of [7], and in [61]. In particular, the optimal distance from the origin for a trap that rotates with a given angular velocity on a ring concentric within the unit disk was shown in [60] to have a highly intricate bifurcation structure.

**Acknowledgements** We gratefully acknowledge the contributions of our collaborators, Andrew Bernoff, Jason Gilbert, Sarafa Iyaniwura, Theodore Kolokolnikov, Alan Lindsay, Colin MacDonald, Wesley Ridgway, Michele Titcombe, and Tony Wong to some of the work described herein.

## References

1. O. Bénichou, R. Voituriez, From first-passage times of random walks in confinement to geometry-controlled kinetics. Phys. Rep. **539**, 225–284 (2014)
2. D. Holcman, Z. Schuss, Time scale of diffusion in molecular and cellular biology, J. of Physics A: Math. Theor. **47**, 173001, (2014)
3. J. Yang, I. Kupka, Z. Schuss, D. Holcman, Search for a small egg by spermatozoa in restricted geometries. J. Math. Biol. **73**, 948–964 (2016)
4. D. Holcman, Z. Schuss, The narrow escape problem. SIAM Rev. **56**, 213–257 (2014)
5. S. Redner, *A Guide to First-Passage Processes* (Cambridge University Press, Cambridge, 2001)

6. R. Metzler, G. Oshanin, S. Redner (eds.), *First-Passage Phenomena and Their Applications* (World Scientific, Singapore, 2014)
7. S. Iyaniwura, T. Wong, M.J. Ward, C. Macdonald, Simulation and optimization of mean first passage time problems in 2-D using numerical embedded methods and perturbation theory. Multiscale Model. Simul. **19**, 1367–1393 (2021)
8. M.J. Ward, W.D. Henshaw, J.B. Keller, Summing logarithmic expansions for singularly perturbed eigenvalue problems. SIAM J. Appl. Math. **53**, 799–828 (1993)
9. M.J. Ward, J.B. Keller, Strong localized perturbations of eigenvalue problems. SIAM J. Appl. Math. **53**, 770–798 (1993)
10. M.J. Ward, Spots, traps, and patches: asymptotic analysis of localized solutions to some linear and nonlinear diffusive systems. Nonlinearity **31**, R189 (2018)
11. S. Iyaniwura, T. Wong, C.B. Macdonald, M.J. Ward, Optimization of the mean first passage time in near-disk and elliptical domains in 2-D with small absorbing traps. SIAM Rev. **63**, 525–555 (2021)
12. W.J.M. Ridgway, A. Cheviakov, Locally and globally optimal configurations of N particles on the sphere with applications in the narrow escape and narrow capture problems. Phys. Rev. E **100**, 042413 (2019)
13. A. Burchard, J. Denzler, On the geometry of optimal windows, with special focus on the square. SIAM J. Math. Anal. **37**, 1800–1827 (2006)
14. E.M. Harrell, P. Kröger, K. Kurata, On the placement of an obstacle or a well so as to optimize the fundamental eigenvalue. SIAM J. Math. Anal. **33**, 240–259 (2001)
15. A. Henrot, On minimization problems for eigenvalues of the Laplacian. J. Evol. Equ. **3**, 443–461 (2003)
16. J. Denzler, Windows of given area with minimal heat diffusion. Trans. Am. Math. Soc. **351**, 569–580 (1999)
17. T. Kolokolnikov, M.S. Titcombe, M.J. Ward, Optimizing the fundamental Neumann eigenvalue for the Laplacian in a domain with small traps. Eur. J. Appl. Math. **16**, 161–200 (2005)
18. D. Coombs, R. Straube, M.J. Ward, Diffusion on a sphere with localized traps: Mean first passage time, eigenvalue asymptotics, and Fekete points. SIAM J. Appl. Math. **70**, 302–332 (2009)
19. F. Paquin-Lefebvre, S. Iyaniwura, M.J. Ward, Asymptotics of the principal eigenvalue of the Laplacian in 2-D periodic domains with small traps. Eur. J. Appl. Math. **33**, 646–673 (2021)
20. J. Gilbert, A. Cheviakov, Global optimization of the mean first passage time for narrow capture problems in elliptic domains. Eur. J. Appl. Math. **34**, 1269–1287 (2023)
21. Y. Chen, C.B. Macdonald, The closest point method and multigrid solvers for elliptic equations on surfaces. SIAM J. Sci. Comp. **37**, A134–A155 (2015)
22. V. Kurella, J.C. Tzou, D. Coombs, M.J. Ward, Asymptotic analysis of first passage time problems inspired by ecology. Bull. Math. Biol. **77**, 83–125 (2015)
23. A. Ismael, F. Vaz, L.N. Vicente, A particle swarm pattern search method for bound constrained global optimization. J. Global Optim. **39**, 197–219 (2007)
24. J. Currie, D.I. Wilson, N. Sahinidis, J. Pinto, OPTI: lowering the barrier between open source optimizers and the industrial MATLAB user. Foundations of computer-aided process operations, **24** (2012)
25. J. Kennedy, *Particle Swarm Optimization*. Encyclopedia of machine learning (2010), pp. 760–766
26. T. Kolokolnikov, M.J. Ward, J. Wei, Spot self-replication and dynamics for the Schnakenburg model in a two-dimensional domain. J. Nonl. Sci **19**, 1–56 (2009)
27. S. Pillay, M.J. Ward, A. Peirce, T. Kolokolnikov, An asymptotic analysis of the mean first passage time for narrow escape problems: Part I: Two-dimensional domains. Multiscale Model. Simul. **8**, 803–835 (2010)
28. S. Iyaniwura, M.J. Ward, Asymptotic analysis for the mean first passage time in finite or spatially periodic 2-D domains with a cluster of small traps. ANZIAM J. **63**, 1–22 (2021)
29. D.S. Grebenkov, A.T. Skvortsov, Mean first-passage time to a small absorbing target in an elongated planar domains. New J. Phys. **22**, 113024 (2020)

30. D.S. Grebenkov, A.T. Skvortsov, Mean first-passage time to a small absorbing target in three-dimensional elongated domains. Phys. Rev. E **105**, 054107 (2022)
31. X. Chen, Y. Oshita, An application of the modular function in nonlocal variational problems. Arch. Rat. Mech. Anal. **186**, 109–137 (2007)
32. D.C. Torney, B. Goldstein, Rates of diffusion-limited reaction in periodic systems. J. Stat. Phys. **49**, 725–750 (1987)
33. A. Cheviakov, M.J. Ward, Optimizing the principal eigenvalue of the Laplacian in a sphere with interior traps. Math. and Comput. Model. **53**, 1394–1409 (2011)
34. S. Condamin, O. Bénichou, M. Moreau, Random walks and Brownian motion: a method of computation for first passage times and related quantities in confined geometries. Phys. Rev. E **75**, 021111 (2007)
35. P.C. Bressloff, Asymptotic analysis of target fluxes in the three-dimensional narrow capture problem. Multiscale Model. Simul. **19**, 612–632 (2021)
36. P.C. Bressloff, Asymptotic analysis of extended two-dimensional narrow capture problems. Proc. Roy. Soc. A: Math. Phys. Eng. Sci. **477**, 20200771 (2021)
37. J. Gilbert, A. Cheviakov, Globally optimal volume-trap arrangements for the narrow-capture problem inside a unit sphere. Phys. Rev. E **99**, 012109 (2019)
38. E.B. Saff, A.B.J. Kuijlaars, Distributing many points on a sphere. Math. Intell. **19**, 5–11 (1997)
39. E.A. Rakhamanov, E.B. Saff, Y.M. Zhou, Minimal discrete energy on the sphere. Math. Res. Lett. **1**, 647–662 (1994)
40. A. Singer, Z. Schuss, D. Holcman, Narrow escape, Part II: The circular disk. J. Stat. Phys. **122**, 465–489 (2006)
41. D.S. Grebenkov, Universal formula for the mean first passage time in planar domains. Phys. Rev. Lett. **117**, 260201 (2016)
42. A.F. Cheviakov, M.J. Ward, R. Straube, An asymptotic analysis of the mean first passage time for narrow escape problems: Part II: The sphere. Multiscale Model. Simul. **8**, 836–870 (2010)
43. A. Cheviakov, D. Zawada, Narrow-escape problem for the unit sphere: Homogenization limit, optimal arrangements of large number of traps, and the N2 conjecture. Phys. Rev. E **87**, 042118 (2013)
44. H.C. Berg, E.M. Purcell, Physics of chemoreception. Biophys. J. **20**, 193–219 (1977)
45. A.E. Lindsay, A.J. Bernoff, M.J. Ward, First passage statistics for the capture of a Brownian particle by a structured spherical target with multiple surface traps. Multiscale Model. Simul. **15**, 74–109 (2017)
46. A.E. Lindsay, A.J. Bernoff, Numerical approximation of diffusive capture rates by planar and spherical surfaces with absorbing pores. SIAM J. Appl. Math. **78**, 266–290 (2018)
47. C.B. Muratov, S.Y. Shvartsman, Boundary homogenization for periodic arrays of absorbers. Multiscale Model. Simul. **7**, 44–61 (2008)
48. A.M. Berezhkovskii, M.I. Monine, C.B. Muratov, S.Y. Shvartsman, Homogenization of boundary conditions for surfaces with regular arrays of traps. J. Chem. Phys. **124**, 036103 (2006)
49. A.J. Bernoff, A.E. Lindsay, D.D. Schmidt, Boundary homogenization and capture time distributions of semi-permeable membranes with periodic patterns of reactive sites. Multiscale Model. Simul. **16**, 1411–1447 (2018)
50. A.E. Lindsay, A.N. Hernandez, B. Quaife, Trapping of planar Brownian motion: Full first passage time distributions by Kinetic Monte-Carlo, asymptotic and boundary integral methods. Multiscale Model. Simul. **20**, 1284–1314 (2022)
51. G. Handy, S.D. Lawley, Revising Berg-Purcell for finite receptor kinetics. Biophys. J. **120**, 2237–2248 (2021)
52. S.D. Lawley, Boundary homogenization for trapping patchy particles. Phys. Rev. E **100**, 032601 (2019)
53. C.E. Plunkett, S.D. Lawley, Boundary homogenization for patchy surfaces trapping patchy particles. J. Chem. Phys. **158**, 094104 (2023)
54. J. Kaye, L. Greengard, A fast solver for the narrow capture and narrow escape problems in the sphere. J. Comput. Phys. **5**, 100047 (2020)

55. R. Stana, G. Lythe, C. Molina-Páris, Diffusion in a disk with inclusion. SIAM J. Appl. Math. **81**, 1287–1302 (2021)
56. D.S. Grebenkov, S.D. Traytak, Semi-analytical computation of Laplacian Green's functions in three-dimensional domains with disconnected spherical boundaries. J. Comput. Phys. **379**, 91–117 (2019)
57. D.S. Grebenkov, Diffusion toward non-overlapping partially reactive spherical traps: Fresh insights into classical problems. J. Chem. Phys. **152**, 244108 (2020)
58. S. Tyagi, Coulomb potentials in two and three dimensions under periodic boundary conditions. J. Chem. Phys. **122**, 014101 (2005)
59. M.C. Kropinski, B.D. Quaife, Fast integral equation methods for the modified Helmholtz equation. J. Comput. Phys. **230**, 425–434 (2011)
60. J.C. Tzou, T. Kolokolnikov, Mean first passage time for a small rotating trap inside a reflective disk. Multiscale Model. Simul. **13**, 231–255 (2015)
61. A.E. Lindsay, J.C. Tzou, T. Kolokolnikov, Optimization of first passage times by multiple cooperating mobile traps. Multiscale Model. Simul. **15**, 920–947 (2017)

# Chapter 11
# Boundary Homogenization for Target Search Problems

**Denis Grebenkov and Alexei Skvortsov**

**Abstract** In this chapter, we describe several approximations in the theory of Laplacian transport near complex or heterogeneously reactive boundaries. This phenomenon, governed by the Laplace operator, is ubiquitous in fields as diverse as chemical physics, hydrodynamics, electrochemistry, heat transfer, wave propagation, self-organization, biophysics, and target search. We overview the mathematical basis and various applications of the effective medium approximation and the related boundary homogenization when a complex heterogeneous boundary is replaced by an effective much simpler boundary. We also discuss the constant-flux approximation, the Fick–Jacobs equation, and other mathematical tools for studying the statistics of first-passage times to a target. Numerous examples and illustrations are provided to highlight the advantages and limitations of these approaches.

## 11.1 Introduction

The diffusion of Brownian particles near complex boundaries is ubiquitous for many natural phenomena and engineering applications. Common examples are heterogeneous catalysis, diffusion-limited aggregation, tracer dispersion over complex canopies in atmosphere and ocean, cell communication via chemical signals, gas exchange in the human lungs and placentas, and many others [1–18]. The mathematics of this process remains unchanged even if the diffusive quantity is continuous (e.g., temperature profile, velocity field, wave amplitude, magnetic field, vorticity) and thus not directly related to any physical particle or Lagrangian marker. In this

D. Grebenkov (✉)
Laboratoire de Physique de la Matière Condensée, CNRS – Ecole Polytechnique, Institut Polytechnique de Paris, 91120 Palaiseau, France
e-mail: denis.grebenkov@polytechnique.edu

A. Skvortsov
Maritime Division, Defence Science and Technology Group, Melbourne VIC 3207, Australia
e-mail: alexei.skvortsov@defence.gov.au



D. Grebenkov et al. (eds.), *Target Search Problems*,
https://doi.org/10.1007/978-3-031-67802-8_11

regard, diffusion is connected to the broad family of phenomena, which are collectively referred to as *Laplacian transport*. The unified analytical framework of the underlying Laplace equation, which also includes the diffusion equation at the steady state, is thus applicable in fields as diverse as chemical physics, acoustics, electrostatics, fluid dynamics, electrochemistry, wetting, etc. [19–30].

The analytical results for diffusive transport near inhomogeneous boundaries are readily available only for simple shapes (plane, sphere, cylinder, etc.). More complex boundary profiles require the application of advanced analytical methods (e.g., conformal mapping, eigenfunction decomposition, perturbation theory, scaling arguments, etc.) and often lead to extensive numerical simulations via Monte Carlo, finite element, or Lagrangian methods. The aim of this review chapter is to show how the conventional framework of the effective medium approximation and boundary homogenization can be applied to target search problems to enable analytical progress even for boundaries of rather complex profiles. We show that this approach gives access not only to some "averaged" properties of the system (e.g., the total flux) but also to the first-passage time (FPT) statistics. The latter plays the central role in target search problems [31–34] as the proxy of efficiency of search strategies [35–37].

Generally speaking, the complexity of the boundary in the context of diffusive transport can be attributed to two independent factors: (i) the geometrical complexity of the boundary profile (spiky, fractal, patchy, etc.), and (ii) the distribution of the absorbing properties on that boundary. For instance, some parts of the boundary can be absorbing and the other be reflecting. This pattern can be rather complex even if the geometrical profile of the boundary is simple (e.g., a flat boundary). The present review deals with both kinds of boundary complexity. Instead of solving the target search problem for a given complex boundary, boundary homogenization aims at replacing such a boundary by an *equivalent simpler one*, for which the target search problem admits an exact simple solution. The review is thus focused on the fundamental question of constructing such an equivalent boundary, and provides theoretical answers and practical ready-to-use recipes.

We begin with an illustrative example. Consider particles diffusing in a layer $0 < y < H$ between two parallel planes separated by distance $H$. At steady state, the particle concentration $C$ within the layer obeys the Laplace equation

$$\Delta C = 0, \tag{11.1}$$

where $\Delta = \partial^2/\partial x^2 + \partial^2/\partial y^2 + \partial^2/\partial z^2$ is the Laplace operator. Assume that the bottom plane $y = 0$ is absorbing (i.e., any particle hitting this plane disappears so that $C = 0$ at $y = 0$), and the concentration of the particles at the top plane $y = H$ is kept constant: $C = C_0$ (i.e., it is a permanent source of particles). One can also think of $C$ as a temperature profile inside a layer, whose two boundaries are kept at constant temperatures 0 and $C_0$. The solution of this trivial problem yields the linear profile of the particle concentration

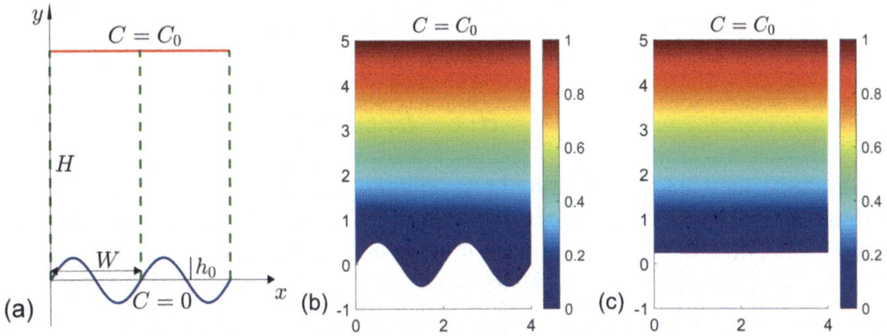


**Fig. 11.1** **a** A schematic view of Laplacian transport from the top (flat) source at $y = H$ (in red) towards an absorbing boundary with sinusoidal profile (in blue), with the period (or wavelength) $W$ and amplitude $h_0$. Dashed vertical lines delimit the periodic cell that is repeated along $x$ coordinate. The system is translationally invariant along $z$ coordinate, which is perpendicular to the figure. **b** A numerical solution of the Laplace equation inside this domain, with reflecting boundary conditions instead of periodic ones on two vertical segments. This numerical solution (obtained by a finite-element method via PDEtool in Matlab) determines the diffusive flux $J$ onto the absorbing boundary, which in turn sets via Eq. (11.4) the offset $\lambda$ (e.g., we get $\lambda \approx 0.26$ for $H = 5$, $W = 2$, and $h_0 = 0.5$). **c** The solution $C = C_0(y - \lambda)/(H - \lambda)$ in a simpler case with a flat absorbing boundary at $y_0 = \lambda$, which yields the same diffusive flux

$$C = \frac{C_0}{H} y = \frac{j_0}{D} y, \tag{11.2}$$

where $j_0 = DC_0/H$ is the particle flux density, and $D$ is the particle diffusivity.

Next we assume that the bottom boundary is subject to a periodic deformation with the period $W \ll H$ (Fig. 11.1a). Near the deformed boundary, the particle concentration is strongly non-uniform, but, for $y \gg W$, the transverse components of the particle flux rapidly disappear and the concentration profile takes the linear form (Fig. 11.1b):

$$C \approx \frac{C_0}{H - \lambda}(y - \lambda), \tag{11.3}$$

where the offset parameter $\lambda$ has the dimension of length and can be positive or negative (note that another convention with $-\lambda$ instead of $\lambda$ in Eq. (11.3) is often used). This parameter aggregates all information about the boundary profile and inhomogeneity of its absorbing properties that is relevant at far field $y \gg W$. For instance, in terms of $\lambda$, the diffusive flux through the system is given by a simple expression

$$\frac{J}{J_0} = \frac{1}{1 - \lambda/H}, \tag{11.4}$$

where $J_0 = j_0\mathcal{A}_0 = DC_0\mathcal{A}_0/H$ is the diffusive flux to a flat absorbing boundary, and $\mathcal{A}_0$ is the cross-sectional area of the periodic cell. At the same time, this relation allows one to determine the offset $\lambda$ from the diffusive flux $J$, which can be found numerically by solving the Laplace equation or via Monte Carlo simulations:

$$\lambda = H(1 - J_0/J). \tag{11.5}$$

Similarly, the far-field solution (11.3) remains valid for a flat bottom boundary with absorbing and reflecting parts (see examples below). If the bottom boundary is fully reflecting, the concentration $C$ is constant, and the diffusive flux $J$ is zero. This situation corresponds to the limit $\lambda = -\infty$. Adding more and more absorbing spots onto the bottom boundary increases the diffusive flux and thus the offset $\lambda$, which thus ranges from $-\infty$ (flat reflecting boundary) to 0 (flat absorbing boundary). Moreover, if the bottom boundary is both deformed and heterogeneous, the far-field solution may correspond to positive $\lambda$.

The linear form (11.3) provides a simple interpretation for the offset $\lambda$: the condition $y = \lambda$ determines the position of the equivalent absorbing boundary that provides the same diffusive flux through the system. The goal of boundary homogenization is to relate $\lambda$ to the geometrical properties of the original complex boundary. As this problem is relevant to various disciplines, the aggregated parameter $\lambda$ bears different names: slip length in the context of superhydrophobic coating [25], blockage coefficient in fluid dynamics [25, 26], grid parameter in electrostatics [27], and displacement length in heat transfer [29, 30]. The mathematical equivalence between steady-state diffusion and electrostatics, both described by the Laplace equation, is particularly helpful for bringing intuitive interpretations [19, 38]. In fact, if $C_0$ is understood as the applied electric potential (voltage) between two metal electrodes located at the top and at the bottom, then $J$ becomes the electric current, while $Z = C_0/J$ is the electric resistance (or impedance) of this system. For instance, if the bottom electrode is flat, one retrieves the classical formula $Z_0 = C_0/J_0 = H/(D\mathcal{A}_0)$ for the resistance of a metal wire of length $H$ and cross-sectional area $\mathcal{A}_0$, while $1/D$ plays the role of electric resistivity. As a consequence, Eq. (11.4) provides another interpretation of the offset $\lambda$ as the relative difference between the electric resistances $Z$ and $Z_0$:

$$\frac{Z_0 - Z}{Z_0} = \frac{\lambda}{H}. \tag{11.6}$$

Moreover, common rules for computing the overall resistance of an electric circuit, composed of sequential and/or parallel connections between its elements, can shed light onto various expressions for the offset $\lambda$ that will be discussed below.

In chemical physics, it is more conventional to relate the offset $\lambda$ to the trapping parameter $\kappa$ that appears in the Robin (or radiation) boundary condition on a partially absorbing (reactive) flat boundary located at some height $y_0$:

$$-D\frac{\partial C}{\partial n} = \kappa C, \tag{11.7}$$

where $\partial/\partial n$ is the normal derivative oriented outwards the domain [39]. The trapping parameter $\kappa$, which has units of meter per second, is also known as surface reactivity, forward reaction constant, permeability, relaxivity, or the inverse of surface resistivity, depending on the application field (see Chap. 4). It can be checked by direct substitution of Eq. (11.3) into Eq. (11.7) that these approaches are equivalent and

$$\kappa = \frac{D}{y_0 - \lambda}. \tag{11.8}$$

While the offset $\lambda$ is uniquely determined by the shape of the bottom boundary and the inhomogeneity of its absorbing parts, the value of the trapping parameter $\kappa$ depends on the *choice* of the location $y_0$ of the effective flat boundary. This ambiguity actually offers some flexibility for boundary homogenization. From the mathematical point of view, the trapping parameter $\kappa$ should be restricted to nonnegative values, with two relevant limits, $\kappa \to 0$ and $\kappa \to \infty$, corresponding to the perfectly reflecting and perfectly absorbing boundary, respectively. When $\lambda < 0$, one can naturally choose $y_0 = 0$. However, if $\lambda > 0$, it is mandatory to choose $y_0 \geq \lambda$ (note that the choice $y_0 = \lambda$ corresponds to an absorbing boundary with $\kappa = \infty$, as discussed above).

For simplicity of notations and more compact formulas, we will employ both parameters $\lambda$ and $\kappa$ interchangeably, keeping in mind that they are related via Eq. (11.8). One of the aims of this chapter is to transfer the analytical results from the different areas of Laplacian transport to the context of particle diffusion and target search problems. Evaluation of $\kappa$ for a given boundary profile is a difficult task that requires solving the Laplace equation in a complex shape domain. Below, we present some analytical results for different types of complex boundaries. In Sect. 11.2, we consider boundary profiles and inhomogeneities that are periodic in $x$ coordinate and translationally invariant along $z$ coordinate, e.g., periodic stripes or riblets. The invariance along $z$ coordinate reduces the original 3D problem to a planar one. In turn, Sect. 11.3 focuses on absorbing spots or deformations, which are periodic in two directions along the surface. Section 11.4 presents a constant-flux approximation, which may offer more accurate results, even though the derived expressions are usually more sophisticated. In Sect. 11.5, we discuss the Fick–Jacobs equation to deal with Laplacian transport in channels of variable cross section and its application for computing the mean first-passage time (MFPT). Section 11.6 concludes this overview by summarizing the main steps and discussing other aspects of Laplacian transport.

## 11.2 Periodic Patterns in Two Dimensions

In this section, we focus on various arrangements of infinitely long stripes or deformations so that the system is invariant along one direction (we call it $z$), and the original three-dimensional problem is thus reduced to a two-dimensional problem in the plane $xy$. Several profiles of stripes and deformations are illustrated in Fig. 11.2.

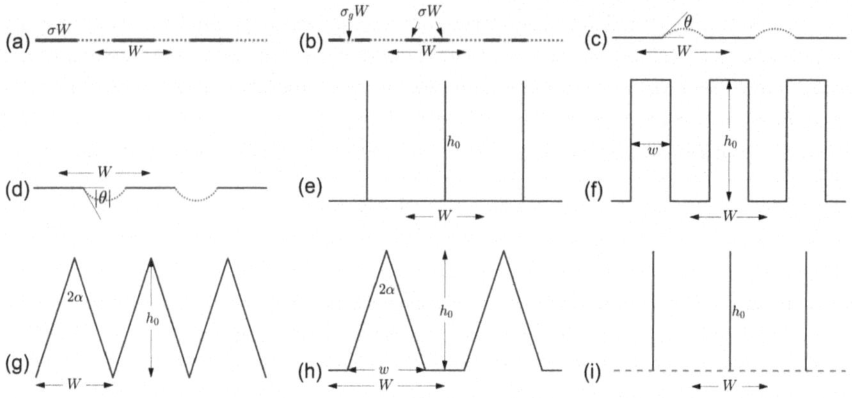


**Fig. 11.2** Various configurations of absorbing stripes or deformations that are infinite along $z$ direction, which is perpendicular to the figure. **a** A flat boundary with a periodic arrangement of absorbing (blue solid) and reflecting (green dashed) stripes, of period $W$ and absorbing fraction $\sigma$. **b** A flat periodic boundary with a pattern of two equal absorbing stripes (blue solid) and two unequal reflecting stripes (green dashed), with the gap $\sigma_g W$ between two absorbing stripes. **c, d** A periodic boundary with flat absorbing stripes (blue solid) and non-flat reflecting parts (green dashed) modeled by circular arcs, with the angle $\theta$ at the joint point between two stripes: **c** upward orientation for $\theta > 0$; **d** downward orientation for $\theta < 0$. **e** An absorbing comb-like boundary. **f** An absorbing boundary with rectangular riblets of width $w$ and height $h_0$. **g** An absorbing boundary of saw-tooth riblets of width $W$ and height $h_0$. **h** An absorbing boundary of riblets with trapezoidal grooves. **i** An absorbing comb-like boundary with reflecting base (green dashed)

For such profiles, there is a wealth of analytical results for $\lambda$ and $\kappa$ due to the conformal invariance of the Laplace equation in two dimensions that enables application of conformal mapping. The general framework for evaluation $\lambda$ (and hence $\kappa$) is given in Refs. [25, 27, 30, 40–43] and references therein.

**Flat Boundaries**

For the flat boundary with a periodic alternating pattern of absorbing and reflecting stripes (Fig. 11.2a), one has [44] (see also [45])[1]

$$\lambda = W \frac{\ln[\sin(\pi\sigma/2)]}{\pi} < 0, \qquad \kappa = \frac{D}{W} \frac{\pi}{\ln[1/\sin(\pi\sigma/2)]}, \tag{11.9}$$

where $W$ is the period of the pattern, $\sigma$ is the fraction of absorbing stripes, and we set $y_0 = 0$ to introduce $\kappa$. As $\sigma \to 0$, the trapping parameter $\kappa$ vanishes logarithmically, $\kappa \propto 1/\ln(1/\sigma) \to 0$, so the heterogeneously absorbing boundary

[1] This solution was obtained in [44] for a periodic cell, in which a single absorbing stripe is at the center; in other words, one solves the Laplace equation $\Delta C(x, y) = 0$ in the half-infinite stripe, $(-W/2, W/2) \times (0, +\infty)$, with $C = 0$ for $|x| < \frac{1}{2}\sigma W$ and $y = 0$, $\partial C/\partial y = 0$ for $\frac{1}{2}\sigma W < |x| < \frac{1}{2} W$ and $y = 0$, $C \to C_0$ as $y \to \infty$, and the additional reflecting boundary condition $\partial C/\partial x = 0$ at two vertical boundaries $|x| = W/2$ and $y > 0$ of the cell. This choice of the periodic cell ensures the periodicity of the solution in the whole upper plane.

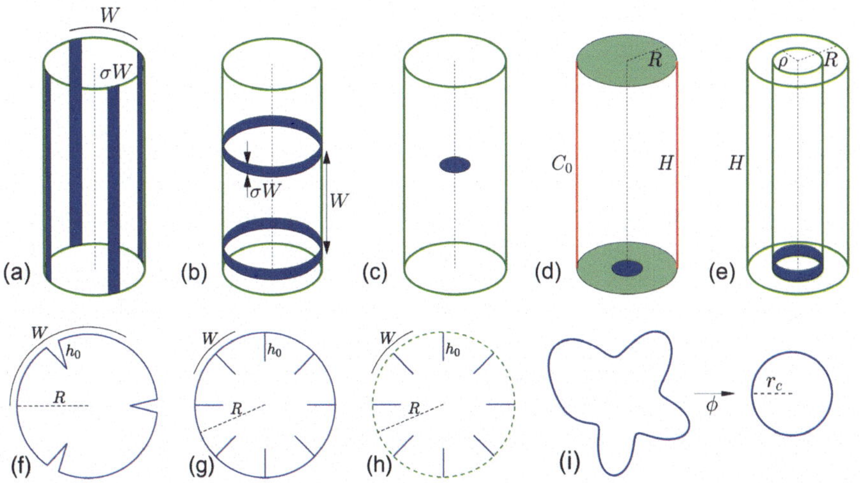


**Fig. 11.3** Cylindrical domains. **a** An infinite reflecting cylinder with absorbing stripes (in blue) oriented parallel to the cylinder axis. **b** An infinite reflecting cylinder with absorbing stripes (in blue) oriented perpendicular to the cylinder axis. **c** A long reflecting cylinder with an absorbing disk (in blue) centered on the cylinder axis. **d** A capped cylinder of height $H$ and radius $R$, with the source located at the cylinder wall and an absorbing disk (in blue), located at the center of the bottom reflecting wall. **e** Two capped coaxial reflecting cylinders of height $H$ and radii $\rho$ and $R$, with an absorbing stripes (in blue) located on the inner cylinder. **f, g** Cross section of an infinite absorbing cylinder with triangular riblets (the cylinder axis is perpendicular to the figure). **h** Cross section of an infinite reflecting cylinder with absorbing spikes. **i** An appropriate conformal map $\phi$ can transform an arbitrary simply connected cross section of an infinite cylinder into an equivalent circular cross section with a conformal radius $r_c$

becomes perfectly reflecting. In the opposite limit $\sigma \to 1$, $\kappa$ diverges as a power-law, $\kappa \propto 1/(1-\sigma)^2 \to \infty$, and the boundary becomes perfectly absorbing.

It is worth noting that the same result is applicable for pressure-driven incompressible Stokes flow in a cylindrical pipe with alternating no-slip and no-stress stripes [46]: when the stripes are parallel to the cylinder axis (Fig. 11.3a), Eq. (11.9) determines the trapping parameter $\kappa$ of the homogenized cylinder; in turn, if the stripes are perpendicular to the cylinder axis (Fig. 11.3b), Eq. (11.9) should be divided by factor 2, see details in [46]. In the context of target search problems, the trapping parameter $\kappa$ can be used to estimate the MFPT to the absorbing stripes on the cylinder from the explicitly known solution for a homogeneous partially absorbing cylinder:

$$T(r) = \frac{R^2 - r^2}{4D} + \frac{R}{2\kappa}, \tag{11.10}$$

where $r$ is the radial coordinate of the starting point. In the case of heterogeneously absorbing cylinder, this is an approximate solution, which is valid far from the boundary, $R - r \gg W$, where $W$ is the period. Moreover, if the starting point is uniformly distributed in the cylinder, the average of $T(r)$ over the volume yields

$$\overline{T} = \frac{1}{\pi R^2} \int_0^{2\pi} d\theta \int_0^R dr\, r\, T(r) = \frac{R^2}{8D} + \frac{R}{2\kappa}\,. \tag{11.11}$$

Exceptionally, the exact expression for the MFPT in the cylinder with absorbing parallel stripes is known [47–49], but it is rather complicated, especially for many stripes. In turn, Eqs. (11.10), (11.11), which we use to illustrate an application of boundary homogenization, provide a rapid and simple way to estimating the MFPT.

For the configurations of stripes with two or more absorbing stripes per period, $\kappa$ cannot be described exclusively by the surface fraction $\sigma$ since the relative position of the absorbing stripes is also important due to their competition for particles (the phenomenon known as diffusional screening or diffusive interaction). The general formula for $\kappa$ is rather cumbersome [50], but, for two equal absorbing stripes per period (not equally spaced), the result has a simple explicit form:

$$\lambda = W\frac{\ln(F)}{2\pi} < 0, \qquad F = \sin^2[\pi(\sigma+\sigma_g)/2] - \sin^2[(\pi\sigma_g)/2], \tag{11.12}$$

where $\sigma$ is the surface fraction of the absorbing stripes, and $\sigma_g$ is the surface fraction of the gap between them (Fig. 11.2b). For $\sigma_g = 0$, we return to the previous formula (11.9), while, in the general case $\sigma_g \neq 0$, the parameter $\sigma_g$ describes the diffusive interaction between the absorbing stripes. We note that the functional form of this interaction is nontrivial: even for $\pi(\sigma+\sigma_g)/2 \ll 1$, one has $\lambda \approx W\ln(\frac{\pi^2}{2}\sigma(\sigma_g+\frac{1}{2}\sigma))/\pi$, i.e., $\lambda$ (and thus $\kappa$) is determined by the product $\sigma(\sigma_g+\frac{1}{2}\sigma)$, which is a signature of strong coupling (two tiny stripes strongly interact even when they are separated by a large distance). Equations (11.9) and (11.12) have been extensively validated numerically with Brownian dynamics simulations, including homogenization of cylindrical surface with different orientations of stripes (longitudinal, transverse, and spiral) [51, 52].

**Non-flat Boundaries**

Next, we consider some examples of non-flat boundaries. We begin with the profile, in which the reflecting parts of the boundary are non-flat. This boundary has a different value of the parameter $\kappa$ because the curved reflecting parts affect the particle ability to reach the absorbing parts of the boundary. To appreciate this effect, we can use the results of Ref. [53], in which the reflecting parts were modeled as circular arcs (see Fig. 11.2c, d). In this case, the formula for $\kappa$ includes a correction factor

$$\frac{\kappa(\theta)}{\kappa(0)} = \frac{3(\pi-\theta)^2}{3\pi^2 - 4\pi\theta + 2\theta^2}, \tag{11.13}$$

where $\theta$ is the angle between reflecting and absorbing parts at their joint, and $\kappa(0)$ is given by Eq. (11.9), in which $\sigma \ll 1$ (dilute limit) was assumed. At $\theta = 0$, one retrieves the flat boundary, while negative angles correspond to the opposite (downwards) orientation of the reflecting arcs. When $\theta$ varies from $-\pi/2$ (half-circle oriented downwards) to $\pi/2$ (half-circle oriented upwards), the correction factor changes from 1.2273 to 0.5, i.e., the boundary becomes less absorbing due obstruction of the absorbing intervals by the reflecting arcs of the boundary. Curiously, the correction factor is larger than 1 for $\theta < 0$, i.e., the inclusion of reflecting arcs oriented downwards enhances the overall reactivity. Indeed, the particle can spend more time in these regions and thus have higher chances to arrive to the absorbing intervals, before returning to the source. Other examples and their correction factors can be found in [54].

An important analytical result for the offset $\lambda$ was obtained for a sinusoidal absorbing boundary of a small amplitude $h_0$ (Fig. 11.1a). In this case, the formula for $\lambda$ can be derived by means of perturbation theory [40]:

$$\lambda \approx \frac{\pi h_0^2}{W} > 0, \quad h_0 \ll W. \tag{11.14}$$

This formula states that the sinusoidal deformation shifts the equivalent boundary upwards, though this shift is small since $\lambda/h_0 \approx \pi h_0/W \ll 1$. The direction of this shift is also understandable: the bottom of the troughs of the deformed boundary is less accessible to the particles, so that most of the particle absorption occurs near the crests (i.e., above the middle line located at $y = 0$), which play the role of "active zones" of absorption (see [17, 55] for this concept). As we can see below, for a small-amplitude deformation of the boundary, the scaling $\lambda \propto (h_0^2/W)$ is rather general: it can be interpreted as the product of the amplitude $h_0$ and a typical slope $\propto h_0/W$ [40]. We stress that this result is perturbative; for instance, substitution of $h_0 = 0.5$ and $W = 2$ from the example in Fig. 11.1 into Eq. (11.14) yields $\lambda \approx 0.39$, whereas a numerical solution of the Laplace equation gave $\lambda \approx 0.26$. Vandembroucq and Roux proposed an improved expression

$$\lambda \approx \frac{2h_0}{\pi} \mathrm{atan}\big(\pi^2 h_0/(2W)\big), \tag{11.15}$$

which is reduced to Eq. (11.14) when $\pi^2 h_0/(2W)$ is small. For our example, this expression gives $\lambda \approx 0.28$, which is much closer to the above numerical value. Note that, if the amplitude $h_0$ was reduced to 0.1, Eq. (11.14) would yield 0.016, which is already close to the numerical solution 0.013. Expectedly, the perturbative expression (11.14) becomes more accurate as $h_0/W$ gets smaller.

For the comb-like absorbing boundary (Fig. 11.2e), there is the exact formula [41]

$$\lambda = W \frac{\ln[\cosh(\pi h_0/W)]}{\pi} > 0, \tag{11.16}$$

where $h_0$ is the height of the riblets. This offset is always positive, i.e., the effective absorbing boundary stands above the base of the comb. This is expected because the spikes efficiently absorb particles and thus increases the diffusive flux $J$, as compared to the flux $J_0$ onto the absorbing base without spikes. For $\pi h_0/W \ll 1$,

Eq. (11.16) leads to a similar scaling as in Eq. (11.14)

$$\lambda \approx \frac{\pi h_0^2}{2W}, \quad h_0 \ll W. \tag{11.17}$$

The difference by factor 2 between these expressions can be attributed to distinct shapes of the sinusoidal profile and the comb structure (note also that $h_0$ was the amplitude of the sine profile, not the height). In contrast, for high riblets $h_0/W \gg 1$, there is a saturation limit

$$\lambda \approx h_0 - W\frac{\ln 2}{\pi}, \quad h_0 \gg W. \tag{11.18}$$

This result implies that most of the particles are absorbed within the distance $\frac{\ln 2}{\pi} W \approx 0.22\, W$ from the tips of the comb-like structure making the rest of the structure irrelevant in this context. This result is the simplest way of illustrating the concept of "active zones" discussed above. As $\lambda$ remains below $h_0$, it is natural to impose the effective partially absorbing boundary at $y_0 = h_0$ (at the tip of the combs), in which case Eq. (11.8) defines the trapping parameter as

$$\kappa = \frac{D}{h_0 - W \ln[\cosh(\pi h_0/W)]/\pi}. \tag{11.19}$$

In two limits $h_0/W \ll 1$ and $h_0/W \gg 1$, one gets $\kappa/D \approx 1/h_0 + \pi/(2W)$ and $\kappa/D \approx (\pi/\ln 2)/W$, respectively.

The direct generalization of the comb-like boundary is a periodic structure with the riblets of a finite thickness. The closed-form solution is only available for high rectangular spikes of height $h_0$ and width $w$ (Fig. 11.2f). If $h_0/W > 0.6$, one gets [56]

$$\lambda \approx h_0 - \frac{W}{2\pi}\mathcal{G}(w/W), \tag{11.20}$$

where $\mathcal{G}(\zeta)$ is the dimensionless function:

$$\mathcal{G}(\zeta) = 2(1-\zeta)\ln(2-\zeta) + \zeta \ln(2\zeta - \zeta^2), \qquad \zeta = w/W \le 1. \tag{11.21}$$

In the limit $w \to 0$, one has $\mathcal{G}(0) = 2\ln 2$, so that Eq. (11.18) for the comb-like boundary is retrieved. In the opposite limit $\zeta = w/W \to 1$, one has $\mathcal{G}(\zeta) \approx (1-\zeta)^2$, i.e., the second term in Eq. (11.20) vanishes, and one gets $\lambda = h_0$ as expected for a flat absorbing boundary at the height $h_0$. As previously, it is natural to impose the effective partially absorbing boundary at $y_0 = h_0$ (at the tips of the riblets), in which case the trapping parameter is

$$\kappa \approx \frac{D}{W}\frac{2\pi}{\mathcal{G}(w/W)}. \tag{11.22}$$

The last family of the 2D profiles that we consider in this section is the saw-tooth boundary (Fig. 11.2g). The profile is characterized by its height $h_0$ and the base width $W$, or by the saw-tooth angle $2\alpha = 2\text{atan}(\frac{1}{2}W/h_0)$. In this case, one gets [57]

$$\lambda = h_0 - \frac{W}{2\pi}\mathcal{S}(\alpha), \tag{11.23}$$

where $\mathcal{S}(\alpha)$ is the dimensionless function of angle $\alpha$:

$$\mathcal{S}(\alpha) = \gamma + 2\ln 2 + \frac{\pi}{\tan\alpha} - \frac{\pi}{\alpha} + \psi\left(1 + \frac{\alpha}{\pi}\right), \tag{11.24}$$

where $\psi(\cdot)$ is the the digamma function, and $\gamma \approx 0.5772$ is the Euler constant. One can check that $\mathcal{S}(\alpha)$ is a monotonously decreasing function of $\alpha$. For $\alpha = 0$ (comb-like boundary), one has $\psi(1) = -\gamma$ and thus $\mathcal{S}(0) = 2\ln 2$, in agreement with Eq. (11.18). For $\alpha = \pi/2$ (flat surface), one gets $\psi(3/2) = -\gamma - 2\ln 2 + 2$, so that $\mathcal{S}(\pi/2) = 0$, as expected. Another explicit result corresponds to $\alpha = \pi/6$ (equilateral triangle), for which $\mathcal{S}(\pi/6) = \pi\sqrt{3}/2 - 3\ln(3)/2$ and thus $\lambda/W = \sqrt{3}/4 + 3\ln(3)/(4\pi)$ [30]. As previously, setting $y_0 = h_0$, one can also introduce the trapping parameter as

$$\kappa = \frac{D}{W}\frac{2\pi}{\mathcal{S}(\alpha)}. \tag{11.25}$$

More complex profiles appear when the saw teeth are separated by a piece of flat boundary, with a period $W$ (the so-called trapezoidal grooves, Fig. 11.2h). The generalized formulas were proposed in Ref. [57], e.g., for high riblets, one has

$$\lambda \approx h_0 - \left[(W - w)\frac{\ln 2}{\pi} + \frac{w}{2\pi}\mathcal{S}(\alpha)\right], \quad h_0/W > 1, \tag{11.26}$$

where $w$ is the width of the foot of the riblet, and $\alpha = \text{atan}(\frac{1}{2}w/h_0)$. Moreover, one can account for rounded wedges at the tips [57]. More numerical and analytical estimates of $\lambda$ for different types of boundaries (piece-wise, pre-fractals, convex, with scalloped and club-like riblets, etc.) can be found in Ref. [57]. We also mention an interesting result established numerically in Ref. [30] regarding the strong non-monotonic dependence of the parameter $\lambda$ on geometrical parameters of $M$-shape profiles.

To conclude this section, we present an example of non-flat boundaries with mixed boundary conditions (absorbing and reflecting). For the comb-like boundary with absorbing spikes and the reflecting base (Fig. 11.2i), the result is [42]

$$\lambda = W\,\frac{\ln[\sinh(\pi h_0/W)]}{\pi}. \tag{11.27}$$

For $\pi h_0/W \gg 1$, the saturation limit, $\lambda \approx h_0 - W\ln(2)/\pi$, remains the same as for the absorbing base (since particles do not reach the bottom of the comb structure, the boundary condition at the base becomes irrelevant). As the comb height $h_0$ decreases, the offset parameter $\lambda$ also decreases and then becomes negative. In the limit $h_0/W \ll 1$, the offset parameter diverges to $-\infty$ logarithmically, and the boundary slowly becomes perfectly reflecting. In this limit, Eqs. (11.9) and (11.27) become identical provided we set $h_0/W = \sigma/2$. The factor 2 can be explained by symmetry argument: a horizontal absorbing interval laying on the reflecting base creates the diffusive flux twice smaller than the flux created by the same vertical interval standing on that base (due to two accessible sides of this interval). These results can be extended to incorporate other shapes of small absorbers by finding an equivalent absorbing interval with the same log-capacity and using it in Eq. (11.9). An effective radiative boundary can be set at $y_0 = h_0$ with the trapping parameter

$$\kappa = \frac{D}{h_0 - W \ln[\sinh(\pi h_0/W)]/\pi}, \tag{11.28}$$

which vanishes very slowly in the limit $h_0 \to 0$.

As earlier in the case of cylinders with absorbing stripes, the above expressions for the trapping parameter $\kappa$, which were derived for a flat boundary, are applicable to curved surfaces. For instance, if the cylinder of radius $R$ contains grooves or riblets that are parallel to the cylinder axis (Fig. 11.3f–h), one can get $\kappa$ by choosing an appropriate expression according to the groove shape, and then substitute it to Eqs. (11.10) and (11.11) for estimating the MFPT. Even though the height $h_0$ of these grooves and the period $W$ are formally expected to be relatively small with respect to $R$ (which plays here the role of $H$ for the flat boundary), such an estimate remains quite accurate for a broad range of parameters. Moreover, there is no smoothness constraint on such a cylindrical surface with grooves or riblets (e.g., one can even consider a comb-like structure). This feature reveals the crucial role of conformal mapping, on which many of the above expressions rely, and thus distinguishes them from other approximations obtained, e.g., via perturbation theory. More generally, for a cylinder of any simply connected cross section $\Omega$ (Fig. 11.3i), one can construct a conformal map from $\Omega$ onto a disk and thus determine its conformal radius $r_c$, from which the MFPT can be estimated by replacing $R$ by $r_c$ in Eq. (11.11) and setting $\kappa = \infty$. It is important to stress the two-step character of this approximation: one first estimates $r_c$ and then evaluates the MFPT for an equivalent disk. In contrast, even though a direct application of conformal mapping to the first-passage problem is possible (see [49] for details), the Poisson equation governing the MFPT is not conformally invariant that brings substantial complications to the exact solution. Despite the simplicity and potential utility of boundary homogenization, its applications are not yet fully explored in the context of target search problems, while systematic investigations on their accuracy and validity range are still missing.

## 11.3 Periodic Patterns in Three Dimensions

When the domain is not translationally invariant along one axis, the original 3D problem is not reducible to a planar one that limits the use of conformal mapping. As a consequence, the analysis of patterns that are periodic in two directions, is much more difficult, with a limited number of available analytical results.

**Flat Boundaries**

Similar to the 2D case, we begin with the trapping parameter of a smooth boundary covered by absorbing and reflecting spots. The first results by Berg and Purcell [60, 61] and by Zwanzig [62] dealt with the particle trapping by small absorbing spots covering a reflecting sphere. The opposite case (a small reflecting spot on the absorbing sphere) was analyzed in [63]. This is still an area of active research [64, 65].

The appeal to electrostatic analogy provides a valuable insight and clear interpretation of the main result on this topic [6, 61]. According to this analogy, the trapping capacity $K$ of an object and its electric capacity (or capacitance) $C$ are related as $K = CD$ [6, 61] (here we employ the convention that the capacity of a sphere of radius $R$ is $4\pi R$). There are many analytical methods for estimating $C$ from the shape of the object [66–69] that enables fruitful translation of many results from electrostatic to diffusion kinetics and target search problems.

We begin with an illustrative example: a sparse periodic array of absorbing disks of radius $a$ on a reflecting surface (Fig. 11.4a). If diffusional screening between disks could be ignored, an estimation of the trapping parameter for such a boundary would be straightforward [70]

$$\kappa = \frac{K}{\mathcal{A}_0} = \frac{CD}{\mathcal{A}_0} = \frac{4aD}{\mathcal{A}_0}, \tag{11.29}$$

where $\mathcal{A}_0$ is the area of the periodic cell, and $C = 4a$ is half of the capacity of a disk in the three-dimensional space $\mathbb{R}^3$ (the factor $1/2$ accounts for the half of the flux to one side of the disk surface). Even though diffusional screening between disks is always present, the formula (11.29) is still valid if $C$ is understood as the capacity

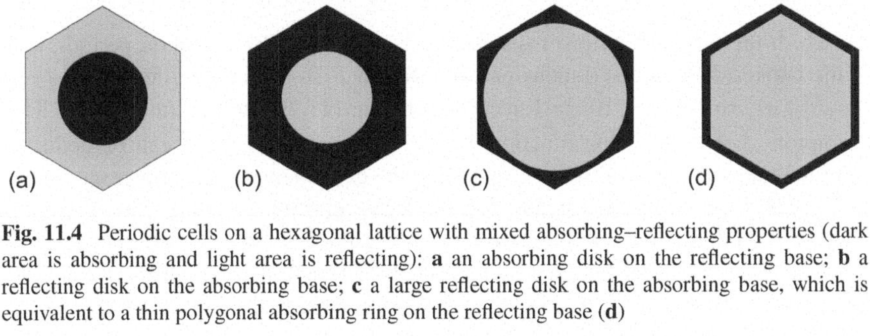


**Fig. 11.4** Periodic cells on a hexagonal lattice with mixed absorbing–reflecting properties (dark area is absorbing and light area is reflecting): **a** an absorbing disk on the reflecting base; **b** a reflecting disk on the absorbing base; **c** a large reflecting disk on the absorbing base, which is equivalent to a thin polygonal absorbing ring on the reflecting base (**d**)

of a single disk in the lattice, which depends on the arrangement of the disks. In the formal limit when the surface fraction of absorbing disks $\sigma$ is equal to 1 (full coverage), one retrieves the infinite capacity of the perfectly conducting plane: $C = \infty$. The calculation of the "renormalized" $C$ (and hence $\kappa$) that accounts for the diffusive interaction between absorbers in the lattice is the main aim of the effective medium approximation in this context. Different analytical methods provide different levels of rigor and accuracy.

The aforementioned electrostatic analogy allows us to reveal some important features of boundary homogenization in three dimensions. The trapping parameter $\kappa$ as function of the surface fraction $\sigma$ of the absorbing part can be strongly dependent on its shape. To illustrate this point, we analyze two complementary configurations: a "spot configuration" (Fig. 11.4a) formed by a hexagonal lattice of absorbing disks ($C = 0$) on the reflecting base ($dC/dy = 0$), and a "mesh configuration" (Fig. 11.4b), which is an inverse of the first one: the same lattice but of the reflecting spots ($dC/dy = 0$) on the absorbing base ($C = 0$). In the limit of small $\sigma$, the first configuration reduces to a sparse periodic arrangement of small disks, so according to Eq. (11.29), one has $\kappa \propto a \propto \sqrt{\sigma}$, while the second configuration transforms to a lattice of flat rings (Fig. 11.4c, d) and $\kappa$ decays much slower: $\kappa \propto 1/\ln(1/\sigma)$ (this follows from the formula for the capacity of a flat ring [71], see also Eq. (11.9) for the stripe configuration). This comparison illustrates that the knowledge of $\sigma$ alone is insufficient for estimating the trapping parameter without additional information on geometrical settings.

It is instructive to pursue the analysis of these two configurations and to deduce explicit formulas for $\kappa$ and then by comparison to appreciate their differences. Our choice of the hexagonal lattice and of the circular shape of absorbing/reflecting spots helps to avoid additional complications related to anisotropy or other asymmetries. To simplify analytical treatment, the periodic cell of this system is often approximated by a circular tube of the same cross section containing a single disk [72].

**Absorbing Disk on Reflecting Base**

For the first configuration (Fig. 11.4a), it is convenient to write the trapping parameter $\kappa$ in the following form [73–75]

$$\kappa = \frac{D}{\sqrt{\mathcal{A}_0}} F_1(\sigma), \tag{11.30}$$

where $\sigma$ is the surface fraction of absorbing area, and $\mathcal{A}_0$ is the area of the periodic cell of the lattice. In turn, the dimensionless function $F_1(\sigma)$ can be written as $F_1(\sigma) = \frac{4}{\sqrt{\pi}}\sqrt{\sigma} f(\sigma)$ to ensure $f(0) = 1$ according to Eq. (11.29) in the limit $\sigma \to 0$. The asymptotic behavior of the function $f(\sigma)$ as $\sigma \to 0$ and $\sigma \to 1$ can be deduced from the known analytical results, including the capacitance approximation.

In 1941, Fock studied the steady flow of electric current down a long circular tube filled with a conducting medium, with a thin circular hole in its center (Fig. 11.3c). In particular, Fock obtained a representation of the function $f(\sigma)$ (and hence $F_1(\sigma)$) as a power-law series of $\sqrt{\sigma}$, evaluated the coefficients of this series in terms of the integrals of Bessel functions, and gave numerical values of the first 12 coefficients [76]. Fock also proposed a simple truncated expression

$$f(\sigma) \approx \frac{1}{1 - 1.41\sigma^{1/2} + 0.34\sigma^{3/2} + 0.07\sigma^{5/2}}. \tag{11.31}$$

This expression provides a very accurate behavior of $f(\sigma)$ as $\sigma \to 0$. Moreover, it ensures the expected divergence of $f(\sigma)$ as $\sigma \to 1$ but fails to yield the correct power-law exponent. This issue was resolved by Leppington and Levine [71], who derived the asymptotic behavior near $\sigma \to 1$:

$$f(\sigma) \approx \frac{1}{2(1 - \sqrt{\sigma})^2} \approx \frac{2}{(1-\sigma)^2}, \quad \sigma \to 1, \tag{11.32}$$

where the second relation was obtained by writing $1 - \sigma = (1 - \sqrt{\sigma})(1 + \sqrt{\sigma}) \approx 2(1 - \sqrt{\sigma})$ as $\sigma \to 1$. Motivated by the relative smallness of the coefficient of the last term in the denominator of Eq. (11.31) and combining two asymptotic results together, one arrives at the following approximation

$$F_1(\sigma) \approx \frac{4}{\sqrt{\pi}} \sqrt{\sigma} \underbrace{\frac{1 + A\sqrt{\sigma} - (A-1)\sigma^{3/2}}{(1-\sigma)^2}}_{\approx f(\sigma)}, \tag{11.33}$$

which is applicable for any $0 < \sigma < 1$. The coefficient $A = 1.41$ is fixed by matching the coefficient in front of $\sqrt{\sigma}$ in the Taylor expansion of the Fock's expression (11.31) as $\sigma \to 0$, whereas the coefficient $A - 1$ in front of $\sigma^{3/2}$ ensures the correct asymptotic behavior (11.32) as $\sigma \to 1$. We stress that this approximation has no fitting parameter. Note that Bernoff et al. derived a similar result by other means [77]. A very similar approximation was earlier proposed by Berezhkovskii et al. by fitting the numerical solution of the Laplace equation [73–75]

$$F_1(\sigma) \approx \frac{4}{\sqrt{\pi}} \sqrt{\sigma} \underbrace{\frac{1 + A\sqrt{\sigma} - B\sigma^2}{(1-\sigma)^2}}_{\approx f(\sigma)}. \tag{11.34}$$

The parameters $A$ and $B$ were estimated from fitting and were shown to depend on the considered lattice, e.g., $A \approx 1.37$ for the hexagonal lattice [73–75]. Note that the second parameter $B$ should be fixed as $A - 1$ to ensure the correct asymptotic behavior (11.32) as $\sigma \to 1$.

One can compare Eq. (11.30) with other results for $\kappa$ found in the literature [77]. The solution of the celebrated Berg–Purcell problem [60] for a reflecting sphere covered by a large number $N \gg 1$ of small absorbing disks of radius $a$ reads in our notations as

$$\kappa_{\mathrm{BP}} = \frac{DaN}{\pi R^2} = \frac{4D}{\pi a} \sigma, \tag{11.35}$$

where $\sigma = \mathcal{N}(a/2R)^2$. As the Berg–Purcell solution ignores diffusional screening between absorbing spots, it is valid only for $\sigma \ll 1$ (the dilute limit). Later Zwanzig heuristically modified the Berg–Purcell result (11.35) to explicitly account for the reduction of the surface area due to absorbing disks [62]:

$$\kappa_{\rm ZW} = \frac{\kappa_{\rm BP}}{1-\sigma}. \tag{11.36}$$

While the diffusive transport towards a spherical boundary is quite different from that on the plane, one can still compare the predictions given by Eqs. (11.36) and (11.30). In the latter case, the surface area of the "periodic cell" can be written as $\mathcal{A}_0 = 4\pi R^2/\mathcal{N}$ so that Eq. (11.30) becomes

$$\kappa \approx \frac{D}{a\sqrt{\pi}} \sqrt{\sigma} F_1(\sigma) = \frac{4D\sigma}{\pi a} f(\sigma). \tag{11.37}$$

As outlined by Bernoff et al. [77], Zwanzig's correction (11.36) yields a slower divergence, $(1-\sigma)^{-1}$ instead of $(1-\sigma)^{-2}$, as $\sigma \to 1$. In turn, Eq. (11.37), which was originally derived by homogenizing the flat boundary, is applicable for the whole range of $\sigma$ even for a sphere.

**Reflecting Disk on Absorbing Base**

For the second configuration (Fig. 11.4b), we also write

$$\kappa = \frac{D}{\sqrt{\mathcal{A}_0}} F_2(\sigma). \tag{11.38}$$

An important insight comes from the identity $\kappa = D/\mathcal{B}$ [26, 28, 78], where $\mathcal{B}$ is the blockage coefficient of a potential flow in a tube with a blocking disk. The value of the blockage coefficient was calculated in [26, 79], yielding the following asymptotic behaviors

$$F_2(\sigma) \approx \frac{\pi^{3/2}}{\ln(2/\sigma)}, \quad \sigma \ll 1, \tag{11.39a}$$

$$F_2(\sigma) \approx \frac{3\pi^{3/2}}{4(1-\sigma)^{3/2}}, \quad \sigma \to 1. \tag{11.39b}$$

Combining two asymptotic results, one arrives at the following interpolation

$$F_2(\sigma) \approx \frac{3\pi^{3/2}}{4(1-\sigma)^{3/2}\ln\big(P + Q\sigma^{-4/3}\big)}, \tag{11.40}$$

with two constants $P$ and $Q$ that follow by matching Eq. (11.40) with the asymptotic relations (11.39a, 11.39b):

$$P = \exp(1) - 2^{4/3} \approx 0.20, \quad Q = 2^{4/3} \approx 2.51. \tag{11.41}$$

While the values of constants $A, B, P, Q$ are affected by the lattice type (see [73–75, 77] for details), the functional form of $F_1$ and $F_2$ remains unchanged. As illustrated on Fig. 11.4c, d, the limit $\sigma \to 0$ is particularly interesting because such a configuration forms a sort of spider web of thin flat absorbing rings on the hexagonal lattice. Moreover, using the capacitance argument, one can replace thin flat stripes by thin absorbing cylinders whose arrangement resembles a graphene layer. While a direct computation of the trapping parameter for such a structure is challenging, Eq. (11.40) can provide a first approximation.

The expression (11.40) for the function $F_2(\sigma)$ can be compared with the results of [78, 80] deduced from fitting the numerical solution:

$$F_2(\sigma) = \frac{3\pi^{3/2} g(\zeta)}{4\zeta^3[1 - \ln(1-\zeta)]}, \tag{11.42}$$

where $\zeta = \sqrt{1-\sigma}$, $g(\zeta) = 1 + 0.6\zeta + 2\zeta^2 - 1.5\zeta^3 - 0.8\zeta^{100}$, and $\sigma$ is the surface fraction of absorbing parts. If Eq. (11.42) may be considered as a benchmark, such a comparison reveals a moderate quality of the interpolation formula (11.40). Moreover, the asymptotic relation (11.39a) is accurate only at very small $\sigma$, whereas next-order corrections are needed for moderately small $\sigma$. One can also compare these results to the capacitance approximation $\kappa = DC/\mathcal{A}_0$ with the capacity $C$ of a flat circular ring, given in [71]. Further investigations could clarify these points and bring more accurate approximations.

**Hindering Effect of a Porous Membrane**

So far we focused on the effect of mixed absorbing/reflecting patterns on the plane at $y = 0$. In a similar way, boundary homogenization can be employed to estimate the hindering effect of pores inside a semi-permeable membrane. Let us consider a flat absorbing boundary at $y = 0$, which is separated from the source at height $H$ by a semi-permeable membrane of thickness $w$, located at some height $h_0$. The presence of the membrane reduces the diffusive flux from the basic value $J_0 = DC_0/H$ without membrane, to [81, 82]

$$J = \frac{C_0}{\frac{H-w}{D} + \frac{w}{D\sigma} + \frac{2}{\kappa_m}}, \tag{11.43}$$

where $\sigma$ is the surface fraction of pores in the membrane, and $\kappa_m$ is given by Eq. (11.30) or Eq. (11.33), depending on the arrangement of pores. As discussed in Sect. 11.1, the ratio $C_0/J$ can be interpreted as the overall diffusive resistance (or impedance) of the system, allowing for a simple physical interpretation: the bulk resistance $(H - w)/D$ of the region without membrane is sequentially added to the resistance of the membrane. The latter has two contributions: the resistance $w/(D\sigma)$ of multiple pores (connected in parallel) and the additional contributions $2/\kappa_m$ accounting for heterogeneous concentration of arrived particles onto the membrane. Expectedly, the overall resistance and thus the flux do not depend on the height $h_0$ due to the sequential addition of resistances. As a consequence, Eqs. (11.5), (11.8) define

$$\lambda = w(1 - 1/\sigma) - \frac{2D}{\kappa_m} < 0, \qquad \kappa = \frac{1}{\frac{w}{D}(1/\sigma - 1) + \frac{2}{\kappa_m}}, \tag{11.44}$$

where the effective radiative boundary was set at $y_0 = 0$. Expectedly, the trapping parameter $\kappa$ decreases when the width $w$ of the membrane increases or the pores vanish ($\sigma \to 0$). Note that the factor 2 in the contribution $2/\kappa_m$ comes from two sides of the membrane; if the membrane lies on the absorbing boundary, $2/\kappa_m$ should be replaced by $1/\kappa_m$.

**Non-flat Boundaries**

There is only a limited number of analytical results for the trapping parameter $\kappa$ for the boundaries with complex 3D profiles (e.g., boundaries with protrusions). Some of these results will be presented below.

Similar to the 2D sinusoidal profile (Fig. 11.1a), one can analyze wavy absorbing surfaces in three dimensions (i.e., a doubly periodic sinusoidal deformation with a square or hexagonal pattern). The problem can be treated by perturbation theory that yields [29]

$$\lambda \approx q h_0^2 / W, \tag{11.45}$$

where $q = 2\pi$ for the square pattern and $q = 4\pi/\sqrt{3}$ for the hexagonal pattern (note a weak dependence on the lattice geometry). These results were validated by numerical simulations [29]. For small values of $h_0/W$ (up to 0.1), the agreement between the asymptotic and numerical results is good, confirming the parabolic scaling near the origin in Eqs. (11.14), (11.45) for both 2D and 3D inhomogeneities.

A lattice of absorbing/reflecting cylinders (pillars) is the natural extension of the model of rectangular riblets considered in Sect. 11.2. For a square lattice of absorbing pillars on the absorbing base (Fig. 11.5a), the trapping parameter was computed and analyzed in [83]. Since the exact expression for $\kappa$ is rather cumbersome, we just mention on two limits.

(i) For thin and tall pillars, one gets

$$\lambda \approx h_0 - W\sqrt{\ln(1/\sigma) - 3/4}, \quad h_0 \gg W, \ \sigma \ll 1, \tag{11.46}$$

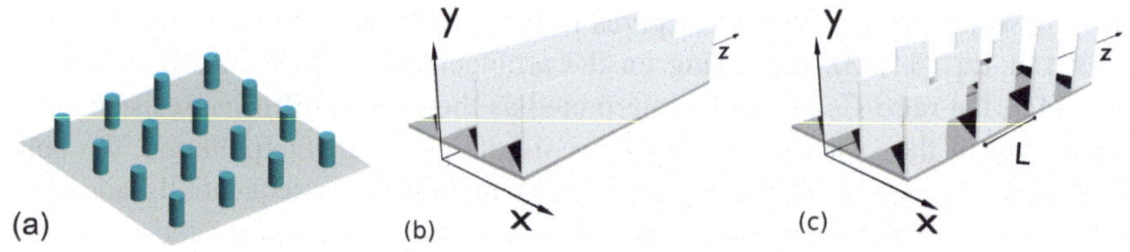


**Fig. 11.5** Examples of irregular boundaries: **a** a square lattice of absorbing pillars on a flat boundary; **b, c** a 3D coating of periodic riblets without or with gaps

where $W$ is the lattice period and $\sigma$ is the surface fraction occupied by the tips of the pillars. Setting an effective radiation boundary at $y_0 = h_0$, one gets

$$\kappa \approx \frac{D}{W} \frac{1}{\sqrt{\ln(1/\sigma) - 3/4}}. \tag{11.47}$$

Similar to the 2D case, this formula is independent of $h_0$ since the particles are mostly absorbed near the top of the pillars.

(ii) For thin and short pillars, one has

$$\lambda \approx h_0 - W^2 \frac{\ln(1/\sigma) - 3/4}{h_0}, \quad \kappa \approx \frac{D}{W} \frac{h_0/W}{\ln(1/\sigma) - 3/4}, \quad h_0 \ll W, \ \sigma \ll 1. \tag{11.48}$$

In this regime, the value of $\lambda$ is affected by the height $h_0$: as the pillar gets shorter, $\lambda$ decreases, as expected. The limit of $h_0 = 0$ is given by Eq. (11.33). In both regimes of high and short pillars, one has $\lambda \to -\infty$ and $\kappa \to 0$ in the limit $\sigma \to 0$ because infinitely thin pillars (needles) are inaccessible to diffusion in three dimensions. This is in sharp contrast to the planar case, for which rectangular riblets of width $w$ were reduced to a comb-like boundary as $\sigma = w/W \to 0$, and the latter was still accessible to diffusing particles; in particular, the offset $\lambda$ and the trapping parameter $\kappa$ remain finite, see Eq. (11.27).

An approximation for the offset $\lambda$ can be derived analytically for a coating made of a periodic system of absorbing riblets with periodic gaps in each riblet (Fig. 11.5b, c). This seems to be one of the simplest realizations of a 3D non-differentiable profile. The result is [84]

$$\lambda \approx \frac{W - 2\lambda_s}{\pi} \ln(\cosh[\pi h_0/(W - 2\lambda_s)]), \tag{11.49}$$

where $W$ is the distance between riblets,

$$\lambda_s = \frac{L}{\pi} \ln[\sin(\pi\sigma_s/2)] < 0, \tag{11.50}$$

$L = s + g$ is the period of the solid-gap structure of an individual riblet, $s$ is the width of the solid part of the riblet (per period), $g = L - s$ is the width of the gap, and $\sigma_s = s/L$. For the case $g = 0$ (no gaps), one has $\sigma_s = 1$, $\lambda_s = 0$, and we return to the 2D solution given by Eq. (11.16).

For a lattice of small absorbers of arbitrary shape on the reflecting plane, an estimation of $\kappa$ can be deduced from the appeal to electrostatic analogy (capacitance arguments), see Eq. (11.29). In fact, a given absorber can be replaced by an equivalent disk, which has the same capacity as the absorber; after that, Eq. (11.30) can be applied. Unfortunately, this approach being physically appealing has a limited range of validity, namely, a very diluted limit (see [84] for details). Another approach is to use relation $K = 1/\mathcal{B}$, where $\mathcal{B}$ is the blockage coefficient for a potential flow in a tube with a blocking soft body of the shape of the protrusion. For different shapes of bodies, expressions for $\mathcal{B}$ are given in [26].

We also mention the perturbative computation of the trapping parameter $\kappa$ for non-spherical axisymmetric boundaries with partial reactivity $\kappa_0$ [85]. In sharp contrast to most earlier studies dedicated to deformations of the plane, these boundaries appear as small deformations of a sphere of radius $R$, which can be written in spherical coordinates as $r(\theta) = R + \delta f(\theta)$, where $\delta \ll R$, and $f(\theta)$ is a dimensionless profile depending on the polar angle $\theta$. If a constant concentration $C_0$ is imposed at infinity, the diffusive flux reads as

$$J = \frac{4\pi DRC_0}{1 + D/(\kappa_0 R)} \left( 1 + B_0 \frac{1 + \frac{2D}{\kappa_0 R}}{1 + \frac{D}{\kappa_0 R}} (\delta/R) + O\big((\delta/R)^2\big) \right), \tag{11.51}$$

where $B_0 = \frac{1}{2}\int_0^\pi f(\theta)\sin(\theta)d\theta$. When $\delta = 0$, one retrieves the diffusive flux onto a partially absorbing sphere found by Collins and Kimball [39], which is further reduced to the Smoluchowski flux $J_0 = 4\pi DC_0 R$ for a perfectly absorbing sphere. Comparison of the diffusive fluxes for $\delta = 0$ and $\delta \neq 0$ allows one to introduce the trapping parameter $\kappa$ of an effective spherical boundary as:

$$\frac{1}{\kappa} = \frac{1}{\kappa_0} - \left( \frac{1}{\kappa_0} + \frac{R}{D} \right) B_0 \frac{1 + \frac{2D}{\kappa_0 R}}{1 + \frac{D}{\kappa_0 R}} (\delta/R) + O\big((\delta/R)^2\big). \tag{11.52}$$

This definition is meaningful only when the right-hand side is positive. One can also introduce the offset $\lambda$, even though the original definition via Eq. (11.8) has to be adapted to this problem.

To conclude this section, we mention two examples of a coating made from mixed absorbing/reflecting parts (reflecting protrusions on the absorbing base). For a hexagonal lattice of cylindrical pits of height $h_0$ inside a reflecting matrix, one has [58, 59]

$$\frac{D}{\kappa} = \frac{h_0}{\sigma} + \frac{\sqrt{\mathcal{A}_0}}{F_1(\sigma)}, \tag{11.53}$$

where $\sigma$ is the absorbing fraction on the plane, $\mathcal{A}_0$ is the area of the periodic cell, and $F_1(\sigma)$ is given by Eq. (11.33). In turn, for a hexagonal lattice of reflecting cylinders (pillars) on the absorbing plane, the result reads

$$\frac{D}{\kappa} = \frac{h_0}{\sigma} + \frac{\sqrt{\mathcal{A}_0}}{F_2(\sigma)}, \tag{11.54}$$

where $F_2(\sigma)$ is given by Eq. (11.40). The structure of Eqs. (11.53), (11.54) is very similar, with two contributions. The first term represents the diffusion resistance of the pit, while the second term accounts for access resistance (a delay due to the time needed for a particle to find an opening in the interface). The relative contributions of these terms depend on the geometrical properties of the boundary profile, and the functional form in the limits $h_0 \to 0$ and $\sigma \to 0$ is very different.

## 11.4 Constant-Flux Approximation

In two previous sections, we mainly focused on absorbing spots or stripes distributed periodically on the plane. The same arguments and often similar results can be derived for curved boundaries such as a sphere (e.g., the Berg–Purcell's problem [60]) or a cylinder (e.g., Stokes flow in a cylindrical pipe with stripes [46]), as discussed earlier. Moreover, the periodicity assumption can also be relaxed by considering random, uniformly distributed arrangements of spots. In fact, specific local variations of the concentration near the absorbing spots are averaged out in the far field, ensuring a similar form for the offset $\lambda$, up to minor changes (in the same way as different lattices yielded different constants in functions $F_1(\sigma)$ and $F_2(\sigma)$ in Sect. 11.3). At the same time, heterogeneous distribution of absorbing spots or their polydispersity (i.e., distribution of their sizes or shapes) can significantly alter the offset $\lambda$, as we illustrated in Sect. 11.3 by considering two complementary configurations of absorbing patterns (Fig. 11.4a,b). In such situations, other approximations may be needed for estimating the diffusive flux and the trapping parameter.

To give an example, we mention the so-called narrow escape problem when a particle diffusing inside a Euclidean domain $\Omega \subset \mathbb{R}^d$ with reflecting boundary $\partial\Omega$ searches for a small hole $\Gamma$ located on that boundary [86, 87]. Many efforts were dedicated to estimating the MFPT $T$ to this hole [47, 49, 63, 65, 77, 88–97]. To get accurate estimates in general domains, one needs to employ elaborate mathematical tools such as conformal mapping (see [49, 87] and references therein) or matched asymptotic analysis (see [98] and Chap. 10 of this book for details). At the same time, boundary homogenization can still be useful even in this setting of a single small absorbing spot. For instance, one of the first results in this field is attributed to Lord Rayleigh who estimated $T \approx |\Omega|/(4aD)$ for a small circular hole of radius $a$ on the reflecting sphere of volume $|\Omega|$. Curiously, this leading-order result can be alternatively deduced by replacing the small absorbing spot on the reflecting sphere by an effective homogeneous partially absorbing sphere with the reactivity $\kappa_{\rm BP}$ given by the Berg–Purcell's formula (11.35) with $N = 1$. In this homogeneous setting, the MFPT can be easily found as

$$\overline{T} = \frac{R^2}{15D} + \frac{R}{3\kappa_{\rm BP}} \tag{11.55}$$

after averaging over the starting point. Substituting $\kappa_{\rm BP}$ from Eq. (11.35), one notices that the second term in Eq. (11.55) is equal to $|\Omega|/(4aD)$ and provides the dominant contribution if $a \ll R$. In other words, despite highly non-uniform distribution of absorbing spots, boundary homogenization still yields the correct result in the leading order. In turn, more advanced techniques are needed for getting next-order corrections or analyzing the dependence on the starting point (see Chap. 10 for details).

In this section, we briefly describe another approximation that allows to deal with non-uniform distributions of absorbing spots. The so-called constant-flux approximation (also known as self-consistent approximation) was proposed by Shoup et al. to calculate the Smoluchowski-type constant for reactions with a small center situated on the otherwise impenetrable surface of a spherical domain [99] (see also earlier works by Keller and Stein [72]). Let us formulate it in a more general setting of diffusion in an unbounded domain $\Omega$, whose reflecting boundary $\partial\Omega$ contains partially absorbing spots $\Gamma$ characterized by reactivity $\kappa_0$. As previously, the steady-state concentration $C$ obeys the Laplace equation, $\Delta C = 0$, with a constant concentration $C_0$ at infinity, and mixed boundary conditions:

$$-D\partial_n C = \kappa_0 C \quad \text{on } \Gamma, \tag{11.56a}$$

$$-D\partial_n C = 0 \quad \text{on } \partial\Omega\backslash\Gamma. \tag{11.56b}$$

The reactivity $\kappa_0$ can either represent finite reaction probability on the spots (see Chap. 4 for details), or result from the previous step of boundary homogenization on $\Gamma$. Note that the limit $\kappa_0 = \infty$ corresponds to the absorbing spots discussed in Sect. 11.3. An exact solution of this boundary value problem is difficult even for simple domains such as a disk or a sphere. In particular, even for planar domains, conformal mapping is not much useful for finite $\kappa_0$ because such a transformation will make the reactivity dependent on the boundary point. The basic idea of the constant-flux approximation consists in replacing the right-hand side of Robin boundary condition (11.56a) by a constant flux density $j_0$. In other words, one aims at approximating the solution $C$ of the original problem by the solution $\hat{C}$ of the modified problem, $\Delta\hat{C} = 0$, with inhomogeneous Neumann boundary condition:

$$-D\partial_n \hat{C} = j_0 \, \mathbb{I}_\Gamma(\boldsymbol{x}) \quad \text{on } \partial\Omega, \tag{11.57}$$

where $\mathbb{I}_\Gamma(\boldsymbol{x})$ is the indicator function of the spots: $\mathbb{I}_\Gamma(\boldsymbol{x}) = 1$ for $\boldsymbol{x} \in \Gamma$, and 0 otherwise. The unknown parameter $j_0$ can be fixed by imposing the self-consistent condition that the Robin boundary condition (11.56a) is satisfied *on average* on $\Gamma$:

$$\kappa_0 \int\limits_\Gamma \hat{C}(\boldsymbol{x}) d\boldsymbol{x} = \int\limits_\Gamma (-D\partial_n \hat{C}) d\boldsymbol{x} = j_0 |\Gamma|, \tag{11.58}$$

where $|\Gamma|$ is the surface area of the spots $\Gamma$. Since $\hat{C}(\boldsymbol{x})$ implicitly depends on $j_0$, this equation can be used to determine $j_0$ (in the limit $\kappa_0 = \infty$, this equation is reduced to $\int_\Gamma \hat{C}(\boldsymbol{x}) d\boldsymbol{x} = 0$, which still determines $j_0$). Even though finding the solution $\hat{C}$ of the modified problem is in general not elementary, it is much simpler than solving the original problem with mixed boundary conditions (11.56).

Since the diffusive flux density of the exact solution, $-D\partial_n C$, is not constant, this approximation fails in a vicinity of the absorbing spot, in the same way as the far-field approximation (11.3) failed near the boundary. However, the approximate solution $\hat{C}$ turns out to be accurate far from the absorbing spots. In particular, the approximation $J = j_0|\Gamma|$ for the diffusive flux is very accurate in many settings. In this light, the constant-flux approximation can be seen as an alternative way of boundary homogenization discussed in Sects. 11.2 and 11.3. The major advantage of the constant-flux approximation is that the diffusive flux density is not uniformly distributed over the boundary $\partial\Omega$, as it was for homogenized boundaries. The accuracy of two methods was compared for absorbing stripes on a cylinder [52].

The original constant-flux approximation by Shoup et al. was later extended and applied to other problems such chemical kinetics of active colloidal particles [100], MFPT in cylindrical and spherical domains [101, 102], and the whole distribution of first-passage times [103–105]. For instance, for diffusion between two capped coaxial cylinders of height $H$ and radii $\rho$ and $R$ (Fig. 11.3e), the MFPT to an absorbing stripe located on the inner cylinder was found in [102]. When the starting point is uniformly distributed, the constant-flux approximation yields

$$T = \frac{H(R^2 - \rho^2)}{\pi D\rho} f, \qquad f = \sum_{n=1}^{\infty} \frac{G_n}{n} \left(\frac{\sin(\pi\sigma n)}{\pi\sigma n}\right)^2, \tag{11.59}$$

with

$$G_n = -\frac{I_1(\pi nR/H)K_0(\pi n\rho/H) + K_1(\pi nR/H)I_0(\pi n\rho/H)}{I_1(\pi nR/H)K_1(\pi n\rho/H) - K_1(\pi nR/H)I_1(\pi n\rho/H)}, \tag{11.60}$$

where $I_\nu(z)$ and $K_\nu(z)$ are the modified Bessel functions of the first and second kind. In turn, using Eq. (11.9) for the trapping parameter $\kappa$ (with the correction factor 1/2 due the perpendicular orientation of stripes, see [46]), one gets a much simpler approximation $f \approx -\ln(\sin(\pi\sigma/2))$. While the simplicity of this expression is advantageous for getting rapid estimates, the constant-flux approximation in Eq. (11.59) is more accurate [52].

Another example of the fruitful use of the constant-flux approximation is the computation of the steady-state reaction rate of diffusion-controlled reactions in sheets [106]. In some applications, the absorbing spot is located on one side of a flat layer (a sheet) between two parallel reflecting boundaries and can be accessed from the lateral sides (Fig. 11.3d). For instance, one can think of a capped cylinder $\{(x, y, z) \in \mathbb{R}^3 \ : \ x^2 + y^2 < R^2,\ 0 < z < H\}$ of height $H$ and radius $R$, with a constant concentration imposed on the cylinder wall and an absorbing disk of radius $\rho$ lying on the bottom. A semi-analytical solution of the Laplace equation, its constant-flux approximation and asymptotic behavior were derived in [106]. In particular, the constant-flux approximation yielded the following asymptotic behavior

$$\frac{J}{J_0} \approx \frac{\pi}{2}\left(\frac{16}{3\pi} + \frac{\rho}{H}\ln\frac{R}{2H}\right)^{-1} \qquad (\rho \ll H \ll R), \tag{11.61}$$

where $J_0 = 4\rho D C_0$ is the diffusive flux on the half-disk in the upper half-space (with $4\rho$ being the half of the capacity of the disk). This ratio determines the offset $\lambda$ and the trapping parameter $\kappa$. Other asymptotic limits were discussed in [106].

## 11.5 Diffusion Inside Channels

Finding buried binding sites by diffusing particles can be seen as another variant of target search problems [58, 59, 107]. In this problem, the binding chemical sites (usually modeled as a finite domain with absorbing boundary) are at the bottom of the deep pits made inside a reflecting porous matrix or at the bottom of a deep reflecting canopy (Fig. 11.6a). The wall of the pit creates the geometrical constraint (an entropy barrier) preventing particles to access the absorbing bottom. This setting is an extension of the example of a semi-circular canopy shown in Fig. 11.2c. In this section, we briefly discuss a powerful approximation via the Fick–Jacobs equation that allows one to reduce the original 2D or 3D problems to an effective one-dimensional problem that can be solved explicitly [108–114]. We start with the planar setting and then briefly mention the 3D case, which is very similar.

**Offset and Trapping Parameter of a Pit**

Let us consider a periodic array of 2D deep pits with reflecting walls and the absorbing bottom (Fig. 11.6b, c). As previously, one can focus on a single periodic cell of width $W$. The Fick–Jacobs equation inside the pit has the form [108–112]

$$\frac{d}{dy}\left[D(y)w(y)\frac{d}{dy}\frac{c(y)}{w(y)}\right] = 0, \quad 0 < y < h_0, \tag{11.62}$$

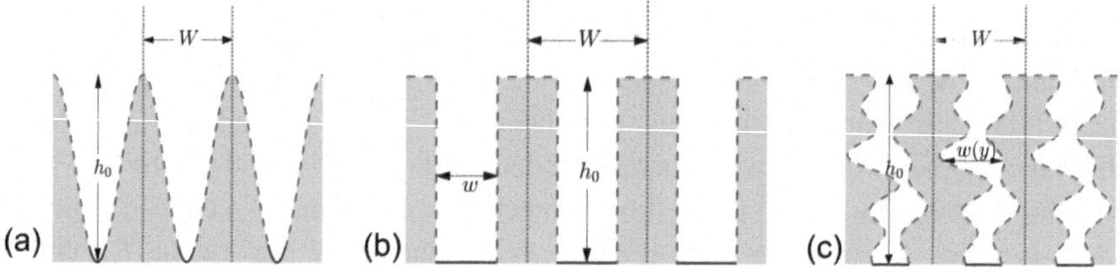


**Fig. 11.6** Arrays of long pits with buried binding sites at the bottom (in blue). **a** Sinusoidal profile. **b** Rectangular profile of width $w$ and height $h_0$, on the period $W$, with a flat absorbing site at the bottom. **c** Smoothly varying profile $w(y)$ of a pit of height $h_0$, with a flat absorbing site at the bottom. Light gray indicates inaccessible solid matrix. The particles arrive from a flat source at $y = H$ located far above the entrances of the pits. Vertical dotted lines delimit one periodic cell

where $w(y)$ is the width of the pit at height $y$, $c(y)$ is the effective one-dimensional concentration, and $D(y)$ is the effective diffusivity determined by $w(y)$. Among numerous approximations for $D(y)$, the model of Kalinay and Percus,

$$D(y) = D\frac{\operatorname{atan}(\frac{1}{2}w'(y))}{\frac{1}{2}w'(y)}, \tag{11.63}$$

is often employed [108–110]. For a smooth profile ($w'(y) \ll 1$), we assume $D(y) = D = const$ in a first approximation [114]. The absorbing bottom of the pit implies the boundary condition $c = 0$, whereas we set $c = c_0 = const$ at some $y = H \gg W$ at the far field. The offset $\lambda$, which is determined by the far-field behavior (11.3) at large $y$, is then evaluated from the explicit solution of Eq. (11.62) as

$$\lambda = h_0 - W\int_0^{h_0} \frac{dy}{w(y)}. \tag{11.64}$$

Setting an effective radiative boundary at $y_0 = h_0$, one gets the trapping parameter

$$\kappa = \frac{D}{W}\left(\int_0^{h_0} \frac{dy}{w(y)}\right)^{-1}. \tag{11.65}$$

If there are discontinuities in the profile $w(y)$ (e.g., near the pit entrance, $y = h_0$, as illustrated in Fig. 11.6b, c), then the Fick–Jacobs equation is not applicable near these points. In this case, one first employs boundary homogenization to estimate the trapping parameter $\kappa$ in the 2D periodic cell, and then imposes a transmission boundary condition for the one-dimensional concentration $c(y)$ (see [108–112] and references therein). This boundary condition results in the additional term in the formula (11.65) for $\kappa$, see [115]. The physical nature of this term (the so-called access resistance) is a delay in particle diffusion due to the time taken by the particle to find the pit entrance.

The above analysis can be carried on in three dimensions for a lattice of long pits by replacing the pit width $w(y)$ by its cross-sectional area $\mathcal{A}(y)$ that leads to a general formula

$$\lambda = h_0 - \mathcal{A}_0\int_0^{h_0} \frac{dy}{\mathcal{A}(y)}, \tag{11.66}$$

where $\mathcal{A}_0$ is the area of the periodic cell. In this way, the trapping parameter $\kappa$ can be analytically derived for a variety of pit profiles (cone, spheroid, spindle, see Refs. [78, 80, 81] for details). We note that the change of the pit profile can modify the functional form of $\kappa(\sigma)$ [26]. Similar to Eq. (11.53), the discontinuity in the pit profile near the opening leads to an additional contribution to $\kappa$ [78, 80, 81].

**Mean First-Passage Time**

In a similar way, the Fick–Jacobs equation can be applied to estimate the MFPT to a small target inside a 2D elongated domain with reflecting boundaries [116].

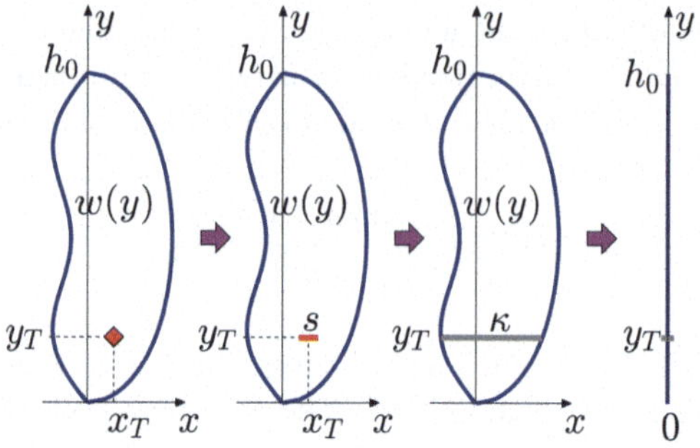


**Fig. 11.7** Three-step transformation of the original MFPT problem inside an elongated planar domain (left) into an effectively one-dimensional problem with semi-permeable semi-absorbing boundary condition (right). A small target of arbitrary shape (here, a red rhombus) is replaced by an equivalent interval (in red) of length $s$ of the same log-capacity, which is then replaced by a semi-permeable semi-absorbing boundary (in gray) with reactivity $\kappa$

The profile $w(y)$ of the domain is assumed to be smooth, slowly changing, but otherwise general, whereas the target is small but of an arbitrary shape. To treat the original 2D problem as one-dimensional (Fig. 11.7), we first replace the absorbing target by an equivalent absorbing interval (i.e., the interval of the same log-capacity). In turn, this absorbing interval is then replaced by a partially absorbing horizontal segment of length $w(y_T)$ and trapping parameter $\kappa$. The latter can be estimated from Eq. (11.12) by thinking of the elongated domain as a stripe of width $w(y_T)$ with reflecting boundaries. We stress that the value of the trapping parameter $\kappa$ depends on the position $y_T$ of the target that enables formulating the MFPT within the effective medium approximation. After that, the Poisson equation for MFPT in the elongated domain reduces to 1D equation that resembles the Fick–Jacobs equation:

$$\frac{D}{w(y)}\frac{d}{dy}\left[w(y)\frac{d\mathcal{T}}{dy}\right] = -1, \tag{11.67}$$

subject to the effective semi-permeable boundary condition at the target location $y_T$:

$$\mathcal{T}(y_T - 0) = \mathcal{T}(y_T + 0), \tag{11.68}$$

$$D\left[\frac{d\mathcal{T}}{dy}(y_T + 0) - \frac{d\mathcal{T}}{dy}(y_T - 0)\right] = \kappa\,\mathcal{T}(y_T). \tag{11.69}$$

This problem was then solved analytically for a variety of profiles $w(y)$ and favorably validated by numerical simulations [116].

This framework has been extended to find the MFPT in 3D elongated domains with axial symmetry [117]. Here, Eq. (11.67) remains the same, with $w(y)$ being replaced by a cross-sectional area $\mathcal{A}(y)$ of the domain. The target is modeled by an equivalent disk and the trapping parameter $\kappa$ is given by Eq. (11.30) via boundary homogenization. We stress that the ideas of boundary homogenization can provide some analytical tools not only for calculating the MFPT but also for evaluating the FPT statistics, in particular, in the long-time limt [118].

## 11.6 Conclusion

Laplacian transport near complex boundaries is relevant in many disciplines, including chemical physics, electrostatics, hydrodynamics, and heat transfer. Despite the long history of intensive research, this field remains very active, with numerous recent advances and applications. In this chapter, we reviewed several approximations that yielded a simplified yet accurate description of this phenomenon. In particular, we discussed the effective medium theory, boundary homogenization, constant-flux approximation and perturbative analysis that allow one to express the diffusive flux and other integral properties of the system in terms of its geometric and absorption parameters. For instance, we described the procedure of boundary homogenization when the complex boundary is replaced by an equivalent flat boundary with Robin boundary condition. The fraction of absorbing parts and the amplitude of boundary profile variations do not need to be small, while the boundary can be rough and spiky. The proposed framework may be useful for a rapid modeling and engineering prototyping before proceeding with extensive computer simulations.

At the same time, many other aspects of Laplacian transport remain unexplored in this chapter: (i) the narrow escape problem and the related matched asymptotic analysis (see Chap. 10); (ii) diffusive search in porous media formed by absorbing/reflecting spherical beads and efficient semi-analytical methods [119–121]; (iii) Laplacian transport towards fractal boundaries [17, 18, 20, 21, 122–125] and its electrochemical measurements [126–132]; (iv) the impact of partial reactivity of complex boundaries, its approximate treatments (including land surveyor approximation [19, 38, 133]), and the spectral theory based on the Dirichlet-to-Neumann operator [22, 134–137] (see Chap. 4); and (v) time-dependent (non-stationary) diffusion in complex media, the distribution of first-passage times, and efficient search strategies [14, 32, 35, 36, 138] (see Chap. 3). We believe that further applications of boundary homogenization and other approximate tools can help uncovering the crucial role of complex geometry and heterogeneity in Laplacian transport and target search problems.

**Acknowledgements** A.T.S thanks Alexander M. Berezhkovskii and Paul A. Martin for useful discussions.

## References

1. D. ben-Avraham, S. Havlin, *Diffusion and Reactions in Fractals and Disordered Systems.* (Cambridge University Press: Cambridge, UK, 2010)

2. Z. Schuss, *Brownian Dynamics at Boundaries and Interfaces in Physics, Chemistry and Biology* (Springer, New York, USA, 2013)
3. A. Bejan, *Shape and Structure, from Engineering to Nature* (Cambridge University Press, Cambridge, UK, 2000)
4. S. Rice, *Diffusion-Limited Reactions* (Elsevier, Amsterdam, the Netherlands, 1985)
5. B.D. Hughes, *Random Walks and Random Environments* (Clarendon Press, Oxford, UK, 1995)
6. P.L. Krapivsky, S. Redner, E. Ben-Naim, *A Kinetic View of Statistical Physics* (Cambridge University Press, New York, USA, 2010)
7. P.R. Nair, M.A. Alam, Dimensionally frustrated diffusion towards fractal adsorber. Phys. Rev. Lett. **99**, 256101 (2007)
8. H. Nepf, M. Ghisalberti, B. White, E. Murphy, Retention time and dispersion associated with submerged aquatic canopies. Water Res. Res. **43**, W04422 (2007)
9. S.L. Edburg, G. Allwine, B. Lamb, D. Stock, H. Thistle, H. Peterson, B. Strom, A simple model to predict scalar dispersion within a successively thinned loblolly pine canopy. J. App. Meteor. Climatol. **49**, 1913–1926 (2010)
10. G.Z. Ramon, E.M.V. Hoek, Transport through composite membranes, part 2: impacts of roughness on permeability and fouling. J. Membr. Sci. **425–426**, 141–148 (2013)
11. M. Felici, M. Filoche, B. Sapoval, Renormalized random walk study of oxygen absorption in the human lung. Phys. Rev. Lett. **92**, 068101 (2004)
12. D.S. Grebenkov, M. Filoche, B. Sapoval, M. Felici, Diffusion-reaction in branched structures: theory and application to the lung acinus. Phys. Rev. Lett. **94**, 050602 (2005)
13. A.S. Serov, C. Salafia, D.S. Grebenkov, M. Filoche, The role of morphology in mathematical models of placental gas exchange. J. Appl. Physiol. **120**, 17–28 (2016)
14. K. Lindenberg, R. Metzler, G. Oshanin (eds.), *Chemical Kinetics: Beyond the Textbook* (World Scientific Press, New Jersey, USA, 2019)
15. D.A. Lauffenburger, J.J. Linderman, *Receptors: Models for Binding, Trafficking, and Signaling* (Oxford University Press, New York, USA, 1993)
16. M. Reva, D.A. DiGregorio, D.S. Grebenkov, A first-passage approach to diffusion-influenced reversible binding: insights into nanoscale signaling at the presynapse. Sci. Rep. **11**, 5377 (2021)
17. M. Filoche, D.S. Grebenkov, J.S. Andrade Jr., B. Sapoval, Passivation of irregular surfaces accessed by diffusion. Proc. Nat. Acad. Sci. USA **105**, 7636 (2008)
18. T.A. Witten Jr., L.M. Sander, Diffusion-limited aggregation, a kinetic critical phenomenon. Phys. Rev. Lett. **47**, 1400 (1981)
19. B. Sapoval, General formulation of Laplacian transfer across irregular surfaces. Phys. Rev. Lett. **73**, 3314 (1994)
20. M. Filoche, B. Sapoval, Transfer across random versus deterministic fractal interfaces. Phys. Rev. Lett. **84**, 5776 (2000)
21. P. Levitz, D.S. Grebenkov, M. Zinsmeister, K.M. Kolwankar, B. Sapoval, Brownian flights over a fractal nest and first passage statistics on irregular surfaces. Phys. Rev. Lett. **96**, 180601 (2006)
22. D.S. Grebenkov, M. Filoche, B. Sapoval, Mathematical basis for a general theory of Laplacian transport towards irregular interfaces. Phys. Rev. E **73**, 021103 (2006)
23. M.Z. Bazant, Exact solutions and physical analogies for unidirectional flows. Phys. Rev. Fluids **1**, 024001 (2016)
24. J.P. Rothstein, Slip on superhydrophobic surfaces. Ann. Rev. Fluid Mech. **42**, 89 (2010)
25. D.G. Crowdy, Frictional slip lengths for unidirectional superhydrophobic grooved surfaces. Phys. Fluids **23**, 072001 (2011)
26. P.A. Martin, A.T. Skvortsov, On blockage coefficients: flow past a body in a pipe. Proc. R. Soc. A **478**, 20210677 (2022)
27. D.P. Hewett, I.J. Hewitt, Homogenized boundary conditions and resonance effects in Faraday cages. Proc. R. Soc. A **472**, 20160062 (2016)

28. P.A. Martin, A.T. Skvortsov, Steady state diffusion in tubular structures: assessment of one-dimensional models. Eur. J. Appl. Math. **34**, 262–279 (2023)
29. M.M. Fyrillas, C. Pozrikidis, Conductive heat transport across rough surfaces and interfaces between two conforming media. Int. J. Heat Mass Transf. **44**, 1789–1801 (2001)
30. M.G. Blyth, C. Pozrikidis, Heat conduction across irregular and fractal-like surfaces. Int. J. Heat Mass Transf. **46**, 1329–1339 (2003)
31. S. Redner, *A Guide to First-Passage Processes*. (Cambridge University Press, 2001)
32. R. Metzler, G. Oshanin, S. Redner (eds.), *First-Passage Phenomena and Their Applications* (World Scientific Press, Singapore, 2014)
33. L.D. Stone, J.O. Royset, A.R. Washburn, *Optimal Search for Moving Targets*. (Springer International Publishing, 2016)
34. J. Masoliver, *Random Processes: First-Passage and Escape*. (World Scientific, 2018)
35. O. Bénichou, C. Loverdo, M. Moreau, R. Voituriez, Intermittent search strategies. Rev. Mod. Phys. **83**, 81–130 (2011)
36. O. Bénichou, R. Voituriez, From first-passage times of random walks in confinement to geometry-controlled kinetics. Phys. Rep. **539**, 225–284 (2014)
37. P.C. Bressloff, J. Newby, Stochastic models of intracellular transport. Rev. Mod. Phys. **85**, 135–196 (2013)
38. B. Sapoval, transport across irregular interfaces: fractal electrodes, membranes and catalysts, in *Fractals and Disordered Systems*, ed. by A. Bunde, S. Havlin. (Springer, Berlin, 1996), pp. 233–261
39. F.C. Collins, G.E. Kimball, Diffusion-controlled reaction rates. J. Colloid Sci. **4**, 425–437 (1949)
40. D. Vandembroucq, S. Roux, Conformal mapping on rough boundaries I: applications to harmonic problems. Phys. Rev. E **55**, 6171–6185 (1997)
41. A.T. Skvortsov, A. Walker, Trapping of diffusive particles by rough absorbing surfaces: boundary smoothing approach. Phys. Rev. E **90**, 023202 (2014)
42. A.T. Skvortsov, A.M. Berezhkovskii, L. Dagdug, Steady-state flux of diffusing particles to a rough boundary formed by absorbing spikes periodically protruding from a reflecting base. J. Chem. Phys. **150**, 194109 (2019)
43. P.A. Martin, On acoustic and electric Faraday cages. Proc. R. Soc. A **470**, 20140344 (2014)
44. J.R. Philip, Flows satisfying mixed no-slip and no-shear conditions. Z. Angew. Math. Phys. **23**, 353–372 (1972)
45. B.. Ya.. Moizhes, Averaged electrostatic boundary conditions for metallic meshes. Zh. Tech. Fiz. **25**, 167–176 (1955). (in Russian)
46. E. Lauga, H.A. Stone, Effective slip in pressure-driven Stokes flow. J. Fluid Mech. **489**, 55–77 (2003)
47. A. Singer, Z. Schuss, D. Holcman, Narrow escape, part II: the circular disk. J. Stat. Phys. **122**, 465 (2006)
48. J.S. Marshall, Analytical solutions for an escape problem in a disc with an arbitrary distribution of exit holes along its boundary. J. Stat. Phys. **165**, 920–952 (2016)
49. D.S. Grebenkov, Universal formula for the mean first passage time in planar domains. Phys. Rev. Lett. **117**, 260201 (2016)
50. A.T. Skvortsov, Mean first passage time for a particle diffusing on a disk with two absorbing traps at the boundary. Phys. Rev. E **102**, 012123 (2020)
51. L. Dagdug, A.M. Berezhkovskii, A.T. Skvortsov, Trapping of diffusing particles by striped cylindrical surfaces. Boundary homogenization approach. J. Chem. Phys. **142**, 234902 (2015)
52. D.S. Grebenkov, R. Metzler, G. Oshanin, L. Dagdug, A.M. Berezhkovskii, A.T. Skvortsov, Trapping of diffusing particles by periodic absorbing rings on a cylindrical tube. J. Chem. Phys. **150**, 206101 (2019)
53. D.G. Crowdy, Slip length for longitudinal shear flow over a dilute periodic mattress of protruding bubbles. Phys. Fluids **22**, 121703 (2010)
54. D.G. Crowdy, Analytical formulae for longitudinal slip lengths over unidirectional superhydrophobic surfaces with curved menisci. J. Fluid Mech. **791**, R7 (2016)

55. B. Sapoval, M. Filoche, K. Karamanos, R. Brizzi, Can one hear the shape of an electrode. I. Numerical study of the active zone in Laplacian transfer. Eur. Phys. J B **9**, 739–753 (1999)
56. S. Richardson, A model for the boundary condition of a porous material. Part 2. J. Fluid Mech. **49**, 327–336 (1971)
57. D.W. Bechert, M. Bartenwerfer, The viscous flow on surfaces with longitudinal ribs. Fluid Mech. **206**, 105–129 (1989)
58. L. Dagdug, A.M. Berezhkovskii, S.M. Bezrukov, G.H. Weiss, Diffusion-controlled reactions with a binding site hidden in a channel. J. Chem. Phys. **118**, 2367–2373 (2003)
59. A.M. Berezhkovskii, A. Szabo, H.-X. Zhou, Diffusion-influenced ligand binding to buried sites in macromolecules and transmembrane channels. J. Chem. Phys. **135**, 075103 (2011)
60. H.C. Berg, E.M. Purcell, Physics of chemoreception. Biophys. J. **20**, 193–219 (1977)
61. H.C. Berg, *Random Walks in Biology*. (Princeton University Press, 1993)
62. R. Zwanzig, Diffusion controlled ligand binding to spheres partially covered by receptors: an effective medium treatment. Proc. Nat. Acad. Sci. U.S.A. **87**, 5856–5857 (1990)
63. L. Dagdug, M.-V. Vazquez, A.M. Berezhkovskii, VYu. Zitserman, Boundary homogenization for a sphere with an absorbing cap of arbitrary size. J. Chem. Phys. **145**, 214101 (2016)
64. G. Handy, S.D. Lawley, Revising Berg-Purcell for finite receptor kinetics. Biophys. J. **120**, 2237–2248 (2021)
65. A.E. Lindsay, A.J. Bernoff, M.J. Ward, First passage statistics for the capture of a Brownian particle by a structured spherical target with multiple surface traps. SIAM Multiscale Model. Simul. **15**, 74–109 (2017)
66. W.R. Smythe, *Static and Dynamic Electricity*, 3rd Ed. (McGraw-Hill, 1958)
67. Y.L. Chow, M.M. Yovanovich, The shape factor of capacitance of a conductor. J. Appl. Phys. **53**, 8470–8475 (1982)
68. O.K. Dudko, A.M. Berezhkovskii, G.H. Weiss, Rate constant for diffusion-influenced ligand binding to receptors of arbitrary shape on a cell surface. J. Chem. Phys. **121**, 1562–1565 (2004)
69. A.M. Berezhkovskii, A.V. Barzykin, Simple formulas for the trapping rate by nonspherical absorber and capacitance of nonspherical conductor. J. Chem. Phys. **126**, 106102 (2007)
70. L. Hill, Effect of rotation on the diffusion-controlled rate of ligand-protein association. Proc. Natl. Acad. Sci. U.S.A. **72**, 4918 (1975)
71. F.G. Leppington, H. Levine, Some axially symmetric potential problems. Proc. Edinburgh Math. Soc. **18**, 55 (1972)
72. K.H. Keller, T.R. Stein, A two-dimensional analysis of porous membrane transport. Math. Biosci. **1**, 421–437 (1967)
73. A.M. Berezhkovskii, Yu.A. Makhnovskii, M.I. Monine, VYu. Zitserman, S.Y. Shvartsman, Boundary homogenization for trapping by patchy surfaces. J. Chem. Phys. **121**, 11390 (2004)
74. A.M. Berezhkovskii, M.I. Monine, C.B. Muratov, S.Y. Shvartsman, Homogenization of boundary conditions for surfaces with regular arrays of traps. J. Chem. Phys. **124**, 036103 (2006)
75. C. B. Muratov S.Y. Shvartsman, Boundary homogenization for periodic arrays of absorbers. Multiscale Model. Simul. **7**, 44–61 (2008)
76. V.A. Fock, A theoretical investigation of the acoustical conductivity of a circular aperture in a wall put across a tube. Dokl. Acad. Nauk SSSR **31**, 875 (1941)
77. A.J. Bernoff, A.E. Lindsay, D.D. Schmidt, Boundary homogenization and capture time distributions of semipermeable membranes with periodic patterns of reactive sites. SIAM Multiscale Model. Simul. **16**, 1411–1447 (2018)
78. A.T. Skvortsov, L. Dagdug, A.M. Berezhkovskii, S.M. Bezrukov, Blockage coefficient of cylindrical blocker and diffusion resistance of membrane channels. Phys. Fluids **35**, 011702 (2023)
79. P.A. Martin, A.T. Skvortsov, Scattering by a sphere in a tube, and related problems. J. Acoust. Soc. Am. **148**, 191 (2020)
80. L. Dagdug, A.T. Skvortsov, A.M. Berezhkovskii, S.M. Bezrukov, Blocker effect on diffusion resistance of a membrane channel. Dependence on the blocker geometry. J. Phys. Chem. B **126**, 6016–6025 (2022)

81. A.T. Skvortsov, L. Dagdug, E.F. Hilder, A.M. Berezhkovskii, S.M. Bezrukov, Permeability and diffusion resistance of porous membranes: analytical theory and its numerical test. J. Chem. Phys. **158**, 054114 (2023)
82. A.T. Skvortsov, L. Dagdug, A.M. Berezhkovskii, I.R. MacGillivray, S.M. Bezrukov, Evaluating diffusion resistance of a constriction in a membrane channel by the method of boundary homogenization. Phys. Rev. E **103**, 012408 (2021)
83. D.S. Grebenkov, A.T. Skvortsov, Diffusion toward a nanoforest of absorbing pillars. J. Chem. Phys. **157**, 244102 (2022)
84. A.T. Skvortsov, D.S. Grebenkov, L. Chan, A. Ooi, Slip length for a viscous flow over spiky surfaces. Europhys. Lett. **143**, 63001 (2023)
85. F. Piazza, D.S. Grebenkov, Diffusion-controlled reaction rate on non-spherical partially absorbing axisymmetric surfaces. Phys. Chem. Chem. Phys. **21**, 25896–25906 (2019)
86. D. Holcman, Z. Schuss, Control of flux by narrow passages and hidden targets in cellular biology. Phys. Progr. Rep. **76**, 074601 (2013)
87. D. Holcman, Z. Schuss, The narrow escape problem. SIAM Rev. **56**, 213–257 (2014)
88. I.V. Grigoriev, Y.A. Makhnovskii, A.M. Berezhkovskii, V.Y. Zitserman, Kinetics of escape through a small hole. J. Chem. Phys. **116**, 9574 (2002)
89. A. Singer, Z. Schuss, D. Holcman, R.S. Eisenberg, Narrow escape, part I. J. Stat. Phys. **122**, 437–463 (2006)
90. A. Singer, Z. Schuss, D. Holcman, Narrow escape, part III: non-smooth domains and Riemann surfaces. J. Stat. Phys. **122**, 491 (2006)
91. Z. Schuss, A. Singer, D. Holcman, The narrow escape problem for diffusion in cellular microdomains. Proc. Nat. Acad. Sci. USA **104**, 16098–16103 (2007)
92. O. Bénichou, R. Voituriez, Narrow-escape time problem: time needed for a particle to exit a confining domain through a small window. Phys. Rev. Lett. **100**, 168105 (2008)
93. S. Pillay, M.J. Ward, A. Peirce, T. Kolokolnikov, An asymptotic analysis of the mean first passage time for narrow escape problems: part I: two-dimensional domains. Multiscale Model. Simul. **8**, 803–835 (2010)
94. A.F. Cheviakov, M.J. Ward, R. Straube, An asymptotic analysis of the mean first passage time for narrow escape problems: part II: the sphere. Multiscale Model. Simul. **8**, 836–870 (2010)
95. A.F. Cheviakov, A.S. Reimer, M.J. Ward, Mathematical modeling and numerical computation of narrow escape problems. Phys. Rev. E **85**, 021131 (2012)
96. S.A. Isaacson, A.J. Mauro, J. Newby, Uniform asymptotic approximation of diffusion to a small target: generalized reaction models. Phys. Rev. E **94**, 042414 (2016)
97. A.J. Bernoff, A.E. Lindsay, Numerical approximation of diffusive capture rates by planar and spherical surfaces with absorbing pores. SIAM J. Appl. Math. **78**, 266–290 (2018)
98. M.J. Ward, J.B. Keller, Strong localized perturbations of eigenvalue problems. SIAM J. Appl. Math. **53**, 770–798 (1993)
99. D. Shoup, G. Lipari, A. Szabo, Diffusion-controlled bimolecular reaction rates. The effect of rotational diffusion and orientation constraints. Biophys. J. **36**, 697 (1981)
100. G. Oshanin, M.N. Popescu, S. Dietrich, Active colloids in the context of chemical kinetics. J. Phys. A: Math. Theor. **50**, 134001 (2017)
101. D.S. Grebenkov, G. Oshanin, Diffusive escape through a narrow opening: new insights into a classic problem. Phys. Chem. Chem. Phys. **19**, 2723–2739 (2017)
102. D.S. Grebenkov, R. Metzler, G. Oshanin, Effects of the target aspect ratio and intrinsic reactivity onto diffusive search in bounded domains. New J. Phys. **19**, 103025 (2017)
103. D.S. Grebenkov, R. Metzler, G. Oshanin, Towards a full quantitative description of single-molecule reaction kinetics in biological cells. Phys. Chem. Chem. Phys. **20**, 16393–16401 (2018)
104. D.S. Grebenkov, R. Metzler, G. Oshanin, Full distribution of first exit times in the narrow escape problem. New J. Phys. **21**, 122001 (2019)
105. D.S. Grebenkov, R. Metzler, G. Oshanin, Distribution of first-reaction times with target regions on boundaries of shell-like domains. New J. Phys. **23**, 123049 (2021)

106. D.S. Grebenkov, D. Krapf, Steady-state reaction rate of diffusion-controlled reactions in sheets. J. Chem. Phys. **149**, 064117 (2018)
107. R. Samson, J.M. Deutch, Diffusion-controlled reaction rate to a buried active site. J. Chem. Phys. **68**, 285–290 (1978)
108. P. Kalinay, J.K. Percus, Extended Fick-Jacobs equation: variational approach. Phys. Rev. E **72**, 061203 (2005)
109. P. Kalinay, J.K. Percus, Corrections to the Fick-Jacobs equation. Phys. Rev. E **74**, 041203 (2006)
110. P. Kalinay, J.K. Percus, Approximations of the generalized Fick-Jacobs equation. Phys. Rev. E **78**, 021103 (2008)
111. M. Mangeat, T. Guérin, D.S. Dean, Dispersion in two dimensional channels – the Fick-Jacobs approximation revisited. J. Stat. Mech. 123205 (2017)
112. M. Mangeat, T. Guérin, D.S. Dean, Dispersion in two-dimensional periodic channels with discontinuous profiles. J. Chem. Phys. **149**, 124105 (2018)
113. A.M. Berezhkovskii, M.A. Pustovoit, S.M. Bezrukov, Diffusion in a tube of varying cross section: Numerical study of reduction to effective one-dimensional description. J. Chem. Phys. **126**, 134706 (2007)
114. A.M. Berezhkovskii, L. Dagdug, S.M. Bezrukov, Range of applicability of modified Fick-Jacobs equation in two dimensions. J. Chem. Phys. **143**, 164102 (2015)
115. L. Dagdug, A.M. Berezhkovskii, VYu. Zitserman, S.M. Bezrukov, Trapping of particles diffusing in two dimensions by a hidden binding site. Phys. Rev. E **103**, 012135 (2021)
116. D.S. Grebenkov, A.T. Skvortsov, Mean first-passage time to a small absorbing target in an elongated planar domain. New J. Phys. **22**, 113024 (2020)
117. D.S. Grebenkov, A.T. Skvortsov, Mean first-passage time to a small absorbing target in three-dimensional elongated domains. Phys. Rev. E **105**, 054107 (2022)
118. D.S. Grebenkov, A.T. Skvortsov, Survival in a nanoforest of absorbing pillars. J. Phys. A: Math. Theor. **56**, 165002 (2023)
119. M. Galanti, D. Fanelli, S.D. Traytak, F. Piazza, Theory of diffusion-influenced reactions in complex geometries. Phys. Chem. Chem. Phys. **18**, 15950–15954 (2016)
120. D.S. Grebenkov, S.D. Traytak, Semi-analytical computation of Laplacian Green functions in three-dimensional domains with disconnected spherical boundaries. J. Comput. Phys. **379**, 91–117 (2019)
121. D.S. Grebenkov, Diffusion toward non-overlapping partially reactive spherical traps: fresh insights onto classic problems. J. Chem. Phys. **152**, 244108 (2020)
122. B.B. Mandelbrot, C.J.G. Evertsz, The potential distribution around growing fractal clusters. Nature **348**, 143 (1990)
123. D.S. Grebenkov, What makes a boundary less accessible. Phys. Rev. Lett. **95**, 200602 (2005)
124. J.S. Andrade Jr., A.D. Araújo, M. Filoche, B. Sapoval, Screening effects in flow through rough channels. Phys. Rev. Lett. **98**, 194101 (2007)
125. A. Rozanova-Pierrat, D.S. Grebenkov, B. Sapoval, Faster diffusion across an irregular boundary. Phys. Rev. Lett. **108**, 240602 (2012)
126. R. de Levie, The influence of surface roughness of solid electrodes on electrochemical measurements. Electrochim. Acta **10**, 113 (1965)
127. S.H. Liu, Fractal model for the ac response of a rough interface. Phys. Rev. Lett. **55**, 529 (1985)
128. T.C. Halsey, Frequency dependence of the double layer impedance at a rough surface. Phys. Rev. A **35**, 3512 (1987)
129. R. de Levie, Fractals and rough electrodes. J. Electroanal. Chem. **281**, 1 (1990)
130. T.C. Halsey, M. Leibig, The double layer impedance at a rough surface. Theoretical results. Ann. Phys. **219**, 109 (1992)
131. T. Pajkossy, Impedance of rough capacitive electrodes. J. Electroanal. Chem. **364**, 111 (1994)
132. E. Chassaing, B. Sapoval, Electrochemical impedance of blocking quasi-fractal 3D electrodes. J. Electrochem. Soc. **141**, 2711 (1994)

133. B. Sapoval, J.S. Andrade Jr., M. Filoche, Catalytic effectiveness of irregular interfaces and rough pores: the land surveyor approximation. Chem. Eng. Sci. **56**, 5011–5023 (2001)
134. D.S. Grebenkov, Scaling properties of the spread harmonic measures. Fractals **14**, 231–243 (2006)
135. D.S. Grebenkov, Analytical representations of the spread harmonic measure. Phys. Rev. E **91**, 052108 (2015)
136. D.S. Grebenkov, Spectral theory of imperfect diffusion-controlled reactions on heterogeneous catalytic surfaces. J. Chem. Phys. **151**, 104108 (2019)
137. D.S. Grebenkov, Paradigm shift in diffusion-mediated surface phenomena. Phys. Rev. Lett. **125**, 078102 (2020)
138. D.S. Grebenkov, R. Metzler, G. Oshanin, Strong defocusing of molecular reaction times results from an interplay of geometry and reaction control. Commun. Chem. **1**, 96 (2018)

# Chapter 12
# Competition of Many Searchers

**Sean D. Lawley**

**Abstract** First passage times (FPTs) are often used to estimate timescales in physical, chemical, and biological processes. FPTs generically describe the time it takes a random "searcher" to find a "target." In many systems, the important timescale is not the time it takes a single searcher to find a target, but rather the time it takes the fastest searcher out of many searchers to find a target. Such fastest FPTs or extreme FPTs result from many searchers competing to find the target and differ markedly from FPTs of single searchers. In this chapter, we review recent results on fastest FPTs. We show how fastest FPTs depend on the mode of stochastic search (including search by diffusion, subdiffusion, superdiffusion, and discrete jumps), the initial searcher distribution, and properties of the spatial domain.

## 12.1 Introduction

Many events in physical, chemical, and biological systems are initiated when a "searcher" finds a "target," which is called a first passage time (FPT) [1]. Depending on the system, the searcher could be, for instance, an ion, protein, cell, or predator, and the target could be a receptor, ligand, cell, or prey. The majority of prior work in this area studies the FPT of a given single searcher. However, the relevant timescale in many systems is the time it takes the fastest searcher out of many searchers to find a target, which is called a fastest or extreme FPT [2–20].

Concretely, let $\tau$ denote the FPT to a target $U_{\text{target}}$ of a single searcher whose stochastic position at time $t \geq 0$ is denoted by $X(t)$. Mathematically,

$$\tau := \inf\{t \geq 0 : X(t) \in U_{\text{target}}\}. \tag{12.1}$$

S. D. Lawley (✉)
Department of Mathematics, University of Utah, Salt Lake City, UT 84112, USA
e-mail: sean.lawley@utah.edu

 
D. Grebenkov et al. (eds.), *Target Search Problems*,
https://doi.org/10.1007/978-3-031-67802-8_12

Note that $U_{\text{target}}$ may be a disjoint union of multiple sets (i.e., the "target" $U_{\text{target}}$ may consist of "multiple targets"). If there are $N \geq 1$ independent and identically distributed (iid) searchers with respective FPTs $\tau_1, \ldots, \tau_N$, then the fastest FPT is

$$T_N := \min\{\tau_1, \ldots, \tau_N\}, \tag{12.2}$$

where $\tau_1, \ldots, \tau_N$ are $N$ iid realizations of $\tau$ in (12.1). More generally, if we set $T_{1,N} := T_N$, then the $k$th fastest FPT is

$$T_{k,N} := \min\left\{\{\tau_1, \ldots, \tau_N\} \setminus \cup_{j=1}^{k-1} \{T_{j,N}\}\right\}, \quad k \in \{1, \ldots, N\}. \tag{12.3}$$

In this chapter, we review recent results on the distribution and statistics of $T_N$ and $T_{k,N}$ for $1 \leq k \ll N$. We consider searchers which move by diffusion in Sect. 12.2, superdiffusion in Sect. 12.3, subdiffusion in Sect. 12.4, and jumps on a discrete network in Sect. 12.5. We conclude in Sect. 12.6 by discussing some closely related problems.

## 12.2 Diffusion

We now study fastest FPTs for diffusive searchers. We start with a simple one-dimensional example.

### 12.2.1 An Introductory Example: Diffusion in One Dimension

Suppose $X = \{X(t)\}_{t \geq 0}$ is a one-dimensional pure diffusion process with diffusivity $D > 0$. Let $\tau$ denote the FPT of $X$ to the origin (i.e., $\tau$ is in (12.1) with $U_{\text{target}} = 0$). Assuming $X(0) = L > 0$, the distribution of $\tau$ is given in terms of the error function,

$$\mathbb{P}(\tau > t) = \operatorname{erf}\Big(\frac{L}{\sqrt{4Dt}}\Big). \tag{12.4}$$

Now, the mean of any nonnegative random variable $Z \geq 0$ is given by the integral of its survival probability, $\mathbb{E}[Z] = \int_0^\infty \mathbb{P}(Z > z)\,\mathrm{d}z$. Hence, the mean time for a single searcher to reach the origin is infinite,

$$\mathbb{E}[\tau] = \int_0^\infty \mathbb{P}(\tau > t)\,\mathrm{d}t = \infty,$$

due to the slow large-time decay of $\mathbb{P}(\tau > t)$ in (12.4),

$$\mathbb{P}(\tau > t) \sim (L/\sqrt{\pi D})t^{1/2} \quad \text{as } t \to \infty, \tag{12.5}$$

where throughout this chapter, $f \sim g$ denotes $f/g \to 1$. Further,

$$\mathbb{E}[T_N] = \int_0^\infty [\mathbb{P}(\tau > t)]^N \, \mathrm{d}t, \tag{12.6}$$

since $\mathbb{P}(T_N > t) = \mathbb{P}(\min\{\tau_1, \dots, \tau_N\} > t) = [\mathbb{P}(\tau > t)]^N$, where we have used that $\tau_1, \dots, \tau_N$ are iid. Therefore, (12.5) and (12.6) imply that $\mathbb{E}[T_N]$ is infinite if $N = 2$, but finite for $N \geq 3$ [21].

To obtain the large $N$ asymptotics of $\mathbb{E}[T_N]$, we combine (12.4) and (12.6) to yield

$$\mathbb{E}[T_N] = \int_0^\infty \Big[\mathrm{erf}\Big(\frac{L}{\sqrt{4Dt}}\Big)\Big]^N \mathrm{d}t \sim \int_0^\varepsilon \Big[\mathrm{erf}\Big(\frac{L}{\sqrt{4Dt}}\Big)\Big]^N \mathrm{d}t \quad \text{as } N \to \infty \tag{12.7}$$

for any $\varepsilon > 0$, since the part of integral from $t = \varepsilon$ to $t = \infty$ vanishes exponentially fast as $N \to \infty$. Using the large $z$ asymptotics of $\mathrm{erf}(z)$, it follows that [2]

$$\mathbb{E}[T_N] \sim \frac{L^2}{4D \ln N} \quad \text{as } N \to \infty. \tag{12.8}$$

In the next two subsections, we review how the result in (12.8) for this simple example extends to much more general scenarios.

### 12.2.2 Higher Spatial Dimensions

The basic result in (12.8) has been extended to bounded two-dimensional and three-dimensional spatial domains using probabilistic methods to prove that the fastest searchers take a direct path to the target [17]. This result is easiest to describe in the case that the spatial domain is a cylinder. Specifically, let the domain $U \subset \mathbb{R}^3$ be a cylinder of radius $r > 0$ and height $h > 0$,

$$U := \big\{(x, y, z) \in \mathbb{R}^3 : x^2 + y^2 < r^2,\ z \in (0, h)\big\}.$$

Suppose the target is a disk of radius $a \in (0, r)$ at the bottom of the cylinder,

$$U_{\text{target}} := \big\{(x, y, 0) \in \mathbb{R}^3 : x^2 + y^2 < a^2\big\}.$$

Suppose that $N \geq 1$ searchers are initially placed at $(0, 0, L)$ with $L \in (0, h)$ and then diffuse in $U$ with diffusivity $D > 0$ and reflecting boundary conditions.

Let $Z_n(t) \in [0, h]$ and $R_n(t) \in [0, r]$ denote the height and radial position of the $n$-th searcher at time $t \geq 0$. The first time that the $n$-th searcher hits the target is

$$\tau_n^{3d} := \inf\{t > 0 : Z_n(t) = 0,\ R_n(t) < a\}, \quad n \in \{1, \ldots, N\}.$$

The first time that the $n$-th searcher hits the bottom of the cylinder (regardless of the radial position) is

$$\tau_n^{1d} := \inf\{t > 0 : Z_n(t) = 0\}, \quad n \in \{1, \ldots, N\}.$$

Hence, the first time that any searcher hits the target and the first time that any searcher hits the bottom of the cylinder are, respectively, given by

$$T_N^{3d} := \min_n\{\tau_n^{3d}\}, \quad T_N^{1d} := \min_n\{\tau_n^{1d}\}.$$

The next result shows that the moments of $T_N^{3d}$ and $T_N^{1d}$ become identical as $N$ grows.

**Theorem 1** (From Ref. [17]) *For any moment $m \geq 1$, we have that*

$$\mathbb{E}[(T_N^{3d})^m] \sim \mathbb{E}[(T_N^{1d})^m] \sim \Big(\frac{L^2}{4D \ln N}\Big)^m \quad \textit{as } N \to \infty. \tag{12.9}$$

We make four comments on Theorem 1. First, the proof of Theorem 1 relies on proving that the path of the first searcher to reach the target is almost a straight line from the initial position to the target. In particular, the fastest searcher out of $N \gg 1$ searchers never leaves a tube of radius $a > 0$ connecting the starting location to the target, and therefore $T_N^{3d} = T_N^{1d}$, and the large $N$ behavior of the moments of $T_N^{1d}$ is already known since $T_N^{1d}$ concerns diffusion in one space dimension [5].

Second, Theorem 1 holds for any fixed target size $a > 0$.

Third, the cylinder is finite (i.e., $r < \infty$ and $h < \infty$). In fact, if $r = h = \infty$, then (12.9) cannot hold since $\mathbb{E}[T_N^{3d}] = \infty$ for all $N \geq 1$. To prove this, note that each searcher has a strictly positive probability of never reaching the target (i.e., $\mathbb{P}(\tau^{3d} = \infty) > 0$) since three-dimensional Brownian motion is transient. Therefore, for any $N \geq 1$, there is a strictly positive probability that no searcher hits the target,

$$\mathbb{P}(T_N^{3d} = \infty) = [\mathbb{P}(\tau^{3d} = \infty)]^N > 0,$$

and therefore $\mathbb{E}[T_N^{3d}] = \infty$ for all $N \geq 1$. This is distinct from the phenomenon where the mean FPT of a single searcher is infinite while the mean FPT of the fastest searcher is finite (see Sect. 12.2.1).

Finally, while Theorem 1 concerns a specialized spatial domain (i.e., a cylinder), the argument can be extended to prove that (12.9) holds in much more general two- and three-dimensional domains by considering a thin tube of length $L$ connecting the starting location to the target (see Theorems 5 and 10 in [17] for a precise statement). However, three restrictive assumptions of this argument are that (i) the searchers all start at a single point, (ii) the domain contains a straight-line path from this single starting location to the target, and (iii) the searchers move by pure diffusion. The next section shows that (12.9) still holds without these three assumptions.

### 12.2.3 A Universal Moment Formula

Looking back to the one-dimensional problem in Sect. 12.2.1, the formula for $\mathbb{P}(\tau > t)$ in (12.4) yields the following short-time behavior of $\mathbb{P}(\tau \leq t)$ on a logarithmic
scale,

$$\lim_{t\to 0+} t \ln \mathbb{P}(\tau \leq t) = -C < 0, \quad \text{where } C := \frac{L^2}{4D} > 0. \tag{12.10}$$

Due to Varadhan's formula from large deviation theory [22, 23], the short-time behavior of the distribution of $\tau$ in (12.10) holds in much greater generality [18]. In these more general scenarios, $D > 0$ is a characteristic diffusivity and $L > 0$ is a certain geodesic distance between the searcher starting location(s) (possibly a distribution on a set of starting locations) and the target that (i) avoids any reflecting obstacles, (ii) includes any spatial variation or anisotropy in diffusivity, and (iii) incorporates any geometry in the case of diffusion on a curved manifold [18]. Further, $L$ is unaffected by deterministic forces on the diffusive searchers (i.e., a drift) or a partially absorbing target [18].

Since the large $N$ behavior of $\mathbb{E}[T_N]$ depends only on the short-time behavior of $\mathbb{P}(\tau \leq t)$ (see (12.7)), the general formula in (12.10) for the short-time behavior of $\mathbb{P}(\tau \leq t)$ for diffusive search suggests that the large $N$ asymptotics of $\mathbb{E}[T_N]$ in (12.8) should extend to the aforementioned general scenarios in which (12.10) holds. The following theorem shows that this is indeed the case, and in fact the result extends to the $m$th moment of the $k$th fastest FPT $T_{k,N}$ for $1 \leq k \ll N$.

**Theorem 2** (From Ref. [18]) *Let $\{\tau_n\}_{n=1}^{\infty}$ be a sequence of iid nonnegative random variables. Assume that*

$$\int_0^{\infty} \left[\mathbb{P}(\tau > t)\right]^N \, \mathrm{d}t < \infty \quad \textit{for some } N \geq 1, \tag{12.11}$$

*and assume that there exists a constant $C > 0$ so that*

$$\lim_{t\to 0+} t \ln \mathbb{P}(\tau \leq t) = -C < 0. \tag{12.12}$$

*Then for any $m \geq 1$ and $k \geq 1$, the $m$th moment of $T_{k,N}$ in* (12.3) *satisfies*

$$\mathbb{E}[(T_{k,N})^m] \sim \left(\frac{C}{\ln N}\right)^m \quad \textit{as } N \to \infty. \tag{12.13}$$

Theorem 2 is a general result that applies to order statistics of any random variable satisfying (12.11)–(12.12). Therefore, Theorem 2 and the universal behavior in (12.10) for diffusive search shows that (12.13) holds with $C = L^2/(4D)$ and the various details in the problem (spatial dimension, drift, domain size, target size, searcher number $k$, etc.) are irrelevant to the leading order statistics of $T_{k,N}$ as $N \to \infty$.

However, the convergence rate of (12.13) is generally quite slow. Hence, it is important to identify how these various details affect the fastest FPT statistics at higher order. The next subsection computes these corrections using extreme value theory.

### 12.2.4 Extreme Value Theory Approach

Extreme value theory is a branch of probability theory and statistics dealing with extreme events in the tails of probability distributions. The theory dates back nearly a century to Fisher, Tippett, Gnedenko [24, 25] and is concerned with determining the distribution of the minima (or maxima) of a large sequence of iid random variables in terms of the short-time distribution of a single random variable. We now apply this theory to FPTs of diffusion.

Looking back again to the simple example in Sect. 12.2.1, the formula in (12.4) yields the following short-time behavior of $\mathbb{P}(\tau \le t)$ on a linear scale (i.e., more information than the logarithmic scale in (12.10)),

$$\mathbb{P}(\tau \le t) \sim A t^p e^{-C/t} \quad \text{as } t \to 0+, \tag{12.14}$$

where $C = L^2/(4D) > 0$, $A = \sqrt{4D/(L^2\pi)}$, and $p = 1/2$. It turns out that this short-time behavior of the survival probability of diffusive search extends to much more general scenarios, where $C = L^2/(4D)$ is the diffusion timescale in (12.10) and $A > 0$ and $p \in \mathbb{R}$ depend on the details of the problem (spatial dimension, drift, target reactivity, etc.). The following theorem gives the distribution and statistics of $T_N$ using the information in (12.14).

**Theorem 3** (From Ref. [19]) *Let $\{\tau_n\}_{n\ge 1}$ be iid and assume that there exist constants $C > 0$, $A > 0$, and $p \in \mathbb{R}$ so that* (12.14) *holds. Then $T_N := \min\{\tau_1, \dots, \tau_N\}$ satisfies*

$$\frac{T_N - b_N}{a_N} \to_{\mathrm{d}} X =_{\mathrm{d}} \text{Gumbel}(0, 1) \quad \text{as } N \to \infty,$$

*where*

$$a_N = \frac{C}{(\ln N)^2}, \quad b_N = \frac{C}{\ln N}\Big(1 + \frac{p \ln(\ln(N))}{\ln N} - \frac{\ln(AC^p)}{\ln N}\Big). \tag{12.15}$$

*If we assume further that $\mathbb{E}[T_N] < \infty$ for some $N \ge 1$, then for each $m \in (0, \infty)$,*

$$\mathbb{E}\left[\left(\frac{T_N - b_N}{a_N}\right)^m\right] \to \mathbb{E}[X^m] \quad \textit{as } N \to \infty, \quad \textit{where } X =_{\mathrm{d}} \mathrm{Gumbel}(0, 1).$$

We note that Theorem 3 can be generalized to describe $T_{k,N}$ for $1 \le k \ll N$ (see Theorems 4 and 5 in [19]).

In Theorem 3 and throughout this chapter, $\to_{\mathrm{d}}$ denotes convergence in distribution [26] and $=_{\mathrm{d}}$ denotes equality in distribution. Further, Gumbel(0, 1) denotes a random variable with a Gumbel distribution with location parameter $b = 0$ and scale parameter $a = 0$. Generally, a random variable $X$ has a Gumbel distribution with location parameter $b \in \mathbb{R}$ and scale parameter $a > 0$ if[1]

$$\mathbb{P}(X > x) = \exp\left[-\exp\left(\frac{x - b}{a}\right)\right], \quad \text{for all } x \in \mathbb{R}. \tag{12.16}$$

Note that if $X =_{\mathrm{d}} \mathrm{Gumbel}(b, a)$, then $\mathbb{E}[X] = b - \gamma a$ where $\gamma \approx 0.5772$ is the Euler–Mascheroni constant, and $\mathrm{Variance}(X) = \frac{\pi^2}{6}a^2$. Theorem 3 thus yields higher order estimates of statistics of $T_N$. In particular,

$$\mathbb{E}[T_N] = b_N - \gamma a_N + o(a_N),$$
$$\mathrm{Variance}(T_N) = \frac{\pi^2}{6}a_N^2 + o(a_N^2).$$

Roughly speaking, Theorem 3 implies that $T_N$ is approximately Gumbel with shape $b_N$ and scale $a_N$.

To illustrate Theorem 3, consider one-dimensional diffusive search as in Sect. 12.2.1, but now suppose there is a constant drift $V > 0$ pushing the searchers toward the target at $x = 0$. That is, the position of a searcher evolves according to the stochastic differential equation,

$$\mathrm{d}X = -V\,\mathrm{d}t + \sqrt{2D}\,\mathrm{d}W, \quad X(0) = L > 0, \tag{12.17}$$

where $W$ is a standard Brownian motion. If $\tau$ is the FPT to the origin (i.e., (12.1) with $U_{\mathrm{target}} = 0$), then the survival probability is

$$\mathbb{P}(\tau > t) = \frac{1}{2}\left[1 + \mathrm{erf}\left(\frac{L - Vt}{\sqrt{4Dt}}\right) - e^{\frac{VL}{D}}\mathrm{erfc}\left(\frac{L + Vt}{\sqrt{4Dt}}\right)\right], \quad t > 0. \tag{12.18}$$

Using (12.18), one can check that (12.14) holds with $C = L^2/(4D)$, $A = \sqrt{1/(C\pi)}e^{LV/(2D)}$, and $p = 1/2$. Hence, Theorem 3 implies that as $N \to \infty$,

[1] A Gumbel distribution is sometimes defined differently by saying that $-X$ has a Gumbel distribution with shape $-b$ and scale $a$ if (12.16) holds.

$$\mathbb{E}[T_N] = \frac{L^2}{4D\ln N}\Big[1 + \frac{\ln(\ln(N))}{2\ln N} - \Big(\frac{LV/D - \ln\pi + 2\gamma}{\ln N}\Big) + o((\ln N)^{-1})\Big]. \tag{12.19}$$

This result shows that increasing $V$ (i.e., increasing the drift toward the target) decreases $\mathbb{E}[T_N]$. This is to be expected, but (12.19) shows that the effect of increasing $V$ only affects $\mathbb{E}[T_N]$ at third order as $N \to \infty$.

We note that the analysis above of how $T_N$ vanishes is all in the limit $N \to \infty$ for a fixed drift, fixed target size, etc. However, it is clear that $T_N$ must diverge if we fix $N \geq 1$ (even a large value of $N$) and take, for example, the size of the target to zero (similarly, $T_N$ diverges if we take the target reactivity to zero or take a repelling drift pushing the searchers away from the target to infinity). For an analysis of the "competing limits" of $N \to \infty$ versus, say, a vanishing target (or an unreactive target or a large repulsive drift), we refer the reader to Reference [27].

### 12.2.5 Uniform Initial Conditions

The analysis above all assumes that the diffusive searchers cannot start arbitrarily close to the target. That is, even in the case where the searchers may have a continuum of possible starting locations (as in Sect. 12.2.3), the support of this set of starting locations is a strictly positive distance away from the closest part of the target. For example, the one-dimensional example in Sect. 12.2.1 could be modified so that the initial distribution of $X$ is uniform on the interval $[a, b]$ as long as $0 < a < b < \infty$. In this case, the moment formula in (12.13) holds with $C = L^2/(4D)$ and $L = a > 0$.

How does the fastest FPT change if searchers can start arbitrarily close to the target? For example, suppose $X$ is initially uniformly distributed on the interval $[0, l]$ with targets at both $x = 0$ and $x = l > 0$ (i.e., $\tau$ is in (12.1) with $U_{\text{target}} = \{0, l\}$). In this case, it was shown back in the first work on fastest FPTs of diffusion that [2]

$$\mathbb{E}[T_N] \sim \frac{\pi l^2}{8D}\frac{1}{N^2} \quad \text{as } N \to \infty, \tag{12.20}$$

which is of course much faster than the $1/\ln N$ decay seen in the sections above for the case that searchers cannot start arbitrarily close to the target.

The much faster decay of $\mathbb{E}[T_N]$ for this problem stems from the following short-time behavior of the distribution of $\tau$ [27],

$$\mathbb{P}(\tau \leq t) \sim At^p \quad \text{as } t \to 0+, \tag{12.21}$$

where $p = 1/2$ and $A = \sqrt{4^2D/(l^2\pi)}$, which is much slower than the short-time decay of $\mathbb{P}(\tau \leq t)$ in (12.10). In fact, rather than just the mean in (12.20), it follows from (12.21) and Theorem 4 that $(AN)^{1/p}T_N$ converges in distribution to a Weibull random variable with unit scale and shape $p$ as $N \to \infty$.

The decay of $\mathbb{E}[T_N]$ for this example slows from $1/N^2$ in (12.20) to $1/N$ if the targets at $x = 0$ and $x = l > 0$ are partially absorbing with reactivity $\kappa \in (0, \infty)$. Specifically, it was shown in [27, 28] that

$$\mathbb{E}[T_N] \sim \frac{l}{2\kappa}\frac{1}{N} \quad \text{as } N \to \infty. \tag{12.22}$$

The behavior in (12.22) follows from the fact that the short-time distribution of $\tau$ satisfies (12.21) with $p = 1$ and $A = 2\kappa/l$ [27]. In fact, rather than just the mean in (12.22), it follows from (12.21) and Theorem 4 that $(2\kappa/l)NT_N$ converges in distribution to a unit rate exponential random variable as $N \to \infty$.

Though we illustrated (12.20) and (12.22) in the case of diffusion in an interval, we emphasize that the $N^{-2}$ decay in (12.20) and the $N^{-1}$ decay in (12.22) are generic results which apply to diffusive searchers which are initially uniformly distributed in the entire domain with either perfectly absorbing targets (the case of (12.20)) or partially absorbing targets (the case of (12.22)) [27, 28].

## 12.3 Superdiffusion

Section 12.2 concerns fastest FPTs of normal diffusion processes. Normal diffusion is marked by a squared displacement that grows linearly in time. However, rather than normal diffusion, an anomalous form of diffusion called superdiffusion has been observed in a variety of physical and biological systems [29]. In this section, we investigate fastest FPTs for a model of superdiffusion called a Lévy flight [30]. We note that a Lévy flight is superdiffusive in the sense that its squared displacement grows superlinearly in time (see (12.26)), even though its *mean* squared displacement is infinite.

### *12.3.1 Lévy Flights*

A Lévy flight with stability index $\alpha \in (0, 2)$ can be obtained from the continuous-time random walk model [29, 31] assuming that the waiting time between jumps of the walker has a finite mean $t_0 \in (0, \infty)$ and the probability density of the jump distance has the following slow power law decay:

$$f(y) \sim \frac{(l_0)^\alpha}{y^{1+\alpha}} \quad \text{as } y \to \infty \text{ for some lengthscale } l_0 > 0. \tag{12.23}$$

The probability density $p(x, t)$ for the position of a Lévy flight satisfies the following space-fractional Fokker–Planck equation [32],

$$\frac{\partial}{\partial t} p = -K(-\Delta)^{\alpha/2} p, \tag{12.24}$$

where $K = (l_0)^\alpha / t_0 > 0$ is the generalized diffusivity and $(-\Delta)^{\alpha/2}$ denotes the fractional Laplacian [33].

A Lévy flight $X$ in $\mathbb{R}^d$ can also be defined as the following random time change (i.e., a subordination) of a Brownian motion [32]:

$$X(t) := B(S(t)) + X(0) \quad t \geq 0, \tag{12.25}$$

where $B = \{B(s)\}_{s\geq 0}$ is a $d$-dimensional Brownian motion with unit diffusivity (i.e., $\mathbb{E}\|B(s)\|^2 = 2ds$ for all $s \geq 0$) and $S = \{S(t)\}_{t\geq 0}$ is an independent, $(\alpha/2)$-stable subordinator with Laplace exponent $\Phi(\beta) = K\beta^{\alpha/2}$ (i.e., $\mathbb{E}[e^{-\beta S(t)}] = e^{-t\Phi(\beta)}$ for all $t \geq 0$, $\beta \geq 0$). Observe that the Brownian scaling (where $=_\mathrm{d}$ denotes equality in distribution),

$$B(s) =_\mathrm{d} s^{1/2} B(1) \quad \text{for all } s > 0,$$

and the $(\alpha/2)$-stable subordinator scaling,

$$S(t) =_\mathrm{d} t^{2/\alpha} S(1) \quad \text{for all } t > 0,$$

imply that the Lévy flight defined in (12.25) satisfies the following scaling:

$$X(t) - X(0) =_\mathrm{d} t^{1/\alpha}(X(1) - X(0)) \quad \text{for all } t > 0. \tag{12.26}$$

Squaring (12.26) implies that the squared displacement of a Lévy flight grows superlinearly in time.

The search efficiency of a Lévy flight has previously been investigated in terms of the FHT of a single searcher to a target in a variety of one-dimensional [34–36] or two-dimensional settings [37]. In the following section, we study the FHT of the fastest searcher of many Lévy flight searchers.

### *12.3.2 Fastest FHTs of Lévy Flights*

Define the first hitting time (FHT) $\tau$ of the Lévy flight $X$ in (12.25) to some target $U_{\text{target}} \subset \mathbb{R}^d$ as in (12.1). We use the term FHT rather than FPT since these two concepts can be distinct for Lévy flights due to their discontinuous sample paths [35, 38–40]. As above, determining the statistics and distribution of $T_{k,N}$ for $1 \leq k \ll N$ in (12.3) requires determining the short-time distribution of $\tau$.

Assuming that $X(0)$ cannot lie in the target $U_{\text{target}}$ (meaning that the intersection of the support of $X(0)$ and the closure of $U_{\text{target}}$ is empty), it was proven in [41] that $\tau$ has the universal short-time distribution,

$$\mathbb{P}(\tau \le t) \sim \rho t \quad \text{as } t \to 0+, \tag{12.27}$$

where $\rho \in (0, \infty)$ is the rate,

$$\rho := \int_0^\infty \mathbb{P}(B(s) + X(0) \in U_{\text{target}}) \frac{\alpha/2}{\Gamma(1-\alpha/2)} \frac{K}{s^{1+\alpha/2}} \, \mathrm{d}s. \tag{12.28}$$

If $X(0) = 0$, then the Gaussianity of $B(s)$ means that $\rho$ can be written as

$$\begin{aligned} \rho &= \int_0^\infty \frac{1}{(4\pi s)^{d/2}} \int_{U_{\text{target}}} \exp\Big(\frac{-\|x\|^2}{4s}\Big) \, \mathrm{d}x \, \frac{\alpha/2}{\Gamma(1-\alpha/2)} \frac{K}{s^{1+\alpha/2}} \, \mathrm{d}s \\ &= 2^\alpha \Gamma\Big(\frac{d+\alpha}{2}\Big) \frac{K}{\pi^{d/2}} \frac{\alpha/2}{\Gamma(1-\alpha/2)} \int_{U_{\text{target}}} \frac{1}{\|x\|^{\alpha+d}} \, \mathrm{d}x. \end{aligned}$$

The following theorem yields the distribution and statistics of $T_N$ in (12.2) from the short-time behavior in (12.27). Analogous results hold for $T_{k,N}$ in (12.3) with $1 \le k \ll N$ (see Theorems 5 and 6 in [27]).

**Theorem 4** (From Ref. [27]) *Let* $\{\tau_n\}_{n\ge1}$ *be an iid sequence of random variables and assume that for some* $A > 0$ *and* $p > 0$*, we have that*

$$\mathbb{P}(\tau_n \le t) \sim At^p \quad \text{as } t \to 0+. \tag{12.29}$$

*Then, the following rescaling of* $T_N := \min\{\tau_1, \ldots, \tau_N\}$ *converges in distribution,*

$$(AN)^{1/p} T_N \to_{\mathrm{d}} \text{Weibull}(1, p) \quad \text{as } N \to \infty.$$

*If we assume further that* $\mathbb{E}[T_N] < \infty$ *for some* $N \ge 1$*, then for each* $m \in (0, \infty)$*,*

$$\mathbb{E}[(T_N)^m] \sim \frac{\Gamma(1+m/p)}{(AN)^{m/p}} \quad \text{as } N \to \infty.$$

Throughout this chapter, $X =_{\mathrm{d}} \text{Weibull}(t, p)$ means that $X$ is a Weibull random variable with scale $t > 0$ and shape $p > 0$, which means

$$\mathbb{P}(X > x) = \exp(-(x/t)^p), \quad x \ge 0. \tag{12.30}$$

Note that Weibull$(1, p)$ is exponential with unit rate if $p = 1$. Therefore, applying Theorem 4 to (12.27) implies that $(\rho N)T_N$ converges in distribution to a unit rate exponential random variable as $N$ grows [41]. Hence, $T_N$ is approximately exponentially distributed with rate $\rho N$. Furthermore, if $\mathbb{E}[T_N] < \infty$ for some $N \ge 1$, then Theorem 4 yields all the moments of $T_N$ for large $N$. In particular,

$$\mathbb{E}[T_N] \sim \sqrt{\text{Variance}[T_N]} \sim \frac{1}{\rho N} \quad \text{as } N \to \infty. \tag{12.31}$$

Analogous results hold for (i) $T_{k,N}$ in (12.3) for any $1 \leq k \ll N$ and (ii) the case that $S$ is any nondeterministic Lévy subordinator [41].

We now contrast (12.31) with the case of normal diffusion. First, the $1/N$ decay in (12.31) is much faster than the $1/\ln N$ decay for normal diffusion processes in Sect. 12.2. Further, fastest FPTs of diffusion depend on the shortest path to the target since the fastest diffusive searchers follow this geodesic [18]. Indeed, fastest FPTs of diffusion are unaffected by changes to the problem outside of this path, such as altering the domain or target size or even the spatial dimension. In contrast, the expression for the rate $\rho$ in (12.28) implies that fastest FHTs of Lévy flights depend on these global properties of the problem. Indeed, the fastest Lévy flights do not take a direct path to the closest part of the target [41].

## 12.4 Subdiffusion

Another form of anomalous diffusion seen in a variety of physical and biological systems is subdiffusion [42–45]. Subdiffusion is marked by a mean-squared displacement that grows according to the following sublinear power law:

$$\mathbb{E}\|X(t) - X(0)\|^2 \propto t^\gamma, \quad \gamma \in (0, 1).$$

### *12.4.1 Subordination*

One common way to model subdiffusion is via a time-fractional Fokker–Planck equation [46], which derives from the continuous-time random walk model with infinite mean waiting times between jumps [31, 47]. Such a time-fractional Fokker–Planck equation describes the distribution of a subdiffusive process $\{X(t)\}_{t\geq 0}$ defined by subordinating a normal diffusive process $\{Y(s)\}_{s\geq 0}$ [48]. Precisely, if $\{Y(s)\}_{s\geq 0}$ is a normal diffusion process satisfying an Itô stochastic differential equation, then

$$X(t) := Y(S(t)), \quad t \geq 0, \tag{12.32}$$

where $S = \{S(t)\}_{t\geq 0}$ is an independent, inverse $\gamma$-stable subordinator. We emphasize that $S$ is an *inverse* stable subordinator in this section (in contrast with Sect. 12.3), which is defined by

$$S(t) := \inf\{s > 0 : \mathfrak{T}(s) > t\},$$

where $\{\mathfrak{T}(s)\}_{s\geq 0}$ is a $\gamma$-stable subordinator.

The definition of $X$ in (12.32) implies that if $\tau$ and $\sigma$ are respective FPTs of $X$ and $Y$ to some target, then the distribution of $\tau$ is

$$\mathbb{P}(\tau \leq t) = \mathbb{E}[F_\sigma(S(t))], \tag{12.33}$$

where $F_\sigma(s) := \mathbb{P}(\sigma \leq s)$ is the cumulative distribution function of the FPT $\sigma$ of the normal diffusion process $Y$. It follows immediately from (12.33) that

$$\mathbb{P}(\tau \leq t) = \int_0^\infty \mathbb{P}(\sigma \leq s) \frac{t}{\gamma s^{1+1/\gamma}} l_\gamma\left(\frac{t}{s^{1/\gamma}}\right) \mathrm{d}s, \tag{12.34}$$

since the probability density that $S(t) = s$ is

$$\frac{\mathrm{d}}{\mathrm{d}s}\mathbb{P}(S(t) \leq s) = \frac{t}{\gamma s^{1+1/\gamma}} l_\gamma\left(\frac{t}{s^{1/\gamma}}\right), \tag{12.35}$$

where $l_\gamma(z)$ is defined by its Laplace transform,

$$\int_0^\infty e^{-rz} l_\gamma(z)\,\mathrm{d}z = e^{-r^\gamma}, \quad \gamma \in (0, 1),\ r \geq 0.$$

### 12.4.2 *Fastest FPTs of Subdiffusion*

By using (12.34), the small $s$ behavior of $\mathbb{P}(\sigma \leq s)$, and the small $z$ behavior of $l_\gamma(z)$, we can develop a theory of fastest FPTs of subdiffusion that is analogous to the theory for normal diffusion in Sect. 12.2. In particular, since $\sigma$ is an FPT of a normal diffusion process, then we have that under very general conditions (see Sect. 12.2.3),

$$\lim_{s \to 0+} s \ln \mathbb{P}(\sigma \leq s) = -L^2/(4D) < 0,$$

where $D$ and $L$ are as in Sect. 12.2.3 for the normal diffusion process $Y$ (note that $Y = Y(s)$ is indexed by "internal time" $s \geq 0$, which is not physical time, but rather has dimension (time)$^\gamma$ and thus $D$ has dimension (length)$^2$(time)$^{-\gamma}$). Then, using (12.34) and the fact that [49, 50]

$$\lim_{z \to 0+} z^{\gamma/(1-\gamma)} \ln l(z) = -(1-\gamma)\gamma^{\gamma/(1-\gamma)} < 0, \tag{12.36}$$

we obtain the following universal behavior of FPTs of subdiffusion modeled by a time-fractional Fokker–Planck equation,

$$\lim_{t \to 0+} t^{\gamma/(2-\gamma)} \mathbb{P}(\tau \leq t) = -C := -(2-\gamma)\gamma^{\frac{\gamma}{2-\gamma}}(L^2/(4D))^{\frac{1}{2-\gamma}} < 0. \tag{12.37}$$

By a result analogous to Theorem 2 (see Theorem 6 in [51]), the short-time behavior on a logarithmic scale in (12.37) implies that the $m$th moment of $T_{k,N}$ in

(12.3) for subdiffusion satisfies (assuming $\mathbb{E}[T_N] < \infty$ for some $N \geq 1$),

$$\mathbb{E}[(T_{k,N})^m] \sim \left(\frac{t_\gamma}{(\ln N)^{2/\gamma-1}}\right)^m \quad \text{as } N \to \infty, \tag{12.38}$$

where $t_\gamma > 0$ is the characteristic subdiffusive timescale,

$$t_\gamma := \left(\gamma^\gamma (2-\gamma)^{2-\gamma} \frac{L^2}{4D}\right)^{1/\gamma} > 0, \quad \gamma \in (0, 1].$$

We can also generalize Theorem 3 to determine higher order corrections to (12.38) and the Gumbel probability distribution of $T_{k,N}$, but we omit these results for brevity (see Theorems 9, 10, and 11 in [51]). Note also that we focused here on the case when searchers cannot start arbitrarily close to the target, but one can also determine the fastest FPTs of subdiffusive searchers which start uniformly in a bounded spatial domain (see the discussion section of [51]).

### *12.4.3 Extreme Statistics: Subdiffusion Is Faster Than Normal Diffusion*

Comparing (12.38) with (12.13) in Sect. 12.2 for normal diffusion yields the counterintuitive result that fastest FPTs of subdiffusion are faster than fastest FPTs of normal diffusion. Concretely, let $\sigma^{\text{norm}}$ and $\tau^{\text{sub}}$ denote FPTs to a target in a bounded domain for normal diffusive and subdiffusive search, respectively. We generally have

$$\mathbb{E}[\sigma^{\text{norm}}] < \mathbb{E}[\tau^{\text{sub}}] = \infty,$$

where $\mathbb{E}[\tau^{\text{sub}}] = \infty$ owes to the slow decay of $\mathbb{P}(\tau^{\text{sub}} > t)$ as $t \to \infty$ [52]. However, if $\mathbb{P}(\sigma^{\text{norm}} > s)$ decays at least as fast as a power law as $s \to \infty$, then (12.34) implies that (see Theorem 8 in [51])

$$\mathbb{E}[T_N^{\text{sub}}] < \infty \quad \text{for sufficiently large } N,$$

where $T_N^{\text{sub}} := \min\{\tau_1^{\text{sub}}, \dots, \tau_N^{\text{sub}}\}$ and $\{\tau_n^{\text{sub}}\}_n$ are iid. Further, (12.38) and (12.13) imply

$$\mathbb{E}[T_N^{\text{sub}}] \ll \mathbb{E}[\Sigma_N^{\text{norm}}] \quad \text{for sufficiently large } N,$$

where $\Sigma_N^{\text{norm}} := \min\{\sigma_1^{\text{norm}}, \dots, \sigma_N^{\text{norm}}\}$ and $\{\sigma_n^{\text{norm}}\}_n$ are iid.

## 12.5 Random Walks on Networks

In Sects. 12.2–12.4, we considered search processes on a continuous state space. We now follow [53] and consider searchers which move on a "network" of discrete states.

### 12.5.1 Random Walk Setup

Each searcher moves on the network according to a continuous-time Markov chain. Hence, the waiting time between jumps is exponentially distributed, but this assumption on the waiting time distribution can be relaxed (see Sect. III in [53] for many searchers and also [54] for the case of a single searcher).

Concretely, let $X = \{X(t)\}_{t \geq 0}$ be a continuous-time Markov chain on a finite or countably infinite state space $I$. The process $X$ is a single searcher and its dynamics are described by its infinitesimal generator matrix $Q = \{q(i, j)\}_{i,j \in I}$ [55]. The off-diagonal entries of $Q$ (i.e., $q(i, j) \geq 0$ for $i \neq j$) are nonnegative and give the rate that $X$ jumps from $i \in I$ to $j \in I$. The diagonal entries of $Q$ are nonpositive (i.e., $q(i, i) \leq 0$ for all $i \in I$) and are chosen so that $Q$ has zero row sums. Assume that $\sup_{i \in I} |q(i, i)| < \infty$ so that $X$ cannot take infinitely many jumps in finite time.

### 12.5.2 Single FPTs

Let $\tau$ in (12.1) be the FPT of $X$ to some target set $U_{\text{target}} \subset I$. Let $\rho = \{\rho(i)\}_{i \in I} = \{\mathbb{P}(X(0) = i)\}_{i \in I}$ denote the initial distribution of $X$. To avoid trivial cases, assume that $X$ cannot start on the target, which means that $U_{\text{target}} \cap \text{supp}(\rho) = \varnothing$, where $\text{supp}(\rho) \subset I$ denotes the support of $\rho$ (i.e., $\text{supp}(\rho) := \{i \in I : \rho(i) > 0\}$).

As in the sections above, determining the distribution of $T_N$ in (12.2) and $T_{k,N}$ in (12.3) requires determining the short-time distribution of $\tau$. Theorem 5 shows that

$$\mathbb{P}(\tau \leq t) \sim \frac{\Lambda}{d!} t^d \quad \text{as } t \to 0+, \tag{12.39}$$

where $d \geq 1$ is the smallest number of jumps that $X$ must take to hit the target and $\Lambda > 0$ is a sum of the products of the jump rates along the shortest path to the target.

To make $d$ and $\Lambda$ in (12.39) precise, define a path $\mathcal{P}$ of length $d \in \mathbb{Z}_{\geq 0}$ from $i_0 \in I$ to $i_d \in I$ to be a sequence of $d + 1$ states in $I$,

$$\mathcal{P} = (\mathcal{P}(0), \ldots, \mathcal{P}(d)) = (i_0, i_1, \ldots, i_d) \in I^{d+1}, \tag{12.40}$$

so that

$$q(\mathcal{P}(k), \mathcal{P}(k+1)) > 0, \quad \text{for } k \in \{0, 1, \dots, d-1\}. \tag{12.41}$$

The condition in (12.41) means that $X$ has a strictly positive probability of following the path $\mathcal{P}$. Assume that there is a path from the support of $\rho$ to the target (otherwise, $\mathbb{P}(\tau = \infty) = 1$ and the problem is trivial).

For a path $\mathcal{P} \in I^{d+1}$, let $\lambda(\mathcal{P})$ be the product of the rates along the path,

$$\lambda(\mathcal{P}) := \prod_{i=0}^{d-1} q(\mathcal{P}(i), \mathcal{P}(i+1)) > 0. \tag{12.42}$$

Let $d_{\min}(I_0, I_1) \in \mathbb{Z}_{\geq 0}$ denote the length of the shortest path from $I_0 \subset I$ to $I_1 \subset I$,

$$d_{\min}(I_0, I_1) := \inf\{d : \mathcal{P} \in I^{d+1}, \mathcal{P}(0) \in I_0, \mathcal{P}(d) \in I_1\}.$$

That is, $d_{\min}(I_0, I_1)$ is the smallest number of jumps that $X$ must take to go from $I_0$ to $I_1$. Define the set of all paths from $I_0$ to $I_1$ with the length $d_{\min}(I_0, I_1)$,

$$\mathcal{S}(I_0, I_1) := \{\mathcal{P} \in I^{d+1} : \mathcal{P}(0) \in I_0, \mathcal{P}(d) \in I_1, d = d_{\min}(I_0, I_1)\}. \tag{12.43}$$

Define

$$\Lambda(\rho, I_1) := \sum_{\mathcal{P} \in \mathcal{S}(\text{supp}(\rho), I_1)} \rho(\mathcal{P}(0))\lambda(\mathcal{P}). \tag{12.44}$$

To explain $\Lambda(\rho, I_1)$, suppose first that $\rho(i_0) = 1$ for some $i_0 \in I$ (i.e., $\text{supp}(\rho) = i_0 = X(0)$). If there is only one path with the minimum number of jumps $d_{\min}(i_0, I_1)$, then $\Lambda(\rho, I_1)$ is the product of the jump rates along this path (i.e., $\lambda(\mathcal{P})$ in (12.42)). If there is more than one shortest path, then $\Lambda(\rho, I_1)$ is the sum of the products of the jump rates along these paths. Finally, if $\rho$ is not concentrated at a single point, then $\Lambda(\rho, I_1)$ merely sums the products of the jump rates along all the shortest paths, where the sum is weighted according to $\rho$. With these definitions in place, we can now give the short-time behavior of the distribution of $\tau$.

**Theorem 5** (From Ref. [53]) *The short-time distribution of $\tau$ satisfies* (12.39), *where*

$$d = d_{\min}(\text{supp}(\rho), U_{\text{target}}) \in \mathbb{Z}_{>0},$$
$$\Lambda = \Lambda(\rho, U_{\text{target}}) > 0.$$

### *12.5.3 Fastest FPTs*

Having determined the short-time probability distribution of a single FPT in Theorem 5, we can immediately determine the limiting distribution and statistics of fastest

FPTs for random walks on networks. In particular, define

$$A = \frac{\Lambda}{d!} > 0,$$

where $d \geq 1$ and $\Lambda > 0$ are as in Theorem 5. Theorem 4 implies that the following rescaling of $T_N$ converges in distribution to a Weibull random variable,

$$(AN)^{1/d} T_N \to_{\mathrm{d}} \mathrm{Weibull}(1, d) \quad \text{as } N \to \infty. \tag{12.45}$$

Furthermore, if $\mathbb{E}[T_N] < \infty$ for some $N \geq 1$, then for each moment $m \in (0, \infty)$,

$$\mathbb{E}[(T_N)^m] \sim \frac{\Gamma(1 + m/d)}{(AN)^{m/d}} \quad \text{as } N \to \infty. \tag{12.46}$$

Compared to a single FPT $\tau$, fastest FPTs $T_N$ are faster and less variable since (12.46) implies that the mean and variance of $T_N$ vanish as $N \to \infty$. Furthermore, fastest FPTs are independent of much of the properties of the network. In particular, (12.45) implies that the large $N$ distribution of $T_N$ is fully determined by $N$, $\Lambda$, and $d$, which depend only on network features along the shortest path(s) from the initial distribution to the target. Similar conclusions hold for $T_{k,N}$ for $1 \leq k \ll N$ (see Theorems 2 and 4 in [53]).

## 12.6 Discussion

In this chapter, we reviewed fastest FPTs out of many searchers for a variety of search processes. We found that such extreme FPTs tend to differ drastically from the behavior of a single FPT. We showed that the statistics and distribution of fastest FPTs are determined by the short-time distribution of a single FPT. Hence, details in the problem (spatial dimension, domain size and geometry, some aspects of searcher dynamics, etc.) can become irrelevant to fastest FPTs in the many searcher limit. These details become irrelevant because the fastest FPTs typically correspond to searchers which take a direct path to the target.

We now briefly mention a few closely related problems. The behavior of fastest searchers mimics so-called "mortal" or "evanescent" searchers [7, 56–61] which have been conditioned to reach the target before a fast inactivation time [53, 62, 63]. Conditioning on not being inactivated effectively filters out searchers which do not take a direct path to the target.

In addition to FPTs, cover times of multiple searchers have also been investigated [64–68]. Cover times measure the speed of exhaustive search and are defined as the first time a searcher(s) comes within a specified "detection radius" of every point in the target region (often the entire spatial domain). For diffusive search, it was recently proven that the asymptotic $L^2/(4D \ln N)$ also holds for the time it takes $N \gg 1$ iid searchers to collectively cover a target if $L$ is the shortest distance from the searcher starting locations to the *farthest* part of the target [67] (see [68] for results on cover times of many discrete random walkers on a network). Another statistic which is related to fastest FPTs was discussed in [69] in terms of first-crossing times for boundary local time processes of individual searchers.

In the case of multiple targets, one can consider the probability that the fastest searcher hits a particular target first (the so-called splitting probability or hitting probability). In the limit of many diffusive searchers, the fastest searcher always hits the nearest target, regardless of target size, domain geometry, drift, etc. [70]. This phenomenon has been shown to offer a way for biological cells to detect the location of the source of a diffusing signal [71, 72] and also offers insight into the choices made by the first deciders in large groups of decision-makers [73, 74].

*Slowest* FPTs have recently found application in ovarian aging and menopause timing [75, 76]. Women are born with $N \in [10^5, 10^6]$ primordial follicles in their ovarian reserve, and this reserve then decays until only about $k = 10^3$ follicles remain which is thought to trigger menopause [77]. Further, there is a long history [78–83] of models that assume that follicles leave the reserve at iid random times $\tau_1, \ldots, \tau_N$. Hence, the menopause age is $T_{N-k,N}$ (using the notation in (12.3)), which means that menopause age is determined by the $k/N \approx 0.2\%$ slowest follicles to leave the reserve. This analysis has been used to offer an explanation for the so-called "wasteful" oversupply of follicles [75] and also to analyze a proposed surgical procedure for delaying menopause [84]. Other instances of apparent "redundancy" in biological systems have been understood in terms of fastest FPTs [7, 10–16].

A possible problem with the analysis of fastest FPTs of diffusive searchers in some scenarios is that the vanishing mean fastest search time,

$$\mathbb{E}[T_N] \sim \frac{L^2}{4D \ln N} \quad \text{as } N \to \infty, \tag{12.47}$$

becomes unphysical. Specifically, if the searchers have a finite maximum speed $v > 0$, then no searcher can reach a target that is distance $L > 0$ from the starting location before time $L/v > 0$. That is, we must have $T_N \geq L/v > 0$, which contradicts (12.47) for sufficiently large $N$. This contradiction stems from the well-known [85] problem of approximating finite speed random walks by diffusion due to the infinite speed of propagation of solutions to the diffusion equation [86, 87]. This discrepancy between finite speed random walks and infinite speed diffusion can often be ignored in many applications, since the discrepancy occurs in the tails of the distribution. However, these tails determine fastest FPTs. For a discussion and analysis of (i) when (12.47) becomes unphysical and (ii) fastest FPTs of run and tumble processes (i.e., finite speed random walks), we refer the reader to Reference [88].

In terms of practical application, an important question concerns the convergence rate of the asymptotic results reviewed in this chapter. Put simply, how large does the number of searchers $N$ need to be for the asymptotics to be accurate? The convergence in (12.47) is generally quite slow, though the "sufficiently large" value of $N$ for which the approximation $\mathbb{E}[T_N] \approx \frac{L^2}{4D \ln N}$ is accurate depends strongly on the specifics of the system. To illustrate in a concrete example, consider three-dimensional diffusive search confined in a sphere of radius $R > 0$ with reflecting boundaries. Suppose the searchers start at the boundary of the sphere and the target is a concentric sphere with radius $\varepsilon R$. If we fix $\varepsilon \in (0, 1)$ (even small a value of $\varepsilon$) and take $N \to \infty$, then the mean fastest search time vanishes according to (12.47) where $L = (1 - \varepsilon)R$. If we instead fix $N$ (even a large value of $N$) and take $\varepsilon \to 0$, then the mean fastest search time diverges according to [27]

$$\mathbb{E}[T_N] \sim \frac{\mathbb{E}[\tau]}{N} \sim \frac{1}{N}\frac{R^2}{3D\varepsilon} \quad \text{as } \varepsilon \to 0, \tag{12.48}$$

where $\mathbb{E}[\tau] \sim R^2/(3D\varepsilon)$ is the diverging mean FPT of a single searcher. Hence, smaller targets require more searchers in order for (12.47) to be accurate. When is the system in the fast escape regime of (12.47) versus the slow escape regime of (12.48)? If we fix $\varepsilon \ll 1$ and increase $N$, (12.48) implies that $\mathbb{E}[T_N]$ initially decreases like $\mathbb{E}[\tau]/N$. However, this rather rapid $1/N$ decay cannot continue indefinitely since it must eventually slow to the $1/\ln N$ decay in (12.47). Since $\mathbb{E}[T_N]$ is monotonically decreasing in $N$, this argument suggests the following simple approximation:

$$\mathbb{E}[T_N] \approx \max\left\{\frac{\mathbb{E}[\tau]}{N}, \frac{L^2}{4D\ln N}\right\}, \tag{12.49}$$

and thus

$$\mathbb{E}[T_N] \approx \frac{L^2}{4D\ln N} \quad \text{if} \quad \frac{\mathbb{E}[\tau]}{N} \ll \frac{L^2}{4D\ln N}. \tag{12.50}$$

The estimates (12.49)–(12.50) can be extended to more general scenarios assuming that $L^2/D \ll \mathbb{E}[\tau] < \infty$ [27]. A more detailed investigation of when various approximations of fastest FPTs of diffusion are valid can be found in Ref. [27].

**Acknowledgements** The author was supported by the National Science Foundation (Grant Nos. DMS-2325258 and DMS-1944574).

## References

1. S. Redner, *A Guide To First-passage Processes*. (Cambridge University Press, 2001)
2. G.H. Weiss, K.E. Shuler, K. Lindenberg, Order statistics for first passage times in diffusion processes. J. Stat. Phys. **31**(2), 255–278 (1983)
3. S.B. Yuste, K. Lindenberg, Order statistics for first passage times in one-dimensional diffusion processes. J. Stat. Phys. **85**(3–4), 501–512 (1996)
4. S.B. Yuste, L. Acedo, Diffusion of a set of random walkers in Euclidean media. First passage times. J. Phys. A Math. Theor. **33**(3), 507 (2000)
5. S.B. Yuste, L. Acedo, K. Lindenberg, Order statistics for $d$-dimensional diffusion processes. Phys. Rev. E **64**(5), 052102 (2001)

6. S. Redner, B. Meerson, First invader dynamics in diffusion-controlled absorption. J. Stat. Mech. **2014**(6), P06019 (2014)
7. B. Meerson, S. Redner, Mortality, redundancy, and diversity in stochastic search. Phys. Rev. Lett. **114**(19), 198101 (2015)
8. S. Ro, Y.W. Kim, Parallel random target searches in a confined space. Phys. Rev. E **96**(1), 012143 (2017)
9. K. Basnayake, Z. Schuss, D. Holcman, Asymptotic formulas for extreme statistics of escape times in 1, 2 and 3-dimensions. J Nonlinear Sci **29**(2), 461–499 (2019)
10. Z. Schuss, K. Basnayake, and D. Holcman. Redundancy principle and the role of extreme statistics in molecular and cellular biology. Phys. Life Rev., 28:52–79, 2019
11. D. Coombs. First among equals: Comment on "Redundancy principle and the role of extreme statistics in molecular and cellular biology" by Z. Schuss, K. Basnayake and D. Holcman. Phys. Life Rev., 28:92–93, 2019
12. S. Redner and B. Meerson. Redundancy, extreme statistics and geometrical optics of Brownian motion: Comment on "Redundancy principle and the role of extreme statistics in molecular and cellular biology" by Z. Schuss, K. Basnayake and D. Holcman. Phys. Life Rev., 28:80–82, 2019
13. I.M. Sokolov, Extreme fluctuation dominance in biology: On the usefulness of wastefulness: Comment on "Redundancy principle and the role of extreme statistics in molecular and cellular biology" by Z. Schuss, K. Basnayake and D. Holcman. Phys. Life Rev, 28:88–91, 2019
14. D.A. Rusakov, L.P. Savtchenko, Extreme statistics may govern avalanche-type biological reactions: Comment on "Redundancy principle and the role of extreme statistics in molecular and cellular biology" by Z. Schuss, K. Basnayake, D. Holcman. Phys. Life Rev, 28:85–87, 2019
15. L. M. Martyushev. Minimal time, Weibull distribution and maximum entropy production principle: Comment on "Redundancy principle and the role of extreme statistics in molecular and cellular biology" by Z. Schuss, K. Basnayake and D. Holcman. Phys. Life Rev., 28:83–84, 2019
16. M. V. Tamm. Importance of extreme value statistics in biophysical contexts: Comment on "Redundancy principle and the role of extreme statistics in molecular and cellular biology" by Z. Schuss, K. Basnayake and D. Holcman. Phys. Life Rev., 28:94–95, 2019
17. S. D. Lawley and J. B. Madrid. A probabilistic approach to extreme statistics of brownian escape times in dimensions 1, 2, and 3. J. Nonlinear. Sci., 1–21, 2020
18. S.D. Lawley, Universal formula for extreme first passage statistics of diffusion. Phys. Rev. E **101**(1), 012413 (2020)
19. S. D. Lawley. Distribution of extreme first passage times of diffusion. J. Math. Biol., 2020
20. J. Morgan, A.E. Lindsay, Modulation of antigen discrimination by duration of immune contacts in a kinetic proofreading model of T cell activation with extreme statistics. PLoS Comput. Biol. **19**(8), e1011216 (2023)
21. K. Lindenberg, V. Seshadri, K. E. Shuler, and G. H. Weiss. Lattice random walks for sets of random walkers. first passage times. J. Stat. Phys., 23:11–25, 1980
22. S.R.S. Varadhan, Diffusion processes in a small time interval. Commun. Pure Appl. Math. **20**(4), 659–685 (1967)
23. S.R.S. Varadhan, On the behavior of the fundamental solution of the heat equation with variable coefficients. Commun. Pure Appl. Math. **20**(2), 431–455 (1967)
24. R. A. Fisher and L. H. C. Tippett. Limiting forms of the frequency distribution of the largest or smallest member of a sample. In Mathematical Proceedings of the Cambridge Philosophical Society, volume 24, pages 180–190. Cambridge University Press, 1928
25. S. Coles. *A*n introduction to statistical modeling of extreme values, volume 208. Springer, 2001
26. P. Billingsley. *C*onvergence of probability measures. John Wiley & Sons, 2013
27. J.B. Madrid, S.D. Lawley, Competition between slow and fast regimes for extreme first passage times of diffusion. J. Phys. A: Math. Theor. **53**(33), 335002 (2020)
28. D.S. Grebenkov, R. Metzler, G. Oshanin, From single-particle stochastic kinetics to macroscopic reaction rates: fastest first-passage time of n random walkers. New J. Phys. **22**(10), 103004 (2020)

29. R. Metzler, J. Klafter, The restaurant at the end of the random walk: recent developments in the description of anomalous transport by fractional dynamics. J. Phys. A: Math. Theor. **37**(31), R161 (2004)
30. A.A. Dubkov, B. Spagnolo, V.V. Uchaikin, Lévy flight superdiffusion: an introduction. Int J Bifurcat Chaos **18**(09), 2649–2672 (2008)
31. E.W. Montroll, G.H. Weiss, Random walks on lattices. II. J Math Phys **6**(2), 167–181 (1965)
32. M. M. Meerschaert and A. Sikorskii. *S*tochastic models for fractional calculus, volume 43. Walter de Gruyter GmbH & Co KG, 2011
33. A. Lischke, G. Pang, M. Gulian, F. Song, C. Glusa, X. Zheng, Z. Mao, W. Cai, M.M. Meerschaert, M. Ainsworth et al., What is the fractional Laplacian? A comparative review with new results. J. Comput. Phys. **404**, 109009 (2020)
34. V.V. Palyulin, A.V. Chechkin, R. Metzler, Lévy flights do not always optimize random blind search for sparse targets. Proc. Natl. Acad. Sci. **111**(8), 2931–2936 (2014)
35. V.V. Palyulin, G. Blackburn, M.A. Lomholt, N.W. Watkins, R. Metzler, R. Klages, A.V. Chechkin, First passage and first hitting times of Lévy flights and Lévy walks. New J. Phys. **21**(10), 103028 (2019)
36. D. Gomez and S. D. Lawley. First hitting time of a one-dimensional Lévy flight to small targets. SIAM J. Appl. Math. **84**, 1140–1162 (2023)
37. J. C. Tzou and L. Tzou. A Counterexample to the Lévy Flight Foraging Hypothesis in the Narrow Capture Framework . *a*rXiv preprint arXiv:2302.13976, 2023
38. T. Koren, M.A. Lomholt, A.V. Chechkin, J. Klafter, R. Metzler, Leapover lengths and first passage time statistics for Lévy flights. Phys. Rev. Lett. **99**(16), 160602 (2007)
39. T. Koren, A.V. Chechkin, J. Klafter, On the first passage time and leapover properties of Lévy motions. Phys. A: Stat. Mech. Appl **379**(1), 10–22 (2007)
40. A. Wardak, First passage leapovers of Lévy flights and the proper formulation of absorbing boundary conditions. J. Phys. A: Math. Theor. **53**(37), 375001 (2020)
41. S.D. Lawley, Extreme statistics of superdiffusive Lévy flights and every other Lévy subordinate Brownian motion. J. Nonlinear Sci. **33**(4), 53 (2023)
42. F.A. Oliveira, R.M.S. Ferreira, L.C. Lapas, M.H. Vainstein, Anomalous diffusion: A basic mechanism for the evolution of inhomogeneous systems. Front. Phys. **7**, 18 (2019)
43. J. Klafter, I.M. Sokolov, Anomalous diffusion spreads its wings. Phys. World **18**(8), 29 (2005)
44. F. Höfling, T. Franosch, Anomalous transport in the crowded world of biological cells. Rep. Prog. Phys. **76**(4), 046602 (2013)
45. E. Barkai, Y. Garini, R. Metzler, Strange kinetics of single molecules in living cells. Phys. Today **65**(8), 29 (2012)
46. R. Metzler, E. Barkai, J. Klafter, Anomalous diffusion and relaxation close to thermal equilibrium: A fractional Fokker-Planck equation approach. Phys. Rev. Lett. **82**(18), 3563 (1999)
47. R. Metzler, J. Klafter, The random walk's guide to anomalous diffusion: a fractional dynamics approach. Phys. Rep. **339**(1), 1–77 (2000)
48. S.D. Lawley, Subdiffusion-limited fractional reaction-subdiffusion equations with affine reactions: Solution, stochastic paths, and applications. Phys. Rev. E **102**(4), 042125 (2020)
49. W. R. Schneider. Stable distributions: Fox function representation and generalization. In *S*tochastic processes in classical and quantum systems, pages 497–511. Springer, 1986
50. E. Barkai, Fractional Fokker-Planck equation, solution, and application. Phys. Rev. E **63**(4), 046118 (2001)
51. S.D. Lawley, Extreme statistics of anomalous subdiffusion following a fractional fokker-planck equation: subdiffusion is faster than normal diffusion. J. Phys. A: Math. Theor. **53**(38), 385005 (2020)
52. S.B. Yuste, K. Lindenberg, Comment on "Mean first passage time for anomalous diffusion." Phys. Rev. E **69**(3), 033101 (2004)
53. S.D. Lawley, Extreme first-passage times for random walks on networks. Phys. Rev. E **102**(6), 062118 (2020)
54. D.S. Grebenkov, L. Tupikina, Heterogeneous continuous-time random walks. Phys. Rev. E **97**(1), 012148 (2018)

55. J.R. Norris, *Markov Chains* (Cambridge University Press, Statistical & Probabilistic Mathematics, 1998)
56. E. Abad, S.B. Yuste, K. Lindenberg, Reaction-subdiffusion and reaction-superdiffusion equations for evanescent particles performing continuous-time random walks. Phys. Rev. E **81**(3), 031115 (2010)
57. E. Abad, S.B. Yuste, K. Lindenberg, Survival probability of an immobile target in a sea of evanescent diffusive or subdiffusive traps: A fractional equation approach. Phys. Rev. E **86**(6), 061120 (2012)
58. E. Abad, S.B. Yuste, K. Lindenberg, Evanescent continuous-time random walks. Phys. Rev. E **88**(6), 062110 (2013)
59. S.B. Yuste, E. Abad, K. Lindenberg, Exploration and trapping of mortal random walkers. Phys. Rev. Lett. **110**(22), 220603 (2013)
60. B. Meerson, The number statistics and optimal history of non-equilibrium steady states of mortal diffusing particles. J. Stat. Mech: Theory Exp. **2015**(5), P05004 (2015)
61. D.S. Grebenkov, J.F. Rupprecht, The escape problem for mortal walkers. J. Chem. Phys.**146**(8), 084106 (2017)
62. J. Ma, M. Do, M.A. Le Gros, C.S. Peskin, C.A. Larabell, Y. Mori, S.A. Isaacson, Strong intracellular signal inactivation produces sharper and more robust signaling from cell membrane to nucleus. PLoS Comp Biol **16**(11), e1008356 (2020)
63. S.D. Lawley, The effects of fast inactivation on conditional first passage times of mortal diffusive searchers. SIAM J. Appl. Math. **81**(1), 1–24 (2021)
64. M. Chupeau, O. Bénichou, R. Voituriez, Cover times of random searches. Nat. Phys. **11**(10), 844–847 (2015)
65. S.N. Majumdar, S. Sabhapandit, G. Schehr, Exact distributions of cover times for n independent random walkers in one dimension. Phys. Rev. E **94**(6), 062131 (2016)
66. J. Dong, W. Han, Y. Wang, X. Chen, L. Huang, Universal cover-time distribution of heterogeneous random walks. Phys. Rev. E **107**(2), 024128 (2023)
67. H. Kim and S. D. Lawley. Cover times of many diffusive or subdiffusive searchers. *S*IAM J Appl Math, 2024
68. H. Kim, S.D. Lawley, Cover times of many random walkers on a discrete network. Phys. Rev. E **109**(1), 014121 (2024)
69. D.S. Grebenkov, Depletion of resources by a population of diffusing species. Phys. Rev. E **105**(5), 054402 (2022)
70. S. Linn, S.D. Lawley, Extreme hitting probabilities for diffusion. J. Phys. A: Math. Theor. **55**(34), 345002 (2022)
71. A.J. Bernoff, A. Jilkine, A. Navarro-Hernández, A.E. Lindsay, Single cell directional sensing from just a few receptor binding events. Biophys. J . **122**(15), 3108–3116 (2023)
72. A. E. Lindsay, A. J. Bernoff, and A. Navarro Hernández. Short-time diffusive fluxes over membrane receptors yields the direction of a signalling source. R. Soc. Open Sci., 10(4):221619, 2023
73. B. Karamched, M. Stickler, W. Ott, B. Lindner, Z.P. Kilpatrick, K. Josic, Heterogeneity improves speed and accuracy in social networks. Phys. Rev. Lett. **125**(21), 218302 (2020)
74. M. Stickler, W. Ott, Z.P. Kilpatrick, K. Josic, B.R. Karamched, Impact of correlated information on pioneering decisions. Phys. Rev. Res. **5**(3), 033020 (2023)
75. S.D. Lawley, J. Johnson, Why is there an "oversupply" of human ovarian follicles? Biol. Reprod. **108**(5), 814–821 (2023)
76. S.D. Lawley, J. Johnson, Slowest first passage times, redundancy, and menopause timing. J. Math. Biol. **86**(6), 1–53 (2023)
77. S.J. Richardson, V. Senikas, J.F. Nelson, Follicular depletion during the menopausal transition: evidence for accelerated loss and ultimate exhaustion. J. Clin. Endocrinol. Metab. **65**(6), 1231–1237 (1987)
78. M.J. Faddy, E.C. Jones, R.G. Edwards, An analytical model for ovarian follicle dynamics. J. Exp. Zool. **197**(2), 173–185 (1976)

79. M.J. Faddy, R.G. Gosden, R.G. Edwards, Ovarian follicle dynamics in mice: a comparative study of three inbred strains and an F1 hybrid. J. Endocrinol. **96**(1), 23–33 (1983)
80. A.N. Hirshfield, Development of follicles in the mammalian ovary. Int. Rev. Cytol. **124**, 43–101 (1991)
81. M.J. Faddy, R.G. Gosden, Physiology: A mathematical model of follicle dynamics in the human ovary. Hum. Reprod. **10**(4), 770–775 (1995)
82. C.E. Finch, T.B.L. Kirkwood, *Chance, development, and aging* (Oxford University Press, USA, 2000)
83. J. Johnson, J.W. Emerson, S.D. Lawley, Recapitulating human ovarian aging using random walks. PeerJ **10**, e13941 (2022)
84. J. Johnson, S. D. Lawley, J. W. Emerson, and K. Oktay. Modeling delay of age at natural menopause with planned tissue cryopreservation and autologous transplantation. Am. J. Obstet. Gynecol., 2024
85. J.B. Keller, Diffusion at finite speed and random walks. Proc. Natl. Acad. Sci. **101**(5), 1120 (2004)
86. R. Kuske, J.B. Keller, Large deviation theory for stochastic difference equations. Eur. J. Appl. Math. **8**(6), 567–580 (1997)
87. D.D. Joseph, L. Preziosi, Heat waves. Rev. Mod. Phys. **61**(1), 41 (1989)
88. S.D. Lawley, Extreme first passage times of piecewise deterministic Markov processes. Nonlinearity **34**(5), 2750 (2021)

# Chapter 13
# First-Passage Times for Random Diffusivity Models

**Yann Lanoiselée and Vittoria Sposini**

**Abstract** Random diffusivity models have gained more and more interest over the last decade as they allow for a rather simple mathematical framework to introduce different kinds of system heterogeneity in the description of diffusion dynamics. In this chapter, we briefly recall the general properties of such models and then provide a detailed description of their first-passage time properties. We start by showing how, through the subordination approach, a formal solution of any first-passage time problem can be defined, starting from the analogous solution of standard Brownian diffusion. We present results on the first-passage time density obtained directly from the solution of the associated Fokker-Planck equation. Eventually, we provide applications to special cases and focus on the statistical properties and asymptotic behaviors of one diffusing diffusivity model. A comparison of such a study with the results from standard (homogeneous) Brownian diffusion is also discussed. To conclude, we highlight possible future directions.

## 13.1 Introduction

The concept of random diffusivity is quite recent in statistical physics and was motivated by the pioneering works of Wang and co-authors [1, 2] in which experiments based on single-particle tracking techniques revealed a new kind of diffusive dynamics. The Mean-Square Displacement (MSD) was shown to be linear in time as expected for Brownian motion. However, the distribution of displacement was shown to depart from the canonical Gaussian distribution and resembled more a Laplace distribution, characterized by exponential tails. Such phenomenology, defined as Brow-

Y. Lanoiselée (✉)
Basque Center for Applied Mathematics (BCAM), Mazarredo Zumarkalea, 14, 48009 Bilbo, Bizkaia, Spain
e-mail: ylanoiselee@bcamath.org

V. Sposini
Department of Physics and Astronomy "G. Galilei", University of Padova, Via F. Marzolo 8, 35131 Padova, Italy
e-mail: vittoria.sposini@unipd.it



D. Grebenkov et al. (eds.), *Target Search Problems*,
https://doi.org/10.1007/978-3-031-67802-8_13

nian (or Fickian) yet non-Gaussian (BnG) diffusion, was later observed in many other experimental and computational systems, highlighting its almost universal behavior. For a detailed list of all the systems displaying this behavior, we refer to recent reviews on the topic [3, 4]. One could be surprised that BnG diffusion has not been discovered before. For us, two main factors are responsible for that, first microscopy techniques have considerably improved in the last two decades. Second, the belief that a linear MSD necessarily means Brownian motion had many research articles solely focused on MSD and neglected the analysis of the displacement distribution.

A proposition for explaining BnG diffusion is that the diffusion coefficient is not necessarily the same for each trajectory accumulated to construct the empirical displacement distribution or even that the diffusion coefficient can vary in time for which the 'diffusivity' word is used. From this idea various models have been developed, the most commonly known as *diffusing diffusivity* models [5].

Interestingly enough, random diffusivity models are directly related to the concept of *superstatistics* introduced by Beck and Cohen in the early 2000s [6–9]. The systems under study cannot be described by a single thermodynamic ensemble because of the distributional character of diffusion coefficients. Therefore the thermodynamic ensemble for the whole system is a superposition of equilibrium ensembles pondered by the probability of diffusion coefficients. While this is true for all models that will be presented below, the physical interpretation of superstatistics is different whether the diffusion coefficient is varying or not. When the diffusion coefficient can explore the whole range of possible diffusivities then each particle can explore entirely the phase space, so the system is ergodic. Quantities averaged along a single trajectory over a sufficiently long time will converge to the ensemble-measured quantity. In turn, when each particle has a fixed diffusion coefficient, then each of them explores a different part of phase space, so the system is fundamentally non-ergodic even though each particle can be described at thermodynamic equilibrium. Both interpretations not only are of great interest from a theoretical point of view but also find real-life applications and have been studied in details [3].

This chapter aims to understand how system heterogeneity, described via random diffusivity models, affects the first-passage properties. Indeed, the study of first-passage problems plays a fundamental role in physics as it provides the mathematical setup for a series of real-life processes such as targeting, translocation, triggering, criticality, and resonance [2]. In general, it is not easy to access the full first-passage time distribution and thus most studies focus on the evaluation of the mean first-passage time. While the latter might be useful to determine a rough time scale of the underlying process, it is well known that even for simple Brownian motion the mean first-passage time is not a reliable parameter to describe the full statistics of the process and it is essential to go beyond [10]. In this regard, we will show that for the class of processes described by random diffusivity models, it is possible to work out an expression for the full first-passage time distribution, allowing us to explore different limiting behaviors and compare our results to the case of standard Brownian motion.

### 13.1.1 General Formulation of Random Diffusivity Models

In general, random diffusivity models can be described starting from the following Langevin equation

$$\frac{dx(t)}{dt} = \sqrt{2D(t)}\,\xi(t), \tag{13.1}$$

where $\xi(t)$ represents a Gaussian noise and $D(t)$ is the stochastically varying, *random diffusivity*, which, depending on the specific model, will display different statistical features. Note that, for the sake of clarity, we focus here on the one-dimensional case; a generalization to higher dimensions is straightforward. The Gaussian noise can a priori have arbitrary correlation structure (see for example [11] for a case with fractional Gaussian noise), however, for first-passage time statistics, we limit ourselves from thereon to the case when $\xi(t)$ is a $\delta$-correlated white noise with zero mean. Note that $\xi(t)$ and $D(t)$ are considered to be independent. Regardless of the features of $D(t)$, one can formally rewrite Eq. (13.1) as

$$x(t) = \int_0^t \sqrt{2D(t')}\,\xi(t')dt', \tag{13.2}$$

from which we can obtain the characteristic function of $x(t)$

$$\Phi(w,t) = \left\langle \overline{\exp\left(iwx(t)\right)} \right\rangle_{D(t)} = \left\langle \overline{\exp\left(iw \int_0^t \sqrt{2D(t')}\,\xi(t')dt'\right)} \right\rangle_{D(t)}, \tag{13.3}$$

where the bar indicates the average over the thermal history ($\xi(t)$) and the angular brackets stand for the average over all the realizations of the random diffusivity ($D(t)$). After computing the average over the thermal noise we get

$$\Phi(w,t) = \left\langle \exp\left(-w^2 \int_0^t D(t')dt'\right) \right\rangle_{D(t)}. \tag{13.4}$$

Once we know the characteristic function of $x(t)$, the Probability Density Function (PDF) of the model can be obtained by inverse Fourier transform, that is

$$P(x,t) = \frac{1}{2\pi} \int_{-\infty}^{\infty} e^{-iwx}\,\Phi(w,t)dw, \tag{13.5}$$

and its moments are given by

$$\langle x^k(t)\rangle = (-i)^k \left(\frac{\partial^k}{\partial w^k}\Phi(w,t)\right)\bigg|_{w=0}. \tag{13.6}$$

In particular, the second moment reads

$$\langle x^2(t)\rangle = 2\langle D(t)\rangle t, \tag{13.7}$$

showing that a linear trend of the mean-squared displacement is guaranteed by assuming an equilibrium initial condition for the diffusivity, such that $\langle D(t)\rangle \equiv \bar{D}$.

It is worth noticing that, due to the relation defined in Eq. (13.4), the main features of any random diffusivity model are solely defined by the properties of the integrated random diffusivity $T(t) = \int_0^t D(t')dt'$, as we will also show below.

For BnG diffusion to converge to normal diffusion at long time thus being at equilibrium, the random diffusivity needs to be a stationary process, such that its distribution $\Pi_s(D)$ is time-independent. Moreover, assuming the diffusivity has an exponential relaxation time with characteristic time $\tau$, the non-Gaussian diffusion is observed only for $t \ll \tau$. In the long-time regime ($t \gg \tau$) the Gaussian distribution is recovered.

During the years a series of different random diffusivity models have been studied. The diffusing diffusivity models are usually introduced to mimic medium heterogeneity; they assume the diffusivity to be itself a diffusing process, such that a second Langevin equation is introduced to define its evolution [5, 12–16]. Switching models are based on the idea that the diffusivity can switch randomly between two (or more) different states [13, 17, 18]. Other models associate the random diffusivity to size fluctuations of the diffusive particles [19, 20]. The integrated diffusivity can also be modeled directly by the Continuous Time Random Integrated Diffusivity process [21]. Once again, we refer to the review papers [3, 4] for a general overview of random diffusivity models for BnG diffusion.

## 13.2 First-Passage in Heterogeneous Systems with Random Diffusivity

In this section we report and discuss the first-passage time properties for random diffusivity models, starting from the results available in the literature [18, 22–28]. In Sects. 13.2.1 and 13.2.2 we present two alternative approaches that can be followed to study the first-passage properties of random diffusivity models. Then, in Sect. 13.3 we report the results for some specific models, discuss the limiting behaviors, and compare them with the ones from standard diffusion in homogeneous media.

### *13.2.1 Derivation Through Subordination Integral*

One way to obtain the first-passage time probability for random diffusivity models is to use the subordination approach. Starting from the general Langevin formulation in

Eq. (13.1), we can rescale time by the diffusivity, $dT = D(t)dt$. The rescaled time is called a *subordinator*, that is a new random path variable $T(t) = \int_0^t D(t')dt'$ which is interpreted as the integrated diffusivity over time. Then, the Langevin equation can be rewritten as

$$\frac{dx(T)}{dT} = \sqrt{2}\,\xi(T). \tag{13.8}$$

This formulation is valid regardless of the geometry of the systems and it suggests that, in this new set of variables, the process is simply a standard Brownian Motion (BM). Then, in these new units, the propagator is simply given by the BM one, $P_{\mathrm{BM}}(x, T)$. Finally, to calculate the PDF in real-time units we need to average over the PDF of the subordinator $Q(T, t)$, so that

$$P(x, t) = \int_0^\infty P_{\mathrm{hom}}(x, T)Q(T, t)dT, \tag{13.9}$$

where $P_{\mathrm{hom}}(x, t)$ refers to the PDF for the homogeneous case, corresponding to standard Gaussian diffusion. This integral is known in the literature as *subordination integral* and it establishes that the probability for the diffusive particle to arrive at position $x$ at time $t$ is given by the probability of being at $T$ on the path at time $t$, multiplied by the probability of being at position $x$ for this path length $T$, summed over all path lengths [14].

Equation (13.9) relates directly the statistics of the particle position to the properties of the integrated diffusivity $T(t)$ and suggests that, by knowing the solution of the first-passage time problem for standard BM and the probability $Q(T, t)$ of the integrated diffusivity, one can immediately obtain the first-passage time probability of the specific random diffusivity model. However, it is often not possible to find a full analytic solution of the integral, in such cases, one can focus on the asymptotic regimes at short and long times [14, 15, 21].

The subordination approach is very general, it can also be used to describe time-fractional diffusion [29, 30] which can be seen as a special case of diffusing diffusivity. Also, a subordinator can itself be subordinated to combine different types of heterogeneity [21].

### *13.2.2 Derivation from the Fokker-Planck Equation for Bounded Domains*

Another way to obtain the first-passage time probability density and all similar statistical properties is to make use of the Fokker-Planck (FP) equation [24]. The FP equation for a particle diffusing in a medium without potential and subject to diffusivity fluctuations describes the joint probability of position and diffusivity as a function of time. When the diffusivity is independent of space and solely depends on time, the FP equation can be written as

$$\frac{\partial P(x, D, t|D_0, t)}{\partial t} = D\Delta P + H\left[D, P, \frac{\partial P}{\partial D}, \frac{\partial^2 P}{\partial D^2}\right], \tag{13.10}$$

where $H\left[D, P, \frac{\partial P}{\partial D}, \frac{\partial^2 P}{\partial D^2}\right]$ is the operator describing the evolution of the diffusivity component. Then, it is useful to make a Laplace transform with respect to diffusivity,

$$\tilde{P}(x, s, t|x_0, D_0) = \int_0^\infty P(x, D, t|x_0, D_0)e^{-sD}dD, \tag{13.11}$$

for which the FP equation becomes

$$\frac{\partial \tilde{P}}{\partial t} + \Delta\frac{\partial \tilde{P}}{\partial s} - \mathcal{L}_{D\to s}\left(H\left[D, P, \frac{\partial P}{\partial D}, \frac{\partial^2 P}{\partial D^2}\right]\right) = 0, \tag{13.12}$$

where $\mathcal{L}_{D\to s}(D)$ denotes the Laplace transform with respect to $D$. A no-flux condition ($\frac{\partial P}{\partial D} = 0$) at $D = 0$ is used to ensure positive diffusivity. The propagator $\tilde{P}$ is decomposed on the complete basis of orthonormal Laplacian eigenfunctions, verifying $\Delta u_n(x) + \lambda_n u_n(x) = 0$ in the domain $\Omega$, with the desired limit condition on the boundary $\delta\Omega$. The quantities $\lambda_n$ are the eigenvalues enumerated by $n = 1, 2, \ldots$ in increasing order. For a review of Laplacian eigenfunctions in different geometries see [31]. Moreover, the orthogonality of eigenfunctions allows one to search the propagator $\tilde{P}$ in the form

$$\tilde{P}(x, s, t|x_0, D_0) = \sum_{n=1}^{\infty} u_n(x)u_n(x_0)p(\lambda_n, s, t|D_0). \tag{13.13}$$

Inserting this into Eq. (13.12) yields a first order partial differential equation

$$\frac{\partial p}{\partial t} - \lambda_n\frac{\partial p}{\partial s} - \mathcal{L}_D\left(H\left[D, p, \frac{\partial p}{\partial D}, \frac{\partial^2 p}{\partial D^2}\right]\right) = 0. \tag{13.14}$$

The solution to this equation gives $p(\lambda_n, s, t|D_0)$.

In an experimental setup, the diffusivity is a hidden variable, so the propagator is averaged over diffusivity, simply by setting $s = 0$. We define

$$\Upsilon(t, \lambda_n|D_0) = p(\lambda_n, s = 0, t|D_0), \tag{13.15}$$

note that this quantity is the Laplace transform of the probability density function of the integrated diffusivity $Q(t, T)$

$$\Upsilon(t, \lambda_n|D_0) = \int_0^\infty e^{-\lambda_n T} Q(T, t)dT. \tag{13.16}$$

Then the propagator is averaged over the initial diffusivity, which is distributed as the stationary one at equilibrium. The stationary diffusivity PDF $\Pi_s(D)$ is obtained by solving

$$H\left[D, \Pi_s, \frac{\partial \Pi_s}{\partial D}, \frac{\partial^2 \Pi_s}{\partial D^2}\right] = 0 \tag{13.17}$$

from which one can obtain the marginal PDF

$$\begin{aligned} P(x, t|x_0) &= \int_0^\infty P(x, t|x_0, D_0)\Pi_s(D_0)dD_0 \\ &= \sum_{n=1}^{\infty} u_n(x)u_n(x_0) \int_0^\infty \Upsilon(t, \lambda_n|D_0)\Pi_s(D_0)dD_0. \end{aligned} \tag{13.18}$$

In the case of diffusion in a homogeneous medium (i.e., Brownian motion) with diffusion coefficient $\bar{D}$ we have

$$\Upsilon_{\text{hom}}(t, \lambda_n) = e^{-\bar{D}\lambda_n t}. \tag{13.19}$$

When the particle undergoes a continuous-time random walk with long stalling periods characterized by an anomalous waiting exponent $0 < \alpha < 1$, one gets $\Upsilon_{\text{frac}}(t; \lambda_n) = E_\alpha(-D_\alpha t^\alpha \lambda_n)$, where $E_\alpha(z)$ is the Mittag-Leffler function, and $D_\alpha$ is the (constant) generalized diffusion coefficient [32, 33].

The survival probability $S(t|x_0)$ of a particle inside the domain is obtained by averaging out the space component from the propagator $P(x, t|x_0)$,

$$S(t|x_0) = \sum_{n=1}^{\infty} u_n(x_0)\Upsilon(t, \lambda_n) \int_\Omega u_n(x)dx. \tag{13.20}$$

Then, the first-passage time density is obtained by taking the time derivative with a minus sign,

$$\rho(t|x_0) = -S'(t|x_0) = -\sum_{n=1}^{\infty} u_n(x_0)\Upsilon'(t, \lambda_n) \int_\Omega u_n(x)dx. \tag{13.21}$$

For some special cases, summation techniques can be used to compute an explicit expression for the infinite sum [34].

Finally, the macroscopic flux of molecules going through the boundary $\delta\Omega$ can be obtained by averaging the first-passage time density over the initial concentration profile $c_0(x_0)$,

$$J(t) = \int_\Omega c_0(x_0)\rho(t|x_0)dx_0. \tag{13.22}$$

### 13.2.3 Half Space

We now consider the case of the diffusion in 1 dimension on the half-space $x \in [0, \infty)$ with a boundary condition at $x = 0$ starting from $x_0$. The propagator for diffusion in an unbounded domain in $\mathbb{R}$ is

$$P_{\mathbb{R}}(x, t|x_0) = \int_{\mathbb{R}} \frac{dq}{2\pi} e^{iq(x-x_0)} \Upsilon(t; |q^2|). \tag{13.23}$$

In this case, one can use the image method to deduce the first-passage time properties. For a perfectly absorbing boundary condition the PDF cancels at the boundary $P_{\mathbb{R}_+}(x = 0, t|x_0) = 0$ which can be obtained by placing a copy of the propagator in free space at the mirror point reflection $\hat{x_0}$ with respect to the reactive boundary at $x = 0$ with a negative sign to cancel the probability at the boundary

$$P_{\mathbb{R}_+}(x, t|x_0) = P_{\mathbb{R}}(x, t|x_0) - P_{\mathbb{R}}(x, t|\hat{x_0}). \tag{13.24}$$

The same technique can be applied in the case of reflecting boundary condition by making the mirror reflection contribution positive thus ensuring no flux at the boundary $\frac{\partial}{\partial x} P_{\mathbb{R}_+}(x, t|x_0)\Big|_{x=0} = 0$. In the absorbing case, the propagator simplifies to

$$P(x, t|x_0) = -\frac{i}{\pi} \int_0^\infty dq e^{iqx} \sin(qx_0) \Upsilon(t; q^2). \tag{13.25}$$

By averaging over $x$ we obtain the survival probability,

$$S(t|x_0) = \int_0^\infty dx \, P(x, t|x_0) = \frac{2}{\pi} \int_0^\infty \frac{dq}{q} \sin(qx_0) \Upsilon(t; q^2). \tag{13.26}$$

Finally, the first-passage time probability is given by

$$\rho(t, x_0) = -\frac{2}{\pi} \int_0^\infty \frac{dq}{q} \sin(qx_0) \Upsilon'(t; q^2). \tag{13.27}$$

## 13.3 Statistical Properties and Asymptotic Behaviors

In this section, we discuss the statistical properties and asymptotic behaviors for a specific diffusing diffusivity model.

### 13.3.1 A Model of Diffusing Diffusivity

A suitable model of diffusing diffusivity is the Cox-Ingersoll-Ross (CIR) model that has been introduced in the finance context to model volatility [35]. The first-passage time statistics of diffusion with CIR diffusivity or some special cases of it have been studied in [22, 24, 26]. The dynamics are described by the coupled Langevin equations

$$\begin{cases} dx(t) &= \sqrt{2D(t)}dW_t^1 \\ dD(t) &= \frac{1}{\tau}(\bar{D} - D(t))dt + \sigma\sqrt{2D(t)}dW_t^2, \end{cases} \tag{13.28}$$

where $dW_t^1$ and $dW_t^2$ denote the increments of two independent Wiener processes. The first equation describes the changes in positions due to thermal fluctuations under a time-dependent diffusivity. The second equation describes the changes of the diffusivity. The first term is a harmonic potential that drags the diffusivity toward its equilibrium value $\bar{D}$ with a correlation time $\tau$. The second term represents diffusivity fluctuations with an amplitude $\sigma$. One can note that the fluctuations depend on the value of the diffusivity itself such that fluctuations are small when the diffusivity is small, thus ensuring non-negative diffusivity. The dimensionless parameter

$$\frac{1}{\nu} = \frac{\mathrm{var}(D)}{\mathrm{mean}(D)^2} = \frac{\sigma^2\tau}{\bar{D}}, \tag{13.29}$$

controls the diffusivity fluctuation strength. This can be seen by rescaling the second Langevin equation in Eq. (13.28) by the average diffusivity $D_t^* = D_t/\bar{D}$ and time by the correlation time $\tau$,

$$dD_t^* = \left(1 - D_t^*\right)d(t/\tau) + \sqrt{\frac{1}{\nu}}\sqrt{2D_t^*}dW_{t/\tau}^2. \tag{13.30}$$

The long-time stationary probability density function of diffusivity is Gamma distributed,

$$\Pi_\infty(D) = \frac{\nu^\nu}{\Gamma(\nu)\bar{D}^\nu}D^{\nu-1}\exp(-\nu D/\bar{D}). \tag{13.31}$$

This model can be solved using the Fokker-Planck equation Eq. (13.10) with

$$H\left[D, P, \frac{\partial P}{\partial D}, \frac{\partial^2 P}{\partial D^2}\right] = \frac{1}{\tau}\frac{\partial}{\partial D}((D - \bar{D})P) + \sigma^2\frac{\partial^2}{\partial D^2}(DP), \tag{13.32}$$

The function $\Upsilon(t, \lambda_n|D_0)$ for this model is known exactly [24],

$$\Upsilon(t, \lambda_n|D_0) = \left( \frac{2\omega e^{-(\omega-1)t/(2\tau)}}{(\omega+1) - (\omega-1)e^{-\omega t/\tau}} \right)^{\nu} \times \exp\left( -\frac{D_0(\omega+1)}{2\sigma^2\tau} \left( 1 - \frac{2\omega}{(\omega+1) - (\omega-1)e^{-\omega t/\tau}} \right) \right), \tag{13.33}$$

where $\omega = \sqrt{1 + 4\sigma^2\tau^2\lambda_n}$.

For diffusion in an infinite domain, at short time, the propagator presents exponential tails, and the shape at the center of the distribution is governed by the strength of the diffusivity fluctuations: It can be peaked when $\frac{1}{\nu} > 1$, purely Laplace when $\frac{1}{\nu} = 1$ and finally Gaussian-like when $\frac{1}{\nu} < 1$. At a time longer than the diffusivity correlation time $t \gg \tau$ the propagator converges to the Gaussian distribution as required for a process at equilibrium.

### 13.3.2 Limiting Behaviors and Comparison with Homogeneous Case

We study the limiting behavior of this diffusing diffusivity model with respect to two quantities. The first one is the disorder strength computed from the stationary distribution of diffusivity in Eq. (13.29), which expresses how large are the fluctuations of diffusivity compared to the mean. The second quantity compares the disorder lengthscale $\sigma\tau$ – which is the average lengthscale explored by the diffusing molecule during the correlation time $\tau$ – to the half-size $R$ of the domain (e.g., the radius in spherical geometries),

$$\mu = \frac{\sigma\tau}{R}. \tag{13.34}$$

When $\mu \ll 1$ with fixed $\nu$ the diffusivity fluctuations are huge but span over very short lengthscales such that they only affect the short-time behavior of the first-passage time density. In turn, when $\mu \gg 1$ diffusivity fluctuations are still relevant even when reaching the target and thus affect the whole distribution of first-passage events.

The function $\Upsilon(t, \lambda_n)$ can be rewritten as a function of $\mu$ and $\nu$,

$$\Upsilon(t, \lambda_n) = \left( \frac{4\omega e^{-(\omega-1)\hat{t}/(2\nu\mu^2)}}{(\omega+1)^2 - (\omega-1)^2 e^{-\omega\hat{t}/(2\nu\mu^2)}} \right)^{\nu}, \tag{13.35}$$

where $\hat{t} = \bar{D}t/R^2$ and $\omega = \sqrt{1 + 4\mu^2R^2\lambda_n}$.

In the case when $1/\nu \to 0$ with $\mu$ fixed, the diffusivity fluctuations are very small but their correlation time is very long. The exponential term $e^{-\omega\hat{t}/(2\nu\mu^2)}$ in Eq. (13.35) can be expanded for a fixed $t$ such that

$$\Upsilon(t, \lambda_n) = \Upsilon_{\text{hom}}(t, \lambda_n) + \mathcal{O}(1/\nu). \tag{13.36}$$

As $1/\nu$ gets smaller, the first-passage time pdf $\rho(t, |x_0)$ converges to that of the homogeneous case, this is expected because when diffusivity fluctuations are vanishingly small the diffusivity becomes constant.

In the limit $\mu \to 0$ with fixed $1/\nu$ one can expand $\omega = \sqrt{1 + 4\mu^2 R^2 \lambda_n} \sim 1 + 2\mu^2 R^2 \lambda_n + \mathbb{O}(\mu^4)$ from which

$$\Upsilon(t, \lambda_n) = \Upsilon_{\text{hom}}(t, \lambda_n) + \mathcal{O}(\mu^2). \tag{13.37}$$

Once again the function $\Upsilon(t, \lambda_n)$ converges to that for homogeneous diffusion in this limit. However, the probability density $\rho(t|x_0)$ remains different from $\rho_{\text{hom}}(t|x_0)$ at short times. This is because $\mu$ appears in Eq. (13.35) as a product with the Laplacian eigenvalues $\lambda_n$ that grows to infinity as $n$ increases. Regardless of the smallness of the parameter $\mu$, there exists an index $n_0$ such that for all $n > n_0$ the above expansion is not applicable. In other words, for any $\mu > 0$, there remain infinitely many terms $\Upsilon'(t, \lambda_n)$ that significantly differ from its homogeneous counterpart. As these terms determine the short-time asymptotic behavior of the probability density function, $\rho(t|x_0)$ exhibits deviations from $\rho_{\text{hom}}(t|x_0)$ at (very) short times for any $\mu > 0$.

For Brownian motion with diffusion coefficient $\bar{D}$, the first-passage time has a broad distribution that can span over several orders of magnitude. A random diffusivity widens the span of the distribution even more. A small diffusivity slows down the first-passage times thus contributing to the tail of the distribution. In turn, a large diffusivity accelerates the first reaction time thus contributing positively to the probability of short reaction times—note that we are working under the assumption of perfectly absorbing targets. The reaction times are thus controlled by two factors. The first factor is whether the path to the target is direct or not just as in the homogeneous case [36, 37]. The second factor is how fast are these paths explored depending on the diffusivity dynamics. The first particles to reach the target take a direct path with high diffusivity, and the last molecules to reach the boundary jiggle around for a long time before reaching the boundary and experiencing mostly low diffusivity. For intermediate reaction time, some more direct paths for homogeneous diffusion can be explored slower due to low diffusivity and conversely, some long paths can be explored faster due to high diffusivity.

### 13.3.3 Short-Time Behavior and Superposition Case

The short-time behavior for this diffusing diffusivity model can be obtained exactly by noting that at times much smaller than the correlation time, $t \ll \tau$, and when $(x - x_0)\sqrt{\nu/(\bar{D}t)} \gg 1$ the diffusivity remains very close to its initial value. The integral representation of the position of a particle undergoing diffusing diffusivity Eq. (13.2) can be simplified by replacing the time-dependent $D(t)$ by its initial value

$D(0)$ as follows

$$x(t) = \int_0^t \sqrt{2D(0)}\,\xi(t')dt'. \tag{13.38}$$

In this limit, each particle experiences a different but unique diffusion coefficient, such that the system is effectively represented as a collection of independent Brownian motions with random but constant diffusion coefficients. The propagator is then a superposition of Brownian propagators,

$$P_{\text{hom}}(x, D(0), t|x_0) = \sum_{n=1}^{\infty} u_n(x)u_n(x_0)\exp(-D(0)\lambda_n t) \tag{13.39}$$

weighted by the initial probability density of the diffusivity, that is,

$$P(x, t|x_0) = \int_0^{\infty} P_{\text{hom}}(x, D(0), t|x_0)\Pi_s(D(0))dD(0), \tag{13.40}$$

from which it is deduced that

$$P(x, t|x_0) = \sum_{n=1}^{\infty} u_n(x)u_n(x_0)\Upsilon_{\text{superpos}}(t, \lambda_n), \tag{13.41}$$

where $\Upsilon_{\text{superpos}}(t, \lambda_n)$ is the Laplace transform of the stationary probability density of the diffusion coefficient,

$$\Upsilon_{\text{superpos}}(t, \lambda_n) = \int_0^{\infty} e^{-\lambda_n D(0)t}\,\Pi_s(D(0))dD(0). \tag{13.42}$$

The first-passage time density is then obtained by averaging out the space component,

$$\rho(x, t|x_0) = -\sum_{n=1}^{\infty} u_n(x_0)\Upsilon'_{\text{superpos}}(t, \lambda_n)\int_{\Omega} u_n(x)dx. \tag{13.43}$$

This description is also perfectly suitable for systems where the diffusion coefficient never changes but is distributed across particles. In this case, the paths to reach the target are always similar to the homogeneous case but are explored at different yet constant 'speeds'. So when considering such a superposition, small diffusion coefficients contribute more to the tails while high diffusion coefficients contribute more to the short-time behavior.

### 13.3.4 Macroscopic Reaction Rate

Smoluchowski obtained the historical macroscopic reaction rate [38] in the case of diffusion from the exterior of a ball of radius $R$. While the macroscopic reaction rate depends on the geometry of the problem, the case of the exterior of the ball is particularly interesting because it explicitly includes the possibility for the diffusing molecules to escape at infinity instead of necessarily reaching the target. As discussed in [24], in 3D spherical coordinates the propagator for unbounded diffusion is

$$P_{\mathbb{R}^3}(r,t|r_0) = \int_0^\infty \frac{\cos(q(r-r_0)) - \cos(q(r+r_0))}{4\pi^2 r r_0} \Upsilon(t,q^2)dq, \tag{13.44}$$

where $r_0$ is the initial distance from the center of the ball (where $r_0 > R$). Note that the coordinates are oriented in the direction of the initial point $x_0 = (r_0, 0, 0)$. The propagator outside the ball can be obtained in the form

$$P(r,t|r_0) = \int_0^\infty \frac{\cos(q(r-r_0)) - \cos(q(r+r_0-2R))}{4\pi^2 r r_0} \Upsilon(t,q^2)dq. \tag{13.45}$$

From this one can deduce the survival probability,

$$S(t,r_0) = \frac{r_0 - R}{r_0} + \frac{2R}{\pi r_0} \int_0^\infty \frac{\sin(q(r_0-R))}{q} \Upsilon(t,q^2)dq, \tag{13.46}$$

where the first term is non-vanishing and corresponds to the probability of escaping at infinity. The first-passage probability density is then

$$\rho(t|r_0) = -\frac{2R}{\pi r_0} \int_0^\infty \frac{\sin(q(r_0-R))}{q} \Upsilon'(t,q^2)dq, \tag{13.47}$$

that is not normalized due to the non-zero probability of escaping to infinity. The macroscopic reaction rate is obtained by averaging Eq. (13.47) with a uniform initial concentration $c_0$,

$$J(t) = 4\pi c_0 R\bar{D}\left(1 - \frac{2R}{\pi\bar{D}} \int_0^\infty \frac{\Upsilon'(t,q^2)}{q^2} dq\right). \tag{13.48}$$

In the case of Brownian motion we have $\Upsilon_{\mathrm{hom}}(t,q^2) = e^{-\bar{D}tq^2}$ from which

$$J_{\mathrm{hom}}(t) = 4\pi c_0 R\bar{D}\left(1 + \frac{R}{\sqrt{\pi\bar{D}t}}\right). \tag{13.49}$$

At long times, the second term vanishes in both Eqs. (13.48) and (13.49) to yield the same stationary reaction rate

$$J_s(t) = 4\pi c_0 R\bar{D}, \tag{13.50}$$

so the diffusing diffusivity only alters the convergence to the stationary state but not its final expression.

In turn for the superposition case or at short time for diffusing diffusivity where $\Upsilon(t, q^2) = (1 + q^2\bar{D}t/\nu)^{-\nu}$ the flux is

$$J_{\text{superpos}}(t) = 4\pi c_0 R\bar{D}\left(1 + \frac{\Gamma(\nu + 1/2)}{\sqrt{\nu}\Gamma(\nu)}\frac{R}{\sqrt{\pi\bar{D}t}}\right). \tag{13.51}$$

This relation is close to Eq. (13.49) for Brownian motion with mean diffusivity $\bar{D}$, in which the time-dependent term is multiplied by the prefactor $\frac{\Gamma(\nu+1/2)}{\sqrt{\nu}\Gamma(\nu)}$ depending only on $\nu$. This prefactor monotonously grows from 0 to 1 as $\nu$ increases from 0 to infinity (the limit $\nu \to \infty$ corresponding to Brownian motion). As a consequence, the dynamic disorder tends to diminish the macroscopic reaction rate. The approximate asymptotic relation in (13.51) does not depend on the disorder scale $\mu$ (which is undefined for the superposition case). According to the superposition approximation, this relation is the lower bound for the flux $J(t)$ corresponding to the limit $\tau \to 0$ or, equivalently, $\mu \to \infty$. In turn, the exact expression in Eq. (13.49) for Brownian motion corresponds to the limit $\mu \to 0$ and thus is close to the upper bound for $J(t)$.

## 13.4 Conclusion and Outlook

In this chapter, we discussed how to obtain the first-passage time properties of random diffusivity models. On the one hand, the subordination approach allows us to define a formal solution to any first-passage time problem, starting directly from the analogous solution of standard Brownian diffusion. On the other hand, we can solve directly the associated FP equations and obtain the first-passage time density. By looking at one specific case of random diffusivity model we have shown that the effect of random diffusivity is to widen the domain of the first-passage time distribution. While the small diffusivity values mostly dominate the tail of the distribution, the fast diffusivities contribute to the short reaction times.

It is important to notice that the formulations introduced here are very general, however, only examples considering particles diffusing in a fixed volume and searching for immobile, perfectly reactive targets have been considered so far. As demonstrated in this chapter, these studies are crucial to understanding what is the general role of the heterogeneity typical of random diffusivity models in targeting phenomena, however, to gain full knowledge it is essential to move to more realistic problem setups. One could start by introducing a more detailed description of the targets and define the role of their aspect ratio and reactivity [39] or study the dynamics of particles near a reactive surface [40, 41]. Alternatively, one could consider different geometries or confining potentials [42] and study for instance the narrow escape problem, which focuses on the mean first-passage time of a diffusion process to a small target on the reflecting boundary of a bounded domain [43]. Another central point is the study of the relevance of multiple searchers in various situations, that is the calculation of the fastest first-passage time of $N$ random walkers [44, 45].

Finally, it would be interesting to understand how much of the methodology developed here could be extended to other models dealing with deterministically or stochastically evolving parameters. In particular, in analogy with the models discussed in this chapter that start from the standard Brownian motion and introduce a random diffusivity, a series of models have been developed in the literature that start from the so-called fractional Brownian motion [46, 47] and introduce different variability in the typical parameters of the model [48–54].

## References

1. B. Wang, S.M. Antony, S.C. Bae, S. Granick, Anomalous yet Brownian process. Proc. Natl. Acad. Sci. USA **106**, 15160 (2009)
2. B. Wang, J. Kuo, S.C. Bae, S. Granick, When Brownian diffusion is not Gaussian. Nat. Mater. **11**, 481 (2012)
3. R. Metzler, Superstatistics and non-Gaussian diffusion. Eur. Phys. J. Spec. Top. **229**, 711–728 (2020)
4. R. Metzler, A.V. Chechkin, Non-Gaussianity in stochastic transport: phenomenology and modelling (2022). arXiv:2204.01048
5. M.V. Chubynsky, G.W. Slater, Diffusing diffusivity: a model for anomalous, yet Brownian, diffusion. Phys. Rev. Lett. **113**, 098302 (2014)
6. C. Beck, E.G.D. Cohen, Superstatistics. Phys. A **322**, 267 (2003)
7. C. Beck, E.G.D. Cohen, H.L. Swinney, From time series to superstatistics. Phys. Rev. E **72**, 056133 (2005)
8. C. Beck, Superstatistical Brownian motion. Prog. Theor. Phys. Suppl. **162**, 29 (2006)
9. S. Abe, C. Beck, E.G.D. Cohen, Superstatistics, thermodynamics, and fluctuations. Phys. Rev. E **76**, 031102 (2007)
10. A. Godec, R. Metzler, First passage time distribution in heterogeneity controlled kinetics: going beyond the mean first passage time. Sci. Rep. **6**, 20349 (2016)
11. W. Wang, A.G. Cherstvy, A.V. Chechkin, S. Thapa, F. Seno, X. Liu, R. Metzler, Fractional Brownian motion with random diffusivity: emerging residual nonergodicity below the correlation time. J. Phys. A Math. Theor. **53**, 474001 (2020)
12. R. Jain, K.L. Sebastian, Diffusion in a crowded, rearranging environment. J. Phys. Chem. B **120**, 3988–3992 (2016)
13. N. Tyagi, B.J. Cherayil, Non-Gaussian Brownian diffusion in dynamically disordered thermal environments. J. Phys. Chem. B **121**, 7204–7209 (2017)
14. A.V. Chechkin, F. Seno, R. Metzler, I.M. Sokolov, Brownian yet non-Gaussian diffusion: from superstatistics to subordination of diffusing diffusivities. Phys. Rev. X. **7**, 021002 (2017)
15. V. Sposini, A.V. Chechkin, F. Seno, G. Pagnini, R. Metzler, Random diffusivity from stochastic equations: comparison of two models for Brownian yet non-Gaussian diffusion. New J. Phys. **20**, 043044 (2018)
16. Y. Lanoiselée, D.S. Grebenkov, A model of non-Gaussian diffusion in heterogeneous media. J. Phys. A Math. Theor. **51**, 145602 (2018)
17. T. Miyaguchi, T. Akimoto, E. Yamamoto, Langevin equation with fluctuating diffusivity: a two-state model. Phys. Rev. E **94**, 012109 (2016)

18. D.S. Grebenkov, A unifying approach to first-passage time distributions in diffusing diffusivity and switching diffusion models. J. Phys. A Math. Theor. **52**, 174001 (2019)
19. F. Baldovin, E. Orlandini, F. Seno, Polymerization induces non-Gaussian diffusion. Front. Phys **7**, 124 (2019)
20. E. Barkai, S. Burov, Packets of diffusing particles exhibit universal exponential tails. Phys. Rev. Lett. **124**, 060603 (2020)
21. Y. Lanoiselée, D.S. Grebenkov, Non-Gaussian diffusion of mixed origins. J. Phys. A Math. Theor. **52**, 304001 (2019)
22. R. Jain, K.L. Sebastian, Diffusing diffusivity: survival in a crowded rearranging and bounded domain. J. Phys. Chem. B **120**, 9215–9222 (2016)
23. N. Tyagi, B.J. Cherayil, Enhancement of target location rates by stochastic modulation of Brownian motion. J. Stat. Mech. Theory Exp. **2018**, 063208 (2018)
24. Y. Lanoiselée, N. Moutal, D.S. Grebenkov, Diffusion-limited reactions in dynamic heterogeneous media. Nat. Commun. **9**, 4398 (2018)
25. R. Jain, K.L. Sebastian, Diffusing diffusivity: Fractional Brownian oscillator model for subdiffusion and its solution. Phys. Rev. E **98**, 052138 (2018)
26. V. Sposini, A.V. Chechkin, R. Metzler, First passage statistics for diffusing diffusivity. J. Phys. A Math. Theor. **52**, 04LT01 (2019)
27. K. Goswami, K.L. Sebastian, Exact solution to the first-passage problem for a particle with a dichotomous diffusion coefficient. Phys. Rev. E **102**, 042103 (2020)
28. D.S. Grebenkov, V. Sposini, R. Metzler, G. Oshanin, F. Seno, Exact first-passage time distributions for three random diffusivity models. J. Phys. A Math. Theor. **54**, 04LT01 (2021)
29. M. Magdziarz, A. Weron, J. Klafter, Equivalence of the fractional Fokker-Planck and Subordinated Langevin Equations: The case of a Time-Dependent Force. Phys. Rev. Lett. **101**, 210601 (2008)
30. N. Leonenko, E. Pirozzi, First passage times for some classes of fractional time-changed diffusions. Stoc. Anal. Appl. **40**(4), 735–763 (2022)
31. D.S. Grebenkov, B.-T. Nguyen, Geometrical structure of Laplacian eigenfunctions. SIAM Rev. **55**(4), 601–667 (2013)
32. K.S. Fa, E.K. Lenzi, Time-fractional diffusion equation with time-dependent diffusion coefficient. Phys. Rev. E **72**, 011107 (2005)
33. D.S. Grebenkov, Subdiffusion in a bounded domain with a partially absorbing-reflecting boundary. Phys. Rev. E **81**, 021128 (2010)
34. D.S. Grebenkov, A physicist's guide to explicit summation formulas involving zeros of Bessel functions and related spectral sums. Rev. Math. Phys. **33**(03), 2130002 (2021)
35. J.C. Cox, J.E. Ingersoll, S.A. Ross, Econometrica **53**, 385–408 (1985)
36. A. Godec, R. Metzler, Universal proximity effect in target search kinetics in the few-encounter limit. Phys. Rev. X. **6**, 041037 (2016)
37. D.S. Grebenkov, R. Metzler, G. Oshanin, Strong defocusing of molecular reaction times results from an interplay of geometry and reaction control. Commun. Chem. **1**, 96 (2018)
38. M. Smoluchowski, Versuch einer mathematischen Theorie der Koagulationskinetic kolloider Lösungen. Z. Phys. Chem. **129**, 129–168 (1917)
39. D.S. Grebenkov, R. Metzler, G. Oshanin, Effects of the target aspect ratio and intrinsic reactivity onto diffusive search in bounded domains. New J. Phys. **19**, 103025 (2017)
40. A.V. Chechkin, I.M. Zaid, M.A. Lomholt, I.M. Sokolov, R. Metzler, Bulk-mediated diffusion on a planar surface: full solution. Phys. Rev. E **86**, 041101 (2012)
41. D.S. Grebenkov, Paradigm shift in diffusion-mediated surface phenomena. Phys. Rev. Lett. **125**, 078102 (2020)
42. Y. Lanoiselée, A. Stanislavsky, D. Calebiro, A. Weron, Temperature and friction fluctuations inside a harmonic potential. Phys. Rev. E **106**, 064127 (2022)
43. D. Holcman, Z. Schuss, The narrow escape problem. SIAM Rev. **56**, 213–257 (2014)
44. Z. Schuss, K. Basnayake, D. Holcman, Redundancy principle and the role of extreme statistics in molecular and cellular biology. Phys. Life Rev. **28**, 52–79 (2019)

45. D.S. Grebenkov, R. Metzler, G. Oshanin, From single-particle stochastic kinetics to macroscopic reaction rates: fastest first-passage time of *N* random walkers. New J. Phys. **22**, 103004 (2020)
46. A.N. Kolmogorov, Wienersche Spiralen und einige andere interessante Kurven im Hilbertschen Raum, C. R. (Dokl). Acad. Sci. URSS (NS) **26**, 115 (1940)
47. B.B. Mandelbrot, J.W. van Ness, Fractional Brownian motions, fractional noises and applications. SIAM Rev. **10**, 422 (1968)
48. D. Molina-García, T.M. Pham, P. Paradisi, C. Manzo, G. Pagnini, Fractional kinetics emerging from ergodicity breaking in random media. Phys. Rev. E **94**, 052147 (2016)
49. A. Maćkała, M. Magdziarz, Statistical analysis of superstatistical fractional Brownian motion and applications. Phys. Rev. **99**, 012143 (2019)
50. M. Balcerek, K. Burnecki, S. Thapa, A. Wyłomańska, A.V. Chechkin, Fractional Brownian motion with random Hurst exponent: accelerating diffusion and persistence transitions. Chaos **32**, 093114 (2022)
51. W. Wang, M. Balcerek, K. Burnecki, A.V. Chechkin, S. Janušonis, J. Ślezak, T. Vojta, A. Wyłomańska, R. Metzler, Memory-multi-fractional Brownian motion with continuous correlations. Phys. Rev. Res. **5**, L032025 (2023)
52. N. Korabel, A. Taloni, G. Pagnini, V. Allan, S. Fedotov, T.A. Weigh, Ensemble heterogeneity mimics ageing for endosomal dynamics within eukaryotic cells. Sci. Rep. **13**, 8789 (2023)
53. M. Balcerek, A. Wyłomańska, K. Burnecki, R. Metzler, D. Krapf, Modelling intermittent anomalous diffusion with switching fractional Brownian motion. New J. Phys. **25**, 103031 (2023)
54. J. Ślezak, R. Metzler, Minimal model of diffusion with time changing Hurst exponent. J. Phys. A Math. Theor. **56**, 35LT01 (2023)

# Chapter 14
# Random Resetting in Search Problems

**Arnab Pal, Viktor Stojkoski, and Trifce Sandev**

**Abstract** By repeatedly returning a search process to a known or random state, random resetting possesses the potential to unveil new trajectories, sidestep potential obstacles, and consequently enhance the efficiency of locating desired targets. In this chapter, we highlight the pivotal theoretical contributions that have enriched our understanding of random resetting within an abundance of stochastic processes, ranging from standard diffusion to its fractional counterpart. We also touch upon the general criteria required for resetting to improve the search process, particularly when the distribution describing the time needed to reach the target is broader compared to a normal one. Building on this foundation, explore real-world applications where resetting optimizes the efficiency of reaching the desired outcome, spanning, topics from home-range search, ion transport to the intricate dynamics of income. Conclusively, the results presented in this chapter offer a cohesive perspective on the multifaceted influence of random resetting across diverse fields.

A. Pal
The Institute of Mathematical Sciences, Chennai 600113, India
e-mail: arnabpal@imsc.res.in

Homi Bhabha National Institute, Mumbai 400094, India

V. Stojkoski
Faculty of Economics, Ss. Cyril and Methodius University, Skopje 1000, Macedonia
e-mail: vstojkoski@eccf.ukim.edu.mk

T. Sandev (✉)
Macedonian Academy of Sciences and Arts, Skopje 1000, Macedonia
e-mail: trifce.sandev@manu.edu.mk

Institute of Physics, Faculty of Natural Sciences and Mathematics, Ss. Cyril and Methodius University, Skopje 1000, Macedonia



D. Grebenkov et al. (eds.), *Target Search Problems*,
https://doi.org/10.1007/978-3-031-67802-8_14

## 14.1 Introduction

*When in doubt, reset!*

Random resetting is a concept that has gained substantial attention in recent years as a novel and somewhat counter-intuitive approach to solving complex search problems. This technique has provided groundbreaking insights into the stochastic nature of search processes whose behavior can often confound traditional models and approaches [1]. In this chapter, we delve into the fascinating interplay between random resetting and search processes. We explore how resetting randomly to a new location [1–11] acts as a mechanism that provides new insights into the dynamics of various stochastic systems, enhancing our ability to predict and control their behavior.

*But why would random resetting improve the modeling of search processes?*

By repeatedly returning the search process to a known or random state, random resetting uncovers new pathways, avoids potential pitfalls, and can lead to an unexpected efficiency in reaching desired targets [1, 2, 12–15]. Consider home-range search, a realm where the influence of random resetting has yielded unprecedented advances in efficiency and understanding [16–18]. In ecological studies, animals often search within a specific home range to find food, mates, or shelter. The home is an important basis since the foraging animals and birds return to their dens and nests respectively to rest, rescue searchers return to their bases to resupply, and drones or patrol aircraft return to their docking stations to recharge or refuel. Traditional models may struggle to accurately represent the stochastic nature of this behavior. The introduction of random resetting, where the search process is repeatedly returned to the "home" of the searcher or a random state, helps to model the realistic unpredictability of search patterns within an animal's (broadly, of a searcher's) home range (see Fig. 14.1). This approach not only simulates the dynamic interplay between an animal's or a searcher's instinctive behavior and random environmental factors, leading to more accurate representations, but also reveals that search with returns at home can outperform free-range search, particularly in conditions of high uncertainty [16].

Now consider the field of economics, specifically in the modeling of individual income dynamics, where random resetting introduces a vital perspective [19–21]. Understanding the time needed for an individual to improve their income is crucial for various economic decisions, from personal financial planning to governmental policy-making [22]. Traditional models, however, may overlook the complex and often abrupt changes in an individual's financial situation, such as losing a job or retirement. By incorporating random resetting, which can mimic these sudden transitions, economists are provided with a more realistic portrayal of the time dynamics involved in income improvement. This approach not only offers a nuanced depiction of income trajectories but also recognizes the stabilizing effect of these "reset" points, acknowledging their real-world relevance.

But the application of random resetting in search problems goes way beyond these two examples. From the intricate pathways of cellular biology [23–26], chemical reactions [26–29], foraging [16, 30–32], nonlinear dynamical systems [33],

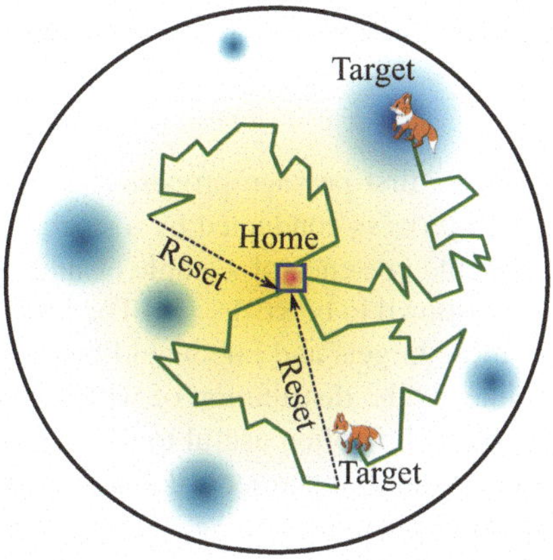


**Fig. 14.1** Schematic for a stochastic search process under resetting in a confined topography in the presence of multiple targets (dark green circles). The searcher (e.g., a prairie fox) looks for targets (which can be food, companion, or other resources) in the vicinity of its home and intermittently returns there to take rest. Following this *spatial reset*, the searcher resumes its search, more likely wandering to a different direction. The search is completed after it has found one or multiple targets in the domain. The central aim here is to understand the behavior of the mean search time under random resetting, in particular, as a function of the frequency of resetting or return

operation research [34] to computer algorithms [35, 36], the principles of random resetting continue to offer fresh insights and innovative solutions. The versatility and adaptability of this concept have allowed researchers and practitioners to tackle previously intractable challenges, forging connections between disparate fields and creating a cohesive understanding of the underlying dynamics of search processes.

In this chapter, we systematically explore the realm of search processes with random resetting. In Sect. 14.2, we introduce a robust mathematical framework, elucidating the core concepts and recent advancements that have shaped our understanding of search under the influence of random resetting. Moving forward, in Sect. 14.3, we delve into examples of search processes under resetting, offering a detailed examination of the intriguing theoretical behaviors and patterns that arise from the interplay between search strategies and random resets. This includes a focus on how random resetting can influence metrics such as first-passage time. Next, in Sect. 14.4, we explore the delicate balance between too much and too little resetting, investigating strategies that maximize efficiency in search processes by optimizing the resetting rate. Then, in Sect. 14.5, we showcase diverse applications of search processes under random resetting, demonstrating the wide-ranging impact of this approach on contemporary science and technology, as well as its practical solutions to real-world challenges. Finally, we synthesize the insights gleaned from our exploration, highlighting both the theoretical underpinnings and practical implications of incorporating random resetting into search processes.

## 14.2 The Renewal Formalism for First-Passage Under Resetting

In this section, we discuss the first-passage properties of a resetting process. Let us consider a generic search process conducted by a stochastic searcher $\mathbf{x}(t)$ in an arbitrary finite domain $\mathcal{D}$ in the presence of a single or multiple targets. The initial condition of the process is assumed to be $\mathbf{x}(0) = \mathbf{x_0}$. The process is completed when the searcher finds one of these targets and one would be interested in the statistics of the stochastic search time $T$, the distribution of which is described by the first-passage time density $f_T(t|\mathbf{x_0})$.[1] It is often useful to introduce $Q_0(t|\mathbf{x_0})$ that defines the survival probability of the underlying process that it has not found any target until time $t$, starting from the initial configuration $\mathbf{x_0}$. Clearly, this quantity is related to the position probability density function, denoted by $P_0(\mathbf{x}, t|\mathbf{x_0}, 0)$, of the searcher in the following way: $Q_0(t|\mathbf{x_0}) = \int_{\mathcal{D}} d\mathbf{x} P_0(\mathbf{x}, t|\mathbf{x_0}, 0)$. In the absence of targets, the searcher always survives; thus, $Q_0(t|\mathbf{x_0}) = 1$. Otherwise, it decays to zero at long times as the searcher eventually finds the target (assuming that the search is being conducted in confined domains). In other words, the search process is completed.

Let us consider that the underlying first-passage process is intermittently stopped and restarted from some pre-selected configuration. Further, one can assume that after each resetting event, the searcher goes back either to a fixed location $\mathbf{x_R}$ or locations drawn from a distribution $P(\mathbf{x_R})$. The waiting times between the resetting events are drawn from a probability density $f_R(t)$. Under this mechanism, the new first-passage time is denoted by $T_r$ which is distributed via the first-passage time density $f_{T_r}(t|\mathbf{x_0})$. The subscript "$r$" is used to indicate the presence of resetting events. Here, $Q_r(t|\mathbf{x_0}; \mathbf{x_R})$ denotes the survival probability in the presence of resetting. In other words, this measures the probability to find the searcher inside the domain $\mathcal{D}$ up to time $t$ given that it had started at $\mathbf{x_0}$ at time zero and experienced multiple resetting to $\mathbf{x_R}$. In this chapter, we assume that the waiting times between resetting are drawn from $f_R(t) = re^{-rt}$ which essentially means that resetting occurs at a rate $r$ and the average waiting time between resetting events is given by $1/r$.

Since each resetting event compels the searcher to start from scratch, the first-passage process under resetting generically belongs to the broad class of stochastic renewal processes. Adapting the mathematical structure from there, one can write a renewal equation for the survival probability $Q_r(t|\mathbf{x_0}; \mathbf{x_R})$ in the following way [1, 2]:

$$Q_r(t|\mathbf{x_0}; \mathbf{x_R}) = e^{-rt} Q_0(t|\mathbf{x_0}) + r \int_0^t d\tau_l e^{-r\tau_l} Q_0(\tau_l|\mathbf{x_R}) Q_r(t - \tau_l|\mathbf{x_0}; \mathbf{x_R}). \quad (14.1)$$

Equation (14.1) has a simple interpretation. The first term on the right-hand side implies that the particle survives till time $t$ without experiencing any reset event. The second term considers the possibility when there are multiple reset events. One can

[1] In this chapter, we have used the notations $f_Z(t)$ and $\langle Z \rangle$ to denote, respectively, the probability density function and the expectation of a non-negative random variable $Z$.

then look at a long trajectory where the last reset event had occurred at time $t - \tau_l$, and after that there has been no reset for the duration $\tau_l$. The probability of this joint event is given by $r d\tau_l e^{-r\tau_l}$. But then this has to be multiplied by $Q_r(t - \tau_l|\mathbf{x_0}; \mathbf{x_R})$, i.e., the probability that the particle survives till time $t - \tau_l$ with multiple reset events. Finally, $Q_0(\tau_l|\mathbf{x_R})$, is the survival probability of the particle for the last non-resetting interval $\tau_l$, starting from $\mathbf{x_R}$. This is a useful formula since the survival properties for the resetting process can be directly understood from the reset-free processes. Notably, the construction of the above equation relied upon the last resetting event, hence this approach is often known as the *last renewal formalism*. Similar to this, a *first renewal formalism* can also be developed for the survival probability. The corresponding renewal equation reads [37]

$$Q_r(t|\mathbf{x_0}; \mathbf{x_R}) = e^{-rt} Q_0(t|\mathbf{x_0}) + r \int_0^t d\tau_f e^{-r\tau_f} Q_0(\tau_f|\mathbf{x_0}) Q_r(t - \tau_f|\mathbf{x_R}; \mathbf{x_R}), \tag{14.2}$$

where we assume that the first resetting occurs at time $\tau_f$, and up to that there was no resetting (with the probability $e^{-r\tau_f}$). Until then the particle survives with $Q_0(\tau_f|\mathbf{x_0})$—the reset free survival probability. For the remaining time interval $t - \tau_f$, the particle survives with many resetting events—the probability of which is assigned by $Q_r(t - \tau_f|\mathbf{x_R}; \mathbf{x_R})$. It is easy to see that both (14.1) and (14.2) are identical (see [1, 37]). Notably, the renewal formalism does not require any particular choice of the underlying dynamics, i.e., the renewal equations hold both for Markovian and non-Markovian anomalous search processes. The only key assumption here is that no memory of the dynamics can be carried forward from one interval to the next. Crucially, as we show below, it is not required to have the expression for the survival probability $Q_0(t|\cdot)$ of the underlying process in the real time domain—remarkably, an expression in the Laplace space is sufficient to extract further information. To see this, let us apply the Laplace transform on both sides of Eq. (14.1) which satisfies

$$\widetilde{Q}_r(s|\mathbf{x_0}; \mathbf{x_R}) = \frac{\widetilde{Q}_0(s + r|\mathbf{x_0})}{1 - r\widetilde{Q}_0(s + r|\mathbf{x_R})}, \tag{14.3}$$

where $\widetilde{\mathcal{G}}(s) = \mathcal{L}_{t\to s}[\mathcal{G}(t)] = \int_0^\infty dt\, e^{-st}\mathcal{G}(t)$ is the Laplace transform of the function $\mathcal{G}(t)$. The first-passage time density $f_{T_r}(t|\mathbf{x_0})$ is the negative derivative of the survival probability in time such that [38, 39]

$$f_{T_r}(t|\mathbf{x_0}; \mathbf{x_R}) = -\frac{\partial Q_r(t|\mathbf{x_0}; \mathbf{x_R})}{\partial t}, \tag{14.4}$$

which essentially provides the statistics of the first-passage time under resetting. In Laplace space, this relation translates to

$$\widetilde{f}_{T_r}(s|\mathbf{x_0}; \mathbf{x_R}) = 1 - s\widetilde{Q}_r(s|\mathbf{x_0}; \mathbf{x_R}), \tag{14.5}$$

where we have assumed the following boundary conditions in time: $Q_r(t = 0|\mathbf{x_0}; \mathbf{x_R}) = 1$, $Q_r(t \to \infty|\mathbf{x_0}; \mathbf{x_R}) < \infty$. The resulting relation (14.5) is quite useful since the moments can be computed readily such as

$$\langle T_r^n(\mathbf{x_0}; \mathbf{x_R})\rangle = (-1)^n \left. \frac{\mathrm{d}^n}{\mathrm{d}s^n} \widetilde{f_{T_r}}(s|\mathbf{x_0}; \mathbf{x_R}) \right|_{s\to 0} . \tag{14.6}$$

For instance, the mean first-passage time (MFPT) under resetting reads (by taking $n = 1$ in Eq. (14.6))

$$\langle T_r\rangle = \frac{\widetilde{Q_0}(r|\mathbf{x_0})}{1 - r\widetilde{Q_0}(r|\mathbf{x_R})} = \frac{1 - \widetilde{f_T}(r|\mathbf{x_0})}{r\widetilde{f_T}(r|\mathbf{x_R})}, \tag{14.7}$$

where we have used two equivalent forms for the mean time under resetting—one in terms of the survival probability and the other in terms of the first-passage time density of the resetting free underlying process. Notation-wise, we have suppressed the dependence of the MFPT on the initial coordinate $\mathbf{x_0}$ and resetting coordinate $\mathbf{x_R}$ for brevity. In the remaining sections, we will mostly assume $\mathbf{x_0} = \mathbf{x_R}$ (unless otherwise stated) without loss of much generality.

## 14.3 Applications to Theoretical First-Passage Models

In this section, we review a few canonical examples of search processes under resetting. The central goal is to show how the formalism developed in the previous section can be applied to all these examples in a unified manner. In doing so, we delve deeper into the ramifications due to resetting on the first-passage statistics. Furthermore, we discuss the search optimization conditions in resetting induced processes.

### *14.3.1 Diffusion with Stochastic Resetting*

Let us consider a diffusion-mediated search in one dimension where a Brownian particle diffuses through the medium starting from $x_0$ at time zero [40]. The motion of the particle can be quantified in terms of the probability density function $P(x, t|x_0)$ for the position $x$ at time $t$, which is given by the diffusion equation

$$\frac{\partial P(x, t|x_0)}{\partial t} = D\frac{\partial^2 P(x, t|x_0)}{\partial x^2}, \tag{14.8}$$

where $D$ is the diffusion constant. The initial condition is given by $P(x, 0|x_0) = \delta(x - x_0)$. The particle is also reset to $x_0 > 0$ intermittently with a rate $r$. We assume that there is an absorbing boundary at the origin, and we are interested in the mean time for the particle to find the boundary. This metric can be computed by employing Eq. (14.7) where we have to use the survival probability of the resetting free process. The latter is well known in the literature (see, e.g., [38]) and is given by

$$Q_0(t|x_0) = \int_0^L dx\, P(x,t|x_0) = \text{erf}\left[\frac{x_0}{\sqrt{4Dt}}\right], \tag{14.9}$$

where $\text{erf}(z) = \frac{2}{\sqrt{\pi}} \int_0^z dy\, e^{-y^2}$ is the error function, from which one can find the following in Laplace space:

$$\widetilde{Q}_0(s|x_0) = \frac{1}{s}\left[1 - e^{-\sqrt{\frac{s}{D}}x_0}\right]. \tag{14.10}$$

The first-passage time density is also straightforward to compute,

$$\widetilde{f}_T(s|x_0) = e^{-\sqrt{\frac{s}{D}}x_0}. \tag{14.11}$$

Substituting either of the above, namely Eq. (14.10) or Eq. (14.11) into Eq. (14.7), gives the MFPT of one-dimensional diffusion under resetting, which was first derived by Evans and Majumdar [2],

$$\langle T_r \rangle = \frac{1}{r}\left[e^{\alpha x_0} - 1\right], \tag{14.12}$$

where $\alpha = \sqrt{\frac{r}{D}}$ is an inverse length that corresponds to the typical distance covered by the particle between two resetting events. Note that $\langle T_r \rangle \sim \frac{1}{r}$ as $r \to 0$, which diverges since the particle hardly experiences any resetting event, and eventually drifts away from the origin. This is no surprise as for simple one-dimensional diffusion $\langle T \rangle = \infty$. In the other extreme limit $r \to \infty$, the mean first-passage time diverges exponentially since the particle almost remains localized around the resetting coordinate. Evidently, this marks the existence for an optimal resetting rate $r^*$, which can be computed by setting

$$\left.\frac{d\langle T_r \rangle}{dr}\right|_{r=r^*} = 0. \tag{14.13}$$

Using the MFPT for diffusion under resetting from Eq. (14.12) in Eq. (14.13) and defining $z^* = \sqrt{\frac{r^*}{D}}x_0$, one obtains the following equation for the optimal resetting rate:

$$\frac{z^*}{2} = 1 - e^{-z^*}. \tag{14.14}$$

Equation (14.14) gives $z^* = 1.5936...$ from which one finds $r^* \sim 2.54\tau_d^{-1}$, where $\tau_d = x_0^2/D$ is the diffusive time scale. In terms of this diffusive time scale, the minimal MFPT becomes $\langle T_{r*}\rangle \sim 1.544\tau_d$. Thus, the minimal MFPT is obtained when resetting is conducted at a diffusive time scale.

#### 14.3.1.1 Diffusion with Resetting in Higher Dimensions

As mentioned above, the formalism can be applied also to a search process that is being conducted in an arbitrary spatial dimension $d$. Consider a diffusive searcher that starts from the initial position $\mathbf{x_0}$ and undergoes stochastic resetting to $\mathbf{x_0}$ with a constant rate $r$. There is a finite size target—an absorbing $d$-dimensional sphere of radius $a$ (with $|\mathbf{x}_0| > a$) centered at the origin. Whenever the searcher reaches the surface of the target sphere, the particle is absorbed. The expression for the survival probability of the diffusive particle starting from $\mathbf{x_0}$ with the absorbing sphere at the origin is given by the following expression in Laplace space (see, e.g., [38]):

$$\widetilde{Q}_0(s|\mathbf{x}_0) = \frac{1}{s} - \frac{1}{s}\frac{R_0^\nu}{a^\nu}\frac{K_\nu(\sqrt{s/D}R_0)}{K_\nu(\sqrt{s/D}a)}, \tag{14.15}$$

where $K_\nu(z)$ is the modified Bessel function of the second kind, with index $\nu = 1 - d/2$. Here, $R_0 = |\mathbf{x}_0|$ is the distance from the resetting position to the target. Using Eq. (14.7), one arrives at the following expression for the MFPT of the searcher to the target sphere [41]:

$$\langle T_r\rangle = \frac{1}{r}\left[\frac{a^\nu}{R_0^\nu}\frac{K_\nu(\sqrt{r/D}a)}{K_\nu(\sqrt{r/D}R_0)} - 1\right]. \tag{14.16}$$

Similar to one-dimensional diffusion, resetting is always useful to render a finite MFPT for such $d$-dimensional diffusive search in unconfined space.

### 14.3.2 *Diffusive Search in a Potential Landscape*

A Brownian particle diffusing in a potential landscape $U(x)$ is described by the Smoluchowski or Fokker-Planck equation

$$\frac{\partial}{\partial t}P(x,t|x_0) = \left(\frac{\partial}{\partial x}U'(x) + D\frac{\partial^2}{\partial x^2}\right)P(x,t|x_0), \tag{14.17}$$

with the initial condition $P(x,0|x_0) = \delta(x - x_0)$. In what follows, we consider different potentials and study the trade-off between the resetting and attraction due to the potential.

#### 14.3.2.1 Linear Potential

Let us first consider a case when the particle starts from $x_0$ (here, we have set $x_0 > 0$ without any loss of generality) and experiences a linear potential $U(x) = k|x|$, the minimum of which is centered at the origin. In addition, the particle is reset to $x_0$ at a rate $r$, and one is interested in the mean time that it takes for the searcher to reach the origin, which is also an absorbing boundary. In this case, the first-passage time density for the underlying process in Laplace space reads [42–44]

$$\widetilde{Q}_0(s|x_0) = \frac{1}{s}\left[1 - e^{-\frac{x_0}{2D}\left(\sqrt{k^2+4Ds}-k\right)}\right]. \tag{14.18}$$

Substituting the above in Eq. (14.7), we find

$$\langle T_r \rangle = \frac{1}{r}\left[e^{\frac{x_0}{2D}\left(\sqrt{k^2+4Dr}-k\right)} - 1\right], \tag{14.19}$$

which in the limit $r \to 0$ becomes $|x_0|/k$. In this case, the optimal resetting rate $r^*$ can be obtained by solving the following transcendental equation [42]:

$$\frac{z^{*2}}{2\sqrt{Pe^2 + z^{*2}}} = 1 - e^{-\left(\sqrt{Pe^2+z^{*2}}-Pe\right)}, \tag{14.20}$$

where $z^* = \sqrt{\frac{r^*}{D}}x_0$ is the rescaled resetting rate and $Pe = kx_0/(2D)$ is the Péclet number. Equation (14.20) asserts that the root $z^*$ depends on the Péclet number [38]. While for small Péclet number ($Pe \ll 1$), the limit is similar to the simple diffusion case, and one has $z^* > 0$, this is not the case with the high Péclet number limit ($Pe \gg 1$) which results in an approximate solution $z^* \approx 0$ (which is also a trivial solution of Eq. (14.20)). This means that as one varies $Pe$, the optimal resetting rate switches between a finite value and zero. More on the physical ground, in the diffusion-dominated regime resetting remains beneficial (resulting in $r^* > 0$), while in the high force gradient limit the searcher is able to find the target quite efficiently making resetting only detrimental ($r^* = 0$) to the search.

#### 14.3.2.2 Harmonic Potential

Let us now consider a Brownian particle in a harmonic potential $U(x) = \frac{1}{2}kx^2$ that starts from $x_0$ and tries to find the target which is located at the origin in the presence of resetting to $x_0$ [3]. The first-passage time of a Brownian particle in a harmonic potential (also known as the Ornstein-Uhlenbeck process) was thoroughly investigated in the past, see for example [45, 46], where the higher-dimensional settings were also discussed. Starting again from the backward Fokker-Planck equation with appropriate boundary conditions in Laplace space, one finds [47, 48]

$$\widetilde{Q}(s|x_0) = \frac{1}{s}\left[1 - \frac{\Gamma\left(\frac{1}{2}+\frac{s}{2k}\right)}{\sqrt{2\pi}} x_0\sqrt{k/D}\,\mathcal{U}\left(\frac{1+s/k}{2}, \frac{3}{2}, \frac{kx_0^2}{2D}\right)\right], \tag{14.21}$$

where $\mathcal{U}(a, b, z)$ is the Tricomi confluent hypergeometric function [49]. The MFPT in this case reads

$$\langle T_r\rangle = \frac{1}{r}\left[\frac{\sqrt{2\pi}}{\Gamma\left(\frac{1}{2}+\frac{r}{2k}\right) x_0\sqrt{k/D}\,\mathcal{U}\left(\frac{1+r/k}{2}, \frac{3}{2}, \frac{kx_0^2}{2D}\right)} - 1\right]. \tag{14.22}$$

The analysis for the optimal resetting rate is similar to the discussion in the previous section. Further studies related to search processes under resetting mediated by logarithmic and power law potential were studied in [50, 51], respectively.

### *14.3.3 Search in Confined Geometry*

Many search processes are often conducted in a confined domain. In such cases, the mean search time usually remains finite and it is not apparent whether resetting can be a useful strategy. In this section, we aim to gain such insights by reviewing some canonical stochastic search processes.

#### 14.3.3.1 Diffusive Search in a Confinement Under Resetting

Consider a Brownian particle, initially located at $x_0$, diffusing in an interval $[0, L]$ in one dimension. The particle can get absorbed by any of these boundaries. In addition, the particle is stochastically reset to the initial position $x_0$ $(0 < x_0 < L)$ and we are interested in the first-passage properties of the particle due to resetting. The probability density $P(x, t|x_0)$ of the underlying process is a classical result and known from the literature [38],

$$P(x, t|x_0) = \frac{2}{L}\sum_{n=1}^{\infty} \psi_n(x_0)\psi_n(x)e^{-k_n t}, \tag{14.23}$$

where $\psi_n(x) = \sin[n\pi x/L]$ are the eigenfunctions and $k_n = n^2\pi^2 D/L^2$ is the rate at which the $n$-th eigenmode $\psi_n(x)$ decays with time. From this one can easily compute the survival probability in Laplace space (see [38])

$$\widetilde{Q}_0(s|x_0) = \frac{1}{s}\left[1 - g_0(x_0, s)\right], \text{ with } g_0(x_0, s) = \frac{\sinh[(L - x_0)\sqrt{s/D}] + \sinh[x_0\sqrt{s/D}]}{\sinh[L\sqrt{s/D}]}. \tag{14.24}$$

The MFPT for this case then can be found by substituting the survival probability into Eq. (14.7),

$$\langle T_r \rangle = \frac{L^2}{4D} \mathcal{G}(\beta, u) \text{ , where } \mathcal{G}(\beta, u) = \frac{1}{\beta^2} \left[ \frac{\cosh(\beta)}{\cosh \beta(1-2u)} - 1 \right], \tag{14.25}$$

where $u = x_0/L$, $\beta = \frac{L}{2}\alpha$ [52, 53]. To find the optimal restart rate, we scale $r^* = 4D\beta^{*2}/b^2$, in terms of $\beta^*$. In this case, one can determine the domain in which resetting expedites the completion of the underlying process: $\mathcal{D} = [(0, u_-) \cup (u_+, 1)]$, where $u_\pm = (5 \pm \sqrt{5})/10$. When $u_- < u < u_+$, the function has a minimum at $\beta^* = 0$, meaning $\langle T_r \rangle$ cannot be reduced by introducing resetting. On the other hand, when $u \in \mathcal{D}$, one can minimize the scaling function $\mathcal{G}(\beta, u)$ at a finite $\beta$ implying that the mean first-passage time is optimized at a finite resetting rate [52]. Similar analysis can also be extended for diffusive search in the presence of a drift [53] and generically in the presence of a potential [54] in a one-dimensional confinement.

One can also consider $d$-dimensional search where a Brownian searcher diffuses between two $d$-dimensional concentric spheres, starting from a distance $|\mathbf{x_0}|$ to the center of sphere, $R_1 < |\mathbf{x_0}| < R_2$ with $R_{1(2)}$ being the radius of the inner (outer) sphere. The process is completed once the searcher hits the inner or outer spherical surface. In addition, the searcher is reset to $|\mathbf{x_0}|$ at a rate $r$. The first-passage properties in the presence of resetting were discussed in [55]. Several extensions were made for a similar setup in the presence of an external potential [56, 57]. First-passage properties under resetting in a two-dimensional circle were studied in [58]. Escape properties of an underdamped particle with resetting from an interval were studied in [59].

#### 14.3.3.2 Anomalous Processes in a Confinement Under Resetting

We consider a particle, initially located at $x_0$, performing a random walk in continuous time within an interval $[0, L]$ in one dimension in the presence of resetting. The walker can get absorbed by any of these boundaries, and one is interested in finding the first-passage time of the walker. We further assume that the walker's characteristic jump distances are small compared to the interval length but the waiting times between the jumps are taken from a distribution $\phi(t)$ [60–62]. The master equation for this continuous time random walk process can be written as [63]

$$\frac{\partial P(x, t|x_0)}{\partial t} = \frac{\overline{\sigma}^2}{2} \int_0^t K(t-t') \frac{\partial^2 P(x, t'|x_0)}{\partial x^2} dt', \tag{14.26}$$

where $K(t)$ is the memory kernel which is related to the waiting time density $\phi(t)$. These two quantities are related to each other in Laplace space such that

$$\widetilde{K}(s) = \frac{s\widetilde{\varphi}(s)}{1 - \widetilde{\varphi}(s)}, \tag{14.27}$$

where $\widetilde{\varphi}(s)$ is the LT of the waiting time density $\phi(t)$. Applying suitable boundary conditions, one can obtain the following expression for the survival probability in the Laplace space,

$$\widetilde{Q}_0(s|x_0) = \frac{1}{s}\left[1 - \frac{\cosh\left(\overline{\alpha}(s)\left(x_0 - \frac{L}{2}\right)\right)}{\cosh\left(\overline{\alpha}(s)\frac{L}{2}\right)}\right], \tag{14.28}$$

where $\overline{\alpha}(s) \equiv \frac{1}{\overline{\sigma}}\sqrt{\frac{2s}{\widetilde{K}(s)}} = \frac{\sqrt{2}}{\overline{\sigma}}\sqrt{\frac{1}{\widetilde{\varphi}(s)} - 1}$. Substituting the above in Eq. (14.7), the MFPT under resetting reads [63]

$$\langle T_r \rangle = \frac{1}{r}\left[\frac{\cosh\left(\frac{\overline{\alpha}(r)L}{2}\right)}{\cosh\left(\overline{\alpha}(r)\left(x_0 - \frac{L}{2}\right)\right)} - 1\right], \tag{14.29}$$

which holds for an arbitrary memory kernel. One can study the MFPT under quite some generalities such as waiting times with finite first and second moments, with finite first moment and diverging second moment and diverging first and second moments. In the former case, one can derive a condition on $x_0/L$ for resetting to be beneficial (this is similar to the diffusion as was discussed in Sect. 14.3.3.1) while in the latter two cases, resetting always expedites the completion marking the existence of a finite optimal resetting rate and serving as an effective cutoff strategy to the long-tailed waiting time density [63].

### 14.3.4 *Fractional Diffusion with Stochastic Resetting*

Consider a subdiffusion process that undergoes stochastic resetting. The setup is similar to previous ones—the walker starts at $x_0 > 0$ at time zero and the process is completed when it finds the boundary at the origin. The corresponding Fokker-Planck equation for the fractional diffusion process is given by [64]

$$\frac{\partial P(x,t|x_0)}{\partial t} = D\frac{d}{dt}\int_0^t \eta(t-t')\frac{\partial^2 P(x,t'|x_0)}{\partial x^2}dt', \tag{14.30}$$

where $D$ is the generalized diffusion coefficient whose dimension will explicitly depend on the form of the memory kernel $\eta(t)$. The memory kernel $\eta(t)$ emanates from the waiting time distribution $\phi(t)$ of the walker between the jumps. This relation, in Laplace space, reads $\widetilde{\phi}(s) = \frac{1}{1+1/\widetilde{\eta}(s)}$. Here we note that for specific choices of $\eta(t)$, such as a power-law form, this is a fractional process. Note that Eq. (14.26) is related to the above equation with the transformation $\widetilde{\eta}(s) \to \widetilde{K}(s)/s$. For this process, the survival probability of a particle starting at $x_0$ at time zero in the absence of resetting reads [65]

$$\widetilde{Q}_0(s|x_0) = \frac{1}{s}\left(1 - e^{-\sqrt{1/[\widetilde{\eta}(s)D]}x_0}\right). \tag{14.31}$$

Using Eq. (14.7) the mean first-passage time becomes

$$\langle T_r \rangle = \frac{1}{r}\left(e^{\sqrt{1/[\widetilde{\eta}(r)D]}x_0} - 1\right), \tag{14.32}$$

which is valid for various choices of the kernel $\eta(t)$. As an illustrative example, we consider the ordinary subdiffusion [66–68] for which $\eta(t) = \frac{t^{\mu-1}}{\Gamma(\mu)}$ so that $\widetilde{\eta}(s) \sim s^{-\mu}$ with $0 < \mu \leq 1$, and the waiting time distribution between the jumps has a power-law decay $\phi(t) \sim t^{-1-\mu}$. Following Eq. (14.31), we find

$$\widetilde{Q}_0(s|x_0) = \frac{1}{s}\left(1 - e^{-\sqrt{s^{\mu}/D}x_0}\right), \tag{14.33}$$

and the corresponding MFPT reads

$$\langle T_r \rangle = \frac{1}{r}\left(e^{\sqrt{r^{\mu}/D}x_0} - 1\right), \tag{14.34}$$

where the generalized diffusion coefficient $D$ has a dimension [length$^2$]/[time$^{\mu}$]; see also [60]. Equation (14.34) is consistent with a more general formula derived with different methods in [69]. For $r \to 0$ and $r \to \infty$, $\langle T_r \rangle$ diverges and in between there is a minimum for an optimal $r^*$, which can be found from the following transcendental equation [65]:

$$1 - e^{-\zeta} = r\frac{d\zeta}{dr}, \quad \zeta = x_0/\sqrt{\widetilde{\eta}(r)D}. \tag{14.35}$$

Equation (14.35) is, in principle, solvable for a given waiting time kernel. In particular, for the subdiffusion case, Eq. (14.35) boils down to $1 - e^{-\zeta} = \frac{\mu\zeta}{2}$. The case $\mu = 1$ reproduces the original diffusion, but otherwise renders a unique solution for the optimal scaled resetting rate $\zeta$ in terms of the subdiffusive index $\mu$ [65]. Since, for a subdiffusive particle, the MFPT to reach the origin (in the absence of resetting) is infinite, it is only natural that resetting will stabilize the system rendering a finite MFPT and consequently, an optimal resetting rate. Graphical representation of the MFPT for subdiffusive search is given in Fig. 14.2 for different values of $\mu$.

### *14.3.5 Heterogeneous Diffusion with Stochastic Resetting*

Consider independent random walkers on a one-dimensional heterogeneous medium, initially located at position $x_0$ and then they search for a target which is located at the origin. The target position defines a bound of the search domain, and the process

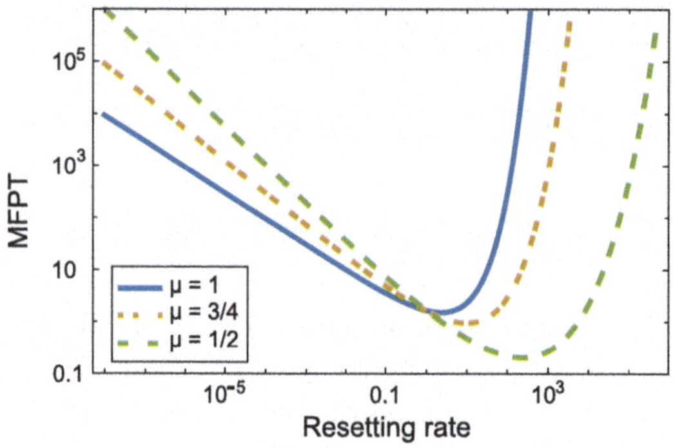


**Fig. 14.2** Mean first-passage time for subdiffusive search process under resetting. MFPT for normal diffusion under stochastic resetting (14.12) with $\mu = 1$ (blue solid line), and MFPT for anomalous diffusion under stochastic resetting (14.34) with $\mu = 3/4$ (red dotted line), and $\mu = 1/2$ (green dashed line). The parameters are set as $D = 1$, $x_0 = 1$. In all the cases, we observe an expedited completion due to resetting

is completed as soon as the target is first detected. The evolution for the probability density function in this case can be written as [70, 71]

$$\frac{\partial P(x,t|x_0)}{\partial t} = \frac{\partial^2}{\partial x^2}[D(x)P(x,t|x_0)] + (A/2-1)\frac{\partial}{\partial x}\left[\frac{\partial D(x)}{\partial x}P(x,t|x_0)\right], \tag{14.36}$$

where $D(x) > 0$ is a space-dependent diffusion coefficient manifesting for the heterogeneous medium and $0 \leq A \leq 2$ is a parameter that takes different values depending on the numerical scheme. For instance, the case of $A = 2$ (Itô convention) is commonly used in finance [72], while $A = 1$ (Stratonovich convention) is popular in physics [73]. The highly anticipated case $A = 0$ (known as isothermal, kinetic, or Hänggi-Klimontovich) also has applications related to Fick's law [74–76]. To mimic the heterogeneous diffusivity, it is often considered $D(x) = D|x|^\nu$ ($\nu < 2$) which has been used extensively to capture the diffusive motion of a particle on fractal objects and diffusion in turbulent media. For different interpretations of the heterogeneous diffusion processes, we refer to [71–77]. For this setup, the survival probability in Laplace space reads [70]

$$\widetilde{Q}_0(s|x_0) = \frac{1}{s} - \frac{2}{\Gamma(b)s}\left[\frac{x_0^{\frac{2-\nu}{2}}}{2-\nu}\sqrt{\frac{s}{D}}\right]^b \times K_b\left(\frac{2x_0^{\frac{2-\nu}{2}}}{2-\nu}\sqrt{\frac{s}{D}}\right), \tag{14.37}$$

where $K_b(z)$ is the modified Bessel function of the second kind and $b = \frac{1-\nu(1-A/2)}{2-\nu} \geq 0$. The MFPT under resetting then can be obtained using Eq. (14.7),

$$\langle T_r \rangle = \frac{1}{r} \left[ \frac{\Gamma(b) \left( \frac{x_0^{\frac{2-\nu}{2}}}{2-\nu} \sqrt{\frac{r}{D}} \right)^{-b}}{2K_b \left( \frac{2x_0^{\frac{2-\nu}{2}}}{2-\nu} \sqrt{\frac{r}{D}} \right)} - 1 \right]. \tag{14.38}$$

Recalling the definition of the Gamma function, we observe from the above that the MFPT is finite for $b > 0$. For the resetting-free process, the underlying first-passage time density has an asymptotic tail $t^{-b-1}$ which is a generalized version of the simple diffusion case. In such cases, the underlying process may have a diverging MFPT depending on $\nu$ and $A$. Similar to the cases discussed above (e.g., diffusion), one can explicitly show that the MFPT becomes finite in the presence of resetting [70, 78].

## 14.4 When Does Resetting Work?

There are an endless variety of ways in which first-passage processes and resetting mechanisms mix and match. Rigorous studies show, as is also evident from the examples above, that for processes like diffusion or subdiffusion, resetting always expedites the search while for the search processes in a confined domain or in the presence of a generic potential landscape, resetting can often be detrimental. A lot of efforts have been spent on finding the physical conditions, which can mark this behavioral transition. To see this, one usually looks into a generic first-passage process (with well-defined first and second moments), adds an infinitesimal resetting rate, and tries to examine its ramifications. Recalling the MFPT under resetting from Eq. (14.7) and expanding it in the following polynomial in the small $r$ limit [13, 53]

$$\langle T_r \rangle = a_0 + a_1 r + a_2 r^2 + a_3 r^3 + \cdots , \tag{14.39}$$

where $\{a_i\}$ are the expansion coefficients. Such kinds of expansions respect a set of postulates, which we state below. By taking $r$ strictly to be zero, we see that $\langle T_0 \rangle = a_0 = \langle T \rangle$, the MFPT of the underlying process in the absence of resetting and is assumed to be positive finite.

*Physical meaning of the coefficients* The coefficients $a_i$, so far defined formally, can be given physical meaning by analyzing $\langle T_r \rangle$ around $r = 0$. Recalling $\widetilde{f_T}(r|\cdot) = \int_0^\infty dt\, e^{-rt}\, f_T(t|\cdot)$ to be the Laplace transformation of the underlying first-passage time density, expanding the same in the small $r$ limit in Eq. (14.7), and furthermore comparing with the Taylor's series expansion of $\langle T_r \rangle$ in Eq. (14.39), we identify

$$a_0 = \langle T \rangle, \ a_1 = -\frac{\langle T^2 \rangle}{2} + \langle T \rangle^2,$$
$$a_2 = \frac{1}{6}\langle T^3 \rangle + \langle T \rangle^3 - \langle T \rangle \langle T^2 \rangle,$$
$$a_3 = -\frac{\langle T^4 \rangle}{4!} + \frac{\langle T^3 \rangle \langle T \rangle}{3} + \frac{\langle T^2 \rangle^2}{4} - \frac{3\langle T^2 \rangle \langle T \rangle^2}{2} + \langle T \rangle^4,$$

and so on, where $\langle T^n \rangle$ is the $n$-th moment of the underlying first-passage time distribution [13, 53]. The moments (hence the coefficients) are explicit functions of the system parameters, and will characterize the transitions.

*A universal criterion* For resetting to reduce the underlying MFPT, one would naturally expect $\langle T_{r\to 0} \rangle < \langle T_{r=0} \rangle$ with the introduction of an infinitesimal resetting rate $r \to 0$. This would guarantee an initial reduction of the MFPT as a function of the resetting rate. Technically, this inequality thus states that in the linear-order expansion of Eq. (14.39), one must have $a_1 < 0$. Rearranging $a_1$ from the above, one finds [13, 79]

$$CV \equiv \frac{\sigma(T)}{\langle T \rangle} > 1, \tag{14.40}$$

where $CV$ is called the the coefficient of variation. This is a measure of statistical dispersion that stands for the ratio between the standard deviation $\sigma(T)$ and the mean $\langle T \rangle$ of the underlying first-passage time $T$. The criterion is completely universal, system independent, and essentially tells that for resetting to be beneficial this measure has to be higher than unity. Note that the $CV$-criterion depends only on the first two statistical metrics (which need to be finite) of the underlying process and not the entire distribution. For stochastic processes, where $CV = \infty$ (broad distribution/heavy tail) or even not well-defined (e.g., Lévy first-passage time distribution for the one-dimensional simple diffusive search), resetting is guaranteed to help and thus this relation becomes redundant. Clearly, for the underlying first-passage processes with narrow distributions (toward a deterministic limit) such that $CV < 1$, resetting can only prolong the completion (see Fig. 14.3). It is worth noting that the $CV = 1$ condition can also be derived from the *inspection paradox* in probability theory [80].

While the $CV$-criterion is the *sufficient* condition for resetting to reduce the MFPT, it is not a *necessary* one. There can be scenarios when the introduction of the resetting rate can first increase the MFPT thus rendering $a_1 > 0$ (equivalently, $CV < 1$) but then the MFPT can be reduced for a finite resetting rate. It turns out that the sign of $a_1$ is not enough to understand such situations and thus, one needs to take into account higher order corrections such as the quantities $a_2$ and $a_3$. In such cases, the $CV$-criterion (14.40) need not be respected yet resetting can be found to be useful (see [53] where a Landau like theory was proposed to understand this phenomena, accompanied by various examples [57, 81]).

*Interpretation of the criterion* The coefficient of variation ($CV$) usually characterizes how broad or narrow a distribution is around its mean and that certainly depends on the underlying search mechanism, search domain, target configurations, and many other intrinsic parameters. Recent single-molecule experiments with colloidal particles, biomolecules, or enzymes have brought deep insights into the first-passage phenomena while looking into various microscopic search mechanisms such as surmounting activation barriers, bottlenecks, gated reactions, transport across channels, facilitated diffusion, and chemical reactions (see, e.g., [82–85] for a review). Many such transport, escape, or activation processes take place in multidimensional energy landscapes with metastable states in the presence of deep kinetic traps and large fluctuating barriers. There, the microscopic searchers tend to spend exceedingly long time and thus the completion time statistics render to heavy-tail, broad distributions [61, 86]. Similar arguments also apply to chemical reactions which are inherently stochastic in nature. Thus the outcome of the reactions are also random—this means various kinds of products can be formed from an enzyme due to the intrinsic fluctuations. Enzymatic catalysis processes often can assume on-pathways or off/parallel-pathways [87, 88]. In the former case, the reactions are usually nearest neighbor and the dwell time distribution is narrower. The latter case however can contain multiple pathways with active catalytic states with relative weights. This gives rise to multiple classes of dwell time distributions which have long, multi-exponential or heavy tails. Thus the natural reaction time will have broader distributions with $CV > 1$. These kinds of scenarios also take place when searchers are macroscopic, e.g., foraging animals, falcons, or even drones [89, 90]. Many such processes take place in unconfined territories in search of a steady supply of nutrients and other essential resources. Indeed, there can be large fluctuations around the average search time due to uncertainty, and lack of cognition and experience, and these can turn out to be deleterious and may often result in death. As shown above, resetting/quenching to the initial configuration for the microscopic searchers, unbinding of the enzymes from the metastable states, or returning to the home for macroscopic searchers can reverse the deleterious effects of large fluctuations in the underlying search time, thus turning a marked drawback into a favorable advantage. Simply put, such *reset kind events* can curtail the long detrimental trajectories and make the process more regular and remarkably efficient. From a broader perspective, whenever we have $CV > 1$ (i.e., worse search conditions for the underlying process), physical scenarios naturally suggest that search with resets/returns may be considered as a useful bet-hedging strategy.

*Optimal resetting rate as an order parameter and resetting transition* The equality $CV = 1$ serves as a sharp boundary for resetting transition. This is not really a thermodynamic phase transition, but mimics the behavioral transition between the phases, namely "resetting-detrimental" and "resetting-beneficial". Assume $CV < 1$—in this case, $\langle T_{r \to 0} \rangle > \langle T_{r=0} \rangle$ and thus the optimal resetting rate $r^*$ that globally minimizes $\langle T_r \rangle$ is fixed at zero. In other words, resetting only delays the completion process. On the other hand, for $CV > 1$, we have $\langle T_{r \to 0} \rangle < \langle T_{r=0} \rangle$ and thus the optimal resetting rate $r^*$ that globally minimizes $\langle T_r \rangle$ takes a non-zero

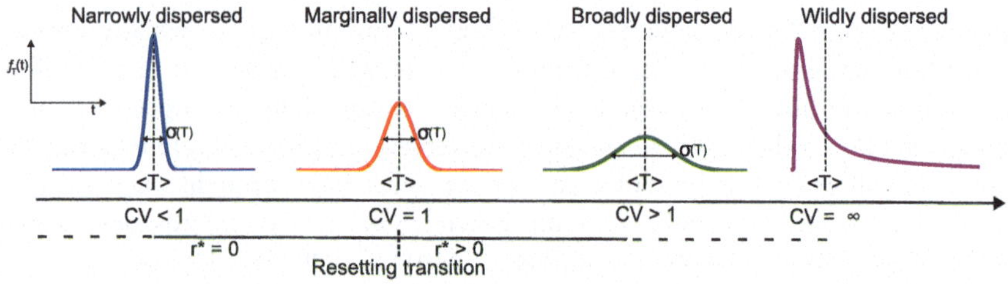


**Fig. 14.3** Schematic illustration of the possible roles of resetting on the underlying first-passage time processes. For the narrowly dispersed distribution, $CV < 1$ and resetting is only detrimental. As $CV$ increases and crosses the boundary $CV = 1$, the effect of resetting becomes more significant. Infinite dispersion $CV = \infty$ may arise when the underlying first-passage time process is characterized by heavy power-law tails (namely $f_T(t) \sim t^{-(1+\gamma)}$ with $0 < \gamma < 2$). Remarkably, in such cases one always observes an overarching gain due to resetting. The interpolation between these behaviors is captured by the optimal resetting rate $r^*$ which plays the role of an order parameter. The transition between $r^* > 0$ and $r^* = 0$ at the phase-separatrix $CV = 1$ is often known as the *resetting transition*

value. Simply put, an optimally reset process can always expedite the completion making resetting beneficial. The transition between $r^* = 0$ to $r^* > 0$ as we vary $CV$ is known as the "resetting transition" and the optimal resetting rate can be regarded as the *order parameter* for such transition in resetting systems (see Fig. 14.3). Since $CV$ is usually a function of the system parameters, one could imagine that there is a control parameter of interest, say $p$, that can be varied to span from $CV < 1$ to $CV > 1$. Henceforth, $CV = 1$ would render a critical $p_c$ in the parameter space where the exact resetting transition will take place. In fact, one can show that near the transition, $r^* \sim |p - p_c|^\beta$, where $\beta = 1$ is a critical exponent that is found to be universal across different reset processes [48, 53]. This is somewhat reminiscent of the continuous phase transition in statistical physics. Furthermore, a Landau-like mean-field approach can be used to probe such universality [53].

## 14.5 Applications of Resetting to Real-World Search Problems

So far, we discussed the theoretical implications of stochastic resetting on search problems. In this section, we shift our perspective and turn our attention to the applications of stochastic resetting in real-world search problems. In particular, we discuss recent developments on applying resetting to home-range search [16, 17], resetting facilitated diffusive transport [81], turnover of chemical reactions [27–29], and income dynamics [22]. This helps us bridge the gap between theoretical research and its tangible applications in real-world scenarios.

### 14.5.1 Home-Range Search

One of the intriguing applications of random resetting in search problems is the home-range search. In this context, a searcher is not indefinitely adrift in a search space but has a reference point—a "home"—to which it can return under certain conditions. This kind of search strategy finds applications in animal foraging behavior, robot exploration, and many areas where search optimization is crucial.

Imagine a searcher that begins at an origin or "home" in a potentially $d$-dimensional search space $\mathcal{D}$. This searcher attempts to locate a target (or multiple targets). For the unconstrained search, the searcher might find the target in a random time $T$, taken from the distribution $f_T(t)$. However, real-world searchers often have limitations or strategies that cause them to return to a starting or known point. In our model, if the searcher does not find the target within a stipulated time $R$ distributed according to $f_R(t) = re^{-rt}$, it retreats back to its home. The time it takes to return, denoted as $\tau(\mathbf{x})$, can vary depending on where the searcher is when the *decision to return home* is made. Once the searcher reaches home, it might rest or re-strategize, staying there for a time span $W$. Post this, the search initiates anew, making the process cyclic [16].

To make this model tractable, we employ two assumptions. First, targets cannot be located during the phases of return and waiting at home. Second, search cycles are isolated events. This means each new search starts with no memory of past endeavors, thus ensuring independence between cycles. The latter fact allows one to utilize the renewal approach as sketched out in Sect. 14.2 and to estimate the MFPT for such home-range-search process which reads [16]

$$\langle T_r \rangle = \underbrace{\frac{\widetilde{Q}_0(r|\mathbf{x_0})}{1 - r\widetilde{Q}_0(r|\mathbf{x_0})}}_{\text{search}} + \underbrace{\frac{r\int_{\mathcal{D}} d\mathbf{x}\ \tau(\mathbf{x})\widetilde{P}_0(\mathbf{x}, r)}{1 - r\widetilde{Q}_0(r|\mathbf{x_0})}}_{\text{return}} + \underbrace{\frac{r\widetilde{Q}_0(r|\mathbf{x_0})}{1 - r\widetilde{Q}_0(r|\mathbf{x_0})}\langle W \rangle}_{\text{home}}, \tag{14.41}$$

where $\tau(\mathbf{x})$ is the home-return time of the searcher starting from the coordinate $\mathbf{x}$ at the time of resetting. Here, $\widetilde{P}_0(\mathbf{x}, r) = \int_0^\infty dt\ e^{-rt} P_0(\mathbf{x}, t)$ is the LT of $P_0(\mathbf{x}, t)$—PDF of the underlying process such that $Q_0(t|\mathbf{x_0}) = \int_{\mathcal{D}} d\mathbf{x}\ P_0(\mathbf{x}, t)$ is the survival probability of the searcher up to time $t$.

#### 14.5.1.1 Return of a Diffusive Searcher with a Constant Velocity

To exemplify Eq. (14.41), let us consider a one-dimensional diffusive search process where the searcher has to find a target placed at $L$. The searcher starts from its home, which is fixed at the origin. Let us assume that upon resetting, the searcher returns to its home with a constant velocity $v_r$ such that $\tau(x) = |x|/v_r$. In this case, elementary calculation yields $\widetilde{P}_0(\mathbf{x}, r) = \frac{1}{\sqrt{4Dr}} \left[ e^{-\sqrt{r/D}|x|} - e^{-\sqrt{r/D}(2L-x)} \right]$ and thus from Eq. (14.41) one finds [16]

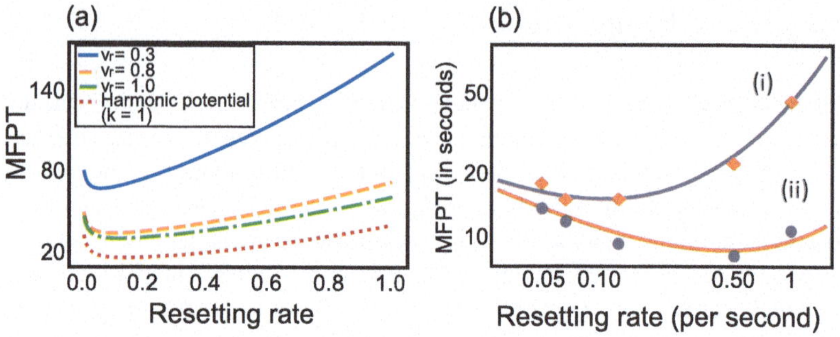


**Fig. 14.4** Behavior of the mean first-passage time of a home-range diffusive searcher as a function of resetting rate under various modes of return. Panel **(a)** shows MFPTs when the searcher returns to its home at a constant velocity or is attracted by a harmonic trap centered at its home. Panel **(b)** shows MFPTs of a silica particle to a target wall (see [17] for the experimental details) under holographic optical tweezer modulated protocols: (i) non-instantaneous constant time return $\tau_0 = 3.79$ s and (ii) instantaneous return $\tau_0 = 0$ respectively. Theoretical predictions (Eq. (14.46), solid lines) are in good agreement with experimental data (symbols). Panel **(b)** is adapted from [17]

$$\langle T_r \rangle = \frac{1}{r}\left(e^{\sqrt{r/D}L} - 1\right) + \frac{L}{v_r}\left(\frac{2\sinh\left(\sqrt{r/D}L\right)}{\sqrt{r/D}L} - 1\right), \tag{14.42}$$

where we have assumed $W = 0$ for simplicity. Note that the first term on the RHS is accounted for the search process and thus identical to the MFPT under instantaneous resetting. The second term on the RHS accounts for the delay due to the finite time return to its home. As $v_r \to \infty$, one recovers the instantaneous limit since the searcher returns to its home almost instantaneously. See Fig. 14.4a which shows the modulation of MFPT under different return velocities.

#### 14.5.1.2 Return of a Searcher Under the Action of a Harmonic Potential

We now consider a return process in which a one-dimensional searcher returns under the influence of a harmonic potential $U(x) = \frac{1}{2}kx^2$, $k > 0$, such that we can write Newton's second law for the return process, $m\ddot{x} = -kx$, where $m$ is the mass, and $\ddot{x}$ is the second derivative of the displacement or the acceleration. This is a deterministic return process and thus the solution of the return path is $x = x_0 \cos\left(\sqrt{\frac{k}{m}}t\right)$, where $x_0$ is the starting location. Using this, one finds the return time to the origin to be $\tau(x) = \frac{\pi}{2}\sqrt{\frac{m}{k}}$, which turns out to be independent of $x$. Plugging this in Eq. (14.41), one finds

$$\langle T_r \rangle = \left(1 + \frac{\pi}{2}\sqrt{\frac{m}{k}}\right)\frac{\widetilde{Q}_0(r|x_0)}{1 - r\widetilde{Q}_0(r|x_0)}, \tag{14.43}$$

which holds for arbitrary underlying search process. In particular, for diffusive process one has [91]

$$\langle T_r \rangle = \left( \frac{1}{r} + \frac{\pi}{2}\sqrt{\frac{m}{k}} \right) \left( e^{\sqrt{r/D}L} - 1 \right). \tag{14.44}$$

When the stiffness of the potential is very high (i.e., $k \to \infty$), one immediately recovers the instantaneous return limit. Otherwise for finite stiffness, it can often facilitate the search compared to the velocity-driven return process (see Fig. 14.4a).

#### 14.5.1.3 Return of a Searcher to Home in Fixed Time Application to Resetting Experiments

We now turn our attention to the case when the searcher returns to its home in fixed time say $\tau_0$. This protocol is quite feasible in single particle experiments since the searcher (say, a silica particle) can be manipulated with great precision by optical traps [17]. In such a case, the MFPT reads

$$\langle T_r \rangle = (1 + \tau_0) \frac{\widetilde{Q}_0(r|x_0)}{1 - r\widetilde{Q}_0(r|x_0)}, \tag{14.45}$$

which again holds for an arbitrary underlying search process. In particular, in the first-passage time experiment [17], the resetting mechanism was conducted stochastically using holographic optical tweezers on a silica particle (which mimics a diffusive dynamics). Following the resetting, the particle was returned to its home (i.e., the initial coordinate) using optical traps in a constant time $\tau_0$ [17]. For a diffusive particle starting from the origin, the above expression for the MFPT to a target set at $L$ simplifies to

$$\langle T_r \rangle = \left( \frac{1}{r} + \tau_0 \right) \left( e^{\sqrt{r/D}L} - 1 \right), \tag{14.46}$$

which was verified against the experimental data to find an excellent fit [17]. See Fig. 14.4b which was adapted from [17].

#### 14.5.1.4 Stochastic Return of the Searcher

Notably, we have assumed that the return to the home is a deterministic process making $\tau(\mathbf{x})$ a deterministic function once $\mathbf{x}$ is fixed. One can also consider a situation when the return processes are stochastic. This is natural in many setups since there will always be random fluctuations that are uncontrollable and thus return protocols need not be absolutely deterministic. In such cases, $\tau(\mathbf{x})$ becomes a stochastic function of the return path, and thus, to emphasize this additional averaging, one needs to replace $\tau(\mathbf{x})$ by $\overline{\tau(\mathbf{x})}$ in Eq. (14.41). Interested readers can take a look at references [47, 92–94] for such scenarios.

### *14.5.2 Resetting Facilitated Diffusive Transport Through Channels*

Diffusion of particles, molecules, or even living microorganisms in confined geometries such as tubes and channels plays a key role across various scales in natural and technological processes [95–97]. A diffusing particle inside a channel crossing a biological membrane can either escape from both sides of the membrane or it may be allowed to escape through one side while the other side simply reflects it back. Key quantities in these processes are the first-passage probabilities and average escape times from the channel (or lifetimes) conditioned to the exit side of the membrane, as well as the overall average lifetime of the particle in the channel [98, 99]. These quantities are ubiquitous in theoretical studies on channel-facilitated transport. Enormous theoretical efforts have been devoted over the years to make the transport inside various channels more efficient. The fact that resetting has the ability to speed up complex search processes naturally compels one to use this as a useful strategy for the channel-facilitated first-passage transport. Indeed this was shown in [81] recently, and we review the results in brief.

The canonical Fick-Jacobs formalism is a key approach to treat diffusion of particles, ions, molecules, or even living microorganisms in confined geometries such as tubes and channels of varying cross-sections [95, 96, 100]. For instance, the effective one-dimensional Fick-Jacobs equation for a diffusing particle, starting at $x_0$, inside a three-dimensional tube of variable radius $R(x)$ and length $L$ can be written as

$$\frac{\partial P(x,t)}{\partial t} = \frac{\partial}{\partial x}\left\{D(x)e^{-\beta_{\mathcal{T}}U(x)}\frac{\partial}{\partial x}\left[e^{\beta_{\mathcal{T}}U(x)}P(x,t)\right]\right\}, \tag{14.47}$$

where the entropic potential is given by $-\beta_{\mathcal{T}}U(x) = \ln(R^2(x)/R^2(x_0))$ for a three-dimensional tube (usually, confinement in higher dimensions gives rise to an effective entropic potential in the reduced dimension [96]). Here, $\beta_{\mathcal{T}} = 1/(k_B\mathcal{T})$ is the inverse temperature with $k_B$ the Boltzmann constant. Replacing $U(x)$ for the three-dimensional tube, the modified equation reads

$$\frac{\partial P(x,t)}{\partial t} = D_\lambda \frac{\partial}{\partial x}\left\{R^2(x)\frac{\partial}{\partial x}\left[\frac{P(x,t)}{R^2(x)}\right]\right\}, \tag{14.48}$$

where $D_\lambda$ is given by the Reguera-Rubí formula $D_\lambda = \frac{D_0}{\sqrt{1+\lambda^2}}$ [101]. We further assume that the tube radius $R(x)$ increases in the axial direction with a constant rate $\lambda$ so that $R(x) = a + \lambda x$, where the $x$-coordinate is measured along the tube axis, and $a$ is the tube radius at $x = 0$. For a conical tube with two absorbing points at

$x = 0$ and $x = L$, the unconditional first-passage time density (in Laplace space) is given by [81]

$$\widetilde{f_T}(s|x_0) = \frac{(\lambda L + a)\sinh(\sqrt{s/D_\lambda}x_0) + a\sinh(\sqrt{s/D_\lambda}(L - x_0))}{(\lambda x_0 + a)\sinh(\sqrt{s/D_\lambda}L)}. \tag{14.49}$$

The first-passage density can then be substituted in Eq. (14.7) to obtain the MFPT of the searcher in the presence of resetting [81]

$$\langle T_r(x_0)\rangle = \frac{(\lambda x_0 + a)\sinh(\alpha_\lambda L) - a\sinh(\alpha_\lambda(L - x_0)) - (\lambda L + a)\sinh(\alpha_\lambda x_0)}{r[a\sinh(\alpha_\lambda(L - x_0)) + (\lambda L + a)\sinh(\alpha_\lambda x_0)]}, \tag{14.50}$$

where $\alpha_\lambda = \sqrt{r/D_\lambda}$. It can further be shown using the $CV$-criterion (14.40) that resetting is going to be beneficial (i.e., an optimal resetting rate $r^* > 0$ would exist) when the following condition is satisfied [81]:

$$(\lambda v + \tilde{a})\left[\lambda(1 + v)(7 - 3v^2) + 15\tilde{a}(1 + v - v^2)\right] \geq 10v(1 - v)\left[\lambda(1 + v) + 3\tilde{a}\right]^2, \tag{14.51}$$

where $\tilde{a} = a/L$ and $v = x_0/L$. Thus, the criterion is fulfilled as long as the particle starts sufficiently close to one of the absorbing boundaries ($v \sim 0$ or $v \sim 1$) but not when it starts out in the center. For starting positions near the center of the tube, increasing the resetting rate increases the MFPT since any trajectory on both sides is going to take the particle closer to the boundary and reset hinders it. But for starting positions closer to the boundaries, resetting decreases the lifetime because there are now many more possible trajectories that are taking the particle away from the boundary and resetting systematically eliminates them [81]. Thus, a carefully navigated resetting strategy (14.51), characterized in terms of the initial position $x_0$ and other system parameters, can indeed facilitate *diffusive transport* of a particle through narrow channels. Likewise, resetting can also be useful for *active transport* and search in a confined arena [102].

### *14.5.3 Turnover of the Gated Chemical Reactions*

Chemical reactions are inherently stochastic in nature wherein the estimation of the reaction-time or the encounter-time between two molecules is often cast as a first-passage time problem [85, 87, 88, 103]. Chemical reactions usually comprise three basic steps: binding (when one reactant $C$ binds another one $R_1$ to form a metastable complex $CR_1$), unbinding or resetting (when the complex $CR_1$ is disassociated from the parent reactants), and catalysis (when a product $P$ is successfully formed). Often, the reactant molecules switch between a reactive and a non-reactive state; and thus

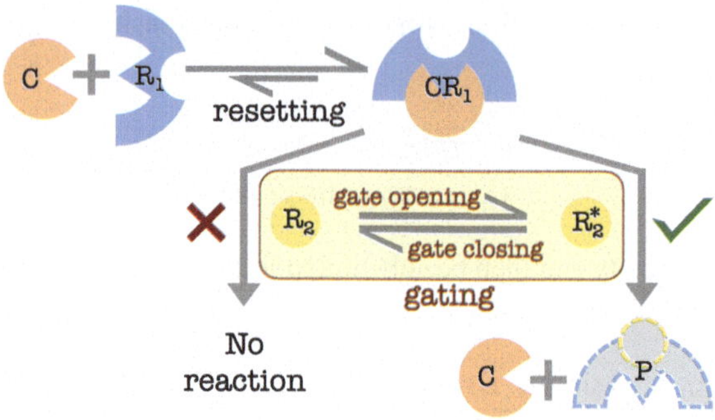


**Fig. 14.5** Scheme for a gated chemical reaction between two reactants $R_1$ and $R_2$, catalyzed by $C$. Initially, the catalyst binds reversibly with $R_1$ to generate $CR_1$, a metastable intermediate: $C + R_1 \rightleftharpoons CR_1$. In the next step, $CR_1$ reacts selectively with $R_2^\star$ (the *reactive* or *open-gate* state of $R_2$) to generate the product $P$ and liberate the catalyst: $CR_1 + R_2^\star \to C + P$. This step can be modeled by gated drift-diffusion, while the unbinding of $C$ from $R_1$ is essentially manifested as resetting. Adapted from [29]

the collisions between the reactants must happen in their reactive states for the completion of a successful product formation (see Fig. 14.5). This is known as "gating" in biochemistry that typically refers to the transition between the open and closed states of an ion channel; and the ions are allowed to flow through the channel only when it is open [104, 105].

Such gated reactions can be illustrated by considering a diffusive transport process confined to the positive semi-infinite space in the presence of a target that randomly switches between an open and a closed state with constant rates. Consider that the target switches from the non-reactive to the reactive state at a rate $\gamma > 0$, and the reverse transition takes place with a constant rate $\beta > 0$. This is a general scenario mimicked by, e.g., a chemical reaction (see Fig. 14.5), where the collisions between reactants [$CR_1$ and $R_2$] lead to the formation of product only when at least one of the reactants is in an activated state [when $R_2$ exists as $R_2^\star$] and not otherwise. Since resetting is an integral part of any consecutive chemical reaction that has a reversible first-step [when $C$ binds to, or unbinds from, $R_1$], the reaction scheme shown in Fig. 14.5 can be modeled as gated diffusion process in a potential landscape with resetting. The mean turnover time of this reaction can then be obtained by casting into the formulation shown above [29, 106],

$$\langle T_r^G \rangle = \frac{e^{\sqrt{\frac{r}{D}}x_0} - 1}{r} + \frac{\beta e^{\sqrt{\frac{r}{D}}x_0}}{\gamma\sqrt{r[r+\gamma+\beta]}}\left[1 + \frac{r e^{-\sqrt{\frac{r+\gamma+\beta}{D}}x_0}}{(\gamma+\beta)}\right]. \tag{14.52}$$

It can be shown explicitly that resetting or unbinding is always going to expedite the turnover process compared to the underlying gated diffusive process [29, 106].

### 14.5.4 Resetting in Income Dynamics

Income mobility describes the dynamic aspects of income within a society. Formally, it quantifies the time it takes for an individual to transition from one income status to another, shedding light on the pace and accessibility of upward mobility in different societies [107–110].

Recently, it was shown that a first-passage under resetting approach to income dynamics can effectively capture micro-level variations and experiences of individuals, a caveat that is often missing in standard income mobility analyses. The idea is that a baseline model for income dynamics called geometric Brownian motion with stochastic resetting allows us to develop a "model"-based view for quantifying the time needed for an individual currently with income $x_0$ to reach *target income* $y$ [19–21, 111–113].

In this application, the particle position $x(t)$ can be thought of as the income of a worker in period $t$ (Fig. 14.6a). This observable grows multiplicatively with a rate $\mu$ and volatility $\sigma$ [114–118] until a random event that occurs with a rate $r$ resets its dynamics [2]. The reset event can be interpreted as a worker that left the job market (for example by retiring, being laid off, or after an injury) and is substituted by another younger worker with a starting income $x_R$ (here, $x_R$ need not be equal to $x_0$) [111]. Hence, the MFPT in srGBM disaggregates the income dynamics to the level of individual workers, offering a more nuanced lens through which we can understand the intricacies of income mobility.

To derive the MFPT in srGBM, we use the Fokker-Planck equation for the distribution of the random income variable $x(t)$ that follows the geometric Brownian motion with $r = 0$. This equation reads [22]

$$\frac{\partial}{\partial t} P(x, t|x_0) = \mathcal{L}(x_0) P(x, t|x_0), \tag{14.53}$$

where $\mathcal{L}(z) = \mu z \frac{\partial}{\partial z} + \frac{\sigma^2}{2} z^2 \frac{\partial^2}{\partial z^2}$ is the generator for the GBM process (following Itô convention) [115, 119]. For Stratonovich and Hänggi-Klimontovich convention of the GBM, we refer to [120, 121]. Using this equation and the general formalism for first-passage, it can be shown that the Laplace form of the solution for the mean first-passage time of a GBM trajectory to reach a finite income target $y$, starting at $x_0$, is [22]

$$\begin{aligned} \widetilde{f_T}(s|x_0) &= \left(\frac{x_0}{y}\right)^{q_1(s)}, \quad x_0 \le y, \\ q_1(s) &= \frac{\sqrt{(\sigma^2 - 2\mu)^2 + 8 s \sigma^2} + (\sigma^2 - 2\mu)}{2\sigma^2} > 0. \end{aligned} \tag{14.54}$$

Substituting the above in Eq. (14.7), one obtains the MFPT of a srGBM trajectory that stochastically resets to $x_R$.

Applying the srGBM framework to study the income dynamics yields enlightening observations. For instance, as the economy experiences a larger growth of income or larger volatility, the MFPT declines, as illustrated in Fig. 14.6b–c. This observation aligns with real-world insights: it is generally easier for workers to traverse the income distribution when the economy booms and there is increased randomness [122].

The stochastic resetting rate $r$ in the srGBM framework provides another layer of depth to our understanding (depicted with red dots in Fig. 14.6b–c). This rate, especially when optimized at $r^*$, can be perceived as a lever that policymakers might tweak to influence the rhythm of income dynamics within an economy. For instance, by calibrating policies around workforce participation—such as the frequency of retirements—authorities could potentially regulate the time it takes for a worker to navigate the income spectrum [123–125]. Two salient features of the optimal resetting rate in srGBM emerge which mirror real-world economic observations. First, when volatility is held constant, a surge in growth ($\mu$) reduces the rate at which the MFPT reaches its minimum. This, for example, could be a result of joint factors leading to economic growth such as efficient qualification programs, or quality foreign investments. Second, for a fixed growth rate, when the randomness in the system is increased, we also observe an increased optimal resetting rate. This, for

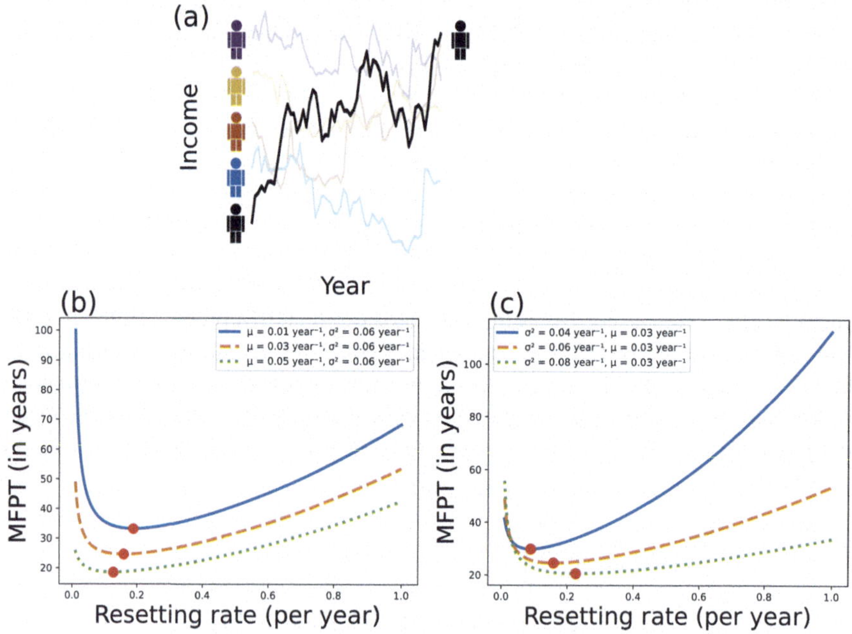


**Fig. 14.6** Resetting in income dynamics. Panel **(a)**: A schematic figure for the income movement of an individual over the years. Panel **(b)**: MFPT (in years) as a function of the resetting rate for various growth rates $\mu$. Panel **(c)**: MFPT (in years) as a function of the resetting rate for various volatilities $\sigma^2$. For panels **(b–c)**: We set $x_0 = x_R = 1$ and set a target $y$ for the individual to double their income (i.e., $y = 2$). Adapted from [22]

example, can be a result of increased intrinsic differences between societal groups when it comes to getting new jobs (e.g., gender or racial differentials).

These theoretical insights have led to an empirical methodology for using the srGBM MFPT in real economies [22, 126]. By integrating the srGBM framework with data, a more textured landscape of income dynamics and economic mobility emerges, paving the way for richer insights and more informed policy decisions. The simplicity of the srGBM MFPT approach paves the way for potential applications in other countries, opening doors to global economic insights.

## 14.6 Discussion

In this chapter, we have shown that stopping a complex search process intermittently only to reset or restart from scratch can often lead to an *accelerated completion*—a counter-intuitive effect that is orthogonal to our general perception. The subject of random resetting has been a topical interest and has seen a wide panorama of applications starting from physics, chemistry, biology, ecology, computer science to economics. By developing a comprehensive framework for the first-passage statistics under resetting, we have applied the formalism to many theoretical first-passage models. In doing so, we discuss in detail the conditions which play a pivotal role in determining whether resetting can indeed be used as a mitigating strategy. Finally, we showcase a few interdisciplinary applications of resetting motivated by the real-world search processes.

It will be quite interesting to study various non-Markovian complex search processes, e.g., geometry-controlled kinetics [127], random walks in fractal geometry [128] under partial [129, 130] and generic resetting mechanism, as well as resetting with random amplitude [131]. Yet another interesting direction that remains quite less explored is the thermodynamical cost of first-passage resetting (see [132–134]). For instance, what will be the energy expenditure for an optimally reset process in comparison to a simply reset process? Will there be a universal thermodynamic trade-off relation at the optimality? Naturally, these questions are fundamental to design resetting strategies in living systems. Needless to say, with the current advent of the experimental studies using laser traps [17, 18, 135], tunable robots [32, 136, 137] or the applications of resetting in quantum search processes [138, 139], queuing theory [34], and material science [140], the field marks the inception of a new era where we expect to see more realistic and trans-disciplinary applications of resetting in complex systems.

**Acknowledgements** AP gratefully acknowledges research support from DST-SERB Start-up Research Grant No. SRG/2022/000080 and the DAE, Government of India. TS acknowledges financial support by the German Science Foundation (DFG, Grant number ME 1535/12-1). TS is supported by the Alliance of International Science Organizations (Project No. ANSO-CR-PP-2022-05). TS is also supported by the Alexander von Humboldt Foundation. We thank Arup Biswas for an illustration.

## References


1. M.R. Evans, S.N. Majumdar, G. Schehr, Stochastic resetting and applications. J. Phys. A: Math. Theor. **53**(19), 193001 (2020)
2. M.R. Evans, S.N. Majumdar, Diffusion with stochastic resetting. Phys. Rev. Lett. **106**(16), 160601 (2011)
3. A. Pal, Diffusion in a potential landscape with stochastic resetting. Phys. Rev. E **91**(1), 012113 (2015)
4. M.R. Evans, S.N. Majumdar, K. Mallick, Optimal diffusive search: nonequilibrium resetting versus equilibrium dynamics. J. Phys. A Math. Theor. **46**(18), 185001 (2013)
5. S. Eule, J.J. Metzger, Non-equilibrium steady states of stochastic processes with intermittent resetting. New J. Phys. **18**(3), 033006 (2016)
6. A. Nagar, S. Gupta, Diffusion with stochastic resetting at power-law times. Phys. Rev. E **93**(6), 060102 (2016)
7. V. Méndez, D. Campos, Characterization of stationary states in random walks with stochastic resetting. Phys. Rev. E **93**(2), 022106 (2016)
8. A. Kumar, A. Pal, Universal framework for record ages under restart. Phys. Rev. Lett. **130**(15), 157101 (2023)
9. L. Kusmierz, S.N. Majumdar, S. Sabhapandit, G. Schehr, First order transition for the optimal search time of Lévy flights with resetting. Phys. Rev. Lett. **113**(22), 220602 (2014)
10. O.L. Bonomo, A. Pal, First passage under restart for discrete space and time: application to one-dimensional confined lattice random walks. Phys. Rev. E **103**, 052129 (2021)
11. S. Belan, Restart could optimize the probability of success in a Bernoulli trial. Phys. Rev. Lett. **120**(8), 080601 (2018)
12. M.R. Evans, S.N. Majumdar, Diffusion with optimal resetting. J. Phys. A Math. Theor. **44**(43), 435001 (2011)
13. A. Pal, S. Reuveni, First passage under restart. Phys. Rev. Lett. **118**(3), 030603 (2017)
14. S. Reuveni, Optimal stochastic restart renders fluctuations in first passage times universal. Phys. Rev. Lett. **116**(17), 170601 (2016)
15. A. Chechkin, I. Sokolov, Random search with resetting: a unified renewal approach. Phys. Rev. Lett. **121**(5), 050601 (2018)
16. A. Pal, Ł Kuśmierz, S. Reuveni, Search with home returns provides advantage under high uncertainty. Phys. Rev. Res. **2**(4), 043174 (2020)
17. O. Tal-Friedman, A. Pal, A. Sekhon, S. Reuveni, Y. Roichman, Experimental realization of diffusion with stochastic resetting. J. Phys. Chem. Lett. **11**(17), 7350–7355 (2020)
18. B. Besga, A. Bovon, A. Petrosyan, S.N. Majumdar, S. Ciliberto, Optimal mean first-passage time for a Brownian searcher subjected to resetting: experimental and theoretical results. Phys. Rev. Research **2**(3), 032029 (2020)
19. V. Stojkoski, P. Jolakoski, A. Pal, T. Sandev, L. Kocarev, R. Metzler, Income inequality and mobility in geometric Brownian motion with stochastic resetting: theoretical results and empirical evidence of non-ergodicity. Philos. Trans. R. Soc. A **380**(2224), 20210157 (2022)
20. X. Gabaix, J.M. Lasry, P.L. Lions, B. Moll, The dynamics of inequality. Econometrica **84**(6), 2071–2111 (2016)
21. D. Vinod, A.G. Cherstvy, W. Wang, R. Metzler, I.M. Sokolov, Nonergodicity of reset geometric Brownian motion. Phys. Rev. E **105**(1), L012106 (2022)
22. P. Jolakoski, A. Pal, T. Sandev, L. Kocarev, R. Metzler, V. Stojkoski, A first passage under resetting approach to income dynamics. Chaos Solitons & Fractals **175**, 113921 (2023)
23. D. Boyer, C. Solis-Salas, Random walks with preferential relocations to places visited in the past and their application to biology. Phys. Rev. Lett. **112**(24), 240601 (2014)
24. É. Roldán, A. Lisica, D. Sánchez-Taltavull, S.W. Grill, Stochastic resetting in backtrack recovery by RNA polymerases. Phys. Rev. E **93**(6), 062411 (2016)
25. A.M. Ramoso, J.A. Magalang, D. Sánchez-Taltavull, J.P. Esguerra, É. Roldán, Stochastic resetting antiviral therapies prevent drug resistance development. Eur. Lett. **132**(5), 50003 (2020)

26. S. Budnar, K.B. Husain, G.A. Gomez, M. Naghibosadat, A. Varma, S. Verma, N.A. Hamilton, R.G. Morris et al., Anillin promotes cell contractility by cyclic resetting of RhoA residence kinetics. Dev. Cell **49**(6), 894–906 (2019)
27. S. Reuveni, M. Urbakh, J. Klafter, Role of substrate unbinding in michaelis-menten enzymatic reactions. Proc. Natl. Acad. Sci. **111**(12), 4391–4396 (2014)
28. T. Rotbart, S. Reuveni, M. Urbakh, Michaelis-Menten reaction scheme as a unified approach towards the optimal restart problem. Phys. Rev. E **92**(6), 060101 (2015)
29. A. Biswas, A. Pal, D. Mondal, S. Ray, Rate enhancement of gated drift-diffusion process by optimal resetting. J. Chem. Phys. **159**(5) (2023)
30. A. Morán, M. Lihoreau, A. Pérez-Escudero, J. Gautrais, Modeling bee movement shows how a perceptual masking effect can influence flower discovery. PLoS Comput. Biol. **19**(3), e1010558 (2023)
31. O. Bénichou, C. Loverdo, M. Moreau, R. Voituriez, Intermittent search strategies. Rev. Mod. Phys. **83**(1), 81 (2011)
32. S. Paramanick, A. Biswas, H. Soni, A. Pal, N. Kumar, Uncovering universal characteristics of homing paths using foraging robots. PRX Life **2**(3), 033007 (2024)
33. A. Ray, A. Pal, D. Ghosh, S.K. Dana, C. Hens, Mitigating long transient time in deterministic systems by resetting. Chaos **31**(1) (2021)
34. O.L. Bonomo, A. Pal, S. Reuveni, Mitigating long queues and waiting times with service resetting. PNAS Nexus **1**(3), pgac070 (2022)
35. M. Luby, A. Sinclair, D. Zuckerman, Optimal speedup of Las Vegas algorithms. Inf. Process. Lett. **47**(4), 173–180 (1993)
36. O. Blumer, S. Reuveni, B. Hirshberg, Stochastic resetting for enhanced sampling. J. Phys. Chem. Lett. **13**(48), 11230–11236 (2022)
37. A. Pal, A. Kundu, M.R. Evans, Diffusion under time-dependent resetting. J. Phys. A Math. Theor. **49**(22), 225001 (2016)
38. S. Redner, *A Guide to First-Passage Processes* (Cambridge University Press, Cambridge UK, 2001)
39. A.J. Bray, S.N. Majumdar, G. Schehr, Persistence and first-passage properties in nonequilibrium systems. Adv. Phys. **62**(3), 225–361 (2013)
40. V. Stojkoski, T. Sandev, L. Kocarev, A. Pal, Autocorrelation functions and ergodicity in diffusion with stochastic resetting. J. Phys. A Math. Theor. **55**(10), 104003 (2022)
41. M.R. Evans, S.N. Majumdar, Diffusion with resetting in arbitrary spatial dimension. J. Phys. A Math. Theor. **47**(28), 285001 (2014)
42. A. Pal, R. Chatterjee, S. Reuveni, A. Kundu, Local time of diffusion with stochastic resetting. J. Phys. A Math. Theor. **52**(26), 264002 (2019)
43. R. Singh, R. Metzler, T. Sandev, Resetting dynamics in a confining potential. J. Phys. A Math. Theor. **53**(50), 505003 (2020)
44. S. Ray, D. Mondal, S. Reuveni, Péclet number governs transition to acceleratory restart in drift-diffusion. J. Phys. A Math. Theor. **52**(25), 255002 (2019)
45. A.N. Borodin, P. Salminen, *Handbook of Brownian Motion-Facts and Formulae* (Springer Science & Business Media, Berlin, 2015)
46. D.S. Grebenkov, First exit times of harmonically trapped particles: a didactic review. J. Phys. A Math. Theor. **48**(1), 013001 (2014)
47. D. Gupta, C.A. Plata, A. Kundu, A. Pal, Stochastic resetting with stochastic returns using external trap. J. Phys. A Math. Theor. **54**(2), 025003 (2020)
48. S. Ahmad, I. Nayak, A. Bansal, A. Nandi, D. Das, First passage of a particle in a potential under stochastic resetting: a vanishing transition of optimal resetting rate. Phys. Rev. E **99**(2), 022130 (2019)
49. V.F. Zaitsev, A.D. Polyanin, *Handbook of Exact Solutions for Ordinary Differential Equations* (CRC Press, London, 2002)
50. S. Ray, S. Reuveni, Diffusion with resetting in a logarithmic potential. J. Chem. Phys. **152**(23) (2020)

51. K. Capała, B. Dybiec, Optimization of escape kinetics by reflecting and resetting. Chaos **33**, 103124 (2023)
52. A. Pal, V.V. Prasad, First passage under stochastic resetting in an interval. Phys. Rev. E **99**(3), 032123 (2019)
53. A. Pal, V.V. Prasad, Landau-like expansion for phase transitions in stochastic resetting. Phys. Rev. Res. **1**(3), 032001 (2019)
54. S. Ahmad, K. Rijal, D. Das, First passage in the presence of stochastic resetting and a potential barrier. Phys. Rev. E **105**(4), 044134 (2022)
55. H. Chen, F. Huang, First passage of a diffusing particle under stochastic resetting in bounded domains with spherical symmetry. Phys. Rev. E **105**(3), 034109 (2022)
56. S. Ahmad, D. Das, Role of dimensions in first passage of a diffusing particle under stochastic resetting and attractive bias. Phys. Rev. E **102**(3), 032145 (2020)
57. S. Ahmad, D. Das, Comparing the roles of time overhead and spatial dimensions on optimal resetting rate vanishing transitions, in Brownian processes with potential bias and stochastic resetting. J. Phys. A Math. Theor. **56**(10), 104001 (2023)
58. A. Chatterjee, C. Christou, A. Schadschneider, Diffusion with resetting inside a circle. Phys. Rev. E **97**(6), 062106 (2018)
59. K. Capała, B. Dybiec, Random acceleration process on finite intervals under stochastic restarting. J. Stat. Mech. **2021**(8), 083216 (2021)
60. R. Metzler, J. Klafter, The random walk's guide to anomalous diffusion: a fractional dynamics approach. Phys. Rep. **339**(1), 1–77 (2000)
61. J. Klafter, I.M. Sokolov, *First Steps in Random Walks: from Tools to Applications* (Oxford University Press, Oxford, 2011)
62. V. Méndez, A. Masó-Puigdellosas, T. Sandev, D. Campos, Continuous time random walks under Markovian resetting. Phys. Rev. E **103**(2), 022103 (2021)
63. V. Méndez, A. Masó-Puigdellosas, D. Campos, Nonstandard diffusion under Markovian resetting in bounded domains. Phys. Rev. E **105**(5), 054118 (2022)
64. I.M. Sokolov, J. Klafter, From diffusion to anomalous diffusion: a century after Einstein's Brownian motion. Chaos **15**(2) (2005)
65. A. Stanislavsky, A. Weron, Optimal non-Gaussian search with stochastic resetting. Phys. Rev. E **104**(1), 014125 (2021)
66. A. Masó-Puigdellosas, D. Campos, V. Méndez, Transport properties and first-arrival statistics of random motion with stochastic reset times. Phys. Rev. E **99**(1), 012141 (2019)
67. J. Masoliver, M. Montero, Anomalous diffusion under stochastic resettings: a general approach. Phys. Rev. E **100**(4), 042103 (2019)
68. Ł Kuśmierz, E. Gudowska-Nowak, Subdiffusive continuous-time random walks with stochastic resetting. Phys. Rev. E **99**(5), 052116 (2019)
69. V. Shkilev, Continuous-time random walk under time-dependent resetting. Phys. Rev. E **96**(1), 012126 (2017)
70. M. Dos Santos, L. Menon Jr., C. Anteneodo, Efficiency of random search with space-dependent diffusivity. Phys. Rev. E **106**(4), 044113 (2022)
71. T. Sandev, V. Domazetoski, L. Kocarev, R. Metzler, A. Chechkin, Heterogeneous diffusion with stochastic resetting. J. Phys. A Math. Theor. **55**(7), 074003 (2022)
72. K. Itô, Stochastic integral. Proc. Imperial Acad. **20**(8), 519–524 (1944)
73. R. Stratonovich, A new representation for stochastic integrals and equations. SIAM J. Control **4**(2), 362–371 (1966)
74. P. Hänggi, Stochastic processes. i. asymptotic behaviour and symmetries. Helv. Phys. Acta **51**, 183–201 (1978)
75. Y.L. Klimontovich, Ito, Stratonovich and kinetic forms of stochastic equations. Phys. A **163**(2), 515–532 (1990)
76. A.G. Cherstvy, A.V. Chechkin, R. Metzler, Anomalous diffusion and ergodicity breaking in heterogeneous diffusion processes. New J. Phys. **15**(8), 083039 (2013)
77. N. Leibovich, E. Barkai, Infinite ergodic theory for heterogeneous diffusion processes. Phys. Rev. E **99**(4), 042138 (2019)

78. S. Ray, Space-dependent diffusion with stochastic resetting: a first-passage study. J. Chem. Phys. **153**(23), 234904 (2020)
79. A. Pal, I. Eliazar, S. Reuveni, First passage under restart with branching. Phys. Rev. Lett. **122**(2), 020602 (2019)
80. A. Pal, S. Kostinski, S. Reuveni, The inspection paradox in stochastic resetting. J. Phys. A Math. Theor. **55**(2), 021001 (2022)
81. S. Jain, D. Boyer, A. Pal, L. Dagdug, Fick–jacobs description and first passage dynamics for diffusion in a channel under stochastic resetting. J. Chem. Phys. **158**(5) (2023)
82. R. Satija, A.M. Berezhkovskii, D.E. Makarov, Broad distributions of transition-path times are fingerprints of multidimensionality of the underlying free energy landscapes. Proc. Natl. Acad. Sci. **117**(44), 27116–27123 (2020)
83. F. Sturzenegger, F. Zosel, E.D. Holmstrom, K.J. Buholzer, D.E. Makarov, D. Nettels, B. Schuler, Transition path times of coupled folding and binding reveal the formation of an encounter complex. Nat. Commun. **9**(1), 4708 (2018)
84. A.L. Thorneywork, J. Gladrow, Y. Qing, M. Rico-Pasto, F. Ritort, H. Bayley, A.B. Kolomeisky, U.F. Keyser, Direct detection of molecular intermediates from first-passage times. Sci. Adv. **6**(18), eaaz4642 (2020)
85. R. Metzler, S. Redner, G. Oshanin, *First-Passage Phenomena and Their Applications*, vol. 35 (World Scientific, Singapore, 2014)
86. J.P. Bouchaud, A. Georges, Anomalous diffusion in disordered media: statistical mechanisms, models and physical applications. Phys. Rep. **195**(4–5), 127–293 (1990)
87. J.R. Moffitt, C. Bustamante, Extracting signal from noise: kinetic mechanisms from a Michaelis-Menten-like expression for enzymatic fluctuations. FEBS J. **281**(2), 498–517 (2014)
88. B.P. English, W. Min, A.M. Van Oijen, K.T. Lee, G. Luo, H. Sun, B.J. Cherayil, S. Kou, X.S. Xie, Ever-fluctuating single enzyme molecules: Michaelis-Menten equation revisited. Nat. Chem. Biol. **2**(2), 87–94 (2006)
89. G.M. Viswanathan, M.G. Da Luz, E.P. Raposo, H.E. Stanley, *The Physics of Foraging: an Introduction to Random Searches and Biological Encounters* (Cambridge University Press, Cambridge, 2011)
90. G.M. Viswanathan, S.V. Buldyrev, S. Havlin, M. Da Luz, E. Raposo, H.E. Stanley, Optimizing the success of random searches. Nature **401**(6756), 911–914 (1999)
91. A.S. Bodrova, I.M. Sokolov, Resetting processes with noninstantaneous return. Phys. Rev. E **101**(5), 052130 (2020)
92. D. Gupta, A. Pal, A. Kundu, Resetting with stochastic return through linear confining potential. J. Stat. Mech. **2021**(4), 043202 (2021)
93. A. Biswas, A. Kundu, A. Pal, Stochasticity in returns can expedite classical first passage under resetting. Phys. Rev. E **110**, L042101 (2024). https://doi.org/10.1103/PhysRevE.110.L042101
94. G. Mercado-Vásquez, D. Boyer, S.N. Majumdar, G. Schehr, Intermittent resetting potentials. J. Stat. Mech. **2020**(11), 113203 (2020)
95. M.H. Jacobs, M. Jacobs, *Diffusion Processes* (Springer, Berlin, 1935)
96. R. Zwanzig, Diffusion past an entropy barrier. J. Phys. Chem. **96**(10), 3926–3930 (1992)
97. Y. Scher, S. Reuveni, Unified approach to gated reactions on networks. Phys. Rev. Lett. **127**(1), 018301 (2021)
98. A. Berezhkovskii, A. Szabo, Time scale separation leads to position-dependent diffusion along a slow coordinate. J. Chem. Phys. **135**(7) (2011)
99. A.M. Berezhkovskii, G. Hummer, S.M. Bezrukov, Identity of distributions of direct uphill and downhill translocation times for particles traversing membrane channels. Phys. Rev. Lett. **97**(2), 020601 (2006)
100. L. Dagdug, A. Berezhkovskii, S.M. Bezrukov, G.H. Weiss, Diffusion-controlled reactions with a binding site hidden in a channel. J. Chem. Phys. **118**(5), 2367–2373 (2003)
101. D. Reguera, J. Rubi, Kinetic equations for diffusion in the presence of entropic barriers. Phys. Rev. E **64**(6), 061106 (2001)

102. G.K. Sar, A. Ray, D. Ghosh, C. Hens, A. Pal, Resetting mediated navigation of active Brownian searcher in a homogeneous topography. Soft Matter **19**, 4502–4518 (2023)
103. S. Iyer-Biswas, A. Zilman, First-passage processes in cellular biology. Adv. Chem. Phys. **160**, 261–306 (2016)
104. P.C. Bressloff, *Stochastic Processes in cell Biology*, vol. 41 (Springer, Cham, 2014)
105. A. Szabo, D. Shoup, S.H. Northrup, J.A. McCammon, Stochastically gated diffusion-influenced reactions. J. Chem. Phys. **77**(9), 4484–4493 (1982)
106. G. Mercado-Vásquez, D. Boyer, Search of stochastically gated targets with diffusive particles under resetting. J. Phys. A Math. Theor. **54**(44), 444002 (2021)
107. M. Jäntti, S.P. Jenkins, in *Handbook of Income Distribution*, vol. 2 (Elsevier, 2015), pp. 807–935
108. V. Stojkoski, Measures of physical mixing evaluate the economic mobility of the typical individual. Chaos Solitons Fractals **180**, 114567 (2024)
109. V. Stojkoski, M. Karbevski, Ergodicity breaking in wealth dynamics: the case of reallocating geometric Brownian motion. Phys. Rev. E **105**(2), 024107 (2022)
110. C.A. La Porta, S. Zapperi, Unraveling the dynamics of wealth inequality and the impact on social mobility and health disparities. J. Phys. Complex. (2023)
111. M. Nirei, W. Souma, in *The Complex Dynamics of Economic Interaction* (Springer, 2004), pp. 161–168
112. S. Aoki, M. Nirei, Zipf's law, Pareto's law, and the evolution of top incomes in the United States. Am. Econ. J. Macroecon. **9**(3), 36–71 (2017)
113. D. Vinod, A.G. Cherstvy, R. Metzler, I.M. Sokolov, Time-averaging and nonergodicity of reset geometric Brownian motion with drift. Phys. Rev. E **106**(3), 034137 (2022)
114. A.G. Cherstvy, D. Vinod, E. Aghion, A.V. Chechkin, R. Metzler, Time averaging, ageing and delay analysis of financial time series. New J. Phys. **19**(6), 063045 (2017)
115. V. Stojkoski, T. Sandev, L. Basnarkov, L. Kocarev, R. Metzler, Generalised geometric Brownian motion: Theory and applications to option pricing. Entropy **22**(12), 1432 (2020)
116. V. Stojkoski, Z. Utkovski, L. Basnarkov, L. Kocarev, Cooperation dynamics in networked geometric Brownian motion. Phys. Rev. E **99**(6), 062312 (2019)
117. J.T. Kemp, L.M. Bettencourt, Statistical dynamics of wealth inequality in stochastic models of growth. Phys. A **607**, 128180 (2022)
118. J.T. Kemp, L.M. Bettencourt, Learning increases growth and reduces inequality in shared noisy environments. PNAS Nexus **2**(4), pgad093 (2023)
119. V. Stojkoski, T. Sandev, L. Kocarev, A. Pal, Geometric Brownian motion under stochastic resetting: a stationary yet nonergodic process. Phys. Rev. E **104**(1), 014121 (2021)
120. T. Sandev, A. Iomin, L. Kocarev, Hitting times in turbulent diffusion due to multiplicative noise. Phys. Rev. E **102**(4), 042109 (2020)
121. T. Sandev, A. Iomin, *Special Functions of Fractional Calculus: Applications to Diffusion and Random Search Processes* (World Scientific, Singapore, 2022)
122. D. Aristei, C. Perugini, The drivers of income mobility in Europe. Econ. Syst. **39**(2), 197–224 (2015)
123. H. Cremer, J.M. Lozachmeur, P. Pestieau, Social security, retirement age and optimal income taxation. J. Public Econ. **88**(11), 2259–2281 (2004)
124. S. Staubli, J. Zweimüller, Does raising the early retirement age increase employment of older workers? J. Public Econ. **108**, 17–32 (2013)
125. J. Sheather, D. Slattery, The great resignation-how do we support and retain staff already stretched to their limit? BMJ **375**(2533) (2021)
126. V. Stojkoski, S. Mitikj, M. Trpkova-Nestorovska, D. Tevdovski, Income mobility and mixing in north macedonia. Tech. rep. (2023)
127. C. Chevalier, B. Meyer, Geometry-controlled kinetics. Nat. Chem. **2**(6), 472–477 (2010)
128. S. Condamin, O. Bénichou, V. Tejedor, R. Voituriez, First-passage times in complex scale-invariant media. Nature **450**(7166), 77–80 (2007)
129. O. Tal-Friedman, Y. Roichman, S. Reuveni, Diffusion with partial resetting. Phys. Rev. E **106**(5), 054116 (2022)

130. C. Di Bello, A.V. Chechkin, A.K. Hartmann, Z. Palmowski, R. Metzler, Time-dependent probability density function for partial resetting dynamics. New J. Phys. **25**(8), 082002 (2023)
131. M. Dahlenburg, A.V. Chechkin, R. Schumer, R. Metzler, Stochastic resetting by a random amplitude. Phys. Rev. E **103**(5), 052123 (2021)
132. A. Pal, S. Reuveni, S. Rahav, Thermodynamic uncertainty relation for systems with unidirectional transitions. Phys. Rev. Res. **3**(1), 013273 (2021)
133. P.S. Pal, A. Pal, H. Park, J.S. Lee, Thermodynamic trade-off relation for first passage time in resetting processes. Phys. Rev. E **108**, 044117 (2023)
134. D. Gupta, C.A. Plata, A. Pal, Work fluctuations and Jarzynski equality in stochastic resetting. Phys. Rev. Lett. **124**(11), 110608 (2020)
135. R. Goerlich, M. Li, L.B. Pires, P.A. Hervieux, G. Manfredi, C. Genet, Experimental test of Landauer's principle for stochastic resetting (2023). arXiv:2306.09503
136. S. Paramanick, A. Pal, H. Soni, N. Kumar, Programming tunable active dynamics in a self-propelled robot. Eur. Phys. J. E **47**(5), 34 (2024)
137. A. Altshuler, O.L. Bonomo, N. Gorohovsky, S. Marchini, E. Rosen, O. Tal-Friedman, S. Reuveni, Y. Roichman, Environmental memory facilitates search with home returns. Phys. Rev. Res. **6**(2), 023255
138. R. Yin, E. Barkai, Restart expedites quantum walk hitting times. Phys. Rev. Lett. **130**(5), 050802 (2023)
139. M. Kulkarni, S.N. Majumdar, First detection probability in quantum resetting via random projective measurements. J. Phys. A: Math. Theor. **56**(38), 385003
140. T. Aquino, T. Le Borgne, J. Heyman, Fluid-solid reaction in porous media as a chaotic restart process. Phys. Rev. Lett. **130**(26), 264001 (2023)

# Chapter 15
# The Distribution of the Maximum of Independent Resetting Brownian Motions

**Alexander K. Hartmann, Satya N. Majumdar, and Grégory Schehr**

**Abstract** The probability distribution of the maximum $M_t$ of a single resetting Brownian motion (RBM) of duration $t$ and resetting rate $r$, properly centered and scaled, is known to converge to the standard Gumbel distribution of the classical extreme value theory. This Gumbel law describes the typical fluctuations of $M_t$ around its average $\sim \sqrt{D/r}\ \ln(r\,t)$ for large $t$ on a scale of $O(1)$. Here we compute the large-deviation tails of this distribution when $M_t = O(t)$ and show that the large-deviation function has a singularity where the second derivative is discontinuous, signaling a dynamical phase transition. Then we consider a collection of independent RBMs with initial (and resetting) positions uniformly distributed with a density $\rho$ over the negative half-line. We show that the fluctuations in the initial positions of the particles modify the distribution of $M_t$. The average over the initial conditions can be performed in two different ways, in analogy with disordered systems: (i) the annealed case where one averages over all possible initial conditions and (ii) the quenched case where one considers only the contributions coming from typical initial configurations. We show that in the annealed case, the limiting distribution of the maximum is characterized by a new scaling function, different from the Gumbel law but the large-deviation function remains the same as in the single particle case. In the quenched case, the limiting (typical) distribution and the large deviation regime remain the same as in the single particle case. However, in the quenched case, an additional scaling regime emerges where $M_t = O(1)$ and the distribution of $M_t$ in this regime is described by a new nontrivial rate function. Our analytical results, both for the typical as well as for the large-deviation regime of

A. K. Hartmann
Institut für Physik, Universität Oldenburg, 26111 Oldenburg, Germany
e-mail: alexander.hartmann@uni-oldenburg.de

S. N. Majumdar
LPTMS, CNRS, Univ. Paris-Sud, Université Paris-Saclay, 91405 Orsay, France
e-mail: satya.majumdar@universite-paris-saclay.fr

G. Schehr (✉)
Sorbonne Université, Laboratoire de Physique Théorique et Hautes Energies, UMR 7589, 4 Place Jussieu, 75252 Paris Cedex 05, France
e-mail: schehr@lpthe.jussieu.fr



D. Grebenkov et al. (eds.), *Target Search Problems*,
https://doi.org/10.1007/978-3-031-67802-8_15

$M_t$, are verified numerically with extremely high precision, down to $10^{-250}$ for the probability density of $M_t$.

## 15.1 Introduction

Consider a random searcher on a line whose dynamics is characterized by a stochastic process $x(t)$, starting say at the origin $x = 0$. Let $M$ denote the location of a fixed target; see Fig. 15.1. How long will the searcher take to find the target? This is a key fundamental question for any search process. A quantity that plays a central role is the survival probability $Q(M, t)$ of the target up to time $t$, i.e., the probability that the target is not found by the searcher up to time $t$. The actual time to find the target is a random variable known as the first-passage time. The probability distribution function (PDF) $P(t_f|M)$ of the first-passage time is very simply related to the survival probability by the relation

$$Q(M, t) = \text{Prob.}\left[t_f \geq t\right] = \int_t^\infty P(t_f|M)\, dt_f, \tag{15.1}$$

which simply follows from the fact that if the target survives up to $t$, it must be found by the searcher only after $t$, i.e., $t_f \geq t$. Taking derivative with respect to $t$ relates the survival probability and the first-passage PDF

$$P(t_f|M) = -\frac{\partial}{\partial t} Q(M, t)\Big|_{t=t_f}. \tag{15.2}$$

Search processes are ubiquitous in nature [1–3]: animals searching for food, proteins searching for a DNA site to bind, chemicals diffusing and searching for other constituents to react, debugging schemes searching for a bug in a computer program,

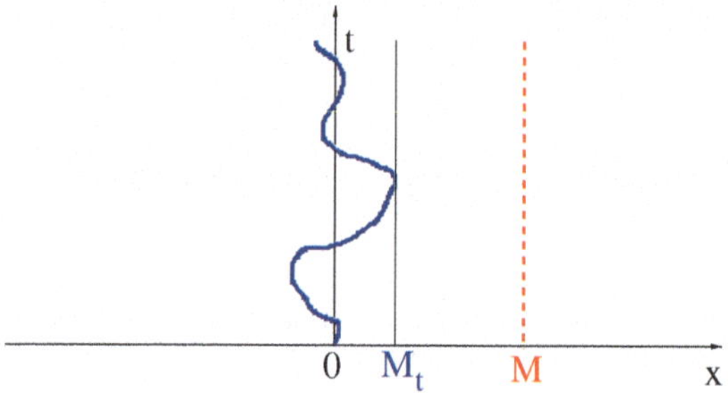


**Fig. 15.1** Schematic trajectory of a stochastic search process $x(t)$, starting from $x(0) = 0$, with global maximal value $M_t$ up to time $t$. A fixed target, shown by the vertical dashed (red) line, is located at $M$. The survival probability $Q(M, t)$ of the target up to time $t$ is exactly the probability of the event $M_t \leq M$, as in Eq. (15.5)

randomized algorithms searching for a global minimum in a high-dimensional landscape and many others. The survival probability and the associated first-passage time distribution is a key concept in random search processes with many applications [4–13]. It can be considered as a cost function for a search process. The efficiency of a search process is enhanced by minimizing the survival probability of the target. Given the ubiquity of search processes, the survival probability $Q(M, t)$ or the associated first-passage PDF $P(t_f|M)$ has been studied for more than 100 years across disciplines: in mathematics, physics, chemistry, biology and computer science [14–19]. In the physics of nonequilibrium many-body systems, the survival probability or "persistence" also appears as a key quantity that characterizes the history dependence of the underlying stochastic systems [18].

Another, a priori unrelated, subject that has also been widely studied with applications ranging from sports, climate science, finance all the way to physics of disordered systems is extreme value statistics (EVS) [20–22]; for a recent pedagogical review on EVS, see Ref. [24]. In EVS, one typically looks at a time series, for example the temperature or rainfall data in a given weather station or the price of a stock, and records the maximum (or minimum) $M_t$ of the process or the time series up to time $t$. The histogram of $M_t$ gives access to its PDF $P(M, t)$. The statistical behavior of $M_t$, encoded in $P(M, t)$, gives important information about the extreme fluctuations and plays a crucial role, e.g., in analyzing data for global warming [25]. When the underlying time series is completely uncorrelated, the EVS is well understood from the classical literature in statistics and mathematics [20, 23] and $P(M, t)$, appropriately centered and scaled, is usually governed by one of the three well-known extreme value distributions, known as the Gumbel, the Fréchet and the Weibull class [24]. However, much less is understood about the EVS when the underlying time series is correlated [24].

These two subjects, namely the survival probability and the EVS, are actually very closely related. To see this, consider again the one-dimensional situation and let the process $x(t)$ represent the motion of the searcher in our underlying time series, with initial value $x(0) = 0$. Let $M_t$ denote the maximum achieved by this process up to time $t$, i.e.,

$$M_t = \max_{0 \le \tau \le t} [\{x(\tau)\}] . \tag{15.3}$$

Then the cumulative distribution function (CDF) of the maximum is defined as

$$\text{Prob.}[M_t \le M] = \int_0^M P(M', t)\, dM', \tag{15.4}$$

where $P(M, t)$ is the PDF of $M_t$. However, if the maximum $M_t$ is less than $M$, this event is equivalent to the probability that all positions of the searcher $\{x(\tau)\}$ up to $t$ must be less than $M$ (see Fig. 15.1). In other words, this is precisely the probability that the target at $M$ is not touched/found by the process up to $t$, i.e., the survival probability $Q(M, t)$ of the target up to time $t$. This establishes a precise and general relation between the EVS and survival probability

$$\text{Prob.}[M_t \leq M] = \int_0^M P(M', t)\, dM' = Q(M, t). \tag{15.5}$$

Thus, if we know the survival probability $Q(M, t)$, the PDF of the maximum $M_t$ is simply obtained from (15.5) by taking a derivative with respect to $M$

$$P(M, t) = \frac{\partial}{\partial M} Q(M, t). \tag{15.6}$$

To summarize, the survival probability $Q(M, t)$ has two arguments $M$ and $t$. Taking a derivative with respect to $M$, as in Eq. (15.6), provides the PDF of the maximum $M_t$ of the process up to time $t$. In contrast, taking the negative of the derivative of $Q(M, t)$ with respect to $t$, as in Eq. (15.2), provides the PDF of the first-passage time to the target at $M$.

So far the discussion has been for any arbitrary random search process in one dimension. We now focus on a specific random search process that has created much interest over the last decade. This is known as the resetting Brownian motion (RBM) [26–28]. In this simple model, the trajectory of the searcher performs an ordinary Brownian motion, starting from its initial position at 0, during a random time interval $\tau$, distributed via $p(\tau)$. At the end of the interval, the trajectory is reset to the starting position and a new random time interval is chosen from $p(\tau)$ during which the searcher undergoes free diffusion and so on. When the interval between resettings is distributed via

$$p(\tau) = r\, e^{-r\,\tau}, \tag{15.7}$$

the model is called Poissonian resetting where the Brownian motion resets to its starting position at a constant $r$ [26]. But other variants, such as "periodic" or "sharp", restart where $p(\tau) = \delta(\tau - T)$ or power-law distributed $p(\tau)$ have also been studied [29–33].

Here we will focus on simple Poissonian resetting with rate $r$, mainly because this model is simple and has a rather nice analytical structure. A nonzero resetting rate $r$ has two major effects [26]: (i) by breaking the detailed balance, it drives the system into a nonequilibrium stationary state where the position distribution is typically non-Gaussian at long times and (ii) in the presence of a target at fixed $M$ and with the RBM starting and resetting at 0, the mean first-passage time (MFPT) to the target, as a function of the resetting rate $r$, achieves a minimum at an optimal value $r^*$. The existence of an optimal $r^*$ makes the RBM an efficient search process [26–28, 30, 32, 34–41] and confirms the scenario that when searching for a target in vain for a while, it is better to restart the search process: the rationale being that the searcher may explore a different pathway searching for the target, thus expediting the search process. However, the existence of an optimal $r^*$ is not always guaranteed. Several models have been studied, where by varying a parameter in the underlying process, the optimal $r^*$ may undergo a phase transition, becoming zero beyond a critical value of the parameter [32, 34, 36, 42–52]. Stochastic resetting has created much interest across fields in the last decade (for recent reviews see Refs. [53–55]). Various observables for diffusion with stochastic resetting have also been measured in experiments using holographic optical tweezers in one- and two-dimensional optical traps [56–59].

The purpose of this chapter is to illustrate the complementary perspectives of the same observable from two different fields, i.e., search processes and the EVS, within the simple model of a resetting Brownian motion as the search process. In Sect. 15.2, we will explore the observable $Q_r(M, t)$ denoting the survival probability of a target placed at $M$, in the presence of a single searcher whose dynamics is governed by the RBM, starting and resetting at the origin with rate $r$. As mentioned earlier, $Q_r(M, t)$ can alternatively be interpreted as the CDF of the maximum $M_t$ of the RBM (without any target) up to time $t$. The Laplace transform (with respect to time $t$) of $Q_r(M, t)$ was already computed exactly in Ref. [26]. However, it turns out that the exact Laplace inversion is difficult, and interesting behaviors of $Q_r(M, t)$ for different scales of $M$ in real time $t$ are actually hidden in the Laplace transform. Treating $\partial_M Q_r(M, t) = P_r(M, t)$ as the PDF of the maximum $M_t$ (see Eq. (15.6)), one can extract the mean $\langle M_t \rangle$ and its variance explicitly and they have been used in many applications. These include computing the statistics of the number of distinct sites visited by a resetting random walker on a one-dimensional lattice [60], computing the mean perimeter and the mean area of the convex hull of a two-dimensional RBM [61] and also calculating the distribution of the time at which the maximum occurs in the stationary state of an RBM [62, 63]. Similarly, higher moments of $M_t$ can also be extracted analytically [64]. However, extracting the full extreme-value distribution $Q_r(M, t)$ for all $M$ from its Laplace transform is highly nontrivial. Our first goal in Sect. 15.2 would be to analyze this Laplace transform carefully to bring out the late time behavior of $Q_r(M, t)$ for all $M$. In particular, treating $Q_r(M, t)$ as the CDF of the maximum $M_t$ brings out new interesting scales of $M_t$. We will see that while the typical fluctuations of $M_t$ are of $O(\ln t)$ for large $t$, there is an interesting large-deviation tail when $M \sim O(t)$ where the associated rate function undergoes a second-order dynamical phase transition. We will analyze $Q_r(M, t)$ in different regions of the $M$-$t$ plane and explore its scaling behaviors.

In Sect. 15.3, we will consider a generalization of the single-particle system to a many-particle scenario. Here we will have $N$ particles distributed uniformly with density $\rho$ on the left half of the origin at $t = 0$, with the target located at $M \geq 0$. Each particle undergoes resetting Brownian motions independently, i.e., the $i$-th particle starts at the initial random position $x_i$ and undergoes diffusion and resetting to its own initial position $x_i$. The process ends when the target is found by any one of the $N$ particles. Let $M_t(i)$ denote the maximum, up to $t$, of the $i$-th particle. Clearly, $M_t(i)$'s are independent random variables. Then the global maximum

$$M_t = \max_{1 \leq i \leq N} [M_t(i)] \tag{15.8}$$

represents the maximum of a set of $N$ independent, but *not identically* distributed variables. The different initial and resetting positions make the variables $M_t(i)$ non-identically distributed. We will see that this has interesting implications for the extreme-value statistics. In addition, the behavior of the survival probability or the EVS depends strongly on how the averaging over the initial positions is performed. If one averages the survival probability over all possible initial conditions with equal weight, this is called *annealed* survival probability, in analogy with disordered systems. In contrast, when one averages only over typical initial configurations (regular equidistant positions with typical interparticle separation $1/\rho$) and ignores the rest, the corresponding survival probability is *quenched*, again in analogy with disordered systems. We will see that the annealed and the quenched survival probabilities have rather different behaviors. This also has interesting significance from the perspective of extreme-value statistics as will be illustrated. Moreover, in the two cases, we will also explore both the typical as well as the large-deviation regimes of the survival probability as a function of $M$ for large $t$. We note that, in Ref. [26], the survival probability for both the annealed and the quenched cases were analyzed, but only for $M = 0$, i.e., when the target is fixed at the origin. However, to understand how the extreme $M_t$ behaves, we need to analyze the survival probability for general $M \geq 0$. Thus, the results for general $M \geq 0$ presented in this chapter are original and go beyond Ref. [26].

In Sect. 15.4, we will verify our analytical predictions via high-precision numerical simulations. Finally, we will conclude with some perspectives in Sect. 15.5.

## 15.2 Single-Particle Case

We start with a single RBM, starting and resetting at the origin with rate $r$, with a target fixed at $M$. Let $M_t$ be the maximum of this RBM up to time $t$; see Fig. 15.1. As mentioned in the introduction, the survival probability $Q_r(M, t)$ of the target up to time $t$ is exactly the CDF of the random variable $M_t$; see Eq. (15.5). Thus $Q_r(M, t)$ has two alternative interpretations. When one thinks of it as the cumulative distribution of the maximum, one thinks $Q_r(M, t)$ as a function of $M$ for a fixed $t$. In contrast, when one thinks of $Q_r(M, t)$ as the survival probability of a fixed target located at $M$, one is more interested in determining how the survival probability decays as a function of time $t$, for fixed $M$. Here, we will therefore analyze $Q_r(M, t)$ in the $M - t$ plane for $M \geq 0$ and $t \geq 0$. In the first case, we will fix $t$ and observe $Q_r(M, t)$ as a function of $M$ where it has the interpretation of a cumulative probability and increases from 0 at $M = 0$ to 1 as $M \to \infty$, for any fixed $t$. In the second case, we will fix $M$ and study how $Q_r(M, t)$ decays as a function of time $t$.

The survival probability $Q_r(M, t)$, or instead its Laplace transform, can be easily computed using the renewal property of the RBM [26, 53]. Let $Q_0(M, t)$ denote the survival probability of an ordinary Brownian motion without resetting. This Brownian motion starts at the origin and diffuses with a diffusion constant $D$ in the presence of a fixed target at $M \geq 0$. This survival probability can be easily computed by solving the diffusion equation with an absorbing boundary condition at $M$, and this well-known result reads [15, 17, 18]

$$Q_0(M,t) = \mathrm{erf}\left(\frac{M}{\sqrt{4Dt}}\right), \tag{15.9}$$

where $\mathrm{erf}(z) = (2/\sqrt{\pi})\int_0^x e^{-u^2}\,du$ is the error function. At late times, it decays as a power law, $Q_0(M,t) \sim M/\sqrt{\pi D t}$. In the presence of resetting with a nonzero rate $r > 0$, one can write the survival probability $Q_r(M,t)$ in terms of $Q_0(M,t)$ via the renewal equation [53]

$$Q_r(M,t) = e^{-rt} Q_0(M,t) + r\int_0^t d\tau\, e^{-r\tau} Q_r(M,t-\tau) Q_0(M,\tau). \tag{15.10}$$

This equation can be interpreted as follows. The first term corresponds to the event when there is no resetting in the full interval $[0,t]$ and the process is then an ordinary Brownian motion that needs to survive up to time $t$. The probability of this event equals the survival probability $Q_0(M,t)$ without resetting times the probability that no resetting occurs in $[0,t]$. The probability of no resetting in time $t$, using Eq. (15.7), is simply $\int_t^\infty p(\tau)d\tau = e^{-rt}$. This then explains the first term in Eq. (15.10). The second term corresponds to the events with one or more resettings. In this case let the last resetting event before $t$ occur at time $t-\tau$. In the interval $[t-\tau,t]$, there is no resetting, preceded by one resetting event at time $t-\tau$: this joint event occurs with probability $p(\tau)d\tau = r\,d\tau e^{-r\tau}$. During this interval $[t-\tau,t]$, the process evolves as a free Brownian motion and its survival probability is simply $Q_0(M,\tau)$. But one needs to ensure also that the process (with resetting) did not cross $M$ during $[0,t-\tau]$, which occurs with a probability $Q_r(M,t-\tau)$. Since these events are independent due to the Markov nature of the process, we take the product of these three probabilities and integrate over $\tau$ from 0 to $t$, leading to the second term in Eq. (15.10). Let us now define the Laplace transform

$$\tilde{Q}_r(M,s) = \int_0^\infty e^{-st}\, Q_r(M,t)\,dt. \tag{15.11}$$

Since Eq. (15.10) has a convolution structure in time, it is natural to take its Laplace transform which then leads to

$$\tilde{Q}_r(M,s) = \frac{\tilde{Q}_0(M,r+s)}{1 - r\tilde{Q}_0(M,r+s)}. \tag{15.12}$$

Using the expression of $Q_0(M,t)$ in Eq. (15.9), one has

$$\tilde{Q}_0(M,s) = \int_0^\infty e^{-st}\, Q_0(M,t)\,dt = \frac{1}{s}\left(1 - e^{-\sqrt{s/D}\,M}\right). \tag{15.13}$$

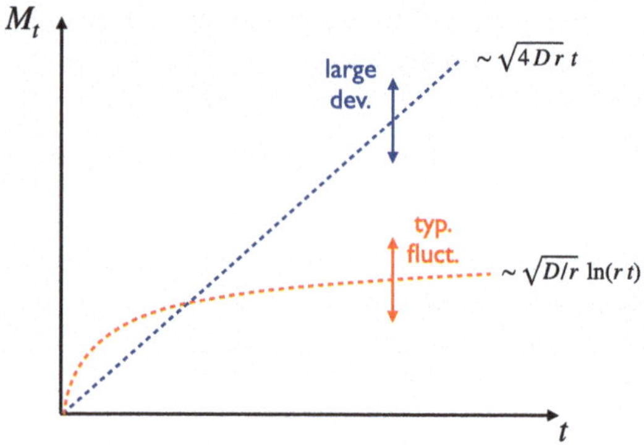


**Fig. 15.2** A schematic representation of the two time scales associated with the fluctuations of the maximum $M_t$ up to time $t$ for a resetting Brownian motion. For a fixed but large $t$, while the typical fluctuations scale as $\ln(rt)$ (shown by the red dashed curve), the large deviations scale linearly with $t$ as shown by the blue dashed line

Substituting this in Eq. (15.12) gives the well-known result [26, 53]

$$\tilde{Q}_r(M,s) = \int_0^\infty e^{-st}\, Q_r(M,t)\,dt = \frac{1 - e^{-\sqrt{\frac{r+s}{D}}\,M}}{s + r\,e^{-\sqrt{\frac{r+s}{D}}\,M}}. \tag{15.14}$$

Inverting this Laplace transform formally, we get

$$Q_r(M,t) = \int_\Gamma \frac{ds}{2\pi i}\, e^{s\,t}\, \frac{1 - e^{-\sqrt{\frac{r+s}{D}}\,M}}{s + r\,e^{-\sqrt{\frac{r+s}{D}}\,M}} \tag{15.15}$$

where $\Gamma$ is a Bromwich contour which has to pass to the right of all singularities of the integrand. In principle, this is an exact expression of the survival probability, or equivalently that of the CDF of $M_t$, valid for all $M$ and all $t$. In the rest of this section, we will analyze the behavior of $Q_r(M,t)$ in different regions of the $M$-$t$ plane; see Fig. 15.2. In the following subsection, we analyze $Q_r(M,t)$ for fixed $t$ as a function of $M$—here the interpretation of $Q_r(M,t)$ is the cumulative distribution of the random variable $M_t$. In the next subsection, we will analyze $Q_r(M,t)$ as a function of $t$ for fixed $M$— this has the interpretation of the survival probability.

### *15.2.1 Probability Distribution of the Maximum*

Here, we interpret $Q_r(M,t)$ as the CDF of the maximum $M_t$ up to time $t$ of the RBM, starting and resetting at the origin with rate $r$. It is also natural to consider the associated PDF $P_r(M,t) = \partial_M Q_r(M,t)$. Using Eq. (15.14), the Laplace transform of the PDF of $M_t$ is then given by

$$\tilde{P}_r(M,s) = \frac{\partial}{\partial M}\tilde{Q}_r(M,s) = \frac{(s+r)^{3/2}}{\sqrt{D}}\,\frac{e^{-\sqrt{\frac{r+s}{D}}\,M}}{(s + r\,e^{-\sqrt{\frac{r+s}{D}}\,M})^2}. \tag{15.16}$$

Therefore $P_r(M,t)$ is given by the Bromwich integral

$$P_r(M,t) = \int_\Gamma \frac{ds}{2\pi i} e^{s\,t} \frac{(s+r)^{3/2}}{\sqrt{D}} \frac{e^{-\sqrt{\frac{r+s}{D}}\,M}}{(s+r\,e^{-\sqrt{\frac{r+s}{D}}\,M})^2}, \tag{15.17}$$

where $\Gamma$ is a Bromwich contour.

We first fix $t$ to be large and consider $P_r(M,t)$ in Eq. (15.17) as a function of $M$. To analyze this integral, we need to examine the singularities of the integrand in the complex $s$-plane and make sure that the vertical Bromwich contour $\Gamma$ is to the right of all the singularities. The integrand clearly has a branch cut at $s = -r$ and also a double pole at $s = s^*$ where $s^*$ is given by the root of the equation

$$s + r\,e^{-\sqrt{\frac{r+s}{D}}\,M} = 0. \tag{15.18}$$

In the limit of large $M$, we expect $|s^*| \ll 1$. Consequently we have, from Eq. (15.18), to leading order for large $M$,

$$s^* = -r\,e^{-\sqrt{\frac{r}{D}}M}. \tag{15.19}$$

We also notice that the integrand in (15.17) admits, for large $t$ and large $M$, a saddle point $s_1$ which is obtained by minimizing $(s\,t - \sqrt{(r+s)/D}\,M)$, leading to

$$s_1 = -r + \frac{1}{4D}\left(\frac{M}{t}\right)^2. \tag{15.20}$$

Therefore, when $s_1 > s^*$, we can deform the Bromwich contour to pass through $s_1$ and pick up the dominant contribution for large $t$—in this case we do not need to consider the contribution from the pole. In contrast, if $s_1 < s^*$, the dominant contribution will come from the pole at $s^*$. From (15.19), we see that for large $M$, the pole $s^*$ is essentially close to zero, $s^* \approx 0$. Thus, from Eq. (15.20), it follows that $s_1 > s^* \approx 0$ if $M > M_c(t)$ where

$$M_c(t) = \sqrt{4\,D\,r\,t}. \tag{15.21}$$

In contrast, when $M < M_c(t)$, the leading contribution comes from the pole. By evaluating the residue at this pole, one gets

$$P_r(M,t) \approx \sqrt{\frac{r}{D}}\,(rt)\,\exp\left[-\sqrt{\frac{r}{D}}\,M - rt\,e^{-\sqrt{\frac{r}{D}}M}\right]. \tag{15.22}$$

Clearly, this is of order $O(1)$ when $M \sim \sqrt{D/r}\,\ln(r\,t)$.

Hence we clearly see that there are two scales that characterize the fluctuations of $M_t$, namely when $M_t \sim \ln(r\,t)$ and $M_t \sim t$. While the first one corresponds to the typical scale of fluctuations of $M_t$, the second one represents large deviations of $M_t$ compared to its typical value. We now consider these two scales of fluctuations separately.

**Typical fluctuations**: In this case, the dominant contribution to the integral in Eq. (15.17) comes from the double pole at $s = s^*$, leading to the asymptotic form of the PDF $P_r(M, t)$ in Eq. (15.22). This can be expressed in a nice scaling form

$$P_r(M,t) \approx \sqrt{\frac{r}{D}}\,\mathcal{F}\left(\sqrt{\frac{r}{D}}\left(M - \sqrt{\frac{D}{r}}\,\ln(r\,t)\right)\right), \tag{15.23}$$

where the scaling function $\mathcal{F}(v)$ has the Gumbel form

$$\mathcal{F}(v) = e^{-v-e^{-v}}, \tag{15.24}$$

which is normalized to unity, $\int_{-\infty}^{\infty} \mathcal{F}(v)\,dv = 1$. From this scaling form (15.23), it is clear that the PDF of $M_t$ is peaked around the typical value $M_t \sim \sqrt{\frac{D}{r}}\,\ln(r\,t)$ and the fluctuations around this typical value is of order $O(1)$ for large $t$. In other words, one can express the random variable $M_t$ as

$$M_t = \sqrt{\frac{D}{r}}\,(\ln(r\,t) + v)\,, \tag{15.25}$$

where $v$ is a $O(1)$ random variable for large $t$, distributed via the Gumbel PDF $\mathcal{F}(v)$.

The result in Eq. (15.25) has a nice physical interpretation [53] in terms of extreme-value statistics (EVS) of independent and identically distributed (IID) random variables [24]. Consider the RBM without the target, starting and resetting at the origin. A typical trajectory of this RBM up to a long time $t$ consists of a large number of intervals separated by resetting events. Since the resetting occurs with rate $r$, the number of such intervals in time $t$ is typically $N_r = t/(1/r) = r\,t$. Within each interval labeled by $i = 1, 2, \ldots, N_r$, there is a local maximum of the displacement denoted by $m_i$. Clearly the global maximum $M_t$ is given by

$$M_t = \max\{m_1, m_2, \ldots, m_{N_r}\}. \tag{15.26}$$

Since the $m_i$'s belong to different resetting intervals, they are clearly independent of each other. To find the PDF of $m_i$, consider the $i$-th interval between resettings. Inside this interval, the motion is purely Brownian and the CDF of the maximum within this interval of duration, say, $\tau$ is given by $\mathrm{erf}\left(\frac{m_i}{\sqrt{4D\tau}}\right)$ as in Eq. (15.9). However, the interval duration $\tau$ itself is exponentially distributed via the PDF $p(\tau) = r\,e^{-r\tau}$.

Averaging over this random duration, one gets the CDF of the maximum within a given interval as

$$\text{Prob.}\,[m_i \leq m] = r \int_0^\infty e^{-r\tau}\,\text{erf}\left(\frac{m}{\sqrt{4D\tau}}\right) d\tau = 1 - e^{-\sqrt{r/D}\,m}, \quad (15.27)$$

where we used Eq. (15.13). Consequently, the PDF of $m_i$ is given by

$$\text{Prob.}\,[m_i = m] = \sqrt{\frac{r}{D}}\,e^{-\sqrt{r/D}\,m}, \qquad m \geq 0. \quad (15.28)$$

Hence, $M_t$ in Eq. (15.26) is the maximum of a set of $N_r = rt$ IID random variables, each distributed via the PDF in Eq. (15.28). Consequently, using the EVS of IID variables each with exponential distribution [24], one would expect that $M_t$ will indeed scale as $\ln N_r = \ln(r\,t)$ with fluctuations of order $O(1)$ around it and in addition, the centered and scaled distribution will have the Gumbel form as in Eq. (15.24).

**Large-deviation regime**: Here we are interested in the large-deviation tail of $P_r(M,t)$ when $M = O(t)$. In this limit, the saddle point $s_1$ is of order $O(1)$ from Eq. (15.20). If $M > M_c(t) = \sqrt{4D\,r}\,t$ given in Eq. (15.21), the saddle point $s_1 > s^* \approx 0$. In this case, deforming the Bromwich contour through the saddle point and performing the steepest descent analysis, we get

$$P_r(M,t) \sim e^{-r\,t - \frac{M^2}{4Dt}}, \quad (15.29)$$

where $\sim$ means the leading order behavior for large $t$ up to pre-exponential subleading factors. Thus so far, we have seen that there are two well-separated scales of fluctuations of $M_t$, respectively of order $O(t)$ and order $O(\ln(r\,t))$, where the PDF takes very different forms given respectively in Eqs. (15.29) and (15.22). One may wonder what happens in the intermediate regime when $\ln t \ll M \ll t$. In this regime, the Gumbel form in Eq. (15.22) continues to hold but one can neglect the term $rt\,e^{-\sqrt{r/D}\,M}$ since $M \gg \ln(r\,t)$. Hence the approximate form of the PDF in this intermediate range reads

$$P_r(M,t) \sim \frac{r^{3/2}}{\sqrt{D}}\,t\,e^{-\sqrt{\frac{r}{D}}M}. \quad (15.30)$$

To summarize, the PDF of the maximum exhibits different behaviors in the three different regimes

$$P_r(M,t) \sim \begin{cases} \sqrt{\frac{r}{D}}\,\exp\left[-\sqrt{\frac{r}{D}}\left(M-\sqrt{\frac{D}{r}}\ln(r\,t)\right)-e^{-\sqrt{\frac{r}{D}}\left(M-\sqrt{\frac{D}{r}}\ln(r\,t)\right)}\right] & \text{for} \quad M = O\,(\ln(r\,t)) \\ \frac{r^{3/2}}{\sqrt{D}}\,t\,e^{-\sqrt{\frac{r}{D}}M}\,, & \text{for} \quad \ln r\,t \ll M \ll r\,t \\ e^{-r\,t-\frac{M^2}{4\,D\,t}}\,, & \text{for} \quad M = O(r\,t). \end{cases} \tag{15.31}$$

While the Gumbel behavior in the first line corresponds to the typical trajectories with a large number of resettings, the last line corresponds to very rare trajectories that have undergone no reset up to time $t$. The intermediate regime corresponds to trajectories with few resets. The right tail of the intermediate regime and the left tail of the last regime can be combined into a single large-deviation form

$$P_r(M,t) \sim e^{-t\,I\left(\frac{M}{t}\right)}, \tag{15.32}$$

where

$$I(z) = \begin{cases} \sqrt{\frac{r}{D}}\,z\,, & z \le z_c = \sqrt{4\,D\,r} \\ r + \frac{z^2}{4D}\,, & z \ge z_c = \sqrt{4\,D\,r}. \end{cases} \tag{15.33}$$

Thus, the second derivative of $I(z)$ is discontinuous at $z = z_c$. In terms of the location of the maximum at time $t$, one sees that there is a "light cone" $M_c(t) = z_c\,t$ in the $M - t$ plane, such that for $M \le M_c(t)$, the typical behavior of the maximum is established, while outside the "light cone" $M \ge M_c(t)$, the distribution of $M$ is still atypical, corresponding to rare trajectories with no resetting. Indeed, this large-deviation form of $P_r(M,t)$ in Eq. (15.33) is exactly identical to the large-deviation form of the position distribution of the RBM [35]. The typical, intermediate and the large-deviation behaviors of $P_r(M,t)$ are summarized schematically in Fig. 15.3.

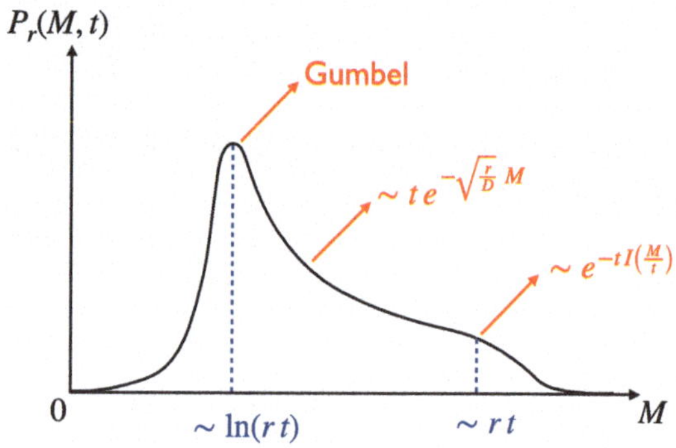


**Fig. 15.3** A schematic plot of the PDF $P_r(M,t)$ of the maximum $M_t$ as a function of $M$, for fixed but large $t$ as in Eqs. (15.31) and (15.32). The two vertical dashed blue lines represent the typical and the large-deviation scales

### 15.2.2 Survival Probability

In this subsection, we consider $Q_r(M,t)$ as a survival probability, with a target fixed at $M$ and analyze it as a function of time $t$. This means traversing the $M-t$ plane in Fig. 15.2 horizontally by increasing $t$ at a fixed $M$.

We first consider the limit of large $t$, with $M$ fixed but not necessarily large. In this case, the dominant behavior of the integral in Eq. (15.15) emerges from the pole of the integrand at $s = s^*$, where we recall from Eq. (15.18) that $s^*$ satisfies

$$s^* + r\, e^{-\sqrt{\frac{r+s^*}{D}}\, M} = 0. \tag{15.34}$$

Evaluating the residue at the pole, we see that, up to pre-exponential factors, $Q_r(M,t)$ decays exponentially for large $t$ as

$$Q_r(M,t) \sim e^{-\theta_r(M)\, t} \quad , \quad \text{where} \quad \theta_r(M) = -s^*. \tag{15.35}$$

It is easy to evaluate $\theta_r(M)$ as a function of $M$ by solving Eq. (15.34) numerically. The asymptotic behaviors of $\theta_r(M)$ for small and large $M$ are given by

$$\theta_r(M) \approx \begin{cases} r - \frac{r^2 M^2}{D} & , \quad M \to 0 \\ \\ r\, e^{-\sqrt{\frac{r}{D}}\, M} & , \quad M \to \infty. \end{cases} \tag{15.36}$$

As the target recedes, i.e., $M$ becomes large, the exponent $\theta_r(M)$ becomes exponentially small, indicating that the target survives longer, as expected. In this large $M$ regime, Eq. (15.35) can be written in the form

$$Q_r(M,t) \sim \exp\left[-rt\, e^{-\sqrt{\frac{r}{D}}\, M}\right] = \exp\left[-\, e^{-\sqrt{\frac{r}{D}}\left(M - \sqrt{\frac{D}{r}}\, \ln(r\, t)\right)}\right]. \tag{15.37}$$

This is exactly the CDF of the Gumbel distribution in Eq. (15.22), associated with the typical fluctuations of the random variable $M_t$.

In the opposite limit $M \to 0$, the exponent $\theta_r(M)$ in Eq. (15.36) approaches a nonzero constant $r$. A priori, this may look a bit strange since one expects that $Q_r(M,t) \to 0$ as $M \to 0$ for any time $t$. Indeed, in Eq. (15.35), we have ignored pre-exponential factors. Taking into account these factors, it is easy to show that, as $M \to 0$,

$$Q_r(M,t) \approx e^{-rt} \frac{M}{\sqrt{\pi D t}}. \tag{15.38}$$

Thus $Q_r(M,t)$ indeed vanishes as $M \to 0$, but the exponent $\theta_r(M=0) = r$ remains finite. Physically, this corresponds to Brownian trajectories that have not reset up to time $t$ and have escaped in the direction opposite to the target. This is because if the trajectory resets even once, the target at $M = 0$ will not survive, since it is close to the resetting position at 0.

So far we have discussed the large $t$ behavior of the survival probability $Q_r(M, t)$ for fixed $M$ and we have seen that when $t = O(e^{\sqrt{\frac{r}{D}}M})$, it acquires the form of the CDF of a Gumbel variable in Eq. (15.37). This corresponds to crossing the curve $M = \sqrt{D/r}\,\ln(r\,t)$ in the $M - t$ plane in Fig. 15.2. One can also analyze the behavior of $Q_r(M, t)$ for intermediate values of $t$, e.g., when $t = O(M)$. In this case, we use the large-deviation behavior of the PDF $P_r(M, t)$ in Eq. (15.32). Integrating over $M$ gives

$$Q_r(M, t) \sim 1 - B\, e^{-t\, I(\frac{M}{t})}, \tag{15.39}$$

where $B$ is an unimportant pre-exponential factor (subleading for large $t$) and the rate function $I(z)$ is given in Eq. (15.33). When $t < M/\sqrt{4D\,r}$, i.e., $z = M/t > z_c = \sqrt{4Dr}$, one gets using Eq. (15.33)

$$Q_r(M, t) \sim 1 - B\, e^{-r\,t - \frac{M^2}{4Dt}} \quad , \quad t < M/\sqrt{4D\,r}. \tag{15.40}$$

In contrast, when $t > M/\sqrt{4D\,r}$, but still $t \ll e^{\sqrt{\frac{r}{D}}M}$, using the first line of (15.33), one gets

$$Q_r(M, t) \approx 1 - B\, e^{-\sqrt{\frac{r}{D}}M}. \tag{15.41}$$

Thus in this regime the survival probability depends only weakly on time through the prefactor $B$. Finally, when $t$ approaches $e^{\sqrt{\frac{r}{D}}M}$, it crosses over to the Gumbel form (15.37). To summarize, for fixed $M$ and as a function of $t$, there are two time scales $t \sim O(M)$ and $t \sim e^{\sqrt{\frac{r}{D}}M}$, corresponding to crossing horizontally the two lines in the $M - t$ plane in Fig. 15.2: the first time scale $t \sim O(M)$ corresponds to the regime of the large deviation of the EVS, while the second one corresponds to the regime of the typical behavior of the EVS.

## 15.3 Multi-particle Case

We now consider $N$ independent RBMs, located initially on the negative half-line with initial positions $\{x_i\}$ $(i = 1, 2, \dots, N)$. The $i$-th particle starts at $x_i$, undergoes independent diffusion and resets to $x_i$ with rate $r$. The initial position $x_i$ of the $i$-th particle is chosen independently from a uniform distribution over $x_i \in [-L, 0]$. We are interested in the thermodynamic limit when $N \to \infty$, $L \to \infty$, with the ratio $N/L = \rho$ fixed. For a fixed set of initial positions, we define $M_t$ as the position of the global maximum of all $N$ particles up to time $t$. Let us denote the CDF of $M_t$ by $Q_r(M, t|\{x_i\}) = \text{Prob.}(M_t \le M)$ for a given set of initial positions $\{x_i\}$'s. This CDF

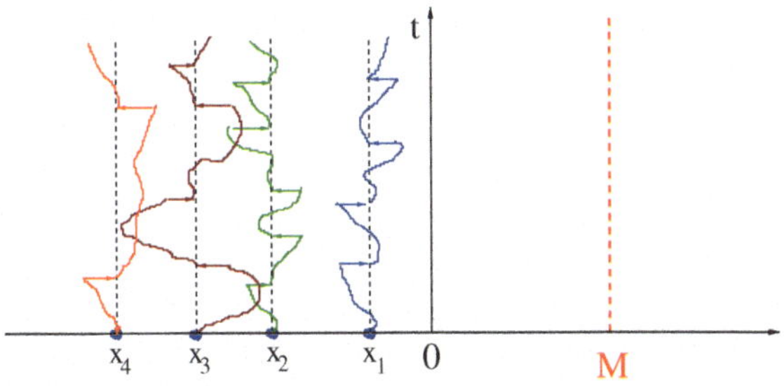


**Fig. 15.4** Schematic trajectories of 4 independent resetting Brownian motions up to time $t$, each starting and resetting at $x_i \leq 0$ ($i = 1, 2,$ 3 and 4). The target is located at $M > 0$ shown by the dashed red vertical line

is also the survival probability of a fixed target located at $M$; see Fig. 15.4. Here, we can interpret this search process as a team-search by $N$ independent walkers. The search terminates when any one of the $N$ walkers finds the target for the first time.

Using the independence of $N$ particles, the CDF for a fixed set of initial positions $\{x_i\}$ is given by

$$Q_r(M, t|\,\{x_i\}) = \prod_{i=1}^{N} Q_r(M - x_i, t), \tag{15.42}$$

where $Q_r(M, t)$ is the single-particle CDF discussed in the previous section. Now we also need to average the survival probability in Eq. (15.42) with respect to the initial positions $\{x_i\}$, drawn independently from a uniform distribution. This averaging can be done in two different ways, namely (i) annealed and (ii) quenched, akin to disordered systems. In fact, this setup has been used to compute the current distribution in nonequilibrium systems of independent stochastic processes, such as for diffusive particles [65–68, 70, 71], for run-and-tumble particles [66, 68, 69] and also for resetting Brownian motions [70]. More recently, the total number of visits to the origin by all particles up to time $t$ (equivalently the local time at the origin for the combined $N$-particle process) has been computed for a similar step-like initial condition [72]. Here, we will use the formalism developed above and adapt it to study the survival probability, or equivalently the CDF of the maximum $M_t$. In the annealed case, one averages $Q_r(M, t|\,\{x_i\})$ in Eq. (15.42) directly, i.e.,

$$Q_{\rm an}(M, t) = \overline{Q_r(M, t|\{x_i\})} = \left[Q_{\rm avg}(M, t)\right]^N, \tag{15.43}$$

where

$$Q_{\rm avg}(M, t) = \lim_{L\to\infty} \frac{1}{L} \int_{-L}^{0} Q_r(M - x, t)\, dx, \tag{15.44}$$

where $\overline{\cdots}$ denotes the averaging over the initial positions $\{x_i\}$. In contrast, in the quenched case, one is interested in the typical initial configurations that dominate the survival probability. One way to extract the contribution from this typical initial configuration is to take the average over the logarithm of $Q_r(M, t|\{x_i\})$ and then re-exponentiate, i.e.,

$$Q_{\rm qu}(M, t) = \exp\left[\overline{\ln Q_r(M, t|\{x_i\})}\right]. \tag{15.45}$$

It turns out that this is equivalent to choosing a specific initial configuration with equispaced points separated by $1/\rho$ [65, 66].

In the two sub-sections below, we discuss the annealed and quenched cases separately.

### *15.3.1 Annealed Case*

In the annealed case, the survival probability or equivalently the CDF $Q_{\rm an}(M, t)$ is given by Eq. (15.43), where the average $\overline{\cdots}$ is over the initial positions $\{x_i\}$s, each drawn independently and uniformly from $x_i \in [-L, 0]$. We first replace $Q_r(M - x_i, t)$ by $1 - (1 - Q_r(M - x_i, t))$ in Eq. (15.42) and average over $\{x_i\}$ as in Eq. (15.43). This gives, using the independence of $x_i$,

$$Q_{\rm an}(M, t) = \left(1 - \frac{1}{L}\int_{-L}^{0} (1 - Q_r(M - x, t))\, dx\right)^N. \tag{15.46}$$

Changing variable $x \to -x_0$ and taking the thermodynamic limit $N \to \infty, L \to \infty$, with the ratio $N/L = \rho$ fixed, we get

$$Q_{\rm an}(M, t) = \exp\left[-\rho \int_0^{\infty} (1 - Q_r(M + x_0, t))\, dx_0\right]. \tag{15.47}$$

We remark that the special case $M = 0$, i.e., the survival probability $Q_{\rm an}(M = 0, t)$ for a target fixed at the origin was analyzed in Ref. [26] with the interesting result that, at long times, $Q_r(M = 0, t)$ decays algebraically as

$$Q_{\rm an}(M = 0, t) \sim t^{-\theta}, \quad \theta = \rho\sqrt{\frac{D}{r}}, \tag{15.48}$$

with a continuously varying exponent $\theta$. Here, our goal is to go beyond $M = 0$, and study the full CDF $Q_r(M, t)$ for all $M$.

To proceed for general $M \geq 0$, we rewrite Eq. (15.47) as

$$Q_{\rm an}(M, t) = e^{-\rho J(M,t)}\,, \text{ where } J(M, t) = \int_M^{\infty} (1 - Q_r(x_0, t))\, dx_0. \tag{15.49}$$

We then define the Laplace transform of $J(M, t)$ with respect to $t$ as

$$\tilde{J}(M,s) = \int_0^\infty e^{-st}\, J(M,t)\, dt. \tag{15.50}$$

Taking the Laplace transform of $J(M,t)$ in Eq. (15.49) and using the single-particle result in Eq. (15.14), we get

$$\tilde{J}(M,s) = \frac{\sqrt{D(r+s)}}{r\,s} \ln\left(1 + \frac{r}{s}\, e^{-\sqrt{\frac{r+s}{D}}M}\right). \tag{15.51}$$

Using the Bromwich inversion formula, we can write $J(M,t)$ formally as

$$J(M,t) = \int_\Gamma \frac{ds}{2\pi i}\, e^{st}\, \frac{\sqrt{D(r+s)}}{r\,s} \ln\left(1 + \frac{r}{s}\, e^{-\sqrt{\frac{r+s}{D}}M}\right), \tag{15.52}$$

where $\Gamma$ is the Bromwich contour.

**Typical regime**: In this case, we anticipate and verify a posteriori that the typical value of $M_t$ scales as $\sqrt{D/r}\,\ln t$ for large $t$ with $O(1)$ fluctuations around it. To extract this typical behavior from the Laplace transform in Eq. (15.52), we rescale $s\,t = \tilde{s}$ and rewrite it as

$$J(M,t) = \sqrt{\frac{D}{r}} \int_\Gamma \frac{d\tilde{s}}{2\pi i}\, \frac{e^{\tilde{s}}}{\tilde{s}} \sqrt{1 + \frac{\tilde{s}}{rt}}\, \ln\left(1 + \frac{r\,t}{\tilde{s}}\, e^{-\sqrt{\frac{r+\frac{\tilde{s}}{t}}{D}}\,M}\right). \tag{15.53}$$

Now we take the $t \to \infty$ limit, $M \to \infty$ limit but keeping $M - \sqrt{D/r}\,\ln(rt) = z$ fixed. One immediately gets in this scaling limit

$$J(M,t) \approx F\left(M - \sqrt{\frac{D}{r}}\,\ln(rt)\right), \tag{15.54}$$

where the scaling function is given by

$$F(z) = \sqrt{\frac{D}{r}} \int_\Gamma \frac{d\tilde{s}}{2\pi i}\, \frac{e^{\tilde{s}}}{\tilde{s}}\, \ln\left(1 + \frac{1}{\tilde{s}}\, e^{-\sqrt{r/D}\,z}\right). \tag{15.55}$$

The scaling function $F(z)$ can be evaluated exactly by expanding the logarithm in the integrand in Eq. (15.55) in a power series in $1/\tilde{s}$ and then performing the Bromwich integral term by term. This gives

$$F(z) = -z + \sqrt{\frac{D}{r}}\left(\gamma_E + \Gamma(0, e^{-\sqrt{\frac{r}{D}}z})\right), \tag{15.56}$$

where $\Gamma(0,u) = \int_u^\infty \frac{e^{-x}}{x}\, dx$ is the incomplete Gamma function and $\gamma_E$ denotes the Euler constant. In particular, the asymptotic behaviors of $F(z)$ are given by

$$F(z) \approx \begin{cases} -z + \sqrt{\frac{D}{r}}\gamma_E \ , & z \to -\infty \\ \sqrt{\frac{D}{r}} e^{-\sqrt{\frac{r}{D}} z} \ , & z \to +\infty. \end{cases} \tag{15.57}$$

Then using Eq. (15.49), we finally obtain our main result

$$Q_{\text{an}}(M, t) \ \sim \ e^{-\rho F\left(M - \sqrt{\frac{D}{r}} \ln(rt)\right)}. \tag{15.58}$$

Thus the CDF of $M_t$ has the final scaling form

$$Q_{\text{an}}(M, t) \approx G_{\text{an}}\left(M - \sqrt{\frac{D}{r}} \ \ln(rt)\right) \tag{15.59}$$

where $G_{\text{an}}(z) = e^{-\rho F(z)} = \exp\left[-\rho\sqrt{\frac{D}{r}}\gamma_E + \rho z - \rho\sqrt{\frac{D}{r}}\,\Gamma\left(0, e^{-\sqrt{\frac{r}{D}} z}\right)\right].$

This function $G_{\text{an}}(z)$ is a new scaling function, different from the standard Gumbel form. Note that in the limit of large $M$, i.e., as $z \to \infty$, using the second line in Eq. (15.57), we get

$$G_{\text{an}}(z) \approx e^{-\rho\sqrt{\frac{D}{r}} e^{-\sqrt{r/D}\, z}}, \tag{15.60}$$

representing the Gumbel-like behavior of the CDF. In the opposite limit when $M \to 0$, i.e., when $z \to -\infty$, we get from the first line of (15.57)

$$Q_{\text{an}}(M \to 0, t) \sim e^{-\rho\sqrt{D/r}\ln(rt)} \sim t^{-\rho\sqrt{r/D}}, \tag{15.61}$$

thus recovering the known result (15.48) of Ref. [26]. As $M$ increases from 0, our main new result in Eq. (15.59) thus interpolates smoothly between $M = 0$ and the typical value of $M \sim \sqrt{D/r}\ \ln t$. Finally, it follows that in this typical regime, the PDF

$$P_{\text{an}}(M, t) = \frac{\partial}{\partial M} Q_{\text{an}}(M, t) \tag{15.62}$$

has the scaling behavior

$$P_{\text{an}}(M, t) \approx G'_{\text{an}}\left(M - \sqrt{\frac{D}{r}} \ \ln(rt)\right), \tag{15.63}$$

where $G'(z) = dG/dz$ is the derivative of the scaling function $G(z)$ in Eq. (15.59).

**Large-deviation regime**: We start from Eq. (15.49), where $J(M, t)$ is given explicitly by Eq. (15.52). Note that the result in Eq. (15.52) is valid for all $M$. For the typical fluctuations, we considered the scaling limit $M \to \infty$, $t \to \infty$ with $M - \sqrt{D/r}\,\ln t = z$ fixed. For the large-deviation regime, when $M$ is very large, one can expand the logarithm in (15.52) and only keep the first term, resulting in

$$J(M,t) \approx \int_{\Gamma} \frac{ds}{2\pi i}\, e^{st}\, \frac{\sqrt{D(r+s)}}{s^2}\, e^{-\sqrt{\frac{r+s}{D}}\,M}. \tag{15.64}$$

The integrand has a double pole at $s = 0$ and a branch cut at $s = -r$. In addition it has a saddle point at $s_1 = -r + (M/t)^2/(4D)$ as in Eq. (15.20). If $s_1 > 0$, the dominant contribution to the integral comes from the saddle point, and evaluating this saddle point one gets the large deviation behavior $J(M,t) \approx e^{-rt - M^2/(4Dt)}$. The condition $s_1 > 0$ translates into $M > M_c(t)$ where

$$M_c(t) = \sqrt{4\,D\,r\,t}. \tag{15.65}$$

When $M < M_c(t)$, the double pole at $s = 0$ dominates and one gets $J(M,t) \approx t\, e^{-\sqrt{\frac{r}{D}}M}$. Thus, in the large-deviation regime when $M = O(t)$, we can combine the behaviors for $M < M_c(t)$ and $M > M_c(t)$ into a single large-deviation form

$$J(M,t) \approx e^{-t\,I\left(\frac{M}{t}\right)} \quad , \quad I(z) = \begin{cases} \sqrt{\frac{r}{D}}\,z, & z < z_c = \sqrt{4\,D\,r} \\ r + \frac{z^2}{4D}, & z > z_c = \sqrt{4\,D\,r}. \end{cases} \tag{15.66}$$

The rate function $I(z)$ is identical to the single particle case in Eq. (15.33) and thus exhibits a second-order dynamical phase transition at $z = \sqrt{4\,D\,r}$. Hence, in this large-deviation regime, using Eqs. (15.49) and (15.66) we get the annealed CDF

$$Q_{\rm an}(M,t) \approx \exp\left[-\rho\, e^{-t\,I\left(\frac{M}{t}\right)}\right] \approx 1 - \rho\, e^{-t\,I\left(\frac{M}{t}\right)}. \tag{15.67}$$

Another alternative way to analyze the large-deviation regime is by probing directly the annealed PDF in Eq. (15.62). Taking the derivative of Eq. (15.49) with respect to $M$ gives the exact expression

$$P_{\rm an}(M,t) = \rho\,[1 - Q_r(M,t)]\; e^{-\rho\, J(M,t)}. \tag{15.68}$$

We now substitute in Eq. (15.68) the large-deviation behavior of $Q_r(M,t)$ in Eq. (15.39) and that of $J(M,t)$ in Eq. (15.66). This leads for large $t$, large $M$ but with $M/t$ fixed, the following large-deviation behavior of $P_{\rm an}(M,t)$ (up to pre-exponential factors),

$$P_{\rm an}(M,t) \sim e^{-t\,I\left(\frac{M}{t}\right)}, \tag{15.69}$$

with the rate function $I(z)$ given in Eq. (15.33).

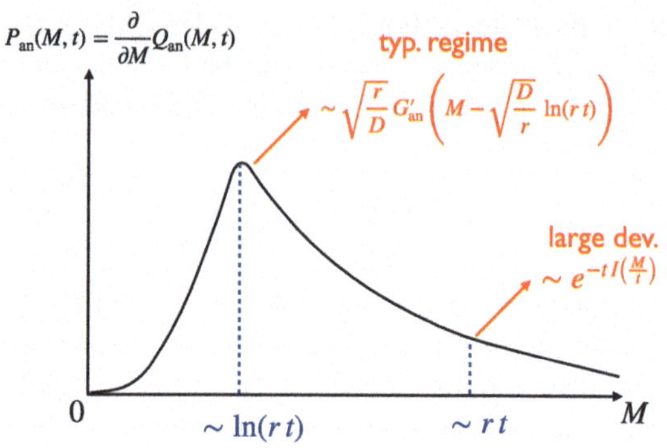


**Fig. 15.5** A schematic plot of the annealed PDF $P_{an}(M, t)$ versus $M$ in Eq. (15.70). The two vertical dashed blue lines represent the typical and the large-deviation scales

Thus, to summarize, we find that in the annealed case, there are essentially two scales of $M$, namely $M \sim \sqrt{D/r}\,\ln(rt)$ characterizing typical fluctuations and $M \sim t$ characterizing atypically large fluctuations. The behavior of the PDF of the maximum, for large $t$, can be summarized as

$$P_{an}(M,t) \approx \begin{cases} G'_{an}\left(M - \sqrt{\frac{D}{r}}\,\ln(rt)\right), & \text{when} \quad M \sim \sqrt{\frac{D}{r}}\,\ln(rt), \\ e^{-t\,I\left(\frac{M}{t}\right)}, & \text{when} \quad M \sim t, \end{cases} \tag{15.70}$$

with $G(z)$ and $I(z)$ given respectively in Eqs. (15.59) and (15.33). A schematic plot of $P_{an}(M, t)$, in the typical and in the large-deviation regimes, is shown in Fig. 15.5.

### *15.3.2 Quenched Case*

In the quenched case, the survival probability or equivalently the CDF $Q_{qu}(M, t)$ is given by Eq. (15.45), where the average $\overline{\cdots}$ is again over the initial positions $\{x_i\}$, each drawn independently and uniformly from $x_i \in [-L, 0]$. Performing this average (by changing $x_i \to -x_i$) and taking the thermodynamic limit gives

$$\begin{aligned} Q_{qu}(M,t) &= \exp\left[\overline{\ln Q_r(M,t|\{x_i\})}\right] = \exp\left[\sum_{i=1}^{N} \int_{-L}^{0} \ln\left[Q_r(M - x_i, t)\right] \frac{dx_i}{L}\right] \\ &= \exp\left[\rho \int_0^\infty \ln\left[Q_r(M + x, t)\right] dx\right]. \end{aligned} \tag{15.71}$$

Shifting $x$ by $M + x$ in the integrand gives

$$Q_{qu}(M,t) = \exp\left[\rho \int_M^\infty \ln\left[Q_r(m,t)\right] dm\right], \tag{15.72}$$

where we recall that $Q_r(m,t)$ is the single-particle survival probability with a target fixed at $m$, and is given by Eq. (15.15), i.e.,

$$Q_r(m,t) = \int_\Gamma \frac{ds}{2\pi i}\, e^{s\,t}\, \frac{1 - e^{-\sqrt{\frac{r+s}{D}}\, m}}{s + r\, e^{-\sqrt{\frac{r+s}{D}}\, m}}. \tag{15.73}$$

Once again, the special case $M = 0$, i.e., the quenched survival probability $Q_{\rm qu}(M = 0, t)$ for a target fixed at the origin was studied in Ref. [26] and it was found to decay exponentially for large $t$ as

$$Q_{\rm qu}(M = 0, t) \sim \exp\left[-\rho\sqrt{r\,D}\,\theta_1\, t\right], \quad \text{with} \quad \theta_1 = 4(1 - \ln 2). \tag{15.74}$$

Note that in Ref. [26], the exponent was $2\,\theta_1$ since the effective density there was $2\rho$. Our goal here is to study $Q_{\rm qu}(M,t)$ for all $M \geq 0$, going beyond the special case $M = 0$.

To make progress for general $M \geq 0$, we first note that for large $t$, the dominant behavior of the integral in Eq. (15.73) is governed by the pole of the integrand and one has, to leading order for large $t$ (ignoring pre-exponential factors),

$$Q_r(m,t) \sim e^{s^*(m)\,t}, \tag{15.75}$$

where the pole $s^*(m)$ satisfies

$$s^*(m) + r\, e^{-\sqrt{\frac{(r+s^*(m))}{D}}\, m} = 0. \tag{15.76}$$

Substituting (15.75) in Eq. (15.72) gives, up to pre-exponential factors,

$$Q_{\rm qu}(M,t) \sim \exp\left[\rho\, t \int_M^\infty s^*(m)\, dm\right]. \tag{15.77}$$

The pole $s^*(m)$ in Eq. (15.76) is necessarily negative. It is then natural to substitute $s^*(m) = -r\,y$ in Eq. (15.77) such that $y \geq 0$. Under this substitution, it follows from Eq. (15.76) that the variable $y$ satisfies

$$y = e^{-\sqrt{\frac{r}{D}(1-y)}\, m}, \quad \text{implying} \quad m = -\sqrt{\frac{D}{r}}\, \frac{\ln y}{\sqrt{1-y}}. \tag{15.78}$$

We now make the change of variable $m \to y$ in Eq. (15.77) and obtain, using (15.78),

$$Q_{\rm qu}(M,t) \sim \exp\left[-\rho\sqrt{r\,D}\, t\, W\left(\sqrt{\frac{r}{D}}\, M\right)\right]. \tag{15.79}$$

Here the function $W(u)$ is given by

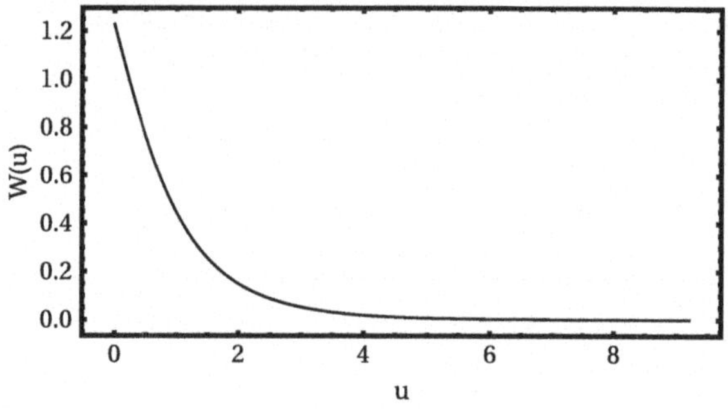


**Fig. 15.6** The function $W(u)$ versus $u$ obtained by eliminating $y^*(u)$ between Eqs. (15.82) and (15.81) using Mathematica. The function $W(u) \to 4(1 - \ln 2) = 1.22741 \cdots$ as $u \to 0$ and decays exponentially as $W(u) \approx e^{-u}$ as $u \to \infty$ as described in Eq. (15.84)

$$W(u) = \int_0^{y^*(u)} \left[ \frac{1}{\sqrt{1-y}} + \frac{y \ln y}{2\,(1-y)^{3/2}} \right] dy, \tag{15.80}$$

where $0 \le y^*(u) \le 1$ is determined in terms of $u$ by solving the equation

$$y^*(u) = e^{-\sqrt{1-y^*(u)}\,u}. \tag{15.81}$$

It turns out that the integral in Eq. (15.80) can be performed explicitly, giving

$$\begin{aligned} W(u) = 4\left(1 - \sqrt{1-y^*}\right) + 4\left[\operatorname{arctanh}\left(\sqrt{1-y^*}\right) - \ln 2\right] \\ + \frac{(2-y^*)\,\ln(y^*)}{\sqrt{1-y^*}}, \end{aligned} \tag{15.82}$$

with $y^* \equiv y^*(u)$ determined from Eq. (15.81) as a function of $u$. From Eq. (15.81), it is easy to derive the asymptotic behavior of $y^*(u)$. One gets

$$y^*(u) \approx \begin{cases} 1 - u^2, & \text{as} \quad u \to 0 \\ e^{-u}, & \text{as} \quad u \to \infty. \end{cases} \tag{15.83}$$

Consequently, the asymptotic behaviors of $W(u)$ in Eq. (15.82) are given by

$$W(u) \approx \begin{cases} 4\,(1 - \ln 2) - u, & \text{as} \quad u \to 0 \\ e^{-u}, & \text{as} \quad u \to \infty. \end{cases} \tag{15.84}$$

A plot of the function $W(u)$, obtained using Mathematica, is provided in Fig. 15.6.

Substituting the $W(u \to 0)$ behavior above in Eq. (15.79), we reproduce the $M \to 0$ limit of $Q_{\text{qu}}(M = 0, t)$ in Eq. (15.74). On the other hand, substituting the large $u$ behavior of $W(u) \approx e^{-u}$ in Eq. (15.79), we see that a natural scaling limit emerges: $M \to \infty$, $t \to \infty$ but with $M - \sqrt{D/r}\,\ln(rt) = z$ fixed. This is indeed the typical behavior of $Q_{\rm qu}(M,t)$.

**Typical regime**: In this typical scaling regime, where $M$ and $t$ are both large with $M - \sqrt{D/r}\,\ln(rt) = z$ fixed, we then get from Eq. (15.79)

$$Q_{\rm qu}(M,t) \sim G_{\rm qu}\left(M - \sqrt{\frac{D}{r}}\,\ln(rt)\right)$$
$$\text{where} \quad G_{\rm qu}(z) = \exp\left[-\rho\sqrt{\frac{D}{r}}\,e^{-\sqrt{r/D}\,z}\right]. \tag{15.85}$$

One should compare this typical scaling behavior in the quenched case with that of the annealed case in Eq. (15.59). While in the annealed case, the scaling function $G_{\rm an}(z)$ is a new function (not of the Gumbel form), in the quenched case the scaling function $G_{\rm qu}(z)$ in Eq. (15.85) has a Gumbel form. Another difference between the two cases is worth remarking. In the annealed case, the very small $M$ behavior ($M \to 0$) is included in the scaling form in Eq. (15.59) corresponding to the $z \to -\infty$ limit. In contrast, in the quenched case, the $M \to 0$ limit in Eq. (15.74) is not part of the scaling function $G_{\rm qu}(z)$ in Eq. (15.85). To see this, we note that substituting $M = 0$, i.e., $z = -\sqrt{\frac{D}{r}}\,\ln(rt)$ in Eq. (15.85) gives $Q_{\rm qu}(0,t) \approx \exp[-\rho\sqrt{D\,r}\,t]$ which is different from the true $M = 0$ behavior in Eq. (15.74). In the quenched case, the behavior of $Q_{\rm qu}(M,t)$, for both small and large $M \sim O(\ln(rt))$, is rather captured by the result in Eq. (15.79) with the function $W(u)$ given explicitly in Eqs. (15.82) and (15.81).

**Large-deviation regime**: In the discussion about typical fluctuations above we probed the CDF $Q_{\rm qu}(M,t)$ as a function of $M$, for fixed large $t$, at the scale $M \sim \sqrt{D/r}\,\ln t$. Let us now investigate the same CDF $Q_{\rm qu}(M,t)$ at the scale of large deviations when $M \sim O(t)$. As in the annealed case, to investigate this large-deviation regime, it is convenient to consider the PDF of $M$, namely

$$P_{\rm qu}(M,t) = \frac{\partial}{\partial M} Q_{\rm qu}(M,t). \tag{15.86}$$

Taking the derivative of Eq. (15.72) with respect to $M$ yields the exact PDF

$$P_{\rm qu}(M,t) = -\rho\,\ln\left[Q_r(M,t)\right]\,\exp\left[\rho\int_M^\infty \ln\left[Q_r(m,t)\right]dm\right], \tag{15.87}$$

where $Q_r(m,t)$ is the single-particle CDF given in Eq. (15.73). In the large-deviation regime when $M \sim O(t)$, we then use the large-deviation result for $Q_r(M,t)$ in Eq. (15.39). Substituting this result in Eq. (15.87) and taking the limit $t \to \infty$, it is straightforward to see that the exponential factor in Eq. (15.87) contributes 1 to leading order and expanding the logarithm in the front, one gets to leading order for large $t$ (ignoring pre-exponential factors as usual)

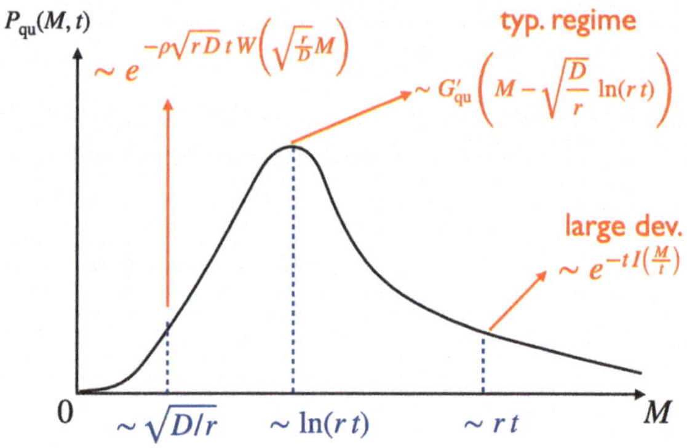


**Fig. 15.7** A schematic plot of the quenched PDF $P_{\rm qu}(M, t)$ versus $M$ in Eq. (15.89). The three vertical dashed blue lines represent the three scales mentioned in Eq. (15.89)

$$P_{\rm qu}(M, t) \sim e^{-t\, I(\frac{M}{t})}, \tag{15.88}$$

where the rate function $I(z)$ is given exactly in Eq. (15.33). The second derivative of the rate function $I(z)$ is discontinuous at $z = z_c = \sqrt{4Dr}$, signaling a second-order phase transition. The result for the quenched case in Eq. (15.88) is identical (up to pre-exponential factors) to that in the annealed case in Eq. (15.69). Thus, on the scale $M \sim O(t)$, both the annealed and the quenched PDFs of $M$, respectively in Eqs. (15.69) and (15.88), are described by an identical large-deviation form that coincides with the single-particle case in Eq. (15.32).

Thus, in the quenched case, the behavior of $Q_{\rm qu}(M, t)$ or equivalently the PDF $P_{\rm qu}(M, t) = \partial_M Q_{\rm qu}(M, t)$ can be summarized as follows (ignoring pre-exponential factors as usual)

$$P_{\rm qu}(M, t) \sim \begin{cases} \exp\left[-\rho\sqrt{r\,D}\,t\,W\left(\sqrt{\frac{r}{D}}\,M\right)\right], & \text{when} \quad M \sim \sqrt{\frac{D}{r}}, \\ G'_{\rm qu}\left(M - \sqrt{\frac{D}{r}}\,\ln(rt)\right), & \text{when} \quad M \sim \sqrt{\frac{D}{r}}\,\ln(rt), \\ e^{-t\,I(\frac{M}{t})}, & \text{when} \quad M \sim t, \end{cases} \tag{15.89}$$

where the functions $W(u)$, $G_{\rm qu}(z)$ and $I(z)$ are given respectively in Eqs. (15.82), (15.85) and (15.33). A schematic plot of the quenched PDF $P_{\rm qu}(M, t)$ is given in Fig. 15.7. Thus, compared to the annealed case in Eq. (15.70), we see that in the quenched case there is an additional scale when $M \sim O(1)$ for large $t$, where the PDF $P_{\rm qu}(M, t)$ behaves differently from the typical regime. This is natural because in the quenched case a scale of $O(1)$ naturally emerges from the initial condition where the particles are equispaced.

## 15.4 Numerical Simulations

To study the problem numerically [73], we start with the **single-particle** case. To obtain a numerical approximation of the cumulative distribution $Q_r(M, t)$ at a fixed given time $t$, we considered an array $\overline{Q}(M, t')$ which is discretized in time with time resolution $\Delta t$, i.e., for $t' = 0, \Delta t, 2\Delta t, \ldots, t$. We applied a discretized version of Eq. (15.10) which reads

$$\overline{Q}(M, t') = e^{-rt'} Q_0(M, t') \\ + (1 - e^{-r\Delta t}) \sum_{\tau=\Delta t, 2\Delta t, \ldots}^{t'} e^{-r(\tau-\Delta t)} \overline{Q}(M, t' - \tau) Q_0(M, \tau) , \qquad (15.90)$$

where the survival probability $Q_0(M, t)$ of an ordinary Brownian motion without resetting is given by Eq. (15.9) and $e^{-r}$ is the probability of experiencing no reset in the unit time span. Thus, each term in the sum represents the case that a particle experiences the last reset at time $t' - \tau$, i.e., $\tau/\Delta t$ "steps" back in time, for all possible values of $\Delta t \leq \tau \leq t'$. This happens with the probability which is a product of the probability $(1 - e^{-r\Delta t})$ for the last reset times the probability $e^{-r(\tau-\Delta t)}$ for experiencing no reset since the last reset, during time $\tau - \Delta t$. For the calculation of $\overline{Q}(M, t')$ one includes the probabilities $\overline{Q}(M, t' - \tau)$ and $Q_0(M, \tau)$ that a particle does not reach position $M$ in both time intervals before and after the last reset, respectively.

To achieve a high numerical accuracy in evaluating very small probabilities like $100^{-100}$ and estimating the sums and differences of probabilities, we used the `mpfr` library [74, 75] with a precision of 1000 binary digits for storing data in arrays and performing arithmetic operations.

To estimate numerically the probability density, we approximate Eq. (15.6) by a finite difference

$$P_r(M, t) \approx \frac{\overline{Q}_r(M + h, t) - \overline{Q}_r(M - h, t)}{2h} . \qquad (15.91)$$

Thus, to investigate $P_r(M, t)$ for a sequence $M = k\Delta M$ ($k = 1, 2, 3, \ldots$) of values of $M$, we actually considered the values $\Delta M - h, \Delta M + h, 2\Delta M - h, 2\Delta M + h, 3\Delta M - h, \ldots$. We tested different spacing values $h$ and found no notable changes when going below $h = 0.01$.

Note that for $\Delta t \to 0$ the factor in front of the sum in Eq. (15.90) becomes $r\Delta t$ and $e^{-r(\tau-\Delta t)} \to e^{-r\tau}$, therefore one obtains Eq. (15.10). One can expect an influence of the discretization but it will be small, because quantities like the survival probability $Q_0(M, \tau)$ and the no-reset probability $e^{-rt'}$ are always exact independently of the choice of $\Delta t$. We have tested this explicitly by calculating the single-particle distribution $P_r(M, t)$ for various values of $\Delta t$, here for the parameter values $t = 1000$, $D = 1$ and $r = 0.1$. The result is shown in Fig. 15.8. For values below $\Delta t = 2$ we

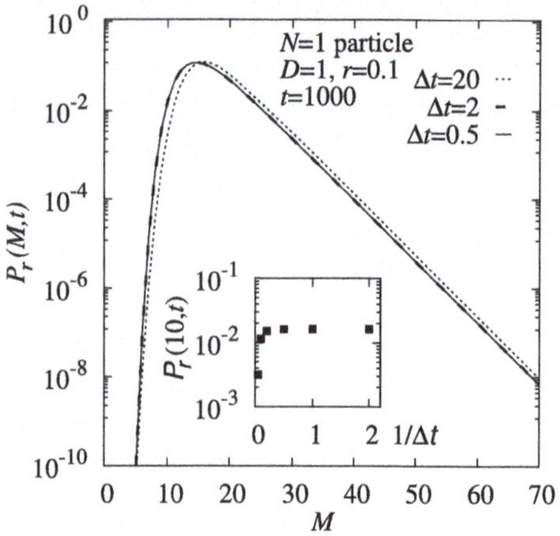


**Fig. 15.8** One-particle distribution $P_r(M, t)$ for $t = 1000$, $D = 1$, $r = 0.1$ and various values $\Delta t$ of the discretization in time. The inset shows how the particular value $P_r(10, t)$ depends on $1/\Delta t$

observed no significant change in the results in Fig. 15.8, so we kept $\Delta t = 1$ throughout our study.

Having tested the dependence on $\Delta t$ of $P_r(M, t)$ in Fig. 15.8, we then fix $\Delta t = 1$ and present the result for the single-particle distribution $P_r(M, t)$ versus $M$ in Fig. 15.9 for $t = 1000$ and compared it to the analytical result in Eq. (15.31). Generally, a very good agreement can be observed, for the left tail, the typical region, the intermediate region and for the large-deviation Gaussian tail. The predicted large-deviation behavior in Eq. (15.33) is confirmed by plotting the numerically obtained rate function $I(z) = -\log(P_r(M, t))/t$ as a function of $z = M/t$ in Fig. 15.10.

For the multi-particle **annealed case** we approximate the average in Eq. (15.44) of the single-particle cumulative function $Q_r(M, t)$ over all possible initial positions in the interval $[-L, 0]$ by a finite sum over $N$ evenly spaced particle positions $x = -(N-1)\delta, -(N-2)\delta, \ldots, \delta, 0$. The particles exhibit a density $\rho = N/L$, i.e., the spacing is $\delta = L/N$. In the present work we use $\delta = 1$. Since starting at position

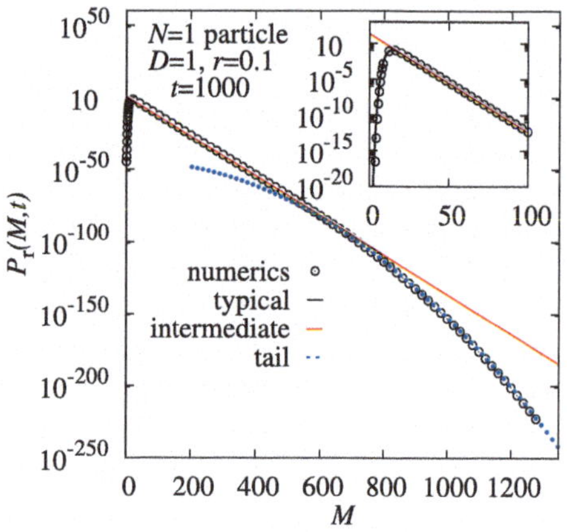


**Fig. 15.9** Numerical one-particle distribution $P_r(M, t)$ for $t = 1000$, $D = 1, r = 0.1$ (symbols) compared to the analytical result in Eq. (15.31) (lines). The inset highlights the range of small values of $M$

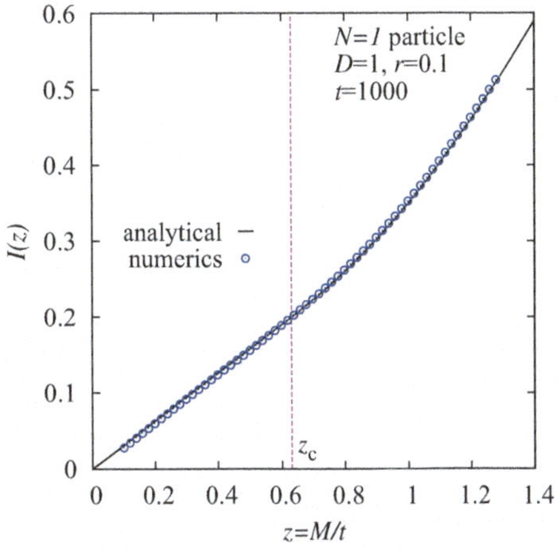


**Fig. 15.10** Numerical result for the one-particle rate function $I(z)$ for $t = 1000$, $D = 1, r = 0.1$ (symbols) compared to the analytical result of Eq. (15.33) (solid line). The blue vertical line provides the location of the critical value $z_c = \sqrt{4Dr} = 0.632456\ldots$

$-i\delta$ and reaching a maximum $M$ is equivalent to start at position 0 with reaching a maximum $M + i\delta$, one obtains

$$Q_{\text{avg}}(M,t) \approx \frac{1}{N}\sum_{i=0}^{N-1} Q_r(M + i\delta, t)\,. \tag{15.92}$$

Note that the spacing of the considered values of $M$ must be such that it is compatible with the needed values $M + i\delta$. For any target value of $M$, we also perform the full calculation for $M - h$ and $M + h$.

The cumulative distribution in Eq. (15.46) for the annealed case is simply $[Q_{\text{avg}}(M,t)]^N$ from Eq. (15.43), therefore $P_{\text{an}}(M,t) = \frac{\partial}{\partial M}[Q_{\text{avg}}(M,t)]^N = N \frac{\partial Q_{\text{avg}}(M,t)}{\partial M}[Q_{\text{avg}}(M,t)]^{N-1}$. The partial derivative is again obtained by taking the numerical difference in Eq. (15.91). The result for the numerical density is then given by

$$P_{\text{an}}(M,t) \approx N\frac{Q_{\text{avg}}(M+h,t) - Q_{\text{avg}}(M-h,t)}{2h}[Q_{\text{avg}}(M,t)]^{N-1}. \tag{15.93}$$

The resulting distribution for $N = 100$ particles in interval $[-(N-1), 0]$ with spacing $\delta = 1/\rho = 1$ is shown in Fig. 15.11. We tested that increasing the number $N$ of particles does not change the result; the additional particles are just too far away. The numerical results confirm very well the analytical results in Eq. (15.70), for the typical region, as well as for the tail which is governed again by the rate function $I(z)$.

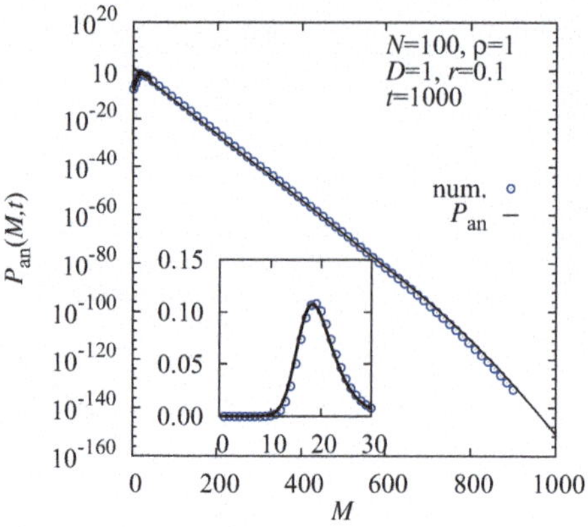


**Fig. 15.11** Numerical result for the annealed distribution $P_{an}(M, t)$ for $t = 1000$, $D = 1$, $r = 0.1$ and $N = 100$ particles (symbols). The lines show the analytical result in Eq. (15.70). The inset highlights the typical region in the first line of Eq. (15.70)

For the multi-particle **quenched case**, we consider just one configuration of particles at positions $x = -(N-1)\delta, -(N-2)\delta, \dots, \delta, 0$. Therefore, the cumulative distribution is given by

$$Q_{qu}(M, t) = \prod_{i=0}^{N-1} Q_r(M + i\delta, t)\,. \tag{15.94}$$

To obtain the density, we take the derivative with respect to $M$ and approximate the derivative of the single-particle cumulative distribution again by a finite difference arriving at

$$P_{qu}(M, t) = \sum_{i=0}^{N-1} \frac{Q_r(M + i\delta + h, t) - Q_r(M + i\delta - h, t)}{2h} \prod_{j=0, j\neq i}^{N-1} Q_r(M + j\delta, t)\,. \tag{15.95}$$

The resulting distribution for the quenched case for $N = 100$ particles in the interval $[-(N-1), 0]$ with density $\rho = 1$ is shown in Fig. 15.12. Also here, the numerical results confirm very well the analytical results in Eq. (15.89), for the typical region, for both tails and for the intermediate region.

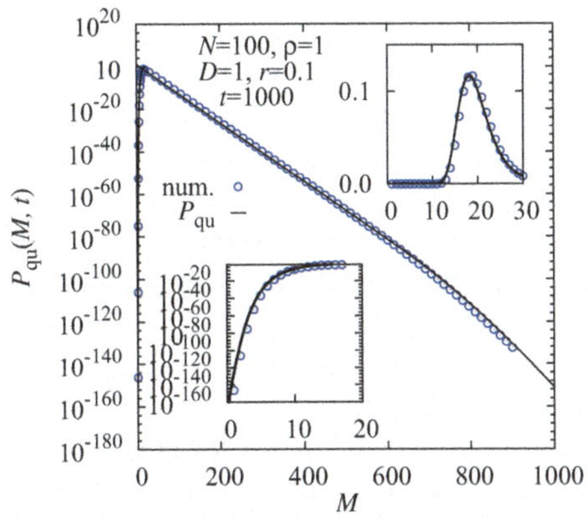


**Fig. 15.12** Numerical result for the quenched distribution $P_{\rm qu}(M,t)$ for $t = 1000$, $D = 1$, $r = 0.1$ and $N = 100$ particles (symbols). The lines show the analytical result in Eq. (15.89). The inset in the upper right corner highlights the typical regime (as given by the second line of Eq. (15.89)). The inset in the lower left corner corresponds to the left tail, when $M \sim \sqrt{D/r}$, as given in the first line of Eq. (15.89)

## 15.5 Conclusion

In this chapter, we first revisited the exact computation of the probability distribution of the maximum $M_t$ of a single resetting Brownian motion (RBM) of duration $t$, starting and resetting at the origin with a constant rate $r$. In this case, it was known that the random variable $M_t$ fluctuates around $\sqrt{D/r}\,\ln(r\,t)$ for large $t$ and the typical scale of fluctuations is of order $O(1)$. The appropriately centered and scaled distribution of $M_t$ approaches at large times to a standard Gumbel distribution. In this chapter, we have analyzed the large-deviation regime of $M_t$ when $M_t \sim O(t)$ and shown that its PDF satisfies a large-deviation form $P_r(M,t) \sim e^{-tI(M/t)}$, where the rate function $I(z)$ has a second-order discontinuity at the critical value $z_c = \sqrt{4Dr}$. This signals a dynamical phase transition in the tail of the distribution of the maximum. Next, we considered a collection of independent RBMs with initial (and resetting) positions uniformly distributed with a density $\rho$ over the negative half-line. In this case, the distribution of the global maximum up to time $t$ depends on how one carries out the average over the initial positions. In the annealed case the stochastic trajectories and the initial positions are averaged simultaneously, while in the quenched case, one averages only over the stochastic trajectories for a fixed initial condition, which is a typical one, i.e., the initial condition that has the largest probability of occurrence. In the annealed case, we showed that the PDF of the maximum, appropriately centered around $\sqrt{D/r}\,\ln(r\,t)$ and scaled by a constant factor, approaches at late times to a new limiting distribution characterized by a scaling function $G'_{\rm an}(z)$, which is different from the well-known Gumbel law for a single particle. Thus, even though the particles are independent, the averaging over the initial positions modifies the typical distribution of the maximum. However, the large-deviation behavior of $M_t$

when $M \sim O(t)$ remains the same as the single particle case. Hence, in the annealed case, the fluctuations of the initial positions modify the typical behavior of the maximum, but not its large-deviation tails. In contrast, for the quenched case, we showed that the PDF of the maximum, centered at $\sqrt{D/r}\,\ln(r\,t)$ and scaled by a constant, does approach the limiting Gumbel form, but with nontrivial large-deviation tails on both sides of the typical value, i.e., both when $M \sim O(1)$ and $M \sim O(t)$. Thus in the quenched case, the fluctuations of the initial positions affect only the large-deviation behavior of $M_t$. Our analytical predictions, both for the typical as well as for the large-deviation regime of $M_t$, have been verified numerically to a very high precision, down to $10^{-250}$ for the PDF of $M_t$.

There are many directions in which this work can be extended. Here we considered independent RBMs distributed initially over the semi-infinite space. Recently, another model of resetting Brownian motions has been introduced where the position of the walkers are simultaneously reset to their initial positions, but they evolve independently between two resetting events [76, 77]. Simultaneous resetting makes the positions of the particles correlated, and it would be interesting to study the distribution of $M_t$ in this correlated case with the semi-infinite initial conditions as studied in this paper for independent RBMs. Another interesting extension would be to study the distribution of the extremes (both the maximum as well as the minimum) for a set of RBMs in higher dimensions with initial positions distributed uniformly over a given region of space.

**Acknowledgements** The authors received support from the *Erwin Schrödinger International Institute for Mathematics and Physics* (ESI) of the University of Vienna, where part of the work was performed. SNM acknowledges the Gay-Lussac Humboldt Research Award (2019), awarded by the Alexander von Humboldt foundation that enabled a visit to the University of Oldenburg where this work was partially done. SNM and GS acknowledge support from ANR Grant No. ANR- 23-CE30-0020-01 EDIPS. The simulations were partially performed at the HPC cluster CARL, located at the University of Oldenburg (Germany) and funded by the DFG through its Major Research Instrumentation Program (INST 184/157-1 FUGG) and the Ministry of Science and Culture (MWK) of the Lower Saxony State.

## References

1. W.J. Bell *Searching Behavior: the behavioral Ecology of Finding Resources* (Chapman and Hall, London, 1991)
2. G.M. Viswanathan, M.G.E. da Luz, E.P. Raposo, H.E. Stanley, *The Physics of Foraging: an Introduction to Random Searches and Biological Encounters* (Cambridge University Press, Cambridge, 2011)
3. O.G. Berg, R.B. Winter, P.H. von Hippel, *Diffusion-driven mechanisms of protein translocation on nucleic acids. 1. Models and theory* Biochemistry-US **20**, 6929 (1981)
4. S. Condamin, O. Bénichou, V. Tejedor, R. Voituriez, J. Klafter, First-passage times in complex scale-invariant media. Nature **450**, 77 (2007)
5. G. Oshanin, K. Lindenberg, H.S. Wio, S. Burlatsky, Efficient search by optimized intermittent random walks. J. Phys. A Math. Theor. **42**, 434008 (2009)

6. P.L. Krapivsky, S.N. Majumdar, A. Rosso, Maximum of N independent Brownian walkers till the first exit from the half-space. J. Phys. A Math. Theor. **43**, 315001 (2010)
7. O. Bénichou, C. Loverdo, M. Moreau, R. Voituriez, Intermittent search strategies. Rev. Mod. Phys. **83**, 81 (2011)
8. B. Meerson, S. Redner, Mortality, redundancy, and diversity in stochastic search. Phys. Rev. Lett. **114**, 198101 (2015)
9. A. Godec, R. Metzler, First passage time distribution in heterogeneity controlled kinetics: going beyond the mean first passage time. Sci. Rep.-UK **6**, 20349 (2016)
10. A. Godec, R. Metzler, Universal proximity effect in target search kinetics in the few-encounter limit. Phys. Rev. X **6**, 041037 (2016)
11. D.S. Grebenkov, G. Oshanin, Diffusive escape through a narrow opening: new insights into a classic problem. Phys. Chem. Chem. Phys. **19**, 2723 (2017)
12. D.S. Grebenkov, R. Metzler, G. Oshanin, Full distribution of first exit times in the narrow escape problem. New J. Phys. **21**, 122001 (2019)
13. G. Mercado-Vásquez, D. Boyer, First hitting times to intermittent targets. Phys. Rev. Lett. **123**, 250603 (2019)
14. W. Feller, *Introduction to Probability Theory and its Applications*, vol. 2 (Wiley, New York, 1966)
15. S. Redner, *A Guide to First-Passage Processes* (Cambridge University Press, Cambridge, UK, 2007)
16. S.N. Majumdar, Persistence in nonequilibrium systems. Curr. Sci. **77**, 370 (1999)
17. S.N. Majumdar, Brownian functionals in physics and computer science. Curr. Sci. **89**, 2076 (2005)
18. A.J. Bray, S.N. Majumdar, G. Schehr, Persistence and first-passage properties in nonequilibrium systems. Adv. Phys. **62**, 225 (2013)
19. R. Metzler, S. Redner, G. Oshanin, *First-Passage Phenomena and Their Applications*, vol. 35 (World Scientific, Singapore, 2014)
20. E.J. Gumbel, *Statistics of Extremes* (Dover, New York, 1958)
21. R.W. Katz, M.B. Parlange, P. Naveau, Statistics of extremes in hydrology. Adv. Water Resour. **25**, 1287 (2002)
22. S.Y. Novak, *Extreme Value Methods with Applications to Finance* (CRC Press, 2011)
23. M.R. Leadbetter, G. Lindgren, H. Rootzén, *Extremes and Related Properties of Random Sequences and Processes* (Springer Science & Business Media, 2012)
24. S.N. Majumdar, A. Pal, G. Schehr, Extreme value statistics of correlated random variables: a pedagogical review. Phys. Rep. **840**, 1 (2020)
25. S. Redner, M.R. Petersen, Role of global warming on the statistics of record-breaking temperatures. Phys. Rev. E **74**, 061114 (2006)
26. M.R. Evans, S.N. Majumdar, Diffusion with stochastic resetting. Phys. Rev. Lett. **106**, 160601 (2011)
27. M.R. Evans, S.N. Majumdar, Diffusion with optimal resetting. J. Phys. A Math. Theor. **44**, 435001 (2011)
28. M.R. Evans, S.N. Majumdar, Diffusion with resetting in arbitrary spatial dimension. J. Phys. A Math. Theor. **47**, 285001 (2014)
29. A. Nagar, S. Gupta, Diffusion with stochastic resetting at power-law times. Phys. Rev. E **93**, 060102 (R) (2016)
30. A. Pal, A. Kundu, M.R. Evans, Diffusion under time-dependent resetting. J. Phys. A Math. Theor. **49**, 225001 (2016)
31. U. Bhat, C. De Bacco, S. Redner, Stochastic search with Poisson and deterministic resetting. J. Stat. Mech. 083401 (2016)
32. A. Pal, S. Reuveni, First passage under restart. Phys. Rev. Lett. **118**, 030603 (2017)
33. I. Eliazar, S. Reuveni, Mean-performance of sharp restart I: statistical roadmap. J. Phys. A Math. Theor. **53**, 405004 (2020)
34. L. Kusmierz, S.N. Majumdar, S. Sabhapandit, G. Schehr, First order transition for the optimal search time of Lévy flights with resetting. Phys. Rev. Lett. **113**, 220602 (2014)

35. S.N. Majumdar, S. Sabhapandit, G. Schehr, Dynamical transition in the temporal relaxation of stochastic processes under resetting. Phys. Rev. E **91**, 052131 (2015)
36. S. Reuveni, Optimal stochastic restart renders fluctuations in first passage times universal. Phys. Rev. Lett. **116**, 170601 (2016)
37. M. Montero, J. Villarroel, Directed random walk with random restarts: The Sisyphus random walk. Phys. Rev. E **94**, 032132 (2016)
38. A. Chechkin, I.M. Sokolov, Directed random walk with random restarts: the Sisyphus random walk. Phys. Rev. Lett. **121**, 050601 (2018)
39. P.C. Bressloff, Diffusive search for a stochastically-gated target with resetting. J. Phys. A Math. Theor. **53**, 425001 (2020)
40. R.G. Pinsky, Diffusive search with spatially dependent resetting. Stoch. Proc. Appl. **130**, 2954 (2020)
41. B. De Bruyne, S.N. Majumdar, G. Schehr, Optimal resetting Brownian bridges via enhanced fluctuations. Phys. Rev. Lett. **128**, 200603 (2022)
42. C. Christou, A. Schadschneider, Diffusion with resetting in bounded domains. J. Phys. A Math. Theor. **48**, 285003 (2015)
43. D. Campos, V. Méndez, Phase transitions in optimal search times: How random walkers should combine resetting and flight scales. Phys. Rev. E **92**, 062115 (2015)
44. S. Belan, Restart could optimize the probability of success in a Bernoulli trial. Phys. Rev. Lett. **120**, 080601 (2018)
45. S. Ray, D. Mondal, S. Reuveni, Péclet number governs transition to acceleratory restart in drift-diffusion. J. Phys. A Math. Theor. **52**, 255002 (2019)
46. S. Ahmad, I. Nayak, A. Bansal, A. Nandi, D. Das, First passage of a particle in a potential under stochastic resetting: a vanishing transition of optimal resetting rate. Phys. Rev. E **99**, 022130 (2019)
47. A. Pal, V.V. Prasad, Landau-like expansion for phase transitions in stochastic resetting. Phys. Rev. Res. **1**, 032001 (2019)
48. A. Pal, V.V. Prasad, First passage under stochastic resetting in an interval. Phys. Rev. E. **99**, 032123 (2019)
49. G. Mercado-Vasquez, D. Boyer, S.N. Majumdar, G. Schehr, Intermittent resetting potentials. J. Stat. Mech. 113203 (2020)
50. G. Mercado-Vasquez, D. Boyer, S.N. Majumdar, Freezing transitions of Brownian particles in confining potentials. J. Stat. Mech. 063203 (2022)
51. G. Mercado-Vasquez, D. Boyer, S.N. Majumdar, Reducing mean first passage times with intermittent confining potentials: a realization of resetting processes. J. Stat. Mech. 093202 (2022)
52. M. Biroli, S.N. Majumdar, G. Schehr, Critical number of walkers for diffusive search processes with resetting. Phys. Rev. E **107**, 064141 (2023)
53. M.R. Evans, S.N. Majumdar, G. Schehr, Stochastic resetting and applications. J. Phys. A **53**, 193001 (2020)
54. A. Pal, S. Kostinski, S. Reuveni, The inspection paradox in stochastic resetting. J. Phys. A Math. Theor. **55**, 021001 (2022)
55. S. Gupta, A.M. Jayannavar, Stochastic resetting: a (very) brief review. Front. Phys. **10**, 789097 (2022)
56. O. Tal-Friedman, A. Pal, A. Sekhon, S. Reuveni, Y. Roichman, Experimental realization of diffusion with stochastic resetting. J. Phys. Chem. Lett. **11**, 7350 (2020)
57. B. Besga, A. Bovon, A. Petrosyan, S.N. Majumdar, S. Ciliberto, Optimal mean first-passage time for a Brownian searcher subjected to resetting: experimental and theoretical results. Phys. Rev. Res. **2**, 032029(R) (2020)
58. B. Besga, F. Faisant, A. Petrosyan, S. Ciliberto, S.N. Majumdar, Dynamical phase transition in the first-passage probability of a Brownian motion. Phys. Rev. E **104**, L012102 (2021)
59. F. Faisant, B. Besga, A. Petrosyan, S. Ciliberto, S.N. Majumdar, Optimal mean first-passage time of a Brownian searcher with resetting in one and two dimensions: experiments, theory and numerical tests. J. Stat. Mech. 113203 (2021)

60. M. Biroli, F. Mori, S.N. Majumdar, Number of distinct sites visited by a resetting random walker. J. Phys. A Math. Theor. **55**, 244001 (2022)
61. S.N. Majumdar, F. Mori, H. Schawe, G. Schehr, Mean perimeter and area of the convex hull of a planar Brownian motion in the presence of resetting. Phys. Rev. E **103**, 022135 (2021)
62. F. Mori, S.N. Majumdar, G. Schehr, Distribution of the time of the maximum for stationary processes. Eur. Lett. **135**, 30003 (2021)
63. F. Mori, S.N. Majumdar, G. Schehr, Time to reach the maximum for a stationary stochastic process. Phys. Rev. E **106**, 054110 (2022)
64. P. Singh, A. Pal, Extremal statistics for stochastic resetting systems. Phys. Rev. E **103**, 052119 (2021)
65. B. Derrida, A. Gerschenfeld, Current fluctuations in one dimensional diffusive systems with a step initial density profile. J. Stat. Phys. **137**, 978 (2009)
66. T. Banerjee, S.N. Majumdar, A. Rosso, G. Schehr, Current fluctuations in noninteracting run-and-tumble particles in one dimension. Phys. Rev. E **101**, 052101 (2020)
67. S. Marbach, Intrinsic fractional noise in nanopores: the effect of reservoirs. J. Chem. Phys. **154**, 171101 (2021)
68. T. Banerjee, R.L. Jack, M.E. Cates, Role of initial conditions in one-dimensional diffusive systems: Compressibility, hyperuniformity, and long-term memory. Phys. Rev E **106**, L062101 (2022)
69. S. Jose, A. Rosso, K. Ramola, Generalized disorder averages and current fluctuations in run and tumble particles. Phys. Rev. E **108**(5), L052601 (2023)
70. C. Di Bello, A.K. Hartmann, S.N. Majumdar, F. Mori, A. Rosso, G. Schehr, Current fluctuations in stochastically resetting particle systems. Phys. Rev. E **108**, 014112 (2023)
71. D.S. Dean, S.N. Majumdar, G. Schehr, Effusion of stochastic processes on a line. J. Stat. Mech. 063208 (2023)
72. I. Burenev, S.N. Majumdar, A. Rosso, Local time of a system of Brownian particles on the line with step-like initial condition. Phys. Rev. E **108**, 064113 (2023)
73. A.K. Hartmann, *Big Practical Guide to Computer Simulations* (World Scientific, Singapore, 2015)
74. MPFR multi precision floating point library. https://www.mpfr.org/
75. L. Fousse, G. Hanrot, V. Lefèvre, P. Pélissier, P. Zimmermann, MPFR: A multiple-precision binary floating-point library with correct rounding. ACM Trans. Math. Software **33**, 13-es (2007)
76. M. Biroli, H. Larralde, S.N. Majumdar, G. Schehr, Extreme statistics and spacing distribution in a Brownian gas correlated by resetting. Phys. Rev. Lett. **130**, 207101 (2023)
77. M. Biroli, H. Larralde, S.N. Majumdar, G. Schehr, Exact extreme, order and sum statistics in a class of strongly correlated system. Phys. Rev. E **109**, 014101 (2024)

# Chapter 16
# Target Search Kinetics for Random Walkers with Memory

**Olivier Bénichou, Thomas Guérin, Nicolas Levernier, and Raphaël Voituriez**

**Abstract** In this chapter, we consider the problem of a non-Markovian random walker (displaying memory effects) searching for a target. We review an approach that links the first passage statistics to the properties of trajectories followed by the random walker in the future of the first passage time. This approach holds in one and higher spatial dimensions, when the dynamics in the vicinity of the target is Gaussian, and it is applied to three paradigmatic target search problems: the search for a target in confinement, the search for a rarely reached configuration (rare event kinetics), or the search for a target in infinite space, for processes featuring stationary increments or transient aging. The theory gives access to the mean first passage time (when it exists) or to the behavior of the survival probability at long times, and agrees with the available exact results obtained perturbatively for examples of weakly non-Markovian processes. This general approach reveals that the characterization of the non-equilibrium state of the system at the instant of first passage is key to derive first-passage kinetics, and provides a new methodology, via the analysis of trajectories after the first-passage, to make it quantitative.

---

O. Bénichou
LPTMC, CNRS/Sorbonne Université, 4, Place Jussieu, Paris Cedex 05, 75252 Paris, France
e-mail: olivier.benichou@sorbonne-universite.fr

T. Guérin (✉)
LOMA, CNRS/University of Bordeaux, 351 Cours De La Libération, 33400 Talence, France
e-mail: thomas.guerin@u-bordeaux.fr

N. Levernier
CINaM, CNRS/Aix Marseille University, Marseille, France
e-mail: nicolas.levernier@univ-amu.fr

R. Voituriez
LJP, CNRS/Sorbonne University, CNRS/Sorbonne Université, Place Jussieu, Paris Cedex 05, 75252 Paris, France
e-mail: raphael.voituriez@upmc.fr

 
D. Grebenkov et al. (eds.), *Target Search Problems*,
https://doi.org/10.1007/978-3-031-67802-8_16

## 16.1 Introduction

How much time does it take for a random walker to go from a point $A$ to a point $B$, or to meet another particle? Answering this question amounts to solving a first passage problem [1]: one has to calculate the first passage time $T$ to reach a "target region". First passage problems appear naturally in many areas of physics, ranging from transport controlled reactions [2, 3] (since a reaction cannot be faster than the time required for the reactants to come into contact), finance [4], polymer translocation [5], or search problems in biophysics (a virus searching for the nucleus of a eukaryotic cell [6], a transcription factor searching for a gene sequence [7], etc.). In this chapter, we consider the generic picture of a random walker of position $\boldsymbol{r}(t)$ evolving with time $t$ in a continuous $d$-dimensional space.

In the last decades, the study of first passage problems has attracted a lot of attention in the literature [1, 8–10]. Historically, results have been derived first for classical random walks (or Brownian motion) on regular lattices (or uniform Euclidean spaces), and then extended to complex media for processes that are memoryless (Markovian), for which the properties of the random walk at time $t + dt$ can be deduced from the knowledge of the position of the random walker only. For such processes, exact analytical tools, such as the renewal equation or backward Fokker-Planck equations, are available as starting points to study first passage properties [1, 11, 12]. However, when a random walker interacts with other variables of a "bath" (which could be internal degrees of freedom, or variables associated with the environment), the position of the random walker only must be described as a non-Markovian process, where memory effects appear when the dynamics of the other variables that interact with the random walker is ignored. A simple example is given by the dynamics of a tagged monomer of a polymer chain. In this case the position of the tagged monomer $\boldsymbol{r}(t + dt)$ at $t + dt$ does not depend on $\boldsymbol{r}(t)$ only but also on the positions of all other monomers of the polymer. This collective dynamics makes the motion of the tagged monomer non-Markovian [13–15].

Existing theoretical approaches to first passage kinetics of non-Markovian processes can be schematically classified as follows. First, for a few examples of specific non-Markovian stochastic processes analytical solutions can be obtained. This is the case of the random acceleration process [16] and of processes generated by telegraphic noises [17–20]. Second, a standard approach was introduced in the field of reactions involving polymers by Wilemski and Fixman [21, 22], and proved to be applicable outside the field of polymer dynamics. This approach consists in considering that the degrees of freedom of the bath are always in a stationary state and will be referred to as a *pseudo-Markovian approach*. This approach has been improved in Ref. [23]. Even if useful [24–28], we will see that this approach can predict wrong scaling laws in some cases (see, e.g., Ref. [29] and below). A third approach consists in developing a perturbation theory for weakly non-Markovian processes [30–36], assumed to be "close" to Brownian motion. These exact results have been so far restricted to one-dimensional processes and necessitate additional hypotheses to be extended to the non-perturbative regime. Last, a lot of work has been performed for

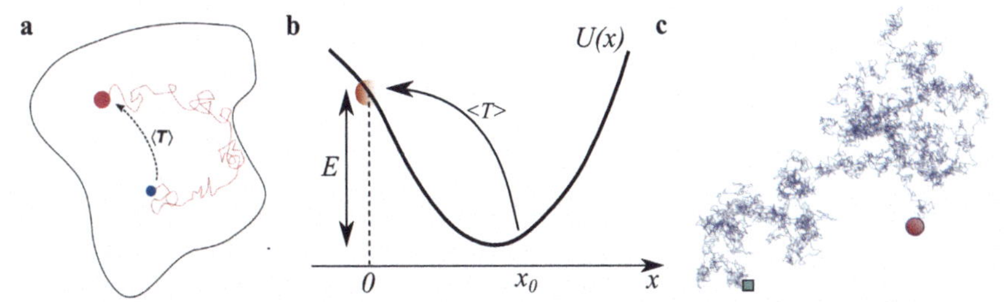


**Fig. 16.1** The three first passage problems that will be investigated in this chapter. **a** Search for a target in confinement. **b** Search, when $x(t)$ is a reaction coordinate, for an energetically costly configuration (rare event kinetics). **c** Search for a target in infinite space

the problem of first passage in infinite space for non-stationary initial conditions, leading to the determination of persistence exponents characterizing the long-time decay of the first passage distribution (see review in [9]).

The goal of this chapter is to review recent works that introduced an alternative non-perturbative theoretical approach. This relies on the analysis of the trajectories followed by the random walker after the first-passage event, and can be developed in any space dimension [37–40]. We will show that this approach provides quantitative results for first passage properties in the three paradigmatic cases represented in Fig. 16.1.

1. In the first situation, shown in Fig. 16.1a, a random walker is looking for a target in a confined domain. In this case the random walker is able to explore the whole volume of the domain, and there is no substantial energy barrier to cross to reach the target. Such geometric confinement can be achieved by hard walls or softer confinement (such as harmonic trapping). The target search can be seen as *limited by entropy* since what limits the search is the exploration of the volume.
2. In the second case, the random walker is assumed to describe a reaction coordinate $x(t)$ moving in an energy landscape $U(x)$ and the target is located at a configuration of high energy [Fig. 16.1b]. The first passage is in this case a *rare event* and is limited by the energy cost to reach the target.
3. In the third case of exploration in infinite space [Fig. 16.1c], where the stationary probability density in the absence of target vanishes $p_s(\boldsymbol{r}) = 0$, first passage properties are quantified by the survival probability, which is characterized by a persistent exponent defining its algebraic decay, and a prefactor. These will be discussed for processes with stationary increments, and for processes with non-stationary initial conditions.

The presentation is based on results of Refs. [37–40]. Note that we will leave aside the case of first passage properties for anomalous transport coming from jumps with broadly distributed waiting times and/or jump lengths, such as in Continuous Time Random Walks or Lévy walks. Strictly speaking, such random walks are also non-Markovian if the waiting times are not exponentially distributed. However, such processes are amenable to Markovian analysis, essentially because renewal equations can be obtained [9, 41–43].

The outline of this chapter is as follows. First, we derive a general equation for the mean first passage time, for non-Markovian non-smooth random processes (Sect. 16.2.1). Then, we show how to predict the mean first passage time for non-Markovian random walkers searching for a target in a large volume (Sect. 16.2.2). Next, we show how the formalism can be adapted to describe the kinetics of rare events in Sect. 16.2.3. Next, we describe the first passage without confinement and describe how to obtain the long-time asymptotics for the survival probability for processes with stationary increments (Sect. 16.3.1). Finally, in Sect. 16.3.2, we show how the formalism can be adapted to calculate persistence exponents in the case of non-stationary initial conditions, such as after a quench.

## 16.2 Mean First Passage Times for Processes Reaching a Stationary State

### *16.2.1 General Formula for the Mean FPT for a Non-Markovian, Non-smooth Random Walker*

We start this section with the derivation of a general relation for the mean first passage time for a non-Markovian random walker $x(t)$ in one dimension, with the hypothesis that $x(t)$ is continuous, non-smooth (so that $\langle \dot{x}^2 \rangle = \infty$ [9], as for overdamped processes), and reaches a stationary state at long time. We denote by $p(x, t)$ its probability density function (PDF) at time $t$, and the stationary PDF is $p_s(x) = \lim_{t\to\infty} p(x, t)$. Both $p$ and $p_s$ are defined in the absence of a target. Now, we call $T$ the first passage time (FPT) to a target at $x = 0$, and $F(T)$ the PDF of FPTs. Our starting point is a "tautological" equation which comes from the following remark: since the process is non-smooth, if the particle is observed at $x$ at $t$ then it means that the particle had already reached the target at some earlier time $\tau$ for the first time, with $\tau < t$. Then, we can write,

$$p(0, t) = \int_0^t d\tau F(\tau) p(0, t|\text{FPT} = \tau), \tag{16.1}$$

where $p(0, t)$ is the probability density to be at $x = 0$ (for the dynamics without absorption to the target) and $p(0, t|\text{FPT} = \tau)$ is the probability density to observe the particle at $x = 0$ at time $t$ given that the FPT is $\tau$ (when the particle is allowed to continue its motion after the FPT). This equation is a generalized "renewal" equation [11]. To the best of our knowledge the fact that one should condition the propagators to the value of the FPT was first noted in Ref. [44]. Here we note that in the Markovian case one could write $p(0, t|\text{FPT} = \tau) = p(0, t - \tau|0)$ so that the above equation would involve a convolution, enabling one to derive a general expression of the FPT

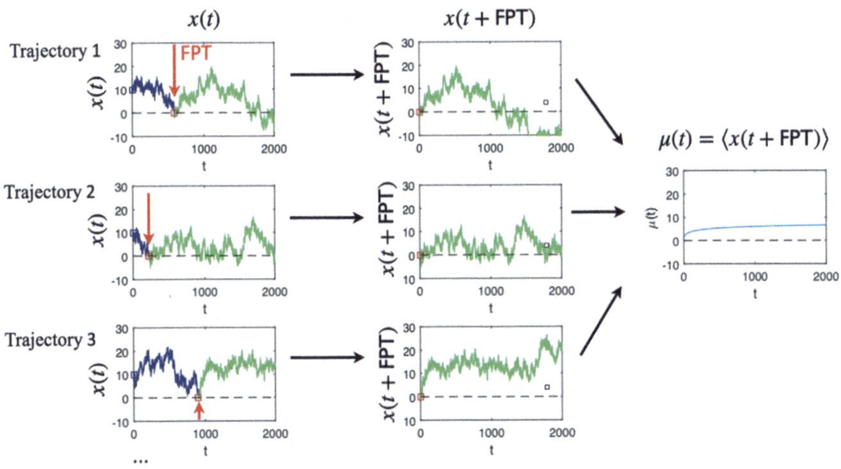


**Fig. 16.2** Construction of the average trajectories in the future of the FPT $\mu(t)$. For each realization of a stochastic trajectory, one measures the FPT (left) and uses this value of time as an initial time for the process in the future of the FPT. Averaging those new trajectories over a large number of realizations, one obtains $\mu(t)$. Here, the trajectories are real examples of a scale invariant subdiffusive process of MSD $\psi(t) = t^{0.72}$

density and its moments (see, e.g., [10, 45]). This is however not possible in the non-Markovian case where the convolution structure is lost and one sees that one has to characterize the probability density conditioned to the fact that a first passage event was observed.

Now, we introduce the process in the future of the FPT, $y_\pi(t) \equiv x(t + \text{FPT})$. The process $y_\pi(t)$ is thus defined from a given stochastic trajectory $x(t)$ by choosing the first passage event as the initial condition, see Fig. 16.2. The suffix $\pi$ is used for historical reasons as $\pi$ is sometimes used to denote the probability with which a part of an extended target is reached for the first time [11, 46], and here $y_\pi$ is linked to the configuration of non-reactive degrees of freedom at the first passage [47]. We define $p_\pi(y, t)$ as the probability density of observing $y_\pi(t) = y$, which reads

$$p_\pi(y, t) = \int_0^\infty d\tau F(\tau) p(y, t + \tau|\text{FPT} = \tau). \tag{16.2}$$

Now we show how to make the mean FPT appear by a few algebraic manipulations. We consider Eq. (16.1), where we substract $p_s(0)$ on both sides, leading to

$$p(0, t) - p_s(0) = \int_0^t d\tau F(\tau)[p(0, t|\text{FPT} = \tau) - p_s(0)] - \int_t^\infty d\tau F(\tau) p_s(0), \tag{16.3}$$

where we have used the fact that $\int_0^\infty d\tau F(\tau) = 1$. Next, we write

$$\int_0^\infty dt \int_t^\infty d\tau F(\tau) = \langle T \rangle, \tag{16.4}$$

which is obtained by changing the order of integration between $t$ and $\tau$. We also note the following equalities:

$$\begin{aligned} \int_0^\infty dt \int_0^t d\tau F(\tau)[p(0,t|\mathrm{FPT}=\tau) - p_s(0)] \\ = \int_0^\infty d\tau \int_0^\infty du\; F(\tau)\,[p(0,\tau+u|\mathrm{FPT}=\tau) - p_s(0)] \end{aligned} \tag{16.5}$$

$$= \int_0^\infty du\,[p_\pi(0,u) - p_s(0)], \tag{16.6}$$

where we have changed the order of between $(t, \tau)$ and used $t = u + \tau$ to obtain Eq. (16.5), and we have used the definition (16.2) to simplify the resulting integral. Using Eqs. (16.4) and (16.6), we see that integrating Eq. (16.3) over $t$ leads to [40]

$$\langle T \rangle p_s(0) = \int_0^\infty dt\,[p_\pi(0,t) - p(0,t)]. \tag{16.7}$$

This equation is general and exact, as soon as $p_s$ exists, for any continuous non-smooth stochastic process, even non-Gaussian. However, at this stage, it is not explicit since we have no information on $p_\pi$.

We therefore consider a two-point generalized version of the renewal equation:

$$p(0,t;x_1,t+t_1) = \int_0^t d\tau F(\tau) p(0,t;x_1,t+t_1|\mathrm{FPT}=\tau). \tag{16.8}$$

Here, $p(0,t;x_1,t+t_1)$ is the joint PDF of $x = 0$ at time $t$ and $x = x_1$ at a later time $t + t_1$. Here $x_1$ and $t_1$ are arbitrarily fixed scalars, with $t_1 > 0$. Multiplying by $x_1$ and integrating over $x_1$ leads to [40]

$$\begin{aligned} \int_0^\infty dt[p_\pi(0,t)\mathbb{E}_\pi(y_\pi(t+t_1)|y_\pi(t)=0) - p(0,t)\mathbb{E}(x(t+t_1)|x(t)=0)] = \\ \langle T \rangle p_s(0)\mathbb{E}_s(x(t_1)|x(t)=0), \end{aligned} \tag{16.9}$$

where the notation $\mathbb{E}(A|E)$ stands for the conditional average of $A$ given that the event $E$ is realized, $\mathbb{E}_\pi$, $\mathbb{E}_s$ and $\mathbb{E}$ denote the average over the respective PDFs $p_\pi$, $p_s$ and $p$. This general equation will be useful to characterize the mean first passage time in two limiting cases: the search for a target for a symmetric random walk in a large volume (Sect. 16.2.2), and the search for a rarely reached event (Sect. 16.2.3).

## 16.2.2 Target Search in a Large Volume

### 16.2.2.1 Explicit Form of the Equations for the Mean FPT

We consider now the more specific case of target search in confinement, when the following property is satisfied: we assume that there is a limit in which $p_s(0) \to 0$, which we call large volume limit, corresponding to boundaries going far away from the target and the initial position of the walker. Note that, in the case of a random walk in an harmonic potential, such a limit is reached when the stiffness of the potential vanishes. We assume that, in this large volume limit, the trajectories of $x(t)$ are not biased (no preferred direction), that $x(t)$ has stationary increments (no aging), and that $x(t)$ is a Gaussian process. Of note this Gaussian hypothesis does not imply that correlations are weak; instead we will see that our theory can deal with processes for which the relaxation time of the increments are formally infinite (such as in the case of the fractional Brownian motion). We also note that there are many physical examples of Gaussian non-Markovian processes, such as the dynamics of tracer particles moving in complex viscoelastic fluids [48, 49], nematic fluids [50], crowded narrow channels [51], or attached to polymers [13, 14].

With these hypotheses, one can show that $x(t)$ is fully characterized as soon as one specifies the Mean Square Displacement (MSD) $\psi(t) = \langle [x(t+\tau) - x(\tau)]^2 \rangle$, since the covariance at times $t, t'$ takes the form

$$\sigma(t, t') = \mathrm{Cov}(x(t), x(t')) = \frac{1}{2} \left[ \psi(|t - t'|) - \psi(t) - \psi(t') \right]. \tag{16.10}$$

We assume that the MSD diverges at long times as $\psi(t) \simeq \kappa t^{2H}$ for $t \to \infty$, with $0 < H < 1$, with $\kappa > 0$ a transport coefficient. The motion at long times can then be either subdiffusive ($H < 1/2$) or superdiffusive ($H > 1/2$), and in all cases the particle is not trapped in the vicinity of the initial point but is free to explore space. Note that $\psi(t)$ is defined in the absence of confinement, so that the behavior $\psi(t) \propto t^{2H}$ is not contradictory with the fact that the "real" MSD must saturate at times for which the random walker reaches the confining boundaries.

In the large volume limit, it is found that all terms in Eqs. (16.7) and (16.9) have a well defined, non-vanishing infinite space limit, except for the term $p_s(0)$. Next, we assume that the trajectories in the future of the first passage event have Gaussian statistics, with the same covariance $\sigma(t, t')$ as the initial process, and an average $\mu(t) = \langle x(t + \mathrm{FPT}) \rangle$. With these approximations, we can calculate the conditional averages entering (16.9) by using existing formulas for conditional averages for Gaussian processes [52], leading to

$$\int_0^\infty \frac{dt}{\sqrt{\psi(t)}} \left\{ [\mu(t+\tau) - \mu(t)K(t,\tau)]\, e^{-\frac{\mu(t)^2}{2\psi(t)}} - x_0\,[1 - K(t,\tau)]\, e^{-\frac{x_0^2}{2\psi(t)}} \right\} = 0, \tag{16.11}$$

with $K(t, \tau) = [\psi(t + \tau) + \psi(t) - \psi(\tau)]/[2\psi(t)]$, while the mean FPT reads

$$\langle T \rangle p_s(0) = \int_0^\infty dt \frac{1}{[2\pi\,\psi(t)]^{1/2}} \left[ e^{-\mu(t)^2/(2\psi(t))} - e^{-x_0^2/(2\psi(t))} \right]. \tag{16.12}$$

Second, the long-time properties of $\mu(t)$ can be obtained. Analyzing Eq. (16.11) for large $\tau$ leads (after some lines of algebra) to the conclusion that

$$\mu(t) \simeq x_0 - \frac{B}{t^{1-2H}}, \qquad (t \to \infty), \tag{16.13}$$

where $B$ is a constant, usually positive, that depends on the behavior of the process at all timescales. This expression clearly shows that for subdiffusive processes ($H < 1/2$) the trajectories in the future of the FPT converge to $x_0$ at long times. This initial position is therefore never forgotten, underlying the strongly non-Markovian nature of the problem. For superdiffusive processes, on the contrary, the trajectory in the future of the FPT typically crosses the target and eventually escapes to $-\infty$. For processes with finite memory, i.e., which become diffusive at long times, $\mu(t)$ reaches a non-vanishing constant. Next, using (16.13), it is clear that the integral (16.12) exists, meaning that the formalism predicts that the mean FPT scales linearly with the volume $V = 1/p_s(0)$ in the large volume limit. This may be compared with the Wilemski-Fixman closure approximation [21, 22]. This approximation was developed in the context of reactions involving polymers, and consists in assuming that the hidden degrees of freedom (the non-reactive monomers in a polymer chain) follow an equilibrium distribution (conditioned to the fact that reactive monomers are in contact). In our framework, this approach translates into the approximation $\mu(t) \simeq 0$, which would incorrectly predict that the mean FPT in (16.12) diverges for $H < 1/3$.

Some predictions of this approach are presented in Fig. 16.3, for various processes, which clearly show that it is much more precise than a pseudo-Markovian approximation. The examples of stochastic processes shown in these figures cover the case of diffusion in a Maxwell viscoelastic fluid [53] and that of the fractional Brownian motion, where $\psi(t)$ is a power law, with infinite memory.

#### 16.2.2.2 Case of Scale Invariant Processes

Let us now consider the case where the process is scale invariant at all times (not only at long times): $\psi(t) = \kappa t^{2H}$. In this case we can show that the solutions of (16.11) satisfy:

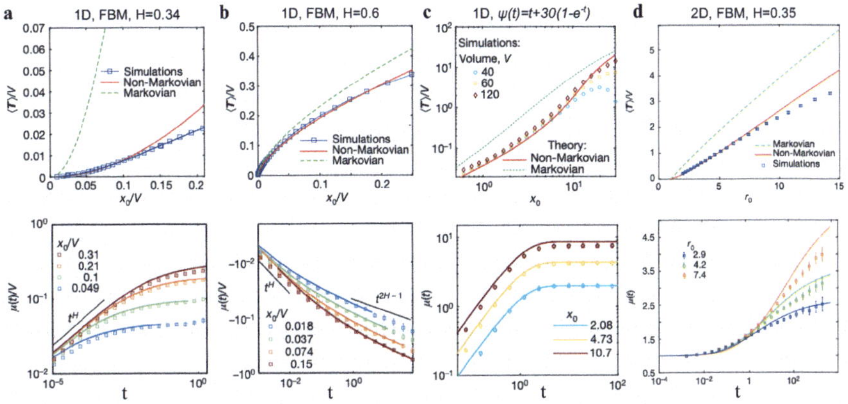


**Fig. 16.3** Test of the non-Markovian for several stochastic processes, upper graphs show the mean FPT, lower graphs display the value of $\mu(t)$. Symbols are the results of numerical simulations; continuous lines are the results of the non-Markovian theory, green dashed lines are the results of the pseudo-Markovian approximation, with $\mu(t) \simeq 0$. The value of the MSD are respectively $\psi(t) = t^{2H}$ for (**a**), (**b**), (**d**) (with $H$ indicated in the graph) and $\psi(t) = 30(1 - e^{-t}) + t$ for (**c**). The examples (**a**), (**b**), (**c**) are one-dimensional (1D), for (**d**) the problem is two-dimensional (2D), with a target radius $a = 1$. Adapted from Ref. [40]

$$\mu(t, x_0) = x_0 \, f_H\left(\frac{t\,\kappa^{1/2H}}{x_0^{1/H}}\right) \tag{16.14}$$

where the function $f_H(u)$ depends only on $H$ and is the solution of (16.11) in the particular case $\psi(t) = t^{2H}$ and $x_0 = 1$. Inserting this scaling into Eq. (16.12) leads to

$$\langle T\rangle/V = \beta_H \frac{x_0^{1/H-1}}{\kappa^{1/2H}}, \qquad \beta_H = \int_0^\infty du \frac{e^{-f_H^2(u)/2u^{2H}} - e^{-1/2u^{2H}}}{\sqrt{2\pi}\,u^H}. \tag{16.15}$$

This means that the scaling with the initial distance is the same as in the Markovian case [45], with however a prefactor that takes into account non-Markovian effects encompassed in the function $f_H$.

#### 16.2.2.3 Generalization to Higher Spatial Dimensions

All the arguments presented above can be adapted in $d$-dimensions. Assuming that the target is a (small) sphere of radius $a$, the problem is spherically symmetric in the large volume limit and it is useful to assume that the initial position is randomly distributed around the target, with a fixed initial distance $r_0$. Then the average trajectory in the future of the FPT can be written as $\mu(t)\hat{\boldsymbol{u}}(\theta)$, where $\theta$ is the angle at which the target is hit for the first time and $\hat{\boldsymbol{u}}$ the unit vector in the direction $\theta$. With these approximations, one obtains the self-consistent equations [40]

$$0 = \int_0^\infty \frac{dt}{\psi(t)^{d/2}} \left[ e^{-\mu(t)^2/[2\psi(t)]} \left( \frac{\mu(t)}{d\,\psi(t)} \left( \mu(t+\tau) - \mu(t) \frac{\sigma(t+\tau,t)}{\psi(t)} \right) + M(t,\tau) \right) \right.$$
$$\left. - e^{-r_0^2/[2\psi(t)]} \left( \frac{r_0}{d\,\psi(t)} \left( r_0 - r_0 \frac{\sigma(t+\tau,t)}{\psi(t)} \right) + M(t,\tau) \right) \right]. \quad (16.16)$$

with $M(t,\tau) = [\sigma(t+\tau,t) - \sigma(t,t)]/\psi(t)$, and $r_0$ is the initial distance to the center of the target. This equation is valid for $d = 2$ and $d = 3$. Note that the radius of the target $a$ enters in the problem by imposing $\mu(t = 0) = a$. The mean FPT to the target is

$$\frac{\langle T \rangle}{V} = \int_0^\infty dt [p_\pi(\boldsymbol{0},t) - p(\boldsymbol{0},t)] = \int_0^\infty dt \frac{e^{-\mu(t)^2/[2\psi(t)]} - e^{-r_0^2/[2\psi(t)]}}{[2\pi\psi(t)]^{d/2}}. \quad (16.17)$$

An example of prediction obtained with this theory is presented in Fig. 16.3d.

### *16.2.3 Rare Event Kinetics*

#### 16.2.3.1 General Formulas

Let us now describe how the formalism can be adapted to the study of rare event kinetics. The problem is now to determine the FPT of a non-Markovian reaction coordinate $x(t)$ to a target at $x = 0$ that is rarely reached, because a large energy barrier has to be crossed. Obviously, equation (16.7) is still valid, and we stress the following key points: (i) first, as long as $x_0$ is not in the close vicinity of the target, $p(0,t)$ is exponentially small (with noise intensity) at all times, (ii) second, the probability $p_\pi(x,t)$ to revisit the target after a time $t$ is exponentially small at long times, but finite at times that immediately follow a FPT event, when $x$ is still close to the target. This suggests that the integral (16.7) is dominated by this short time contribution, where $p_\pi$ can be replaced by its value obtained by considering the (Gaussian) linearized dynamics around the target point. Let us assume that the trajectories of $x(t)$, conditioned to $x(0) = 0$, are Gaussian with covariance $\sigma$ given by (16.10) and average $m_s(t)$. Then, we approximate the dynamics after the first passage by a Gaussian dynamics, with same covariance $\sigma$, and mean $\mu(t)$. With these approximations, the estimate (16.7) becomes [38]

$$\langle T \rangle p_s(0) \simeq \int_0^\infty dt \, e^{-\mu(t)^2/[2\psi(t)]}/[2\pi\psi(t)]^{1/2}, \quad (16.18)$$

and the self-consistent equation (16.9) for $\mu(t)$ becomes [38]

$$\int_0^\infty dt \frac{e^{-\mu(t)^2/(2\psi(t))}}{\psi(t)^{1/2}} \left\{ \mu(t+\tau) - \mu(t) \frac{\psi(t+\tau) + \psi(t) - \psi(\tau)}{2\psi(t)} - m_s(\tau) \right\} = 0. \tag{16.19}$$

These equations are obtained from Eqs. (16.7) and (16.9) by neglecting the terms proportional to $p(0, t)$, since, according to the hypothesis (ii), this propagator is exponentially small at all times. The above equation suggests a two-step strategy to obtain $\langle T \rangle$. The first step consists in characterizing the static quantity $p_s(0)$ ; for equilibrium systems one obtains $p_s(0) \propto e^{-E/k_B T}$ and in particular $\langle T \rangle$ follows an Arrhenius-like law. The second step consists in analyzing the dynamics of $x(t)$ in the vicinity of the target to deduce $\mu(t)$. We note that this theory holds for general equilibrium and non-equilibrium systems as well. For equilibrium systems, for which one can write a generalized Langevin equation linearized around the target point, one has

$$\int_0^t dt' \dot{x}(t') K(t-t') = F + \xi(t), \qquad \langle \xi(t)\xi(t') \rangle = k_B T K(|t-t'|), \tag{16.20}$$

where $F = -\partial_z U(z)|_{z=0}$ is the opposite of the slope of the potential at the target and $K$ is a memory friction kernel. Analyzing this equation leads to the following form of the fluctuation-dissipation theorem

$$m_s(t) = \frac{F\psi(t)}{2k_B T}. \tag{16.21}$$

Inserting this expression in Eqs. (16.19) and (16.18) leads to the determination of $\langle T \rangle$ as a function of the local slope $F$ of the potential near the target, the thermal energy $k_B T$, the MSD characterizing the dynamics $\psi(t)$, and the equilibrium probability density $p_s(0)$ at the target. As in all rare event theories the initial state does not matter since it is forgotten much faster than the time to reach the target. For the Markovian (diffusive) case with $\psi(t) \propto t$, there is an obvious solution $\mu(t) = m_s(t)$. For non-Markovian variables, this relation does not hold and the future trajectory $\mu(t)$ reflects the state of the non-reactive degrees of freedom at the FPT.

#### 16.2.3.2 Scale Invariant Processes and Link with Pickands' Constants

In the case of a MSD with power-law scaling at short times, with $\psi(t) = \kappa t^{2H}$, where $0 < H < 1$, Eq. (16.19) predicts that $\mu$ takes the scaling form

$$\mu(t) = \frac{k_B T}{F} f\left(t/t^*\right), \qquad t^* = \left(\frac{k_B T}{\sqrt{\kappa}\,|F|}\right)^{1/H}, \tag{16.22}$$

and the mean FPT reads

$$\langle T \rangle p_s(0) = \frac{A_H (k_B T)^{\frac{1-H}{H}}}{|F|^{\frac{1-H}{H}} \kappa^{\frac{1}{2H}}}, \qquad A_H = \int_0^\infty du \, \frac{e^{-\frac{f^2(u)}{2u^{2H}}}}{\sqrt{2\pi} u^H}, \tag{16.23}$$

where $f$ satisfies

$$\int_0^\infty dt \, \frac{e^{-f(t)^2/(2t^{2H})}}{t^H} \left\{ f(t+\tau) - f(t) \frac{(t+\tau)^{2H} + t^{2H} - \tau^{2H}}{2t^{2H}} + \frac{\tau^{2H}}{2} \right\} = 0. \tag{16.24}$$

The formula (16.23) provides an explicit asymptotic relation for the mean FPT, as a function of the subdiffusion coefficient $\kappa$, the local force $F$ and the temperature $k_B T$, and $A_H$ depends only on $H$. This result is compatible with the mathematical results of Pickands [54] obtained when $x(t)$ is a Gaussian process (whereas here we only assume that $x(t)$ is locally Gaussian, in the vicinity of the target), if one identifies

$$A_H = 2^{1/\alpha} \mathcal{H}_\alpha, \qquad \alpha = 2H, \tag{16.25}$$

where $\mathcal{H}_\alpha$ are Pickands' constants. Our theory therefore provides approximations of $\mathcal{H}_\alpha$. Of note, the formalism presented here can be analyzed in the limit $\varepsilon = H - 1/2 \to 0$, leading to

$$A_H = 2 + 4\varepsilon(\gamma_e - \ln 2) + \mathcal{O}(\varepsilon^2), \tag{16.26}$$

where $\gamma_e$ is Euler's constant. This result agrees with the calculation of Ref. [33].

#### 16.2.3.3 Example of Flexible Chain Reaching a Long Extension

As an example of application of our formalism, we consider the "simple" problem of determining the mean time for a flexible chain to reach a given large extension. This problem plays a key role in the quantification of the mechanisms of constraint release that determine the rheological properties of untangled star polymers [55, 56], and is also involved in ligand adhesion kinetics when mediated by flexible linkers [57]; it has been reconsidered in Ref. [58]. We assume here that the polymer chain can be described by an harmonic chain of $N$ phantom beads with friction drag $\gamma$ linked by springs of stiffness $k$, whose dynamics satisfies

$$\dot{x}_i = \tau_0^{-1}(x_{i+1} - 2x_i + x_{i-1}) + f_i(t), \quad \langle f_i(t) f_j(t') \rangle = 2\frac{l_0^2}{\tau_0} \delta(t-t')\delta_{ij}, \tag{16.27}$$

where $l_0 = \sqrt{k_B T/k}$ is the typical bond length, and $\tau_0 = \gamma/k$ the typical relaxation time of a single bond. The first monomer is fixed, $x_1 = 0$, and we study the mean time $\langle T \rangle$ for the second polymer end $x(t) = x_N(t)$ to reach a threshold value $x = z$.

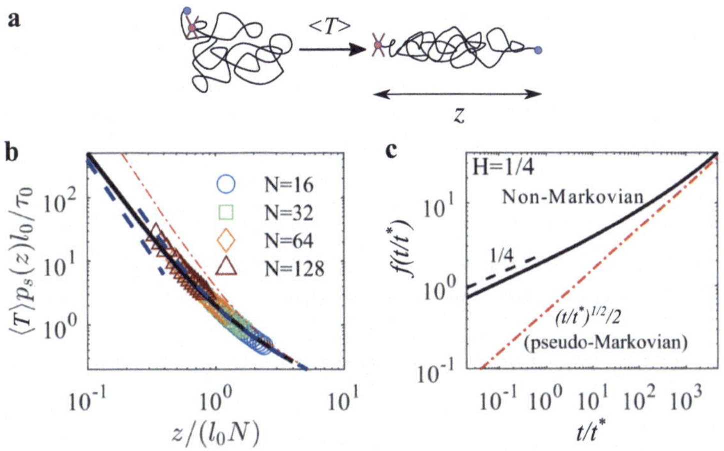


**Fig. 16.4** **a** Sketch of the problem: considering an attached flexible polymer, we compute the first time that a large extension $z$ is reached. **b** Scaled mean FPT as a function of the rescaled variables $z/(l_0 N)$. Symbols: simulation data of Ref. [58]. Black line: non-Markovian theory. Dashed blue line: asymptotic results (16.28) of the theory. Red semi-dashed line: pseudo-Markovian approximation, with $\mu \simeq m_s$. **c** Rescaled average trajectory after the FPT $f(t/t^*)$ [see Eq. (16.22)] obtained by solving numerically (16.24) for $H = 1/4$. The dashed red line would be the future trajectory by assuming equilibrium at initial time. Adapted from Ref. [38]

The energy at fixed $z$ is given by $E = kz^2/(2N)$, and we assume $E \gg k_B T$, so that first-passage events to $z$ are rare.

Using the formalism above, one readily finds that the rescaled mean FPT $\langle T \rangle p_s(z)$ can be expressed as a function of the parameter $z/N$. This scaling is checked by comparing with the data of Ref. [58] on Fig. 16.4b, and agrees well with the numerical solution of the above equations. Furthermore, one can derive explicit asymptotic laws [38]

$$\langle T \rangle p_s(z) \simeq \frac{\tau_0}{l_0} \begin{cases} 0.39(Nl_0/z)^3 & (l_0\sqrt{N} \ll z \ll Nl_0) \\ \frac{Nl_0}{z}\left[1 + \left(\frac{Nl_0}{z}\right)^2\right] & (Nl_0 \ll z) \end{cases} . \tag{16.28}$$

For this problem, a general solution in the weak noise limit (small temperature limit at fixed $N$) can be found [formula (10.117) of Ref. [59]]. This result however gives only the leading order term when $Nl_0 \ll z$, meaning that this exact weak-noise approach does not capture the impact of memory effects in all the regimes. The result for very long chains in the first line of (16.28) is obtained by making use of the fact that the monomer motion is subdiffusive at short times, with $H = 1/4$. The prefactor is actually found to be eight times larger than predicted by the pseudo-Markovian approximation, indicating strong memory effects. The importance of these memory effects can be seen directly from the reactive trajectories $\mu(t)$ in Fig. 16.4c where one compares $\mu(t)$ and $m_s(t)$ in the regime of infinitely long chains: one clearly sees that $\mu(t)$ at short times displays a different scaling with $t$ as compared to $m_s(t)$, meaning that *the folding dynamics after a first passage* [$\mu(t)$] *is infinitely slower than the folding dynamics* $m_s(t)$ *starting from an equilibrium state* (where the chain is conditioned to be stretched). This means that the return probability $p_\pi(z, t)$ is largely overestimated by the pseudo-Markovian approximation.

## 16.3 Survival Probability of Unconfined Compact Non-markovian Random Walks

### *16.3.1 Stochastic Processes with Stationary Increments*

Let us now focus on the case of an unconfined random walk, for which one cannot define a stationary PDF. Here, we focus on non-smooth Gaussian symmetric random walks with stationary increments, with MSD $\psi(t) \simeq \kappa t^{2H}$ at long times, with $0 < H < 1$. In this case, for compact random walks (satisfying $dH < 1$), it is well known that the mean time to reach a target is infinite, and the first passage statistics is characterized by the probability of not having reached the target (survival probability, $S(t)$) at long times,

$$S(t) \underset{t\to\infty}{\simeq} S_0/t^\theta, \tag{16.29}$$

where $\theta$ is known as a *persistence exponent* and $S_0$ is a prefactor. In one dimension, $\theta = 1 - H$ can be obtained [60–62], and its generalization to $d$-dimension is $\theta = 1 - dH$ (see Ref. [43] and below). This law is valid for compact processes, for which the probability to reach a target, even pointlike, in infinite space is equal to one.

At long times, the PDF of the position $p(\boldsymbol{r}, t)$ decays as a power-law,

$$p(\boldsymbol{0}, t) \underset{t\to\infty}{\simeq} K/t^{dH}, \tag{16.30}$$

where, in the case of Gaussian processes, $K = 1/[2\pi\kappa]^{d/2}$. Let us briefly describe how to obtain the prefactor $S_0$ in (16.29). We start with the generalized renewal equation (16.1) (adapted to the $d$-dimensional case), which we integrate over $t \in ]0, A[$ to obtain

$$\begin{aligned}\int_0^A dt p(\boldsymbol{0}, t) &= \int_0^A dt \int_0^t d\tau F(\tau) p(\boldsymbol{0}, t|\text{FPT} = \tau)\\ &= \int_0^A du \int_0^{A-u} d\tau F(\tau) p(\boldsymbol{0}, \tau + u|\text{FPT} = \tau),\end{aligned} \tag{16.31}$$

where we have used $t = \tau + u$. As before, we consider the PDF of the position $p_\pi(\boldsymbol{x}, t)$ at a time $t$ after the first passage event, given by Eq. (16.2) (generalized to higher spatial dimensions), so that for any $A > 0$

$$\int_0^A dt[p_\pi(\boldsymbol{0},t)-p(\boldsymbol{0},t)] = \int_0^A dt \int_{A-t}^{\infty} d\tau F(\tau) p(\boldsymbol{0},\tau+t|\text{FPT}=\tau). \quad (16.32)$$

The right-hand side can be estimated by assuming that, at long times, one has the following decoupling approximation:

$$p(\boldsymbol{0},(u+v)A|\text{FPT}=uA) \underset{A\to\infty}{\simeq} p_\pi(\boldsymbol{0},vA) \sim K/(vA)^{dH}. \quad (16.33)$$

Using $t = uA$ and $\tau = vA$ in (16.32) and the above approximation, we obtain

$$\int_0^A dt[p_\pi(\boldsymbol{0},t)-p(\boldsymbol{0},t)] = A^{1-dH-\theta} \int_0^1 du \int_{1-u}^{\infty} dv \frac{K S_0 \theta}{u^{\theta+1} v^{dH}}. \quad (16.34)$$

Now, for $A \to \infty$, we see that the left-hand side is nothing but the value of the mean FPT for large volumes, rescaled by the volume. Since its value is finite for large $A$, we deduce that the right-hand side of the above equation must also reach a finite value for large $A$. This leads to [39]

$$\theta = 1 - dH, \qquad S_0 \simeq \frac{\sin(\pi dH)}{\pi K} \times \lim_{V\to\infty} \frac{\langle T \rangle}{V}, \quad (16.35)$$

where $K$ is defined in Eq. (16.30). For non-Markovian processes, this result requires the "long time decoupling approximation" (16.33). However, relaxing this approximation (by introducing a scaling function $G(u/v)$ in (16.33)) would only change the results by a scaling factor, while the dependence of $S_0$ on the initial distance to the target would be taken into account in $\langle T \rangle / V$.

Equation (16.35) indicates that the prefactor $S_0$ is directly proportional to the mean FPT for the same random walker in confinement, thus enabling one to calculate $S_0$ by the same methods that were used in confinement. Figure 16.5 compares our predictions for $S_0$ with simulations of various processes. In fact, Eq. (16.35) is very general: it holds for compact processes with and without memory, for any spatial dimension (including fractal spaces), for perfect or imperfect reactivity. We could also use this formula for other forms of confinement, as provided by considering the resetting of the random walker with some resetting rate. The above formula (16.35) would then link the prefactor of the survival probability in absence of resetting to the mean FPT to the target in presence of resetting for low resetting rates, provided that one replaces $1/V$ by the stationary probability to be at the target in presence of resetting. This can be checked directly by using the formulas present in Ref. [63].

Last, we note that the above equation gives an information on the first passage time density even in confinement at intermediate times when the boundaries are not reached. Hence, although the mean FPT does not characterize alone the full distribution (which is not exponential), the above formula indicates that the mean FPT provides valuable informations on the shape of the FPT density.

### 16.3.2 Non-stationary Initial Conditions: Survival Probability After a Quench

In this last section, we discuss the case of non-stationary initial conditions. This could typically occur for example after a sharp change of temperature of the whole system at initial time (a quench). For such non-stationary initial conditions, a classical problem is that of the determination of persistence exponents $\theta$ [9], which can be non-trivial even in simple models, with rare exact results [64–66]. Here, we focus on Gaussian processes that are transiently aging: we assume that the PDF of $x(t + t_1)$, conditioned to the value of $x(t)$, reaches for large $t$ a finite steady state value that does not depend on $t$. An example of such processes is provided by the dynamics of polymers or interfaces exposed to a temperature quench at $t = 0$, in which case the covariance takes the form

$$\sigma(t, t') \propto T_0(t^{2H} + t'^{2H}) + (1 - T_0)(t + t')^{2H} - |t - t'|^{2H}, \qquad (16.36)$$

where $T_0$ is the temperature at past times $t < 0$ (not to be confounded with the first passage time $T$), while the temperature of the dynamics is taken as unity for $t > 0$. Of note, the temperature $T_0$ before the quench can be lower or larger than the temperature $T_0 = 1$ of the dynamics for $t > 0$. This covariance function (16.36) is relevant for the

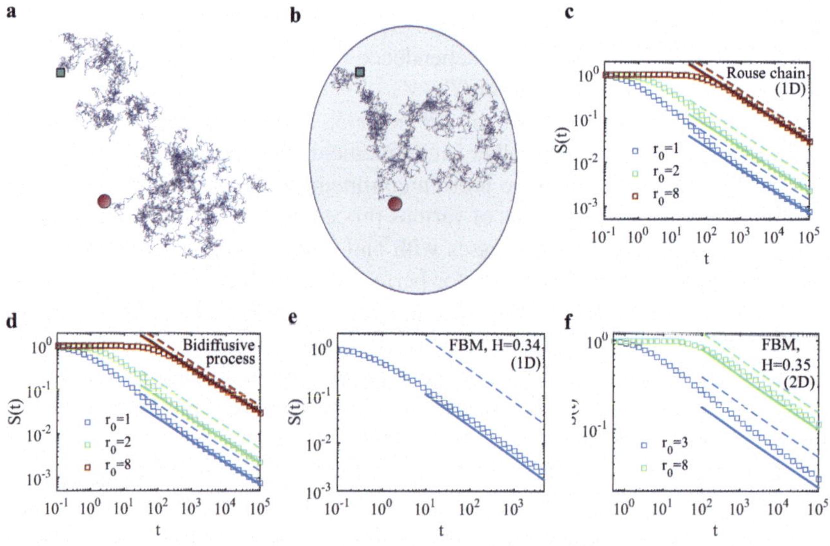


**Fig. 16.5** **a**, **b** Sketch of two first passage problems, without (**a**) and with (**b**) confinement for the same stochastic process. In case (**a**) the mean FPT to reach the target is infinite, but it is finite in case (**b**) and can be used to characterize the long-time decay of the survival probability in problem (**a**). **c**–**f** figures showing simulation results for $S(t)$ compared to the prediction obtained with (16.35) and the present formalism, for various stochastic processes indicated in each sub-figure. Adapted from Ref. [39]

dynamics of several models of interfaces and polymer dynamics [61]. For example $H = 1/4$ corresponds to a Rouse chain and to Edwards-Wilkinson interface, $H = 3/8$ to the Mullins-Herring interface (or semiflexible chains), other values of $H$ can be obtained for macromolecules with fractal (hyperbranched) architecture [67].

There exists a fairly general method to study persistence exponents, which consists in defining a stationary process $X(\tau) \equiv x(e^\tau)/\langle x(e^\tau)^2\rangle^{1/2}$ (where $\tau = \ln t$), and analyzing the first passage properties of $X$ within the independent interval approximation. This method applies to processes that are aging at all times, such as the random acceleration process [68, 69] or systems in which the dynamics occurs at zero temperature [64–66, 70–75]. However, it appears that this method is not applicable to transiently aging processes such as those described by the covariance (16.36). Technically speaking, the reason of this failure is that the obtained process $X(\tau)$ is not smooth, so that the independent interval approximation cannot be applied to analyze its first passage properties. The only available methods to study persistence exponents are perturbative methods (around Markovian processes) and the derivation of upper and lower bounds [61].

Here, we may establish a link between the long time behavior of trajectories after a first passage event and the persistence exponents. Indeed, assuming that $\theta < 1 - dH$, we may use Eq. (16.34) (which is still valid here) and see that it can be satisfied only if

$$p_\pi(\boldsymbol{0}, t) - p(\boldsymbol{0}, t) \underset{t\to\infty}{\sim} \frac{1}{t^{dH+\theta}}. \tag{16.37}$$

Here, we focus on processes with zero mean (in practice, this is realized by assuming that $x(t = 0) = 0$, and considering first-passage events only for $t > \Delta$ for a fixed $\Delta > 0$). Hence, we assume that the trajectories after a first passage event are Gaussian, with zero mean, and covariance $\sigma_\pi(t, t')$ (for each spatial coordinate). In this case, $p_\pi(\boldsymbol{0}, t) \simeq 1/[2\pi\sigma_\pi(t, t)]^{d/2}$, so that Eq. (16.37) can be satisfied only if

$$\sigma_\pi(t, t) - \sigma(t, t) \underset{t\to\infty}{\sim} t^{2H-\theta}. \tag{16.38}$$

This means that *the calculation of the exponent $\theta$ amounts to that of the covariance $\sigma_\pi(t, t')$ of the trajectories in the late future after a first-passage event.*

A procedure to find the persistence exponent then consists in writing a self-consistent equation for $\sigma_\pi$ (as was done to determine the average trajectories $\mu(t)$), followed by an analysis of this equation at long times to identify $\theta$ via Eq. (16.38). This procedure is done in Ref. [37] and leads to a linear integral equation of the form

$$\int_0^1 K_\theta(u, v)\left[z_\theta(u) - z_\theta(1)\left(1 - \frac{(2H-\theta)(1-u)}{2}\right)\right] du = f_\theta(v), \tag{16.39}$$

where $\sigma_\pi(t, t') - \sigma(t, t') \simeq t^{2H-\theta} z(t/t')$ at long times, and $K_\theta$, $f_\theta$ can be calculated in terms of $\sigma$. Finding the value of $\theta$ for which $z$ does not diverge then leads to the prediction of the value of $\theta$. This leads to the results presented in Fig. 16.6. There, one

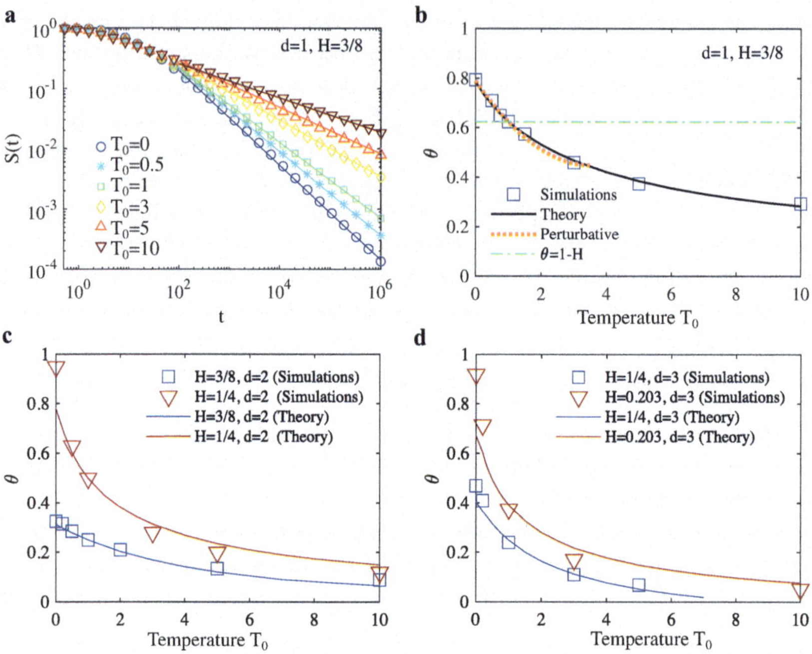


**Fig. 16.6** Persistence for "quenched fBM", whose covariance follows Eq. (16.36). **a** Example of survival probabilities for interface dynamics with $H = 3/8$ at different temperatures. The slopes of the continuous lines is the value of $\theta$ predicted by our approach. **b**, **c**, **d**: Systematic comparison of $\theta$ as measured in simulations versus theoretical values for different $T_0$ and (**b**) $d = 1$, (**c**) $d = 2$ and (**d**) $d = 3$ spatial dimensions. In (**b**), the perturbation results are Eq. (16.40). Adapted from Ref. [37]

clearly sees that $\theta$ depends on the temperature $T_0$. It is found that the theory captures quantitatively the dependence of the persistence exponents on the temperature quench for all these models. Of note, perturbative analysis of the problem for small values of $\varepsilon = H - 1/2$ can be conducted and yields

$$\theta = 1 - H - 2(\sqrt{2} - 1)(1 - T_0)\varepsilon + \{a_1(1 - T_0)[a_2 + (1 - T_0)]\}\varepsilon^2 + \mathcal{O}\left(\varepsilon^3\right), \tag{16.40}$$

where analytical expressions of $a_1$ and $a_2$ can be obtained, with numerical estimates $a_1 \simeq 1.77$, $a_2 \simeq 1.28$. Interestingly, in the particular cases $T_0 = 0$ and $T_0 = 1$, the first order terms coincide with the exact first order solution of Ref. [61], which points towards the exactness of this approach at this order.

The dependence on the temperature quench shows that the exponents that quantify the kinetics of first passage to a target are significantly modified if the system is prepared with non-stationary initial conditions. Of note, this result holds in space dimensions $d \geq 1$, and in particular for $d = 3$. In the context of chemical reactions, this means that the kinetics of reaction involving complex macromolecules could be significantly modified by a change of initial conditions, obtained either by a temperature quench or by imposing a constraint, such as a geometric confinement or an external field, that would be relaxed at $t = 0$. The above method then allows to quantify the kinetics of such reactions.

## 16.4 Conclusion

As a summary, this chapter has reviewed an approach that allows for a quantitative determination of the first passage properties of non-Markovian random walkers to target sites in various geometries and space dimensions. The originality of the approach is that it relies on the analysis of the trajectories of the random walker after the first passage event, as documented in references [37–40]; this has allowed in recent years for new developments in paradigmatic examples of target search problems. This approach was first applied to prototypical problems of target search within confined spaces; it was next extended to the case of rare events kinetics, and finally to the case of infinite geometries, with the refined determination of the survival probability and persistence of processes featuring stationary increments or transient aging. A key point to assess the importance of memory effects lies in the fact that for non-Markovian dynamics, trajectories after the first passage event differ from those originating from an equilibrium state. This discrepancy comes from the fact that the degrees of freedom of the bath are not in an equilibrium state at the instant of the first passage, as reported in specific examples in Refs. [47, 67, 76–79]. The approach that we have presented shows that one does not need to describe explicitly the distribution of the degrees of freedom of the bath; the position of the random walker is in principle sufficient. Interestingly, this approach agrees with the available exact results obtained perturbatively for examples of weakly non-Markovian processes. Of note, similar arguments were used recently to predict the shape of the distribution of first passage times [80]. Finally, these series of studies reveal that the characterization of the non-equilibrium state of the system at the instant of first passage is key to derive first-passage kinetics, and they provide a new methodology, via the analysis of trajectories after the first-passage, to make it quantitative.

## References

1. S. Redner, *A Guide to First- Passage Processes* (Cambridge University Press, Cambridge, UK, 2001)
2. D. ben Avraham, S. Havlin, *Diffusion and Reactions in Fractals and Disordered Systems* (Cambridge University Press, Cambridge, UK, 2000)
3. O.G. Berg, P.H. von Hippel, Diffusion-controlled macromolecular interactions. Annu. Rev. Biophys. Biophys. Chem. **14**, 131–160 (1985)

4. R. Chicheportiche, J.-P. Bouchaud, Some applications of first-passage ideas to finance, in *First-Passage Phenomena and Their Applications* (World Scientific, 2014), pp. 447–476
5. V.V. Palyulin, T. Ala-Nissila, R. Metzler, Polymer translocation: the first two decades and the recent diversification. Soft Matt. **10**, 9016–9037 (2014)
6. A.-T. Dinh, T. Theofanous, S. Mitragotri, A model for intracellular trafficking of adenoviral vectors. Biophys. J. **89**, 1574–88 (2005)
7. M. Coppey, O. Bénichou, R. Voituriez, M. Moreau, Kinetics of target site localization of a protein on DNA: a stochastic approach. Biophys. J. **87**, 1640–1649 (2004)
8. R. Metzler, S. Redner, G. Oshanin, *First-Passage Phenomena and Their Applications* (World Scientific, 2014)
9. A.J. Bray, S.N. Majumdar, G. Schehr, Persistence and first-passage properties in nonequilibrium systems. Adv. Phys. **62**, 225–361 (2013)
10. O. Bénichou, R. Voituriez, From first-passage times of random walks in confinement to geometry-controlled kinetics. Phys. Rep. **539**, 225–284 (2014)
11. N. Van Kampen, *Stochastic Processes in Physics and Chemistry*, 1st edn. (Elsevier Science, Amsterdam, 1992)
12. C. Gardiner, *Handbook of Stochastic Methods for Physics, Chemistry and the Natural Sciences*, 2nd edn. (Springer, Belin, 1985)
13. D. Panja, Anomalous polymer dynamics is non-Markovian: memory effects and the generalized Langevin equation formulation. J. Stat. Mech. Theor. Exp. **2010**, P06011 (2010)
14. D. Panja, Generalized Langevin equation formulation for anomalous polymer dynamics. J. Stat. Mech. Theor. Exp. **2010**, L02001 (2010)
15. J.T. Bullerjahn, S. Sturm, L. Wolff, K. Kroy, Monomer dynamics of a wormlike chain. Europhys. Lett. **96**, 48005 (2011)
16. D. Bicout, T. Burkhardt, Absorption of a randomly accelerated particle: gambler's ruin in a different game. J. Phys. A Math. Gen. **33**, 6835–6841 (2000)
17. J. Masoliver, K. Lindenberg, B.J. West, First-passage times for non-Markovian processes. Phys. Rev. A **33**, 2177 (1986)
18. P. Hänggi, P. Talkner, First-passage time problems for non-Markovian processes. Phys. Rev. A **32**, 1934 (1985)
19. J. Masoliver, K. Lindenberg, B.J. West, First-passage times for non-Markovian processes: Correlated impacts on bound processes. Phys. Rev. A **34**, 2351 (1986)
20. J. Masoliver, K. Lindenberg, B.J. West, First-passage times for non-Markovian processes: Correlated impacts on a free process. Phys. Rev. A **34**, 1481 (1986)
21. G. Wilemski, M. Fixman, Diffusion-controlled intrachain reactions of polymers. 1. Theory. J. Chem. Phys. **60**, 866–877 (1974)
22. G. Wilemski, M. Fixman, Diffusion-controlled intrachain reactions of polymers. 2. Results for a pair of terminal reactive groups. J. Chem. Phys. **60**, 878–890 (1974)
23. I.M. Sokolov, Cyclization of a polymer: first-passage problem for a non-Markovian process. Phys. Rev. Lett. **90**, 080601 (2003)
24. J. Kappler, F. Noé, R.R. Netz, Cyclization and relaxation dynamics of finite-length collapsed self-avoiding polymers. Phys. Rev. Lett. **122**, 067801 (2019)
25. A. Dua, B. Cherayil, The dynamics of chain closure in semiflexible polymers. J. Chem. Phys. **116**, 399–409 (2002)
26. D. Campos, V. Méndez, Two-point approximation to the Kramers problem with coloured noise. J. Chem. Phys. **136**, 074506 (2012)
27. C. Hyeon, D. Thirumalai, Kinetics of interior loop formation in semiflexible chains. J. Chem. Phys. **124**, 104905 (2006)
28. P. Debnath, B.J. Cherayil, Dynamics of chain closure: approximate treatment of nonlocal interactions. J. Chem. Phys. **120**, 2482–9 (2004)
29. L.P. Sanders, T. Ambjörnsson, First passage times for a tracer particle in single file diffusion and fractional Brownian motion. J. Chem. Phys. **136**, 175103 (2012)
30. K.J. Wiese, S.N. Majumdar, A. Rosso, Perturbation theory for fractional Brownian motion in presence of absorbing boundaries. Phys. Rev. E **83**, 061141 (2011)

31. M. Delorme, K.J. Wiese, Maximum of a fractional Brownian motion: analytic results from perturbation theory. Phys. Rev. Lett. **115**, 210601 (2015)
32. M. Delorme, K.J. Wiese, Perturbative expansion for the maximum of fractional Brownian motion. Phys. Rev. E **94**, 012134 (2016)
33. M. Delorme, A. Rosso, K.J. Wiese, Pickands' constant at first order in an expansion around Brownian motion. J. Phys. A Math. Theor. **50**, 16LT04 (2017)
34. T. Sadhu, M. Delorme, K.J. Wiese, Generalized arcsine laws for fractional Brownian motion. Phys. Rev. Lett. **120**, 040603 (2018)
35. K.J. Wiese, First passage in an interval for fractional Brownian motion. Phys. Rev. E **99**, 032106 (2019)
36. M. Arutkin, B. Walter, K.J. Wiese, Extreme events for fractional Brownian motion with drift: theory and numerical validation. Phys. Rev. E **102**, 022102 (2020)
37. N. Levernier, T.V. Mendes, O. Bénichou, R. Voituriez, T. Guérin, Everlasting impact of initial perturbations on first-passage times of non-Markovian random walks. Nat. Commun. **13**, 5319 (2022)
38. N. Levernier, O. Bénichou, R. Voituriez, T. Guérin, Kinetics of rare events for non-Markovian stationary processes and application to polymer dynamics. Phys. Rev. Res. **2**, 012057 (2020)
39. N. Levernier, M. Dolgushev, O. Bénichou, R. Voituriez, T. Guérin, Survival probability of stochastic processes beyond persistence exponents. Nat. Commun. **10**, 1–7 (2019)
40. T. Guérin, N. Levernier, O. Bénichou, R. Voituriez, Mean first-passage times of non-Markovian random walkers in confinement. Nature **534**, 356–359 (2016)
41. B.D. Hughes, *Random Walks and Random Environments* (Oxford Science publications, Oxford, 1995)
42. B. Meyer, C. Chevalier, R. Voituriez, O. Bénichou, Universality classes of first-passage-time distribution in confined media. Phys. Rev. E **83**, 051116 (2011)
43. N. Levernier, O. Bénichou, T. Guérin, R. Voituriez, Universal first-passage statistics in aging media. Phys. Rev. E **98**, 022125 (2018)
44. A.E. Likthman, C.M. Marques, First-passage problem for the Rouse polymer chain: an exact solution. Europhys. Lett. **75**, 971–977 (2006)
45. S. Condamin, O. Bénichou, V. Tejedor, R. Voituriez, J. Klafter, First-passage times in complex scale-invariant media. Nature **450**, 77–80 (2007)
46. S. Condamin, V. Tejedor, R. Voituriez, O. Bénichou, J. Klafter, Probing microscopic origins of confined subdiffusion by first-passage observables. Proc. Natl. Acad. Sci. USA **105**, 5675–5680 (2008)
47. T. Guérin, O. Bénichou, R. Voituriez, Non-Markovian polymer reaction kinetics. Nat. Chem. **4**, 568–573 (2012)
48. T. Mason, K. Ganesan, J. Van Zanten, D. Wirtz, S. Kuo, Particle tracking microrheology of complex fluids. Phys. Rev. Lett. **79**, 3282 (1997)
49. T.G. Mason, D. Weitz, Optical measurements of frequency-dependent linear viscoelastic moduli of complex fluids. Phys. Rev. Lett. **74**, 1250 (1995)
50. T. Turiv, I. Lazo, A. Brodin, B.I. Lev, V. Reiffenrath, V.G. Nazarenko, O.D. Lavrentovich, Effect of collective molecular reorientations on Brownian motion of colloids in nematic liquid crystal. Science **342**, 1351–1354 (2013)
51. Q.-H. Wei, C. Bechinger, P. Leiderer, Single-file diffusion of colloids in one-dimensional channels. Science **287**, 625–627 (2000)
52. M.L. Eaton, *Multivariate Statistics, A Vector Space Approach*, vol. 53 (Institute of Mathematical Statistics, Beachwood, Ohio, USA, 1983)
53. M. Grimm, S. Jeney, T. Franosch, Brownian motion in a Maxwell fluid. Soft Matt. **7**, 2076–2084 (2011)
54. J. Pickands, Asymptotic properties of the maximum in a stationary Gaussian process. Trans. Am. Math. Soc. **145**, 75–86 (1969)
55. S. Milner, T. McLeish, Parameter-free theory for stress relaxation in star polymer melts. Macromolecules **30**, 2159–2166 (1997)

56. S. Milner, T. McLeish, Reptation and contour-length fluctuations in melts of linear polymers. Phys. Rev. Lett. **81**, 725 (1998)
57. C. Jeppesen, J.Y. Wong, T.L. Kuhl, J.N. Israelachvili, N. Mullah, S. Zalipsky, C.M. Marques, Impact of polymer tether length on multiple ligand-receptor bond formation. Science **293**, 465–468 (2001)
58. J. Cao, J. Zhu, Z. Wang, A.E. Likhtman, Large deviations of Rouse polymer chain: first passage problem. J. Chem. Phys. **143**, 20 (2015)
59. Z. Schuss, *Theory and Applications of Stochastic Processes: an Analytical Approach*, vol. 170 (Springer Science & Business Media, 2009)
60. M. Ding, W. Yang, Distribution of the first return time in fractional Brownian motion and its application to the study of on-off intermittency. Phys. Rev. E **52**, 207 (1995)
61. J. Krug, H. Kallabis, S.N. Majumdar, S.J. Cornell, A.J. Bray, C. Sire, Persistence exponents for fluctuating interfaces. Phys. Rev. E **56**, 2702–2712 (1997)
62. G. Molchan, Maximum of a fractional Brownian motion: probabilities of small values. Commun. Math. Phys. **205**, 97–111 (1999)
63. A. Pal, S. Reuveni, First passage under restart. Phys. Rev. Lett. **118**, 030603 (2017)
64. M. Poplavskyi, G. Schehr, Exact persistence exponent for the 2 D-Diffusion equation and related Kac polynomials. Phys. Rev. Lett. **121**, 150601 (2018)
65. B. Derrida, V. Hakim, V. Pasquier, Exact first-passage exponents of 1D domain growth: relation to a reaction-diffusion model. Phys. Rev. Lett. **75**, 751 (1995)
66. I. Dornic, Universal Painlevé VI probability distribution in Pfaffian persistence and Gaussian first-passage problems with a sech-Kernel (2018). arXiv:1810.06957
67. M. Dolgushev, T. Guérin, A. Blumen, O. Bénichou, R. Voituriez, Contact kinetics in fractal macromolecules. Phys. Rev. Lett. **115**, 208301 (2015)
68. T. Burkhardt, Semiflexible polymer in the half-plane and statistics of the integral of a Brownian curve. J. Phys. A Math. Gen. **26**, L1157–L1162 (1993)
69. G. De Smedt, C. Godreche, J. Luck, Partial survival and inelastic collapse for a randomly accelerated particle. Europhys. Lett. **53**, 438 (2001)
70. B. Derrida, A. Bray, C. Godreche, Non-trivial exponents in the zero temperature dynamics of the 1D Ising and Potts models. J. Phys. A: Math. Gen. **27**, L357 (1994)
71. S. Majumdar, C. Sire, Survival probability of a gaussian non-Markovian process: application to the T=0 dynamics of the Ising model. Phys. Rev. Lett. **77**, 1420–1423 (1996)
72. S. Majumdar, C. Sire, A. Bray, S.J. Cornell, Nontrivial exponent for simple diffusion. Phys. Rev. Lett. **77**, 2867–2870 (1996)
73. Derrida, V. Hakim, Zeitak, Persistent spins in the linear diffusion approximation of phase ordering and zeros of stationary gaussian processes. Phys. Rev. Lett. **77**, 2871–2874 (1996)
74. A. Watson, Persistence pays off in defining history of diffusion. Science **274**, 919–920 (1996)
75. T. Newman, W. Loinaz, Critical dimensions of the diffusion equation. Phys. Rev. Lett. **86**, 2712 (2001)
76. N. Levernier, M. Dolgushev, O. Bénichou, A. Blumen, T. Guérin, R. Voituriez, Non-Markovian closure kinetics of flexible polymers with hydrodynamic interactions. J. Chem. Phys. **143**, 204108 (2015)
77. O. Bénichou, T. Guérin, R. Voituriez, Mean first-passage times in confined media: from Markovian to non-Markovian processes. J. Phys. A Math. Theor. **48**, 163001 (2015)
78. T. Guérin, O. Bénichou, R. Voituriez, Reactive conformations and non-Markovian kinetics of a Rouse polymer searching for a target in confinement. Phys. Rev. E **87**, 032601 (2013)
79. T. Guérin, O. Bénichou, R. Voituriez, Reactive conformations and non-Markovian cyclization kinetics of a Rouse polymer. J. Chem. Phys. **138**, 094908 (2013)
80. Y. Sakamoto, T. Sakaue, First passage time statistics of non-Markovian random walker: Dynamical response approach. Phys. Rev. Res. **5**, 043148 (2023)

# Chapter 17
# Non-Markovian and Collective Search Strategies

**Hugues Meyer and Heiko Rieger**

**Abstract** Agents searching for a target can improve their efficiency by memorizing where they have already been searching or by cooperating with other searchers and using strategies that benefit from collective effects. This chapter reviews such concepts: non-Markovian and collective search strategies. We start with the first-passage properties of continuous non-Markovian processes and then proceed to the discrete random walker with 1-step and n-step memory. Next, we discuss the auto-chemotactic walker, a random walker that produces a diffusive chemotactic cue which the walker tries to avoid. Then ensembles of agents searching for a single target are discussed, whence the search efficiency may also comprise metabolic costs in addition to the first-passage time. We consider the first-passage properties of ensembles of chemotactic random walkers and then the pursuit problem, in which searchers (or hunters / predators) see the mobile target over a certain distance. Evasion strategies of single or many targets are also elucidated. Finally, we review collective foraging strategies comprising many searchers and many immobile targets. We finish with an outlook on future research directions comprising yet unexplored search strategies of immune cells and swarm robotics.

## 17.1 Introduction

Understanding search processes, either solitary or collectively, is important in many disciplines like biology, physics computer science, search and rescue, and robotics. A review about animal foraging strategies can be found in the books by Sumpter [70]

H. Meyer · H. Rieger (✉)
Theoretical Physics and Center for Biophysics, Saarland University, Campus E2 6, 66123 Saarbrücken, Germany
e-mail: Heiko.Rieger@uni-saarland.de

H. Meyer
e-mail: Hugues.Meyer@uni-saarland.de

H. Rieger
Leibniz Institute for New Materials INM, Campus D1 1, 66123 Saarbrücken, Germany

 
D. Grebenkov et al. (eds.), *Target Search Problems*,
https://doi.org/10.1007/978-3-031-67802-8_17

or Viswanathan et al. [80]. In this chapter, we will focus on the question: How can a single searcher use its memory to optimize the search process and how can multiple searchers act collectively, by communication or interaction, to optimize the search process?

## 17.2 Non-Markovian Search Strategies

Most dynamical processes are not Markovian: their further evolution depends on their past history and not only their instantaneous state. More precisely, the time-evolution of a small set of observables of a system depends a priori on the values they have previously taken. Of course, any system whose microscopic dynamics is Markovian can be considered as Markovian by simply using the microscopic degrees of freedom as variables of interest. Less extremely, theoretical methods grouped under the name of Markovian embedding procedures have been developed to find sets of observables that are guaranteed to be Markovian. However, the observables derived in this way are not necessarily relevant variables in the context of search processes, where one is mostly interested in the time-evolution of the positions of searchers, which can hence be described as non-Markovian variables.

### *17.2.1 Non-Markovian Random Walk Models*

Many natural systems are equipped with memory and use it to optimize their search efficiency. It can take various forms, either by explicitly remembering the past locations or by modifying the environment to leave clues of a previous passage. Self-interacting random walks, e.g., on a lattice can be defined by nearest neighbour jump rates that depend on a weight function $w_i(n)$ that depends on the number of times $n$ that the walker has visited the site $i$ before. Introducing a 'potential' $V(n)$ by $w(n) = \exp(-V(n))$ one can interpret the process as representing a walker depositing a signal that, in turn, modifies the local energy landscape $V$ experienced by the walker. In contrast to the auto-chemotactic walker—involving a diffusing chemical clue—discussed below the deposited signal is here assumed to be static and permanent. $V(n)$ could be positive (attraction) or negative (repulsion), and $V(n) = -\beta\theta(n)$ with $\beta \to \infty$ corresponds to the self-avoiding random walk [3, 43]. Such self-interacting or reinforced random walks have been considered in [3, 10, 15, 17, 18, 34, 35, 53, 55, 58, 59, 64, 75]. Another prominent example of non-Markovian random walks is the elephant walk [65].

First-passage time properties of reinforced random walks have been considered in [5] and for non-Markovian polymer reaction kinetics in [23]. A general formalism to analyze first-passage time properties of non-Markovian processes based only on the properties of the time-dependent mean square displacement has been presented in [24]. In this chapter, we are mainly interested in search strategies, which are specific choices of the parameters of the random walk, and their implications on the mean first-passage times (MFPT)—with the ultimate goal to optimize them. This has explicitly been done for random walks with $n$-step memory and the auto-chemotactic walk.

### *17.2.2 Lattice Random Walk with 1-Step Memory*

Many discrete-time models of random walkers are defined such that particles account for their last previous step to determine their next move. Before even mentioning first-passage properties, we should first emphasize that the dynamics of such systems are already significantly impacted by the 1-step memory. In Ref. [63], Sadjadi et al. studied a two-dimensional model where a particle chooses at each time step either to continue in the same direction with a certain probability or to take a turn with an angle drawn from a probability distribution $R(\theta)$. The diffusive properties of such a particle were analytically calculated and shown to depend in a non-trivial way on $R(\theta)$. Most importantly, the sign of the quantity $\mathcal{R} = \int_{-\pi}^{\pi} d\theta \cos\theta R(\theta)$ was shown to determine whether the dynamics is super-diffusive ($\mathcal{R} > 0$) or sub-diffusive ($\mathcal{R} < 0$).

As the dynamical properties of searchers greatly influence their first-passage time properties, it is reasonable to expect short-term memory to significantly impact search efficiency. In Ref. [74], Tejedor et al. have shown that searchers with short-term memory can already minimize the typical time to find a randomly located target by a substantial amount. The model consists of a random walker evolving on a $d$-dimensional cubic lattice of length $L$ whose probability to continue in the same direction as the one of the previous step is favoured to all other directions. Noting $\epsilon$ the difference between the probabilities of the favoured direction and all other directions, it is shown that the mean first-passage time to reach a randomly located target in a system with periodic boundary conditions can be calculated analytically as a function of $\epsilon$. Most interestingly, this function is shown to admit a minimum that depends on the system size. In particular, in the limit of large systems, it is shown that the optimal MFPT scales proportionally to $V = L^d$ while it scales as $V \ln V$ for a blind random walk.

Remarkably, optimal search strategies of agents that are trained by machine learning turn out to resemble closely a persistent random walk: In [51] the searcher was a lattice random walker equipped with a counter that memorized the number of consecutive straight steps that the searcher performed before the current time. Reinforced learning was applied to learn the optimal probability to turn left or right conditioned on the number of consecutive straight steps $n$ before, with the result being a sigmoidal function of $n$ jumping from zero to non-zero at a specific $n_{\mathrm{opt}}$. This search strategy is essentially equivalent to a persistent random walk with an optimal persistence length. Similarly, in [30], the searcher was an active Brownian particle that can switch between mobility modes defined by zero or non-zero activity, corresponding to zero or non-zero Péclet number and hence between zero and non-zero persistence. Reinforced learning techniques were applied to learn the optimal time intervals to switch between the two mobility modes. It turned out that the MFPT to find a randomly located target is minimized by a specific switch rate giving rise to a specific effective (large scale) persistence length.

### 17.2.3 Lattice Random Walk with n-Step Memory

As memory becomes longer, one can expect a searcher to be more and more efficient to reach a target as it can in principle avoid more easily to return to previously visited locations. In any system there is however a strict lower bound on the mean first-passage time to hit a randomly located target. In a lattice setting, for example, the most efficient search strategy for a single walker consists in scanning each lattice site one by one without returning twice on any site. In such a case, the mean first-passage time is given by $T_\infty = V/2$ where $V$ is the total number of sites. One can therefore wonder about the gain in efficiency using *long-term* memory and ask about the number of steps in memory required to reach an MFPT as close as possible to the strict lower bound $T_\infty$.

This question was raised and answered in Ref. [46] by Meyer and Rieger, who formalized it in the following way. Consider a walker on a $d$-dimensional lattice that decides where to jump at the next step based on the knowledge of its $n$ previous steps noted $\mathbf{e}_{i_0}, \ldots, \mathbf{e}_{\mathbf{i_{n-1}}}$. Formally, it jumps to position $r + \mathbf{e}_k$ from position $\mathbf{r}$ with probability $p(\mathbf{e}_k|\mathbf{e}_{i_0}, \ldots, \mathbf{e}_{\mathbf{i_{n-1}}})$. Using a backward master equation, it is possible to show that the mean first-passage time to reach a randomly located target can be calculated exactly from the conditional probabilities $p(\mathbf{e}_k|\mathbf{e}_{i_0}, \ldots, \mathbf{e}_{\mathbf{i_{n-1}}})$, provided periodic boundary conditions. This implies that the mean first-passage time can be minimized by optimizing the values of these probabilities. Such an optimization problem can quickly become computationally expensive as the number of variables scales as $z^n$, where $z$ is the coordination number of the lattice (e.g., $z = 4$ for a square 2-dimensional lattice), and the critical step in computing the mean first-passage time consists in inverting a matrix of size $z^n \times z^n$. Hence, as $n$ gets larger, not only the number of variables grow exponentially, but also the computing time to evaluate the objective function. Since, the cases $n = 1, 2, 3$ were numerically optimized and show that for $n > 1$, the optimal strategies consist in the concatenation of blocks of steps of length larger than $n$ which can be constructed from the knowledge of the last $n$ steps. In particular, these blocks are not mirror-symmetric, indicating that the left–right symmetry should be broken in order to be efficient in the search. Finally, the optimal strategies lead to a mean first-passage time equal to $V$ for $n = 1, 3V/4$ for $n = 2$, and $2V/3$ for $n = 3$, as $V \to \infty$, suggesting the conjecture that the optimal MFPT for an arbitrary value of $n$ scales as $\left(1 + \frac{1}{n}\right)\frac{V}{2}$.

### 17.2.4 Auto-chemotactic Walker

A natural realization of systems with explicit environmental memory is observed in organisms leaving chemical trails along their path. While the deposition of pheromones by ants is often mentioned as the typical example of such a phenomenon, trail-leaving systems are observed in many other contexts and at various scales. At the micron scale, some immune cells as well as a few amoeba species are known to perform auto-chemotaxis, where the cue is produced and emitted by the organisms themselves. While the theoretical description of auto-chemotaxis in the continuum for a large number of particles has been introduced in the 1970s via the Keller–Segels equations [31], more recent articles have investigated particle-based models in order to study the dynamics and collective behaviours of trail-leaving systems on the level of individual particles.

The overall dynamics of a trail-leaving particle highly depends on the behaviour of the trail itself. In Ref. [34], Kranz et al. considered a model of an active Brownian particle leaving a finite-width trail which neither diffuses nor decays, which can be considered as an infinite memory timescale. In addition, the trail exerts a repelling force on the particle itself by favouring its director vector to be perpendicular to the trail. The effective dynamics of the particle, namely its effective rotational and translational diffusion constants, is then mainly determined by the quantity $\Omega\tau = k\chi/(\pi R v_0^2)$, where $k$ is the deposition rate of the trail, $\chi$ is the sensitivity of the particle to the trail, and $v_0$ is the active speed of the particle. In particular, a phase transition is observed at $\Omega\tau > 2$, where the rotational diffusion constant diverges and the system transitions from a diffusive behaviour to a state where the particle performs very confined loops and is practically trapped by its own trail in a small region of space. This regime is of course unfavourable for a search process as space exploration is considerably hindered.

If the chemical cue starts diffusing and/or decaying, one enters the realm of auto-chemotactic particles. A time- and space-continuous model of such objects was introduced in [72]. There, the particle is again modelled as a point-like active Brownian particle whose velocity unit vector experiences a torque generated by the chemotactic field, i.e., the gradient of the concentration field. In the weak chemotactic coupling limit, the effective dynamics of the walker is simply diffusive and evolves according to a Langevin equation parameterized by an effective diffusion constant. For a strong attractive coupling (i.e., for a particle attracted by the trail), circular trajectories are also observed, but not in the case of chemo-repulsion. A similar dependence of the dynamics of an auto-chemo-repulsive particle was reported in Ref. [46], where a simpler lattice model was introduced, c.f. Fig. 17.1a. There a walker jumps from a lattice site to a neighbouring one with a probability favouring sites with low values of the concentration field defined on the lattice. Then it adds a fixed amount to this field at the location it has jumped to. The walker is shown to have a persistence length highly dependent on the diffusion constant of the concentration field, which effectively acts as a pushing and aligning force on the walker (see Fig. 17.1a for a snapshot of the emerging concentration). It turns out that the MFPT has a pronounced minimum for

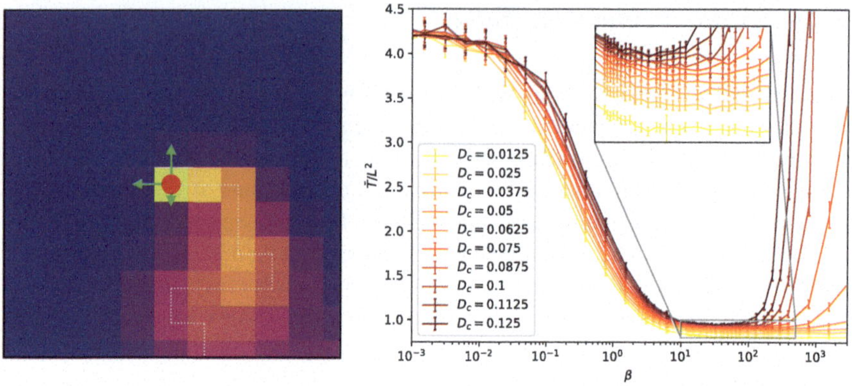


**Fig. 17.1** Snapshot of an auto-chemotactic walk as defined in [46]. Left panel: The red circle indicates the position of the walker while the dotted line tracks its path and the colour gets darker for sites visited further in the past. The background colour codes for the concentration field where darker colour represents lower values, which determines the probability to jump to one of the neighbouring sites at the next step as indicated by the size of the green arrows. Right panel: Mean first-passage time $\overline{T}$ of a single auto-chemotactic walker to reach a randomly located target, normalized by the system size, as a function of the coupling constant $\beta$ between the walker and the chemotactic field for various values of the diffusion constant $D_c$ of the chemotactic field. From [46]

a specific coupling constant between the walker and the chemotactic field as shown in Fig. 17.1b. In [46], it was also shown that the main effect of increasing $\beta$ was to increase the persistence length of the auto-chemotactic walker, implying that the optimal coupling constant corresponds to an optimal persistence length.

### 17.2.5 Miscellaneous

Other non-Markovian searches have been studied in recent literature, often motivated by experimental observation of biological systems. In the realm of systems that remember previously visited locations, it is, e.g., observed in some animal species, such as bats, monkeys, or even humans, that the probability of visiting new sites depends on the total number of sites already visited, and the probability to choose a specific site among all visited sites depends on the number of previous visits on that site [11, 78]. It is shown that this dependence can be fine-tuned for the search to allow frequent returns to preferred sites while enabling an extensive scan of all sites of the domain, which leads to a good balance between energy consumption and risk management. On a smaller scale, similar conditional behaviours leading to long-term memory effects are also observed and modelled. In [28], the hopping of two species of molecular motors along a couple of microtubules is considered. Each species can travel along a single direction, and each motor can either bind or unbind to a filament, or hop along it, and the probability of binding to a filament changes whether the last motors that have been on that portion of the filament were of the same species or not. Transport properties and in particular the efficiency of bidirectional transport are shown to be greatly influenced by such a conditional binding probability.

More explicit non-Markovian searches have also been investigated, under the form of Generalized Langevin Equations (GLE) where memory is directly encoded in the equation of motion of the searcher via a memory kernel. While it is known that the overall dynamics, in particular first-passage properties, are highly sensitive to the form of the memory kernel of the GLE [25], it was shown numerically that exponentially decaying kernels are optimally efficient in the context of composite searches, i.e., searches which alternate between intensive and extensive searches, for spatially correlated targets [33].

Finally, searches with environmental memory like the auto-chemotactic walk were introduced in recent years. In particular, the Sokoban walk [9] considers a walker on a lattice where obstacles are placed. The walker can push the obstacle with a certain probability, provided that there is no other obstacle behind it. An experimental realization of a similar system using bristle robots and light movable obstacles was presented in [2] and showed that the effective memory created via the trails of the robot through the obstacles allows for a more efficient search for randomly located targets.

## 17.3 Collective Search Strategies

Multiple searchers will generally find a single target faster than a single one, see [52, 61], and the process can be collectively optimized by utilizing interactions or communications between them. The standard, against which any optimization by collectivism should be measured, is the FPT (to reach a single target) distribution of a number $N$ of non-interacting random searchers, which is given by the distribution of the minimum of FPTs of $N$ individual searchers. Naively, one can expect that the MFPT of $N$ searchers is proportional to $1/N$ times the MFPT of a single searcher [52, 61], but the ensemble-averaged realization-dependent first-passage time exhibits quite a non-trivial and sometimes counter-intuitive behaviour and was shown to actually decay much slower as $1/\log N$ for independently diffusing walkers [21, 22, 38, 39, 81].

A simple strategy for $N$ random searchers to improve their search efficiency for single or many targets would be to avoid clustering and to achieve a homogeneous distribution over the search area. A theoretical model for random walkers interacting via long-range repulsion (e.g., Coulomb potential) has been proposed in [73] and it has been found that the search efficiency can indeed be substantially increased by this interaction: the MFPT shows a pronounced minimum for a specific interaction strength, analogous to the charge in the Coulomb interactions. The observation that the MFPT depends non-monotonously on the interaction strength is plausible since a too strong repulsion decreases the effective diffusion constant of each searcher.

### 17.3.1 Chemotactic Searchers

A similar strategy underlies random searchers with chemotactically repulsive interactions [47]: each searcher produces a chemotactic clue that diffuses into the environment and tells other searchers to turn away. This behaviour is modelled by incorporating the gradient of the concentration field of the chemotactic clue into the equations of motion of each searcher: on a lattice by a concentration (difference) dependence of the transition probabilities [47] and in continuous space by a concentration (gradient) torque in the equation of motion for the angle of, e.g., active Brownian particles (Wysocki and Rieger, unpublished). It has been found that in both cases the MFPT can be substantially reduced by a specific choice of the strength of the interaction between the particle and the chemotactic clue. It should noted that the chemotactic clue can already improve the search efficiency of a single searcher since the self-produced concentration field leads to a preference to move in the same direction, away from the clue, and thus increases effectively the persistence length of the searcher—which has already the potential to improve and optimize the search efficiency [74]. For multiple searchers, it has been shown that in addition to the single particle effect the MFPT can be decreased further, a manifestation of a collective effect.

In the aforementioned example, the search efficiency measured by the MFPT alone is a monotonous function of the number of searchers simply decreasing with $N$. This would imply that it would always be better to involve as many searchers as possible to have the most efficient search. From a searchers perspective, this would be unrealistic since the cost of a search is not given by the search time alone but usually also contains a metabolic contribution: each searcher consumes energy to move. Incorporating this metabolic cost into the measure for search efficiency leads to a well-defined optimal number of searchers, $N_{opt}$ (Meyer and Rieger, to be published). However, it should be noted that for large enough diffusion constants of the concentration field, auto-chemorepulsive particles can self-organize and form macroscopic bands travelling at a constant speed as the particle density increases, due to the effective alignment interaction between particles mediated by the concentration field. Such structures are deleterious for search processes as particles are confined in a small region of space and thus do not scan space efficiently. A detailed analysis of this phenomenon is presented in [47], where phase diagrams are reported.

Note that auto-chemoattraction is also relevant for collective search processes. In fact, it can be shown to be an optimal collective search strategy when accounting for various contributions to the search efficiency [60]. Consider a set of $N$ particles evolving according to Brownian dynamics driven by the gradient of a potential $Z$ of the form $Z(x_1, \ldots, x_N) = \zeta(x_1) \ldots \zeta(x_N)$ where $x_i$ is the position of particle $i$. A cost function defined as the weighted contributions of the elapsed time, the kinetic energy, and the collisions between particles until all particles have found a target is minimized if and only if the function $\zeta$ and the particle density field satisfy the Keller–Segel equations for auto-chemoattraction [31]. This is particularly relevant in the context of scouting, occurring when a first particle has found the target, stops moving, and recruits other particles guided by the cue it produces.

### 17.3.2 *Chemotactic Searcher–Target Attraction*

We have discussed so far how a repulsive (auto)chemotactic searcher–searcher interaction can improve search efficiency. In many search problems, it is the target that emits a signal or chemical clue received by the searcher, mostly attractive like bacteria searched by macrophages or rabbits hunted by dogs, or repulsive as an evasion strategy of the target. Typically the searcher, or predator, moves along the steepest gradient of the chemical concentration to ultimately find its ejection source. Likewise the prey (or the target, for example, another microbe) 'smells' a secreted chemical from the advancing predator and tries to escape by moving along in the opposite direction of its maximal gradient. For such a situation, a discrete chemotactic predator–prey model was proposed in [67], in which the prey secrets a diffusing chemical which is sensed by the predator and vice versa. Two dynamical states corresponding to catching and escaping were identified and it is shown that steady hunting is unstable. A subsequent work analyzed the efficiency of such a chemotactic pursuit and compared blind search with temporal and spatial gradient sensing [45]. Although strategies based on the temporal or spatial sensing of chemotactic gradients are significantly more efficient than unguided migration, a "blind search" turned out to work surprisingly well, in particular if the searcher is fast and directionally persistent. This result underpins the importance of the persistent random walk with an optimally chosen persistence length (c.f. Sect. 17.2.2) as an efficient search strategy in comparison with chemotactic gradient search.

More complex searcher–target interactions emerge for instance in the immune systems, where, e.g., T cell migration is essential for T cell responses and an efficient immune response to virus-infected or tumour cells. It allows for the detection of cognate antigen at the surface of antigen-presenting cells and for interactions with other cells involved in the immune response. Although appearing random, growing evidence suggests that T cell motility patterns are strategic and governed by mechanisms that are optimized for both the activation stage of the cell and for environment-specific cues [36].

Another example is the neutrophil recruitment from blood to extravascular sites of infectious tissue damage, which is a hallmark of early innate immune responses. Once outside the vessel, individual neutrophils often show extremely coordinated chemotaxis and cluster formation reminiscent of the swarming behaviour of insects. Such a coordinated chemotaxis over several millimetres cannot be achieved by diffusion alone: with a diffusion constant of $D \approx 10^{-10}\,\mathrm{m}^2/\mathrm{s}$ for small signalling molecules diffusive signalling would take hours which is incompatible with the observed time scales in neutrophil swarming. In [37] it was revealed how local events are propagated over long-range distances, and how auto-signalling produces coordinated, self-organized neutrophil-swarming behaviour that isolates the wound or infectious site from surrounding viable tissue. These local events constitute a so-called signalling relay: the first neutrophil searcher reaching the target emits a signal that diffuses into the environment. When other neutrophils close by detect this signal they also start to produce the signal and so forth (c.f. Fig. 1 in the Supplementary Information of [37]). Similar swarming behaviour has been observed in T cells [19]. The dynamics of such diffusive signalling relays, which are also relevant in developmental biology and microbial consortia, has been analyzed in [16] and it was shown that it gives rise to signalling waves propagating with a density and diffusion constant dependent velocity $v$. With a grain of salt, this propagation velocity can be interpreted as a measure of the efficiency of the observed collective search strategy.

It should be noted that a vast literature exists on chemotaxis among a single group of individuals (i.e., without a searcher–target scenario), leading to agglomeration and pattern formation in the attractive case, mostly based on the Patlak–Keller–Segel model, for which we refer to [56] for a review.

### *17.3.3 The Pursuit Problem: Collective Chase*

In the pursuit problem, the searcher (or hunter / predator) can 'see' the target (or prey), which either moves deterministically or stochastically. In an analytically solvable model for $N$ non-interacting predators hunting a single randomly moving prey it was assumed that the predators always move in the direction towards the prey, which is assumed to be always visible [6]. This is neither the most efficient pursuit strategy for the predator nor the most efficient escape strategy for the prey, but this simple model allowed for the analytical computation of the mean capture time. It turns out that depending on the number of chasers, the mean capture time as a function of the prey's diffusion coefficient can be monotonically increasing, decreasing, or attaining a minimum at a finite value [6]. From the hunter's perspective, possibly more relevant than the mean capture is the energy spent by the hounds to capture the prey (c.f. chemotactic search above). Then an optimal speed and number of chasers exist and depend on each chaser's baseline power consumption [6]. The extension of the model to the case in which the stochastic prey sees its hunters was discussed in [69]. The evasion strategy of the target consisted in a deterministic contribution to the stochastic equation of motion, which averages over repulsive forces exerted by the hunter, each weighted by an exponential decay with distance from the hunter (Fig. 17.2).

To analyze efficient group chase strategies, interactions between predators have to be considered (analogously for escape strategies of groups of prey, to be discussed in the next section), which results in a many-body problem that in most cases can only be tackled numerically, see for instance [4, 26, 27, 29, 42, 62, 71, 77]. An active matter approach [68] was taken in [4], where two groups of self-propelled particles were considered, each with intra-group alignment as in the Vicsek model for flocking [76] and repulsion (to avoid collision). Chase and escape were described by inter-group interactions, attraction (for predators) or repulsion (for preys) from the nearest

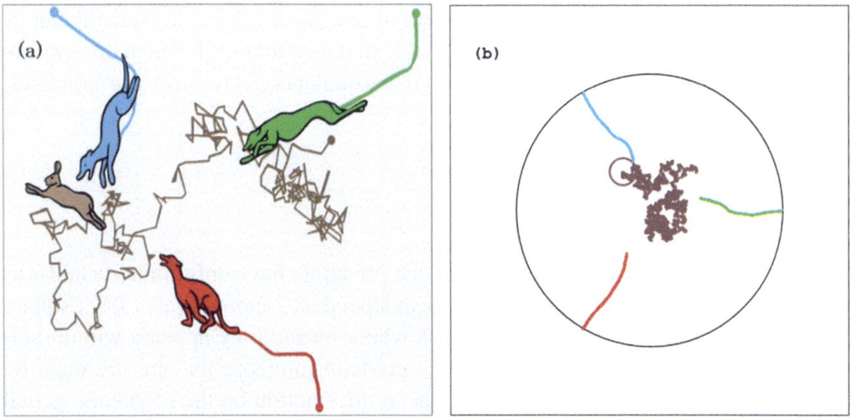


**Fig. 17.2** **a** Sketch of the group hunting model studied in [69]. **a** The hare/prey diffuses according to simple Brownian dynamics while the hounds/predators run at a constant speed and adjust their direction to always aim at the prey. **b** Typical trajectories of the model studied in [69] for three hunters and one target starting from the circular boundary and the circle centre, respectively. From [69]

particles of the opposite group. It was found that the capture time depends non-monotonously on the effective range of the predator–prey interaction and could be maximized in favour of the prey.

A more biology-inspired approach modelling group chase tactics was taken in [27], considering in particular the challenge for predator groups to catch a faster prey: For the predators or chasers the model contained equations of motion describing that each predator chases the closest target. The collective chasing strategy included a soft repulsion between the chasers (to avoid clustering of the predators emerging frequently when all predators move towards the same spot, the location of the prey) and a prediction of their target's position. The escape strategy of the prey was mainly defined by moving in the free direction that is furthest away from all the chasers within the sensitivity range, but also contained a stochastic component describing seemingly erratic escape behaviour frequently observed in nature. A panic parameter was introduced that depends on the distance between the escaper and its nearest chaser and controls this erratic behaviour. In nature, faster prey are frequently captured by predators when running into obstacles or confining walls. Consequently, the model also contained rules for escape behaviour at walls: At the wall confining the simulation area, the escaper aligns its velocity to the wall. If the escaper can slip away between the nearest two chasers, then it returns to the field.

It was found that repulsive interactions between the predators are important for capture efficiency and can lead to optimal groups of chasers that can catch a faster prey. It was also observed that emergent behaviour can occur: with certain parameters, the chasers have the chance to encircle (encage) their prey. If chasers use the prediction method to forecast their targets' position, their efficiency increases, and this can also overcome great delay times in the chasers' reactions to prey movement. Finally, erratic escape behaviour of prey in the form of stochastic zigzag movement can be advantageous for the prey when there is a delay in the system [27], which brings us to the discussion of target evasion strategies—in particular collective evasion strategies.

### *17.3.4 Evasion Strategies*

A variant of the pursuit problem in which the predators have only limited sight was studied in [54]: here the hunters perform independent random walks on a square lattice with V sites and start a direct chase whenever the prey appears within their sighting range. The prey is caught when a predator jumps to the site occupied by the prey. When the prey is blind, i.e., it has no information on the predators' actual positions (its sighting range is zero), the best strategy is to stay still [13, 48, 49]. Contrastingly, in [54], the efficiency of a minimal-effort evasion strategy was analyzed according to which the prey tries to avoid encounters with the predators by making a hop only when any of the predators appears within its sighting range; otherwise, the prey stays still. It was shown that if the sighting range of such a lazy prey is equal to 1 lattice spacing, at least 3 predators are needed in order to catch the prey on a square lattice. In this situation, a simple asymptotic relation $\ln P_{ev}(t) \propto (N/V)^2 \ln P_{imm}(t)$ between the survival probabilities of an evasive and an immobile prey was established. Hence, when the density $\rho = N/V$ of the predators is low, $\rho \ll 1$, the lazy evasion strategy leads to the spectacular increase of the survival probability [54]. A single predator hunting a herd of prey, both with limited sight, was also considered in [66], including explicitly the self-volume of the prey restraining their dynamics on the lattice. It was found that the prey's sighting range dominates their life expectancy and the predator profits more from a bad eye-sight of the prey than from his own good eye-sight.

Collective evasion strategies, i.e., collective behavioural patterns of groups of prey when attacked by a predator, can be observed in animal swarms, herds, and flocks, with the primary function to protect members when attacked by predators. One way to reduce the predator's chance of making a successful kill is to confuse the predator, which may be accomplished by collective evasion behaviour. In [82], such a behavioural pattern was modelled by self-propelled particles (in discrete time) choosing their direction of motion to a) achieve flocking by rules for medium-range alignment, short-range collision avoidance, and long-range attraction and b) escape rules that are activated when a (single) predator is in sighting range. These escape rules can either be selfish or cooperative, the latter implying a 'patience' zone in which escape is not activated although the predator is in sight. The (single) predator is also modelled as a self-propelled particle and chooses at each time step one prey randomly within its attack zone having a fan shape whose centre axis is in the direction from the predator to its previous target and whose spread angle is fixed. The predator tries to approach directly the target at each moment, but its ability to change its movement direction is in general inferior to the ability of the prey. It was shown that the mean first capture time is maximized for a specific value of the width of the 'patience' zone, underpinning the efficiency of cooperative evasion strategies.

A more minimalistic approach, but essentially in the same spirit, was taken in [14]: The movement of prey particles and a single predator were described by (over-damped) equations of motion in which the velocity changes were determined by repulsive and attractive forces (decaying algebraically with distance). Prey–prey interactions comprised an attractive long-range part (leading to flocking) and a repulsive short-range part (avoiding collisions), prey–predator interaction was repulsive, and predator–prey interaction was an average over attractive forces from each prey. The model predicts the formation of a stable donut-like arrangement of the prey positions around the predator in certain parameter ranges, an emergent collective behaviour also observed in nature (c.f. Fig. 17.3). The predator can be 'trapped' at the centre of the prey swarm while the prey forms a concentric annulus where the repulsion exerted by the predator cancels out owing to the symmetry. If the predator strength $c$, the coupling parameter between the average predator–prey attractive forces and the velocity change of the predator, is sufficiently small, the swarm will escape the predator. If $c$ is increased past a threshold, the predator becomes more 'focused' and less 'confused', resulting in 'chasing dynamics' which can lead to very

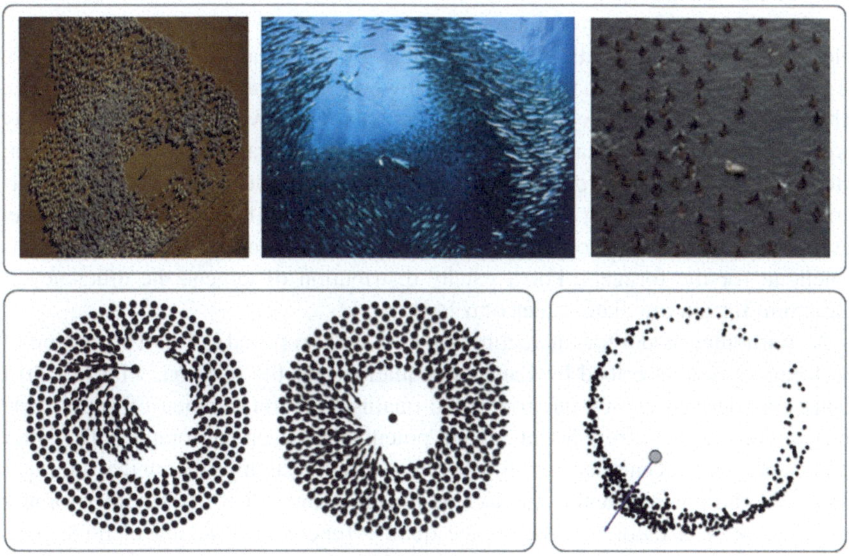

**Fig. 17.3** Upper panel: Pictures of the evasion strategy of groups of sheep (left), sardines (middle), and ducks (right) in the presence of a predator (always in the centre of the emerging 'donut' as observed in nature. From [14]. Lower panel: Simulation snapshots of a group evasion strategy model studied in [14]. Left: first-order model (i.e., overdamped equations of motion for the particle velocities) with two different predator–prey attraction strengths. Right: second-order model (i.e., equations of motion for the accelerations). From [14]

complex periodic or chaotic behaviour [14]. Thus for a 'weak' predator, the swarm is able to escape the predator completely.

### 17.3.5 Collective Foraging

In contrast to 'hunting' discussed in the previous sections, which involved mobile prey, we focus here on collective 'foraging', meaning many searchers and many immobile targets, like plant food for herbivores. In contrast to the mathematical problem for $N$ searchers and $M$ targets [32, 57], foraging involves the consumption / removal of a target once it is found and usually also the renewal of targets, such that the target density remains approximately constant. Thus, the foraging process can reach a stationary state, in which the foraging efficiency, measured by targets found per time, is of interest. In particular when the distribution of targets is 'patchy' (i.e., randomly distributed high-density clusters of targets in an environment with otherwise low target density) it appears plausible that communication between the searchers could be useful.

This hypothesis was scrutinized in [44] where a model of a group of interacting foragers was introduced to study the optimal strategy to search for heterogeneously distributed targets. In the model, foragers move on a square lattice containing immobile but regenerative targets. At any instant, a forager is able to detect only those targets that happen to be in the same site. However, the foragers are allowed to have information about the state of other foragers. A forager who has not detected any target walks towards the nearest location, where another forager has detected a target, with a probability $\exp(-\alpha d)$, where $d$ is the distance between the foragers and $\alpha$ is a parameter characterizing the propensity of the foragers to aggregate. The model reveals that neither overcrowding ($\alpha \to 0$) nor independent searching ($\alpha \to \infty$) is beneficial for the foragers. For a patchy distribution of targets, the efficiency is maximum for intermediate values of $\alpha$ [44].

A continuum model for interacting foragers was proposed in [7]: The motion of the foragers was described by Langevin equations for the position, with two drift terms: one derived from a environmental quality function (amount of grass, prey, etc.) and one derived from an interaction potential between the foragers. The latter just spreads with a Gaussian kernel with width $\sigma$ the local information that a forager has about the environmental quality. To illustrate the effect of this inter-forager communication a theoretical landscape quality function consisting of three non-normalized Gaussian functions (centred at different target positions) was considered and the mean first arrival time of foragers at the targets was evaluated. It was shown that the so-defined search could be minimized for a specific value of the interaction or communication range $\sigma$: the number of individuals from which a given searcher receives a signal will typically increase with the interaction range. When this scale is too small, individuals receive too little information (no information when $\sigma = 0$) and thus exhibit low search efficiency. Similarly, interaction ranges that are too large lead to individuals being overwhelmed with information from all directions, also resulting in inefficient searches [7]. As an application the model was applied to the acoustic communication among Mongolian gazelles, for which data are available, searching for good habitat areas. The key observation of the model and the application is that, in general, intermediate communication distances optimize search efficiency in terms of time and quality.

## 17.4 Summary and Outlook

As soon as many searchers and / or many targets are involved in a target problem, collective search or evasion strategy is advantageous—either for the searcher or the evader. A few fundamental facts about collective searches are known and have been analyzed quantitatively in the past, as we reviewed here: for many agents searching randomly (i.e., blindly) one or many randomly distributed targets, a repulsive interaction between them is advantageous, for instance via chemotaxis; in case random searchers are supposed to swarm to a specific randomly located spot over long distances (e.g., neutrophils to a wound) a signalling relay is useful to overcome the limits of diffusion; for group hunting (i.e., many searchers with limited visibility) repulsion between the hunters and prey trajectory prediction is advantageous; for group evasion (i.e., escape behaviour of prey seeing the predator) attractive interaction between prey and a mixture of deterministic and random movement in the presence of a predator is advantageous; for group foraging searchers should communicate the local target density to others nearby.

Nature is of course more sophisticated than this and developed many more collective search strategies which are largely unexplored. For instance, the immune system is one arena where searching (and killing) is of utmost importance. T cells change their migratory behaviour in different environments—like in lymph nodes or tissue [36]. There are striking similarities between the strategies of immune cells to find pathogens and strategies of ant colonies to forage for food and it has been hypothesized that the ability of searchers to sense and adapt to dynamic targets and environmental conditions enhances search effectiveness through adjustments to movement and communication patterns [50]. Other immune cells, like T helper cells, B cells, dendritic cells, macrophages, etc., are involved in an efficient immune response, too, interacting with each other and altering migratory and secretory patterns.

Inspired by such natural examples a plentitude of complex search strategies could be imagined: searchers adapting their migration behaviour to a spatially and/or dynamically changing environment or target distribution, different types of searchers communicating in different ways and with different ranges, mobile or immobile bystander agents that influence or relay information about target distribution or type, and much more. Similarly, sophisticated escape strategies could be imagined, like switching between flocking, group splitting, and dispersion to confuse an approaching predator, long-range communication or the involvement of a bystander providing information about predator movements, etc. Moreover, one could expect significant advantages of search and escape strategies that involve the memory of the searchers or the moving targets (the escapers) like in the non-Markovian search processes that we reviewed at the beginning of this chapter.

More complex search and hunting strategies than those reviewed here have been developed in social animals during evolution. Four levels of increasing complexity characterize group hunting [79]. The lowest level of complexity, similarity, is characterized by individuals directing their behaviours at a single target, as we reviewed in Sect. 17.3.3; however, these actions lack temporal and spatial coordination. When the behaviours of the group align temporally, it is known as synchrony. Coordination represents a further increase in complexity as individuals relate their actions temporally and spatially to the actions of the other group members. Finally, the highest level of complexity is collaboration. When collaborating, individuals not only coordinate their behaviours, but also perform a variety of complementary behaviours directed at the same target, which may require understanding the others' role in completing the goal. Examples for the highest levels of complexity in group hunting among social animals have been reviewed in [79].

Swarm intelligence observed in nature often stimulated the design of analogous strategies in artificial systems like robots [8, 12]. For instance, sophisticated foraging strategies were implemented in swarms of unmanned aerial vehicles (UAV) or drones, which have become more and more accessible and are now increasingly utilized across a variety of domains, including exploration of unknown territories and monitoring of hazardous environments. In its simplest form, a swarm of UAVs is characterized by a large number of homogeneous individuals (agents) with local communication, sensing, and actuation capabilities, and its swarm behaviour can be designed by employing three main conceptual frameworks [12]: (i) collective decision making, in which UAVs aim at achieving consensus on a collective strategy, in order to maximize the performance of the swarm; (ii) spatially organized behaviour, in which UAVs aggregate according to a set of spatial constraints to arrange themselves following specific patterns, chains, or structures; and (iii) navigation-based behaviour, in which each UAV follows a set of simple behavioural rules to steer the whole swarm and cooperatively explore an environment. The latter is mostly used for search and coverage tasks and is frequently designed by exploiting biologically inspired approaches [1].

In [1], swarm behaviour of UAVs was modelled considering and comparing two different paradigms, namely biological behaviour, which mimics social animal meta-heuristics, and computational behaviour, which considers enhancements exploiting UAVs information technology. For instance, computational flocking involved different rules for movement in three different zones: 'separate' in the near zone (corresponding to short-range inter-searcher repulsion), 'align' with neighbours in the medium zone (corresponding to the Vicsek model), and 'cohere' in the far field zone (corresponding to long-range attraction). Note that similar movement rules were applied to groups of prey/escapers in some of the models for evasion strategies reviewed in Sect. 17.3.4. Computational 'stigmergy' (meaning agents leaving trails in the environment to coordinate other's movement, like chemotaxis) involves 'digital pheromones', which are maintained in a virtual space, called pheromone map, and can have an instant diffusion, to immediately propagate the environmental information to nearby UAV. Pheromones models can include both attractive and repulsive pheromones, to mark detected targets and visited areas, respectively. The various possible strategies are then usually optimized by using evolutionary algorithms or machine learning.

Finally, we would like to mention that chemotaxis and swarming are not only ubiquitous in biological systems but have recently been found in synthetic active matter, like active colloids, too [20, 40, 41]. This offers the possibility to explore collective search strategies experimentally, which would pave the way for novel applications, as for instance in (nano)medicine.

## References

1. A.L. Alfeo, M.G.C.A. Cimino, G. Vaglini, Enhancing biologically inspired swarm behavior: metaheuristics to foster the optimization of UAVs coordination in target search. Comput. Oper. Res. **110**, 34–47 (2019)
2. A. Altshuler, O.L. Bonomo, N. Gorohovsky, S. Marchini, E. Rosen, O. Tal-Friedman, S. Reuveni, Y. Roichman, Environmental memory facilitates search with home returns. Phys. Rev. Res **6**, 023255 (2024). https://doi.org/10.1103/PhysRevResearch.6.023255; arXiv:2306.12126
3. D.J. Amit, G. Parisi, L. Peliti, Asymptotic behavior of the "true" self-avoiding walk. Phys. Rev. B **27**(3), 1635–1645 (1983)
4. L. Angelani, Collective predation and escape strategies. Phys. Rev. Lett. **109**(11), 118104 (2012)
5. A. Barbier-Chebbah, O. Bénichou, R. Voituriez, Self-interacting random walks: aging, exploration, and first-passage times. Phys. Rev. X **12**(1), 011052 (2022)
6. D. Bernardi, B. Lindner, Run with the Brownian hare, hunt with the deterministic hounds. Phys. Rev. Lett. **128**(4), 040601 (2022)
7. K. Bhattacharya, T. Vicsek, Collective foraging in heterogeneous landscapes. J. R. Soc. Interface **11**(100), 20140674 (2014)
8. E. Bonabeau, M. Dorigo, G. Theraulaz, *Swarm Intelligence: From Natural to Artificial Systems* (Oxford University Press, 1999)
9. O.L. Bonomo, S. Reuveni, Loss of percolation transition in the presence of simple tracer-media interactions. Phys. Rev. Res. **5**(4), L042015 (2023)
10. D. Boyer, J.C.R. Romo-Cruz, Solvable random-walk model with memory and its relations with Markovian models of anomalous diffusion. Phys. Rev. E **90**(4), 042136 (2014)
11. D. Boyer, C. Solis-Salas, Random walks with preferential relocations to places visited in the past and their application to biology. Phys. Rev. Lett. **112**(24), 240601 (2014)
12. M. Brambilla, E. Ferrante, M. Birattari, M. Dorigo, Swarm robotics: a review from the swarm engineering perspective. Swarm Intell. **7**(1), 1–41 (2013)
13. A.J. Bray, S.N. Majumdar, R.A. Blythe, Formal solution of a class of reaction-diffusion models: reduction to a single-particle problem. Phys. Rev. E **67**(6), 060102 (2003)
14. Y. Chen, T. Kolokolnikov, A minimal model of predator "swarm interactions". J. R. Soc. Interface **11**(94), 20131208 (2014)
15. B. Davis, Reinforced random walk. Probab. Theory Relat. Fields **84**(2), 203–229 (1990)
16. P.B. Dieterle, J. Min, D. Irimia, A. Amir, Dynamics of diffusive cell signaling relays. Elife **9**, e61771 (2020)
17. A. Dussutour, V. Fourcassié, D. Helbing, J.-L. Deneubourg, Optimal traffic organization in ants under crowded conditions. Nature **428**(6978), 70–73 (2004)

18. J.G. Foster, P. Grassberger, M. Paczuski, Reinforced walks in two and three dimensions. New J. Phys. **11**(2), 23009 (2009)
19. J.L. Galeano Niño, S.V. Pageon, S.S. Tay, F. Colakoglu, D. Kempe, J. Hywood, J.K. Mazalo, J Cremasco, M.A. Govendir, L.F. Dagley, K. Hsu, S. Rizzetto, J. Zieba, G. Rice, V. Prior, G.M. O'Neill, R.J. Williams, D.R. Nisbet, B. Kramer, A.I. Webb, F. Luciani, M.N. Read, M. Biro, Cytotoxic T cells swarm by homotypic chemokine signalling. Elife **9**, e56554 (2020)
20. J. Grauer, H. Löwen, A. Be'er, B. Liebchen, Swarm hunting and cluster ejections in chemically communicating active mixtures. Sci. Rep. **10**(1), 5594 (2020)
21. D.S. Grebenkov, Depletion of resources by a population of diffusing species. Phys. Rev. E **105**(5), 54402 (2022)
22. D.S. Grebenkov, R. Metzler, G. Oshanin, From single-particle stochastic kinetics to macroscopic reaction rates: fastest first-passage time of N random walkers. New J. Phys. **22**(10), 103004 (2020)
23. T. Guérin, O. Bénichou, R. Voituriez, Non-Markovian polymer reaction kinetics. Nat. Chem. **4**(7), 568–573 (2012)
24. T. Guérin, N. Levernier, O. Bénichou, R. Voituriez, Mean first-passage times of non-Markovian random walkers in confinement. Nature **534**(7607), 356–359 (2016)
25. P. Hanggi, P. Talkner, First-passage time problems for non-Markovian processes. Phys. Rev. A **32**(3), 1934–1937 (1985)
26. T. Iwama, M. Sato, Group chase and escape with some fast chasers. Phys. Rev. E **86**(6), 067102 (2012)
27. M. Janosov, C. Virágh, G. Vásárhelyi, T. Vicsek, Group chasing tactics: how to catch a faster prey. New J. Phys. **19**(5), 53003 (2017)
28. R. Jose, L. Santen, Self-organized lane formation in bidirectional transport by molecular motors. Phys. Rev. Lett. **124**(19), 198103 (2020)
29. A. Kamimura, T. Ohira, *Group Chase and Escape* (Springer, 2020)
30. H. Kaur, T. Franosch, M. Caraglio, Adaptive active Brownian particles searching for targets of unknown positions. Mach. Learn. Sci. Technol. **4**(3), 35008 (2023)
31. E.F. Keller, L.A. Segel, Model for chemotaxis. J. Theor. Biol. **30**(2), 225–234 (1971)
32. S.M.J. Khadem, S.H.L. Klapp, R. Klages, Search efficiency of discrete fractional Brownian motion in a random distribution of targets. Phys. Rev. Res. **3**(2), 23169 (2021)
33. A. Klimek, R.R. Netz, Optimal non-Markovian composite search algorithms for spatially correlated targets. Europhys. Lett. **139**(3), 32003 (2022)
34. W.T. Kranz, A. Gelimson, K. Zhao, G.C.L. Wong, R. Golestanian, Effective dynamics of microorganisms that interact with their own trail. Phys. Rev. Lett. **117**(3), 038101 (2016)
35. W.T. Kranz, R. Golestanian, Trail-mediated self-interaction. J. Chem. Phys. **150**(21), 214111 (2019)
36. M.F. Krummel, F. Bartumeus, A. Gérard, T cell migration, search strategies and mechanisms. Nat. Rev. Immunol. **16**(3), 193–201 (2016)
37. T. Lämmermann, P.V. Afonso, B.R. Angermann, J.M. Wang, W. Kastenmüller, C.A. Parent, R.N. Germain, Neutrophil swarms require LTB4 and integrins at sites of cell death in vivo. Nature **498**(7454), 371–375 (2013)
38. S.D. Lawley, Distribution of extreme first passage times of diffusion. J. Math. Biol. **80**(7), 2301–2325 (2020)
39. S.D. Lawley, Universal formula for extreme first passage statistics of diffusion. Phys. Rev. E **101**(1), 12413 (2020)
40. B. Liebchen, H. Löwen, Synthetic chemotaxis and collective behavior in active matter. Acc. Chem. Res. **51**(12), 2982–2990 (2018)
41. B. Liebchen, D. Marenduzzo, M.E. Cates, Phoretic interactions generically induce dynamic clusters and wave patterns in active colloids. Phys. Rev. Lett. **118**(26), 268001 (2017)
42. Y. Lin, N. Abaid, Collective behavior and predation success in a predator-prey model inspired by hunting bats. Phys. Rev. E **88**(6), 062724 (2013)
43. N. Madras, G. Slade, *The Self-avoiding Walk* (Springer Science & Business Media, 2013)

44. R. Martinez-Garcia, J.M. Calabrese, T. Mueller, K.A. Olson, C. López, Optimizing the search for resources by sharing information: Mongolian gazelles as a case study. Phys. Rev. Lett. **110**(24), 248106 (2013)
45. C. Metzner, On the efficiency of chemotactic pursuit - comparing blind search with temporal and spatial gradient sensing. Sci. Rep. **9**(1), 14091 (2019)
46. H. Meyer, H. Rieger, Optimal Non-Markovian search strategies with n-step memory. Phys. Rev. Lett. **127**(7), 070601 (2021)
47. H. Meyer, H. Rieger, Alignment interaction and band formation in assemblies of autochemorepulsive walkers. Phys. Rev. E **108**(3), 034604 (2023)
48. M. Moreau, G. Oshanin, O. Bénichou, M. Coppey, Pascal principle for diffusion-controlled trapping reactions. Phys. Rev. E **67**(4), 045104 (2003)
49. M. Moreau, G. Oshanin, O. Bénichou, M. Coppey, Lattice theory of trapping reactions with mobile species. Phys. Rev. E **69**(4), 046101 (2004)
50. M.E. Moses, J.L. Cannon, D.M. Gordon, S. Forrest, Distributed adaptive search in T cells: lessons from ants. Front. Immunol. **10** (2019)
51. G. Muñoz-Gil, A. López-Incera, L.J. Fiderer, H.J. Briegel, Optimal foraging strategies can be learned (2023)
52. I. Nayak, A. Nandi, D. Das, Capture of a diffusive prey by multiple predators in confined space. Phys. Rev. E **102**(6), 062109 (2020)
53. A. Ordemann, E. Tomer, G. Berkolaiko, S. Havlin, A. Bunde, Structural properties of self-attracting walks. Phys. Rev. E **64**(4), 046117 (2001)
54. G. Oshanin, O. Vasilyev, P.L. Krapivsky, J. Klafter, Survival of an evasive prey. Proc. Natl. Acad. Sci. **106**(33), 13696–13701 (2009)
55. H.C. Ottinger, The generalised true self-avoiding walk-a model with continuously variable exponent $\nu$. J. Phys. A. Math. Gen. **18**(7), L363 (1985)
56. K.J. Painter, Mathematical models for chemotaxis and their applications in self-organisation phenomena. J. Theor. Biol. **481**, 162–182 (2019)
57. V.V. Palyulin, V.N. Mantsevich, R. Klages, R. Metzler, A.V. Chechkin, Comparison of pure and combined search strategies for single and multiple targets. Eur. Phys. J. B **90**, 1–16 (2017)
58. L. Peliti, L. Pietronero, Random walks with memory. La Riv. del Nuovo Cim. **10**(6), 1–33 (1987)
59. R. Pemantle, A survey of random processes with reinforcement. Probab. Surv. **4**(none), 1–79 (2007)
60. A. Pezzotta, M. Adorisio, A. Celani, Chemotaxis emerges as the optimal solution to cooperative search games. Phys. Rev. E **98**(4), 42401 (2018)
61. S. Redner, P.L. Krapivsky, Capture of the lamb: diffusing predators seeking a diffusing prey. Am. J. Phys. **67**(12), 1277–1283 (1999)
62. P. Romanczuk, I.D. Couzin, L. Schimansky-Geier, Collective motion due to individual escape and pursuit response. Phys. Rev. Lett. **102**(1), 010602 (2009)
63. Z. Sadjadi, M.R. Shaebani, H. Rieger, L. Santen, Persistent-random-walk approach to anomalous transport of self-propelled particles. Phys. Rev. E **91**(6), 62715 (2015)
64. V.B. Sapozhnikov, Self-attracting walk with $\nu < 1/2$. J. Phys. A. Math. Gen. **27**(6), L151 (1994)
65. G.M. Schütz, S. Trimper, Elephants can always remember: exact long-range memory effects in a non-Markovian random walk. Phys. Rev. E **70**(4), 045101 (2004)
66. M. Schwarzl, A. Godec, G. Oshanin, R. Metzler, A single predator charging a herd of prey: effects of self volume and predator - prey decision-making. J. Phys. A Math. Theor. **49**(22), 225601 (2016)
67. A. Sengupta, T. Kruppa, H. Löwen, Chemotactic predator-prey dynamics. Phys. Rev. E **83**(3), 031914 (2011)
68. M.R. Shaebani, A. Wysocki, R.G. Winkler, G. Gompper, H. Rieger, Computational models for active matter. Nat. Rev. Phys. **2**(4), 181–199 (2020)
69. M. Su, D. Bernardi, B. Lindner, Pursuit problem with a stochastic prey that sees its chasers. New J. Phys. **25**(2), 23033 (2023)
70. D.J.T. Sumpter, *Collective Animal Behavior* (Princeton University Press, 2011)

71. A. Surendran, M.J. Plank, M.J. Simpson, Spatial structure arising from chase-escape interactions with crowding. Sci. Rep. **9**(1), 14988 (2019)
72. J. Taktikos, V. Zaburdaev, H. Stark, Modeling a self-propelled autochemotactic walker. Phys. Rev. E **84**(4), 41924 (2011)
73. N.P. Tani, A. Blatt, D.A. Quint, A. Gopinathan, Optimal cooperative searching using purely repulsive interactions. J. Theor. Biol. **361**, 159–164 (2014)
74. V. Tejedor, R. Voituriez, O. Bénichou, Optimizing persistent random searches. Phys. Rev. Lett. **108**(8), 88103 (2012)
75. B. Tóth, Self-interacting random motions, in *The European Congress of Mathematics* ed, by C. Casacuberta, R.M. Miró-Roig, J. Verdera, S. Xambó-Descamps (Birkhäuser, Basel, 2001), pp. 555–564
76. T. Vicsek, A. Czirók, E. Ben-Jacob, I. Cohen, O. Shochet, Novel type of phase transition in a system of self-driven particles. Phys. Rev. Lett. **75**(6), 1226–1229 (1995)
77. R. Vidal, O. Shakernia, H.J. Kim, D.H. Shim, S. Sastry, Probabilistic pursuit-evasion games: theory, implementation, and experimental evaluation. IEEE Trans. Robot. Autom. **18**(5), 662–669 (2002)
78. O. Vilk, D. Campos, V. Méndez, E. Lourie, R. Nathan, M. Assaf, Phase transition in a non-Markovian animal exploration model with preferential returns. Phys. Rev. Lett. **128**(14), 148301 (2022)
79. J. Vincent, *Cooperative Hunting* (Springer International Publishing, Cham, 2020), pp. 1–4
80. G.M. Viswanathan, M.G.E. da Luz, E.P. Raposo, H.E. Stanley, *The Physics of Foraging* (Cambridge University Press, 2011)
81. G.H. Weiss, K.E. Shuler, K. Lindenberg, Order statistics for first passage times in diffusion processes. J. Stat. Phys. **31**, 255–278 (1983)
82. M. Zheng, Y. Kashimori, O. Hoshino, K. Fujita, T. Kambara, Behavior pattern (innate action) of individuals in fish schools generating efficient collective evasion from predation. J. Theor. Biol. **235**(2), 153–167 (2005)

# Chapter 18
# Modeling Ionic Flow Between Small Targets: Insights from Diffusion and Electro-Diffusion Theory

**Frédéric Paquin-Lefebvre and David Holcman**

**Abstract** The flow of ions through permeable channels causes voltage drop in physiological nanodomains such as synapses, dendrites and dendritic spines and other protrusions. How the voltage changes around channels in these nanodomains has remained poorly studied. We focus this chapter on summarizing recent efforts in computing the steady-state current, voltage and ionic concentration distributions based on the Poisson–Nernst–Planck equations as a model of electro-diffusion. We first consider the spatial distribution of an uncharged particle density and derive asymptotic formulas for the concentration difference by solving Laplace's equation with mixed boundary conditions. We study a constant particle injection rate modeled by a Neumann flux condition at a channel represented by a small boundary target, while the injected particles can exit at one or several narrow patches. We then discuss the case of two species (positive and negative charges) and take into account motions due to both concentration and electrochemical gradients. The voltage resulting from charge interactions is calculated by solving Poisson's equation. We show how deep an influx diffusion propagates inside a nanodomain, for populations of both uncharged and charged particles. We estimate the concentration and voltage changes in relations with geometrical parameters and quantify the impact of membrane curvature.

F. Paquin-Lefebvre · D. Holcman (✉)
Group of Data Modeling and Computational Biology, IBENS, École Normale Supérieure – Université PSL, 75005 Paris, France
e-mail: david.holcman@ens.fr

F. Paquin-Lefebvre
e-mail: paquin@bio.ens.psl.eu

D. Holcman
Department of Applied Mathematics and Theoretical Physics and Churchill College, University of Cambridge, Cambridge CB3 0WA, UK

 
D. Grebenkov et al. (eds.), *Target Search Problems*,
https://doi.org/10.1007/978-3-031-67802-8_18

## 18.1 Introduction

The flow of particles (molecules or ions) entering through narrow channels on the cellular membrane regulates the concentrations but also the potential equilibrium in excitable cells such as neurons [1, 2]. While the selection of particles going through a channel and the amplitude of the particle current inside a single channel has been well characterized over the past 50 years, how entering particles evolve inside the cells in the microdomain around the single channel remains less explored. However, these regions can already contain neighboring channels that could be activated to respond as soon as possible to the first entrance. In particular, this could influence locally the membrane potential, modulated by the influx and efflux of ions across these neighboring channels [3, 4]. These local influxes modify constantly the local ionic concentrations and the electric field in small nanoregions, difficult to access experimentally.

In some cases involving small compartments such as cilia [5], dendritic spines [6, 7] and glial protrusions, a local difference of concentration generated by the influx through a channel could be much larger than the background ion concentration. For instance, the calcium concentration inside a dendrite is of the order of a few 10–100 nM at rest [8], while calcium influx through channels can lead to a concentration ranging from 1 to 100 of $\mu$M, a factor of at least 10. These concentration differences cause a local electric field and voltage drop that triggers the opening of neighboring channels or receptors [9] (Fig. 18.1c). However the size of this region as well as the exact mechanism of ionic regulation in these regions near the channel exits remains unclear.

This chapter summarizes recent efforts in modeling and computing the voltage and the dynamics of ions at the channel exit in microdomains containing other regulatory channels that influence the concentration of entering ions. The simplest modeling approach is to consider that ions are simply diffusing and then account for possible electrical charge interactions using an electro-diffusion model [10, 11]. Electro-diffusion models are described by the Poisson–Nernst–Planck equations, which consist in coupling the Fokker–Planck description for the density with Poisson's equation for the voltage. One way to simplify the static charge case is to use classical Boltzmann solutions, for which ionic densities are exponential functions of the voltage, within Poisson's equation, thereby resulting in a single nonlinear model equation. This approach, known as the Poisson–Boltzmann theory, is appropriate to describe properties of electrolytes near one or two planar membranes [12–14] and can serve to introduce spatial scales, such as the Debye length, quantifying local charge interactions. However, ionic densities given by Boltzmann solutions cannot account for sudden localized ionic fluxes resulting from channels and pump dynamics, which is exactly the paradigm of our study: ions enter through one window, spread inside a microdomain (such as a ball and a cylinder) and leave through another ensemble of target windows (Fig. 18.2a, b).

The chapter is organized as follows. In Sect. 18.2, we introduce briefly the electro-diffusion modeling based on the Poisson–Nernst–Planck equations. In Sect. 18.3 we present, in the diffusion limit (neglecting electrochemical gradients), the computation of the concentration and the flow between two channels for a single species by solving Laplace's equation. Then in Sect. 18.4 we consider two ionic species, positive and negative, that can interact. We compute the voltage profile using the Poisson–Nernst–Planck equations. We present some asymptotic analyses and the resulting biophysical formulas revealing the role of geometrical parameters on the voltage dynamics. Finally, we briefly conclude in Sect. 18.5, summarizing the asymptotic and simulation difficulties and discuss some perspectives and implications in biophysics.

## 18.2 Modeling Ionic Fluxes Exiting a Channel Inside a Domain

Voltage changes in excitable cells such as neurons rely on currents generated by ionic channels opening and closing [1]. At the cellular level, voltage dynamics in neuronal cells is well characterized generating depolarization, bursts of activity or action potentials, as revealed by patch-clamp measurements or voltage-sensitive dyes [15]. However, voltage dynamics due to ionic transport in small micro-compartments (Fig. 18.1a, b) is much less known due to spatio-temporal challenges, often below the diffraction limit, thus preventing direct experimental access.

In the absence of direct measurement, modeling and numerical simulations provide instructive insights on possible physiological mechanisms underlying voltage changes. By neglecting concentration changes, voltage can be modeled by the classical cable theory [16]. The 1-D voltage propagation within a neuron has provided a theoretical foundation for electrophysiology [17]. However this theory breaks down when applied to small neuronal compartments of the scale of 100 nm to 1 μm, like dendritic spines, synaptic terminals or small neuronal processes such as filopodia or glial protrusions, because they assume spatial ionic homogeneity [18]. Indeed, voltage changes can be generated by a concentration gradient in neuronal nanodomains [7]. How voltage is regulated and spatially extends within such domains remain unclear: ionic concentration gradients may well be key elements, although they should be studied by further accounting for the local electric field generated by these gradients.

To study voltage changes at a subcellular level, the Poisson–Nernst–Planck (PNP) mean-field equations are used to model electro-diffusion, where diffusive ions can interact and generate an electric field. This description accounts for the geometry in computing the flow of electrical current. For a single charge density $c(\boldsymbol{x}, t)$ of valence $z$ within a domain $\Omega$ and with $v(\boldsymbol{x}, t)$ measuring the voltage, the Nernst–Planck equation formulates as [20]

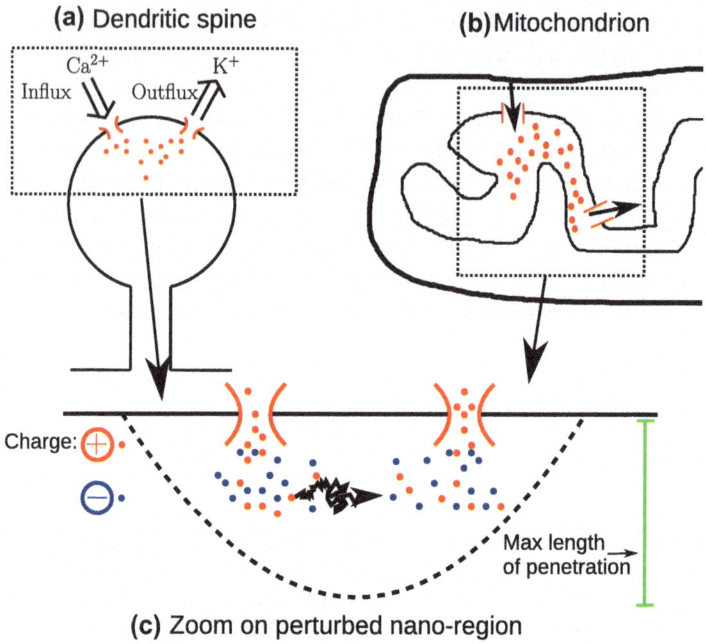


**Fig. 18.1 Channel dynamics and voltage in neuronal nanodomains. a** Dendritic spine receiving a calcium ion influx. **b** Inner and outer membranes of a mitochondrion, with the inner membrane curvature thought to regulate voltage and calcium ion intake [19]. **c** Formation of a voltage nanodomain in the vicinity of two ion channels (cf. [9])

$$\frac{\partial c(\boldsymbol{x},t)}{\partial t} = D\left(\Delta c(\boldsymbol{x},t) + \frac{ze}{k_B\mathcal{T}}\nabla(c(\boldsymbol{x},t)\nabla v(\boldsymbol{x},t))\right), \quad \boldsymbol{x} \in \Omega, \tag{18.1}$$

with diffusion coefficient $D$, Boltzmann constant $k_B$, temperature $\mathcal{T}$ and electron charge $e$. The ratio $k_B\mathcal{T}/e$ is known as the thermal voltage, and its value is approximately 26 mV (Table 18.1). If the membrane $\partial\Omega$ is impermeable to the passage of ions then no-flux boundary condition holds, yielding

$$D\left(\frac{\partial c(\boldsymbol{x},t)}{\partial \boldsymbol{n}} + \frac{ze}{k_B\mathcal{T}}c(\boldsymbol{x},t)\frac{\partial v(\boldsymbol{x},t)}{\partial \boldsymbol{n}}\right) = 0, \quad \boldsymbol{x} \in \partial\Omega, \tag{18.2}$$

where $\partial/\partial\boldsymbol{n}$ is the outward normal derivative. The voltage $v(\boldsymbol{x},t)$ is then the solution of Poisson's equation

$$\Delta v(\boldsymbol{x},t) = -\frac{z\mathcal{F}}{\varepsilon\varepsilon_0}c(\boldsymbol{x},t), \quad \boldsymbol{x} \in \Omega, \tag{18.3a}$$

with Faraday constant $\mathcal{F}$, as well as relative and vacuum permittivities $\varepsilon, \varepsilon_0$ (values for physical constants are summarized in Table 18.1). Finally, the voltage outward normal

**Table 18.1 Model parameters and physical constants**

| Parameter | Symbol | Value |
|---|---|---|
| Domain length scale | $R$ | 500 nm |
| Window radius | $A$ | 10 nm |
| Inter-window distance | $L$ | [25, 200] nm |
| Diffusion coefficient | $D_{p,m}$ | $200\,\mu\text{m}^2\text{s}^{-1}$ |
| Concentration | $C_0$ | 200 mM |
| Current amplitude | $I$ | 100 pA |
| Electron charge | $e$ | $1.60 \times 10^{-19}$ C |
| Boltzmann constant | $k_B$ | $1.38 \times 10^{-23}\,\text{J}\,\text{K}^{-1}$ |
| Temperature | $\mathcal{T}$ | 298 K |
| Avogadro constant | $N_A$ | $6.02 \times 10^{23}\,\text{mol}^{-1}$ |
| Relative permittivity | $\varepsilon$ | 78.4 |
| Vacuum permittivity | $\varepsilon_0$ | $8.85 \times 10^{-12}\,\text{C}\,\text{V}^{-1}\,\text{m}^{-1}$ |
| Faraday constant | $\mathcal{F} = eN_A$ | $96485.33\,\text{C}\,\text{mol}^{-1}$ |
| Thermal voltage | $k_B\mathcal{T}/e$ | 25.68 mV |
| Dimensionless current (Eq. (18.44)) | $J = \frac{IR}{\mathcal{F}\pi A^2 D_p C_0}$ | 41.24 |
| Coupling strength (Poisson's Eq. (18.43)) | $\lambda = \frac{e^2 R^2 N_A C_0}{k_B \mathcal{T} \varepsilon\varepsilon_0}$ | $2.71 \times 10^5$ |

derivative must be proportional to the surface charge density, which is Poisson's compatibility condition, thereby yielding

$$\frac{\partial v(\boldsymbol{x},t)}{\partial \boldsymbol{n}} = -\frac{z\mathcal{F}}{\varepsilon\varepsilon_0|\partial\Omega|}\int_\Omega c(\boldsymbol{x},t)d\boldsymbol{x}\,, \quad \boldsymbol{x} \in \partial\Omega\,. \tag{18.3b}$$

For such a class of non-electroneutral electro-diffusion models, charges accumulate near the boundary [20, 21], and high membrane curvature as yielded by cusp or funnel-shaped boundaries plays a significant role in regulating internal voltage changes [22, 23]. However to reduce voltage effects caused by large charge imbalances, more realistic electro-diffusion models should incorporate multiple charges, with global (but not necessarily local) electroneutrality condition. Interestingly for an immobile excess of negative charges at the center of a ball domain [24], voltage effects were found to extend on spatial scales much longer than the Debye length defined for two monovalent charges as [12]

$$\lambda_D = \sqrt{\frac{\varepsilon\varepsilon_0 k_B \mathcal{T}}{2e^2 N_A C_0}}\,. \tag{18.4}$$

For a background ionic density $C_0$ of hundreds of millimolars, $\lambda_D$ is less than one nanometer.

One goal of this chapter is to present how non-local electroneutrality, with spatial charge imbalance coming from a localized flux of positive ions on narrow boundary patches, can affect the voltage inside the nanodomain (Sect. 18.4). This analysis could be applied to calcium or sodium ions entering neuronal nanocompartments via ionic channels, disturbing locally the voltage distribution and causing the formation of a voltage drop nanodomain, and finally leading to the activation of nearby potassium channels from which excess of positive charges will exit [9] (Fig. 18.1c).

## 18.3 Influx Diffusion Steady-State Models for a Single Species

How far inside a domain can an influx of Brownian particles, entering through a narrow window, perturb the steady-state bulk concentration when they can escape through a neighboring window? We summarize in this section the properties of a single-species steady-state diffusion system, with voltage and electrochemical effects neglected. We first consider a general bounded domain with two narrow windows on its boundary (Fig. 18.2a). Local concentration differences are quantified using asymptotic formulas that link and mix together current and geometrical parameters (Sect. 18.3.1). In Sect. 18.3.2, we recall how to estimate the size of nanodomains where the concentration is spatially perturbed. A measure of the perturbation is characterized by the length of penetration (introduced in [25]), computed as the maximum distance from the boundary of the flow line going from the center of the influx window, where particles are injected, to the center of the neighboring window where they are expelled (Fig. 18.2c). Finally, we will see how a collection of exit windows (Sect. 18.3.3) with Dirichlet absorbing boundary conditions (Fig. 18.2b) can affect the concentration distribution.

### *18.3.1 Entry and Exit of Particles Mediated by Neumann Flux Boundary Conditions*

The geometry to consider is a bounded domain $\Omega$ with two small patches on the boundary $\partial\Omega$. The two patches $\partial\Omega_{A_i}$ for $i = 1, 2$, correspond to narrow circular windows of identical radius $A$, each centered at $\boldsymbol{x}_i$. The steady-state concentration distribution $c(\boldsymbol{x})$ for a single species diffusing (with diffusion coefficient $D$) inside $\Omega$ is influenced by Neumann flux boundary conditions, modeling the injection of a particle current of amplitude $I$ through the entry point $\partial\Omega_{A_1}$ and the exit through $\partial\Omega_{A_2}$ with an identical current $-I$ of opposite sign. The remaining boundary part $\partial\Omega_r$ is reflecting and the full domain size is such that the isoperimetric ratio, given by $|\partial\Omega|/|\Omega|^{2/3}$, is of order $O(1)$. By then setting the valence $z$ to zero (to neglect electrochemical effects) within Eq. (18.1), we find that at steady state the density

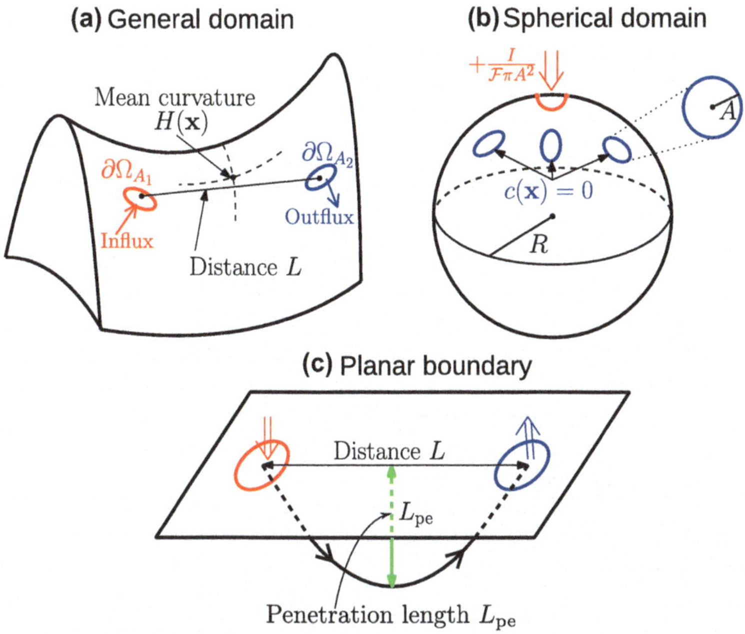


**Fig. 18.2 3-D geometrical domain types. a** An arbitrary domain with general mean curvature function $H(\boldsymbol{x})$. **b** A ball domain of radius $R$. **c** For two nearby narrow windows, we can approximate the boundary by a tangent plane

$c(\boldsymbol{x})$ is the solution of Laplace's equation,

$$D\Delta c(\boldsymbol{x}) = 0\,, \qquad \boldsymbol{x} \in \Omega\,, \tag{18.5a}$$

with boundary conditions

$$D\frac{\partial c(\boldsymbol{x})}{\partial \boldsymbol{n}} = \frac{I}{\mathcal{F}\pi A^2}\,, \quad \boldsymbol{x} \in \partial\Omega_{A_1}\,, \quad D\frac{\partial c(\boldsymbol{x})}{\partial \boldsymbol{n}} = -\frac{I}{\mathcal{F}\pi A^2}\,, \quad \boldsymbol{x} \in \partial\Omega_{A_2}\,, \tag{18.5b}$$

$$\frac{\partial c(\boldsymbol{x})}{\partial \boldsymbol{n}} = 0\,, \quad \boldsymbol{x} \in \partial\Omega_r\,, \tag{18.5c}$$

where $\mathcal{F}$ is the Faraday constant. We fix the total concentration by choosing the constant $C_0$ as the mean density:

$$\frac{1}{|\Omega|}\int_{\Omega} c(\boldsymbol{x})d\boldsymbol{x} = C_0\,. \tag{18.5d}$$

The question at hand is to study the difference of concentration $c(\boldsymbol{x}_1) - c(\boldsymbol{x}_2)$ with respect to the model parameters and the local window organization. For an almost convex or spherical domain, we consider that the local radius of curvature $R$ of the boundary where the two windows are located is such that $R \approx \sqrt{|\partial\Omega|}$. The windows are much smaller than the total boundary area, with $\sqrt{|\partial\Omega_{A_i}|} \ll \sqrt{|\partial\Omega|}$ yielding $A \ll R$. In that case, the concentration difference depends on the current $I$ and various geometrical parameters, and is given by

$$c(\boldsymbol{x}_1) - c(\boldsymbol{x}_2) = \frac{2I}{\mathcal{F}\pi AD}\left(1 - \frac{H(\boldsymbol{x}_1) + H(\boldsymbol{x}_2)}{8} A \log\left(\frac{A}{R}\right) + o\left(\frac{A}{R}\right)\right), \quad (18.6)$$

where $H(\boldsymbol{x}_i)$ is the mean curvature computed at the center $\boldsymbol{x}_i$ of the narrow windows [25]. At leading order in $A/R$, the difference increases linearly with the current amplitude $I$, while it is inversely proportional to the diffusion coefficient $D$ and the window radius $A$. Another key parameter is the Euclidean distance $L = \|\boldsymbol{x}_1 - \boldsymbol{x}_2\|$ between the two window centers, but its influence is restricted to higher order terms of the expansion (18.6). This expression shares similarity with the classical narrow escape theory [26] with multiple targets, where the mean escape time is only affected by the targets' geometrical organization via third-order terms [27].

A brief derivation of formula (18.6) uses the Neumann Green's function $G_s(\boldsymbol{x}; \boldsymbol{y})$ with singularity $\boldsymbol{y}$ on the boundary $\partial\Omega$. This Green's function is the solution of the boundary value problem

$$D\Delta G_s(\boldsymbol{x}; \boldsymbol{y}) = \frac{R}{|\Omega|}, \ \boldsymbol{x} \in \Omega, \ \boldsymbol{y} \in \partial\Omega, \quad D\frac{\partial G_s(\boldsymbol{x}; \boldsymbol{y})}{\partial \boldsymbol{n}} = R\delta(\boldsymbol{x} - \boldsymbol{y}), \ \boldsymbol{x}, \boldsymbol{y} \in \partial\Omega, \quad (18.7)$$

with $\int_\Omega G_s(\boldsymbol{x}; \boldsymbol{y}) d\boldsymbol{x} = 0$, and its behavior near the singular diagonal $\boldsymbol{x} = \boldsymbol{y}$ is given explicitly by [26]

$$G_s(\boldsymbol{x}; \boldsymbol{y}) \approx \frac{1}{D}\left(\frac{R}{2\pi \|\boldsymbol{x} - \boldsymbol{y}\|} - \frac{RH(\boldsymbol{y})}{4\pi} \log\left(\frac{\|\boldsymbol{x} - \boldsymbol{y}\|}{R}\right) + O(1)\right), \quad (18.8)$$

where $H(\boldsymbol{y})$ is the mean surface curvature in $\boldsymbol{y}$ on the boundary. Green's identity given by

$$\begin{aligned}\int_\Omega D\left(G_s(\boldsymbol{x}; \boldsymbol{y})\Delta c(\boldsymbol{x}) - c(\boldsymbol{x})\Delta G_s(\boldsymbol{x}; \boldsymbol{y})\right) d\boldsymbol{x} = \\ \int_{\partial\Omega} D\left(G_s(\boldsymbol{x}; \boldsymbol{y})\frac{\partial c(\boldsymbol{x})}{\partial \boldsymbol{n}} - u\frac{\partial G_s(\boldsymbol{x}; \boldsymbol{y})}{\partial \boldsymbol{n}}\right) d\boldsymbol{x}\end{aligned} \quad (18.9)$$

is used for the substitution of (18.5) and (18.7) yielding

$$c(\boldsymbol{y}) = \frac{1}{|\Omega|}\int_\Omega c(\boldsymbol{x}) d\boldsymbol{x} + \frac{I}{\mathcal{F}\pi A^2 R}\left(\int_{\partial\Omega_{A_1}} G_s(\boldsymbol{x}; \boldsymbol{y}) d\boldsymbol{x} - \int_{\partial\Omega_{A_2}} G_s(\boldsymbol{x}; \boldsymbol{y}) d\boldsymbol{x}\right). \quad (18.10)$$

Thus the density difference between the two narrow windows is given by

$$c(\boldsymbol{x}_1) - c(\boldsymbol{x}_2) = \frac{I}{\mathcal{F}\pi A^2 R}\left(\int_{\partial\Omega_{A_1}} G_s(\boldsymbol{x};\boldsymbol{x}_1)d\boldsymbol{x} + \int_{\partial\Omega_{A_2}} G_s(\boldsymbol{x};\boldsymbol{x}_2)d\boldsymbol{x} - \int_{\partial\Omega_{A_2}} G_s(\boldsymbol{x};\boldsymbol{x}_1)d\boldsymbol{x} - \int_{\partial\Omega_{A_1}} G_s(\boldsymbol{x};\boldsymbol{x}_2)d\boldsymbol{x}\right). \tag{18.11}$$

The integrals around the singularity of Green's function are then evaluated directly using expansion (18.8) with radial coordinate $r = \|\boldsymbol{x} - \boldsymbol{x}_i\|$:

$$\begin{aligned}\int_{\partial\Omega_{A_i}} G_s(\boldsymbol{x};\boldsymbol{x}_i)d\boldsymbol{x} &\approx \frac{2\pi}{D}\int_0^A \left(\frac{R}{2\pi r} - \frac{RH(\boldsymbol{x}_i)}{4\pi}\log\left(\frac{r}{R}\right) + O(1)\right) r dr\,, \\ &\approx \frac{RA}{D}\left(1 - \frac{H(\boldsymbol{x}_i)}{4}\log\left(\frac{A}{R}\right) + O\left(\frac{A}{R}\right)\right).\end{aligned} \tag{18.12}$$

The mixed integral terms can be incorporated within the remainder term since

$$\int_{\partial\Omega_{A_i}} G_s(\boldsymbol{x};\boldsymbol{x}_j)d\boldsymbol{x} \approx \pi A^2 G_s(\boldsymbol{x}_i;\boldsymbol{x}_j)\,. \tag{18.13}$$

Hence by substituting (18.12) and (18.13) within (18.11), we recover the two-term asymptotic formula given above in Eq. (18.6).

#### 18.3.1.1 Density–Current Distribution in a Spherical Domain

When the domain $\Omega$ consists of a ball of radius $R$, we can extend expansion (18.6) with an additional term involving the distance parameter $L = \|\boldsymbol{x}_1 - \boldsymbol{x}_2\|$. By using the mean curvature function $H(\boldsymbol{x}) = 1/R$ and the exact solution for Green's function $G_s(\boldsymbol{x};\boldsymbol{y})$ on the sphere given by [27]

$$\begin{aligned}G_s(\boldsymbol{x};\boldsymbol{y}) = \frac{R}{D}\Bigg(&\frac{1}{2\pi\|\boldsymbol{x}-\boldsymbol{y}\|} - \frac{1}{4\pi R}\log\left(\frac{\|\boldsymbol{x}-\boldsymbol{y}\|^2}{2R^2} + \frac{\|\boldsymbol{x}-\boldsymbol{y}\|}{R}\right) \\ &+\frac{1}{4\pi R}\left(\log(2) - \frac{9}{5}\right)\Bigg),\end{aligned} \tag{18.14}$$

with $\|\boldsymbol{x}\| = \|\boldsymbol{y}\| = R$ and $\boldsymbol{x} \neq \boldsymbol{y}$, we obtain the three-term asymptotic expansion [25]

$$\begin{aligned}c(\boldsymbol{x}_1) - c(\boldsymbol{x}_2) = \frac{2I}{\mathcal{F}\pi AD}\Bigg(&1 - \frac{A}{4R}\log\left(\frac{A}{R}\right) + \Bigg(\frac{1}{8} - \frac{R}{2L} \\ &+\frac{1}{4}\log\left(\frac{L^2}{2R^2} + \frac{L}{R}\right)\Bigg)\frac{A}{R} + O\left(\left(\frac{A}{R}\right)^2\right)\Bigg).\end{aligned} \tag{18.15}$$

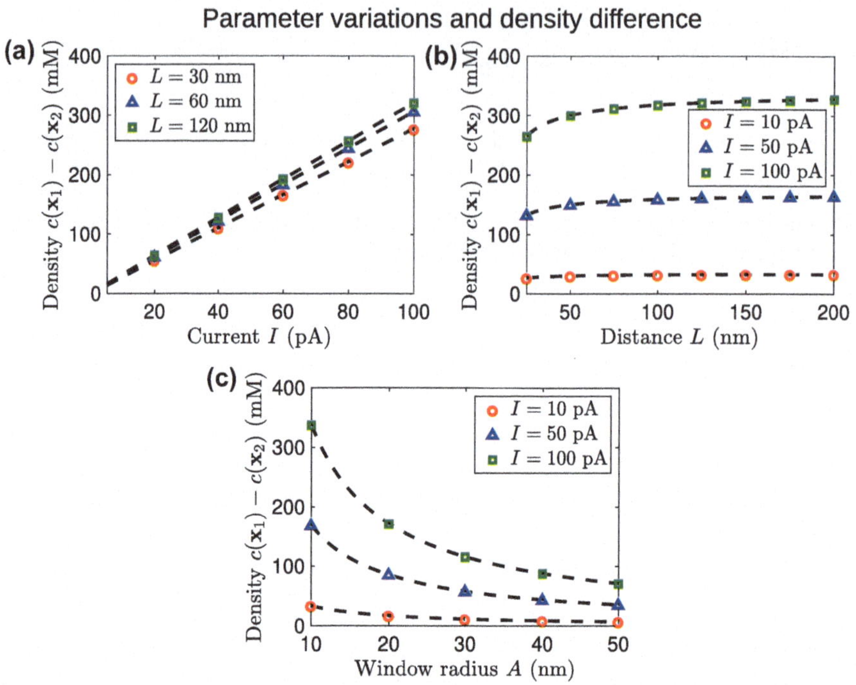


**Fig. 18.3 Spherical domain case.** Density difference as a function of the current amplitude $I$ (**a**), the distance $L$ (**b**) and the narrow radius $A$ (**c**). The three sets of symbols (circles, squares and triangles) indicate numerical solutions obtained with COMSOL, while asymptotic solutions are computed from Eq. (18.15) and are shown by black dashed curves. Model parameters are $R = 500$ nm, $A = 10$ nm, $C_0 = 200$ mM, $D = 200\,\mu\text{m}^2/\text{s}$. In **c** each window is centered at a Pole, thus $L = 2R = 1\,\mu$m. Values for other physical parameters are given in Table 18.1

Numerical simulations are used to explore how geometrical parameters can affect the concentration differences due to influx diffusion (Fig. 18.3). We solve Eq. (18.5) using COMSOL [28] on a spherical ball domain (see Appendix 2 for a mesh decomposition description). As predicted by formula (18.15), the density drop $c(\boldsymbol{x}_1) - c(\boldsymbol{x}_2)$ increases linearly with the current $I$ (Fig. 18.3a) and behaves as the reciprocal of the window radius $A$ (Fig. 18.3c). Finally, increasing the inter-window distance $L$ has an effect on the density drop only when the windows are nearby, with $L \sim O(A)$ (Fig. 18.3b).

### 18.3.2 Density–Current Relation in an Infinite Half-Space and the Notion of Penetration Length

To measure how deep inside a medium an influx current can perturb the steady-state concentration, the penetration length $L_{\text{pe}}$ was recently introduced based on model (18.5). The penetration length $L_{\text{pe}}$ is the maximal distance from the boundary achieved by trajectories following the gradient field lines and connecting the two windows at their centers. Thus it is associated with paths of the steepest concentration descent. Explicit solutions for the concentration gradient $\nabla c(\boldsymbol{x})$ are in general not available for arbitrary domains $\Omega$. However, using a tangent plane approximation (on a smooth surface), a boundary layer analysis in the vicinity of two nearby (almost tangent) windows maps the solutions of model (18.5) onto the ones of Laplace's equation in the infinite half-space, with a planar boundary. This is illustrated below starting from a spherical ball domain.

The procedure is as follows: first, we use spherical coordinates $(r, \phi, \theta)$ to parameterize the narrow windows, with radius $r < R$, azimuth $\phi$ and colatitude $\theta$, yielding

$$\partial\Omega_{A_j} = \left\{ (R, \phi, \theta) \,\middle|\, (\theta - \theta_j)^2 + \sin^2(\theta_j)(\phi - \phi_m)^2 \le \left(\frac{A}{R}\right)^2 \right\}, \quad j = 1, 2. \tag{18.16}$$

The two windows are aligned on the same azimuth $\phi_m$, and since they can be at most tangent the colatitudes satisfy $\theta_2 - \theta_1 > 2A/R$ . By then defining $\theta_m = (\theta_1 + \theta_2)/2$ as the midpoint colatitude coordinate, we can introduce a set of local Cartesian coordinates $(\eta, s_1, s_2)$ [27],

$$\eta = \frac{R - r}{A}, \quad s_1 = \frac{R}{A}\sin(\theta_m)(\phi - \phi_m), \quad s_2 = \frac{R}{A}(\theta - \theta_m), \tag{18.17}$$

with $0 \le \eta < \infty$ and $-\infty < s_1, s_2 < \infty$. With these new coordinates, we can map the narrow windows defined in Eq. (18.16) onto the plane $\eta = 0$ (Fig. 18.4),

$$\partial\Omega_{A_j} = \left\{ (0, s_1, s_2) \,\middle|\, \rho_i^2 \le 1 \right\}, \quad \rho_j = \sqrt{\left(\frac{\sin(\theta_j)}{\sin(\theta_m)} s_1\right)^2 + \left(s_2 + (-1)^{j+1}\frac{L}{2A}\right)^2}. \tag{18.18}$$

for $j = 1, 2$. The inner density variable is defined by

$$C(\eta, s_1, s_2) = \frac{1}{C_0} c\left(R - \eta A, \phi_m + \frac{A}{R\sin(\theta_m)} s_1, \theta_m + \frac{A}{R} s_2\right), \tag{18.19}$$

and by substituting within the model (18.5) we find that at leading order (in terms of the ratio $A/R \ll 1$) it is the solution in the infinite half-space of Laplace's equation

$$\left(\frac{\partial^2}{\partial \eta^2}+\frac{\partial^2}{\partial s_1^2}+\frac{\partial^2}{\partial s_2^2}\right)C(\eta, s_1, s_2)=0\,,\quad \eta>0,\quad -\infty<s_1,s_2<\infty. \tag{18.20a}$$

Furthermore, the corresponding boundary conditions on the plane $\eta = 0$ are

$$-\left.\frac{\partial C}{\partial \eta}\right|_{\eta=0}=\frac{I}{\mathcal{F}\pi AC_0D},\quad \rho_1\le 1 \quad -\left.\frac{\partial C}{\partial \eta}\right|_{\eta=0}=-\frac{I}{\mathcal{F}\pi AC_0D},\quad \rho_2\le 1\,, \tag{18.20b}$$

$$\left.\frac{\partial C}{\partial \eta}\right|_{\eta=0}=0\,,\quad \rho_1>1 \text{ and } \rho_2>1\,, \tag{18.20c}$$

and the far-field condition is $C(\eta, s_1, s_2) \to 1$ as $\sqrt{\eta^2+s_1^2+s_2^2}\to\infty$. The solution of this boundary value problem (Eq. (18.20)) has an integral form [29],

$$C(\eta, s_1, s_2)=1+\frac{I}{\mathcal{F}\pi AC_0D}\int_0^\infty e^{-m\eta}\frac{J_1(m)}{m}\left(J_0\left(m\rho_1\right)-J_0\left(m\rho_2\right)\right)dm \tag{18.21}$$

where $J_n(x)$ is the Bessel function of order $n$. When calculating the difference between the entry and exit points, we obtain

$$\begin{aligned} c(\boldsymbol{x}_1)-c(\boldsymbol{x}_2)&=C_0C\left(0,0,-\frac{L}{2A}\right)-C_0C\left(0,0,\frac{L}{2A}\right)\\ &=\frac{2I}{\mathcal{F}\pi AD}\left(1-\int_0^\infty J_1(m)J_0\left(m\frac{L}{A}\right)dm\right)\,, \end{aligned} \tag{18.22}$$

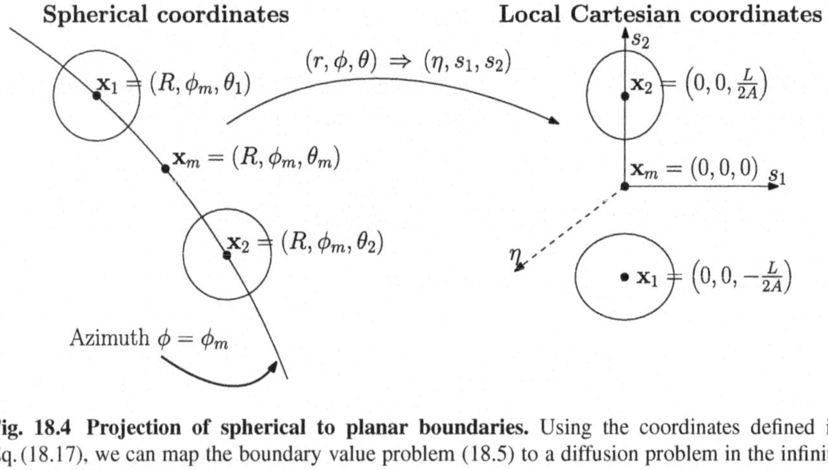


**Fig. 18.4 Projection of spherical to planar boundaries.** Using the coordinates defined in Eq. (18.17), we can map the boundary value problem (18.5) to a diffusion problem in the infinite half-space (Eq. (18.20))

and interestingly for well-spaced windows, i.e. with $L/A \gg 1$, the integral term vanishes and we recover the leading-order term from Eq. (18.6). Finally, from the time-dependent trajectories $C(t) = (\eta(t), s_1(t), s_2(t))^T$ solutions of the dynamical system

$$\frac{d}{dt}C(t) = -\begin{pmatrix} \frac{\partial}{\partial \eta}C(\eta(t), s_1(t), s_2(t)) \\ \frac{\partial}{\partial s_1}C(\eta(t), s_1(t), s_2(t)) \\ \frac{\partial}{\partial s_2}C(\eta(t), s_1(t), s_2(t)) \end{pmatrix}, \quad C(0) = \begin{pmatrix} 0 \\ 0 \\ -\frac{L}{2A} \end{pmatrix}, \tag{18.23}$$

with starting point at the center of the influx receiving window, we define the penetration length as $L_{\text{pe}} = A\eta_{\text{pe}}$, with $\eta_{\text{pe}}$ the maximal distance from the planar boundary along $C(t)$,

$$\eta_{\text{pe}} = \max\left\{\eta(t) \,\middle|\, \frac{d}{dt}C(t) = -\nabla_{(\eta,s_1,s_2)}C\left(C(t)\right) \text{ with } C(0) = \left(0, 0, -\frac{L}{2A}\right)^T\right\}. \tag{18.24}$$

Because of symmetry, we can reduce (18.23) to a 2-D dynamical system on the half-plane $s_1 = 0$ and $\eta > 0$ that is given by

$$\begin{pmatrix} \frac{d\eta}{dt} \\ \frac{ds_2}{dt} \end{pmatrix} = \frac{I}{\mathcal{F}\pi AC_0D}\begin{pmatrix} \int_0^\infty e^{-m\eta}J_1(m)\left(J_0\left(m\left(\frac{L}{2A}+s_2\right)\right) - J_0\left(m\left(\frac{L}{2A}-s_2\right)\right)\right)dm \\ \int_0^\infty e^{-m\eta}J_1(m)\left(J_1\left(m\left(\frac{L}{2A}+s_2\right)\right) + J_1\left(m\left(\frac{L}{2A}-s_2\right)\right)\right)dm \end{pmatrix}, \tag{18.25a}$$

with starting point

$$\begin{pmatrix} \eta(0) \\ s_2(0) \end{pmatrix} = \begin{pmatrix} 0 \\ -\frac{L}{2A} \end{pmatrix}. \tag{18.25b}$$

This reduced initial value problem can be solved numerically (Fig. 18.5). It allows to study how the current intensity $I$ and the inter-window distance $L$ affect the penetration length $L_{\text{pe}}$. Interestingly, an influx current can penetrate on ranges from 50 to 200 nm, independent of the current amplitude $I$ (Fig. 18.5b) while being linearly proportional to the distance $L$ between the narrow windows (Fig. 18.5c). The discrepancy observed when $L$ is large between the penetration lengths calculated by numerically integrating the initial value problem (18.25), and directly from COMSOL simulation results, results from the membrane curvature. This aspect is neglected at leading order within the boundary layer analysis (Eq. (18.20)).

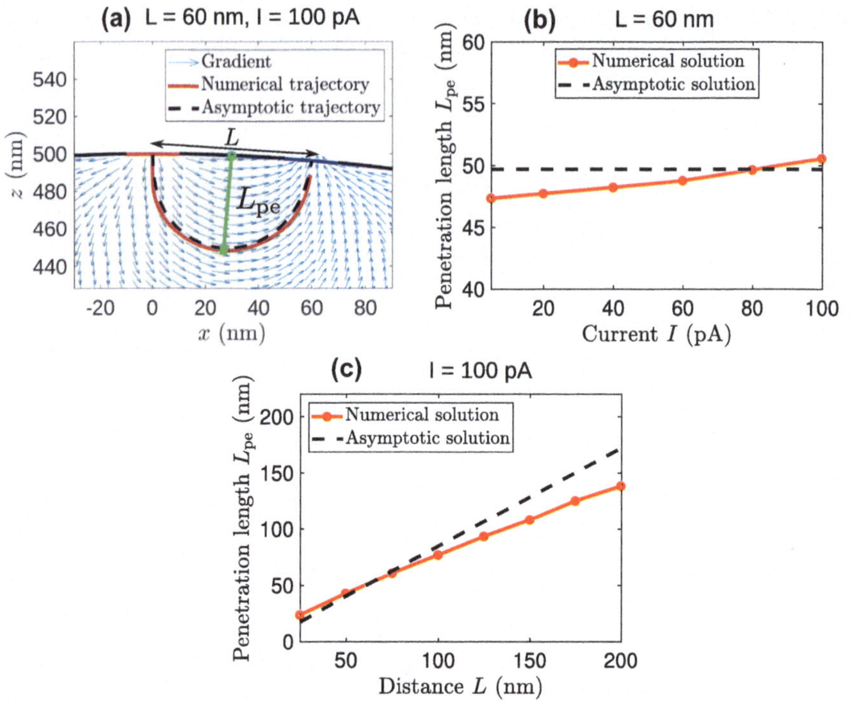


**Fig. 18.5 Field lines and penetration length. a** Example of trajectories tangent to the field lines (blue arrows) and connecting the entry and exit windows, with distance and current parameters $L = 60$ nm and $I = 100$ pA. The red trajectory follows a gradient calculated with COMSOL [28], while the black dashed trajectory is computed by numerically integrating the initial value problem (18.25). Finally the green line indicates the depth of penetration $L_{\text{pe}}$. **b** Penetration length $L_{\text{pe}}$ versus the influx magnitude $I$, with $L = 60$ nm. **c** Penetration length $L_{\text{pe}}$ versus the distance $L$, with $I = 100$ pA. Other parameter values are the same as in Fig. 18.3

### *18.3.3 Concentration Distribution with Multiple Absorbing Windows*

An alternative model to (18.5) consists of diffusing particles that are expelled from the domain through a collection of exit sites each with absorbing boundary conditions (Fig. 18.2b). The boundary $\partial\Omega$ is now divided into a reflecting part $\partial\Omega_r$, a single narrow influx window $\partial\Omega_{A_1}$, and $N-1$ narrow absorbing windows $\partial\Omega_{A_j}$ for $j = 2, \ldots, N$. All the windows $\partial\Omega_{A_j}$ are identical disks of radius A. By denoting the union of all exit sites as $\partial\Omega_a = \cup_{j=2}^{N} \partial\Omega_{A_j}$, the influx diffusion model (18.5) is as follows:

$$D\Delta c(\boldsymbol{x}) = 0\,, \qquad \boldsymbol{x} \in \Omega\,, \tag{18.26a}$$

with boundary conditions

$$D\frac{\partial c(\boldsymbol{x})}{\partial \boldsymbol{n}} = \frac{I}{\mathcal{F}\pi A^2}, \quad \boldsymbol{x} \in \partial\Omega_{A_1}, \quad c(\boldsymbol{x}) = 0, \quad \boldsymbol{x} \in \partial\Omega_a, \tag{18.26b}$$

$$\frac{\partial c(\boldsymbol{x})}{\partial \boldsymbol{n}} = 0, \quad \boldsymbol{x} \in \partial\Omega_r. \tag{18.26c}$$

Since $|\partial\Omega_{A_j}| \ll |\partial\Omega|$, the concentration $c(\boldsymbol{x}_1)$ at the center of the influx window [25] can be computed as

$$c(\boldsymbol{x}_1) = \frac{I}{\mathcal{F}\pi AD}\left(1 + \frac{\pi}{4(N-1)} - \left(H(\boldsymbol{x}_1) + \frac{1}{(N-1)^2}\sum_{i=2}^{N} H(\boldsymbol{x}_i)\right)\frac{A}{4}\log\left(\frac{A}{R}\right) + O\left(\frac{A}{R}\right)\right), \tag{18.27}$$

where the mean curvature $H(\boldsymbol{x}_j)$ is evaluated at the center $\boldsymbol{x}_j$ of window $\partial\Omega_{A_j}$. The derivation uses the classical Weber's solution [30] to approximate the outward flux at each exit site,

$$D\frac{\partial c(\boldsymbol{x})}{\partial \boldsymbol{n}} = \frac{K_j}{\sqrt{A^2 - \|\boldsymbol{x} - \boldsymbol{x}_j\|^2}}, \quad \text{for} \quad \|\boldsymbol{x} - \boldsymbol{x}_j\| < A, \quad j \neq 1, \tag{18.28}$$

where the constant $K_j$ controls the strength of the exit flux through the boundary $\partial\Omega_{A_j}$. By integrating (18.28) over each window area $\partial\Omega_{A_j}$, we get that the total outflux $\Phi_j$, in units of mol/s, is given by

$$\Phi_j = \int_{\partial\Omega_{A_j}} D\frac{\partial c(\boldsymbol{x})}{\partial \boldsymbol{n}} dx = 2\pi K_j \int_0^A \frac{rdr}{\sqrt{A^2 - r^2}} = 2\pi A K_j, \quad j \neq 1. \tag{18.29}$$

The contribution from all exiting fluxes must compensate the total influx, $\Phi_1 = I/\mathcal{F}$ (compatibility condition), thus linking together the unknown constants $K_j$,

$$\sum_{j=2}^{N} K_j = -\frac{I}{\mathcal{F}2\pi A}. \tag{18.30}$$

For a general domain $\Omega$, this formula involves the mean curvature function $H(\boldsymbol{x})$ [25],

$$\Phi_j = \frac{-I}{(N-1)\mathcal{F}}\left(1 - \frac{A\log(A/R)}{\pi(N-1)}\sum_{i=2}^{N}(H(\boldsymbol{x}_i) - H(\boldsymbol{x}_j)) + O\left(\frac{A}{R}\right)\right), \quad j \neq 1. \tag{18.31}$$

#### 18.3.3.1 Explicit Solution on a Spherical Domain

For a ball of radius $R$, we use the exact Green's function (18.14) to derive a three-term asymptotic expansion for the concentration at the influx location [25],

$$
\begin{aligned}
c(\boldsymbol{x}_1) = \frac{I}{(N-1)\mathcal{F}\pi AD}\Bigg(N-1+\frac{\pi}{4}-\frac{NA}{4R}\log\left(\frac{A}{R}\right)+\Bigg(\frac{N+1}{8}-\frac{\log(2)}{4} \\
+\frac{1}{N-1}\sum_{i=2}^{N}\sum_{j=i+1}^{N}\left(\frac{R}{L_{ij}}-\frac{1}{2}\log\left(\frac{L_{ij}^2}{2R^2}+\frac{L_{ij}}{R}\right)\right) \\
-\sum_{j=2}^{N}\left(\frac{R}{L_{1j}}-\frac{1}{2}\log\left(\frac{L_{1j}^2}{2R^2}+\frac{L_{1j}}{R}\right)\right)\Bigg)\frac{A}{R}+O\left(\frac{A^2}{R^2}\right)\Bigg),
\end{aligned}
\tag{18.32}
$$

revealing the role of the windows' geometrical organization on the maximal concentration drop across the domain. For a single exit window, this formula reduces to

$$
\begin{aligned}
c(\boldsymbol{x}_1) = \frac{I}{\mathcal{F}\pi AD}\Bigg(1+\frac{\pi}{4}-\frac{A}{2R}\log\left(\frac{A}{R}\right)+\Bigg(\frac{3}{8}-\frac{\log(2)}{4} \\
-\frac{R}{L}+\frac{1}{2}\log\left(\frac{L^2}{2R^2}+\frac{L}{R}\right)\Bigg)\frac{A}{R}+O\left(\left(\frac{A}{R}\right)^2\right)\Bigg).
\end{aligned}
\tag{18.33}
$$

Next, for the total outflux $\Phi_j$ through each exit window, the second term of the expansion (18.31) vanishes since the mean curvature is constant on spherical domains. An additional term of order $O(A/R)$ yields the formula below [25],

$$
\begin{aligned}
\Phi_j = -\frac{I}{(N-1)\mathcal{F}}\Bigg(1+\Bigg(\sum_{\substack{i=2\\ i\neq j}}^{N}\Bigg(\frac{R}{L_{1j}}-\frac{R}{L_{1i}}-\frac{R}{L_{ij}} \\
-\frac{1}{2}\log\left(\frac{2R^2L_{1j}^2+4R^3L_{1j}}{\left(L_{1i}^2+2RL_{1i}\right)\left(L_{ij}^2+2RL_{ij}\right)}\right)\Bigg) \\
+\frac{2}{(N-1)}\sum_{i=2}^{N}\sum_{k=i+1}^{N}\left(\frac{R}{L_{ik}}-\frac{1}{2}\log\left(\frac{L_{ik}^2}{2R^2}+\frac{L_{ik}}{R}\right)\right)\Bigg)\frac{2A}{\pi R}+O\left(\left(\frac{A}{R}\right)^2\right)\Bigg), \quad j\neq 1.
\end{aligned}
\tag{18.34}
$$

Finally, numerical results illustrating how the relative position with respect to the influx window can affect the strength of the exit fluxes are shown in Fig. 18.6. In a spherical domain with three narrow windows (Fig. 18.6a), for which only the distance $L_{13}$ between the first (receiving the influx) and third windows is allowed to vary, Eq. (18.34) with $N = 3$ is used to compute the flux ratio $\Phi_3/\Phi_2$ between the third and second windows,

$$
\frac{\Phi_3}{\Phi_2} = \frac{1-\frac{2A}{\pi R}\left(\frac{R}{L_{12}}-\frac{R}{L_{13}}+\frac{1}{2}\log\left(\frac{L_{13}^2+2RL_{13}}{L_{12}^2+2RL_{12}}\right)\right)+O\left(\left(\frac{A}{R}\right)^2\right)}{1+\frac{2A}{\pi R}\left(\frac{R}{L_{12}}-\frac{R}{L_{13}}+\frac{1}{2}\log\left(\frac{L_{13}^2+2RL_{13}}{L_{12}^2+2RL_{12}}\right)\right)+O\left(\left(\frac{A}{R}\right)^2\right)}.
\tag{18.35}
$$

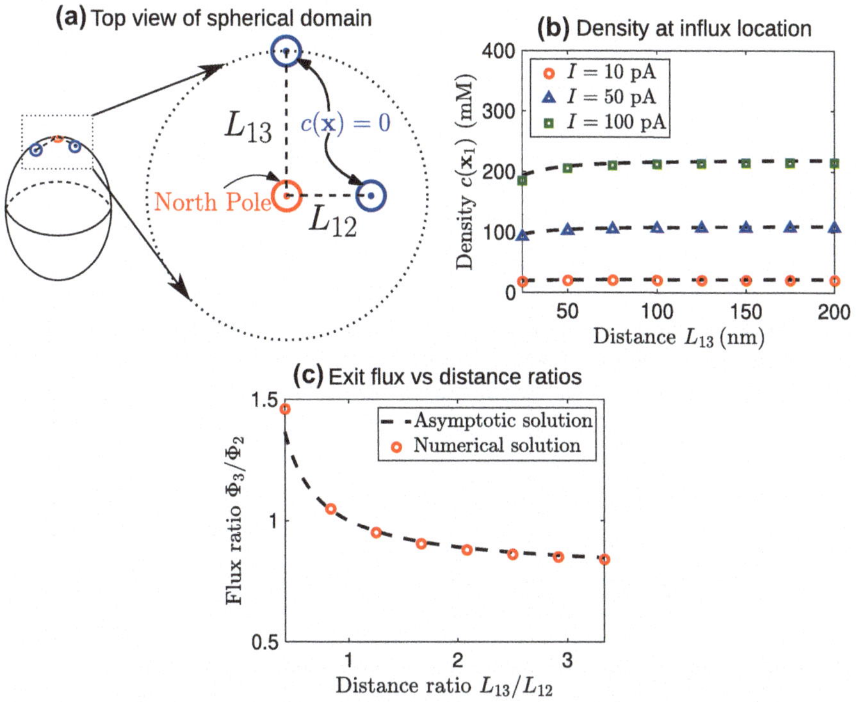


**Fig. 18.6 Splitting fluxes between two exit sites. a** Top view of a spherical domain with $N = 3$ narrow windows. $L_{12}$ and $L_{13}$ denote the distances from the influx window to the exit sites. **b** Density $c(\boldsymbol{x}_1)$ at influx location versus $L_{13}$. The three sets of symbols indicate the COMSOL numerical solutions, while asymptotic solutions are computed from Eq. (18.32) and are shown by black dashed curves. **c** Flux ratio $\Phi_3/\Phi_2$ versus distance ratio, showing both numerical (red circles) and asymptotic (Eq. (18.35), black dashed curve) solutions. Here we have $L_{12} = 60$ nm, with other parameters the same as in Fig. 18.3

Comparing this against ratios numerically computed with COMSOL [28], we conclude that exit sites located near the influx receiving window contribute more to the escape dynamics than the farther ones (Fig. 18.6c).

## 18.4 Concentration and Voltage Distributions in a Two-Charge Electro-Diffusion Model

We consider here a two-species steady-state model with positive $c_p(\boldsymbol{x})$ and negative $c_m(\boldsymbol{x})$ monovalent charges, and voltage $v(\boldsymbol{x})$. We first consider an arbitrary domain $\Omega$ with mean membrane curvature function $H(\boldsymbol{x})$ and characteristic length scale $R$. The boundary $\partial\Omega$ is smooth and is punctured by two narrow windows receiving and emitting an electrical current composed of positive charges only. The membrane is impermeable to negative charges. The positive and negative charge densities are each solution of the Nernst–Planck equation:

$$\begin{cases} D_p\left(\Delta c_p(\boldsymbol{x}) + \frac{e}{k_B \mathcal{T}}\nabla(c_p(\boldsymbol{x})\nabla v(\boldsymbol{x}))\right) = 0\,, & \boldsymbol{x} \in \Omega\,, \\ D_m\left(\Delta c_m(\boldsymbol{x}) - \frac{e}{k_B \mathcal{T}}\nabla(c_m(\boldsymbol{x})\nabla v(\boldsymbol{x}))\right) = 0\,, & \boldsymbol{x} \in \Omega\,, \end{cases} \tag{18.36a}$$

with possibly distinct diffusion coefficients $D_p, D_m$. The voltage $v(\boldsymbol{x})$ is the solution of Poisson's equation, here given by

$$\Delta v(\boldsymbol{x}) = -\frac{\mathcal{F}}{\varepsilon\varepsilon_0}\left(c_p(\boldsymbol{x}) - c_m(\boldsymbol{x})\right)\,, \quad \boldsymbol{x} \in \Omega\,. \tag{18.36b}$$

The boundary $\partial\Omega$ is divided into a reflective part $\partial\Omega_r$, and two narrow windows $\partial\Omega_{A_1}$ and $\partial\Omega_{A_2}$ receiving and emitting positive charge electrical currents. The windows are identical disks of radius $A$ centered at $\boldsymbol{x}_1$ and $\boldsymbol{x}_2$ on the boundary. For the positive species $c_p(\boldsymbol{x})$, we have

$$D_p\left(\frac{\partial c_p}{\partial \boldsymbol{n}} + \frac{e}{k_B\mathcal{T}}c_p\frac{\partial v}{\partial \boldsymbol{n}}\right) = 0\,, \quad \boldsymbol{x} \in \partial\Omega_r\,, \tag{18.36c}$$

$$D_p\left(\frac{\partial c_p}{\partial \boldsymbol{n}} + \frac{e}{k_B\mathcal{T}}c_p\frac{\partial v}{\partial \boldsymbol{n}}\right) = \frac{I}{\mathcal{F}\pi A^2}\,, \quad \boldsymbol{x} \in \partial\Omega_{A_1}\,, \tag{18.36d}$$

$$D_p\left(\frac{\partial c_p}{\partial \boldsymbol{n}} + \frac{e}{k_B\mathcal{T}}c_p\frac{\partial v}{\partial \boldsymbol{n}}\right) = -\frac{I}{\mathcal{F}\pi A^2}\,, \quad \boldsymbol{x} \in \partial\Omega_{A_2}\,, \tag{18.36e}$$

while no-flux condition holds everywhere on the boundary for the negative charge,

$$D_m\left(\frac{\partial c_m}{\partial \boldsymbol{n}} - \frac{e}{k_B\mathcal{T}}c_m\frac{\partial v}{\partial \boldsymbol{n}}\right) = 0\,, \quad \boldsymbol{x} \in \partial\Omega\,. \tag{18.36f}$$

Moreover global electroneutrality, with average positive and negative densities equal to $C_0$, yields the constraints

$$\frac{1}{|\Omega|}\int_\Omega c_p(\boldsymbol{x})d\boldsymbol{x} = C_0\,, \quad \frac{1}{|\Omega|}\int_\Omega c_m(\boldsymbol{x})d\boldsymbol{x} = C_0\,, \tag{18.36g}$$

and therefore to satisfy the compatibility condition of Poisson's equation (Eq. (18.36b)), we impose that the total surface charge vanishes at the boundary

$$\frac{\partial v}{\partial \boldsymbol{n}} = 0\,, \quad \boldsymbol{x} \in \partial\Omega\,. \tag{18.36h}$$

To compensate for the localized fluxes of positive charges and due to electrochemical forces, the negative ion steady-state distribution must also become spatially heterogeneous. Thus, we can quantify the resulting density differences between the entry and exit points (see Sect. 18.3), in relation with current and geometrical parameters. Moreover, asymptotic analysis allows to compute the density and voltage differences as (cf. Appendix 1)

$$c_p(\boldsymbol{x}_1) - c_p(\boldsymbol{x}_2) = c_m(\boldsymbol{x}_1) - c_m(\boldsymbol{x}_2) = \frac{I}{\mathcal{F}\pi A D_p}\left(1 - \frac{H(\boldsymbol{x}_1) + H(\boldsymbol{x}_2)}{8}A\log\left(\frac{A}{R}\right) + o\left(\frac{A}{R}\right)\right), \tag{18.37a}$$

while the voltage is the logarithm of the ratio between the entry $c_m(\boldsymbol{x}_1)$ and exit $c_m(\boldsymbol{x}_2)$ negative charge densities,

$$v(\boldsymbol{x}_1) - v(\boldsymbol{x}_2) = \frac{k_B\mathcal{T}}{e}\log\left(\frac{C_0 + \frac{I}{2\mathcal{F}\pi A D_p}\left(1 - \frac{H(\boldsymbol{x}_1)+H(\boldsymbol{x}_2)}{8}A\log\left(\frac{A}{R}\right) + O\left(\frac{A}{R}\right)\right)}{C_0 - \frac{I}{2\mathcal{F}\pi A D_p}\left(1 - \frac{H(\boldsymbol{x}_1)+H(\boldsymbol{x}_2)}{8}A\log\left(\frac{A}{R}\right) + O\left(\frac{A}{R}\right)\right)}\right), \tag{18.37b}$$

where $H(\boldsymbol{x}_j)$ is the mean curvature calculated at the center $\boldsymbol{x}_j$, $j = 1, 2$. These formulas show that the membrane curvature can modulate the voltage [22, 23, 31]. Interestingly the deviation magnitudes predicted by formulas Eqs. (18.37a) and (18.6) (diffusion only) relate by a factor one-half. To conclude, less significant density differences are achieved with electro-diffusion than for diffusion.

When an absorbing condition $c_p(\boldsymbol{x}) = 0$ is imposed on the exit window $\partial\Omega_{A_2}$, an alternative formula for the voltage drop is obtained by considering the following ansatz:

$$v(\boldsymbol{x}_1) - v(\boldsymbol{x}_2) = \frac{k_B\mathcal{T}}{e}\log\left(1 + \gamma\left(\frac{IR}{\mathcal{F}D_p\pi A^2 C_0}\right)^{\alpha}\frac{A}{R}F\left(\frac{A}{R}\right)\right), \tag{18.38}$$

where $F\left(\frac{A}{R}\right)$ is defined as

$$F\left(\frac{A}{R}\right) = 1 + \frac{\pi}{4} - \frac{H(\boldsymbol{x}_1) + H(\boldsymbol{x}_2)}{4}A\log\left(\frac{A}{R}\right) + O\left(\frac{A}{R}\right). \tag{18.39}$$

Here the exponent $\alpha$ and the parameter $\gamma$ are found by numerical fit.

### 18.4.1 Electro-Diffusion with Two Species in a Spherical Domain

For the case of a unit ball domain, an additional term to the asymptotic formulas (18.37) can be obtained that reveals the role of the relative window positions on the voltage. The asymptotic expansion of $F(A/R)$ (Eq. (18.39)) leads to

$$F\left(\frac{A}{R}\right) = 1 - \frac{A}{4R}\log\left(\frac{A}{R}\right) + \left(\frac{1}{8} - \frac{R}{2L} + \frac{1}{4}\log\left(\frac{L^2}{2R^2} + \frac{L}{R}\right)\right)\frac{A}{R} + O\left(\left(\frac{A}{R}\right)^2\right), \tag{18.40}$$

and the density differences are of the same order for both species (cf. Appendix 1),

$$c_p(\boldsymbol{x}_1) - c_p(\boldsymbol{x}_2) = c_m(\boldsymbol{x}_1) - c_m(\boldsymbol{x}_2) = \frac{I}{\mathcal{F}\pi A D_p}F\left(\frac{A}{R}\right), \tag{18.41a}$$

while the voltage is given by

$$v(\boldsymbol{x}_1) - v(\boldsymbol{x}_2) = \frac{k_B \mathcal{T}}{e} \log \left( \frac{C_0 + \frac{I}{2\mathcal{F}\pi A D_p} F\left(\frac{A}{R}\right)}{C_0 - \frac{I}{2\mathcal{F}\pi A D_p} F\left(\frac{A}{R}\right)} \right). \tag{18.41b}$$

Formula (18.41b) predicts nonlinear voltage effects deviating from Ohm's law at large current amplitude, with those nonlinear effects also observed numerically (Fig. 18.7c). Moreover, we report voltage spatial variations on ranges of size of 10–25 nm (Fig. 18.7a). This range is well beyond the Debye length $\lambda_D$, obtained from classical electrolyte theory [12], which for two monovalent ions leads to $\lambda_D = \sqrt{\frac{\varepsilon\varepsilon_0 k_B \mathcal{T}}{2e^2 N_A C_0}} \approx 0.7$ nm (parameter values are from Table 18.1).

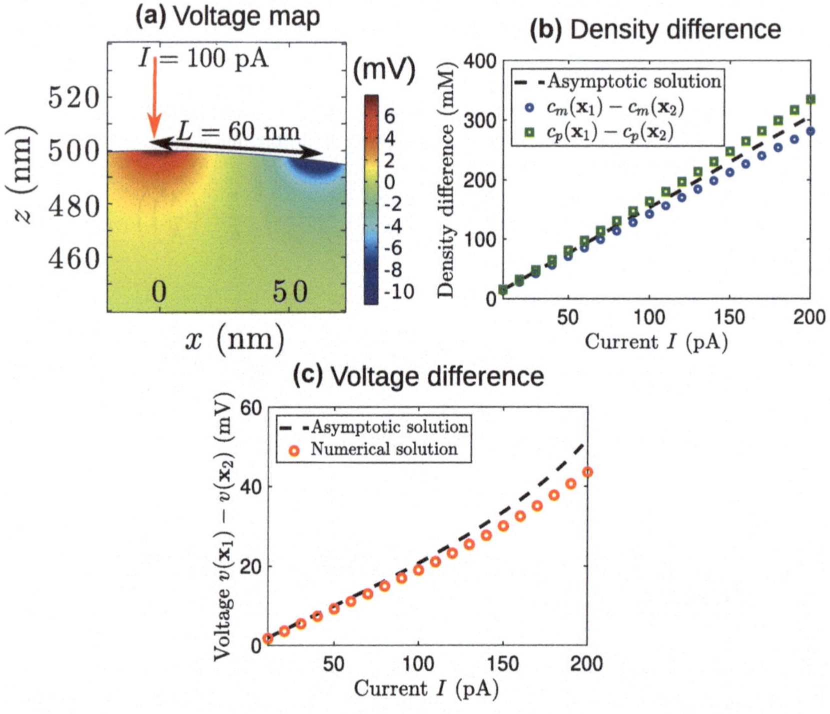


**Fig. 18.7 Nonlinear voltage–current relations. a** 2-D numerical voltage map computed with COMSOL [28], with distance and current parameters $L = 60$ nm and $I = 100$ pA. **b** Comparison between positive (green squares) and negative (blue circles) charge concentration differences, as a function of the current magnitude $I$. The black dashed curve indicates the asymptotic solution (18.41a). **c** Voltage drop versus the current magnitude $I$, showing both numerical (red circles) and asymptotic (Eq. (18.41b), black dashed curve) solutions. In **b–c** the distance is set to $L = 60$ nm, with other parameter values taken from Table 18.1

## 18.5 Conclusions and Perspectives

In this chapter, we presented models and analyses of voltage dynamics and ion density distribution after their entry through an ionic channel, their motion inside the bulk of a cell, until their eventual extrusion by a neighboring pump. In the diffusion approximation with pumps or channels modeled as small windows, an influx of ions generates a local difference of concentration between the source and a neighboring target window. Computing local ionic concentration differences allows to quantify the extension of nanodomains where voltage distribution triggers the opening of neighboring channels [9]. The mathematical problem was first formulated with Laplace's equation in Sect. 18.3, which neglects electrochemical effects. Asymptotic formulas for the concentration difference between two narrow windows, each with a Neumann flux boundary condition (Sect. 18.3.1), reveal that it depends at first order on the window size $A$, the current intensity $I$ and the diffusion coefficient $D$, while at second order it also depends on the mean curvature computed at the center of each window. The third-order terms involve the geometrical organization of all the windows via the explicit solution of Green's function. These computations and formulas resemble those from the classical narrow escape theory [26, 27, 32–38] describing the average escape time of a stochastic particle from a domain with small absorbing targets on an otherwise reflective boundary. But while the mathematical formulation is similar as it also involves a mixed boundary value problem, with Poisson's instead of Laplace's equations, the underlying biophysical modeling differs. Indeed, narrow escape theory provides time scales to quantify biochemical reaction rates, and applies to a single Brownian particle with a starting point inside a microdomain, without any external influx of particles. In Sect. 18.3.1, instead of exit times we modeled steady-state ionic densities and evaluated local concentration changes between narrow windows receiving and emitting particle fluxes.

In addition, the present approach was used to measure how far from the channel the interior concentration can be perturbed (Sect. 18.3.2). For that goal, we used the penetration length introduced in [25]. This length scale is constructed using a field lines approach, by measuring the maximal distance from the boundary for trajectories connecting the flux receiving and emitting windows. The penetration length is independent of the current intensity $I$ of the field, but depends linearly on the distance $L$ between the windows. In Sect. 18.3.3 we discussed the case of multiple exit windows with Dirichlet absorbing boundary conditions. Numerical and asymptotic results confirmed how exit fluxes and their magnitudes are mediated by their relative positions with respect to the entry site.

Finally, the electric field generated by two charged ions was modeled using electro-diffusion theory in Sect. 18.4, where the localized flux current was composed of positive charges only. From density difference formulas, we obtained voltage–current laws emphasizing the role of geometrical parameters and membrane curvature on voltage dynamics. We also reported interior voltage simulation results suggesting the formation of a nanodomain, with voltage effects extending well-beyond the Debye screening length from the classical theory of electrolytes [12].

To conclude, in this chapter we briefly summarized diffusion and electro-diffusion modeling approaches to investigate concentration and voltage spatial changes near the opening of voltage-gated channels. These local variations are behind the formation of voltage nanodomains [9], which are involved in controlling channel open probabilities in dendrites and small protrusions of neuronal cells [7]. By focusing on steady-state problems, we were able to derive simple biophysical laws relating the geometrical organization of channels to ionic density and voltage effects. However these results do not provide information about return time scales to resting potential following a sudden ion influx through a channel. Such a study extension to time-dependent processes will be tackled somewhere else.

**Acknowledgements** F. P.-L. gratefully acknowledges the support from the Foundation ARC through a postdoctoral fellowship (award No ARCPDF12020020001505). D. H. was supported by the European Research Council (ERC) under the European Union's Horizon 2020 Research and Innovation Program (grant agreement No 882673), and ANR AstroXcite.

## Appendix 1: Asymptotic Analysis for the Poisson–Nernst–Planck Model

In this appendix, we recall the asymptotic analysis for the two-charge Poisson–Nernst–Planck model (18.36). This analysis is used to derive the density and voltage difference formulas (18.37) and (18.41). First, the characteristic length scale $R$ serves to non-dimensionalize the equation. Upon introducing the dimensionless spatial coordinates $\tilde{\boldsymbol{x}} = \frac{\boldsymbol{x}}{R}$ within the domain $\tilde{\Omega} = \Omega/R$, with new density $\tilde{c}_p$, $\tilde{c}_m$ and voltage $\tilde{v}$ variables given by

$$\tilde{c}_p(\tilde{\boldsymbol{x}}) = \frac{c_p(\tilde{\boldsymbol{x}}R)}{C_0}, \quad \tilde{c}_m(\tilde{\boldsymbol{x}}) = \frac{c_m(\tilde{\boldsymbol{x}}R)}{C_0}, \quad \tilde{v}(\tilde{\boldsymbol{x}}) = \frac{e}{k_B\mathcal{T}} v(\tilde{\boldsymbol{x}}R), \tag{18.42}$$

we can reformulate system (18.36) as

$$\Delta\tilde{c}_p + \nabla\left(\tilde{c}_p\nabla\tilde{v}\right) = 0, \quad \Delta\tilde{c}_m - \nabla\left(\tilde{c}_m\nabla\tilde{v}\right) = 0, \quad \Delta\tilde{v} = -\lambda(\tilde{c}_p - \tilde{c}_m), \tag{18.43}$$

with $\tilde{\boldsymbol{x}} \in \tilde{\Omega}$ that is subject to the boundary conditions

$$\begin{aligned} &\frac{\partial\tilde{c}_p}{\partial\boldsymbol{n}} = J, \quad \tilde{\boldsymbol{x}} \in \partial\tilde{\Omega}_{a_1}, \quad \frac{\partial\tilde{c}_p}{\partial\boldsymbol{n}} = -J, \quad \tilde{\boldsymbol{x}} \in \partial\tilde{\Omega}_{a_2}, \\ &\frac{\partial\tilde{c}_p}{\partial\boldsymbol{n}} = 0, \quad \tilde{\boldsymbol{x}} \in \partial\tilde{\Omega}_r, \quad \frac{\partial\tilde{c}_m}{\partial\boldsymbol{n}} = \frac{\partial\tilde{v}}{\partial\boldsymbol{n}} = 0, \quad \tilde{\boldsymbol{x}} \in \partial\tilde{\Omega}. \end{aligned} \tag{18.44}$$

Here the narrow windows $\partial\tilde{\Omega}_{a_1}$ and $\partial\tilde{\Omega}_{a_2}$ have a radius $a = A/R$ and are separated by a distance $l = \|\tilde{\boldsymbol{x}}_1 - \tilde{\boldsymbol{x}}_2\| = L/R$, while $J$ and $\lambda$ are two dimensionless parameters defined as

$$J = \frac{IR}{\mathcal{F}\pi A^2 D_p C_0}, \quad \lambda = \frac{e^2R^2C_0N_A}{k_B\mathcal{T}\varepsilon\varepsilon_0}. \tag{18.45}$$

The asymptotic analysis is valid in the following parameter regime:

$$a \ll 1, \quad J \sim O(1) \ \text{ and } \ \lambda \gg 1, \quad \text{yielding} \quad \frac{J}{\sqrt{\lambda}} \ll 1, \tag{18.46}$$

which holds for an electrical current $I$ and a background concentration $C_0$ of the order of hundreds of picoamperes and millimolars (cf. Table 18.1). To remove the large coupling parameter $\lambda$ from Poisson's equation, we introduce the coordinate $\boldsymbol{\xi} = \sqrt{\lambda}\tilde{\boldsymbol{x}}$ within the stretched domain $\Omega_{\boldsymbol{\xi}} = \sqrt{\lambda}\tilde{\Omega}$, and the new variables

$$C_p(\boldsymbol{\xi}) = \tilde{c}_p(\boldsymbol{\xi}/\sqrt{\lambda}), \quad C_m(\boldsymbol{\xi}) = \tilde{c}_m(\boldsymbol{\xi}/\sqrt{\lambda}), \quad \mathcal{V}(\boldsymbol{\xi}) = \tilde{v}(\boldsymbol{\xi}/\sqrt{\lambda}), \tag{18.47}$$

we find

$$\Delta C_p + \nabla\left(C_p \nabla \mathcal{V}\right) = 0, \quad \boldsymbol{\xi} \in \Omega_{\boldsymbol{\xi}}, \tag{18.48a}$$
$$\Delta C_m - \nabla\left(C_m \nabla \mathcal{V}\right) = 0, \quad \boldsymbol{\xi} \in \Omega_{\boldsymbol{\xi}}, \tag{18.48b}$$
$$\Delta \mathcal{V} = -(C_p - C_m), \quad \boldsymbol{\xi} \in \Omega_{\boldsymbol{\xi}}, \tag{18.48c}$$

subject to the boundary conditions

$$\frac{\partial C_p}{\partial \boldsymbol{n}} = \frac{J}{\sqrt{\lambda}}, \quad \boldsymbol{\xi} \in \partial\Omega_{\boldsymbol{\xi}_1}, \quad \frac{\partial C_p}{\partial \boldsymbol{n}} = -\frac{J}{\sqrt{\lambda}}, \quad \boldsymbol{\xi} \in \partial\Omega_{\boldsymbol{\xi}_2}, \tag{18.48d}$$
$$\frac{\partial C_p}{\partial \boldsymbol{n}} = 0, \quad \boldsymbol{\xi} \in \partial\Omega_{\boldsymbol{\xi}_r}, \quad \frac{\partial C_m}{\partial \boldsymbol{n}} = \frac{\partial \mathcal{V}}{\partial \boldsymbol{n}} = 0, \quad \boldsymbol{\xi} \in \partial\Omega_{\boldsymbol{\xi}}, \tag{18.48e}$$

and constraints

$$\frac{1}{|\Omega_{\boldsymbol{\xi}}|}\int_{\Omega_{\boldsymbol{\xi}}} C_p(\boldsymbol{\xi})d\boldsymbol{\xi} = 1, \quad \frac{1}{|\Omega_{\boldsymbol{\xi}}|}\int_{\Omega_{\boldsymbol{\xi}}} C_m(\boldsymbol{\xi})d\boldsymbol{\xi} = 1. \tag{18.48f}$$

The parameter regime (18.46) justifies a regular asymptotic expansion in terms of the small ratio $\frac{J}{\sqrt{\lambda}} \ll 1$:

$$C_m(\boldsymbol{\xi}) = C_e(\boldsymbol{\xi}) + C_{p1}(\boldsymbol{\xi})\frac{J}{\sqrt{\lambda}} + O\left(\frac{J^2}{\lambda}\right), \tag{18.49a}$$

$$C_m(\boldsymbol{\xi}) = C_e(\boldsymbol{\xi}) + C_{m1}(\boldsymbol{\xi})\frac{J}{\sqrt{\lambda}} + O\left(\frac{J^2}{\lambda}\right), \tag{18.49b}$$

$$\mathcal{V}(\boldsymbol{\xi}) = \mathcal{V}_1(\boldsymbol{\xi})\frac{J}{\sqrt{\lambda}} + O\left(\frac{J^2}{\lambda}\right). \tag{18.49c}$$

Upon substituting (18.49) within (18.48), we can derive a series of boundary value problems at each order in $J/\sqrt{\lambda}$. At leading order, we get

$$\Delta C_e = 0, \quad \boldsymbol{\xi} \in \Omega_{\boldsymbol{\xi}}, \quad \frac{\partial C_e}{\partial \boldsymbol{n}} = 0, \quad \boldsymbol{\xi} \in \partial\Omega_{\boldsymbol{\xi}}, \quad \frac{1}{|\Omega_{\boldsymbol{\xi}}|}\int_{\Omega_{\boldsymbol{\xi}}} C_e(\boldsymbol{\xi})d\boldsymbol{\xi} = 1, \tag{18.50}$$

whose solution is $C_e(\boldsymbol{\xi}) = 1$, while collecting terms of order $O(J/\sqrt{\lambda})$ yields

$$\Delta C_{p1} + \nabla\left(C_e \nabla \mathcal{V}_1\right) = 0, \quad \boldsymbol{\xi} \in \Omega_{\boldsymbol{\xi}}, \tag{18.51}$$
$$\Delta C_{m1} - \nabla\left(C_e \nabla \mathcal{V}_1\right) = 0, \quad \boldsymbol{\xi} \in \Omega_{\boldsymbol{\xi}}, \tag{18.52}$$
$$\Delta \mathcal{V}_1 = -(C_{p1} - C_{m1}), \quad \boldsymbol{\xi} \in \Omega_{\boldsymbol{\xi}}, \tag{18.53}$$

with the boundary conditions

$$\frac{\partial C_{p1}}{\partial \boldsymbol{n}} = 1\,, \quad \boldsymbol{\xi} \in \partial\Omega_{\boldsymbol{\xi}_1}\,, \quad \frac{\partial C_{p1}}{\partial \boldsymbol{n}} = -1\,, \quad \boldsymbol{\xi} \in \partial\Omega_{\boldsymbol{\xi}_2}\,, \quad \frac{\partial C_{p1}}{\partial \boldsymbol{n}} = 0\,, \quad \boldsymbol{\xi} \in \partial\Omega_{\boldsymbol{\xi}_r}\,, \tag{18.54}$$

$$\frac{\partial C_{m1}}{\partial \boldsymbol{n}} = 0\,, \quad \boldsymbol{\xi} \in \partial\Omega_{\boldsymbol{\xi}}\,, \quad \frac{\partial \mathcal{V}_1}{\partial \boldsymbol{n}} = 0\,, \quad \boldsymbol{\xi} \in \partial\Omega_{\boldsymbol{\xi}}\,, \tag{18.55}$$

and global electroneutrality constraints

$$\frac{1}{|\Omega_{\boldsymbol{\xi}}|}\int_{\Omega_{\boldsymbol{\xi}}} C_{p1}(\boldsymbol{\xi})d\boldsymbol{\xi} = \frac{1}{|\Omega_{\boldsymbol{\xi}}|}\int_{\Omega_{\boldsymbol{\xi}}} C_{m1}(\boldsymbol{\xi})d\boldsymbol{\xi} = 0\,. \tag{18.56}$$

By adding and subtracting Eqs. (18.51) and (18.52) we find that, since $C_e = 1$ is constant,

$$\Delta\left(C_{p1} + C_{m1}\right) = 0\,, \quad \Delta(C_{p1} - C_{m1}) - 2(C_{p1} - C_{m1}) = 0\,, \quad \boldsymbol{\xi} \in \Omega_{\boldsymbol{\xi}}\,. \tag{18.57}$$

Then we use Green's function $\mathcal{G}_s(\boldsymbol{\xi};\boldsymbol{\eta})$ solution of

$$\Delta\mathcal{G}_s(\boldsymbol{\xi};\boldsymbol{\eta}) = \frac{\sqrt{\lambda}}{|\Omega_{\boldsymbol{\xi}}|}\,, \quad \boldsymbol{\xi} \in \Omega_{\boldsymbol{\xi}}\,, \boldsymbol{\eta} \in \partial\Omega_{\boldsymbol{\xi}}\,, \tag{18.58}$$

$$\frac{\partial\mathcal{G}_s(\boldsymbol{\xi};\boldsymbol{\eta})}{\partial \boldsymbol{n}} = \sqrt{\lambda}\delta(\boldsymbol{\xi} - \boldsymbol{\eta})\,, \quad \boldsymbol{\xi}, \boldsymbol{\eta} \in \partial\Omega_{\boldsymbol{\xi}} \tag{18.59}$$

that behaves near the singular diagonal $\boldsymbol{\xi} = \boldsymbol{\eta}$ as

$$\mathcal{G}_s(\boldsymbol{\xi};\boldsymbol{\eta}) \approx \frac{\sqrt{\lambda}}{2\pi\|\boldsymbol{\xi} - \boldsymbol{\eta}\|} - \sqrt{\lambda}\frac{\mathcal{H}(\boldsymbol{\eta})}{4\pi}\log\left(\frac{\|\boldsymbol{\xi} - \boldsymbol{\eta}\|}{\sqrt{\lambda}}\right) + O(1)\,, \tag{18.60}$$

with mean boundary curvature function $\mathcal{H}(\boldsymbol{\eta})$. We can solve (18.57) as follows using Green's identity

$$C_{p1}(\boldsymbol{\eta}) + C_{m1}(\boldsymbol{\eta}) = \frac{1}{\sqrt{\lambda}}\left(\int_{\partial\Omega_{\boldsymbol{\xi}_1}} \mathcal{G}_s(\boldsymbol{\xi};\boldsymbol{\eta})d\boldsymbol{\xi} - \int_{\partial\Omega_{\boldsymbol{\xi}_2}} \mathcal{G}_s(\boldsymbol{\xi};\boldsymbol{\eta})d\boldsymbol{\xi}\right)\,, \tag{18.61a}$$

$$\begin{aligned} C_{p1}(\boldsymbol{\eta}) - C_{m1}(\boldsymbol{\eta}) = \frac{1}{\sqrt{\lambda}}\Bigg(&\int_{\partial\Omega_{\boldsymbol{\xi}_1}} \mathcal{G}_s(\boldsymbol{\xi};\boldsymbol{\eta})d\boldsymbol{\xi} - \int_{\partial\Omega_{\boldsymbol{\xi}_2}} \mathcal{G}_s(\boldsymbol{\xi};\boldsymbol{\eta})d\boldsymbol{\xi} \\ &-2\int_{\Omega_{\boldsymbol{\xi}}} \mathcal{G}_s(\boldsymbol{\xi};\boldsymbol{\eta})\left(C_{p1}(\boldsymbol{\xi}) - C_{m1}(\boldsymbol{\xi})\right)d\boldsymbol{\xi}\Bigg)\,. \end{aligned} \tag{18.61b}$$

Upon setting $\boldsymbol{\eta} = \boldsymbol{\xi}_1$ and $\boldsymbol{\xi}_2$ within Eq. (18.61a), we evaluate the surface integrals as [25]

$$C_{p1}(\boldsymbol{\xi}_1) + C_{m1}(\boldsymbol{\xi}_1) = \sqrt{\lambda}aF_1(a)\,, \quad C_{p1}(\boldsymbol{\xi}_2) + C_{m1}(\boldsymbol{\xi}_2) = -\sqrt{\lambda}aF_2(a)\,, \tag{18.62}$$

where $F_i(a)$ is the following asymptotic expansion:

$$F_i(a) = 1 - \sqrt{\lambda}\frac{\mathcal{H}(\boldsymbol{\xi}_i)}{4}a\log(a) + O(a)\,, \quad i = 1, 2\,. \tag{18.63}$$

Doing the same for Eq. (18.61b) with $\eta = \xi_1$ yields

$$C_{p1}(\xi_1) - C_{m1}(\xi_1) = \sqrt{\lambda} a F_1(a) - \frac{2}{\sqrt{\lambda}} \int_{\Omega_\xi} G_s(\xi;\eta) \left(C_{p1}(\xi) - C_{m1}(\xi)\right) d\xi \quad (18.64)$$

which further approximates as

$$\begin{aligned} C_{p1}(\xi_1) - C_{m1}(\xi_1) &= \sqrt{\lambda} a F_1(a) \\ &- \frac{2}{\sqrt{\lambda}} \left(C_{p1}(\xi_1) - C_{m1}(\xi_1)\right) \lambda^{\frac{3}{2}} \frac{a^2}{2} \left(1 + O(a \log(a))\right), \end{aligned} \quad (18.65)$$

and gets finally solved as

$$C_{p1}(\xi_1) - C_{m1}(\xi_1) = \frac{\sqrt{\lambda} a F_1(a)}{1 + \lambda a^2 (1 + O(a \log(a)))}. \quad (18.66)$$

We use symmetry arguments to derive the result for $\eta = \xi_2$,

$$C_{p1}(\xi_2) - C_{m1}(\xi_2) = -\frac{\sqrt{\lambda} a F_2(a)}{1 + \lambda a^2 (1 + O(a \log(a)))}, \quad (18.67)$$

and solving Eqs. (18.62), (18.66) and (18.67) for the individual perturbations and neglecting terms of order $O(a \log(a))$ in the denominators, we find,

$$C_{p1}(\xi_1) = \frac{\sqrt{\lambda} a F_1(a)}{2} \left(1 + \frac{1}{1 + \lambda a^2}\right), \quad C_{p1}(\xi_2) = -\frac{\sqrt{\lambda} a F_2(a)}{2} \left(1 + \frac{1}{1 + \lambda a^2}\right), \quad (18.68)$$

$$C_{m1}(\xi_1) = \frac{\sqrt{\lambda} a F_1(a)}{2} \left(1 - \frac{1}{1 + \lambda a^2}\right), \quad C_{m1}(\xi_2) = -\frac{\sqrt{\lambda} a F_2(a)}{2} \left(1 - \frac{1}{1 + \lambda a^2}\right). \quad (18.69)$$

Finally, upon calculating the positive and negative dimensionless density differences between the flux receiving and emitting windows, we obtain

$$C_p(\xi_1) - C_p(\xi_2) = Ja \left(\frac{F_1(a) + F_2(a)}{2}\right) \left(1 + \frac{1}{1 + \lambda a^2}\right) + O\left(\frac{J^2}{\lambda}\right), \quad (18.70)$$

$$C_m(\xi_1) - C_m(\xi_2) = Ja \left(\frac{F_1(a) + F_2(a)}{2}\right) \left(1 - \frac{1}{1 + \lambda a^2}\right) + O\left(\frac{J^2}{\lambda}\right), \quad (18.71)$$

but since $\lambda \gg 1$ we can neglect the ratio $1/(1 + \lambda a^2)$, yielding the same formula for both positive and negative charges

$$C_p(\xi_1) - C_p(\xi_2) = C_m(\xi_1) - C_m(\xi_2) = Ja \left(\frac{F_1(a) + F_2(a)}{2}\right) + O\left(\frac{J^2}{\lambda}\right). \quad (18.72)$$

We multiply by the background density $C_0$ to recover the original density variables, and in terms of the original spatial coordinate $\boldsymbol{x} \in \Omega$ this yields

$$c_p(\boldsymbol{x}_1) - c_p(\boldsymbol{x}_2) = c_m(\boldsymbol{x}_1) - c_m(\boldsymbol{x}_2) = \frac{I}{\mathcal{F}\pi A D_p} F\left(\frac{A}{R}\right) + O\left(C_0 \frac{J^2}{\lambda}\right). \quad (18.73)$$

For a general domain $\Omega$ with length scale $R$ and mean boundary curvature function $H(\boldsymbol{x}) = \frac{\sqrt{\lambda}}{R}\mathcal{H}(\boldsymbol{x}\sqrt{\lambda}/R)$, we have the expansion $F(a) = F(A/R)$ defined as

$$F\left(\frac{A}{R}\right) = 1 - \frac{H(\boldsymbol{x}_1) + H(\boldsymbol{x}_2)}{8} A \log\left(\frac{A}{R}\right) + O\left(\frac{A}{R}\right), \quad (18.74)$$

while for a ball of radius $R$ we obtain [25]

$$F\left(\frac{A}{R}\right) = 1 - \frac{A}{4R}\log\left(\frac{A}{R}\right) + \left(\frac{1}{8} - \frac{R}{2L} + \frac{1}{4}\log\left(\frac{L^2}{2R^2} + \frac{L}{R}\right)\right)\frac{A}{R} + O\left(\left(\frac{A}{R}\right)^2\right). \quad (18.75)$$

**Boltzmann Solution and Voltage Difference Formula**

We derive the voltage–current map from the negative charge density, which satisfies the classical Boltzmann distribution,

$$C_m(\xi) = \frac{|\Omega_\xi|}{\int_{\Omega_\xi} e^{\mathcal{V}(\xi)} d\xi} e^{\mathcal{V}(\xi)}, \quad \xi \in \Omega_\xi, \quad (18.76)$$

and thus we find that the voltage drop between the flux receiving and emitting windows is given by the log of the ratio of negative charge densities

$$\mathcal{V}(\boldsymbol{x}_1) - \mathcal{V}(\boldsymbol{x}_2) = \log\left(\frac{C_m(\xi_1)}{C_m(\xi_2)}\right). \quad (18.77)$$

We then substitute the regular asymptotic expansion,

$$\mathcal{V}(\xi_1) - \mathcal{V}(\xi_2) = \log\left(\frac{1 + C_{m1}(\xi_1)\frac{J}{\sqrt{\lambda}} + O\left(\frac{J^2}{\lambda}\right)}{1 + C_{m1}(\xi_2)\frac{J}{\sqrt{\lambda}} + O\left(\frac{J^2}{\lambda}\right)}\right), \quad (18.78)$$

and after performing a Taylor expansion in terms of $J/\sqrt{\lambda} \ll 1$ small, and using the expressions for $C_{m1}(\xi_1)$ and $C_{m1}(\xi_2)$ given in Eq. (18.69), we finally obtain

$$\mathcal{V}(\xi_1) - \mathcal{V}(\xi_2) = \log\left(\frac{1 + \frac{Ja}{2}F(a)\left(1 - \frac{1}{1+\lambda a^2}\right)}{1 - \frac{Ja}{2}F(a)\left(1 - \frac{1}{1+\lambda a^2}\right)}\right) + O\left(\frac{J^2}{\lambda}\right). \quad (18.79)$$

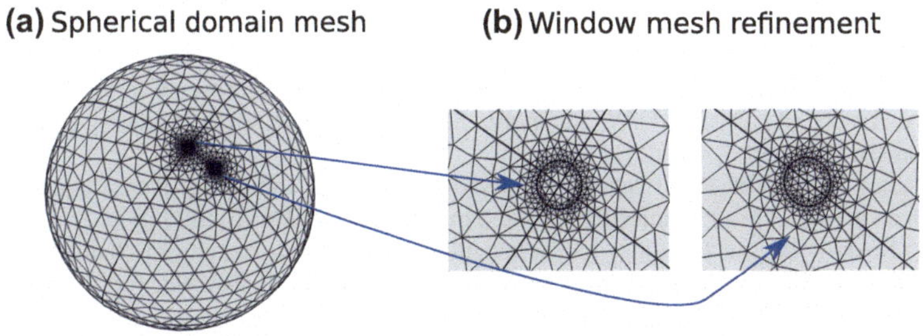


**Fig. 18.8 Spherical domain mesh. a** The domain mesh is selected to be *fine*. **b** The narrow window mesh is locally refined to be *extremely fine*

To restore the original dimensional variables we multiply by the thermal voltage $k_B\mathcal{T}/e$, and upon dropping high-order terms and neglecting the ratio $1/(1+\lambda a^2)$ (since $\lambda \gg 1$), we obtain the nonlinear voltage map

$$v(\boldsymbol{x}_1) - v(\boldsymbol{x}_2) = \frac{k_B\mathcal{T}}{e}\log\left(\frac{C_0 + \frac{I}{2\mathcal{F}\pi A D_p}F\left(\frac{A}{R}\right)}{C_0 - \frac{I}{2\mathcal{F}\pi A D_p}F\left(\frac{A}{R}\right)}\right). \tag{18.80}$$

The expansion $F(A/R)$ is as given in Eqs. (18.74)–(18.75) depending whether $\Omega$ is an arbitrary domain with curvature function $H(\boldsymbol{x})$ or a sphere of radius $R$.

## Appendix 2: Domain Mesh and COMSOL Simulations

The *Coefficient Form PDE* module from COMSOL Multiphysics version 5.2a [28] is used to numerically solve the boundary value problems. All simulations are carried on 3-D spherical ball domains discretized with a free tetrahedral mesh that is *extremely fine* on the narrow absorbing targets and *fine* elsewhere on the domain (Fig. 18.8). The predefined mesh preferences from COMSOL, which control the maximum and minimum element sizes, vary from *extremely coarse* to *extremely fine*.

## References

1. B. Hille, *Ionic Channels of Excitable Membranes* (Sinauer Associates, Sunderland USA, 2001)
2. R.S. Eisenberg, From structure to function in open ionic channels. J. Membrane Biol. **171**, 1 (1999)
3. F. Bezanilla, The voltage sensor in voltage-dependent ion channels. Physiol. Rev. **80**, 555 (2000)
4. F. Bezanilla, How membrane proteins sense voltage. Nat. Rev. Mol. Cell. Biol. **9**, 323 (2008)
5. G. Fain, *Sensory Transduction* (Oxford University Press, Oxford UK, 2019)
6. R. Yuste, *Dendritic Spines* (The MIT Press, Cambridge USA, 2010)

7. D. Holcman, R. Yuste, The new nanophysiology: regulation of ionic flow in neuronal subcompartments. Nat. Rev. Neurosci. **16**, 685 (2015)
8. M. Segal, Dendritic spines and long-term plasticity. Nat. Rev. Neurosci. **6**, 277 (2005)
9. R.W. Turner, A. Dustin, G.W. Zamponi, Signaling complexes of voltage-gated calcium channels. Channels **5**, 440 (2011)
10. Z. Schuss, B. Nadler, R.S. Eisenberg, Derivation of Poisson and Nernst-Planck equations in a bath and channel from a molecular model. Phys. Rev. E **64**, 036116 (2001)
11. S.Y. Noskov, W. Im, B. Roux, Ion permeation through the $\alpha$-hemolysin channel: theoretical studies based on Brownian dynamics and Poisson-Nernst-Planck electrodiffusion theory. Biophys. J. **87**, 2299 (2004)
12. C.R. Safinya, J.E. Radler, *Handbook of Lipid Membranes: Molecular, Functional, and Materials Aspects* (CRC Press, Boca Raton USA, 2021)
13. D. Ben-Yaakov, D. Andelman, D. Harries, R. Podgornik, Beyond standard Poisson-Boltzmann theory: ion-specific interactions in aqueous solutions. J. Phys.: Condens. Matter **21**, 424106 (2009)
14. I. Borukhov, D. Andelman, H. Orland, Adsorption of large ions from an electrolyte solution: a modified Poisson-Boltzmann equation. Electrochim. Acta **46**, 221 (2000)
15. J. Cartailler, T. Kwon, R. Yuste, D. Holcman, Deconvolution of voltage sensor time series and electro-diffusion modeling reveal the role of spine geometry in controlling synaptic strength. Neuron **97**, 1126 (2018)
16. H.C. Tuckwell, *Introduction to Theoretical Neurobiology. Volume 1: Linear Cable Theory and Dendritic Structure* (Cambridge University Press, Cambridge UK, 1988)
17. W. Rall, Electrophysiology of a dendritic neuron model. Biophys. J. **2**, 145 (1962)
18. N. Qian, T.J. Sejnowski, An electro-diffusion model for computing membrane potentials and ionic concentrations in branching dendrites. Spines Axons. Biol. Cyber. **62**, 1 (1989)
19. D.M. Wolf, M. Segawa, K. Kondadi, R. Anand, S.T. Bailey, A.S. Reichert, A.M. van der Bliek, D.B. Shackelford, M. Liesa, O.S. Shirihai, Individual cristae within the same mitochondrion display different membrane potentials and are functionally independent. EMBO J. **38**, e101056 (2019)
20. J. Cartailler, Z. Schuss, D. Holcman, Analysis of the Poisson-Nernst-Planck equation in a ball for modeling the voltage-current relation in neurobiological microdomains. Phys. D Nonlinear Phenom. **339**, 39 (2017)
21. J. Cartailler, Z. Schuss, D. Holcman, Electrostatics of non-neutral biological microdomains. Sci. Rep. **7**, 11269 (2017)
22. J. Cartailler, Z. Schuss, D. Holcman, Geometrical effects on nonlinear electrodiffusion in cell physiology. J. Nonlinear Sci. **27**, 1971 (2017)
23. J. Cartailler, D. Holcman, Steady-state voltage distribution in three-dimensional cusp-shaped funnels modeled by PNP. J. Math. Biol. **79**, 155 (2019)
24. A. Tricot, I.M. Sokolov, D. Holcman, Modeling the voltage distribution in a non-locally but globally electroneutral confined electrolyte medium: applications for nanophysiology. J. Math. Biol. **82**, 31 (2021)
25. F. Paquin-Lefebvre, D. Holcman, Modelling and asymptotic analysis of the concentration difference in a nanoregion between an influx and outflux diffusion across narrow windows. Proc. R. Soc. A. **477**, 20210501 (2021)
26. D. Holcman, Z. Schuss, *Stochastic Narrow Escape in Molecular and Cellular Biology* (Springer, New York, New York USA, 2015)
27. A.F. Cheviakov, M.J. Ward, R. Straube, An asymptotic analysis of the mean first passage time for narrow escape problems: part II: the sphere. Multiscale Model. Simul. **8**, 836 (2010)
28. *COMSOL Multiphysics* (2016). www.comsol.fr/release/5.2a, Version 5.2a
29. H.S. Carslaw, J.C. Jaeger, *Conduction of Heat in Solids* (Oxford University Press, Oxford UK, 1986)
30. J. Crank, *The Mathematics of Diffusion* (Oxford University Press, Oxford UK, 1975)
31. J. Cartailler, D. Holcman, Electrical transient laws in neuronal microdomains based on electro-diffusion. Phys. Chem. Chem. Phys. **20**, 21062 (2018)

32. D. Holcman, Z. Schuss, Escape through a small opening: receptor trafficking in a synaptic membrane. J. Statist. Phys. **117**, 975 (2004)
33. Z. Schuss, A. Singer, D. Holcman, The narrow escape problem for diffusion in cellular microdomains. Proc. Natl. Acad. Sci. **104**, 16098 (2007)
34. O. Bénichou, R. Voituriez, Narrow-escape time problem: time needed for a particle to exit a confining domain through a small window. Phys. Rev. Lett. **100**, 168105 (2008)
35. S. Pillay, M.J. Ward, A. Peirce, T. Kolokolnikov, An asymptotic analysis of the mean first passage time for narrow escape problems: part i: two-dimensional domains. Multiscale Model. Simul. **8**, 803 (2010)
36. A.F. Cheviakov, D. Zawada, Narrow-escape problem for the unit sphere: homogenization limit, optimal arrangements of large numbers of traps, and the $N^2$ conjecture. Phys. Rev. E **87**, 042118 (2013)
37. D.S. Grebenkov, Universal formula for the mean first passage time in planar domains. Phys. Rev. Lett. **117**, 260201 (2016)
38. D.S. Grebenkov, R. Metzler, G. Oshanin, Full distribution of first exit times in the narrow escape problem. New J. Phys. **21**, 122001 (2019)

# Chapter 19
# Target Search by Active Particles

**Urna Basu, Sanjib Sabhapandit, and Ion Santra**

**Abstract** Active particles, which are self-propelled nonequilibrium systems, are modeled by overdamped Langevin equations with colored noise, emulating the self-propulsion. In this chapter, we present a review of the theoretical results for the target search problem of these particles. We focus on the three best known models, namely, run-and-tumble particles, active Brownian particles, and direction reversing active Brownian particles, which differ in their self-propulsion dynamics. For each of these models, we discuss the first-passage and survival probabilities in the presence of an absorbing target. We also discuss how resetting helps the active particles find targets in a finite time.

## 19.1 Introduction

Survival of most species depends on successful search operations. In the animal world, life begins with a successful search for a mate and thereupon sperm cells searching for an oocyte to fertilize [1]. Similarly, in the plant kingdom, the propagation of species depends on pollination which is nothing but a random search process. Sustenance of animals requires foraging and finding suitable shelters [2]. Even within our body, the DNA-binding proteins use random search to locate their specific targets to bind on DNA [3]. Apart from these fundamental biological processes, search processes are also an indispensable part of human society—searching for information on the web, searching for lost items, and trying to find the right word during a conversation are part of our daily lives. Random search optimization algorithms play a crucial

U. Basu (✉)
S. N. Bose National Centre for Basic Sciences, Kolkata 700106, India
e-mail: urna@bose.res.in

S. Sabhapandit
Raman Research Institute, Bengaluru 560080, India
e-mail: sanjib@rri.res.in

I. Santra
Institute for Theoretical Physics, Georg-August-Universität Göttingen, 37073 Göttingen, Germany
e-mail: ion.santra@theorie.physik.uni-goettingen.de

 
D. Grebenkov et al. (eds.), *Target Search Problems*,
https://doi.org/10.1007/978-3-031-67802-8_19

role in computer science—finding the ground state of a complex system, combinatorial optimization problems, cryptography, artificial intelligence applications such as machine learning, natural language processing all depend on randomized search algorithms.

Despite the diversity, a commonality unites most search problems; see the special issue Ref. [4] for a comprehensive list of diverse search problems. Typically, the searcher "moves" stochastically, exploring the space and the process culminates upon finding the target. The underlying commonality allows us to use minimal stochastic models for such random search processes, the paradigmatic examples being the random walk and its continuous counterpart, the Brownian motion. Due to their inherent stochasticity, the methods of statistical physics are quite useful to study these random search models [5, 6].

One of the most important observables, in the context of search processes, is the time required to find a target. Mathematically, the time at which a stochastic searcher hits a target for the first time is known as the first-passage time. This time $t$ is evidently a random quantity, whose probability density function (PDF) $F(a,t)$ also depends on the separation $a$ between the starting position and the location of the target. Another related quantity is the survival probability $Q(a,t) = \int_t^\infty F(a,t')\,dt'$, which denotes the probability that the searcher does not hit the target up to time $t$. Therefore, the mean first-passage time, i.e., the average time to find the target, is given by

$$T(a) = \int_0^\infty t\,F(a,t)\,dt = \int_0^\infty Q(a,t)\,dt. \tag{19.1}$$

For a Brownian particle moving in one dimension with diffusion constant $D$, the first-passage time PDF is given by the Lévy–Smirnov distribution,

$$F(a,t) = \frac{1}{\sqrt{4\pi D}}\frac{a}{t^{3/2}}\exp\left(-\frac{a^2}{4Dt}\right). \tag{19.2}$$

The corresponding survival probability is given by [7]

$$Q(a,t) = \text{erf}\left(\frac{a}{\sqrt{4Dt}}\right). \tag{19.3}$$

At large time, $Q(a,t) \sim t^{-\alpha}$ with $\alpha = 1/2$, where $\alpha$ is known as the persistence exponent. For $N$ independent searchers $\alpha = N/2$, and consequently, $F_N(a,t) \sim t^{-(N/2+1)}$. Thus, the mean first-passage time is finite only if there are three or more searchers ($N \geq 3$).[1]

It turns out that the searching process becomes drastically faster if the particle intermittently returns to the starting point and restarts the search [9, 10]. Let $Q^{(r)}(a,t)$ denote the survival probability for a Brownian particle in the presence of such a

[1] Note that, $\langle t^{-1}\rangle \equiv \int_0^\infty t^{-1}\,F(a,t)\,dt$ is sometimes used to characterize search efficiency when $T(a)$ diverges [8].

stochastic resetting with a rate $r$. Its Laplace transform is related to the Laplace transform of $Q(a, t)$ by [10],

$$\tilde{Q}^{(r)}(a, s) = \frac{\tilde{Q}(a, r+s)}{1 - r\tilde{Q}(a, r+s)}, \tag{19.4}$$

where $\tilde{Q}^*(a, s) = \int_0^\infty Q^*(a, t)\, e^{-st}\, dt$ denotes Laplace transform. The corresponding mean first-passage time is nothing but

$$T_r = \tilde{Q}^{(r)}(a, 0) = \frac{\tilde{Q}(a, r)}{1 - r\tilde{Q}(a, r)} = -\frac{1}{r} + \frac{1}{r}\exp\left(a\sqrt{r/D}\right), \tag{19.5}$$

which is finite for any non-zero resetting rate. It diverges for both small and large $r$: as a power law $\sim r^{-1/2}$ as $r \to 0$, and stretched exponentially as $r \to \infty$, with a minimum at an optimal resetting rate satisfying $r^*(a^2/D) = 2.539\ldots$.

The Brownian search is a Markov process. This is not true for most living organisms, which often have a finite memory, manifested in their persistent motion along their body axis. For example, living organisms ranging from bacteria at the microscopic scale to an ant or the birds, at the macroscopic scale, move persistently for a certain time before changing the direction substantially. These types of self-propelled motion are often referred to as *active motion* and the agents displaying such motion are referred to as *active particles*. Such motion has also been realized in synthetic systems like Janus particles [11–13]. For example, when a polystyrene spherical bead with one side coated with platinum is immersed in hydrogen peroxide solution, the platinum side catalyzes the hydrogen peroxide to oxygen and water, causing a self-propulsion of the bead by diffusiophoresis [14]. These artificial active particles, including nano/microscopic machines, have a diverse range of potential applications in biomedical target search [15, 16]. Target search by such non-Markovian processes is expected to exhibit richer behavior compared to the simple Brownian search, which is the subject of this chapter.

The self-propelled active motions, despite having complex and varied underlying mechanisms, can be broadly categorized into a few classes depending on their microscopic swimming strategies. In the minimal statistical models, describing the dynamics of typical Janus particles or microscopic organisms, the time evolution of the position vector $\boldsymbol{r}(t)$ of a single active particle can be described by the overdamped Langevin equation, $\dot{\boldsymbol{r}}(t) = v_0\, \hat{\boldsymbol{n}}(t)$, where the different swimming strategies are encoded by the different stochastic dynamics of the unit vector $\hat{\boldsymbol{n}}(t)$, indicating the orientation of the body axis. Perhaps the most well-known model is the so-called run-and-tumble particle (RTP), where the active particle moves in a straight line before intermittently changing its direction $\hat{\boldsymbol{n}}$ in a discrete manner. The orientation $\hat{\boldsymbol{n}}$ undergoes a continuous change via a rotational diffusion for an active Brownian particle (ABP), whereas a direction reversing active Brownian particle (DRABP) undergoes an intermittent complete reversal of direction in addition to the continuous ABP dynamics. The speed $v_0$ is constant in all these models.

In this chapter, we discuss the survival probability and first-passage time PDF for RTP, ABP and DRABP in the presence of fixed targets. We further explore how stochastic resetting optimizes the search for these systems. We conclude with a summary and some open problems.

## 19.2 Run-and-Tumble Particles

The run-and-tumble particle (RTP) is one of the first theoretical models of active particles, developed to mimic the overdamped motion of the *E. coli* bacteria [17, 18] that consists of alternating run-and-tumble phases. In the run phase, the bacterium moves almost in a straight line at a constant speed, whereas in the tumble phase, it reorients randomly, resulting in a change of the direction of motion. The typical duration of a tumbling phase is much shorter than that of a run phase. The RTP model assumes the tumbles to be instantaneous events occurring at a constant rate $\gamma$. Consequently, the position $\boldsymbol{r}$ and the unit vector $\hat{\boldsymbol{n}}$, indicating the direction of motion of an RTP, evolve by

$$\dot{\boldsymbol{r}}(t) = v_0\, \hat{\boldsymbol{n}}(t), \quad \text{and } \hat{\boldsymbol{n}} \to \hat{\boldsymbol{n}}' \text{ with rate } \gamma, \tag{19.6}$$

where the speed $v_0$ remains constant during the run phases. In other words, the duration $\tau$ of each run phase is drawn independently from the exponential distribution $\rho(\tau) = \gamma \exp(-\gamma\tau)$. Note that, while the time evolution of the position $\boldsymbol{r}(t)$ is non-Markovian by itself, the combined $(\boldsymbol{r}(t), \boldsymbol{n}(t))$ process is Markov.

The first-passage time PDF has been computed exactly for a one-dimensional RTP in the presence of a fixed absorbing boundary [19]. In one dimension, the unit vector $\hat{\boldsymbol{n}}$ reduces to a dichotomous random variable $\sigma = \pm 1$, and Eq. (19.6) simplifies to

$$\dot{x}(t) = v_0 \sigma(t), \tag{19.7}$$

where $\sigma \to -\sigma$ at a rate $\gamma$. In the following, we present the main steps of the derivation of the survival probability, which was obtained in [19].

### *19.2.1 Survival Probability*

The survival probability $Q_\sigma(x, t)$, which denotes the probability that the RTP has not crossed the absorbing boundary at the origin up to time $t$, starting from an initial position $x > 0$ and orientation $\sigma$, follows a backward Fokker–Planck equation [20]. To obtain the BFPE, it is convenient to consider a time interval $[0, t + \Delta t]$, separated into two parts $[0, \Delta t]$ and $[\Delta t, t + \Delta t]$. In the first interval, either $\sigma \to -\sigma$ with probability $\gamma \Delta t$, or the particle moves a distance $\sigma v_0 \Delta t$ ballistically with probability $(1 - \gamma \Delta t)$. An RTP starting with a negative velocity at the origin gets immediately absorbed. On the other hand, an RTP starting with a positive velocity does not get absorbed immediately and survives with a probability $1-\gamma\Delta t$. Thus, using the temporal homogeneity of the $(x,\sigma)$ process, we have

$$Q_\sigma(x,t+\Delta t)=(1-\gamma\Delta t)Q_\sigma(x+v_0\sigma\Delta t,t)+\gamma\Delta t\,Q_{-\sigma}(x,t),\quad x\in(0,\infty),$$
$$Q_-(0,t)=0,\quad\text{and}\quad Q_+(0,t+\Delta t)=(1-\gamma\Delta t)Q_+(v_0\Delta t,t).$$

Taking a Taylor series expansion of the above equation and taking the $\Delta t\to 0$ limit, we get the backward Fokker–Planck equation [20]

$$\frac{\partial Q_\sigma(x,t)}{\partial t}=v_0\sigma\frac{\partial Q_\sigma(x,t)}{\partial x}-\gamma\Big[Q_\sigma(x,t)-Q_{-\sigma}(x,t)\Big] \tag{19.8}$$

and the boundary conditions $Q_-(0,t)=0$ and $[\partial_t-v_0\partial_x+\gamma]Q_+(0,t)=0$. Since a particle starting at infinity cannot cross the origin at any finite time, we have another set of boundary conditions $Q_\sigma(x\to\infty,t)=1$ for all times. The initial conditions are evidently $Q_\sigma(x,0)=1$.

Taking a Laplace transform of Eq. (19.8), defined by $\tilde{Q}_\sigma(s)=\int_0^\infty dt\, e^{-st}Q_\sigma(t)$, we get

$$\mathcal{L}_\sigma\tilde{Q}_\sigma(x,s)=1+\gamma\tilde{Q}_{-\sigma}(x,s)\quad\text{with}\ \mathcal{L}_\sigma\equiv s+\gamma-v_0\sigma\frac{\partial}{\partial x}, \tag{19.9}$$

where the boundary conditions translate to $\tilde{Q}_-(0,s)=0$, $\mathcal{L}_+\tilde{Q}_+(0,s)=1$, and $\tilde{Q}_\sigma(x\to\infty,s)=1/s$. Operating $\mathcal{L}_{-\sigma}$ on both sides of the above equation, we get

$$\left[v_0^2\frac{\partial^2}{\partial x^2}-s(s+2\gamma)\right]Q_\sigma(x,s)=-(s+2\gamma). \tag{19.10}$$

Solving the above equation and using the boundary conditions, we get

$$\tilde{Q}_+(x,s)=\frac{1}{s}\left(1-\frac{(s+\gamma)-\lambda}{\gamma}e^{-\lambda x/v_0}\right),\quad \tilde{Q}_-(x,s)=\frac{1}{s}(1-e^{-\lambda x/v_0}), \tag{19.11}$$

where $\lambda=\sqrt{s(s+2\gamma)}$. It is immediately evident from the above equations that the mean first-passage times, $T_\pm(x)=\tilde{Q}_\pm(x,0)$, is infinite.

Since the Laplace transforms of the first-passage time PDF $\tilde{F}_\pm(x,s)=\int_0^\infty F_\pm(x,t)\,e^{-st}\,dt$ are related to the survival probability by $\tilde{Q}_\pm(x,s)=[1-\tilde{F}_\pm(x,s)]/s$, the Laplace transforms $\tilde{F}_\pm(x,s)$ are readily available from the above equations. Inverting the Laplace transform yields

$$F_+(x,t) = \frac{\gamma e^{-\gamma t}}{t + x/v_0}\left[\frac{x}{v_0} I_0\left(\gamma\sqrt{t^2 - (x/v_0)^2}\right) + \frac{1}{\gamma}\frac{\sqrt{t - x/v_0}}{\sqrt{t + x/v_0}} I_1\left(\gamma\sqrt{t^2 - (x/v_0)^2}\right)\right]\Theta(t - x/v_0), \tag{19.12}$$

$$F_-(x,t) = \frac{\gamma e^{-\gamma t}\, x/v_0}{\sqrt{t^2 - (x/v_0)^2}} I_1\left(\gamma\sqrt{t^2 - (x/v_0)^2}\right)\Theta(t - x/v_0) + e^{-\gamma t}\delta\left(t - x/v_0\right), \tag{19.13}$$

where $I_n(z)$ denotes the $n$th order modified Bessel function of the first kind [21].

Note that the minimum possible value of the first-passage time is $t = x/v_0$ which is achieved when the particle starts with a negative velocity, i.e., $\sigma = -1$ does not tumble before reaching the target. The Dirac-delta function appearing in $F_-(x,t)$ corresponds to these deterministic trajectories which occur with a probability $e^{-\gamma t}$. There is no analogous term for $F_+(x,t)$ as a particle starting with a positive velocity, i.e., $\sigma = 1$ must undergo at least one tumbling in order to reach the target at the origin. For all the trajectories undergoing at least one tumbling event, the first-passage time $t > x/v_0$. The Heaviside Theta function $\Theta(t - x/v_0)$ appearing in Eqs. (19.12) and (19.13) indicates the contributions from these trajectories.

Evidently, the above expressions for the first-passage time PDF for an RTP are very different from that of a Brownian particle. However, in the limit $\gamma \to \infty$ and $v_0 \to \infty$, while keeping $v_0^2/\gamma = 2D_{\mathrm{RT}}$ finite, the above equations reduce to the first-passage time PDF for the Brownian motion given by Eq. (19.2) with $D_{\mathrm{RT}}$ being the diffusion coefficient.

For any finite $\gamma$ and $v_0$, the large-time $t \gg x/v_0, \gamma^{-1}$ asymptotic behavior comes out to be

$$F_-(x,t) \simeq \frac{\sqrt{\gamma}/v_0}{\sqrt{2\pi}}\frac{x}{t^{3/2}}, \quad F_+(x,t) \simeq \frac{\sqrt{\gamma}/v_0}{\sqrt{2\pi}}\frac{\bar{x}}{t^{3/2}}, \tag{19.14}$$

where $\bar{x} = x + v_0/\gamma$. A particle starting with a positive velocity, i.e., $\sigma = 1$ typically moves a distance $v_0/\gamma$ before the first tumbling event. Since the particle moves with a negative velocity immediately after the first tumbling event, the first-passage time PDF $F_+(x,t)$ is equivalent to $F_-(\bar{x},t)$ for large $t$.

Of particular interest is the special case $x = 0^+$, i.e., when the initial position of the particle is infinitesimally close to the target. Obviously, if the initial velocity is negative, i.e., $\sigma = -1$, then the particle gets absorbed immediately. Consequently, $F_-(0,t) = \delta(t)$, which is also obtained from Eq. (19.13) in the $x = 0$ limit. The corresponding survival probability $Q_-(0,t) = 0$, which is, in fact, one of the boundary conditions on Eq. (19.8). On the other hand, a particle starting from $x = 0^+$ with a positive velocity still survives for a finite time, characterized by the PDF $F_+(0,t) = t^{-1}e^{-\gamma t}I_1(\gamma t)$ obtained from Eq. (19.12) by setting $x = 0$. The corresponding survival probability in this case is given by $Q_+(0,t) = e^{-\gamma t}\left[I_0(\gamma t) + I_1(\gamma t)\right]$. Therefore choosing the initial velocity direction $\sigma = \pm 1$ with equal probability 1/2, we get the total survival probability $Q(0, t) = [Q_+(0, t) + Q_-(0, t)]/2$ as

$$Q(0, t) = \frac{1}{2} e^{-\gamma t} \big[ I_0(\gamma t) + I_1(\gamma t) \big]. \tag{19.15}$$

In the one-dimensional model of RTPs discussed so far, the velocity direction $\sigma = \pm 1$ switches at each tumbling event, which occurs with a rate $\gamma$. One can also think of a different version of an RTP, where at each tumbling event occurring with rate $\gamma$, the new velocity direction is chosen independently from $\sigma = \pm 1$ with equal probability $1/2$. Therefore, in a small time interval $dt$, the probability that the RTP reverses its velocity direction is $\gamma dt/2$, resulting in a waiting time PDF $\rho(\tau) = (\gamma/2)\, e^{-\gamma\tau/2}$. For such an RTP, the survival probability Eq. (19.15) becomes

$$Q(0, t) = \frac{1}{2} e^{-\gamma t/2} \big[ I_0(\gamma t/2) + I_1(\gamma t/2) \big]. \tag{19.16}$$

Note that the constant speed $v_0$ can be thought of as drawing the speed from a PDF $W(v) = \delta(v - v_0)$ after each tumbling event.

It turns out that Eq. (19.16) is universal [22] in the sense that it remains valid for a much wider class of RTPs with any arbitrary speed PDF $W(v)$, as a consequence of the Sparre Andersen theorem for one-dimensional discrete-time random walks with symmetric and continuous jump distributions [23]. Naturally, this generalized RTP also describes the dynamics of the $x$ component of a $d$-dimensional RTP, and Eq. (19.16) remains valid for the corresponding survival probability [22].

### *19.2.2 RTP with Resetting*

The presence of stochastic resetting expedites the target search process for an RTP, similar to a Brownian particle. In this section, we discuss the behavior of the first-passage properties of an RTP starting from an initial position $x_0 > 0$, in the presence of an absorbing boundary at $x = 0$, undergoing stochastic resetting with rate $r$. Since the state space of the RTP consists of $(x, \sigma)$, we need to specify the resetting protocol for both $x$ and $\sigma$. Here we consider the resetting protocol, when at each resetting event, the position $x(t) \to x_0$ and the velocity direction $\sigma(t) \equiv \sigma' \to \sigma''$ with probability $p_{\sigma',\sigma''}$.

We assume that the RTP starts with the equilibrium state of the velocity orientation, where $\sigma = \pm 1$ with equal probability $1/2$. Our goal is to find the corresponding survival probability $Q^{(r)}(x_0, t)$, i.e., the probability that starting from the position $x_0$, the RTP has not crossed the target up to time $t$. To this end, it is useful to start with the last renewal equation for $Q^{(r)}_{\sigma',\sigma}(x_0, t)$, which denotes the probability that starting from the state $\{x(0) = x_0, \sigma(0) = \sigma\}$, the RTP has not crossed the target (absorbing boundary) up to time $t$ and $\sigma(t) = \sigma'$. Evidently,

$$Q^{(r)}(x_0,t) = \frac{1}{2}\sum_{\sigma,\sigma'} Q^{(r)}_{\sigma',\sigma}(x_0,t). \tag{19.17}$$

To construct the renewal equation for $Q^{(r)}_{\sigma',\sigma}(x_0,t)$, we consider two scenarios: (i) there is no resetting during the interval $(0,t)$, which has a probability $e^{-rt}$ and (ii) there is at least one resetting and the last reset occurs within the time interval $(t-\tau, t-\tau+d\tau)$ with a probability $r d\tau$. The probability that there is no reset during the remaining time is $e^{-r\tau}$. Suppose $\sigma(t-\tau) = \sigma''$ and during the reset $\sigma'' \to \sigma'''$ with probability $p_{\sigma'',\sigma'''}$. Summing over all possible $\sigma''$, we get the renewal equation as

$$\begin{aligned} Q^{(r)}_{\sigma',\sigma}(x_0,t) &= e^{-rt} Q_{\sigma',\sigma}(x_0,t) \\ &\quad + r\int_0^t d\tau\, e^{-r\tau} \sum_{\sigma'',\sigma'''} Q_{\sigma',\sigma'''}(x_0,\tau)\, Q^{(r)}_{\sigma'',\sigma}(x_0,t-\tau)\, p_{\sigma'',\sigma'''}, \end{aligned} \tag{19.18}$$

where $Q_{\sigma',\sigma}(x_0,t) \equiv Q^{(0)}_{\sigma',\sigma}(x_0,t)$ denotes the survival probability in the absence of resetting. Taking a Laplace transform in the above equation, $\tilde{Q}^*_{\sigma',\sigma}(x_0,s) = \int_0^\infty Q^*_{\sigma',\sigma}(x_0,t)\, e^{-st}\, dt$, we get

$$\tilde{Q}^{(r)}_{\sigma',\sigma}(x_0,s) = \tilde{Q}_{\sigma',\sigma}(x_0,r+s) + r \sum_{\sigma'',\sigma'''} \tilde{Q}_{\sigma',\sigma'''}(x_0,r+s)\, \tilde{Q}^{(r)}_{\sigma'',\sigma}(x_0,s)\, p_{\sigma'',\sigma'''}. \tag{19.19}$$

This is a system of four linear equations and can be solved easily to obtain $\tilde{Q}^{(r)}_{\sigma',\sigma}(x_0,s)$ in terms of $\tilde{Q}_{\sigma',\sigma}(x_0,s)$. Summing over $\{\sigma,\sigma'\}$, and using Eq. (19.17), we get

$$\tilde{Q}^{(r)}(x_0,s) = \frac{\tilde{Q}(x_0,r+s) + \frac{r}{2}\left(\sum_{\sigma,\sigma'} \sigma\sigma'\, p_{\sigma,\sigma'}\right) \mathbf{det}[\tilde{Q}(x_0,r+s)]}{1 - r\sum_{\sigma,\sigma'} p_{\sigma,\sigma'}\, \tilde{Q}_{\sigma,\sigma'}(x_0,r+s) + r^2 \mathbf{det}[p] \times \mathbf{det}[\tilde{Q}(x_0,r+s)]}, \tag{19.20}$$

where $[p]$ and $[\tilde{Q}]$ denote $2\times 2$ matrices, given by

$$[p] = \begin{bmatrix} p_{+,+} & p_{+,-} \\ p_{-,+} & p_{-,-} \end{bmatrix} \quad \text{and} \quad [\tilde{Q}] = \begin{bmatrix} \tilde{Q}_{+,+} & \tilde{Q}_{+,-} \\ \tilde{Q}_{-,+} & \tilde{Q}_{-,-} \end{bmatrix}. \tag{19.21}$$

It is noteworthy that Eq. (19.20) is very different from its passive counterpart given by Eq. (19.4) and depends explicitly on how the active degree of freedom is reset. Evidently, the mean first-passage time can be obtained by setting $s = 0$ in Eq. (19.20),

$$T^{(r)}(x_0) = \frac{\tilde{Q}(x_0, r) + \frac{r}{2}\Big(\sum_{\sigma,\sigma'} \sigma\sigma'\, p_{\sigma,\sigma'}\Big)\mathbf{det}[\tilde{Q}(x_0, r)]}{1 - r\sum_{\sigma,\sigma'} p_{\sigma,\sigma'}\, \tilde{Q}_{\sigma,\sigma'}(x_0, r) + r^2\mathbf{det}[p] \times \mathbf{det}[\tilde{Q}(x_0, r)]}. \tag{19.22}$$

Note that, the presence of multiple resetting options for the orientation of an RTP makes the above expression of the MFPT quite different from its Brownian counterpart, given by a much simpler form in Eq. (19.5).

The Laplace transforms $\tilde{Q}_{\sigma,\sigma'}(x_0, s)$ can be obtained using a backward Fokker–Planck equation approach [24], similar to the one discussed in the previous section,

$$\begin{aligned}
\tilde{Q}_{+,+}(x_0, s) &= \frac{1}{s(s+2\gamma)}\left[s + \gamma - (s + \gamma - \lambda)e^{-\lambda x_0/v_0}\right],\\
\tilde{Q}_{+,-}(x_0, s) &= \frac{\gamma}{s(s+2\gamma)}\left[1 - e^{-\lambda x_0/v_0}\right],\\
\tilde{Q}_{-,+}(x_0, s) &= \frac{1}{s(s+2\gamma)}\left[\gamma - \frac{(\gamma+s)(\gamma+s+\lambda)}{\gamma}e^{-\lambda x_0/v_0}\right],\\
\tilde{Q}_{-,-}(x_0, s) &= \frac{s+\gamma}{s(s+2\gamma)}\left[1 - e^{-\lambda x_0/v_0}\right].
\end{aligned} \tag{19.23}$$

In the following, we consider two distinct resetting protocols for $\sigma$. In the first case [24], $\sigma$ either reverses sign with probability $\mu_1$ or remains unchanged with probability $1 - \mu_1$ at reset, where $0 \le \mu_1 \le 1$. This corresponds to $p_{\sigma,\sigma'} = \mu_1\delta_{\sigma',-\sigma} + (1-\mu_1)\delta_{\sigma',\sigma}$, and the mean first-passage time, given by Eq. (19.22), simplifies to

$$T^{(r)}_{\mu_1} = -\frac{1}{r} + \frac{1}{r + 2\gamma - \kappa}\left[\frac{2\gamma}{r}e^{\kappa x_0/v_0} - \frac{(2\mu_1 - 1)(\kappa - r)}{2(\mu_1 r + \gamma)}\right], \tag{19.24}$$

with $\kappa = \sqrt{r(r+2\gamma)}$. The mean first-passage time shows a qualitatively different behavior depending on whether $x_0$ is smaller or larger than the persistence length scale, $\ell = v_0\gamma^{-1}$, of the RTP, as shown in Fig. 19.1a and b, respectively. Though in both cases, $T^{(r)}_{\mu_1}$ is finite for all $r > 0$ and has a minimum at an optimum value $r^*$, the optimal resetting rate $r^*$ is a non-monotonic function of $\mu_1$ for $x_0 < \ell_a$, while it decreases monotonically with $\mu_1$ for $x_0 > \ell_a$ [see Fig. 19.1c].

On the other hand, for the second case, a new velocity direction is chosen from $[1, -1]$ with a certain probability independent of the previous orientation. This corresponds to $p_{\sigma,\sigma'} = \mu_2\delta_{\sigma,1} + (1-\mu_2)\delta_{\sigma',-1}$, where the mean first-passage time is given by

$$T^{(r)}_{\mu_2} = -\frac{1}{r} + \frac{2\gamma\, e^{\kappa x_0/v_0} - (2\mu_2 - 1)(\kappa - r)}{2r[\gamma - \mu_2(\kappa - r)]}. \tag{19.25}$$

The mean first-passage time $T^{(r)}_{\mu_2}$ is finite for all $r > 0$ and has a minimum at an optimum value $r^*(\mu_2)$, which is shown in Fig. 19.2a and b for $x_0 < \ell_a$ and $x_0 >$

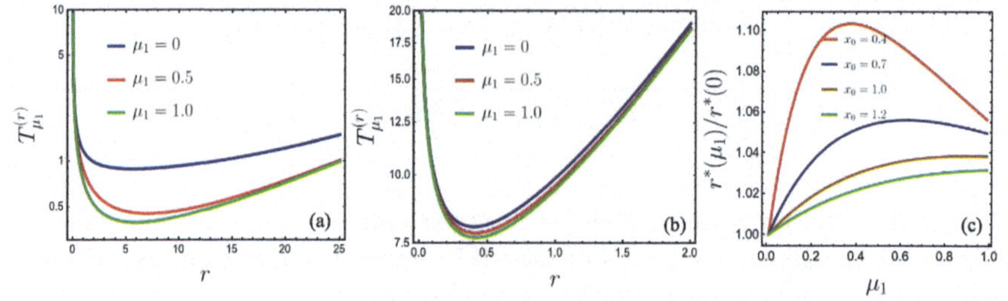

**Fig. 19.1** Plots for protocol I with $\gamma = 1$ and $v_0 = 1$. **a** shows the MFPT versus $r$ for $x_0 = 0.1 < \ell_a = 1$; **b** shows the MFPT versus $r$ for $x_0 = 1.1 > \ell_a$; **c** shows the normalized optimal resetting rate as a function of $\mu_1$, which demonstrates a non-monotonic behavior for $x_0 < \ell_a$ and a monotonically decreasing behavior for $x_0 > \ell_a$

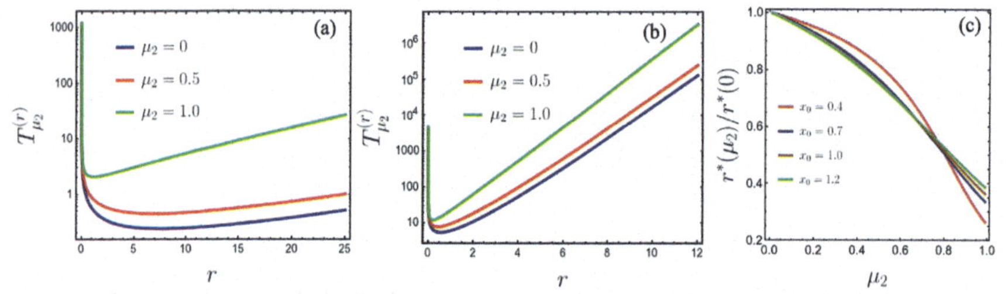

**Fig. 19.2** Plots for protocol II with $\gamma = 1$ and $v_0 = 1$. **a** shows the MFPT versus $r$ for $x_0 = 0.1 < \ell_a = 1$; **b** shows the MFPT versus $r$ for $x_0 = 1.1 > \ell_a$; **c** shows the normalized optimal resetting rate as a function of $\mu_2$, which always exhibits a monotonically decreasing behavior

$\ell_a$, respectively. The optimal resetting rate $r^*(\mu_2)$ decreases monotonically with $\mu_2$ [see Fig. 19.2c]. In the Brownian limit $\gamma \to \infty$, $v_0 \to \infty$ with $v_0^2/\gamma = 2D_{RT}$, both Eqs. (19.24) and (19.25) converge to Eq. (19.5).

Note that, the two resetting protocols become equivalent for $\mu_1 = \mu_2 = 1/2$ for our chosen initial condition. This special case corresponds to a complete resetting protocol, with both the position and the orientation being reset to their initial conditions. In fact, in this case, $p_{\sigma,\sigma'} = 1/2$, for all $(\sigma, \sigma')$, and Eq. (19.22) becomes identical to the first equality of Eq. (19.4).

Several variants of this one-dimensional run-and-tumble model have also been studied which show interesting first-passage properties. For example, exact results for the mean first-passage time for an RTP with space-dependent velocity and tumbling rate, in the presence of an external potential on an interval, were obtained in [25]. The survival probability has also been obtained for run-and-tumble motion with position- and orientation-dependent tumbling rates [26] and with an external confining potential [27]. RTPs with partially absorbing boundaries were studied in [28, 29] using the encounter-based approach to study diffusion-mediated surface phenomena [30–33]. Recently, it was shown that the mean first-passage time of a chiral RTP, with discrete orientation space, moving on a finite two-dimensional region, exhibits a minimum at an optimum value of chirality [34]. The effect of the initial condition on the mean first-passage time of an RTP in the presence of two absorbing boundaries has been explored recently, using a perturbative analysis, in the small Péclet limit [35]. The first-passage properties of an RTP, with discrete internal states and moving on a lattice, were studied in [36], and it was shown that activity facilitates the return to the origin. The effect of resetting on the mean first-passage time of an RTP moving diffusively in an anharmonic external potential has been studied in [37].

## 19.3 Active Brownian Particles

Not all microorganisms exhibit a run-and-tumble motion. For example, some mutants of *E. coli* (CheC497), unlike a wild type, change its direction of motion continuously [17]. A similar motion is also observed for typical Janus particles [14]. An "active Brownian particle" models such motion, where the orientation vector of the particle undergoes a rotational Brownian motion on a unit sphere. The two-dimensional model is the simplest case, where the orientation vector $\hat{\mathbf{n}} = (\cos\theta, \sin\theta)$ and the position $\mathbf{r}(t) = (x(t), y(t))$ evolve according to

$$\dot{x}(t) = v_0 \cos\theta(t), \quad \dot{y}(t) = v_0 \sin\theta(t), \quad \text{and } \dot{\theta}(t) = \sqrt{2D_R}\,\zeta(t), \tag{19.26}$$

where $\zeta(t)$ is a Gaussian white noise with zero mean and correlations $\langle\eta(t)\eta(t')\rangle = \delta(t-t')$. The rotational diffusion constant $D_R$ sets the active timescale $D_R^{-1}$, separating the short- and long-time dynamical regimes. The presence of activity leads to the emergence of a strongly anisotropic and non-diffusive behavior in the short-time regime $t \ll D_R^{-1}$. On the other hand, in the long-time regime $t \gg D_R^{-1}$, the dynamics Eq. (19.26) converges to a Brownian motion with an effective diffusion coefficient $D_{\mathbf{eff}} = D_{\mathbf{AB}} = v_0^2/(2D_R)$ [38]. Naturally, the first-passage properties also show different behaviors in the short-time and long-time regimes, which we discuss below.

### *19.3.1 Survival Probability*

The active nature of the short-time dynamics of an ABP leads to strongly anisotropic first-passage properties. These effects are best understood if one looks at the survival probabilities along and orthogonal to the initial orientation of the ABP [39]. In the following, we choose the initial orientation $\theta(0) = 0$, and look at the survival probabilities of two effective one-dimensional processes, parallel and orthogonal to the initial orientation, separately: (i) along the $x$-direction with the $y$-axis as an absorbing boundary and (ii) along the $y$-direction with $x$-axis as an absorbing boundary.

Let $Q_x(x_0, t)$ denote the probability that the ABP, starting from the $x = x_0$ line, has not crossed the absorbing boundary at $x = 0$. Similarly, $Q_y(y_0, t)$ denotes the survival probability for the $y$-component, starting from the $y = y_0$ line with an absorbing boundary at $y = 0$. Although it is straightforward to write the corresponding backward Fokker–Planck equations, it is hard to solve them analytically with appropriate boundary conditions. However, one can predict the leading order behaviors of $Q_x(x_0, t)$ and $Q_y(y_0, t)$ by analyzing Eq. (19.26) in the short-time and long-time limits separately.

Since at late times $t \gg D_R^{-1}$, the active Brownian motion converges to a usual Brownian motion, one expects

$$Q_x(x_0, t) \sim \frac{1}{\sqrt{4D_{\text{eff}}t}}, \quad Q_y(y_0, t) \sim \frac{1}{\sqrt{4D_{\text{eff}}t}}. \tag{19.27}$$

On the other hand, starting from $\theta(0) = 0$, at short times $t \ll D_R^{-1}$, the orientation angle $\theta(t) \sim O(\sqrt{D_R t}) \ll 1$. Consequently, $\cos\theta(t) \simeq 1$ and $\sin\theta(t) \simeq \theta(t)$. The motion of the ABP in the short-time regime can then be described by the effective Langevin equations,

$$\dot{x} \simeq v_0, \quad \dot{y} \simeq v_0\,\theta(t) \tag{19.28}$$

where $\theta(t)$ undergoes a Brownian motion with diffusion constant $D_R$.

The above effective dynamics implies that, in this short-time regime, $x(t)$ stays positive with probability close to unity, implying $Q_x(x_0, t) \simeq 1$. In the $x_0 \to 0$ limit, the crossover from this ballistic behavior to the diffusive behavior Eq. (19.27) can be captured by the scaling form

$$Q_x(x_0 \to 0, t) = \mathcal{F}_x\,(t\,D_R)\,, \tag{19.29}$$

where the scaling function follows

$$\mathcal{F}_x(u) \sim \begin{cases} 1 & \text{for} \quad u \ll 1, \\ u^{-1/2} & \text{for} \quad u \gg 1. \end{cases} \tag{19.30}$$

To extract the short-time behavior of the survival probability $Q_y(y_0, t)$ it is convenient to take another time derivative of $\dot{y}(t)$ in Eq. (19.28), which yields

$$\ddot{y}(t) \approx \sqrt{2\,v_0^2 D_R}\;\zeta(t), \tag{19.31}$$

and $\zeta(t)$ is the white noise defined in Eq. (19.26). The above equation is nothing but the Langevin equation for the well-known "random acceleration process" (RAP), which is one of the simplest non-Markovian processes [40]. The first-passage properties of the RAP have been studied extensively in the literature, and the survival probability has been computed in the $y_0 \to 0$ limit [41]. This result, translated to the case of the short-time regime of an ABP, gives

$$Q_y(y_0, t) \simeq \frac{2^{5/6}\Gamma(-4/3)}{3^{2/3}\pi\Gamma(3/4)}\left(\frac{y_0 D_R}{v_0}\right)^{1/6}(t D_R)^{-1/4} . \tag{19.32}$$

Clearly, the above result implies that the $y$ component, i.e., the component orthogonal to the initial orientation, of the ABP shows a persistence exponent $\alpha_y = 1/4$ in the short-time regime. The crossover from this anomalous persistence behavior to the diffusive behavior Eq. (19.27) at late times suggests the following scaling: form

$$Q_y(y_0, t) = \left(\frac{y_0 D_R}{v_0}\right)^{1/6} \mathcal{F}_y(t D_R) . \tag{19.33}$$

The scaling function $\mathcal{F}_y(u)$ has the limiting behaviors

$$\mathcal{F}_y(u) \sim \begin{cases} u^{-1/4} & \text{for} \quad u \ll 1 \\ u^{-1/2} & \text{for} \quad u \gg 1. \end{cases} \tag{19.34}$$

Note that, due to the $t^{-1/2}$ tail of the survival probability, the mean first-passage time for both the $x$ and $y$ components of ABP diverge, similar to the Brownian motion. Figure 19.3 illustrates the scaling behaviors for $Q_x(x_0, t)$ and $Q_y(y_0, t)$ predicted in Eqs. (19.29) and (19.33) which show the expected crossovers near $D_R t \sim 1$.

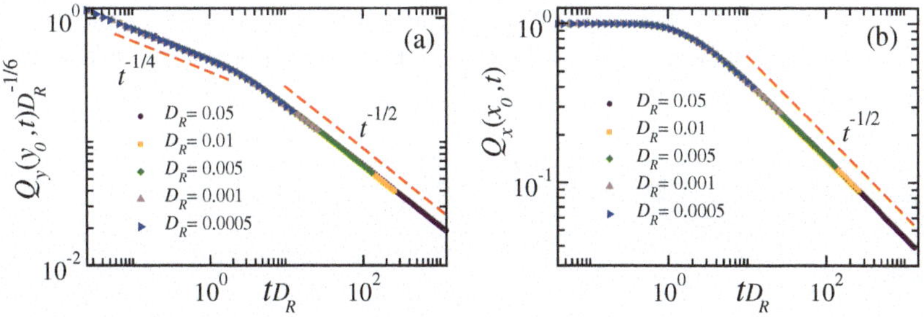


**Fig. 19.3** Survival probability of an ABP: **a** Scaling collapse of $Q_y(y_0, t)$ according to Eq. (19.33) for different values of $D_R$, using data obtained from numerical simulations. The dashed lines indicate the analytically predicted behaviors for the short-time and long-time regimes. Here we have used $v_0 = 1$ and $y_0 = 0.1$. **b** Scaling collapse of $Q_x(x_0, t)$ according to Eq. (19.29) for different values of $D_R$, using data obtained from numerical simulations. The dashed line indicates the diffusive $t^{-1/2}$ decay. Here $v_0 = 1$ and $x_0 = 0.1$

### 19.3.2 ABP with Resetting

Similar to the RTP, the presence of resetting also helps the ABP to find the target in a finite time. Since, for an ABP, the motion orthogonal to the initial orientation shows a much richer behavior compared to the motion along the initial orientation, we concentrate on the first-passage properties in the presence of resetting for the orthogonal component only. The simplest resetting protocol is when the position and orientation are both reset to their initial values. As before, we set the initial orientation $\theta(0) = 0$, so that the $y$-component represents the orthogonal motion. For this resetting protocol with a constant resetting rate $r$ and resetting position $y = y_0$, the last renewal equation takes the form

$$Q^{(r)}(y_0, t) = e^{-rt} Q(y_0, t) + r \int_0^t d\tau e^{-r\tau} Q^{(r)}(y_0, t - \tau) Q(y_0, \tau) \qquad (19.35)$$

where $Q^{(r)}(y_0, t)$ denotes the survival probability of the $y$-component in the presence of an absorbing wall at $y = 0$. Taking a Laplace transform of the above equation, we get the mean first-passage time

$$T_y = \frac{\tilde{Q}(y_0, r)}{1 - r\tilde{Q}(y_0, r)}, \qquad (19.36)$$

where $\tilde{Q}(y_0, r)$ denotes the Laplace transform of $Q(y_0, t)$. In general, it is hard to calculate $T_y$ explicitly. However, for $D_R \ll r$ we can use the effective RAP picture to compute the MFPT. For an RAP starting at rest from a position $y_0$, the Laplace transform of the survival probability $\tilde{Q}(y_0, s)$, for an absorbing boundary at the origin, is known exactly [41],

$$\tilde{Q}(y_0, s) = \frac{1}{s}\left(1 - p\left(\frac{y_0 r^{3/2}}{v_0\sqrt{D_R}}\right)\right), \qquad (19.37)$$

where

$$p(w) = \int_0^\infty \frac{dz}{z^{5/3}} e^{-wz} \mathrm{Ai}(z^{-2/3}) \left[1 + \frac{1}{4\sqrt{\pi}} \Gamma\left(-\frac{1}{2}, \frac{2}{3z}\right)\right], \qquad (19.38)$$

where $Ai(z)$ denotes the Airy function [21].

Thus for an ABP, being reset to initial position $y_0$, with velocity $\theta = 0$ at rate $r$, the mean first passage, using Eq. (19.36) is given by

$$T_y = \left(\frac{y_0}{v_0\sqrt{D_R}}\right)^{2/3} \mathcal{H}\left(\frac{y_0 r^{3/2}}{v_0\sqrt{D_R}}\right), \quad \text{where } \mathcal{H}(w) = \frac{1 - p(w)}{w^{2/3} p(w)}. \qquad (19.39)$$

While the integral in Eq. (19.38) is hard to evaluate exactly, the limiting behavior of the function $p(w)$ can be found. For small $w$, the integral is dominated by the large $z$ behaviors of the integrand. A naive large $z$ approximation of the integrand results in a divergence from around $z = 0$. However, such a divergence is absent (to the leading order) for the function $p'(w)$. Hence, we first evaluate $p'(w)$ with the large $z$ approximation of the integrand and then integrate with respect to $w$ to obtain

$$p(w) = 1 - \frac{3^{5/6} w^{1/6}}{2^{1/6}\Gamma(1/3)} - \frac{3^{1/3} w^{2/3}\Gamma(1/3)}{4\Gamma(2/3)} + \cdots \quad \text{as } w \to 0, \tag{19.40}$$

where the integrand constant is fixed by the condition $p(0) = 1$. For large $w$, the integral Eq. (19.38) is dominated by the small $z$ behavior of the integrand, i.e., $\mathrm{Ai}(z^{-2/3})\left[1 + \frac{1}{4\sqrt{\pi}}\Gamma\left(-\frac{1}{2}, \frac{2}{3z}\right)\right] \to e^{-\frac{2}{3z}} z^{1/6}/(2\sqrt{\pi})$ to the leading order, which results in

$$p(w) \to \sqrt{\frac{3}{8}} e^{-\sqrt{8w/3}} \quad \text{as } w \to \infty. \tag{19.41}$$

Hence, $\mathcal{H}(w)$ diverges as $w^{-1/2}$ as $w \to 0$ and $e^{\sqrt{8w/3}}$ as $w \to \infty$. Consequently, the mean first-passage time in the presence of resetting diverges as a power law ($T_y \sim r^{-3/4}$) as $r \to 0$ and stretched exponentially ($\ln T_y \sim r^{3/4}$) as $r \to \infty$. The function $p(w)$ is minimum at $w^* = 0.358\ldots$, and, consequently, the mean first-passage time $T_y$ has a minimum at the optimal resetting rate given by $\left[y_0/(v_0\sqrt{D_R})\right]^{2/3} r^* = 0.504\ldots$. The scaling behavior in Eq. (19.39), along with the asymptotic behaviors, Eqs. (19.40) and (19.41), is illustrated in Fig. 19.4.

Recently, the first-passage properties of an ABP confined within a circular absorbing boundary were investigated perturbatively around the passive limit [42], which was shown to depend more strongly on the magnitude of the propulsion velocity, compared to the rotational diffusion coefficient. First-passage properties of an ABP in the presence of a Lorentz force and stochastic resetting were investigated in [43],

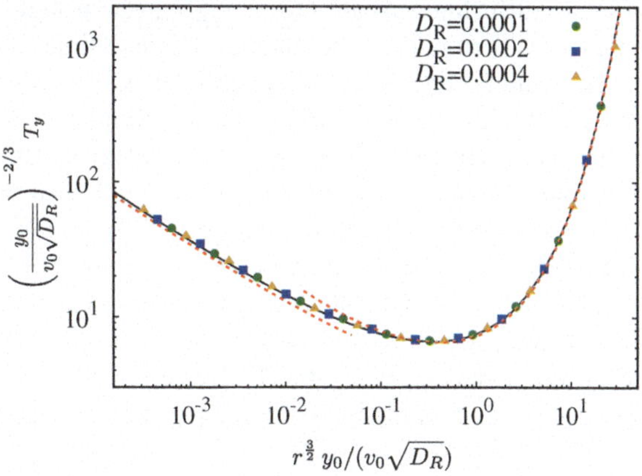


**Fig. 19.4** ABP under reset: Scaled MFPT corresponding to the $y$-component in the short-time regime $D_R \ll r$. The symbols indicate the data obtained from numerical simulations. The solid black line indicates the predicted scaling function Eq. (19.39), while the dashed red lines correspond to the asymptotic expressions Eqs. (19.41) and (19.40), of the scaling functions for small and large $w$

where it was shown that, in the presence of an inhomogeneous magnetic field, an active particle takes longer to reach a target compared to a passive particle.

## 19.4 Direction Reversing Active Brownian Particles

The two most common models of active motion discussed so far, namely, run-and-tumble and active Brownian particles, exhibit qualitatively different types of motion—the RTP exhibits intermittent, discrete changes of orientation, while the orientation for an ABP changes continuously. There exists a class of microorganisms, which shows an ABP-like motion with intermittent directional reversals. Examples include soil bacteria like Myxococcus xanthus [44–47], Pseudomonas putida [48, 49], Pseudomonas citronellolis [50], marine bacteria like Pseudoalteromonas haloplanktis, and other bacteria like Shewanella putrefaciens [51, 52]. A minimal statistical model to describe such motion is the so-called direction reversing active Brownian particle (DRABP) model [53], which in two dimensions evolves by the Langevin equations

$$\dot{x}(t) = v_0\sigma(t)\cos\theta(t), \quad \dot{y}(t) = v_0\sigma(t)\sin\theta(t), \quad \text{and} \quad \dot{\theta}(t) = \sqrt{2D_R}\,\zeta(t), \tag{19.42}$$

where $\sigma(t)$ is a dichotomous noise which alternates between $\pm 1$ with a rate $\gamma$ and $\zeta(t)$ is a Gaussian white noise with zero mean and $\langle\zeta(t)\zeta(t')\rangle = \delta(t-t')$. The continuous evolution of $\theta(t)$ gives rise to an ABP-like motion between two successive reversal events, indicated by $\sigma \to -\sigma$ transitions.

The most interesting behavior is observed when the reversal timescale $\gamma^{-1}$ is much smaller than the rotational timescale $D_R^{-1}$. This leads to the emergence of three distinct dynamical regimes, namely, a short-time regime $t \ll \gamma^{-1}, D_R^{-1}$, an intermediate-time regime $\gamma^{-1} \ll t \ll D_R^{-1}$, and a long-time regime $t \gg \gamma^{-1}, D_R^{-1}$. While in the long-time regime the DRABP behaves like a Brownian particle with an effective diffusion constant $D_{\text{eff}} = \frac{v_0^2}{2(D_R+2\gamma)}$, a strongly non-diffusive and anisotropic behavior is seen in the short and intermediate-time regimes [53].

The complex dynamical behavior of DRABP also leads to non-trivial first-passage properties with signatures of anisotropy. Similar to ABP, this anisotropic behavior is best characterized by looking at the survival probabilities along and orthogonal to the initial orientation. To this end, we look at the survival probabilities along the $x$- and $y$-directions separately, by choosing $\theta(0) = 0$ and $\sigma(0) = 1$.

Let $Q_x(x_0, t)$ denote the probability that the DRABP, starting from the $x = x_0$ line, has not crossed the absorbing boundary at $x = 0$. Similarly, $Q_y(y_0, t)$ denotes the survival probability for the $y$-component, starting from the $y = y_0$ line with an absorbing boundary at $y = 0$. The long-time diffusive behavior of DRABP implies that the survival probabilities follow Eq. (19.27) with $D_{\text{eff}} = \frac{v_0^2}{2(D_R+2\gamma)}$. On the other hand, in the short-time regime, the behavior of the DRABP is similar to that of an ABP, with $Q_x(x_0, t) \simeq 1$ and $Q_y(y_0, t)$ given by Eq. (19.32).

The interplay of the reversal and rotational diffusion leads to a novel first-passage behavior in the intermediate regime, which we discuss below. In this regime, the dichotomous noise $\sigma(t)$ emulates a Gaussian white noise $\xi(t)$ with zero mean and autocorrelation $\langle \xi(t)\xi(t')\rangle = \gamma^{-1}\delta(t-t')$. Thus the dynamics of DRABP in this regime is described effectively by the Langevin equations,

$$\dot{x}(t) = v_0\xi(t), \quad \dot{y}(t) = v_0\xi(t)\theta(t). \tag{19.43}$$

Clearly, the dynamics along the $x$-component is just a Brownian motion, leading to the survival probability given by Eq. (19.27) with an effective diffusion coefficient $v_0^2/(2\gamma)$. The $y$-component, on the other hand, undergoes a diffusion where the noise strength $\theta$ itself undergoes an independent Brownian motion.

The survival probability $Q_y(y_0, t)$, i.e., the probability that a particle starting with a $y(0) = 0$, and an initial orientation $\theta_0 = 0$ has not crossed the $y = 0$ line till time $t$, is given by

$$Q_y(y_0, t) = \int_0^\infty \mathrm{d}y\; P(y, t; y_0), \tag{19.44}$$

where $P(y, t; y_0)$ is the marginal PDF of the $y$-component in the presence of an absorbing wall at $y = 0$, starting from the initial position $y_0$. The corresponding forward Fokker–Planck equation for $\mathcal{P}(y, \theta, t)$, i.e., the probability that $y(t) = y$ and $\theta(t) = \theta$ read

$$\frac{\partial}{\partial t}\mathcal{P}(y, \theta, t) = \frac{v_0^2\theta^2}{2\gamma}\frac{\partial^2}{\partial y^2}\mathcal{P}(y, \theta, t) + D_R\frac{\partial^2}{\partial \theta^2}\mathcal{P}(y, \theta, t), \tag{19.45}$$

with the initial condition $\mathcal{P}(y, \theta, 0) = \delta(y - y_0)\delta(\theta)$ and boundary conditions $\mathcal{P}(y, \theta, t) \to 0$ as $\theta \to \pm\infty$ and $\mathcal{P}(0, \theta, t) = \mathcal{P}(\infty, \theta, t) = 0$. Note that, we have suppressed the initial position dependence for notational convenience. Making the change of variable $\frac{v_0^2 t}{2\gamma} \to t$ and taking a sine-Laplace transform, defined by

$$\tilde{\mathcal{P}}(k, \theta, s) = \int_0^\infty dt\; e^{-st}\int_0^\infty dy\; \sin(ky)\,\mathcal{P}(y, \theta, t), \quad \text{with } k > 0, \tag{19.46}$$

the Fokker–Planck equation (19.45) reduces to

$$\Lambda^2\frac{\mathrm{d}^2}{\mathrm{d}\theta^2}\tilde{\mathcal{P}}(k, \theta, s) - (s + \theta^2k^2)\tilde{\mathcal{P}}(k, \theta, s) = -\sin(ky_0)\delta(\theta), \tag{19.47}$$

with $\Lambda^2 = 2\gamma D_R/v_0^2$. For $\theta \neq 0$, the above ordinary differential equation has two independent solutions given by $D_{-q}\left(\theta\sqrt{2k/\Lambda}\right)$ and $D_{-q}\left(-\theta\sqrt{2k/\Lambda}\right)$, where

$q = \frac{1}{2}(1 + \frac{s}{k\Lambda})$ and $D_\nu(z)$ denotes the parabolic cylinder function [21]. The solutions $D_{-q}\left(\mp\theta\sqrt{2k/\Lambda}\right) \to \infty$ as $\theta \to \pm$, respectively. Therefore, using the boundary conditions $\tilde{\mathcal{P}}(k,\theta,s) \to 0$ for $\theta \to \pm\infty$, and the continuity of $\tilde{\mathcal{P}}(k,\theta,s)$ at $\theta = 0$, we get $\tilde{\mathcal{P}}(k,\theta,s) = aD_{-q}\left(|\theta|\sqrt{2k/\Lambda}\right)$. Fixing the constant $a$ by using the discontinuity of the derivatives of $\tilde{\mathcal{P}}(k,\theta,s)$ across $\theta = 0$, i.e., $(\mathrm{d}\tilde{\mathcal{P}}/\mathrm{d}\theta)_{0^+} - (\mathrm{d}\tilde{P}/\mathrm{d}\theta)_{0^-} = -\sin(ky_0)/\Lambda^2$, we get the exact solution

$$\tilde{\mathcal{P}}(k,\theta,s) = \frac{2^{\frac{q}{2}}\sin(ky_0)}{\sqrt{8\pi k\Lambda^3}}\Gamma\left(\frac{q}{2}\right)D_{-q}\left(|\theta|\sqrt{\frac{2k}{\Lambda}}\right). \tag{19.48}$$

Since only $P(y,t)$ is needed to obtain the survival probability, we integrate over $\theta$ to get the sine-Laplace transform of $P(y,t;y_0)$,

$$\tilde{P}(k,s) = \frac{2\sin(ky_0)}{s+k\Lambda}\,{}_2\tilde{F}_1\left(1,\frac{q+1}{2},\frac{q+2}{2},-1\right)\Gamma\left(\frac{q+2}{2}\right), \tag{19.49}$$

where ${}_2\tilde{F}_1(a,b,c,z)$ denotes the regularized hypergeometric function which is analytic for all values of $a$, $b$, $c$ and $z$. From Eq. (19.49), it is straightforward to identify the singularities of $\tilde{P}(k,s)$ on the complex $s$-plane, all of which lie on the negative real $s$-axis: $s_n = -k\Lambda(4n+1)$ with $n = 0, 1, 2, \cdots$ where $s_0$ comes from the prefactor $(s+\Lambda k)^{-1}$ while $s_{n>0}$ are obtained from the singularities $q_n = -2(n+1)$ of $\Gamma\left(\frac{q+2}{2}\right)$. The inverse Laplace transform of Eq. (19.49) can then be expressed as $\hat{P}(k,t) = \sum_{n=0}^{\infty} e^{s_n t} R_n$, where $R_n$ denotes the residue of $\tilde{P}(k,s)$ at $s = s_n$ and can be obtained exactly (see Supplementary material of [53]). Using the exact form of $R_n$, we get

$$\hat{P}(k,t) = \sqrt{2}\sin(ky_0)e^{-k\Lambda t}\sum_{n=0}^{\infty}\binom{-\frac{1}{2}}{n}e^{-4nk\Lambda t} = \frac{\sin(ky_0)}{\sqrt{\cosh(2k\Lambda t)}}. \tag{19.50}$$

Using the trigonometric identity $2\sin(ky)\sin(ky_0) = [\cos(k(y-y_0)) - \cos(k(y+y_0))]$ in the inverse sine-transform $P(y,t;y_0) = (2/\pi)\int_0^\infty dk\ \sin(ky)\hat{P}(k,t)$, it is straightforward to see that the PDF of the position has a scaling form

$$P(y,t;y_0) = \frac{1}{4\Lambda t}\left[f\left(\frac{y-y_0}{4\Lambda t}\right) - f\left(\frac{y+y_0}{4\Lambda t}\right)\right],$$

where the scaling function $f(z)$ can be evaluated exactly,

$$f(z) = \frac{1}{\pi}\int_0^\infty d\kappa\ \frac{\cos(\kappa z)}{\sqrt{\cosh(\kappa/2)}} = \frac{1}{\sqrt{2\pi^3}}\Gamma\left(\frac{1}{4}+iz\right)\Gamma\left(\frac{1}{4}-iz\right). \tag{19.51}$$

Thus, the survival probability, given by Eq. (19.44), also has a scaling form,

$$Q_y(y_0, t) = g\left(\frac{y_0}{v_0 t}\sqrt{\frac{\gamma}{8D_R}}\right), \tag{19.52}$$

where the scaling function $g(z_0)$ is given by

$$g(z_0) = \int_0^\infty dz\,[f(z - z_0) - f(z + z_0)] = 2\int_0^{z_0} dz\, f(z). \tag{19.53}$$

The long-time behavior of the survival probability can be easily extracted by noting that $g(z_0) = 2z_0 f(0) + O(z_0^2)$ for $z_0 \ll 1$. Thus, we have

$$Q_y(y_0, t) \approx \frac{\Gamma(1/4)^2}{2\pi^{3/2}}\sqrt{\frac{\gamma}{D_R}}\frac{y_0}{v_0 t} \quad \text{for} \quad \frac{y_0}{v_0 t} \ll \sqrt{\frac{D_R}{\gamma}} \ll 1. \tag{19.54}$$

Thus the survival probability in the direction orthogonal to the initial orientation exhibits a non-monotonic decay exponents, increasing from an initial $\alpha = 1/4$ to $\alpha = 1$ at around $t \sim O(\gamma^{-1})$, and then decreasing to $\alpha = 1/2$ at $t \sim O(D_R^{-1})$ as shown in Fig. 19.5.

Note that, the exact first-passage PDF $F_y(y_0, t) = -\partial_t Q_y(y_0, t)$ can be easily computed from Eq. (19.52),

$$F_y(y_0, t) = \frac{y_0\sqrt{2}\gamma^{3/2}}{v_0^3 t^2\sqrt{D_R}} f\left(\frac{y_0}{v_0 t}\sqrt{\frac{\gamma}{8D_R}}\right). \tag{19.55}$$

In terms of the scaled time $\tau = v_0 t\sqrt{\frac{8D_R}{\gamma}}$,

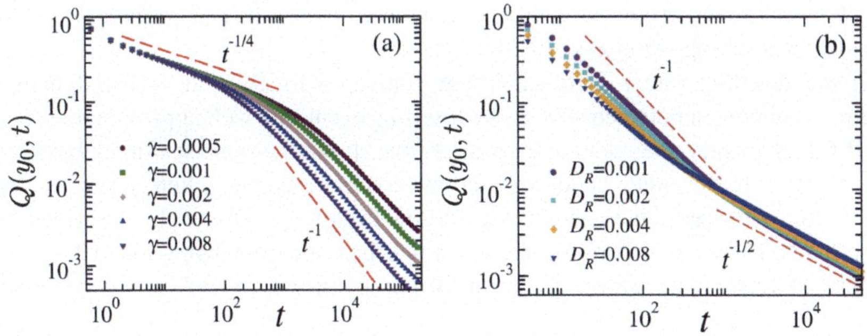


**Fig. 19.5** Survival probability of the $y$-component of the DRABP: **a** shows the crossover from the short-time $\sim t^{-1/4}$ behavior to the intermediate time $\sim t^{-1}$ behavior for $D_R = 10^{-5}$ and $y_0 = 0.001$. **b** shows the crossover from the intermediate time $\sim t^{-1}$ to long-time $\sim t^{-1/2}$ decay for $\gamma = 1$ and $y_0 = 0.1$

$$Q_y(\tau) = g(1/\tau) = 2\int_0^{1/\tau} dz f(z) = 2\int_\tau^\infty \frac{dz}{z^2} f(1/z), \tag{19.56}$$

where $f(z)$ is defined in Eq. (19.51).

**Survival probability along the direction of the initial orientation** The survival probability along the direction of the initial orientation, i.e., the probability that a particle starting from $x_0 > 0$ with an initial orientation $\theta_0 = 0$ has not crossed the line $x = 0$ up to time $t$ is, however, less interesting. At very short times ($t \ll \gamma^{-1}, D_R^{-1}$), the trajectories undergo none or very few reversals, as a result the particle starting with the initial orientation $\theta_0 = 0$ always moves away from the line $x = 0$. Due to this the particle almost always survives and the corresponding survival probability remains unity. In the intermediate- and long-time regime, the particle exhibits diffusive motion as a result of which the survival probability falls off as $t^{-1/2}$.

### 19.4.1 DRABP with Resetting

Similar to RTP and ABP, the introduction of stochastic resetting helps a DRABP in finding a target in a finite amount of time. In the following, we discuss the first-passage properties of a DRABP along the direction orthogonal to the initial orientation, which shows a very rich behavior even in the absence of resetting. We choose initial conditions $\theta(0) = 0$ and $\sigma(0) = 1$, and investigate the survival probability of the $y$-component, starting from $y = y_0$, in the presence of an absorbing wall at $y = 0$. At each resetting event, the position and orientation are both reset to their initial values at a constant rate $r$. In this case, the renewal equation for the survival probability $Q^{(r)}(y_0, t)$ is the same as Eq. (19.35). Consequently the mean first-passage time can also be determined from the Laplace transform of the survival probability in the absence of resetting using Eq. (19.36).

If the resetting rate $r$ is much larger compared to $D_R$ and $\gamma$, the dynamics of the $y$-component between successive resetting events are well approximated by an RAP. Consequently, the mean first-passage time shows the same scaling behavior as Eq. (19.39). On the other hand, a different scenario emerges when $\gamma \gg r \gg D_R$, where the $y$-process, between two resetting events, is effectively described by Eq. (19.43). Hence, to determine the mean first-passage time, using Eq. (19.36), we require the Laplace transform of the survival probability Eq. (19.52), which is given by

$$\tilde{Q}(y_0, r) \equiv \int_0^\infty e^{-rt} Q(y_0, t)\, dt = \frac{1}{r}\left[1 - p\left(\frac{r y_0}{v_0}\sqrt{\frac{\gamma}{8D_R}}\right)\right], \tag{19.57}$$

where

$$p(w) = 2\int_0^\infty e^{-w/z} f(z)\, dz. \tag{19.58}$$

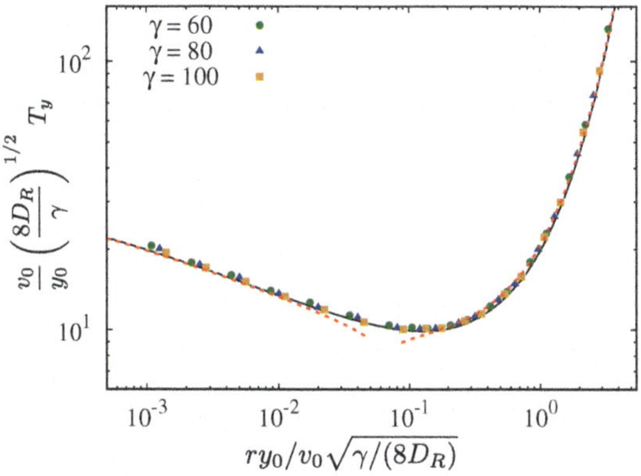


**Fig. 19.6** DRABP under reset: Scaled MFPT corresponding to the $y$-component in the intermediate regime $D_R \ll r \ll \gamma$. The symbols indicate the data obtained from numerical simulations. The solid black line indicates the predicted scaling function Eq. (19.60), while the dashed red lines correspond to the asymptotic expressions Eqs. (19.62) and (19.61), of the scaling functions for small and large $w$

The MFPT, in turn, has a scaling form,

$$T_y = \frac{y_0}{v_0}\sqrt{\frac{\gamma}{8D_R}}\,\mathcal{G}\left(\frac{r y_0}{v_0}\sqrt{\frac{\gamma}{8D_R}}\right), \tag{19.59}$$

with the scaling function

$$\mathcal{G}(w) = \frac{1 - p(w)}{w\, p(w)}. \tag{19.60}$$

This analytical scaling form is compared with numerical simulations in Fig. 19.6, where a perfect scaling collapse of the MFPT for different values of $\gamma$ is observed.

Although it is hard to find closed-form expressions for $p(w)$, its behavior for small and large $w$ can be extracted. To find the leading small-$w$ behavior, it is convenient first to consider the second derivative $p''(w) = 2w^{-1}\int_0^\infty F_1(z, w)\, f(z)\, dz$, where $F_1(z, w) = we^{-w/z}/z^2$ is a Fréchet FDT. In the limit $w \to 0$, most of the weight of $F_1(z, w)$ is concentrated around $z = w$, where $f(w)$ to the leading order can be replaced by $f(0)$. Hence, $p''(w) \approx 2f(0)/w$ to the leading order, where we have used the normalization $\int_0^\infty F_1(z, w)\ \mathrm{dz} = 1$. This becomes more apparent when one enlarges the region of integration near $z = 0$ by the change of variable $z = e^u$, which yields $p''(w) = 2w^{-1}\int_{-\infty}^\infty G(u - \ln w)\, f(e^u)\, du$, where $G(x) = e^{-x-e^{-x}}$ is the well-known Gumbel distribution. In the limit $w \to 0$, the distribution $G(u - \ln w)$ is peaked around $u = \ln w \ll 0$, where $f(e^u)$ is almost flat, and to the leading order, can be approximated by its limiting value $f(0)$ for $u \to -\infty$. Therefore, using $\int_{-\infty}^\infty G(x)\, dx = 1$, for small $w$, to the leading order, we have $p''(w) \approx 2f(0)/w$. Integrating twice, to the leading order, we get $p(w) \approx 2f(0)\ \mathrm{w} \ln w + 1$, where the integration constant 1 is fixed by the exact result $p(0) = 1$. In fact, one can go beyond this heuristic argument and find $p(w)$ for small $w$, to any order, systematically. We refer to [54], where, from exact calculation, we found the integral $-p'(w) = -a_1 \ln w + b_1 + c_1 w + d_1 w^2 + e_1 w^2 \ln w + h_1 w^3 + O(w^4)$, where $a_1 = 2f(0) \approx 3.34\ldots$, $b_1 \approx -6.93\ldots$, $c_1 \approx 22.25\ldots$, $d_1 \approx 13.47\ldots$, $e_1 \approx 28.71\ldots$, and $h_1 = -60.29\ldots$. Integrating once, we get

$$p(w) = 1 + \left(a_1 - \frac{e_1 w^2}{3}\right) w \ln w - (a_1 + b_1) w - \frac{c_1 w^2}{2} + \left(\frac{e_1}{9} - \frac{d_1}{3}\right) w^3 - \frac{h_1 w^4}{4} + O(w^5), \tag{19.61}$$

where, once again, we fixed the integration constant using $p(0) = 1$.

Due to the presence of the essential singularity $e^{-w/z}$ in Eq. (19.58), the behavior of the integral for large $w$ is dominated by large-$z$ behavior of $f(z)$. Hence, using the asymptotic behavior $f(z) = e^{-\pi z}\sqrt{\frac{2}{\pi z}}\left[1 + (8z)^{-2} + \cdots\right]$ for $z \gg 1$, in Eq. (19.58) and performing the integral, we get

$$p(w) = \frac{e^{-2\sqrt{\pi w}}}{\sqrt{2\pi}}\left[4 + \frac{\pi}{16w} + \frac{\sqrt{\pi}}{32w^{3/2}} + O\left(w^{-2}\right)\right]. \tag{19.62}$$

Therefore, the mean first-passage time in the presence of resetting diverges logarithmically ($T_y \sim -\ln r$) as $r \to 0$ and stretched exponentially ($\ln T_y \sim \sqrt{r}$) as $r \to \infty$. At an optimal resetting rate given by $\sqrt{\gamma/(8D_R)}\,(y_0/v_0)\, r^* = 0.2656\ldots$, $T_y$ is a minimum. The predicted asymptotic behaviors, Eqs. (19.61) and (19.62), of the scaling function are also compared with the scaled MFPT, obtained from numerical simulations, in Fig. 19.6, which shows an excellent match.

## 19.5 Conclusion and Outlook

In this chapter, we have reviewed recent theoretical results and reported some new findings about the first-passage properties of self-propelled/active particles. The dynamics of these particles are modeled by overdamped Langevin equations with colored noises, the correlation times of the noise being a measure of the activity. We discuss the first-passage and survival probabilities of the three most well-known active particle models, namely, RTP, ABP, and DRABP, in the presence of an absorbing target. While all these processes eventually become diffusive at times much larger than the active timescale, leading to the familiar $\sim t^{-3/2}$ tail of the first-passage time (PDF), the presence of the activity gives rise to a set of interesting scaling behavior and non-trivial persistence exponents. We also review how stochastic resetting expedites the search process by an active particle. Because of the presence of an additional internal orientation, these particles exhibit a rich first-passage behavior in the presence of resetting. In particular, we obtain interesting non-trivial functional dependence of the mean first-passage time on the resetting rate $r$ for the different models.

The advancement in single-particle tracking experiments provides an excellent opportunity to extract model parameters for real systems. The theoretical results, reviewed here, would then aid in the design of artificial micro/nanobots that are increasingly employed in targeted drug delivery to remote intracellular sites [55].

Apart from the potential applications, there are several open questions which are of interest from the theoretical point of view. The exponential waiting time between two successive tumblings or directional reversals, considered here, is an idealization of the real phenomena [49], and it would be interesting to study how the first-passage properties are altered when the waiting times are selected from PDFs that are non-Poissonian [56–58] or the activity dynamics is heterogenous [59, 60]. Active particles quite often have a "chirality," which refers to a constant drift in the evolution of the orientation [61, 62]. The first-passage properties of chiral active particles are virtually unexplored. For ABP and DRABP, we considered a complete reset, i.e., both position and orientation being reset to the initial condition. It would be interesting to consider more general resetting protocols, such as partial resetting [63, 64] and non-instantaneous resetting [65–67].

A particularly challenging question is how the first-passage properties change in higher dimensions. For example, considering specific target zones (say, a circular hole at the origin) in two dimensions, instead of a line target, which has been discussed in this review. This is particularly hard for active particles as the absorbing boundary conditions typically involve the orientation of the particle as well, unlike their passive Brownian counterpart [68]. Last but not the least, considering that the active particles, modeling living organisms, are typically found in groups or colonies with competition for resources, exploring the first-passage properties of interacting active particles, especially in the presence of passive crowders [69, 70], could be a promising avenue for future research.

## References

1. M. Ikawa, N. Inoue, A.M. Benham, M. Okabe et al., J. Clin. Investig. **120**, 984 (2010)
2. G.M. Viswanathan, M.G. Da Luz, E.P. Raposo, H.E. Stanley, *The Physics of Foraging: An Introduction to Random Searches and Biological Encounters* (Cambridge University Press, 2011)
3. J. Chen, Z. Zhang, L. Li, B.C. Chen, A. Revyakin, B. Hajj, W. Legant, M. Dahan, T. Lionnet, E. Betzig et al., Cell **156**, 1274 (2014)
4. M.G. Da Luz, A. Grosberg, E.P. Raposo, G.M. Viswanathan, J. Phys. A Math. Theor. **42**, 430301 (2009)
5. O. Bénichou, C. Loverdo, M. Moreau, R. Voituriez, Rev. Mod. Phys. **83**, 81 (2011)
6. A.J. Bray, S.N. Majumdar, G. Schehr, Adv. Phys. **62**, 225 (2013)
7. S. Redner, *A Guide to First-passage Processes* (Cambridge University Press, 2001)
8. V.V. Palyulin, A.V. Chechkin, R. Metzler, Proc. Nat. Acad. Sci. **111**, 2931 (2014)
9. M.R. Evans, S.N. Majumdar, J. Phys. A Math. Theor. **44**, 435001 (2011)

10. M.R. Evans, S.N. Majumdar, Phys. Rev. Lett. **106**, 160601 (2011)
11. Z. Yang, D.A. Aarts, Y. Chen, S. Jiang, P. Hammond, I. Kretzschmar, H.J. Schneider, M. Shahinpoor, A.H. Muller, C. Xu et al., *Janus Particle Synthesis, Self-assembly and Applications* (Royal Society of Chemistry, 2012)
12. C. Bechinger, R. Di Leonardo, H. Löwen, C. Reichhardt, G. Volpe, G. Volpe, Rev. Mod. Phys. **88**, 045006 (2016)
13. A. Walther, A.H. Muller, Chem. Rev. **113**, 5194 (2013)
14. J.R. Howse, R.A. Jones, A.J. Ryan, T. Gough, R. Vafabakhsh, R. Golestanian, Phys. Rev. Lett. **99**, 048102 (2007)
15. B.J. Nelson, I.K. Kaliakatsos, J.J. Abbott, Annu. Rev. Biomed. Eng. **12**, 55 (2010)
16. J. Wang, W. Gao, ACS Nano **6**(7), 5745 (2012)
17. H.C. Berg, D.A. Brown, Nature **239**(5374), 500 (1972)
18. P.S. Lovely, F. Dahlquist, J. Theor. Biol. **50**, 477 (1975)
19. K. Malakar, V. Jemseena, A. Kundu, K. Vijay Kumar, S. Sabhapandit, S.N. Majumdar, S. Redner, A. Dhar, J. Stat. Mech. **2018**, 043215 (2018)
20. H. Risken, *The Fokker-Planck Equation: Methods of Solution and Applications*. Springer Series in Synergetics (Springer, Berlin Heidelberg, 2012)
21. *NIST Digital Library of Mathematical Functions*. https://dlmf.nist.gov/, Release 1.1.11 of 2023-09-15. F.W.J. Olver, A.B. Olde Daalhuis, D.W. Lozier, B.I. Schneider, R.F. Boisvert, C.W. Clark, B.R. Miller, B.V. Saunders, H.S. Cohl, M.A. McClain, eds
22. F. Mori, P. Le Doussal, S.N. Majumdar, G. Schehr, Phys. Rev. Lett. **124**, 090603 (2020)
23. E. Sparre Andersen, Math. Scand. **2**, 195 (1954)
24. M.R. Evans, S.N. Majumdar, J. Phys. A Math. Theor. **51**, 475003 (2018)
25. L. Angelani, R. Di Leonardo, M. Paoluzzi, Eur. Phys. J. E **37**, 1 (2014)
26. P. Singh, S. Sabhapandit, A. Kundu, J. Stat. Mech. **2020**, 083207 (2020)
27. A. Dhar, A. Kundu, S.N. Majumdar, S. Sabhapandit, G. Schehr, Phys. Rev. E **99**, 032132 (2019)
28. P.C. Bressloff, J. Phys. A Math. Theor. **56**, 435004 (2023)
29. P.C. Bressloff, J. Stat. Mech. **2023**, 043208 (2023)
30. D.S. Grebenkov, Phys. Rev. Lett. **125**, 078102 (2020)
31. D.S. Grebenkov, Phys. Rev. E **102**, 032125 (2020)
32. Z. Benkhadaj, D.S. Grebenkov, Phys. Rev. E **106**, 044121 (2022)
33. D.S. Grebenkov, J. Chem. Phys. **158**, 214111 (2023)
34. R. Mallikarjun, A. Pal, Physica A **622**, 128821 (2023)
35. S.A. Iyaniwura, Z. Peng, SIAM J. Appl. Math. **84**, 1079–1095 (2024). arXiv:2310.04446
36. S. Jose, J. Stat. Mech. **2022**, 113208 (2022)
37. A. Scacchi, A. Sharma, Mol. Phys. **116**, 460 (2018)
38. I. Santra, U. Basu, S. Sabhapandit, J. Phys. A Math. Theor. **55**, 385002 (2022)
39. U. Basu, S.N. Majumdar, A. Rosso, G. Schehr, Phys. Rev. E. **98**, 062121 (2018)
40. M.C. Wang, G.E. Uhlenbeck, Rev. Mod. Phys. **17**, 323 (1945)
41. T. Burkhardt, J. Phys. A Math. Gen. **26**, L1157 (1993)
42. F. Di Trapani, T. Franosch, M. Caraglio, Phys. Revi. E **107**, 064123 (2023)
43. I. Abdoli, A. Sharma, Soft Matter **17**, 1307 (2021)
44. Y. Wu, A.D. Kaiser, Y. Jiang, M.S. Alber, Proc. Natl. Acad. Sci. **106**, 1222 (2009)
45. S. Thutupalli, M. Sun, F. Bunyak, K. Palaniappan, J.W. Shaevitz, J.R. Soc, Interface **12**, 20150049 (2015)
46. S. Leonardy, I. Bulyha, L. Sögaard-Andersen, Mol. Biosyst. **4**, 1009 (2008)
47. G. Liu, A. Patch, F. Bahar, D. Yllanes, R.D. Welch, M.C. Marchetti, S. Thutupalli, J.W. Shaevitz, Phys. Rev. Lett. **122**, 248102 (2019)
48. C.S. Harwood, K. Fosnaugh, M. Dispensa, J. Bacteriol. **171**, 4063 (1989)
49. M. Theves, J. Taktikos, V. Zaburdaev, H. Stark, C. Beta, Biophys. J. **105**, 1915 (2013)
50. B.L. Taylor, D. Koshland Jr., J. Bacteriol. **119**, 640 (1974)
51. J.E. Johansen, J. Pinhassi, N. Blackburn, U.L. Zweifel, Å. Hagström, Aquat. Microb. Ecol. **28**, 229 (2002)
52. G.M. Barbara, J.G. Mitchell, F.E.M.S. Microbiol, Ecol. **44**, 79 (2003)

53. I. Santra, U. Basu, S. Sabhapandit, Phys. Rev. E **104**, L012601 (2021)
54. I. Santra, U. Basu, S. Sabhapandit, J. Phys. A Math. Theor. **55**, 414002 (2022)
55. C. Xin, D. Jin, Y. Hu, L. Yang, R. Li, L. Wang, Z. Ren, D. Wang, S. Ji, K. Hu et al., ACS Nano **15**, 18048 (2021)
56. R. Großmann, F. Peruani, M. Bär, New J. Phys. **18**, 043009 (2016)
57. F. Detcheverry, Phys. Rev. E **96**, 012415 (2017)
58. V. Zaburdaev, S. Denisov, J. Klafter, Rev. Mod. Phys. **87**, 483 (2015)
59. S.M.J. Khadem, N.H. Siboni, S.H.L. Klapp, Phys. Rev. E **104**, 064615 (2021)
60. E. Lemaitre, I.M. Sokolov, R. Metzler, A.V. Chechkin, New J. Phys. **25**, 013010 (2023)
61. B. Liebchen, D. Levis, EPL **139**, 67001 (2022)
62. S. Das, U. Basu, J. Stat. Mech. **2023**, 063205 (2023)
63. I. Santra, U. Basu, S. Sabhapandit, J. Stat. Mech. **2020**, 113206 (2020)
64. V. Kumar, O. Sadekar, U. Basu, Phys. Rev. E **102**, 052129 (2020)
65. D. Gupta, C.A. Plata, A. Kundu, A. Pal, J. Phys. A Math. Theor. **54**, 025003 (2020)
66. I. Santra, S. Das, S.K. Nath, J. Phys. A Math. Theor. **54**, 334001 (2021)
67. M. Radice, Phys. Rev. E **104**, 044126 (2021)
68. F. Spitzer, in *Random Walks, Brownian Motion, and Interacting Particle Systems: A Festschrift in Honor of Frank Spitzer*, ed. by R. Durrett, H. Kesten (Birkhäuser Boston, MA, 1991)
69. A. Biswas, J. Cruz, P. Parmananda, D. Das, Soft Matter **16**, 6138 (2020)
70. M. Khatami, K. Wolff, O. Pohl, M.R. Ejtehadi, H. Stark, Sci. Rep. **6**, 37670 (2016)

# Chapter 20
# Deterministic Pursuit of Randomly Moving Targets

**Davide Bernardi and Benjamin Lindner**

**Abstract** In this chapter, we explore several mathematical models describing the pursuit of an object moving randomly in a one- or two-dimensional space. Our central focus lies in analytically calculating the mean capture time within these models, treating it as a mean first passage time. In certain special cases, we can derive exact expressions, which we subsequently utilize to develop approximations valid for the general scenarios. We examine how this statistical measure is influenced by both the diffusion coefficient of the target and the number of chasers involved. Additionally, we explore strategies for optimizing the pursuit from the hunters' perspective. Looking ahead, we consider potential generalizations and extensions of the model.

## 20.1 From Hunting Hounds and Chasing Pirates to Targeting Objects in Microbiology

Pursuit and evasion challenge living organisms across a multitude of domains. In our early years, we encounter a first form of this challenge while engaging in games such as *tag*. Indeed, the act of pursuing a target is a ubiquitous endeavor that permeates a wide spectrum of both animal and human activities. It stands as a fundamental concern in the realms of terrestrial, aquatic, aerial, and even space exploration, spanning both civilian and military contexts throughout the human history. Importantly, this predicament also manifests itself in biology, where pursuits for sustenance or prey occur at various scales, spanning from bacterial chemotaxis to the hunting behaviors of lions pursuing zebras.

A simple pursuit problem that is often used as an instructive example in differential calculus is that of a dog (or *hound*) chasing a hare that runs on a straight line with

D. Bernardi (✉)
Department of Physics and Astronomy, University of Padova, Via Marzolo 8, 35131 Padova, Italy
e-mail: davide.bernardi@unipd.it

B. Lindner
Physics Department, Humboldt University Berlin, Newtonstr. 15, 12489 Berlin, Germany
e-mail: benjamin.lindner@physik.hu-berlin.de

 
D. Grebenkov et al. (eds.), *Target Search Problems*,
https://doi.org/10.1007/978-3-031-67802-8_20

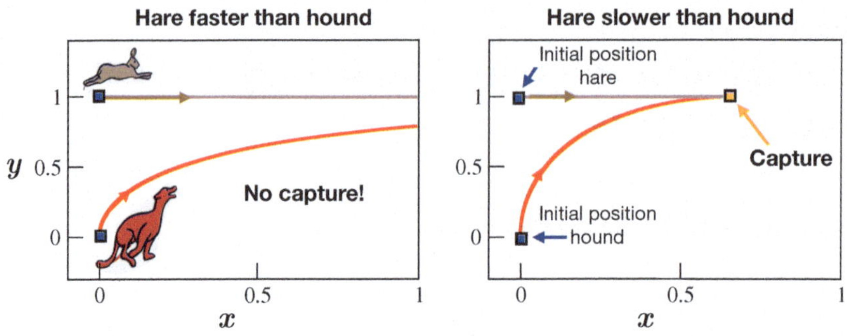


**Fig. 20.1** Classical pursuit problem. A hare (brown) is running along a horizontal line with constant speed. A dog (red) starting at the origin pursues the hare with fixed speed, constantly adjusting its direction such that it always points to the hare. If the hare is faster than the dog, it will never be caught by the dog, which will stay behind the hare forever; if it is slower than the dog, it will be captured within a finite time (the location of capture is indicated by the orange square on the right)

constant speed. The hound also maintains a constant speed, but it constantly adjusts its direction to align with their direct connection (see Fig. 20.1). The problem consists in finding a differential equation for the curve drawn by the hound in the 2d space, solving this equation, and finally determining the capture time (the capture can occur only if the hound is faster than the hare). Two trajectories (such curves are called radiodromes) are shown in Fig. 20.1 for the two cases of a fast hare (left, no capture) and a slow hare (right, capture in a finite time), respectively. Explicit formulas for the trajectories and (if finite) capture time can be found in textbooks on the topic [1].

Of course, there are other interpretations of the same mathematical problem, the most famous one is that of a pirate ship pursuing a merchant vessel, formulated and solved by the French mathematician Pierre Bouguer in the early eighteenth century. Various generalizations have been studied, for instance the case of several pursuers, the chase in spatial dimensions other than two, the question of the optimal escape strategy, several pursuers that pursue each other (n-bug problem) and many more (see [1, 2] for a review of the classical problems of pursuit theory).

If we turn to microscopic objects, relevant for instance in cell biology, we encounter other types of chases in which the target (an ion, a molecule, an organelle, or an entire cell) moves in a quasi random fashion due to thermal fluctuations or other noise sources. There is a large class of models that employ a random searcher, which, especially in combination with a random resetting [3–16], can also perform a surprisingly successful pursuit, or rather search, since in these cases the chaser typically moves randomly and the target is fixed. However, here we want to focus on a searcher that really pursues the object actively and systematically, as it is, for instance, the case for certain bacteria executing chemotaxis. This kind of detemininistic pursuit is substantially different from another class of encounter problems between randomly diffusing searchers and targets, which are the topic of many studies [17–22].

In nanorobotics, several mechanisms are being studied to endow microswimmers with the capability of detecting and chasing stochastically moving targets by taxis

[23–31]. Questions of this nature, along with others like bacterial chemotaxis, have spurred research into pursuit models incorporating stochastic elements [32–44]. Common to these kind of approaches, is that they are mostly too complicated to be explored analytically. Simple 2d and 1d models of pursuit of a Brownian particle may not be the most realistic descriptions of an artificial microswimmer or of the goal-directed movement of *Escherichia coli*. However, they allow for some theoretical insight [45, 46], which is the topic of this chapter.

## 20.2 A Tractable Pursuit Problem with a Stochastically Moving Target

In many instances of the pursuit-and-escape problem, particularly in biology, the target may not move entirely deterministically. As a starting point, we can take the limit of a completely stochastic prey and ask the question: how long does it take to hunt down a Brownian particle? We start here with a specific scenario of one hunter pursuing one diffusing prey in two spatial dimensions. Several generalizations, such as the case of multiple pursuers and other spatial dimensions, are discussed afterwards.

### *20.2.1 How Long Does a Hound Need to Hunt Down a Brownian Particle?*

Our foundational model is shown in Fig. 20.2: A Brownian particle, approximated by a circle of radius $R$, diffuses with diffusion coefficient $D$ in a 2d space. The (point-like) hound pursues the particle with speed $v_0$, always directing its velocity toward the hare. If $\mathbf{X}$ and $\mathbf{Y}$ are the vectors indicating the particle's and chaser's position, respectively, the equations of motion read:

$$\dot{\mathbf{X}} = \sqrt{2D}\boldsymbol{\xi}(t), \quad \dot{\mathbf{Y}} = v_0 \frac{\mathbf{X}-\mathbf{Y}}{||\mathbf{X}-\mathbf{Y}||}. \tag{20.1}$$

In Eq. (20.1), the dot indicates a derivative with respect to time, and the components of the vector $\boldsymbol{\xi}(t)$ are zero-mean independent Gaussian white noise processes, the mean and autocorrelation of which are defined by $\langle \xi_i(t) \rangle = 0$ and $\langle \xi_i(t)\xi_j(t') \rangle = \delta_{ij}\delta(t-t')$, respectively.

We let the system start at initial positions $\mathbf{X}_0$ and $\mathbf{Y}_0$ and monitor the distance between the positions:

$$d(t) = \sqrt{(X_1(t) - Y_1(t))^2 + (X_2(t) - Y_2(t))^2}. \tag{20.2}$$

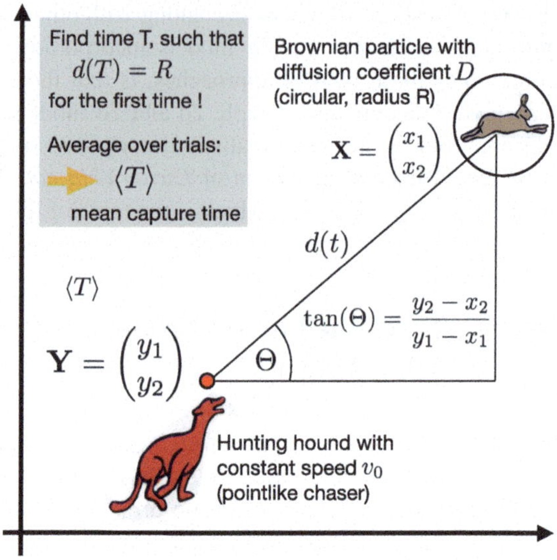


**Fig. 20.2** Pursuit of a Brownian particle. The hare (brown) is approximated as a circle of radius $R$ and performs a Brownian motion with diffusion coefficient $D$. The hound (red) pursues the particle under the angle $\Theta$ (determined by the connecting line between pursuer and pursuee) and constant speed $v_0$. We focus on the average capture time, i.e. the time at which the hound reaches the circle (the distance between the chaser and the center of the particle falls below $R$)

If this distance becomes smaller than the radius, the Brownian hare has been hunted down and the simulation is stopped. We measure the elapsed time $T$, which is the *capture time* (CT). We can repeat this numerical experiment many times and average over these repetitions (trials) to determine $\langle T \rangle$, the mean CT.

To reduce the parameters in the system, we rotate the coordinate system such that $X_2(0) = 0$, measure distances in terms of the initial distance $d_0 = d(0)$ and rescale time as $\tau = v_0 t / d_0$, leading to the following system of equations:

$$\frac{\mathrm{d}\boldsymbol{x}}{\mathrm{d}\tau} = \sqrt{2\hat{D}}\boldsymbol{\xi}(\tau); \qquad \frac{\mathrm{d}\boldsymbol{y}}{\mathrm{d}\tau} = \frac{\boldsymbol{x} - \boldsymbol{y}}{||\boldsymbol{x} - \boldsymbol{y}||}, \tag{20.3}$$

with initial conditions $y_1(0) = y_2(0) = x_2(0) = 0$ and $x_1(0) = 1$ and the new noise parameter $\hat{D} = D/(v_0 d_0)$. These stochastic differential equations can be transformed to polar coordinates [47] for the distance vector, i.e., $x_1 - y_1 = r\cos(\Theta)$ and $x_2 - y_2 = r\sin(\Theta)$, yielding the much simpler problem

$$\frac{\mathrm{d}\Theta}{\mathrm{d}\tau} = \sqrt{2D}\xi_\Theta(t), \quad \frac{\mathrm{d}r}{\mathrm{d}\tau} = \frac{D}{r} - 1 + \sqrt{2D}\xi_r(t). \tag{20.4}$$

In these equations $\xi_\Theta(t)$ and $\xi_r(t)$ are two uncorrelated Gaussian noise sources of unit intensity and the initial conditions are $\Theta(0) = 0$ and $r(0) = 1$. Here and in the following we drop the hat on the new parameter $D$ for notational convenience (which simply corresponds to setting $d_0 = 1, v_0 = 1$).

We note that the first term in the second equation is the so-called Stratonovich drift, which acts as a repulsion term near zero thus preventing the radius to take unphysical negative values. In terms of the new distance variable $r(t)$, it becomes clear that the capture problem is essentially one-dimensional (the angle variable $\Theta$ has no effect on the dynamics of the distance between hound and hare). Starting at $r(0) = 1$, we ask for the time $T$ it takes $r(T)$ to reach $R$ from above. Put differently, the capture problem reduces to a first-passage-time problem in the nonlinear potential $U(r; D)$ (the parametric dependence of the potential on the noise intensity is explicitly indicated):

$$\frac{\mathrm{d}r}{\mathrm{d}\tau} = -U'(r; D) + \sqrt{2D}\xi_r(t), \quad U(r; D) = r - D\ln(r). \tag{20.5}$$

The mean CT then follows from a standard formula for the mean first-passage-time from $r = 1$ to $r = R$ [47] and can be fortunately still significantly simplified:

$$\begin{aligned}\langle T\rangle &= \frac{1}{D}\int_R^1 \mathrm{d}y e^{U(y;D)/D}\int_y^\infty \mathrm{d}z\, e^{-U(z;D)/D}\\ &= \frac{1}{D}\int_R^1 \mathrm{d}y\frac{e^{y/D}}{y}\int_y^\infty \mathrm{d}z\, ze^{-z/D} = 1 - R + D\ln\left(\frac{1}{R}\right).\end{aligned} \tag{20.6}$$

Interestingly, the potential depends on the noise intensity, and different values of $D$ lead to qualitatively different situations as illustrated in Fig. 20.3.

For weak noise ($D < R$), the potential's minimum at $r_{\mathrm{min}} = D$ is to the left of the absorbing boundary at $r = R$—here the first passage time is dominated by the linear part of the potential. In this case, the full probability density of the CT can be well approximated by an inverse Gaussian—the first passage problem is identical to that of an overdamped Brownian particle in a constant field of gravity hitting the ground for the first time after being released from a fixed height, a problem that was solved

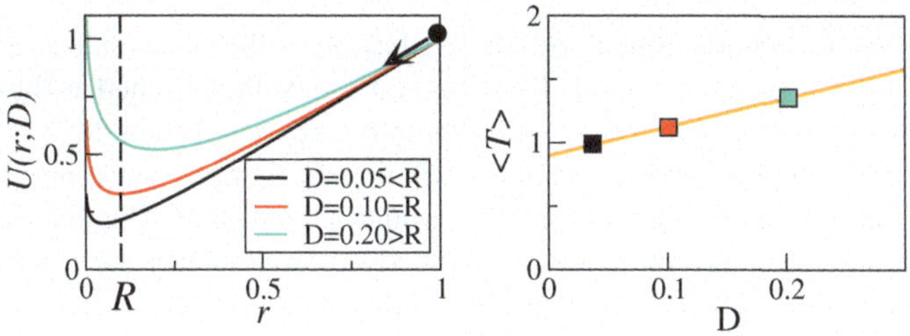


**Fig. 20.3** Effective potential (left) and mean capture time for one hound chasing a Brownian particle in 2d (right). Different noise levels lead to different shapes of the potential. In particular, the potential's minimum can be either beyond the absorbing boundary ($D = 0.05$, black), right on it ($D = 0.1$, red) or to its right ($D = 0.20$, turquoise). On increasing $D$, the potential becomes more nonlinear, which leads to a higher mean CT

by Erwin Schrödinger more than hundred years ago [48], and that we also encounter once more in the next subsection.

One might be tempted to think that for $r_{\min} = D > R$ the potential shape leads to a Kramers escape problem, and that CTs can become exponentially long with a high barrier [the difference $U(R) - U(D)$]. However, this is not the case because $D$ influences both the potential barrier and the intensity of the fluctuations needed to escape over it. This is not good news for a very active hare—although a high diffusion coefficient increases the CT (and prolongs the hare's survival), the increase of the CT with $D$ is only linear. Below, we will also encounter cases with more hunters, in which a higher diffusion coefficient may even shorten the hare's life.

### *20.2.2 Even Simpler: Hunting Brownian Particles on a Line*

Before moving on to more complicated scenarios, in which the hare is chased in two spatial dimensions by multiple hounds, it is instructive to take one step back and consider the simpler situation, that is, hare and hounds are constrained to move along a line. In this case, it turns out that the full CT distribution can be computed explicitly.

If the system is one-dimensional, the number of hounds can be two at most. We first consider a symmetric initial condition with the two hounds placed at $y_{1(2)}(0) = \pm 1$. With these conventions, the equations defining the system's evolution take a particularly simple form

$$\begin{aligned} \dot{x} &= \sqrt{2D}\xi(t) \\ \dot{y}_1 &= v_h \frac{x - y_1}{|x - y_1|} = -v_h \\ \dot{y}_2 &= v_h \frac{x - y_2}{|x - y_2|} = +v_h, \end{aligned} \tag{20.7}$$

where the condition for capture is $\min_{i=1,2}\{|x - y_i|\} < R$. It is easy to realize from Eq. (20.7) that the hounds' trajectories do not depend on the noise realization, and are $y_{1(2)}(t) = 1 \mp v_h t$ (see Fig. 20.4a). The situation can be thus framed as the diffusion of a Brownian particle with moving absorbing boundaries. When there is only one chasing hound, our problem is equivalent to finding the first-passage time distribution of a Brownian diffusion with constant drift through a fixed point. This classic problem was solved a century ago by using the method of images [48]: the solution to the diffusion equation with the same initial conditions $P_{s,1}(x, 0) = \delta(x)$ but natural boundary conditions is combined with a "negative image" of it, the initial conditions of which are chosen such that the absorbing condition, $P_{s,1}(1 - R - v_h t, t) = 0$, is fulfilled:

$$P_{s,1}(x, t) = \frac{1}{\sqrt{4\pi Dt}} \left[ e^{-\frac{x^2}{4Dt}} - e^{\frac{x_s v_h}{D}} e^{-\frac{(2x_s - x)^2}{4Dt}} \right], \tag{20.8}$$

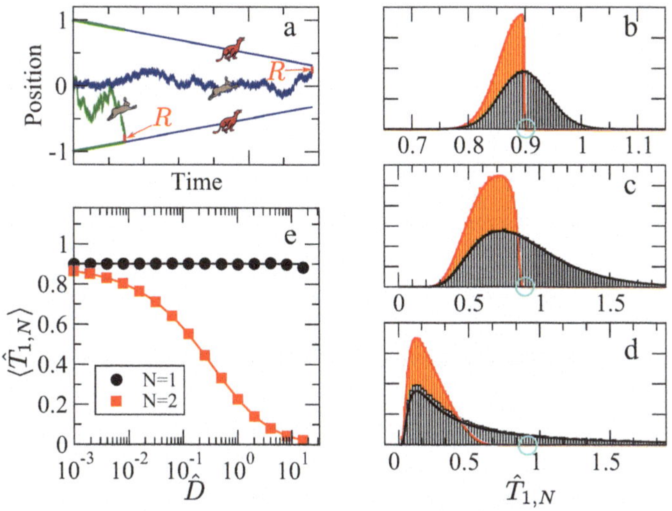


**Fig. 20.4** Chasing a random target on a line. Two typical trajectories of the hare (wiggly green and blue lines) with capture by the left and right hound, respectively **(a)**. The right column shows histograms from simulations (black for one hunter, orange for two hunters) compared to theory, Eq. (20.9) for one hound (black line) and Eq. (20.16) for two hounds (red line), for different noise intensities $D = 0.001$ **(b)**, $D = 0.064$ **(c)**, $D = 1.024$ **(d)**. Mean CT with one hound (black) and two (red) as function of noise intensity **(e)**. Lines are theory Eqs. (20.17) and (20.18), respectively; symbols depict results of simulations

where $x_s = 1 - R$. The CT distribution is the probability flux through the absorbing boundary. Since the probability density approaches zero at the boundary, the probability flux is the spatial derivative of the probability density multiplied by $-D$. The first passage time density turns out to be

$$\rho_{\mathrm{CT},1}(t) = -D\frac{\partial}{\partial x}P_{s,1}(x,t)\bigg|_{x=x_s-v_h t} = \frac{x_s}{\sqrt{4\pi D t^3}}\exp\left(-\frac{(x_s - v_h t)^2}{4Dt}\right). \quad (20.9)$$

This distribution is known as "inverse Gaussian". The name is due to the fact that this is the distribution of the time needed for a Brownian particle to reach a fixed level, which is the "inverse problem" of the distribution of the particle's position at a fixed time, which is a Gaussian distribution [49].

This method can be used to solve the case that two chasing hounds are present. Here, however, there are two absorbing boundaries that need to be taken care of at the same time. When one "virtual probability point source" is added to cancel the excess probability at one boundary, the absorbing boundary condition on the other side is affected. The problem can be solved by adding an infinite series of images, in which each term on one side cancels the residual effect of the previous one of the opposite side. Hence, we make the ansatz:

$$P_{s,2}(x,t) = P_0(x,t) + \sum_{k=1}^{\infty}(-1)^k\big(P_k^+(x,t) + P_k^-(x,t)\big), \tag{20.10}$$

where

$$P_0(x,t) = \frac{1}{\sqrt{4\pi Dt}}\exp\left(-\frac{x^2}{4Dt}\right) \tag{20.11}$$

$$P_k^{\pm}(x,t) = \frac{1}{\sqrt{4\pi Dt}}\exp\left(\frac{x_s v_h \lambda_k}{D}\right)\exp\left(-\frac{(x \mp \kappa_k x_s)^2}{4Dt}\right) \tag{20.12}$$

and $\kappa_k$, $\lambda_k$ are integer numbers to be determined such that

$$P_{k+1}^+(x_s - v_h t, t) = P_k^-(x_s - v_h t, t) \tag{20.13}$$

$$P_{k+1}^-(-(x_s - v_h t), t) = P_k^+(-(x_s - v_h t), t), \tag{20.14}$$

and $P_0^+ = P_0^- = P_0(x,t)$. The conditions in Eqs. (20.13) and (20.14) are satisfied if

$$\begin{aligned} \kappa_{k+1} &= \kappa_k + 2 \\ \lambda_{k+1} &= \lambda_k + \kappa_k + 1, \end{aligned} \tag{20.15}$$

with $\kappa_0 = \lambda_0 = 0$. These recursion relations are solved by $\kappa_k = 2k$; $\lambda_k = k^2$. Owing to the perfect symmetry of the problem, the CT density is twice the probability flux through one of the two barriers. After some manipulations, the CT density is [45]

$$\begin{aligned} \rho_{\mathrm{CT},2}(t) &= -2D\left[\frac{\partial P_0(x,t)}{\partial x} + \sum_{k=1}^{\infty}(-1)^k\left(\frac{\partial P_k^+(x,t)}{\partial x} + \frac{\partial P_k^-(x,t)}{\partial x}\right)\right]_{x=x_s-v_h t} \\ &= \frac{1}{\sqrt{4\pi Dt^3}}\sum_{k=-\infty}^{\infty}(-1)^k e^{\frac{x_s v_h k^2}{D}} e^{-\frac{(x_s(1-2k)-v_h t)^2}{4Dt}}\Big[x_s(1-2k) - v_h t\Big]. \end{aligned} \tag{20.16}$$

The shape of the CT distribution is crucially determined by the noise intensity. When the noise intensity is small (for instance, $D = 0.001$), the CT distribution in the case of a single chasing hound is nearly symmetrical (Fig. 20.4b, black line and gray histogram) and it is centered around the deterministic CT $x_s/v_h = (1-R)/v_h$ (Fig. 20.4b, blue circle). When $N = 2$ chasing hounds are present, the CT distribution has a peculiar left-skewed shape (Fig. 20.4b, red line and orange histogram) and sharply drops at the deterministic CT, which, in this case, is also the maximum CT, i.e. the time at which the distance between the two chasers is equal to the particle's diameter. Increasing the noise intensity ($D = 0.064$) shifts the peak of both distributions to lower CTs (Fig. 20.4c). A further increase to strong noise ($D = 1.024$) makes both distributions have a peak at low CTs and a positive skew (Fig. 20.4d). There is, however, an important difference: while the right tail of the $N = 1$ case grows to larger and larger values, the right tail of the $N = 2$ is constrained to remain below the deterministic time. The intuitive explanation is as follows. For $N = 1$, a strong noise can push the hare either toward the only chaser or away from it; in case of an extremely large excursion, the CT can grow, in principle, arbitrarily large. Interestingly, the mean CT (Fig. 20.4e, black line and circles) stays constant regardless of $D$, as it can be shown by computing

$$\left\langle \hat{T}_{1,1} \right\rangle = \int_0^\infty \mathrm{d}t\, t\, \rho_{\mathrm{CT},1}(t) = \frac{1-R}{v_h}. \tag{20.17}$$

When $N = 2$, the noise can only drive the hare toward either chasing hound. Hence, the noise can only accelerate the capture. As a consequence, the mean CT decreases with the noise intensity. The explicit formula for the mean CT is hard to interpret at first sight:

$$\begin{aligned} \left\langle \hat{T}_{1,2}(\hat{D}) \right\rangle &= \hat{D}\left[\left(\sum_{k=-\infty}^{\infty} (-1)^k \mathcal{B}\left(\frac{x_c}{\hat{D}}, k\right)\right) - 1\right]; \\ \mathcal{B}(\alpha, k) &= \frac{\alpha(2k-1)^2 + 1 - 2k}{|2k-1|} e^{\alpha k^2} \operatorname{erfc}(\sqrt{\alpha}|k|) - 2\sqrt{\frac{\alpha}{\pi}}. \end{aligned} \tag{20.18}$$

However, its numerical evaluation (red line in Fig. 20.4e; squares are simulations of the equation of motion) shows that the CT is indeed a decreasing function of $D$.

This procedure can be generalized to the case when the initial starting distance of the two hounds is not the same, i.e. we set $y_1(0) = 1$ and $y_2(0) = -\alpha$. Defining $x_a = y_1(0) - R$, $x_b = y_2(0) - R$, $\beta = x_a + x_b$, we find that the time-dependent probability density of the hare's position has the same form as Eq. (20.10), but here

$$\hat{P}_k^+(x, t) = F\Big(2\beta k - 2x_b - x, t\Big) G\big(x_a, x_b\big) - F\Big(2\beta k - x, t\Big) \hat{G}\big(x_a, x_b\big), \tag{20.19}$$

with

$$F(x, t) = \frac{1}{\sqrt{4\pi Dt}} \exp\left(\frac{-x^2}{4Dt}\right); \tag{20.20}$$

$$G(x, y) = \exp\left(\frac{v_h(xq_k + yp_k)}{D}\right); \tag{20.21}$$

$$\hat{G}(x, y) = \exp\left(\frac{v_h(xq_k + yp_{k+1})}{D}\right) \tag{20.22}$$

and

$$q_k = 2k^2 - k; \qquad p_k = 2k^2 - 3k + 1 \tag{20.23}$$

and $\hat{P}_k^-$ is obtained from $\hat{P}_k^-$ with the substitution $x_a \to x_b$, $x_b \to x_a$, $-x \to +x$.

Since the symmetry around the origin is broken, the CT distribution is now the sum of the probability flux at $x = x_a - v_h t$ and $x = -x_b + v_h t$:

$$\rho_{\mathrm{CT},2}(t) = D\left(\left.\frac{\partial P_{s,2}(x,t)}{\partial x}\right|_{x=-x_b+v_h t} - \left.\frac{\partial P_{s,2}(x,t)}{\partial x}\right|_{x=x_a-v_h t}\right)$$

$$= J_0(t) + \sum_{k=1}^{\infty} J_k(t, x_a - v_h t) - J_k(t, v_h t - x_b) \tag{20.24}$$

where

$$J_0(t) = \frac{(x_b - v_h t)}{2t} F(-x_b + v_h t, t) + \frac{(x_a - v_h t)}{2t} F(x_a - v_h t, t) \tag{20.25}$$

and

$$\begin{aligned} J_k(t,x) = \frac{1}{2t}\Big[&(2\beta k - 2x_b - x)F\big(2\beta k - 2x_b - x, t\big)G\big(x_a, x_b\big) \\ &- (2\beta k - x)F\big(2\beta k - x, t\big)\hat{G}\big(x_a, x_b\big) \\ &- (2\beta k - 2x_a + x)F\big(2\beta k - 2x_a + x, t\big)G\big(x_b, x_a\big) \\ &+ (2\beta k + x)F\big(2\beta k + x, t\big)\hat{G}\big(x_b, x_a\big)\Big]. \end{aligned} \tag{20.26}$$

Note that Eqs. (20.24) to (20.26) hold for $0 < t < t_{\max} = (x_a + x_b)/2v_h$. Figure 20.5 shows that the CT distribution can take a peculiar shape with a broad and asymmetric peak when one of the two chasers is at a starting distance that is twice as large as the other's (this is the case for both $\alpha = 1/2$ and $\alpha = 2$).

### 20.2.3 Many Chasers for One Particle, One Particle for Many Chasers: Approximating the Mean Capture Time

Moving back to the more interesting two-dimensional world, we could now ask what happens if multiple chasers are added to the model. We make the simplifying assumption that all chasers start at the same distance $\ell$ from the target and equidistant from each other, and that they move with the same speed $v_0$. The equations become:

$$\frac{\mathrm{d}\mathbf{X}}{\mathrm{d}t} = \sqrt{2D}\boldsymbol{\xi}(t); \qquad \frac{\mathrm{d}\mathbf{Y}_n}{\mathrm{d}t} = v_0 \frac{\mathbf{X} - \mathbf{Y}_n}{||\mathbf{X} - \mathbf{Y}_n||}, \tag{20.27}$$

where $\mathbf{Y}_n(t)$ is the position of the $n$th chaser and the chase ends whenever any chaser reaches the capture distance $R$. The CT is then defined as

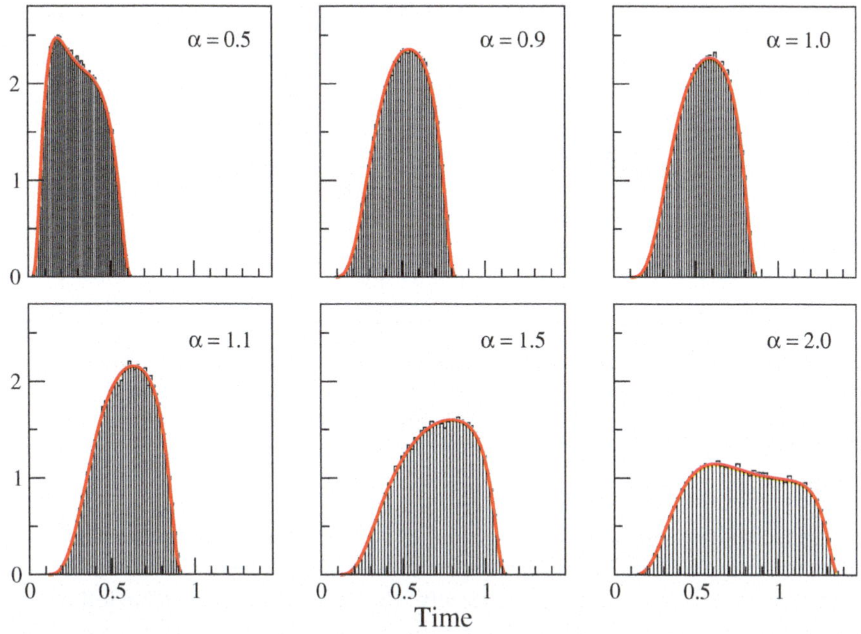


**Fig. 20.5** Capture time distribution with asymmetric starting positions. Here, the second hound starts at a distance $\alpha$ from the hare, and the first hound's starting distance is one. The noise is set here to the intermediate value $D = 0.128$. Results of simulations are shown by black histograms, theory (Eqs. (20.24–20.26)) by red lines

$$T_{2,N} = \min \left\{ t \mid \min_{n=1\ldots N} \{ ||\boldsymbol{X}(t) - \boldsymbol{Y}_n(t)|| \} \leq R\,\ell \right\}. \tag{20.28}$$

As explained above, two parameters can be eliminated by measuring space in units of the initial distance $\ell$ and time in units of $\ell/v_0$. By doing so, Eq. (20.41) becomes note that $\xi(t) = \sqrt{v_0/\ell}\xi(\tau)]$

$$\frac{\mathrm{d}\mathbf{x}}{\mathrm{d}\tau} = \sqrt{2\hat{D}}\boldsymbol{\xi}(\tau); \qquad \frac{\mathrm{d}\mathbf{y}_n}{\mathrm{d}\tau} = \frac{\mathbf{x} - \mathbf{y}_n}{||\mathbf{x} - \mathbf{y}_n||}. \tag{20.29}$$

The system is now fully specified by three parameters: the diffusion constant $\hat{D}$, the number of hounds $N$, and the rescaled capture distance $0 < R < 1$. The rescaled CT will be indicated as $\hat{T}_{2,N}$. In the following, we set $R = 0.1$ and systematically vary $\hat{D}$ and $N$. Results do not change qualitatively if $R$ is not chosen too close to the less interesting extreme cases $R = 0$ or $R = 1$, which correspond to impossible or immediate capture, respectively.

The chase for the Brownian particle cannot be reduced to an effectively 1d problem when multiple chasers are present. In fact, the diffusion of the particle occurs in

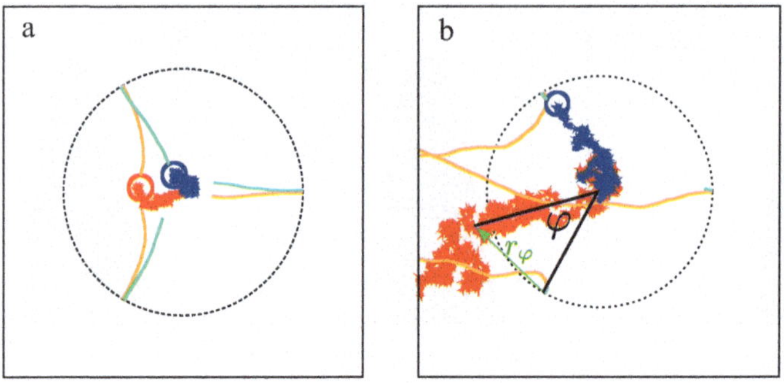


**Fig. 20.6** Examples of typical trajectories of the system at weak (**a**) and strong (**b**) noise. Two realizations are shown. In the first (second) trial, the particle's trajectory is shown in blue (red) and the chasers' trace in cyan (orange). At strong noise (**b**), the chasers' traces for the "ensnared" trajectory (cyan) are so short that they are barely visible

a 2d space with moving absorbing boundaries with no specific symmetries. Furthermore, the position of the boundaries depends on the particle's path. Finding an exact solution in the general case seems extremely challenging. However, in the following, a series of approximations will lead to a theoretical description that well captures the general picture.

In the weak-noise limit, two typical trajectories of the hare are depicted in Fig. 20.6a in red and blue, respectively. In both cases, one of the $N = 3$ chasers (trajectories shown in orange and cyan, respectively) reaches the diffusing particle when the latter has not moved far from the starting position. In this limit, the chasers act as a sort of tightening noose, and this picture becomes increasingly accurate when more chasers are added. Qualitatively, this situation is similar to that of the hare being trapped between two chasers on a line. Hence, in the weak-noise limit we will approximate the mean CT in 2d with $N$ chasers with Eq. (20.18), i.e. the theory for the 1d scenario with two chasers.

In the strong-noise limit, the hare's average displacement is much larger than that of the hounds, which, when $\hat{D} \to \infty$, appear motionless (this argument can be made precise by rescaling time as $\tau' = \hat{D}\tau$). In this limit, the typical trajectory of the system can fall into two categories: (i) "ensnared" trajectories (Fig. 20.6b, blue trajectory), that run from the origin into one of the nearly immobile chasers; (ii) "escaped" trajectories (Fig. 20.6b, red trajectory), that slip through the chasers and exit the ring on which the hounds' initial positions are (Fig. 20.6b, black dotted line). For ensnared trajectories, we expect Eq. (20.18), the 1d theory for the two-chaser case, to be still an acceptable approximation. In regard to escaped trajectories, we anticipate that when the Brownian particle diffuses away, the chasers will tend to group together and the closest pursuer will matter the most. We assume that this situation can be described by the theory for the Brownian particle pursued by a single chaser, discussed in Sect. 20.2.1. The mean CT of this part of the ensemble will then be given by

$$\left\langle \hat{T}_{2,1}(r_\varphi, \hat{D}) \right\rangle = r_\varphi - R + D \ln\left(\frac{r_\varphi}{R}\right). \qquad (20.30)$$

Equation (20.30) is Eq. (20.6), in which the initial distance $r(0) = r_\varphi$ is set to the distance from the nearest chaser to the position where the particle exits from the ring with radius $1 + R$ (Fig. 20.6b, green arrow). Of course, $r_\varphi$ depends on the particle's exit angle $\varphi$ (Fig. 20.6b, black lines). This approximation will tend to overestimate the mean CT because a strong noise fluctuation can push the particle toward a different chaser than the closest one. Averaging over the two kinds of trajectories yields

$$\left\langle \hat{T}_{2,N}(\hat{D}) \right\rangle \approx p_N \left\langle\left\langle \hat{T}_{2,1}(r_\varphi, \hat{D}) \right\rangle\right\rangle_\varphi + (1 - p_N) \left\langle \hat{T}_{1,2}(\hat{D}) \right\rangle, \qquad (20.31)$$

where $p_N$ represents the probability to observe an escape when $N$ hounds are present. For simplicity, the average over $\varphi$ assumes a uniform distribution in the interval $(0, \pi/N)$. The angular diffusion makes this approximation rather inaccurate especially at the right end of the interval, where the absorbing boundary is. Computing $p_N$ exactly is a hard problem even in the limit $\hat{D} \to \infty$, because the absorbing boundary, although static, has a complex shape with no circular symmetry. In the following, however, we show how rather crude simplifications lead to an estimate that shows a reasonable agreement with numerical simulations. Figure 20.7a displays $p_N$ for several choices of $R$ in a logarithmic scale. All points fall close to the straight line, which implies that $p_N \approx p_1^N$, at least for moderate values of $N$. How can we interpret this result? Suppose that only one chaser is present, then

$$p_1 = 1 - \frac{\omega_1}{\Omega}, \qquad (20.32)$$

where $\frac{\omega_1}{\Omega}$ is the fraction of trajectories that start from the origin and hit the only "sitting" hound before hitting the blue escape boundary (Fig. 20.7b, orange line). If we add a second hound, the fraction of escaped trajectories is now

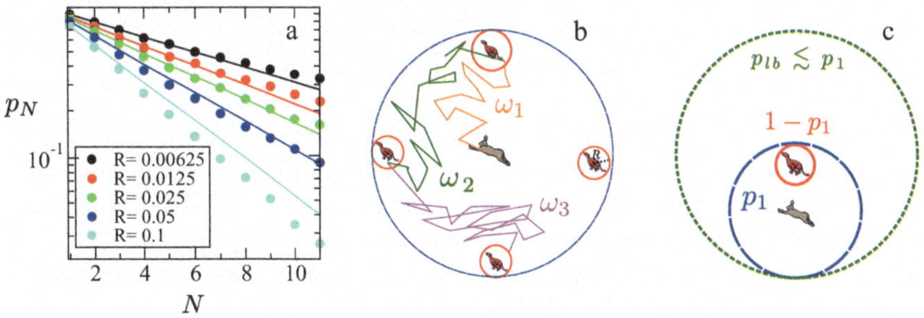


**Fig. 20.7** Approximations to the probability of "escaping" trajectories

$$p_2 = 1 - 2\frac{\omega_1}{\Omega} + \frac{\omega_2}{\Omega}. \tag{20.33}$$

The reasoning is as follows: we need to subtract $\omega_1$ trajectories that run through the first hound (regardless of the second) and $\omega_1$ that run through the second hound (regardless of the first). However, the $\omega_2$ trajectories that run through *both* hounds (Fig. 20.7b, green line) have been double-counted and thus need to be added back. This line of reasoning can be extended to the case of $N$ hounds, which leads to

$$p_N = 1 - \sum_{k=1}^{N} (-1)^{k-1} \binom{N}{k} \frac{\omega_k}{\Omega}, \tag{20.34}$$

in which the alternating sum is in the same spirit of the Wiener-Rice formula for level crossings [50, 51]. Finally, if we assume that hounds are so far from each other that the probability of diffusing from one hound to the other is the same as diffusing from the origin to one of the hounds, and that it is independent of how many hounds the trajectory has already been through, then $\omega_k \approx \omega_{k-1}\frac{\omega_1}{\Omega} = \omega_1^k/\Omega^{k-1}$. Inserting this into Eq. (20.34) leads to

$$p_N = 1 - \sum_{k=1}^{N} (-1)^{k-1} \binom{N}{k} \frac{\omega_k}{\Omega} \approx 1 - \sum_{k=1}^{N} (-1)^{k-1} \binom{N}{k} \left(\frac{\omega_1}{\Omega}\right)^k = \left(1 - \frac{\omega_1}{\Omega}\right)^N = p_1^N, \tag{20.35}$$

which is what we observe in Fig. 20.7a. Finally, we can estimate $p_1$ by replacing the actual exit boundary (Fig. 20.7c, blue dashed line) with a larger boundary with radius $2 + R$ and centered on the hound (Fig. 20.7c, green dotted line). This new boundary encompasses the actual one. Thus, the probability of reaching it before hitting the hound is a lower bound, $p_{1,\text{lb}} \leq p_1$. Interestingly, it turns out that $p_{1,\text{lb}}$ underestimates $p_1$ typically by 10% or less. Since the capture and exit boundaries (Fig. 20.7c red and green lines, respectively) are concentric, classic results on spherical Brownian motion apply [52], and we obtain a final expression for the escape probability,

$$p_N \approx p_1^N \approx p_{1,\text{lb}}^N = \left[\frac{\ln(R)}{\ln[R/(2+R)]}\right]^N. \tag{20.36}$$

It is worth emphasizing that all approximations that led to Eq. (20.36) are valid for strong noise and moderate values of $N$. In particular, when $N$ reaches the value $N_{\text{max}} = \lceil \pi/\arcsin R \rceil$, the distance between the hounds' initial positions becomes smaller than $2R$. Hence the hound ring is effectively inescapable ($p_{N \geq N_{\text{max}}} = 0$). This latter fact holds also for any noise intensity, since the distance between any two chasers can never grow in the model studied here.

Equation (20.36) can be inserted into Eq. (20.31), which applies to the strong noise limit. In principle, if we could estimate $p_N$ for arbitrary noise levels, we could use Eq. (20.31) to describe the general case. When the noise intensity is diminished,

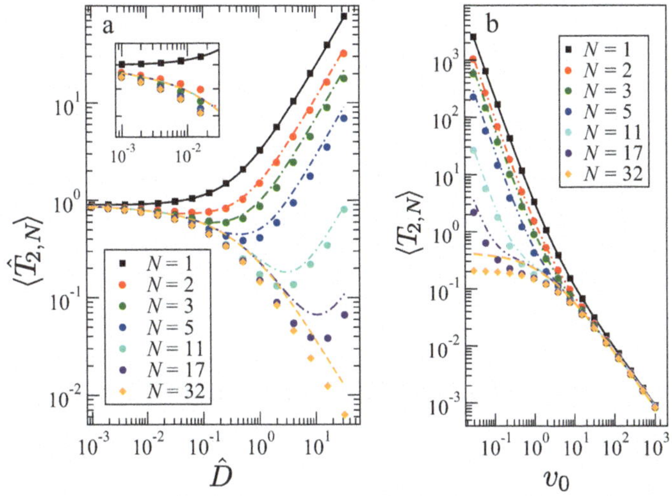


**Fig. 20.8** The mean time needed by a pack of $N$ hounds to catch a Brownian particle in 2d as a function of the particle's diffusion coefficient $\hat{D}$ (**a**) or as a function of the hounds' speed $v_0$ (**b**). In the first case, time is measured in units of $\ell/v_0$ (see text). The inset in (**a**) is a closeup of the weak-noise range. Symbols are simulation results, while lines represent the approximation Eq. (20.31) (in **a**) or the latter formula in combination with the rescaling Eq. (20.37) (in **b**)

$p_N$ must decrease and eventually vanish as $\hat{D} \to 0$. As a crude replacement for this, we set $p_N = 0$ whenever the entire right hand side of Eq. (20.31) falls below $\left\langle \hat{T}_{1,2}(\hat{D}) \right\rangle$. In other words, we use the theory for the ensnared case as a lower bound for the general case, which must be true at weak noise. When $p_N = 0$, Eq. (20.31) indeed reduces to the small-noise approximation Eq. (20.18).

This latter rudimentary approximation yields a theory that describes the general picture in a rather satisfying way, as Fig. 20.8a shows. Here, the mean CT measured in numerical simulations is plotted as a function of $\hat{D}$ for several choices of $N$ (symbols in various colors as in the legend) and compared to the theory (lines of corresponding color). Remarkably, the mean CT displays a minimum at a nonzero noise intensity for all values of $N$ except when $N = 1$ and $N = N_{\max} = 32$, for which it is monotonically increasing or decreasing, respectively. The single chaser case was solved exactly in Sect. 20.2.1, in which we also discussed why the mean CT is an increasing function of $\hat{D}$. The other extreme case of $N = N_{\max} = 32$ is the scenario in which the ring is inescapable and only ensnared trajectories are possible. The picture is qualitatively similar to the 1d trapped scenario of Sect. 20.2.2. The theory correctly predicts the decreasing behavior of the mean CT, although it underestimates it to a little extent at strong noise levels because the nonlinear noise-induced drift is neglected. When $1 < N < N_{\max}$, the two monotonic behaviors that correspond to escaped and ensnared trajectory are weighted according to $p_N$ as in Eq. (20.31), which results in the minimum at a nonzero level. The intuitive explanation is as follows. The probability of escaping the ring stays negligible over a range of weak noise intensities, in which the mean CT keeps decreasing as the noise is increased. However, once some "critical" noise level is reached, the chances of escaping the ring become relevant and the mean CT bends up, driven by trajectories that take a long time to get caught. We note that the probability of escaping saturates at a value, estimated by Eq. (20.36), that is strictly smaller than one.

From the standpoint of the hare, the findings of Fig. 20.8a can be summarized as follows: if there are multiple chasers, moving at random can only prolong the survival time if the diffusion coefficient can go beyond a certain critical level, which increases with the number of chasers. Otherwise, the hare is better off staying still.

### 20.2.4 Switching Sides: The Perfect Brownian-Particle Hunter

In the final part of this section, we will switch sides and consider the problem from the opposite point of view, and ask what is the optimal way of choosing the chasers' parameters. To this end, we revert to the original units as given in Eq. (20.41). The mean CT as a function of the chasers' speed $v_0$ is

$$\left\langle T_{2,N}(v_0)\right\rangle = \frac{\ell}{v_0}\left\langle \hat{T}_{2,N}\left[\frac{D}{\ell v_0}\right]\right\rangle, \tag{20.37}$$

which is plotted with simulation results in Fig. 20.8b. The mean CT is a decreasing function of both $N$ and $v_0$, which corresponds to the intuition that many faster chasers should get the job done quicker than fewer, slower ones. In the limit of very large speed, all curves converge to the deterministic limit $\sim \ell(1-R)/v_0$. In the limit of vanishing speed the mean CT saturates at a finite level only when $N = N_{\max} = 32$ and the hare is fully encircled, otherwise it diverges.

Figure 20.8b affirms that, if time is the only issue, increasing the number and speed of hounds is always beneficial, a somewhat unsurprising result. However, there are other less trivial takeaways. For instance, Fig. 20.8b also shows that the benefit of adding more chasers depends on how fast they are. If the speed and number of hounds is constrained in some way, the "best" way to choose $v_0$ and $N$ may be possibly different than the rather obvious "as fast and as many as possible". For instance, in this regard it is natural to take into account that the chasers need energy to move. If we assume a linear drag force $\boldsymbol{F} = -\gamma \mathrm{d}\boldsymbol{Y}_n/\mathrm{d}t$, the total energy expended by the chasers is

$$\langle E_N\rangle = \left\langle \sum_{n=1}^{N} \int \mathrm{d}\boldsymbol{Y}_n \cdot \gamma \frac{\mathrm{d}\boldsymbol{Y}_n}{\mathrm{d}t}\right\rangle = N\gamma v_0^2 \left\langle T_{2,N}\right\rangle, \tag{20.38}$$

which is shown in Fig. 20.9a for $\gamma = 1$. For any given $N$, the expended energy is, of course, an increasing function of the speed $v_0$. Interestingly, all curves roughly

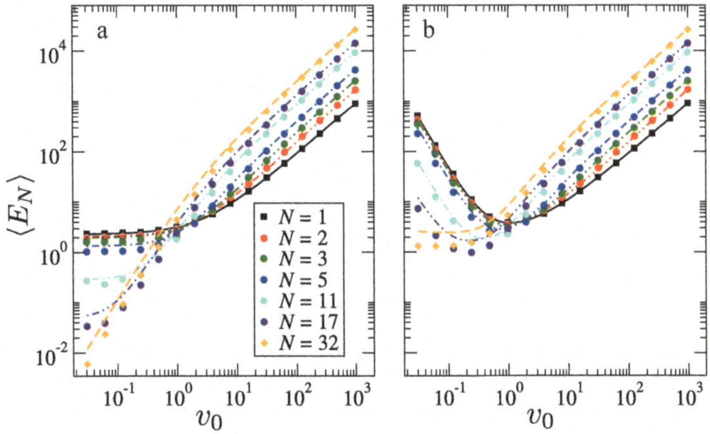


**Fig. 20.9** Mean energy spent by $N$ chasers to catch the Brownian particle as a function of the speed. In the first scenario (**a**), each pursuer is subject to a linear drag force. In the second scenario (**b**), an additional baseline power consumption is introduced, so that a global minimum appears at a nonzero speed. Lines are theory according to Eq. (20.38) (in **a**) and (20.39) (in **b**) where we use our approximation for the mean CT Eqs. (20.31) and (20.37) while symbols use Eqs. (20.38) and (20.39) with the mean CTs measured in stochastic simulations

intersect at $v_0 \approx D = 1$, which implies that around this value the total energy spent by the chasers is essentially independent of the number of hounds. At higher speeds, $\langle E_N \rangle$ increases with $N$ and when $v_0 \to \infty$ it is $\langle E_N \rangle \sim N\ell v_0(1 - R)$. Since the mean CT depends weakly on $N$ when hounds are fast, it is more efficient to employ only one. The opposite holds when chasers are slow. In the limit $v_0 \to 0$, $\langle E_1 \rangle \to D \ln(1/R)$ for $N = 1$. For $2 \leq N < N_{\max}$, $\langle E_N \rangle \sim N p_1^N \, \mathrm{D} \langle \ln(r_\varphi/R) \rangle_\varphi$, which is a decreasing function of $N$ (note that the term with the average over $\varphi$ also decreases with $N$). Finally, when $N = N_{\max}$, $E_{N_{\max}}(v_0 \to 0) \to 0$. In other words, the global minimum is achieved by surrounding the hare with motionless hounds and waiting for the hare to hop into them, the proverbial free lunch!

Altogether, Fig. 20.9a implies that, in general, employing many slow hounds is more efficient. Beside the somewhat artificial situation leading to the global optimum ($N = N_{\max}, v_0 = 0$), it is worth recalling that mean CT diverges when $v_0 \to 0$, which makes this limit theoretically fascinating but of little practical relevance.

It stands to reason that waiting should have a cost. It is natural to model this cost through a constant power drain $\beta$ independent of the chasers' motion, which accounts for the baseline metabolism, the power necessary to keep each hound functional. This additional term changes Eq. (20.38) into

$$\langle E_N \rangle = N(\gamma v_0^2 + \beta) \langle T_{2,N} \rangle . \tag{20.39}$$

Figure 20.9b shows how the new term changes the picture (here $\beta = 0.2$). At large $v_0$, the drag force (the cost of running) dominates. Hence, there is little change on the right side of the picture. At low speeds, however, all curves bend up because long CTs have now an energetic cost. As a consequence, a minimum (an energetically optimal speed) appears for all values of $N$ a finite value of $v_0$. The optimal $v_0$ and $N$ depend on $\beta$. Clearly, increasing $\beta$ makes $v_0$ shift to larger values.

## 20.3 Outlook: Open Problems in the Pursuit of Stochastic Targets

This chapter has hopefully convinced the reader that pursuit problems with randomly moving target offer some interesting challenges in the theory of stochastic processes and some elementary ones can be tackled in the framework of first-passage-time problems. Our explicit results for the capture statistics may be applicable in nanorobotics, where simple scenarios with random targets can easily be implemented. They may be also useful in the analysis of mesoscopic motion of organelles in cells that is shaped by random fluctuations.

There are many potential extensions of the simple situations we have considered. First of all, we may consider a chase in a three-dimensional space, relevant for instance in cellular biology. The chase could also take place in a potential landscape (experienced by both the pursuer and the pursuee or by only one of them) and there could be not only several hunters but also several preys. The pursuers could also be subject to fluctuations and to heterogeneity—a population of hunters with distributed speeds, for instance, is realistic in many applications. The target may not move with a simple Brownian motion but anomalously—the effect on the CT statistics is difficult to estimate but certainly worth further exploration.

Another kind of problem emerges when we take into account that often the pursuer is also a pursuee! Especially, in biology predators at the intermediate levels of the food chain are prey to others. What are the conditional CTs in models of three or more agents? The methods introduced in this chapter may be applicable to all these interesting problems.

We would like to finally briefly discuss a special extension that we have recently explored together with a co-author [46]: Here the target is subject to repelling interactions with the pursuing agents—the hare can effectively see the hounds and thus might be able to more efficiently escape from the pursuer than a purely randomly moving ('blind') hare. We introduce an evasion term into the dynamics of the hare of the form

$$\frac{\mathrm{d}\mathbf{X}}{\mathrm{d}t} = \sqrt{2D}\boldsymbol{\xi}(t) + \boldsymbol{E}, \tag{20.40}$$

$$\boldsymbol{E} = \frac{v_e}{N}\sum_{n=1}^{N} g(r_n)\frac{\boldsymbol{X}-\boldsymbol{Y}_n}{\|\boldsymbol{X}-\boldsymbol{Y}_n\|}, \quad g(r_n) = \exp\left[-\frac{r_n - R}{S}\right], \tag{20.41}$$

whereas the dynamics of the hounds are unchanged. The velocity of the hare is here influenced by the distances to the hounds with closer pursuers making a stronger impact to the effective repulsion (the parameter $S$ corresponds to an effective 'viewing range'). For this kind of model the mean CT can still be calculated analytically for one hound only in the one-and two-dimensional cases. For this case as well as for a few hounds chasing the hare one finds that if the evasion term is sufficiently strong, it prolongs the mean escape dramatically, in particular if the noise intensity is weak. Here, however, it is also important to pay attention to the kind of statistics considered: the drastic increase in the mean CT is due to the rare events that the hare has slipped through the net of its pursuers and goes off driven by the invasion term—such rare realizations of the process can significantly enlarge the mean CT. However, if one focuses on the *typical* CT as reflected, e.g., by the median of the escape time distribution a strong enlargement is not noticeable anymore. We finally mention that the determination of the critical value of the evasion term, $v_e$, is a nontrivial affair, discussed in detail in Ref. [46].

In conclusion, pursuit-and-escape models with stochastic targets remain an exciting subject of research in the theory of stochastic processes with many open problems.

**Acknowledgements** We would like to thank Meng Su for a fruitful collaboration on the topic discussed in the outlook.

## References

1. P.J. Nahin, *Chases and Escapes: The Mathematics of Pursuit and Evasion* (Princeton University Press, 2007)
2. R. Isaacs, *Differential Games: A Mathematical Theory with Applications to Warfare and Pursuit, Control and Optimization* (Wiley, New York, 1965)
3. M.A. Lomholt, T. Ambjörnsson, R. Metzler, Optimal target search on a fast-folding polymer chain with volume exchange. Phys. Rev. Lett. **95**, 260603 (2005)
4. G. Oshanin, H.S. Wio, K. Lindenberg, S.F. Burlatsky, Intermittent random walks for an optimal search strategy: one-dimensional case. J. Phys. Condens. Matter **19**, 065142 (2007)
5. M.R. Evans, S.N. Majumdar, Diffusion with stochastic resetting. Phys. Rev. Lett. **106**, 160601 (2011)
6. C. Mejía-Monasterio, G. Oshanin, G. Schehr, First passages for a search by a swarm of independent random searchers. J. Stat. Mech. Theory Exp. **2011**, P06022 (2011)
7. L. Kusmierz, S.N. Majumdar, S. Sabhapandit, G. Schehr, First order transition for the optimal search time of Lévy flights with resetting. Phys. Rev. Lett. **113**, 220602 (2014)
8. V.V. Palyulin, A.V. Chechkin, R. Metzler, Lévy flights do not always optimize random blind search for sparse targets. Proc. Natl. Acad. Sci. **111**, 2931 (2014)
9. A. Pal, S. Reuveni, First passage under restart. Phys. Rev. Lett. **118**, 030603 (2017)
10. A. Falcón-Cortés, D. Boyer, L. Giuggioli, S.N. Majumdar, Localization transition induced by learning in random searches. Phys. Rev. Lett. **119**, 140603 (2017)
11. A. Pezzotta, M. Adorisio, A. Celani, Chemotaxis emerges as the optimal solution to cooperative search games. Phys. Rev. E **98**, 042401 (2018)
12. J. Noetel, V.L.S. Freitas, E.E.N. Macau, L. Schimansky-Geier, Optimal noise in a stochastic model for local search. Phys. Rev. E **98**, 022128 (2018)

13. A. Chechkin, I.M. Sokolov, Random search with resetting: a unified renewal approach. Phys. Rev. Lett. **121**, 050601 (2018)
14. G. Mercado-Vásquez, D. Boyer, First hitting times to intermittent targets. Phys. Rev. Lett. **123**, 250603 (2019)
15. M.R. Shaebani, R. Jose, L. Santen, L. Stankevicins, F. Lautenschläger, Persistence-speed coupling enhances the search efficiency of migrating immune cells. Phys. Rev. Lett. **125**, 268102 (2020)
16. M. Dahlenburg, A.V. Chechkin, R. Schumer, R. Metzler, Stochastic resetting by a random amplitude. Phys. Rev. E **103**, 052123 (2021)
17. S. Redner, P.L. Krapivsky, Capture of the lamb: diffusing predators seeking a diffusing prey. Am. J. Phys. **67**, 1277 (1999)
18. S. Redner, *A Guide to First-Passage Processes* (Cambridge University Press, Cambridge, UK, 2001)
19. S.B. Yuste, G. Oshanin, K. Lindenberg, O. Bénichou, J. Klafter, Survival probability of a particle in a sea of mobile traps: a tale of tails. Phys. Rev. E **78**, 021105 (2008)
20. G. Oshanin, O. Vasilyev, P.L. Krapivsky, J. Klafter, Survival of an evasive prey. Proc. Natl. Acad. Sci. **106**, 13696 (2009)
21. F. Le Vot, S.B. Yuste, E. Abad, D.S. Grebenkov, First-encounter time of two diffusing particles in confinement. Phys. Rev. E **102**, 032118 (2020)
22. F. Le Vot, S.B. Yuste, E. Abad, D.S. Grebenkov, First-encounter time of two diffusing particles in two- and three-dimensional confinement. Phys. Rev. E **105**, 044119 (2022)
23. R. Golestanian, T.B. Liverpool, A. Ajdari, Propulsion of a molecular machine by asymmetric distribution of reaction products. Phys. Rev. Lett. **94**, 220801 (2005)
24. Y. Hong, N.M.K. Blackman, N.D. Kopp, A. Sen, D. Velegol, Chemotaxis of nonbiological colloidal rods. Phys. Rev. Lett. **99**, 178103 (2007)
25. L. Zhang et al., Artificial bacterial flagella: fabrication and magnetic control. Appl. Phys. Lett. **94**, 64107 (2009)
26. S. Sengupta, M.E. Ibele, A. Sen, Fantastic voyage: designing self-powered nanorobots. Angew. Chem. Int. Ed. **51**, 8434 (2012)
27. M. You, C. Chen, L. Xu, F. Mou, J. Guan, Intelligent micro/nanomotors with taxis. Acc. Chem. Res. **51**, 3006 (2018)
28. M.N. Popescu, W.E. Uspal, C. Bechinger, P. Fischer, Chemotaxis of active Janus nanoparticles. Nano Lett. **18**, 5345 (2018). (PMID: 30047271)
29. S. Palagi, P. Fischer, Bioinspired microrobots. Nat. Rev. Mater. **3**, 113 (2018)
30. M. Medina-Sánchez, V. Magdanz, M. Guix, V. Fomin, O. Schmidt, Adv. Funct. Mater. **28**, 1707228 (2018)
31. L. Alvarez, M. Fernandez-Rodriguez, A. Alegria, Reconfigurable artificial microswimmers with internal feedback. Nat. Commun. **12**, 4762 (2021)
32. J.M. Dobbie, Search for an avoiding target. SIAM J. Appl. Math. **28**, 72 (1975)
33. P. Bernhard, A.-L. Colomb, G.P. Papavassilopoulos, Rabbit and hunter game: two discrete stochastic formulations. Comput. Math. Appl. **13**, 205 (1987)
34. A. Bonato, D. Mitsche, X. Pérez-Giménez, P. Praíat, A probabilistic version of the game of zombies and survivors on graphs. Theor. Comput. Sci. **655**, 2 (2016). Special Issue on Theory and Applications of Graph Searching Problems
35. F. Simard, J. Desharnais, F. Laviolette, General cops and robbers games with randomness. Theor. Comput. Sci. **887**, 30 (2021)
36. R. Vidal, O. Shakernia, H.J. Kim, D.H. Shim, S. Sastry, Probabilistic pursuit-evasion games: theory, implementation, and experimental evaluation. IEEE Trans. Rob. Autom. **18**, 662 (2002)
37. P. Romanczuk, I.D. Couzin, L. Schimansky-Geier, Collective motion due to individual escape and pursuit response. Phys. Rev. Lett. **102**, 010602 (2009)
38. A. Kamimura, T. Ohira, Group chase and escape. New J. Phys. **12**, 053013 (2010)
39. L. Angelani, Collective predation and escape strategies. Phys. Rev. Lett. **109**, 118104 (2012)
40. T. Iwama, M. Sato, Group chase and escape with some fast chasers. Phys. Rev. E **86**, 067102 (2012)

41. Y. Lin, N. Abaid, Collective behavior and predation success in a predator-prey model inspired by hunting bats. Phys. Rev. E **88**, 062724 (2013)
42. M. Janosov, C. Virágh, G. Vásárhelyi, T. Vicsek, Group chasing tactics: how to catch a faster prey. New J. Phys. **19**, 053003 (2017)
43. A. Surendran, M. Plank, M. Simpson, Spatial structure arising from chase-escape interactions with crowding. Sci. Rep. **9**, 14988 (2019)
44. S. Goh, R.G. Winkler, G. Gompper, Noisy pursuit and pattern formation of self-steering active particles. New J. Phys. **24**, 093039 (2022)
45. D. Bernardi, B. Lindner, Run with the Brownian hare, hunt with the deterministic hounds. Phys. Rev. Lett. **128**, 040601 (2022)
46. M. Su, D. Bernardi, B. Lindner, Pursuit problem with a stochastic prey that sees its chasers. New J. Phys. **25**, 023033 (2023)
47. C.W. Gardiner, *Handbook of Stochastic Methods* (Springer-Verlag, Berlin, 1985)
48. E. Schrödinger, Zur Theorie der Fall- und Steigversuche an Teilchen mit Brownscher Bewegung. Phys. Z. **16**, 289 (1915)
49. M.C.K. Tweedie, Inverse statistical variates. Nature **155**, 453 (1945)
50. T. Verechtchaguina, I.M. Sokolov, L. Schimansky-Geier, First passage time densities in non-Markovian models with subthreshold oscillations. EPL **73**, 691 (2006)
51. T. Schwalger, Mapping input noise to escape noise in integrate-and-fire neurons: a level-crossing approach. Biol. Cybern. **115**, 539 (2021)
52. J.G. Wendel, Hitting spheres with Brownian motion. Ann. Probab. **8**, 164 (1980)

# Chapter 21
# Conservative Versus Non-conservative Diffusion Toward a Target in a Networked Environment

**Ernesto Estrada**

**Abstract** The networked nature of complex systems determines the way in which "information" navigates the system from a source to a target. This navigation is governed by the lack of central controllers and the fact that every individual entity ignores the global structure of the system. Consequently, targeted shortest-path searches are almost automatically excluded in these systems, leaving the more blind diffusive processes as the main mechanism for navigating complex networks. Here we show that non-conservative diffusion has some advantages over the "classical" (conservative) diffusion for searching a target in a network. The non-conservative nature of the diffusion process is given by the possibility that the network "communicates" with the environment in which it is embedded. We use analytical and computational methods to show that non-conservative diffusion uses trajectories which are more prone to find a target than the conservative one. We illustrate the existence of this mechanisms in systems as varied as traffic in urban environments, volume transmission in the brain, and communication through online social networks.

## 21.1 Introduction

"The Target Problem" seems to be an easy problem when it deals with searching strategies toward an efficient target identification on a networked environment. This environment can be represented by the discrete space created by a graph $G = (V, E)$, in which a set of vertices $V$ are interconnected by pairs forming a set of edges $E$ [25]. Therefore, once a target has been identified in the graph, e.g., the vertex $w \in V$, our task can be thought to be reduced to find the shortest topological path connecting our current location, e.g., the vertex $v \in V$, with the target. A path refers to a sequence of different vertices and edges between two corresponding vertices. Among all existing paths between two vertices, the one having the minimum length, in terms of the number of edges in the path, is the shortest path. There are several

E. Estrada (✉)
Institute for Cross-Disciplinary Physics and Complex Systems (IFISC), CSIC-UIB, Palma de Mallorca, Spain
e-mail: estrada@ifisc.uib-csic.es



D. Grebenkov et al. (eds.), *Target Search Problems*,
https://doi.org/10.1007/978-3-031-67802-8_21

algorithms for finding the shortest path between two vertices in a graph [48, 92, 100]. The Bellman–Ford–Moore algorithm [7, 40, 77] allows to find the shortest paths from the source vertex to all other vertices in a graph in which edges can be weighted by positive numbers. The algorithm has a time complexity of $O\left(V \cdot E\right)$, which because $\#E \geq \#V - 1$ in connected graphs, makes the complexity relatively large, i.e., in general larger, than $O\left(V^2\right)$. The most popular algorithm for finding shortest paths is the Dijkstra algorithm [20], which finds the shortest paths from the source vertex to all other vertices in the graph with time complexity $O\left(V^2\right)$. Finally, the Floyd–Warshall algorithm [39, 93] finds the shortest paths between all pairs of vertices in a graph in $O\left(V^3\right)$ allowing weights in the edges, which can be positive or negative.

Let us now see the target problem from the perspective of a single vertex in a network representation of a complex system [31]. Think, for instance, about a neuron which is interconnected to others in a human brain composed of about $10^{10}$ neuronal cells and $10^{14}$ interconnections [64]. Can this single neuron find the shortest path to a target neuron in this network? If the neuron is going to use a routing/navigation process to find the shortest path to a specific target it needs to have a "global knowledge about the network topology" [50]. This is exactly what the previously mentioned algorithms use: global information about the network. It is hard to digest that every single neuron has a map of the 86 billion neurons in the brain. In spite of this, there are authors who believe that [68] "the shortest path plays an important role in the information transmission of a brain network, and it is a very important measure to describe the internal structure of the brain network". The claim is mainly based on the apparent fact that "the shortest path can transmit the information more quickly and reduce brain consumption" [68]. For such a thing to be possible, it is necessary that there exists a central controller with "global knowledge about the network topology" [50] to direct the information through the shortest paths connecting pairs of brain regions. But even in this "mystical" scenario, the use of the shortest path could be inefficient from an energetic point of view. Tomasi et al. [90] have found experimentally that "a higher degree of connectivity was associated with nonlinear increases in metabolism". That is, the more connected a vertex is, the higher its energy consumption. It is evident that a highly connected vertex supports many shortest paths crossing it—a vertex of degree $k$ may support up to $k\left(k - 1\right)/2$ shortest paths of length two between its nearest neighbors, apart from the rest. Therefore, it is not necessarily true that the use of shortest paths "reduces brain consumption" as claimed in [68].

How is then possible that information finds its way from a specific source to a specific target in such complex networked environments like a human brain? Diffusion may be a plausible solution. At the end, a diffusive process can transmit information between a source and a target without any "knowledge about global network topology" [50]. This strategy will also avoid the problem of increasing glucose consumption because it "prevents particles or messages from taking shortest paths" [50]. Diffusion is ubiquitous in nature and in many man-made systems, even in those in which it seems to be counterintuitive [72]. For instance, when navigating a city at rush hour, drivers in general know the shortest topological paths between origin–target pairs. However, it has been seen statistically that they frequently avoid such paths for several reasons [3]. The first is that such routes contain, with high probability, highly interconnected intersections. Therefore, traveling through shortest topological paths necessarily would imply more complicated maneuvers and more waiting time due to traffic signals, jams, etc. The second is that in navigating between origin–target pairs, drivers frequently use their own "cognitive maps", which allow them "establishing locations, understanding distances between locations, comprehending the direction of one location from another, linking locations in sequence, and transferring knowledge from the mental arena to the surrounding physical environment" [49]. We can think that every driver in the same city has his own cognitive map to go from one place to another, particularly at rush hour. Indeed, Golledge and Gärling [49] have found, in travel-related literature, more than 20 different strategies used by drivers to find their routes. Their search includes reports in fields such as travel behavior, operations research, transport geography, and behavioral travel modeling. Such mental algorithms include strategies like: using the "longest leg first", or using the "shortest leg first", selecting the route that has "fewest turns", or "fewest lights or stop signs", or "fewest obstacles or obstructions", etc. If we assign randomly and independently these algorithms to the drivers in a city at rush hour, what we will observe are patterns similar to those of a diffusive motion [3, 61, 98].

When we talk about diffusion we typically talk about a series of different physical phenomena, which include classical conservative diffusion, non-conservative diffusion, and anomalous diffusion (sub- and superdiffusion) [18, 46, 60, 66, 84]. By conservative we mean that the amount of diffusive particles is constant in the graph, while in a non-conservative diffusion the number of diffusive particles can change with time. The reason for the latest is that some particles are created/annihilated in the graph, maybe because they escape to the environment in which this graph is embedded. Here, we study the similarities and differences between conservative and non-conservative diffusion on graphs. We discover here that the non-conservative diffusive strategy present certain advantages for a single vertex in a network to send information to a target in a faster way relative to the conservative one. The main reason behind this advantage of non-conservative diffusion over the conservative one resides in the trajectories that diffusive particles follow in both processes. These trajectories are found here by means of a geometrization of the graphs using the Euclidean distances induced by the diffusive processes. We find here that non-conservative diffusive particles follow trajectories involving low-connected vertices, which coincide with the bypasses recently found to play an important role in complex networks navigation [33], while the conservative ones can follow trajectories more similar to those of the shortest topological paths. We finalize this chapter with some examples of complex systems in which such non-conservative diffusive strategies are used.

## 21.2 The Setting

Here we only consider simple, undirected and unweighted graphs $G = (V, E)$, where $V = \{v_1, \ldots, v_n\}$ is the set of vertices with $\#V = n$, and $E = \{(v_i, v_j) \mid v_i, v_j \in V\}$ is the set of edges with $\#E = m$. We use indistinctly the terms graphs and networks to refer to $G$, although the term "network" is mainly reserved to the skeleton of complex systems in the real world. The terms vertex and node are also used indistinctly.

A walk of length $l$ in a graph is a sequence of (not necessarily different) vertices $v_1, v_2, \ldots, v_l, v_{l+1}$ such that for each $i = 1, 2, \ldots l$ there is an edge from $v_i$ to $v_{i+1}$. The walk is known as closed if $v_{l+1} = v_1$. If all the vertices and edges of the walk are different we say that it corresponds to a path. Among all the paths that connect a pair of vertices, the one having the minimum length is the shortest path. The shortest path between two vertices is a distance. A graph for which there is at least a path between every pair of vertices is said to be connected. We will consider here only connected graphs.

Two vertices are adjacent in the graph if they share a common edge. Then the square symmetric matrix $A$, whose entries are defined as

$$A_{i,j} = \begin{cases} 1 & (i, j) \in E \\ 0 & (i, j) \notin E, \end{cases} \tag{21.1}$$

represents the adjacency between pairs of vertices in the graph and it is known as the adjacency matrix.

The following is a very well-known result (see, for instance, [8]):

**Theorem 1** *Let $G$ be a simple connected graph. Then, the $(i, j)$ entry of the $l$th power of $A$, $\left(A^l\right)_{ij}$, counts the number of walks of length $l$ between the vertices $i$ and $j$.* □

The number of vertices adjacent to a vertex $i$ is known as the degree of $i$ and denoted by $k_i$. The diagonal matrix $K$ of the degree of the vertices in $G$ is known as the degree matrix. Then, the matrix $L = K - A$ is defined as the graph Laplacian matrix of $G$, see further for a "first-principles" definition.

## 21.3 Normal Diffusion

Let us consider a simple undirected graph as the one illustrated in Fig. 21.1 in which there is a "concentration" $C_i(0) = C_i(t = 0) \in \mathbb{R}_+ \cup \{0\}$ of a given "item" at the node $i$ at a time $t = 0$.

At every edge $(i, j)$ there will be a gradient $C_i(0) - C_j(0)$ of concentrations. As time evolves, there is a "movement" of the "items" from high concentration to low concentration. Let us pick a vertex of $G$, let say $A$, and analyze how the concentration of the items at this vertex changes in time. Because $A$ is connected to $B$ and to $C$ we

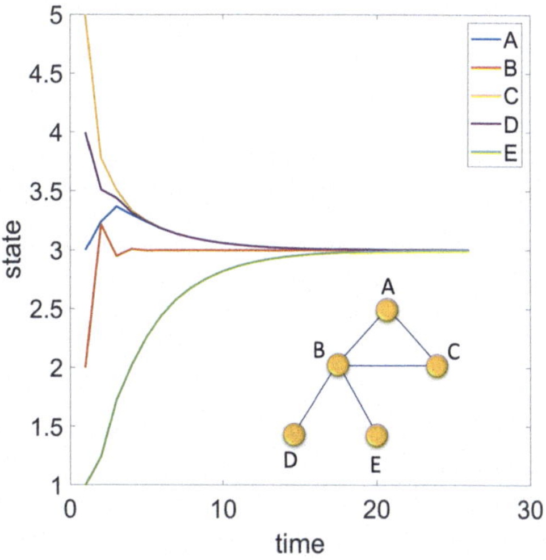


**Fig. 21.1** Illustration of the time evolution of the standard diffusion on the graph illustrated in the inset, when the initial concentrations at the vertices are given by $C^0 = [3, 2, 5, 4, 1]^T$ for the vertices $V = \{A, B, C, D, E\}$

have that the change of concentration at $A$ is proportional to the sum of the gradients with both $B$ and $C$:

$$\frac{dC_A(t)}{dt} = \gamma \left[ (C_B(t) - C_A(t)) + (C_C(t) - C_A(t)) \right], \tag{21.2}$$

where $\gamma$ is a proportionality coefficient known as diffusivity or diffusion coefficient [46]. We can rearrange this equation for the node $A$ as

$$\frac{dC_A(t)}{dt} = \gamma \left[ C_B(t) + C_C(t) - 2C_A(t) \right], \tag{21.3}$$

where $C_B(t) + C_C(t)$ is the sum of the concentrations at the nearest neighbors of $A$ and $2C_A(t)$ is the concentration at $A$ multiplied by the number of nearest neighbors it has, i.e., its degree. Let us hereafter consider $\gamma = 1$. Then, we can write generically the following equation:

$$\dot{C}_i(t) = \left( \sum_{(j,i) \in E} C_j(t) \right) - k_i C_i(t), \tag{21.4}$$

where $k_i$ is the degree of vertex $i$. It is worth mentioning here that this equation is a particular case of the so-called "master equation", which is considered in other chapters of this book.

Let us now define the following.

**Definition 2** Let $\mathcal{H} := \ell^2(V)$ be a Hilbert space and let $f \in \mathcal{H}$ be a function. Then, the adjacency operator $\mathcal{A}$ of $G$ is defined as [70]

$$(\mathcal{A}f)(v) := \sum_{(w,v)\in E} f(w). \tag{21.5}$$

Notice that if $G$ is infinite but locally finite, i.e., $k_i < \infty$ for all vertices of $G$, the adjacency operator is a bounded self-adjoint operator in $\mathcal{H}$ [70]. If the graph is finite the adjacency operator coincides with the adjacency matrix $A$ of the graph.

**Definition 3** Let $\mathcal{H} := \ell^2(V)$ be a Hilbert space and let $f \in \mathcal{H}$ be a function. Then, the degree operator $\mathcal{K}$ of $G$ is defined as

$$(\mathcal{K}f)(v) := k_v f(v). \tag{21.6}$$

If the graph is finite the degree operator is realized by the diagonal matrix $K$ of vertices degrees of the graph.

Let $\dot{C}(t) := \left[\dot{C}_{v_1}(t) \ldots \dot{C}_{v_n}(t)\right]^T$ and let $C(t) := \left[C_{v_1}(t) \ldots C_{v_n}(t)\right]^T$. Then, we can write the set of diffusion equations for all the vertices of the graph in the following matrix-vector form:

$$\begin{aligned}\dot{C}(t) &= AC(t) - KC(t)\\ &= (A - K)C(t)\\ &= -LC(t),\end{aligned} \tag{21.7}$$

where $L := K - A$ is the Laplacian of the graph, which can be formally defined as

**Definition 4** Let $\mathcal{H} := \ell^2(V)$ be a Hilbert space and let $f \in \mathcal{H}$ be a function. The Laplacian operator $\mathcal{L}$ of $G$ is defined as [53]

$$(\mathcal{L}f)(v) := \sum_{(w,v)\in E} (f(w) - f(v)). \tag{21.8}$$

When the graph is finite, then $L$ is the matrix representation of this operator. □

We then have the diffusion equation of the graph as

$$\dot{C}(t) = -LC(t), \tag{21.9}$$

where $L := K - A$ is the graph Laplacian and $K$ is a diagonal matrix of vertex degrees, and where the initial condition is $C(0) = C^0 \in \mathbb{R}_+^{n\times 1} \cup \{0\}$.

### 21.3.1 Some Properties of L

Here we state some of the general properties of the graph Laplacian matrix $L$ (see [54, 55, 73, 76]).

**Definition 5** Let $m$ be the number of edges and $n$ be the number of vertices. Then the incidence matrix $\nabla$ is the $m \times n$ matrix given by

$$\nabla_{v,e} = \begin{cases} -1, & e = (v, w) \text{ and } v > w \\ 1, & e = (v, w) \text{ and } v < w \\ 0, & \text{otherwise.} \end{cases} \tag{21.10}$$

**Lemma 6** *The Laplacian matrix of the graph L is given by*

$$L = \nabla^T \nabla, \tag{21.11}$$

*such that the incidence matrix can be considered as a gradient matrix of a graph where concentrations are arbitrarily fixed at the nodes.* □

The following describes some of the general properties of the Laplacian.

**Theorem 7** *The Laplacian matrix L of a graph* □

1. is a positive semidefinite matrix, indeed $f^T L f = \sum_{(v,w)\in E} (f(v) - f(w))^2 \geq 0$;
2. always has an eigenvalue equal to 0;
3. the multiplicity of the zero eigenvalue is equal to the number of connected components of the graph.

## 21.3.2 Conservative Nature of the Graph Diffusion

**Definition 8** A diffusion process is said to be conservative if the number of diffusive particles is constant along the time. That is, if and only if $\mathbf{1}^T C(t) = \mathbf{1}^T C^0$ for any $0 \leq t \leq \infty$. Otherwise the process is said to be non-conservative □

**Lemma 9** *The diffusion (21.9) on a connected graph is always conservative.* □

***Proof*** The solution of the diffusion equation is

$$C(t) = e^{-tL} C^0. \tag{21.12}$$

Let us take the sum of the entries of $C(t)$ at an arbitrary time $t$,

$$\mathbf{1}^T C(t) = \mathbf{1}^T e^{-tL} C^0 \tag{21.13}$$

and let us expand the matrix exponential in its Taylor series

$$\mathbf{1}^T C(t) = \mathbf{1}^T C^0 - t\mathbf{1}^T L C^0 + \frac{t^2}{2!}\mathbf{1}^T L^2 C^0 + \cdots + \left(\frac{(-1)^k t^k}{k!}\right)\mathbf{1}^T L^k C^0 + \cdots. \tag{21.14}$$

Because $L$ is positive semidefinite, we have that $\mathbf{1}^T L = 0$ and so $\left(\mathbf{1}^T L\right) L^{k-1} = 0$, so that $\mathbf{1}^T C(t) = \mathbf{1}^T C^0$ for any $t$. □

In plain words, the previous result means that at any time $t$ the sum of concentrations of particles at the nodes of the graph is exactly the same, independently of $t$.

Let us now see what is the steady state of the diffusive dynamics in a graph.

**Lemma 10** *Let* $C_i^0$ *be the state of the vertex* $i$ *at time* $t = 0$. *Then, when* $t \to \infty$ *the solution of the diffusion equation of the graph converges to* $C_i\left(t \to \infty\right) = \frac{1}{n}\sum_{i=1}^{n} C_i^0$, *i.e., to the average of the state of nodes in the initial condition. The rate of convergence of the diffusion is dictated by the smallest nontrivial eigenvalue of* $L$. □

***Proof*** Let $0{=}\mu_1 < \mu_2 \leq \cdots \leq \mu_n$ be the eigenvalues of $L$ and let $\varphi_i$ the orthonormalized eigenvector associated with $\mu_i$. Then, we can write

$$\begin{aligned} C\left(t\right) &= e^{-tL}C^0 \\ &= \varphi_1 e^{-t\mu_1}\left(\varphi_1^T C^0\right) + \varphi_2 e^{-t\mu_2}\left(\varphi_2^T C^0\right) + \cdots + \varphi_n e^{-t\mu_n}\left(\varphi_n^T C^0\right). \end{aligned} \tag{21.15}$$

When $t$ is sufficiently large, we have

$$\begin{aligned} \lim_{t\to\infty} C\left(t\right) &= \varphi_1 e^{-t\mu_1}\left(\varphi_1^T C^0\right) \\ &= \frac{1}{\sqrt{n}}\mathbf{1}\left(\frac{1}{\sqrt{n}}\mathbf{1}^T C^0\right) \\ &= \mathbf{1}\left(\frac{1}{n}\sum_{i=1}^{n} C_i^0\right), \end{aligned} \tag{21.16}$$

which proves the first part of the result. Because $\mu_2$ is the second smallest eigenvalue of $L$, it will dictate the rate of convergence of the process. □

## 21.3.3 Intuition of the Conservative Diffusion on Graphs

We start here by writing an iterative form of the diffusion dynamics on graphs in discrete time:

$$C_i\left(r+1\right) = C_i\left(r\right) - \varepsilon C_i\left(r\right)k_i + \varepsilon \sum_{(j,i)\in E} C_j\left(r\right), \tag{21.17}$$

where $\varepsilon > 0$ is the step size. If we introduce the following matrix $P := I - \varepsilon L$, which was proposed to be called the Perron matrix by Olfati-Saber et al. [79], we can write the discrete-time diffusion dynamics as

$$C\left(r+1\right) = PC\left(r\right). \tag{21.18}$$

This formulation allows us to interpret what is happening in the diffusive process on a graph in a step-by-step basis. That is, if the "concentration" of an item at the vertex $i$ at time step $r$ is equal $C_i(r)$, then the node $i$ will transfer the amount $\varepsilon C_i(r)$ to each of its nearest neighbors at time step $r+1$, at the same time that it will receive $\varepsilon C_j(r)$ from each of its $j$th nearest neighbors. This means that at every time step there is a trade-off in which pairs of connected vertices interchange concentrations between each other, but such a process goes in the direction of lower concentration.

Let us explain this with an example based on the graph illustrated in Fig. 21.1. Let us consider the following initial condition: $C^0 = [3, 2, 5, 4, 1]^T$. That is, the vertex $A$, which is connected to vertices $B$ and $C$, has an initial concentration of $C_A^0 = 3$. At the next time step, $r = 1$, $A$ will transfer to $B$ the amount $\varepsilon C_A^0 = 3\varepsilon$ and will receive from $B$ the amount $\varepsilon C_B^0 = 2\varepsilon$. This means that the net balance of $A$ for its trade-off with $B$ is $C_{A \rightleftharpoons B}(1) = -3\varepsilon + 2\varepsilon = -\varepsilon$, which implies that $A$ has transferred more to $B$ than what it has received from that vertex. If we do the same analysis for the trade-off between $A$ and $C$, which has initial concentration 5, we have $C_{A \rightleftharpoons C}(1) = -3\varepsilon + 5\varepsilon = 2\varepsilon$, clearly indicating that $A$ receives more from $C$ than what it transfers to that node. Because $A$ is connected only to $B$ and $C$, we have $C_A(1) = C_A^0 + C_{A \rightleftharpoons B}(1) + C_{A \rightleftharpoons C}(1) = 3 - \varepsilon + 2\varepsilon = 3 + \varepsilon$, indicating that in the first step $A$ increases its concentration in spite of the fact that it has transferred more to $B$ than what it has received from it, but such a loss is compensated by the gains from $C$ which has a high initial concentration. The process continues with these "negotiations" of concentrations between pairs of connected vertices so that the steady state is eventually reached as illustrated in Fig. 21.1. This kind of negotiation is the main reason why this process is known in engineering as the "consensus protocol" (see, for instance, [74]).

## 21.4 Non-conservative Diffusion

Let us consider the following matrix: $\mathscr{L}_\chi := \chi I - A$, which was first analyzed in [44] (see also [45]) and we propose here to call it the Lerman–Ghosh Laplacian of a graph. Let us then consider the following diffusive process:

$$\dot{C}(t) = -(\chi I - A)\, C(t) = -\mathscr{L}_\chi C(t). \tag{21.19}$$

At the local level, the concentration of items at a vertex $i$ at time $t$ is given by

$$\dot{C}_i(t) = \left( \sum_{(i,j) \in E} C_j(t) \right) - \chi C_i(t). \tag{21.20}$$

We now analyze the solution of this dynamics. Let $\lambda_1 > \lambda_2 \geq \cdots \geq \lambda_n$ be the eigenvalues of the adjacency matrix of a simple, connected graph $G$, and let $\psi_j$ be the (orthonormalized) eigenvector associated with $\lambda_j$. Then, we have the following result.

**Theorem 11** *Let $\mathcal{L}_\chi = \chi I - A$ be the Lerman–Ghosh Laplacian and let*

$$\dot{C}(t) = -\mathcal{L}_\chi C(t) \quad (21.21)$$

*with initial condition $C(t) = C_0$. Then,*

$$\lim_{t\to\infty} C(t) = \begin{cases} \left(\psi_1^T C^0\right)\psi_1 e^{t(\lambda_1-\chi)} = \infty \text{ for } \chi < \lambda_1 \\ \left(\sum_j C_j^0 \psi_1(j)\right)\psi_1 \text{ for } \chi = \lambda_1 \\ \left(\psi_1^T C^0\right)\psi_1 e^{-t(\chi-\lambda_1)} = 0 \text{ for } \chi > \lambda_1. \end{cases} \quad (21.22)$$

***Proof*** The solution of the diffusion equation is given by

$$C(t) = e^{-t(\chi I - A)} C^0, \quad (21.23)$$

which can be written as

$$C(t) = e^{t(\lambda_1-\chi)}\left(\psi_1^T C^0\right)\psi_1 + e^{t(\lambda_2-\chi)}\left(\psi_2^T C^0\right)\psi_2 + \cdots + e^{t(\lambda_n-\chi)}\left(\psi_n^T C^0\right)\psi_n. \quad (21.24)$$

Then, when $\chi < \lambda_1$, we have

$$\lim_{t\to\infty} C(t) = e^{t(\lambda_1-\chi)}\left(\psi_1^T C^0\right)\psi_1, \quad (21.25)$$

which diverges as $t \to \infty$.

If $\chi = \lambda_1$ we have that the first term of Eq. (21.24) is zero, and the rest are negative, such that

$$\lim_{t\to\infty} C(t) = \left(\psi_1^T C^0\right)\psi_1, \quad (21.26)$$

which indicates that the solution is proportional to the entries of the eigenvector $\psi_1$ associated with the spectral radius $\lambda_1$ of $A$. This eigenvector was introduced by Bonacich [9–11] as a centrality index of the vertices in a graph and it is nowadays known as the eigenvector centrality. Therefore, the current framework provides a dynamics interpretation of this centrality index in terms of the concentration reached by a vertex at the steady state of a non-conservative diffusion controlled by the Lerman–Ghosh Laplacian matrix when $\chi = \lambda_1$. Because the second smallest eigenvalue of $(\lambda_1 I - A)$ is $\lambda_1 - \lambda_2$, it determines the rate of convergence of the diffusive process. Notice that if $C^0 = \psi_1$, then the diffusion process is conservative because

$$\mathbf{1}^T C(t) = \left[\mathbf{1}^T \psi_1\right] = \mathbf{1}^T C^0 \quad . \quad (21.27)$$

Finally, if $\chi > \lambda_1$ then

$$\lim_{t\to\infty} C(t) = e^{-t(\chi-\lambda_1)}\left(\psi_1^T C^0\right)\psi_1, \tag{21.28}$$

which goes to zero as $t \to \infty$.

**Corollary 12** *Let G be a regular graph, i.e., a graph in which all vertices have the same degree. Then, the diffusion model (21.19) with $\chi = \lambda_1$ is conservative for any initial condition, where $\lim_{t\to\infty} C_i(t) = \frac{1}{n}\sum_j C_j^0$ for all $j \in V$. If the graph is not regular, the diffusion process with $\chi = \lambda_1$ and $C^0 \neq \psi_1$ is non-conservative. This is because $\psi_1(j) < 1$ and then $\lim_{t\to\infty} C(t) = \left(\sum_j C_j^0 \psi_1(j)\right)\psi_1 < \frac{1}{n}\sum_j C_j^0$ such that $0 < C(t \to \infty) < C^0 < \infty$. The exception is when $\psi_1(j) = \frac{1}{\sqrt{n}}$ for all $j \in V$, which is the case of the regular graph.* □

**Remark 13** The diffusion process (21.19) is non-conservative for all values of $\chi$ and initial conditions $C^0$, except for the specific case in which $\chi = \lambda_1$ and $C^0 = \psi_1$. □

In Fig. 21.2, we illustrate the three examples of non-conservative diffusion modeled by Eq. (21.19) using the Lerman–Ghosh Laplacian for the graph and initial condition of Fig. 21.1 with $\chi < \lambda_1$ (a), $\chi = \lambda_1$ (b) and $\chi > \lambda_1$ (c). Notice that when $\chi = \lambda_1$ and $C^0 \neq \psi_1$ the process can be said to be quasi-conservative as the number of particles when $t \to \infty$ is just a bit smaller than at $t = 0$.

### 21.4.1 On the Intuition of the Non-conservative Diffusion

In the discrete-time setting Eq. (21.19) indicates that the resulting concentration at every vertex of the graph at the discrete-time step $r + 1$ is given by

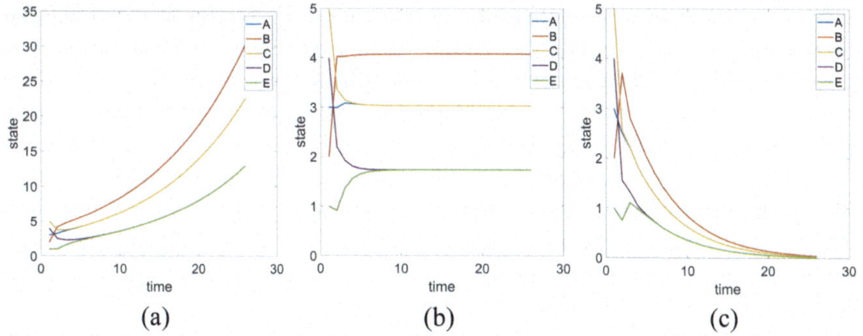


**Fig. 21.2** Illustration of the non-conservative diffusion process on the graph illustrated in Fig. 21.1 for $\chi < \lambda_1$ (**a**), $\chi = \lambda_1$ (**b**), and $\chi > \lambda_1$ (**c**)

$$C_i(r+1) = C_i(r) - \varepsilon\chi C_i(r) + \varepsilon \sum_{(j,i)\in E} C_j(r). \tag{21.29}$$

If we introduce the following matrix $Q := I - \varepsilon(\chi I - A) = (1 - \varepsilon\chi) I + A$, then we can write the discrete-time diffusion dynamics as

$$C(r+1) = QC(r). \tag{21.30}$$

In order to gain intuition on what is happening let us start by considering that a graph is connected to a given reservoir in which there could be an infinite reserve of the item to be diffused across the graph. Every vertex of the graph is connected to the reservoir by means of a semi-edge, which has one endpoint at the vertex, and the other one is freely dangling in the reservoir. We then consider a two-step diffusion process as follows. First every vertex $i$ expels $\varepsilon\chi C_i(r)$ to the reservoir, where $\chi$ is a constant such that $\varepsilon\chi \leq \max_i C_i^0$ to guarantee that no vertex is left with a negative concentration. This first step creates a pool of the item being diffused in the reservoir which adds to the existing reserve of items in the reservoir (in case it exists). In the second step of the process, every vertex sucks from the reservoir an amount equal to $\varepsilon \sum_{(j,i)\in E} C_j(r)$.

For instance, let us consider the graph of Fig. 21.1, which has $\lambda_1 \approx 2.343$, and let us use $\chi = 3$ and $\varepsilon = 0.25$ as before. Then, vertex $A$ expels $\varepsilon\chi C_A^0 = 2.25$ of its initial concentration to the reservoir. The amounts expelled by the other vertices are 1.5, 3.75, 3.0, 0.75. In total, the vertices have expelled 11.25 of its initial total concentrations that sum 15. At every vertex remains a concentration equal to 0.75, 0.50, 1.25, 1.0, 0.25, respectively, as can be seen in Fig. 21.3a. In the second step, vertex $A$ has to suck $\varepsilon \sum_{(j,A)\in E} C_j^0 = 1.75$, while the others will suck 3.25, 1.25, 0.50, 0.50, respectively, for a total of 7.25. This amount to be sucked from the reservoir is smaller than the one expelled previously to it. Therefore, in the reservoir there will remain $11.25 - 7.25 = 4$, and the sum of the concentrations at the nodes will be 11 instead of the 15 that there was initially. This is illustrated in Fig. 21.3b. The repetition of this process makes that the vertices get empty as time evolves because in every time step there are more "items" deposited in the reservoir than the ones extracted from it. Notice that in this case the reservoir at $t = 0$ may be empty. Here, the reservoir is just a virtual space used to gain an intuition on how the Kerman–Ghosh Laplacian operates over a diffusive process on the graph. More physically realistic scenarios can be adapted according to specific situations. It is worth mentioning that this description suggests that a reservoir can be introduced as an additional sink or source node to the graph, which is connected to all other nodes. It has been shown that accounting for asymmetric transition rates (from and to the reservoir) requires the use of transition matrices, see, e.g., [52] and references therein. We show here that the Lerman–Ghost Laplacian represents an interesting alternative way to describe sink/source nodes in graphs.

If $\chi < \lambda_1$ the amount $\varepsilon\chi C_i(r)$ expelled by the vertices to the reservoir is smaller than the amount $\varepsilon \sum_{(j,i)\in E} C_j(r)$ that the vertex has to suck from the reservoir. For instance, if we use $\chi = 1$, the vertices will expel in total 3.75 to the reservoir as

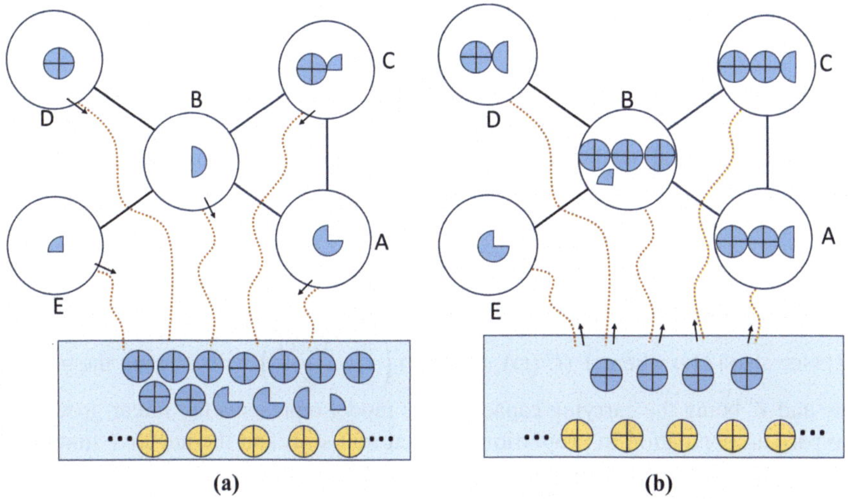


**Fig. 21.3** Schematic representation of the non-conservative diffusive process when $\chi > \lambda_1$

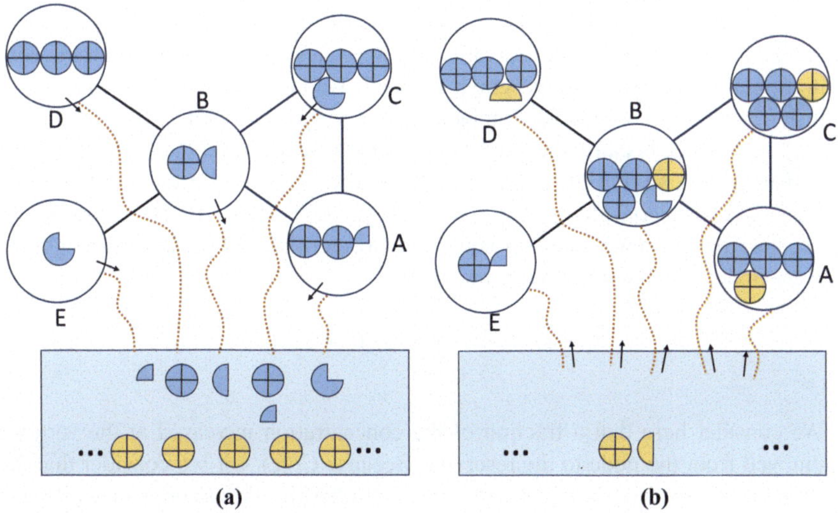


**Fig. 21.4** Schematic representation of the non-conservative diffusive process when $\chi < \lambda_1$

illustrated in Fig. 21.4a. However, the vertices have to extract 7.25 from the reservoir, which means they will take the 3.75 previously expelled plus 3.5 from the reserve in the reservoir. Here the total amount at the vertices at $t = 1$ is 19 (see Fig. 21.4b) instead of the amount equal to 15 which existed at $t = 0$. The consequence of the repetition of this process is that the concentrations at the vertices grow to infinity as time evolves due to the fact that vertices extract more "items" from the reservoir than the ones they previously expel to it.

## 21.5 Logistic Diffusion

It is certainly unrealistic to think about systems in which a non-conservative diffusion blows up when $t \rightarrow \infty$. This situation is avoided in the so-called reaction–diffusion models, where we allow that the vertices create or annihilate items to avoid that their concentrations go to zero or to infinity:

$$\dot{C}(t) = -LC(t) + f(C(t)). \tag{21.31}$$

An example of this model is the Fisher–Kolmogorov–Petrovsky–Piskunov [38, 62] (see also [21]) where $f(C(t)) = \alpha C(t)\left(1 - \frac{C(t)}{\mathscr{C}}\right)$ with $\alpha$ being the growth rate and $\mathscr{C}$ being the carrying capacity. The model represents the linear growth of the particle population in proportion to its current size, and the negative quadratic one has the effect of curtailing the items population growth.

Let us here consider a general model of the form

$$\dot{C}(t) = -(\chi I - A)C(t) + f(C(t)), \tag{21.32}$$

such that although it is non-conservative it reaches a steady state in which the concentration at every vertex in the graph is equal. We consider the non-conservative diffusion with Lerman–Ghosh Laplacian, $\gamma = 1$, $\chi < \lambda_1$ and $C(0) = C^0$ with $0 \leq C_i^0 < 1$. Then, because the graphs considered here are never the trivial graph we have that $\lambda_1 > 0$, so that by setting $\chi = 0$ we always guarantee the condition that $0 = \chi < \lambda_1$. If we consider $f(C(t)) = 0$, the model $\dot{C}(t) = AC(t)$ always blows up because the adjacency operator sums the concentrations of the newest neighbors of every node, which continuously increases

$$\dot{C}_i(t) = \sum_{(j,i)\in E} A_{ij} C_j(t). \tag{21.33}$$

We consider here that a fraction of the concentration increased at the vertex $i$ is removed from the node to the reservoir. Because $C_i(t) < 1$ we consider that the fraction to be removed is equal to $C_i(t)$ multiplied by the amount in which the concentration has increased:

$$\dot{C}_i(t) = \sum_{(j,i)\in E} A_{ij} C_j(t) - C_i(t) \sum_{(j,i)\in E} A_{ij} C_j(t). \tag{21.34}$$

Obviously, we can regroup the terms to obtain

$$\dot{C}_i(t) = (1 - C_i(t)) \sum_{(j,i)\in E} A_{ij} C_j(t), \tag{21.35}$$

which is nothing else than the logistic differential equation [6] (for applications see also [16, 56, 63, 99]).

The logistic model can be rewritten (see [1]) as

$$\frac{1}{1-C_i(t)}\frac{dC_i(t)}{dt}=\sum_{(j,i)\in E}A_{ij}\left(1-e^{-(-\log(1-C_j(t)))}\right), \tag{21.36}$$

which is equivalent to

$$\frac{dy_i(t)}{dt}=\sum_{(j,i)\in E}A_{ij}f\left(y_j(t)\right), \tag{21.37}$$

where $y_i(t):=g(C_i(t))=-\log(1-C_i(t))\in[0,\infty]$, $f(y):=1-e^{-y}=g^{-1}(y)$.

Lee et al. [65] have considered for a given $C^0$ an approximate solution $\hat{C}(t)$ to the logistic equation, which is given by $\hat{C}(t)=f\left(\hat{y}(t)\right)$, where $\hat{y}(t)$ is the solution of the following linearized version of the previous nonlinear equation:

$$\frac{d\hat{y}(t)}{dt}=A\text{diag}(1-C_0)\,\hat{y}(t)+b(C_0), \tag{21.38}$$

where $\text{diag}(1-C_0)$ represents a diagonal matrix whose main diagonal entries are given by the values of $1-C_0$ and the rest of entries are zero. Here, $\hat{y}_0=g\left(C^0\right)$ and $b(C):=C+(1-C)\log(1-C)$, which can be considered as a reaction–diffusion equation on the new variable $\hat{y}(t)$. Let $a\preceq b$ means that $a(i)\leq b(i)$for all entries $i=1,\ldots,n$ of the two vectors $a$ and $b$. We then reproduce the following result.

**Theorem 14** ([65]) *For any t it is true that $C(t)\preceq\hat{C}(t)$ when they have the same initial conditions, and the solution of the linearized logistic model is* □

$$\begin{aligned}\hat{y}(t)&=e^{(t-t_0)A\text{diag}\left(1-C^0\right)}g\left(C^0\right)\\&+\sum_{k=0}^{\infty}\frac{(t-t_0)^{k+1}}{(k+1)!}\left[A\text{diag}\left(1-C^0\right)\right]^kAb\left(C^0\right),\end{aligned} \tag{21.39}$$

which reduces to

$$\hat{y}(t)=g\left(C^0\right)+\left[e^{tA\text{diag}\left(1-C^0\right)}-I\right]\text{diag}\left(1-C^0\right)^{-1}C^0, \tag{21.40}$$

when $C^0\prec\mathbf{1}$, and to

$$\hat{y}(t)=g\left(C^0\right)+\sum_{k=0}^{\infty}\frac{t^{k+1}}{(k+1)!}\left[A\text{diag}\left(1-C^0\right)\right]^kAC^0 \tag{21.41}$$

when $C_i^0\in\{0,1\}$.

**Remark 15** Here we show that both solutions when $C^0 \prec \mathbf{1}$, and when $C_i^0 \in \{0, 1\}$ are the same. For that, let us consider $D = \text{diag}\left(1 - C^0\right)$ in the solution for $C_i^0 \in \{0, 1\}$. Then,

$$\begin{aligned}\hat{y}(t) &= g\left(C^0\right) + t\sum_{k=0}^{\infty}\frac{[tAD]^k}{(k+1)!}AC^0 \\ &= g\left(C^0\right) + tE_{1,2}(A)\,AC^0 \\ &= g\left(C^0\right) + t\left(\frac{e^{tAD} - I}{tAD}\right)AC^0 \\ &= g\left(C^0\right) + \left(\frac{e^{tAD} - I}{D}\right)C^0,\end{aligned} \tag{21.42}$$

which is the solution for $C^0 \prec \mathbf{1}$. We have used the definition of the Mittag-Leffler matrix function:

$$E_{\alpha,\beta}(M) := \sum_{k=0}^{\infty}\frac{M^k}{\Gamma(\alpha k + \beta)}, \tag{21.43}$$

where $M$ is a matrix, which for $\alpha = 1$ and $\beta = 2$ is

$$E_{1,2}(M) := \sum_{k=0}^{\infty}\frac{M^k}{\Gamma(k+2)} = \sum_{k=0}^{\infty}\frac{M^k}{(k+1)!} = \frac{e^M - I}{M}, \tag{21.44}$$

where $I$ is the identity matrix. This specific Mittag-Leffler matrix function [5, 29, 43] is also known as the $\Psi$-matrix function (see Sect. 10.7.4 in [57]). Notice that $E_{1,1}(M) = \exp(M)$.

Hereafter, we consider the realistic scenario in which the initial concentration of the item under study at one given vertex $i$ is very small $C_i^0 \ll 1$, and zero elsewhere. In this scenario, $D \approx I$ and the solution of the approximate logistic diffusion equation is □

$$\hat{y}(t) \approx e^{tA}C^0 - \left(C^0 + \log\left(1 - C^0\right)\right), \tag{21.45}$$

and

$$\hat{x}(t) = 1 - e^{-\hat{y}(t)}. \tag{21.46}$$

For instance, we illustrate in Fig. 21.5 the temporal evolution of this diffusion model on the simple graph of Fig. 21.1 with both the approximate and exact solutions.

Obviously,

$$\lim_{t\to\infty}\hat{y}(t) = e^{t\lambda_1}\left(\psi_1^T C^0\right)\psi_1 - \left(C^0 + \log\left(1 - C^0\right)\right) = \infty, \tag{21.47}$$

such that

$$\lim_{t \to \infty} \hat{x}(t) = \mathbf{1}, \tag{21.48}$$

implying that the process is non-conservative as $\mathbf{1}^T C^0 = C_i^0 \ll 1$, and $\mathbf{1}^T C(\infty) = n$.

## 21.6 Sending "Information" to a Target

In order to compare the two diffusion processes considered here, conservative and non-conservative ones, we select a graph formed by two cliques, which are interconnected by a path. Although these graphs are closely related to the so-called dumbbell graphs, they are not exactly the same, so we propose here to call them weight-lifting (WL) graphs. In Fig. 21.6, we illustrate the WL graph with two cliques of 10 vertices interconnected by a path of the same size.

This kind of graphs is a good example of a difficult networked environment in which a particle in one of the cliques has to diffuse to find a target in the other clique. While it is relatively easy for a diffusive particle to navigate one of the cliques, it is more difficult to find the only path communicating the two cliques to move to the other. The WL graph built here has 10 vertices in each clique and 10 vertices in the path connecting them. Thus, we will call it $WL(10)$. In Fig. 21.7a, we illustrate the temporal evolution of the conservative diffusion of a concentration initially located in a clique vertex $i$ different from the one connecting the clique to the path, such that $C_i^0 = 0.001$ and $C_j^0 = 0$ for all $j \neq i$. For the sake of comparison, we illustrate in Fig. 21.7b the same process in an Erdős–Rényi random graph [22] having the same number of vertices and edges as the $WL(10)$ one. That is, this is a graph in which the difficulties inherent to the structure of the WL graph have been changed by an environment in which the particle can navigate in an easier way, although having the same number of vertices and edges. As can be seen in Fig. 21.7, the conservative diffusion takes about 100 times more time to reach the steady state in the dumbbell graph than in its random analogous.

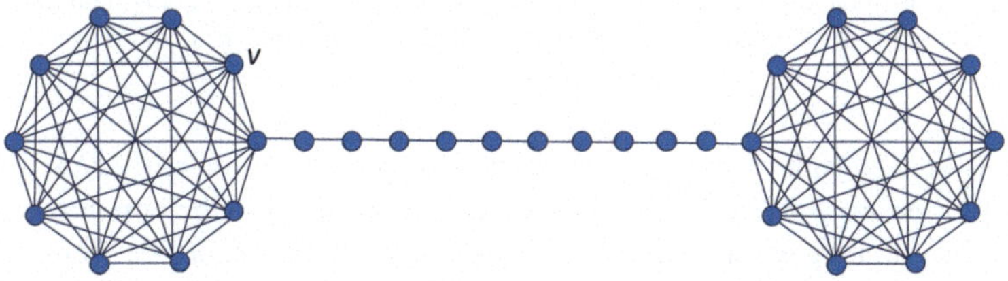


**Fig. 21.5** Illustration of the time evolution of the logistic diffusion on the graph of Fig. 21.1 using the approximation Eq. (21.45) instead of the exact solution of Eq. (21.40). Here $C_A^0 = 0.001$ and $C_i^0 = 0$ for $i = \{B, C, D, E\}$

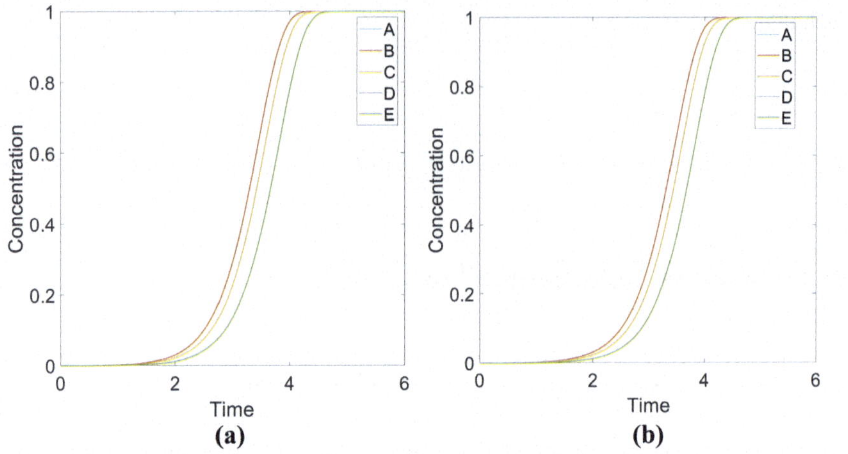


**Fig. 21.6** Illustration of a WL graph formed by two cliques of 10 vertices interconnected by a path of the same size. A vertex which is in the clique but not in the path is denoted by $v$

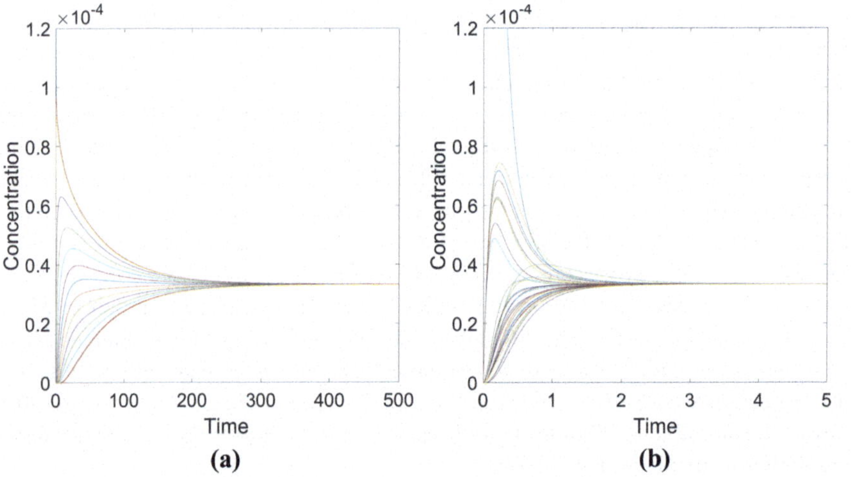


**Fig. 21.7** Time evolution of conservative diffusion dynamics on a WL graph (**a**) with two cliques of 10 vertices interconnected by a linear chain of 10 vertices, and in the random graph (**b**)—of the Erdős–Rényi type—having the same number of vertices and edges like the graph in (**a**)

Let us now consider what happens with the non-conservative diffusion. To consider a process in which a steady state is reached, we will use the non-conservative logistic diffusion equation on graphs using the approximation of Lee–Tenneti–Eun and the same initial condition as before.

It is remarkable to notice that the non-conservative diffusion only takes about two times more time to reach the steady state in the WL graphs than in its random analogous. This is illustrated in Fig. 21.8. If we compare this non-conservative process

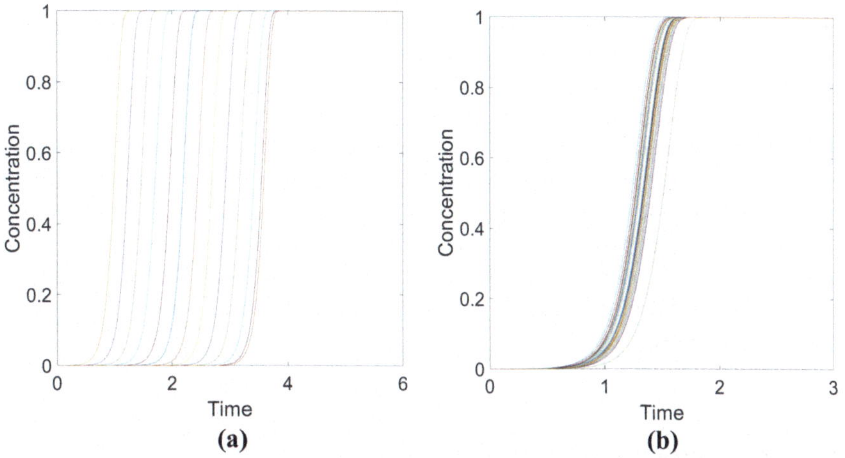


**Fig. 21.8** Time evolution of non-conservative logistic diffusion dynamics on a WL graph (**a**) with two cliques of 10 vertices interconnected by a linear chain of 10 vertices, and in the random graph (**b**)—of the Erdős–Rényi type—having the same number of vertices and edges like the graph in (**a**)

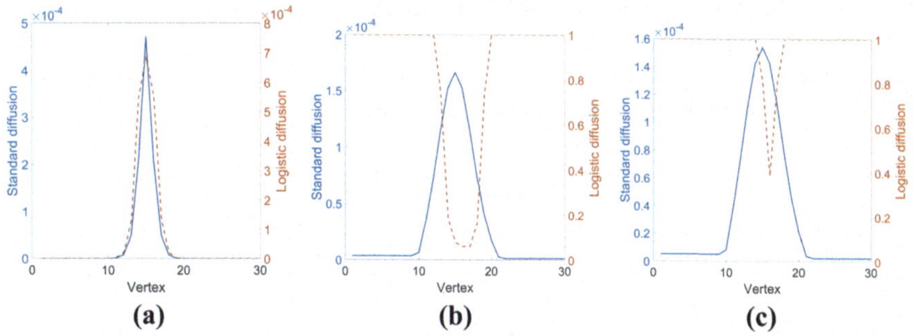


**Fig. 21.9** Illustration of the time evolution of the diffusion through the $WL(10)$ graph with 30 vertices. The initial concentration is $C_i^0 = 0.001$ at a vertex close to the center of the linear chain connecting the two cliques. The dynamics is controlled by the standard diffusion equation (solid blue line) and by the logistic diffusion (broken red line). The plots are for $t = 0.5$ (**a**), $t = 3.0$ (**b**), and $t = 3.5$ (**c**)

with the conservative one in the WL graph we observe that the first is about 100 times faster than the second one.

In Fig. 21.9, we illustrate the differences in the way in which the concentration of items is diffused from a vertex of the WL graph to the rest of the vertices. In this case, we selected a vertex of the linear chain which is located close to the center (the linear chain has 10 vertices, and the vertex selected has four vertices to the right and five to the left). At very short times, as illustrated in Fig. 21.9a, it is already seen that the logistic diffusion spreads faster the concentration of items to the nearest neighbors of the central vertex than the standard diffusion. The main fundamental difference between the two dynamics occurs when the time advances. The standard diffusion evolves by flattening the peak of concentration at the central node while spreading the concentration from the closest to the farthest vertices (see Fig. 21.9b). The logistic diffusion progresses by extracting concentration from the central vertex and filling the two cliques at its expense. That is, it evolves by emptying the vertex in which the whole concentration is initially located and its nearest neighbors to fill the two cliques. Once the cliques are filled, as illustrated in Fig. 21.9c, the central vertex and its nearest neighbors are filled again. This mechanism is possible only because the process is non-conservative and the vertices can "extract" concentration from the reservoir. The result of the process is that it is faster than the standard diffusion as we have seen before.

## 21.7 Trajectories of "Items" to a Target

Both the conservative classical diffusion model and the non-conservative ones studied here have solutions in which the term depending on the structure of the network is $\exp(\beta M)$, where $\beta$ is a parameter usually depending on time and diffusion coefficient and where $M = \{-L, A\}$. Hereafter we will consider only the case $\beta = 1$ for the sake of simplicity. Let us focus on a couple of vertices designated by $v$ and $w$. Then, at a given time the concentration at these nodes are

$$C_v(t) = \sum_j (\exp(M))_{vj} C_j^0 \tag{21.49}$$

and

$$C_w(t) = \sum_j (\exp(M))_{wj} C_j^0. \tag{21.50}$$

Then, let

$$\mathscr{F}_{vw|C_j^0=\delta_{jv}} = C_{v|C_j^0=\delta_{jv}}(t) - C_{w|C_j^0=\delta_{jv}}(t), \tag{21.51}$$

be the diffusive flow from $v$ to $w$ in a graph when the initial concentration is totally located at the vertex $v$, $C_0(j) = \delta_{j,v}$, where $\delta_{i,j}$ is the Kronecker delta. Let us define the same in the other direction when the initial concentration is completely located at the vertex $w$,

$$\mathscr{F}_{wv|C_j^0=\delta_{jw}} = C_{w|C^0=\delta_{jw}}(t) - C_{v|C_j^0=\delta_{jw}}(t). \tag{21.52}$$

### 21.7.1 Inducing a Geometric Embedding

**Definition 16** Let $\mathscr{F}_{vw|C_j^0=\delta_{jv}}$ and $\mathscr{F}_{wv|C_j^0=\delta_{jw}}$ as defined previously. Then, let us define the sum of the gradients between the two nodes in both directions, which is given by □

$$\mathscr{D}_{vw}(M) := \mathscr{F}_{vw|C_j^0=\delta_{jv}} + \mathscr{F}_{wv|C_j^0=\delta_{jw}} = (\exp(M))_{vv} + (\exp(M))_{ww} - 2(\exp(M))_{vw}. \tag{21.53}$$

We then have the following result [23, 51, 81].

**Proposition 17** *$\mathscr{D}_{vw}(M)$ is a squared Euclidean distance between the pairs of vertices $v$ and $w$.* □

Let us consider that $M = U^T \Lambda U$ where $\Xi$ is a diagonal matrix of eigenvalues of $M$ and $U$ is an orthogonal matrix of eigenvectors. Then, if $\varphi_i$ is the $i$th column of $U^T$, we have

$$\begin{aligned}\mathscr{D}_{vw}(M) &= (\varphi_v - \varphi_w)^T e^{\Lambda} (\varphi_v - \varphi_w) \\ &= \left(e^{\Lambda/2}\varphi_v - e^{\Lambda/2}\varphi_w\right)^T \left(e^{\Lambda/2}\varphi_v - e^{\Lambda/2}\varphi_w\right) \\ &= (x_v - x_w)^T (x_v - x_w) \\ &= \|x_v - x_w\|^2.\end{aligned} \tag{21.54}$$

**Definition 18** Let $s = \mathrm{diag}\left(e^M\right)$. Then, we define the diffusion Euclidean Distance Matrix (EDM) as the matrix

$$\mathscr{M} = s\mathbf{1}^T + \mathbf{1}s^T - 2e^M. \tag{21.55}$$

An EDM is called [4, 59, 67, 87] spherical EDM or circum-EDM if the points $p^1, p^2, ..., p^n$ that generate the EDM lie on a hypersphere. Then, in the case of a graph, the points represent the vertices of the graph. If the corresponding EDM $\mathscr{M}$ of the graph is circum-EDM it will immediately imply that the vertices of the graph can be embedded on the surface of a hypersphere such that the length of the chord separating two vertices $v$ and $w$ on the sphere is given by the corresponding Euclidean distance, which is the entry $v, w$ of $\mathscr{M}$. We now prove that $\mathscr{M}$ is indeed circum-Euclidean. □

**Proposition 19** *$\mathscr{M}$ is a circum-Euclidean distance matrix.* □

***Proof*** We follow Gower [51] who proved that the points that generate the matrix $M$ lie on the surface of a hypersphere if and only if $\mathbf{1}^T M^{-1} \mathbf{1} \neq 0$. Thus, we have to prove that $M^{-1}$ exists. We first use the Sherman–Morrison–Woodbury formula (see p. 50 in [47]) to have

$$\begin{aligned}
\mathscr{M}^{-1} &= \left(\left(-2e^{M} + \mathbf{1}s^{T}\right) + s\mathbf{1}^{T}\right)^{-1} \\
&= -\frac{1}{2}e^{-M} - \frac{\left(e^{-M}\mathbf{1}s^{T}e^{-M}\right)}{4-2s^{T}e^{-M}\mathbf{1}} \\
&\quad - \frac{\left(-\frac{1}{2}e^{-A} - \frac{\left(e^{-M}\mathbf{1}s^{T}e^{-M}\right)}{4-2s^{T}e^{-M}\mathbf{1}}\right) s\mathbf{1}^{T} \left(-\frac{1}{2}e^{-M} - \frac{\left(e^{-M}\mathbf{1}s^{T}e^{-M}\right)}{4-2s^{T}e^{-M}\mathbf{1}}\right)}{1+\mathbf{1}^{T}\left(-\frac{1}{2}e^{-M} - \frac{\left(e^{-M}\mathbf{1}s^{T}e^{-M}\right)}{4-2s^{T}e^{-M}\mathbf{1}}\right)s} \\
&= -\frac{1}{2}e^{-M} - \frac{ab}{4-2\varepsilon} - \frac{\left(-\frac{1}{2}b^{T} - \frac{ac}{4-2\varepsilon}\right)\left(-\frac{1}{2}a^{T} - \frac{db}{4-2\varepsilon}\right)}{1+\left(-\frac{1}{2}\varepsilon - \frac{dc}{4-2\varepsilon}\right)},
\end{aligned} \tag{21.56}$$

from which we get

$$\mathscr{M}^{-1} = -\frac{1}{2}e^{-A} + \frac{caa^{T} - (\varepsilon-2)\,b^{T}a^{T} - (\varepsilon-2)\,ab + db^{T}b}{2\left(cd - (\varepsilon-2)^{2}\right)}, \tag{21.57}$$

where $a = e^{-A}\mathbf{1}$, $b = s^{T}e^{-M}$, $c = s^{T}e^{-M}s$, $d = \mathbf{1}^{T}e^{-M}\mathbf{1}$, and $\varepsilon = s^{T}e^{-M}\mathbf{1}$. We still need to prove that the denominator is not zero: $cd - (\varepsilon-2)^{2} \neq 0$. It is easy to see that $c \geq \varepsilon$, with equality if and only if $s = \mathbf{1}$ which only happen if the graph is trivial, which is excluded here as the graphs are connected. Therefore,

$$-\varepsilon^{2} + cd + 4\varepsilon - 4 > -\varepsilon^{2} + (d+4)\,\varepsilon - 4 > 0, \tag{21.58}$$

where the last inequality is obvious from the roots of the quadratic equation.

We now obtain $\mathbf{1}^{T}\mathscr{M}^{-1}\mathbf{1}$, as

$$\mathbf{1}^{T}\mathscr{M}^{-1}\mathbf{1} = -\frac{1}{2}d - \frac{d\varepsilon}{4-2\varepsilon} - \frac{\left(-\frac{1}{2}\varepsilon - \frac{dc}{4-2\varepsilon}\right)\left(-\frac{1}{2}d - \frac{d\varepsilon}{4-2\varepsilon}\right)}{1+\left(-\frac{1}{2}\varepsilon - \frac{dc}{4-2\varepsilon}\right)} \tag{21.59}$$

$$= \frac{2d}{cd - (\varepsilon-2)^{2}}. \tag{21.60}$$

Because $d = \mathbf{1}^T e^{-M}\mathbf{1}$ is the sum of all the entries of $e^{-M}$ and $M \neq 0$ (because the graph is not trivial), we have that $d > 0$ and because we have proved before that $cd - (\varepsilon - 2)^2 > 0$, we have proved that $\mathscr{M}$ is a nonsingular spherical EDM. □

**Remark 20** For properties of the circum-EDM $\mathscr{M}$ when $M = A$, the reader is referred to [30, 36, 37]. For analysis of the circum-EDM $\mathscr{M}$ when $M = L$, see [17, 82]. □

### *21.7.2 Diffusive Trajectories*

In order to analyze the trajectories of the diffusive particles in a graph, we need to convert the distances generated in the previous section into geodesics which can be traversed by these particles. That is, we need a geometrization of the graph. A graph is geometrized if we consider every edge $e = vw$ in $E$ as a compact one-dimensional manifold with boundary $\partial e = v \cup w$. We then assign to $e = vw$ the metric $\mathbb{L}_{vw}(M)$, such that $\tilde{e}_{vw}(M) \cong [0, \mathscr{D}_{vw}(M)]$ if $e \in E$ or zero otherwise. In this way, we transform the network into a space which is locally compact, complete, and geodetic [15, 71].

Technically, what we need is to create the corresponding distance matrix $\mathscr{D}(M)$ and then multiply it in an entrywise way by $A$, creating a weighted adjacency matrix $W$ in which every weight corresponds to the length of the corresponding edge. We now can obtain the weighted shortest paths (SP) connecting every pair of vertices. While the unweighted shortest paths correspond to the topological ones, those based on $\mathscr{D}(-L)$ correspond to the conservative diffusive SP, and the ones based on $\mathscr{D}(A)$ are the non-conservative diffusive SP, also known as communicability SP.

In Fig. 21.10, we illustrate the shortest topological (STP) and diffusive (conservative (SCP) and non-conservative (SNCP)) paths between a pair of vertices in a random rectangular graph. As expected, the STP crosses the vertices with the largest degree in the graph. However, the SNCP goes by a longer path to connect the same pair of origin–destination vertices, but in a way that avoids those highly connected vertices, not only in terms of their degrees but in terms of their general cliquishness. Somehow unexpected is the fact that the SCP is very similar to the STP, as it also crosses the most connected vertices in the graph.

To understand the differences between the conservative and non-conservative diffusion trajectories, let us first consider the non-conservative case. The non-conservative diffusion trajectory between two vertices corresponds to the path with the minimum sum of the corresponding distances $\mathscr{D}_{vw}(A)$ along the path. The distance between individual pairs of vertices can be written as

$$\mathscr{D}_{vw}(A) = \left(1 + \frac{(A^2)_{vv}}{2!} + \frac{(A^3)_{vv}}{3!} + h.o\right) + \left(1 + \frac{(A^2)_{ww}}{2!} + \frac{(A^3)_{ww}}{3!} + h.o\right) - 2\left(A_{vw} + \frac{(A^2)_{vw}}{2!} + \frac{(A^3)_{vw}}{3!} + h.o\right),$$

which can be written as

$$\mathscr{D}_{vw}(A) = \left(1 + \frac{k_v}{2} + \frac{t_v}{3} + h.o\right) + \left(1 + \frac{k_w}{2} + \frac{t_w}{3} + h.o\right) - 2\left(P_{vw}^1 + \frac{P_{vw}^2}{2!} + \frac{P_{vw}^3}{3!} + h.o\right),$$

where $k_i$ and $t_i$ are the degree and the number of triangles of $i$, and $P_{ij}^l$ is the number of paths of length $l$ between $i$ and $j$. Therefore, $\mathbb{L}_{vw}(A)$ avoids the vertices with the largest degree, largest number of triangles, etc., i.e., it avoids the vertices of largest cliquishness. Let us now see what happens to $\mathscr{D}_{vw}(-L)$ along the path. We can write

$$\mathscr{D}_{vw}(-L) = \left(1 - \frac{k_v}{2} - \frac{k_v^3}{6} - \frac{\sum_{(j,v)\in E} k_j}{6} + \frac{t_v}{6} + h.o\right) + \left(1 - \frac{k_w}{2} - \frac{k_w^3}{6} - \frac{\sum_{(j,w)\in E} k_j}{6} + \frac{t_w}{6} + h.o\right) - 2\left(-L_{vw} + \frac{(L^2)_{vw}}{2!} - \frac{(L^3)_{vw}}{3!} + h.o\right),$$

indicating that $\mathbb{L}_{vw}(-L)$ does not necessarily avoid the vertices with high cliquishness as it has been the case of the example in Fig. 21.10.

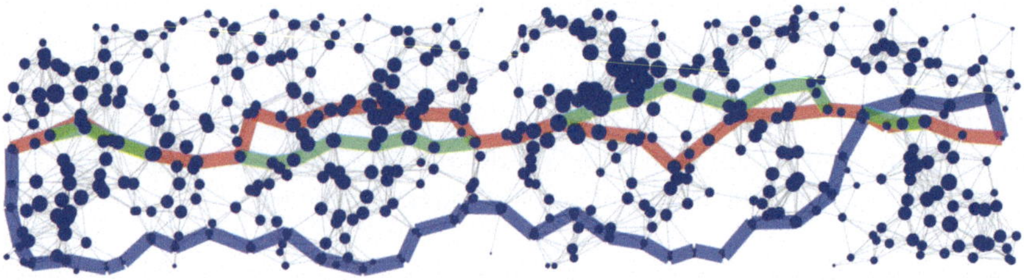

**Fig. 21.10** Illustration of the diffusive paths between two vertices in a random rectangular graph. Blue: shortest non-conservative path (SNCP); Red: shortest conservative path (SCP); Green: shortest topological path (STP)

### 21.7.3 Where to Find Non-conservative Diffusion in Graphs

We have seen in this chapter that the non-conservative diffusion has some advantages relative to conservative diffusion for finding a target on a networked environment. Namely, the steady state in the non-conservative process is reached much faster than in the conservative one by tracing trajectories that avoids the potentially most connected vertices. Then, it is not surprising to find real-world systems in which non-conservative diffusion is present, some of which are described below.

The most intuitive of these examples is the one of traffic at rush hour that we have described in the Introduction. But, where does the "non-conservative" part come from? The number of cars departing from one intersection is not necessarily the same that arrives at the next one. The reason is that a few cars will end up their trajectory in the street leg between two intersections due to the existence of parking spaces in them. In a similar way, some cars can emerge from these spaces, such that the number of cars can increase/decrease from one vertex to another.

A second illustrative example is the diffusive neurotransmission by means of chemical synapses in neuronal systems [78, 91]. Nowadays it is well established that chemical synapses do not only occur by the wiring intercellular communication in which two neurons interchange neurotransmitters [2]. Apart from this conservative process, neurons and brain regions also communicate by means of a volume transmission (VT), which uses the extracellular fluid filling channels of the extracellular space (ECS) and the cerebrospinal fluid filling ventricular space and sub-arachnoidal space [2, 13, 41, 85, 86, 94, 95]. That is, some amounts of the neurotransmitters are spilled over to the ECS in the perisynaptic region and from there it can be retaken by other neurons, or they can even be transported by blood and/or cerebrospinal fluid over long distances where they can also be retaken. From the point of view of the network of neuronal or brain regions, the process is clearly non-conservative, although at the organ level it is of course a conservative process. Although not treated here, the process in which neurotransmitters, or other substances, are captured at longer distances from its origin by navigating across the reservoir allowing for long-range displacements resembles the mechanism of Levy flights in the setting of continuous diffusion, which is treated in this book. The current approach will be extended in the near future to consider these spatial nonlocalities as well as temporal ones.

A third example of networked non-conservative diffusion is the communication in social media [69, 80, 97] like Twitter or Facebook where a user can post a message which can be read by her followers, but also (if not constrained by the user) by non-followers, all of whom can propagate such information to others [88]. By constructing the network of followers, the propagation of information is non-conservative because part of the information leaves the network and some other information is subsequently received by the users from outside the network.

Other areas in which non-conservative diffusion on networks is present include, for instance, diffusive processes in ecology [12, 14, 83], diffusion of microplastics and other materials in the ocean [58] where the locations can be recorded by means of interconnected nodes, among others.

## 21.8 Conclusions

When referring to transmission principles in the brain, Tognoli and Scott-Kelso [89] stated that they "do not scale well upward from simple "channels" of synaptic interactions to the larger and more complex web of evolved brains. Thus, it is without surprise that the brain betrays an essential communicational etiquette: its parts do not behave in a sequential—one-talks-at-a-time—manner". This statement can be extended to most complex systems where the sender of information does not necessarily know the global topology of the network in which it is embedded.

The lack of a sequential one-talks-at-a-time communication is a characteristic feature of diffusive processes in which the sender of information spreads it across all potential routes to find its target(s). We have seen here that even a conservative diffusive way in which one-talks-to-everybody is not the most efficient way of reaching potential targets in networked environments. The discovery presented here by first time that the non-conservative diffusion is more efficient in communicating a source and a target in a network is somehow surprising. On the light of this discovery, it is then clear why some new ways of social communication, such as online social networks, are so efficient in spreading information. They certainly use non-conservative diffusive ways of finding their targets. However, we should not forget that this process may be more costly energetically than a conservative one, particularly when diffusive particles need to be constantly created at the nodes of the network.

In the case of conservative diffusion, we have previously extended the mathematical framework to consider quantum systems via the Schrödinger equation without potential on graphs [26]. In that case, we considered a quantum walker on a graph which can hop not only to nearest neighbors but also to distant ones. The current framework of non-conservative diffusion with Lerman–Ghosh Laplacian can be mathematically extended to consider quantum non-conservative diffusion. The analysis of this theoretical scenario opens new avenues for the study of quantum processes on graphs. Other plausible frameworks for extending the current approaches are the consideration of biased Laplacians [27, 42, 75] or its combination with centrality-based stochastic resetting of random walks [96].

The current work also opens new avenues that should be explored to better understand the ways in which source-to-target communication occurs in complex systems. For instance, anomalous diffusion on graphs described by means of the $d$-path Laplacians [24, 28, 34, 35] and its combination with fractional temporal derivatives [19], are all conservative processes. The extension of $d$-path Laplacians to $d$-path Lerman–Ghosh Laplacians seems to be a necessary extension of the current work. This will allow us to investigate, for instance, if the non-conservative anomalous diffusion is more efficient than its conservative analogous in finding a target on a graph. Similarly, the Lerman–Ghosh Laplacians can be used in degree-biased advection–diffusion models on undirected graphs/networks [75] to simulate realistic scenarios in complex systems. Another avenue is to investigate non-conservative diffusion in metaplexes [32], where not only the particles diffuse across a discrete space, but also in a continuous one existing inside the nodes. All in all, we hope this chapter helps the readers in their particular "Target Problems" for understanding complex systems.

**Acknowledgements** The author thanks Project OLGRA (PID2019-107603GB-I00) funded by Spanish Ministry of Science and Innovation and the Maria de Maeztu project CEX2021-001164-M funded by the MCIN/AEI/10.13039/501100011033.

## References

1. L. Abadias, G. Estrada-Rodriguez, E. Estrada, Fractional-order susceptible-infected model: definition and applications to the study of covid-19 main protease. Fract. Calc. Appl. Anal. **23**(3), 635–655 (2020)
2. L.F. Agnati, D. Guidolin, M. Guescini, S. Genedani, K. Fuxe, Understanding wiring and volume transmission. Brain Res. Rev. **64**(1), 137–159 (2010)
3. M. Akbarzadeh, E. Estrada, Communicability geometry captures traffic flows in cities. Nat. Hum. Behav. **2**(9), 645–652 (2018)
4. A. Alfakih, A remark on the faces of the cone of Euclidean distance matrices. Linear Algebra Appl. **414**(1), 266–270 (2006)
5. F. Arrigo, F. Durastante, Mittag-Leffler functions and their applications in network science. SIAM J. Matrix Anal. Appl. **42**(4), 1581–1601 (2021)
6. N. Bacaër, N. Bacaër, Verhulst and the logistic equation (1838). *A Short History of Mathematical Population Dynamics* (2011), pp. 35–39
7. R. Bellman, On a routing problem. Q. Appl. Math. **16**(1), 87–90 (1958)
8. C. Berge, *The Theory of Graphs* (Courier Corporation, 2001)
9. P. Bonacich, Factoring and weighing approaches to clique identification. J. Math. Sociol. **92**, 1170–1182 (1971)
10. P. Bonacich, Power and centrality: a family of measures. Am. J. Sociol. **92**(5), 1170–1182 (1987)
11. P. Bonacich, Some unique properties of eigenvector centrality. Soc. Netw. **29**(4), 555–564 (2007)
12. S.R. Borrett, B.C. Patten, Structure of pathways in ecological networks: relationships between length and number. Ecol. Model. **170**(2–3), 173–184 (2003)
13. D.O. Borroto-Escuela, M. Perez De La Mora, P. Manger, M. Narváez, S. Beggiato, M. Crespo-Ramírez, G. Navarro, K. Wydra, Z. Díaz-Cabiale, A. Rivera et al., Brain dopamine transmission in health and Parkinson's disease: modulation of synaptic transmission and plasticity through volume transmission and dopamine heteroreceptors. Front. Synaptic Neurosci. **10**, 20 (2018)
14. Å. Brännström, D.J. Sumpter, Coupled map lattice approximations for spatially explicit individual-based models of ecology. Bull. Math. Biol. **67**(4), 663–682 (2005)
15. M.R. Bridson, A. Haefliger, *Metric Spaces of Non-positive Curvature*, vol. 319 (Springer Science & Business Media, 2013)
16. P. Brogueira, J.D. de Deus, Evolution equation for soft physics at high energy. J. Phys. G: Nucl. Part. Phys. **37**(7), 075006 (2010)
17. R.R. Coifman, S. Lafon, Diffusion maps. Appl. Comput. Harmon. Anal. **21**(1), 5–30 (2006)
18. J. Crank, *The Mathematics of Diffusion* (Oxford University Press, 1979)
19. F. Diaz-Diaz, E. Estrada, Time and space generalized diffusion equation on graph/networks. Chaos, Solitons & Fractals **156**, 111791 (2022)
20. E.W. Dijkstra, A note on two problems in connexion with graphs, in *His Life, Work, and Legacy*, ed. by E.W. Dijkstra (2022), pp.287–290
21. M. El-Hachem, S.W. McCue, W. Jin, Y. Du, M.J. Simpson, Revisiting the fisher-kolmogorov-petrovsky-piskunov equation to interpret the spreading-extinction dichotomy. Proc. R. Soc. A **475**(2229), 20190378 (2019)
22. P. Erdős, A. Rényi, On random graphs i. Publ. Math. Debrecen **6**(290–297), 18 (1959)

23. E. Estrada, The communicability distance in graphs. Linear Algebra Appl. **436**(11), 4317–4328 (2012)
24. E. Estrada, Path Laplacian matrices: introduction and application to the analysis of consensus in networks. Linear Algebra Appl. **436**(9), 3373–3391 (2012)
25. E. Estrada, *The Structure of Complex Networks: Theory and Applications* (Oxford University Press, USA, 2012)
26. E. Estrada, d-path Laplacians and quantum transport on graphs. Mathematics **8**(4), 527 (2020)
27. E. Estrada, 'hubs-repelling' Laplacian and related diffusion on graphs/networks. Linear Algebra Appl. **596**, 256–280 (2020)
28. E. Estrada, Path Laplacians versus fractional Laplacians as nonlocal operators on networks. New J. Phys. **23**(7), 073049 (2021)
29. E. Estrada, The many facets of the Estrada indices of graphs and networks. SeMA J. **79**(1), 57–125 (2022)
30. E. Estrada, Every nonsingular spherical Euclidean distance matrix is a resistance distance matrix. Linear Algebra Appl. **656**, 198–209 (2023)
31. E. Estrada, What is a complex system, after all? Founda. Sci. 1–28 (2023)
32. E. Estrada, G. Estrada-Rodriguez, H. Gimperlein, Metaplex networks: influence of the exo-endo structure of complex systems on diffusion. SIAM Rev. **62**(3), 617–645 (2020)
33. E. Estrada, J. Gómez-Gardeñes, L. Lacasa, Network bypasses sustain complexity. Proc. Natl. Acad. Sci. **120**(31), e2305001120 (2023)
34. E. Estrada, E. Hameed, N. Hatano, M. Langer, Path Laplacian operators and superdiffusive processes on graphs. i. one-dimensional case. Linear Algebra Appl. **523**, 307–334 (2017)
35. E. Estrada, E. Hameed, M. Langer, A. Puchalska, Path Laplacian operators and superdiffusive processes on graphs. ii. two-dimensional lattice. Linear Algebra Appl. **555**, 373–397 (2018)
36. E. Estrada, N. Hatano, Communicability angle and the spatial efficiency of networks. SIAM Rev. **58**(4), 692–715 (2016)
37. E. Estrada, M. Sanchez-Lirola, J.A. De La Peña, Hyperspherical embedding of graphs and networks in communicability spaces. Discret. Appl. Math. **176**, 53–77 (2014)
38. R.A. Fisher, The wave of advance of advantageous genes. Ann. Eugen. **7**(4), 355–369 (1937)
39. R.W. Floyd, Algorithm 97: shortest path. Commun. ACM **5**(6), 345 (1962)
40. L.R. Ford, *Network Flow Theory* (Rand Corporation Santa Monica, CA, 1956)
41. K. Fuxe, D.O. Borroto-Escuela, A. Tarakanov, W.R. Fernandez, P. Manger, A. Rivera, K. van Craenenbroeck, K. Skieterska, Z. Diaz-Cabiale, M. Filip et al., Understanding the balance and integration of volume and synaptic transmission. Relevance for psychiatry. Neurol. Psych. Brain Res. **19**(4), 141–158 (2013)
42. L.V. Gambuzza, M. Frasca, E. Estrada, Hubs-attracting Laplacian and related synchronization on networks. SIAM J. Appl. Dyn. Syst. **19**(2), 1057–1079 (2020)
43. R. Garrappa, M. Popolizio, Computing the matrix Mittag-Leffler function with applications to fractional calculus. J. Sci. Comput. **77**(1), 129–153 (2018)
44. R. Ghosh, K. Lerman, Parameterized centrality metric for network analysis. Phys. Rev. E **83**(6), 066118 (2011)
45. R. Ghosh, K. Lerman, T. Surachawala, K. Voevodski, S. Teng, Non-conservative diffusion and its application to social network analysis. J. Complex Netw. **12**(1), cnae006 (2024). https://doi.org/10.1093/comnet/cnae006
46. D.T. Gillespie, E. Seitaridou, *Simple Brownian Diffusion: An Introduction to the Standard Theoretical Models* (Oxford University Press, USA, 2013)
47. G.H. Gloub, C.F. Van Loan, *Matrix Computations*, 3rd edn. (Johns Hopkins Universtiy Press, 1996)
48. B. Golden, Shortest-path algorithms: a comparison. Oper. Res. **24**(6), 1164–1168 (1976)
49. R.G. Golledge, T. Gärling, Cognitive maps and urban travel, in *Handbook of Transport Geography and Spatial Systems* (Emerald Group Publishing Limited, 2004), pp. 501–512
50. J. Goñi, A. Avena-Koenigsberger, N. Velez de Mendizabal, M.P. van den Heuvel, R.F. Betzel, O. Sporns, Exploring the morphospace of communication efficiency in complex networks. PLoS ONE **8**(3), e58070 (2013)

51. J.C. Gower, Properties of Euclidean and non-Euclidean distance matrices. Linear Algebra Appl. **67**, 81–97 (1985)
52. D.S. Grebenkov, L. Tupikina, Heterogeneous continuous-time random walks. Phys. Rev. E **97**(1), 012148 (2018)
53. A. Grigor'yan, *Introduction to Analysis on Graphs*, vol. 71. (American Mathematical Society, 2018)
54. R. Grone, R. Merris, The Laplacian spectrum of a graph ii. SIAM J. Discret. Math. **7**(2), 221–229 (1994)
55. R. Grone, R. Merris, V. Sunder, The Laplacian spectrum of a graph. SIAM J. Matrix Anal. Appl. **11**(2), 218–238 (1990)
56. A. Hernando, A. Plastino, Scale-invariance underlying the logistic equation and its social applications. Phys. Lett. A **377**(3–4), 176–180 (2013)
57. N.J. Higham, *Functions of Matrices: Theory and Computation* (SIAM, 2008)
58. A. Isobe, S. Iwasaki, K. Uchida, T. Tokai, Abundance of non-conservative microplastics in the upper ocean from 1957 to 2066. Nat. Commun. **10**(1), 417 (2019)
59. G. Jaklič, J. Modic, On Euclidean distance matrices of graphs. Electr. J. Linear Algebra **26**, 574–589 (2013)
60. M.S. Keil, Local to global normalization dynamic by nonlinear local interactions. Physica D **237**(6), 732–744 (2008)
61. S.-S. Kim, M. Chung, Y.-K. Kim, Urban traffic prediction using congestion diffusion model, in *2020 IEEE International Conference on Consumer Electronics-Asia (ICCE-Asia)*, pp. 1–4 (IEEE, 2020)
62. A.N. Kolmogorov, I.G. Petrovsky, N.S. Piskunov, A study of the equation of diffusion with increase in the quantity of matter, and its application to a biological problem. Moscow Univ. Bull. Math. **1**, 1–25 (1937)
63. B. Kooi, M. Boer, S. Kooijman, On the use of the logistic equation in models of food chains. Bull. Math. Biol. **60**(2), 231–246 (1998)
64. J. Kwapień, S. Drożdż, Physical approach to complex systems. Phys. Rep. **515**(3–4), 115–226 (2012)
65. C.-H. Lee, S. Tenneti, D.Y. Eun, Transient dynamics of epidemic spreading and its mitigation on large networks, in *Proceedings of the Twentieth ACM International Symposium on Mobile Ad Hoc Networking and Computing* (2019), pp. 191–200
66. E. Lenzi, R. Mendes, C. Tsallis, Crossover in diffusion equation: anomalous and normal behaviors. Phys. Rev. E **67**(3), 031104 (2003)
67. C.-K. Li, T. Milligan, M. Trosset, Euclidean and circum-Euclidean distance matrices: characterizations and linear preservers. Electr. J. Linear Algebra **20**, 739–752 (2010)
68. J. Liu, M. Li, Y. Pan, W. Lan, R. Zheng, F.-X. Wu, J. Wang et al., Complex brain network analysis and its applications to brain disorders: a survey. Complexity (2017)
69. L. Liu, J. Tang, J. Han, S. Yang, Learning influence from heterogeneous social networks. Data Min. Knowl. Disc. **25**, 511–544 (2012)
70. M. Măntoiu, S. Richard, R.T. de Aldecoa, Spectral analysis for adjacency operators on graphs, in *Annales Henri Poincaré*, vol. 8. (Springer, 2007), pp. 1401–1423
71. S. Markvorsen, Minimal webs in Riemannian manifolds. Geom. Dedicata. **133**, 7–34 (2008)
72. N. Masuda, M.A. Porter, R. Lambiotte, Random walks and diffusion on networks. Phys. Rep. **716**, 1–58 (2017)
73. R. Merris, A survey of graph laplacians. Linear Multilinear Algebra **39**(1–2), 19–31 (1995)
74. M. Mesbahi, M. Egerstedt, *Graph Theoretic Methods in Multiagent Networks* (Princeton University Press, 2010)
75. M. Miranda, E. Estrada, Degree-biased advection-diffusion on undirected graphs/networks. Math. Modell. Nat. Phen. **17**, 30 (2022)
76. B. Mohar, Y. Alavi, G. Chartrand, O. Oellermann, The Laplacian spectrum of graphs. Graph Theory Combin. Appl. **2**(871–898), 12 (1991)
77. E.F. Moore, The shortest path through a maze, in *Proceedings of the International Symposium on the Theory of Switching* (Harvard University Press, 1959), pp. 285–292

78. C. Nicholson, Diffusion and related transport mechanisms in brain tissue. Rep. Prog. Phys. **64**(7), 815 (2001)
79. R. Olfati-Saber, J.A. Fax, R.M. Murray, Consensus and cooperation in networked multi-agent systems. Proc. IEEE **95**(1), 215–233 (2007)
80. L.A. Overbey, B. Greco, C. Paribello, T. Jackson, Structure and prominence in twitter networks centered on contentious politics. Soc. Netw. Anal. Min. **3**, 1351–1378 (2013)
81. I.J. Schoenberg, Metric spaces and positive definite functions. Trans. Am. Math. Soc. **44**(3), 522–536 (1938)
82. C. Scott, E. Mjolsness, Graph diffusion distance: properties and efficient computation. PLoS ONE **16**(4), e0249624 (2021)
83. D. Scott, The determination and use of thermodynamic data in ecology. Ecology **46**(5), 673–680 (1965)
84. I.M. Sokolov, Models of anomalous diffusion in crowded environments. Soft Matter **8**(35), 9043–9052 (2012)
85. E. Sykova, Extrasynaptic volume transmission and diffusion parameters of the extracellular space. Neuroscience **129**(4), 861–876 (2004)
86. K.H. Taber, R.A. Hurley, Volume transmission in the brain: beyond the synapse. J. Neuropsychiatry Clin. Neurosci. **26**(1), iv–4 (2014)
87. P. Tarazaga, T. Hayden, J. Wells, Circum-Euclidean distance matrices and faces. Linear Algebra Appl. **232**, 77–96 (1996)
88. I. Taxidou, P.M. Fischer, Online analysis of information diffusion in Twitter. In *Proceedings of the 23rd International Conference on World Wide Web* (2014), pp. 1313–1318
89. E. Tognoli, J.S. Kelso, Enlarging the scope: grasping brain complexity. Front. Syst. Neurosci. **8**, 122 (2014)
90. D. Tomasi, G.-J. Wang, N.D. Volkow, Energetic cost of brain functional connectivity. Proc. Natl. Acad. Sci. **110**(33), 13642–13647 (2013)
91. J. Tønnesen, S. Hrabětová, F.N. Soria, Local diffusion in the extracellular space of the brain. Neurobiol. Dis. **177**, 105981 (2023)
92. X.Z. Wang, The comparison of three algorithms in shortest path issue, in *Journal of Physics: Conference Series*, vol. 1087. (IOP Publishing, 2018), p. 022011
93. S. Warshall, A theorem on boolean matrices. J. ACM **9**(1), 11–12 (1962)
94. K. Wiencke, A. Horstmann, D. Mathar, A. Villringer, J. Neumann, Dopamine release, diffusion and uptake: a computational model for synaptic and volume transmission. PLoS Comput. Biol. **16**(11), e1008410 (2020)
95. H. Xiong, E. Lacin, H. Ouyang, A. Naik, X. Xu, C. Xie, J. Youn, B.A. Wilson, K. Kumar, T. Kern et al., Probing neuropeptide volume transmission in vivo by simultaneous near-infrared light-triggered release and optical sensing. Angew. Chem. **134**(34), e202206122 (2022)
96. K. Zelenkovski, T. Sandev, R. Metzler, L. Kocarev, L. Basnarkov, Random walks on networks with centrality-based stochastic resetting. Entropy **25**(2), 293 (2023)
97. A. Zeng, C.H. Yeung, Predicting the future trend of popularity by network diffusion. Chaos: Interdiscip. J. Nonlinear Sci. **26**(6) (2016)
98. C. Zhang, H. Bai, S. Wang, C. Xie, Review of urban traffic congestion formation and diffusion mechanism, in *CICTP* (2018), pp. 2376–2385
99. H. Zhang, X. Zhao, X. Yu, L. Liu, S. Ma, Complex network characteristics and evolution research of software architecture, in *2016 IEEE Advanced Information Management, Communicates, Electronic and Automation Control Conference (IMCEC)*, pp. 1785–1788 (IEEE, 2016)
100. L. Zhao, J. Zhao, Comparison study of three shortest path algorithm, in *2017 International Conference on Computer Technology, Electronics and Communication (ICCTEC)*. (IEEE, 2017), pp. 748–751

# Chapter 22
# Searching Through Cellular Landscapes

**Elena F. Koslover**

**Abstract** Target search processes by molecular components underlie a broad variety of cellular functions, from signal processing to sorting and export, to transcriptional regulation. These processes take place on an immensely complex landscape that encompasses both the geometry of intracellular structures and the localized interactions that modulate molecular motion and encounter times. The subcellular world includes subdomains with different dimensionalities, such as 1D filaments, and 2D sheets, as well as more complex morphologies such as networks and trees. Here, we summarize cellular applications of mathematical models for geometry-dependent kinetics. In addition, we discuss how the energetic landscape emerging from transient interactions within cellular condensates, and with extended filamentous and membrane structures, can serve to accelerate some searchers while filtering out others. The landscape picture provides a unifying framework linking molecular and cellular search processes, together with the mathematics of random walks in complex spatial architectures.

## 22.1 Introduction

The intracellular world provides a complicated landscape through which particles ranging from ions and molecules to micron-scale organelles must navigate and search. The ability of cellular components to find specific target sites within the cell underlies a vast array of dynamical processes and reactions, contributing to nearly all cellular functions. For instance, metabolism relies on substrates finding their enzymes, with the kinetics of several metabolic enzymes known to push the diffusive search limit [1]. Transcriptional regulation requires transcription factors to find their way into the nucleus [2–5] and onto specific DNA binding sites [6–8]. Loading of nascent proteins into the endoplasmic reticulum (ER) [9] and mitochondria [10, 11] requires RNA and/or proteins to encounter specific pores and transport complexes on the organelle surface. Many signal processing pathways rely on

E. F. Koslover (✉)
University of California, San Diego, La Jolla, CA 92093, USA
e-mail: ekoslover@ucsd.edu



D. Grebenkov et al. (eds.), *Target Search Problems*,
https://doi.org/10.1007/978-3-031-67802-8_22

encounters between kinases and their substrates that hail from different regions of the cell [12].

All of these processes involve a search by a molecular particle for its binding partner. The dynamics of the particle itself can vary depending on its mode of transport within the cell. Small particles moving over short length scales exhibit random motions that can be treated as an effective diffusion. Such diffusion-like motion has been observed both in single-particle trajectories [13–17], and in bulk fluorescence recovery and photoactivated spreading measurements, which track the spatial evolution of a concentration profile [16, 18–20]. Notably, the seemingly diffusive behavior of intracellular particles is thought to be driven primarily by broadly distributed active forces rather than thermal agitation, with overall mobility greatly reduced upon ATP depletion [21, 22]. Heterogeneity and crowding of structures within the cytoplasm contribute to setting the effective particle diffusivity, as well as underlying the anomalous non-Gaussian shape of particle step-size distributions observed in living cells [23, 24].

One consequence of intracellular crowding is the steep dependence of the diffusion coefficient on particle size, with particles larger than approximately 50 nm experiencing greatly reduced diffusion [14, 25]. For large particles (such as RNA-protein complexes or vesicles), diffusion is too slow to enable efficient search over the cellular scale. Such particles tend to undergo active processive motion driven by molecular motors [26] that carry cargos over cytoskeletal highways [27, 28]. The distribution of run-lengths for processively moving particles tends to exhibit a broad power-law tail, so that actively driven cargo can be modeled as undergoing Lévy-flights [29–31]. Many intracellular components engage in intermittent transport, switching between processive and diffusive-like motion. For such particles, the rate of target encounter depends on the rates of transition between motility states [32–34], as well as the spatial organization of the cytoskeletal highways for active transport [35–37]. Here, we focus specifically on intracellular search processes by particles that exhibit effectively diffusive motion, with the role of active transport explored in more detail in other chapters.

The landscape navigated by searchers within the cell is defined by both the spatial architecture of subcellular regions and the pattern of local interactions with other cellular components. Important features of the landscape can arise from the morphology of membrane-enclosed domains (such as the endoplasmic reticulum or mitochondrial networks), which limit the search space for particles confined within them [38–40], while also serving as obstacles for cytoplasmic particles [41, 42]. In addition to such direct barriers, particles also undergo transient weak interactions with cellular components that can both slow and accelerate the search [43]. Such interactions are thought to play a key functional role in the selective recruitment of particles by liquid-like localized condensates within the cell [44–46]. In this chapter we explore how the intracellular landscape guides and modulates the target search process, considering both the general principles of target search on complex landscapes and their application to specific functionally important systems in cell biology.

## 22.2 Biological Functions for Intracellular Target Search

Any short-range interaction between subcellular components requires them to first encounter each other within the intracellular environment. In cases where the components are present in sufficiently low copy number, such encounters require a spatial target search process that can limit reaction rates. Described here are several categories of interactions where search times may play an important role.

**Escape from Compartments**
Escape from subcellular domains is one biologically important category of target search processes. This process falls within the purview of 'narrow-escape' problems—searching for small targets located on a domain boundary—which have been extensively explored mathematically [47–50]. An example of such narrow escape is the release of ions from the endoplasmic reticulum (ER) or mitochondria into the cytoplasm during signaling events, wherein the ions must find narrow channels in the organelle membrane [51–54]. Within the ER, proteins targeted for secretion accumulate at ER exit sites (ERES) [55, 56], which requires them to find structures thought to be gated by very narrow (few nm) necks [57] while moving diffusively through stacked sheets or tubular networks [15, 58, 59]. Proteins that regulate transcription must first enter the nucleus, which requires finding and passing through a pore in the nuclear membrane [5]. Conversely, proteins and RNAs must also leave the nucleus by encountering the same nuclear pores from the inner side.

In these examples, the molecular components are thought to find their exit and entry sites by random search rather than long-range sensing or directed delivery to the target of interest. Notably, pores and channels in organelle membranes often involve molecular filters that can aid or hinder the target search process by interacting with molecules approaching sufficiently close to them. Such filters can include the disordered peptide tails surrounding the cytoplasmic side of the nuclear pore [60], or the cloud of charged residues thought to funnel $Ca^{2+}$ ions into $IP_3R$ channels on the ER membrane [61].

**Assembly of Molecular Complexes**
Another category of intracellular target search processes encompasses the assembly and nucleation of multi-component structures. This includes, for instance, the assembly of transcription factor complexes on enhancer sites within genomic DNA, an important step for initiation and regulation of gene expression [62, 63]. Many transcription factors are thought to search through the genome by a combination of 3D diffusion and 1D sliding while non-specifically bound to the DNA backbone, a dimensional reduction which has been shown to greatly accelerate the search [6, 7, 64]. Other multi-protein assembly processes include recruitment of proteins to focal adhesions [65] and the nucleation and growth of cytoskeletal filaments [66]. The monomers or individual protein components of these structures must search along the cytoplasm and/or plasma membrane to encounter a partner for nucleation or to join an existing complex.

In certain cases, the search process for assembly is thought to be facilitated by the nearby presence of phase-segregated liquid droplets that can aid in the recruitment of specific binding partners to the complexes. For instance, phosphorylated Nephrin receptors guide N-WASP and NCK proteins to form condensates at the inner surface of the plasma membrane, which then serve to recruit actin nucleation factor Arp2/3 and trigger actin assembly [67]. Similarly, a liquid-like assembly of pericentriolar proteins helps accelerate the nucleation and growth of microtubule asters at centrosomes [68]. Within the nucleus, condensates that assemble at enhancer sites may speed up the recruitment of transcriptional cofactors for regulating gene expression [69, 70], while those forming at double-strand breaks help speed the arrival of DNA repair proteins [71].

**Intracellular Timing and Proofreading**
In certain intracellular target search systems, the fastest search times may not necessarily be optimal for function. In particular, the target search process can be utilized as a timer for regulating dynamic pathways. The endocytic system, for instance, is thought to control the duration of signaling events initiated by EGFR and other receptors at the plasma membrane. The duration is modulated by the time it takes for endosomes containing the receptors to move across the cell to the perinuclear zone and encounter lysosomes or phosphatases localized on the ER membrane, triggering deactivation [72, 73]. Slowing the target search process for these organelles can thus allow for a longer lasting persistent signal. In other cases, rapid degradation of signal molecules may narrow the distribution of search times through an obstacle-laden environment, allowing for more precise timing of the search process as a whole, without speeding up the arrival rate of individual molecules [74].

Extended intracellular transport times can also serve as a spatial proofreading mechanism [75]. When the location at which a molecule becomes activated and the location where it binds downstream effectors are spatially separated, there is an increased time period available for deactivation before reaching the target. Consequently, enhanced accuracy can be achieved in terms of allowing only the 'correct' (more slowly deactivated) molecules to reach the effector. This mechanism may underlie selective glycosylation pathways that modulate protein quality control in the ER [76] and the Golgi [77], as well as the propagation of kinase pathways that initiate at the plasma membrane [12]. More complex kinetic control can be achieved via multi-layered structures (as in rRNA processing in the nucleolus [78]), where specific enzymes are localized to specific layers. The accuracy of the multi-step reaction or assembly process is then expected to depend on the molecular target search rate within each layer compared to the timescale of moving through to another layer.

In all of these cases, the search by molecular reactants for their binding partners serves as a timer mechanism for regulating their functional activity. Optimization of these target search systems would require maximizing the precision of search (in terms of both timing and which molecules are successful), rather than necessarily maximizing the search speed.

## 22.3 The Role of Dimensionality

Given the complexity of the intracellular landscape and particle dynamics within live cells, it can be convenient to simplify both by treating them as self-similar fractals, defined by their effective dimensionality [80, 81]. The dimension of transport ($d_w$) is defined by the scaling of the mean-squared displacement (MSD) with respect to time, according to MSD $\sim t^{2/d_w}$. Diffusive particles (such as molecular-scale components in the cytoplasm [13–17]) have $d_w = 2$, while larger cytoplasmic particles that exhibit subdiffusion have been observed to move with a dimensionality of $d_w \approx 2.7$ [24, 82–84], and genomic loci can go up to $d_w \approx 5$ [83, 84]. Active motor-driven particles have a transport dimension in the range $1 < d_w < 2$, with the lower limit corresponding to unidirectional processive motion [30, 85, 86].

The dimension of the domain ($d_f$) describes how the 'volume' (or $N$, number of sites that can be visited) scales with the spatial extent ($R$), according to $N \sim R^{d_f}$. The cellular landscape of confinement and interactions defines the domain dimension. Depending on the subcellular structures considered, the relevant dimension can range from $d_f = 1$ (for tubular regions), through fractional dimensionalities [87], to $d_f = 2$ (on membranes) and $d_f = 3$ (in bulk cytoplasm). While prefactors are highly variable, the scaling exponents relating displacement, time, and number of sites sampled tend to be consistent between individual particles, and to remain robust under various biochemical and pharmacalogical perturbations [83, 88, 89].

The mean target search time is known to obey distinct scaling laws depending on the domain and transport dimensions $d_f$ and $d_w$ [81]. When $d_f < d_w$, particles exhibit compact search, wherein each visited region of the domain is fully explored before moving to the next region. In this regime, search times are dependent on the initial starting position of a particle relative to its target (even at long distances), the distribution of search times is not well-defined by a single timescale [90, 91], and the average search rate across the domain scales super-linearly with the target concentration [81, 92, 93]. By contrast, in the non-compact regime ($d_f > d_w$), the searching particle leaves behind many gaps in each region it visits, the search time becomes independent of starting position for sufficiently large distances from the target, and most particles reach the domain boundary before they ever find their target. This is the regime of classical bulk kinetics, where search rates are linearly proportional to concentration.

For diffusing particles ($d_w = 2$), the search rate (defined as $k = 1/\langle T\rangle$, where $\langle T\rangle$ is the mean first passage time averaged over all starting positions) can be summarized for different domain dimensions [92, 94]:

$$k \sim \begin{cases} Dc^2, & d_f = 1 \\ Dc/\log[(ca^2)^{-1}], & d_f = 2 \\ Dac, & d_f = 3 \end{cases} \tag{22.1}$$

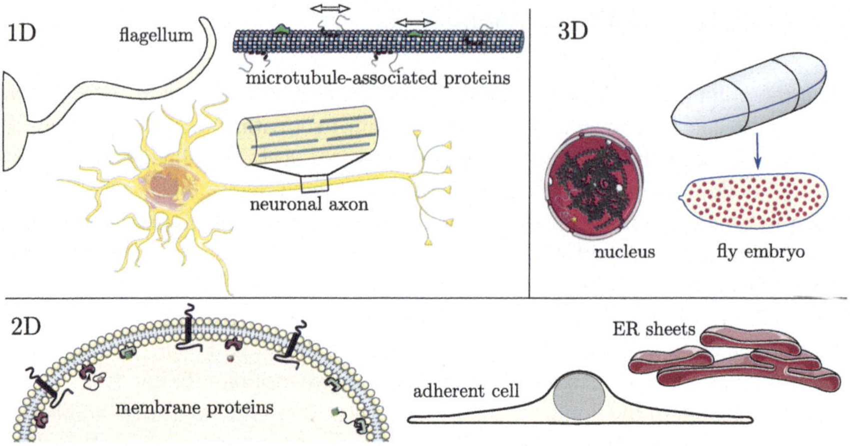


**Fig. 22.1** Sample cellular systems with different effective dimensionalities: 1D: one-dimensional systems, including narrow projections and sliding along cytoskeletal filaments. 2D: two-dimensional systems, including integral and peripheral membrane proteins, flat adherent cells, and sheet structures in the ER. 3D: three-dimensional bulk systems, including the nucleus and multinucleated *Drosophila* embryos [79]

where $c$ is the target concentration, and $a$ the target size. It should be noted that this search rate can be substantially faster than the reaction rate. In many biologically relevant cases, successful reactions constitute only a small fraction of the encounter events [95, 96]. Due to the differing dependence on reactant proximity, the dimensionality of the system affects the extent to which particles lose track of their binding partner after an unsuccessful encounter. Confinement to 2D has been shown to increase the frequency of rebinding events, allowing the same enzyme to proceed through multiple sequential reactions [97]. This switch from distributive to processive kinetics can qualitatively alter enzyme function by suppressing bistability and ultrasensitivity in multi-step reaction pathways [97–99].

In the usual regime where targets are small and sparse, 3D search can be very slow. Cells have thus developed mechanisms to effectively reduce the dimensionality as well as the volume through which particles must search for their targets. Such reduction of dimensionality can be achieved by confining particles within topologically isolated domains (Fig. 22.1), by attaching them to a 2D membrane surface or a 1D filament, or by enabling weak interactions with these cellular structures that allow for a combination of bulk 3D motion and sliding along a lower dimension. These systems can be thought of as search over an energy landscape with broad flat valleys or narrow channels extending from the target. Below we provide a brief overview of intracellular dimensions as well as examples of dimensional reduction for accelerating target search.

**3D Search**

Three-dimensional search is the default for processes occurring in bulk cytoplasm as well as inside globular organelles (such as the nucleus, peroxisomes, and autophagosomes). Thus, the encounter rate for two particles in the cytoplasm is expected to scale in proportion to their concentrations, as in the usual mass-action kinetics. The typical timescale for a ligand to find its enzyme in a spherical domain of radius $R$ can be estimated as $\tau \approx R^3/(12Da)$, where $D$ is the average diffusivity of both ligand and enzyme, and $a$ is the radius of each [100]. Using a typical cytoplasmic protein diffusivity of order $D \approx 10\,\mu\text{m}^2/\text{s}$, and molecular size $a \approx 2\,\text{nm}$, the time for two specific proteins to meet is on the order of $\tau \approx 0.03\,\text{s}$ inside a 200 nm vesicle, $\tau \approx 4\,\text{s}$ in a 1 μm bacterial cell, $\tau \approx 1\,\text{h}$ in a 10 μm mammalian fibroblast, and a whopping $\tau \approx 100$ years in a 1 mm frog egg. Clearly, large cells require high copy numbers of enzymes to enable ligands to encounter them on reasonable timescales.

Encounter times can be accelerated by organization of the subcellular environment. This can include confining specific groups of enzymes within sub-micron-sized organelles (such as endosomes and carboxysomes) [101], or placing multiple enzymes in spatial proximity by binding to a scaffold [102, 103]. The non-compact nature of 3D search implies that reaction products and intermediates tend to quickly escape into the bulk rather than remaining to interact with a nearby enzyme, even if that nearby binding site is only a few nanometers away [104]. Nevertheless, clustering of many enzymes into a compact aggregate can allow for the effective channeling of intermediates, passed directly to their next target site with a low probability of escape [105]. Such clustering has been observed in a number of metabolic pathways, including purine [106] and pyrimidine [107] biosynthesis.

**2D Search**

Two-dimensional search is relevant for particles associated with or embedded within cellular membranes. Examples include encounters between transmembrane receptors and their effector proteins [108] (e.g., G-protein-coupled receptor signaling [109]) and newly synthesized proteins in the perinuclear ER membrane which must find punctate ER exit sites (ERES) for export [55]. Additional examples of 2D search occur when particles are confined in a geometry with one dimension much shorter than the others, so long as the target itself is of comparable size to the shortest dimension. For example, adherent cells grown on a 2D substrate tend to spread out into very flat structures, so that the search process for a protein or vesicle to encounter the surface of the nucleus can be considered effectively two-dimensional.

A common intracellular scenario involves a cytoplasmic protein or small molecule that must find a target localized to a 2D membrane surface. Some examples include the search for import channels on the ER membrane by nascent proteins bound to signal recognition particles [110], and the assembly of actin monomers at nucleation sites formed by transmembrane nephrin proteins [67, 111]. Such encounters can be accelerated through dimensional reduction, by allowing a multi-step search process where the aqueous particle first associates non-specifically with the membrane, and then slides along the membrane to reach the target [100, 112, 113]. Because the rate of sliding to the target along the surface scales logarithmically with target size $a$, while the rate of purely 3D search scales linearly with the target size, the multi-step process can greatly accelerate search rates for small targets. In the case of irreversible binding to the surface, the acceleration in search time is approximated by [100]

$$\frac{\tau_{2D}}{\tau_{3D}} \approx 6\left(\frac{a}{R_{\text{mem}}}\right)\left(\frac{D_{2D}}{D_{3D}}\right)\left(\frac{R_{\text{mem}}^3}{R_{\text{dom}}^3}\right)\log\left(\frac{R_{\text{mem}}}{a}\right), \tag{22.2}$$

where $R_{\text{mem}} \gg a$ is the size of the membranous structure, $R_{\text{dom}}$ is the size of the entire 3D domain, and $D_{2D}$, $D_{3D}$ are the membrane and bulk diffusion coefficients, respectively. For the case of weak, reversible binding and very small targets, the particle has a chance to equilibrate throughout the entire domain before encountering the target. The search acceleration is then given by a similar expression with the substitution $R_{\text{dom}}^3 \to 2\pi R_{\text{mem}}^2 d e^{E_A/k_b T}$, where $d$ is the interaction distance, $E_A$ the interaction energy for the membrane, and $k_b T$ is the thermal energy [114].

Enhancement in encounter rates by such dimensional reduction from 3D to 2D target search has been proposed to contribute to several intracellular signaling systems. Specifically, the process by which cytoplasmic proteins find nuclear pores is thought to include a 2D surface-sliding step while interacting with a patch of FG motifs surrounding the pore on the outer surface of the nuclear envelope [115]. At the plasma membrane, recruitment of the nucleotide exchange factor Sos from the cytoplasm to the membrane allows the activation of the Ras signaling pathway [116], in part by speeding up the encounter rates between Sos and Ras proteins [117].

**1D Search**

One-dimensional search can be used to describe the encounter rates of particles attached to intracellular filaments. For instance, the motor proteins kinesin and dynein have both been shown to exhibit a weakly bound diffusive state during which they slide along the microtubule surface [118, 119]. A variety of other proteins, such as the regulatory microtubule-coating protein tau [120, 121] and the centromere-associated kinase Aurora-B [122], also slide diffusively along the microtubule track. Target search via this mode of motion enables proteins to find the microtubule plus and minus ends [123], to coalesce into structures coating microtubule regions [124], and to rapidly activate microtubule-associated substrates [122]. Confinement within narrow domains can also lead to effectively 1D search processes in extended cellular regions such as fungal hyphae [125, 126], neuronal projections [127], and cilia or flagella [128] (Fig. 22.1). For the latter case, the slow transport time (scaling quadratically with length) for diffusing kinesin motors to reach the flagellar base has been implicated as a timing mechanism in the regulation of flagellar length [128, 129].

Acceleration of search through dimensional reduction from 3D to 1D can be achieved through weak interaction with a filament. A classic example is the facilitated diffusion of DNA-binding proteins, which have been shown to bind non-specifically and slide along the DNA chain while probing for their target sequence [130, 131]. The target search of such proteins involves an interplay of 1D sliding, 3D microhops between nearby DNA segments, and 3D macrohops between genomically distal but spatially proximal sequences [6, 7, 64]. The target search rate is maximized when proteins spend roughly equal amounts of time in 3D and 1D search [132–134]. In conditions where DNA is dilute, such facilitated diffusion allows proteins to find their targets orders of magnitude faster than through 3D search alone [6, 135]. In physiological conditions, the process can be approximately described as an effectively diffusive search for an extended 'antenna' around the target site, followed by rapid sliding to the target [7, 134].

A similar acceleration through nonspecific sliding along a 1D scaffold has been proposed for the activity of a microtubule-associated kinase, Aurora-B, which plays a role in spindle organization and cytokinesis [122]. Weak electrostatic nonspecific interaction with the microtubule scaffold was shown to enhance the rate at which the kinase could phosphorylate further copies of itself, as well as other substrates on the microtubule. Here, again, reduction of dimensionality from a 3D to a 1D system accelerates functionally important kinetic processes within the cell.

## 22.4 Subcellular Geometry for Target Search

The geometry of subcellular structures plays a crucial role in defining the landscape over which particles search for their targets. Here, we explore the effect of common geometrical motifs, including sheets, tubular networks, and trees, on intracellular target search processes.

### *22.4.1 Searching Through Sheets and Stacks*

A common structural motif in eukaryotic cells is a stack of membranous sheets or disks. A sheet consists of two bilayer membranes sandwiching an aqueous lumen to form a flat, topologically enclosed domain. Examples of such structures are found in the perinuclear ER [136, 143], the specialized outer segments of photoreceptor cells [138], and the thylakoid disk stacks within chloroplasts [144] (Fig. 22.2a). Lamellar structures are also formed by mitochondrial cristae—invaginated folds of the inner mitochondrial membrane [137, 145, 146]. Target search processes that are affected by the sheet architecture can involve proteins moving along the membrane surface, particles in the hollow lumen of sheets, and cytoplasmic or organelle matrix particles that must navigate around sheet-like obstacles.

**ER Sheets**

In the ER, membrane proteins targeted for the cell surface or aqueous proteins targeted for secretion are cotranslationally inserted into the perinuclear sheets by ER-associated ribosomes. These proteins then undergo folding and several rounds of quality control, during which they interact with ER-resident chaperone proteins [147]. Under stress conditions, the quality control and protein degradation

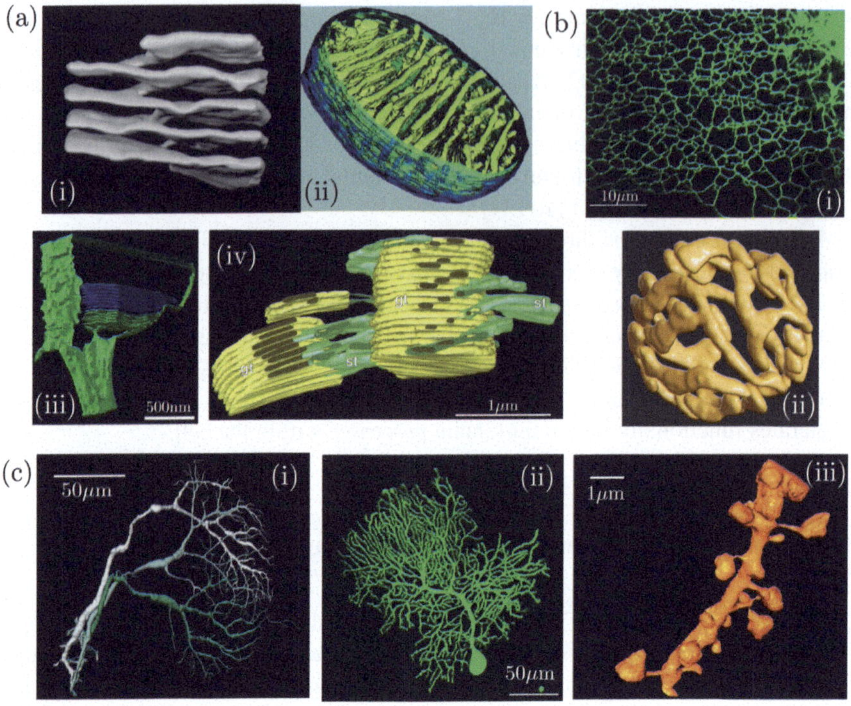


**Fig. 22.2** Common motifs of intracellular structures. **a** Flat sheets, including (i) spiral ramps in the peripheral ER [136], (ii) mitochondrial cristae [137], (iii) outer segment of photoreceptor [138], and (iv) grana and stroma thylakoids in plant cell [139]. **b** Tubular networks, including (i) peripheral ER network in mammalian cell (image courtesy of L. Westrate) and (ii) mitochondrial network in yeast cell (image courtesy of M. Viana). **c** Trees, including (i) dendritic tree of *Drosophila* optical sensory neuron [140], (ii) mouse Purkinje cell [141], and (iii) dendritic spines from mouse pyramidal neuron [142]. Images are adapted with permission from the references cited

machinery localizes to specific subdomains within the ER, necessitating a search process by nascent proteins to assemble in those regions [148]. After successful processing, both ER membrane and luminal proteins must search out punctate ER exit sites (ERES) in order to be packaged and released to proceed along the secretory pathway [56]. In addition, the accumulation of unfolded proteins in the ER can trigger signaling through the 'unfolded protein response' pathway, which involves a number of search steps, including the recruitment of activated ATF6 proteins to the ERES and clustering of IRE1 proteins on the ER membrane [149, 150].

One type of search process in sheets involves a luminal particle searching for a small target on the bounding membrane. This includes ions exiting through small pores, such as $Ca^{2+}$ release from the ER lumen [51, 54], or ER luminal proteins searching for the entry pore into the ERES. In the limit where the target extent is comparable to the sheet height, this process corresponds to effectively 2D target search, and the search rate scales logarithmically with target size [151]. However, channel pores tend to be on the order of 1 nm in size, while ER sheets have a height of about 50 nm [152]. In the limit of small, sparse targets, the search rate is similar (up to a logarithmic correction factor) to the 3D search process for an isolated absorbing target in empty space [151].

Long-range target search within a stack depends additionally on the dynamics of moving between distinct sheets. In the ER, consecutive sheets have been shown to be connected by spiral dislocations analogous to parking-garage ramps [136] (Fig. 22.2a.i). Winding around these ramps, rather than searching for direct pores connecting the sheets, allows for more efficient diffusivity of particles in the perpendicular direction across the stack, scaling in proportion to $\rho h^2$ where $h$ is the separation between sheets and $\rho$ the area density of the dislocations [153].

**Mitochondrial Cristae**

Mitochondrial cristae form lamellar structures (Fig. 22.2a.ii) which are partially open at one end, allowing their lumen to be continuous with the inter-membrane space along the surface of the mitochondrion [137]. However, the tubular cristae junctions where this lumen connects with the rest of the inter-membrane space are quite narrow (on the order of 30 nm [146]) and gated by a variety of proteins that enable cristae to remain partially isolated [154]. Such limitations on passage through the junctions enable individual cristae to maintain distinct electrical potentials across the inner mitochondrial membrane, implying that ions in the inter-membrane space cannot flow rapidly between different cristae [155].

Computational models of protein diffusion in the inner mitochondrial membrane indicate that the broad distribution of escape times from cristae junctions is expected to lead to slow subdiffusive motion over a range of intermediate timescales [146]. Proteins in the inner mitochondrial membrane include subunits of ATP synthase and the electron transport chain, which are responsible for maintaining the metabolic function of mitochondria. Measurements of spreading dynamics for an ATP synthase subunit (ATP6) indicate that it does not move far from the mitochondrial nucleoid producing it, even on timescales of several cell divisions [156]. This observation implies that inner membrane proteins tend to remain compartmentalized within mitochondria, potentially allowing for selective fission or mitophagy to isolate mutant nucleoids away from the mitochondrial population [157]. Perturbation of cristae junction morphology increases the rate of ATP6 spreading throughout the mitochondria [156]. This effect could be explained in part by enhanced search rates for inner membrane proteins to find transitions between distinct cristae.

### 22.4.2 Searching Through Tubular Networks

Interconnected networks of tubules form another common structural motif in the intracellular world. Such reticulated structures are found in the peripheral ER, which forms a lattice-like network of membrane-enclosed tubules stretching across the cell [158–160] (Fig. 22.2b.i). The membrane and lumen of the peripheral ER tubules are contiguous with the sheet-like structures found in the perinuclear region. ER tubules are also sometimes interspersed with peripheral sheet regions; the extent of such regions can be tuned by changing expression levels of ER-shaping morphogen proteins [152]. Search processes on ER networks include membrane and luminal proteins encountering ERES [40, 56] and contact sites with other organelles recruiting ER-resident molecular machinery [161, 162].

In addition to the ER, mitochondria fuse into extensive network structures in some cell types. Budding yeasts tend to exhibit highly connected mitochondrial networks localized to the cell surface [38] (Fig. 22.2b.ii). Mammalian stem-cells, by contrast, contain 3D mitochondrial networks that have been shown to balance at the percolation transition between fragmented and connected structures [163]. Perturbations of mitochondrial fission and fusion proteins can substantially alter the connectivity of the network [38, 163]. Environmental conditions also modulate network structure, with mitochondrial hyperfusion observed under a variety of stress conditions (starvation, UV exposure) [164, 165], and fragmentation of the network occurring in nutrient-rich environments [166]. Within the mitochondria, the ability of proteins to find regions adjoining specific nucleoids may aid in the complementation and rescue of mutated mtDNA [157].

**Computing First Passage Times in Networks**

The simplest picture of target search through reticulated cellular structures involves treating the network tubules as effectively 1D edges connected at point-like junctions. Diffusing particles move along the edges, randomly selecting which edge to enter when passing through a junction. The dynamics of such particles can be treated as a heterogeneous continuous time random walk [167] between Markovian states, where each state $i$ consists of trajectories that start at node $i$ at time 0 and have passed no other nodes. The average dwell time in each state $Q_i$ and the splitting probability $P_{ik}$ of the next passing to state $k$ can be found as simple functions of the particle diffusivity $D$ and the edge lengths $\ell_{ij}$ of the network [168]:

$$P_{ik} = \frac{1/\ell_{ik}}{\sum 1/\ell_{ij}}, \quad Q_i = \frac{1}{2D} \cdot \frac{\sum \ell_{ij}}{\sum 1/\ell_{ij}}, \tag{22.3}$$

where the summations are over all neighboring nodes. The splitting probabilities can be used to define a weighted graph Laplacian on the network [169], enabling calculation of the mean first passage time from node $i$ to $j$ according to

$$\tau_i = \left[ (\mathbf{I} - \mathbf{P})^{-1} \cdot \mathbf{Q} \right]_i, \tag{22.4}$$

where $\mathbf{I}$ is the identity matrix, and the $j$th row and column are removed from all matrices. Higher order moments of the target search distribution can similarly be computed as a function of edge lengths on the network [168].

Calculation of the global mean first-passage time (GMFPT, averaged over all starting positions) provides a convenient measure of the diffusive accessibility of different network regions, and the target-averaged GMFPT (TA-GMFPT) gives a search time metric across the entire structure. Typical search times for a membrane protein with $D \approx 1\,\mu m^2/s$ across the peripheral ER network of a kidney cell are on the order of an hour [40], while particles with the same diffusivity can search through the much smaller yeast mitochondrial network in approximately a minute [39].

#### 22.4.2.1 Network Connectivity and Target Search

The search time on a network depends on both the network topology (connectivity between nodes) and its spatial embedding (edge lengths and node positions). Intuitively, directed transport that relies on the shortest path between each pair of nodes must be accelerated by adding connections between network nodes. For diffusing particles, however, the addition of 'shortcuts' that connect distant nodes across the network can actually increase the TA-GMFPT by effectively trapping the particle inside long edges [170]. When the network nodes are spread out evenly over space, the optimal network architecture for fastest search times approximates a lattice of nearest-neighbor contacts [170].

The negative effect of distant shortcuts on search efficiency can also be viewed as interfering with dimensional reduction. The dimension of a spatial network can differ from the dimension of the space in which it is embedded, with long-range shortcuts potentially increasing the former [172]. A relevant example for intracellular transport involves production in a central perinuclear region of particles that must reach targets on the cell periphery. If we consider the average of the MFPT to each target, it can be shown that a network structure with direct connections from the center to each

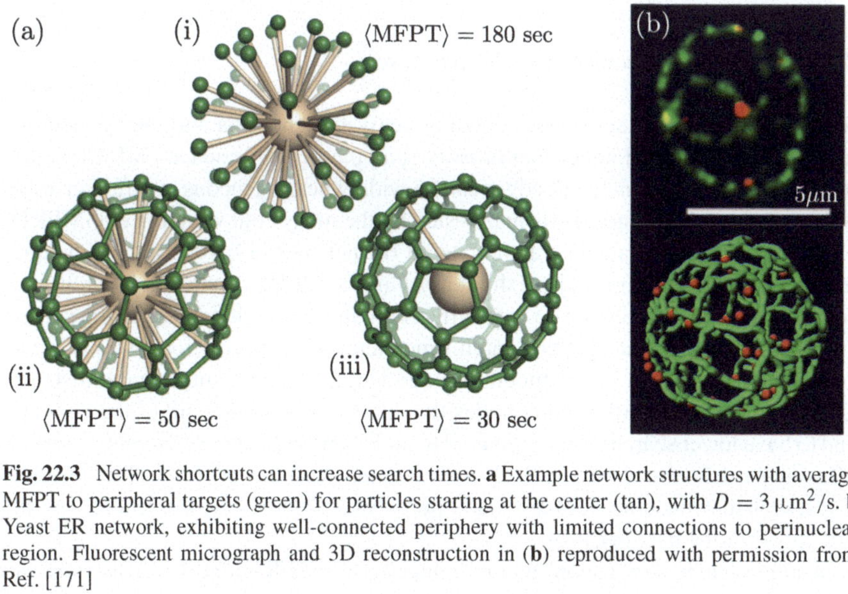


**Fig. 22.3** Network shortcuts can increase search times. **a** Example network structures with average MFPT to peripheral targets (green) for particles starting at the center (tan), with $D = 3\,\mu m^2/s$. **b** Yeast ER network, exhibiting well-connected periphery with limited connections to perinuclear region. Fluorescent micrograph and 3D reconstruction in (**b**) reproduced with permission from Ref. [171]

target is suboptimal (Fig. 22.3). In such a network, particles that enter the wrong edge must backtrack to the center prior to finding their target. A more efficient architecture is one where peripheral targets are connected by a lattice-like network along the cell surface, with only a few connections leading back to the central source. Such a morphology funnels particles toward the surface and reduces the search to an effectively 2D process. Intriguingly, the peripheral ER in yeast cells exhibits the characteristic features of this optimal structure—a well-connected cortical lattice of tubules, linked to the perinuclear region by very sparse connections.

For a lattice-like planar network (as in the peripheral ER of adherent mammalian cells or the surface-bound yeast mitochondria), the TA-GMFPT increases with the total edge length of tubules and decreases for networks with greater connectivity [39]. Lattice networks with missing edge connectivity can be mapped to percolation systems, which have been studied in many contexts [173]. Random walks on networks close to the percolation transition are known to exhibit universal scaling behaviors in metrics such as the effective spatial diffusivity [80, 173]. In particular, the diffusive TA-GMFPT has been shown to be approximated by [39] $T \sim \left[L^2/(L+\Gamma)\right]^{\mu}$, where $L$ is the total edge length, $\Gamma$ is the cyclomatic loop number [174], and $\mu \approx 1.3$ is a scaling exponent for diffusivity in 2D percolation lattices [173].

Notably, perturbation of proteins responsible for mitochondrial fusion and fission can reduce the network connectivity, raising the calculated search times on the networks due to the lower loop number [38]. The reaction rates for simulated particle pairs on the network are decreased substantially in the less well-connected networks, but only for sufficiently sparse particles. Thus, organelle network architecture has the potential to modulate reaction rates between low-copy-number proteins and nucleoids in the mitochondrial network.

#### 22.4.2.2 Heterogeneous Network Structures

Intracellular organelle networks, including yeast mitochondria and the ER, tend to have evenly distributed nodes that broadly span the available space [38]. However, the connectivity of the network edges can be widely heterogeneous in different areas of the cell. Such heterogeneity is manifested by the many-fold variation of GMFPTs to different target nodes within an individual ER network [39, 40]. Notably, heterogeneous network structures also exhibit asymmetric MFPTs, so that the search time from one node to another can be very different than the search time when the source and target are switched. Global search times tend to be lowest for nodes that are well-connected, with a large amount of edge length within a small graph-distance neighborhood [40]. Furthermore, nodes that are located in the bulk of the network tend to have lower search times compared to nodes on the periphery. Notably, pairs of diffusing particles are also more likely to encounter each other at such well-connected nodes, such that reactions between sparse particles are more likely to occur in specific 'hot-spots' of the network [40].

In addition to heterogeneous tubule connectivity, the ER network can also contain non-tubular structures, including peripheral and perinuclear sheet-like structures or variable-size junction nodes. Although the search time for individual junctions is not correlated to their size, the presence of expanded junction regions does raise the overall global search times in the network, due to increased waiting times for exiting the junctions [40]. Overexpression of the ER tubule-forming morphogen Rtn3 enables an extreme modification of network architecture to form expanded bubble-like regions with some remaining connectivity between them. This structural perturbation has been shown to increase protein and ion delivery times across the network, and to reduce the magnitude of local $Ca^{2+}$ release from the ER [175].

An additional putative source of network heterogeneity comes from variable tubule widths. Narrowing of ER tubules through overexpression of Rtn4 has been shown to reduce transport rates of luminal proteins across the network and to also hinder $Ca^{2+}$ release [176]. One possible explanation for the slowed transport times is a reduction in active flows that have been proposed to drive observed processive movements of luminal particles in the ER [177]. An alternate explanation could involve variability in width along individual tubules, resulting in localized partial constrictions. Diffusion through tubules of variable width can be treated as motion over a 1D free energy landscape, with locally wide regions corresponding to entropic traps and constrictions serving as barriers between them [178, 179]. The effective diffusivity through a tube with periodically varying radii is given by $D_{\text{eff}} = D_0/\left[\langle A\rangle\langle 1/A\rangle\right]$, where $A$ is the cross-sectional area [180]. Overall narrowing of a tubule that already has high variability in $A$ could decrease the effective diffusion coefficient and slow delivery rates of particles across the network.

Overall, the morphology of organelle networks helps shape the landscape within which luminal and membrane particles search for their binding partners, thereby playing an important role in determining search rates.

### *22.4.3 Searching Through Trees and Spines*

An alternate geometry for intracellular search processes consists of the highly-branched tree-like networks formed by neuronal dendrites and axons [181]. These structures lack loops and generally have a root node at the soma where all mRNA synthesis and much of the protein synthesis take place [182, 183]. Distal regions of the tree contain important signaling elements—pre-synaptic boutons in axons and post-synaptic spines in dendrites—necessitating long-range transport from the soma. The highly branched structure of dendritic arbors, in particular, has been explored in the context of organelle, protein, and mRNA delivery [184–187]. Perturbations to dendritic morphology are associated with a number of neuropathologies, including Alzheimer's disease and schizophrenia [188].

Target search processes on dendritic arbors include the delivery of newly translated cytoplasmic and membrane proteins to pre-synaptic and post-synaptic structures, as well as mRNA molecules to distal translation outposts [182]. Furthermore, renewal of critical organelles, including distally stationed mitochondria [187, 189] and Golgi outposts [190], also necessitates delivery through the arbor. Given the extended lengths of neurites, long-range delivery from the soma generally relies on motor-driven transport of organelles, vesicles, or RNA-protein particles. However, short-range diffusive target search may contribute to the distribution of proteins produced at local translation factories [182] as well as the recycling and downstream signaling of activated receptors [191].

**Tree Morphology**
The diffusion of particles along dendritic trees is governed by several morphological features, including the widths of the dendritic tubes, their branching patterns, and the presence of spine structures branching off the dendrite. In some cell types, individual dendrites have tapering widths, with the radius narrowing toward the distal end of the branch [192]. This effect leads to a biased diffusion of membrane-bound and aqueous particles, equivalent to a tilted landscape that leads to preferential motion toward the wider end [186]. Furthermore, dendritic radii become narrower at each branching junction, with the relative widths of daughter $(r_1, r_2)$ and parent $(r_0)$ branches often described by a scaling law of the form: $r_0^\alpha = r_1^\alpha + r_2^\alpha$. The value of $\alpha$ varies between different neuronal cell types, with $\alpha = 2$ and $\alpha = 2.7$ reported for sensory dendrites [187, 193] and motor neuron axons [194], respectively. This relationship implies that daughter branches have higher surface-area-to-volume ratio compared to the parent branch, creating a bias for membrane proteins to move preferentially into more distal branches, and increasing their distal density relative to aqueous particles [184]. Notably, for proteins with a fixed decay rate, enhanced diffusive entry into a wider branch can increase the steady-state protein density in that branch. As a result, matching wider branch radii to longer branches allows for a more uniform distribution of proteins over the tree, and equivalently requires lower total protein production to supply a minimum quantity at each distal tip [184]. A similar effect is observed for the distribution of actively moving organelles, where the branch width must be proportional to the 'bushiness' of the subtree in order to allow an equitable distribution of organelles in an asymmetric tree [187].

Both the total edge length of a tree and the timescales for diffusive search tend to scale exponentially with the depth (number of branching levels) in the tree. For a hierarchical tree, formed by $N$ rounds of replacing each edge with a triskelion of identical edges, the MFPT to traverse between the two most distant nodes scales as $\tau \sim 6^N \sim L^{\log 6/\log 3} \sim L^{1.63}$ [195], where $L$ is the total edge length. The global MFPT to the center of such a fractal structure also exhibits the same scaling [196]. For a binary tree, the MFPT from the root node to a distal tip scales as $\tau \sim 2^N \sim L$ [197]. Overall, arranging a given length of neurite into a tree-like morphology allows for more rapid diffusive search than would be expected for a linear structure (where $\tau \sim L^2$).

For a binary tree, the search time-scaling also depends on the bias parameter $p$, which gives the probability of a particle entering the parent edge when passing a junction. In particular, for $0 < p < 1/2$, the MFPT for a particle at the distal tip to reach the root (soma) scales exponentially with the tree depth $N$, according to $\tau \sim 2^N \sim L$ [186]. By contrast, the case where $p = 1/2$ gives $\tau \sim N^2 \sim \log L$ and higher values of $p$ give $\tau \sim N$ [186]. Purely unbiased motion corresponds to $p = 1/3$, but the narrower radii of daughter dendrites are expected to give rise to higher values of this bias parameter. In particular, for a tree structure wherein the cross-sectional area is conserved across a junction ($r_0^2 = r_1^2 + r_2^2$), a cytoplasmic protein should exhibit $p = 1/2$ whereas a membrane-bound protein would have $p \approx 0.4$. Consequently, the MFPT for membrane-bound proteins to reach the soma would scale exponentially with tree depth, whereas that of aqueous proteins would scale only quadratically.

Dendritic arbor depths are reduced in the course of aging and Alzheimer's progression [198, 199], thus leading to shorter return times to the soma, reduced gradients [186], and potentially altered balance of cytoplasmic and membrane proteins in the neurite. These examples highlight how the morphology of the neuronal cell defines a geometric landscape that can govern diffusive search times for different types of proteins.

**Dendritic Spines**

An additional morphological feature of dendritic arbors are the narrow-necked bulbous projections called dendritic spines. These spines are associated with synaptic regulation and plasticity, reconfiguring their shape and size in response to synaptic input [142, 200]. They serve as effective traps for molecules diffusing along the dendrite, reducing the effective diffusivity of membrane proteins along the dendritic branch by a factor that depends on the overall surface area of each spine and the spacing between them [201]. An uneven distribution of spine spacing gives rise to subdiffusive motion along the dendrite over intermediate times [202]. The subdiffusive movement of membrane proteins on seconds timescales and the reduced diffusivity at longer times as a function of spine density has been observed experimentally in several neuronal cell types [203, 204].

The narrow neck of the spine limits both the rate of entry and the rate of exit from the spine head, allowing for effective compartmentalization between individual spine structures [205]. As the exit time from the spine head scales quadratically with neck length, longer necks imply that ions within an individual spine can be pumped into ER storage or out to the extracellular space before ever exchanging with the rest of the dendrite [206]. The spine neck width exemplifies a trade-off between narrow necks that promote compartmentalization and limit escape of proteins within the spine head and wide necks that speed delivery of new proteins into the head [205]. Notably, the balance between these two effects implies that the overall rate at which proteins within the main dendrite can find a distal target is not expected to depend strongly on the neck geometry [201]. Nevertheless, the trapping effect of dendrites slows search rates through the dendritic tree, and decreased spine densities (as observed in Alzheimer's, schizophrenia, and other neuronal disorders) are expected to speed up the rates at which diffusive proteins can traverse the arbor [186].

## 22.5 Landscapes for Intracellular Search

Confinement within membrane-enclosed organelle and cell geometries, as discussed in Sect. 22.4, can be thought of as an extreme form of intracellular landscape with very high barriers preventing transition out of the confined domain. In general, particles within the cell move over a rough energy landscape with many finite wells and barriers that modulate their preferred locations and the transition rates between them.

The kinetics of search over an energy landscape have a long history of exploration in the context of molecular reactions [207, 208]. Protein folding [209] and cluster rearrangements [210] are traditionally described as thermal motion over a complex high-dimensional landscape describing molecular configuration. Such landscapes can be coarse-grained into a discrete set of basins (energy wells) [207], with rapid equilibration within a basin and transition rates between neighboring basins scaling exponentially with the height of the barrier between them, according to the classic Kramer's rate theory [211, 212]. Such 'basin-hopping' descriptions of the system convert first passage times over a continuous landscape into a set of discrete hops over an interconnected network of states. Analogous approaches, wherein coarse-grained locally equilibrated states are extracted from molecular dynamics trajectories, are termed Markov State Models [213, 214]. In both cases, transitions between connected states are assumed to have constant rates, independent of past states occupied by the system. The rate of finding a target state (e.g., the correctly folded conformation of a protein) can then be found using the usual approach for search times over a network (see Eq. 22.4).

The structure of the network and its transitions define the dynamic behavior of the molecular system when searching for its target states. For example, the landscape can be structured as a funnel, guiding the system toward a preferred basin and allowing for search times much faster than would be expected from purely random exploration [215]. Alternately, a fractal landscape encompassing a broad spectrum of deep wells and tall barriers can confer glassy behavior and non-ergodic dynamics with diverging average transition times between clusters of states [216, 217]. Kinetic traps and barriers can arise, for example, from topological constraints that prohibit self-intersection of a polymer, and prevent it from sampling an equilibrium distribution of configurations [218]. On a more coarse-grained scale, the search for a particular configuration on such a landscape cannot be described in terms of a single well-defined rate, violating the usual framework of transition state theory [217].

The complex intracellular environment offers a landscape guiding the movement of particles over micrometer rather than nanometer distances. A key distinction between cell-scale versus molecular-scale landscapes is that many interactions within a cell are expected to be relatively localized. The landscape is defined locally by binding interactions with various cellular structures, or by electrostatic or hydrophobic effects that might repel a particle from specific regions. However, screening in the intracellular environment gives rise to a very short Debye length ($\ell_d < 1\,\text{nm}$ [219]), implying that interactions between charged particles are substantial only at a short range. Consequently, the funnel-like landscapes that can substantially speed up protein folding and other molecular search processes [207] are not a common feature of cytoplasmic transport. Instead, cellular particles transition between locally flat landscape regions defined by the location of other structures with which they interact. The funneling role of a tilted landscape is instead fulfilled by ATP-burning, motor-driven transport processes which enable processive motion toward a target [220]. In addition, particles that bind to long-range cellular structures (including membranes and filaments) can experience intermediate-range interactions mediated by the mechanics of the substrate itself (see Sect. 22.5.3).

### 22.5.1 Search Acceleration with Condensates

To organize their interior, cells rely not only on confinement within membrane-bound compartments but also on the formation of local condensates, which can recruit or exclude specific molecules [221]. In many cases, these condensates exhibit the hallmark features of phase-separated liquid droplets, including a spherical shape, ability to fuse upon contact, and high mobility of particles within the droplet [222, 223]. Such liquid-like condensates have been observed in the cytoplasm (including P-granules, stress granules, and purinosomes [224]), the plasma membrane (lipid-ordered and disordered domains) [225], and the nucleus (repair foci, silencing foci, nucleosomes [226]). Other foci form more solid-like aggregates, with phase-segregation of polymers containing multiple binding sites resulting in aggregation of the bound proteins [224, 226–228].

In both cases, the foci form a localized microenvironment that serves as a broad (micron-scale) well in the energy landscape experienced by molecules that undergo weak interactions with the foci components (Fig. 22.4). In addition to the energetic preference for entering into the condensate, the diffusivity of molecules within it is generally altered from its value in the surrounding medium, either due to transient binding with stationary sites in a solid condensate or to increased viscosity in a liquid droplet. A key distinction between the two types of condensates is that the binding strength (and hence the energetic well depth) in a solid condensate is directly related

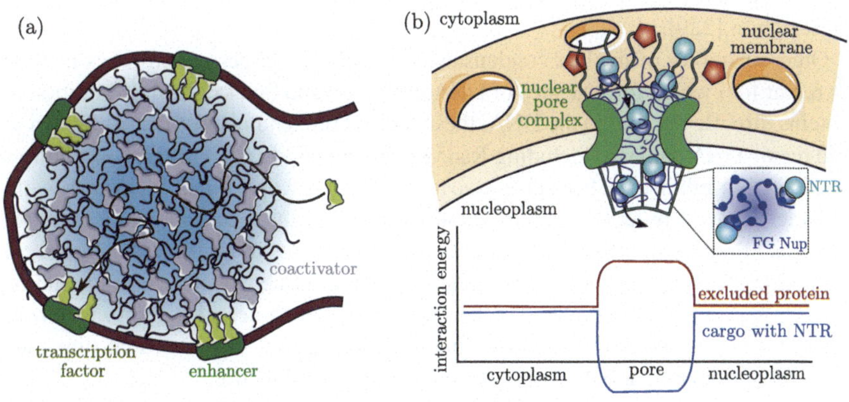


**Fig. 22.4** Examples of intracellular landscapes arising from weak transient interactions. **a** Liquid condensates formed at super-enhancer sites may speed assembly of transcription factor complexes. **b** Unstructured domains in the nuclear pore complex form reversible hydrogel that selectively enhances passage of NTR-shielded proteins [79]

to the effective diffusivity; in a liquid droplet, the two parameters can be considered independently of one another [44].

One of the key functional roles proposed for intracellular condensates is the acceleration of diffusion-limited reactions to targets sequestered within the focus. On the one hand, the energy well surrounding the target serves as a primitive single-step funnel that increases the search rate. On the other hand, decreased diffusivity within the focus slows down the search process for a small target within it. For a sufficiently deep well, the search time to hit a target of radius $r_0$ within a spherical droplet of radius $r_f$, for a target confined within an overall cellular or nuclear domain of radius $r_n$, is given by [44]

$$\tau \approx \frac{r_f^3}{3D_f r_0} + \frac{r_n^3}{3D_n r_f}, \tag{22.5}$$

where $D_f$ is diffusivity within the focus and $D_n$ is the diffusivity outside it. This corresponds to two sequential processes, where the searcher first finds the focus itself and then finds the target within it. The search time is minimized for a focus of size $r_f^* = (\frac{1}{3} r_0 r_n^3 D_f / D_n)^{1/4}$. For physiological parameters relevant for the dynamics of DNA repair proteins, this optimal size is $r_f^* \approx 120\,\text{nm}$, comparable to the observed size of DNA repair foci within the nucleus, and the presence of the condensate is estimated to speed up the search process by approximately 6-fold [44].

The formation of condensates within the nucleus has been proposed to enhance the kinetics of several biological processes critical for genomic maintenance and expression. A variety of transcription factors, as well as RNA polymerase itself, contains low-complexity domains that engage in weak interactions and form condensates at transcriptional hubs [70]. This process allows a few transcription factors bound to enhancer elements on the DNA to rapidly recruit a pool of polymerase molecules that are then held locally available for transcription [229]. Phosphorylation of the polymerase decreases its affinity for the condensate and releases it to initiate active transcription [229].

Other examples of nuclear condensates that enhance target search times include the repair foci which form at sites of DNA double-strand breaks [226]. In addition to proteins directly binding to the DNA, these foci recruit a number of factors known to help repair DNA damage, including Rad52 [71] and p53 [230]. Periodic oscillations in p53 concentration have been proposed to limit the growth of repair foci via Ostwald ripening [231], a process wherein small droplets shrink while larger ones grow. These oscillations stabilize the formation of condensates at multiple damage sites, maintaining foci of an optimal size to enable rapid recruitment of other repair proteins. Overall, the ability of cytoplasmic and nuclear proteins to condense into localized foci gives rise to wells in the intracellular landscape that reduce search times for recruited factors to reach the targets of their biological function.

### 22.5.2 Filtration at Pore Entrance

Weak interactions with localized structures also play an important role in intracellular search processes that involve particles passing through narrow pores. In these cases, a protein, ion, or metabolite must enter a channel in order to enter or exit a membrane-enclosed structure. The necessity of entering a narrow channel and diffusing through to the other side can be formulated as an entropic barrier on an effective energy landscape [49, 50, 178, 179]. Repulsive or attractive interactions with structures at the channel entrance further modulate this landscape to set the typical escape time [49, 50].

One classic example is the transport of $Ca^{2+}$ ions out of the endoplasmic reticulum, which serves as a cellular calcium reservoir and releases the ions in response to a variety of signals [232]. Passage out of the ER requires each $Ca^{2+}$ ion to find a channel pore that is only about 1 nm across. The search can be thought of as a multi-step process wherein the ion samples nearby regions of the membrane several times before wandering back to the bulk of the ER lumen; as a consequence, the search rate scales linearly with target radius rather than with its area [54, 114]. The search is thought to be accelerated still further by negatively charged residues surrounding the channel entrance. These residues form a calcium-permeable 'sponge' that funnels ions toward the channel [61]. The resulting 'Donnan potential' well extends for a distance of 5−10 nm [61], transforming the search into a sequential two-step process where the ion must first encounter this larger target and then quickly finds the pore while confined within the well.

Another example of target search through narrow filters arises in the context of translocation through a nuclear pore (Fig. 22.4b). The nuclear pore complex contains proteins with large, unstructured repeating regions (FG domains) that form a hydrogel to selectively gate passage through the pore into the nucleus [5, 60]. These unstructured domains interact weakly with each other and with nuclear transport receptors (NTRs). The NTRs form a shield for activated transcription factors, allowing them to pass efficiently through the nuclear pore to enter the nucleus [60]. In the 'selective phase' model, a hydrogel of self-cohesive FG domains forms a sieve that excludes sufficiently large inert particles, while enabling rapid entry of particles smaller than the mesh size [233]. At the same time, NTRs surrounding a protein can compete for binding to the FG domains, locally dissolving the hydrogel and enabling passage [60, 234]. In essence, such a model relies on the FG domains to form localized energy wells that enhance the transit rate of NTR-shielded proteins as compared to inert particles, which may experience a flat landscape or steric barriers, depending on their size.

Simplified mathematical models of nuclear pore transport consider the dynamics of a particle diffusing in one dimension, while transiently binding and unbinding from locally anchored polymeric structures (FG domains). The landscape experienced by such a particle resembles a series of overlapping harmonic wells, with the binding strength setting the well depth. The effective diffusivity of the particle depends both on the tether flexibility and on the rate of unbinding [235]. As in the case of higher dimensional condensates (discussed in the previous section), the rate at which a particle reaches the boundary of this landscape depends non-monotonically on the binding strength [45, 235]. In particular, very weak binding reduces selectivity (as compared to inert non-interacting particles), while very strong binding hinders the ability of the particle to hop between tethers. The latter effect can be mitigated by allowing particles to directly transition from one tether to another without unbinding, significantly enhancing selectivity [235]. An alternate model describes passage through the pore as stochastically gated diffusion, where particles must hop between local confining domains, but can only do so when they are in an unbound state [46]. Confinement into a series of smaller subdomains was shown to greatly increase the MFPT [46], implying that the elastic flexibility of FG domains could play an important role in selective transport by increasing the confinement range of bound particles. Intriguingly, binding to elastic structures has been shown to increase transit rates for multiple sterically interacting particles [236], implying a possible non-trivial effect from crowding of proteins on their passage times through a nuclear pore.

Transit through a pore, whether an ion channel or the more complex nuclear pore complex, constitutes an important intracellular target search problem, with the search rates accelerated or reduced by the molecular interaction landscape at the entrance and within the pore. Because a distinct landscape is presented to different molecules, the search times can vary widely, allowing these pores to serve as effective filters for sorting cellular components.

### *22.5.3 Substrate-Mediated Landscapes*

Intracellular particles moving along membranes and filaments can experience complicated landscapes defined by the substrate itself (e.g., lipid composition of the membrane, sequence of a DNA strand), as well as interactions with other particles propagated through the substrate. A classic example of membrane-mediated interactions involves transmembrane proteins whose hydrophobic domains are mismatched to the thickness of the membrane. Such proteins cause local deformation of the membrane bilayer thickness, which extends over a few nanometer length-scale $\lambda$, followed by a slight overshoot to the opposite sign deformation, before decaying to zero [237]. Two proteins approaching close to each other experience a slight repulsive potential at intermediate distances, and an attractive potential at separations below $4\lambda$ where the overlap of their deformation field causes less overall perturbation of the membrane [237–239]. As a result, proteins with similar transmembrane thickness experience attractive wells in the energy landscape drawing them close together (Fig. 22.5a).

Membrane thickness-mediated interactions are thought to aid in the recruitment of proteins to ER exit sites. Binding to specific cargo receptors may localize a few proteins with extended transmembrane domain length, allowing other proteins with similar thickness to selectively accumulate at the site [55]. Furthermore, these interactions may contribute to the assembly of chemoreceptors, allowing for nucleation

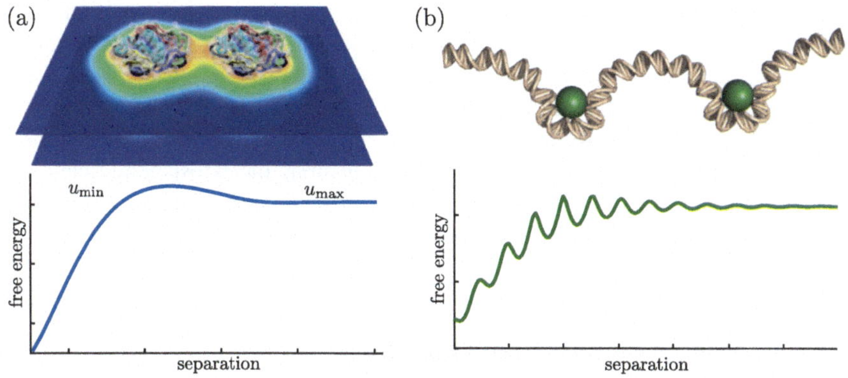


**Fig. 22.5** Substrate-mediated landscapes arise for **a** integral membrane proteins with mismatched thickness of the transmembrane domain [237] and **b** DNA-bending proteins [240]. Rates of encounter between sliding particles can be slowed or accelerated by landscape shape. Figure **a**, top is reprinted from Ref. [237], copyright 2016 by the American Physical Society

followed by recruitment to existing clusters even at relatively low chemoreceptor concentration [238]. The target search dynamics of new receptors arriving to a cluster are accelerated by the attractive well in the energy landscape induced by membrane deformation.

Similarly, interactions between curvature-inducing proteins propagate along a membrane and allow for short-range attraction mediated by membrane deformation. Such interactions allow for the efficient coalescence of arc-shaped FtsZ proteins into tightly compact rings during bacterial cytokinesis [241]. Curvature-based interactions have also been shown to yield nanometer-range attraction between BAR-domain proteins, enabling their rapid assembly into linear aggregates [242, 243].

A distinct substrate-mediated landscape arises for proteins sliding along a DNA strand under tension. Many DNA-binding proteins induce bending of the helix, and placing two such bends close to each other can reduce the energetic cost of deforming the strand against tension [244], causing an overall DNA-mediated attraction between the proteins (Fig. 22.5b). Furthermore, due to the twist of the DNA strand, the interaction landscape exhibits periodic hills and wells, with a separation corresponding to the 10 basepair pitch of the helix [240, 245]. This landscape speeds up the encounter time between proteins sliding along the DNA at low tensions and slows down encounters at high tensions, with optimal tension on the order of 0.5 pN minimizing the search time between two sliding proteins [240].

Cytoskeletal fibers may also serve as a substrate for sliding proteins, which experience a landscape propagated through the fiber structure. A number of microtubule-binding proteins, including kinesin-1 and kinesin-4 motors as well as non-motile tau and EB1, have been shown to alter the conformation of the microtubule lattice [246].

The associated changes in lattice compaction and/or introduction of defects can propagate over several micrometers and affect the binding strength of other microtubule-associated proteins [247–249]. Similarly, actin-binding proteins such as cofilin, filamin, and myosin motors have been shown to induce propagating conformational changes in actin filaments which in turn serve to attract other proteins to specific filament regions [250]. For proteins walking or sliding along the cytoskeletal filaments, such cooperative allosteric effects may form a landscape that drives biased motion and accelerates the formation of coats and aggregates on the filament [121].

## 22.6 Perspectives and Open Questions

The intracellular world provides a uniquely complex landscape for target search processes over a multitude of spatiotemporal scales. Calculations of search time shed light on cellular processes ranging from signaling (via ion release [54, 251], nuclear translocation [46], or escape from dendritic branches [186]) to maintenance (protein replenishment at synapses [201, 252] or in a mitochondrial network [38]), to transcription (binding to DNA target sites [6, 7]) and secretion (exit from the ER [40]). These systems have motivated studies of how target search processes can be modulated by different geometries [39, 186, 206, 251], by dimensional reduction [100, 113, 253], by transitions between states with varying transport properties [32, 46], by finite lifetimes [74], and large searcher populations [254].

The energy landscape picture provides a crucial conceptual framework for understanding the search behavior of passive particles in a complex environment. Unlike original applications in molecular dynamics, cellular landscapes tend to be low-dimensional and broadly flat, with localized features defined by relatively short-range interactions (compared to the distances that must be covered by searching particles). Searching through cellular landscapes thus requires not only hopping between well-defined barrier-separated wells, but also substantial periods of diffusive transport across broad valleys in one, two, or three dimensions. These landscapes are defined by both the geometry of the system and the ability of the particle to interact with membranes and filaments within the cell. The intracellular landscape thus motivates new investigations into how landscape structure governs search times. Furthermore, the active nature of the cellular environment gives rise to a number of questions that remain poorly understood.

One set of open questions centers on the dynamic nature of cellular landscapes. Many organelle and filament structures rearrange over timescales of seconds to minutes, potentially altering the search times of particles confined within or bound to these structures. For example, the tubules of the endoplasmic reticulum tend to shorten as a result of constitutive membrane tension, causing dynamic rearrangement of junctions, balanced by sporadic growth events wherein new tubules are pulled out and joined into the network [58, 255]. These dynamics lead to substantial alteration of network structure over a period of approximately a minute. For monomeric luminal and membrane proteins diffusing within the ER, these rearrangements have been shown to have little effect on search times through the network [40]. However, for larger assemblies, such as the ER exit sites, network dynamics are likely to substantially affect the rate at which they move through the network [256]. Mitochondrial populations in cells also create dynamic networks, ranging from fragmented 'social networks' with transient interactions [257, 258] to highly fused yet dynamic physical networks [38, 163]. The rate at which disconnected network components fuse and break up again is expected to greatly modulate the ability of individual particles to find targets within the mitochondrial population. This rate depends on both cellular transport processes that bring distant mitochondria into proximity with each other and the fusion and fission frequencies. The quantitative effect of network dynamics on search times within mitochondrial populations remains currently unexplored.

In addition, the cell is full of active transport processes that can directly or indirectly guide the motion of searching particles. For example, slow cytoplasmic flows can enhance mixing and positioning of cellular components in oocytes, embryos, fungal hyphae, and crawling neutrophils [125, 259–261]. Localized flows arising from hydrodynamic entrainment by passing motor-driven vesicles have been hypothesized to contribute to the motion of particles in axons [262]. Similarly, short-range flows engendered by the sweeping motion of actively plucked microtubules have been suggested to contribute to the transport of passive particles in the cytoplasm [23]. Furthermore, many cellular particles switch between periods of directed motion (through coupling to molecular motors) and passive diffusion in the cytoplasm. While several studies have explored the delivery times of particles that experience both directed transport and diffusive periods [32, 37, 252], the overall effects of active driving on search times over a landscape are not well understood. In particular, the optimal coupled spatial organization of both flows (or highways for active transport) and local interaction landscapes remains an open question.

Finally, studies of target search within a living cell inevitably need to be considered in the context of biological function. The average search time for an individual particle is not always the most relevant metric, nor is it necessarily minimized for optimal biological function. In many signaling processes, for instance, positive feedback at the target implies that the arrival time for the first of many searchers governs the rate of signal propagation [254]. The behavior of such extreme statistics on different landscape structures thus forms an important avenue for future work. In other cases, high particle densities imply that only local search between nearby targets and searchers is relevant, limiting the importance of the global landscape structure [38]. Still other situations, such as those involving spatial kinetic proofreading [75] or signal timing through delayed propagation [72, 74], may require the cell to optimize not for the fastest target search but for a narrowly bounded intermediate search time. The breadth of cellular functions thus makes it necessary to explore a variety of metrics for search processes through time and space, going far beyond the classic mean first passage time. The complex, dynamic landscape structure, the presence of active driving forces, and the existence of a broad range of search timescales makes the intracellular world an ideal playground for applying and developing the mathematical framework for target search problems.

**Acknowledgements** This work was partially funded by the US National Science Foundation (Grant No. 1848057) and by the Chan Zuckerberg Initiative.

## References

1. A. Bar-Even, E. Noor, Y. Savir, W. Liebermeister, D. Davidi, D.S. Tawfik, R. Milo, The moderately efficient enzyme: evolutionary and physicochemical trends shaping enzyme parameters. Biochemistry **50**, 4402–4410 (2011)
2. A. Cook, F. Bono, M. Jinek, E. Conti, Structural biology of nucleocytoplasmic transport. Annu. Rev. Biochem. **76**, 647–671 (2007)
3. S.R. Wente, M.P. Rout, The nuclear pore complex and nuclear transport. Cold Spring Harb. Perspect. Biol. **2**, a000562 (2010)
4. W.A. Herbst, K.C. Martin, Regulated transport of signaling proteins from synapse to nucleus. Curr. Opin. Neurobiol. **45**, 78–84 (2017)
5. C. Strambio-De-Castillia, M. Niepel, M.P. Rout, The nuclear pore complex: bridging nuclear transport and gene regulation. Nat. Rev. Mol. Cell. Bio. **11**, 490–501 (2010)
6. S.E. Halford, J.F. Marko, How do site-specific DNA-binding proteins find their targets? Nucleic Acids Res. **32**, 3040–3052 (2004)
7. E.F. Koslover, M.A.D. de la Rosa, A.J. Spakowitz, Theoretical and computational modeling of target-site search kinetics in vitro and in vivo. Biophys. J. **101**, 856–865 (2011)
8. O. Pulkkinen, R. Metzler, Distance matters: the impact of gene proximity in bacterial gene regulation. Phys. Rev. Lett. **110**, 198101 (2013)
9. Y. Nyathi, B.M. Wilkinson, M.R. Pool, Co-translational targeting and translocation of proteins to the endoplasmic reticulum. Biochimica et Biophysica Acta (BBA)-Mol. Cell Res. **1833**, 2392–2402 (2013)
10. X.G. Arceo, E.F. Koslover, B.M. Zid, A.I. Brown, Mitochondrial mRNA localization is governed by translation kinetics and spatial transport. PLoS Comput. Biol. **18**, e1010413 (2022)
11. F. Devaux, G. Lelandais, M. Garcia, S. Goussard, C. Jacq, Posttranscriptional control of mitochondrial biogenesis: spatio-temporal regulation of the protein import process. FEBS Lett. **584**, 4273–4279 (2010)
12. B.N. Kholodenko, Cell-signalling dynamics in time and space. Nat. Rev. Mol. Cell Bio. **7**, 165–176 (2006)
13. B.R. Daniels, B.C. Masi, D. Wirtz, Probing single-cell micromechanics in vivo: the microrheology of C. Elegans developing embryos. Biophys. J. **90**, 4712–4719 (2006)
14. F. Etoc, E. Balloul, C. Vicario, D. Normanno, D. Liße, A. Sittner, J. Piehler, M. Dahan, M. Coppey, Non-specific interactions govern cytosolic diffusion of nanosized objects in mammalian cells. Nat. Mater. **17**, 740–746 (2018)
15. Y. Sun, Z. Yu, C.J. Obara, K. Mittal, J. Lippincott-Schwartz, E.F. Koslover, Unraveling trajectories of diffusive particles on networks. Phys. Rev. Res. **4**, 023182 (2022)
16. M. Sittewelle, S.J. Royle, Passive diffusion accounts for the majority of intracellular nanovesicle transport. Life Sci Alliance **7** (2024)
17. A.N. Kapanidis, S. Uphoff, M. Stracy, Understanding protein mobility in bacteria by tracking single molecules. J. Mol. Biol. **430**, 4443–4455 (2018)
18. N.W. Goehring, D. Chowdhury, A.A. Hyman, S.W. Grill, FRAP analysis of membrane-associated proteins: lateral diffusion and membrane-cytoplasmic exchange. Biophys. J. **99**, 2443–2452 (2010)
19. P.D. Calvert, W.E. Schiesser, E.N. Pugh Jr., Diffusion of a soluble protein, photoactivatable GFP, through a sensory cilium. J. Gen. Physiol. **135**, 173–196 (2010)

20. R.G. Sadovsky, S. Brielle, D. Kaganovich, J.L. England, Measurement of rapid protein diffusion in the cytoplasm by photo-converted intensity profile expansion. Cell Rep. **18**, 2795–2806 (2017)
21. B.R. Parry, I.V. Surovtsev, M.T. Cabeen, C.S. O'Hern, E.R. Dufresne, C. Jacobs-Wagner, The bacterial cytoplasm has glass-like properties and is fluidized by metabolic activity. Cell **156**, 183–194 (2014)
22. M. Guo, A.J. Ehrlicher, M.H. Jensen, M. Renz, J.R. Moore, R.D. Goldman, J. Lippincott-Schwartz, F.C. Mackintosh, D.A. Weitz, Probing the stochastic, motor-driven properties of the cytoplasm using force spectrum microscopy. Cell **158**, 822–832 (2014)
23. P. Witzel, M. Götz, Y. Lanoiselée, T. Franosch, D.S. Grebenkov, D. Heinrich, Heterogeneities shape passive intracellular transport. Biophys. J. **117**, 203–213 (2019)
24. T.J. Lampo, S. Stylianidou, M.P. Backlund, P.A. Wiggins, A.J. Spakowitz, Cytoplasmic RNA-protein particles exhibit non-Gaussian subdiffusive behavior. Biophys. J. **112**, 532–542 (2017)
25. A.S. Verkman, Solute and macromolecule diffusion in cellular aqueous compartments. Trends Biochem. Sci. **27**, 27–33 (2002)
26. R.D. Vale, The molecular motor toolbox for intracellular transport. Cell **112**, 467–480 (2003)
27. M. Burute, L.C. Kapitein, Cellular logistics: unraveling the interplay between microtubule organization and intracellular transport. Ann. Rev. Cell Dev. Bi **35**, 29–54 (2019)
28. L. Balabanian, A.R. Chaudhary, A.G. Hendricks, Traffic control inside the cell: microtubule-based regulation of cargo transport. Biochemist **40**, 14–17 (2018)
29. K. Chen, B. Wang, S. Granick, Memoryless self-reinforcing directionality in endosomal active transport within living cells. Nat. Mater. **14**, 589–593 (2015)
30. S. Fedotov, N. Korabel, T.A. Waigh, D. Han, V.J. Allan, Memory effects and Lévy walk dynamics in intracellular transport of cargoes. Phys. Rev. E **98**, 042136 (2018)
31. M.S. Song, H.C. Moon, J.-H. Jeon, H.Y. Park, Neuronal messenger ribonucleoprotein transport follows an aging Lévy walk. Nat. Commun. **9**, 1–8 (2018)
32. O. Bénichou, C. Loverdo, M. Moreau, R. Voituriez, Intermittent search strategies. Rev Mod Phys **83**, 81 (2011)
33. A. Godec, R. Metzler, Active transport improves the precision of linear long distance molecular signalling. J. Phys. A : Math. Theor. **49**, 364001 (2016)
34. S.S. Mogre, E.F. Koslover, Multimodal transport and dispersion of organelles in narrow tubular cells. Phys. Rev. E **97**, 042402 (2018)
35. D. Ando, N. Korabel, K.C. Huang, A. Gopinathan, Cytoskeletal network morphology regulates intracellular transport dynamics. Biophys. J. **109**, 1574–1582 (2015)
36. A.E. Hafner, H. Rieger, Spatial organization of the cytoskeleton enhances cargo delivery to specific target areas on the plasma membrane of spherical cells. Phys. Biol. **13**, 066003 (2016)
37. A.E. Hafner, H. Rieger, Spatial cytoskeleton organization supports targeted intracellular transport. Biophys. J. **114**, 1420–1432 (2018)
38. M.P. Viana, A.I. Brown, I.A. Mueller, C. Goul, E.F. Koslover, S.M. Rafelski, Mitochondrial fission and fusion dynamics generate efficient, robust, and evenly distributed network topologies in budding yeast cells. Cell Syst. **10**, 287-297.e5 (2020)
39. A.I. Brown, L.M. Westrate, E.F. Koslover, Impact of global structure on diffusive exploration of organelle networks. Sci. Rep. **10**, 1–13 (2020)
40. Z.C. Scott, K. Koning, M. Vanderwerp, L. Cohen, L.M. Westrate, E.F. Koslover, Endoplasmic reticulum network heterogeneity guides diffusive transport and kinetics. Biophys J. (2023)
41. H. Li, S.-X. Dou, Y.-R. Liu, W. Li, P. Xie, W.-C. Wang, P.-Y. Wang, Mapping intracellular diffusion distribution using single quantum dot tracking: compartmentalized diffusion defined by endoplasmic reticulum. J. Am. Chem. Soc. **137**, 436–444 (2015)
42. A. Sabri, X. Xu, D. Krapf, M. Weiss, Elucidating the origin of heterogeneous anomalous diffusion in the cytoplasm of mammalian cells. Phys. Rev. Lett. **125**, 058101 (2020)
43. P.A. Vasquez, B. Walker, K. Bloom, D. Kolbin, N Caughman, R. Freeman, M. Lysy, C. Hult, K.A. Newhall, M. Papanikolas, et al., The power of weak, transient interactions across biology: a paradigm of emergent behavior. Physica D 133866 (2023)

44. M.L. Heltberg, J. Miné-Hattab, A. Taddei, A.M. Walczak, T. Mora, Physical observables to determine the nature of membrane-less cellular sub-compartments. eLife **10**, e69181 (2021)
45. A. Zilman, S. Di Talia, B.T. Chait, M.P. Rout, M.O. Magnasco, Efficiency, selectivity, and robustness of nucleocytoplasmic transport. PLoS Comput. Biol. **3**, e125 (2007)
46. P.C. Bressloff, Stochastically gated diffusion model of selective nuclear transport. Phys. Rev. E **101**, 042404 (2020)
47. O. Bénichou, R. Voituriez, Narrow-escape time problem: time needed for a particle to exit a confining domain through a small window. Phys. Rev. Lett. **100**, 168105 (2008)
48. Z. Schuss, A. Singer, D. Holcman, The narrow escape problem for diffusion in cellular microdomains. Proc. Natl. Acad. Sci. **104**, 16098–16103 (2007)
49. D.S. Grebenkov, G. Oshanin, Diffusive escape through a narrow opening: new insights into a classic problem. Phys. Chem. Chem. Phys. **19**, 2723–2739 (2017)
50. D.S. Grebenkov, R. Metzler, G. Oshanin, Full distribution of first exit times in the narrow escape problem. New J. Phys. **21**, 122001 (2019)
51. R. Thul, M. Falcke, Release currents of IP3 receptor channel clusters and concentration profiles. Biophys. J. **86**, 2660–2673 (2004)
52. K. Bentele, M. Falcke, Quasi-steady approximation for ion channel currents. Biophys. J. **93**, 2597–2608 (2007)
53. L. Bruno, G. Solovey, A.C. Ventura, S. Dargan, S.P. Dawson, Quantifying calcium fluxes underlying calcium puffs in Xenopus laevis oocytes. Cell Calcium **47**, 273–286 (2010)
54. Z. Yang, E.F. Koslover, Diffusive exit rates through pores in membrane-enclosed structures. Phys. Biol. **20**, 026001 (2023)
55. N. Borgese, Getting membrane proteins on and off the shuttle bus between the endoplasmic reticulum and the Golgi complex. J. Cell Sci. **129**, 1537–1545 (2016)
56. L.M. Westrate, M.J. Hoyer, M.J. Nash, G.K. Voeltz, Vesicular and uncoated Rab1-dependent cargo carriers facilitate ER to Golgi transport. J Cell Sci. **133**, jcs239814 (2020)
57. A.V. Weigel, C.-L. Chang, G. Shtengel, C.S. Xu, D.P. Hoffman, M. Freeman, N. Iyer, J. Aaron, S. Khuon, J. Bogovic et al., ER-to-Golgi protein delivery through an interwoven, tubular network extending from ER. Cell **184**, 2412–2429 (2021)
58. L. Westrate, J. Lee, W. Prinz, G. Voeltz, Form follows function: the importance of endoplasmic reticulum shape. Annu. Rev. Biochem. **84**, 791–811 (2015)
59. S. Wang, H. Tukachinsky, F.B. Romano, T.A. Rapoport, Cooperation of the ER-shaping proteins atlastin, lunapark, and reticulons to generate a tubular membrane network. eLife, **5**, e18605 (2016)
60. H.B. Schmidt, D. Görlich, Transport selectivity of nuclear pores, phase separation, and membraneless organelles. Trends Biochem. Sci. **41**, 46–61 (2016)
61. R. Mejía-Alvarez, C. Kettlun, E. Ríos, M. Stern, M. Fill, Unitary Ca2+ current through cardiac ryanodine receptor channels under quasi-physiological ionic conditions. J. Gen. Physiol. **113**, 177–186 (1999)
62. F. Spitz, E.E. Furlong, Transcription factors: from enhancer binding to developmental control. Nat. Rev. Genet. **13**, 613–626 (2012)
63. M. Levine, C. Cattoglio, R. Tjian, Looping back to leap forward: transcription enters a new era. Cell **157**, 13–25 (2014)
64. M.A. Lomholt, B. van den Broek, S.-M.J. Kalisch, G.J. Wuite, R. Metzler, Facilitated diffusion with DNA coiling. Proc. Natl. Acad. Sci. **106**, 8204–8208 (2009)
65. S. Schumacher, R. Vazquez Nunez, C. Biertümpfel, N. Mizuno, Bottom-up reconstitution of focal adhesion complexes. FEBS J. **289**, 3360–3373 (2022)
66. J. Kraus, R. Alfaro-Aco, B. Gouveia, S. Petry, Microtubule nucleation for spindle assembly: one molecule at a time. Trends Biochem. Sci. (2023)
67. P. Li, S. Banjade, H.-C. Cheng, S. Kim, B. Chen, L. Guo, M. Llaguno, J.V. Hollingsworth, D.S. King, S.F. Banani et al., Phase transitions in the assembly of multivalent signalling proteins. Nature **483**, 336–340 (2012)
68. J.B. Woodruff, B.F. Gomes, P.O. Widlund, J. Mahamid, A. Honigmann, A.A. Hyman, The centrosome is a selective condensate that nucleates microtubules by concentrating tubulin. Cell **169**, 1066–1077 (2017)

69. K. Wagh, D.A. Garcia, A. Upadhyaya, Phase separation in transcription factor dynamics and chromatin organization. Curr. Opin. Struc. Biol. **71**, 148–155 (2021)
70. B.R. Sabari, A. Dall'Agnese, A. Boija, I.A. Klein, E.L. Coffey, K. Shrinivas, B.J. Abraham, N.M. Hannett, A.V. Zamudio, J.C. Manteiga et al., Coactivator condensation at super-enhancers links phase separation and gene control. Science **361**, eaar3958 (2018)
71. J. Miné-Hattab, M. Heltberg, M. Villemeur, C. Guedj, T. Mora, A.M. Walczak, M. Dahan, A. Taddei, Single molecule microscopy reveals key physical features of repair foci in living cells. eLife **10**, e60577 (2021)
72. J. Bakker, M. Spits, J. Neefjes, I. Berlin, The EGFR odyssey-from activation to destruction in space and time. J. Cell Sci. **130**, 4087–4096 (2017)
73. J. Neefjes, M.M. Jongsma, I. Berlin, Stop or go? Endosome positioning in the establishment of compartment architecture, dynamics, and function. Trends Cell Biol. **27**, 580–594 (2017)
74. J. Ma, M. Do, M.A. Le Gros, C.S. Peskin, C.A. Larabell, Y. Mori, S.A. Isaacson, Strong intracellular signal inactivation produces sharper and more robust signaling from cell membrane to nucleus. PLoS Comput. Biol. **16**, e1008356 (2020)
75. V. Galstyan, K. Husain, F. Xiao, A. Murugan, R. Phillips, Proofreading through spatial gradients. eLife **9**, e60415 (2020)
76. A.I. Brown, E.F. Koslover, Design principles for the glycoprotein quality control pathway. PLoS Comput. Biol. **17**, e1008654 (2021)
77. A. Jaiman, M. Thattai, Golgi compartments enable controlled biomolecular assembly using promiscuous enzymes. eLife **9**, e49573 (2020)
78. M. Feric, N. Vaidya, T.S. Harmon, D.M. Mitrea, L. Zhu, T.M. Richardson, R.W. Kriwacki, R.V. Pappu, C.P. Brangwynne, Coexisting liquid phases underlie nucleolar subcompartments. Cell **165**, 1686–1697 (2016)
79. Parts of the figure were drawn using Servier Medical Art, provided by Servier, licensed under a Creative Commons Attribution 3.0 unported license
80. D. Ben-Avraham, S. Havlin, *Diffusion and Reactions in Fractals and Disordered Systems* (Cambridge University Press, Cambridge, UK, 2000)
81. O. Bénichou, C. Chevalier, J. Klafter, B. Meyer, R. Voituriez, Geometry-controlled kinetics. Nat. Chem. **2**, 472 (2010)
82. I.M. Tolić-Nørrelykke, E.-L. Munteanu, G. Thon, L. Oddershede, K. Berg-Sørensen, Anomalous diffusion in living yeast cells. Phys. Rev. Lett. **93**, 078102 (2004)
83. S.C. Weber, A.J. Spakowitz, J.A. Theriot, Bacterial chromosomal loci move subdiffusively through a viscoelastic cytoplasm. Phys. Rev. Lett. **104**, 238102 (2010)
84. S. Yu, J. Sheats, P. Cicuta, B. Sclavi, M. Cosentino Lagomarsino, K.D. Dorfman, Subdiffusion of loci and cytoplasmic particles are different in compressed Escherichia coli cells. Commun. Biol. **1**, 176 (2018)
85. D. Arcizet, B. Meier, E. Sackmann, J.O. Rädler, D. Heinrich, Temporal analysis of active and passive transport in living cells. Phys. Rev. Lett. **101**, 248103 (2008)
86. R. Benelli, M. Weiss, From sub-to superdiffusion: fractional Brownian motion of membraneless organelles in early C. Elegans embryos. New J. Phys. **23**, 063072 (2021)
87. M.A. Aon, B. O'rourke, S. Cortassa, The fractal architecture of cytoplasmic organization: scaling, kinetics and emergence in metabolic networks. Mol. Cell. Biochem. **256**, 169–184 (2004)
88. I. Golding, E.C. Cox, Physical nature of bacterial cytoplasm. Phys. Rev. Lett. **96**, 098102 (2006)
89. S.C. Weber, A.J. Spakowitz, J.A. Theriot, Nonthermal ATP-dependent fluctuations contribute to the in vivo motion of chromosomal loci. Proc. Natl. Acad. Sci. **109**, 7338–7343 (2012)
90. A. Godec, R. Metzler, Universal proximity effect in target search kinetics in the few-encounter limit. Phys. Rev. X **6**, 041037 (2016)
91. D.S. Grebenkov, R. Metzler, G. Oshanin, Strong defocusing of molecular reaction times results from an interplay of geometry and reaction control. Commun. Chem. **1**, 96 (2018)
92. O. Bénichou, R. Voituriez, From first-passage times of random walks in confinement to geometry-controlled kinetics. Phys. Rep. **539**, 225–284 (2014)

93. S. Condamin, O. Bénichou, V. Tejedor, R. Voituriez, J. Klafter, First-passage times in complex scale-invariant media. Nature **450**, 77–80 (2007)
94. F. Le Vot, S. Yuste, E. Abad, D.S. Grebenkov, First-encounter time of two diffusing particles in two-and three-dimensional confinement. Phys. Rev. E **105**, 044119 (2022)
95. D. Drenckhahn, T.D. Pollard, Elongation of actin filaments is a diffusion-limited reaction at the barbed end and is accelerated by inert macromolecules. J. Biol. Chem. **261**, 12754–12758 (1986)
96. S.S. Mogre, J.R. Christensen, C.S. Niman, S.L. Reck-Peterson, E.F. Koslover, Hitching a ride: mechanics of transport initiation through linker-mediated hitchhiking. Biophys. J. **118**, 1357–1369 (2020)
97. S.M. Abel, J.P. Roose, J.T. Groves, A. Weiss, A.K. Chakraborty, The membrane environment can promote or suppress bistability in cell signaling networks. J. Phys. Chem. B **116**, 3630–3640 (2012)
98. K. Takahashi, S. Tănase-Nicola, P.R. Ten Wolde, Spatio-temporal correlations can drastically change the response of a MAPK pathway. Proc. Natl. Acad. Sci. **107**, 2473–2478 (2010)
99. P.J. Mlynarczyk, R.H. Pullen, S.M. Abel, Confinement and diffusion modulate bistability and stochastic switching in a reaction network with positive feedback. J. Chem. Phys. **144** (2016)
100. D.S. Grebenkov, R. Metzler, G. Oshanin, Search efficiency in the Adam-Delbrück reduction-of-dimensionality scenario versus direct diffusive search. New J. Phys. **24**, 083035 (2022)
101. A. Küchler, M. Yoshimoto, S. Luginbühl, F. Mavelli, P. Walde, Enzymatic reactions in confined environments. Nat. Nanotechnol. **11**, 409–420 (2016)
102. H. Lee, W.C. DeLoache, J.E. Dueber, Spatial organization of enzymes for metabolic engineering. Metab. Eng. **14**, 242–251 (2012)
103. L.K. Langeberg, J.D. Scott, Signalling scaffolds and local organization of cellular behaviour. Nat. Rev. Mol. Cell Bio. **16**, 232–244 (2015)
104. P. Bauler, G. Huber, T. Leyh, J.A. McCammon, Channeling by proximity: the catalytic advantages of active site colocalization using Brownian dynamics. J. Phys. Chem. Lett. **1**, 1332–1335 (2010)
105. M. Castellana, M.Z. Wilson, Y. Xu, P. Joshi, I.M. Cristea, J.D. Rabinowitz, Z. Gitai, N.S. Wingreen, Enzyme clustering accelerates processing of intermediates through metabolic channeling. Nat. Biotechnol. **32**, 1011–1018 (2014)
106. S. An, R. Kumar, E.D. Sheets, S.J. Benkovic, Reversible compartmentalization of de novo purine biosynthetic complexes in living cells. Science **320**, 103–106 (2008)
107. I. Chitrakar, D.M. Kim-Holzapfel, W. Zhou, J.B. French, Higher order structures in purine and pyrimidine metabolism. J. Struct. Biol. **197**, 354–364 (2017)
108. D.A. Lauffenburger, J. Linderman, *Receptors: Models for Binding, Trafficking, and Signaling* (Oxford University Press, Oxford, UK, 1996)
109. C.J. Brinkerhoff, J.S. Choi, J.J. Linderman, Diffusion-limited reactions in G-protein activation: unexpected consequences of antagonist and agonist competition. J. Theor. Biol. **251**, 561–569 (2008)
110. D. Akopian, K. Shen, X. Zhang, S.-O. Shan, Signal recognition particle: an essential protein-targeting machine. Annu. Rev. Biochem. **82**, 693–721 (2013)
111. T. Wiegand, A.A. Hyman, Drops and fibers-how biomolecular condensates and cytoskeletal filaments influence each other. Emerging Top. Life Sci. **4**, 247 (2020)
112. G. Adam, M. Delbrück, Reduction of dimensionality in biological diffusion processes. Struct. Chem. Mol. Biol. **198**, 198–215 (1968)
113. D. Axelrod, M. Wang, Reduction-of-dimensionality kinetics at reaction-limited cell surface receptors. Biophys. J. **66**, 588–600 (1994)
114. H.C. Berg, E.M. Purcell, Physics of chemoreception. Biophys. J. **20**, 193–219 (1977)
115. R. Peters, Translocation through the nuclear pore complex: selectivity and speed by reduction-of-dimensionality. Traffic **6**, 421–427 (2005)
116. A. Aronheim, D. Engelberg, N. Li, N. Al-Alawi, J. Schlessinger, M. Karin, Membrane targeting of the nucleotide exchange factor Sos is sufficient for activating the Ras signaling pathway. Cell **78**, 949–961 (1994)

117. W.Y.C. Huang, S.G. Boxer, J.E. Ferrell Jr., Membrane localization accelerates association under conditions relevant to cellular signaling. Proc. Nat. Acad. Sci. **121**(10), e2319491121 (2024)
118. J.R. Cooper, L. Wordeman, The diffusive interaction of microtubule binding proteins. Curr. Opin. Cell Biol. **21**, 68–73 (2009)
119. G. Bhabha, G.T. Johnson, C.M. Schroeder, R.D. Vale, How dynein moves along microtubules. Trends Biochem. Sci. **41**, 94–105 (2016)
120. T. Scholz, E. Mandelkow, Transport and diffusion of Tau protein in neurons. Cell. Mol. Life Sci. **71**, 3139–3150 (2014)
121. M.H. Hinrichs, A. Jalal, B. Brenner, E. Mandelkow, S. Kumar, T. Scholz, Tau protein diffuses along the microtubule lattice. J. Biol. Chem. **287**, 38559–38568 (2012)
122. M. Noujaim, S. Bechstedt, M. Wieczorek, G.J. Brouhard, Microtubules accelerate the kinase activity of Aurora-B by a reduction in dimensionality. PLoS ONE **9**, e86786 (2014)
123. Y. Shin, Y. Du, S.E. Collier, M.D. Ohi, M.J. Lang, R. Ohi, Biased Brownian motion as a mechanism to facilitate nanometer-scale exploration of the microtubule plus end by a kinesin-8. Proc. Natl. Acad. Sci. **112**, E3826–E3835 (2015)
124. V. Siahaan, J. Krattenmacher, A.A. Hyman, S. Diez, A. Hernández-Vega, Z. Lansky, M. Braun, Kinetically distinct phases of tau on microtubules regulate kinesin motors and severing enzymes. Nat. Cell Biol. **21**, 1086–1092 (2019)
125. C. Lin, M. Schuster, S.C. Guimaraes, P. Ashwin, M. Schrader, J. Metz, C. Hacker, S.J. Gurr, G. Steinberg, Active diffusion and microtubule-based transport oppose myosin forces to position organelles in cells. Nat. Commun. **7**, 11814 (2016)
126. S.S. Mogre, J.R. Christensen, S.L. Reck-Peterson, E.F. Koslover, Optimizing microtubule arrangements for rapid cargo capture. Biophys. J. **120**, 4918–4931 (2021)
127. T.L. Blasius, N. Reed, B.M. Slepchenko, K.J. Verhey, Recycling of kinesin-1 motors by diffusion after transport. PLoS ONE **8**, e76081 (2013)
128. A. Chien, S.M. Shih, R. Bower, D. Tritschler, M.E. Porter, A. Yildiz, Dynamics of the IFT machinery at the ciliary tip. eLife **6**, e28606 (2017)
129. W.F. Marshall, The flagellar length control system: exploring the physical biology of organelle size. Phys. Biol. **20**, 021001 (2023)
130. Y. Wang, R.H. Austin, E.C. Cox, Single molecule measurements of repressor protein 1D diffusion on DNA. Phys. Rev. Lett. **97**, 048302 (2006)
131. J. Gorman, E.C. Greene, Visualizing one-dimensional diffusion of proteins along DNA. Nat. Struct. Mol. Biol. **15**, 768–774 (2008)
132. M. Slutsky, L.A. Mirny, Kinetics of protein-DNA interaction: facilitated target location in sequence-dependent potential. Biophys. J. **87**, 4021–4035 (2004)
133. M.A.D. de la Rosa, E.F. Koslover, P.J. Mulligan, A.J. Spakowitz, Dynamic strategies for target-site search by DNA-binding proteins. Biophys. J. **98**, 2943–2953 (2010)
134. T. Hu, A.Y. Grosberg, B. Shklovskii, How proteins search for their specific sites on DNA: the role of DNA conformation. Biophys. J. **90**, 2731–2744 (2006)
135. R.B. Winter, O.G. Berg, P.H. Hippel, Diffusion-driven mechanisms of protein translocation on nucleic acids. 3. The Escherichia coli lac repressor-operator interaction: kinetic measurements and conclusions. Biochemistry **20**, 6961–6977 (1981)
136. M. Terasaki, T. Shemesh, N. Kasthuri, R.W. Klemm, R. Schalek, K.J. Hayworth, A.R. Hand, M. Yankova, G. Huber, J.W. Lichtman et al., Stacked endoplasmic reticulum sheets are connected by helicoidal membrane motifs. Cell **154**, 285–296 (2013)
137. T. Frey, C. Renken, G. Perkins, Insight into mitochondrial structure and function from electron tomography. BBA - Bioenerg. **1555**, 196–203 (2002)
138. S. Volland, L.C. Hughes, C. Kong, B.L. Burgess, K.A. Linberg, G. Luna, Z.H. Zhou, S.K. Fisher, D.S. Williams, Three-dimensional organization of nascent rod outer segment disk membranes. Proc. Natl. Acad. Sci. **112**, 14870–14875 (2015)
139. J.R. Austin, L.A. Staehelin, Three-dimensional architecture of grana and stroma thylakoids of higher plants as determined by electron tomography. Plant Physiol. **155**, 1601–1611 (2011)

140. K.M. Boergens, C. Kapfer, M. Helmstaedter, W. Denk, A. Borst, Full reconstruction of large lobula plate tangential cells in Drosophila from a 3D EM dataset. PLoS ONE **13**, e0207828 (2018)
141. Y. Sakai, H. Kassai, H. Nakayama, M. Fukaya, T. Maeda, K. Nakao, K. Hashimoto, H. Sakagami, M. Kano, A. Aiba, Hyperactivation of mTORC1 disrupts cellular homeostasis in cerebellar Purkinje cells. Sci. Rep. **9**, 2799 (2019)
142. J. Tønnesen, U.V. Nägerl, Dendritic spines as tunable regulators of synaptic signals. Front Psychiatry **7**, 101 (2016)
143. L.K. Schroeder, A.E. Barentine, H. Merta, S. Schweighofer, Y. Zhang, D. Baddeley, J. Bewersdorf, S. Bahmanyar, Dynamic nanoscale morphology of the ER surveyed by STED microscopy. J. Cell Biol. **218**, 83–96 (2019)
144. M. Pribil, M. Labs, D. Leister, Structure and dynamics of thylakoids in land plants. J. Exp. Bot. **65**, 1955–1972 (2014)
145. Y.-F. Jiang, S.-S. Lin, J.-M. Chen, H.-Z. Tsai, T.-S. Hsieh, C.-Y. Fu, Electron tomographic analysis reveals ultrastructural features of mitochondrial cristae architecture which reflect energetic state and aging. Sci. Rep. **7**, 45474 (2017)
146. V.M. Sukhorukov, J. Bereiter-Hahn, Anomalous diffusion induced by cristae geometry in the inner mitochondrial membrane. PLoS ONE **4**, e4604 (2009)
147. B.M. Adams, M.E. Oster, D.N. Hebert, Protein quality control in the endoplasmic reticulum. Protein J. **38**, 317–329 (2019)
148. E.M. Sontag, W.I. Vonk, J. Frydman, Sorting out the trash: the spatial nature of eukaryotic protein quality control. Curr. Opin. Cell Biol. **26**, 139–146 (2014)
149. C. Hetz, K. Zhang, R.J. Kaufman, Mechanisms, regulation and functions of the unfolded protein response. Nat. Rev. Mol Cell Bio. **21**, 421–438 (2020)
150. L.T. Kischuck, A.I. Brown, Tube geometry controls protein cluster conformation and stability on the endoplasmic reticulum surface. Soft Matter **19**, 6771–6783 (2023)
151. D.S. Grebenkov, D. Krapf, Steady-state reaction rate of diffusion-controlled reactions in sheets. J. Chem. Phys. **149**, 064117 (2018)
152. Y. Shibata, T. Shemesh, W.A. Prinz, A.F. Palazzo, M.M. Kozlov, T.A. Rapoport, Mechanisms determining the morphology of the peripheral ER. Cell **143**, 774–788 (2010)
153. G. Huber, M. Wilkinson, Terasaki spiral ramps and intracellular diffusion. Phys. Biol. **16**, 065002 (2019)
154. B. Gottschalk, C.T. Madreiter-Sokolowski, W.F. Graier, Cristae junction as a fundamental switchboard for mitochondrial ion signaling and bioenergetics. Cell Calcium **101**, 102517 (2022)
155. D.M. Wolf, M. Segawa, A.K. Kondadi, R. Anand, S.T. Bailey, A.S. Reichert, A.M. van der Bliek, D.B. Shackelford, M. Liesa, O.S. Shirihai, Individual cristae within the same mitochondrion display different membrane potentials and are functionally independent. EMBO J. **38**, e101056 (2019)
156. C. Jakubke, R. Roussou, A. Maiser, C. Schug, F. Thoma, D. Bunk, D. Hörl, H. Leonhardt, P. Walter, T. Klecker et al., Cristae-dependent quality control of the mitochondrial genome. Sci. Adv. **7**, eabi8886 (2021)
157. H. Hoitzing, I.G. Johnston, N.S. Jones, What is the function of mitochondrial networks? A theoretical assessment of hypotheses and proposal for future research. BioEssays **37**, 687–700 (2015)
158. J.R. Friedman, J.R. DiBenedetto, M. West, A.A. Rowland, G.K. Voeltz, Endoplasmic reticulum-endosome contact increases as endosomes traffic and mature. Mol. Biol. Cell **24**, 1030–1040 (2013)
159. M.S. Tikhomirova, A. Kadosh, A.J. Saukko-Paavola, T. Shemesh, R.W. Klemm, A role for endoplasmic reticulum dynamics in the cellular distribution of microtubules. P. Natl. Acad. Sci. **119**, e2104309119 (2022)
160. H.T. Perkins, V.J. Allan, T.A. Waigh, Network organisation and the dynamics of tubules in the endoplasmic reticulum. Sci. Rep. **11**, 16230 (2021)

161. C. Raiborg, E.M. Wenzel, H. Stenmark, ER-endosome contact sites: molecular compositions and functions. EMBO J. **34**, 1848–1858 (2015)
162. S. Aoyama-Ishiwatari, Y. Hirabayashi, Endoplasmic reticulum-mitochondria contact sites-emerging intracellular signaling hubs. Front. Cell Dev. Biol. **9**, 1040 (2021)
163. N. Zamponi, E. Zamponi, S.A. Cannas, O.V. Billoni, P.R. Helguera, D.R. Chialvo, Mitochondrial network complexity emerges from fission/fusion dynamics. Sci. Rep. **8**, 1–10 (2018)
164. A.S. Rambold, B. Kostelecky, N. Elia, J. Lippincott-Schwartz, Tubular network formation protects mitochondria from autophagosomal degradation during nutrient starvation. P. Natl. Acad. Sci. **108**, 10190–10195 (2011)
165. D. Tondera, S. Grandemange, A. Jourdain, M. Karbowski, Y. Mattenberger, S. Herzig, S. Da Cruz, P. Clerc, I. Raschke, C. Merkwirth et al., SLP-2 is required for stress-induced mitochondrial hyperfusion. EMBO J. **28**, 1589–1600 (2009)
166. M. Liesa, O.S. Shirihai, Mitochondrial dynamics in the regulation of nutrient utilization and energy expenditure. Cell Metab. **17**, 491–506 (2013)
167. D.S. Grebenkov, L. Tupikina, Heterogeneous continuous-time random walks. Phys. Rev. E **97**, 012148 (2018)
168. Z.C. Scott, A.I. Brown, S.S. Mogre, L.M. Westrate, E.F. Koslover, Diffusive search and trajectories on tubular networks: a propagator approach. Eur. Phys. J. E **44**, 80 (2021)
169. N. Masuda, M.A. Porter, R. Lambiotte, Random walks and diffusion on networks. Phys. Rep. **716**, 1–58 (2017)
170. G. Gounaris, E. Katifori, Braess's paradox analog in physical networks of optimal exploration. Phys. Rev. Lett. **133**(6), 067401 (2024)
171. C.-H. Wu, S.-C. Lee, C.-W. Wang, Viral protein targeting to the cortical endoplasmic reticulum is required for cell-cell spreading in plants. J. Cell Biol. **193**, 521–535 (2011)
172. L. Daqing, K. Kosmidis, A. Bunde, S. Havlin, Dimension of spatially embedded networks. Nat. Phys. **7**, 481–484 (2011)
173. D. Stauffer, A. Aharony, *Introduction to Percolation Theory* (CRC Press, 1994)
174. M. Barthélemy, Spatial networks. Phys. Rep. **499**, 1–101 (2011)
175. C.C. Crapart, Z.C. Scott, T. Konno, A. Sharma, P. Parutto, D.M.D. Bailey, L.M. Westrate, E. Avezov, E.F. Koslover, Luminal transport through intact endoplasmic reticulum limits the magnitude of localized Ca2+ signals. Proc. Nat. Acad. Sci. **121**(13), e2312172121 (2024)
176. T. Konno, P. Parutto, C.C. Crapart, V. Davì, D.M.D. Bailey, M.A. Awadelkareem, C. Hockings et al., Endoplasmic reticulum morphology regulation by RTN4 modulates neuronal regeneration by curbing luminal transport. Cell Rep. **43**(7) (2024)
177. D. Holcman, P. Parutto, J.E. Chambers, M. Fantham, L.J. Young, S.J. Marciniak, C.F. Kaminski, D. Ron, E. Avezov, Single particle trajectories reveal active endoplasmic reticulum luminal flow. Nat. Cell Biol. **20**, 1118 (2018)
178. D. Reguera, J. Rubi, Kinetic equations for diffusion in the presence of entropic barriers. Phys. Rev. E **64**, 061106 (2001)
179. R. Zwanzig, Diffusion past an entropy barrier. J. Phys. Chem. **96**, 3926–3930 (1992)
180. L. Dagdug, A.M. Berezhkovskii, V.Y. Zitserman, S.M. Bezrukov, Effective diffusivity of a Brownian particle in a two-dimensional periodic channel of abruptly alternating width. Phys. Rev. E **103**, 062106 (2021)
181. G.A. Ascoli, D.E. Donohue, M. Halavi, NeuroMorpho. Org: a central resource for neuronal morphologies. J. Neurosci. **27**, 9247–9251 (2007)
182. Y. Fonkeu, N. Kraynyukova, A.-S. Hafner, L. Kochen, F. Sartori, E.M. Schuman, T. Tchumatchenko, How mRNA localization and protein synthesis sites influence dendritic protein distribution and dynamics. Neuron **103**, 1109–1122 (2019)
183. C.E. Holt, K.C. Martin, E.M. Schuman, Local translation in neurons: visualization and function. Nat. Struct. Mol. Biol. **26**, 557–566 (2019)
184. F. Sartori, A.-S. Hafner, A. Karimi, A. Nold, Y. Fonkeu, E.M. Schuman, T. Tchumatchenko, Statistical laws of protein motion in neuronal dendritic trees. Cell Rep. **33** (2020)

185. A.H. Williams, C. O'donnell, T.J. Sejnowski, T. O'leary, Dendritic trafficking faces physiologically critical speed-precision tradeoffs. eLife **5**, e20556 (2016)
186. R. Jose, L. Santen, M.R. Shaebani, Trapping in and escape from branched structures of neuronal dendrites. Biophys. J. **115**, 2014–2025 (2018)
187. E.J. Donovan, A. Agrawal, N. Liberman, J.I. Kalai, A.J. Adler, A.M. Lamper, H.Q. Wang, N.J. Chua, E.F. Koslover, E.L. Barnhart, Dendrite architecture determines mitochondrial distribution patterns in vivo. Cell Rep. **43**(5) (2024)
188. V.A. Kulkarni, B.L. Firestein, The dendritic tree and brain disorders. Mol. Cell. Neurosci. **50**, 10–20 (2012)
189. A. Agrawal, E.F. Koslover, Optimizing mitochondrial maintenance in extended neuronal projections. PLoS Comput. Biol. **17**, e1009073 (2021)
190. A.C. Horton, B. Rácz, E.E. Monson, A.L. Lin, R.J. Weinberg, M.D. Ehlers, Polarized secretory trafficking directs cargo for asymmetric dendrite growth and morphogenesis. Neuron **48**, 757–771 (2005)
191. E.M. Petrini, J. Lu, L. Cognet, B. Lounis, M.D. Ehlers, D. Choquet, Endocytic trafficking and recycling maintain a pool of mobile surface AMPA receptors required for synaptic potentiation. Neuron **63**, 92–105 (2009)
192. M. Ferrante, M. Migliore, G.A. Ascoli, Functional impact of dendritic branch-point morphology. J. Neurosci. **33**, 2156–2165 (2013)
193. M. Liao, X. Liang, J. Howard, The narrowing of dendrite branches across nodes follows a well-defined scaling law. Proc. Natl. Acad. Sci. **118**, e2022395118 (2021)
194. D.B. Chklovskii, A. Stepanyants, Power-law for axon diameters at branch point. BMC Neurosci. **4**, 1–7 (2003)
195. S. Redner, *A Guide to First-passage Processes* (Cambridge University Press, Cambridge, UK, 2001)
196. Y. Lin, B. Wu, Z. Zhang, Determining mean first-passage time on a class of treelike regular fractals. Phys. Rev. E **82**, 031140 (2010)
197. Y.-P. Förster, L. Gamberi, E. Tzanis, P. Vivo, A. Annibale, Exact and approximate mean first passage times on trees and other necklace structures: a local equilibrium approach. J. Phys. A-Math. Theor. **55**, 115001 (2022)
198. D.L. Dickstein, C.M. Weaver, J.I. Luebke, P.R. Hof, Dendritic spine changes associated with normal aging. Neuroscience **251**, 21–32 (2013)
199. J. Tsai, J. Grutzendler, K. Duff, W.-B. Gan, Fibrillar amyloid deposition leads to local synaptic abnormalities and breakage of neuronal branches. Nat. Neurosci. **7**, 1181–1183 (2004)
200. C. Sala, M. Segal, Dendritic spines: the locus of structural and functional plasticity. Physiol. Rev. **94**, 141–188 (2014)
201. P. Bressloff, B. Earnshaw, Diffusion-trapping model of receptor trafficking in dendrites. Phys. Rev. E **75**, 041915 (2007)
202. P.C. Bressloff, B.A. Earnshaw, M.J. Ward, Diffusion of protein receptors on a cylindrical dendritic membrane with partially absorbing traps. SIAM J. Appl. Math. **68**, 1223–1246 (2008)
203. F. Santamaria, S. Wils, E. De Schutter, G.J. Augustine, Anomalous diffusion in Purkinje cell dendrites caused by spines. Neuron **52**, 635–648 (2006)
204. F. Santamaria, S. Wils, E. De Schutter, G.J. Augustine, The diffusional properties of dendrites depend on the density of dendritic spines. Eur. J. Neurosci. **34**, 561–568 (2011)
205. C.M. Simon, I. Hepburn, W. Chen, E. De Schutter, The role of dendritic spine morphology in the compartmentalization and delivery of surface receptors. J. Comput. Neurosci. **36**, 483–497 (2014)
206. A. Biess, E. Korkotian, D. Holcman, Diffusion in a dendritic spine: the role of geometry. Phys. Rev. E **76**, 021922 (2007)
207. D. Wales, *Energy Landscapes: Applications to Clusters, Biomolecules and Glasses*. Cambridge Molecular Science (Cambridge University Press, Cambridge, UK, 2003)
208. D.J. Wales, Exploring energy landscapes. Annu. Rev. Phys. Chem. **69**, 401–425 (2018)

209. J.N. Onuchic, Z. Luthey-Schulten, P.G. Wolynes, Theory of protein folding: the energy landscape perspective. Annu. Rev. Phys. Chem. **48**, 545–600 (1997)
210. D.J. Wales, J.P. Doye, M.A. Miller, P.N. Mortenson, T.R. Walsh, Energy landscapes: from clusters to biomolecules. Adv. Chem. Phys. **115**, 1–111 (2000)
211. P. Hänggi, P. Talkner, M. Borkovec, Reaction-rate theory: fifty years after Kramers. Rev. Mod. Phys. **62**, 251 (1990)
212. H.A. Kramers, Brownian motion in a field of force and the diffusion model of chemical reactions. Physica **7**, 284–304 (1940)
213. J.D. Chodera, F. Noé, Markov state models of biomolecular conformational dynamics. Curr. Opin. Struc. Biol. **25**, 135–144 (2014)
214. J.-H. Prinz, H. Wu, M. Sarich, B. Keller, M. Senne, M. Held, J.D. Chodera, C. Schütte, F. Noé, Markov models of molecular kinetics: generation and validation. J. Chem. Phys. **134** (2011)
215. P.E. Leopold, M. Montal, J.N. Onuchic, Protein folding funnels: a kinetic approach to the sequence-structure relationship. Proc. Natl. Acad. Sci. **89**, 8721–8725 (1992)
216. H.J. Hwang, R.A. Riggleman, J.C. Crocker, Understanding soft glassy materials using an energy landscape approach. Nat. Mater. **15**, 1031–1036 (2016)
217. X. Hu, L. Hong, M. Dean Smith, T. Neusius, X. Cheng, J.C. Smith, The dynamics of single protein molecules is non-equilibrium and self-similar over thirteen decades in time. Nat. Phys. **12**, 171–174 (2016)
218. E. Lieberman-Aiden, N.L. Van Berkum, L. Williams, M. Imakaev, T. Ragoczy, A. Telling, I. Amit, B.R. Lajoie, P.J. Sabo, M.O. Dorschner et al., Comprehensive mapping of long-range interactions reveals folding principles of the human genome. Science **326**, 289–293 (2009)
219. T. Ando, J. Skolnick, Crowding and hydrodynamic interactions likely dominate in vivo macromolecular motion. P. Natl. Acad. Sci. **107**, 18457–18462 (2010)
220. N.J.L. Alamilla, I.S. Holek, Reconstructing the free-energy landscape associated to molecular motors processivity. Biophys. Chem. **167**, 16–25 (2012)
221. A.S. Lyon, W.B. Peeples, M.K. Rosen, A framework for understanding the functions of biomolecular condensates across scales. Nat. Rev. Mol. Cell Bio. **22**, 215–235 (2021)
222. C.P. Brangwynne, C.R. Eckmann, D.S. Courson, A. Rybarska, C. Hoege, J. Gharakhani, F. Jülicher, A.A. Hyman, Germline P granules are liquid droplets that localize by controlled dissolution/condensation. Science **324**, 1729–1732 (2009)
223. A.A. Hyman, C.A. Weber, F. Jülicher, Liquid-liquid phase separation in biology. Annu Rev Cell Dev Bi **30**, 39–58 (2014)
224. C.P. Brangwynne, P. Tompa, R.V. Pappu, Polymer physics of intracellular phase transitions. Nat. Phys. **11**, 899–904 (2015)
225. S.A. Shelby, S.L. Veatch, The membrane phase transition gives rise to responsive plasma membrane structure and function. Cold Spring Harb. Perspect Biol. a041395 (2023)
226. J. Miné-Hattab, A. Taddei, Physical principles and functional consequences of nuclear compartmentalization in budding yeast. Curr. Opin. Cell Biol. **58**, 105–113 (2019)
227. M. Prouteau, R. Loewith, Regulation of cellular metabolism through phase separation of enzymes. Biomolecules **8**, 160 (2018)
228. S. Alberti, A.A. Hyman, Biomolecular condensates at the nexus of cellular stress, protein aggregation disease and ageing. Nat. Rev. Mol. Cell Bio. **22**, 196–213 (2021)
229. M. Boehning, C. Dugast-Darzacq, M. Rankovic, A.S. Hansen, T. Yu, H. Marie-Nelly, D.T. McSwiggen, G. Kokic, G.M. Dailey, P. Cramer et al., RNA polymerase II clustering through carboxy-terminal domain phase separation. Nat. Struct. Mol. Biol. **25**, 833–840 (2018)
230. Y.-H. Wang, T.L. Ho, A. Hariharan, H.C. Goh, Y.L. Wong, N.S. Verkaik, M.Y. Lee, W.L. Tam, D.C. van Gent, A.R. Venkitaraman et al., Rapid recruitment of p53 to DNA damage sites directs DNA repair choice and integrity. P. Natl. Acad. Sci. **119**, e2113233119 (2022)
231. M.S. Heltberg, A. Lucchetti, F.-S. Hsieh, D.P.M. Nguyen, S.-H. Chen, M.H. Jensen, Enhanced DNA repair through droplet formation and p53 oscillations. Cell **185**, 4394–4408 (2022)
232. G. Santulli, R. Nakashima, Q. Yuan, A.R. Marks, Intracellular calcium release channels: an update. J. Physiol. **595**, 3041–3051 (2017)

233. K. Ribbeck, D. Görlich, Kinetic analysis of translocation through nuclear pore complexes. EMBO J. **20**, 1320–1330 (2001)
234. S. Frey, D. Görlich, A saturated FG-repeat hydrogel can reproduce the permeability properties of nuclear pore complexes. Cell **130**, 512–523 (2007)
235. L. Maguire, M. Stefferson, M.D. Betterton, L.E. Hough, Design principles of selective transport through biopolymer barriers. Phys. Rev. E **100**, 042414 (2019)
236. B. Fogelson, J.P. Keener, Transport facilitated by rapid binding to elastic tethers. SIAM J. Appl. Math. **79**, 1405–1422 (2019)
237. O. Kahraman, P.D. Koch, W.S. Klug, C.A. Haselwandter, Bilayer-thickness-mediated interactions between integral membrane proteins. Phys. Rev. E **93**, 042410 (2016)
238. C.A. Haselwandter, N.S. Wingreen, The role of membrane-mediated interactions in the assembly and architecture of chemoreceptor lattices. PLoS Comput. Biol. **10**, e1003932 (2014)
239. A. Shrestha, O. Kahraman, C.A. Haselwandter, Mechanochemical coupling of lipid organization and protein function through membrane thickness deformations. Phys. Rev. E **105**, 054410 (2022)
240. E.F. Koslover, A.J. Spakowitz, Twist-and tension-mediated elastic coupling between DNA-binding proteins. Phys. Rev. Lett. **102**, 178102 (2009)
241. R. Shlomovitz, N. Gov, Membrane-mediated interactions drive the condensation and coalescence of FtsZ rings. Phys. Biol. **6**, 046017 (2009)
242. M. Simunovic, A. Srivastava, G.A. Voth, Linear aggregation of proteins on the membrane as a prelude to membrane remodeling. P. Natl. Acad. Sci. **110**, 20396–20401 (2013)
243. M. Simunovic, P. Bassereau, G.A. Voth, Organizing membrane-curving proteins: the emerging dynamical picture. Curr. Opin. Struc. Biol. **51**, 99–105 (2018)
244. J. Rudnick, R. Bruinsma, DNA-protein cooperative binding through variable-range elastic coupling. Biophys. J. **76**, 1725–1733 (1999)
245. S. Kim, E. Broströmer, D. Xing, J. Jin, S. Chong, H. Ge, S. Wang, C. Gu, L. Yang, Y.Q. Gao et al., Probing allostery through DNA. Science **339**, 816–819 (2013)
246. K.J. Verhey, R. Ohi, Causes, costs and consequences of kinesin motors communicating through the microtubule lattice. J Cell. Sci. **136**, jcs260735 (2023)
247. E. Muto, H. Sakai, K. Kaseda, Long-range cooperative binding of kinesin to a microtubule in the presence of ATP. J. Cell Biol. **168**, 691–696 (2005)
248. R.A. Cross, Microtubule lattice plasticity. Curr. Opin. Cell Biol. **56**, 88–93 (2019)
249. S.S. Wijeratne, S.A. Fiorenza, A.E. Neary, R. Subramanian, M.D. Betterton, Motor guidance by long-range communication on the microtubule highway. P. Natl. Acad. Sci. **119**, e2120193119 (2022)
250. K. Tokuraku, M. Kuragano, T.Q. Uyeda, Long-range and directional allostery of actin filaments plays important roles in various cellular activities. Int. J. Mol. Sci. **21**, 3209 (2020)
251. D. Holcman, Z. Schuss, Control of flux by narrow passages and hidden targets in cellular biology. Rep. Prog. Phys. **76**, 074601 (2013)
252. J.M. Newby, P.C. Bressloff, Directed intermittent search for a hidden target on a dendritic tree. Phys. Rev. E **80**, 021913 (2009)
253. A.M. Berezhkovskii, A.V. Barzykin, Search for a small hole in a cavity wall by intermittent bulk and surface diffusion. J. Chem. Phys. **136** (2012)
254. Z. Schuss, K. Basnayake, D. Holcman, Redundancy principle and the role of extreme statistics in molecular and cellular biology. Phys. Life Rev. **28**, 52–79 (2019)
255. C. Lin, Y. Zhang, I. Sparkes, P. Ashwin, Structure and dynamics of ER: minimal networks and biophysical constraints. Biophys. J. **107**, 763–772 (2014)
256. L. Stadler, K. Speckner, M. Weiss, Diffusion of exit sites on the endoplasmic reticulum: a random walk on a shivering backbone. Biophys. J. **115**, 1552–1560 (2018)
257. M. Picard, C. Sandi, The social nature of mitochondria: implications for human health. Neurosci. & Biobehav. Rev. **120**, 595–610 (2021)
258. J.M. Chustecki, D.J. Gibbs, G.W. Bassel, I.G. Johnston, Network analysis of Arabidopsis mitochondrial dynamics reveals a resolved tradeoff between physical distribution and social connectivity. Cell Syst. **12**, 419–431 (2021)

259. M.E. Quinlan, Cytoplasmic streaming in the Drosophila oocyte. Annu. Rev. Cell Dev. Bi **32**, 173–195 (2016)
260. R. Illukkumbura, T. Bland, N.W. Goehring, Patterning and polarization of cells by intracellular flows. Curr. Opin. Cell Biol. **62**, 123–134 (2020)
261. E.F. Koslover, C.K. Chan, J.A. Theriot, Cytoplasmic flow and mixing due to deformation of motile cells. Biophys. J. **113**, 2077–2087 (2017)
262. M. Mussel, K. Zeevy, H. Diamant, U. Nevo, Drag of the cytosol as a transport mechanism in neurons. Biophys. J. **106**, 2710–2719 (2014)

# Chapter 23
# How to Find Targets That Are Always Hidden: Nucleosome-Covered DNA and Pioneer Transcription Factors

**Anupam Mondal and Anatoly B. Kolomeisky**

**Abstract** Association of protein transcription factors (TFs) to specific DNA sites in promoter regions is the fundamental event that initiates gene regulation. However eukaryotic DNA is mostly covered by nucleosomes, creating obstacles for TFs to access their targets. Nature solved this problem by utilizing a special class of protein molecules, known as pioneer TFs, that are capable to enter chromatin structures and reach DNA sites hidden by nucleosomes. However, the microscopic origin of the underlying processes remains not well understood. In this chapter, we present a brief review of the current understanding of the molecular mechanisms of the target search by pioneer TFs. It is argued that it is not enough for TFs to explore only nucleosome breathing, and they might also need to directly displace the nucleosome native contacts, substituting them with equivalent contacts between nucleosomal DNA and pioneer TFs. Using this idea, a comprehensive discrete-state stochastic framework is developed to evaluate the protein target search phenomena on nucleosome-embedded DNA.

## 23.1 Introduction

The information stored in cellular DNA is crucial for the proper functioning and development of living organisms [1–3]. It is located in specific DNA segments, often referred to as "genes" or "target sites". These are small patches of nucleotide sequences that encode the synthesis of RNA molecules during transcription, which eventually leads to the synthesis of proteins during translation. The first step in all

A. Mondal
Department of Chemistry, Center for Theoretical Biological Physics, Rice University, Houston, TX 77005, USA
e-mail: am275@rice.edu

A. B. Kolomeisky (✉)
Department of Chemistry, Department of Chemical and Biomolecular Engineering, Department of Physics and Astronomy, Center for Theoretical Biological Physics, Rice University, Houston, TX 77005, USA
e-mail: tolya@rice.edu



D. Grebenkov et al. (eds.), *Target Search Problems*,
https://doi.org/10.1007/978-3-031-67802-8_23

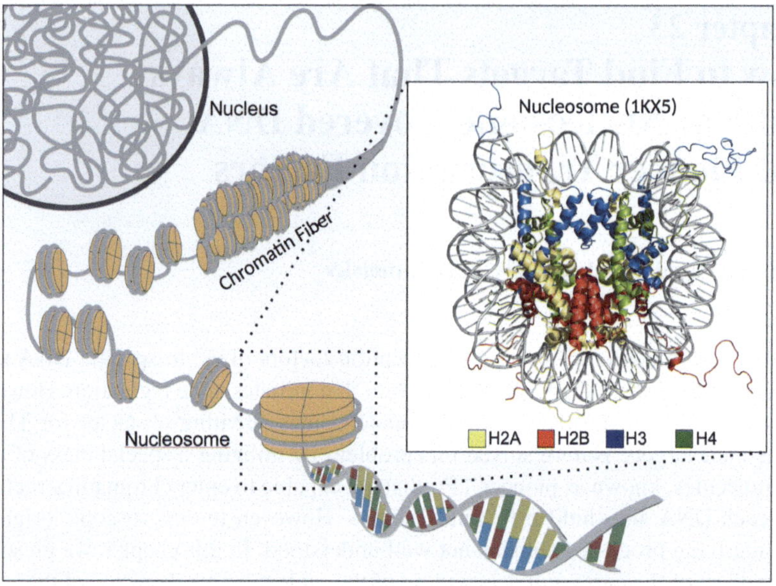


**Fig. 23.1** The genomic DNA is compactly packed in the form of a chromatin fiber inside a cell nucleus, where the fundamental building block of chromatin is the nucleosome (shown in the inset). A single nucleosome comprises ~150 bp of DNA wrapped around a histone octameric core protein. Copyright OIST (Okinawa Institute of Science and Technology Graduate University)

these processes, however, is locating the target sequences amidst a sea of non-specific DNA sites by several classes of protein molecules, which are known as transcription factors (TFs) [2, 4, 5]. The association of TFs to their specific targets might be complicated in eukaryotes because DNA is hierarchically packaged and compacted in the form of dense chromatin structures to fit them into small cellular volumes [2, 3] (see Fig. 23.1). The basic organizational unit of chromatin is the nucleosome, which is composed of an octameric histone core protein complex that contains two copies of the four histone core protein domains, H2A, H2B, H3, and H4, around which ~ 150 base pairs (bp) of DNA is tightly wrapped into ~1.65 superhelical turns [6] (Fig. 23.1). The structure of the nucleosome core particle has been resolved by X-ray crystallography with 2.8 Å resolution (PDB ID: 1KX5, see Fig. 23.1), revealing that such tight association of the nucleosomal DNA with histone proteins might severely restrain the accessibility of the target DNA sites for protein TFs [7, 8]. This raises a fundamental question of how cells manage to simultaneously fulfill their two most important functions of transcription regulation and compact genome organization.

Recent experimental studies [9, 10] have shown that nucleosome structures are highly mobile rather than static objects, and the histone–DNA contacts are often disrupted spontaneously to adopt alternative conformations that are different from the canonical crystal structure. This results in a highly dynamic landscape for the regulation of DNA target genes. One of the most important components of nucle-

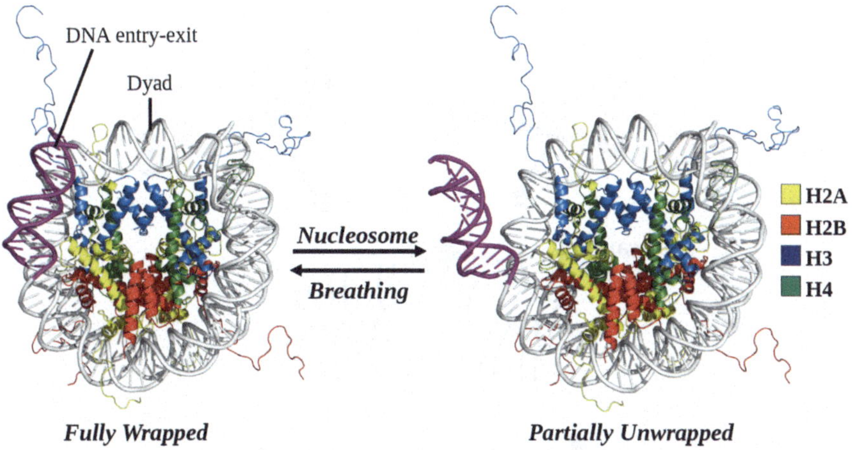


**Fig. 23.2** A stretch of DNA (∼15 bp of DNA, colored in magenta) in the fully wrapped (FW) nucleosome structure detaches from the histone surface at one entry-exit DNA site and forms a partially unwrapped (PU) nucleosome conformation. The reversible transitions between conformations with different degrees of unwrapping define the nucleosome breathing dynamics. Both the FW and PU nucleosome structures shown here are obtained from cryo-EM structures with PDB IDs: 6ESF and 6ESH, respectively. The missing histone residues are modeled using the crystal structure 1KX5. The nucleosomal DNA is shown in grey and the histone octamers are indicated by four different colors

osome transformations is the "breathing dynamics" or "site exposure" in which a segment of nucleosomal DNA can spontaneously unwrap and rewrap starting from one end in the entry-exit DNA site [10–15] (see Fig. 23.2), resulting in several different conformational states: fully wrapped (FW) and a variety of partially unwrapped (PU). A recent cryo-EM study has characterized distinct conformations of partially unwrapped structures with varying lengths of liberated DNA segments [10]. Explicit measurements of the wrap-unwrap kinetics of DNA at the nucleosome termini suggest that the transitioning between FW and PU conformations occur on a millisecond time scale under physiological conditions [12, 14]. This rapid partial unfolding of the nucleosomal DNA complex, therefore, provides a window of opportunity for TFs to efficiently access their target DNA sites located on the outer DNA stretches. However, the DNA sites that are located within the nucleosome interior region seemed to be not easily accessible, raising a question of how deeply buried genes that are almost always hidden can be activated.

Nature resolves this problem by utilizing a special class of proteins that are called pioneer TFs. They have the ability to efficiently bind their DNA target motifs, even in inaccessible regions, and subsequently create states for regulatory processes to proceed [16–20]. Several experimental studies suggested that pioneer TFs trigger diverse transcriptional programs [21], and this leads to opening silent chromatin regions [22, 23]. Direct interactions with nucleosomes through a winged-helix DNA binding domain of pioneer TF that resembles linker histone, probably allows the DNA

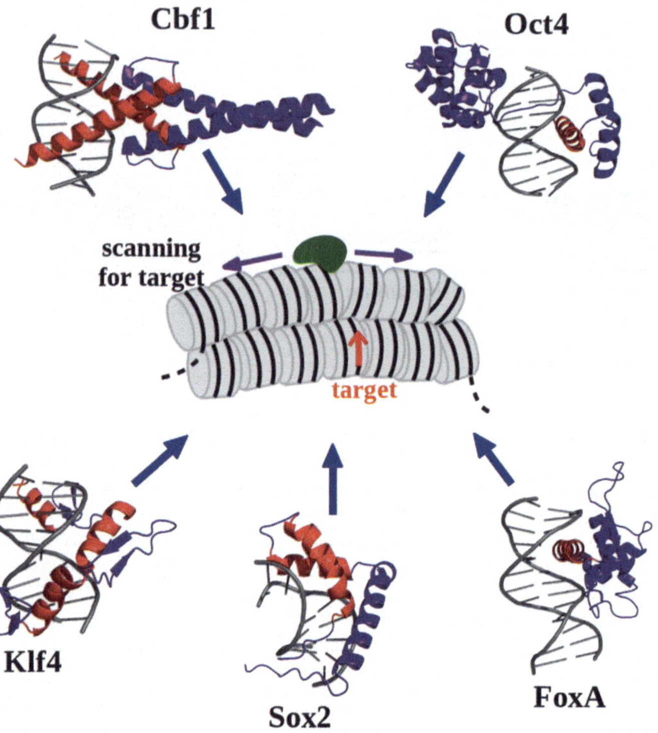


**Fig. 23.3** Schematic representation of the DNA target search by pioneer TFs on compact chromatin. Cartoon representations of the 3D structures of several pioneer TFs: Cbf1 (PDB-7SSA, partly shown), Klf4 (PDB-2WBU), Sox2 (PDB-1GT0), Oct4 (PDB-3L1P), and FoxA (PDB-1VTN) in complexes with DNA containing canonical motifs. The DNA binding domains of pioneer TFs are colored in red and the DNA motifs are shown in black colors

motifs to partially open and gain access to chromatin [24, 25]. Examples of pioneer TFs include Sox2, Oct4, Klf4, Cbf1, FoxA, and several other protein families, which are able to invade chromatin and activate target genes on nucleosome-covered DNA regions (see Fig. 23.3). In this connection, it is worth mentioning that for binding of single-stranded DNA binding proteins a similar kinetic regulation could be identified [26] that can be understood in terms of an effective free-energy picture [27]. The proper functioning of pioneer TFs, therefore, is essential because their malfunctioning can create defects in large-scale chromatin structures and compromise human health [17]. Indeed, the mutations in pioneer TFs are found in many forms of cancer [17, 28]. Despite their critical roles, the molecular mechanisms of the functioning of pioneer TFs remain poorly understood.

The target search associated with the non-pioneer (normal) TFs on naked DNA has been extensively investigated in the last 40 years by a wide spectrum of experimental and theoretical approaches [29–46]. Tremendous efforts have been devoted to understanding the microscopic details of how these proteins can rapidly find their target

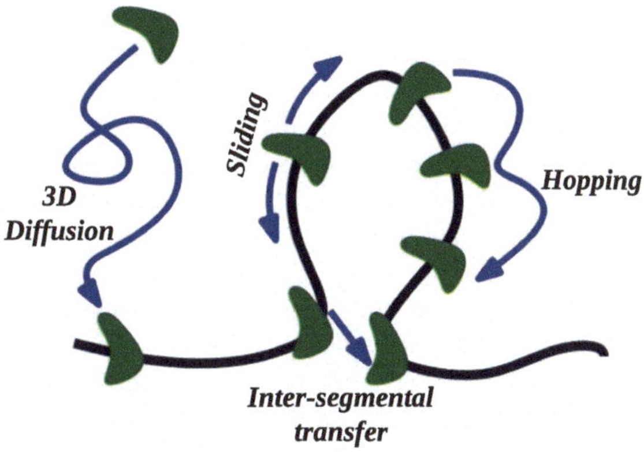


**Fig. 23.4** Schematic representation of the facilitated diffusion mechanism of target search on naked DNA. The protein molecule might adopt different dynamic modes: sliding, hopping, and inter-segmental transfer

sites among an astronomically large genomic background of non-specific DNA sites. It was suggested that normal TFs might accelerate their search by a combination of 3D diffusion in the bulk and 1D diffusion on the DNA, a mechanism known as "facilitated diffusion" [30, 35–37, 39, 40]. In this approach, normal TFs alternate between different modes of translocation for scanning their target sites as shown in Fig. 23.4. When attached non-specifically to DNA, TFs can perform sliding dynamics along the DNA contour during which they either stochastically move between adjacent DNA bases in forward and backward directions or rotate around the helical major grooves of DNA [31, 43] (see Fig. 23.4). When detached from DNA, they diffuse in the solution and then re-associate to a nearby DNA site after some time [43]. Furthermore, when the two distant DNA sites are nearby, proteins can jump from one segment to another non-contiguous segment through inter-segmental transfer [44, 47].

The crucial aspect of all the proposed mechanisms of target search is that the experiments to test them have been performed mostly under in vitro conditions where DNA molecules are not obstructed and considered free in solutions. However, the situation inside real in vivo cellular environment is much more complex since approximately three-fourths of all the eukaryotic DNA are bound by nucleosomes [48], apparently preventing other proteins from easily accessing their target sites. Although few studies have attempted to theoretically understand the target search phenomena of normal TFs in real cellular systems [40, 49–51], the microscopic details for the binding of pioneer TFs to their target sites on nucleosome-covered DNA segments are still not understood at all.

Recent experiments have shown that non-pioneer and pioneer TFs interact differently with naked and nucleosomal DNA [52]. More specifically, it has been shown that while the association rates are similar for both normal and pioneer TFs on both naked and nucleosome-covered DNA, the dissociation rates of pioneer TFs are much slower from the nucleosome-covered region compared to canonical non-pioneer TFs, while the situation is completely reversed for the unbinding dynamics from naked DNA segments. These observations stimulated several theoretical investigations that attempted to present a quantitative picture of how pioneer TFs find their targets on nucleosome [53, 54]. A specific model has been proposed [53], which is based on three different assumptions: (i) pioneer TFs can associate with closed nucleosomes, but normal TFs are unable to do that, (ii) both pioneer and normal TFs can reach their targets by sliding along the transiently opened DNA segments that appear due to nucleosome breathing motion and (iii) rewrapping of nucleosomal DNA displaces the normal TFs off into the solution, whereas pioneer TFs remain bound on the nucleosome surface and wait for the nucleosomal DNA to unwrap again and resume its search.

These theoretical studies were able to estimate the target search dynamics of pioneer TFs by utilizing experimentally measured transition rates. However, although they provided a reasonable microscopic picture of the binding of pioneer TF on exposed DNA segments, the obtained quantitative estimates of the target search times for deeply buried DNA sites are on the order of $10^5$ s or several hours [53], which are too slow for activating genes inside living cells. Experiments also suggested that unwrapping/rewrapping of nucleosomal DNA occurs with very high frequency [12, 14] (on the millisecond time scale), and the unwrapping probability decreases exponentially with the distance far away from the nucleosome ends [55]. This indicates that although the target sites that are close to nucleosome ends can be reached via this approach, the deeply buried DNA sites are not accessible by this mechanism.

In this chapter, we would like to briefly discuss a complementary theoretical method to explain the mechanism of target search processes by pioneer TFs. It is based on a recently proposed idea of an interaction-compensation mechanism in which the pioneer TF that comes to DNA covered by nucleosome weakens some hydrogen bonds between histone protein and DNA and replaces them with equivalent DNA-pioneer non-specific contacts [18, 56] (as illustrated in Fig. 23.5). The structural similarity between the DNA binding domain of pioneer TF and the linker histone-like protein stimulated this idea [18]. These interactions would allow the pioneer TFs to slide along the nucleosome-covered DNA regions more efficiently and it also helps to quickly find any DNA target sites covered by the nucleosome core particle. By exploring this idea, we develop a theoretical framework to analyze how pioneer TFs search for nucleosome-covered DNA sites through a discrete-state stochastic

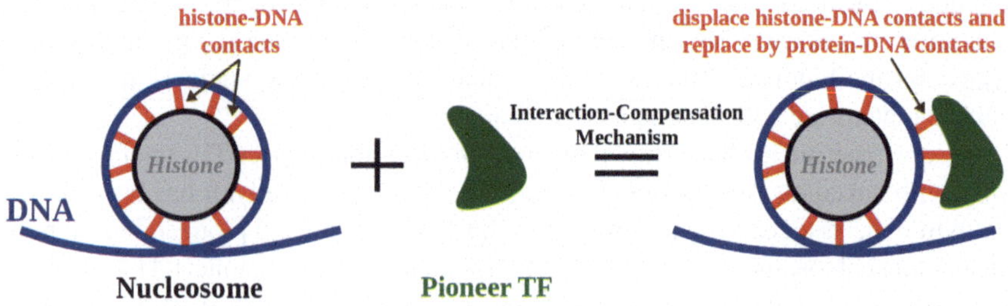


**Fig. 23.5** Schematic representation of the interaction-compensation mechanism in which the pioneer TF displaces some histone-DNA contacts and it substitutes them with similar DNA-Pioneer TF contacts

description that takes into account the most relevant physical-chemical processes in the system.

## 23.2 Discrete-State Stochastic Model of the Protein Target Search on Nucleosomal and Naked DNA

In order to understand how pioneer TFs utilize the interaction-compensation mechanism to invade nucleosome-covered DNA sites and how their search dynamics differ from normal TFs, we consider a system consisting of linker DNA segments attached to both ends of a single nucleosome particle, as shown in Fig. 23.6a. The nucleosomal DNA (grey-colored) is fully wrapped around the histone core octamer and linker DNAs (black-colored) are flanked around it. The DNA molecule is viewed as a lattice of $L$ sites, where each DNA site corresponds to about 10 bps. This is the average size of the region of protein-DNA interactions (effective target size). Out of a total of $L$ sites, the sites starting from 1 to $l - 1$ are the first linker region followed by nucleosome-covered DNA region of size $\Delta$ that starts from $n = l$ and ends at $n = l + \Delta - 1$, and the remaining sites are the second linker region (see Fig. 23.6b).

It is assumed that initially (at $t = 0$) the protein molecule starts from the bulk solution which is labeled as a state 0 (Fig. 23.6). From the solution, the protein molecule can associate with either linker DNA or nucleosome-covered DNA region. Still, the corresponding rates differ depending on which segment it will land. The association rate to any of the linker DNA segments is equal to $k_{on}$ (per unit site), while the corresponding association rate per unit site on the nucleosome-covered DNA region is $k'_{on}$ (see Fig. 23.6b). Similarly, the non-specifically bound protein dissociates into the solution with rates $k_{off}$ and $k'_{off}$ from linker DNA and nucleosome-covered DNA segments, respectively. In addition, the protein can slide along the linker DNA segment with rate $u$, while the sliding rate along the nucleosome-covered DNA region is equal to $u'$ (Fig. 23.6b). The target site is positioned at the site $m$ ($1 \leq m \leq L$), and the search process ends when the protein molecule reaches the target site for the first time.

The transition rates for binding and unbinding of pioneer and non-pioneer TFs have been measured experimentally for both naked and nucleosome-covered DNA, and we utilize them in our theoretical calculations [52, 57, 58]. It is found that from the solution the association rates to the nucleosome-covered DNA sites for both non-pioneer and pioneer TFs are comparable but typically low, i.e., $k'_{on}$ (non-pioneer) $\simeq k'_{on}$ (pioneer), whereas their binding rates on the naked DNA sites are comparable and much faster, i.e., $k_{on}$ (non-pioneer) $\simeq k_{on}$ (pioneer). This is easy to understand because the searching protein must separate the nucleosome-covered part from the DNA, and this requires energy. Therefore, it is easier to associate with naked DNA segments.

However, the dissociation rates of pioneer and normal (non-pioneer) TFs differ significantly. The non-pioneer TF quickly dissociates from the nucleosome-covered

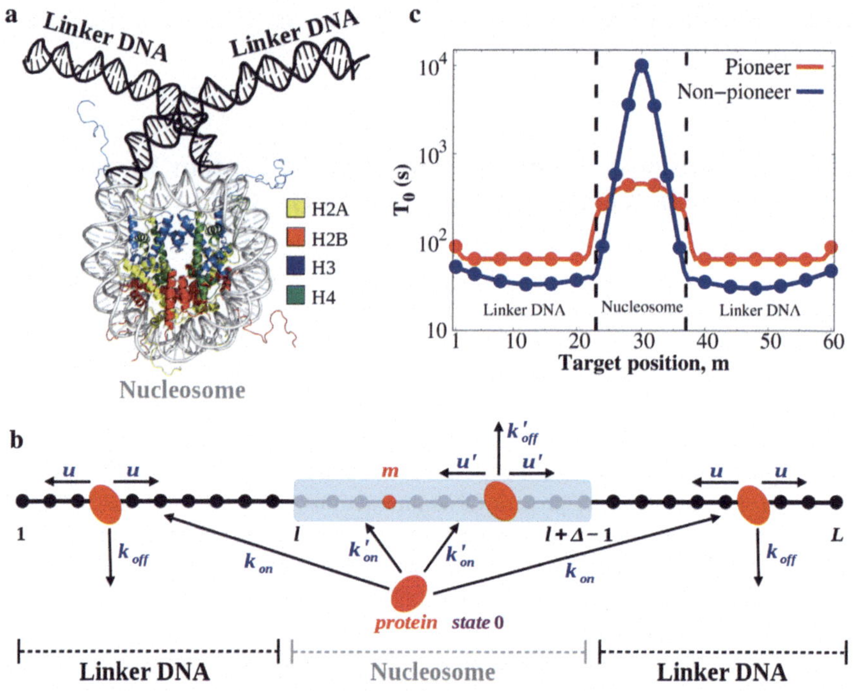


**Fig. 23.6** **a** Schematic view of the DNA-nucleosome system consisting of a single nucleosome particle with attached linker DNAs. **b** Discrete-state stochastic model of the target search on DNA. The nucleosome-covered DNA sites are grey-colored, and the remaining sites are for two linker DNA regions. All the possible transitions are indicated by arrows. The search process continues until the protein (red ellipse) finds its DNA target site (red circle) located at the site $m$. **c** Mean search times as a function of the target position along the DNA chain for pioneer and non-pioneer TFs. Analytical results are presented by solid lines and symbols are from Monte Carlo simulations. The error bars are defined as standard errors and the associated error bars are smaller than the symbols sizes. The following parameters were used in calculations: $L = 60$, $l = 23$, $\Delta = 15$, and all other transition rates are given in Table 23.1. Black vertical dashed lines are drawn at $m = 23$ and $m = 37$ to determine the boundaries of two linker DNA segments and the nucleosome region

DNA segments, but the pioneer TF stays much longer on covered DNA regions, i.e., $k'_{off}$ (non-pioneer) $\gg$ $k'_{off}$ (pioneer). This behavior can be explained by the structural properties of pioneer TFs that exhibit similarity with the linker histone-like domains [18] (see Fig. 23.3). Thus, pioneer TFs most probably interact with the nucleosome in the following way. While interacting with the closed nucleosome, the pioneer TF displaces some contacts between histone and DNA and substitutes them with its own DNA binding domain (see Fig. 23.5). These newly created contacts compensate for lost interactions with DNA. As a result, this interaction-compensation mechanism makes the non-specifically bound protein–DNA complex energetically more stable, explaining the low dissociation rates of pioneer TFs. A recent experimental study found that a basic helix-loop-helix motif of a budding yeast pioneer

**Table 23.1** Transition rates for non-pioneer and pioneer TFs adopted from Refs. [52, 57, 58]

| TF | $u$ (s$^{-1}$) | $u'$ (s$^{-1}$) | $k_{\text{on}}$ (s$^{-1}$) | $k'_{\text{on}}$ (s$^{-1}$) | $k_{\text{off}}$ (s$^{-1}$) | $k'_{\text{off}}$ (s$^{-1}$) |
|---|---|---|---|---|---|---|
| non-pioneer | 10 | 1 | $10^{-2}$ | $10^{-4}$ | $10^{-2}$ | 1 |
| pioneer | 1 | 10 | $10^{-2}$ | $10^{-4}$ | 1 | $10^{-2}$ |

TF enables efficient nucleosome invasion [59]. Such interactions, therefore, not only increase long residence times on nucleosomes for pioneer TFs but also allow them to successfully invade any nucleosome-covered DNA sites. Furthermore, the situation is reversed for the unbinding rates from naked DNA segments. The pioneer TF dissociates very quickly while the non-pioneer TF stays much longer, yielding $k_{off}$ (pioneer) $\gg k_{off}$ (non-pioneer). The possible explanation is that the associated linker histone-like domain present in the pioneer TF tends to lower the non-specific protein–DNA interactions, thus helping the protein to easily dissociate from the linker DNA segment.

In addition, it is postulated that the interaction-compensation mechanism enables the pioneer TFs to slide faster along the nucleosome-covered DNA segment compared to any non-pioneer TF (i.e., $u'$ (pioneer) $\gg u'$ (non-pioneer)), while the complete opposite behavior can be observed on free DNA segments, i.e., $u$ (non-pioneer) $\gg u$ (pioneer). This leads to different transition rates on naked and nucleosome-covered DNA segments for both pioneer and non-pioneer TF, as shown in Table 23.1.

To analyze the dynamics of the protein target search, we use a method of first-passage probabilities that have been successfully employed to understand the molecular mechanisms of protein-DNA interactions [4, 5, 38, 44]. For the model presented in Fig. 23.6b, one can introduce a function $F_n(t)$, which is defined as the probability for the protein to reach the target site $m$ at time $t$ for the first time if at $t = 0$ the protein starts at the site $n$ ($n = 1, 2, \ldots, L$). We also define $F_0(t)$ as the first-passage probability of the protein reaching the target site at time $t$, if at $t = 0$ the protein starts in the bulk solution (state 0). To evaluate these probabilities, backward master equations that describe the temporal evolution of these quantities can be utilized [60],

$$\frac{dF_n(t)}{dt} = u\big[F_{n-1}(t) + F_{n+1}(t)\big] + k_{off}F_0(t) - (2u + k_{off})F_n(t) \tag{23.1}$$

for linker DNA regions with $1 < n < l$ or $l + \Delta - 1 < n < L$, and

$$\frac{dF_n(t)}{dt} = u'\big[F_{n-1}(t) + F_{n+1}(t)\big] + k'_{off}F_0(t) - (2u' + k'_{off})F_n(t) \tag{23.2}$$

for the nucleosome-covered DNA region with $l \leq n \leq l + \Delta - 1$. The physical interpretation of these two expressions is as follows: The first two terms on the right-hand side depict the probabilities of the protein moving unbiasedly from site $n$ to the adjacent sites $n - 1$ and $n + 1$ with a 1D diffusion rate $u$ (Eq. 23.1) or $u'$ (Eq. 23.2). The third term encompasses the possibility of the protein diffusing into the bulk solution from site $n$ with a rate $k_{off}$ (Eq. 23.1) or $k'_{off}$ (Eq. 23.2). The reverse mechanism, where the protein is equally likely to reach site $n$ from its adjacent sites and bulk solution, is represented by the last term, and the associated rate is $-(2u + k_{off})$ (for Eq. 23.1) or $-(2u' + k'_{off})$ (for Eq. 23.2). It is important to note that despite the widespread use of an alternative approach, namely the forward Chapman-Kolmogorov equation, which describes measurable quantities directly as functions of the measured (real) time, employing the backward master equation for modeling the target search dynamics of proteins might be more advantageous. This is because the backward approach reconstructs past events and is preferentially used to predict the first-passage times of reaching the target DNA site. Furthermore, it might better correspond to experiments that measure these processes. Additional details can be found elsewhere [61].

The dynamic processes associated with the boundaries of the DNA chain ($n = 1$ and $n = L$) are slightly different, leading to

$$\frac{dF_1(t)}{dt} = uF_2(t) + k_{off}F_0(t) - (u + k_{off})F_1(t), \tag{23.3}$$

$$\frac{dF_L(t)}{dt} = uF_{L-1}(t) + k_{off}F_0(t) - (u + k_{off})F_L(t). \tag{23.4}$$

In addition, the dynamics in the bulk solution ($n = 0$) is given by

$$\frac{dF_0(t)}{dt} = k_{on}\sum_{n=1}^{l-1} F_n(t) + k'_{on}\sum_{n=l}^{l+\Delta-1} F_n(t) + k_{on}\sum_{n=l+\Delta}^{L} F_n(t) - \left[(L-\Delta)k_{on} + \Delta k'_{on}\right]F_0(t). \tag{23.5}$$

Finally, the initial condition is $F_n(t) = \delta_{n,m}(t) = \begin{cases} 0, & \text{if } n \neq m \\ 1, & \text{if } n = m \end{cases}$, which means that if initially (at $t = 0$) the searching protein is already positioned at the target site $m$, then the search process ends immediately.

The evolution equations for probabilities can be solved explicitly by using Laplace transformations, $\widetilde{F_n}(s) \equiv \int_0^\infty e^{-st}F_n(t)dt$, leading to simpler algebraic expressions,

$$(s + 2u + k_{off})\widetilde{F_n}(s) = u\left[\widetilde{F_{n-1}}(s) + \widetilde{F_{n+1}}(s)\right] + k_{off}\widetilde{F_0}(s), \tag{23.6}$$

$$(s + 2u' + k'_{off})\widetilde{F_n}(s) = u'\left[\widetilde{F_{n-1}}(s) + \widetilde{F_{n+1}}(s)\right] + k'_{off}\widetilde{F_0}(s), \tag{23.7}$$

$$(s + u + k_{off})\widetilde{F_1}(s) = u\widetilde{F_2}(s) + k_{off}\widetilde{F_0}(s), \tag{23.8}$$

$$(s + u + k_{off})\widetilde{F_L}(s) = u\widetilde{F_{L-1}}(s) + k_{off}\widetilde{F_0}(s), \tag{23.9}$$

$$\left(s + (L - \Delta)k_{on} + \Delta k'_{on}\right)\widetilde{F}_0(s) = k_{on}\sum_{n=1}^{l-1}\widetilde{F}_n(s) + k'_{on}\sum_{n=l}^{l+\Delta-1}\widetilde{F}_n(s) + k_{on}\sum_{n=l+\Delta}^{L}\widetilde{F}_n(s), \tag{23.10}$$

with the initial condition $\widetilde{F}_m(s) = 1$ (Laplace transformation of the Kronecker delta function is equal to 1). These equations can be solved exactly for all ranges of parameters by looking for solutions in the form

$$\widetilde{F}_n(s) = A_1 y^n + A_2 y^{-n} + B, \tag{23.11}$$

where the parameters $A_1$ and $A_2$ can be obtained by utilizing the boundary conditions which produce complex expressions for $A_1, A_2$ and are not given here, but the explicit form of $y$ and $B$ are given by

$$y = \frac{(s + 2u + k_{off}) - \sqrt{(s + 2u + k_{off})^2 - 4u^2}}{2u}, \tag{23.12}$$

$$B = \frac{k_{off}}{s + k_{off}}\widetilde{F}_0(s), \tag{23.13}$$

if $n$ is the site in the linker DNA region, i.e., $1 \leq n \leq l - 1$ or $l + \Delta \leq n \leq L$; or

$$\widetilde{F}_n(s) = A'_1 x^n + A'_2 x^{-n} + B', \tag{23.14}$$

where the complex expressions of $A'_1$ and $A'_2$ can be determined from boundary conditions and the expressions for $x$ and $B'$ are given by,

$$x = \frac{(s + 2u' + k'_{off}) - \sqrt{(s + 2u' + k'_{off})^2 - 4u'^2}}{2u'} \tag{23.15}$$

$$B' = \frac{k'_{off}}{s + k'_{off}}\widetilde{F}_0(s), \tag{23.16}$$

if $n$ is the site in the nucleosome-covered region of the DNA, i.e., $l \leq n \leq l + \Delta - 1$. The final expressions for $\widetilde{F}_0(s)$ are quite bulky but can be written explicitly, providing explicit estimates for mean search times [53, 60],

$$T_0 \equiv \int_0^{\infty} t F_0(t) dt = -\left.\frac{\partial \widetilde{F}_0(s)}{\partial s}\right|_{s=0}. \tag{23.17}$$

## 23.3 Pioneer TFs Search Faster for Nucleosome-Covered DNA Target Sites

To understand the protein target search phenomena in real cellular environments, we selected the parameters that mimic realistic conditions in living cells. For instance, we choose the size of the nucleosome-covered DNA segment as $\Delta = 15$ because each DNA site corresponds to 10 bp in our theoretical model and ~150 bps of DNA is wrapped in a single nucleosome. In addition, the average length of the linker DNA between neighboring nucleosomes, known as nucleosome repeat length (NRL), is estimated to be close to 200 bps [62, 63]. We then set the total DNA length as $L = 60$, which approximately accommodates a single nucleosome with two linker DNA regions flanked around it, covering up to 600 bps of total DNA length.

Figure 23.6c presents the results of our calculations for target search times for pioneer and non-pioneer TFs, which are also supported by our Monte Carlo computer simulations. One can see that when the target site is located on the linker DNA regions ($m = 1, \ldots, l - 1$ or $m = l + \Delta, \ldots, L$), the non-pioneer TFs search faster than pioneer TFs. However, a completely opposite behavior is observed on the nucleosome-covered DNA segment ($m = l, \ldots, l + \Delta - 1$). In this region, despite the fact that the overall search times is higher for both types of TFs, the pioneer TFs find their target sites significantly faster than non-pioneer TFs. Interestingly, the theoretical predictions of the search times for pioneer TFs on nucleosome-covered segments (~10 min) are consistent with the recent kinetic measurements of search times for GAGA pioneer TF (GAF) in live cells [64], and this gives additional support to the proposed theoretical framework.

To quantify this effect of search facilitation, we define a dimensionless parameter

$$\alpha = \frac{T_{0,\text{ non-pioneer}}}{T_{0,\text{ pioneer}}} \tag{23.18}$$

which specifies how pioneer TFs are faster to locate target sites in comparison with non-pioneer TFs, and the result is shown in Fig. 23.7 with varying positions of the histone starting site and the target site. One can see that the acceleration in the search time for pioneer TF can be achieved when the target site is positioned on the region covered by the histone protein. Specifically, the larger effect in the search facilitation is observed for deeply buried DNA sites for which the acceleration can reach up to two times faster and the effect is stronger when the nucleosome is positioned at the end of the DNA chain. We argue that this is the result of the interaction-compensation mechanism which allows the pioneer TFs to explore a large fraction of the covered DNA site more efficiently than any other normal TFs. The main reason is that the lower dissociation rate of pioneer TFs on nucleosome-covered DNA segments allows these proteins to stay longer on DNA and the faster sliding rate permits them to invade nucleosome-covered targets more efficiently compared to non-pioneer TFs. It seems that nature optimized pioneer TFs for the specific task of searching on the nucleosomal DNA, but not on the naked DNA.

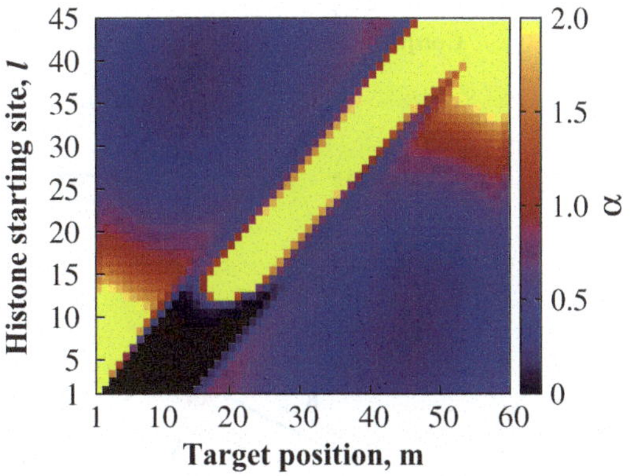


**Fig. 23.7** Acceleration of the search times on nucleosome-covered DNA region for the pioneer TFs in comparison with the non-pioneer TFs for varying positions of the core histone and the target site. The following parameters are used in calculations: $L = 60$, $\Delta = 15$ and all other transition rates are given in Table 23.1

## 23.4 The Role of Nucleosome Breathing Dynamics in Target Search

Now it is crucial to understand if adding the nucleosome breathing dynamics would also influence the target search dynamics. For this purpose, we extended the original discrete-state stochastic model by considering a system as shown in Fig. 23.8a. It is assumed that the nucleosome can be found in one of two major conformations: fully wrapped (FW) or partially unwrapped (PU). In the FW conformation, the DNA molecule (colored grey) is tightly wrapped around the histone protein (blue circle), whereas a DNA segment of length $l_B$ can dissociate off the histone surface in the PU conformation and the remaining $L - l_B$ DNA sites are still tightly bound to the nucleosome (see Fig. 23.8a). The transition rate to unwrap the part of the DNA chain is equal to $k_{open}$, while the rewrapping rate is $k_{close}$. These transition rates have already been measured ($k_{open} \simeq 210\ \text{s}^{-1}$ and $k_{close} \simeq 370\ \text{s}^{-1}$) in recent single-molecule experiments [14], and we utilize them in our calculations.

The reversible transitions between FW and PU conformations specify the nucleosome breathing dynamics in our theoretical model. It is important to note that we only consider asymmetric breathing, i.e., unwrapping/wrapping only from one nucleosome end, which is supported by multiple experimental observations [13, 65, 66]. The corresponding discrete-state stochastic model of the protein search dynamics is presented in Fig. 23.8b. In this description, the transition rates for the searching protein in the two nucleosome conformations are the same as described in Sect. 23.2 on both naked DNA and nucleosome-covered DNA segments. Note also that the search process within this framework is accomplished when the protein reaches the target site positioned at the site $m$ ($1 \leq m \leq L$) in either FW or PU conformations for the first time.

To explicitly analyze the target search dynamics coupled with nucleosome breathing, we define $F_n^{FW}(t)$ as the probability density function for the protein to reach the target site $m$ at time $t$ for the first time if at $t = 0$ the protein starts at the site $n$ (where $n = 1, 2, \ldots, L$) on the FW nucleosome conformation. A similar first-passage prob-

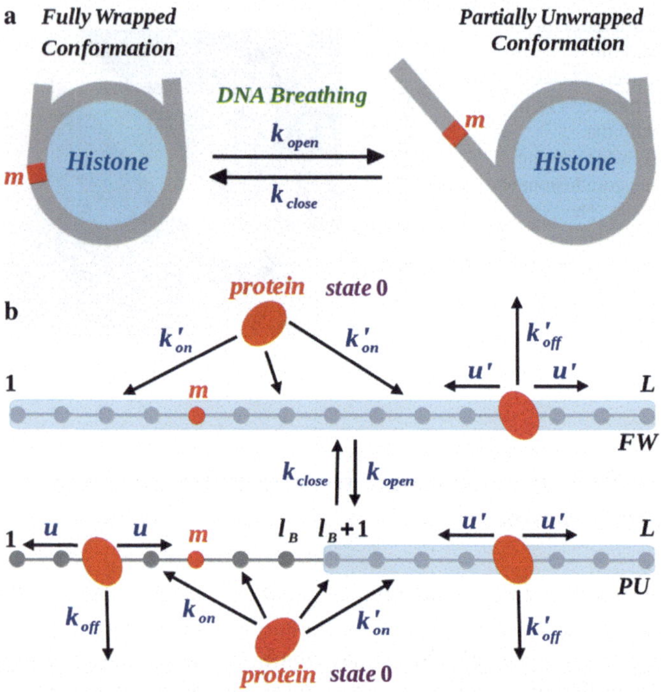


**Fig. 23.8** **a** Schematic view of the nucleosome breathing. The red patch denotes the location of the target at site $m$. **b** Discrete-state stochastic model of the protein target search coupled with nucleosome breathing. The transition rates between FW and PU conformations are given by $k_{open}$ and $k_{close}$ and all other transition rates are indicated by black arrows. Panel B is reprinted with permission from Ref. [67]. Copyright 2023 American Chemical Society

ability density function for the PU conformation can be specified as $F_n^{PU}(t)$. We also define $F_0^{FW}(t)$ $\left(F_0^{PU}(t)\right)$ as the first-passage probability density function of the protein to reach the target site at time $t$ if at $t = 0$ the protein starts in the surrounding bulk solution (state 0) and the nucleosome exists in any conformation. Then the temporal evolution of these probabilities can be described by a set of backward master equations [67],

$$\frac{dF_n^{FW}(t)}{dt} = u'\left[F_{n-1}^{FW}(t) + F_{n+1}^{FW}(t)\right] + k'_{off}F_0^{FW}(t) + k_{open}F_n^{PU}(t) - (2u' + k'_{off} + k_{open})F_n^{FW}(t), \quad (23.19)$$

for the FW conformation with $1 < n < L$; and

$$\frac{dF_n^{PU}(t)}{dt} = u\left[F_{n-1}^{PU}(t) + F_{n+1}^{PU}(t)\right] + k_{off}F_0^{PU}(t) + k_{close}F_n^{FW}(t) - (2u + k_{off} + k_{close})F_n^{PU}(t), \quad (23.20)$$

for the naked DNA segment in the PU conformation ($1 < n \leq l_B$); and

$$\frac{dF_n^{PU}(t)}{dt} = u'\left[F_{n-1}^{PU}(t) + F_{n+1}^{PU}(t)\right] + k'_{off} F_0^{PU}(t) + k_{close} F_n^{FW}(t) - (2u' + k'_{off} + k_{close}) F_n^{PU}(t), \quad (23.21)$$

for the wrapped segment in the PU conformation ($l_B < n < L$). The dynamics at the extremities of the DNA chain ($n = 1$ and $n = L$) in two nucleosome conformations are slightly different,

$$\frac{dF_1^{FW}(t)}{dt} = u' F_2^{FW}(t) + k'_{off} F_0^{FW}(t) + k_{open} F_1^{PU}(t) - (u' + k'_{off} + k_{open}) F_1^{FW}(t), \quad (23.22)$$

$$\frac{dF_L^{FW}(t)}{dt} = u' F_{L-1}^{FW}(t) + k'_{off} F_0^{FW}(t) + k_{open} F_L^{PU}(t) - (u' + k'_{off} + k_{open}) F_L^{FW}(t), \quad (23.23)$$

$$\frac{dF_1^{PU}(t)}{dt} = u F_2^{PU}(t) + k_{off} F_0^{PU}(t) + k_{close} F_1^{FW}(t) - (u + k_{off} + k_{close}) F_1^{PU}(t), \quad (23.24)$$

$$\frac{dF_L^{PU}(t)}{dt} = u' F_{L-1}^{PU}(t) + k'_{off} F_0^{PU}(t) + k_{close} F_L^{FW}(t) - (u' + k'_{off} + k_{close}) F_L^{PU}(t). \quad (23.25)$$

In addition, the search dynamics in the bulk solution ($n = 0$) are governed by

$$\frac{dF_0^{FW}(t)}{dt} = k_{open} F_0^{PU}(t) + k'_{on} \sum_{n=1}^{L} F_n^{FW}(t) - \left[k_{open} + L k'_{on}\right] F_0^{FW}(t), \quad (23.26)$$

and

$$\frac{dF_0^{PU}(t)}{dt} = k_{close} F_0^{FW}(t) + k_{on} \sum_{n=1}^{l_B} F_n^{PU}(t) + k'_{on} \sum_{n=l_B+1}^{L} F_n^{PU}(t) - \left[k_{close} + l_B k_{on} + (L - l_B) k'_{on}\right] F_0^{PU}(t). \quad (23.27)$$

Furthermore, the initial conditions are the following, $F_n^{FW}(t) = \delta_{n,m}(t)$ or $F_n^{PU}(t) = \delta_{n,m}(t)$, which means that if the searching protein is already at the target site $m$ at $t = 0$ in any of the nucleosome conformations (FW or PU), the search process ends immediately.

Following the similar procedures as described in Sect. 23.2, the above equations can be transformed into simple algebraic equations through Laplace transformations, $\left(\widetilde{F_n^{FW}}(s) \equiv \int_0^\infty e^{-st} F_n^{FW}(t)dt \text{ and } \widetilde{F_n^{PU}}(s) \equiv \int_0^\infty e^{-st} F_n^{PU}(t)dt\right)$, which can be exactly solved for all ranges of parameters. Because we are interested in the mean search times to locate the target site, when the protein starts the search process from the bulk solution with DNA in the FW conformation, the mean search times are given by

$$T_0^{FW} \equiv \int_0^\infty t F_0^{FW}(t)dt = -\frac{\partial \widetilde{F_0^{FW}}(s)}{\partial s}\Big|_{s=0}. \tag{23.28}$$

Similarly, starting from the bulk solution when DNA is in the PU conformation,

$$T_0^{PU} \equiv \int_0^\infty t F_0^{PU}(t)dt = -\frac{\partial \widetilde{F_0^{PU}}(s)}{\partial s}\Big|_{s=0}. \tag{23.29}$$

For our theoretical calculations, it is also convenient to consider the mean search time averaged over the two initial conformational states [67],

$$T_0 = \frac{k_{open}}{k_{open} + k_{close}} T_0^{PU} + \frac{k_{close}}{k_{open} + k_{close}} T_0^{FW}. \tag{23.30}$$

In Fig. 23.9, we present theoretical predictions for mean search times at different target locations, and the theoretical results are also supplemented with Monte Carlo computer simulations. For our specific calculations, we choose the length of the liberated DNA segment as $l_B = 7$ in the PU conformation, however, other values of $l_B$ do not alter the predictions of our theoretical analysis. Using experimentally determined transition rates, we compare the search dynamics of both non-pioneer

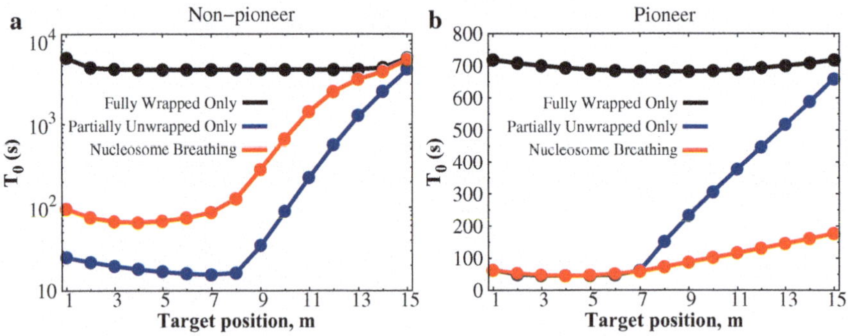


**Fig. 23.9** Mean search times as a function of target position for **a** non-pioneer and **b** pioneer TFs. Three different scenarios are considered: when the nucleosome is always fully wrapped (black curves), always partially unwrapped (blue curves) and the nucleosome is breathing (red curves). Solid lines are from analytical calculations and symbols are from Monte Carlo computer simulations. The following parameters are used in calculations: $L = 15$, $l_B = 7$, $k_{open} = 210$ s$^{-1}$, $k_{close} = 370$ s$^{-1}$ and all other transition rates are given in Table 23.1. The figure is reprinted with permission from Ref. [67]. Copyright 2023 American Chemical Society

and pioneer TFs for three different scenarios: when the nucleosome is always fully wrapped (black curves), always partially unwrapped (blue curves), and exhibiting breathing dynamics between these two conformations (red curves). One can see that non-pioneer TFs achieve the fastest target search when the DNA is always partially unwrapped from the histone surface and the slowest search is observed for always fully wrapped DNA (Fig. 23.9a). Considering the breathing dynamics in the system, the search time for non-pioneer TFs exhibits an expected intermediate behavior (Fig. 23.9a). However, a surprising result is observed for the mean search times of pioneer TFs (see Fig. 23.9b). Here the fastest search is observed for the DNA sites that are always hidden for the system with nucleosome breathing, while the search times coincide with the always partially unwrapped system for the DNA sites that fluctuate during the breathing event. This result is fully unexpected and it suggests that somehow pioneer TFs can take advantage of nucleosome breathing dynamics. The target search for sites hidden deep inside the nucleosome is facilitated by nucleosome breathing. Strikingly, our theoretical calculations for the search time of hidden DNA sites ($\sim$75 – 200 s) fully agree with the recent experimental evaluation of search times for GAF pioneer TFs in live *Drosophila* cells [64].

Our theoretical approach can also be utilized to evaluate if there is an optimal breathing dynamics that leads to the fastest search and if experimentally measured values of transition rates correspond to these conditions. The results of the calculations are presented in Fig. 23.10 together with experimental observations for wrapping/unwrapping transition rates. As one can see (Fig. 23.10b), the fastest search for pioneer TFs is achieved for fast breathing dynamics, i.e., for large $k_{open}$ and $k_{close}$, independently of the position of the target. This agrees well with the experimentally observed rates of nucleosome wrapping/unwrapping. However, for non-pioneer TFs

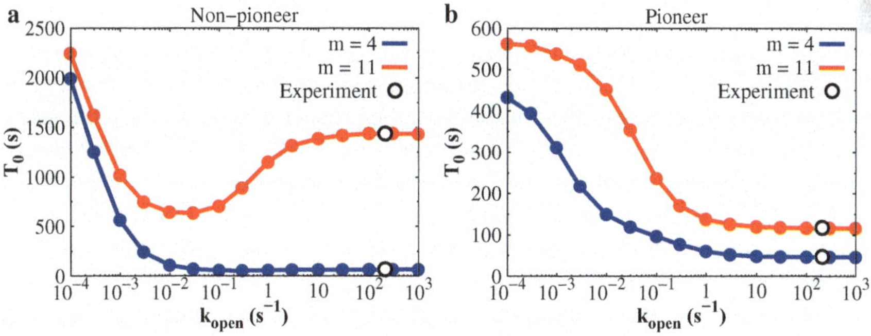


**Fig. 23.10** Mean search times as a function of the nucleosome unwrapping rate $k_{open}$ at two different target positions $m = 4$ and $m = 11$ for **a** non-pioneer and **b** pioneer TFs. While varying $k_{open}$, $k_{close}$ is changed according to the fixed equilibrium ratio $K_{eq} = \frac{k_{open}}{k_{close}} = 0.56$ as obtained in experimental measurements [14]. Solid lines are from analytical calculations and symbols are from Monte Carlo computer simulations. The black circle highlights the experimentally determined $k_{open}$ [14]. The following parameters are used in calculations: $L = 15$, $l_B = 7$ and all other transition rates are given in Table 23.1

(see Fig. 23.10a) larger opening/closing rates do not always correspond to the fastest target search. These arguments again emphasize that pioneer TFs explore nucleosome breathing to optimize their functioning.

To understand the microscopic picture behind the fast target search kinetics of pioneer TFs for deeply buried DNA sites, we present the following arguments. From the bulk solution, the dominating path for the pioneer TF is to land first on the liberated DNA segment in the PU conformation rather than the covered DNA region. This is because the association rate on the naked DNA segment ($k_{on} \simeq 10^{-2}$ s$^{-1}$) is much faster than the corresponding association rate on the nucleosome-covered DNA segment ($k'_{on} \simeq 10^{-4}$ s$^{-1}$, see Table 23.1). Once associated with the unwrapped segment, the protein molecules might enter the nucleosome-covered segment via sliding. Consequently, to reach the DNA sites further buried deep inside the nucleosome ($m = l_B + 1, \ldots, L$), the protein needs to walk towards the other end of the nucleosome. The pioneer TF performs this task very efficiently because of its very low dissociation rates ($k'_{off} \simeq 10^{-2}$ s$^{-1}$) as well as relatively fast sliding rates ($u' \simeq 10$ s$^{-1}$). This is a result of the interaction-compensation mechanism as explained above. However, a higher dissociation rate of the non-pioneer TF ($k'_{off} \simeq 1$ s$^{-1}$) and also slower sliding rates ($u' \simeq 1$ s$^{-1}$) stimulates more frequent detachments into the solution during the sliding. Every time it dissociates the non-pioneer factor has to re-enter the liberated segment again to access the desired buried sites. The effect is stronger when the target site is located at the end of the DNA chain. This explains the surprisingly large kinetic efficiency of pioneer TFs over the non-pioneer TFs for the targets especially located in the always histone-covered nucleosome regions.

## 23.5 Conclusions

A concept of an interaction-compensation mechanism has been introduced to explain the ability of pioneer TFs to enter chromatin regions that are heavily covered by nucleosomes. According to this mechanism, the pioneer TF can displace the native contacts between histone and DNA and replace them with similar contacts between DNA and pioneer TF. Utilizing this idea, we developed a discrete-state stochastic framework to analyze the protein target search on nucleosomal DNA. It takes into account all relevant physical-chemical processes in the system, allowing us to evaluate the mean search times via explicit first-passage probability calculations. From our analysis, several important dynamic features of the search dynamics of pioneer TFs have been identified. It is found that pioneer TFs significantly accelerate their search dynamics compared to normal TFs when the target sites are positioned on the nucleosome-covered regions, while their search dynamics on linker DNA regions are not as efficient. We also investigated the role of nucleosome breathing dynamics on the target search dynamics of pioneer TFs. By exploring a similar stochastic approach, it is shown that, surprisingly, the pioneer TFs might take advantage of nucleosome breathing dynamics and efficiently find those DNA sites that are always hidden inside the nucleosome. A possible microscopic picture to explain this effect is also proposed. It is argued that pioneer TFs mostly enter the nucleosome-covered region through the DNA segment that rapidly fluctuates between closed and open states during the breathing event. Then pioneer TFs utilize the interaction-compensation mechanism to efficiently reach any buried targets inside the nucleosome. This mechanism facilitates the search for hidden DNA targets by pioneer TFs several orders of magnitude faster than any non-pioneer TFs. Our theoretical analysis is fully tested and supported by extensive Monte Carlo computer simulations as well as recent experimental observations.

Although our presented theoretical framework provides a reasonable microscopic picture of invading nucleosome-covered DNA sites by pioneer TFs, it is important to discuss its limitations. In our theoretical model, we neglected the DNA sequence heterogeneity [42, 68, 69] and assumed uniform association, dissociation, and sliding rates for transcription factor search dynamics. Furthermore, the structural details of nucleosomal DNA [70] in its fully wrapped and partially unwrapped conformations are also not taken into account in our model. Despite these limitations, using experimentally derived rate constants (that implicitly consider the above-mentioned factors depending on the experimental conditions), our theoretical approach still provides a clear physical-chemical picture of complex regulatory processes that can be quantitatively tested using advanced experimental and theoretical techniques.

**Acknowledgements** The work was supported by the Welch Foundation (C-1559), by the NSF (CHE-2246878), and by the Center for Theoretical Biological Physics sponsored by the NSF (PHY-2019745).

## References

1. R.R. Sinden, *DNA Structure and Function* (Academic Press, San Diego USA, 1994)
2. R. Phillips, J. Kondev, J. Theriot, H. Garcia, *Physical Biology of the Cell* (Garland Science, Boca Raton USA, 2012)
3. H. Lodish, A. Berk, C.A. Kaiser, M. Krieger, M.P. Scott, A. Bretscher, H. Ploegh, P. Matsudaira, *Molecular Cell Biology* (W. H. Freeman, New York USA, 2007)
4. A.A. Shvets, M.P. Kochugaeva, A.B. Kolomeisky, Mechanisms of protein search for targets on DNA: theoretical insights. Molecules **23**, 2106 (2018)
5. J. Iwahara, A.B. Kolomeisky, Discrete-state stochastic kinetic models for target DNA search by proteins: theory and experimental applications. Biophys. Chem. **269**, 106521 (2021)
6. K. Luger, A.W. Mäder, R.K. Richmond, D.F. Sargent, T.J. Richmond, Crystal structure of the nucleosome core particle at 2.8 Å resolution. Nature **389**, 251 (1997)
7. C.T. Harbison, D.B. Gordon, T.I. Lee, N.J. Rinaldi, K.D. Macisaac, T.W. Danford, N.M. Hannett, J. Tagne, D.B. Reynolds, J. Yoo et al., Transcriptional regulatory code of a eukaryotic genome. Nature **431**, 99 (2004)
8. R. Pique-Regi, J.F. Degner, A.A. Pai, D.J. Gaffney, Y. Gilad, J.K. Pritchard, Accurate inference of transcription factor binding from DNA sequence and chromatin accessibility data. Genome Res. **21**, 447 (2011)
9. A. Miyagi, T. Ando, Y.L. Lyubchenko, Dynamics of nucleosomes assessed with time-lapse high-speed atomic force microscopy. Biochemistry **50**, 7901 (2011)

10. S. Bilokapic, M. Strauss, M. Halic, Histone octamer rearranges to adapt to DNA unwrapping. Nat. Struct. Mol. Biol. **25**, 101 (2018)
11. G. Li, J. Widom, Nucleosomes facilitate their own invasion. Nat. Struct. Mol. Biol. **11**, 763 (2004)
12. G. Li, M. Levitus, C. Bustamante, J. Widom, Rapid spontaneous accessibility of nucleosomal DNA. Nat. Struct. Mol. Biol. **12**, 46 (2005)
13. H.S. Tims, K. Gurunathan, M. Levitus, J. Widom, Dynamics of nucleosome invasion by DNA binding proteins. J. Mol. Biol. **411**, 430 (2011)
14. J. Kim, J. Lee, T.H. Lee, Lysine acetylation facilitates spontaneous DNA dynamics in the nucleosome. J. Phys. Chem. B **119**, 15001 (2015)
15. A. Mondal, S.K. Mishra, A. Bhattacherjee, Nucleosome breathing facilitates cooperative binding of pluripotency factors Sox2 and Oct4 to DNA. Biophys. J. **121**, 4526 (2022)
16. M. Iwafuchi-Doi, K.S. Zaret, Pioneer transcription factors in cell reprogramming. Genes Dev. **28**, 2679 (2014)
17. M. Iwafuchi-Doi, K.S. Zaret, Cell fate control by pioneer transcription factors. Development **143**, 1833 (2016)
18. K.S. Zaret, J.S. Carroll, Pioneer transcription factors: establishing competence for gene expression. Genes Dev. **25**, 2227 (2011)
19. K.S. Zaret, S.E. Mango, Pioneer transcription factors, chromatin dynamics, and cell fate control. Curr. Opin. Genet. Dev. **37**, 76 (2016)
20. M. Slattery, T. Zhou, L. Yang, A.C. Dantas Machado, R. Gordân, R. Rohs, Absence of a simple code: how transcription factors read the genome. Trends Biochem. Sci. **39**, 381 (2014)
21. K.S. Zaret, Pioneer transcription factors initiating gene network changes. Annu. Rev. Genet. **54**, 367 (2020)
22. L.A. Cirillo, C.E. McPherson, P. Bossard, K. Stevens, S. Cherian, E.Y. Shim, K.L. Clark, S.K. Burley, K.S. Zaret, Binding of the winged-helix transcription factor HNF3 to a linker histone site on the nucleosome. EMBO J. **17**, 244 (1998)
23. L.A. Cirillo, F.R. Lin, I. Cuesta, D. Friedman, M. Jarnik, K.S. Zaret, Opening of compacted chromatin by early developmental transcription factors HNF3 (FoxA) and GATA-4. Mol. Cell **9**, 279 (2002)
24. A. Soufi, M.F. Garcia, A. Jaroszewicz, N. Osman, M. Pellegrini, K.S. Zaret, Pioneer transcription factors target partial DNA motifs on nucleosomes to initiate reprogramming. Cell **161**, 555 (2015)
25. M. Iwafuchi-Doi, G. Donahue, A. Kakumanu, J.A. Watts, S. Mahony, B.F. Pugh, D. Lee, K.H. Kaestner, K.S. Zaret, The pioneer transcription factor FoxA maintains an accessible nucleosome configuration at enhancers for tissue-specific gene activation. Mol. Cell **62**, 79 (2016)
26. K. Pant, R.L. Karpel, M.C. Williams, Kinetic regulation of single DNA molecule denaturation by T4 gene 32 protein structural domains. J. Mol. Biol. **327**, 571 (2003)
27. T. Ambjörnsson, R. Metzler, Coupled dynamics of DNA breathing and of proteins that selectively bind to single-stranded DNA. Phys. Rev. E Stat. Nonlin. Soft Matter Phys. **72**, 030901 (2005)
28. K.M. Jozwik, J.S. Carroll, Pioneer factors in hormone-dependent cancers. Nat. Rev. Cancer **12**, 381 (2012)
29. O.G. Berg, R.B. Winter, P.H. von Hippel, Diffusion-driven mechanisms of protein translocation on nucleic acids. 1. Models and theory. Biochemistry **20**, 6929 (1981)
30. P.H. von Hippel, O.G. Berg, Facilitated target location in biological systems. J. Biol. Chem. **264**, 675 (1989)
31. P.C. Blainey, G. Luo, S.C. Kou, W.F. Mangel, G.L. Verdine, B. Bagchi, X.S. Xie, Nonspecifically bound proteins spin while diffusing along DNA. Nat. Struct. Mol. Biol. **16**, 1224 (2009)
32. J. Elf, G.W. Li, X.S. Xie, Probing transcription factor dynamics at the single-molecule level in a living cell. Science **316**, 1191 (2007)
33. G.M. Clore, Exploring translocation of proteins on DNA by NMR. J. Biomol. NMR **51**, 209 (2011)

34. J. Iwahara, G.M. Clore, Detecting transient intermediates in macromolecular binding by paramagnetic NMR. Nature **440**, 1227 (2006)
35. S.E. Halford, J.F. Marko, How do site-specific DNA-binding proteins find their targets? Nucleic Acids Res. **32**, 3040 (2004)
36. L. Mirny, M. Slutsky, Z. Wunderlich, A. Tafvizi, J. Leith, A. Kosmrlj, How a protein searches for its site on DNA: the mechanism of facilitated diffusion. J. Phys. A: Math. Theor. **42**, 434013 (2009)
37. A.B. Kolomeisky, Physics of protein-DNA interactions: mechanisms of facilitated target search. Phys. Chem. Chem. Phys. **13**, 2088 (2011)
38. A. Veksler, A.B. Kolomeisky, Speed-selectivity paradox in the protein search for targets on DNA: is it real or not? J. Phys. Chem. B **117**, 12695 (2013)
39. M.A. Lomholt, B. van den Broek, S.M. Kalisch, G.J. Wuite, R. Metzler, Facilitated diffusion with DNA coiling. Proc. Natl. Acad. Sci. U. S. A. **106**, 8204 (2009)
40. D. Gomez, S. Klumpp, Facilitated diffusion in the presence of obstacles on the DNA. Phys. Chem. Chem. Phys. **18**, 11184 (2016)
41. M.P. Kochugaeva, A.A. Shvets, A.B. Kolomeisky, How conformational dynamics influences the protein search for targets on DNA. J. Phys. A: Math. Theor. **49**, 444004 (2016)
42. A.A. Shvets, A.B. Kolomeisky, Sequence heterogeneity accelerates protein search for targets on DNA. J. Chem. Phys. **143**, 245101 (2015)
43. A. Mondal, A. Bhattacherjee, Searching target sites on DNA by proteins: role of DNA dynamics under confinement. Nucleic Acids Res. **43**, 9176 (2015)
44. A. Mondal, A. Bhattacherjee, Understanding the role of DNA topology in target search dynamics of proteins. J. Phys. Chem. B **121**, 9372 (2017)
45. A.W. Lau, B.D. Hoffman, A. Davies, J.C. Crocker, T.C. Lubensky, Microrheology, stress fluctuations, and active behavior of living cells. Phys. Rev. Lett. **91**, 198101 (2003)
46. J. Ma, M. Do, M.A. Le Gros, C.S. Peskin, C.A. Larabell, Y. Mori, S.A. Isaacson, Strong intracellular signal inactivation produces sharper and more robust signaling from cell membrane to nucleus. PLoS Comput. Biol. **16**, e1008356 (2020)
47. A. Mondal, Sangeeta, A. Bhattacherjee, Torsional behaviour of supercoiled DNA regulates recognition of architectural protein Fis on minicircle DNA. Nucleic Acids Res. **50**, 6671 (2022)
48. E. Segal, Y. Fondufe-Mittendorf, L. Chen, A. Thåström, Y. Field, I.K. Moore, J.P. Wang, J. Widom, A genomic code for nucleosome positioning. Nature **442**, 772 (2006)
49. C. Felipe, J. Shin, Y. Loginova, A.B. Kolomeisky, The effect of obstacles in multi-site protein target search with DNA looping. J. Chem. Phys. **152**, 025101 (2020)
50. E.F. Koslover, M.A. Díaz de la Rosa, A.J. Spakowitz, Theoretical and computational modeling of target-site search kinetics in vitro and in vivo. Biophys. J. **101**, 856 (2011)
51. K. Mondal, S. Chaudhury, A theoretical study of the role of bulk crowders on target search dynamics of DNA binding proteins. J. Stat. Mech. **2020**, 093204 (2020)
52. B.T. Donovan, H. Chen, C. Jipa, L. Bai, M.G. Poirier, Dissociation rate compensation mechanism for budding yeast pioneer transcription factors. Elife **8**, e43008 (2019)
53. A. Mondal, S.K. Mishra, A. Bhattacherjee, Kinetic origin of nucleosome invasion by pioneer transcription factors. Biophys. J. **120**, 5219 (2021)
54. S.K. Mishra, A. Bhattacherjee, How do nucleosome dynamics regulate protein search on DNA? J. Phys. Chem. B **127**, 5702 (2023)
55. K.J. Polach, J. Widom, Mechanism of protein access to specific DNA sequences in chromatin: a dynamic equilibrium model for gene regulation. J. Mol. Biol. **254**, 130 (1995)
56. A. Mondal, A.B. Kolomeisky, Role of nucleosome sliding in the protein target search for covered DNA sites. J. Phys. Chem. Lett. **14**, 7073 (2023)
57. Y. Luo, J.A. North, M.G. Poirier, Single molecule fluorescence methodologies for investigating transcription factor binding kinetics to nucleosomes and DNA. Methods **70**, 108 (2014)
58. Y. Luo, J.A. North, S.D. Rose, M.G. Poirier, Nucleosomes accelerate transcription factor dissociation. Nucleic Acids Res. **42**, 3017 (2014)
59. B.T. Donovan, H. Chen, P. Eek, Z. Meng, C. Jipa, S. Tan, L. Bai, M.G. Poirier, Basic helix-loop-helix pioneer factors interact with the histone octamer to invade nucleosomes and generate nucleosome-depleted regions. Mol. Cell **83**, 1251 (2023)

60. C. Felipe, J. Shin, A.B. Kolomeisky, How pioneer transcription factors search for target sites on nucleosomal DNA. J. Phys. Chem. B **126**, 4061 (2022)
61. P. Schuster, *Stochasticity in Processes Fundamentals and Applications to Chemistry and Biology* (Springer, AG Switzerland, 2016)
62. A.K. Singh, F. Mueller-Planitz, Nucleosome positioning and spacing: from mechanism to function. J. Mol. Biol. **433**, 166847 (2021)
63. A. Valouev, S.M. Johnson, S.D. Boyd, C.L. Smith, A.Z. Fire, A. Sidow, Determinants of nucleosome organization in primary human cells. Nature **474**, 516 (2011)
64. X. Tang, T. Li, S. Liu, J. Wisniewski, Q. Zheng, Y. Rong, L.D. Lavis, C. Wu, Kinetic principles underlying pioneer function of GAGA transcription factor in live cells. Nat. Struct. Mol. Biol. **29**, 665 (2022)
65. Y. Chen, J.M. Tokuda, T. Topping, S.P. Meisburger, S.A. Pabit, L.M. Gloss, L. Pollack, Asymmetric unwrapping of nucleosomal DNA propagates asymmetric opening and dissociation of the histone core. Proc. Natl. Acad. Sci. U. S. A. **114**, 334 (2017)
66. A. Mondal, A.B. Kolomeisky, Why are nucleosome breathing dynamics asymmetric? J. Phys. Chem. Lett. **15**, 422 (2024)
67. A. Mondal, C. Felipe, A.B. Kolomeisky, Nucleosome breathing facilitates the search for hidden DNA sites by pioneer transcription factors. J. Phys. Chem. Lett. **14**, 4096 (2023)
68. J. Culkin, L. de Bruin, M. Tompitak, R. Phillips, H. Schiessel, The role of DNA sequence in nucleosome breathing. Eur. Phys. J. E Soft Matter **40**, 106 (2017)
69. L. de Bruin, M. Tompitak, B. Eslami-Mossallam, H. Schiessel, Why do nucleosomes unwrap asymmetrically? J. Phys. Chem. B **120**, 5855 (2016)
70. R. Rohs, S.M. West, A. Sosinsky, P. Liu, R.S. Mann, B. Honig, The role of DNA shape in protein-DNA recognition. Nature **461**, 1248 (2009)

# Chapter 24
# Exploring the Benefits of DNA-Target Search With Antenna

**Lucas Hedström and Ludvig Lizana**

**Abstract** The most common gene regulation mechanism is when a protein binds to a regulatory sequence to change RNA transcription. However, these sequences are short relative to the genome length, so finding them poses a challenging search problem. This chapter presents two mathematical frameworks capturing different aspects of this problem. First, we study the interplay between diffusional flux through a target where the searching proteins get sequestered on DNA far from the target because of non-specific interactions. From this model, we derive a simple formula for the optimal protein-DNA unbinding rate, maximizing the particle flux. Second, we study how the flux flows through a target on a single antenna with variable length. Here, we identify a non-trivial logarithmic correction to the linear behavior relative to the target size proposed by Smoluchowski's flux formula.

## 24.1 Introduction

Cells use designated proteins called transcription factors (TFs) to regulate gene expression. These TFs adjust the expression levels of genes in response to external and internal triggers such as changed nutrient or toxin concentrations, viral attacks, the physical presence of other cells, or damaged DNA. Mechanistically, TFs regulate genes by binding to short operator sequences typically located a few base pairs downstream of the gene's transcription start site from where they may hinder or help RNA polymerases to transcribe the gene into messenger RNA. However, because DNA is so long, TFs face a considerable challenge in finding these operator sites. And in some cases, they must find them quickly to counteract critical changes. Also, once found, they must stay bound long enough to influence RNA transcription (or return sufficiently often).

For illustrative purposes, let's discuss the TF search problem in *Escherichia coli* bacteria (E. coli) using actual numbers. E. coli encodes their environments in concentrations of about 300 TF types that regulate the expression levels of

L. Hedström · L. Lizana (✉)
Integrated Science Lab, Department of Physics, Umeå University, SE-901 87, Umeå, Sweden
e-mail: ludvig.lizana@umu.se

 
D. Grebenkov et al. (eds.), *Target Search Problems*,
https://doi.org/10.1007/978-3-031-67802-8_24

approximately 4,200 genes [1]. Since copy numbers typically are low, $\sim 10^2$ per TF species (geometric average), and operator sites are scattered across a $5 \cdot 10^6$ basepair (bp) long DNA, each TF must scan about 50,000 bps per second to achieve second-scale response times. Translated into a one-dimensional diffusion constant, this corresponds to $D_{\mathrm{1D}} \sim 5 \cdot 10^4$ bp$^2$sec$^{-1}$, which is not too far from with *in vitro* single-particle tracking measurements on the LacI repressor yielding $D_{\mathrm{1D}}^{\mathrm{LacI}} = (4.0 \pm 0.9) \cdot 10^5$ bp$^2$sec$^{-1}$[2]. However, this value appears to be about ten times higher *in vivo* [2], indicating that linear diffusion cannot be the primary search mechanism. The standard explanation for this discrepancy is that LacI diffuses along short DNA segments and occasionally transfers between segments through the cytoplasm. Since the apparent diffusion constant represents a weighted average between two diffusion mechanisms, one slow (DNA) and one fast (cytoplasm), it should exceed $D_{\mathrm{1D}}^{\mathrm{LacI}}$.

Next, to estimate realistic binding constants, we note that $10^2$ proteins correspond to approximately $c^{\mathrm{TF}} = 0.1$ micromolar ($\mu$M) intracellular concentration (1 protein in E. coli's volume is $1.6 \cdot 10^{-3}$ $\mu$M [3]). This suggests that TF-operator binding constants must be $K_{\mathrm{op}}^{\mathrm{TF}} \sim 0.01$ $\mu$M to ensure a 90% occupation probability.[1] This binding constant corresponds the binding energy $\epsilon_{\mathrm{op}}^{\mathrm{TF}} = -18.4\ k_B T$ if using $\epsilon = k_B T \ln(K_{\mathrm{op}}^{\mathrm{TF}}/[1\mathrm{M}])$ [4] ($k_B T$ is thermal energy), or $\epsilon_{\mathrm{op}}^{\mathrm{TF}} = -11.4$ kcal mol$^{-1}$ at room temperature ($k_B T = 0.62$ kcal mol$^{-1}$). As discussed below, $\epsilon_{\mathrm{op}}^{\mathrm{TF}}$ is almost four times smaller than the binding strength $\epsilon_{\mathrm{DNA}}^{\mathrm{TF}}$ associated with unspecific DNA binding.

As discussed above and explored in many theoretical models, TFs use a mixed strategy to find the designated target binding site (e.g., [2, 5–14] and many more). This strategy combines one-dimensional diffusional sliding along the DNA chain and three-dimensional jumps between DNA segments. By restricting a portion of the TF's search to several one-dimensional segments, the binding rate is effectively enhanced compared to a purely three-dimensional search. It improves because it is enough to find the correct DNA piece holding the target rather than diffusing directly into a nano-sized target from the bulk (a 10 bp target is only 2.7 nm long, 1 bp = 0.27 nm). Also, making three-dimensional jumps helps de-correlate the search, thus reducing repetitive visits to the same DNA sites, which characterizes a purely one-dimensional diffusive search. This combined process relies on a weak non-specific affinity between TFs and DNA, much weaker than binding to an operator site ($K_{\mathrm{op}}^{\mathrm{TF}}/K_{\mathrm{DNA}}^{\mathrm{TF}} \sim 10^6$, see[2]).

But how strong should the non-specific interaction be to minimize the search time? It should be long enough to ensure the target gets found if associated in it's vicinity but weak enough to allow TFs to quickly examine many segments and not get stuck. This problem is similar to an inspector looking for defective items (e.g., food) and wanting to know how much time should be invested in each one to find

[1] Assuming first-order binding kinetics, the binding probability is $p_b = c_{\mathrm{TF}}/(c_{\mathrm{TF}} + K_{\mathrm{op}}^{\mathrm{TF}})$, or $K_{\mathrm{op}}^{\mathrm{TF}} = c_{\mathrm{TF}}(1 - p_b)/p_b$. Using that $c_{\mathrm{TF}} = 0.1$ $\mu$M and $p_b = 0.9$ gives $K_{\mathrm{op}}^{\mathrm{TF}} = 1.1 \cdot 10^{-8}$ M.

[2] Using $\epsilon_{\mathrm{DNA}}^{\mathrm{TF}} \approx -4.8 k_B T$ for unspecific TF-DNA binding [4] and $\epsilon_{\mathrm{op}}^{\mathrm{TF}} = -18.4 k_B T$, gives $K_{\mathrm{op}}^{\mathrm{TF}}/K_{\mathrm{DNA}}^{\mathrm{TF}} = e^{(\epsilon_{\mathrm{DNA}}^{\mathrm{TF}} - \epsilon_{\mathrm{op}}^{\mathrm{TF}})/(k_B T)} \approx 8 \cdot 10^5 \approx 10^6$.

as many faulty items as possible in an enormous pile of candidate items. If going through them too quickly, there is a significant risk that faulty items go undetected. But if inspecting too slowly, there is not enough time to go through the pile.

A similar issue arises in the so-called speed-stability paradox. Here, TFs scan the DNA sequence for specific binding motifs [15, 16]. If the TF often pauses to check if it is on top of the target sequence, in other words, being very sensitive to the local base pair structure, it will find it with a probability of passing over the target. But, on the other hand, frequent pauses lead to a vanishingly small diffusion constant and long search times to distant targets.

This chapter explores how the TF search times (or binding frequencies) change with the length of the DNA stretches they examine before detaching. We call these segments antennae and consider a setup with millions of them, where only one harbors the designated target. We envision the target representing a gene regulatory sequence much shorter than the antenna itself. We open this chapter with a simple base case and calculate the flux through a single target lacking antenna. Next, we solve analytically a mathematical model capturing the rich interplay between unspecific DNA-TF binding of an extensive DNA segment pool and target-search times. Finally, we consider the search to a target on a single antenna and show that the search time has a non-trivial logarithmic correction with respect to its length.

## 24.2 Binding Rates to a Single Target

To define a reference case, imagine a target with radius $a$ floating in a cellular volume $V$ and a single searching protein. The protein randomly samples every target-sized 3D volume to find the target. If the trails are independent and the target volume fraction is $\phi = a^3/V$, the probability of finding the target after $n$ attempts is

$$\phi(1-\phi)^n, \tag{24.1}$$

where the average number of trials is $\langle n \rangle$ is

$$\langle n \rangle = \frac{1}{\phi}. \tag{24.2}$$

Next, assuming the protein explores each subvolume by diffusion, it spends approximately the time $\tau = a^2/D_{3\mathrm{D}}$ in each one, where $D_{3\mathrm{D}}$ is the diffusion constant. The total search time $T$ thus becomes $T = \langle n \rangle \times \tau$, or

$$T = \frac{\tau}{\phi} = \frac{V}{D_{3\mathrm{D}}a}. \tag{24.3}$$

By inverting this relationship, we obtain the corresponding stationary rate, or flux $J$

$$J = D_{3\mathrm{D}}ac, \tag{24.4}$$

where $c = 1/V$ indicates a uniform protein concentration. Apart from a factor of $4\pi$, this equation is the famed Smoluchowski rate derived in numerous settings (e.g., in this issue[3]).

Let's estimate $J$ from actual E. coli data. Using that $V = 1(\mu\text{m})^3$, $c = 100/V$ (100 proteins), $a = 10$ nm (few base pairs wide target), and $D_{3\text{D}} = 5(\mu\text{m})^2\text{sec}^{-1}$ (nano-sized protein complex), gives the binding frequency $J = 0.5\ \text{sec}^{-1}$ or $T = 2$ seconds between rebinding events.

Next, we make the target-search problem more complex by letting the target be a small part of a DNA segment in a large pool of other segments to which proteins may bind and stay sequestered for significant durations.

## 24.3 Target-Binding Rates in a Pool of Disconnected Antennae

As discussed in the Introduction, TFs search for target sequences by combining 3D diffusion and 1D sliding, where this mechanism builds on a non-specific DNA-protein attraction. However, there is a problem inherent with this setup. If proteins associate too strongly with DNA, they may spend significant time searching in the wrong places. On the other hand, if they happen to be close enough ("close" being within sliding distance or "antenna length"), the proteins will detect it with almost certainty because they have a chance to diffuse over the target several times. Therefore, sifting through many DNA segments is critical to shortening the search time. But this requires lowering the unspecific binding energy. And if too weak, proteins start missing the target even if binding to the correct DNA segment because they dissociate too quickly. This poses an optimization problem that depends on the sliding or antenna length. Or, put differently, the optimal time spent searching each DNA piece. Below, we formulate a simple mathematical theory that allows us to explore this problem and calculate analytically the steady-state flux through a DNA-target as a function of the unbinding rate $k_{\text{off}}$, which is proportional to the inverse of the unspecific binding time. We use E. coli as a primary example for our theory, where DNA permeates the entire cell volume. However, it applies equally to a cell nucleus in a mammalian cell that harbors the DNA (bacteria lack nuclei).

Consider the model sketched in Fig. 24.1. To the left, we depict the cell interior with DNA (solid line), surrounding DNA-binding proteins (filled circles), and a target (open circle). To the right, we illustrate an effective model suitable for analytical treatment. It shows DNA segments as unfilled circles or "nodes" ($i = 1, \ldots, N$) that are dynamically coupled through the cell volume ("bulk"); The nodes are similar to the "DNA blobs" introduced in [17]. The arrows in Fig. 24.1 highlight that proteins may bind and unbind DNA from and to the bulk. We associate these arrows with on and off rates, denoted $k_{\text{on}}$ and $k_{\text{off}}$.

[3] M. Lomholt and R. Metzler Chap. 25.

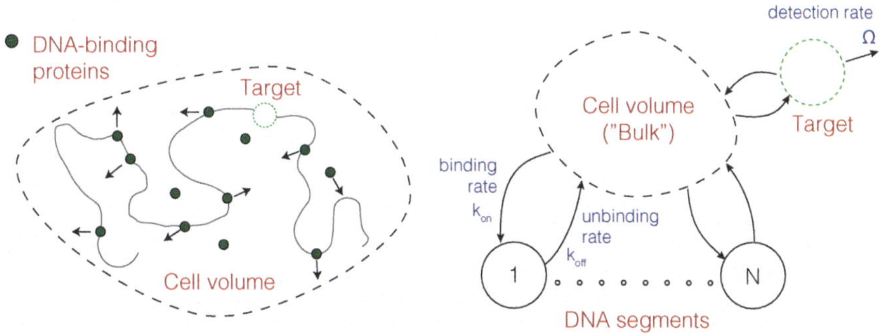


**Fig. 24.1** (left) Schematic cell interior where DNA (solid line) fills the cell volume (dashed). The figure also illustrates DNA-binding proteins (filled circles) diffusing in the cell volume and interacting with DNA to find a designated target sequence (open circle). (right) Effective DNA-target search model. The nodes $1, \ldots, N$ represent disconnected DNA segments (antennae) that are dynamically coupled via the cell volume ("Bulk"). The target node (dashed circle) is connected to the bulk like all other DNA nodes but has an additional detection or target-finding rate $\Omega$

The model omits proteins that translocate directly between nodes, so-called inter-segmental transfer [6]. This implies we treat the DNA segments as spatially scattered antennae that lack structural correlations related to DNA's 3D folding (e.g., as predicted by Hi-C experiments [18, 19] or power-law distance decays in traditional polymer models).

In addition to the DNA segments and the bulk, the figure shows the target as a separate node (dashed circle, far right). It has the same $k_{\text{on}}$ and $k_{\text{off}}$ as the other nodes but has also a target-detection rate $\Omega$. Given $k_{\text{off}}$ and $\Omega$, the probability of finding the target sequence once a protein binds the correct node is

$$\frac{\Omega}{\Omega + k_{\text{off}}}. \tag{24.5}$$

This equation shows that the detection probability drops to zero if $k_{\text{off}}$ is large. This suggests that $k_{\text{off}}$ should be small to increase the detection probability. But if it becomes too low, the protein flux into the target reduces because most proteins stay bound to the other DNA nodes.

Below, we calculate how the flux depends on $k_{\text{off}}$ by recasting Fig. 24.1 into a system of coupled equations for how the protein concentration in each node $c_i(t)$ changes over time $t$. Specifically, we are interested in the steady-state flux $J_\infty = J(t \to \infty)$ through the target and how it changes with $k_{\text{off}}$. We define $J_\infty$ as

$$J_\infty = c_{\text{target}}(t \to \infty)\Omega \equiv \bar{c}_{\text{target}}\Omega. \tag{24.6}$$

To calculate $\bar{c}_{\text{target}}$, we first an write equation for $c_{\text{target}}(t)$,

$$\frac{dc_{\mathrm{target}}(t)}{dt} = k_{\mathrm{on}} c_{\mathrm{bulk}}(t) - (k_{\mathrm{off}} + \Omega) c_{\mathrm{target}}(t), \tag{24.7}$$

where $c_{\mathrm{bulk}}(t)$ denotes the bulk protein concentration. Next, we write corresponding equations for each node $i = 1, \ldots N$, all following a similar structure:

$$\begin{aligned} \tfrac{dc_1}{dt} &= k_{\mathrm{on}} c_{\mathrm{bulk}}(t) - k_{\mathrm{off}} c_1(t) \\ &\vdots \\ \tfrac{dc_N}{dt} &= k_{\mathrm{on}} c_{\mathrm{bulk}}(t) - k_{\mathrm{off}} c_N(t). \end{aligned} \tag{24.8}$$

By summing these equations, we may write a single equation for the protein concentration on DNA, $c_{\mathrm{DNA}}(t)$, as

$$\frac{dc_{\mathrm{DNA}}(t)}{dt} = \frac{d}{dt}\big(c_1(t) + \ldots + c_N(t)\big) = k_{\mathrm{on}} N c_{\mathrm{bulk}}(t) - k_{\mathrm{off}} c_{\mathrm{DNA}}(t). \tag{24.9}$$

In steady-state, Eqs. (24.9) and (24.7) gives

$$\bar{c}_{\mathrm{target}} = \frac{k_{\mathrm{on}}}{k_{\mathrm{off}} + \Omega} \bar{c}_{\mathrm{bulk}}, \quad \bar{c}_{\mathrm{DNA}} = \frac{N k_{\mathrm{on}}}{k_{\mathrm{off}}} \bar{c}_{\mathrm{bulk}}. \tag{24.10}$$

Next, we relate $\bar{c}_{\mathrm{target}}$ to the cell's total protein concentration $c$. Because of mass conservation, the proteins must either be sequestered on DNA, diffusing in the bulk, or attached to the target node.

$$c = c_{\mathrm{target}}(t) + c_{\mathrm{DNA}}(t) + c_{\mathrm{bulk}}(t) \approx c_{\mathrm{DNA}}(t) + c_{\mathrm{bulk}}(t). \tag{24.11}$$

We motivate the last step by noting that the target segment is a vanishingly small fraction compared to the rest of the genome ($\ll 10^{-4}$ in bacteria). Using this equation and that $\bar{c}_{\mathrm{DNA}}$ from Eq. (24.10), we obtain

$$\bar{c}_{\mathrm{bulk}} = \frac{c}{1 + \frac{N k_{\mathrm{on}}}{k_{\mathrm{off}}}}, \tag{24.12}$$

and thus

$$\bar{c}_{\mathrm{target}} = \frac{k_{\mathrm{on}}}{k_{\mathrm{off}} + \Omega} \frac{c}{1 + \frac{N k_{\mathrm{on}}}{k_{\mathrm{off}}}}, \tag{24.13}$$

Using this expression to calculate the steady-state flux $J_\infty = \bar{c}_{\mathrm{target}} \Omega$, we obtain

$$J_\infty = \Omega \bar{c}_{\mathrm{target}} = \frac{1}{1 + \frac{k_{\mathrm{off}}}{\Omega}} \frac{k_{\mathrm{on}} c}{1 + \frac{N k_{\mathrm{on}}}{k_{\mathrm{off}}}}. \tag{24.14}$$

We note that each one of the factors in $J_\infty$ has intuitive interpretations. The first factor, $1/(1 + k_{\mathrm{off}}/\Omega)$, is the target node's reaction-limited contribution (or binding probability, Eq. (24.5)). When $\Omega$ is small, this factor drops to zero, indicating that the target requires a high revisiting frequency to reach short search times. The second factor, $k_{\mathrm{on}}c/(1 + Nk_{\mathrm{on}}/k_{\mathrm{off}})$ represents the reduced (Smoluchowski) diffusion flux ($\propto k_{\mathrm{on}}c$) due to the proteins' being unspecifically bound elsewhere on DNA.

Interestingly, the two factors in Eq. (24.14) have an opposing dependence on $k_{\mathrm{off}}$. This suggests that there exists an optimal $k^*_{\mathrm{off}}$ that maximizes the target flux $J^{\mathrm{max}}_{\infty} = J_{\infty}(k_{\mathrm{off}} = k^*_{\mathrm{off}})$. We explore this optimum in the following section.

### 24.3.1 Finding the Optimal $k_{off}$ that Maximizes Target-Binding Rates $J_\infty$

Before doing any analytical calculations, we plot the flux $J_\infty$ to see how it changes with the off-rate $k_{\mathrm{off}}$ for a few detection rates $\Omega$ keeping $Nk_{\mathrm{on}}$ ($= \bar{k}_{\mathrm{on}}$) fixed (Fig. 24.2, left). Besides the distinct maximum at $k^*_{\mathrm{off}}$, we note two regimes. These may be understood as follows. For small $k_{\mathrm{off}}$, the flux increases linearly as $J_\infty \sim k_{\mathrm{off}}$ (dashed line) in all cases. Here, most proteins remain sequestered on DNA for significant times, where the bulk concentration becomes low, $\bar{c}_{\mathrm{bulk}} \approx ck_{\mathrm{off}}/\bar{k}_{\mathrm{on}}$. This results in an $\Omega$-independent flux $J_\infty \approx k_{\mathrm{on}}\bar{c}_{\mathrm{bulk}} = (c/N)\, k_{\mathrm{off}}$, thus limited by $k_{\mathrm{off}}$.

This contrasts with the other extreme when $k_{\mathrm{off}}$ is large. Here, most of the proteins reside in the bulk, $\bar{c}_{\mathrm{bulk}} \approx c(1 - \bar{k}_{\mathrm{on}}/k_{\mathrm{off}})$, and the flux becomes $J_\infty \approx (\Omega/k_{\mathrm{off}})\, k_{\mathrm{on}}c$. In this case, proteins frequently find the correct segment but stay only for a short

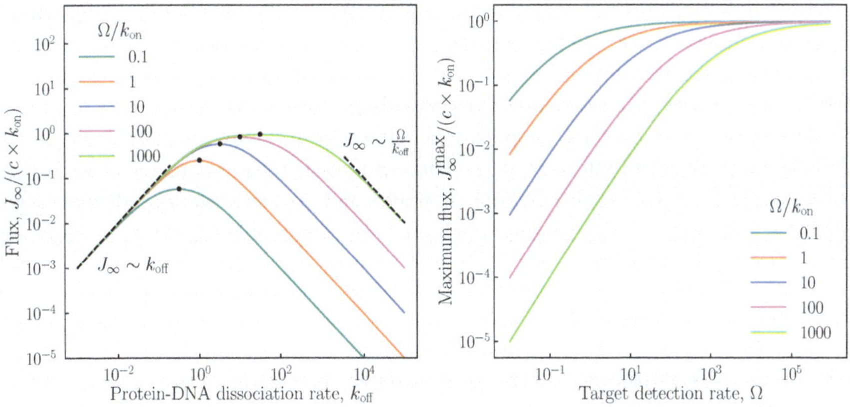


**Fig. 24.2** (left) Target flux $J_\infty$ (Eq. (24.14)) versus the rate $k_{\mathrm{off}}$ of detaching from DNA for five detection rates $\Omega$. We plot asymptotic behaviors for small and large $k_{\mathrm{off}}$ as dashed lines. For large $k_{\mathrm{off}}$, we only plot the asymptote for $\Omega/k_{\mathrm{on}} = 1000$ for clarity. The filled black circles indicate optimal points ($k^*_{\mathrm{off}}$, $J^{\mathrm{max}}_\infty$) calculated from Eqs. (24.16) and (24.17). (right) Maximal target flux versus target-detection rate $\Omega$ for five on rates $\Omega/k_{\mathrm{on}}$ (Eq. (24.17)). We note that the curves saturate for large $\Omega$, indicating that increasing the detection rate further does not lead to significantly higher flux

while, and the limiting rate is $\Omega$; the probability of finding the target is $\Omega/k_{\text{off}}(1 - \mathcal{O}(\Omega/k_{\text{off}}))$. We indicate the decay $\Omega/k_{\text{off}}$ as a dashed line.

Between these limiting cases resides an optimal $k^*_{\text{off}}$ that maximizes the flux. To find this optimum, we calculate $\frac{dJ_\infty}{dk_{\text{off}}} = 0$ using Eq. (24.14), yielding

$$\frac{\Omega k_{\text{on}} c\,(\Omega \bar{k}_{\text{on}} + k^2_{\text{off}})}{(k_{\text{off}} + \bar{k}_{\text{on}})^2(\Omega + k_{\text{off}})^2} = 0. \tag{24.15}$$

Solving this equation analytically gives (we omit the negative solutions)

$$k^*_{\text{off}} = \sqrt{\bar{k}_{\text{on}}\Omega} = \sqrt{N k_{\text{on}}\Omega}. \tag{24.16}$$

which gives the maximal rate $J^{\max}_\infty = J_\infty(k_{\text{off}} = k^*_{\text{off}})$

$$J^{\max}_\infty = \frac{k_{\text{on}} c}{\left(1 + \sqrt{\frac{N k_{\text{on}}}{\Omega}}\right)^2}. \tag{24.17}$$

We marked the optimal points $(k^*_{\text{off}}, J^{\max}_\infty)$ by symbols (•) in Fig. 24.2 (left). We also note that the plateau to the right optimal point becomes broader with increasing $\Omega$. We interpret this as if $k_{\text{off}}$ increases until the optimal point $k^*_{\text{off}}$, increasing it more does not lead to a larger $J^{\max}_\infty$ because the flux is controlled by $\Omega$. But eventually, $k_{\text{off}}$ grows beyond $\Omega$, and the flux starts to decline as there is a significant chance of missing the target.

We also plotted how the maximal flux $J^{\max}_\infty$ changes with the target-finding rate $\Omega$ for varying on-rates $N k_{\text{on}}$ in Fig. 24.2 (right). While $J^{\max}_\infty$ ranges over several orders of magnitude for small $\Omega$, it approaches the diffusion-limited Smoluchowski rate $J^{\max}_\infty \simeq k_{\text{on}} c$ as the target becomes fully absorbing ($\Omega \to \infty$).

In biology, $k_{\text{off}}$ is inversely proportional to the protein-DNA-residence time, making this parameter difficult to change. While it is conceivable that $k_{\text{off}}$ may change due to the binding of small molecules that alter the protein's binding affinity (e.g., allolactose binding to the Lac repressor), it is more likely that $k_{\text{off}}$ changes occur on evolutionary time scales.

## 24.4 Target-Binding Rates to a Single Antenna with Varying Length

The preceding discussion concerned diffusional flux through a target in the presence of many DNA segments that weakly bind the searching proteins. Here, we return to the single-antenna case and investigate how Smoluchowski's reaction rate through a target changes with DNA segment, or antenna, length $\lambda$. Specifically, we will study a

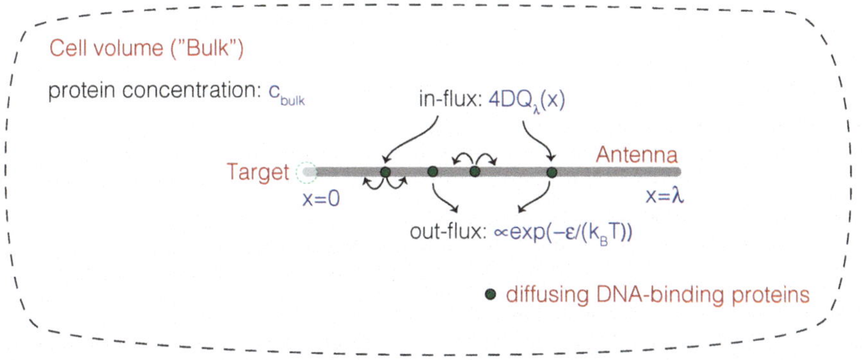


**Fig. 24.3** Schematic illustrating the single-antenna search problem. The antenna remains immersed (and immobile) in a large volume with constant protein concentration (filled circles). These proteins may bind to the antenna and diffuse while staying associated. After some time, they unbind to the bulk. We put the target at the antenna's left end ($x = 0$). We assume the influx $4DQ_\lambda(x)$ and the outflux $\propto e^{-\epsilon/(k_BT)}$ are in steady-state

case analytically and demonstrate that the flux has a non-trivial logarithmic correction $J_\lambda \propto \lambda/\log\lambda$ when protein-antenna interactions are strong [20].

To model this scenario, consider a volume with constant particle density and a suspended antenna with some length $\lambda$ (Fig. 24.3). At one of its ends resides a target, which we represent as a small absorbing sphere. Like in the previous sections, we are interested in the steady-state conditions, where there is a balance between the diffusional flux from the volume onto the antenna and through the target. But unlike before, particles may bind to the target directly from the bulk without first associating with the antenna and then sliding (albeit this is a small contribution relative to the flux coming from the antenna).

Next, the antenna is associated with an interaction energy $\epsilon$ (or binding constant $K$) to particles diffusing in the surrounding volume. For now, we treat this interaction as very strong ($\epsilon \to \infty$ or $K \to 0$), implying that the particles never detach once associated with the antenna. However, at the end of this section, we relax this condition.

### *24.4.1 Perfectly Absorbing Antenna*

To define the system's geometry, we use a 3D fcc lattice with coordinates $\mathbf{r} = (x, y, z)$, where $x, y, z$ are multiples of the lattice spacing $l_0$. The lattice spans symmetrically to infinity in all directions, where we place the antenna in the origin and let it extend along the positive $x$-axis from $\mathbf{r}_0 = (0, 0, 0)$ to $\mathbf{r}_\lambda = (\lambda, 0, 0)$. To simplify notation, we introduce linear coordinates $x$, where $x = 0$ and $x = \lambda$ are the antenna's two ends, and $\bar{c}(x) = c(x, t \to \infty)$ is the effective linear particle density in steady-state. Note that the antenna has a finite thickness that equals the lattice spacing $l_0$, turning the antenna into a rod ($l_0$ is also the target's diameter).

Far from the antenna, we assume the bulk concentration is constant $c_{\text{bulk}}$. We also assume the diffusion constant $D$ on the antenna and in the bulk are identical, that is $D = (l_0^2/6) \times k_{\text{jump}}$ (admittedly, in actual cells, the diffusion constant may be orders of magnitude smaller). Below, we put the jump rate $k_{\text{jump}} = 1$ as we are interested in the flux-dependence on varying antenna lengths rather than diffusion rates. Also, we replaced $\lambda/l_0 \to \lambda$ to simplify notation ($l_0$ could represent one base pair).

To mathematically set up the problem, we formulate a one-dimensional diffusion equation for $\bar{c}(x)$. In the strong-binding limit, we do not allow particles to escape the antenna back to the bulk. Therefore, the flux into every inner antenna site from the four (out of six) adjacent bulk sites is $4 \times DQ_\lambda(x)$, where $Q_\lambda(x)$ is the density at those adjacent sites. This density interpolates the densities in the bulk $c_{\text{bulk}}$ and on the antenna $\bar{c}(x)$.

If approximating $x$ as a continuous variable, we model the particle density $\bar{c}(x)$ as a diffusion equation with a linear line source term $-4 \times DQ_\lambda(x)$, representing the influx. That is,

$$D\frac{d^2\bar{c}(x)}{dx^2} = -4DQ_\lambda(x), \quad 0 \le x \le \lambda \tag{24.18}$$

The absorbing condition at the target ($x = 0$) and flux boundary condition at the other end ($x = \lambda$) are

$$\bar{c}(x = 0) = 0, \tag{24.19}$$

$$D\left(\frac{d\bar{c}(x)}{dx}\right)_{x=\lambda} = 5DQ_\lambda(x = \lambda) \tag{24.20}$$

To get the last condition, we note that the flux at the antenna ends slightly differs from the flux into the inner nodes because the ends have five adjacent bulk nodes instead of four. This gives $5 \times DQ_\lambda(x = \lambda)$.

As a last step, we define the Smoluchowski rate through the target as

$$J_\lambda = -D\left(\frac{d\bar{c}(x)}{dx}\right)_{x=0}, \tag{24.21}$$

which is the quantity of interest.

To calculate, $J_\lambda$, we must find $Q_\lambda(x)$. As demonstrated in [20], this is a challenging and technical problem. However, it follows approximately

$$Q_\lambda(x) = \frac{c_{\text{bulk}}}{1 + \frac{1}{\pi}(\operatorname{arcsinh}(x) + \operatorname{arcsinh}(\lambda - x))}. \tag{24.22}$$

Plotting this expression, we note that it is constant along most of the antenna, apart from the ends (Fig. 24.4). Therefore, if the antenna is long, we may approximate $Q_\lambda(x)$ by its average

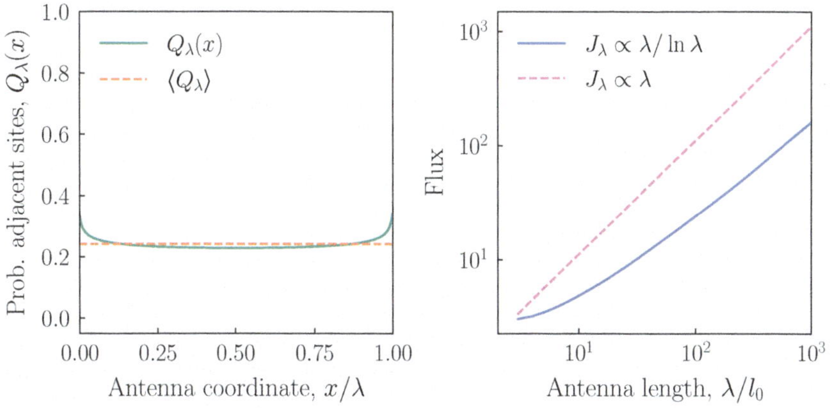


**Fig. 24.4** (left) Particle density on sites adjacent to the antenna $Q_\lambda(x)$. The lines represent: Eq. (24.22) with $c_{\text{bulk}} = 1$ (solid); antenna average Eq. (24.24) (dashed). (right) Flux through the target predicted from the theory Eq. (24.25) (solid) and Smoluchowski's linear prediction (dashed)

$$\langle Q_\lambda \rangle = \frac{1}{\lambda} \int_0^\lambda Q_\lambda(x)dx \simeq \frac{A}{1 + \frac{2}{\pi} \ln \lambda}, \quad \lambda \gg 1, \tag{24.23}$$

where $A \approx 1.06$. By using $Q_\lambda(x) \approx \langle Q_\lambda \rangle$ in Eq. (24.18), thus removing the explicit $x$-dependence in the source term, and using the boundary conditions in Eqs. (24.19) and (24.20) (and $d^2\bar{c}(x)/dx^2 = 4\langle Q_\lambda \rangle$) to determine constants, we obtain

$$\bar{c}(x) = \langle Q_\lambda \rangle (-2x^2 + (4\lambda + 5)x). \tag{24.24}$$

Using this expression in Eq. (24.21) to calculate the steady-state flux $J_\lambda$ through the target gives

$$J_\lambda \simeq -2\pi AD \frac{\lambda}{\ln \lambda}, \quad \lambda \gg 1. \tag{24.25}$$

This result highlights the logarithmic correction $[\ln(\lambda)]^{-1}$ in $J_\lambda$. As we portray in Fig. 24.4, this correction substantially lowers the flux for long antennae relative to the linear increase $J \sim \lambda$ (Eq. (24.4)), representing Smoluchowski's prediction.

### *24.4.2 Partially Absorbing Antenna*

To generalize these results to a partially absorbing antenna, we modify the diffusion equation Eq. (24.18). In particular, we change the right-hand side by adding an outflux term to the source

$$D\frac{d^2\bar{c}(x)}{dx^2} = -4D\left(\langle Q_\lambda\rangle - e^{-\epsilon/(k_BT)}\bar{c}(x)\right). \tag{24.26}$$

One of the boundary conditions in Eq. (24.20) also changes. While the absorbing condition at $x = 0$ stays intact, the flux condition at $x = \lambda$ modifies to

$$D\left(\frac{d\bar{c}(x)}{dx}\right)_{x=\lambda} = 5D\left(\langle Q_\lambda\rangle - e^{-\epsilon/(k_BT)}\bar{c}(x)\right)_{x=\lambda}. \tag{24.27}$$

A relatively lengthy calculation gives that the leading $\lambda$ behavior is [20]

$$J_\lambda \simeq J_0 \times \left[1 + \frac{\langle Q_\lambda\rangle e^{\epsilon/(2k_BT)}}{3J_0}\tanh\left(\frac{2\lambda}{e^{\epsilon/(2k_BT)}}\right)\right], \tag{24.28}$$

where we introduced the shorthand $J_0$ denoting the standard Smoluchowski flux when the antenna length is zero, $J_{\lambda\to 1}$ (i.e., one single absorbing site in the 3D fcc lattice).

This expression allows us to explore $J_\lambda$ for varying $\epsilon$, keeping $\lambda$ large but fixed. First, we note that when $\epsilon \to \infty$, we recover our previous expression by expanding $\tanh\xi \approx \xi - \xi^3/3 + \ldots$

$$J_\lambda \simeq J_0 \times \frac{\langle Q_\lambda\rangle e^{\epsilon/(2k_BT)}}{3J_0}\left[\frac{2\lambda}{e^{\epsilon/(2k_BT)}} - \frac{1}{3}\left(\frac{2\lambda}{e^{\epsilon/(2k_BT)}}\right)^3 + \ldots\right] \tag{24.29}$$

where

$$|J_\lambda| \sim \langle Q_\lambda\rangle\lambda \propto \frac{\lambda}{\ln\lambda}. \tag{24.30}$$

Furthermore, plotting Eq. (24.28) reveals two noteworthy observations (Fig. 24.5). First, $J_\lambda$ increases rapidly for low to intermediate $\epsilon$. To find this behavior analytically, we note that the hyperbolic tangent approaches unity when $2\lambda/(e^{\epsilon/(2k_BT)}) \gg 1$. Using this approximation gives an exponential dependence on $\epsilon$:

$$|J_\lambda| \simeq \frac{\langle Q_\lambda\rangle}{3}e^{\epsilon/(2k_BT)}. \tag{24.31}$$

We plot this approximation as dashed lines alongside Eq. (24.28). We note that the match is better for longer antenna lengths.

Second, if instead plotting Eq. (24.28) as a function of antenna length for a few fixed energies, we note that the $J_\lambda$-curves start to cluster and the flux becomes less and less dependent on $\epsilon$ (Fig. 24.5, right). This suggests that there is some characteristic $\epsilon_c$, above which increasing it more does not lead to a significantly larger flux, where the limiting maximal flux is

$$J_\lambda^{\max} = \lim_{\epsilon\to\infty}|J_\lambda| \propto \lambda/\ln\lambda. \tag{24.32}$$

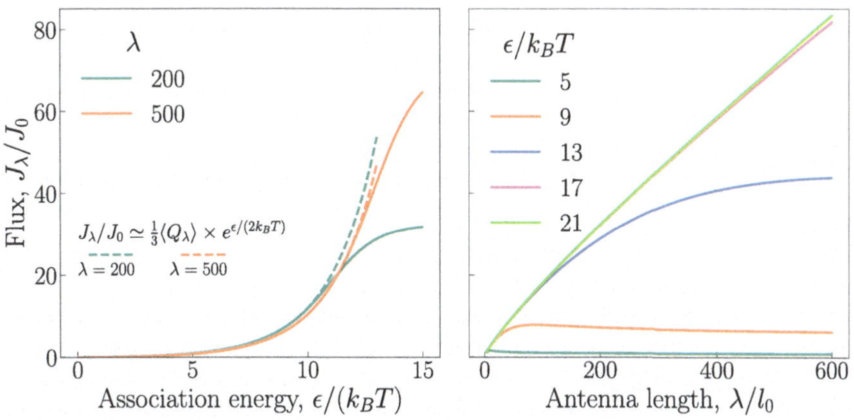


**Fig. 24.5** (left) Target flux versus particle-antenna association energy for two different antenna lengths. The solid lines represent Eq. (24.28), and the dashed lines highlight the approximation in Eq. (24.31). (right) Target flux with increasing antenna lengths for five different particle-antenna association energies. As in the left panel, the solid lines represent Eq. (24.28). Note that the lines cluster for increasing $\epsilon$, suggesting that there is a limiting $J_\lambda$ that does not depend on $\epsilon$

## 24.5 Discussion and Closing Remarks

The most common gene regulation mechanism is when a protein binds to a regulatory sequence to increase or decrease RNA transcription rates. However, these sequences are short relative to the genome length, so finding them poses a demanding search problem. This chapter presents two mathematical frameworks capturing different aspects of this problem. First, we studied the interplay between diffusional flux through a target where the searching proteins get sequestered on DNA out-of-reach from the target because of non-specific interactions. However, lowering these DNA-protein interactions increases the chance of missing the target, even if the proteins are close because of fleeting contacts. To make the model analytically solvable, we treated DNA as an ensemble of a disconnected antenna, thereby omitting structural correlations in 3D appearing naturally in polymer models or Hi-C experiments [18, 19]. We found that the optimal binding rate to be $(Nk_{\text{on}}\Omega)^{1/2}$ ($N$ is the number of segments, $k_{\text{on}}$ is the protein association rate, and $\Omega$ is the target-detection rate). Next, we studied the particle flux flows through a single antenna when changing its length. We found a non-trivial logarithmic correction to the linear behavior suggested by Smoluchowski's flux formula.

The mathematical theories presented here rest on several assumptions. Below, we make a few remarks.

The model in Fig. 24.1 divides DNA into $N$ disconnected segments, where we calculate the maximal flux $J_\infty^{\max}$ as a function of $k_{\text{off}}$ keeping $N$ fixed. Keeping $N$ fixed also means keeping the antenna length $\lambda$ fixed as $N \propto 1/\lambda$. However, under actual conditions, $k_{\text{off}}$ sets the proteins' DNA-residence time, making $\lambda$ and $k_{\text{off}}$ dependent variables. Using the 1D diffusion constant, we can estimate the antenna (or sliding) length as $\lambda \propto (D_{1D}/k_{off})^{1/2}$. Thus, changing $k_{off}$ in principle modifies $N$ too. This observation leads to a more complicated mathematical theory that we leave as an open problem for interested readers to develop.

Next, we assumed that the antenna was straight. This assumption is realistic for transcription factors with sliding lengths shorter than DNA's persistence length $\approx 150$ bp [21]. However, several empirical studies report sliding lengths in the same order or longer (e.g., [22]). This situation suggests considering a coiled rather than a straight antenna. As demonstrated in a suite of papers [11, 12], it is possible to formulate a target-search model for a coiled antenna following elegant but relatively intuitive arguments. Let's first consider the straight antenna. Smoluchowski's formula says the flux through a small sphere with radius $a$ is $J_{3D} \sim a$, where $a \sim \lambda$ for a straight antenna. Next, like Sect. 24.4 in this chapter, we balance the bulk flux into the sphere with the particle flux a $J_{1D}$ onto the antenna inside the sphere. This flux sets the linear protein concentration on the antenna far from the target. Finally, knowing that the concentration vanishes at the absorbing target position, we may estimate the incoming linear flux $J$ using Fick's law. Now to the critical point in the argument: in steady-state, the fluxes on all scales must be in balance, thus

$$J_{3D} = J_{1D} = J. \tag{24.33}$$

This condition gives similar relationships as we derived in Sect. 24.2, from where it is possible to calculate the optimal $\lambda$ that minimizes the search. To generalize these arguments to a coiled antenna, it suffices to note that the sphere's radius $a$ is not linear in $\lambda$. Instead, because the antenna now is a coiled polymer, we associate $a$ with the polymer's radius of gyration, i.e., $a \sim \lambda^{\nu}$, where $\nu = 1/2$ for a Gaussian coil. The rest of the analysis remains the same, and the optimal $\lambda$ now depends on $\nu$.

Also, we did not explore time-dependent transients associated with the particle flux or effects of initial conditions. But in E.coli, it seems "distance matters" [23], where search times depend critically on where TFs are produced and how long they must diffuse to reach their designated targets. Some papers argue that this process drives genes to cluster spatially [24, 25].

Finally, we remark that the on-rate in Eq. (24.9) easily generalizes to more complex DNA-node configurations than in Fig. 24.1. If rewriting the total on-rate as

$$\bar{k}_{on} = k_{on}N = k_{on}e^{\log N} \equiv k_{on}e^{\Delta G/(k_B T)}, \tag{24.34}$$

we may identify the free energy $\Delta G$ associated with the entropy of the node structures. For instance, the nodes may be wired into a complex 3D interaction network with scale-dependent community partitions [26, 27]. By including network configurations as free energy offers a framework to formulate effective target-search theories embracing space-filling fractal networks complementing previous efforts [14, 28–30].

Even if the DNA-target search problem is close to five decades old [5, 31, 32], it still attracts new researchers, theoretical and empirical, across disciplines. We hope this chapter inspired future work exploring some unknown aspects of DNA-search processes or similar constrained search problems in other areas.

## References


1. R. Milo, R. Phillips, *Cell Biology by the Numbers* (Garland Science, New York, 2015)
2. J. Elf, G.-W. Li, X.S. Xie, Probing transcription factor dynamics at the single-molecule level in a living cell. Science **316**, 1191–1194 (2007)
3. R. Phillips, J. Kondev, J. Theriot, H. Garcia, *Physical Biology of the Cell* (Garland Science, Boca Raton, 2012)
4. K. Sneppen, *Models of Life* (Cambridge University Press, Cambridge, 2014)
5. P.H. von Hippel, O. Berg, Facilitated target location in biological systems. J. Biol. Chem. **264**, 675–678 (1989)
6. M.A. Lomholt, B. van den Broek, S.-M.J. Kalisch, G.J. Wuite, R. Metzler, Facilitated diffusion with DNA coiling. Proceed. Nat. Acad. Sci. **106**, 8204–8208 (2009)
7. P. Hammar, P. Leroy, A. Mahmutovic, E.G. Marklund, O.G. Berg, J. Elf, The lac repressor displays facilitated diffusion in living cells. Science **336**, 1595–1598 (2012)
8. O. Bénichou, C. Loverdo, M. Moreau, R. Voituriez, Intermittent search strategies. Rev. Mod. Phys. **83**, 81 (2011)
9. M. Klein, T.J. Cui, I. MacRae, C. Joo, M. Depken, Skipping and sliding to optimize target search on protein-bound DNA and RNA. BioRxiv, pp. 2020–06 (2020)
10. L. Mirny, M. Slutsky, Z. Wunderlich, A. Tafvizi, J. Leith, A. Kosmrlj, How a protein searches for its site on DNA: the mechanism of facilitated diffusion. J. Phys. A: Math. Theoret. **42**, 434013 (2009)
11. T. Hu, A.Y. Grosberg, B. Shklovskii, How proteins search for their specific sites on dna: the role of DNA conformation. Biophys. J. **90**, 2731–2744 (2006)
12. L. Hu, A.Y. Grosberg, R. Bruinsma, Are DNA transcription factor proteins Maxwellian demons? Biophys. J. **95**, 1151–1156 (2008)
13. A.B. Kolomeisky, Physics of protein-DNA interactions: mechanisms of facilitated target search. Phys. Chem. Chem. Phys. **13**, 2088–2095 (2011)
14. L. Hedström, L. Lizana, Modelling chromosome-wide target search. New J. Phys. **25**, 033024 (2023)
15. O. Bénichou, Y. Kafri, M. Sheinman, R. Voituriez, Searching fast for a target on DNA without falling to traps. Phys. Rev. Lett. **103**, 138102 (2009)
16. M. Slutsky, L.A. Mirny, Kinetics of protein-DNA interaction: facilitated target location in sequence-dependent potential. Biophys. J. **87**, 4021–4035 (2004)
17. M. Bauer, R. Metzler, In vivo facilitated diffusion model. PloS One **8**, e53956 (2013)
18. E. Lieberman-Aiden, N.L. Van Berkum, L. Williams, M. Imakaev, T. Ragoczy, A. Telling, I. Amit, B.R. Lajoie, P.J. Sabo, M.O. Dorschner et al., Comprehensive mapping of long-range interactions reveals folding principles of the human genome. Science **326**, 289–293 (2009)
19. J.-M. Belton, R.P. McCord, J.H. Gibcus, N. Naumova, Y. Zhan, J. Dekker, Hi-C: A comprehensive technique to capture the conformation of genomes. Methods **58**, 268–276 (2012)
20. O.A. Vasilyev, L. Lizana, G. Oshanin, Smoluchowski rate for diffusion-controlled reactions of molecules with antenna. J. Phys. A: Math. Theoret. **50**, 264004 (2017)
21. H.G. Garcia, P. Grayson, L. Han, M. Inamdar, J. Kondev, P.C. Nelson, R. Phillips, J. Widom, P.A. Wiggins, Biological consequences of tightly bent DNA: the other life of a macromolecular celebrity. Biopolymers: Original Res. Biomol. **85**, 115–130 (2007)

22. E.G. Marklund, A. Mahmutovic, O.G. Berg, P. Hammar, D. van der Spoel, D. Fange, J. Elf, Transcription-factor binding and sliding on DNA studied using micro-and macroscopic models. Proceed. Nat. Acad. Sci. **110**, 19796–19801 (2013)
23. O. Pulkkinen, R. Metzler, Distance matters: the impact of gene proximity in bacterial gene regulation. Phys. Rev. Lett. **110**, 198101 (2013)
24. G. Kolesov, Z. Wunderlich, O.N. Laikova, M.S. Gelfand, L.A. Mirny, How gene order is influenced by the biophysics of transcription regulation. Proc. Nat. Acad. Sci. **104**, 13948–13953 (2007)
25. M. Di Stefano, A. Rosa, V. Belcastro, D. di Bernardo, C. Micheletti, Colocalization of coregulated genes: a steered molecular dynamics study of human chromosome 19. PLoS Comput. Biol. **9**, e1003019 (2013)
26. D. Bernenko, S.H. Lee, P. Stenberg, L. Lizana, Mapping the semi-nested community structure of 3d chromosome contact networks. PLOS Comput. Biol. **19**, e1011185 (2023)
27. A. Holmgren, D. Bernenko, L. Lizana, Mapping robust multiscale communities in chromosome contact networks. Sci. Rep. **13**, 12979 (2023)
28. J. Smrek, A.Y. Grosberg, Facilitated diffusion of proteins through crumpled fractal DNA globules. Phys. Rev. E **92**, 012702 (2015)
29. O. Bénichou, C. Chevalier, B. Meyer, R. Voituriez, Facilitated diffusion of proteins on chromatin. Phys. Rev. Lett. **106**, 038102 (2011)
30. M. Nyberg, T. Ambjörnsson, P. Stenberg, L. Lizana, Modeling protein target search in human chromosomes. Phys. Rev. Res. **3**, 013055 (2021)
31. G. Adam, M. Delbrück, A. Rich, N. Davidson, *Structural Chemistry and Molecular Biology* (San Francisco, USA, 1968)
32. A.D. Riggs, H. Suzuki, S. Bourgeois, lac repressor-operator interaction: I. equilibrium studies. J. Mol. Biol. **48**, 67–83 (1970)

# Chapter 25
# Target Search on DNA—Effect of Coiling

**Michael A. Lomholt and Ralf Metzler**

**Abstract** Some proteins can find their targets on DNA faster than by pure diffusion in the three-dimensional cytoplasm, through the process of facilitated diffusion: They can loosely bind to DNA and temporarily slide along it, thus being guided by the DNA molecule itself to the target. This chapter examines this process in mathematical detail with a focus on including the effect of DNA coiling on the search process.

## 25.1 Introduction

Already in 1968 Adam and Delbrück [1] proposed that diffusion-limited biological processes could be aided by the mechanism of dimensional reduction. A prime example of this are proteins searching for a specific location on a long DNA molecule, typically megabases long. In 1970 Riggs et al. [2] observed the very fast association of the lac repressor protein to its specific operator site on DNA, and Richter and Eigen suggested in 1974 [3] that it was exactly by exploiting the possibility of one-dimensional diffusion along the DNA that the lac repressor could be so fast. Detailed mathematical modelling of this facilitated diffusion process, during which the protein can switch between three-dimensional diffusion in water and one-dimensional diffusion along the DNA, has been refined in a number of articles by Berg and various coauthors [4–8]. For a more contemporary work on dimensional reduction, we refer the reader to [9].

The present chapter examines the effect of DNA coiling on the target search process of proteins searching for a target on DNA. The sliding of DNA-binding proteins

M. A. Lomholt (✉)
PhyLife—Center for Physical Life Science, University of Southern Denmark, Campusvej 55, 5230 Odense M, Denmark
e-mail: mlomholt@sdu.dk

R. Metzler
Institute of Physics & Astronomy, University of Potsdam, Karl-Liebknecht-Str 24/25, Haus 28, 14476 Potsdam-Golm, Germany
e-mail: rmetzler@uni-potsdam.de

 
D. Grebenkov et al. (eds.), *Target Search Problems*,
https://doi.org/10.1007/978-3-031-67802-8_25

was demonstrated in single DNA experiments for both single-strand DNA binders [10] and double-strand binders [11, 12]. Berg and Blomberg suggested already in 1977 [5] that coiling would speed up the target localisation by removing some of the correlations that would be present for a straight DNA molecule. The speed-up by DNA coiling has been confirmed experimentally by Gowers and Halford exploiting supercoiling of DNA in [13], and later by optical tweezers experiments by van den Broek et al. [14]. The theory of the effect of DNA coiling that is presented in this chapter is focused on modelling the latter experiment. It builds on the facilitated diffusion model by Berg and Ehrenberg [7] for straight DNA, and it constitutes a significantly more detailed version of the mathematical theory presented in [15].

The account begins with a re-derivation of Smoluchowski's 1916 result for diffusion-limited reactions with spherical targets [16] in Sect. 25.2. This is the target localisation rate that dimensional reduction helps in surpassing. Section 25.2 also introduces, for this simple case, some of the mathematical apparatus that the remaining sections use. Next the facilitated diffusion process is presented from the point of view of diffusion along the DNA in Section 25.3. The specific theory for straight and coiled DNA then follows in Sects. 25.4 and 25.5. In Sect. 25.6, the mechanism of intersegmental transfers is briefly discussed before ending with conclusions in Sect. 25.7. Additionally, the Matlab scripts used for calculating target localisation rates and creating the figures in this chapter can be found at [17].

## 25.2 Smoluchowski Result

We will first revisit the case of diffusing proteins reacting with a spherical target at the origin as first studied by Smoluchowski in 1916 [16]. To find out how fast a protein can find the target, we will study the diffusion-limited binding to the target. Our dynamical quantity is the three-dimensional volume density $n = n(r, t)$. We assume spherical symmetry so the density only depends on the radial distance $r$ to the origin and time $t$. The dynamics follows the diffusion equation

$$\frac{\partial}{\partial t} n(r, t) = D_{3d} \frac{1}{r^2} \frac{\partial}{\partial r} \left( r^2 \frac{\partial}{\partial r} n(r, t) \right), \tag{25.1}$$

where $D_{3d}$ is the protein diffusion constant. The proteins react with the target placed at the origin whenever they are within a distance $b$ of it (measured from the centre of the protein to the centre of the target). This means that we have an absorbing boundary condition at $r = b$,

$$n(r = b, t) = 0. \tag{25.2}$$

Far away from the target we will assume that the density approaches a constant value $n_{bulk}$, meaning that we have the second boundary condition

$$\lim_{r\to\infty} n(r,t) = n_{\text{bulk}}. \tag{25.3}$$

To solve the diffusion equation we Laplace-transform the density $n(r,t)$,

$$n(r,u) = \int_0^\infty dt\, e^{-ut} n(r,t). \tag{25.4}$$

Note that we indicate a Laplace-transformed quantity by a change of the variable to Laplace time $u$. When Laplace-transforming the diffusion equation, it is converted into the ordinary differential equation

$$un(r,u) - n_{\text{bulk}} = D_{\text{3d}} \frac{1}{r^2}\frac{\partial}{\partial r}\left(r^2 \frac{\partial}{\partial r} n(r,u)\right), \tag{25.5}$$

where we assumed that initially the concentration is constant $n(r, t = 0) = n_{\text{bulk}}$. The general solution to this equation is

$$n(r,u) = \frac{n_{\text{bulk}}}{u} + \frac{A(u)}{r} e^{\sqrt{u/D_{\text{3d}}}\,r} + \frac{B(u)}{r} e^{-\sqrt{u/D_{\text{3d}}}\,r}, \tag{25.6}$$

where $A$ and $B$ are integration constants that can depend on $u$. These can be identified via the two boundary conditions, leading to the unique selection

$$n(r,u) = \frac{n_{\text{bulk}}}{u}\left(1 - \frac{b}{r} e^{-\sqrt{u/D_{\text{3d}}}(r-b)}\right). \tag{25.7}$$

The current density along the radial direction is $-D_{\text{3d}}\partial n/\partial r$. The current $j$ of proteins into the target is minus this current times the surface area. In Laplace space, we find

$$\begin{aligned} j(u) &= 4\pi b^2 D_{\text{3d}} \left.\frac{\partial n(r,u)}{\partial r}\right|_{r=b} \\ &= 4\pi D_{\text{3d}} b n_{\text{bulk}} \left(\frac{1}{u} + \frac{b}{\sqrt{D_{\text{3d}} u}}\right). \end{aligned} \tag{25.8}$$

Transforming back to real time, this gives

$$j(t) = 4\pi D_{\text{3d}} b n_{\text{bulk}} \left(1 + \frac{b}{\sqrt{\pi D_{\text{3d}} t}}\right). \tag{25.9}$$

From this solution for the current into the target, we see that it reaches the stationary value $j_{\text{stat}} = 4\pi D_{\text{3d}} b n_{\text{bulk}}$ at times $t \gg b^2/D_{\text{3d}}$. With inverse mobility $1/D_{\text{3d}}$ and with the square of the target radius $b$, this characteristic time increases, as expected.

Let us for the moment assume that we are in the eventual stationary situation with a constant flux of proteins into the target. Then at any given moment, there is a probability density $j_{\text{stat}}$ of a protein arriving at the target. If we look at the waiting time $\tau$ before the next protein arrives, then this will be exponentially distributed with rate parameter $j_{\text{stat}}$, i.e., the average waiting time reads

$$\langle \tau \rangle = j_{\text{stat}}^{-1}. \tag{25.10}$$

This equation converts between the macroscopic view of the equations for the density of proteins and the microscopic view of the arrival of the first protein to the target. When $\langle \tau \rangle \gg b^2/D_{3\text{d}}$, i.e., when $n_{\text{bulk}} \ll b^{-3}$, then the transient contribution (the second term) in Eq. (25.9) is unimportant for the statistics of the protein binding. In the following, we will assume that we are at such dilute concentrations of proteins such that we only need to study the stationary regime to know the binding statistics. Furthermore, we will remove the dependence on concentration by defining the rate constant for the binding reaction as

$$k_{\text{on}} \equiv \frac{j_{\text{stat}}}{n_{\text{bulk}}} = 4\pi D_{3\text{d}} b. \tag{25.11}$$

This is indeed Smoluchowski's result [16].

The experimental motivation for developing another model than the above for DNA binding proteins comes from in vitro experiments on the Lac repressor [2] where a value $k_{\text{on}} \approx 10^{10}/[(\text{mol/l}) \cdot \text{sec}]$ was measured. We can compare this with Smoluchowski's theory for spherical targets by estimating its prediction for $k_{\text{on}}$. For a typical diameter of a transcription factor of $2R = 5\,\text{nm}$ by Stokes formula, we obtain $D_{3\text{d}} = 6\pi\eta R \approx 10^2 \mu\text{m}^2/\text{s}$ where we used the viscosity of water $\eta \approx 10^{-3}\,\text{Pa} \cdot \text{s}$. If we say the protein has to situate itself on the DNA within a precision matching the length of one base-pair, it corresponds to a target of size $2b \approx 0.3\,\text{nm}$. Consequently, we find $k_{\text{on}} \approx 0.2\,\mu\text{m}^3/\text{s} \approx 10^8/[(\text{mol/l}) \cdot \text{sec}]$. That some proteins can operate faster than this upper diffusion limit for spherical targets motivates the facilitated diffusion model in the next section. We note that the considerations here apply to in vitro situations. In vivo conditions will be different. For instance, a much reduced value for $D_{3\text{d}}$, compared with the estimate in water above, was found inside living *Escherichia coli* in [18].

## 25.3 Facilitated Diffusion Along DNA

The faster search can be explained by the fact that proteins can bind loosely ("non-specifically" [19–21]) to the DNA, and then slide along the DNA diffusively, searching for the target. To model this process mathematically, we will formulate it as a diffusion process along the DNA. Thus our central dynamic quantity $n(x, t)$ is now the density of loosely bound proteins per length of the DNA at position $x$, where $x$ is the distance along the DNA contour with the origin defined to be at the target. We label the diffusion constant along the DNA $D_{1\text{d}}$ to distinguish it from the diffusion

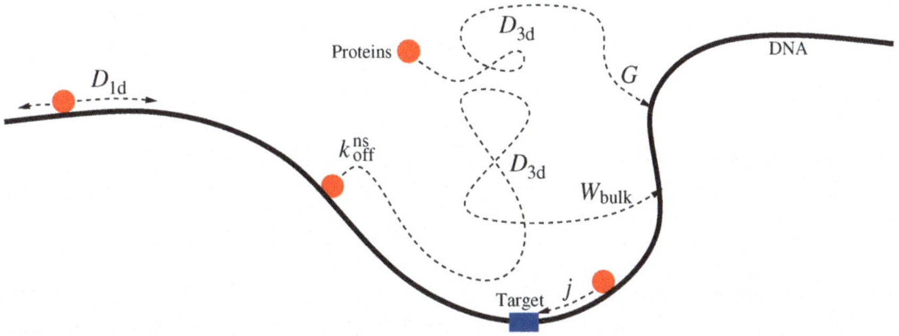


**Fig. 25.1** Schematic illustration of the different terms in Eq. (25.12)

constant $D_{3d}$ for unbound proteins. The diffusion equation governing $n(x, t)$ then has several extra terms as compared with Eq. (25.1) (see Fig. 25.1 for an illustration of this model):

$$\frac{\partial n(x,t)}{\partial t} = \left(D_{1d}\frac{\partial^2}{\partial x^2} - k_{off}^{ns}\right) n(x,t) - j(t)\delta(x) + G(x,t)$$
$$+k_{off}^{ns}\int_{-\infty}^{\infty} dx' \int_0^t dt'\, W_{bulk}(x - x', t - t')n(x', t'). \qquad (25.12)$$

Here $k_{off}^{ns}$ is the dissociation rate from the DNA of the non-specifically bound proteins; $j(t)$ is, as in the previous section, the number of proteins per time binding to the target site located at $x = 0$ and they are thus removed; $W_{bulk}(x - x', t - t')$ is the joint probability density that a protein rebinds on the DNA at the point $x$ and at the time $t$ given that it has fallen of the DNA at $x'$ at time $t'$. The remaining term, $G(x, t)$, accounts for the proteins that bind to the DNA at time $t$ and position $x$ without previously having been bound between time zero and time $t$. We will argue later that details of $G(x, t)$ (such as if it is uniform in $x$) will not matter for dilute protein concentrations. We are assuming that any protein diffusing across the target will be absorbed by it, imposing the condition $n(x = 0, t) = 0$. This condition determines the current $j(t)$ of proteins binding to the target. We note that in this model the proteins can only find the target by first binding loosely to the DNA, and then diffusing one-dimensionally along the DNA into the target.

In the above, we made a number of assumptions to simplify the interplay between one-dimensional diffusion along the DNA and diffusion in the bulk around the DNA: We assumed that the DNA is long enough such that we can take the end points to infinity when considering the diffusion to the target, i.e., we can integrate to infinity in the convolution integral containing $W_{bulk}$. This assumption should hold, if the absorption at the target only depletes proteins at a distance from the target that is small compared with the distance to the end points. In the next section we will introduce an effective sliding length that provides a length scale for this depletion. Another assumption is that there is no dependence between excursion events to the bulk. A static coiled configuration of the DNA could lead to such dependence, since two excursions following each other could bring the protein back and forth between two parts of the DNA that are far away from each measured along the DNA with high probability. The assumption should hold for straight or quickly fluctuating DNA though. Assuming we are in either of these situations, the kernel $W_{\text{bulk}}(x-x', t-t')$ is taken to be homogeneous in space and time i.e., the distance and duration of excursions in the bulk do not depend on the initial location or time. Additionally, we assume that the diffusion along the DNA is homogeneous, i.e., it does not depend on the local sequence of nucleotides. See for instance [22, 23] for discussions of how sequence heterogeneity could be a barrier for fast search, but also for a mechanism with which proteins can circumvent this barrier. Finally, we assume that the targets' extension along the DNA length is short compared with typical sliding lengths along the DNA. This implies that the possibility that the protein can bind to the DNA directly on top of the target via three-dimensional diffusion in the bulk does not matter for the search time, and we therefore consider only the possibility that the protein can find the target by sliding along the DNA.

As in the previous section, we want to find the stationary flux into the target $j_{\text{stat}}$, and we will use the same technique of Laplace transforming, replacing time $t$ with the Laplace variable $u$. Furthermore, we will Fourier-transform along the DNA:

$$n(q, u) = \int_{-\infty}^{\infty} dx\, e^{-iqx} n(x, u), \tag{25.13}$$

similarly indicating the transform by replacing the position coordinate $x$ with wavenumber $q$. Applying the two transforms, Eq. (25.12) becomes

$$\frac{1}{W(q, u)} n(q, u) - n(q, t = 0) = G(q, u) - j(u), \tag{25.14}$$

where

$$\frac{1}{W(q, u)} = u + D_{1d} q^2 + k_{\text{off}}^{\text{ns}} [1 - W_{\text{bulk}}(q, u)]. \tag{25.15}$$

To extract the long time behaviour, we rewrite Eq. (25.14) to arrive at

$$n(q, u) = n^{\text{ns}}(q, u) - j(u) W(q, u), \tag{25.16}$$

where we have introduced

$$n^{\text{ns}}(q, u) = W(q, u)[G(q, u) + n(q, t = 0)]. \tag{25.17}$$

$n^{\text{ns}}(x, t)$ is the density of proteins on the DNA in the absence of the sink at the target. For this situation we would expect equilibration to occur, such that eventually we will have a constant density $n_{\text{eq}}^{\text{ns}}$, since without the sink there is no inhomogeneity along the DNA. In the following, we will neglect all dependence on the initial distribution of proteins by assuming we are in this equilibrium limit of loose binding to the DNA.

Due to the linearity of our equations we expect this assumption to hold in the dilute limit, since the equilibration time should be independent of the protein concentration, while the search time will increase as concentration is lowered [10]. In the Laplace domain this means that $n^{\text{ns}}(x, u) \sim n^{\text{ns}}_{\text{eq}}/u$, where $\sim$ indicates that we are looking at the behaviour at long times, which by Tauberian theorems [24] is equivalent to studying the limit of $u$ going to zero. Equations (25.8) and (25.9) provide an example of this, where it can be found that the term that dominates at small $u$ converts to a term that dominates at large $t$, and vice versa for the other term.

We can find the current $j(u)$ by inverse Fourier-transforming Eq. (25.16) at $x = 0$ using the condition of a sink at the target, $n(x = 0, t) = 0$, to arrive at

$$j(u) = \frac{n^{\text{ns}}(x = 0, u)}{W(x = 0, u)}. \tag{25.18}$$

To proceed we will assume that the value $W(x = 0, u = 0)$ is finite and non-zero, which will be the case for the three-dimensional diffusion problems we study here with the bulk space stretching far in all directions, leading to long tailed distributions for the bulk excursions. With this assumption we find the limiting behaviour at small $u$: $j(u) \sim k_{1\text{d}} n^{\text{ns}}_{\text{eq}}/[uW(x = 0, u = 0)]$, i.e., that the current becomes stationary at long times with value

$$j_{\text{stat}} \sim k_{1\text{d}} n^{\text{ns}}_{\text{eq}}, \tag{25.19}$$

where $k_{1\text{d}} = 1/W(x = 0, u = 0)$, or explicitly

$$k_{1\text{d}}^{-1} = \int_{-\infty}^{\infty} \frac{dq}{2\pi} \frac{1}{D_{1\text{d}} q^2 + k^{\text{ns}}_{\text{off}}(1 - \lambda_{\text{bulk}}(q))}, \tag{25.20}$$

with the distribution $\lambda_{\text{bulk}}(x) = W_{\text{bulk}}(x, u = 0)$ of the relocation lengths along the DNA contour of the three-dimensional bulk excursions.

To proceed and calculate the integral in Eq. (25.20), we need to know $\lambda_{\text{bulk}}(q)$. There is, however, a situation we can discuss already without a detailed form of $\lambda_{\text{bulk}}(q)$: If the density $\lambda_{\text{bulk}}(x)$ has its probability mass spread out sufficiently such that a protein is unlikely to rebind to a point where it was previously bound to the DNA, then for the purpose of calculating the integral in Eq. (25.20) we can effectively consider $\lambda_{\text{bulk}}(q)$ to vanish for $q \neq 0$ (it will always be unity for $q = 0$ by normalisation). We remark that the integrand will be singular at $q = 0$. But the overall three-dimensional geometry of the bulk means that the rate of target finding will be finite as in Sect. 25.2. Thus the singularity will be soft enough for the integral to converge (see the discussion above Eq. (25.32) for the explicit case of straight DNA). Assuming we are in this situation, where $\lambda_{\text{bulk}}(q)$ does not contribute significantly to the integral, we discard $\lambda_{\text{bulk}}(q)$ and then integrate analytically to find

$$k_{1\text{d}} = 2\sqrt{D_{1\text{d}} k^{\text{ns}}_{\text{off}}}. \tag{25.21}$$

If we multiply this result on both sides with $n_{\mathrm{eq}}^{\mathrm{ns}}$ it can be rewritten as

$$j_{\mathrm{stat}} = 2l_{\mathrm{sl}}k_{\mathrm{off}}^{\mathrm{ns}}n_{\mathrm{eq}}^{\mathrm{ns}}. \tag{25.22}$$

To understand this expression, first note that $k_{\mathrm{off}}^{\mathrm{ns}}n_{\mathrm{eq}}^{\mathrm{ns}}$ is the rate of unbinding events per length of the DNA. In the equilibrated situation, this is equivalent to the rate of rebinding events per length. If such a rebinding happens within a distance of $l_{\mathrm{sl}}$ on either side of the target, then the protein will be likely to find the target, thus resulting in Eq. (25.22) for the rate of target localisation.

In the previous section, we preferred to express results in terms of $k_{\mathrm{on}} = j_{\mathrm{stat}}/n_{\mathrm{bulk}}$, where $n_{\mathrm{bulk}}$ was the density per volume in the bulk, i.e., the density of non-bound proteins in the water around the DNA. At equilibrium of non-specific binding the two concentrations are related by the non-specific binding constant per length of DNA, which is $K_{\mathrm{ns}} = n_{\mathrm{eq}}^{\mathrm{ns}}/n_{\mathrm{bulk}}$. Using this, we can write

$$k_{\mathrm{on}} = k_{\mathrm{1d}}K_{\mathrm{ns}}. \tag{25.23}$$

Note that if a significant fraction of the proteins are bound to the DNA, it becomes necessary to distinguish the bulk concentration of proteins $n_{\mathrm{bulk}}$ from the total amount of proteins per volume $n_{\mathrm{total}}$. If $l_{\mathrm{DNA}}$ is the length of the DNA divided by the volume of the surrounding water, then the concentrations are related by

$$n_{\mathrm{total}} = n_{\mathrm{bulk}} + l_{\mathrm{DNA}}^{\mathrm{total}}n_{\mathrm{eq}}^{\mathrm{ns}} = (1 + l_{\mathrm{DNA}}^{\mathrm{total}}K_{\mathrm{ns}})n_{\mathrm{bulk}}. \tag{25.24}$$

In terms of $n_{\mathrm{total}}$, the target localisation rate can therefore be written as

$$j_{\mathrm{stat}} = \frac{k_{\mathrm{on}}}{1 + K_{\mathrm{ns}}l_{\mathrm{DNA}}^{\mathrm{total}}}n_{\mathrm{total}}. \tag{25.25}$$

In the following two sections, we will aim at calculating $k_{\mathrm{on}}$ first for straight and then coiled DNA.

## 25.4 Straight DNA

In this section, we will obtain $W_{\mathrm{bulk}}$ for the case when the DNA is completely straight. This can be achieved, e.g., by stretching the DNA in an optical tweezers setup [14, 25] or in microfluidic setups [26]. We can then consider the problem as a diffusion problem around a cylinder, which is a problem that is well known in the literature; see for instance [27–29]. Here we provide a derivation tailored towards obtaining $W_{\mathrm{bulk}}(x, t)$. We first determine the probability density $P$ for a single protein to be at a position in space at a certain time $t$. In this cylindrical geometry, the DNA is situated along the $x$-axis, and the protein will interact with it when it is at a distance $r = r_{\mathrm{int}}$ from the $x$-axis (see Fig. 25.2). For the directions orthogonal to the $x$-axis,

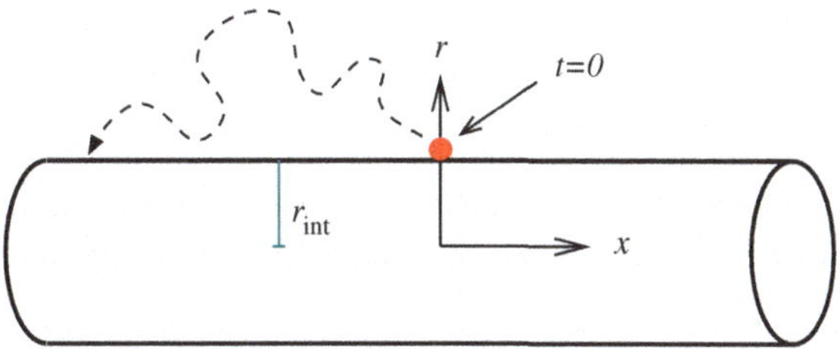


**Fig. 25.2** Schematic of the bulk excursion of a protein being released at the cylindrical DNA-interaction surface. This process is modelled by the diffusion equation (25.26) and the boundary condition (25.27)

we use polar coordinates with the radial coordinate $r$. Since we are only interested in the value of $x$ at which the protein rebinds to the DNA, we will drop the dependence on the polar angle and $P$ is then rotationally symmetric around the $x$-axis. Thus we have $P = P(x, r, t)$, and the three-dimensional diffusion equation can be written as

$$\frac{\partial P}{\partial t} = D_{3\text{d}} \left( \frac{\partial^2}{\partial x^2} + \frac{1}{r} \frac{\partial}{\partial r} r \frac{\partial}{\partial r} \right) P, \tag{25.26}$$

with the diffusion constant $D_{3\text{d}}$. The boundary condition we use at $r = r_{\text{int}}$ can be expressed as the probability flux density in the radial direction,

$$- D_{3\text{d}} \frac{\partial P}{\partial r} \bigg|_{r=r_{\text{int}}} = - \frac{k_{\text{on}}^{\text{ns}}}{2\pi r_{\text{int}}} P \bigg|_{r=r_{\text{int}}} + \frac{\delta(x)}{2\pi r_{\text{int}}} \delta(t). \tag{25.27}$$

On the right-hand side, the first term represents the probability rate of the protein loosely binding to the DNA per surface area of the cylinder. $k_{\text{on}}^{\text{ns}}$ is a rate constant with units of area per time. The second term represents the protein release from the boundary at $x = 0$ at time $t = 0$. Note that consistency with non-specific equilibration, $n_{\text{eq}}^{\text{ns}} k_{\text{off}}^{\text{ns}} = n_{\text{bulk}} k_{\text{on}}^{\text{ns}}$, means that $k_{\text{on}}^{\text{ns}} = K_{\text{ns}} k_{\text{off}}^{\text{ns}}$. That the protein enters the bulk via the boundary condition means that we will take the bulk to be empty initially: $P(x, r, t = 0) = 0$. Additionally we assume that $P$ vanishes infinitely far away from the spatial origin at all times.

To solve Eq. (25.26), we Laplace-transform the time $t$ as usual and also Fourier-transform along the $x$-axis again to convert the derivatives with respect to $x$ to factors of $iq$, i.e.,

$$uP(q, r, u) = D_{3\text{d}} \left( -q^2 + \frac{1}{r} \frac{\partial}{\partial r} r \frac{\partial}{\partial r} \right) P(q, r, u). \tag{25.28}$$

The general solution of this equation is given in terms of modified Bessel functions $I_n$ and $K_n$ as

$$P(q, r, u) = A I_0(\bar{q} r) + B K_0(\bar{q} r), \tag{25.29}$$

where $\bar{q} = \sqrt{q^2 + u/D_{3\text{d}}}$. The boundary condition that $P$ vanishes infinitely far from the origin tells us that $A = 0$. Inserting the term with $B$ into the remaining boundary condition (25.27) gives

$$B = \frac{1}{k_{\rm on}^{\rm ns} K_0(\bar{q} r_{\rm int}) + 2\pi r_{\rm int} D_{\rm 3d} \bar{q} K_1(\bar{q} r_{\rm int})}. \tag{25.30}$$

Now we have arrived at a result that we can use to calculate $W_{\rm bulk}$, since it is simply given by $W_{\rm bulk}(x, t) = k_{\rm on}^{\rm ns} P(x, r = r_{\rm int}, t)$, i.e., the probability density that the protein binds at the boundary at position $x$ and time $t$. This produces the result

$$W_{\rm bulk}^{\rm cyl}(q, u) = \left(1 + \frac{2\pi D_{\rm 3d} \bar{q} r_{\rm int} K_1(\bar{q} r_{\rm int})}{k_{\rm on}^{\rm ns} K_0(\bar{q} r_{\rm int})}\right)^{-1}. \tag{25.31}$$

The superscript cyl indicates that this is the solution when the DNA is forming a straight cylinder.

We are now in a position where we can evaluate the integral in Eq. (25.20) numerically, since we can obtain $\lambda_{\rm bulk}$ from $W_{\rm bulk}^{\rm cyl}$ simply by setting $u = 0$, i.e., $\bar{q} = |q|$. The numerical result for $k_{\rm on}$ is shown as a function of $k_{\rm on}^{\rm ns}$ in Fig. 25.3, together with two asymptotic approximations. The asymptotic expression for small $k_{\rm on}^{\rm ns}$ is based on Eqs. (25.21) and (25.23) and is limited by the non-specific binding to DNA. This limit holds when $k_{\rm on}^{\rm ns} \ll D_{\rm 3d}$. In the opposite limit, $k_{\rm on}^{\rm ns} \gg D_{\rm 3d}$, the detailed form of $W_{\rm bulk}^{\rm cyl}(q, u = 0)$ at small values of $|q|$ becomes important. To obtain an expression for $k_{\rm on}$ in this limit, we use that the asymptotic behaviour of the modified Bessel functions at small arguments are $K_1(x) \approx 1/x$ and $K_0(x) \approx -\ln(x)$. Thus at small $|q|$, $W_{\rm bulk}^{\rm cyl}(q, u = 0)$ will deviate from unity only by a small term that decays slowly as $-1/\ln(|q| r_{\rm int})$. If we ignore this logarithmic factor and examine the behaviour of the denominator of the integrand in Eq. (25.20), we see that the slow logarithmic behaviour becomes dominant at $|q|$ below $1/l_{\rm sl}^{\rm eff}$ where

$$l_{\rm sl}^{\rm eff} = \sqrt{\frac{k_{\rm on}^{\rm ns}}{(2\pi D_{\rm 3d})}} l_{\rm sl}. \tag{25.32}$$

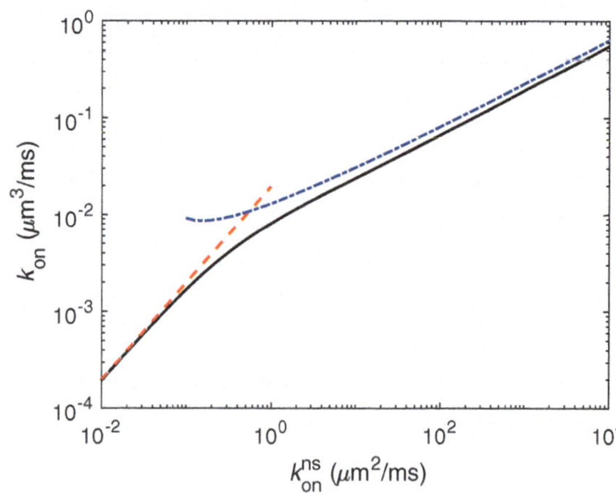


**Fig. 25.3** Plot of $k_{\rm on}$ as a function of $k_{\rm on}^{\rm ns}$ for $D_{\rm 3d} = 0.1\,\mu{\rm m}^2/{\rm ms}$, $D_{\rm 1d} = 10^{-4}, \mu{\rm m}^2/{\rm ms}$, $r_{\rm int} = 3$ nm and $k_{\rm off}^{\rm ns} = 1\,{\rm ms}^{-1}$. The black solid line is the full numerical result based on Eqs. (25.20), (25.23) and (25.31), while the red dashed line matching at low $k_{\rm on}^{\rm ns}$ is the result in Eqs. (25.21) and (25.23) and the blue dash-dotted line is the result in Eq. (25.33)

If we simply replace the last term in the denominator by its asymptotic value at $q = 1/l_{sl}^{eff}$, i.e., we set $1 - \lambda_{bulk}(q) = 2\pi D_{3d}/[k_{on}^{ns} \ln(l_{sl}^{eff}/r_{int})]$ then we can perform the integral and obtain

$$k_{on} \sim \frac{4\pi D_{3d} l_{sl}^{eff}}{[\ln(l_{sl}^{eff}/r_{int})]^{1/2}}. \tag{25.33}$$

This is the approximate asymptotic expression plotted for large $k_{on}^{ns}$ in Fig. 25.3. The crossover to this asymptotic behaviour occurs around $k_{on}^{ns} \approx 2\pi D_{3d}$, but the final convergence only sets in very slowly as $k_{on}^{ns}$ increases due to the involved logarithmic behaviours.

In the limit when $k_{on}^{ns} \gg D_{3d}$ the proteins that fall of the DNA will have a tendency to rebind immediately, thus performing what is in effect just a tiny hop on the spot. The tiny probability that the protein avoids such a hop must be proportional to $D_{3d}/k_{on}^{ns}$ by dimensional arguments. Therefore the number of hops and thus the time spent loosely bound on the DNA scale with $k_{on}^{ns}/D_{3d}$. The effective distance covered by one-dimensional diffusion along the DNA while sliding and hopping on the spot will therefore increase by a factor proportional to $\sqrt{k_{on}^{ns}/D_{3d}}$ since the displacement in Brownian motion increases by the square root of time. With this reasoning, an interpretation becomes evident of $l_{sl}^{eff}$ as the effective sliding length resulting from combining sliding events connected by quick rebindings. By comparing Eq. (25.33) with the Smoluchowski result (25.11), this reasoning also provides an interpretation of $l_{sl}^{eff}$ as an increased target size (sometimes referred to as an antenna [30]) with $1/[\ln(l_{sl}^{eff}/r_{int})]^{1/2}$ entering as a geometrical factor due to the cylindrical shape of the target.

Finally, we note that the results of Eqs. (25.20) and (25.31) match the result obtained by Berg and Ehrenberg in 1982 [7] for infinitely long straight DNA.

## 25.5 Coiled DNA

We now turn to the calculation of $k_{on}$ when the DNA is coiled. To reach an analytical expression, we assume that the DNA has a sufficiently high persistence length, such that locally around the target the DNA can effectively be thought of as a number of cylindrical straight rods passing close to each other (see left part of Fig. 25.4). The idea behind this is that we can solve the diffusion problem for a single straight rod (see right part of Fig. 25.4), and then we can combine the solutions by assuming independent interactions. In this independent interactions approach, the individual solutions are combined by letting the actual binding event be the binding event that happens first among the multiple rods. This means that we can calculate the "survival" probability of the protein not having bound anywhere as simply the product of all the probabilities for each rod that it has not yet bound to that individual rod.

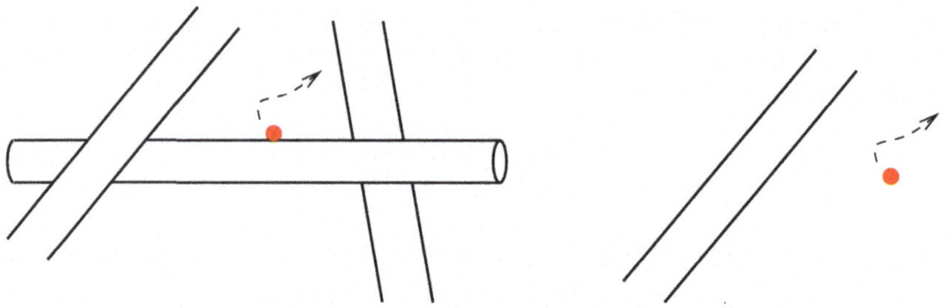

**Fig. 25.4** Left: The DNA is considered made up of locally straight rods with the protein diffusing among them. Right: The diffusion problem is solved for each rod individually and then combined assuming independent interactions

We call $P_{\mathrm{surv}}^{\mathrm{foreign}}(t)$ the survival probability that the protein has not yet bound to any foreign straight rod ignoring the existence of the (non-foreign) straight rod that it unbound from. With the assumption of independent interactions, we can write

$$P_{\mathrm{surv}}^{\mathrm{foreign}}(t) = \left[1 - J_{\mathrm{single}}(t)\right]^N, \qquad (25.34)$$

where $N$ is the, for the moment finite, number of foreign straight rods and $J_{\mathrm{single}}(t)$ is the probability that it has reacted with a particular foreign rod ignoring the existence of all the others. Assuming we have found $J_{\mathrm{single}}(t)$ and thereby $P_{\mathrm{surv}}^{\mathrm{foreign}}(t)$, we will then calculate $W_{\mathrm{bulk}}$ as

$$W_{\mathrm{bulk}}(q,t) = W_{\mathrm{bulk}}^{\mathrm{cyl}}(q,t) P_{\mathrm{surv}}^{\mathrm{foreign}}(t) \quad (q \neq 0), \qquad (25.35)$$

where $W_{\mathrm{bulk}}^{\mathrm{cyl}}$ is the solution found in the last section when the DNA was a single straight rod. Here we have again used the assumption of independent interactions to calculate the probability that it returns to the straight rod it unbound from, i.e., we multiply the solution without foreign rods with the probability that it has not yet bound to any foreign rod. Also, we have assumed that binding to a foreign rod leads the protein so far away from where it unbound that these foreign rod-bindings do not contribute significantly to the integral of Eq. (25.20), and we have therefore discarded them in Eq. (25.35). This is the same kind of approximation that was used previously when arriving at the reaction-limited result in Eq. (25.21). We have indicated the approximation by the condition $q \neq 0$ in Eq. (25.35), since these excursions, no matter how long they tend to be, will always contribute at $q = 0$ to $W_{\mathrm{bulk}}(q,t)$.

The single foreign rod binding probability $J_{\mathrm{single}}$ will be found by solving the usual three-dimensional diffusion equation, Eq. (25.26), but now with the initial condition that the protein can start anywhere in the bulk relative to the rod. This means that the initial condition is a uniform concentration, $P(x,r,t=0) = 1/V$, with $V$ being a constant. For finite $V$ this implies that $P(x,r,t=0)$ is not normalised. However, $V$ will later be taken to infinity in a way where it will represent the volume of the bulk water, thus restoring proper normalisation of probability when this limit is taken. After a Laplace transformation of time $t$, we get

$$uP(r,u) - \frac{1}{V} = D_{3d}\frac{1}{r}\frac{\partial}{\partial r}r\frac{\partial}{\partial r}P(r,u), \tag{25.36}$$

where the $x$ dependence has been left out since the problem is uniform along the $x$ axis. The general solution to this equation is

$$P(r,u) = \frac{1}{uV} + AI_0(\sqrt{u/D_{3d}}r) + BK_0(\sqrt{u/D_{3d}}r), \tag{25.37}$$

with the integration constants $A$ and $B$ that may depend on $u$. As one of the boundary conditions, we take that the probability density $P$ should approach the constant $1/V$ far from the DNA. In Laplace space this boundary condition can be written as $\lim_{r\to\infty} P(r,u) = 1/(uV)$. Thus we have that $A = 0$. For the boundary condition on the DNA surface we employ Eq. (25.27) without the release from the surface, i.e.,

$$-D_{3d}\frac{\partial P}{\partial r}\bigg|_{r=r_{\rm int}} = -\frac{k_{\rm on}^{\rm ns}}{2\pi r_{\rm int}}P\bigg|_{r=r_{\rm int}}. \tag{25.38}$$

We then obtain

$$P(r,u) = \frac{1}{uV} - \frac{k_{\rm on}^{\rm ns}K_0(\sqrt{u/D_{3d}}r)/(uV)}{k_{\rm on}^{\rm ns}K_0(\sqrt{u/D_{3d}}r_{\rm int}) + 2\pi r_{\rm int}\sqrt{D_{3d}u}K_1(\bar{q}r_{\rm int})}. \tag{25.39}$$

We can now find the probability of having reacted with a single rod, $J_{\rm single}(t) = \int_0^t Lk_{\rm on}^{\rm ns}P(r_{\rm int},t')dt'$, which in Laplace space becomes

$$J_{\rm single}(u) = \frac{Lk_{\rm on}^{\rm ns}}{u^2V}\left(1 + \frac{k_{\rm on}^{\rm ns}K_0(\sqrt{u/D_{3d}}r_{\rm int})}{2\pi\sqrt{uD_{3d}}r_{\rm int}K_1(\sqrt{u/D_{3d}}r_{\rm int})}\right)^{-1}. \tag{25.40}$$

This puts us in a position to combine the results for each rod via Eq. (25.34), and take the limit of $N$, $L$, and $V$ growing to infinity with $L/V = l_{\rm DNA}/N$ for constant $l_{\rm DNA}$. This results in

$$P_{\rm surv}^{\rm foreign}(t) = \exp\left[-J_{\rm cap}(t)\right], \tag{25.41}$$

where $J_{\rm cap}(t)$ is the inverse Laplace transform of

$$J_{\rm cap}(u) = \frac{k_{\rm on}^{\rm ns}l_{\rm DNA}}{u^2}\left(1 + \frac{k_{\rm on}^{\rm ns}K_0(\sqrt{u/D_{3d}}r_{\rm int})}{2\pi\sqrt{uD_{3d}}r_{\rm int}K_1(\sqrt{u/D_{3d}}r_{\rm int})}\right)^{-1}. \tag{25.42}$$

We can evaluate the inverse Laplace transform of $J_{\rm cap}(u)$ numerically and thereby obtain $P_{\rm surv}^{\rm foreign}(t)$. However, to obtain $W_{\rm bulk}(q,u)$ we would have to perform a Laplace transform on top of that. To reduce the number of transforms, i.e., integrals, that need to be performed, we will use a convenient approximation for $J_{\text{cap}}(t)$. To argue for this approximation we first note that for $k_{\text{on}}^{\text{ns}} \gg D_{3\text{d}}$, we can divide the $u$-dependence of $J_{\text{cap}}(u)$ into the three regimes

$$J_{\text{cap}}(u) \approx \begin{cases} k_{\text{on}}^{\text{ns}} l_{\text{DNA}} u^{-2} & , \; u^{-1} \ll \frac{D_{3\text{d}} r_{\text{int}}^2}{(k_{\text{on}}^{\text{ns}})^2} \\ 2\pi l_{\text{DNA}} r_{\text{int}} \sqrt{D_{3\text{d}}} u^{-3/2} & , \; \frac{D_{3\text{d}} r_{\text{int}}^2}{(k_{\text{on}}^{\text{ns}})^2} \ll u^{-1} \ll \frac{r_{\text{int}}^2}{D_{3\text{d}}} \\ \dfrac{4\pi l_{\text{DNA}} D_{3\text{d}}}{u^2 \ln(D_{3\text{d}}/(u r_{\text{int}}^2))} & , \; u^{-1} \gg \frac{r_{\text{int}}^2}{D_{3\text{d}}} \end{cases} . \tag{25.43}$$

When the regimes are well separated in this way, with decaying $u$ dependence, they can each be Laplace-inverted independently to find

$$J_{\text{cap}}(t) \approx \begin{cases} k_{\text{on}}^{\text{ns}} l_{\text{DNA}} t & , \; t \ll \frac{D_{3\text{d}} r_{\text{int}}^2}{(k_{\text{on}}^{\text{ns}})^2} \\ 4 l_{\text{DNA}} r_{\text{int}} \sqrt{\pi D_{3\text{d}} t} & , \; \frac{D_{3\text{d}} r_{\text{int}}^2}{(k_{\text{on}}^{\text{ns}})^2} \ll t \ll \frac{r_{\text{int}}^2}{D_{3\text{d}}} \\ \dfrac{4\pi l_{\text{DNA}} D_{3\text{d}} t}{\ln(D_{3\text{d}} t / r_{\text{int}}^2)} & , \; t \gg \frac{r_{\text{int}}^2}{D_{3\text{d}}} \end{cases} , \tag{25.44}$$

where Tauberian theorems [24] were applied in the last regime. When $k_{\text{on}}^{\text{ns}} \gg D_{3\text{d}}$, the first regime will be very brief and the probability of being captured by a foreign rod will only accumulate to $l_{\text{DNA}} r_{\text{int}}^2 D_{3\text{d}} / k_{\text{on}}^{\text{ns}}$. This will be a very small number, and therefore we will ignore this regime. In the last regime we will approximate the slowly varying logarithm with a constant. Thus we will assume $J_{\text{cap}}(t) \approx k_{\text{cap}} t$ in this regime where the precise value of $k_{\text{cap}}$ will be fixed later, but it will be somewhere around $l_{\text{DNA}} D_{3\text{d}}$. Combining the last two regimes we see that a function that gives the correct approximate behaviour in each is

$$P_{\text{app}}(t) = \exp[-J_{\text{app}}(t)] \quad \text{with} \quad J_{\text{app}}(t) = 4 l_{\text{DNA}} r_{\text{int}} \sqrt{\pi D_{3\text{d}} t} + k_{\text{cap}} t, \tag{25.45}$$

and this is the function we will replace $P_{\text{surv}}^{\text{foreign}}$ with. To fix $k_{\text{cap}}$ we choose the condition that we want the same probability that the protein returns to the non-foreign rod for both of $P_{\text{surv}}^{\text{foreign}}$ and $P_{\text{app}}$. This gives the following equation for $k_{\text{cap}}$

$$\int_0^\infty dt \; W_{\text{bulk}}^{\text{cyl}}(q = 0, t) \left[ P_{\text{surv}}^{\text{foreign}}(t) - P_{\text{app}}(t) \right] = 0, \tag{25.46}$$

which can be handled numerically. This equation involves two Laplace inversions and an integral over these for each numerical guess at a solution, but we have a good guess by using $l_{\text{DNA}} D_{3\text{d}}$ as a starting value for $k_{\text{cap}}$. See Fig. 25.5 for a comparison of $J_{\text{cap}}(t)$ and $J_{\text{app}}(t)$. The numerical Laplace inversions were handled using the Talbot and Euler algorithms presented in [31].

The $\sqrt{t}$ part of $P_{\text{app}}$ can be rewritten as a Laplace transform in the following way:

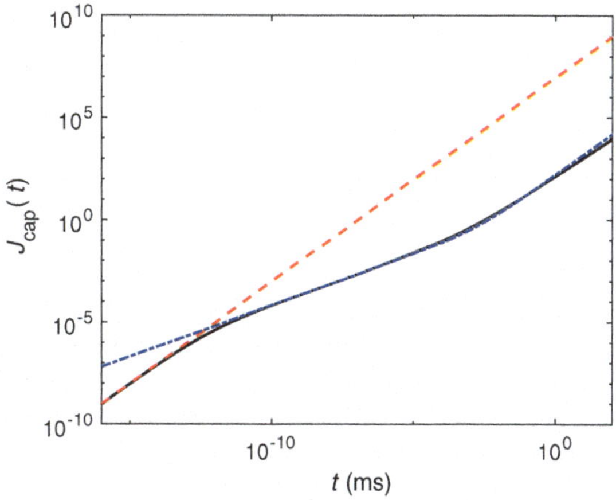


**Fig. 25.5** Plot of $J_{\text{cap}}(t)$ as a function of time $t$ for $D_{3d} = 0.1$, $\mu\text{m}^2/\text{ms}$, $r_{\text{int}} = 3$ nm and $k_{\text{on}}^{\text{ns}} = 10^4$, $\mu\text{m}^2/\text{ms}$. The black solid line is the result of numerically Laplace-inverting Eq. (25.42), while the red dashed line matching at low $t$ is the result for the first regime in Eq. (25.44) and the blue dash-dotted line is the approximation $J_{\text{app}}(t)$ given in Eq. (25.45)

$$P_{\text{app}}(t) = \int_0^\infty ds\; e^{-(s+k_{\text{cap}})t} \frac{a}{2\sqrt{\pi s^3}} e^{-a^2/(4s)}, \tag{25.47}$$

where $a = 4\sqrt{\pi} r_{\text{int}} l_{\text{DNA}} \sqrt{D_{3d}}$. This is very convenient when it is inserted in Eq. (25.35) instead of $P_{\text{surv}}^{\text{foreign}}$ and Laplace-transformed to obtain $W_{\text{bulk}}(q, u)$. Then one finds

$$W_{\text{bulk}}(q, u) = \int_0^\infty ds\; \frac{a\, e^{-a^2/(4s)}}{2\sqrt{\pi s^3}} W_{\text{bulk}}^{\text{cyl}}(q, u + s + k_{\text{cap}}). \tag{25.48}$$

With this expression the number of involved integrals has been reduced to a single one, and the expression is now ready to be used inside the integral of Eq. (25.20). Along the way in the derivation of the expression, we assumed $D_{3d} \ll k_{\text{on}}^{\text{ns}}$. But note that in the opposite limit $D_{3d} \gg k_{\text{on}}^{\text{ns}}$ the expression is consistent with the vanishing of $\lambda_{\text{bulk}}(q)$ for $q \neq 0$ and thus it extrapolates to the result in Eq. (25.21) in that limit.

To get some analytic insight into the effects of the coiling or presence of foreign DNA, we can apply approximations similar to the ones taken in the $D_{3d} \ll k_{\text{on}}^{\text{ns}}$ regime of the straight rod. Thus we approximate $W_{\text{bulk}}^{\text{cyl}}$ by setting $q = 1/l_{\text{sl}}^{\text{eff}}$, and furthermore we approximate $k_{\text{cap}} = D_{3d} l_{\text{DNA}}$. Using this we can write

$$\begin{aligned} 1 - \lambda_{\text{bulk}}(q) &\approx \int_0^\infty ds\; \frac{a\, e^{-a^2/(4s)}}{2\sqrt{\pi s^3}} [1 - W_{\text{bulk}}^{\text{cyl}}(q = 1/l_{\text{sl}}^{\text{eff}}, u = s + k_{\text{cap}})] \\ &= C_1 + C_2. \end{aligned} \tag{25.49}$$

In the second line we split the integral over $s$ into two parts, where $C_1$ represents the integral from $s = 0$ to $s = D_{3d}/r_{\text{int}}^2$ and $C_2$ the remaining integral to infinity.

To obtain an approximation for $C_1$, we assume that the dominant contribution of the integral occurs for $s \ll k_{\text{cap}} + D_{3d}/(l_{\text{sl}}^{\text{eff}})^2$ and thus we simply ignore the $s$ dependence in $W_{\text{bulk}}^{\text{cyl}}$. This gives

$$
\begin{aligned}
C_1 &\approx \int_0^{D_{3d}/r_{\rm int}^2} ds\, \frac{a\, e^{-a^2/(4s)}}{2\sqrt{\pi s^3}} [1 - W_{\rm bulk}^{\rm cyl}(q = 1/l_{\rm sl}^{\rm eff}, u = k_{\rm cap})] \\
&\approx 1 - W_{\rm bulk}^{\rm cyl}(q = 1/l_{\rm sl}^{\rm eff}, u = k_{\rm cap}) \\
&\approx -\frac{2\pi D_{3d}}{k_{\rm on}^{\rm ns} \ln\left(\sqrt{(l_{\rm sl}^{\rm eff})^{-2} + l_{\rm DNA} r_{\rm int}}\right)}.
\end{aligned}
\tag{25.50}
$$

For the approximation in the second line we used $a^2 \ll D_{3d}/r_{\rm int}^2$ to extend the integral back to infinity again, and normalisation of the factor in front of the square brackets to perform the integration. In the third line we used the asymptotic expansion of the modified Bessel functions at small arguments, as we did when deriving Eq. (25.33).

To obtain an approximation for $C_2$, we do the opposite as for $C_1$ and ignore instead $1/l_{\rm sl}^{\rm eff}$ and $k_{\rm cap}$ for the remainder of the integral. This gives

$$
\begin{aligned}
C_2 &\approx \int_{D_{3d}/r_{\rm int}^2}^{\infty} ds\, \frac{a\, e^{-a^2/(4s)}}{2\sqrt{\pi s^3}} [1 - W_{\rm bulk}^{\rm cyl}(q = 0, u = s)] \\
&\approx \int_{D_{3d}/r_{\rm int}^2}^{\infty} ds\, \frac{a}{2\sqrt{\pi s^3}} \frac{1}{1 + k_{\rm on}^{\rm ns}/(2\pi r_{\rm int}\sqrt{D_{3d} s})} \\
&= 4 r_{\rm int}^2 l_{\rm DNA} \frac{2\pi D_{3d}}{k_{\rm on}^{\rm ns}} \ln\left(1 + \frac{k_{\rm on}^{\rm ns}}{2\pi D_{3d}}\right).
\end{aligned}
\tag{25.51}
$$

For the approximation in the second line, we used $a^2 \ll D_{3d}/r_{\rm int}^2$ to discard the exponential and we used that the asymptotic expansion of the two modified Bessel functions are identical for $\bar{q} r_{\rm int} \gg 1$ to discard these functions too.

Finally, setting $1 - \lambda_{\rm bulk}(q) = C_1 + C_2$ we can perform the $q$ integral in Eq. (25.20) and reach the approximate expression

$$
k_{\rm on} \sim 4\pi D_{3d} l_{\rm sl}^{\rm eff} \sqrt{\frac{k_{\rm on}^{\rm ns}}{2\pi D_{3d}} (C_1 + C_2)}. \tag{25.52}
$$

In Fig. 25.6 the ratio of $k_{\rm on}$ for the coiled and straight DNA is plotted as a function of $l_{\rm DNA}^{1/2}$ for both using Eq. (25.48) in Eq. (25.20) and for the approximate result of Eq. (25.52) divided by the asymptotic straight approximation in Eq. (25.33). The approximate result shows clear deviations, but it qualitatively captures the steep increase in $k_{\rm on}$ when $l_{\rm DNA}$ becomes denser than $1/(l_{\rm sl}^{\rm eff})^2$.

To give a feeling for the quantity $l_{\rm DNA}$, we provide an estimate of it in the special case of the Worm Like Chain model. Consider a point on the DNA a contour distance $s$ from the target. According to [32] the probability that this point happens to be in a tiny volume $dV$ around the target can be approximated by the expression

$$
j_{\rm M}(s) dV = \left(\frac{3}{4\pi |s| \ell_{\rm p}}\right)^{3/2} \exp\left(-\frac{\beta \ell_{\rm p}^2}{|s|^2}\right) dV, \tag{25.53}
$$

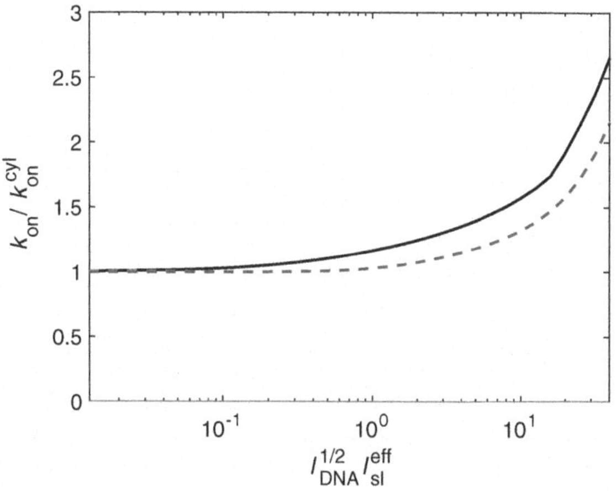


**Fig. 25.6** Plot of $k_{\mathrm{on}}/k_{\mathrm{on}}^{\mathrm{cyl}}$ as a function of $\sqrt{l_{\mathrm{DNA}}} l_{\mathrm{sl}}^{\mathrm{eff}}$ for $D_{3\mathrm{d}} = 0.1\,, \mu\mathrm{m}^2/\mathrm{ms}$, $D_{1\mathrm{d}} = 10^{-4}\,, \mu\mathrm{m}^2/\mathrm{ms}$, $r_{\mathrm{int}} = 3\,\mathrm{nm}$, $k_{\mathrm{on}}^{\mathrm{ns}} = 10^4\,, \mu\mathrm{m}^2/\mathrm{ms}$ and $k_{\mathrm{off}}^{\mathrm{ns}} = 1\,\mathrm{ms}^{-1}$ (giving $l_{\mathrm{sl}}^{\mathrm{eff}} = 1.26\,, \mu\mathrm{m}$). $k_{\mathrm{on}}^{\mathrm{cyl}}$ represents the value for straight DNA, i.e., for $l_{\mathrm{DNA}} = 0$. The black solid line is the numerical result obtained using Eq. (25.48) in Eq. (25.20), while the red dashed line is the result in Eq. (25.52) divided by the result in Eq. (25.33). The value of $k_{\mathrm{on}}^{\mathrm{ns}}$ has been chosen to be far inside the regime $k_{\mathrm{on}}^{\mathrm{ns}} \gg D_{3\mathrm{d}}$

where $\ell_{\mathrm{p}}$ is the persistence length of the DNA and $\beta \approx 8$. Integrating this over the full DNA contour we obtain the average amount of DNA in $dV$, i.e., we obtain the DNA density

$$l_{\mathrm{DNA}} = \int_{-L_1}^{L_2} j_{\mathrm{M}}(s) ds, \tag{25.54}$$

where $L_1$ and $L_2$ are the lengths of the DNA before and after the target. If we take the limit of infinity long DNA, i.e., $L_1$ and $L_2$ being infinite, we find

$$l_{\mathrm{DNA}} = \left(\frac{3}{4\pi}\right)^{3/2} \beta^{-1/4}\Gamma(1/4)\,\ell_{\mathrm{p}}^{-2} \approx 0.25\,\ell_{\mathrm{p}}^{-2} \tag{25.55}$$

using $\beta \approx 8$. For a persistence length around $\ell_{\mathrm{p}} \approx 50\,\mathrm{nm}$ [33] this provides us with the length scale of $l_{\mathrm{DNA}}^{-1/2} \approx 100\,\mathrm{nm}$ for the distances to nearby DNA segments.

## 25.6 Intersegmental Transfers

Some proteins, such as the Lac repressor [8], are able to bind to more than one DNA strand simultaneously. As a consequence these proteins can move directly from one segment of DNA to another, without having to unbind from the DNA and diffuse temporarily in the bulk water. This mechanism is called an intersegmental transfer [8], and it can be incorporated in the diffusion equation along the DNA, Eq. (25.12), by introducing the extra term

$$\frac{\partial n(x,t)}{\partial t} = \left( D_{1d}\frac{\partial^2}{\partial x^2} - k_{\rm off}^{\rm ns} \right) n(x,t) - j(t)\delta(x) + G(x,t)$$
$$+ k_{\rm off}^{\rm ns} \int_{-\infty}^{\infty} dx' \int_0^t dt'\, W_{\rm bulk}(x-x', t-t') n(x', t')$$
$$-k_{\rm intra} \left( n(x,t) - \int_{-\infty}^{\infty} dx'\, \lambda_{\rm intra}(x-x') n(x', t) \right). \tag{25.56}$$

Here $k_{\rm intra}$ is the rate at which intersegmental transfers are occurring, and $\lambda_{\rm intra}(x)$ is the probability density of distances the protein transfers measured along the DNA. We here again assume independence between transfers, i.e., that the dynamics of the conformation of the coil is fast compared with the rate of the transfers $k_{\rm intra}$, to simplify the modelling considerably. By applying the same mathematical steps as in Sect. 25.3 we find that the rate $k_{1d}$ can now be expressed as

$$k_{1d}^{-1} = \int_{-\infty}^{\infty} \frac{dq}{2\pi} \frac{1}{D_{1d}q^2 + k_{\rm off}^{\rm ns}(1-\lambda_{\rm bulk}(q)) + k_{\rm intra}(1-\lambda_{\rm intra}(q))}. \tag{25.57}$$

We conclude from this that the intersegmental transfers will increase $k_{1d}$ and result in faster target search, since the extra term contributes positively in the denominator of the integrand. Thus increasing $k_{\rm intra}$ will always result in faster target search. This is opposed to increasing $k_{\rm off}^{\rm ns}$, since here there will be the opposing effects of increasing $k_{1d}$ but lowering the density of loosely bound proteins $n_{\rm ns}^{\rm eq}$ on the DNA.

There are two limits in which we can simplify the form of the extra term in Eq. (25.57). One occurs when the transfers are typically very long, so that they do not lead the protein to a place it has already visited. In this case, we can make the approximation $\lambda_{\rm intra}(q) \approx 0$ that we have previously used for $\lambda_{\rm bulk}$ also. This limit was studied in [34], and it was found there that the intersegmental transfers can have a significant impact with respect to lowering search times. Another relatively simple limit, studied in [35], emerges if the polymer that the protein diffuses along is a freely fluctuating coil with low persistence length (i.e., for DNA with a persistence length of around 150 base pairs [33], this would require a coarse-graining in a sufficiently long chain). In this case the distribution $\lambda_{\rm intra}(x)$ has a power law tail at large $x$, i.e., $\lambda_{\rm intra}(x) \propto |x|^{-1-\alpha}$ ($|x|$ large), where for instance $\alpha = 0.5$ for Gaussian chains. Motions with power law tails for jumping distances with $0 < \alpha < 2$ are called Lévy flights. If the power law tail sets in at distances short enough such that the motion is dominated by one-dimensional diffusion along the chain at these distances, then it will only be the power law tail of the intersegmental transfer term that contributes to the integral in Eq. (25.57), and we can approximate $k_{\rm intra}(1-\lambda_{\rm intra}(q)) \approx D_{\rm L}|q|^{\alpha}$

with $D_L$ being a diffusion constant for the Lévy flight. This $q$-scaling means that the corresponding convolution term in Eq. (25.56) can be written as a fractional derivative. See [35] for more on this including a study of optimal search strategies, i.e., values of $k_{\rm off}^{\rm ns}$ that minimise search time in this limit. We note that this example of a Lévy flight involves power-law distributed jumps in the chemical space measured along the DNA. In real space, the intersegmental transfers cover a short distance.

## 25.7 Conclusion

The mathematical theory presented in this chapter provides a way to make predictions about the effect of coiling of DNA for target search by proteins. For proteins with only one DNA binding site, the effect of coiling is quantified through the concentration $l_{\rm DNA}$ of DNA around the local segment of DNA with the target. However, the theory relies on many assumptions that limit the applications. These include the following: the DNA molecule is long and has a large persistence length, no correlations with previous bulk excursions (possibly through dynamics of the DNA), and dilute protein concentration. Thus there are many relevant alternatives of modelling the effect of DNA coiling on target search that has not been covered in this chapter. Just a few examples are scaling regimes when the persistence length cannot be assumed to be long [30], effect of large-scale heterogeneous network structure [36, 37], or effect of the fractal nature of chromatin [38].

Sequence heterogeneity was mentioned in Sect. 25.3 as a potential barrier for fast search. However, combined with looping, sequence heterogeneity in the form of auxiliary sites within looping distance of the target can also help in accelerating search [39]. Without looping such sites will act as traps, and this introduces additional long-time scales in the search process. Modelling such traps will mean that the assumption of non-specific equilibration and exponentially distributed search time employed in this chapter will have to be examined carefully [23].

We finish by noting that more recent studies have shown that at low chemical concentrations chemical rate constants (i.e., inverse mean search times) may become insufficient. Instead, even in quite generic geometries, repeated first-passage times may be significantly different from each other [40, 41], with characteristic times spanning several orders of magnitude [42, 43]. Given that transcription factors in living cells may reach such low numbers (ten or few tens in the entire cell), it will be of interest to explore the full span of relevant target localisation times, in supercomputing studies [44] or experimentally. In such scenarios of low copy numbers the bursty protein production of a DNA-binding protein encoded in one gene leads to strong concentration fluctuations at the target gene, in situations when the gene-gene distance is short [45].

Facilitated diffusion was demonstrated to occur both in vitro and in living cells. Single-DNA experiments unveiled contributions of DNA coiling to the localisation dynamics of DNA-binding proteins to their specific target site on the DNA. In the future it will be of interest to follow individual proteins from their production [46] across the cell [47], to their eventual binding to the DNA.

## References

1. G. Adam, M. Delbrück, Structural chemistry and molecular biology, (Freeman, San Francisco, 1968) Chap. Reduction of Dimensionality in Biological Diffusion Processes
2. A.D. Riggs, S. Bourgeois, M. Cohn, The lac repressor-operator interaction. J. Mol. Biol. **53**, 401 (1970)
3. P.H. Richter, M. Eigen, Diffusion controlled reaction rates in spheroidal geometry: application to repressor-operator association and membrane bound enzymes. Biophys. Chem. **2**, 255 (1974)
4. O.G. Berg, C. Blomberg, Association kinetics with coupled diffusional flows. Special application to the lac repressor–operator system. Biophys. Chem. **4**, 367 (1976)
5. O.G. Berg, C. Blomberg, Association kinetics with coupled diffusion. An extension to coiled-chain macromolecules applied to the lac repressor-operator system. Biophys. Chem. **7**, 33 ( 1977)
6. O.G. Berg, R.B. Winter, P.H. von Hippel, Diffusion-driven mechanisms of protein translocation on nucleic acids. 1. Models and theory. Biochemistry **20**, 6929 (1981)
7. O.G. Berg, M. Ehrenberg, Association kinetics with coupled three- and one-dimensional diffusion. Chain-length dependence of the association rate of specific DNA sites. Biophys. Chem. **15**, 41 (1982)
8. P.H. von Hippel, O.G. Berg, Facilitated target location in biological systems. J. Biol. Chem. **264**, 675 (1989)
9. D.S. Grebenkov, R. Metzler, G. Oshanin, Search efficiency in the Adam-Delbrück reduction-of-dimensionality scenario versus direct diffusive search. New. J. Phys. **24**, 083035 (2022)
10. I. Sokolov, R. Metzler, K. Pant, M. Williams, Target search of $n$ sliding proteins on a DNA. Biophys. J. **89**, 895 (2005)
11. Y.M. Wang, R.H. Austin, E.C. Cox, Single molecule measurements of repressor protein 1d diffusion on DNA. Phys. Rev. Lett. **97**, 048302 (2006)
12. P. Hammar, P. Leroy, A. Mahmutovic, E.G. Marklund, O.G. Berg, J. Elf, The lac repressor displays facilitated diffusion in living cells. Science **336**, 1595 (2012)
13. D.M. Gowers, S.E. Halford, Protein motion from non-specific to specific DNA by three-dimensional routes aided by supercoiling. EMBO J. **22**, 1410 (2003)
14. B. van den Broek, M.A. Lomholt, S.-M.J. Kalisch, R. Metzler, G.J.L. Wuite, How DNA coiling enhances target localization by proteins. Proc. Natl. Acad. Sci. USA **105**, 15738 (2008)
15. M.A. Lomholt, B. van den Broek, S.-M. Kalisch, G.J.L. Wuite, R. Metzler, Facilitated diffusion with DNA coiling. Proc. Natl. Acad. Sci. USA **106**, 8204 (2009)
16. M. von Smoluchowski, Drei Vorträge über Diffusion, Brownsche Molekularbewegung und Koagulation von Kolloidteilchen. Phys. Z. **17**, 557 (1916)
17. https://github.com/mlomholt/coiledDNA (2023)
18. J. Elf, G.-W. Li, X.S. Xie, Probing transcription factor dynamics at the single-molecule level in a living cell. Science **316**, 1191 (2007)
19. M. Ptashne, A. Gann, *Genes and Signals* (Cold Spring Harbor Laboratory Press, Plainview, NY, 2002)
20. G.D. Stormo, D.S. Fields, Specificity, free energy and information content in protein-DNA interactions. Trends Biochem. Sci. **23**, 109 (1998)
21. N.E. Buchler, U. Gerland, T. Hwa, On schemes of combinatorial transcription logic. Proc. Natl. Acad. Sci. USA **100**, 5136 (2003)
22. M. Slutsky, L.A. Mirny, Kinetics of protein-DNA interaction: facilitated target location in sequence-dependent potential. Biophys. J. **87**, 4021 (2004)

23. O. Bénichou, Y. Kafri, M. Sheinman, R. Voituriez, Searching fast for a target on DNA without falling to traps. Phys. Rev. Lett. **103**, 138102 (2009)
24. B.D. Hughes, *Random Walks and Random Environments*, vol. 1 (Oxford University Press, Oxford, 1995)
25. K. Pant, R.L. Karpel, I. Rouzina, M.C. Williams, Mechanical measurement of single-molecule binding rates: kinetics of DNA helix-destabilization by T4 gene 32 protein. J. Mol. Biol. **336**, 851 (2004)
26. J. Krog, M. Alizadehheidari, E. Werner, S.K. Bikkarolla, J.O. Tegenfeldt, B. Mehlig, M.A. Lomholt, F. Westerlund, T. Ambjörnsson, Stochastic unfolding of nanoconfined DNA: experiments, model and Bayesian analysis. J. Chem. Phys. **149**, 215101 (2018)
27. H.S. Carslaw, J.C. Jaeger, *Conduction of Heat in Solids* (Oxford University Press, London, GB, 1959)
28. A.V. Chechkin, I.M. Zaid, M.A. Lomholt, I.M. Sokolov, R. Metzler, Bulk-mediated surface diffusion along a cylinder: propagators and crossovers. Phys. Rev. E **79**, 040105(R) (2009)
29. A.V. Chechkin, I.M. Zaid, M.A. Lomholt, I.M. Sokolov, R. Metzler, Effective surface motion on a reactive cylinder of particles that perform intermittent bulk diffusion. J. Chem. Phys. **134**, 204116 (2011)
30. T. Hu, A.Y. Grosberg, B.I. Shklovskii, How proteins search for their specific sites on DNA: the role of DNA conformation. Biophys. J. **90**, 2731 (2006)
31. J. Abate, W. Whitt, A unified framework for numerically inverting Laplace transforms. INFORMS J. Comput. **18**, 408 (2006)
32. L. Ringrose, S. Chabanis, P.-O. Angrand, C. Woodroofe, A.F. Stewart, Quantitative comparison of DNA looping in vitro and in vivo: chromatin increases effective DNA flexibility at short distances. EMBO J. **18**, 6630 (1999)
33. G.S. Manning, The persistence length of DNA is reached from the persistence length of its null isomer through an internal electrostatic stretching force. Biophys. J. **91**, 3607 (2006)
34. M. Sheinman, Y. Kafri, The effects of intersegmental transfers on target location by proteins. Phys. Biol. **6**, 016003 (2009)
35. M.A. Lomholt, T. Ambjörnsson, R. Metzler, Optimal target search on a fast-folding polymer chain with volume exchange. Phys. Rev. Lett. **95**, 260603 (2005)
36. M. Nyberg, T. Ambjörnsson, P. Stenberg, L. Lizana, Modeling protein target search in human chromosomes. Phys. Rev. Res. **3**, 013055 (2021)
37. L. Hedström, L. Lizana, Modelling chromosome-wide target search. New J. Phys. **25**, 033024 (2023)
38. O. Bénichou, C. Chevalier, B. Meyer, R. Voituriez, Facilitated diffusion of proteins on chromatin. Phys. Rev. Lett. **106**, 038102 (2011)
39. M. Bauer, E.S. Rasmussen, M.A. Lomholt, R. Metzler, Real sequence effects on the search dynamics of transcription factors on DNA. Sci. Rep. **5**, 10072 (2015)
40. C. Mejía-Monasterio, G. Oshanin, G. Schehr, First passages for a search by a swarm of independent random searchers. J. Stat. Mech. **2011**, P06022 (2011)
41. T. Mattos, C. Mejía-Monasterio, R. Metzler, G. Oshanin, First passages in bounded domains: when is the mean first passage time meaningful? Phys. Rev. E **86**, 031143 (2012)
42. A. Godec, R. Metzler, Universal proximity effect in target search kinetics in the few encounter limit. Phys. Rev. X **6**, 041037 (2016)
43. D. Grebenkov, R. Metzler, G. Oshanin, Strong defocusing of molecular reaction times: geometry and reaction control. Commun. Chem. **1**, 96 (2018)
44. J. Ma, M. Do, M.A.L. Gros, C.S. Peskin, C.A. Larabell, Y. Mori, S.A. Isaacson, Strong intracellular signal inactivation produces sharper and more robust signaling from cell membrane to nucleus. PLoS Comp. Biol. **16**, e1008356 (2020)
45. O. Pulkkinen, R. Metzler, Distance matters: the impact of gene proximity in bacterial gene regulation. Phys. Rev. Lett. **110**, 198101 (2013)
46. J. Yu, J. Xiao, X. Ren, K. Lao, X.S. Xie, Probing gene expression in live cells, one protein molecule at a time. Science **311**, 1600 (2006)
47. C.D. Rienzo, V. Piazza, E. Gratton, F. Beltram, F. Cardarelli, Probing short-range protein Brownian motion in the cytoplasm of living cells. Nat. Commun. **5**, 5891 (2014)

# Chapter 26
# First-Reaction Times at the Terminal Target Site in Molecular Relay Races

**Denis Grebenkov, Ralf Metzler, and Gleb Oshanin**

**Abstract** In diverse processes encountered in molecular biology, e.g. in intracellular signal transduction or in cell-to-cell communication, the search for the terminal reaction centre often necessitates a series of intermediate steps performed in a due order—a cascade of sequential diffusion-controlled reactions involving so-called "messengers". In a typical situation, a distant cell sends a "messenger" molecule that diffuses in the extracellular medium seeking for a special binding site on the surface of another cell, where it initiates a special kind of a reaction. This process produces the second messenger right at the inner surface of this cell. The second messenger diffuses now within the cell seeking, e.g. the entrance to the nuclear pore. Diffusing through the pore, it penetrates into the nucleus and then searches diffusively for the terminal binding site located on the DNA. In this chapter, we discuss the form of the probability density function (PDF) of the first time instant at which the terminal reaction event in such a reaction cascade takes place. We present a general formalism for the derivation of the PDF and formally exact expressions, which permit to draw some far-reaching conclusions about the behaviour of the cumulants of the PDF and its short- and long-time tails. We also provide some illustrative simple examples—diffusion to a series of multiple ordered targets in unbounded one- and two-dimensional domains—in which the final results have lucid and instructive forms.

D. Grebenkov (✉)
Laboratoire de Physique de la Matière Condensée (PMC), CNRS/Ecole Polytechnique/Institut Polytechnique de Paris, 91120 Palaiseau, France
e-mail: denis.grebenkov@polytechnique.edu

R. Metzler
Institute of Physics and Astronomy, University of Potsdam, 14476 Potsdam-Golm, Germany
e-mail: rmetzler@uni-potsdam.de

G. Oshanin
Laboratoire de Physique Théorique de la Matière Condensée, CNRS/Sorbonne Université, 4 Place Jussieu, 75252 Paris, France
e-mail: gleb.oshanin@sorbonne-universite.fr

 
D. Grebenkov et al. (eds.), *Target Search Problems*,
https://doi.org/10.1007/978-3-031-67802-8_26

## 26.1 Introduction

Many reaction-diffusion processes occurring in cellular systems follow a more complicated chemical pathway than common reactions in purely chemical containers [1, 2]. In the simplest settings, the latter reactions often[1] only necessitate an encounter of two chemically active molecules converting them into a desired reaction product or finding diffusively, in case of ligands, specific binding sites. Different kinetic aspects of such basic reactions were studied in detail (see [4, 5] for a review). The situation becomes more complicated when the reactions are not perfect such that not every first encounter leads to a reaction event; in such a situation multiple encounters may be needed, interspersed with diffusive excursions in the bulk, in order to ensure a successful chemical conversion [6]. As a consequence, the first-reaction times in such systems may not be exhaustively well characterised by a single time scale and may have a significant scatter—the mean first-reaction time may be bigger than the typical value of this random variable by several orders of magnitude. Focusing on the situation in which a single diffusive ligand seeks an immobile binding site in a bounded domain, the form of the full probability density function (PDF) of the first-reaction time and its effective broadness are discussed in this book (see Chap. 3).

In signal transduction pathways or in cell-to-cell communications, a molecule cannot physically meet the terminal target, because they are separated by natural barriers, i.e. the plasma membranes, which are often impermeable to many molecular species such that the molecule simply cannot pass through them. In order to circumvent these impenetrable steric constraints, Nature invented more sophisticated scenarios. In particular, the signal transduction has to proceed in an indirect way, via an ordered sequence of intermediate reaction events occurring at different positions in space. Only after all of these events take place in a due order, the terminal reaction event may happen. An elementary example of such a pathway, resembling a "relay race" in human competitions, is realised when one cell passes a "message" to another one (Fig. 26.1a). In this case, a cell secrets a messenger (a protein or another molecule produced by a sending cell) and releases it into the extracellular space. The messenger diffuses there until it finds an immobile receptor on the cellular membrane of another cell, and binds to it. Upon binding to this intermediary target, the first messenger affects its activity and triggers a conformational change of the receptor, producing the "primary effector", which stimulates a synthesis of the second messenger, now directly inside the second cell. In this way, the boundary is crossed. The second messenger diffuses within the cell itself and activates an intra-cellular process, after it binds to a specific terminal target site. In other words, the chemical pathway to the terminal reaction event involves an ordered sequence of intermediate ligand-binding events. Such "indirect" pathways, sometimes much more complicated

[1] We note that some reactions involved in industrial processes may be, of course, much more complicated; they may require, e.g. the presence of a catalyst—a third partner which promotes a reaction (see, e.g. [3] and references therein), or consist of many consecutive stages.

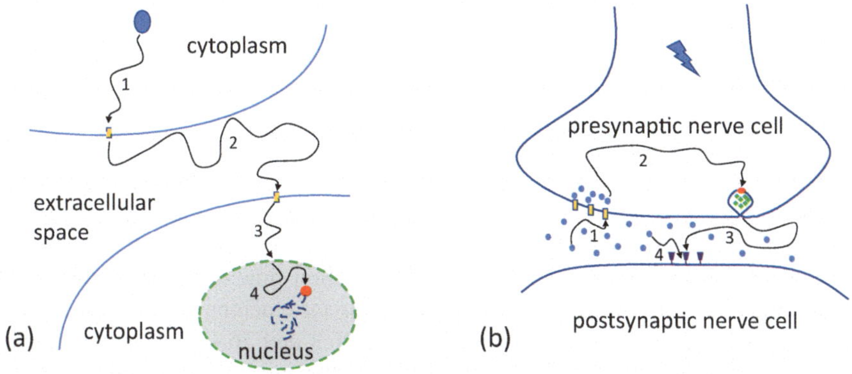


**Fig. 26.1** Two biological examples of a molecular relay race. (**a**) In a cell-to-cell communication, a first messenger is released inside the cytoplasm (or from an organelle, or from a nucleus) and diffuses towards a receptor on the plasma membrane; the second messenger is then released on the outer part of the plasma membrane and diffuses through the extracellular space towards a receptor on the plasma membrane of another cell; the third messenger is released on the inner part of that membrane and diffuses towards a nuclear pore; finally, the last messenger searches for a specific binding site on the DNA strand. (**b**) In signal transmission in the nervous system, the arriving action potential opens calcium channels (in yellow) on the presynaptic membrane and allows for the diffusive uptake of calcium ions (blue circles); after entering the presynaptic cell, each calcium ion searches for a calcium-sensing protein (in red) sitting on the vesicle with neurotransmitters (in green); the arrival of calcium ions onto that protein triggers a vesicle opening and the consequent release of neurotransmitters that diffuse through the synaptic cleft towards receptors on the postsynaptic nerve cell; the latter act as ligand-gated ion channels that open to calcium ions; the induced depolarisation of the membrane generates an action potential that propagates through the neuron

ones, are ubiquitous in cellular biology, and present an efficient way to control the signal transduction by activating or inhibiting intermediate targets [1, 2].

Another instance of a biological process, in which the terminal reaction event is conditioned on a multi-step passage through some intermediary targets, is the signal transmission between neurons [7–9]. Here, the first step consists in opening the calcium channels on the surface of the presynaptic bouton by an electrical impulse in one neuron and the consequent influx of calcium ions (Fig. 26.1b). Their subsequent diffusion towards specific sensor proteins is the second step that allows to launch, in case of a successful reaction event, the neurotransmitters into the synaptic cleft. In the third step, the neurotransmitters seek diffusively specific receptors on the postsynaptic spine boundary of the second neuron. In the fourth step, calcium ions from the synaptic cleft search for the opened calcium channels on the postsynaptic boundary to induce an action potential in the second neuron.

In yet another setting, a diffusive signalling molecule is released in a bulbous head of the dendritic spine and first needs to find an escape window—yet another intermediate target site, on the otherwise impermeable boundary. This escape represents the standard setup of the so-called Narrow Escape Problem [10–22], which is also discussed in the present book. Surmounting an entropy barrier [20] the molecule then enters into a narrow neck and thus has a chance to eventually reach the terminal target site, which is the parent dendrite to which the neck is connected via its opposite extremity. Note that in such a cascade of reactions, the reactive particles diffuse at different stages in domains of different spatial dimensions—from three-dimensional to effectively one-dimensional ones.

In this Chapter, summarising the results presented in our earlier paper [23], we address the question how to calculate the full PDF of the time instant at which the terminal reaction event in such a sequential reaction cascade takes place. We present some general results for the PDF of the first-reaction time at the terminal target site in Sect. 26.2, in which we also discuss their far-reaching consequences and most important characteristic features. In Sects. 26.3 and 26.4 we provide simple illustrative examples for diffusion in one- and two-dimensional unbounded systems, respectively, in which the predicted behaviour is made apparent.

## 26.2 Model and General Results

Consider a sequential cascade of reactions in the sense described above, which consists of $N$ intermediate processes. At the $j$-th stage ($j = 1, 2, \dots, N$), a particle—the $j$-th messenger—is launched at a certain time instant at position $\boldsymbol{x}_{0,j}$ and searches diffusively, with the diffusion coefficient $D_j$, for some target, which can be either immobile and located at a fixed position $\boldsymbol{x}_j$ or diffuse itself (see, e.g. [24, 25]). The former case is indeed realised in most intra-cellular transduction pathways, while the latter may occur at the initial stage of the cell-to-cell communication process, when a messenger secreted by a cell seeks a binding site on the surface of another cell.

The reaction with the target may be *imperfect* (see, e.g. [26, 27]) and happen upon an encounter with the target only with a finite probability $0 < p_j < 1$ (defining the so-called intrinsic reaction rate $\kappa_j < \infty$). This is the most common case in which there is a barrier against the reaction, such that the first encounter with the target does not necessarily lead to a successful reaction event [6]. In this case, the messenger will keep on diffusing and meeting the target again and again until the barrier is crossed and a successful reaction event takes place. Evidently, the meaningful property here is the first-reaction time. In turn, in the absence of such a reaction barrier, the reaction is called *perfect*, with the reaction probability $p_j = 1$ ($\kappa_j = \infty$), such that a successful reaction event occurs upon the first encounter with the target. In this case, the first-reaction time is actually the first-passage time to the corresponding target. In both cases, we denote the random duration of the $j$-th stage of the reaction cascade as $\tau_j$, which is the time-interval between the instant when the messenger is launched and the instant at which the successful reaction event happens. At this very time instant, the $j$-th stage terminates and the $(j + 1)$-th stage is initiated.

The PDF of the duration $\tau_j$, denoted as $\Psi_j(t)$, depends on the $j$-th messenger's starting point, the geometry and the composition of the environment, in which the diffusion takes place, as well as the target's location (or its diffusion), the reaction probability at this stage, the spatial dimension and the diffusion coefficient. We also note parenthetically that in most of realistic situations the spatial domains in which the search processes take place are bounded, and consequently, the reaction events are certain to occur. Therefore, $\Psi_j(t)$ is normalised. In Chap. 3 of this book, we discuss the form of the PDF $\Psi_j(t)$ for a simple example of a diffusive ligand seeking an immobile, small imperfect binding site located in the centre of a spherical bounded domain.

Generically, for a diffusive search in bounded domains, $\Psi_j(t)$ exhibits four distinct temporal regimes, as will be discussed in more detail below (see also Chap. 3):

(i) A short times, $\Psi_j(t)$ shows a sharp exponential rise terminating at the most probable value of the first-reaction time $\tau_{mp}$, at which the PDF attains its maximal value. We will comment on the functional form of this regime below;
(ii) $\Psi_j(t)$ shows a slow power-law descent up to another characteristic time scale $\tau_c$, at which the particle first engages with the boundaries and realises that it moves in a bounded domain;
(iii) There is an extended plateau region beyond $\tau_c$ which stretches up to the mean first-reaction time $\overline{\tau}$. This regime, in which all values of the first-reaction times between $\tau_c$ and $\overline{\tau}$ are almost equally probable, is more pronounced when the corresponding reaction probability is smaller. Interestingly, in the case of a spherical domain, around 70 per cent of all reaction events take place up to $\overline{\tau}$ [28];
(iv) The plateau is followed by an ultimate long-time exponential cut-off.

In general, a calculation of $\Psi_j(t)$ is quite a difficult task. For several examples of domains with a complicated geometry, a thorough discussion of different temporal regimes and their relevance can be found in [21, 29–34].

The above-mentioned cascade of diffusion-reaction processes proceeds until the random moment in time when the $N$-th messenger successfully reacts with the terminal, $N$-th target. We are interested in what follows in the form of the PDF $H(t)$ of the total duration

$$\mathcal{T} = \tau_1 + \tau_2 + \ldots + \tau_N \tag{26.1}$$

of the relay race process, from the initial release of the first messenger to the terminal successful reaction of the $N$-th messenger with the $N$-th target.

Despite the fact that each subsequent reaction event is conditioned on the successful accomplishment of the previous one, in our natural settings, the durations of the intermediate stages are *independent*, such that $\mathcal{T}$ is the sum of independent (though not necessarily equally distributed) random variables. Consequently, the desired PDF $H(t)$ can be formally written as the $N$-fold convolution

$$H(t) = \int_0^t dt_{N-1} \int_0^{t_{N-1}} dt_{N-2} \ldots \int_0^{t_2} dt_1 \prod_{j=1}^{N} \Psi_j(t_j - t_{j-1}), \quad t_N = t, \;\; t_0 = 0. \tag{26.2}$$

Denoting the moment-generating function of the time span $\tau_j$ as

$$\Phi_j(p) = \int_0^\infty dt\, e^{-pt}\, \Psi_j(t), \tag{26.3}$$

one realises that the form in Eq. (26.2) ensures that the generating function of $\mathcal{T}$, which we denote as $H(p)$, is simply the product of the generating functions of $\tau_j$:

$$H(p) = \prod_{j=1}^{N} \Phi_j(p). \tag{26.4}$$

This is a fairly general result, which is a consequence of the statistical independence of the first-reaction times at the intermediate stages. As such, it permits us to make the following, quite general statements:

(i) Expression (26.4) signifies that the $n$-th order cumulant of the total time $\mathcal{T}$ is simply the sum of the cumulants of the first-reaction times $\tau_j$ at the intermediate stages of the relay race, *if all of them exist*. In particular, we have

$$\overline{\mathcal{T}} = \overline{\tau_1} + \overline{\tau_2} + \ldots + \overline{\tau_N}, \tag{26.5}$$

where the overbar denotes averaging with respect to the corresponding PDF, and

$$\mathrm{Var}(\mathcal{T}) = \mathrm{Var}(\tau_1) + \mathrm{Var}(\tau_2) + \ldots + \mathrm{Var}(\tau_N), \tag{26.6}$$

with Var being the variance of the corresponding first-reaction time. The same holds for higher-order cumulants. We note, however, that even in bounded domains the moments (and hence, the cumulants) of first-passage or first-reaction times are often not representative of the actual behaviour [35, 36] because their values are supported by anomalously long searching trajectories, corresponding to the long-$t$ (right) tails of the PDF. As a consequence, the moments or cumulants alone do not tell us much, if anything, about a typical behaviour observed in most of realisations of the process.

(ii) Expressions (26.2) and (26.4) allow us to draw some general conclusions about the short-$t$ and long-$t$ asymptotic tails of the PDF $H(t)$. As evidenced in Refs. [21, 29–34], the short-$t$ behaviour of $\Psi_j(t)$, i.e. an exponential rise to the maximal value, is supported by the so-called "direct" trajectories [37–40], which in $d$-dimensional systems go straight from the starting point to the target site, and the reaction event takes place upon the first encounter with the target. These short-$t$ asymptotics are remarkably universal and, regardless of the actual spatial dimension of the system, they follow

$$\Psi_j(t) \sim \exp\left(-\frac{|\boldsymbol{x}_{0,j} - \boldsymbol{x}_j|^2}{4D_j t}\right), \quad t \to 0, \tag{26.7}$$

where $|\boldsymbol{x}_{0,j} - \boldsymbol{x}_j|$ is the shortest distance from the starting point to the target (the "geometry control", see [28, 39]), and the sign "$\sim$" denotes the leading (singular) behaviour (up to power-law corrections) in the corresponding limit. We note that the reaction probability enters only the amplitude of this asymptotic form. Expectedly, the leading singular behaviour of $\Psi_j(t)$ which stems from the direct trajectories, has the same form as that of the Lévy-Smirnov density (see, e.g. [41]), which describes the first-passage time density to a target on the half-line.

Correspondingly, the large-$p$ behaviour of the moment-generating function at the $j$-th stage follows

$$\Phi_j(p) \sim \exp\left(-|\boldsymbol{x}_{0,j} - \boldsymbol{x}_j|\sqrt{p/D_j}\right), \quad p \to \infty, \tag{26.8}$$

such that we have, in virtue of Eq. (26.4),

$$H(p) \sim \exp\left(-\sqrt{Tp}\right), \quad T = \left(\sum_{j=1}^{N} \frac{|\boldsymbol{x}_{0,j} - \boldsymbol{x}_j|}{\sqrt{D_j}}\right)^2, \quad p \to \infty. \tag{26.9}$$

Inverting the latter expression, we find that the PDF $H(t)$ of the terminal reaction event exhibits the following short-$t$ behaviour in the leading order,

$$H(t) \sim \exp\left(-\frac{T}{4t}\right), \quad t \to 0, \tag{26.10}$$

which has again the Lévy-Smirnov form. Note that the characteristic time $T$ in the latter expression, and hence, $H(t)$ itself are controlled by the stages with the largest distance between the starting point and the position of the target, or stages with the smallest diffusion coefficient.

In turn, if all search processes are diffusive and proceed in bounded domains, each individual $\Psi_j(t)$ decays exponentially in the long-$t$ limit,

$$\Psi_j(t) \sim \exp\left(-\lambda_{1,j} t\right), \quad t \to \infty, \tag{26.11}$$

where $\lambda_1$ is the smallest eigenvalue of the diffusion operator describing the $j$-th stage. This eigenvalue is often close to the inverse of the corresponding mean first-reaction time $\overline{\tau_j}$ [21, 29–34]. Performing the Laplace transform, inserting the result into Eq. (26.4) and then inspecting the contributions of different terms, one finds that $H(t)$ obeys

$$H(t) \sim \exp\left(-\frac{t}{\tau_{max}}\right), \quad t \to \infty, \tag{26.12}$$

where

$$\tau_{max} = \max\{\lambda_{1,1}^{-1}, \lambda_{1,2}^{-1}, \ldots, \lambda_{1,N}^{-1}\} \approx \max\{\overline{\tau_1}, \overline{\tau_2}, \ldots, \overline{\tau_N}\} \tag{26.13}$$

denotes the maximum of the mean first-reaction times at the intermediate stages. Therefore, the long-$t$ decay of the PDF $H(t)$ in a relay race is not controlled by the sum of all mean first-reaction times, but merely by the largest of them. This signifies that, quite intuitively, the decay of $H(t)$ is controlled by the slowest search process. This substantiates to some extent our remark about the significance of moments and cumulants made at the end of the previous paragraph: both short-$t$ and long-$t$ behaviours of the PDF cannot in general be inferred from the behaviour of the low-order moments.

(iii) Expressions (26.2) and (26.4) are valid not only for standard diffusive motion but also for anomalous diffusion; indeed, there exists ample evidence that at least the short-time dynamics of particles in the cellular cytoplasm does not follow ordinary Fickian diffusion. As a matter of fact, particles move in a complex crowded environment, filled with impermeable organelles, filaments and proteins, which impose steric constraints on the dynamics [42, 43]. As a consequence, the diffusion of the particle in the interior part of the cell often takes place effectively in a tortuous labyrinthine spatial domain. A thorough discussion of this issue can be found in Chap. 22. Moreover, the successive processes need not be of identical nature: in fact, sub-diffusive, diffusive, and super-diffusive processes can be combined and alternated with each other. As we have already remarked, the target sites can be mobile themselves and also the search for a target site on each stage can proceed in spaces of different spatial dimensions, as happens, e.g. in the above-mentioned example of the transmission of signalling ions or proteins between neurons [7–9] or, more generally, in the Adam-Delbrück scenario of reactions in complex biophysical systems [34, 44].

(iv) A natural reset of memory that happens with reactions occurring at intermediate steps within the cascade renders expressions (26.2) and (26.4) applicable to non-Markovian processes with a memory, such as, e.g. a fractional Brownian motion. We will comment more on this issue at the end of Sect. 26.3.

(v) In some pathways, a launch of the $(j+1)$-th messenger after the arrival of the preceding one to its corresponding target site does not occur instantaneously. Sometimes, it is necessary to produce first the corresponding effector that will trigger a synthesis of the $(j+1)$-th messenger. This process of synthesis may require some random time. In this case, expression (26.2) is to be appropriately complemented to include such intermediate stages but will still maintain the form of a convolution. We note, however, that these purely chemical processes usually do not involve a diffusive transport stage and hence, happen at much shorter time scales.

(vi) Expressions (26.2) and (26.4) may also be appropriately generalised to take into account a finite lifetime of searchers—an issue that is relevant for many biological systems [45–47]. For mortal searchers one studies the first-reaction times conditioned on the survival of the diffusing particles. These effects can be particularly relevant for morphogen gradient formation (see, e.g. [48]), in which complex signal transduction pathways involving extracellular diffusion, receptor binding and transcriptional effectors for intra-cellular signalling are at play.

(vii) In many important pathways, at certain stages not a single messenger is launched but rather $M$ of them (with $M$ being often substantially larger than unity). It happens, for example, when a ligand diffusing in the extracellular medium binds to a receptor located on the exterior part of the cell's surface, which subsequently opens a gated channel and permits some number of ions to enter the cell. This situation with a so-called "amplified" signal has been well-studied in the past (see, e.g. [30, 32, 34, 49–54]) such that the corresponding form of $\Psi_j(t)$ is known in great detail (see, e.g. Chap. 3). In particular, it is known that the "direct" trajectories provide the most important contribution in this case. When the $j$-th stage is realised by $M$ messengers, the first-reaction time $\tau_j$ should be replaced by the minimum $\tau_j^{(M)}$ among $M$ independent realisations of $\tau_j$. The probability density of $\tau_j^{(M)}$ can be explicitly written in terms of $\Psi_j(t)$ in time domain, namely, $\Psi_j^{(M)}(t) = M\Psi_j(t)[\int_t^{\infty} dt'\Psi_j(t')]^{M-1}$. In turn, the moment-generating function $\Phi_j^{(M)}(p)$ is a convolution of multiple $\Phi_j(p)$ so that it is not straightforward to appreciate the effect of the "amplified" signal onto the relay race. Curiously, if the above "parallel" search is easily described in time domain (via $\Psi_j(t)$), the "sequential" search in the relay race admits a simpler description in Laplace domain (via $\Phi_j(p)$).

In Sects. 26.3 and 26.4 below we complement some of our general statements about the first-reaction times in the relay race processes by providing exact solutions obtained for simple exactly solvable situations. Namely, we consider a cascade of *perfect* reactions occurring in one- and two-dimensional unbounded systems with messengers performing standard Brownian motion. A salient feature of these situations is that here even the first moment of any intermediate stage does not exist, such that $H(t)$ will not possess even the first moment. We note, as well, that the results for imperfect reactions are also available and can be found, together with the discussion of the effects of a finite reactivity in the original paper [23]. These results, however, have rather cumbersome forms even in the one-dimensional case such that we do not present them here.

## 26.3 One-Dimensional Systems

Consider a one-dimensional system that is infinite in both directions, with immobile targets placed at positions $x_j$, $j = 1, 2, \ldots, N$. The positions $x_j$ are fixed and not ordered in any way. The first messenger starts at the origin (in our previous notations, $x_{0,1} = 0$) and seeks diffusively, with the diffusion coefficient $D_1$, the target site located at $x_1$. We suppose, for simplicity, that the other targets are "invisible" for the first messenger. As soon as this messenger arrives at $x_1$ for the first time, the second messenger is launched from this position. This messenger diffuses with the diffusion coefficient $D_2$ and seeks for the target at position $x_2$, not seeing any other target. The relay race continues until the $N$-th messenger, that starts at $x_{N-1}$ arrives at position $x_N$.

The first-passage time PDF for each individual stage has the form of the Lévy-Smirnov density

$$\Psi_j(t) = \frac{|x_j - x_{j-1}|}{\sqrt{4\pi D_j t^3}} \exp\left(-\frac{(x_j - x_{j-1})^2}{4D_j t}\right), \tag{26.14}$$

and its moment-generating function $\Phi_j(p)$ is

$$\Phi_j(p) = \exp\left(-|x_j - x_{j-1}|\sqrt{p/D_j}\right). \tag{26.15}$$

Consequently, the moment-generating function of the terminal reaction event is given for any $p$ by

$$H(p) = \exp\left(-\sqrt{Tp}\right), \tag{26.16}$$

where $T$ is defined in Eq. (26.9). Inverting the latter result, we arrive at the following expression for the desired PDF $H(t)$,

$$H(t) = \frac{\sqrt{T}}{\sqrt{4\pi t^3}} \exp\left(-\frac{T}{4t}\right), \tag{26.17}$$

which is valid for any value of $t$. One can see that, remarkably, the PDF $H(t)$ of the terminal reaction event preserves the form of the individual components, corresponding to the intermediate stages in the reaction cascade and has exactly the Lévy-Smirnov form. This circumstance is, of course, not accidental and we comment on it below.

We close this Section with the following remark. While the long-$t$ behaviour of $H(t)$ depends on the details of the process under study, some general statements can be made about the short-$t$ forms of $H(t)$ for arbitrary Gaussian processes (Markovian or non-Markovian, such as, e.g. fractional Brownian motion). In a recent paper [55], which used the "geometrical-optics" arguments in order to determine the short-time forms of the first-passage density, it was shown that for an arbitrary Gaussian process whose MSD obeys $\langle x^2(t)\rangle = 2D_j t^\alpha$, one has

$$\Psi_j(t) \sim \exp\left(-\frac{(x_j - x_{j-1})^2}{4D_j t^\alpha}\right), \quad t \to 0, \tag{26.18}$$

where $D_j$ is the "generalised" diffusion coefficient (with dimension length$^2$/time$^\alpha$), and $\alpha$ is the anomalous diffusion exponent (e.g. $\alpha = 2H$ with $H \in (0, 1)$ being the Hurst index in case of fractional Brownian motion). Performing the Laplace transform of the expression (26.18) using the steepest-descent method, we find that the large-$p$ asymptotics of the generating function $\Phi_j(p)$ obeys (up to the pre-exponential dependence on $p$)

$$\Phi_j(p) \sim \exp\left(-c_1 \left(\frac{(x_j - x_{j-1})^2}{D_j}\right)^{\frac{1}{\alpha+1}} p^{\frac{\alpha}{\alpha+1}}\right), \quad p \to \infty, \tag{26.19}$$

where $c_1$ is a computable constant. As a consequence, an analogue of Eq. (26.4) reads

$$H(p) \sim \exp\left(-c_1 \left(T_\alpha p\right)^{\frac{\alpha}{\alpha+1}}\right), \quad p \to \infty, \tag{26.20}$$

where

$$T_\alpha = \left(\sum_{j=1}^{N} \left(\frac{(x_j - x_{j-1})^2}{D_j}\right)^{\frac{1}{\alpha+1}}\right)^{\frac{\alpha+1}{\alpha}}. \tag{26.21}$$

Let us dwell on expressions (26.19) and (26.20) from a broader perspective. If we were to interpret these expressions as valid for arbitrary $p$, it will mean that both the intermediate PDF $\Psi_j(t)$ and the final result $H(t)$ are simply (heavy-tailed) one-sided Lévy stable laws of order $\nu = \alpha/(\alpha + 1)$ (see, e.g. [56]). It is expedient now to recall that the Lévy-Smirnov density is, in fact, also a member of this family—a one-sided Lévy stable law with $\alpha = 1$ (and $\nu = 1/2$). In other words, in the case with standard Brownian motion studied above, the random variable $\mathcal{T}$ is the sum of independent, identically distributed random variables with a one-sided Lévy stable distribution with $\alpha = 1$, which is valid for any $t$. Therefore, it is not a surprise at all that the PDF of the terminal reaction event in Eq. (26.17) is also a stable one-sided law with $\alpha = 1$.

In a more general case with the PDF in Eq. (26.18) with $\alpha \neq 1$, expression (26.20) is valid only for $p \to \infty$, such that this interpretation is only meaningful to describe the asymptotic behaviour of $H(t)$ in the limit $t \to 0$. Using the short-$t$ asymptotic form of the one-sided Lévy stable law (see, e.g. [56]), we find that $H(t)$ obeys

$$H(t) \sim \exp\left(-c_2 (T_\alpha/t)^\alpha\right), \quad t \to 0, \tag{26.22}$$

where $c_2$ is a computable constant. We note that the time-dependence in Eq. (26.22) is exactly the same as the one in the individual $\Psi_j(t)$ in Eq. (26.18), which is the consequence of the fact that here we also deal with a one-sided stable law. In other words, the short-time asymptotics of the PDF of the time of the terminal reaction event preserves the form of the individual components, but now the dependence on the distances and the generalised diffusion coefficients has a more complicated form as compared to the standard diffusion case $\alpha = 1$, as a consequence of a more complicated law of addition of stable variables. Setting $\alpha = 1$ in Eq. (26.22), one of course recovers the same dependence on these parameters as that in the exponential from Eq. (26.17).

In the same vein, one can analyse how molecular crowding of the cytoplasm or reversible binding to immobile sites affect the cascade reactions in the anomalous diffusion model of continuous-time random walks with heavy-tailed waiting time distributions [57–59]. In this case, the PDF $\Psi_j(t)$ can be obtained as a solution of the fractional diffusion equation. In analogy to the fractional Brownian motion considered above, the short-time behaviour of $\Psi_j(t)$ admits a stretched-exponential form resembling Eq. (26.18); in contrast, the stretched-exponential decay (26.22) is replaced by a power-law decay due to anomalously long stalling peroids. We stress that this analysis is not limited to one-dimensional settings, see [60–62] for details.

## 26.4 Two-Dimensional Systems

We consider next such a cascade of diffusion-controlled reactions in a two-dimensional system, which is unbounded in all directions. We suppose that the immobile targets, which are labelled sequentially by the index $j = 1, 2, 3, ..., N$, are discs of radii $R_j$ and their centres appear at given fixed positions $\boldsymbol{x}_j$. To somewhat shorten the notation, we introduce the distances $\rho_j = |\boldsymbol{x}_j - \boldsymbol{x}_{j-1}|$ between the targets and assume that typically $\rho_j \gg R_j$. As before, the first messenger starts at the origin and seeks diffusively the first target, paying no attention to other targets which are "invisible" to it. As soon as it touches the first target, the second messenger is launched from the centre of the first target and searches diffusively the second target. The process repeats until the $N$-th messenger arrives to the $N$-th target.

In two dimensions, the moment-generating function of the first-passage time by a diffusive particle (with diffusion coefficient $D_j$) to a perfectly reactive circular target with radius $R_j$, starting its motion at distance $\rho_j$ apart from the target's centre, has the rather compact form (see, e.g. [31])

$$\Phi_j(p) = \frac{K_0\left(\rho_j\sqrt{p/D_j}\right)}{K_0\left(R_j\sqrt{p/D_j}\right)}, \tag{26.23}$$

where $K_\nu(z)$ is the modified Bessel function of the second kind [63]. However, the inverse Laplace transform of expression (26.23) cannot be written in terms of common special functions (see also Chap. 27). In turn, one can employ a suitable integral representation. To this end, it is convenient to represent the inverse Laplace transform as the Bromwich integral and to exploit the regularity of the $K_\nu(z)$ in order to change the integration variable and to deform the integration contour in the complex plane. These operations allow one to express the corresponding $\Psi_j(t)$ as the integral over the positive real half-axis (see, e.g. [64] for more details and extensions),

$$\Psi_j(t) = -\frac{2}{\pi t}\int_0^\infty dz\, z\, e^{-z^2}\mathrm{Im}\, A\left(zR_j/\sqrt{D_j t}, \rho_j/R_j\right), \tag{26.24}$$

where Im denotes the imaginary part, and

$$A(z,r) = \frac{(J_0(z)+iY_0(z))(J_0(zr)-iY_0(zr))}{J_0^2(z)+Y_0^2(z)}. \tag{26.25}$$

Here $J_\nu(z)$ and $Y_\nu(z)$ are the Bessel functions of the first and second kind, respectively [63]. Following [64], we first generalise expression (26.24) for the reaction-cascade case. In virtue of Eq. (26.4), the moment-generating function of the first-passage time to the terminal reaction event in a cascade reads

$$H(p) = \prod_{j=1}^{N} \frac{K_0\left(\rho_j\sqrt{p/D_j}\right)}{K_0\left(r_j\sqrt{p/D_j}\right)}. \tag{26.26}$$

Since the product of ratios of modified Bessel functions of the second kind has no poles, one can apply the same technique as above to get the following result:

$$H(t) = -\frac{2}{\pi t}\int_0^\infty dz\, z\, e^{-z^2}\mathrm{Im}\left(\prod_{j=1}^{N} A\left(zR_j/\sqrt{D_j t}, \rho_j/R_j\right)\right). \tag{26.27}$$

Equation (26.27) is exact for any $t$ and any values of the parameters, and generalises the Bromwich-type representation of the PDF of first-reaction times for the problem with a single perfect target (see, e.g. [31]) to the case of a cascade of diffusion-controlled reactions with an array of perfect targets. In Fig. 26.2 we depict $H(t)$ from Eq. (26.27) as a function of $t$ in the case of a single intermediate target. Note that it is possible to construct an analogous expression for imperfect targets [23], but calculations in this case become essentially more involved.

The short-time behaviour of the PDF in Eq. (26.27) can be obtained by analysing the large-$p$ limit of the moment-generating function $H(p)$ in Eq. (26.26). In this limit, $K_\nu(z) \sim e^{-z}\sqrt{\pi/(2z)}$ [63] and hence

$$H(t) \sim \frac{\left(\prod_{j=1}^{N}(R_j/\rho_j)\right)^{1/2}\sqrt{T}}{\sqrt{4\pi t^3}}\exp\left(-\frac{T}{4t}\right), \quad t \to 0, \tag{26.28}$$

where now

$$T = \left(\sum_{j=1}^{N}\frac{\rho_j - R_j}{\sqrt{D_j}}\right)^2 \tag{26.29}$$

is a characteristic time scale, which only slightly differs from our result (26.9) above including the dependence on the targets' radii $R_j$. Therefore, the short-time tail of $H(t)$ in two-dimensional systems with perfect reactions has the same time-dependence as the (one-dimensional) Lévy-Smirnov density, and differs from it only by the dependence of the prefactor and of the characteristic time on the system parameters. As we remarked above, this is not a mere coincidence but a consequence of the fact that the short-$t$ tails in various geometries are entirely supported by the essentially one-dimensional trajectories, which go straight from the starting point to the target.

The analysis of the long-time behaviour of $H(t)$ is generally much more subtle and here the integral representation (26.27) appears to be particularly useful. Following [64] we notice that the long-time limit corresponds to $z \to 0$, for which

$$A(z,r) \sim 1 - B(z)\ln r, \qquad B(z) \sim \frac{1}{\ln(1/z) + \ln 2 - \gamma - \pi i/2}, \tag{26.30}$$

where $\gamma$ is the Euler-Mascheroni constant. Substituting these expressions into Eq. (26.27) and taking advantage of some standard arguments (see [64]) we find the following long-$t$ asymptotic relation

$$H(t) \sim -\frac{1}{\pi t}\mathrm{Im}\left(\prod_{j=1}^{N}\left(1 - B_j(t)\ln(\rho_j/R_j)\right)\right), \tag{26.31}$$

where

$$B_j(t) = \frac{1}{\ln(\sqrt{D_j t}/R_j) + \ln 2 - \gamma - \pi i/2}. \tag{26.32}$$

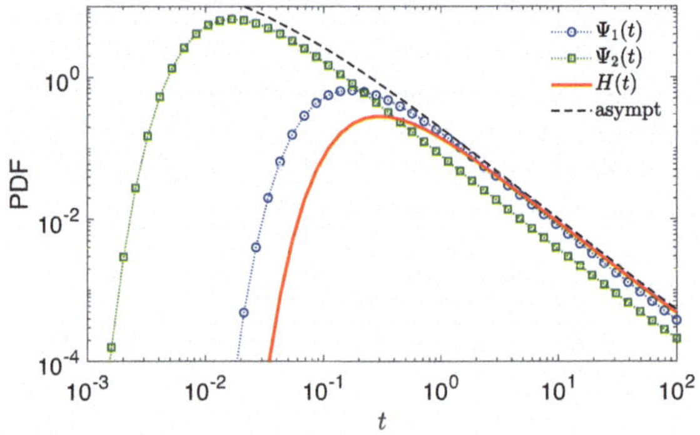


**Fig. 26.2** PDFs $\Psi_1(t)$ and $\Psi_2(t)$ of two FPTs $\tau_1$ and $\tau_2$ in two dimensions with $R_1 = R_2 = 1$, $\rho_1 = \rho_2 = 2$, $D_1 = 1$ and $D_2 = 10$ (circles and squares), and the PDF $H(t)$ of $\tau = \tau_1 + \tau_2$ (solid line). These curves were obtained by numerical integration of Eq. (26.27). The dashed line indicates the long-time approximation (26.31)

Figure 26.2 illustrates the high quality of this approximation.

Expression (26.31) contains, however, an imaginary part of a rational function, which somewhat obscures its actual time dependence, and it can thus be instructive to calculate this imaginary part explicitly. To this end, we rewrite $B_j(t)$ formally as

$$B_j(t) = 2\frac{\ln(4e^{-2\gamma} D_j t/R_j^2) + \pi i}{\ln^2(4e^{-2\gamma} D_j t/R_j^2) + \pi^2}, \tag{26.33}$$

which implies that $B_j(t)$ tends to zero as $t \to \infty$. Expanding then the product in Eq. (26.31) in inverse powers of this small parameter and keeping only the leading-order terms we find

$$H(t) \sim \frac{2}{t}\sum_{j=1}^{N} \frac{\ln(\rho_j/R_j)}{\ln^2(4e^{-2\gamma} D_j t/R_j^2) + \pi^2}, \qquad t \to \infty, \tag{26.34}$$

which defines the long-time limit explicitly. This remarkable asymptotic form is valid for arbitrary values of $\rho_j$, $R_j$ and $D_j$. Note, as well, that the overall long-$t$ decay of $H(t)$ obeys $H(t) \sim 1/(t \ln^2(t))$, i.e. is the same as of the parental distributions $\Psi_j(t)$. Therefore, $H(t)$ is normalised but does not possess even the first moment.

## 26.5 Conclusion

We have discussed the problem of determining the full probability density function of the time instant of the terminal reaction event in a cascade of consequent diffusion-controlled ligand-binding processes, as encountered, e.g. in cell-to-cell communications or in intra-cellular signal transduction pathways. The salient feature of such pathways is that a chemically active particle—often called the first "messenger" or a "signal"—which initiates the process, cannot physically meet the terminal target site because they are separated by impermeable boundaries, e.g. the plasma membranes. Instead, the process proceeds by a series of intermediate stages when a corresponding messenger engages with an intermediate binding site on a boundary, producing the next messenger on the other side of this boundary, which permits to cross it in an indirect way. Two examples of numerous cascade pathways were discussed in the Introduction and depicted in Fig. 26.1.

The processes taking place at the intermediate stages are conditioned on a successful termination of the previous stage, but their durations (the first-reaction times) are random variables which are statistically independent from each other. This observation permits us to represent the first-reaction time density of the terminal reaction event as the convolution of the first reaction-time densities at the intermediate stages. This is a fairly general representation, which is valid for an arbitrary number of the intermediate stages, any type of random or deterministic motion of messengers and also for any effective dimensionality of space; for realistic pathways some of the stages may take place in three-dimensional domains, others on two-dimensional surfaces, while third ones occur in effectively one-dimensional setups (e.g. in nuclear pores or in axons). Concurrently, diffusion of messengers, if it takes place in very crowded environments, may be anomalous, at least at sufficiently short times. As well, our general result is valid for situations in which the signal is "amplified" at certain stages, i.e. not a single messenger is launched at a certain stage but many of them. We note parenthetically that the first reaction-time PDFs in various geometries, which mimic different stages of signal transduction pathways, have been intensively studied within the recent years and many important results have been obtained.

On this basis, we have drawn here some important conclusions about the statistical properties of the first reaction-time at the terminal target site. Apart from some rather trivial ones, such as, e.g. an observation that the cumulant of arbitrary order of the terminal first-reaction time is simply the sum of the cumulants of the first reaction times at the intermediate stages, we made several statements about the short-time and long-time asymptotic behaviours of the first reaction-time PDF for diffusion in bounded domains. Namely, we argued that the short-time tail of the probability density function of the first reaction-times at the terminal target site generically has the Lévy-Smirnov form, in which the distance entering this function is simply the sum, over all stages, of the geodesic distances between the starting points and the locations of the corresponding intermediate binding sites. This statement therefore generalises previous claims about the importance of the so-called direct paths. Furthermore, we have shown that the long-time tails are always exponential functions of time with the characteristic decay time being equal to the longest mean first-reaction time among such times for all the stages. Such a conclusion may be counter-intuitive at first glance—one may rather expect this characteristic time to be equal to the sum of the mean first-reaction times.

Lastly, we discussed the behaviour in one- and two-dimensional unbounded domains with an array of intermediate target sites, for which case we have derived explicit forms of the probability density function of the first reaction-time.

## References

1. B. Alberts et al., *Molecular Biology of the Cell*, 6th edn. (Garland, New York, 2015)
2. D.P. Snustad, M.J. Simmons, *Principles of Genetics*, 7th edn. (John Wiley, New York, NY, 2015)
3. G. Oshanin, A. Blumen, Kinetic description of diffusion-limited reactions in random catalytic media. J. Chem. Phys. **108**, 1140–1147 (1998)
4. K. Lindenberg, R. Metzler, G. Oshanin (eds.), *Chemical Kinetics: Beyond the Textbook* (World Scientific Publishers Europe, London, 2019)
5. E. Gudowska-Nowak, K. Lindenberg, R. Metzler, Preface: Marian Smoluchowski's 1916 paper—a century of inspiration. J. Phys. A Math. Theor. **50**, 380301 (2017)
6. D.S. Grebenkov, Diffusion-controlled reactions: an overview. Molecules **28**, 7570 (2023)
7. D. Holcman, E. Korkotian, M. Segal, Calcium dynamics in dendritic spines: modeling and experiments. Cell Calcium **37**, 467 (2005)

8. A. M. Berezhkovskii, A. V. Barzykin V. Yu. Zitserman, Escape from cavity through narrow tunnel, J. Chem. Phys. **130**, 245104 (2009)
9. M. Reva, D.A. DiGregorio, D.S. Grebenkov, A first-passage approach to diffusion-influenced reversible binding: insights into nanoscale signaling at the presynapse. Sci. Rep. **11**, 5377 (2021)
10. I.V. Grigoriev, Y.A. Makhnovskii, A.M. Berezhkovskii, V.Y. Zitserman, Kinetics of escape through a small hole. J. Chem. Phys. **116**, 9574 (2002)
11. D. Holcman, Z. Schuss, Escape through a small opening: receptor trafficking in a synaptic membrane. J. Stat. Phys. **117**, 975 (2004)
12. O. Bénichou, R. Voituriez, Narrow-escape time problem: time needed for a particle to exit a confining domain through a small window. Phys. Rev. Lett. **100**, 168105 (2008)
13. A.F. Cheviakov, A.S. Reimer, M.J. Ward, Mathematical modeling and numerical computation of narrow escape problems. Phys. Rev. E **85**, 021131 (2012)
14. G. Oshanin, M. Tamm, O. Vasilyev, Narrow-escape times for diffusion in microdomains with a particle-surface affinity: mean-field results. J. Chem. Phys. **132**, 235101 (2010)
15. O. Bénichou, D.S. Grebenkov, P. Levitz, C. Loverdo, R. Voituriez, Optimal reaction time for surface-mediated diffusion. Phys. Rev. Lett. **105**, 150606 (2010)
16. O. Bénichou, D.S. Grebenkov, P. Levitz, C. Loverdo, R. Voituriez, Mean first-passage time of surface-mediated diffusion in spherical domains. J. Stat. Phys. **142**, 657 (2011)
17. J.-F. Rupprecht, O. Bénichou, D.S. Grebenkov, R. Voituriez, Kinetics of active surface-mediated diffusion in spherically symmetric domains. J. Stat. Phys. **147**, 891 (2012)
18. A. Hafner, H. Rieger, Spatial cytoskeleton organization supports targeted intracellular transport. Biophys. J. **114**, 1420 (2018)
19. A. Hafner, H. Rieger, Spatially inhomogeneous search strategies in Ref. [4], Ch. 12 (2019)
20. D.S. Grebenkov, G. Oshanin, Diffusive escape through a narrow opening: new insights into a classic problem. Phys. Chem. Chem. Phys. **19**, 2723 (2017)
21. D.S. Grebenkov, R. Metzler, G. Oshanin, Full distribution of first exit times in the narrow escape problem. New J. Phys. **21**, 122001 (2019)
22. Z. Yang, E.F. Koslover, Diffusive exit rates through pores in membrane-enclosed structures. Phys. Biol. **20**, 026001 (2023)
23. D.S. Grebenkov, R. Metzler, G. Oshanin, A molecular relay race: sequential first-passage events to the terminal reaction centre in a cascade of diffusion-controlled processes. New J. Phys. **23**, 093004 (2021)
24. F. Le Vot, S.B. Yuste, E. Abad, D.S. Grebenkov, First-encounter time of two diffusing particles in confinement. Phys. Rev. E **102**, 032118 (2020)
25. F. Le Vot, S.B. Yuste, E. Abad, D.S. Grebenkov, First-encounter time of two diffusing particles in two- and three-dimensional confinement. Phys. Rev. E **105**, 044119 (2022)
26. O. Bénichou, M. Moreau, G. Oshanin, Kinetics of stochastically gated diffusion-limited reactions and geometry of random walk trajectories. Phys. Rev. E **61**, 3388 (2000)
27. D.S. Grebenkov, Imperfect diffusion-controlled reactions, in Ref. [4], Ch. 8 (2019)
28. D.S. Grebenkov, R. Metzler, G. Oshanin, Strong defocusing of molecular reaction times results from an interplay of geometry and reaction control. Commun. Chem. **1**, 96 (2018)
29. O. Bénichou, R. Voituriez, From first-passage times of random walks in confinement to geometry-controlled kinetics. Phys. Rep. **539**, 225 (2014)
30. D.S. Grebenkov, R. Metzler, G. Oshanin, From single-particle stochastic kinetics to macroscopic reaction rates: fastest first-passage time of *N* random walkers. New J. Phys. **22**, 103004 (2020)
31. D.S. Grebenkov, R. Metzler, G. Oshanin, Towards a full quantitative description of single-molecule reaction kinetics in biological cells. Phys. Chem. Chem. Phys. **20**, 16393 (2018)
32. S.D. Lawley, Distribution of extreme first passage times of diffusion. J. Math. Biol. **80**, 2301 (2020)
33. D.S. Grebenkov, R. Metzler, G. Oshanin, Distribution of first-reaction times with target regions on boundaries of shell-like domains. New J. Phys. **23**, 123049 (2021)

34. D.S. Grebenkov, R. Metzler, G. Oshanin, Search efficiency in the Adam-Delbrück reduction-of-dimensionality scenario versus direct diffusive search. New J. Phys. **24**, 083035 (2022)
35. C. Mejía-Monasterio, G. Oshanin, G. Schehr, First passages for a search by a swarm of independent random searchers. J. Stat. Mech. P06022 (2011)
36. T.G. Mattos, C. Mejía-Monasterio, R. Metzler, G. Oshanin, First passages in bounded domains: when is the mean first passage time meaningful? Phys. Rev. E **86**, 031143 (2012)
37. L. Mirny, M. Slutsky, Z. Wunderlich, A. Tafvizi, J. Leith, A. Kosmrlj, How a protein searches for its site on DNA: the mechanism of facilitated diffusion. J. Phys. A Math. Theor. **42**, 434013 (2009)
38. O. Pulkkinen, R. Metzler, Distance matters: the impact of gene proximity in bacterial gene regulation. Phys. Rev. Lett. **110**, 198101 (2013)
39. A. Godec, R. Metzler, Universal proximity effect in target search kinetics in the few encounter limit. Phys. Rev. X **6**, 041037 (2016)
40. A. Godec, R. Metzler, First passage time distribution in heterogeneity controlled kinetics: going beyond the mean first passage time. Sci. Rep. **6**, 20349 (2016)
41. S. Redner, *A Guide to First Passage Processes* (Cambridge University Press, Cambridge, 2001)
42. S.R. McGuffee, A.H. Elcock, Diffusion, crowding & protein stability in a dynamic molecular model of the bacterial cytoplasm. PLoS Comput. Biol. **6**, e1000694 (2010)
43. S.K. Ghosh, A.G. Cherstvy, D.S. Grebenkov, R. Metzler, Anomalous, non-Gaussian tracer diffusion in crowded two-dimensional environments. New J. Phys. **18**, 013027 (2016)
44. G. Adam, M. Delbrück, Reduction of dimensionality in biological diffusion processes, in *Structural Chemistry and Molecular Biology*, ed. by A. Rich, N. Davidson (San Francisco, CA: W. H. Freeman & Co, 1968)
45. B. Meerson, S. Redner, Mortality, redundancy, and diversity in stochastic search. Phys. Rev. Lett. **114**, 198101 (2015)
46. D.S. Grebenkov, J.-F. Rupprecht, The escape problem for mortal walkers. J. Chem. Phys. **146**, 084106 (2017)
47. V. Balakrishnan, E. Abad, T. Abil, J.J. Kozak, First-passage properties of mortal random walks: Ballistic behavior, effective reduction of dimensionality, and scaling functions for hierarchical graphs. Phys. Rev. E **99**, 062110 (2019)
48. S.B. Yuste, E. Abad, K. Lindenberg, Reaction-subdiffusion model of morphogen gradient formation. Phys. Rev. E **82**, 061123 (2010)
49. G.H. Weiss, K.E. Shuler, K. Lindenberg, Order statistics for first passage times in diffusion processes. J. Stat. Phys. **31**, 255 (1983)
50. K. Basnayake, A. Hubl, Z. Schuss, D. Holcman, Extreme narrow escape: shortest paths for the first particles among n to reach a target window. Phys. Lett. A **382**, 3449 (2018)
51. Z. Schuss, K. Basnayake, D. Holcman, Redundancy principle and the role of extreme statistics in molecular and cellular biology. Phys. Life Rev. **28**, 52–79 (2019)
52. S.D. Lawley, J.B. Madrid, A probabilistic approach to extreme statistics of Brownian escape times in dimensions 1, 2, and 3. J. Nonlinear Sci. **30**, 1207–1227 (2020)
53. S.N. Majumdar, A. Pal, G. Schehr, Extreme value statistics of correlated random variables: a pedagogical review. Phys. Rep. **840**, 1 (2020)
54. D.S. Grebenkov, Depletion of resources by a population of diffusing species. Phys. Rev. E **105**, 054402 (2022)
55. B. Meerson, G. Oshanin, Geometrical optics of large deviations of fractional Brownian motion. Phys. Rev. E **105**, 064137 (2022)
56. G. Schehr, P. Le Doussal, Extreme value statistics from the real space renormalization group: Brownian motion, Bessel processes and continuous time random walks. J. Stat. Mech. P01009 (2010)
57. H. Scher, E.W. Montroll, Anomalous transit-time dispersion in amorphous solids. Phys. Rev. B **12**, 2455 (1975)
58. R. Metzler, J. Klafter, The random walk's guide to anomalous diffusion: a fractional dynamics approach. Phys. Rev. **339**, 1 (2000)

59. R. Metzler, J. Klafter, The restaurant at the end of the random walk: recent developments in the description of anomalous transport by fractional dynamics. J. Phys. A **37**, R161 (2004)
60. S.B. Yuste, K. Lindenberg, Subdiffusive target problem: survival probability. Phys. Rev. E **76**, 051114 (2007)
61. D.S. Grebenkov, Searching for partially reactive sites: analytical results for spherical targets. J. Chem. Phys. **132**, 034104 (2010)
62. D.S. Grebenkov, Subdiffusion in a bounded domain with a partially absorbing-reflecting boundary. Phys. Rev. E **81**, 021128 (2010)
63. M. Abramowitz, I.A. Stegun (eds.), *Handbook of Mathematical Functions with Formulas, Graphs, and Mathematical Tables* (Dover, Washington D.C., 1983)
64. D.S. Grebenkov, Statistics of boundary encounters by a particle diffusing outside a compact planar domain. J. Phys. A Math. Theor. **54**, 015003 (2021)

# Chapter 27
# Size-Distribution of Polymer-Free Voids in Frozen Polymer Solutions

**Sergei F. Burlatsky and Gleb Oshanin**

**Abstract** In this chapter we address a question about a functional form of the distribution of sizes of polymer-free voids in frozen solutions of long polymer chains. We show that when the polymers can be considered as statistically independent, the problem can be solved exactly by mapping it onto the Target Problem. We present exact results for Gaussian chains in three- and two-dimensional systems. For chains of arbitrary conformations (but still independent), the probability of polymer-free voids is determined using some heuristic arguments. We show that the fact that all the monomers are linked into the chains causes a significant departure of the distribution of sizes of polymer-free voids from the Poisson distribution, and generically leads to a very uneven spatial distribution of the polymer fraction.

## 27.1 Introduction

In this chapter we are concerned with the following question: Consider a polymer solution comprising *long* linear polymer chains, each composed of many repeating subunits—the monomers. The chains are assumed to be immobile or very slowly moving such that their conformations do not appreciably change within some observation time. One may expect, on intuitive grounds, that due to the fact that the monomers are linked into a long chain and therefore, their spatial positions are very strongly correlated, a polymer fraction should be quite unevenly distributed in space and characterized, in particular, by rather large spatial regions completely devoid of polymer subunits. On the contrary, for short chains comprising only a few monomers, one expects a much more uniform coverage of space. Then, it is legitimate to inquire

S. F. Burlatsky
United Technologies Research Center, East Hartford, CT 06108, USA
e-mail: BurlatSF@utrc.utc.com

G. Oshanin (✉)
Laboratoire de Physique Théorique de la Matière Condensée (UMR CNRS 7600), CNRS, Sorbonne Université, 4 place Jussieu, 75252 Paris Cedex 05, France
e-mail: gleb.oshanin@sorbonne-universite.fr

 
D. Grebenkov et al. (eds.), *Target Search Problems*,
https://doi.org/10.1007/978-3-031-67802-8_27

what is a typical size and, desirably, the full probability density function of sizes of polymer-free voids in case of long chains. The importance of this problem was discussed in the book by de Gennes [1] and a typical size of such voids has been estimated.

On the basis of our earlier papers [2, 3], we proceed to show in this chapter that if all the chains can be considered as statistically independent, the problem of finding the probability density function of sizes of the polymer-free voids becomes exactly solvable. Moreover, it maps, in fact, onto the standard Target Problem (see, e.g., [4–6]), with the salient aspect being that in our case one is interested in finding the "target-size" dependence of the target survival probability, rather than the commonly studied "time" (i.e., the chain's length in our context) dependence. Our analysis shows that an intuitive expectation that the polymer fraction has a much more uneven spatial distribution than the one for a solution of monomers is indeed correct and the probability of having a polymer-free void of radius $R$ is much higher than this probability in a frozen solution of monomers present at the same density as the polymer fraction.

We note that the assumption that polymers can be considered as some independent paths is not completely unrealistic. Indeed, polymer chains appear as independent Gaussian coils in dense solutions (melts), in which the volume fraction of polymers exceeds some threshold value (the $c^*$-theorem due to Flory [1]). In dilute solutions, in which the polymers are well-separated from each other, the chains can evidently be considered as independent, but their conformations depend largely on the quality of the embedding solvent. Chains are swollen, as compared to Gaussian shapes, in good solvents due to excluded-volume effects. Conversely, in $\theta$-solvents the isolated polymers have Gaussian shapes due to a compensation of the excluded-volume interactions. We will focus mostly on the Gaussian case, in which an exact analysis can be carried out, but will also present a heuristic generalization for the case when (independent) chains have arbitrary conformations.

## 27.2 The Model and Basic Notations

Consider a $d$-dimensional volume $V_d$ with $K$ randomly placed immobile polymer chains, each containing $N + 1$ monomers. Consequently, the case $N = 0$ corresponds to a solution of monomers. We also note parenthetically that in many practically interesting situations the number $N$ of subunits in a chain can also be a random variable, as it happens, e.g., for polydisperse polymer solutions. We do not discuss this case here although the analysis presented below can be readily generalized over this case too, as it was done in [7] for the Target Problem involving particles with randomly distributed life-times (see also Chap. 7).

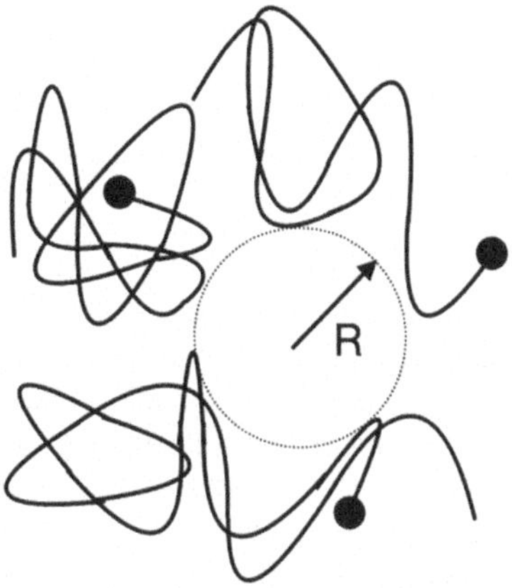


**Fig. 27.1** A sketch of a solution of immobile polymer chains with a polymer-free void of radius $R$

Next, we tag one of the end-points of the polymers (denoted by thick filled circles in Fig. 27.1) and label the monomers along each chain by the index $n$. Respectively, the vector $\mathbf{R}_{k,n}$ denotes the position of the $n$th monomer in the $k$th chain. Let $I(\mathbf{r})$ be a Boolean function such that $I(\mathbf{r}) = 1$ if $|\mathbf{r}| > R$ (i.e., $\mathbf{r}$ is outside of a $d$-dimensional sphere of radius $R$ centered at the origin, see Fig. 27.1) and equals zero otherwise. Then, the product

$$I_R = \prod_{k=1}^{K} \prod_{n=0}^{N} I\left(\mathbf{R}_{k,n}\right) \tag{27.1}$$

is the indicator function of the event that neither monomer of neither chain appears at a distance less than $R$ from the origin. Being averaged over all possible arrangements of polymer chains (we denote in what follows such an average by angle brackets, $\langle \ldots \rangle$), the indicator function in Eq. (27.1) defines the so-called "survival" probability (in the language of the Target Problem) or, in our context, the probability of having a spherical polymer-free void of radius $R$ in a frozen polymer solution. Respectively, the probability density function $P(R)$ is given by

$$P(R) = -\frac{d\langle I_R \rangle}{dR}. \tag{27.2}$$

Evidently, the above probability density function is normalized, because $\langle I_R \rangle = 1$ for $R = 0$ and $\langle I_R \rangle = 0$ for $R = \infty$.

Assuming that all chains are statistically independent from each other, we can write the averaged value of $I_R$ in a factorized form. Supposing next that the tagged end-points are uniformly distributed throughout the volume $V_d$, this factorized form writes

$$
\begin{aligned}
\langle I_R \rangle &= \prod_{k=1}^{K} \left\langle \prod_{n=0}^{N} I\left(\mathbf{R}_{k,n}\right) \right\rangle \\
&= \left[ \frac{1}{V_d} \int_{V_d} d\mathbf{R}_0 \left\langle \prod_{n=0}^{N} I\left(\mathbf{R}_n\right) \right\rangle_0 \right]^K ,
\end{aligned} \tag{27.3}
$$

where the angle brackets with the subscript 0 denote averaging over all possible conformations of a single chain with its end-point fixed at $\mathbf{R}_0$. Note that since the averaging is now effectively reduced to a single-chain problem, likewise it happens in the standard Target Problem [4–6], we drop the subscript $k$.

The next step consists, in a usual way, in adding and subtracting unity in the expression in brackets in the right-hand-side of Eq. (27.3), and then in splitting the integration over the entire volume $V_d$ into the integrations over the interior part of the sphere and the exterior one, denoted by $V'_d$. In doing so, we get

$$
\begin{aligned}
\langle I_R \rangle &= \left[ 1 - \frac{1}{V_d} \int_{V_d} d\mathbf{R}_0 \left\langle 1 - \prod_{n=0}^{N} I\left(\mathbf{R}_n\right) \right\rangle_0 \right]^K \\
&= \left[ 1 - \frac{1}{V_d} \int_{|\mathbf{R}_0|<R} d\mathbf{R}_0 \left\langle 1 - \prod_{n=0}^{N} I\left(\mathbf{R}_n\right) \right\rangle_0 \right. \\
&\quad \left. - \frac{1}{V_d} \int_{V'_d} d\mathbf{R}_0 \left\langle 1 - \prod_{n=0}^{N} I\left(\mathbf{R}_n\right) \right\rangle_0 \right]^K .
\end{aligned} \tag{27.4}
$$

In the inner domain $|\mathbf{R}_0| < R$, the product $\prod_{n=0}^{N} I\left(\mathbf{R}_n\right) \equiv 0$ (because in this domain of integration $\mathbf{R}_0$ is located within the sphere) and hence, the first integral is just equal to the volume of the sphere,

$$
\int_{|\mathbf{R}_0|<R} d\mathbf{R}_0 \left\langle 1 - \prod_{n=0}^{N} I\left(\mathbf{R}_n\right) \right\rangle_0 = v_d R^d , \tag{27.5}
$$

where $v_d$ is the volume of a unit $d$-dimensional sphere, $v_d = \pi^{d/2}/\Gamma(d/2+1)$, $\Gamma(x)$ being the Gamma function. In the outer domain, the indicator function $1 - \prod_{n=0}^{N} I\left(\mathbf{R}_n\right)$ is equal to 1 if at least one of the monomers appears with the sphere, and is equal to zero, otherwise. In consequence, the averaged value of this indicator function,

$$
\left\langle 1 - \prod_{n=0}^{N} I\left(\mathbf{R}_n\right) \right\rangle_0 = P_R(\mathbf{R}_0), \tag{27.6}
$$

defines the probability $P_R(\mathbf{R}_0)$ of the event that for a chain composed of $N+1$ monomers with its tagged end-point fixed at $\mathbf{R}_0$, at least one of the monomers appears within the sphere of radius $R$ centered at the origin.

Eventually, we turn to the thermodynamic limit $V_d \to \infty$ and $K \to \infty$ while keeping fixed their ratio $c = K/V_d$—the volume number density of tagged chains' ends. In this limit, $\langle I_R \rangle$ becomes

$$\langle I_R \rangle = \exp\left(-v_d c R^d\right) \exp\left(-c \int_{V_d'} d\mathbf{R}_0\, P_R(\mathbf{R}_0)\right). \tag{27.7}$$

Before we proceed to the analysis of explicit forms of $\langle I_R \rangle$, it may be instructive to make the following comments: The first multiplier in Eq. (27.7), i.e., $\exp(-v_d c R^d)$, defines the probability of having a spherical void of radius $R$ that is completely free of the tagged *end-points* of polymers. (Note that the second, "untagged" end-point may be present in this spherical volume). For $N = 0$, i.e., for the solution of immobile, randomly placed monomers, this is the only existing term that decays with $R$ very rapidly, as $\exp(-R^d)$, signifying that large monomer-free voids are very improbable. In turn, the second multiplier is the most interesting point here that emerges due to the fact that the monomers are all arranged into long polymer chains. The expression in the exponential in this term is evidently a function of $R$ and $N$, but its form is unclear a priori and represents, in fact, the main challenge for the analysis. Concurrently, it is intuitively clear that the expression in the exponential in the second multiplier in Eq. (27.7) grows with $R$ slower than $R^d$, and therefore, the ultimate large-$R$ behavior of $\langle I_R \rangle$ should be completely dominated by the first multiplier. On the other hand, this multiplier contains a small parameter $c$ and is independent of $N$, such that for intermediate values of $R$ the second multiplier may potentially become the dominant one and this intermediate regime should be more pronounced the larger $N$ is.

## 27.3 Gaussian Coils in Three- and Two-Dimensional Systems

For ideal polymers, one supposes that $N$ and $n$ are continuous variables and exploit the equivalence between a Gaussian coil and a Brownian motion trajectory, in order to calculate the integral in the exponential in the second multiplier exactly. This is done (see [2, 3]) by representing $P_R(\mathbf{R}_0)$ in terms of the solution of the diffusion equation subject to the adsorbing boundary condition imposed at the surface of a $d$-dimensional sphere of radius $R$. This procedure is well-documented for the standard Target Problem [4–6], and even for somewhat more complicated settings, e.g., for the above-mentioned situation with spontaneously evanescent Brownian motions [7] or for particles with a fluctuating reactivity, which switches at random between reactive and non-reactive states [8]. Following this procedure, one eventually arrives at the general relation

$$\int_{V_d'} d\mathbf{R}_0\, P_R(\mathbf{R}_0) = \int_0^N dn\, \kappa_n(R), \tag{27.8}$$

where $\kappa_n(R)$ is defined as the current of diffusive particles through the surface of an immobile $d$-dimensional sphere of radius $R$ and, therefore, depends on both $n$ and $R$. In standard chemical kinetics, it often suffices to know only the large-$n$ (long-time) forms of $\kappa_n(R)$. In particular, in three dimensions, $\kappa_n(R)$ attains a limiting value as $n \to \infty$, which is the so-called Smoluchowski constant [9] (see also [10]). In two dimensions, $\kappa_n(R)$ does not approach a constant value as $n \to \infty$ but rather is a slowly (logarithmically) decreasing function of $n$ in the large-$n$ limit (see, e.g., [11] and Chap. 26). Most of analyses take into account these leading asymptotic forms only. For our purposes here this knowledge is insufficient, because the sub-dominant in time terms may have a stronger dependence on $R$ than the leading ones.

### *27.3.1 Three-Dimensional Systems*

In three-dimensional systems $\kappa_n(R)$ has a compact exact form, which is valid for any $n$ (see, e.g., [9, 10]) :

$$\kappa_n(R) = 4\pi DR\left(1 + \frac{R}{\sqrt{\pi Dn}}\right), \quad D = a^2/6, \tag{27.9}$$

where $a$ is the size of a monomer. Consequently, $\langle I_R \rangle$ in Eq. (27.7) is given explicitly by

$$\langle I_R \rangle = \exp\left(-\text{const}_1\, \phi\, a^2 R - \text{const}_2\, \phi \frac{a}{N^{1/2}} R^2 - \nu_3\, c\, R^3\right), \tag{27.10}$$

where $\phi = cN$ defines the density of the polymer fraction, while $\text{const}_1$ and $\text{const}_2$ are computable numerical constants.

We now discuss the exact expression (27.10), focusing on the limit $N \gg 1$ in which the density of the polymer fraction exceeds by far the density of the chains. Therefore, in this limit the fact that monomers are not placed independently of each other but are all arranged in long chains should have the strongest consequences on the form of $\langle I_R \rangle$. As it could be expected on intuitive grounds, the crucial parameter in Eq. (27.10) is $R_c = aN^{1/2}$—the averaged end-to-end distance of polymer chains. This parameter distinguishes between two different regimes: In the limit $R \gg R_c$, the third term in the exponential in Eq. (27.10) is the dominant one, such that

$$\langle I_R \rangle \simeq \exp\left(-\nu_3\, c\, R^3\right), \quad R \to \infty. \tag{27.11}$$

Note that the prefactor in front of $R^3$ is $c$, i.e., the number density of the tagged ends of the chains, which is much smaller for $N \gg 1$ than the number density $\phi$ of the polymer fraction. This signifies that on such scales the chains (comprising many monomers) are simply seen as point-like particles, and the true fraction of space occupied by polymers is irrelevant. At short and intermediate scales, $R < R_c$, it is the first term in the exponential in Eq. (27.10), which appears to be the dominant one. Hence, for such values of $R$ we have

$$\langle I_R \rangle \simeq \exp\left(-\text{const}_1\, \phi\, a^2 R\right), \quad R < R_c. \tag{27.12}$$

The function in Eq. (27.12), which highlights the effects of correlations between the monomers, vanishes with $R$ at a much slower rate than the function in Eq. (27.11). This signifies, substantiating the claim made in the Introduction, that at such scales having a polymer-free void in a polymer solution is much more probable than in a solution of monomers, present at the same density. Moreover, Eq. (27.12) suggests that for $N \to \infty$ the polymer-free void with the *maximal* size has the Gumbel distribution. We finally remark that the second term in the exponential in Eq. (27.10) is comparable to the first and the third terms only in the vicinity of $R = R_c$ and is much less than both of them away from this point.

### *27.3.2 Two-Dimensional Systems*

In two dimensions, the explicit form of $\kappa_n(R)$ valid for arbitrary $n$ is not known. However, its Laplace-transform is given by a rather compact expression

$$\kappa_\lambda(R) = \int_0^\infty dn e^{-\lambda n} \kappa_n(R) = 2\pi R \sqrt{\frac{D^*}{\lambda}} \frac{K_1(\sqrt{\lambda/D^*}R)}{K_0(\sqrt{\lambda/D^*}R)}, \quad D^* = a^2/4, \tag{27.13}$$

where $K_\nu$ is the modified Bessel function, but in the $n$ (or time) domain only the large-$n$ behavior of $\kappa_n(R)$ is well-established (see, e.g., [11] and Chap. 26). In this limit, $\kappa_n(R)$ does not approach a constant value, in contrast to its counterpart in three-dimensional systems (see Eq. (27.9)), but rather slowly (logarithmically) decreases to zero. In the leading, in this asymptotic limit, order, one has

$$\kappa_n(R) \simeq \frac{4\pi D^*}{\ln\left(4D^* n/R^2\right)}, \quad n \to \infty. \tag{27.14}$$

We will comment on the origin of such a behavior in the next Section. For the analysis of $\langle I_R \rangle$, the above expression is applicable for small values of $R$ only and in order to access the behavior for larger values of $R$ we have to resort to either numerical inversion of Eq. (27.13) or to some approximations. A rather accurate approximate result for $\kappa_n(R)$ for $R < R_c$ has been derived in [3]. Multiplying both sides of Eq. (27.13) by $K_0(\sqrt{\lambda/D^*}R)$ and noticing that the product of the two Laplace-transformed functions on the left-hand-side can be regarded as the Laplace transform of the convolution of these function in the $n$ domain, it was shown that $\kappa_n(R)$ obeys the following integral equation

$$\int_0^N dn\, \kappa_n(R) \frac{\exp\left(-R^2/(4D^*(N-n))\right)}{N-n} = 4\pi D^* \exp\left(-R^2/(4D^*N)\right). \quad (27.15)$$

Assuming that $\kappa_n(R)$ is a slowly varying function of $n$ (and actually it is), it was realized that $\kappa_n(R)$ can be approximated as

$$\kappa_n(R) \approx 4\pi D^* \frac{\exp\left(-R^2/(4D^*n)\right)}{E_1\left(R^2/(4D^*n)\right)}, \quad (27.16)$$

where $E_1$ denotes the exponential integral. In the limit $n \to \infty$, expression (27.16) yields the asymptotic form in Eq. (27.14), as one may verify directly. Further on, comparing the expression (27.16) and the numerical solution of the integral equation (27.15), it was shown in [3] that Eq. (27.16) agrees very well with the latter for $R$ up to $R_c$ when the asymptotic form in Eq. (27.14) becomes invalid.

On the other hand, within the opposite limit $R \gg R_c$ Eq. (27.16) predicts an incorrect behavior. Namely, one finds from Eq. (27.16) that $\kappa_n(R) \simeq \pi R^2/n$ for $R \gg an^{1/2}$. In reality, $\kappa_n(R)$ increases with $R$ at a slower rate. The correct behavior can be simply accessed from the Laplace-transformed expression (27.13). Letting $R \to \infty$, and inverting the resulting expression one finds (see also [11])

$$\kappa_n(R) \simeq \sqrt{\frac{4\pi D^*}{n}} R, \quad R \to \infty, \quad (27.17)$$

which has, in fact, exactly the same form (apart from the factor $R$) as the Smoluchowski constant in one-dimensional systems. The latter expression implies that the integral of the Smoluchowski constant over $n$ grows with $R$ only linearly, i.e., at a slower rate than the Poisson term $\mu_2 c R^2$ in the exponential in the first multiplier in Eq. (27.7).

Therefore, the overall behavior of $\langle I_R \rangle$ is as follows: At scales $R < R_c$, the probability $\langle I_R \rangle$ obeys

$$\langle I_R \rangle \simeq \exp\left(-\pi a^2 c \int_0^N dn \frac{\exp(-R^2/a^2 n)}{E_1(R^2/a^2 n)}\right), \quad R < R_c, \quad (27.18)$$

which simplifies for very small values of $R$ ($R \to 0$ or $N \gg 1$) to give a very weak, logarithmic dependence on $R$ :

$$\langle I_R \rangle \simeq \exp\left(-\frac{\pi a^2 \phi}{\ln(a^2 N/R^2)}\right). \quad (27.19)$$

Within the opposite limit, i.e., for $R > R_c$, we have

$$\langle I_R \rangle \simeq \exp\left(-\pi c R^2\right), \quad R > R_c. \quad (27.20)$$

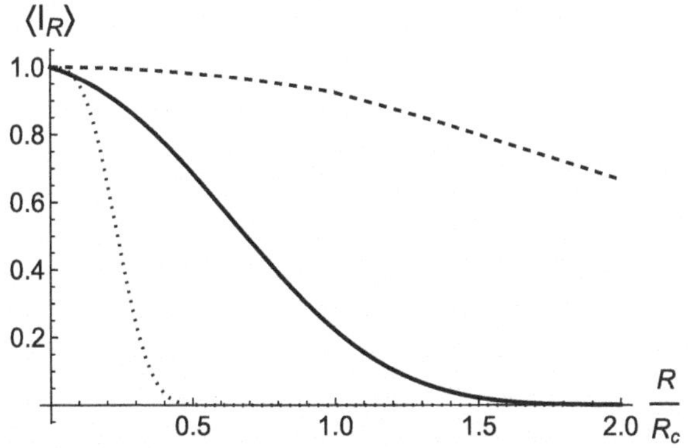


**Fig. 27.2** Probability $\langle I_R \rangle$ in Eq. (27.7) versus the size $R$ of a polymer-free void measured in units of the end-to-end distance $R_c$. $N = 100$, $c = 10^{-3}$ and $a = 1/2$. The upper (dashed) curve depicts $\langle I_R \rangle$ obtained via a numerical solution of Eq. (27.15). The solid curve in the middle corresponds to the result in three dimensions, Eq. (27.10). The dotted curve in the bottom corresponds to a three-dimensional solution of *monomers* present at density $\phi = cN = 0.1$, i.e., the same density as the polymer fraction in the two above curves. In this case $\langle I_R \rangle$ is defined by Eq. (27.11) for any $R$ with $c$ replaced by $\phi$

Hence, in two-dimensional systems the scenario is exactly the same as in three-dimensional systems in the sense that at scales $R$ which are smaller than the averaged end-to-end distance of chains, the decay of $\langle I_R \rangle$ is dominated by the effects of the monomers' arrangement into long polymer chains, while at longer scales the chains are seen as point-like particles present with density $c$, which entails a faster decay in Eq. (27.20).

In Fig. 27.2 we depict $\langle I_R \rangle$ as function of $R$ for three-dimensional and two-dimensional polymer solutions in which the polymer fraction has the density $\phi = 0.1$, as well as, for a comparison, for a three-dimensional solution of immobile monomers present at the same density $\phi$. We observe that $\langle I_R \rangle$ in two dimensions (e.g., for chains adsorbed on a solid surface) appears well above the two other curves. This signifies that here rather large polymer-free voids are probable even for $R$ that exceeds $R_c$, which is the consequence of a very weak (weaker than a power-law) dependence of the Smoluchowski constant on $R$ at intermediate scales, and also of a somewhat weaker decrease of $\langle I_R \rangle$ for $R > R_c$ (compare Eqs. (27.20) and (27.11)). In fact, very large polymer-free voids are indeed apparent in a numerically generated polymer substrate (see Fig. 1 in [3]). In three dimensions, the arrangement of monomers into the chains has a weaker effect on the size of polymer-free voids than in the two-dimensional case, but still $\langle I_R \rangle$ is much greater for any $R$ than the corresponding probability of voids in a solution of monomers.

## 27.4 Non-Gaussian Conformations of Polymers

The case of polymers having non-Gaussian conformations is particularly interesting and realistic. In this Section, we present some arguments which permit to generalize the above analysis over the case of non-ideal (albeit still assumed to be independent) polymer chains. Such a generalization is based on the following geometrical interpretation of the integral $\int_0^N dn\kappa_n(R)$: for Brownian motion this integral defines the averaged, over all realizations, volume of the so-called Wiener "sausage" (see, e.g., [10, 12, 13]); that being, a volume (or an area for diffusion in two dimensions, see Fig. 27.3) swept by a diffusing particle of radius $R$ within time $t$.

A non-trivial point is that the visited parts of space are counted only once; if the particle revisits some already visited area or only some fraction of it, the revisited region does not add any new contribution to the Wiener "sausage". This geometrical subtlety is most apparent for a discrete-space counterpart of the Wiener sausage—a number $S$ of *distinct* sites visited by an $N$ step random walk on a lattice [14]. This random variable increases by unity each time instant when the walker steps on a site that it has not visited before, and maintains the same integer value if this site has been already visited. The mean values of the volume of the Wiener sausage and of the number of distinct visited sites are evidently monotonically growing functions of time and moreover, both have the same time-dependence in systems of the same spatial dimension $d$. Here, the value of $d$ is crucial; For $d < 2$ (e.g., in one

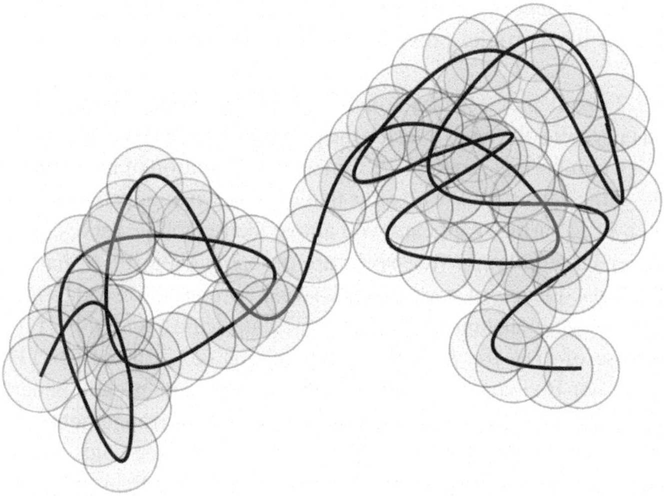

**Fig. 27.3** Area swept up to time $t$ by a particle of radius $R$ that moves randomly on a plane. Thick solid curve represents a given particle's trajectory in a continuous time. Shaded circles are the "snapshots" of the particle taken at some discrete time moments. The area of the Wiener "sausage" for this given realization is the area delimited by the curves that envelope all the snapshots taken at the consecutive time moments, in the limit when the time-step tends to zero

dimension), the mean volume of the Wiener sausage, which also defines the mean maximal extent (range) of trajectories, grows in proportion to $\sqrt{t}$. This means that in one dimension each trajectory very compactly explores the space revisiting many times already visited regions. For $d = 2$, such an oversampling is only marginal and the mean volume of the Wiener sausage grows as $t/\ln(t)$, i.e., slightly slower than linearly with time. This is precisely why the Smoluchowski constant $\kappa_n(R)$, which can be thought of as a density of newly visited sites for a trajectory of length $n$, decreases as the logarithm of $n$ (see Eq. (27.14)) and being integrated over $n$ gives $N/\ln(N)$. Finally, for $d > 2$, the mean volume of the Wiener sausage, and also the mean number of distinct visited sites both grow linearly with time. In this case, each trajectory explores the space in a non-compact way leaving within the already visited regions of space multiple unvisited domains.

The next question is why the spatial dimension 2 is so special. The point is that each given trajectory of the Brownian motion is a self-similar fractal, and its fractal dimension $d_f$ is equal to 2. That being, the end-to-end distance $R_c$ of Brownian motion trajectory scales with time as $t^{2/d_f}$ with $d_f = 2$ in a system of any spatial dimension. If we were to deal with a fractal object with $d_f \neq 2$, in most cases[1] we would encounter a similar scenario as above with the exception that the border-line dimension is $d = d_f$ [16, 17]. More specifically, for $d < d_f$ the mean volume of the Wiener sausage or of the number of distinct sites visited will grow with time as $t^{d/d_f}$, in the marginal case $d = d_f$ one will find $t/\ln(t)$ and eventually, for $d > d_f$ a linear growth with time will take place. Evidently, for $d_f = 2$ we recover the above mentioned growth laws.

Turning to the model under study, we constrain our analysis to the case when the fractal dimension of polymer chains of length $N$ obeys $d_f < d$. Hence, we expect that trivially $\int_0^N dn\, \kappa_n(R) \sim N$, but the most interesting question is how the prefactor in this scaling law depends on $R$. To find the answer, we evoke the concept of blobs [1] and suppose, in a standard fashion, that the chain consists of $M$ blobs of radius $R$, $M = N/L$ where $L$ is the number of monomers in each blob, $L \sim (R/a)^{d_f}$ [1]. Then, the mean volume of the Wiener sausage, up to a numerical factor, is proportional to the number of blobs and their volume,

$$\int_0^N dn\, \kappa_n(R) \sim M R^d = N a^{d_f} R^{d-d_f}. \tag{27.21}$$

In this expression $R$ appears to the power $\gamma = d - d_f$, which is the so-called co-dimension—the dimension of a fractal object with dimensionality $d_f$ embedded in a $d$-dimensional space. Note that the prediction in Eq. (27.21) agrees well with the result found for Gaussian coils in three dimensions, and also for systems of

[1] With an exception, e.g., of an irrelevant for the current analysis situation when fractal trajectories with $d_f \neq 2$ are generated by continuous-time processes with temporal trapping with a broad distribution of trapping times (see [15]). In this case the number of steps, i.e., the length of a trajectory is itself a random variable.

orientationally disordered rigid rod-like molecules, for which the mean volume of the Wiener volume can be calculated analytically to give $\int_0^N dn\, \kappa_n(R) \sim NaR^{d-1}$ [18].

Therefore, the probability of having a polymer-free void of radius $R$ can be written down as

$$\langle I_R \rangle \simeq \exp\left(-\nu_d c R^d\right) \exp\left(-\phi a^{d_f} R^{d-d_f}\right), \tag{27.22}$$

where the first multiplier defines, as above, the probability of having a void of radius $R$ that is free of the polymers' tagged ends, while the second multiplier appears due to the connectivity of monomers into polymer chains. As one may check directly, the term in the exponential in the first multiplier is dominant for scales that exceed the end-to-end distance $R_c = aN^{1/d_f}$ of the chain, while the second multiplier governs the dependence of $\langle I_R \rangle$ on $R$ in the opposite limit. Note that the dependence on $R$ becomes progressively weaker, the closer $d_f$ to $d$ is, and is the weakest for $d_f = d$, as we have observed already for Gaussian coils in two-dimensional systems. Finally, we note that for swollen coils (self-avoiding walks), for which $d_f = 1/\nu$ where $\nu = 3/(d+2)$ (such that $\nu = 3/4$ and $\nu = 3/5$ in two and three dimensions, respectively) is the Flory exponent, one finds from Eq. (27.22) that the mean size of a polymer-free void in three dimensions obeys $\langle R \rangle \sim a\phi^{-3/4}$ in the limit $N \gg 1$. This is precisely the result obtained in [1] using some scaling arguments and a dimensional analysis.

## 27.5 Conclusions and Perspectives

To conclude, in the present chapter we addressed a question about a functional form of the distribution of sizes of polymer-free voids in a solution of long immobile polymer chains. We have shown that in the case when the latter can be considered as statistically independent, the problem can be solved exactly by mapping it onto the Target Problem, which explains why this chapter is presented in this book. We note that the assumption of a statistical independence of chains is not unrealistic and, in fact, this case is realized in dense polymer melts, in which according to the Flory theorem the chains behave as independent Gaussian coils. Evidently, the chains can also be considered as independent in dilute solutions, in which they are well-isolated from each other. Conformations of such isolated chains depend on the solvent; they are swollen in good solvents and are Gaussian coils in $\theta$-solvents [1].

The exact solutions have been obtained for ideal (Gaussian) chains in three- and two-dimensional systems. For chains having arbitrary conformations (but still independent from each other), the probability $\langle I_R \rangle$ of having a polymer-free void of radius $R$ has been determined using some heuristic arguments. We have shown that, not counter-intuitively, the fact that all the monomers are linked into long polymer chains causes a significant departure of the form of $\langle I_R \rangle$ from the Poisson one, $\exp(-\nu_d c R^d)$, where $\nu_d$ is the volume of a $d$-dimensional sphere of unit radius and $c$ is here the density of independently placed monomers. On scales less than the end-to-end distance of a polymer chain, the probability $\langle I_R \rangle$ decreases with $R$ much slower than the above Poisson form, e.g., as $\exp(-R)$ in three dimensions and even slower in two-dimensional systems. At larger scales, $\langle I_R \rangle$ has the Poisson form but the density which enters this expression is not the overall density of the polymer fraction, but a much smaller quantity—the density of the polymers' tagged ends. This signifies that on such scales the chains comprising many monomers are seen as point-like particles. For chains with arbitrary conformations characterized by a fractal dimension $d_f$, which are all embedded in a $d$-dimensional solution, the intermediate-$R$ decrease of $\langle I_R \rangle$ is again much slower than the Poisson form and is progressively slower the closer $d_f$ is to $d$. At larger scales, the Poisson form again provides the dominant dependence on $R$ with, however, the density equal to the density of tagged ends of chains. Overall, our results show that a polymer fraction in a frozen solution is very unevenly distributed in space, as compared to the solution of monomers, and is therefore characterized by much larger regions completely devoid of polymers.

The analysis presented in this chapter concerned the case of homopolymers, in which all the monomers are supposed to have the same size. An interesting question is to define $\langle I_R \rangle$ for solutions of heteropolymers containing subunits of two very different sizes. The second interesting problem, which we have already mentioned, is to determine $\langle I_R \rangle$ for polydisperse polymer solutions. Clearly enough, one may expect completely different behavior of $\langle I_R \rangle$ depending on whether the size-distribution of polymer chains is a monotonically decreasing (e.g., exponential) function of its length, as it happens in case a polymer solution obtained via a polymerization process [1], or is peaked at some large most probable value.

Another possible direction of research concerns search processes—the subject which is amply discussed in the present book. In typical settings, one or many searchers perform random motion searching for a single or multiple, mobile or immobile targets. The random motion can be simply diffusive, as it often happens in systems of biophysical interest (see, e.g., [19, 20]), or follow some artificially designed algorithms. Among others, this can be a search with a stochastic resetting [21–24], or the so-called intermittent random motion [25–28]. In the former algorithm, a searcher is instantaneously reset at a random time moment to its initial position and then the search process starts anew. In the latter, the searcher's random motion consists of randomly alternating fast relocation stages, during which the searcher does not perceive a target and therefore is not capable to identify it, and slow diffusive (or even subdiffusive) tours of a random duration in which the target can be eventually found. For both algorithms, most analyses concentrated on finding such values of parameters, i.e., an optimal reset rate or an optimal relative duration of the relocation/diffusion stages, which provide a minimal value of the mean first-passage time to the target [21–26]. In [27, 28] a different strategy has been proposed—to minimize the probability that a target is not found up to a fixed prescribed time moment. To this end, the parameters of the intermittent motion were tuned to ensure that the mean number of distinct sites visited by a searcher up to this time instant is maximal, giving therefore the target the least possible chances to remain undetected.

The problem discussed in the present chapter suggests yet another measure of the efficiency of a given motion algorithm for a search by multiple particles. Namely, how thoroughly, for a given kind of motion, the searchers explore the space leaving the unvisited domains to be minimal in their sizes. As evidenced by our model-dependent results, (for other motion algorithms the form of Eq. (27.22) may be different), if the consecutive positions of a searcher are correlated, a Brownian search appears to be the worst possible scenario (especially in two dimensions). Ballistic trajectories with $d_f = 1$, on the contrary, provide the narrowest size-distribution of the unvisited areas. In general, it seems advantageous to shorten the trajectories and to increase the number of searchers, which is however not always possible. On the other hand, search with a stochastic reset or an intermittent search, which both reset the correlations between the consecutive positions at some time moments, may turn to be more beneficial and more thoroughly exploring the space, upon an appropriate optimization of parameters. Mathematically, this requires finding the solution of the Target Problem for these kinds of motion.

**Acknowledgements** The authors acknowledge helpful discussions with Olivier Bénichou and Maxim Dolgushev.

## References

1. P.-G. de Gennes, *Scaling Concepts in Polymer Physics* (Cornell University Press, Ithaca and London, 1979)
2. S.F. Burlatsky, G. Oshanin, Diffusion-controlled reactions with polymers. Phys. Lett. A **145**, 61–65 (1990)
3. G. Oshanin, A. Mogutov, S.F. Burlatsky, A two-dimensional model of trapping reactions with Gaussian coils. Phys. Lett. A **149**, 55–59 (1990)
4. S. Redner, K. Kang, Kinetics of the "scavenger" reaction. J. Phys. AâŁ¯: Math. Gen. **17**, L451–L455 (1984)
5. M. Tachiya, Theory of diffusion-controlled reactions: formulation of the bulk reaction rate in terms of the pair probability. Radiat. Phys. Chem. **21**, 167–175 (1983)
6. S.F. Burlatsky, A.A. Ovchinnikov, Effect of reactant-fluctuation density on the kinetics of recombination, multiplication, and trapping processes. Zh. Eksp. Teor. Fiz. **92**, 1618–1634 (1987). [Sov. Phys. JETP **65**, 908–917 (1987)]
7. S.B. Yuste, J.J. Ruiz-Lorenzo, K. Lindenberg, The target problem with evanescent subdiffusive traps. Phys. Rev. E **74**, 046119 (2006)
8. O. Bénichou, M. Moreau, G. Oshanin, Kinetics of stochastically gated diffusion-limited reactions and geometry of random walk trajectories. Phys. Rev. E **61**, 3388–3406 (2000)
9. M. Smoluchowski, Versuch einer matematischen Theorie der Koagulationskinetik kolloider Lösungen. Z. Phys. Chem. **92**U, 129–168 (1918)
10. G. Oshanin, M. Moreau, S.F. Burlatsky, Models of chemical reactions with participation of polymers. Adv. Colloid Interface Sci. **49**, 1–49 (1994)
11. K. Uchiyama, The expected area of the Wiener sausage swept by a disc. Stoch. Process. Appl. **123**, 191–211 (2013)
12. M.D. Donsker, S.R.S. Varadhan, Asymptotics for the Wiener sausage. Comm. Pure Appl. Math. **28**, 525–565 (1975)
13. https://en.wikipedia.org/wiki/Wiener_sausage
14. B.D. Hughes, *Random Walks and Random Environments* (Clarendon Press, Oxford, 1995)
15. S.B. Yuste, J. Klafter, K. Lindenberg, Number of distinct sites visited by a subdiffusive random walker. Phys. Rev. E **77**, 032101 (2008)

16. D. ben-Avraham, S. Havlin, *Diffusion and Reactions in Fractals and Disordered Systems* (Cambridge University Press, Cambridge, 2000)
17. L. Régnier, M. Dolgushev, S. Redner, O. Bénichou, Universal exploration dynamics of random walks. Nat. Comm. **14**, 618 (2023)
18. J. Klafter, A. Blumen, G. Zumofen, in *Molecular Dynamics in Restricted Geometries* ed. by J. Klafter, J.M. Drake (Wiley, New York, 1989)
19. D.S. Grebenkov, R. Metzler, G. Oshanin, From single-particle stochastic kinetics to macroscopic reaction rates: fastest first-passage time of $N$ random walkers. New J. Phys. **22**, 103004 (2020)
20. D.S. Grebenkov, R. Metzler, G. Oshanin, Search efficiency in the Adam-Delbrück reduction-of-dimensionality scenario versus direct diffusive search. New J. Phys. **24**, 083035 (2022)
21. M.R. Evans, S.N. Majumdar, Diffusion with stochastic resetting. Phys. Rev. Lett. **106**, 160601 (2011)
22. M.R. Evans, S.N. Majumdar, Diffusion with optimal resetting. J. Phys. A: Math. Theor. **44**, 435001 (2011)
23. M.R. Evans, S.N. Majumdar, G. Schehr, Stochastic resetting and applications. J. Phys. A: Math. Theor. **53**, 193001 (2020)
24. S. Gupta, A.M. Jayannavar, Stochastic resetting: a (very) brief review. Front. Phys. **10**, 789097 (2022)
25. O. Bénichou, M. Coppey, M. Moreau, P.-H. Suet, R. Voituriez, Optimal search strategies for hidden targets. Phys. Rev. Lett. **94**, 198101 (2005)
26. O. Bénichou, C. Loverdo, M. Moreau, R. Voituriez, Intermittent search strategies. Rev. Mod. Phys. **83**, 81–128 (2011)
27. G. Oshanin, H.S. Wio, K. Lindenberg, S.F. Burlatsky, Intermittent random walks for an optimal search strategy: one-dimensional case. J. Phys.: Condens. Matter **19**, 065142 (2007)
28. G. Oshanin, H.S. Wio, K. Lindenberg, S.F. Burlatsky, Efficient search by optimized intermittent random walks. J. Phys. A: Math. Theor. **42**, 434008 (2009)

# Chapter 28
# Target Search Processes During Horizontal Gene Transfer in Bacteria

**Rinat Arbel-Goren and Joel Stavans**

**Abstract** In the absence of sex as in higher organisms such as mammals and plants, bacteria engage in horizontal gene transfer processes in order to acquire new beneficial traits and drive their genetic diversity in one stroke. Often, exogenous DNA integrates into the genome of a host at unique sites that are found within physiologically relevant times, among millions of possible sites on the host's genome. Here we address the question of what are the search mechanisms that allow these unique sites to be found efficiently, within the complex interior of a bacterial cell. We review recent advances in the field, enabled by the ability to follow in real time search processes at the level of individual molecules within single cells.

## 28.1 Introduction

As opposed to higher organisms, bacteria do not have sex. Instead, they rely on a set of mechanisms known collectively as horizontal (or lateral) gene transfer (HGT) [1, 2], in order to generate genetic diversity beyond the long, gradual improvements on fitness afforded by the (vertical) alterations to their genomes that occur as they divide by binary fission [3]. Indeed, comparative genomics has revealed that microbial evolution is dominated by episodic HGT events during which exogenous DNA is introduced into bacterial cells and integrated into their genome, allowing cells to acquire in one shot new genes and traits that contribute to adaptation and survival, such as antibiotic resistances and pathogenicity [4]. These massive additions of genetic information take place in microbial communities, including the human microbiome, the rhizosphere of plant roots, biofilms and many other natural habitats.

R. Arbel-Goren · J. Stavans (✉)
Department of Physics of Complex Systems, Weizmann Institute of Science, 7610001 Rehovot, Israel
e-mail: joel.stavans@weizmann.ac.il

R. Arbel-Goren
e-mail: rinat.goren@weizmann.ac.il

 
D. Grebenkov et al. (eds.), *Target Search Problems*,
https://doi.org/10.1007/978-3-031-67802-8_28

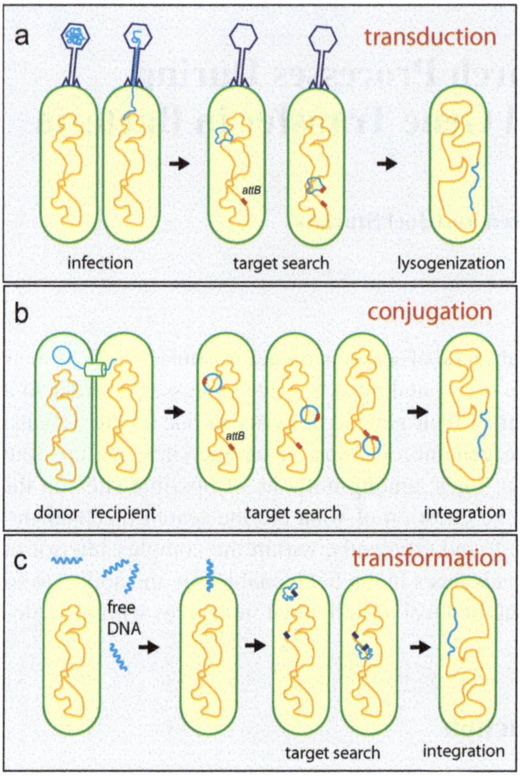


**Fig. 28.1** Mechanisms of horizontal gene transfer. **a** Bacteriophage-mediated transfer of genetic material (blue), sometimes referred to as transduction (specialized and generalized), to an infected recipient cell. The bacterial chromosome is depicted in orange and the attachment site *attB* in red. **b** Conjugation is a process requiring cell to cell contact via either cell surface pili or adhesins, through which DNA (blue) is transferred from a donor cell to a recipient cell. The attachment site *attB* where integration takes place is shown (red). **c** Transformation is the uptake, integration, and functional expression of naked fragments of extracellular DNA (blue). Common to these three processes is a target search for an integration site along the bacterial chromosome (orange)

There are essentially three HGT mechanisms: transduction, conjugation, and transformation (Fig. 28.1). Transduction is mediated by bacteriophages, viruses that infect bacterial cells; conjugation requires instead both donor and recipient cells. Lastly, transformation is the uptake of DNA tracts from the extracellular environment. We note that other mechanisms for conveying exogenous DNA into cells have been discovered, including transfer by phage derived agents, outer membrane vesicles, and nanotubes [2, 5].

Common to all mechanisms and their successful completion requires integration of the exogenous DNA into the host chromosome, which usually takes place at unique sites. This poses a considerable hurdle: even when not replicated, bacterial genomes are typically millions of base pairs long.

What are the physical mechanisms that allow the integration site to be found efficiently among the many possible, within physiologically relevant times? Here we review recent advances in tackling this question. The ability to follow processes at the level of single molecules in vivo in individual cells has proved invaluable in studying intracellular search processes, particularly in the challenging case of bacteria. In this review we focus on dynamical, large-scale aspects of search processes rather than on molecular scales.

## 28.2 Search Processes During Horizontal Gene Transfer

### *28.2.1 Search During Transduction*

A search during transduction takes place during the establishment of lysogeny after infection of bacterial cells by temperate bacteriophages. Temperate bacteriophages are viruses that infect bacterial cells and have two possible developmental fates after infecting a bacterial cell. They can follow the *lytic* pathway, whereby they use cell resources to multiply and create progeny virions that lyse the cell and go on to infect others. Alternatively, they can follow the *lysogenic* pathway and integrate their genome into the host's, replicating passively with the latter [6]. The lysis-lysogeny decision depends on a number of factors including the number of phages infecting a cell (multiplicity of infection). Usually, integration occurs at a unique 15-bp-long attachment site (*attB*), which is located among ~5 million possible address sites, as many as base pairs in a bacterial genome, within physiologically relevant timescales. Recognition of *attB* is mediated by a phage-encoded site-specific recombinase, which can excise and integrate the viral chromosome into and out of the chromosome of its host [7].

The search for an integration site along the chromosome of an *E. coli* cell has been studied in vivo at the level of single molecules in the case of the celebrated, temperate bacteriophage lambda [8]. The lambda genome is ~50 kbp long, much smaller than the *E. coli* chromosome ($5 \times 10^6$ bp). Therefore, it was a surprise when the experiments revealed that instead of an active search, lambda DNA remained mostly confined near the entry point into a cell following infection (Fig. 28.1a), primarily in the polar regions of a cell, or at the midpoint where incipient poles are formed. Indeed, the mean squared displacement of the lambda DNA reached a plateau value within the bacterial cell, compatible with a confinement radius of ~120 nm, below optical resolution. The encounter between the 15-bp-long target sequence on the chromosome and the recombination site on the viral genome is facilitated by the directed motion of bacterial DNA, generated during chromosome replication and segregation. This motion is known to be an out-of-equilibrium, metabolically-driven process, with chromosomal loci moving sub-diffusively through a viscoelastic cytoplasm [9, 10]. The typical timescale for the encounter of lambda DNA and *attB* was found to be of the order of the cell's doubling time under the conditions of the

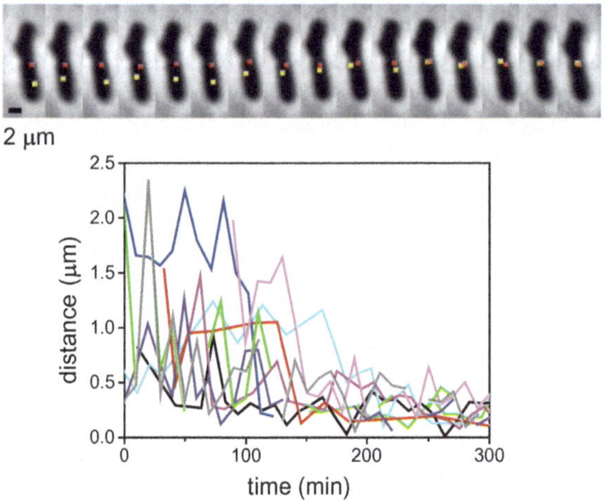


**Fig. 28.2** Target location and encounter of lambda phage DNA with its integration site on an *E. coli* chromosome. Top: Snapshots taken every 10 min after infection show the ParB-RFP locus (red) marking the attachment site on the phage DNA and the ParB-yGFP locus (yellow) marking the *attB* site on the phage DNA. ParB-RFP and the ParB-yGFP loci are separated by ~12 kbp after integration. Bottom: Distance on the focal plane between the attachment site *attB* and phage DNA for individual events. Adapted from [8]

experiments, as shown by a plot of the distance between lambda DNA and *attB* for various events in living cells (Fig. 28.2). Evidence for phage DNA motion spanning the whole cell, in addition to the localized motion mode has also been reported [11]. Preference for either of the modes depended on where along the cell the phage DNA was imported, and the type of motion was not correlated with the outcome of the lysis-lysogeny decision. One may anticipate this search mechanism to be relevant for other temperate phages and bacterial cells, though this notion remains to be tested in the future.

### *28.2.2 Search During Conjugation*

Visualization of different aspects of conjugation in single bacterial cells has been reported. The direct transfer of DNA from donor to recipient *E. coli* cells was observed to occur over long distances, reaching 12 μm (the characteristic cell length is 2 μm), and the subsequent integration of the transferred DNA was observed to be highly efficient, reaching ~96% [12]. The transfer of F-plasmid as single-stranded DNA into a cell, and its subsequent completion to double stranded form was observed to take place with specific timing and subcellular localization [13].

However, target search processes were not followed in these studies. Integrative and conjugative elements (ICEs) are a diverse group of mobile genetic elements found in both Gram-positive and Gram-negative bacteria. Individual ICEs have been found to bestow a wide range of phenotypes on their hosts, including resistance to antibiotics. These DNA elements sharing a common life cycle vary in size from tens to hundreds of kbps and retain the ability to excise and to transfer by conjugation (Fig. 28.1b). Typically the recognition of the target sequence for integration is mediated at the molecular level by an ICE-encoded site-specific recombinase [14].

Recently, the dynamics of target search following conjugation by ICEB*s*1 (21 kbp) for an integration site along the chromosome of recipient *B. subtilis* cells was studied in vivo [15]. The experiments revealed that ICEB*s*1 and the chromosomal *attB* integration site displayed different behavior in their motion during the search process (Fig. 28.3a). The dynamics of *attB* was confined to well-defined regions along the

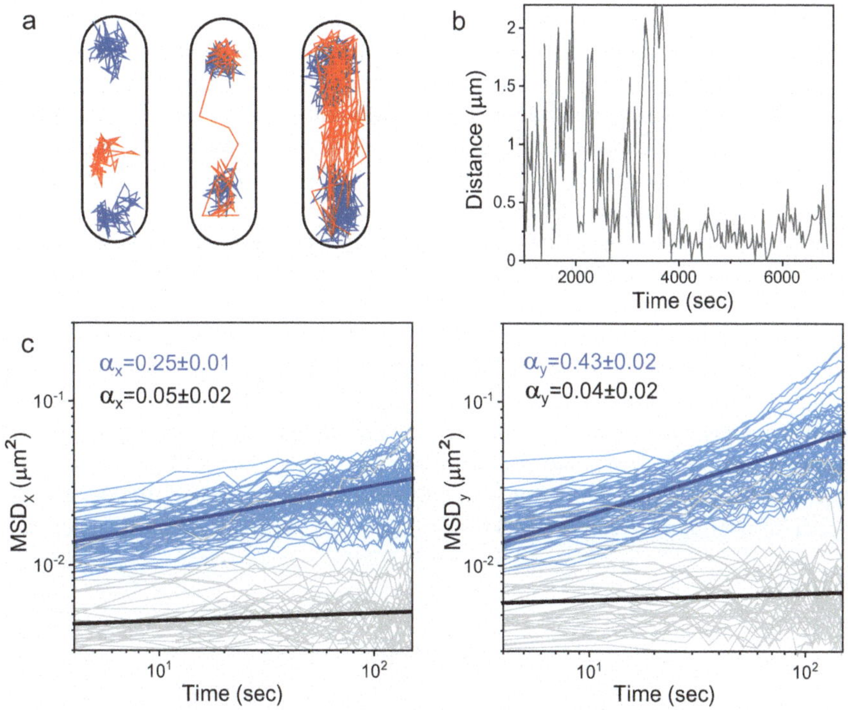


**Fig. 28.3** Dynamics of search by integrative conjugative elements (ICEs) for their integration sites *attB* along the chromosomes of *B. subtilis* cells. **a** Two-dimensional projections of the instantaneous positions of *attB* sites (blue) and ICE*Bs*1 (red) during target search. **b** Distance between an ICE*Bs*1 and the *attB* site into which it eventually integrates. The data correspond to the event shown in the rightmost search process in panel (**a**). **c** Anisotropic mean squared displacement of *attB*. Time-averaged mean squared displacement in the transversal (left panel) and longitudinal (right panel) directions ($MSD_x$ and $MSD_y$ respectively) of individual trajectories of *attB* (grey traces). The value of the exponents $\alpha$ defined by $MSD \propto Dt^{\alpha}$ corresponding to the mean MSD over all trajectories (thick lines) are given in each panel. Grey: *attB* locus $MSD_x$ and $MSD_y$ in ATP and GTP-depleted cells. Adapted from [15]

cell's main axis, with two loci (indicative of a partly replicated chromosome) found typically at 1/4 and 3/4 along the longitudinal axis, compatible with their position along the chromosomal contour [16]. In contrast, ICEB*s*1 exhibited a variety of dynamical behaviors, ranging from compact local explorations, to fast motion spanning the cell length within seconds-long intervals. Three illustrative examples are shown in Fig. 28.3a. Eventual integration into one of the *attB* sites in the rightmost panel is quantified in a plot of the two-dimensional Euclidean distance between the ICEB*s*1 and the *attB* in Fig. 28.3b.

To characterize the motion of *attB* integration sites, the mean squared displacements were calculated. Such experiments showed three salient features (Fig. 28.3c): (i) the dynamics were sub-diffusive, as for chromosomal loci in *E. coli* [9]. Subdiffusivity was attributed to the elasticity of the intracellular medium, as evidenced by the behavior of velocity-velocity correlation functions [15]. Importantly, the mean displacement corresponding to the longest times probed in these measurements is much smaller than the dimensions of a cell, so the rod-shape cell boundaries do not account for subdiffusivity; (ii) the dynamics were anisotropic, with fluctuations along the longitudinal main axis of the cell being different from those in the transversal (radial) directions. Anisotropic fluctuations have only been previously observed in a very different system [17]; (iii) the dynamics of both *attB* and ICE are driven by non-equilibrium, metabolic fluctuations that fluidize the chromosome, as observed in other bacteria [15, 18]. This was shown by measurement of the mean squared displacement of the *attB* locus in cells in which metabolic activity was inhibited (Fig. 28.3c).

The behavior of ICEB*s*1 was rather different. Its highly heterogeneous dynamics led to a distribution of positions peaked at the cell center, with a moderate overlap with the distribution of positions of *attB* sites. This led to the suggestion that brush density there is reduced, as expected between the two nascent nucleoids of incipient daughter cells (Fig. 28.4), and is indicative of an entropic price for ICEs to penetrate into the respective bottlebrush structure of the chromosome [19]. Despite the moderate overlap between the two distributions, the target location efficiency was

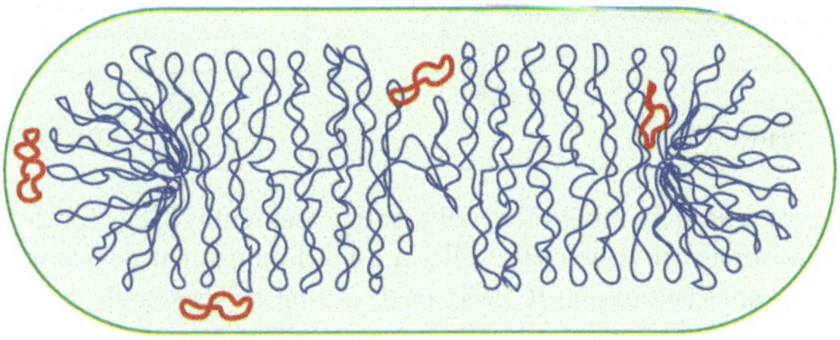

**Fig. 28.4** Bottlebrush structure model of bacterial chromosomes. Chromosomes consist of topologically-isolated, plectonemic loops (blue), extruded by structural maintenance of chromosomes (SMC) proteins from a central core spanning the main cell axis [19]. Free DNA such as ICEs, plasmids or other exogenous DNA (red) can be confined to poles, carry out compact explorations within the bottlebrush or be entropically evicted from the brush [21]

observed to be very high. We point out that *B. subtilis* cells that grow on surfaces display growth in chains, which facilitates cell-to-cell transfer of ICEs, when the latter do not attain integration [20]. It is noteworthy from an evolutionary perspective that different solutions for site-specific search mechanisms were found for transduction and conjugation.

### 28.2.3 Search During Transformation

To date, no experimental studies have been carried out following search processes at the single-molecule level in individual cells in the case of transformation, even though single-cell observations of transformation indicate that the process is one of high efficiency [22]. As opposed to both transduction and conjugation processes, in which the imported DNA encodes for site-specific recombination factors that recognize less than 30 bp sequences on both the host chromosome and the searcher DNA molecule, target search following transformation is mediated by the host's RecA protein [23]. The recognition of homology between DNA tracts in both the imported and genomic DNA depends on the length of homology and the similarity between the sequences [24, 25]. Importantly, RecA polymerizes on DNA substrates, stretching them by $\sim$50%. It has been pointed out that this stretching facilitates enormously the location of homology [26], and the analogy of this antenna effect to Brouwer's fixed point theorem from algebraic topology has been pointed out [27]. We hypothesize that RecA polymerization will affect intracellular motion of exogenous DNA following transformation, in contrast to motion of free DNA such as mini-plasmids [18]. While target search processes following transformation have not been characterized explicitly, a number of single-cell studies have illuminated certain ingredients relevant to the search for homology and eventual integration. Real-time visualization of transformation in competent *Streptococcus pneumoniae* has been reported [28]. By tracking the localization of fluorescent fusions to RecA in *Streptococcus pneumoniae*, it has been observed that dynamic RecA filaments form on internalized single-stranded DNA, as well as emanate from replication forks, implicating replication forks as substrates or landing pads for the formation of homologous recombination intermediates, together with presynaptic filaments on imported single-stranded DNA [29]. This mechanism accounts for the cell doubling time as the characteristic scale for finding homology.

While not forming a proper part of HGT, the target search for a duplicated sequence that takes place during the repair of DNA lesions by homologous recombination may share features in common with target search processes following transformation. The RecA-mediated repair of double stranded breaks in DNA was also followed in individual *E. coli* cells in vivo [30]. Evidence was found that homology search is completed in $\sim$10 min, and is mediated by a highly dynamic RecA-ssDNA nucleoprotein filament that can stretch throughout the whole cell. The authors proposed that it is this architectural feature that effectively limits the homology search to one dimension, greatly speeding up the location of homology, and allowing the RecA-ssDNA nucleoprotein filament to probe many different sequences along the chromosome in parallel. A more recent study has followed the dynamics of RecA nucleoprotein filaments following the induction of double-strand breaks [31]. This study was carried out in live *Caulobacter* cells, implicated RecN, a structural maintenance of chromosome (SMC)-like protein in the translocation of RecA-ssDNA nucleoprotein filaments. Their motion was observed to be directional, repeatedly crossing the cell from pole to pole rapidly. This motion was non-equilibrium, enabled by the ATP consumption of RecN, putting on more precise grounds previous proposals about the fluidization of the bacterial cytoplasm by metabolic processes [18]. Whether RecN is functionally active following transformation or participates also in the motion of ICEs following conjugation remains to be determined.

### 28.2.4 Contrasts with Target Search by Transcription Factors

The search of integration sites during HGT and that of transcription factors for their cognate sites share some commonalities, but there are significant differences as well. Since the pioneering work of Riggs and collaborators who demonstrated that the Lac repressor can locate its binding sites more efficiently than the diffusion-limited rate in three-dimensions would allow [32], it has been appreciated that transcription factors carry out their search by three mechanisms dubbed collectively as facilitated diffusion [33, 34]: (i) hopping in which the protein undergoes a sequence of binding-unbinding events; (ii) one-dimensional diffusion or sliding, mediated by weak non-specific interactions to the DNA substrate, and (iii) intersegmental transfer, whereby the transcription factor transfers to distal DNA sites by looped DNA intermediates. Evidence for facilitated diffusion mechanisms has been observed in vivo [35], consistent with modeling approaches based on a generalized Berg-von Hippel model [36]. Some of these mechanisms may have counterparts during target search in HGT. For example, the sub-diffusive search and its transient binding of ICEs to the chromosome can be viewed as hopping. Similar behavior is expected from exogenous DNA imported by transformation. The antenna effect whereby transcription factors can scan DNA by transient non-specific one-dimensional diffusion before dissociating from a chromosome can be loosely compared to the effect provided by the ~50% extension of single-stranded DNA as substrate for RecA polymerization. This extension greatly facilitates the homology search, making it much more robust to relative displacements between the presynaptic complex and a chromosomal segment [26, 27]. Note that without extension, out-of-registry misalignment of both by only one base pair suffices to destroy homology. Extension of the single-stranded DNA substrate by RecA in the formation of a presynaptic complex, together with a short-range sliding mechanism [37], and the large persistence length of presynaptic complexes ~800 nm allow for multiple contacts between the complex and different (distal) segments of the chromosome. Evidence for multiple contacts has been obtained in single-molecule both in vitro [38] and in vivo experiments in individual cells in the case of double-stranded break repair [30]. Multiple contacts may be relevant for the search by transcription factors in situations involving for example DNA looping [39].

It is also noteworthy that whereas transcription factors can move relatively unhindered through the chromosomal mesh due to their small size, imported exogenous DNA in HGT can have lengths of tens of thousands of base pairs (e.g., lambda phage, ICEs and more), and their three-dimensional conformations are expected to be larger than transcription factors. Their motion is thus expected to be considerably more affected by the crowded interior of bacterial cells, in particular by the polymer mesh of the bacterial chromosome [40].

## 28.3 Outlook and Future Directions

Despite tremendous progress, much remains to be done and we conclude with a list of unsolved issues that in our view should push the field forward. From an experimental point of view, following RecA-mediated homology search in the case of HGT may reveal different search strategies and mechanisms beyond those involved in the repair of DNA lesions [30, 31]. For example, exogenous DNA imported by transformation is independently moving and much smaller than DNA segments adjacent to a double-strand break, which form part of the long chromosome itself. Furthermore, it is not clear that SMC-like proteins such as RecN will be involved in translocating RecA filaments as during recombinational repair.

Just as the intracellular diffusion of transcription factors depends on their size [41], the mobility of ICEs may also exhibit size-dependent mobility. ICEs range from $\sim$12 to $\sim$500 kbp [42, 43]. We anticipate that this mobility will depend strongly on the mesh size $\xi$ of the bacterial nucleoid, regarded as a polymer network. The mesh size has been measured recently to be $\xi \sim 50$ nm in the case of *E. coli* [40]. Consequently, we also anticipate that the dynamics of the search process will depend on ICE size. The interplay between ICE size inside a cell and the effective mesh size imposed by the polymeric nature of the chromosome may impose different dynamics of search.

Despite the differences between transduction, conjugation, and transformation discussed above, extant experimental evidence points to highly efficient search and target location, which is typically completed within a generation time timescale. In all cases, the search processes take place within the crowded, complex interior of a bacterial cell. One may therefore wonder whether search processes have undergone a process of optimization during evolution, and if yes, what has been optimized. Somewhat counterintuitively, sub-diffusive behavior appears to be beneficial for target search processes [44].

One possible direction is to investigate optimality as a function of the sub-diffusive character of the intracellular medium, as parameterized by the Hurst exponent $H$, which obeys $\alpha = 2H$. Whereas $H = 1$ corresponds to ballistic motion, sub-diffusive motion is characterized by $H < 0.5$. Trajectories with $H < 0.5$ display antipersistent behavior, such as that measured for chromosomal sites in different bacterial

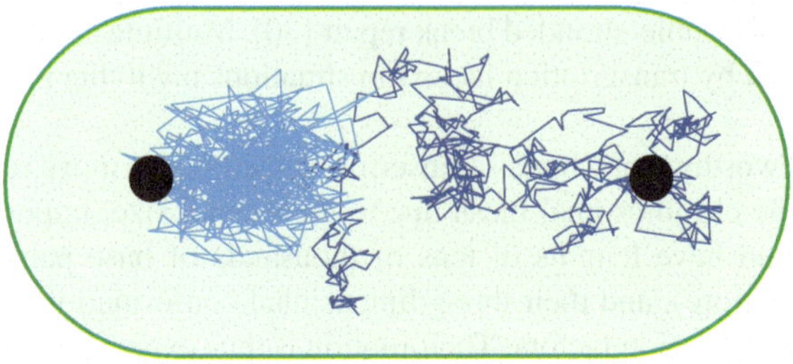

**Fig. 28.5** Simulated two-dimensional sub-diffusive, fractional Brownian motion trajectories (generated with the Matlab function wfbm), with mean squared displacements characterized by different values of the Hurst exponent $H = 0.1$ (light blue) and $H = 0.4$ (dark blue). Black circles represent hypothetical targets such as *attB* sites in a bacterial cell

species [9, 15]. We show simulated two-dimensional trajectories corresponding to two values of the Hurst exponent in Fig. 28.5.

Note that the symmetric character of both trajectories differs significantly from asymmetric trajectories as those measured for *attB* sites in *B. subtilis* [9, 15]. Furthermore, the simulated trajectories assume a homogeneous medium, whereas the non-homogeneous structure of folded chromosomes and their distribution within a cell impose a heterogeneous environment for diffusion. First steps into this direction have already been taken, with search studied as function of the value of Hurst exponents [45].

Experimental studies point to a close connection between chromosomal structure and the target search process [8, 15, 31]. The observed compact local explorations interspersed by multiple pole-to-pole traversals observed in both target search by ICEs after conjugation, as well as in the repair of double strand breaks are reminiscent of models of intermittent search strategies [46], and optimal search by stochastic resetting [47]. However this analogy needs to be put in a firmer ground and closer interaction between experiment and theory is called for. A promising alternative direction to the top-bottom in vivo experimental approaches to study search processes within cells, is a bottom-up in vitro approach recently proposed to study chromosomes [48]. This approach is based on taking genome-sized DNA isolated from live cells and subsequently adding purified DNA-organizing elements, followed by encapsulation in cell-sized containers within microfluidic chips. The advantages of such an approach are clear: it allows for the possibility of varying experimental parameters in a controlled way, and following the consequences of such variations.

Lastly, anisotropic target search mechanisms may prove advantageous in physical or biological contexts in which search processes take place in heterogeneous media or in which anisotropy is structurally built in [15, 17]. This constitutes unchartered territory from the theoretical point of view. Further theoretical exploration is needed.

**Acknowledgements** The research was funded by grant ISF 950/23 from the Israel Science Foundation. We thank useful discussions with G. Teza, R. Metzler, D. Mukamel and J. Marko.

## References


1. E. Koonin, Horizontal gene transfer: Essentiality and evolvability in prokaryotes, and roles in evolutionary transitions, F1000Research **5**, 1805 (2016)
2. B. Arnold, I. Huang, W. Hanage, Horizontal gene transfer and adaptive evolution in bacteria. Nat. Rev. Microbiol. **20**, 206 (2022)
3. R. Lenski, Convergence and divergence in a long-term experiment with bacteria. Am. Nat. **190**, S57 (2017)
4. W. Li, G. Zhang, Detection and various environmental factors of antibiotic resistance gene horizontal transfer. Environ. Res. **212**, 113267 (2022)
5. G. Liu, L.E. Thomsen, J.E. Olsen, Antimicrobial-induced horizontal transfer of antimicrobial resistance genes in bacteria: a mini-review. J. Antimicrob. Chemother. **77**, 556 (2022)
6. A.B. Oppenheim, O. Kobiler, J. Stavans, D.L. Court, S. Adhya, Switches in bacteriophage lambda development. Annu. Rev. Genet. **39**, 409 (2005)
7. A. Landy, The $\lambda$ integrase site-specific recombination pathway, Mob. DNA III **91** (2015)
8. A. Tal, R. Arbel-Goren, N. Costantino, D.L. Court, J. Stavans, Location of the unique integration site on an *Escherichia coli* chromosome by bacteriophage lambda DNA in vivo. Proc. Natl. Acad. Sci. USA **111**, 7308 (2014)
9. S.C. Weber, A.J. Spakowitz, J.A. Theriot, Bacterial chromosomal loci move subdiffusively through a viscoelastic cytoplasm. Phys. Rev. Lett. **104**, 238102 (2010)
10. Y. Zhang, O.K. Dudko, First-passage processes in the genome. Annu. Rev. Biophys. **45**, 117 (2016)
11. Q. Shao, A. Hawkins, L. Zeng, Phage DNA dynamics in cells with different fates. Biophys. J. **108**, 2048 (2015)
12. A. Babić, A.B. Lindner, M. Vulić, E.J. Stewart, M. Radman, Direct visualization of horizontal gene transfer. Science **319**, 1533 (2008)
13. A. Couturier, C. Virolle, K. Goldlust, A. Berne-Dedieu, A. Reuter, S. Nolivos, Y. Yamaichi, S. Bigot, C. Lesterlin, Real-time visualisation of the intracellular dynamics of conjugative plasmid transfer. Nat. Commun. **14**, 294 (2023)
14. C.M. Johnson, A.D. Grossman, Integrative and conjugative elements (ICEs): what they do and how they work. Annu. Rev. Genet. **49**, 577 (2015)
15. R. Arbel-Goren, D. McKeithen-Mead, S.A. Voglmaier, I. Afremov, G. Teza, A. Grossman, J. Stavans, Target search by an imported conjugative DNA element for a unique integration site along a bacterial chromosome during horizontal gene transfer. Nucleic Acids Res. **51**, 3116 (2023)
16. X. Wang, P.M. Llopis, D.Z. Rudner, Organization and segregation of bacterial chromosomes. Nat. Rev. Genet. **14**, 191 (2013)
17. P. Dieterich, O. Lindemann, M.L. Moskopp, S. Tauzin, A. Huttenlocher, R. Klages, A. Chechkin, A. Schwab, Anomalous diffusion and asymmetric tempering memory in neutrophil chemotaxis. PLoS Comput. Biol. **18**, e1010089 (2022)
18. B.R. Parry, I.V. Surovtsev, M.T. Cabeen, C.S. O'Hern, E.R. Dufresne, C. Jacobs-Wagner, The bacterial cytoplasm has glass-like properties and is fluidized by metabolic activity. Cell **156**, 183 (2014)
19. T. Le, M. Imakaev, L. Mirny, M. Laub, High-resolution mapping of the spatial organization of a bacterial chromosome. Science **342**, 731 (2013)
20. A. Babic, M.B. Berkmen, C.A. Lee, A.D. Grossman, Efficient gene transfer in bacterial cell chains. mBio **2**, e00027 (2011)
21. C. Planchenault et al., Intracellular positioning systems limit the entropic eviction of secondary replicons toward the nucleoid edges in bacterial cells. J. Mol. Biol. **432**, 745 (2020)
22. L. Deng, C. Wang, X. Zhang, W. Yang, H. Tang, X. Chen, S. Du, X.-Y. Chen, Cell-to-cell natural transformation in *Bacillus subtilis* facilitates large scale of genomic exchanges and the transfer of long continuous DNA regions. Nucleic Acids Res. **51**, 3820 (2023)
23. E.C. Greene, DNA sequence alignment during homologous recombination. J. Biol. Chem. **291**, 11572 (2016)

24. D. Sagi, T. Tlusty, J. Stavans, High fidelity of RecA-catalyzed recombination: a watchdog of genetic diversity. Nucleic Acids Res. **34**, 5021 (2006)
25. A. Mani, I. Braslavsky, R. Arbel-Goren, J. Stavans, Caught in the act: The lifetime of synaptic intermediates during the search for homology on DNA. Nucleic Acids Res. **38**, 2036 (2010)
26. K. Klapstein, T. Chou, R. Bruinsma, Physics of RecA-mediated homologous recognition. Biophys. J. **87**, 1466 (2004)
27. I. Braslavsky, J. Stavans, Application of algebraic topology to homologous recombination of DNA. iScience **4**, 64 (2018)
28. I. Mortier-Barrière, P. Polard, N. Campo, Direct visualization of horizontal gene transfer by transformation in live pneumococcal cells using microfluidics. Genes **11**, 675 (2020)
29. C.H.G. Johnston, R. Hope, A.L. Soulet, M. Dewailly, D. De Lemos, P. Polard, The RecA-directed recombination pathway of natural transformation initiates at chromosomal replication forks in the pneumococcus. Proc. Nat. Acad. Sci. USA **120**, e2213867120 (2023)
30. J. Wiktor, A.H. Gynnå, P. Leroy, J. Larsson, G. Coceano, I. Testa, J. Elf, RecA finds homologous DNA by reduced dimensionality search. Nature **597**, 426 (2021)
31. A. Chimthanawala, J.J. Parmar, S. Kumar, K.S. Iyer, M. Rao, A. Badrinarayanan, SMC protein RecN drives RecA filament translocation for in vivo homology search. Proc. Natl. Acad. Sci. USA **119** (2022)
32. A. Riggs, S. Bourgeois, M. Cohn, The Lac repressor-operator interaction: III. Kinetic studies. J. Mol. Biol. **53**, 401 (1970)
33. S.E. Halford, J.F. Marko, How do site-specific DNA-binding proteins find their targets? Nucleic Acids Res. **32**, 3040 (2004)
34. P. H. von Hippel, O.G. Berg, Facilitated target location in biological systems. J. Biol. Chem. **264**, 675 (1989)
35. J. Elf, G.W. Li, X.S. Xie, Probing transcription factor dynamics at the single-molecule level in a living cell. Science **316**, 1191 (2007)
36. M. Bauer, R. Metzler, In vivo facilitated diffusion model. PLoS ONE **8**, e53956 (2013)
37. K. Ragunathan, C. Liu, T. Ha, RecA filament sliding on DNA facilitates homology search. eLife **1**, e00067 (2012)
38. A.L. Forget, S.C. Kowalczykowski, Single-molecule imaging of DNA pairing by RecA reveals a three-dimensional homology search. Nature **482**, 423 (2012)
39. M. Bauer, E. Rasmussen, M. Lomholt, R. Metzler, Real sequence effects on the search dynamics of transcription factors on DNA. Sci. Rep. **5**, 10072 (2015)
40. Y. Xiang, I. Surovtsev, Y. Chang, S. Govers, R. Parry, J. Liu, C. Jacobs-Wagner, Interconnecting solvent quality, transcription, and chromosome folding in *Escherichia coli*. Cell **184**, 3626 (2021)
41. N. Bellotto, J. Agudo-Canalejo, R. Colin, R. Golestanian, G. Malengo, V. Sourjik, Dependence of diffusion in *Escherichia coli* cytoplasm on protein size, environmental conditions, and cell growth. eLife **11**, e82654 (2022)
42. R. Wozniak, M. Waldor, Integrative and conjugative elements: mosaic mobile genetic elements enabling dynamic lateral gene flow. Nat. Rev. Microbiol. **8**, 552 (2010)
43. J. Cury, M. Touchon, E.P.C. Rocha, Integrative and conjugative elements and their hosts: composition, distribution and organization. Nucleic Acids Res. **45**, 8943 (2017)
44. G. Guigas, M. Weiss, Sampling the cell with anomalous diffusion–the discovery of slowness. Biophys. J. **94**, 90 (2008)
45. S. Khadem, S. Klapp, R. Klages, Search efficiency of discrete fractional Brownian motion in a random distribution of targets. Phys. Rev. Res. **3**, 23169 (2021)
46. O. Bénichou, C. Loverdo, M. Moreau, R. Voituriez, Intermittent search strategies. Rev. Mod. Phys. **83**, 81 (2011)
47. M.R. Evans, S.N. Majumdar, G. Schehr, Stochastic resetting and applications. J. Phys. A: Math. Theor. **53**, 193001 (2020)
48. A. Birnie, C. Dekker, Genome-in-a-box: building a chromosome from the Bottom Up. ACS Nano **15**, 111 (2020)

# Chapter 29
# Optimal Random Searches: Lévy Walks and Animal Movement

**Marcos G. E. da Luz, Gandhimohan M. Viswanathan, and Ernesto P. Raposo**

**Abstract** We review the general process of random search and discuss its main ingredients, potential goals, as well as ways to quantity search efficiency. We address Lévy walk search models, with emphasis on how they are widely applicable to simple yet concrete phenomena. Moreover, since search strategies can alter encounter rates, it is natural to expect that how animals move must have a bearing on ecological and evolutionary phenomena. We discuss some examples of empirical studies in the latter context. We also present a theoretical analysis of an idealized model in one and two spatial dimensions. A different open question relates to the extent to which the empirically observed Lévy walk patterns are related to the optimization of the search efficiency conferred by Lévy walks, since the latter fact does not logically imply the former. A major development in this context was a study of *Drosophila* larvae that showed the existence of intrinsic neurophysiological mechanisms for generating patterns similar to Lévy walks. Finally, as a digression, we briefly comment on how search is inherent in human nature.

## 29.1 What is a Random Search?

To find her again
Upon this my dreams are depending
Through the dark

M. G. E. da Luz
Departamento de Física, Universidade Federal do Paraná, 81531-980 Curitiba-PR, Brazil
e-mail: luz@fisica.ufpr.br

G. M. Viswanathan (✉)
Department of Physics & National Institute of Science and Technology of Complex Systems, Federal University of Rio Grande do Norte, 59078-900 Natal-RN, Brazil
e-mail: gandhi@fisica.ufrn.br

E. P. Raposo
Laboratório de Física Teórica e Computacional, Departamento de Física, Universidade Federal de Pernambuco, Recife-PE 50670-901, Brazil
e-mail: ernesto.raposo@ufpe.br

 
D. Grebenkov et al. (eds.), *Target Search Problems*,
https://doi.org/10.1007/978-3-031-67802-8_29

I sense the pounding of her heart
Next to mine
She's the sweetest love I could find
So I guess I'll be hunting
high and low...high...
There's no end to the lengths I'll go to ...
high and low...high...
—*Hunting high and low*, A-ha (1985)

One only searches for something when one does not know precisely where this something (the target of the search) is located. Indeed, full knowledge about its location implies simply going straight to the target, and in a perfect world, along the best possible trajectory, say, optimizing distance, time, energy, etc. Therefore, full information (denoted here as $I = 1$) ideally leads to a global minimization of the cost ($C_{\text{global min}}$) to get to the target. Such a perfect information context characterizes a fully deterministic process. If the reward of finding the target is $R$ and the efforts for succeeding in this task is represented by $C$, we can define a searching efficiency as

$$\eta = \frac{R}{C}. \tag{29.1}$$

Conceivably, the highest possible value for $\eta$ is then $\eta_{\text{global max}} = R/C_{\text{global min}}$.

Nonetheless, even in the case $I = 1$, achieving $\eta_{\text{global max}}$ might not be a simple task. For example, the famous traveling salesman problem (TSP) consists of determining the shortest route allowing to visit a set of cities connected by a specified collection of roads (all known, so $I = 1$), Fig. 29.1. But since the TSP, a combinatorics riddle, is NP-hard (a concept from computational complexity theory [1]), the exact optimal pathway can be extremely difficult to ascertain.

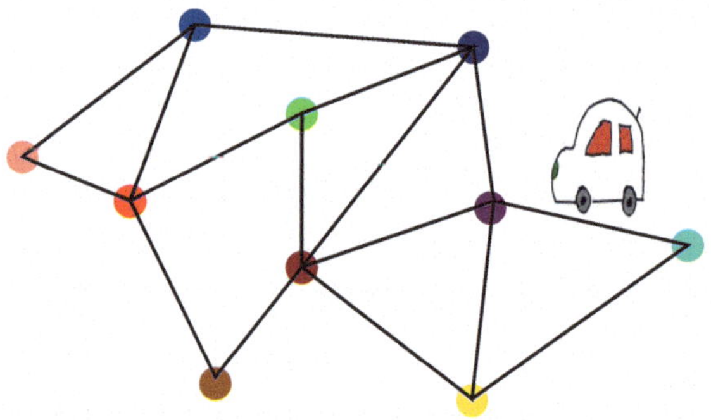

**Fig. 29.1** In the classical traveling salesman problem (TSP), given a set of cities (here vertices) and roads connecting them (here edges), the salesperson should find the best possible path in order to minimize the total distance traveled while visiting all the cities. The cities plus roads form a weighted graph $\mathcal{G}_w$, whose weight of each edge is proportional to the length of the associated road. Thus, the TSP corresponds to find the least weighted Hamiltonian cycle of $\mathcal{G}_w$, which sets a NP-hard problem [1]

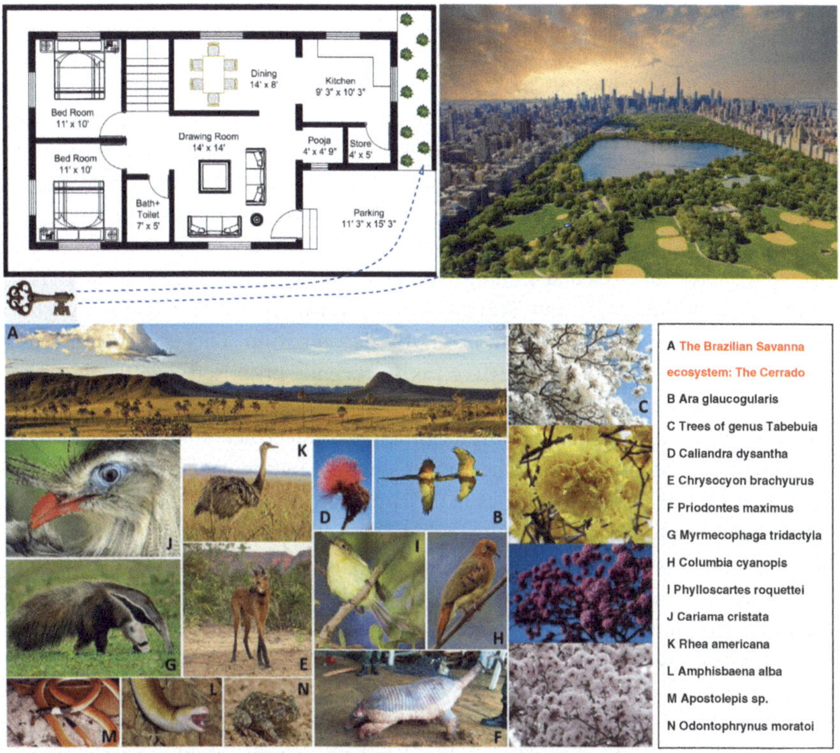


**Fig. 29.2** There are different types of random searches, each presenting specific conditions and thus demanding distinct procedures for a positive outcome. Consider, for instance, the context of a single versus repeated events, or a multitude of events. In the former, the way to proceed for finding, say, a lost key would depend on the search landscape, e.g., whether confined to one's own home, or else perhaps comprising a relatively large region, like in Central Park in New York city. On the other hand, in the latter, when studying the long run outcomes of biological searching in large ecological biosystems having great diversity of species in a rich environment—e.g., the huge Brazilian Savanna (the Cerrado)—the interest is in the global results arising from many realizations of the encounter processes. In these cases, it makes sense to employ stochastic analysis and models to determine the best strategies. Central Park photo: Ingus Kruklitis (creator), Getty Images/iStockphot (credit). Brazilian Cerrado biodiversity: adapted from Fig. 2 of Ref. [4], with the kind permission from the authors

Conversely, if the search agent, or just the searcher, is uncertain about the target position, then there is no other option than to rely in some sort of "trial and error" protocol. In this scenario, the degree of lack of information (i.e., $I < 1$) about the target's locations delimits the kind of strategy (i.e., procedure) to be followed by the searcher. As an example, if Alice has lost her key somewhere in her home and Bob in Central Park in New York city (Fig. 29.2)—with Bob remembering all the particular sections he has explored after a morning in the Park, but not the precise paths he took from one to another—it is safe to say that $I_B < I_A$ and the Alice and

Bob strategies should be different. Certainly, Bob's search strategy should be based on a more tiring extensive exploration approach, and thus $C_{\mathrm{Bob}} > C_{\mathrm{Alice}}$.

Actually, lower $I$'s tend to imply less effective strategies, since less clues are available and thus more generalist they ought to be, this in principle increasing $C$. Hence, almost always $C(I)$ is a decreasing function of $I$ (for few interesting exceptions, refer to [2]). From this, it might not be a surprise that for a total lack of information and given that certain conditions $\mathfrak{C}$ are observed, optimal strategies are potentially universal, i.e., leading to the best outcomes in any case subjected to $\mathfrak{C}$.

So far we have discussed searching, nevertheless without specifying the overall goal. In our example, Alice and Bob are seeking for their lost key, so they wish to find an unique target, indeed a common day-by-day situation. This usually requires a dedicated action (a one-shot scheme), e.g., as the recently proposed specific algorithm to search and rescue missing people in the sea [3]. However, equally important is when the aim is to find many targets and thus the efficiency can be accounted for in terms of the long run final outcome. This is the typical situation in large ecosystems, specially in rather complex and diverse ones, as in the Brazilian savannas [4], the Cerrado (see Fig. 29.2), and in Madagascar [5]. Given that $I < 1$, the best strategies may contrast in the single versus multiple targets scenarios. This has a parallel (although with many distinctions) with formal game theory [6], where the optimal strategies to play just once and to play repeatedly a game can differ.

For a multiple searching of $N$ targets, provided that $N$ is large enough and we can ascribe to each target an average reward $R$, the best strategy is that yielding the lowest $\langle C \rangle = \sum_n C_n/N$, with $C_n$ the cost o finding the $n$-th target. Such "mean field" approach [7, 8] obviously should be modified, becoming more involved, if the targets are somehow distinct and the rewards $R_n$ cannot be approximated by a mean value [9], or the searcher cannot afford a specific $C_n$ to be too high [8, 10, 11]. For the former, an example is when one needs to find targets of different types, so that $\eta$ is not gauged only in terms of quantity but also in diversity of targets [9]. The latter is the case, e.g., if a not top predator shall avoid a hunting risk strategy, i.e., one resulting in high $\langle C \rangle$ but for various reasons [12, 13] leading to strong fluctuations for the $C_n$'s. So, in particular occasions $n$, it may take very long times and/or energy to get a prey (leading to prohibited $C_n$), causing the forager to perish due to starvation.

The central point is that for $I < 1$ and if the interest is to look for multiple similar targets (with $R_n = R$), efficient strategies should have some probabilistic component. In fact, since $I < 1$ the concrete configurations of the successive searching events $n$ tend, in a greater or lesser degree, to differ from each other and a specific and "deterministic" search rule—say, always goes straight $\ell$ meters, if it does not find a target, turn right an angle of exact $\theta$ degrees, and then repeat the process—could work well in few instances $n$, but not in most of them. Moreover, as $I \to 0$, one should expect the mentioned probabilistic component to become more and more crucial. We stress that the actual panorama described here is very common among living organisms, motivating the development of the so-called foraging theory [14–16]. Once biological encounters are recurrent activities, the searching strategies followed by the individuals (when guided by natural instinct, consequence of evolutionary traits [17, 18]) are those related to foraging for numerous targets. This encompasses a large number of biological processes, perhaps the first crossing our minds being predators looking for their prey in different habitats. However, this is far from exhausting all the possibilities, with some other being proteins diffusing and locating DNA strands, T-cells circulating and then detecting and killing cellular pathogens, insects going after flowers, and males seeking fertile females, to cite just a few examples.

The function $\eta$ in Eq. (29.1) in principle relates only to finding the target, but not pursuing and eventually acquiring it (eating, storing, etc). For instance, the predator first needs to locate a prey, but after that, there is the whole chasing dynamics, where not always the predator ends up capturing the pray. Of course, it would be possible to define, once detected, the efficiency to obtain the target, $\eta_{\text{obt}}$. But if the finding and acquisition of a target are relatively independent, one could incorporate the average probability of securing a target as an effective extra term directly into the cost function $C$ [19, 20]. This may not be feasible in more convoluted situations [21], as recently reported in the behavior of malaria parasite while trying to invade the blood vessels of the host [22], demanding a more elaborated theoretical description. The contexts where $\eta_{\text{obt}}$ needs to be treated separately will not be addressed in the present contribution.

Still regarding the target features, they can be stationary or mobile [8, 10, 23], their availability can vary in time [7] and as already mentioned, they might have distinct characteristics, e.g., levels of attractiveness to the searcher [9]. The targets can be detected by the searchers through various types of cues, such as olfactory, auditory, or visual information (see references cited in [15] as well as the review in [20]). As for the searching arenas (environments), they may be homogeneous or disordered [19, 24], perhaps even fractal [2, 24], or yet to display a patchy structure [18, 21]. One would expect random searches to take place in our usual three-dimensional space. This is actually the case for many animal species in the ocean [25], or in the historical context of hunting for submarines during the World War II [26]. But much more often the search effectively takes place on two dimensions, along the planet surface or in particular layers in the sea. However, it is possible to define the problem to arbitrary $d$-dimensions, with $d = 1$ or even for $d \gg 3$, the latter finding applications in the digital realm as the searching in large scale databases [27].

In this way, as it should be clear from all that have been discussed above, a rather simplistic answer to the question in the title of the present section could be: random search corresponds to any instance of looking for one or more targets when their exact location is not known with 100% confidence ($I < 1$). But we emphasize that this is an oversimplification for the countless situations we may find under such extremely broad definition. When modeling random searches, trying to mathematically determine the best strategies, many ingredients need to be included in the analysis, such as the skills of the searcher, characteristics of the environment, properties of the targets and, of course, the value of $I$ and how the partial information (if any) can be used.

So, on the one hand an unique universal optimal universal searching strategy is not attainable. This is specially true in biology due to the huge diversity of foraging behavior observed in nature [28–30]. On the other hand, there is a whole class of animal foraging behavior in which the specificities of the underlying processes are considerably straightforward [11, 15, 18, 31, 32], specially because $I \approx 0$. Therefore, our focus in the following will be on random search in the context of animal foraging, discussing a family of models describing these simpler, but still realistic instances.

Crucially, such models are able to capture a good portion of the key mechanisms associated with more complex random searches, with the great advantage of not being so complicated as to render the theoretical framework intractable or to make it too difficult to interpret the ensuing dynamics.

### *29.1.1 A Prototype General Random Search with $I = 0$*

In order to properly formalize a certain class of random searches, it is necessary to clearly establish the conditions $\mathfrak{C}$ under which the associated dynamics takes place. Those which hereafter we choose to focus posses two very positive features, namely, are pertinent to address a wide range of actual foraging situations and are amenable to a well-grounded mathematical treatment. They can be summarized as the following (for a more detailed presentation see, e.g., [15, 18, 33]).

Consider a search where the targets, all akin, are *a priori* randomly distributed in space and there is zero knowledge about their positions. The searcher can detect the targets only when it is within some finite distance, $r_v$, from them. Further, assume that after finding a target, the searcher does not acquire relevant knowledge about how to locate the next one. Under such assumptions, there is no reason to adopt any deterministic repetitive rule of locomotion. Hence, until a target comes within $r_v$, the searcher could move according to some stochastic process. There is a large variety of possible movement types. Examples include Brownian motion, fractional Brownian motion, Lévy walks, correlated random walks, correlated Lévy walks, and so forth. The specific kind of movement should not be really determinant or relevant if the targets are plentiful [7, 11], i.e., their density $\rho$ is high. Indeed, any or no strategy will do when times are plentiful. Conversely, the mobility characteristics are expected to play a very important role in determining $\eta$ at low $\rho$'s.

Supposing the above conditions $\mathfrak{C}$ and small $\rho$, one of the most studied mobility patterns in the context of random searches (particularly in biology) is the Lévy walk. The motivations for this interest are twofold. First, it is a very plastic random walk, e.g., the variation of a single parameter yields different degrees of diffusionness (next section), which can be relevant in evolutionary arguments [18, 34]. Second, strong empirical, numerical, and analytical results point to the existence of optimal strategies [15, 32, 35]. These facts have supported a widely studied model [15], believed to display many of the essential elements of an effective random search (refer to Fig. 29.3):

(1) If there are targets sites located within a range $r_v$ , then the searcher moves on a straight line to the closest one.

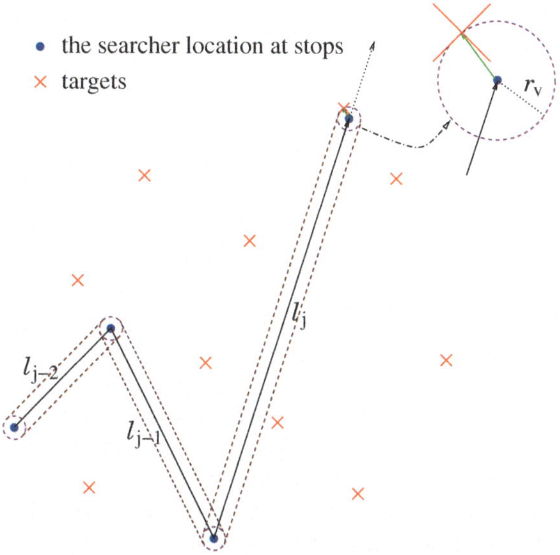


**Fig. 29.3** An illustrative example of the application of rules (1) and (2) of the model in Sect. 29.1.1. At step $j-2$ of the trajectory, the searcher chooses a random direction from a uniform distribution and a distance $\ell_{j-2}$, drawn from a PDF $p$, to walk in a straight line. Then, along the way the searcher keeps scanning its vicinity within a distance $r_v$ (represented by a circle of radius $r_v$). Not finding a target, the searcher repeats the procedure for the step $j-1$. Once more it does not spot a target. Finally, after selecting a direction and a distance $\ell_j$, somewhere along the traveled distance it detects as target. Thus, the searcher truncates the original path, going directly toward the identified target (as shown in the blow up)

(2) Otherwise, the searcher chooses a direction $\theta \in [0, 2\pi)$ from a uniform distribution and a distance $\ell_j$ from a probability density function (PDF) $p(\ell_j)$. It then continuously moves with constant velocity to the new point, while it looks for any target—within the range of detection $r_v$—along the way. If no target is detected, the searcher stops after traversing the distance $\ell_j$ and a new direction and distance are selected. If a target is identified, the motion stops and step (1) is implemented.

(3) The rules (1) and (2) are applied uninterruptedly until reaching the halting requirement. This could be, for example, (a) that a certain number of targets $N$ is found, (b) or else that a certain total distance $L_{\text{total}}$ is traveled.

The above simple model of random search already leads to interesting and even counter-intuitive results [15, 18], to be discussed later. Regarding the efficiency function $\eta$, let us consider the reward for getting a single target as $R$ and the cost of acquiring it as $C = c\, L^{(p)}$, where $L^{(p)}$ the distance traveled for so and $c$ a constant. The superscript $p$ reminds us that the mentioned distance in fact depends on $p$, thus on the exact searching strategy. For either $N$ or $L_{\text{total}}$ above large enough, we can properly define an average for the set of $L_n^{(p)}$'s, with $n$ standing for the $n$-th target found. For the halting criteria (a) and (b), we have, respectively, $\eta = N\, R/(c \sum_{n=1}^{N} L_n^{(p)})$ and $\eta \approx J\, R/(c \sum_{n=1}^{J} L_n^{(p)})$, where $J$ is an integer such that $L_{\text{total}} \approx \sum_{n=1}^{J} L_n^{(p)}$. Now, approximating $\sum_{n=1}^{M} L_n^{(p)} \approx M\, \langle L^{(p)} \rangle$, for either (a) or (b) it reads

$$\eta \approx \frac{R}{c} \frac{1}{\langle L^{(p)} \rangle}. \tag{29.2}$$

Therefore, the maximization of $\eta$ accounts for the PDF $p$ which minimizes $\langle L^{(p)} \rangle$. In this way, the framework for establishing optimal strategies becomes mathematically well defined.

In the aforementioned random search problem, there is a well-defined lower bound for the smallest step length. However, one could imagine the situation in which this lower bound goes to zero, in which case the limit trajectory might have properties similar to a Wiener (or Lévy) process. Such trajectories are fractal, so that the total distance traveled may be infinite. To avoid such an unrealistic situation, one can use the first passage time, instead of the distance traveled, to compute the efficiency. In our formulation, this issue does not arise. Nevertheless, first passage times are discussed in other chapters of this book, as well as briefly in Sect. 29.2.2.

Lastly, concerning the distribution $p(\ell)$, if it has a finite variance, then the resulting motion in the absence of targets is Brownian. Indeed, in this case the central limit theorem applies and the diffusion is normal, with the mean squared displacement growing linearly in time. When targets are added to the mix, some steps are truncated due to the encounters, consequently decreasing the mean step size $\langle \ell_j \rangle$. If, however, the $p(\ell)$ does not have finite variance, then the motion in the absence of targets does not correspond to Brownian motion and the diffusion is not normal. This is the situation for Lévy walks, which provide the best random search strategies whenever the conditions $\mathfrak{C}$ explained in this section are observed.

## 29.2 Lévy Walks

Lévy processes are those whose underlying PDFs $p(\ell)$ of relevant quantities $\ell$ are drawn from $\alpha$-stable Lévy distributions (refer to the Appendix for a brief overview). However, for large values of $\ell$, such PDFs are well described by simpler inverse power laws in the form $p \sim 1/\ell^{\alpha+1} = 1/\ell^{\mu}$. In this way, it is common to model many phenomena related to Lévy flights and Lévy walks directly based on power-law distributions. This is often the case in random search theory.

A Lévy flight is a random walk where at each time step $j$ the searcher "disappears" at the old position and "reappears" at a distance $\ell_j$ away, with $\ell_j$ distributed according to a normalizable power-law tailed distribution with infinite variance. So the steps are somehow "instantaneous". On the other hand, a more realistic type of motion is the Lévy walk, which is closely related to Lévy flights. In a Lévy walk

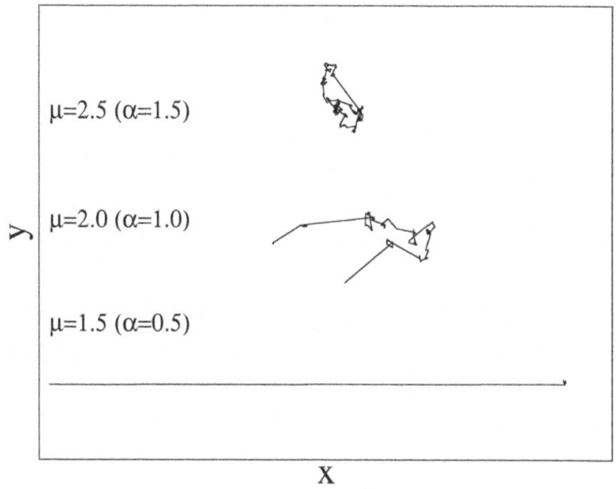


**Fig. 29.4** Examples of some Lévy walks (in 2D) for different values of the power law exponent $\mu$, Eq. (29.3). The relation between $\mu$ and the exponent $\alpha$ of the associated $\alpha$-stable Lévy distribution is also indicated

the displacement from the old to the new position is undertaken at a finite, typically constant, velocity.

Consider a distribution with an asymptotic power law tail of the form

$$p(\ell) \sim 1/\ell^{\mu}, \quad 1 < \mu < 3 \,. \tag{29.3}$$

For the exponent $\mu$ in this range, the distribution has infinite variance. The generalized central limit theorem for such distributions then guarantees that the sums of the corresponding independent and identically distributed random variables must converge to the Lévy $\alpha$-stable distribution, with Lévy index $\alpha = \mu - 1$ (see the Appendix). Examples of walks from Eq. (29.3) are shown in Fig. 29.4.

Note that real animals (the foragers) cannot possibly perform actual mathematical Lévy walks. Indeed, real trajectories are not made of perfect straight line segments. Nor are power law tails ever exact in the natural world. So when one speaks of Lévy walks by animals, what is meant is that the behavior is much more similar to Lévy than to other kinds of stochastic locomotion, like Brownian motion. This can be identified in two specific ways. First, the step lengths are heavy-tailed or fat-tailed. Second, the trajectories have superdiffusive properties, namely, the mean squared displacement grows not linearly in time, instead considerably faster.

It is important to note that the projection of a $d$-dimensional Lévy walk or flight onto a $d-1$-dimensional space is itself a Lévy walk. Nonetheless, the projection of a $d$-dimensional Lévy distribution onto a lower dimensional space is not in general a Lévy distribution, although it will still be asymptotically power law tailed. This fact, usually overlooked, should be taken into consideration in certain situations when investigating experimental data (see, e.g., [22] and refs. therein).

### 29.2.1 Random Search Optimization Via Lévy Walks in the Context of Animal Foraging

In the 1990s, empirical findings [31] about how animals move led to efforts to understand the mathematical properties of Lévy walks in foraging. In 1999, analytical and numerical results showed [32] that Lévy walks with $\mu = 2$ can optimize uninformed random searches for sparse, randomly distributed, revisitable targets when the searching restarts near the previously visited target (so, under the scope of the conditions $\mathfrak{C}$ in Sect. 29.1.1). This optimization occurs because such motion maximizes the encounter rate by neither always going to far away targets, nor always staying close to the previous target, but rather exploiting both options in a "Goldilocks" manner.

The original 1999 paper attracted considerable attention and to some extent caused surprise. The possibility that such "exotic" motion might be possible in biological and ecological contexts was, at that time, considered counter-intutitive to many. Since then, these results have not only been confirmed but have become widely accepted [15, 33, 34].

### 29.2.2 Lévy Searches: A Powerful but a Technical Challenging Theoretical Framework

Whereas the theory of Brownian motion is fairly straightforward to generalize from one to higher dimensions, the same cannot be said of other subtler types of random movement. Lévy walks and flights, in particular, have some counter-intuitive features that have led to mistakes in the literature (see below). As an example, one of the most counter-intuitive properties of Lévy flights is that the mean value of moments of lower order do not converge, so that if they are estimated "on the fly" during the motion, they will behave strangely. For instance, for Lévy index $\alpha < 1$ the mean flight step does not converge and will appear to increase with time (or distance traveled) if it is estimated along a trajectory. Slow convergence or divergence of statistically defined quantities makes it more difficult to develop intuition about the motion. Indeed, it took several years for the results of 1999 [32] to be reproduced.

A recent example of a more serious miscalculation concerns the question of whether the limit $\alpha \to 1$ (or $\mu \to 2$) is optimal for searches with revisits in dimensions larger than 1, assuming that the density $\rho$ of the uniformly distributed targets also goes to zero. These results for Lévy walks have been rigorously established many years ago for the case of $d = 1$ [36, 37].

During the years 2010–2020, it became widely accepted that inverse square Lévy walks (i.e., $\mu \approx 2$) can optimize random searches in any dimension (see also next section for a comparison between Lévy searches in $d = 1$ and $d = 2$). The arguments used to reach this conclusion had a more qualitative "flavor", closer in style to the kind of reasoning used to find universality classes and key exponents in the study of critical phenomena (continuous phase transitions). For example, microscopic details were neglected and "coarse-graining" of the motion was used to estimate the quantities of interest. Thus, most of the analytical and numerical studies focused on the large scale behavior of the search (more important to determine the process encounter rates) at the expense of the small scale behavior (eventually, relevant to infer the exact mechanisms inducing the correct optimal behavior). It should be emphasized that this point of view is not a problem as far as the determination of $\eta_{\text{opt}}$ is concerned: for the interesting aspects of Lévy walks and flights rely precisely on their large scale characteristics. Still, these subtleties led to misunderstandings.

The original 1999 study [32] and many subsequent works (see references cited in Chap. 10 of [15]) had proposed that the 1D outcomes generalized to higher dimensions. Then, in 2021 a paper entitled "Inverse square Lévy walks are not optimal search strategies for d $\geq$ 2," was published [38]. As suggested by the provocative title, it claimed that the inverse square Lévy walks ($\alpha \to 1, \mu \to 2$) were not optimal. The misconception in such work were mainly due to two distinct points. The first, and certainly less problematic, was rather technical. Indeed, the theoretical methods used in [38] are more appropriate for Brownian motion, so they cannot be directly transferred to Lévy walks and flights. Nonetheless, this can be overcome as properly demonstrated in [35, 39] (see also [40]). The second was methodological. The analysis in [38] was based on the concept of first passage time (FPT), i.e., the very first time a random walk reaches a determined distance from its starting point. This *per se* does not constitute a problem since it is a well-known procedure to compute many important quantities associated to random walks in general (including Lévy's). The FPT model constructed in [38] involved a large number of parameters, so displaying a large parameters space $\Gamma$. But a very subtle aspect of the framework was that the region $\Gamma_{\mathfrak{C}}$ of biological interest—at least for random foraging, representing our conditions $\mathfrak{C}$—was indeed small. Moreover, to reach $\Gamma_{\mathfrak{C}}$ intricate limits for the main equations in [38] were necessary (see, e.g., the renormalization group approach in [35]). Inadvertently, it was assumed in [38] that the results valid in most of $\Gamma$ would trivially extended to $\Gamma_{\mathfrak{C}}$. After proper analytic and numerical calculations in [35, 39], it was established that this was not the case, with the optimality findings (but not necessarily all other claims) in [15, 32] holding true.

To completely solve the problem in general $d$-dimensions is mathematically very challenging. Presently, there is no mathematically rigorous analytical solution for it. However, random search has been addressed numerically and by means of analytic approximations and the results are fully consistent with the optimality holding in any dimensions [35, 39]. So, it turns our that the inverse square Lévy walk search for the assumptions $\mathfrak{C}$ is always the best strategy regardless of the spatial dimensions after all.

In a nutshell, in higher dimensions what happens is that in the limit of zero target density and in the case of revisiting targets (very usual in realistic foraging), the search restarts arbitrarily close to the previously visited relative to the next closest target. So the effects of the curvature of the detection horizon disappear around the target, analogously to how the earth appears flat when seen from very near the ground. Consequently, the dynamics in $d$-dimensions is very similar to that in 1D and the well-established theory for one dimension directly applies [36, 37]. A full explanation is given in [35].

## 29.3 Comparing a Simple Lévy Search Model in One and Two Dimensions

To better illustrate some of the points made in the previous section, we will now discuss the significant similarities and differences between Lévy searches in $d = 1$ (1D) and $d = 2$ (2D) spatial dimensions. For this purpose, we will consider idealized situations, which are highly convenient for our discussion.

In 1D, if we assume that the searcher cannot jump over a target without detecting it, then it is sufficient to define the interval $L$ between consecutive targets as the searching arena, with a target at each end. On the other hand, if in 2D the searcher keeps looking for targets in a limited habitat, we can suppose the process taking place in a finite square region of area $A$. To compare with the 1D case, we set $A = L \times L$ with periodic boundary conditions. The targets are randomly (Poisson) distributed in 2D.

The searcher has a detection distance range $r_v$. Further, the walks' step lengths follow a power law distribution as in Eq. (29.3), but with a lower cutoff $\ell_0$ to allow $p(\ell)$ to be normalizable. As already mentioned, due to the generalized central limit theorem, the statistical behavior for the power law tailed $p(\ell)$ with exponent $\mu$ is similar to those for the Lévy $\alpha$-stable distribution with $\alpha = \mu - 1 \in (0, 2]$. In both cases, one obtains Lévy walks.

The direction of the steps is chosen randomly, which means that in 1D the searcher goes forward or backwards with equal probability, and that in 2D an angle is chosen from an uniform distribution in the interval $[0, 2\pi)$.

A key statistical quantity of interest in a random search [18], is the efficiency $\eta$, Eq. (29.1). In the present context, for the searcher always starting at a distance $x_0$ from the last target found, $\eta$ is given as the number of targets located during the search path divided by the total distance traversed. Heuristic arguments like those in [15] suggest that the statistical behavior of the optimal efficiency with respect to $\alpha$ should not depend strongly on the dimensionality of the search space (see Sects. 29.2.1 and 29.2.2). Conceivably, if the random search model defines each step as rectilinear irrespective of the landscape dimension, then the full 2D search trajectory in principle can be built as a concatenation of 1D segments delimited by turning points. Indeed, in both 1D and 2D low-density regimes the efficiency $\eta$ as a function of $\alpha$ displays a maximum around $\alpha = 1$, for searchers starting very close to the last visited target (small $x_0$. This is the so-called nondestructive search. In contrast, one obtains as optimal the ballistic limit of $\alpha \to 0$ when targets are initially very distant to the searcher (large $x_0$, destructive search), see, e.g., [41].

These results can be understood as follows. In the destructive searches the maximum efficiency in the scarce regime is achieved in the large-step ballistic limit

$\alpha \to 0$, which favors the encounter of faraway targets in just one or few steps. Conversely, in nondestructive searches a compromise is reached between large ($\alpha \to 0$) and small ($\alpha \to 2$) steps to find, respectively, far and nearby targets leads to a maximum efficiency for $\alpha \approx 1$. However, it is also interesting to note (see, e.g., [41]) that in 2D, the efficiency $\eta$ develops a "plateau" when $\alpha \lesssim 0.5$ and $x_0$ is small. This indicates that, differently from 1D searches, no significant relative gain in the 2D search $\eta$ takes place in this $\alpha$ range, corresponding to usually very long single steps.

In contrast with the statistical trend of the search efficiency, the probabilities of the searcher to return to the last target found, i.e., the closer one ($P_0$) and to encounter faraway targets ($P_L$), as function of $\alpha$, behave quite differently in 1D and 2D. Indeed, as shown in Fig. 29.5, whereas in 1D one always has $P_0 \geq P_L$, an interesting crossover takes place in 2D. There is a separation in the search regimes, with $P_0 > P_L$ ($P_L > P_0$) for higher (lower) $\alpha$'s depending on the initial distance to the last target found.

By inspecting Fig. 29.5a, we see that in 1D the closer (the last found) target is always encountered with higher probability when compared to the finding of the

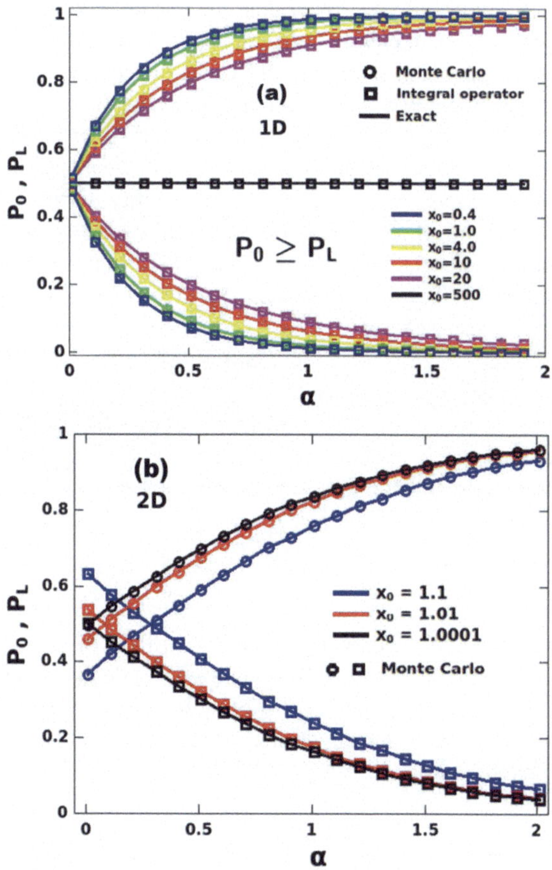


**Fig. 29.5** **a** Probabilities to find the closest ($P_0$) and distant ($P_L$) targets as function of the exponent $\alpha = \mu - 1$ of the power-law $p(\ell)$, Eq. (29.3). **a** Random searches in $d = 1$ (1D), for $L = 10^3$, $\ell_0 = 0.2$, $r_v = 0$, and various $x_0$. We see nice agreement between Monte Carlo (circles), integral operator, an analytical-numerical method (squares), and exact (solid lines) results. **b** Random searches in $d = 2$ (2D), for $10^6$ randomly distributed targets, $L \times L = 10^{10}$, $\ell_0 = 0.1$, $r_v = 1$, and various $x_0$. Monte Carlo results are shown in symbols and solid lines are guide to the eye. $P_0$ and $P_L$ in 2D are depicted by circles and squares, respectively, and display an interesting crossover not present in 1D. Adapted from [41]

faraway target. Nevertheless, a remarkable change in the probabilities $P_0$ and $P_L$ is observed in 2D searches, Fig. 29.5b. For $x_0$ greater than a certain value, a crossover between $P_0$ and $P_L$, absent in 1D, does occur in 2D but. This crossover separates the 2D searching into two regimes: $P_0 < P_L$ for lower $\alpha$ and $P_0 > P_L$ for higher $\alpha$ (we emphasize that $P_0 \geq P_L$ for any $x_0$ and $\alpha$ in 1D).

The phenomenological origin of this lies in the fact that in 2D the searcher can actually pass very close to a target (at a distance $\gtrsim r_v$) without detecting it. Consider, for instance, the very first step of a searcher starting at a small distance $x_0 \gtrsim r_v$ from the last visited target. If in 2D the searcher does not sort the step direction within the correct angle range, then it will miss the detection circle of radius $r_v$ around the closest target and, in the large-$\ell$ limit of small $\alpha$, the searcher will end up very far away from the starting point already in the very beginning of the search dynamics.

We observe that the probability of missing the closest target in 2D tends to one as the searcher's detection distance $r_v \to 0$. This highlights the relevance of the parameter $r_v$ in 2D, in contrast with the 1D case. Therefore, depending on the values of $x_0$ and $\alpha$, the probability of finding the last visited (so closer) target in a 2D search path can actually be lower than the one of encountering any other target, $P_0 < P_L$. Such argument clearly does not apply in 1D since the searcher cannot miss a target by jumping over it without detection—invariably $P_0 \geq P_L$ for any $x_0 \leq L/2$ and $\alpha$, in agreement with the results in Fig. 29.5a.

Figure 29.5 shows results for three distinct approaches. On the one hand, the integral operator method consists of a statistical physics calculation in which a power-law (or Lévy) kernel drives the probability density for the walker departing from position $x_0$ in a 1D absorbing interval of size $L$ to be at a given site after some number of steps [36, 37] (see, also, [41]). In this approach the search efficiency and probability to find the farthest target read [36, 37], respectively,

$$\eta^{-1}(x_0) = [(\mathbb{1} - \mathcal{L})^{-1}\langle|\ell|\rangle](x_0), \tag{29.4}$$

$$P_L(x_0) = [(\mathbb{1} - \mathcal{L})^{-1} p_L](x_0). \tag{29.5}$$

Above, the integral operator $\mathcal{L}$ is defined as

$$[\mathcal{L}f(x')](x) = \int_0^L p(x - x') f(x') dx',$$

with the kernel generally given by the PDF of step lengths $p(\ell)$. Further, $\mathbb{1}$ is the identity operator and $\langle|\ell|\rangle(x_0)$ and $p_L(x_0)$ denote, respectively, the mean step length to find a target in the 1D absorbing interval and the probability of encountering the target at $x = L$ in a single step (starting from $x = x_0$). For the mean step length, we explicitly write

$$\langle|\ell|\rangle(x_0) = \int_{-x_0}^{0} |\ell| p(\ell) d\ell + \int_0^{L-x_0} \ell p(\ell)$$

$$+ x_0 \int_{-\infty}^{-x_0} p(\ell)d\ell + (L - x_0) \int_{L-x_0}^{\infty} p(\ell)d\ell, \quad (29.6)$$

where in the first (last) two integrals a boundary target is not (is) found. Since for power-law and Lévy PDFs $p(\ell)$ of step lengths the inverse operator $(\mathbb{1} - \mathcal{L})^{-1}$ in Eqs. (29.4)–(29.5) is not known analytically, for the results in Fig. 29.5 we express it in a matrix form upon discretization of the 1D search interval. For so, we consider $x = j\Delta x$, with integer $j$ and $\Delta x/L \ll 1$ (for the results in Fig. 29.5 we set $\Delta x/L = 2 \times 10^{-4}$).

By taking the limit $\ell_0 \to 0$ in Eq. (29.5) for the lower cutoff step length of the power-law PDF or for the scale parameter of the Lévy PDF, one gets the exact result (see [36, 37] for details)

$$P_L(x_0, \alpha) = f(x_0, \alpha) \left(\frac{x_0}{L}\right)^{\alpha/2}, \quad (29.7)$$

where

$$f(x_0, \alpha) = \frac{2\,{}_2F_1(\alpha/2, 1 - \alpha/2; 1 + \alpha/2; x_0/L)}{\alpha B(\alpha/2, \alpha/2)}, \quad (29.8)$$

in which ${}_2F_1$ and $B$ denote, respectively, hypergeometric and beta functions.

Lastly, Fig. 29.5 also displays comparisons of results obtained using the above approaches with those from Monte Carlo simulations. In the latter, the search walker is prompted to follow the search rules described in Sect. 29.1.1. The averages are taken over $10^5$ search walks (up to $10^6$ in some runs).

## 29.4 Random Search and Movement Ecology: Some Examples

There are many interesting aspects related to the application of random search theory to biological problems and movement ecology. For instance, it may establish connections between searching behavior, sensory systems, cognitive processes, and ecological and evolutionary trends [18, 42]. Indeed, it has been argued [42] that directed movement and navigation are hallmarks of living systems, and that search is a fundamental navigational task, without which survival and reproduction would not be possible. Such a line of thought leads to fascinating discussions about the possibility of integrating the study of biological mechanisms of perception [43] (especially those associated with locomotion [44]) with ecological and evolutionary phenomena [42]. A recent example in this direction can be found in a paper aptly titled "The Scent of Fear Makes Sea Urchins Go Ballistic" [45].

A comprehensive investigation using GPS data of terrestrial animals has shed light into the problem of movement inside a complex environment [46]. The premise is that animals that move through landscapes must simultaneously find sufficient resources to grow and reproduce while also minimizing encounters with predators (see, e.g., [8, 10, 21]). If one assumes that encounter rates are determined by the mean correlation distance, then this trade-off should be reflected in the observed trajectories. In [46], more than $10^3$ individuals from more than 60 species of terrestrial mammals have been surveyed. The main finding has been that the "directed motion" (which is a measure of directional correlation) of predators was almost an order of magnitude larger than that of prey of comparable size. This impressive result shows how the mere need to search (in this case for prey) can drastically alter the movement comportment in metapopulations.

As a last illustration, the size of the large jumps and whether or not an animal moves in a Brownian or Lévy manner can have a much broader influence on their survival and reproduction than just the obvious fitness in finding food and mates. Indeed, a work [47] using GPS data of more than 800 individuals from across more than 50 species found that animals in low human footprint regions moved to an extent 2 or 3 times larger compared to animals in high human footprint regions. One possible explanation is that superdiffusive animals either need to adapt their behavior to smaller or more fragmented areas (in many cases something not feasible) or else end up being expelled or suffering extinction in regions of high human footprint. In this way, preservation and conservation measures should take into account the actual movement traits of the species in a given endangered ecosystem. We can further speculate and argue that the issue of how animals move and search are important to ecosystem functioning and species survival.

### 29.4.1 The Lévy Flight Foraging Hypothesis

We now return to the main topic of this chapter, viz. Lévy walks, and its close connection with efficient random searches. It is now firmly established that Lévy walks and akin superdiffusive movements can yield, comparatively, very high—perhaps globally optimal—encounter rates between organisms in specific circumstances (refer to the conditions $\mathfrak{C}$ in Sect. 29.1.1). In this context, the Lévy flight foraging hypothesis, sometimes also known simply as the Lévy flight hypothesis [33], holds that natural selection [18] would have led to adaptations to allow organisms to exploit the mathematical properties of Lévy walks. It is fascinating that now we humans are starting to learn mathematics from the animal kingdom, an example being the application of foraging behavior to solve problems in combinatorial optimization; a famous example being the "bees algorithm" [48].

Some claims are no longer in debate and can be considered as established. In particular, high-quality empirical studies have determined beyond reasonable doubt that organisms really do move in ways similar to Lévy walks. Specifically, Lévy walk movement patterns have been observed for seabirds, reindeer, jackals, dinoflagellates, spider monkeys, sharks, fruit flies, honey bees, sharks, bony fish, sea turtles, and even penguins. We refer the interested reader to ref. [15]. As previously mentioned, a second fact is that Lévy walks can increase encounter rates and search efficiency. Still we can ask, is there really a connection between these two indeed seemingly related facts?

It is important to recognize that the above is not a merely pedantic or meaningless question. For instance, patterns resembling Lévy walk trajectories can arise as a side effect or byproduct of special properties of the search environment—for details, see Chap. 13 of [15]. For two decades the puzzle had thus remained unsolved as to whether some (but obviously not all) of the observed Lévy flight patterns were due to endogenous dynamics generated by neural mechanisms that evolved as adaptions via natural selection, or else always due to environmental or other factors.

An important step in answering this crucial question was a study of *Drosophila* in which the larvae were cutoff from sensory and other information [49]. By shutting off the sensory pathways of the organisms, the involved research team found remarkable and highly convincing evidence that Lévy walk patterns can be generated by internal neurophysiological mechanisms. The work demonstrated that there are innate neural mechanisms for generating superdiffusive movement such as seen in Lévy walks. In fact, given that the larvae in the investigation were moving without sensory input, one can be confident that the motion is not due to environmental factors, hence ruling out epiphenomena. These results attracted a lot of attention, even inspiring articles in newspapers and popularization of science-like journals, as Quanta Magazine [50].

There is now broad support for the view that adaptations for Lévy walks and similar movement have indeed evolved. However, it remains an open question whether the explanations are mutually exclusive or whether there are multiple causes of Lévy-like behavior. Hopefully, future experiments and theoretical developments will clarify this point as well, as usual in the (many times random) search for the scientific truth.

### *29.4.2 Final Few Further Situations*

As previously emphasized, it is impossible for a single framework, even if extremely broad, to account for the huge diversity of foraging comportment found in the wild [28–30]. Thus, we shall end the present section addressing a couple of extra situations, which very appropriately exemplifies the richness of factors influencing the optimization of random search.

There can be different ways in which a forager should look and capture resources. This is particularly true in the predator-prey "game". In fact, we have explicitly mentioned that the approach presented in this contribution is more appropriate in those cases in which the efficiency $\eta_{\mathrm{obt}}$ to obtain the target does not require a special theoretical treatment. However, in some instances the adequate final step in acquiring the target, hence related to $\eta_{\mathrm{obt}}$, somehow cannot be dissociated from the searching process itself. For example, many predators have an ambushing conduct. First, they search for a hot spot, say, a pond. Then they hide, waiting for other animals to show up to drink. More broadly, the ambushing belongs to the so-called saltatory patterns of movement [51, 52], alternating between a phase in which the predator, in more detail, looks for prey (LP) and a phase where the goal is only to relocate to new territories (RT). Actually, there exists a class of models aimed to investigate such intermittent search strategies [53]. Interestingly, the searching times are minimized by assuming a continuous normal diffusion random walk for LP and a fast ballistic motion for RT—for a full account see [53] and the references therein.

The discussions so far have assumed a single walker. However, collective searching is also common in nature (refer, e.g., to [54]). In such context, we can further separate the problem into two possibilities: a "follow the leader" stratagem from a truly group behavior. In the first, usually there is an alpha individual leading the foraging. Then, even when certain specific constraints are observed, like to avoid a too strong spreading of the group (say, for safety reasons), essentially are the actions of the leader—imitate by the others—which determine the final combined outcomes. In this case, Lévy strategies still can be highly profitable [55].

The panorama might be distinct in the second case, with a paradigmatic instance being the run-and-tumble (RT) type of locomotion of bacteria (for these organisms, and more generally for self-propelled particles, see the review in [56]). This motion is characterized by random periods of persistent swimming (runs), alternated by abrupt changes of direction (tumbles). One may argue this is not different from sorting a step length $\ell_j$, from $P(\ell_j)$, followed by an angular reorientation. However, the RT dynamics results from flagella synchronization mechanisms, still not well understood, in eukaryotes, which may or may not originate from chemotaxis [57]. Hence, the degree of information $I$ can vary depending on the environment, bacteria species and a bit surprising, colony generation once bacteria can mutate to better fit their medium [58]. Moreover, the individuals strongly interact, conceivably leading to an assembled swarming aimed to find resource-rich regions [59]. All these factors result in a very diverse movement dynamics [60], certainly going much beyond the assumptions $\mathfrak{C}$ of the Lévy flight hypothesis.

## 29.5 Epilogue: The Human Search

Ultimately, the scientific endeavor shall aim to understand—and when possible, control—the general processes and laws that determine the actual world around us. So, sometimes, if we are lucky, a certain topic we choose to investigate turns out to regulate a vast number of natural phenomena; such is the case with random search. Much less common, however, is when this very subject matter also strongly relates to our everyday activities; and again, this is true for random search.

Although not addressed in dept in this contribution, it should be clear that searching is not only ubiquitous in biology but also in our social realm as well. After all, what is browsing the World Wide Web? In fact, pursuing has always been present in the history of humankind, as seen in our spread across the planet Earth, seeking new lands to inhabit. Eventually (and perhaps unconsciously) we even have followed Lévy patterns during the journey [61]—as hypothesized for the discovery of the African island of Madagascar (see Figs. 7.2–7.5 of [15]).

Search is not limited to the need to find targets. Indeed, beyond the known rational and material significance, the unknown to be found appeals directly to our imagination. We are often looking for something, and more frequently than not, this something is rather intangible: our place in the world, our dreams, our curiosity—or yet the search for the truth. It is not an accident that many our great idealized heroes are always searching, with their strength invariably emerging from the hope of succeeding. Often they search alone, by themselves,

> But you, brave and adept from this day on ...
> there's hope that you will reach your goal ...
> the journey that stirs you now is not far off.
> —*The Odyssey*, Homer (circa 8th century BC)

but with the determination of trying hard,

> ...
> To strive, to seek, to find, and not to yield.
> —*Ulysses*, Alfred, Lord Tennyson (1842)

knowing that the quest to find the best search strategy is itself a search, so coming back full circle.

> Think you're escaping and run into yourself.
> Longest way round is the shortest way home.
> —*Ulysses* , James Joyce (1922)

**Acknowledgements** This work was supported by the Brazilian agency CNPq (grants 304532/2019-3 (MGEL), 302414/2022-3 (GMV) and 305062/2017-4 (EPR)).

## Appendix: The $\alpha$-stable Lévy Distributions and Their Power-Law Asymptotics

Suppose the Lévy index $\alpha \in (0, 2]$ and the asymmetry $\beta \in [-1, 1]$, scale $\sigma > 0$ and shift $u \in (-\infty, +\infty)$ as the parameters. Then, one defines the $\alpha$-stable Lévy distributions as the following PDF of $\ell$ (with sign$(t)$ the sign of $t$),

$$p(\ell; \alpha, \beta, \sigma, u) = \frac{1}{2\pi} \int_{-\infty}^{\infty} dt \; \exp[\phi(t; \alpha, \beta, \sigma, u)] \; \exp[-i \ell t], \qquad (29.9)$$

with the characteristic function $\phi$ given by

$$\phi(t;\alpha,\beta,\sigma,u) = \begin{cases} itu - |\sigma\, t|^{\alpha}\Big(1 - i\,\beta\ \mathrm{sign}[t]\ \tan[\frac{\pi}{2}\alpha]\Big), & \text{for } \alpha \neq 1 \\ itu - |\sigma\, t|\ \Big(1 + i\,\beta\ \mathrm{sign}[t]\ \frac{2}{\pi}\ \ln[|t|]\Big), & \text{for } \alpha = 1\,. \end{cases} \tag{29.10}$$

Generally $u$ and $\sigma$ can be factored in the above formulas by partially reincorporating them into the variable $\ell$, hence for sake of discussion we can assume $u = 0$ and $\sigma = 1$. For the resulting $p(\ell; \alpha, \beta)$, analytical expressions based on simple functions are known only for three particular sets of $\alpha$ and $\beta$: the Gaussian ($\alpha = 2$) and Cauchy ($\alpha = 1$) when $\beta = 0$, hence the symmetric cases. The third is the Lévy distribution ($\alpha = 1/2$) for the maximally asymmetric case $\beta = 1$. For other $\alpha$'s and $\beta$'s, in principle $p(x; \alpha, \beta)$ can be cast in terms of Fox-$H$ functions, whose computations rely on complex integrals of the Mellin-Barnes type. Special situations can also be represented as a sum of hypergeometric and/or $G$ functions.

We point out that, via a complex analytic continuation $p(\ell) \to p(z)$, recent analyses [62] have put forward strong arguments pointing to the nonexistence of other closed elementary expressions for the $\alpha$-stable Lévy distributions, excepting the ones mentioned above.

Considering that (i) the exact involved formulas for $p$, Eqs. (29.9) and (29.10), are often too cumbersome to work with, and (ii) to calculate the efficiency of random searches, only the large step length regime of $p(\ell)$ really matters [15, 35], a simplification is possible. Usually we can consider an abridged version of $p$, provided it properly fits $p(\ell)$ when $\ell \gg 1$. To obtain heuristically such a suitable approximation (for a throughout and rigorous treatment see, e.g., [63]), consider $\beta = u = 0, \sigma = 1$ and $\ell > 0$, so that

$$p(\ell;\alpha,\beta=0) = \frac{1}{2\pi}\int_{-\infty}^{\infty} dt\ \exp[-|t|^{\alpha}]\ \exp[-i\ell t]. \tag{29.11}$$

As previously mentioned, if $\alpha = 2$ one directly finds that $p(\ell; 2, 0) = \exp[-(\ell/2)^2]/(2\sqrt{\pi})$, the Gaussian distribution, whereas for $\alpha = 1$ one gets $p(\ell; 1, 0) = 1/(\pi\ (1 + \ell^2))$, the Cauchy distribution. Notably, for large values of $\ell$ the Cauchy goes as $p(\ell, 1, 0) \sim 1/\ell^2$, or since $\alpha = 1$ as $p(\ell, \alpha = 1, 0) \sim 1/\ell^{1+\alpha}$.

If in Eq. (29.11) $\ell$ is large, the complex exponential $\exp[-i\ell t]$ rapidly oscillates as $|t|$ varies. Furthermore, we have an exponentially decaying function $\exp[-|t|^{\alpha}]$ in the integrand of Eq. (29.11)—faster for greater $\alpha$'s. These two trends cause cancelations almost everywhere as we integrate in $t$. Thence, for $\ell \gg 1$ there must be a fine-tuning in Eq. (29.11) leading to a scaling of $p$ with $\lambda$, for $p$ decreasing as $\ell$ increases (more strongly for larger $\alpha$'s). We recall this general trait concurs with the asymptotics of the Cauchy distribution.

Keeping the above in mind, let us rewrite Eq. (29.11) as (implementing trivial changes of variable)

$$p(\ell;\alpha,0) = \frac{1}{2\pi\,\ell}\int_{-\infty}^{\infty} dt\ \exp\Big[-\frac{|t|^{\alpha}}{\ell^{\alpha}}\Big]\exp[-it] = \frac{1}{\pi\,\ell}\int_{0}^{\infty} dt\ \exp\Big[-\frac{t^{\alpha}}{\ell^{\alpha}}\Big]\cos[t]$$

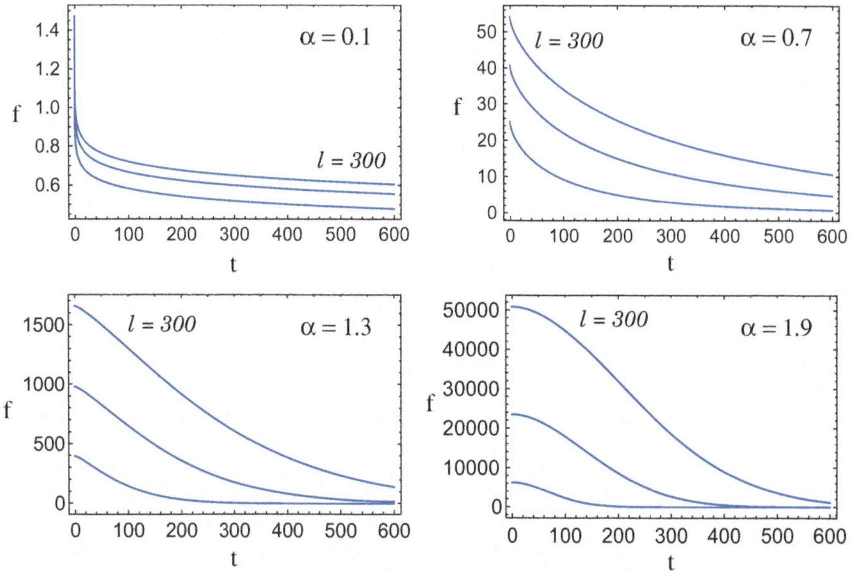


**Fig. 29.6** Plots of $f(t; \ell, \alpha)$ versus $t$ for distinct $\alpha$. The values of $\ell$ are 300 (indicated), 200, and 100. Notice that for a given $\alpha$, for any $t$ the function $f$ monotonically decreases with $\ell$, nevertheless with the overall shapes being fairly similar

$$= \frac{1}{\ell^{\alpha+1}} \frac{1}{\pi} \int_0^\infty dt \, \cos[t] \, f(t; \ell, \alpha),$$

$$f(t; \ell, \alpha) = \exp\big[-\frac{t^\alpha}{\ell^\alpha}\big] \, \ell^\alpha. \tag{29.12}$$

For large $\ell$, the shape of $f$ changes considerably with $\alpha$ but not with $\ell$. Indeed, for fixed $\alpha$, graphs of $f$ for increasing $\ell$'s are fairly blow ups of each other, with little distortion—illustrations in Fig. 29.6. Once these rescalings are essentially washed out by the integration of $I = \int_0^\infty dt \, \cos[t] \, f(t; \ell, \alpha)$, we have just a mild dependence of $I$ on $\ell$, so that in first order $I \sim g(\alpha)$. Thus, we finally reach the following conclusion (extended to arbitrary $\beta$): for a non-Gaussian case, i.e., $\alpha < 2$, $p(\ell; \alpha < 2, \beta)$ is a "fat tailed" distribution. More concretely, for large values of $\ell$, we have asymptotically the power-law $p \sim 1/|\ell|^\mu$, where $\mu = \alpha + 1$ with $\mu \in (1, 3]$.

## References


1. S. Arora, B. Barak, *Computational Complexity: A Modern Approach* (Cambridge University Press, Cambridge, 2009)
2. A.S. Ferreira, E.P. Raposo, G.M. Viswanathan, M.G.E. da Luz, The influence of the environment on Lévy random search efficiency: Fractality and memory effects. Phys. A **391**, 3234

(2012)
3. M. Serra, P. Sathe, I. Rypina, A. Kirincich, S.D. Ross, P. Lermusiaux, A. Allen, T. Peacock, G. Haller, Search and rescue at sea aided by hidden flow structures. Nat. Commun. **11**, 2525 (2020)
4. G. Damasco, C. Fontes, R. Françoso, R. Haidar, The Cerrado biome: A forgotten biodiversity hotspot. Front. Young Minds **6**, 22 (2018)
5. A. Antonelli, et al., Madagascar's extraordinary biodiversity: Evolution, distribution, and use. Science **378**, eabf0869 (2022)
6. M.J. Osborne, A. Rubinstein, *A Course in Game Theory* (MIT University Press, Cambridge, MA, 1994)
7. F.R. Rusch, M.E. Wosniack, E.P. Raposo, G.M. Viswanathan, M.G.E. da Luz, Transient dynamics in a nonequilibrium superdiffusive reaction-diffusion process: Nonequilibrium random search as a case study. Phys. Rev. E **102**, 012126 (2020)
8. C.L. Faustino, L.R. da Silva, M.G.E. da Luz, E.P. Raposo, G.M. Viswanathan, Search dynamics at the edge of extinction: Anomalous diffusion as a critical survival state. EPL **77**, 30002 (2007)
9. M.E. Wosniack, E.P. Raposo, G.M. Viswanathan, M.G.E. da Luz, Efficient search of multiple types of targets. Phys. Rev. E **92**, 062135 (2015)
10. C.L. Faustino, M.L. Lyra, E.P. Raposo, G.M. Viswanathan, M.G.E. da Luz, The universality class of random searches in critically scarce environments. EPL **97**, 50005 (2012)
11. S.A. Sotelo-López, M.C. Santos, E.P. Raposo, G.M. Viswanathan, M.G.E. da Luz, Conditions under which a superdiffusive random-search strategy is necessary. Phys. Rev. E **86**, 031133 (2012)
12. C. Zimmer, M. Boos, N. Poulin, A. Gosler, O. Petit, J.P. Robin, Evidence of the trade-off between starvation and predation risks in ducks. PLoS ONE **6**, e22352 (2011)
13. A. Charalabidis, F.-X. Dechaume-Moncharmont, S. Petit, D.A. Bohan, Risk of predation makes foragers less choosy about their food. PLoS ONE **12**, e0187167 (2017)
14. D.W. Stephens, J.R. Krebs, *Foraging Theory* (Princeton University Press, Princeton, NJ, 2019)
15. G.M. Viswanathan, M.G.E. da Luz, E.P. Raposo, H.E. Stanley, *The Physics of Foraging: An Introduction to Random Searches and Biological Encounters* (Cambridge University Press, Cambridge, 2011)
16. V. Méndez, D. Campos, F. Bartumeus, *Stochastic Foundations in Movement Ecology* (Springer, Berlin, 2014)
17. D.L. Kramer, Foraging behavior, in *Evolutionary Ecology: Concepts and Case Studies*, C.W. Fox, D.A. Roff, and D.J. Fairbairn (Eds). (Oxford Academic, Oxford, 2001), p. 232
18. M.E. Wosniack, M.C. Santos, E.P. Raposo, G.M. Viswanathan, M.G.E. da Luz, The evolutionary origins of Lévy walk foraging. PLoS Comput. Biol. **13**, e1005774 (2017)
19. M.C. Santos, G.M. Viswanathan, E.P. Raposo, M.G.E. da Luz, Optimization of random searches on defective lattices networks. Phys. Rev. E **77**, 041101 (2008)
20. G.M. Viswanathan, E.P. Raposo, M.G.E. da Luz, Lévy flights and superdiffusion in the context of biological encounters and random searches. Phys. Life Rev. **5**, 133 (2008)
21. B.B.S. Niebuhr, M.E. Wosniack, M.C. Santos, E.P. Raposo, G.M. Viswanathan, M.G.E. da Luz, M. Pie, Survival in patchy landscapes: The interplay between dispersal, habitat loss and fragmentation. Sci. Rep. **5**, 11898 (2015)
22. P. Formaglio, M.E. Wosniack, R.M. Tromer, J.G. Polli, Y.B. Matos, H. Zhong, E.P. Raposo, M.G.E. da Luz, R. Amino, Plasmodium sporozoite search strategy to locate hotspots of blood vessel invasion. Nat. Commun. **14**, 2965 (2023)
23. F. Bartumeus, J. Catalan, U. L. Fulco, M. L. Lyra, and G. M. Viswanathan, Optimizing the encounter rate in biological interactions: Lévy versus Brownian strategies. Phys. Rev. Lett. **88**, 097901 (2002); Erratum: Phys. Rev. Lett. **89**, 109902 (2002)
24. M.F. Shlesinger, J. Klafter, Lévy walks versus Lévy flights, in *On Growth and Form*, H.E. Stanley, N. Ostrowsky (Eds). (Nijhoff, Dordrecht, 1986), p. 279
25. S. Andrzejaczek *et al*, Diving into the vertical dimension of elasmobranch movement ecology. Sci. Adv. **8**, eabo1754 (2022)
26. M.F. Shlesinger, Random searching. J. Phys. A **42**, 434001 (2009)

27. Z. Li, H. Ning, L. Cao, T. Zhang, Y. Gong, and T. S. Huang, Learning to search efficiently in high dimensions, in *Advances in Neural Information Processing Systems 24 (NIPS 2011)*, J. Shawe-Taylor, R. Zemel, P. Bartlett, F. Pereira, K.Q. Weinberger (Eds). (Currant Associates Inc., 2011), p. 1710
28. F. Provenza, J. Villalba, C. Cheney, S. Werner, Self-organization of foraging behaviour: From simplicity to complexity without goals. Nutr. Res. Rev. **11**, 199 (1998)
29. M.E. Gilmour, J.A. Castillo-Guerrero, A.B. Fleishman, S. Hernández-Vázquez, H.S. Young, S.A. Shaffer, Plasticity of foraging behaviors in response to diverse environmental conditions. Ecosphere **9**, e02301 (2018)
30. J.D. Davidson, A.E. Hady, Foraging as an evidence accumulation process. PLoS Comput. Biol. **15**, e1007060 (2019)
31. G.M. Viswanathan, V. Afanasyev, S.V. Buldyrev, E.J. Murphy, P.A. Prince, H.E. Stanley, Lévy flight search patterns of wandering albatrosses. Nature (London) **381**, 413 (1996)
32. G.M. Viswanathan, S.V. Buldyrev, S. Havlin, M.G.E. da Luz, E.P. Raposo, H.E. Stanley, Optiziming the success of random searches. Nature (London) **401**, 911 (1999)
33. M.G.E. da Luz, E.P. Raposo, G.M. Viswanathan, And yet it optimizes - Comment on "Liberating Lévy walk research from the shackles of optimal foraging." Phys. Life Rev. **14**, 94 (2015)
34. D.W. Sims, N.E. Humphries, N. Hu, V. Medan, J. Berni, Optimal searching behaviour generated intrinsically by the central pattern generator for locomotion. ELife **8**, e50316 (2019)
35. L.G.P. Caramês, Y.B. Matos, F. Bartumeus, C.G. Bezerra, T. Macrì, M.G.E. da Luz, E.P. Raposo, G.M. Viswanathan, Lévy walkers inside spherical shells with absorbing boundaries: Towards settling the optimal Lévy walk strategy for random searches. Phys. Rev. E **106**, 054147 (2022)
36. S.V. Buldyrev, S. Havlin, A.Y. Kazakov, M.G.E. da Luz, E.P. Raposo, H.E. Stanley, G.M. Viswanathan, Average time spent by a Lévy flyer on an interval with absorbing boundaries. Phys. Rev. E. **64**, 041108 (2001)
37. S.V. Buldyrev, M. Gitterman, S. Havlin, A.Y. Kazakov, M.G.E. da Luz, E.P. Raposo, H.E. Stanley, G.M. Viswanathan, Properties of Lévy flights on an interval with absorbing boundaries. Phys. A **302**, 148 (2001)
38. N. Levernier, J. Textor, O. Bénichou, R. Voituriez, Inverse square Lévy walks are not optimal search strategies for $d \geq 2$. Phys. Rev. Lett. **124**, 080601 (2020)
39. S.V. Buldyrev, E.P. Raposo, F. Bartumeus, S. Havlin, F.R. Rusch, M.G.E. da Luz, G.M. Viswanathan, Comment on "Inverse square Lévy walks are not optimal search strategies for $d \geq 2$." Phys. Rev. Lett. **126**, 048901 (2021)
40. N. Levernier, J. Textor, O. Bénichou, R. Voituriez, Reply to the "Comment on 'Inverse square Lévy walks are not optimal search strategies for $d \geq 2$'". Phys. Rev. Lett. **126**, 048902 (2021)
41. J.R. Colaço, H.A. Araújo, M.G.E. da Luz, G.M. Viswanathan, F. Bartumeus, E.P. Raposo, Effect of the search space dimension for the finding of close and faraway targets in random searches. Phys. Rev. E **106**, 034124 (2022)
42. A.M. Hein, F. Carrera, D.R. Brumley, R. Stocker, S.A. Levin, Natural search algorithms as a bridge between organisms, evolution, and ecology. Proc. Natl. Acad. Sci. U.S.A. **113**, 9413 (2016)
43. J. Lange, M. Lappe, A Model of biological motion perception from configural form cues. J. Neurosci. **26**, 2894–2906 (2006)
44. D.H.F. Chang, N.F. Troje, Characterizing global and local mechanisms in biological motion perception. J. Vis. **9**, 1–10 (2009)
45. J.F. Pagès, F. Bartumeus, J. Romero, T. Alcoverro, The scent of fear makes sea urchins go ballistic. Mov. Ecol. **9**, 50 (2021)
46. M.J. Noonan, et al., The search behavior of terrestrial mammals. BioRxiv, 2022.12.31.521874 (2023)
47. M.A. Tucker et al., Moving in the Anthropocene: Global reductions in terrestrial mammalian movements. Science **359**, 466 (2018)
48. D.T. Pham, A. Ghanbarzadeh, E. Koc, S. Otri, S. Rahim, M. Zaidi, *The Bees Algorithm, Technical Note, Manufacturing Engineering Centre* (Cardiff University, UK, 2005)

49. D.W. Sims, N.E. Humphries, N. Hu, V. Medan, J. Berni, Optimal searching behaviour generated intrinsically by the central pattern generator for locomotion. ELife **8**, e50316 (2019)
50. L. Drew, Random search wired into animals may help them hunt, Quanta Magazine (2020). https://www.quantamagazine.org/random-search-wired-into-animals-may-help-them-hunt-20200611/
51. W.J. O'Brien, H.I. Browman, B.I. Evans, Search strategies of foraging animals. Am. Sci. **78**, 152–160 (1990)
52. D.L. Kramer, R.L. McLaughlin, The behavioral ecology of intermittent locomotion. Am. Zool. **41**, 137–153 (2001)
53. O. Bénichou, C. Loverdo, M. Moreau, R. Voituriez, Intermittent search strategies. Ref. Mod. Phys. **83**, 81–129 (2011)
54. K. Bhattacharya, T. Vicsek, Collective foraging in heterogeneous landscapes. J. R. Soc. Interface **11**, 20140674 (2014)
55. M.C. Santos, E.P. Raposo, G.M. Viswanathan, M.G.E. da Luz, M G E, Can collective searches profit from Lévy walk strategies? J. Phys. A **42**, 434017 (2009)
56. M.E. Cates, J. Tailleur, Motility-induced phase separation. Annu. Rev. Condens. Matter Phys. **6**, 219–44 (2015)
57. M. Polin, I. Tuval, K. Drescher, J.P. Gollub, R.E. Goldstein, Chlamydomonas swims with two gears in a eukaryotic version of run-and-tumble locomotion. Science **325**, 487–490 (2009)
58. M.E. Wosniack, M.G.E. da Luz, L.S. Schulman, Punctuated equilibrium as an emergent process and its modified thermodynamic characterization. J. Theor. Biol. **412**, 113–122 (2017)
59. D. Breoni, F.J. Schwarzendahl, R. Blossey, H. Löwen, A one-dimensional three-state run-and-tumble model with a cell cycle. Eur. Phys. J. E **45**, 83 (2022)
60. M. Sidortsov, Y. Morgenstern, A. Be'er, Role of tumbling in bacterial swarming. Phys. Rev. E **96**, 022407 (2017)
61. D.A. Raichlen, B.M. Wood, A.D. Gordon, A.Z.P. Mabulla, F.W. Marlowe, H. Pontzer, Evidence of Lévy walk foraging patterns in human hunter-gatherers. Proc. Nat. Acad. Sci. **111**, 728–733 (2014)
62. E.C. Rocha, M.G.E. da Luz, E.P. Raposo, G.M. Viswanathan, Why Lévy $\alpha$-stable distributions lack general closed-form expressions for arbitrary $\alpha$. Phys. Rev E **100**, 010103(R) (2019)
63. J.P. Nolan, *Univariate Stable Distributions: Models for Heavy Tailed Data* (Springer, Cham, 2020)

# Chapter 30
# Olfactory Search

**Antonio Celani and Emanuele Panizon**

*If you do not expect the unexpected you will not find it, for it is not to be reached by search or trail*

*Heraclitus*

**Abstract** The task of olfactory search is ubiquitous in nature and in technology, from animals in the quest of food or of a mating partner to robots searching for the source of hazardous fumes in a chemical plant. Here, we focus on the algorithmic approach to this task: we systematically review the different olfactory search strategies. Special emphasis is given to the formal description as a Partially Observable Markov Decision Process, which allows the computation of optimal actions and helps in clarifying the relationships between several effective heuristic search strategies.

## 30.1 Introduction

Olfactory navigation and search are the processes by which organisms harness their sense of smell to navigate their environment, locate resources, and communicate with conspecifics. Olfaction, the sense of smell, serves as a cornerstone of survival and reproduction for a multitude of species across the animal kingdom, from microscopic bacteria to large mammals, transcending the boundaries of size and habitat [1].

The olfactory sensory organs allow to perceive and interpret chemical cues, whether to detect the scent of prey, find a mate, identify food sources, or avoid danger where visual and auditory cues may fall short or remain unreliable [2].

A. Celani (✉) · E. Panizon
The Abdus Salam International Center for Theoretical Physics ICTP, Str. Costiera, 11, 34151 Trieste, Italy
e-mail: celani@ictp.it

E. Panizon
e-mail: epanizo1@ictp.it



D. Grebenkov et al. (eds.), *Target Search Problems*,
https://doi.org/10.1007/978-3-031-67802-8_30

For some animals, like rodents, olfactory navigation is indispensable for exploring maze-like environments, remembering spatial layouts, and locating hidden rewards. For others, such as moths, it allows to track the scent of a potential mate across vast distances. Marine animals use their olfactory sense to migrate across oceans to their natal streams for spawning. These diverse examples highlight the broad spectrum of olfactory navigation applications in the natural world [3].

Several challenges must be faced by organisms that rely on olfaction for their navigation tasks. Odor cues are dispersed by turbulence in the air or water, diffused by the environment, and mixed with other scent sources, leading to ever-changing odor landscapes. Navigating this complexity requires an impressive array of adaptations, from the intricate structure of an insect's antennae to the advanced neural system of a canine's brain [4].

In addition to natural organisms, the principles of olfactory navigation have far-reaching applications in the development of autonomous systems and robotics. Robots equipped with odor sensors can mimic the abilities of natural navigators, allowing them to track the scent of chemical leaks in industrial settings or search for survivors in disaster zones. Understanding the mechanisms and challenges of olfactory navigation in nature contributes significantly to the development of these technologies. Vice versa, understanding the algorithmic principles of olfactory navigation can shed light on biological search processes.

A comprehensive exploration of olfactory navigation and search would span biology, sensory physiology, behavioral ecology, neurobiology, fluid dynamics, and computational science. More humbly, in this chapter, we focus on the algorithmic view of olfactory search. After a brief review of the basic biological and physical facts that form the backdrop of our discussion, we will move quickly into the world of search strategies. A hopefully useful map to navigate this subject is given in Fig. 30.1.

## 30.2 The Physics and Biology of Olfactory Search

Pheromones, specialized chemical compounds, serve as an astonishing example of long-range communication among various organisms. One of the most striking examples of this phenomenon can be observed in Lepidoptera, a vast order of insects that includes moths. These organisms display an exceptional ability to detect and respond to sex pheromones released by potential mates.

In Lepidoptera, and many other species, the communication of sex pheromones is vital for successful reproduction. Most Lepidoptera can detect and be consistently drawn to calling females from considerable distances, sometimes spanning several hundred meters. What makes this achievement all the more impressive is that females release their pheromone messages into the turbulent atmospheric surface layer: males face the challenging task of deciphering the source's location from a signal that is not only attenuated by the environment but also garbled and mixed with various other olfactory stimuli.

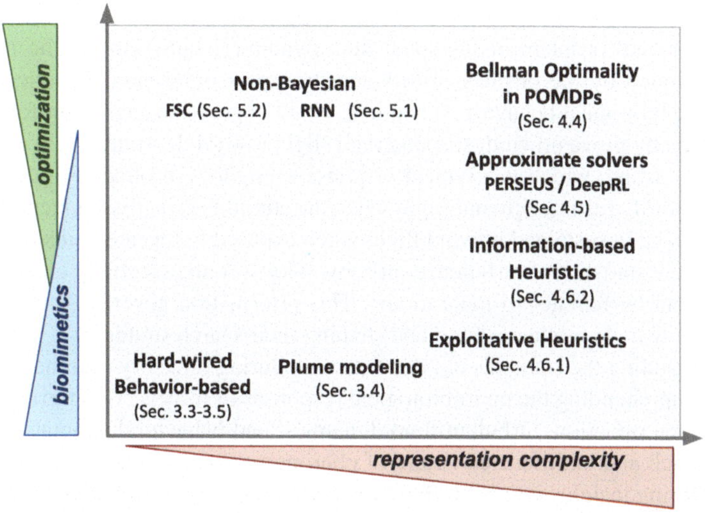


**Fig. 30.1** Concept map of the main algorithmic approaches to olfactory search discussed here and displayed on an abstract plane. On the $y$-axis search strategies are evaluated on how much they are grounded on principles of optimization or, conversely, rooted in biomimetics. The $x$-axis discriminates depending on the level of task-representation required by the agent: on the leftmost side, algorithms are purely reactive, while on the rightmost side, full spatial maps and Bayesian inference are employed

The pheromone communication system in moths, and other insects, operates under strong evolutionary pressure. Adult moths from certain families, such as Saturniidae and Bombycidae, have relatively short lifespans as adults. These moths primarily allocate their brief adult lives to reproduction, having stored up energy in the form of lipids during their larval stages. As a result, natural selection has shaped their olfactory systems to be exquisitely sensitive to pheromones. Just a few pheromone molecules coming into contact with a male moth's antenna are sufficient to alert the insect and trigger a change in its cardiac frequency. Even low concentrations of a few hundred molecules per cubic centimeter can elicit specific behavioral responses that precede flight.

Apart from the intensity of the pheromone signal, its quality and time-course play crucial roles. In terms of quality, the signal is typically a blend of two or more chemical compounds. Discrimination is achieved not by the components themselves but by different combinations and ratios within the mixture. Precision in discrimination is essential for finding a compatible mate.

The time-course of the pheromone signal is another crucial aspect. Turbulence in the environment has a profound impact, distorting the pheromone signal significantly. This leads to irregular and intermittent fluctuations in the concentration of pheromone at distances far from the source. The turbulent atmosphere causes the signal to feature alternating bursts of pheromone and clean-air periods, with a wide range of durations.

Characterizing the properties of odor detections in turbulent flows is a complex and fundamental problem in statistical fluid dynamics. Furthermore, the intermittency generated by the physics of turbulent transport is essential for eliciting the appropriate biological behavior. When insects are exposed to steady, uniform stimuli, they briefly move upwind, suspend their flight toward the source, and engage in crosswind casting, which is a typical response to the loss of olfactory cues. Males may temporarily resume upwind flight when the stimulus is increased incrementally or may engage in sustained upwind flight when exposed to repeated pulses. In these instances, the statistics of turbulence-airborne odor stimuli essentially constitute the message sent by females to male moths. This information governs their behavior, dictating how male moths respond and shaping their search strategies.

Understanding the statistics of odor detection during olfactory searches is essential for comprehending the neurobiological responses of insects. This intricate interplay between olfaction, turbulent fluid dynamics, and behavioral adaptations highlights the rich and multifaceted nature of pheromone communication in the natural world, offering insights into evolution, sensory biology, and the challenges faced by organisms in their quest for reproductive success.

The reader interested in the biological and physical mechanisms at work in olfactory search can find excellent introductions elsewhere [3–5]. The characterization of the dynamical odor landscape in a turbulent environment can be found in [6].

## 30.3 A Day at the Zoo

The olfactory search problem spans diverse research fields, from biology to robotics, where it is commonly known as Odor Source Localization (OSL). Its applications are broad, ranging from detecting hazardous gas leaks in power plants to gaining insights into animal behavior. Given the variety of perspectives applied to the olfactory search in recent decades, it is perhaps unsurprising that the methods and algorithms presented in the scientific literature often lack a coherent structure.

In this section, we aim at summarizing—without any pretense of an exhaustive review—a wide array of methods proposed and analyzed to create efficient search strategies from a heuristic perspective. We will showcase examples of search strategies constructed with well-defined, hard-wired behaviors, which, although sometimes involving parameter inference, are not the outcome of an optimization process.

In the subsequent sections, we will introduce a formal framework in which we will cast the olfactory search problem in the formal language of optimal decision-making in the presence of uncertainty, i.e., the theory of Partially Observed Markov Decision Processes, detailed in Sect. 30.4.2. The latter will be the primary focus of this chapter.

Readers who wish to take a deep dive into the robotic aspects of OSL can refer to [7] or to the more recent reviews in [8, 9].

### 30.3.1 Zero Information

An extreme case of the search process is the one occurring under conditions of zero information, where the probability of encountering new odor signals is so low that it can be neglected.

One potential approach in such cases involves exhaustive searches, a concept that has also been explored in the context of naval warfare [10]. These strategies produce trajectories that attempt at a systematic coverage of all available space, either by gradually expanding in spiral-like patterns or by traversing parallel lines of fixed width. It is intriguing to compare the findings of this approach with the spiraling search patterns that emerge in the realm of "optimal" searches, as discussed in Sect. 30.4.1.

In a more biologically inspired context, it has been suggested that certain animals, like albatrosses, employ Lévy flights [11] as a means to maximize the area they explore. However, the relative effectiveness of Lévy flights compared to potentially biased Brownian motion remains a topic of ongoing debate [12–14]. Connected to the notion of Lévy flights, an intermittent dual-mode strategy, combining fast-moving exploration and slow-moving detection, has been recognized as a widespread feature in target searches conducted under conditions of minimal information [15].

### 30.3.2 Smooth Odor Landscapes

Conversely, at the opposite end of the observational spectrum, we encounter what are often termed "tactic" methods of motion [9]. Notably, these tactics do not necessitate an extended memory of previous observations, but rather rely on the continuous measurement of relevant field gradients, such as chemical concentration (chemotaxis), wind direction (anemotaxis), variations across the spatial sensor array (tropotaxis), or combinations thereof.

These tactic strategies appear in the natural world across a wide range of scales, from bacteria to ants and rodents. However, they typically falter in the presence of turbulence, where the environmental landscape is far from being smooth. A comparative review of tactic animal behaviors, contrasted with more complex strategies, is given in [3]. Wadhwa and Berg [16] explore the mechanical aspects of motility in bacteria, shedding light on how chemotaxis is accomplished.

### 30.3.3 Behavior-Inspired Search Algorithms

Expanding upon these tactical models of motion, researchers have introduced a diverse array of biomimetic algorithmic search patterns. These patterns often consist of straightforward, diagrammatic instructions, following the "if-then-else" format, which aim to replicate or imitate behaviors observed in nature. These algorithms rely on a series of pre-programmed behaviors, with the searcher switching between them based on immediate sensory inputs and/or predefined timing rules.

Some of these algorithms are designed to mimic purely reactive "taxis"-like movements. For instance, there are Braitenberg-style tropotactic robots, equipped with two independent motors connected to sensors positioned at different locations. These robots utilize differences in signal intensity at the sensors to realign themselves toward or away from the signal source. In a more direct emulation of natural behavior, Holland and Melhuish [17] introduced an "E. coli algorithm" for a model run-and-tumble bacterium, while Sabelis and Schippers [18] explored anemotactic strategies, i.e., based on local wind direction, within odor plumes featuring variable wind directions. More recently, a tropotactic robot equipped with displaced gas sensors was developed [19], exploiting the spatial fluctuations of the signal.

Solutions based on hard-wired behaviors have also been proposed to address the challenge of non-continuous signals, such as those presented by odor in turbulent environments. Typically, the initial efforts aimed at replicating the cast-and-surge behavior observed in moths, a behavior that had been increasingly scrutinized in controlled experimental settings [20–22].

The first generation of reactive methods, as proposed by [23], involved alternations between upwind surges upon detecting odor and back-and-forth flights when no odor was detected. Building upon this fundamental concept, [24] introduced and compared several variants, including those with increased casting ranges. Interestingly, the introduction of varying casting ranges, dependent on the last detection, introduced the need for an internal clock and, consequently, a form of memory. This advancement expanded these algorithms beyond purely reactive approaches.

A more recent variation of these algorithms is the cast-and-surge algorithm developed by [25], where the alternating pattern of increasing casting range and surging motion demonstrated remarkable efficiency in reaching the source within (model) turbulent streams on a 2D square grid. Shigaki et al. [26] introduced a behavioral model based on sequences of surges, zigzags, and loops, optimizing the timescale of transitions through experimentation with robotic setups.

In a different vein, [27] proposed the alternation of surging with spiraling and/or increased casting. They compared the performance of these methods with the more complex Infotaxis (refer to Sect. 30.4.6.2) in experimental setups. Additional examples of diagram-based search in robotics can be found in [28], where a strategy that involved delayed turns and a discrete set of different motion modes was used to aid a robot in navigating maze-like environments, leveraging "a combination of upwind search, detection, and knowledge of gross fluid dynamics."

### *30.3.4 Building a Model of the Odor Plume—And Using It*

The algorithms discussed above primarily operate at a local level in their response, relying on pure reactivity or, at most, incorporating a rudimentary internal counter to modulate their search behavior. However, at a more advanced conceptual level, there are algorithms that involve some form of plume modeling. These "cognitive" methods are fundamentally grounded in a pre-constructed understanding of the mechanism that produces odor encounters. As signals accumulate over time, this model becomes increasingly defined, and the estimated source location converges to a smaller region. Therefore, the searcher faces two concurrent challenges: updating and refining the plume model and navigating within it.

It is important to remark that in this section, we will present methods that rely on plume modeling through techniques like parameter fitting or Bayesian inference. However, the navigation aspect is handled through hard-wired behaviors. In the sequel (Sects. 30.4.4 and 30.4.6), we will explore techniques where the navigation strategies are inextricably linked with the information-gathering process.

The initial step toward imbuing spatial "awareness" in the agent involves constructing a real-time plume distribution based on parameter inference from the signals received by the agent as it moves through the environment. One of the early instances of such plume modeling can be found in [29], where the agent maintains a gas distribution model that is continuously fitted in real-time and subsequently used for odor source localization. In [30], a combination of plume modeling through Hidden Markov States and behavior switching diagrams is applied for underwater Chemical Plume Tracing. Bayesian methods have also been employed for estimating chemical concentrations in water, as seen in [31], where inference was utilized to construct and update a source-likelihood map. This model incorporated inputs from concentration levels and fluid flow velocity, although it focused primarily on the "mapping" aspect of the problem, neglecting the search strategy.

Furthermore, [32] presents a robotic OSL algorithm based on particle filters. When odor signals are detected, the searcher engages in predefined exploratory behavior and employs the input signal to update the estimated source locations.

### *30.3.5 Modeling Search Strategies from Experimental Data*

Recent advancements in real-time insect tracking technology allow to record and analyze experimental insect trajectories under various wind conditions. These trajectories have served as the foundation for developing behavioral models that elucidate the responses of insects to odor traces within turbulent flows. While these models may not explicitly represent search behavior, they hold considerable relevance for comparison, particularly when juxtaposed with heuristic or hard-wired algorithms.

In [33], experimental trajectories of walking Drosophila melanogaster were utilized to infer a behavioral model describing their responses to odor inputs within wind chambers. By incorporating adaptation and temporal filtering, the researchers proposed two-mode models where the history of the odor signal is transformed into ON/OFF activations. Various models were suggested, with the two modes corresponding to either an increase or decrease in ground speed accompanied by an opposing decrease or increase in turn rates. Another model variant involved different modulations dependent on the fly heading.

A subsequent work by [34] introduced a "neural circuit"-based model for navigation, incorporating inputs such as heading, wind direction, and odor, and simulating the fly response through layers of neural-inspired computations.

In [35], flies were observed as they responded to a smoke plume that attracted them. The flies exhibited a limited set of discernible actions, including "stop," "walk," and "turn" (either upwind or downwind). The authors successfully quantified the statistics of these actions in conjunction with the local smoke concentration, employing a straightforward model based on transition rates between these actions.

Kadakia et al. [36] expanded upon the analysis of Drosophila trajectories, constructing a more effective model that incorporated directional cues related to odor motion.

## 30.4 The Optimality Perspective of Olfactory Search

The preceding section aimed at providing a non-exhaustive overview of the extensive efforts, spanning several decades, in the pursuit of efficient solutions to the olfactory search problem. What these proposals share is the reliance on well-defined sequences of behaviors, crafted through intuition, ingenuity, and common sense. While their performance arises from the precise combination of these features, none of these methods claim to be, nor are they, inherently "optimal".

A distinct approach exists that frames olfactory search as a quest for an "optimal" behavior within a rigorously defined formal framework, and this approach constitutes the primary focus of this section. The quest for *optimal* search strategies can be marked by a departure from human intuition. In optimal strategies, the selection of actions is inherently balanced between *exploration*, aimed at reducing the uncertainty of source location, and *exploitation*, directed at closing in on the likely source position.

Given the stochastic nature of signaling and the uncertainty regarding source location, olfactory search naturally falls within the domain of Partially Observable Markov Decision Processes (POMDPs), as detailed in Sect. 30.4.2. Within the framework of POMDPs, we redefine the concept of spatial maps associated with "cognitive" approaches using probabilistic density maps referred to as *beliefs* representing the "perfect" encoding of past histories through Bayes' updates.

This section delves into methods that aim to (approximately) achieve optimality, as expounded in Sect. 30.4.4, while simultaneously addressing their strengths and limitations in the context of olfactory search. Intriguingly, the redefinition of the problem in terms of beliefs has served as a crucial starting point for the formulation and comprehension of several heuristic methods (Sect. 30.4.6). These heuristic methods, valued for their simplicity and practicality, continue to be some of the most widely applied techniques in olfactory search.

### 30.4.1 The Special Case of Zero Information

Before describing the general framework of POMDPs, we discuss one of the extreme ends of the olfactory search problem. This is the case where the odor is so rarefied that no new olfactory signal becomes available to the agent during the search. The task is then to find a hidden source which is distributed according to a probability density function $b(s)$, typically peaked at the initial position of the search.

In two or three dimensions, it is expected that the optimal strategy relies on exhaustive searches by non-self-crossing paths, such as outward-going spiraling. While it is known how heuristic strategies perform in this case, finding an (approximate) optimal solution even for 2D grid-worlds requires approximate methods (to be described in the next sections).

A special case, however, is the case of one-dimensional search, where this problem has been widely studied under the name *Linear Search Problem* (LSP) [37]. On a line, any search strategy effectively amounts to a sequence of turning points $x_0 = 0, x_1, \ldots, x_n, \ldots$. The searcher starts in 0, then moves straight to $x_1$, reverses direction and after crossing the origin reaches $x_2$, reverses again, and so on and so forth until the source is found. We can define the expected cost of a strategy as the space traveled to reach the source located in $s$. This strategy cost reads

$$C = s\mathbf{1}(x_0 \leq s < x_1) + (|s| + 2x_1)\mathbf{1}(x_2 \leq s < x_0) + (s + 2x_1 + 2|x_2|)\mathbf{1}(x_1 \leq s < x_0) + \cdots \tag{30.1}$$

where $\mathbf{1}(\cdot)$ is the indicator function, equal to one if the argument is true and zero otherwise, and we assumed that odd turning points are positive and even ones negative. If we define $B(x) = \int_{-\infty}^{x} b(s)\mathrm{d}s$, it can be shown by requiring $x_n$ to be a stationary point of the average cost, i.e., $\partial C/\partial x_n = 0$, that the turning points which minimize the strategy cost must satisfy the following recursion for jump lengths:

$$|x_{n+1} - x_n| = \frac{1 - |B(x_n) - B(x_{n-1})|}{b(x_n)}. \tag{30.2}$$

Notably, for a few selected functional forms for $b_0(s)$, an analytical solution to the recursion can be obtained, which produces the optimal search strategy. If the starting distribution falls as a power law $b(s) \sim s^{-\alpha}$, for example, the resulting sequence of turns is geometric $|x_n| \sim \rho^n$, where $\rho$ satisfies the relation

$$1 + \rho = \frac{1}{\alpha}\left(1 + \rho^{\alpha}\right). \tag{30.3}$$

For a general $b(s)$, the recursion cannot be solved in continuous space. On an infinite linear grid, however, it can always be solved up to any small approximation error [38]. This is possible since one can always truncate the distribution on a finite segment, leading to a controllable error, and then solve the bounded, finite-horizon task with standard methods of Dynamic Programming.

### 30.4.2 Partially Observable Markov Decision Processes

The olfactory search task falls in the wider framework of Partially Observable Markov Decision Processes (POMDPs). For a comprehensive discussion of POMDPs, we refer readers to [39]. In this section, we will introduce the minimal elements required to establish a language to address algorithms within the field of olfactory search.

In POMDPs, the task is composed by a state space $S$, the set of actions $A$ and observations $O$, and a reward function $R : S \times A \to \mathbb{R}$. The agent at each time step selects an action $a \in A$ to move from its state $s \in S$ according to a transition probability $Pr(s'|s, a)$, which defines the physical environment. As it moves, it perceives an observation $o \in O$ and receives a reward $R(s, a)$ (discounted in time by a factor $\gamma$). Since we are interested in minimizing the search time, a common reward function is to assign a positive reward of 1 at the end of the task, when the source is found.

The task is then defined as an optimization problem, where the aim of the agent is to construct a *policy*, i.e., the search strategy in our case, that maximizes the expected cumulative reward. In POMPDs, the searcher does not have a complete knowledge of its state $s$—which in the olfactory search case includes both the position of the agent and of the source. As a consequence, the policy in principle should map the whole history of past observations and actions $h_t = \{a_0, o_0, a_1, o_1, \ldots, a_t, o_t\}$ to the probability of selecting an action at time $t + 1$, i.e., $\pi(a|h_t)$. The quantity to maximize is the expected total reward $J$

$$J = \mathbb{E}[R_{tot}] = \mathbb{E}\left[\sum_{t=0}^{\infty} \gamma^t R(s_t, a_t)\right] = \mathbb{E}[\gamma^{T-1}], \tag{30.4}$$

where $T$ is the arrival time to the source. Explicitly constructing a policy on past histories is almost always unfeasible, since histories grow exponentially over time. To solve this issue, several methods are available to construct tractable object which can in general be considered *memories*, i.e., abstract representation of the past histories in a possibly reduced space. This can be obtained by exploiting the full knowledge of the observation model, as in Bayes' propagation of beliefs, but also in a model-free framework.

### 30.4.3 Beliefs and Cognitive Spatial Maps

A formal solution to the above problem of encoding past histories of observations is through Bayes' propagation of belief. The agent constructs and evolves a *belief* $b(s)$ which is a probability distribution over the state space $S$, here assumed to be discrete and finite. The belief lives in the (possibly very highly dimensional) simplex $b \in \Delta^{|S|-1}$ and is a sufficient statistic of the past history of observations and actions.

For navigation tasks, and in particular for olfactory search, the belief is a probabilistic spatial map that encodes the probability that the target is at a given location.

The policy $\pi(a|b)$ is a probability distribution over actions defined for any belief. Beliefs are updated according to Bayes' rule:

$$b_{o,a}(s') = \mathcal{T}^{o,a}(b) = \frac{Pr(o|s',a)\sum_s Pr(s'|s,a)b(s)}{\sum s, s' Pr(o|s',a)\sum_s Pr(s'|s,a)b(s)}. \tag{30.5}$$

Typically in the olfactory search literature the transitions—encoded in $Pr(s'|s,a)$—are taken to be deterministic, expressing the approximation that the searcher has perfect control of its own movements. While this is not a necessary assumption, it allows to clear the stage from secondary issues other than the sensory/memory aspects. Eq. (30.5) defines the operator of belief update $\mathcal{T}^{o,a}$. Note that for the agent to construct the evolution of belief it is necessary to know the observation model $P(o|s',a)$. Moreover, the agent must have an initial belief $b_0(s)$, whose initialization will influence the performance and behavior of the searcher.

A relevant quantity is the *value* $V_\pi(b)$ of a policy, which is the expected reward that an agent with belief $b$ will obtain following a policy $\pi$. In our case $V_\pi(b) = \mathbb{E}_{a\sim\pi}[\gamma^{T-1}|b_0 = b]$.

#### 30.4.3.1 Models of Plume

We have now to construct a model for the observation $P(o|s,a)$. To do so, we rely on results on concentration fields in the presence of a turbulent flow to produce a stationary probability distribution of odor encounters at any point in space. The advection-diffusion equation for the concentration field of an odor source with emission rate $R$ and a turbulent flow with mean wind $V\hat{y}$ reads

$$\partial_t c + V\partial_x c = D\nabla^2 c + R\delta(\mathbf{r}) - c/\tau, \tag{30.6}$$

where $c$ is the concentration field and $\tau$ the lifetime of the odor particle. Eq. (30.6) can be solved for its stationary solution in arbitrary dimensions. In particular, the solution in two dimensions is

$$c(\mathbf{r}) = \frac{R}{2\pi D} e^{\frac{-V(y-y_0)}{2D}} K_0\left(\frac{|\mathbf{r}-\mathbf{r}_0|}{\lambda}\right), \tag{30.7}$$

where $K_0$ is the zeroth-order modified Bessel function and $\lambda = \sqrt{D\tau/(1+\frac{V^2\tau}{4D})}$. From the mean concentration, one can obtain the average number of detections during a time $\Delta t$: $h(\mathbf{r}) = c(\mathbf{r})D\Delta t 4\pi l$, $l$ being the characteristic size of the searcher. In this case, therefore, we can extract the number of detections $o$ using a Poisson distribution with average number $h$: $P(o|s,a) = \frac{h^o \exp(-h)}{o!}$. While this is an obvious over-simplification, this model is very useful in model-based techniques.

The model above, however, can only reproduce the mean rate of detection as a function of the position. Clearly, it will neglect higher space- and time-correlations. Unfortunately, while a full understanding of these properties remains a challenging and fundamental problem in statistical fluid dynamics [6], correlations are increasingly seen as a key ingredient of many animal responses to odor cues. For example, it was shown that the *Drosophila* flies respond with an upwind motion only to *pulsed* carbon dioxide stimuli [40].

Nonetheless, the performance of searchers optimized in a stochastic environment can be put to the test against realistic turbulent flow obtained via numerical simulations, or via imaging of experimental experiments, such as smoke puffs in wind tunnels [35] or fluorescent dyes in water [41, 42] or in simple particle-based models for dynamical plumes in the presence of wind, as in [43] or [25].

### 30.4.4 Optimality and Bellman Equation

We define $V^*$ as the value function of the *optimal* policy $\pi^*$, which maximizes the expected total reward. $V^*$ satisfies the Bellman equation

$$V^*(b) = \max_{a \in A} \left[ \sum_s b(s) R(s, a) + \gamma \sum_{o \in O} Pr(o|b, a) V^*(b_{o,a}) \right] \qquad (30.8)$$

where $Pr(o|b, a)$ is the probability of observing $o$, given the distribution $b$. If the optimal value $V^*$ is known, the optimal policy can be easily obtained by $\pi^*(b) = \arg\max_a \left[ \sum_s b(s) R(s, a) + \gamma \sum_{o \in O} Pr(o|b, a) V^*(b_{o,a}) \right]$.

It will be sometimes more convenient to use a related quantity, the optimal *quality* function of the belief-action pair $Q^*(b, a)$. This function is related to the (optimal) value function by

$$Q^*(b, a) = \sum_s b(s) R(s, a) + \gamma \sum_{o \in O} P(o|b, a) V^*(b_{o,a}). \qquad (30.9)$$

$Q^*(b, a)$ is the optimal discounted expected return after taking action $a$ and subsequently following the optimal policy.

When the observation corresponds to the source location, the POMDP actually reduces to a fully observable MDP. This case is just the specific instance where the belief is concentrated in a single point $s$, $b(s) = \delta(s)$. Since there is no need for further exploration the optimal solution is in general trivial and the trajectory is given by the shortest path to the source. In the following, we will refer to this case with $V_{\mathrm{MDP}}(s) = V^*(b = \delta(s))$ and the optimal action as $\Pi_{\mathrm{MDP}}(s)$, following the convention in [44].

### 30.4.5 Approximate Solvers

The Bellman optimality equation, Eq. (30.8), can in principle be solved by an iterative process, which is guaranteed to converge on the optimal value function. Introducing the Bellman operator $\mathcal{B}(\cdot)$, defined on the space of the value functions of beliefs,

$$\mathcal{B}(V) = \max_{a \in A} \left[ \sum_{s} b(s) R(s, a) + \gamma \sum_{o \in O} Pr(o|b, a) V(b_{o,a}) \right], \tag{30.10}$$

one has that Bellman operators are contractions, so that starting from any initial guess $V_0(b)$, one could iteratively construct $V_1(b) = \mathcal{B}(V_0)$, $V_2(b) = \mathcal{B}(V_1)$, and so on, that must converge to a fixed point. This *value iteration* process is of practical use when the state space is discrete and relatively small. The high dimensionality and the continuity of the belief space of a POMDP, however, make the update step over all the belief space intractable even for the smallest state spaces $S$. In the following section, we will describe two different approaches, where this "dimensionality curse" is tackled either by point-based methods in belief space or by a functional approximation via Artificial Neural Networks (ANN).

#### 30.4.5.1 DeepRL Solver with Bellman Error Minimizations

The Bellman equation, Eq. (30.8), holds for the optimal value function $V^*$. For any other function of the belief, the l.h.r and r.h.s. of the equation will have a mismatch called Bellman Error. Recently [45] proposed to find the optimal policy using value functions parametrized by a Neural Network with weights $w$, $V(b; w)$, and rephrasing the optimization task as finding the $w^*$ that minimize of the residual Bellman Error. This error reads

$$\mathcal{L}(w) = \mathbb{E}_b \left[ \max_{a \in A} \left( \sum_{s} b(s) R(s, a) + \gamma \sum_{o \in O} Pr(o|b, a) V(b_{o,a}; w) \right) - V(b; w) \right]^2 . \tag{30.11}$$

In principle, the expectation could be taken over the whole space of beliefs, since the residual error of the true optimal value function $V^*(b)$ is zero everywhere. However, $V(b; w)$ is an approximation and the expectation in [46] runs over the beliefs $b$ effectively visited by the searcher. The optimization of $w$ is done using the functional $\mathcal{L}(w)$ as a "loss function" to be minimized through standard stochastic gradient descent, via back-propagation methods.

This method was first proposed in [45] and then further exploited in [46] to solve the case of stochastic plumes in $1D$ and $2D$. The results showed that it converges to solutions outperforming other heuristic methods and is competitive with other methods to reach optimized solutions [47].

#### 30.4.5.2 Approximate Value Iteration with Perseus

The Perseus algorithm [48] is based on *point-based value iteration*, where the value function is parameterized by a finite number of vectors in the belief space. At each iteration step $n$, Perseus performs value iteration on a large sample of beliefs, using a finite set of vectors $\mathcal{A}^n$. For each belief, it approximates the value function as a piecewise linear, convex function:

$$V(b) = \max_{\alpha \in \mathcal{A}^n} b \cdot \alpha, \tag{30.12}$$

where $\cdot$ indicates the dot product over all states $s$. The approximation is then plugged into the iteration—i.e., in the last term of Eq. (30.8)—where it can be written in terms of *propagation* vectors $g^i_{a,o} = \sum_{s' \in S} Pr(o|s', a) Pr(s'|s, a) \alpha_i(s')$. The vectors $g^i_{a,o}$ encode the evolution of the belief as a consequence of action $a$ and observation $o$ and need to be computed and stored only once, after constructing the belief set. Each iteration can then be shown to consist in *backup* operations where the vectors $\alpha$ are updated,

$$\alpha'_a = R(s, a) + \gamma \sum_o \arg\max_{g^i_{a,o}} b \cdot g^i_{a,o}. \tag{30.13}$$

Since each $\alpha$-vector is associated to an action, each belief is mapped into the action of its corresponding maximizing $\alpha$. The order of the backup operation matters, and it has been found that performing it in order of decreasing Bellman error offers a better convergence rate [47].

The Perseus algorithm was applied to solve the olfactory search task of a odor plume in a turbulent flow [47], where its efficiency is compared to several other heuristics, as discussed in the following sections. Rigolli et al. [49] utilized the same algorithm to study the *alternation* model of olfactory search. There, the search process involves two different modes of odor sampling, mimicking the rodent behavior that alternates between sniffing on the ground while moving, or stopping to sniff the air. Since odor puffs travel differently close to the surface, where flow is laminar, than farther in the air the agent essentially can probe the environment through two different communication channels (in POMDP terms, two different functions $Pr_x(o|s, a)$, where $x \in \{\text{ground, air}\}$).

#### 30.4.5.3 Parameterizing the Belief Space

A different approach to tackle the curse of dimensionality of the belief space was used by [50] for the task of tracking a surface-bound odor trail. There, the posterior is propagated through standard Bayes updates, but it is parameterized by a mixture of $k$ Gaussian basis function. Using this reduced basis, the authors are able to use the SARSA Reinforcement Learning algorithm to learn the quality-function $Q_\pi$ and successfully reproduce the characteristic crisscrossing patterns of trail searching.

### 30.4.6 *Bayesian Heuristics*

The complexity of solving the Bellman optimality equation for POMDPs has spurred efforts to produce effective algorithms based on spatial beliefs, without necessarily solving any optimization task. These methods are generally called "heuristics" in the language of POMDPs (not to be confused with the usage of the same term in robotics).

Some of the following methods can also be recast in terms of approximation of the value- or quality-functions $V(b)$, or the quality-function $Q(b, a)$, in terms of their equivalents for the case of full observability, i.e., $V_{\mathrm{MDP}}(s)$ and $Q_{\mathrm{MDP}}(a, s)$ and the relative optimal solution $\Pi_{\mathrm{MDP}}(s)$.

#### 30.4.6.1 Simple Heuristics

A selection of the simpler heuristics can be found in [44], and their performance in a virtual environment (and against more advanced methods) can be evaluated using the codes freely available in the OTTO package [51].

The simplest method, called *Most Likely State* (MLS) [52], prescribes that the action to be taken given a belief $b$ is simply to move in the direction where the probability to find the source is highest. In other terms $Q(b, a)$ is replaced by $Q_{\mathrm{MDP}}(\arg\max_s b(s), a)$. A common variation relies on Thompson sampling, which relaxes the requirement of moving toward the most-likely state only. In Thompson sampling, the agent samples a position for the source with probability given by $b$ and then selects the optimal action for that location. This is equivalent to selecting actions randomly following the expected optimal actions $\pi_{\mathrm{Thompson}}(a|b) = \sum_s b(s)\mathbf{1}\,(a = \Pi_{\mathrm{MDP}}(s))$.

Other methods include Action Voting—which is equivalent to selecting the most probable action as defined for the expected-Thompson algorithm—and QMDP, where the policy is constructed assigning each action to the corresponding average $Q_{\mathrm{MDP}}$ value, weighted by the belief: $\pi_{\mathrm{QMDP}}(b) = \arg\max_a \sum_s b(s) Q_{\mathrm{MDP}}(s, a)$.

While relatively efficient at a very low computational cost, all these methods are essentially exploitative since they do not consider, neither explicitly nor implicitly, the benefit of performing actions to reduce uncertainty. Moreover, they are known to suffer from a sort of "Buridan's ass syndrome" when the belief landscape has multiple maxima and the agent is equidistant to many of them. Since observations are stochastic, the relative intensity of the maxima can fluctuate rapidly in time and the resulting back-and-forth motion can lead to severe performance loss. A possible solution to this is to permit time-persistent action-selection [53], at the cost of introducing additional hyper-parameters to the algorithm.

#### 30.4.6.2 Information-Seeking Methods

As mentioned, the previous methods do not correctly address the exploration part of the delicate exploration/exploitation balance required for optimality. In this section, we will deal with information-seeking methods, which introduce an explicit push for exploration.

Several heuristics have been proposed which consider the (expected) effect of actions on the future beliefs. In particular, this can be achieved by rewarding the agent with the minimization of the expected entropy of the belief $b(s)$ at each step. While this method was first proposed as the "uncertainty reducing" mode in the Dual mode control [44, 52], it is most commonly referred to by the name "Infotaxis" after its more recent application to the olfactory task in a turbulent medium [54].

The Infotaxis heuristic is based on taking actions that minimize the uncertainty on the source location. Such uncertainty is measured by the Shannon entropy of its distribution, i.e., of the belief, as $H(b) = -\sum_s b(s) \log_2 b(s)$. At each time, Infotaxis selects the action that maximizes the *expected reduction of the belief's entropy* $G(b, a)$:

$$G(b, a) = H(b) - \sum_o Pr(o|b, a) H(b_{o,a}). \tag{30.14}$$

Contrary to the previous methods, Infotaxis is biased toward exploration against exploitation. In particular, it has no incentive to move once it has removed all uncertainty over its position and tends to err on the side of curiosity-driven behavior. The Dual mode control [44, 52] counteracts this deficiency by imposing a mode switch: above a certain threshold on the entropy, actions are selected to minimize uncertainty, whereas below it the behavior switches to any of the exploitative heuristics, such as Thompson sampling, QMDP, or Action Voting.

A recent variant, called Space-Aware Infotaxis [45], addresses this issue differently, reformulating the quantity to minimize. The new objective is

$$G_{SAI}(b, a) = -\sum_o Pr(o|b, a) \left( \mathrm{e}^{H(b_{o,a})} + D(b_{o,a}) \right), \tag{30.15}$$

where $D(b) = \sum_s b(s)||x - x_a||$ and the state $s = (x, x_a)$ contains the positions of the searcher $x$ and of the source $x_a$. The second term measures the expected distance between the agent and the source, given $b$. Its minimization, i.e., the exploitation of the current belief $b$, counter-balances the information-gathering aspect of Infotaxis. Note that the exponentiation of the Shannon entropy, without the addition of the distance term, in principle does not modify the policy of Infotaxis since any ordering in the actions is preserved by it. However, it allows for a more direct interpretation, since it can be shown that $\mathrm{e}^{H(b)}$ is strictly related to the minimum expected time needed to explore a distribution $b$ when arbitrary jumps are allowed.

The uncertainty about the position of the source is not the only one that can be addressed. In their original paper [52] stipulate that there is a "benign entropy", where the uncertainty is between states which however share the same (MDP) optimal action. They therefore propose a Dual Mode algorithm based on the entropy of action, instead of state. Given $\phi(a) = \sum_s b(s)\mathbf{1}\,(a = \Pi_{\mathrm{MDP}}(s))$ to measure the relative probability to be in a state $s$ where the optimal action is $a$, the algorithm acts to minimize the entropy $H(\phi)$.

Yet another variant, called Entrotaxis, uses the entropy on "predictive measurement distribution as opposed to the entropy of the expected posterior" [55]. A similar approach to SAI, albeit with quite a different formulation, comes from [56], which modifies the inference process to reduce the requirements for "space-processing capabilities" and proposes a "mapless search scheme". The searcher keeps in memory only a finite number $M$ of detections alongside their previous position, which are used to construct a posteriori probabilities of the source location via Gaussian mixtures. The searcher then moves accordingly to a target which combines the entropy of source distribution with an explicit term which favors exploitation.

### *30.4.7 Comparison Between Heuristics and Optimal Strategies*

Most of the strategies discussed so far, both derived by purely heuristic considerations and from (approximate) optimization, have a long history in robotics and in the olfactory search literature.

A general, qualitative comparison can already be drawn by looking at the typical shapes of trajectories during the search processes. The simplest heuristics were hard-wired to produce behavioral patterns like surges and casts, mimicking natural behavior. The choice of these features, handpicked by human intuition, seems to have been validated by the following research. In a more complex form, the alternation of casts and surges is still recognizable at all levels of algorithm complexity. Infotaxis, and its variants, produce it, as do simpler heuristics such as Thompson sampling or QMDP. Moreover, they are clearly present for optimal strategies.

While this qualitative comparison has always been the subject of discussion, a consistent, comparative performance was only recently tackled. Loisy and Eloy [45], where the SAI heuristic was proposed, demonstrated its superior performance with respect to "Vanilla" Infotaxis, a result further confirmed in [47]. In the latter work, the resulting strategies from Perseus and several heuristics were tested: while Perseus outperformed all heuristics, the much simpler QMDP and Thompson sampling were relatively efficient at large and small emission rates, respectively. Different approaches to solve the POMDPs were also compared in [46].

## 30.5 Perspectives

Belief update allows for the complete encoding of the past odor history into the spatial maps $b$. However, it requires the knowledge of the observation model, which is not always accessible. Moreover, the biological plausibility that an animal can perform Bayesian computations is a subject of debate. Interestingly, it is possible to construct and/or optimize different mappings of past histories into coded memories. Given that the agent has the specific goal of searching a task, it may very well be that the relevant information from the past history can be compressed in a much lower-dimensional space than the belief.

One can imagine to replace the belief $b$ and its update operator $\mathcal{T}^{o,a}$—otherwise uniquely determined by Bayes' rule—with a different encoding by an abstract vector $\mu$ and a "memory-update operator" $\mu' = \mathcal{T}_{\omega}^{a,o}\mu$. As in the previous case, the goal of the learning will be to produce an optimal policy, which now depends on the instantaneous observation and on the current memory state $\pi(a|\mu, o)$.

It is still possible to exploit Reinforcement Learning techniques in POMDPs on these "memory"-spaces, even though typically these involve the Policy Gradient theorem and not explicitly the solution of the Bellman equation. The optimization problem is now effectively divided into two inextricably joint optimization problems, the $\theta$ parameters of the action-selection $\pi(a|o, \mu; \theta)$ and the $\omega$ parameters for the memory-propagation $\mathcal{T}_{\omega}^{a,o}$. How to construct and update such memories is still an open problem. Possible approaches were outlined in [58] and more recently in [57]. Here, we will showcase two different approaches which have been recently applied to the olfactory search problem.

### *30.5.1 Recurrent Neural Networks as Memory Models*

An approach to construct a memory-based process is to use Recurrent Neural Networks (RNN) to encode the memory transition $\mathcal{T}_{\omega}^{a,o}$. Within this method $\mu$ is a vector in an abstract $\mathcal{R}^d$ space, where generally $d \ll |S|$. Singh et al. [43] propose a method to optimize the movement of an agent in an odor plume transported in different turbulent-like conditions of wind, in a model-free context. Analyzing the features of the internal representation $\mu$, they recognize the emergence of diverse behavioral "modes" which closely resemble a diagram-like behavior. A similar approach was also adopted in robotics, where [59] used a Long/short-term memory RNN (LSTM) in combination with a Deterministic Policy Gradient algorithm to optimize the movement of an autonomous underwater vehicle in a deep-sea turbulent environment.

### 30.5.2 Finite-State Controller as Minimal Memories

Recently [60] proposed to encode past observations in a discrete set of "memories", using a Finite-State Controller (FSC) [61]. In this method, the space $S$ is "multiplied" by the number $M$ of memories. At each time the agent is in a spatial position $s$ and in a "memory" state $\mu \in \{1, 2, \ldots, M\}$. The policy, however, depends only on the current memory and the instantaneous observation. The optimization process requires to optimize both the policy $\pi(a|\mu, o; \theta)$ *and* the memory transition $\mu' = \mathcal{T}_{\omega}^{a,o}\mu$. The resulting policy is clearly sub-optimal with respect to policies which depend on the full belief—after all, the belief space is much larger than a finite, discrete set—but it is shown to be robust and very efficient given its very limited computational burden. The optimization process is achieved using the model-based version of the Policy Gradient theorem but it could be in principle be obtained also in a model-*free* setting, where the gradient is stochastically evaluated by experience only, making it relevant for applications in real turbulent flows.

## 30.6 Conclusions

Olfactory search is the prototypical example of a POMDP that is very hard to solve. The difficulty arises from the large size of the unobservable state space, i.e., the set of all possible locations of the odor source, and from the paucity of information cues that are available to the searcher when it is far away from the target.

In this chapter, we aimed at offering a principled review of some algorithmic approaches to olfactory search. The scope of the methods here described goes well beyond this specific example and naturally extends to more general search processes, including those discussed in this book.

## References

1. T.D. Wyatt, *Pheromones and Animal Behavior* (Cambridge University Press, Cambridge, England, 2003)
2. R.T. Cardé, M.A. Willis, Navigational strategies used by insects to find distant, wind-borne sources of odor. J. Chem. Ecol. **34**, 854–866 (2008)
3. K.L. Baker, M. Dickinson, T.M. Findley, D.H. Gire, M. Louis, M.P. Suver, J.V. Verhagen, K.I. Nagel, M.C. Smear, Algorithms for olfactory search across species. J. Neurosci. **38**(44), 9383–9389 (2018)
4. G. Reddy, V.N. Murthy, M. Vergassola, Olfactory sensing and navigation in turbulent environments. Annu. Rev. Conden. Ma. P. **13**, 191–213 (2022)
5. J. Murlis, J.S. Elkinton, R.T. Carde, Odor plumes and how insects use them. Annu. Rev. Entomol. **37**(1), 505–532 (1992)
6. A. Celani, E. Villermaux, M. Vergassola, Odor landscapes in turbulent environments. Phys. Rev. X **4**(4), 041015 (2014)

7. G. Kowadlo, R.A. Russell, Robot odor localization, a taxonomy and survey. Int. J. Robot. Res. **27**(8), 869–894 (2008)
8. B. Ristic, A. Skvortsov, A. Gunatilaka, A study of cognitive strategies for an autonomous search. Inform. Fusion **28**, 1–9 (2016)
9. T. Jing, Q.H. Meng, H. Ishida, Recent progress and trend of robot odor source localization. IEEJ Trans. Electr. Electron. Eng. **16**(7), 938–953 (2021)
10. C. Champagne, Search theory, agent-based simulation, and u-boats in the Bay of Biscay, in *Proceedings of the Winter Simulation Conference, (WSC'03)*, vol. 1, pp. 991–998. IEEE (2003)
11. G.M. Viswanathan, V. Afanasyev, S.V. Buldyrev, E.J. Murphy, P.A. Prince, H.E. Stanley, Lévy flight search patterns of wandering albatrosses. Nature **381**(6581), 413–415 (1996)
12. M.A. Lomholt, K. Tal, R. Metzler, J. Klafter, Lévy strategies in intermittent search processes are advantageous. Proc. Natl. Acad. Sci. U.S.A. **105**(32), 11055–11059 (2008)
13. V.V. Palyulin, A.V. Checkin, R. Metzler, Lévy flights do not always optimize random blind search for sparse targets. Proc. Natl. Acad. Sci. U.S.A. **111**(8), 2931–2936 (2014)
14. N. Levernier, J. Textor, O. Bénichou, R. Voituriez, Inverse square Lévy walks are not optimal search strategies for d $\geq$ 2. Phys. Rev. Lett. **124**(8), 080601 (2020)
15. O. Bénichou, C. Loverdo, M. Moreau, R. Voituriez, Intermittent search strategies. Rev. Mod. Phys. **83**(1), 81 (2011)
16. N. Wadhwa, H.C. Berg, Bacterial motility, machinery and mechanisms. Nat. Rev. Microbiol. **20**(3), 161–173 (2022)
17. O. Holland, C. Melhuish, Some adaptive movements of animats with single symmetrical sensors. From Animals to Animats **4**(6), 55–64 (1996)
18. M. Sabelis, P. Schippers, Variable wind directions and anemotactic strategies of searching for an odour plume. Oecologia **63**, 225–228 (1984)
19. D. Martinez, O. Rochel, E. Hugues, A biomimetic robot for tracking specific odors in turbulent plumes. Auton. Robots **20**, 185–195 (2006)
20. J. Kennedy, Zigzagging and casting as a programmed response to wind-borne odour: a review. Phys. Entomol. **8**(2), 109–120 (1983)
21. K. Ryohei, S. Naoko, S. Tatsuaki, Self-generated zigzag turning of bombyx mori males during pheromone-mediated upwind walking. Zool. Sci. **9**(3), 515–527 (1992)
22. L. Kuenen, R.T. Carde, Strategies for recontacting a lost pheromone plume, casting and upwind flight in the male gypsy moth. Phys. Entomol. **19**(1), 15–29 (1994)
23. E. Kramer, A tentative intercausal nexus and its computer model on insect orientation in windborne pheromone plumes, in *Insect Pheromone Research. New Directions*, pp. 232–247. Springer (1997)
24. J.H. Belanger, M.A. Willis, Biologically-inspired search algorithms for locating unseen odor sources, in *Proceedings of the IEEE Symposium on Intelligent Control (ISIC'98) and IEEE Symposium on Computational Intelligence in Robotics and Automation (CIRA'98)*, pp. 265–270. IEEE (1998)
25. E. Balkovsky, B.I. Shraiman, Olfactory search at high Reynolds number. Proc. Natl. Acad. Sci. U.S.A. **99**(20), 12589–12593 (2002)
26. S. Shigaki, T. Sakurai, N. Ando, D. Kurabayashi, R. Kanzaki, Time-varying moth-inspired algorithm for chemical plume tracing in turbulent environment. IEEE Robot. Autom. Lett. **3**(1), 76–83 (2017)
27. N. Voges, A. Chaffiol, P. Lucas, D. Martinez, Reactive searching and infotaxis in odor source localization. PLoS Comput. Biol. **10**(10), e1003861 (2014)
28. R.A. Russell, Tracking chemical plumes in constrained environments. Robotica **19**(4), 451–458 (2001)
29. H. Ishida, T. Nakamoto, T. Moriizumi, Remote sensing and localization of gas/odor source and distribution using mobile sensing system, in *Proceedings of International Solid State Sensors and Actuators Conference (Transducers' 97)*, vol. 1, pp. 559–562. IEEE (1997)
30. J.A. Farrell, S. Pang, W. Li, R. Arrieta, Chemical plume tracing experimental results with a REMUS AUV, in *Oceans 2003*. (IEEE Cat. No. 03CH37492), vol. 2, pp. 962–968. IEEE (2003)

31. S. Pang, J.A. Farrell, Chemical plume source localization. IEEE T. Syst. Man. Cy. B **36**(5), 1068–1080 (2006)
32. J.G. Li, Q.H. Meng, Y. Wang, M. Zeng, Odor source localization using a mobile robot in outdoor airflow environments with a particle filter algorithm. Auton. Robots **30**, 281–292 (2011)
33. E. Álvarez-Salvado, A.M. Licata, E.G. Connor, M.K. McHugh, B.M. King, N. Stavropoulos, J.D. Victor, J.P. Crimaldi, K.I. Nagel, Elementary sensory-motor transformations underlying olfactory navigation in walking fruit-flies. Elife **7**, e37815 (2018)
34. A.M. Matheson, A.J. Lanz, A.M. Medina, A.M. Licata, T.A. Currier, M.H. Syed, K.I. Nagel, A neural circuit for wind-guided olfactory navigation. Nat. Comm. **13**(1), 4613 (2022)
35. M. Demir, N. Kadakia, H.D. Anderson, D.A. Clark, T. Emonet, Walking drosophila navigate complex plumes using stochastic decisions biased by the timing of odor encounters. Elife **9**, e57524 (2020)
36. N. Kadakia, M. Demir, B.T. Michaelis, B.D. DeAngelis, M.A. Reidenbach, D.A. Clark, T. Emonet, Odour motion sensing enhances navigation of complex plumes. Nature **611**(7937), 754–761 (2022)
37. S. Alpern, S. Gal, *The Theory of Search Games and Rendezvous*, vol. 55. (Springer, 2006)
38. M.Y. Kao, M.L. Littman, Algorithms for informed cows, in *Proceedings of the AAAI 1997 Workshop on Online Search*, vol. 1997 (1997)
39. L.P. Kaelbling, M.L. Littman, A.R. Cassandra, Planning and acting in partially observable stochastic domains. Artif. Intell. **101**(1–2), 99–134 (1998)
40. D. Zocchi, E.S. Ye, V. Hauser, T.F. O'Connell, E.J. Hong, Parallel encoding of CO2 in attractive and aversive glomeruli by selective lateral signaling between olfactory afferents. Curr. Biol. **32**(19), 4225-4239.e7 (2022)
41. J. Duplat, C. Innocenti, E. Villermaux, A nonsequential turbulent mixing process. Phys. Fluids **22**(3) (2010)
42. M. Kree, J. Duplat, E. Villermaux, The mixing of distant sources. Phys. Fluids **25**(9) (2013)
43. S.H. Singh, F. van Breugel, R.P. Rao, B.W. Brunton, Emergent behaviour and neural dynamics in artificial agents tracking odour plumes. Nat. Mach. Intell. **5**(1), 58–70 (2023)
44. J.L. Fernández, R. Sanz, R.G. Simmons, A.R. Diéguez, Heuristic anytime approaches to stochastic decision processes. J. Heuristics **12**, 181–209 (2006)
45. A. Loisy, C. Eloy, Searching for a source without gradients, how good is infotaxis and how to beat it. Proc. R. Soc. A. **478** (2022)
46. A. Loisy, R.A. Heinonen, Deep reinforcement learning for the olfactory search POMDP, a quantitative benchmark. Eur. Phys. J. E **46**(3), 17 (2023)
47. R.A. Heinonen, L. Biferale, A. Celani, M. Vergassola, Optimal policies for Bayesian olfactory search in turbulent flows. Phys. Rev. E **107**(5), 055105 (2023)
48. M.T. Spaan, N. Vlassis, Perseus, Randomized point-based value iteration for POMDPs. J. Artif. Intell. **24**, 195–220 (2005)
49. N. Rigolli, G. Reddy, A. Seminara, M. Vergassola, Alternation emerges as a multi-modal strategy for turbulent odor navigation. eLife **11**, e76989 (2022). https://doi.org/10.7554/eLife.76989
50. G. Reddy, B.I. Shraiman, M. Vergassola, Sector search strategies for odor trail tracking. Proc. Natl. Acad. Sci. U.S.A. **119**(1), e2107431118 (2022)
51. A. Loisy, C. Eloy, OTTO: a Python package to simulate, solve and visualize the source-tracking pomdp. J. Open Source Softw. **7**(74), 4266 (2022)
52. A.R. Cassandra, L.P. Kaelbling, J.A. Kurien, Acting under uncertainty, Discrete Bayesian models for mobile-robot navigation, in *Proceedings of IEEE/RSJ Conference on Intelligent Robots and Systems (IROS'96)*, vol. 2, pp. 963–972. IEEE (1996)
53. D.J. Russo, B. Van Roy, A. Kazerouni, I. Osband, Z. Wen et al., A tutorial on Thompson sampling. Found. Trends Mach. Learn. **11**(1), 1–96 (2018)
54. M. Vergassola, E. Villermaux, B.I. Shraiman, 'infotaxis' as a strategy for searching without gradients. Nature **445**, 406–409 (2007)
55. M. Hutchinson, H. Oh, W.H. Chen, Entrotaxis6 as a strategy for autonomous search and source reconstruction in turbulent conditions. Inform. Fusion **42**, 179–189 (2018)

56. J.B. Masson, Olfactory searches with limited space perception. Proc. Natl. Acad. Sci. U.S.A. **110**(28), 11261–11266 (2013)
57. T. Tottori, T.J. Kobayashi, Memory-limited partially observable stochastic control and its mean-field control approach. Entropy **24**(11) (2022). https://doi.org/10.3390/e24111599
58. N. Meuleau, L. Peshkin, K.E. Kim, L.P. Kaelbling, Learning finite-state controllers for partially observable environments. arXiv preprint arXiv:1301.6721 (2013)
59. H. Hu, S. Song, C.P. Chen, Plume tracing via model-free reinforcement learning method. IEEE T. Neur. Net. Lear. **30**(8), 2515–2527 (2019)
60. K.V.B. Verano, E. Panizon, A. Celani, Olfactory search with finite-state controllers. Proc. Natl. Acad. Sci. U.S.A. **120**(34), e2304230120 (2023)
61. E.A. Hansen, Solving POMDPs by searching in policy space. arXiv preprint arXiv:1301.7380 (2013)

# Chapter 31
# From Features to Foraging: Current Developments in Visual Search by Human Observers

**Todd S. Horowitz**

**Abstract** Visual search is a ubiquitous behavior among sighted organisms. The study of visual search in humans has a rich history in cognitive psychology and allied fields going back over a century. Visual search paradigms have been used to probe the nature of the visual system for decades. However, these studies were predominantly characterized by a single known target appearing with 50% prevalence. Partly driven by the scientific challenge of doing work that is both theoretically relevant and applicable to real world search challenges such as medical image perception and airport baggage screening, researchers have been expanding the parameters of search experiments. Key developments in the field include the study of the effects of prevalence, individual differences, and "hybrid search" for multiple potential targets. We are starting to consider visual search as a branch of foraging, opening up the possibility for collaboration with population ecology and other disciplines that study search.

## 31.1 Introduction

In this volume, "the Target Problem" is addressed from a variety of different perspectives. The problem is conceived in terms of "agents" searching for "targets". The agents can be particles, molecules, drugs, or predators; the targets can be DNA binding sites, organs, or prey animals. In this chapter, I will discuss human agents searching for a variety of visual targets, from letters of the alphabet and simple shapes to fruit to tumors. In cognitive psychology and allied disciplines, this topic is known as "visual search".

---

T. S. Horowitz (✉)
National Cancer Institute, National Institutes of Health, 9609 Medical Center Drive, Rockville, MD 20850, USA
e-mail: todd.horowitz@nih.gov



D. Grebenkov et al. (eds.), *Target Search Problems*,
https://doi.org/10.1007/978-3-031-67802-8_31

### *31.1.1 The Search for Features*

While investigations of visual search date back to at least the work of Kingsley [1], most contemporary research can be traced to three papers published nearly a half-century ago: a pair of papers by Schneider and Shiffrin [2, 3], presenting a set of experiments and rigorous modeling outlining a distinction between controlled and automatic modes of search, and Treisman and Gelade's seminal paper presenting the influential Feature Integration Theory (FIT, [4]). Treisman and Gelade's paper, in particular, set the tone for the next two decades of work on visual search. FIT was designed to reconcile functional specialization [5], the discovery that the visual cortex is organized into spatially separated populations of neurons that specialized in processing basic visual features, such as color, orientation, or motion, with the unitary nature of perceptual experience. Treisman's insight was that attention could serve as a unifying process: features were initially processed separately, but bound into a perceptual object by the act of focusing attention at a location in perceptual space. This theory had very specific implications for visual search: targets that were defined by these basic visual features should be detected easily, while finding targets that were defined by combinations of features should be difficult. More specifically, *feature* targets (e.g., the only red item in an array of green items) should be found by parallel processing (corresponding to Schneider and Shiffrin's automatic mode), meaning that the time to detect such a target would be independent of the number of items to search through. In contrast, finding a *conjunction* target (e.g., a red vertical bar among green vertical and red horizontal bars) would require attending to each item serially in order to conjoin the features (corresponding to Schneider and Shiffrin's controlled mode), leading to search times that increased linearly with the number of items to process.

Using visual search performance as a diagnostic for basic visual features became almost a cottage industry following the publication of FIT. The criterion that the time to find the target should be independent of the number of items is widely used, but should be viewed as necessary but not sufficient for feature status; Treisman proposed a number of converging criteria [6, 7]. For a review of the cumulative findings of this enterprise, see Wolfe and Horowitz [8]. Briefly, while the visual system is a complex and sophisticated apparatus capable of detecting and identifying a rich array of stimuli, our "search engine" is much more limited than its internet counterparts. We cannot search for any arbitrary visual property. Instead, search is limited to a small number (perhaps two dozen) of "guiding features" that can capture or direct attention.

While I do not mean to imply that the scientific debates around basic features and the nature of visual search have been settled [9–11], the intent of the rest of this chapter is to place some guideposts for new directions in the study of visual search.

### 31.1.2 Methods and Terminology

Since many readers of this volume may not be familiar with vision science methods, I will provide a concise introduction to a "classic" visual search experiment here, along with relevant terminology. A useful example is the work of Palmer et al. [12], which was designed to provide benchmark behavioral data for three types of visual search: feature search, conjunction search, and *spatial configuration* search (see below).

In these experiments, the *observer* (our agent, a human experimental subject, also called a *participant*) would sit in front of a computer screen and see a randomly generated display resembling the images in Fig. 31.1. The observer would be instructed to press one key on the keyboard if they saw a target, and another if they did not see a target; this is the *task*. Observers were instructed to respond as quickly and accurately as possible. The time from display onset to response is the *response time*. Each instance of a display being presented and a response recorded is termed a *trial*. Observers in these experiments performed 4000 trials in each of the three experimental *conditions* (i.e., feature search, conjunction search, and spatial configuration search). Targets were presented on 50% of trials. The number of items on the screen, termed the visual *set size*, was chosen from the set (3, 6, 12, 18). So, for example, a *target-present trial* with a set size of 3 would have one target and two *distractors* (non-target items), and a *target-absent trial* of set size 3 would have three distractors.

As noted above, feature search is when the target is distinguished by the presence of a unique feature. Consider the top left panel of Fig. 31.1. Here the target is a red vertical bar, presented in an array of green vertical bars. The non-target items (green verticals) are termed *distractors*. You can probably find the target easily; indeed you likely noticed the target before even being told that it was the target. This phenomenon is known as "pop-out", because the subjective impression is that the target effortlessly pops out of the display. Objectively, this is typically demonstrated by plotting the response time by set size function (Fig. 31.1, lower panel, blue triangles). For a true feature search, this function is a theoretically flat line,[1] indicating that processing time is independent of the set size.

In conjunction search, the target is defined by a combination of features present separately in the distractors. So in the top middle panel of Fig. 31.1, the target is again a red vertical bar, but now the distractors are green vertical bars and red horizontal bars. Here the reaction time by set size function for target-present trials has a positive slope (the "search slope") of 9.2 ms/item.

Spatial configuration search is a special kind of conjunction search in which targets differ from distractors only in the relative spatial arrangement of their components. Here, for example, the target "2" and the distractor "5"s are mirror images of each other. Note the steeper search slope of 42.1 ms/item. Indeed, spatial configuration searches were called "serial searches" for a long time because they reliably produce

[1] Recent studies have demonstrated a shallow logarithmic curve, indicative of exhaustive parallel processing, see [13].

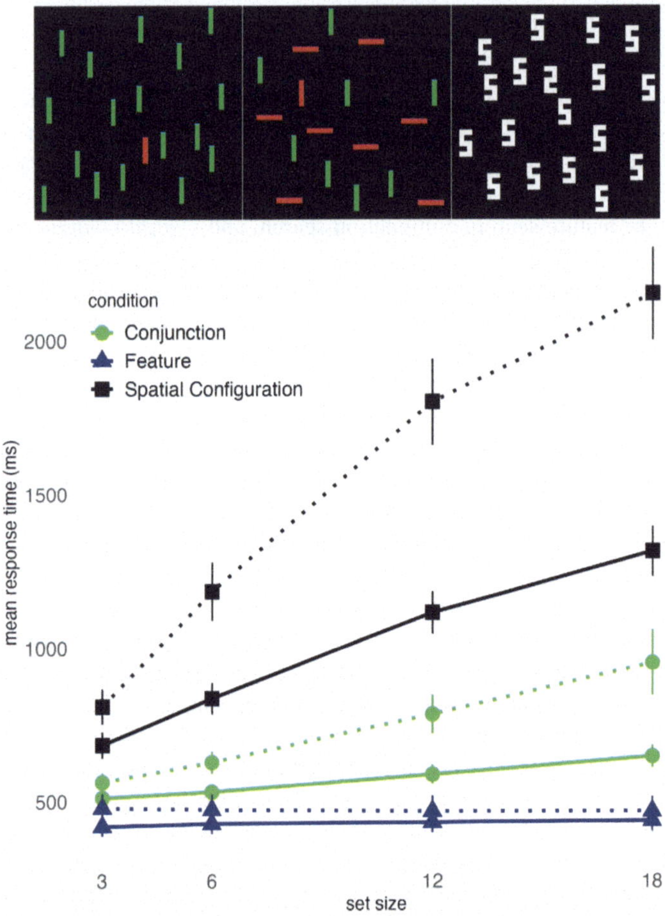


**Fig. 31.1** Top panels present illustrations of three classic visual search tasks. On the left, the target was a red vertical rectangle among green vertical rectangles (feature search). In the middle, the target was a red vertical rectangle among green vertical and red horizontal rectangles (conjunction search). On the right, the target was a digital 2 among digital 5s (spatial configuration search). Bottom panel: Typical response time data as a function of visual set size for correct trials, data replotted from [12] (Data downloaded from https://search.bwh.harvard.edu/new/data_set_files.html). Blue triangles denote feature search, green circles denote conjunction search, and black squares denote spatial configuration search. Solid lines indicate target-present data, dotted lines target absent data. Error bars represent the standard error of the mean

steep slopes and other signatures of serial search (e.g., [14, 15]), but that term is theoretically loaded so I prefer the more descriptive "spatial configuration" [16].

Target-absent slopes are almost always steeper than target-present slopes (unless target prevalence is high, see Sect. 31.2.1). The ratio of target-absent to target-present slopes is typically described as being around 2:1 for conjunction and spatial configuration searches, consistent with the assumption of sampling without replacement [17]; however, empirical slope ratios tend to be significantly greater than 2 [16].

In the experiments described above, observers are typically asked to respond "as quickly and accurately as possible," because the investigator is interested in how fast the task can be correctly performed. However, human performance is constrained by the speed-accuracy tradeoff function (SATF, [18]). In these studies, we generally assume that observers are operating at near-asymptotic accuracy, so that all of the scientifically relevant variability occurs in the response time (speed) domain. Nevertheless, observers typically make a non-trivial number of errors. These errors tend to covary with response time, so observers may be operating at different points along the SATF, leading to underestimates of response time in more difficult conditions.

Other experiments operate in the accuracy domain, limiting exposure to the stimulus but giving observers as much time as they want to respond. Much of this work [19, 20] is underpinned by Signal Detection Theory [21], which allows researchers to transform raw accuracy measures into estimates of discriminability (the ability to tell targets and distractors apart) and criterion (overall bias to respond "present" or "absent"). There are also methodologies to measure the exposure time needed to achieve a fixed accuracy level [22] or to measure the SATF directly [23].

## 31.2 Current Issues

In this section, I discuss a number of scientific topics currently being debated in the visual search literature. Many of these issues are driven by a quest to make a bridge between the highly controlled, often abstract experimental procedures of cognitive psychology and the messiness of the real world. "The real world" sometimes refers to specific, socially important search contexts (e.g., airport baggage screening [24], lifeguarding [25], or radiology [26]), and sometimes to ecological validity (i.e., search tasks that animals or humans might carry out in everyday life [27, 28]). This research is inspired by the idea of "Pasteur's Quadrant" [29], that scientific research is at its best when it is both a quest for a fundamental understanding of nature and an enterprise that casts light on problems of major social import [30].

### *31.2.1 Prevalence*

In the classic visual search paradigms described above, targets are present on 50% of trials (50% *target prevalence*). This is a compromise between data collection efficiency (experimenters are typically interested in the process of finding a target, not the process of deciding that there is no target, but see [31]) and keeping observers honest. Recently, however, researchers have become interested in the effects of target prevalence on search behavior. This interest comes out of practical interests in topics such as cancer screening (e.g., mammography or lung cancer screening), where prevalence is on the order of 1% [32, 33], and airport baggage screening, where prevalence is unknown, but presumably extremely low [34].

Target prevalence exerts systematic effects on search behavior [35]. As prevalence decreases, false negatives (or misses) increase, while false positives (false alarms) decrease [36, 37]. In signal detection theory terms, this amounts to a criterion shift; signal strength is unaffected, but more evidence is needed to classify an item as a target when targets are rare. In the response time domain, "yes" responses (both true positives and false positives) are unaffected by prevalence, but "no" responses are markedly faster at low prevalence and slower at high prevalence. Wolfe and van Wert [38] propose a dual-criterion model to explain this pattern. In their account, focal attention is directed sequentially to potential target items. For each attended item, a target/non-target decision must be made, and the threshold to classify something as a target varies with prevalence. After each decision, an additional decision must be made as to whether to abandon further search and declare the trial to be "target-absent". The threshold for this decision also varies with prevalence, such that observers are willing to quit early when targets are rare, leading to fast "no" response times, and reluctant to quit when targets are frequent, leading to slow "no" response times.

The dual-criterion model suggests that higher false positive rates at high prevalence are due to changes in the first criterion (classification); at high prevalence, observers are actually more likely to see ambiguous stimuli as targets. However, Peltier and Becker [39] argue that false positives also come from guesses. On some trials, observers may stop searching early (hitting the quitting threshold), and use combined information about overall target prevalence and how much of the display they have already searched to make an educated guess as to whether or not there is a target in the unsearched portion of the display.

Meanwhile, Trueblood et al. [40] have been able to partition the classification criterion more finely. Using the Diffusion Decision Model [41], they were able to separate two types of biases, response expectancy (overall bias towards "yes" or "no" responses) and stimulus evaluation biases (bias in how a stimulus is perceived). They showed that novices and experts showed different types of criterion shifts in response to changes in prevalence in a hematopathology classification task (blast cells vs. non-blast). For novice observers, prevalence affected the response expectancy bias; since they are less familiar with the stimuli and the task, they are more heavily influenced by the overall base rate environment. However, expert observers primarily showed shifts in the stimulus evaluation bias parameter, suggesting that the prevalence rate influenced how they perceived the blood cells in question. Since this was strictly speaking not a search task, this experiment provides insight into the first criterion of the Wolfe and van Wert model [38].

One interesting problem is how the brain computes prevalence. In laboratory studies with artificial search tasks, the observer has to learn the prevalence as they go through the experiment, with some studies estimating that behavior is governed by prevalence over the last 30–50 trials [38, 42]. It can be difficult to influence behavior by cueing observers about prevalence on a trial-by-trial basis [43, 44], and observers may come to the laboratory with a "sticky" assumption that prevalence is 50% [45]. On the other hand, in more realistic tasks, expert observers clearly come in with expectations based on their prior experience in the real world [46]. Zhang and Houpt [47] applied insights from decision theory to the problem, noting that observers tend to behave differently when their expectations of probability are based on explicit instruction (e.g., "In this block, 35% of trials will have targets") rather than experience. In line with findings from the decision literature, they found that criterion shifts are more extreme (i.e., more conservative at low prevalence and more liberal at high prevalence) when observers are explicitly instructed about prevalence (e.g., "10% of the trials will have a target") than when they have to estimate prevalence from their experience.

## *31.2.2 Multiple Targets*

In the "classic" visual search paradigm, as popularized by Treisman and Gelade [4], while the number of distractors may vary (according to the set size), there is a single target in memory, and a single target in the search array. For example, the observer might be asked to find the letter "T" in an array of "L" distractors. Once she finds a single target, she can respond, and then go on to the next trial. The target may be absent or present on a given trial, but for a block of dozens or hundreds of trials, she is looking for Ts, and there will be zero or one T in the array. This design was well suited for the enterprise of testing for "basic" or "guiding" features. However, more recent work has focused on the role of memory and quitting strategies in search. As a result, experimental paradigms featuring multiple potential targets (only one of which might be present in the array) or multiple instances of targets available to be found have become popular.

These paradigms also address the drive to be more relevant to the "real world". In applied search contexts ranging from detecting product defects [48] to lung cancer screening [49], the image or search array can potentially include multiple targets. Furthermore, observers are often on the lookout for more than one type of target. Consider the airport baggage screener, who has to be alert for guns, knives, and improvised explosive devices [50].

### 31.2.2.1 Hybrid Search

Hybrid search refers to an important topic of research in the visual search domain that had lain dormant for several decades. As I noted at the start of the chapter, one of the seminal influences on modern visual search research was a pair of papers by Schneider and Shiffrin [2, 3] in which they orthogonally manipulated the visual set size and the memory set size (i.e., the number of potential targets) from 1 to 4 items. In these experiments, the stimuli were alphanumeric characters, and observers might search for digits among consonants, consonants among consonants, or digits among digits. They found that response time was a linear function of both visual and memory set size. They proposed a serial model in which each item from the memory set is compared sequentially to each item in the visual set before moving on to the next memory set item.

While these papers were quite influential over the last few decades, these findings were not followed up on until recently. Wolfe [51] noted that as the memory set grew larger, the Schneider and Shiffrin [2] model would predict that visual search would become untenably slow. Instead, Wolfe predicted that memory search time would increase logarithmically as memory set sizes exceeded the four-item limit of the Schneider and Shiffrin studies. He trained observers to search for arbitrary pictures of objects (e.g., bowling balls, fruit, furniture, flowers). Visual set size was varied from 1 to 16 items, but the memory sets were extended as far as 100 items. He found that response times remained linear as a function of visual set size, but increased logarithmically as a function of the memory set size. Why is memory search so much more efficient than visual search? One mechanism might be to simultaneously compare all or some of the items in the visual array to a subset of the items in the memory set. However, Drew and Wolfe [52] were able to show that this is not the case by restricting visual search to a single item at a time and obtaining the same logarithmic functions. Instead, the data are consistent with a "carwash" model [53, 54]. A carwash is a serial pipeline where one car enters at a time, but you don't have to wait for the car in front of you to finish being washed before you can drive your car in. In this model, items from the visual array enter the system serially (thus the linear response time by visual set size functions), but multiple items can be matched against the current target or category template at the same time, leading to the logarithmic response time by memory set functions.

While Wolfe [51] used a random assortment of specific pictures as stimuli, real-world searches tend to be for categories of targets, rather than specific individual exemplars. So a radiologist looks for lesions that have known characteristics, rather than specific lesions, just as a grocery shopper might be looking for fruits, rather than a specific, known lemon. It turns out that search time is logarithmic as a function of the number of categories, just as it is for individual targets [55], suggesting that targets are evaluated for category membership using a process similar to (but slower than) that for comparison to a specific target template. Observers can also switch which memory set they are relying on as the context changes [56], for example when moving from the produce aisle to the cereal aisle of a grocery store, you can partition your search so that you don't compare cereal boxes to vegetables.

Target templates, the internal neural "definitions" of what the observer is looking for, have typically been assumed to reside in visual working memory [57, 58]. However, visual working memory is usually held to have a limited capacity of approximately four items [59, 60]. The hybrid search experiments described above demonstrate that observers can maintain memory sets that exceed this limit, in some cases by orders of magnitude. These findings clearly demonstrate that items in long-term memory stores can be used to guide search [61]. This means that we have to incorporate insights from the study of long-term memory in order to understand hybrid search.

Object recognition, the process of identifying an object that you are seeing as something that you have seen before, proceeds by two pathways, the relatively automatic, rapid *familiarity* pathway, which yields the subjective impression that "I have seen this thing before, but I don't know when", and the controlled, slower *recollection* pathway [62]. While it would seem likely that the surprisingly efficient search through large memory sets utilizes the familiarity pathway, Guild et al. [63] found that the logarithmic memory search function remained when they introduced manipulations that should have reduced the putative familiarity signal, including presenting mostly target-absent trials, and requiring observers to rely on words rather than images to define the targets. These experiments suggest that hybrid search is served by the recollection pathway.

#### 31.2.2.2 Foraging

Foraging is a fairly broad term that describes an agent searching for resources in the environment that are characterized by some spatial and temporal distribution. One might argue that classic visual search tasks are a very narrow, perhaps even degenerate, subset of foraging, where the "resource" is red vertical bars and there is only one target, spatially distributed across the computer screen, temporally distributed with 50% probability per trial. And indeed, cognitive theorists have made this argument: Klein and MacInnes [64], for instance, argued that inhibition of return (the phenomenon that attention seems to be biased against returning to recently attended locations, see [65]) serves as a "foraging facilitator in visual search". Whereas Klein and MacInnes asked observers to search for "Waldo" in images from the popular *Where's Waldo?*[2] series of picture books, Gilchrist, et al. [66] scattered a variable number of film canisters across the floor of a room and asked participants to find the one canister with a marble in it, a task which required physically reaching out to shake the canister. They found that this foraging task produced data similar to visual search, including search times that increased linearly with the number of canisters, and a 2:1 absent/present slope ratio. The main difference to classic visual search results was that people were less likely to revisit previously examined items when participants had to walk around a room and reach out to pick up film canisters than when they could simply move their eyes around a computer screen (cf. [67]).

These studies were inspired by the metaphor of foraging, but strictly tied to existing questions in the attention and visual search literature. More recently, cognitive psychologists have been designing studies informed by the study of animals foraging for food in population ecology studies. Wolfe [68] began an effort to see how to apply mathematical accounts of optimal foraging behavior (derived from population biology, e.g., [69]) to human visual foraging behavior. Wolfe's studies were designed around a "berry picking" metaphor. Observers were presented with an "aerial view" intended to depict a large area featuring "berries" (bluish and reddish

---

[2] Also known as *Where's Wally?* in the UK.

circles of various sizes) and green "leaves". They could select a sector of the overall field and zoom down to the "patch" level where they could collect the berries using the computer mouse. The difficulty of collecting berries, the perceptual ease of discriminating "good" berries from "bad" berries, and so forth were experimental variables to be manipulated. The key dependent variable was how much time the observer spent in each patch before moving on to the next patch. Marginal Value Theorem (MVT, [70]) predicts that foragers will move on to a new patch when the instantaneous rate of return in the current patch drops below the average rate for the entire field. MVT describes human behavior well in a number of studies from Wolfe's group [71, 72]. Deviations from optimality can also be informative. For example, Cain et al. [45] studied foraging in a more conventional visual search task where observers searched for T-shaped targets among L-shaped distractors against a background of randomly-generated grayscale clouds. Unlike Wolfe's berry patch structure, they used a standard trial structure, and varied the number of targets per trial using a geometric distribution. They applied the Potential Value Theorem [73], a variant of MVT adapted for situations with discrete target probabilities, and found near-optimal behavior, though observers tended to behave as if the target distribution were more uniform than it actually was, giving insight into the Bayesian priors observers bring to laboratory experiments.

Kristjánsson et al. [74] devised a foraging paradigm inspired by a different problem in the population biology literature. Work on how selection processes shape both the protective coloration of prey species and the perceptual and cognitive apparatus of predator species has revealed that when multiple types of prey are available, predators tend to repeatedly select the same prey type in "runs", rather than randomly switching between prey types. Kristjánsson and colleagues set out to test whether human observers would behave similarly. In their paradigm, the stimulus was an array of colored shapes on a black background presented on an iPad screen, and observers "foraged" by tapping on the desired targets. The arrays were constructed according to the rules for feature and conjunction search described in Sect. 31.1. In the feature condition, targets could be red or green, with blue and yellow distractors (or vice versa), while in the conjunction condition, targets were green circles and red squares among red circles and green squares (or vice versa). Unlike classic visual search, here half of the stimuli on each trial were targets. Consistent with the predictions from population biology, they found that in the feature condition, observers switched easily between the two target types, but in the conjunction condition, characterized by more complex stimuli and less salient targets, observers tended to engage in runs, exhaustively searching for one type of target before switching to the other type. The authors suggest that this behavioral difference is due to the higher cognitive demands of conjunction search; in feature search, the target templates are simpler, so it is easy to maintain two active templates simultaneously, but it is better to hold a single template in working memory when the templates are more complex. Wolfe et al. [75] also found that foraging targets were selected in runs, using pictures of complex objects rather than colored shapes. They did not find exhaustive search, probably because their stimuli were even more complex than two-feature conjunctions, and there were up to five target types. They found that runs ended somewhat more quickly than would be predicted on the basis of MVT. Wolfe et al. suggested instead that target switching behavior may be opportunistic; while on a run of searching for Target A, an observer might find their gaze or cursor very close to an exemplar of Target B, and find it easier to simply switch to the other target type.

Jóhannesson et al. [76] compared behavior in the Kristjánsson et al. [74] paradigm using both finger-tapping responses (as in the original) and gaze responses (i.e., the observer was asked to fixate on each target, rather than tapping them). They found that the response mode changed the results such that switching behavior was much more likely in the gaze-response condition than in the finger-response condition. This result may indicate that motor control of the fingers is more cognitively taxing than moving the eyes, or possibly that finger tapping requires more precisely localization. Either way, this finding illustrates how foraging studies bring our attention (so to speak) to factors beyond the stimulus that can modulate search behaviors. Prpic et al. [77] introduced another factor when they devised a 3D version of this foraging paradigm, using the Unity3D game engine. Here, observers explored a virtual reality environment, playing as a squirrel searching for nuts, making the foraging metaphor more literal. The nuts varied in shape (walnuts vs. acorns) and color (red, blue, purple, yellow), allowing the experiments to construct feature and conjunction search versions of the experiment. Again, observers behaved similarly in the conjunction and feature conditions, switching frequently rather than engaging in runs.

Another line of research that provides insight into foraging behavior is work on the Multiple Item Localization paradigm (MILO, [78]). In the MILO task, there is a sequence of targets. In the simplest version of the paradigm, a touchscreen display contains an array of billiard balls, each labeled with a number. The task is simply to tap on the billiard balls in numerical order. In some experiments, the order could be alphabetical [79], or more complex cognitive constructs such as the precedence order in chess [80]. This paradigm has elements of both hybrid search, in that observers must keep a target sequence in memory, and foraging, in that there are multiple targets in the visual display that must be exhaustively captured. The advantage of MILO is that it allows experimenters to measure the temporal context of search. For example, if we remove the billiard balls as they are tapped (the *vanish* condition), this should theoretically make search easier because the observer doesn't have to remember which items they already dealt with. However, when we compare this condition to one where the billiard balls remain on the screen (the *remain* condition), we see very little difference in response time, suggesting that observers have very good retrospective memory for locations that they have responded to. This only applied to locations, not objects, because if we set the balls in motion around the screen, we now see a huge advantage for the vanish condition [81].

MILO can also be used to tap prospective cognition. An interesting feature of MILO is that all items serve as both targets and distractors. If I am searching for the "4", the "5" is a distractor, but I know it will become a target as soon as I have located the "4". If I come across the "5" while searching for the "4", do I remember that location? We can test this by randomizing the location of all upcoming targets whenever the observer taps a target. So after I tap the "4", the "5" moves to a new location. In this *shuffle* condition, response times are again elevated compared to a non-shuffle control, indicating that observers do plan ahead [78, 79].

The development of the foraging task represents an exciting new development in the study of visual search. The foraging task adds a number of new experimental dimensions to the classic visual search paradigm. In addition to manipulating the visual features that define or distinguish targets and distractors, the visual set size, the memory set size [82], the difficulty of travel between patches, the spatial distribution of targets, and the distribution of target values, etc., are all task parameters that can be explored. This is a richer experimental design space that includes classic visual search and hybrid search as special cases. As Kristjánsson et al. [74] showed, findings from the classic search literature can be profitably imported into this higher dimensional space of foraging studies. Foraging tasks also provide a much richer set of outcome variables that can help give us a more nuanced picture of selective attention [83]. Foraging paradigms also broaden the applicability of visual search to other applied contexts, such as satellite imagery [71, 84].

### *31.2.3 Individual Differences*

One interesting finding to come out of Kristjánsson et al. [74] was that four of their sixteen participants showed no differences between the feature and conjunction conditions, maintaining the random selection of the feature condition even in the more difficult conjunction condition. The authors speculated that these individuals were "super-foragers", because they could hold multiple complex target templates in working memory simultaneously. In a follow-up study [85], they replicated their basic findings, and looked for correlations between super-forager behavior and measures of individual attentional function and working memory capacity, but found none. In both studies, the sample size of sixteen observers (including the four "super-foragers") was probably orders of magnitude too small to observe any meaningful correlations. Nevertheless, the fact that they tried to look at different patterns of performance represents a new trend in cognitive psychology in general and the study of visual search in particular.

Traditionally, experimental psychologists have been interested in searching for effects that will generalize across individuals, rather than looking at differences between individuals. Indeed, in classical psychophysics, acquiring precise data often requires an individual observer to run thousands of trials, which means that even undergraduate samples won't do; many studies rely on 3-4 dedicated observers, often including authors (e.g., [86]). The visual search experiments that I have been describing typically have used 10-30 observers. As a result, our experimental paradigms are generally not sensitive to individual differences [87]. Individual differences were seen as the domain of personality psychology (e.g., [88]). However, this attitude is starting to change. The Jóhannesson et al. [85] study is an example of a class of experiments which look for correlations between search and performance on other tasks thought to indicate relevant abilities, such as working memory [89–91], with mixed results, as sample sizes still tend to be rather small in these studies. An emerging set of studies has tried to characterize the relationship among a larger set of cognitive tasks thought to involve attention, in order to get a better idea whether attention is one faculty or multiple [92–94]. These studies involve somewhat larger sample sizes (i.e., N > 200), and require each observer to complete a battery of 10-20 tasks. Factor analysis techniques are then used to elucidate relationships among the different tasks. Huang et al. [92] concluded that a single latent factor, which they dubbed the General Attention factor, or *a*, underlies performance on a set of 9 "primary attention paradigms", including spatial configuration visual search and feature search. In contrast, Treviño et al. [93] suggested that three factors were required to explain attention: Capacity, search, and sustained attention. Skogsberg et al. [94] identified four factors: Global attention, sustained attention, transient attention, and spatiotemporal attention. Unfortunately, these three studies all used a different selection of tasks, and only Huang et al. and Treviño et al. included visual search tasks. This is a very promising line of work, but is in its initial stages. The rapid increase in the acceptability and feasibility of online studies in recent years [95, 96], accelerated by the demands imposed by the COVID-19 pandemic, means that these types of studies, which require much larger sample sizes than traditional laboratory studies, will be easier to conduct in the future.

A major impetus for studying individual differences is the needs of applied search fields [97]. If you work with professional visual searchers (e.g., radiologists, airport screeners), they are highly interested in tests that will allow them to predict which potential trainees would be good searchers and which should look elsewhere for a career. This is another area in which efforts are promising, but still in the early stages [98].

## 31.3 Conclusions

Visual search is a core cognitive capacity for any organism that has a visual system, but especially for humans. Visual search is ubiquitous in our daily lives [99], from groping for our coffee cup in the morning to navigating our way to work and back, to finding where the dog left her favorite toy. This is not to mention radiology and pathology, airport baggage screening, looking for drowning swimmers, scanning an aircraft fuselage for cracks, and other specialized variants of visual search that can help save lives.

Every academic paper and grant proposal on the topic for decades has started with some version of the previous paragraph. In the last decade or so, however, scientific work in the field has actually started to live up to these grand pronouncements, moving our experimental designs closer to the "real world" while retaining the rigor of the controlled laboratory experiment. Moving beyond the default settings of 50% prevalence, single targets, and group averages represents real methodological and theoretical progress.

Perhaps most importantly for the audience of this volume, conceptualizing visual search as a branch of foraging allows cognitive psychologists access to new mathematical tools (e.g., MVT and its successors, [73]) to help understand their data, and provides a scaffolding for building bridges to other disciplines and other conceptions of “the target problem”.

## References

1. H.L. Kingsley, Search; a function intermediate between perception and thinking. Psychol. Monogr. Gen. Appl. **35**, 16 (1926)
2. W. Schneider, R.M. Shiffrin, Controlled and automatic human information processing: i. detection, search, and attention. Psychol. Rev. **84**, 1 (1977)
3. R. M. Shiffrin, W. Schneider, Controlled and automatic human information processing: ii. perceptual learning, automatic attending, and a general theory. Psychol. Rev. **84**, 127 (1977)
4. A.M. Treisman, G. Gelade, A feature-integration theory of attention. Cognit. Psychol. **12**, 97 (1980)
5. S. Zeki, J.D. Watson, C.J. Lueck, K.J. Friston, C. Kennard, R.S. Frackowiak, A direct demonstration of functional specialization in human visual cortex. J. Neurosci. Off. J. Soc. Neurosci. **11**, 641 (1991)
6. A.M. Treisman, Perceptual grouping and attention in visual search for features and for objects. J. Exp. Psychol. Hum. Percept. Perform. **8**, 194 (1982)
7. A.M. Treisman, J. Souther, Search asymmetry: a diagnostic for preattentive processing of separable features. J. Exp. Psychol. Gen. **114**, 285 (1985)
8. J.M. Wolfe, T.S. Horowitz, Five factors that guide attention in visual search. Nat. Hum. Behav. **1**, 0058 (2017)
9. H. Chang, R. Rosenholtz, Search performance is better predicted by tileability than presence of a unique basic feature. J. Vis. **16**, 1 (2016)
10. J. Hulleman, C.N.L. Olivers, The impending demise of the item in visual search. Behav. Brain Sci. **40**, e132 (2017)
11. A. Kristjánsson, Reconsidering visual search. i-Percept. **6**, 2041669515614670 (2015)
12. E.M. Palmer, T.S. Horowitz, A. Torralba, J.M. Wolfe, What are the shapes of response time distributions in visual search? J. Exp. Psychol. Hum. Percept. Perform. **37**, 58 (2011)
13. S. Buetti, D.A. Cronin, A.M. Madison, Z. Wang, A. Lleras, Towards a better understanding of parallel visual processing in human vision: evidence for exhaustive analysis of visual information. J. Exp. Psychol. Gen. **145**, 672 (2016)
14. J.R. Bergen, B. Julesz, Parallel versus serial processing in rapid pattern discrimination. Nature **303**, 696 (1983)
15. H. Egeth, D. Dagenbach, Parallel versus serial processing in visual search: further evidence from subadditive effects of visual quality. J. Exp. Psychol. Hum. Percept. Perform. **17**, 551 (1991)
16. J.M. Wolfe, What can 1 million trials tell us about visual search? Psychol. Sci. **9**, 33 (1998)
17. T.S. Horowitz, Revisiting the variable memory model of visual search. Vis. Cogn. **14**, 668 (2006)
18. R.P. Heitz, The speed-accuracy tradeoff: history, physiology, methodology, and behavior. Front. Neurosci. **8** (2014)
19. M.P. Eckstein, J.P. Thomas, J. Palmer, S.S. Shimozaki, A signal detection model predicts the effects of set size on visual search accuracy for feature, conjunction, triple conjunction, and disjunction displays. Percept. Psychophys. **62**, 425 (2000)
20. P. Verghese, Visual search and attention: a signal detection theory approach. Neuron **31**, 523 (2001)

21. N.A. Macmillan, C.D. Creelman, *Detection Theory: A User's Guide* (Lawrence Erlbaum Associates, Mahwah, NJ, 2005)
22. J.L. Zacks, R.T. Zacks, Visual search times assessed without reaction times: a new method and an application to aging. J. Exp. Psychol. Hum. Percept. Perform. **19**, 798 (1993)
23. A.M. Giordano, B. McElree, M. Carrasco, On the automaticity and flexibility of covert attention: a speed-accuracy trade-off analysis. J. Vis. **9**, 30 (2009)
24. H.J. Godwin, T. Menneer, K.R. Cave, S. Helman, R.L. Way, N. Donnelly, The impact of relative prevalence on dual-target search for threat items from airport X-ray screening. Acta Psychol. (Amst.) **134**, 79 (2010)
25. L.K. Lanagan-Leitzel, E. Skow, C.M. Moore, Great expectations: perceptual challenges of visual surveillance in lifeguarding. Appl. Cogn. Psychol. **29**, 425 (2015)
26. L. Williams, A. Carrigan, W. Auffermann, M. Mills, A. Rich, J. Elmore, T. Drew, The invisible breast cancer: experience does not protect against inattentional blindness to clinically relevant findings in radiology. Psychon. Bull. Rev. (2020)
27. M. Bella-Fernández, M. Suero Suñé, B. Gil-Gómez de Liaño, Foraging behavior in visual search: a review of theoretical and mathematical models in humans and animals. Psychol. Res. (2021)
28. P. Chassy, F. Gobet, Visual search in ecological and non-ecological displays: evidence for a non-monotonic effect of complexity on performance. PLoS ONE **8**, e53420 (2013)
29. D.R. Stokes, *Pasteur's Quadrant: Basic Science and Technological Innovation* (Brookings Institution Press, Washington, DC, 1997)
30. J.M. Wolfe, Rethinking the basic-applied dichotomy. Cogn. Res. Princ. Implic. **1**, 1 (2016)
31. M.M. Chun, J.M. Wolfe, Just say no: how are visual searches terminated when there is no target present? Cognit. Psychol. **30**, 39 (1996)
32. C.I. Lee et al., National performance benchmarks for screening digital breast tomosynthesis: update from the breast cancer surveillance consortium. Radiology **307**, e222499 (2023)
33. The national lung screening trial research team, results of initial low-dose computed tomographic screening for lung cancer. N. Engl. J. Med. **368**, 1980 (2013)
34. J.M. Wolfe, D.N. Brunelli, J. Rubinstein, T.S. Horowitz, Prevalence effects in newly trained airport checkpoint screeners: trained observers miss rare targets, too. J. Vis. **13**, 33 (2013)
35. T.S. Horowitz, Prevalence in visual search: from the clinic to the lab and back again. Jpn. Psychol. Res. **59**, 65 (2017)
36. J.M. Wolfe, T.S. Horowitz, N.M. Kenner, Rare items often missed in visual searches. Nature **435**, 439 (2005)
37. J.M. Wolfe, T.S. Horowitz, M.J. van Wert, N.M. Kenner, S.S. Place, N. Kibbi, Low target prevalence is a stubborn source of errors in visual search tasks. J. Exp. Psychol. Gen. **136**, 623 (2007)
38. J.M. Wolfe, M.J. van Wert, Varying target prevalence reveals two dissociable decision criteria in visual search. Curr. Biol. **20**, 121 (2010)
39. C. Peltier, M.W. Becker, Target-present guessing as a function of target prevalence and accumulated information in visual search. Atten. Percept. Psychophys. **79**, 1064 (2017)
40. J.S. Trueblood, Q. Eichbaum, A.C. Seegmiller, C. Stratton, P. O'Daniels, W.R. Holmes, Disentangling prevalence induced biases in medical image decision-making. Cognition **212**, 104713 (2021)
41. C.N. White, R.A. Poldrack, Decomposing bias in different types of simple decisions. J. Exp. Psychol. Learn. Mem. Cogn. **40**, 385 (2014)
42. K. Ishibashi, S. Kita, J.M. Wolfe, The effects of local prevalence and explicit expectations on search termination times. Atten. Percept. Psychophys. **74**, 115 (2012)
43. K. Ishibashi, S. Kita, Probability cueing influences miss rate and decision criterion in visual searches. i-Percept. **5**, 170 (2014)
44. J.S.-H. Lau, L. Huang, The prevalence effect is determined by past experience, not future prospects. Vis. Res. **50**, 1469 (2010)
45. M.S. Cain, E. Vul, K. Clark, S.R. Mitroff, A Bayesian optimal foraging model of human visual search. Psychol. Sci. **23**, 1047 (2012)

46. K.K. Evans, R.L. Birdwell, J.M. Wolfe, If you don't find it often, you often don't find it: why some cancers are missed in breast cancer screening. PLoS ONE **8**, e64366 (2013)
47. H. Zhang, J.W. Houpt, Exaggerated prevalence effect with the explicit prevalence information: the description-experience gap in visual search. Atten. Percept. Psychophys. **82**, 3340 (2020)
48. R. Yu, A.H.S. Chan, Visual search time in detection tasks with multiple targets: considering change of the effective stimulus field area. Int. J. Ind. Ergon. **43**, 328 (2013)
49. K.S. Berbaum, E.A. Krupinski, K.M. Schartz, R.T. Caldwell, M.T. Madsen, S. Hur, A.T. Laroia, B.H. Thompson, B.F. Mullan, E.A. Franken Jr., Satisfaction of search in chest radiography 2015. Acad. Radiol. (2015)
50. X. Liu, A. Gale, T. Song, Detection of terrorist threats in air passenger luggage: expertise development, in *2007 41st Annual IEEE International Carnahan Conference on Security Technology* (2007), pp. 301–306
51. J.M. Wolfe, Saved by a log: how do humans perform hybrid visual and memory search? Psychol. Sci. **23**, 698 (2012)
52. T. Drew, J.M. Wolfe, Hybrid search in the temporal domain: evidence for rapid, serial logarithmic search through memory. Atten. Percept. Psychophys. **76**, 296 (2014)
53. C.M. Moore, J.M. Wolfe, in *The Limits of Attention: Temporal Constraints in Human Information Processing*, ed. by K. Shapiro (Oxford University Press, 2001), pp. 178–198
54. J.M. Wolfe, Moving towards solutions to some enduring controversies in visual search. Trends Cogn. Sci. (2003)
55. C.A. Cunningham, J.M. Wolfe, The role of object categories in hybrid visual and memory search. J. Exp. Psychol. Gen. **143**, 1585 (2014)
56. S.E.P. Boettcher, T. Drew, J.M. Wolfe, Hybrid search in context: how to search for vegetables in the produce section and cereal in the cereal aisle. Vis. Cogn. **21** (2013)
57. D.J.K. Barrett, O. Zobay, Attentional control via parallel target-templates in dual-target search. PLoS ONE **9**, e86848 (2014)
58. C.N.L. Olivers, J. Peters, R. Houtkamp, P.R. Roelfsema, Different states in visual working memory: when it guides attention and when it does not. Trends Cogn. Sci. **15**, 327 (2011)
59. N. Cowan, The magical number 4 in short-term memory: a reconsideration of mental storage capacity. Behav. Brain Sci. **24**, 87 (2001)
60. N. Cowan, E.M. Elliott, J. Scott Saults, C.C. Morey, S. Mattox, A. Hismjatullina, A.R.A. Conway, On the capacity of attention: its estimation and its role in working memory and cognitive aptitudes. Cognit. Psychol. **51**, 42 (2005)
61. T. Drew, S.E.P. Boettcher, J.M. Wolfe, Searching while loaded: visual working memory does not interfere with hybrid search efficiency but hybrid search uses working memory capacity. Psychon. Bull. Rev. **23**, 201 (2016)
62. L. Jacoby, A process dissociation framework: separating automatic from intentional uses of memory. J. Mem. Lang. **30**, 513 (1991)
63. E.B. Guild, J.M. Cripps, N.D. Anderson, N. Al-Aidroos, Recollection can support hybrid visual memory search. Psychon. Bull. Rev. **21**, 142 (2013)
64. R.M. Klein, W.J. MacInnes, Inhibition of return is a foraging facilitator in visual search. Psychol. Sci. **10**, 346 (1999)
65. M.I. Posner, Y. Cohen, Components of visual orienting. Atten. Perform. **10**, 531 (1984)
66. I.D. Gilchrist, A. North, B.M. Hood, Is visual search really like foraging? Perception **30**, 1459 (2001)
67. I.D. Gilchrist, M. Harvey, Refixation frequency and memory mechanisms in visual search. Curr. Biol. CB **10**, 1209 (2000)
68. J.M. Wolfe, When is it time to move to the next raspberry bush? Foraging rules in human visual search. J. Vis. **13**, 10 (2013)
69. G.H. Pyke, H.R. Pulliam, E.L. Charnov, Optimal foraging: a selective review of theory and tests. Q. Rev. Biol. **137** (1977)
70. E.L. Charnov, Optimal foraging, the marginal value theorem. Theor. Popul. Biol. **9**, 129 (1976)
71. K.A. Ehinger, J.M. Wolfe, When is it time to move to the next map? Optimal foraging in guided visual search. Atten. Percept. Psychophys. **78**, 2135 (2016)

72. J. Zhang, X. Gong, D. Fougnie, J.M. Wolfe, How humans react to changing rewards during visual foraging. Atten. Percept. Psychophys. **79**, 2299 (2017)
73. O. Olsson, J.S. Brown, The foraging benefits of information and the penalty of ignorance. Oikos **112**, 260 (2006)
74. A. Kristjánsson, O.I. Jóhannesson, I.M. Thornton, Common attentional constraints in visual foraging. PLOS ONE **9**, e100752 (2014)
75. J.M. Wolfe, A.M. Aizenman, S.E.P. Boettcher, M.S. Cain, Hybrid foraging search: searching for multiple instances of multiple types of target. Vision Res. **119**, 50 (2016)
76. O.I. Jóhannesson, I.M. Thornton, I.J. Smith, A. Chetverikov, A. Kristjánsson, Visual foraging with fingers and eye gaze. i-Percept. **7**, 2041669516637279 (2016)
77. V. Prpic, I. Kniestedt, E. Camilleri, M.G. Maureira, A. Kristjánsson, I.M. Thornton, A serious game to explore human foraging in a 3D environment. PLoS ONE **14**, e0219827 (2019)
78. I.M. Thornton, T.S. Horowitz, The multi-item localization (MILO) task: measuring the spatiotemporal context of vision for action. Percept. Psychophys. **66**, 38 (2004)
79. I.M. Thornton, T.S. Horowitz, Searching through alternating sequences: working memory and inhibitory tagging mechanisms revealed using the MILO task. i-Percept. **11**, 1 (2020)
80. T. Zammit, Location Memory in Visual Search. The Multi-Item Localization (MILO) Task: A Tool for Investigation (Dissertation, University of Malta, 2017)
81. T.S. Horowitz, I.M. Thornton, Objects or locations in vision for action? Evidence from the MILO task. Vis. Cogn. **16**, 486 (2008)
82. J.M. Wolfe, M.S. Cain, A.M. Aizenman, Guidance and selection history in hybrid foraging visual search. Atten. Percept. Psychophys. **81**, 637 (2019)
83. T. Kristjánsson, I.M. Thornton, A. Chetverikov, A. Kristjánsson, Dynamics of visual attention revealed in foraging tasks. Cognition **194**, 104032 (2020)
84. K.A. Ehinger, J.M. Wolfe, Foraging in satellite imagery: when is it time to move to the next map? J. Vis. **15**, 759 (2015)
85. O.I. Jóhannesson, A. Kristjánsson, I.M. Thornton, Are foraging patterns in humans related to working memory and inhibitory control? Jpn. Psychol. Res. **59**, 152 (2017)
86. S.P. McKee, L. Welch, Sequential recruitment in the discrimination of velocity, J. Opt. Soc. Am. A Opt. Image Sci. Vis. **2**, 243 (1985)
87. C. Hedge, G. Powell, P. Sumner, The reliability paradox: why robust cognitive tasks do not produce reliable individual differences. Behav. Res. Methods **50**, 1166 (2018)
88. S.D. Gosling, V.S.Y. Kwan, O.P. John, A dog's got personality: a cross-species comparative approach to personality judgments in dogs and humans. J. Pers. Soc. Psychol. **85**, 1161 (2003)
89. M.J. Kane, B.J. Poole, S.W. Tuholski, R.W. Engle, Working memory capacity and the top-down control of visual search: exploring the boundaries of "executive attention". J. Exp. Psychol. Learn. Mem. Cogn. **32**, 749 (2006)
90. A. Muhl-Richardson, H.J. Godwin, M. Garner, J.A. Hadwin, S.P. Liversedge, N. Donnelly, Individual differences in search and monitoring for color targets in dynamic visual displays. J. Exp. Psychol. Appl. **24**, 564 (2018)
91. C. Peltier, M.W. Becker, Working memory capacity predicts selection and identification errors in visual search. Perception **46**, 109 (2017)
92. L. Huang, L. Mo, Y. Li, Measuring the interrelations among multiple paradigms of visual attention: an individual differences approach. J. Exp. Psychol. Hum. Percept. Perform. **38**, 414 (2012)
93. M. Treviño, X. Zhu, Y.Y. Lu, L.S. Scheuer, E. Passell, G.C. Huang, L.T. Germine, T.S. Horowitz, How do we measure attention? using factor analysis to establish construct validity of neuropsychological tests. Cogn. Res. Princ. Implic. **6**, 51 (2021)
94. K. Skogsberg, M. Grabowecky, J. Wilt, W. Revelle, L. Iordanescu, S. Suzuki, A relational structure of voluntary visual-attention abilities. J. Exp. Psychol. Hum. Percept. Perform. **41**, 761 (2015)
95. L. Germine, K. Nakayama, B.C. Duchaine, C.F. Chabris, G. Chatterjee, J.B. Wilmer, Is the web as good as the lab? Comparable performance from web and lab in cognitive/perceptual experiments. Psychon. Bull. Rev. **19**, 847 (2012)

96. L. Germine, K. Reinecke, N.S. Chaytor, Digital neuropsychology: challenges and opportunities at the intersection of science and software. Clin. Neuropsychol. **33**, 271 (2019)
97. A.T. Biggs, M.R. Kramer, S.R. Mitroff, Using cognitive psychology research to inform professional visual search operations. J. Appl. Res. Mem. Cogn. **7**, 189 (2018)
98. M.R. Kramer, C.L. Porfido, S.R. Mitroff, Evaluation of strategies to train visual search performance in professional populations. Curr. Opin. Psychol. **29**, 113 (2019)
99. M.P. Eckstein, Visual search: a retrospective. J. Vis. **11**, 14 (2011)

**SPRINGER NATURE**

# GPSR Compliance

*The European Union's (EU) General Product Safety Regulation (GPSR) is a set of rules that requires consumer products to be safe and our obligations to ensure this.*

*If you have any concerns about our products, you can contact us on ProductSafety@springernature.com*

In case Publisher is established outside the EU, the EU authorized representative is:

Springer Nature Customer Service Center GmbH
Europaplatz 3
69115 Heidelberg, Germany

The manufacturer's authorised representative in the EU is Springer Nature Customer Service Centre GmbH, Europaplatz 3, 69115 Heidelberg, Germany. If you have any concerns regarding our products, please contact ProductSafety@springernature.com

Printed and bound by CPI Group (UK) Ltd, Croydon, CR0 4YY
01/10/2026
02225910-0020

Target Search Problems